# Clinical Anatomy for Medical Students

## By the Same Author

### Atlas of Clinical Anatomy

A full-color atlas including 55 surface anatomy
photographs and 326 illustrations, which is
coordinated with the following text

### Gross Anatomy Dissector

A completely illustrated dissecting room manual,
which is coordinated with the textbook *Clinical
Anatomy for Medical Students* and with *Atlas of
Clinical Anatomy*

The other texts by Dr. Snell are *Clinical and
Functional Histology for Medical Students; Clinical
Neuroanatomy for Medical Students; Clinical
Embryology for Medical Students*, Third Edition;
and *An Atlas of Normal Radiographic Anatomy*
(with Alvin C. Wyman, M.D.).

# Clinical Anatomy for Medical Students

## Third Edition

**Richard S. Snell, M.D., Ph.D.**

*Professor and Chairman, Department of Anatomy*
*George Washington University School of Medicine and Health Sciences,*
*Washington, D.C.*

Little, Brown and Company
Boston/Toronto

Copyright © 1973, 1981, 1986 by Little, Brown and
Company (Inc.)

Third Edition

*Third Printing*

Library of Congress catalog card No. 85-81112

ISBN 0-316-802174

Printed in the United States of America

DON


Asian reprint: Medical Sciences International, Ltd., 1981

Greek edition: Medical Books, Costas Litsas, 1986

Italian edition: USES Edizioni Scientifiche Firenze, 1984

Japanese edition: Medical Sciences International, Ltd., 1983

Portuguese edition: Medsi Editora Médica e Científica Ltda,
1984

Spanish edition: Nueva Editorial Interamericana S. A. de
C. V., 1984

To My Students—Past, Present, and Future

# Contents

# Preface

This book is written by an anatomist who is a physician to provide preclinical medical students with the basic knowledge of anatomy necessary for clinical practice. While it is appreciated that during their preclinical years few students know in which areas they will eventually wish to specialize after taking their M.D. degree, this book attempts to offer a core of anatomical knowledge that will serve physicians throughout their professional life. For those areas of the body that are commonly diseased, the anatomy is dealt with in detail; other areas of the body are covered more superficially.

In this Third Edition great emphasis has again been placed on surface anatomy and surface markings, since the majority of practicing physicians seldom explore tissues to any depth beneath the skin. Photographs of living subjects have been added for further help in this study. Normal radiographs of most regions of the body have again been included, with a few substitutions made to improve quality. Examples of CT (computerized tomographic) scans of the head and body and sonograms have also been added. Labelled black and white photographs of cross-sectional anatomy of the head,

neck, and trunk have been included to stimulate students to think in terms of three-dimensional anatomy, which is so important in the interpretation of CT scans and sonograms.

The practical application of anatomical facts to clinical medicine is stressed throughout the book, in the form of Clinical Notes. Clinical Problems that require anatomical knowledge for their solution are presented at the end of each chapter. Both the Clinical Notes and the Clinical Problems have been brought up to date. Since many medical schools require students to take the National Board Part I Examination, examples of National Board type questions are given at the end of each chapter. The answers to the Clinical Problems and the National Board Type Questions will be found at the back of the book.

In this edition, many of the simple colored illustrations have been redrawn to improve accuracy and understanding. Illustrations summarizing the nerve and blood supplies of regions have been retained as well as overviews of the distribution of cranial nerves. Tables summarizing the attachments of muscles, their nerve supply, and their function have also been included.

Again I say to the medical student: The first day that you look at or place your hand on a patient, you require a basic knowledge of anatomy to interpret your observations. It is in the Anatomy Department that you learn the basic medical vocabulary that you will carry with you throughout your professional career and that will enable you to converse with your colleagues. Anatomy can be a boring subject; clinical anatomy is fascinating.

The writing of this book would not have been possible without the benefit of the works of anatomists and physicians too numerous to mention, and I gratefully acknowledge their assistance.

I thank the many medical students, clinical colleagues, and friends who have made valuable suggestions regarding the preparation of this new edition. I am greatly indebted to the late Dr. Alvin C. Wyman, Clinical Professor of Radiology at the George Washington University School of Medicine, for the loan of radiographs that have been reproduced in different sections of the book. I am also grateful to Dr. David O. Davis, who has supplied me with many examples of CT scans. In this connection I also thank the University's Audiosvisual Department for excellent photographic work.

As in the past, I wish to express sincere thanks to Mrs. Terry Dolan and Mrs. Virginia Childs for the earlier preparation of artwork and to Myra Feldman for the very fine new art in this edition.

To the librarians of the George Washington University School of Medicine thanks are due for their continued help in procuring much needed reference material. Special appreciation goes to Betty Hodge and Barbara Chambers for their skill and patience in typing the manuscript.

Finally, to the staff of Little, Brown and Company I express deep gratitude for their continued enthusiasm and unfailing assistance throughout the preparation of this book.

R. S. S.

# Clinical Anatomy for Medical Students

# 1. Introduction

## DESCRIPTIVE ANATOMICAL TERMS

### Terms Related to Position

Anatomy is the study of the structure of the body and the relationship of its constituent parts to each other. All descriptions of the human body are based on the assumption that the person is standing erect, with the upper limbs by the sides and the face and palms of the hands directed forward (Fig. 1-1). This is the so-called *anatomical position*. The various parts of the body are then described in relation to certain imaginary planes.

The *median sagittal plane* is a vertical plane passing through the center of the body, dividing it into equal right and left halves (Fig. 1-1). Planes situated to one or the other side of the median plane and parallel to it are termed *paramedian*. A structure situated nearer to the median plane of the body than another is said to be *medial* to the other. Similarly, a structure that lies farther away from the median plane than another is said to be *lateral* to the other.

*Coronal planes* are imaginary vertical planes at right angles to the median plane (Fig. 1-1). *Horizontal* or *transverse* planes are at right angles to both the median and coronal planes (Fig. 1-1).

The terms *anterior* and *posterior* are used to indicate the front or back of the body, respectively (Fig. 1-1); so that to describe the relationship of two structures, one is said to be anterior or posterior to the other insofar as it is closer to the anterior or posterior body surface.

In describing the hand, the terms *palmar* and *dorsal* surfaces are used in place of anterior and posterior, and in describing the foot, the terms *plantar* and *dorsal surfaces* are used instead of lower and upper surfaces (Fig. 1-1). The terms *proximal* and *distal* describe the relative distances from the roots of the limbs; for example, the arm is proximal to the forearm and the hand is distal to the forearm.

The terms *superficial* and *deep* denote the relative distances of structures from the surface of the body, and the terms *superior* and *inferior* denote levels relatively high or low with reference to the upper and lower ends of the body.

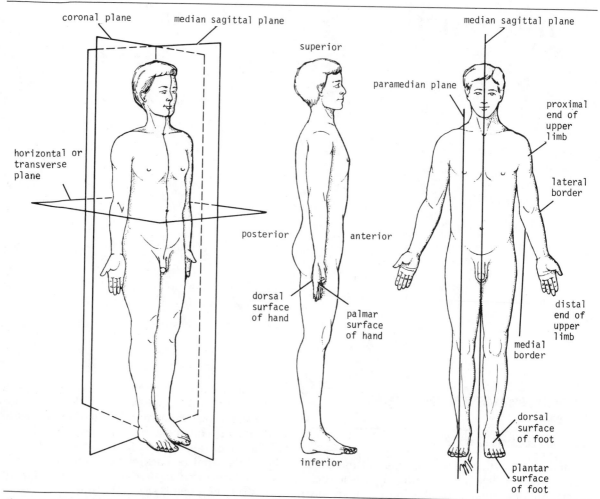

**Fig. 1-1.** Anatomical terms used in relation to position. Note that subjects are standing in anatomical position.

The terms *internal* and *external* are used to describe the relative distance of a structure from the center of an organ or cavity; for example, the *internal carotid artery* is found inside the cranial cavity and the *external carotid artery* is found outside the cranial cavity.

The term *ipsilateral* refers to the same side of the body; for example, the left hand and left foot are *ipsilateral*. *Contralateral* refers to opposite sides of the body; for example, the left *biceps brachii muscle* and the right *rectus femoris muscle* are contralateral.

The *supine* position of the body is lying on the back. The *prone* position is lying face downward.

## Terms Related to Movement

The site where two or more bones come together is known as a *joint*. Some joints have no movement (sutures of skull), some have only slight movement (superior tibiofibular joint), and some are freely movable (shoulder joint).

*Flexion* is a movement that takes place in a sagittal plane. For example, flexion of the elbow joint

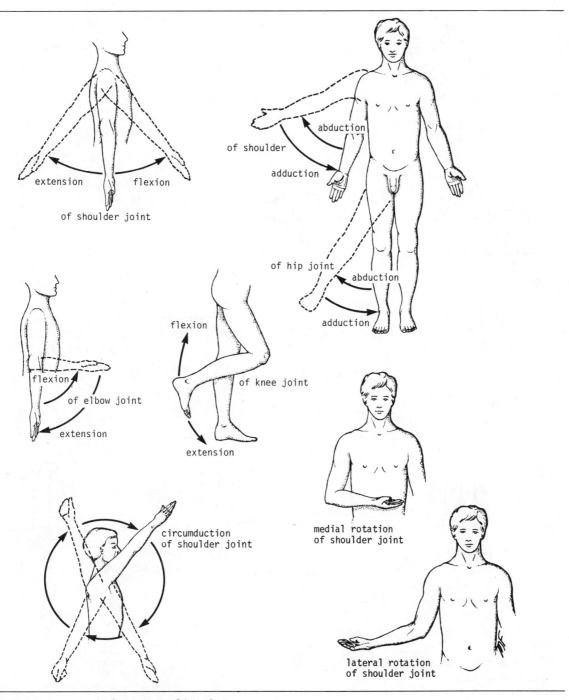

**Fig. 1-2. Some anatomical terms used in relation to movement. Note difference between flexion of elbow and knee.**

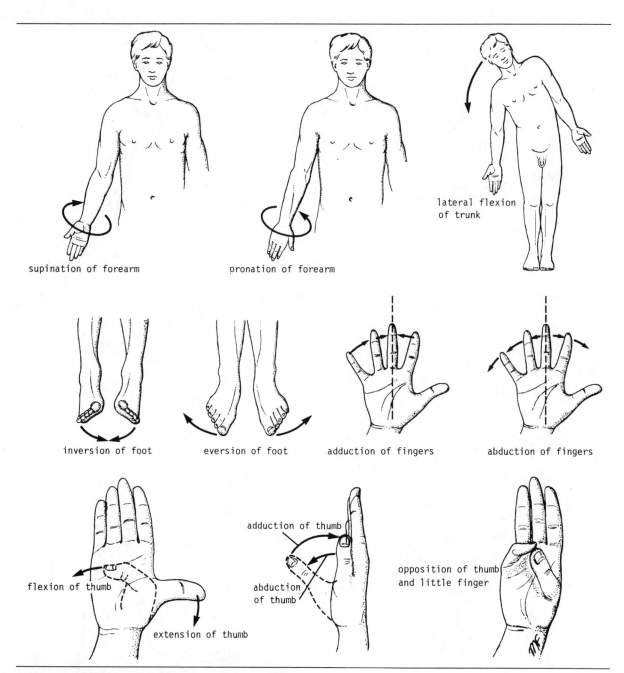

**Fig. 1-3. Additional anatomical terms used in relation to movement.**

approximates the anterior surface of the forearm to the anterior surface of the arm. It is usually an anterior movement, but it is occasionally posterior, as in the case of the knee joint (Fig. 1-2). *Extension* means straightening the joint and usually takes place in a posterior direction (Fig 1-2). *Lateral flexion* is a movement of the trunk in the coronal plane (Fig. 1-3).

*Abduction* of a limb is the movement away from the midline of the body in the coronal plane (Fig. 1-2). *Adduction* of a limb is the movement toward the body in the coronal plane (Fig. 1-2). In the fingers and toes, abduction is applied to the spreading of these structures, and adduction is applied to the drawing together of these structures (Fig. 1-3). The movements of the thumb (Fig. 1-3), which are a little more complicated, are described on page 514.

*Rotation* is the term applied to the movement of a part of the body around its long axis. *Medial rotation* is the movement that results in the anterior surface of the part facing medially; *lateral rotation* is the movement that results in the anterior surface of the part facing laterally. *Pronation of the forearm* is a medial rotation of the forearm in such a manner that the palm of the hand faces posteriorly (Fig. 1-3). *Supination of the forearm* is a lateral rotation of the forearm from the pronated position, so that the palm of the hand comes to face anteriorly (Fig. 1-3).

*Circumduction* is the combination in sequence of the movements of flexion, extension, abduction, and adduction (Fig. 1-2).

*Protraction* is to move forward, *retraction* is to move backward (used to describe the forward and backward movement of the jaw at the temporomandibular joints).

*Inversion* is the movement of the foot so that the sole faces in a medial direction (Fig. 1-3). *Eversion* is the opposite movement of the foot so that the sole faces in a lateral direction (Fig. 1-3).

## SOME BASIC ANATOMICAL STRUCTURES

### Skin

The skin is divided into two distinct parts, the superficial part, the *epidermis*, and the deep part, the *dermis* (Fig. 1-4). The epidermis is a stratified epithelium whose cells become flattened as they mature and rise to the surface. On the palms of the hands and the soles of the feet, the epidermis is extremely thick, to withstand the wear and tear that occurs in these regions. In other areas of the body, for example on the anterior surface of the arm and forearm, it is thin. The dermis is composed of dense connective tissue containing many blood vessels, lymphatic vessels, and nerves. It shows considerable variation in thickness in different parts of the body, tending to be thinner on the anterior than the posterior surface. It is thinner in women than in men. The dermis of the skin is connected to the underlying deep fascia or bones by the *superficial fascia*, otherwise known as *subcutaneous tissue.*

In the dermis the bundles of collagen fibers are mostly arranged in parallel rows. A surgical incision through the skin made along or between these rows causes the minimum of disruption of the collagen, and the wound heals with the minimum of scar tissue. On the other hand, an incision made across the rows of collagen disrupts and disturbs it, resulting in the massive production of fresh collagen and the formation of a broad ugly scar. The direction of the rows of collagen is known as the *lines of cleavage* (Langer's lines), and they tend to run longitudinally in the limbs and circumferentially in the neck and trunk (Fig. 1-5).

The skin over joints always folds in the same place, the *skin creases* (Fig. 1-6). At these sites the skin is thinner than elsewhere and is firmly tethered to underlying structures by strong bands of fibrous tissue.

The appendages of the skin are the *nails, hair follicles, sebaceous glands,* and *sweat glands.*

The *nails* are keratinized plates on the dorsal surfaces of the tips of the fingers and toes. The proximal edge of the plate is the *root of the nail* (Fig. 1-6). With the exception of the distal edge of the plate, the nail is surrounded and overlapped by folds of skin known as the *nail folds*. The surface of skin covered by the nail is the *nail bed* (Fig. 1-6).

*Hairs* grow out of *follicles*, which are invaginations of the epidermis into the dermis (Fig. 1-4). The follicles lie obliquely to the skin surface, and their expanded extremities, called the *hair bulbs,* penetrate to the deeper part of the dermis. Each hair bulb is concave at its end, and the concavity is occupied by vascular connective tissue, the *hair pa*

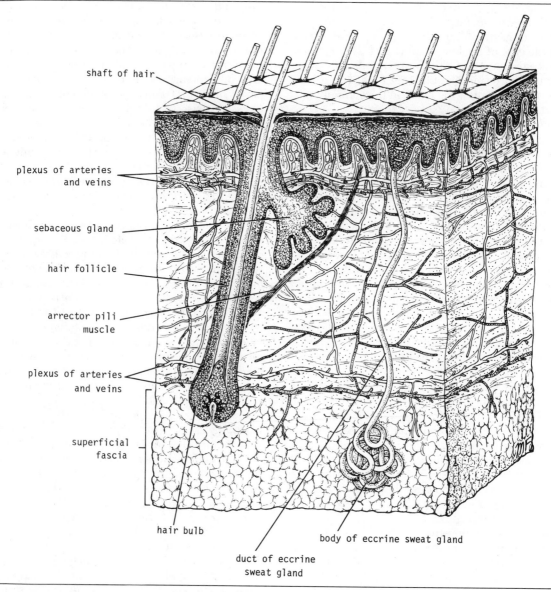

shaft of hair

plexus of arteries
and veins

sebaceous gland

hair follicle

arrector pili
muscle

plexus of arteries
and veins

superficial
fascia

hair bulb

duct of eccrine
sweat gland

body of eccrine sweat gland

**Fig. 1-4. General structure of skin and its relationship to the superficial fascia. Note that hair follicles extend down into the deeper part of the dermis or even into the superficial fascia, while sweat glands extend deeply into the superficial fascia.**

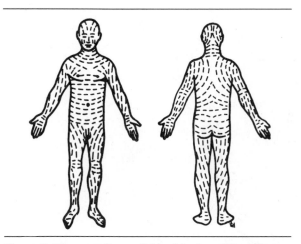

**Fig. 1-5. Cleavage lines of the skin (modified from Last).**

**Fig. 1-6. The various skin creases on palmar surface of hand and anterior surface of wrist joint. Relationship of nail to other structures of finger is also shown.**

*pilla.* A band of smooth muscle, the *arrector pili,* connects the undersurface of the follicle to the superficial part of the dermis (Fig. 1-4). The muscle is innervated by sympathetic nerve fibers, and its contraction causes the hair to move into a more vertical position; it also compresses the sebaceous gland and causes it to extrude some of its secretion. The pull of the muscle also causes dimpling of the skin surface, the so-called *gooseflesh.* Hairs are distributed in various numbers over the whole surface of the body except the lips, the palms of the hands, the sides of the fingers, the glans penis and clitoris, the labia minora and the internal surface of the labia majora, and the soles and sides of the feet and the sides of the toes.

*Sebaceous glands* pour their secretion, the sebum, onto the shafts of the hairs as they pass up through the necks of the follicles. They are situated on the sloping undersurface of the follicles and lie within the dermis (Fig. 1-4). *Sebum* is an oily material that helps to preserve the flexibility of the emerging hair. It also oils the surface epidermis around the mouth of the follicle.

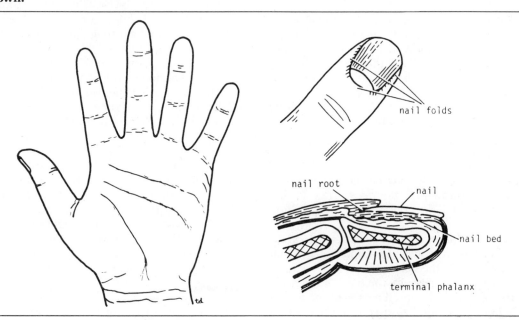

nail folds

nail root

nail

nail bed

terminal phalanx

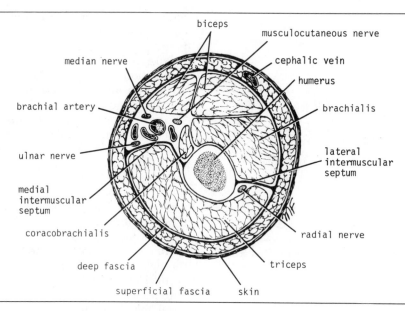

biceps

musculocutaneous nerve

median nerve

cephalic vein

humerus

brachial artery

brachialis

ulnar nerve

lateral intermuscular septum

medial intermuscular septum

coracobrachialis

radial nerve

deep fascia

triceps

superficial fascia

skin

**Fig. 1-7. Section through distal third of right arm, to show arrangement of superficial and deep fascia. Note how fibrous septa extend between groups of muscles dividing arm up into fascial compartments.**

*Sweat glands* are long, spiral, tubular glands distributed over the surface of the body except the red margins of the lips, the nail beds, and the glans penis and clitoris (Fig. 1-4). They extend through the full thickness of the dermis and their extremities may lie in the superficial fascia. The sweat glands are therefore the most deeply penetrating structures of all the epidermal appendages.

## Fasciae

The fasciae of the body may be divided into two types, the *superficial* and the *deep.* They lie between the skin and the underlying muscles and bones.

The *superficial fascia*, or subcutaneous tissue, is a mixture of loose areolar and adipose tissue that unites the dermis of the skin to the underlying deep fascia (Fig. 1-7). In the scalp, the back of the neck, the palms of the hands, and the soles of the feet, it contains numerous bundles of collagen fibers that hold the skin firmly to the deeper structures. In the

eyelids, auricle of the ear, penis and scrotum, and clitoris, it is devoid of adipose tissue.

The *deep fascia* is a membranous layer of connective tissue that invests the muscles and other deep structures (Fig. 1-7). In the neck it forms well-defined layers, which may play an important role in determining the path taken by pathogenic organisms during the spread of infection. In the thorax and abdomen it is merely a thin film of areolar tissue covering the muscles and aponeuroses. In the limbs it forms a definite sheath around the muscles and other structures, holding them in place. Fibrous septa extend from the deep surface of the membrane, between the groups of muscles, and in many places divide up the interior of the limbs into compartments (Fig. 1-7). In the region of joints the deep fascia may be considerably thickened to form restraining bands called *retinacula* (Fig. 1-8). Their function is to hold underlying tendons in position or to serve as pulleys around which the tendons may move.

## Muscle

There are three types of muscle: *skeletal, smooth,* and *cardiac.*

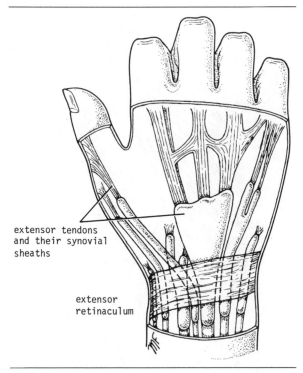

**Fig. 1-8. Extensor retinaculum on posterior surface of wrist holding underlying tendons of extensor muscles in position.**

extensor tendons
and their synovial
sheaths

extensor
retinaculum

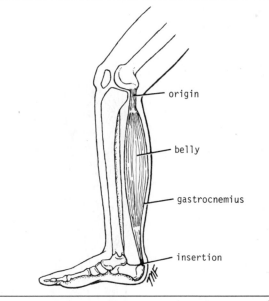

origin

belly

gastrocnemius

insertion

**Fig. 1-9. Origin, insertion, and belly of gastrocnemius muscle.**

## SKELETAL MUSCLE

Skeletal muscles are the muscles that produce the movements of the skeleton; they are sometimes called *voluntary muscles* and are made up of striped muscle fibers. A skeletal muscle has two or more attachments. The attachment that moves the least is referred to as the *origin*, and that which moves the most, as the *insertion* (Fig. 1-9). Under varying circumstances the degree of mobility of the attachments may be reversed, and therefore the terms *origin* and *insertion* are interchangeable.

The fleshy part of the muscle is referred to as its *belly* (Fig. 1-9). The ends of a muscle are attached to bones, cartilage, or ligaments by cords of fibrous tissue called *tendons* (Fig. 1-10). Occasionally, flattened muscles are attached by a thin but strong sheet of fibrous tissue called an *aponeurosis* (Fig. 1-10). A *raphe* is an interdigitation of the tendinous ends of fibers of flat muscles (Fig. 1-10).

### Internal Structure of Skeletal Muscle

The muscle fibers are bound together with delicate areolar tissue, which is condensed on the surface to form a fibrous envelope, the *epimysium*. The individual fibers of a muscle are arranged either parallel or oblique to the long axis of the muscle (Fig. 1-11). Since a muscle shortens by one-third to one-half its resting length when it contracts, then it follows that muscles whose fibers run parallel to the line of pull will bring about a greater degree of movement as compared with those whose fibers run obliquely. Examples of muscles with parallel arranged fibers (Fig. 1-11) are the *sternocleidomastoid*, the *rectus abdominis*, and the *sartorius*.

Muscles whose fibers run obliquely to the line of pull are referred to as *pennate muscles* (they resemble a feather) (Fig. 1-11). A *unipennate muscle* is one in which the tendon lies along one side of the

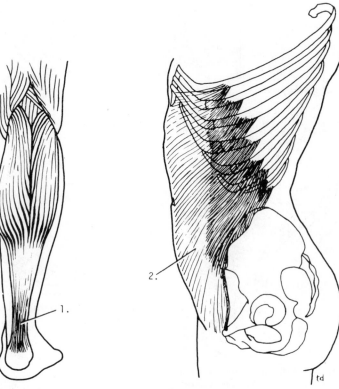

Common tendon for the insertion
of the gastrocnemius and
soleus muscles

External oblique aponeurosis

Raphe of mylohyoid muscles

**Fig. 1-10. Examples of (1) a tendon, (2) an aponeurosis, and (3) a raphe.**

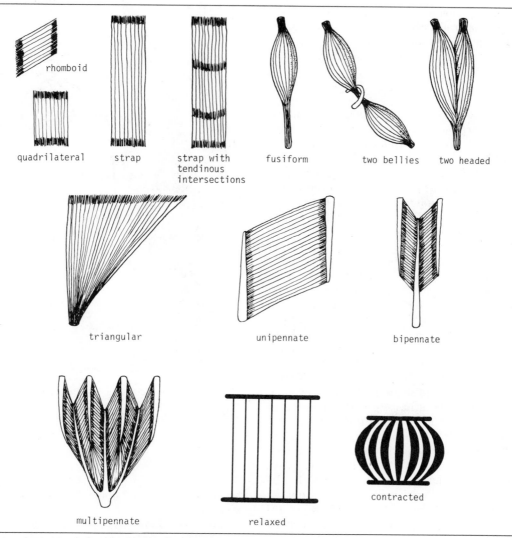

**Fig. 1-11. The different forms of internal structure of skeletal muscle. A relaxed and a contracted muscle are also shown; note how the muscle fibers, on contraction, shorten by one-third to one-half of their resting length. Note also how the muscle swells.**

muscle and the muscle fibers pass obliquely to it (e.g., *extensor digitorum longus*). A *bipennate muscle* is one in which the tendon lies in the center of the muscle and the muscle fibers pass to it from two sides (e.g., *rectus femoris*). A *multipennate* muscle (1) may be arranged as a series of bipennate muscles lying alongside one another (e.g., acromial fibers of the *deltoid*) or (2) may have the tendon lying within its center and the muscle fibers passing to it from all sides, converging as they go (e.g., *tibialis anterior*).

For a given volume of muscle substance, pennate

muscles have many more fibers as compared with muscles with parallel arranged fibers, and they are therefore more powerful; in other words, range of movement has been sacrificed to strength.

## Muscle Tone and Muscle Action

A *motor unit* consists of a motor neuron in the anterior gray horn or column of the spinal cord and all the muscle fibers it supplies (Fig. 1-12). In a large buttock muscle, such as the *gluteus maximus*, where fine control is unnecessary, a given motor neuron may supply as many as 200 muscle fibers. In contrast, in the small muscles of the hand or the extrinsic muscles of the eyeball, where fine control is required, one nerve fiber supplies only a few muscle fibers.

While resting, every skeletal muscle is in a partial state of contraction. This condition is referred to as *muscle tone*. Since muscle fibers are either fully contracted or relaxed, there being no intermediate stage, it follows that a few muscle fibers within a muscle are fully contracted all the time. To bring about this state and to avoid fatigue, different groups of motor units, and thus different groups of muscle fibers, are brought into action at different times. This is accomplished by the asynchronous discharge of nervous impulses in the motor neurons in the anterior gray horn of the spinal cord.

Basically, muscle tone is dependent on the integrity of a simple monosynaptic reflex arc composed of two neurons in the nervous system (Fig. 1-13). The degree of tension in a muscle is detected by sensitive sensory endings called *muscle spindles* and *tendon spindles* (Fig. 1-13). The nervous impulses travel in the afferent neurons that enter the spinal cord. There, they synapse with the motor neurons situated in the anterior gray horn, which, in turn, send impulses down their axons to the muscle fibers (Fig. 1-13). Should the afferent or efferent pathways of this simple reflex arc be cut, the muscle would immediately lose its tone and become flaccid. A flaccid muscle on palpation feels like a mass of dough and has completely lost its resilience. It quickly atrophies and becomes reduced in volume. It is important to realize that the degree of activity of the motor anterior horn cells, and therefore the degree of muscle tone, depends on the summation of the nerve impulses received

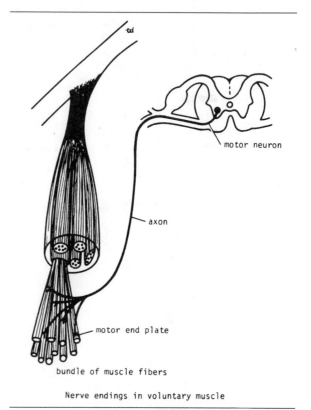

**Fig. 1-12. Components of a motor unit.**

by these cells from other neurons of the nervous system.

Muscle movement is accomplished by bringing into action increasing numbers of motor units and at the same time reducing the activity of the motor units of muscles that will oppose or antagonize the movement. When the maximum effort is required, all the motor units of a muscle are thrown into action.

It is important to understand that all movements are the result of the coordinated action of many muscles. However, to understand a muscle's action it is necessary to study it individually.

A muscle may work in the following ways: as (1) a prime mover, (2) an antagonist, (3) a fixator, and (4) a synergist.

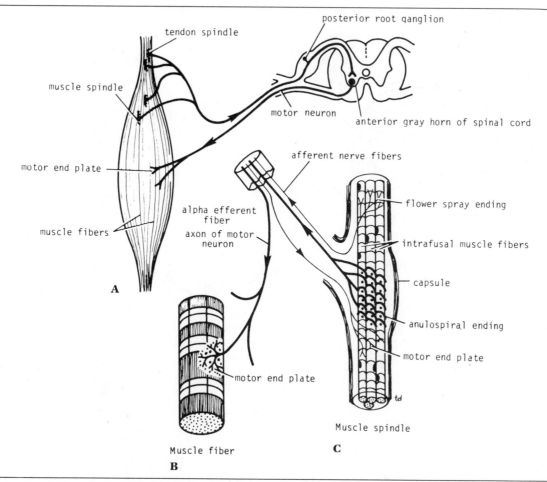

Fig. 1-13. (A) Simple reflex arc consisting of afferent neuron arising from muscle spindles and tendon spindles and efferent neuron whose cell body lies in anterior gray horn of spinal cord. (B) Axon from motor neuron ending on muscle fiber at motor end plate. (C) Structure of muscle spindle.

## Prime Mover

A muscle is a prime mover when it is the chief muscle or member of a chief group of muscles responsible for a particular movement. For example, the quadriceps femoris is a prime mover in the movement of extending the knee joint (Fig. 1-14).

## Antagonist

Any muscle that opposes the action of the prime mover is an antagonist. For example, the *biceps femoris* opposes the action of the quadriceps femoris when the knee joint is extended (Fig. 1-14). Before a prime mover can contract, there must be

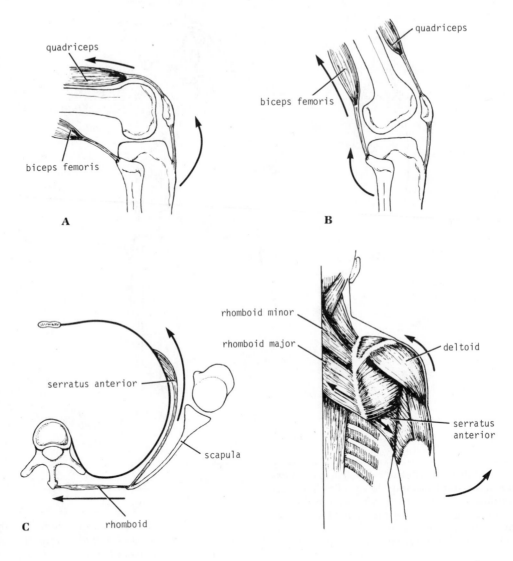

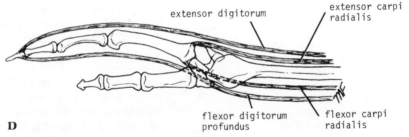

equal relaxation of the antagonist muscle; this is brought about by nervous reflex inhibition.

## Fixator

This is a muscle that contracts isometrically to stabilize the origin of the prime mover so that it may act efficiently. For example, the muscles attaching the shoulder girdle to the trunk contract as fixators to allow the *deltoid* to act on the shoulder joint (Fig. 1-14).

## Synergist

There are many examples in the body where the prime-mover muscle crosses a number of joints before it reaches the joint at which its main action takes place. To prevent unwanted movements in an intermediate joint, groups of muscles called synergists contract and stabilize the intermediate joints. For example, the flexor and extensor muscles of the carpus contract to fix the wrist joint, and this allows the long flexor and extensor muscles of the fingers to work efficiently (Fig. 1-14).

It should be understood that these are terms applied to the action of a particular muscle during a particular movement; many muscles can act as a prime mover, an antagonist, a fixator, or a synergist, depending on the movement to be accomplished.

## *Nerve Supply of Skeletal Muscles*

The nerve trunk to a muscle is a mixed nerve, about 60 percent being motor and 40 percent, sensory, and it also contains some sympathetic autonomic fibers. The nerve enters the muscle at about the midpoint on its deep surface, often near the margin; the place of entrance is known as the *motor point*. This arrangement allows the muscle to move with the minimum interference with the nerve trunk.

The *motor fibers* are of two types: the larger *alpha fibers* derived from large cells in the anterior gray horn, and the smaller *gamma fibers* derived from smaller cells in the spinal cord. Each fiber is myelinated and ends by dividing into many branches, each of which ends on a muscle fiber at the *motor end plate* (Fig. 1-13). Each muscle fiber has at least one motor end plate; the longer fibers possess more.

The *sensory fibers* are myelinated and arise from specialized sensory endings lying within the muscle or tendons called *muscle spindles* or *tendon spindles*, respectively. These endings are stimulated by tension in the muscle, which may occur during active contraction or by passive stretching. The function of these sensory fibers is to convey to the central nervous system information regarding the degree of tension of the muscles. This is essential for the maintenance of muscle tone and body posture and for carrying out coordinated voluntary movements.

The *sympathetic fibers* are nonmyelinated and pass to the smooth muscle in the walls of the blood vessels supplying the muscle. Their function is to regulate the blood flow to the muscles.

## SMOOTH MUSCLE

Smooth muscle consists of long, spindle-shaped cells closely arranged in bundles or sheets. In the tubes of the body it provides the motive power for propelling the contents through the lumen. In the digestive system it also causes the ingested food to be thoroughly mixed with the digestive juices. A wave of contraction of the circularly arranged fibers passes along the tube, milking the contents onward. By their contraction the longitudinal fibers pull the wall of the tube proximally over the contents. This method of propulsion is referred to as *peristalsis*.

In storage organs such as the urinary bladder or the uterus, the fibers are irregularly arranged and interlaced with one another. Their contraction is slow and sustained and brings about the expulsion of the contents of the organs. In the walls of the

**Fig. 1-14. The different types of muscle action. (A) Quadriceps femoris extending knee as a prime mover and biceps femoris acting as antagonist. (B) Biceps femoris flexing knee as a prime mover and quadriceps acting as antagonist. (C) Muscles around shoulder girdle fixing scapula so that movement of abduction can take place at shoulder joint. (D) Flexor and extensor muscles of carpus acting as synergists and stabilizing carpus so that long flexor and extensor tendons can flex and extend fingers.**

blood vessels the smooth muscle fibers are arranged circularly, and they serve to modify the caliber of the lumen.

Depending on the organ, smooth muscle fibers may be made to contract by local stretching of the fibers, by nerve impulses from autonomic nerves, or by hormonal stimulation.

## CARDIAC MUSCLE

Cardiac muscle consists of striated muscle fibers that branch and unite with each other. It is found in the myocardium of the heart. Its fibers tend to be arranged in whorls and spirals, and they have the property of spontaneous and rhythmical contraction. Specialized cardiac muscle fibers form the *conducting system of the heart.*

Cardiac muscle is supplied by autonomic nerve fibers that terminate in the nodes of the conducting system and in the myocardium.

# Joints

The site where two or more bones come together, whether or not there is movement between them, is called a *joint.* Joints are classified according to the tissues that lie between the bones: fibrous joints, cartilaginous joints, and synovial joints.

## FIBROUS JOINTS

The articulating surfaces of the bones are joined by fibrous tissue (Fig. 1-15), and thus very little movement is possible. The degree of movement depends on the length of the collagen fibers uniting the bones. The sutures of the vault of the skull and the inferior tibiofibular joints are examples of fibrous joints.

## CARTILAGINOUS JOINTS

Cartilaginous joints may be divided into two types, primary and secondary. A *primary cartilaginous joint* is one in which the bones are united by a plate or bar of hyaline cartilage. Thus, the union between the *epiphysis* and the *diaphysis* of a growing bone and that between the first rib and the manubrium sterni are examples of such a joint. No movement is possible.

A *secondary cartilaginous joint* is one in which

the bones are united by a plate of fibrocartilage, and the articular surfaces of the bones are covered by a thin layer of hyaline cartilage. Examples are the *intervertebral joints* (Fig. 1-15) and the *symphysis pubis.* The amount of movement possible is dependent on the physical qualities of the fibrocartilage. (See Pelvic Joints, p. 328.)

## SYNOVIAL JOINTS

The articular surfaces of the bones are covered by a thin layer of hyaline cartilage separated by a joint cavity (Fig. 1-15). This arrangement permits a great degree of freedom of movement. The cavity of the joint is lined by *synovial membrane,* which extends from the margins of one articular surface to those of the other. The synovial membrane is protected on the outside by a tough fibrous membrane referred to as the *capsule* of the joint. The articular surfaces are lubricated by a viscous fluid called *synovial fluid.* In certain synovial joints, for example, in the knee joint, discs or wedges of fibrocartilage are interposed between the articular surfaces of the bones. These are referred to as articular discs.

*Fatty pads* are found in some synovial joints lying between the synovial membrane and the fibrous capsule or bone. Examples are found in the hip (Fig. 1-15) and knee joints.

The degree of movement in a synovial joint is limited by the shape of the bones participating in the joint, the coming together of adjacent anatomical structures (for example, the thigh against the anterior abdominal wall on flexing the hip joint), and the presence of fibrous *ligaments* uniting the bones. Most ligaments lie outside the joint capsule, but in the knee some important ligaments, the *cruciate ligaments,* lie within the capsule (Fig. 1-17).

## TYPES OF SYNOVIAL JOINTS

Synovial joints may be classified according to the arrangement of the articular surfaces and the types of movement that are possible.

1. **PLANE JOINTS.** In these joints, the apposed articular surfaces are flat or almost flat, and this permits the bones to slide upon one another. Examples of plane joints are the sternoclavicular and acromioclavicular joints (Fig. 1-16).

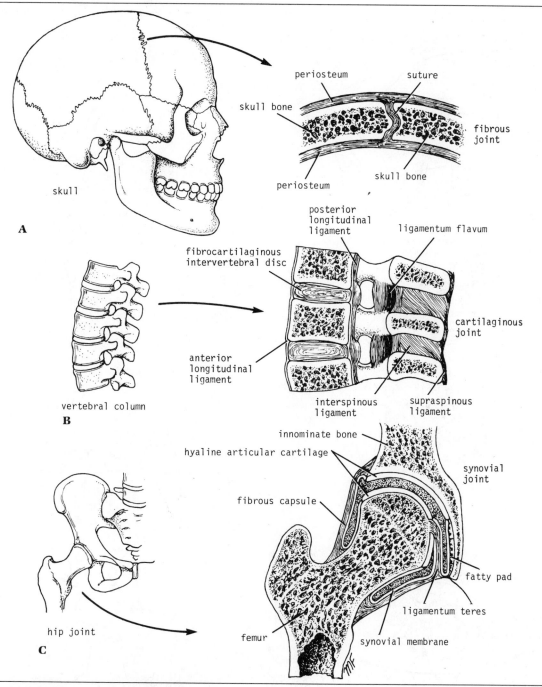

**Fig. 1-15. Examples of three types of joints: (A) fibrous joint (coronal suture of skull), (B) cartilaginous joint (joint between two lumbar vertebral bodies), and (C) synovial joint (hip joint).**

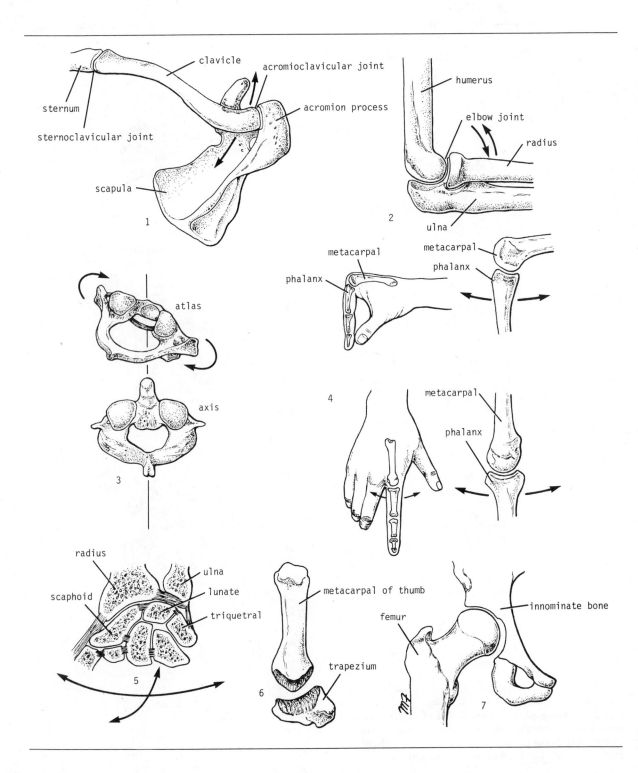

clavicle

acromioclavicular joint

sternum

acromion process

sternoclavicular joint

scapula

1

humerus

elbow joint

radius

ulna

2

atlas

axis

3

metacarpal

phalanx

metacarpal

phalanx

4

metacarpal

phalanx

radius

ulna

scaphoid

lunate

triquetral

5

metacarpal of thumb

trapezium

6

femur

innominate bone

7

2. **HINGE JOINTS.** These joints resemble the hinge on a door, so that flexion and extension movements are possible. Examples of hinge joints are the elbow, knee, and ankle joints (Fig. 1-16).

3. **PIVOT JOINTS.** In these joints, there is a central bony pivot surrounded by a bony-ligamentous ring (Fig. 1-16). In this type of joint, rotation is the only movement possible. The atlanto-axial and superior radioulnar joints are good examples.

4. **CONDYLOID JOINTS.** These joints have two distinct convex surfaces that articulate with two concave surfaces. The movements of flexion, extension, abduction, and adduction are possible together with a small amount of rotation. The metacarpophalangeal joints or knuckle joints are good examples (Fig. 1-16).

5. **ELLIPSOID JOINTS.** In these joints, there is an elliptical convex articular surface that fits into an elliptical concave articular surface. The movements of flexion, extension, abduction, and adduction can take place, but rotation is impossible. The wrist joint is a good example (Fig. 1-16).

6. **SADDLE JOINTS.** In these joints, the articular surfaces are reciprocally concavoconvex and resemble a saddle on a horse's back. These joints permit flexion, extension, abduction, adduction, and rotation. The best example of this type of joint is the carpometacarpal joint of the thumb (Fig. 1-16).

7. **BALL-AND-SOCKET JOINTS.** In these joints, a ball-shaped head of one bone fits into a socket-like concavity of another. This arrangement permits very free movements, including flexion, extension, abduction, adduction, medial rotation, lateral rotation, and circumduction. The shoulder and hip joints are good examples of this type of joint (Fig. 1-16).

**Fig. 1-16. Examples of the different types of synovial joints: (1) plane joints (sternoclavicular and acromioclavicular joints), (2) hinge joint (elbow joint), (3) pivot joint (atlanto-axial joint), (4) condyloid joint (metacarpophalangeal joint), (5) ellipsoid joint (wrist joint), (6) saddle joint (carpometacarpal joint of the thumb), and (7) ball-and-socket joint (hip joint).**

## JOINT STABILITY

The stability of a joint depends on three main factors: (1) the shape, size, and arrangement of the articular surfaces, (2) the ligaments, and (3) the tone of the muscles around the joint.

### Articular Surfaces

The ball-and-socket arrangement of the hip joint (Fig. 1-17) and the mortise arrangement of the ankle joint are good examples of how bone shape plays an important role in joint stability. There are other examples of joints, however, in which the shape of the bones contributes little or nothing to the stability; for example, the acromioclavicular joint, the calcaneocuboid joint, and the knee joint.

### Ligaments

*Fibrous ligaments* will prevent excessive movement in a joint (Fig. 1-17), but if the stress is continued excessively long, then fibrous ligaments stretch. For example, the ligaments of the joints between the bones forming the arches of the feet will not by themselves support the weight of the body. Should the tone of the muscles that normally support the arches become impaired by fatigue, then the ligaments will stretch and the arches will collapse, producing *flat feet.*

*Elastic ligaments*, on the other hand, return to their original length after stretching. The elastic ligaments of the auditory ossicles play an active part in supporting the joints and assisting in the return of the bones to their original position after movement.

### Muscle Tone

In most joints, muscle tone is the major factor controlling stability. For example, the muscle tone of the short muscles around the shoulder joint keeps the hemispherical head of the humerus in the shallow glenoid cavity. Without the action of these muscles, very little force would be required to dislocate this joint. The knee joint is very unstable without the tonic activity of the quadriceps femoris muscle. The joints between the small bones forming the arches of the feet are largely supported by

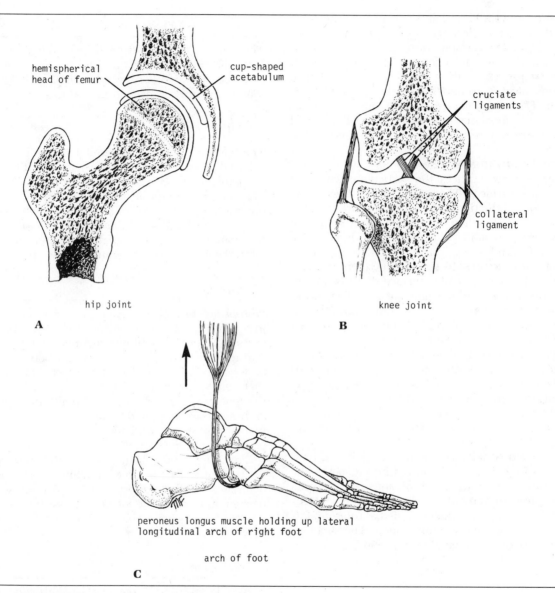

hemispherical
head of femur

cup-shaped
acetabulum

hip joint

**A**

cruciate
ligaments

collateral
ligament

knee joint

**B**

peroneus longus muscle holding up lateral
longitudinal arch of right foot

arch of foot

**C**

Fig. 1-17. The three main factors responsible for
stabilizing a joint. (A) Shape of articular surfaces,
(B) ligaments, and (C) muscle tone.

the tone of the muscles of the leg, whose tendons are inserted into the bones of the feet (Fig. 1-17).

## NERVE SUPPLY OF JOINTS

The capsule and ligaments receive an abundant sensory nerve supply. The blood vessels receive autonomic sympathetic fibers. The cartilage covering the articular surfaces possesses only a few nerve endings near its edges. Overstretching of the capsule and ligaments produces reflex contraction of muscles around the joint; excessive stretching produces pain. The stretch receptors in the capsule and ligaments are continually sending proprioceptive information up to the central nervous system, keeping it informed of the position of the joints. This supplements the information passing to the nervous system from the muscle and tendon spindles, helps to maintain postural tone, and coordinates voluntary movements.

The sympathetic fibers control the blood supply to the joint.

**Hilton's Law.** A nerve supplying a joint also supplies the muscles moving the joint and the skin over the insertions of these muscles.

## Ligaments

A ligament is a cord or band of connective tissue uniting two structures. Commonly found in association with joints, ligaments are of two types: The majority are composed of dense bundles of collagen fibers and are unstretchable under normal conditions (e.g., *iliofemoral ligament* of the hip joint and the *collateral ligaments* of the elbow joint. The second type is composed largely of elastic tissue and can therefore regain its original length after stretching (e.g., *ligamentum flavum* of the vertebral column and the *calcaneonavicular ligament* of the foot).

## Bursae

A bursa is a lubricating device consisting of a closed fibrous sac lined with a delicate smooth membrane. Its walls are separated by a film of viscous fluid. Bursae are found wherever tendons rub against bones, ligaments, or other tendons. They are commonly found close to joints where the skin rubs against underlying bony structures, e.g., the

*prepatellar bursa* (Fig. 1-18). Occasionally, the cavity of a bursa communicates with the cavity of a synovial joint. For example, the *suprapatellar bursa* communicates with the knee joint (Fig. 1-18), and the *subscapularis bursa* communicates with the shoulder joint.

## Synovial Sheath

A synovial sheath is a tubular bursa that surrounds a tendon. The tendon invaginates the bursa from one side so that the tendon becomes suspended within the bursa by a *mesotendon* (Fig. 1-18). The mesotendon enables blood vessels to enter the tendon along its course. In certain situations, where the range of movement is extensive, the mesotendon disappears or remains in the form of narrow threads, the *vincula* (e.g., the long flexor tendons of the fingers and toes).

## Blood Vessels

Blood vessels are of three types: arteries, veins, and capillaries (Fig. 1-19).

The *arteries* convey blood from the heart and distribute it to the various tissues of the body by means of their *branches* (Fig. 1-19 and 20). The smallest arteries, less than 0.1 mm in diameter, are referred to as *arterioles*. The union of branches of arteries is called an *anastomosis*. There are no valves in arteries.

*Anatomical end arteries* (Fig. 1-20) are vessels whose terminal branches do not anastomose with branches of arteries supplying adjacent areas. *Functional end arteries* are vessels whose terminal branches do anastomose with those of adjacent arteries, but the caliber of the anastomosis is insufficient to keep the tissue alive should one of the arteries become occluded.

The *veins* are vessels that convey blood back to the heart; many of them possess valves. The smallest veins are called *venules* (Fig. 1-20). The smaller veins, or *tributaries*, unite to form larger veins, which commonly join with one another to form *venous plexuses*. Medium-sized deep arteries are often accompanied by two veins, one on each side, called *venae comitantes*.

Veins leaving the gastrointestinal tract do not go directly to the heart, but converge on the *portal vein*; this enters the liver and breaks up again into

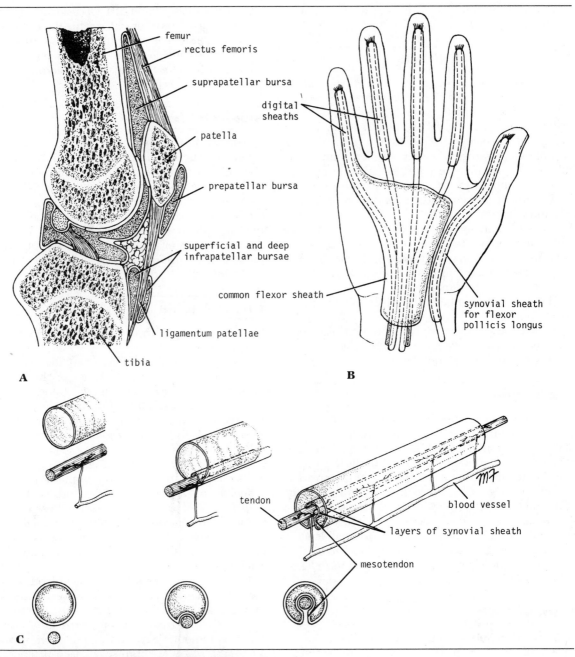

**Fig. 1-18. (A) Four bursae related to front of knee joint. Note that suprapatellar bursa communicates with cavity of joint. (B) Synovial sheaths around long tendons of fingers. (C) How tendon indents synovial sheath during development, and how blood vessels reach tendon through mesotendon.**

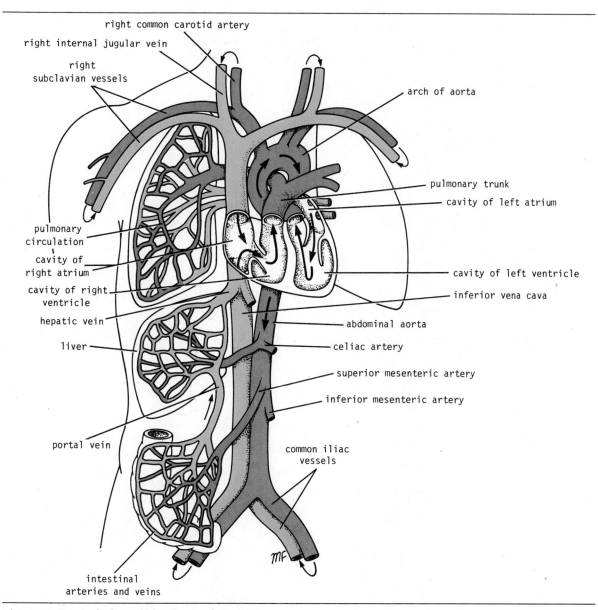

right common carotid artery

right internal jugular vein

right subclavian vessels

arch of aorta

pulmonary trunk

cavity of left atrium

pulmonary circulation

cavity of right atrium

cavity of right ventricle

hepatic vein

liver

cavity of left ventricle

inferior vena cava

abdominal aorta

celiac artery

superior mesenteric artery

inferior mesenteric artery

portal vein

common iliac vessels

intestinal arteries and veins

**Fig. 1-19. General plan of blood vascular system.**

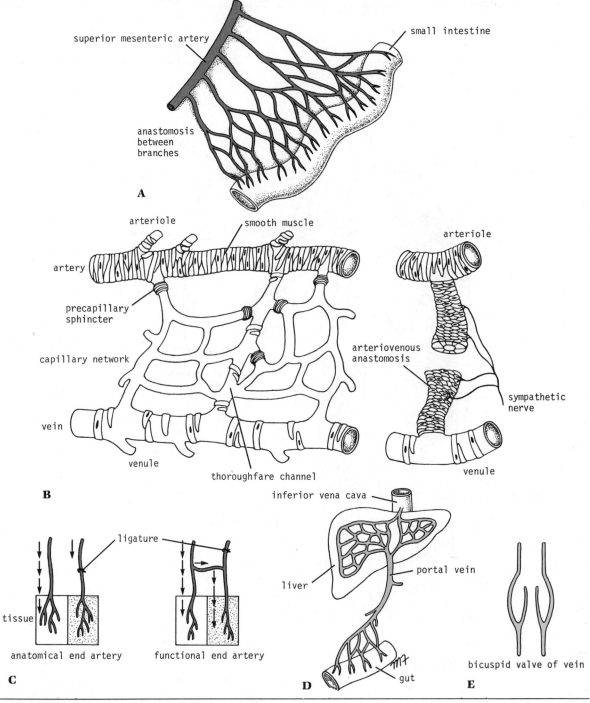

**Fig. 1-20. The different types of blood vessels and their methods of union. (A) Anastomosis between branches of the superior mesenteric artery. (B) A capillary network and an arteriovenous anastomosis. (C) Anatomical end artery and functional end artery. (D) A portal system. (E) Structure of bicuspid valve in a vein.**

veins of diminishing size, which ultimately join capillary-like vessels, termed *sinusoids*, in the liver (Fig. 1-20). A *portal system* is thus a system of vessels interposed between two capillary beds.

The *capillaries* are microscopic vessels in the form of a network connecting the arterioles to the venules (Fig. 1-20).

In some areas of the body, principally the tips of the fingers and toes, there are direct connections between the arteries and veins without the intervention of capillaries. The sites of such connections are referred to as *arteriovenous anastomoses* (Fig. 1-20).

## Lymphatic System

*Lymph* is the name given to tissue fluid once it has entered a lymph vessel. *Lymph capillaries* are the network of fine vessels draining lymph from the tissues. The lymph capillaries are in turn drained by *small lymph vessels*, which unite to form *large lymph vessels*. Lymph vessels have a beaded appearance due to the presence of numerous *valves* along their course.

Lymph ultimately drains into the bloodstream, but before it does so, it passes through at least one lymph node and often through several. The lymph vessels that carry lymph to a lymph node are referred to as *afferent* vessels (Fig. 1-21); those that transport it away from a node are *efferent* vessels. The lymph reaches the bloodstream at the root of the neck by large lymph vessels called the *right lymphatic duct* and the *thoracic duct* (Fig. 1-21).

## Nervous System

The nervous system is divided into two main parts, the *central nervous system*, consisting of the brain and spinal cord, and the *peripheral nervous system*, consisting of the cranial and spinal nerves and their associated ganglia.

The central nervous system is composed of large numbers of nerve cells and their processes, supported by specialized tissue called *neuroglia*. The *neuron* is the name given to the nerve cell and all its processes. The long processes of a nerve cell are called *axons*, or *nerve fibers* (Fig. 1-24).

The interior of the central nervous system is organized into gray and white matter. *Gray matter* consists of nerve cells and the proximal portions of

their processes embedded in neuroglia. *White matter* consists of nerve fibers embedded in neuroglia.

In the peripheral nervous system the cranial and spinal nerves are seen on dissection to be cords of grayish white color. They are made up of bundles of nerve fibers supported by delicate areolar tissue.

There are 12 pairs of *cranial nerves* that leave the brain and pass through foramina in the skull. There are 31 pairs of *spinal nerves* that leave the spinal cord and pass through intervertebral foramina in the vertebral column (Figs. 1-22 and 1-23). The spinal nerves are named according to the regions of the vertebral column with which they are associated: 8 *cervical*, 12 *thoracic*, 5 *lumbar*, 5 *sacral*, and 1 *coccygeal*. Note that there are 8 cervical nerves and only 7 cervical vertebrae and that there are 1 coccygeal nerve and 4 coccygeal vertebrae.

Each spinal nerve is connected to the spinal cord by two *roots*, the *anterior root* and the *posterior root* (Figs. 1-23 and 1-24). The anterior root consists of bundles of nerve fibers carrying nerve impulses away from the central nervous system (Fig. 1-24). Such nerve fibers are called *efferent* fibers. Those efferent fibers that go to skeletal muscle and cause them to contract are called *motor fibers*. Their cells of origin lie in the anterior gray horn of the spinal cord.

The posterior root consists of bundles of nerve fibers that carry impulses to the central nervous system and are called *afferent* fibers (Fig. 1-24). Since these fibers are concerned with conveying information about sensations of touch, pain, temperature, and vibrations, they are called *sensory fibers*. The cell bodies of these nerve fibers are situated in a swelling on the posterior root called the *posterior root ganglion* (Figs. 1-23 and 1-24).

At each intervertebral foramen the anterior and posterior roots unite to form a spinal nerve (Fig. 1-23). Here, the motor and sensory fibers become mixed together, so that a spinal nerve is made up of a mixture of motor and sensory fibers (Fig. 1-24). On emerging from the foramen, the spinal nerve divides into a large *anterior ramus* and a smaller *posterior ramus*. The posterior ramus passes posteriorly around the vertebral column to supply the muscles and skin of the back (Figs. 1-23 and 1-24). The *anterior ramus* continues anteriorly to supply the muscles and skin over the anterolateral body wall and all the muscles and skin of the limbs.

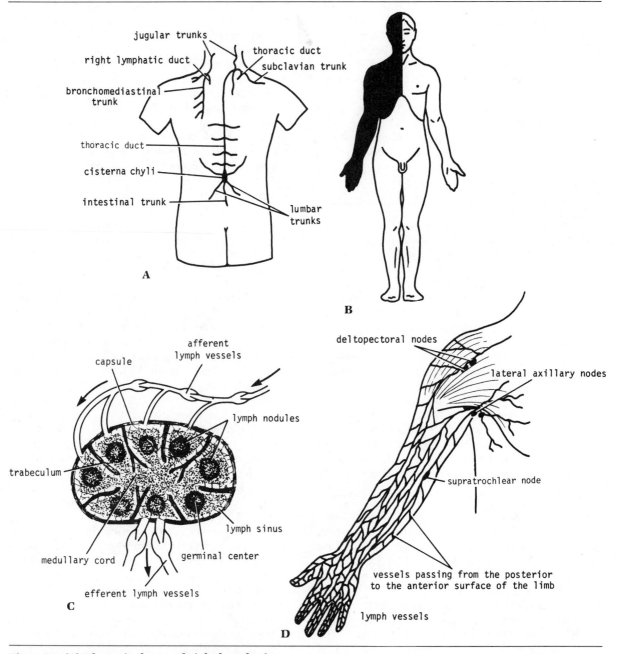

**Fig. 1-21. (A)** Thoracic duct and right lymphatic duct and their main tributaries. **(B)** The areas of body drained into thoracic duct (clear) and right lymphatic duct (black). **(C)** General structure of a lymph node. **(D)** Lymph vessels and nodes of upper limb.

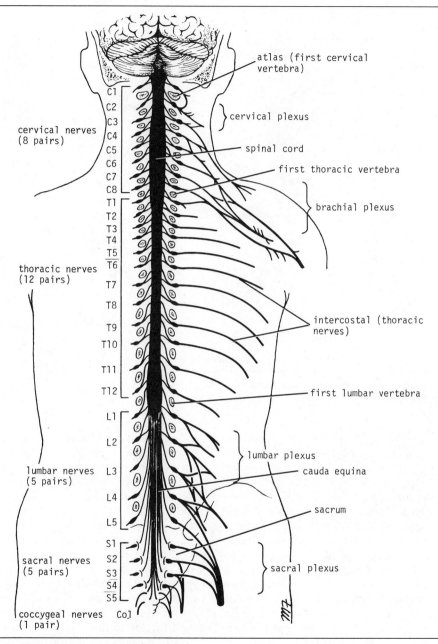

atlas (first cervical
vertebra)

cervical plexus

spinal cord

first thoracic vertebra

brachial plexus

intercostal (thoracic
nerves)

first lumbar vertebra

lumbar plexus

cauda equina

sacrum

sacral plexus

cervical nerves
(8 pairs)

C1
C2
C3
C4
C5
C6
C7
C8

thoracic nerves
(12 pairs)

T1
T2
T3
T4
T5
T6
T7
T8
T9
T10
T11
T12

lumbar nerves
(5 pairs)

L1
L2
L3
L4
L5

sacral nerves
(5 pairs)

S1
S2
S3
S4
S5

coccygeal nerves
(1 pair)

Co1

**Fig. 1-22. Brain, spinal cord, spinal nerves, and
plexuses of limbs.**

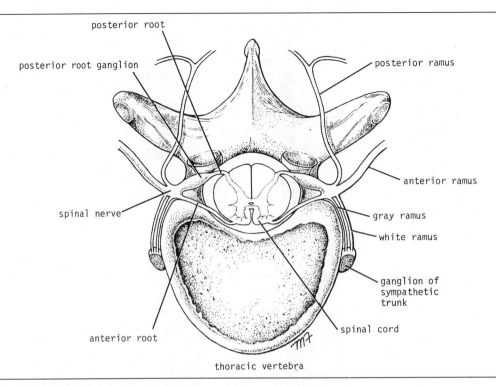

posterior root

posterior root ganglion

posterior ramus

anterior ramus

gray ramus

white ramus

spinal nerve

ganglion of
sympathetic
trunk

anterior root

spinal cord

thoracic vertebra

**Fig. 1-23. The association between spinal cord,
spinal nerves, and sympathetic trunks.**

At the root of the limbs the anterior rami join
one another to form complicated *nerve plexuses*
(Fig. 1-22). At the root of the arms are the *cervical*
and *brachial plexuses*, and at the root of the legs,
the *lumbar* and *sacral plexuses*.

It is important to realize that the classic division
of the nervous system into central and peripheral
parts is purely artificial and one of descriptive convenience, since the processes of the neurons pass
freely between the two. For example, a motor neuron located in the anterior gray horn of the first
thoracic segment of the spinal cord gives rise to an
axon that passes through the anterior root of the
first thoracic nerve (Fig. 1-25), through the brachial plexus, travels down the arm and forearm in
the ulnar nerve, and finally reaches the motor end
plates on several muscle fibers of a small muscle of
the hand—a total distance of about 90 cm (3 feet).

To take another example: Consider the sensation
of touch felt on the lateral side of the little toe. This
area of skin is supplied by the first sacral segment

of the spinal cord (S1). The fine terminal branches
of the sensory axon, called *dendrites*, leave the sensory organs of the skin and unite to form the axon
of the sensory nerve. The axon passes up the leg in
the sural nerve (Fig. 1-25) and then in the tibial
and sciatic nerves to the lumbosacral plexus. It
then passes through the posterior root of the first
sacral nerve to reach the cell body in the posterior
root ganglion of the first sacral nerve. The central
axon now enters the posterior white column of the
spinal cord and passes up to the *nucleus gracilis* in
the *medulla oblongata*—a total distance of about
1½ m (5 feet). Thus, a single neuron extends from
the little toe to the inside of the skull.

Both these examples illustrate the extreme length
of a single neuron.

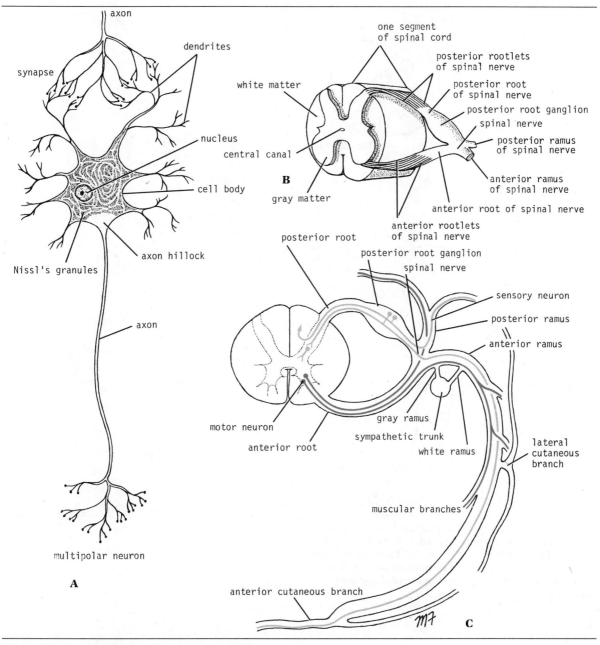

**Fig. 1-24. (A) Multipolar motor neuron with connector neuron synapsing with it. (B) Section through thoracic segment of spinal cord with spinal roots and posterior root ganglion. (C) Cross section of thoracic segment of spinal cord, showing roots, spinal nerve, and anterior and posterior rami and their branches.**

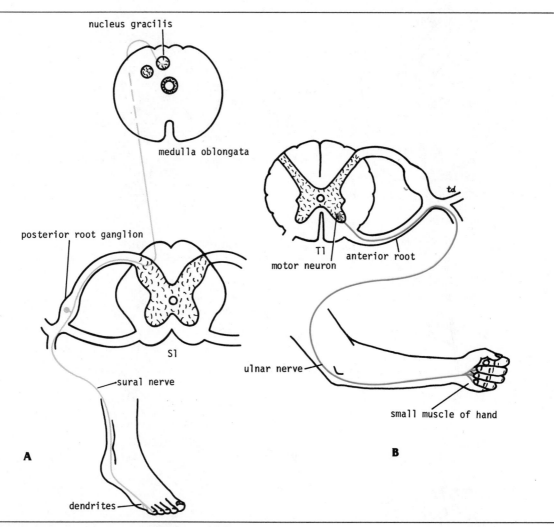

**Fig. 1-25. Two neurons that pass from central to peripheral nervous system. (A) Afferent neuron that extends from little toe to brain. (B) Efferent neuron that extends from anterior gray horn of first thoracic segment of spinal cord to small muscle of hand.**

## Autonomic Nervous System

The autonomic nervous system is the part of the nervous system concerned with the innervation of involuntary structures such as the heart, smooth muscle, and glands throughout the body. It is distributed throughout the central and peripheral nervous system. The autonomic system may be divided into two parts, the *sympathetic* and the *parasympathetic*, and in both parts there are afferent and efferent nerve fibers.

The activities of the sympathetic part of the autonomic system prepare the body for an emergency. It accelerates the heart rate, causes constriction of the peripheral blood vessels, and raises the blood pressure. The sympathetic part of the autonomic system brings about a redistribution of the blood so that it leaves the areas of the skin and intestine and becomes available to the brain, heart, and skeletal muscle. At the same time it inhibits

**Fig. 1-26. General arrangement of somatic part of nervous system (on left) compared with autonomic part of nervous system (on right).**

peristalsis of the intestinal tract and closes the sphincters.

The activities of the parasympathetic part of the autonomic system aim at conserving and restoring energy. They slow the heart rate, increase peristalsis of the intestine and glandular activity, and open the sphincters.

## THE SYMPATHETIC PART OF THE AUTONOMIC SYSTEM

### Efferent Nerve Fibers

The gray matter of the spinal cord, from the first thoracic segment to the second lumbar segment, possesses a lateral horn, or column, in which are located the cell bodies of the sympathetic connector neurons (Fig. 1-26). The myelinated axons of these cells leave the spinal cord in the anterior nerve roots and then pass via the *white rami communicantes* to the *paravertebral ganglia* of the *sympathetic trunk* (Figs. 1-23, 1-26, and 1-27). The connector cell fibers are called preganglionic as

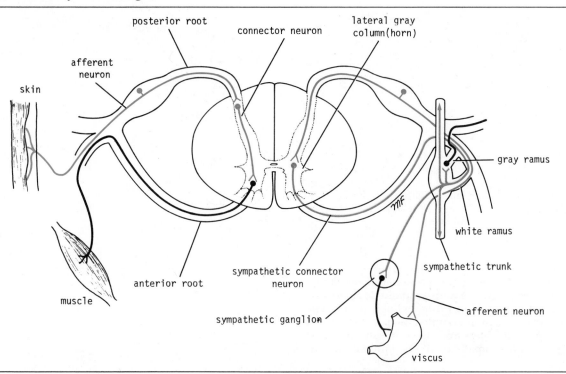

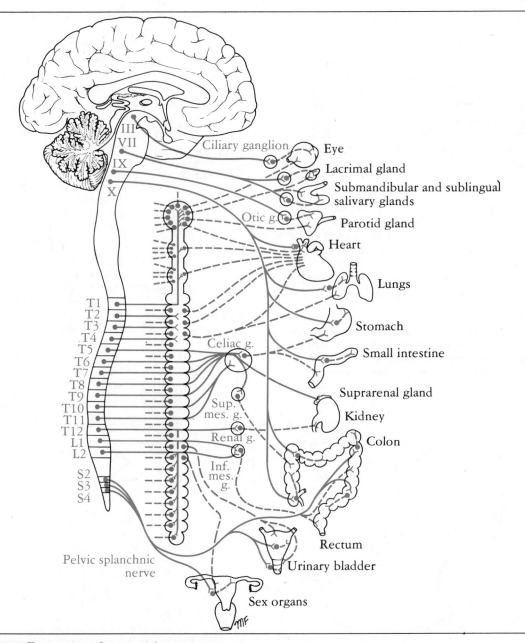

**Fig. 1-27. Efferent part of autonomic nervous system. Preganglionic parasympathetic fibers are shown in solid blue, postganglionic parasympathetic fibers, in interrupted blue. Preganglionic sympathetic fibers are shown in solid red, postganglionic sympathetic fibers, in interrupted red.**

they pass to a peripheral ganglion. Once the preganglionic fibers reach the ganglia in the sympathetic trunk, they may pass to the following destinations:

1. They may terminate in the ganglion they have entered by synapsing with an excitor cell in the ganglion (Fig. 1-26). A *synapse* may be defined as the site where two neurons come into close proximity but not into anatomical continuity. The gap between the two neurons is bridged by a neurotransmitter substance, *acetylcholine*. The axons of the excitor neurons leave the ganglion and are nonmyelinated. These postganglionic nerve fibers now pass to the thoracic spinal nerves as *gray rami communicantes* and are distributed in the branches of the spinal nerves to supply the smooth muscle in the walls of blood vessels, the sweat glands, and the arrector pili muscles of the skin.

2. Those fibers entering the ganglia of the sympathetic trunk high up in the thorax may travel up in the sympathetic trunk to the ganglia in the cervical region, where they synapse with excitor cells (Figs. 1-26 and 1-27). Here again, the postganglionic nerve fibers leave the sympathetic trunk as gray rami communicantes, and most of them join the cervical spinal nerves. Many of the preganglionic fibers entering the lower part of the sympathetic trunk from the lower thoracic and upper two lumbar segments of the spinal cord travel down to ganglia in the lower lumbar and sacral regions, where they synapse with excitor cells (Fig. 1-27). The postganglionic fibers leave the sympathetic trunk as gray rami communicantes that join the lumbar, sacral, and coccygeal spinal nerves.

3. The preganglionic fibers may pass through the ganglia on the thoracic part of the sympathetic trunk without synapsing. These myelinated fibers form the *splanchnic nerves* (Fig. 1-27), of which there are three. The *greater splanchnic nerve* arises from the fifth to the ninth thoracic ganglia, pierces the diaphragm, and synapses with excitor cells in the ganglia of the celiac plexus. The *lesser splanchnic nerve* arises from the tenth and eleventh ganglia, pierces the diaphragm, and synapses with excitor cells in the ganglia of the lower part of the celiac plexus. The *lowest splanchnic nerve* (when present)

arises from the twelfth thoracic ganglion, pierces the diaphragm, and synapses with excitor cells in the ganglia of the renal plexus. The splanchnic nerves are therefore composed of preganglionic fibers. The postganglionic fibers arise from the excitor cells in the peripheral plexuses previously noted and are distributed to the smooth muscle and glands of the viscera. A few preganglionic fibers traveling in the greater splanchnic nerve end directly on the cells of the suprarenal medulla. These medullary cells may be regarded as modified sympathetic excitor cells.

## Afferent Nerve Fibers

The afferent myelinated nerve fibers travel from the viscera through the sympathetic ganglia without synapsing (Fig. 1-26). They enter the spinal nerve via the white rami communicantes and reach their cell bodies in the posterior root ganglion of the corresponding spinal nerve. The central axons then enter the spinal cord and may form the afferent component of a local reflex arc. Others may pass up to higher autonomic centers in the brain.

## THE PARASYMPATHETIC PART OF THE AUTONOMIC NERVOUS SYSTEM

### Efferent Nerve Fibers

The connector cells of this part of the system are located in the brain and the sacral segments of the spinal cord (Fig. 1-27). Those in the brain form parts of the nuclei of origin of cranial nerves 3, 7, 9, and 10, and the axons emerge from the brain contained in the corresponding cranial nerves.

The sacral connector cells are found in the gray matter of the second, third, and fourth sacral segments of the cord. These cells are not sufficiently numerous to form a lateral gray horn, as do the sympathetic connector cells in the thoracolumbar region. The myelinated axons leave the spinal cord in the anterior nerve roots of the corresponding spinal nerves. They then leave the sacral nerves and form the *pelvic splanchnic nerves*.

All the efferent fibers described so far are preganglionic, and they synapse with excitor cells in peripheral ganglia, which are usually situated close to the viscera they innervate. The cranial preganglionic fibers relay in the *ciliary, pterygopalatine, submandibular,* and *otic ganglia* (Fig. 1-27). The preganglionic fibers in the pelvic splanchnic nerves relay in ganglia in the *pelvic plexuses.* Characteristically, the postganglionic fibers are nonmyelinated and are of relatively short length as compared with sympathetic postganglionic fibers.

### Afferent Nerve Fibers

The afferent myelinated fibers travel from the viscera to their cell bodies located either in the sensory ganglia of the cranial nerves or in the posterior root ganglia of the sacrospinal nerves. The central axons then enter the central nervous system and take part in the formation of local reflex arcs, or pass to higher centers of the autonomic nervous system.

It is important to realize that the afferent component of the autonomic system is in actual fact identical to the afferent component of somatic nerves and forms part of the general afferent segment of the entire nervous system. The nerve endings in the autonomic afferent component may not be activated by such sensations as heat or touch, but rather by stretch or lack of oxygen. Once the afferent fibers gain entrance to the spinal cord or brain, they are thought to travel alongside, or are mixed with, the somatic afferent fibers.

## Mucous Membranes

*Mucous membrane* is the name given to the lining of organs or passages that communicate with the surface of the body. A mucous membrane consists essentially of a layer of epithelium supported by a layer of connective tissue, the *lamina propria.* Smooth muscle, called the *muscularis mucosa,* is sometimes present in the connective tissue. A mucous membrane may or may not secrete mucus on its surface.

## Serous Membranes

*Serous membranes* line the cavities of the trunk and are reflected onto the mobile viscera lying within these cavities (Fig. 1-28). They consist of a smooth layer of mesothelium supported by a thin layer of connective tissue. The serous membrane lining the wall of the cavity is referred to as the *parietal layer,* and that covering the viscera is called the *visceral*

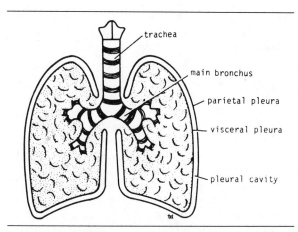

trachea

main bronchus

parietal pleura

visceral pleura

pleural cavity

**Fig. 1-28. Arrangement of pleura within thoracic cavity. Note that under normal conditions pleural cavity is slitlike space, parietal and visceral layers of pleura being separated by a small amount of serous fluid.**

*layer.* The narrow slitlike interval that separates these layers forms the *pleural, pericardial,* and *peritoneal cavities* and contains a small amount of serous liquid, the *serous exudate.* The serous exudate lubricates the surfaces of the membranes and allows the two layers to slide readily on each other.

The mesenteries, omenta, and serous ligaments are described in other chapters of this book.

### *Nerve Supply*

The parietal layer of a serous membrane is developed from the somatopleure and is richly supplied by spinal nerves. It is therefore sensitive to all common sensations such as touch and pain. The visceral layer is developed from the splanchnopleure and is supplied by autonomic nerves. It is insensitive to touch and temperature, but very sensitive to stretch.

## Bone

Bone is a living tissue capable of changing its structure as the result of the stresses to which it is subjected. Like other connective tissues, it consists of cells, fibers, and matrix. It is hard because of the calcification of its extracellular matrix and pos-

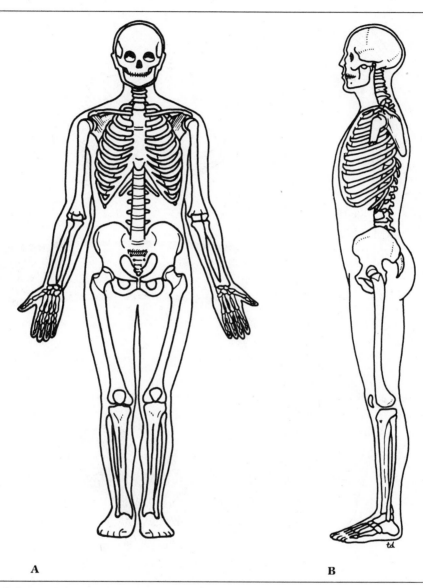

A                                    B

**Fig. 1-29. The skeleton. (A) Anterior view. (B) Lateral view.**

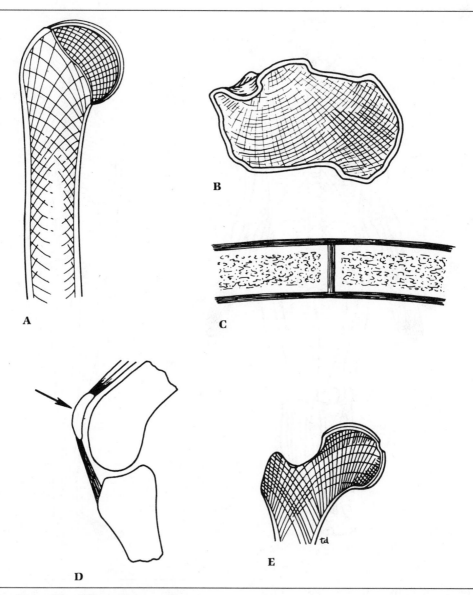

Fig. 1-30. Sections of the different types of bones.
(A) Long bone (humerus). (B) Irregular bone (calcaneum). (C) Flat bone (two parietal bones separated by the sagittal suture). (D) Sesamoid bone
(patella). (E) Note arrangement of trabeculae to act
as struts to resist both compression and tension
forces in upper end of femur.

sesses a degree of elasticity due to the presence of organic fibers. Bone has a protective function; the skull and vertebral column, for example, protect the brain and spinal cord from injury; the sternum and ribs protect the thoracic and upper abdominal viscera (Fig. 1-29). It serves as a lever, as seen in the long bones of the limbs. It is an important storage area for calcium salts. It houses and protects within its cavities the delicate blood-forming bone marrow.

Bone exists in two forms: *compact* and *cancellous.* Compact bone appears as a solid mass; cancellous bone consists of a branching network of trabeculae (Fig. 1-30). The trabeculae are arranged in such a manner as to resist the stresses and strains to which the bone is exposed.

## CLASSIFICATION OF BONES

Bones may be classified regionally or according to their general shape. The regional classification is summarized in Table 1-1. Bones may have the following general shapes: (1) long and short bones, (2) irregular bones, (3) flat bones, and (4) sesamoid bones.

The *long and short bones* are found in the limbs; the *flat and irregular bones*, in the skull, vertebral column, and the limb girdles. The *sesamoid bones* are small nodules of bone that are found in certain tendons where they rub over bony surfaces. The greater part of a sesamoid bone is buried in the tendon and the free surface is covered with articular cartilage. The function of a sesamoid bone is to reduce friction; it may also alter the direction of pull of a tendon. The largest sesamoid bone is the patella, which is situated in the tendon of the quadriceps femoris. Other examples are found in the tendons of flexor pollicis brevis and flexor hallucis brevis.

### *Surface Markings of Bones*

The surfaces of bones show various markings or irregularities. Where bands of fascia, ligaments, tendons, or aponeuroses are attached to bone, the surface is raised or roughened. These roughenings are not present at birth. They appear at puberty and become progressively more obvious during adult life. The pull of these fibrous structures

Table 1-1. Regional Classification of Bones

| Region of skeleton | Number of bones |
|---|---|
| Axial skeleton | |
|   Skull | |
|     Cranium | 8 |
|     Face | 14 |
|     Auditory ossicles | 6 |
|   Hyoid | 1 |
|   Vertebrae | 26 |
|   Sternum | 1 |
|   Ribs | 24 |
| Appendicular skeleton | |
|   Shoulder Girdles | |
|     Cavicle | 2 |
|     Scapula | 2 |
|   Upper extremities | |
|     Humerus | 2 |
|     Radius | 2 |
|     Ulna | 2 |
|     Carpals | 16 |
|     Metacarpals | 10 |
|     Phalanges | 28 |
|   Pelvic girdle | |
|     Innominate (hip) bone | 2 |
|   Lower extremities | |
|     Femur | 2 |
|     Patella | 2 |
|     Fibula | 2 |
|     Tibia | 2 |
|     Tarsals | 14 |
|     Metatarsals | 10 |
|     Phalanges | 28 |
| | 206 |

causes the periosteum to be raised and new bone to be deposited beneath.

In certain situations the surface markings are large and are given special names. Some of the more important markings are summarized in Table 1-2.

## BONE MARROW

Bone marrow occupies the marrow cavity in long and short bones and the interstices of the cancellous bone in flat and irregular bones. At birth, the

Table 1-2. Surface Markings of Bones

| Bone marking | Example |
| --- | --- |
| Linear elevation | |
| Line | Superior nuchal line of the occipital bone |
| Ridge | The medial and lateral supracondylar ridges of the humerus |
| Crest | The iliac crest of the innominate bone |
| Rounded elevation | |
| Tubercle | Pubic tubercle |
| Protuberance | External occipital protuberance |
| Tuberosity | Greater and lesser tuberosities of the humerus |
| Malleolus | Medial malleolus of the tibia and lateral malleolus of the fibula |
| Trochanter | Greater and lesser trochanters of the femur |
| Sharp elevation | |
| Spine or spinous process | Ischial spine, spine of vertebra |
| Styloid process | Styloid process of temporal bone |
| Expanded ends for articulation | |
| Head | Head of humerus, head of femur |
| Condyle (knuckle-like process) | Medial and lateral condyles of femur |
| Epicondyle (a prominence situated just above condyle) | Medial and lateral epicondyles of femur |
| Small flat area for articulation | |
| Facet | Facet on head of rib for articulation with vertebral body |
| Depressions | |
| Notch | Greater sciatic notch of innominate bone |
| Groove or sulcus | Bicipital groove of humerus |

Table 1-2. (continued)

| Bone marking | Example |
| --- | --- |
| Fossa | Olecranon fossa of humerus, acetabular fossa of innominate bone |
| Openings | |
| Fissure | Superior orbital fissure |
| Foramen | Infraorbital foramen of the maxilla |
| Canal | Carotid canal of temporal bone |
| Meatus | External acoustic meatus of temporal bone |

marrow of all the bones of the body is red and hematopoietic. This blood-forming activity gradually lessens with age, and the red marrow is replaced by yellow marrow. At 7 years of age, yellow marrow begins to appear in the distal bones of the limbs. This replacement of marrow gradually moves proximally, so that by the time the person becomes adult, the red marrow is restricted to the bones of the skull, the vertebral column, the thoracic cage, the girdle bones, and the head of the humerus and femur.

All bone surfaces, other than the articulating surfaces, are covered by a thick layer of fibrous tissue called the *periosteum*. The periosteum has an abundant vascular supply, and the cells on its deeper surface are osteogenic. The periosteum is particularly well united to bone at sites where muscles, tendons, and ligaments are attached to bone. Bundles of collagen fibers known as Sharpey's fibers extend from the periosteum into the underlying bone. The periosteum receives a rich nerve supply and is very sensitive.

## Development of Bone

Bone is developed by two methods: (1) membranous and (2) endochondral. In the first method the bone is developed directly from a connective tissue membrane; in the second, a cartilaginous model is first laid down and is later replaced by bone. For details of the cellular changes involved, a textbook of histology or embryology should be consulted.

The bones of the vault of the skull are developed rapidly by the membranous method in the embryo, and this serves to protect the underlying developing brain. At birth, small areas of membrane persist between the bones. This is important clinically since it allows the bones a certain amount of mobility, so that the skull can undergo molding during its descent through the female genital passages.

The long bones of the limbs are developed by endochondral ossification. It is a slow process and not completed until the eighteenth to twentieth year or even later. The center of bone formation found in the shaft of the bone is referred to as the *diaphysis*, the centers at the ends of the bone as the *epiphyses*. The plate of cartilage at each end, lying between the epiphysis and diaphysis in a growing bone, is called the *epiphyseal plate*. The *metaphysis* is that part of the diaphysis that abuts onto the epiphyseal plate.

## Cartilage

Cartilage is a form of connective tissue in which the cells and fibers are embedded in a gel-like matrix, the latter being responsible for its firmness and resilience. Except on the exposed surfaces in joints, it is covered by a fibrous membrane called the *perichondrium*. There are three types of cartilage: hyaline, fibrous, and elastic.

*Hyaline cartilage* has a high proportion of amorphous matrix that has the same refractive index as the fibers embedded in it. Throughout childhood and adolescence it plays an important part in the growth in length of long bones (epiphyseal plates are composed of hyaline cartilage). It has a great resistance to wear and covers the articular surfaces of nearly all synovial joints. It is incapable of repair when fractured; the defect is filled with fibrous tissue.

*Fibrocartilage* has a large number of collagen fibers embedded in a small amount of matrix. It is found in the discs within joints (e.g., the temporomandibular joint, sternoclavicular joint, knee joint) and on the articular surfaces of the clavicle and mandible. If damaged, it repairs itself slowly in a manner similar to fibrous tissue elsewhere. Joint discs have a poor blood supply and therefore do not repair themselves if damaged.

*Elastic cartilage* possesses large numbers of elastic fibers embedded in matrix. As would be expected, it is very flexible and is found in the *auricle of the ear*, the *external auditory meatus*, the *auditory tube*, and the *epiglottis*. If damaged, it repairs itself with fibrous tissue.

Hyaline cartilage and fibrocartilage tend to calcify or even ossify in later life.

## Effects of Sex, Race, and Age on Structure

Descriptive anatomy tends to concentrate on a fixed descriptive form. Medical personnel must always remember that there are sexual and racial differences and that the body's structure and function change as a person grows and ages.

The adult male tends to be taller than the adult female and has longer legs; his bones are bigger and heavier and his muscles are larger. He has less subcutaneous fat, which makes his appearance more angular. His larynx is larger and his vocal cords are longer, so that his voice is deeper. He has a beard and coarse body hair. He possesses axillary and pubic hair, the latter extending to the region of the umbilicus.

The adult female tends to be shorter than the adult male and has smaller bones and less bulky muscles. She has more subcutaneous fat, fat accumulations in the breasts, buttocks, and thighs giving her a more rounded appearance. Her head hair is finer and her skin is smoother in appearance. She has axillary and pubic hair, but the latter does not extend up to the umbilicus. The adult female has larger breasts and a wider pelvis than the male. She has a wider carrying angle at the elbow, which results in a greater lateral deviation of the forearm on the arm.

Until the age of approximately 10 years, boys and girls grow at about the same rate. Around 12 years, boys often start to grow faster than girls, so that most males reach adulthood taller than females.

Puberty begins between ages 10 and 14 in girls and between 12 and 15 in boys. In the girl at puberty, the breasts enlarge and the pelvis broadens. At the same time, a boy's penis, testes, and scrotum enlarge, and in both sexes axillary and pubic hair appear.

Racial differences may be seen in the color of the skin, hair, and eyes, and in the shape and size of the eyes, nose, and lips. Africans and Scandinavians tend to be tall, due to long legs, whereas Orientals tend to be short, with short legs. The heads of central Europeans and Orientals also tend to be round and broad.

After birth and during childhood, the bodily functions become progressively more efficient, reaching their maximum degree of efficiency during young adulthood. During late adulthood and old age, many bodily functions become less efficient.

## CLINICAL NOTES

### DESCRIPTIVE ANATOMICAL TERMS

It is important for medical students to have a sound knowledge and understanding of the basic anatomical terms. With the aid of a medical dictionary, you will find that understanding anatomical terminology greatly assists you in the learning process. For example, when studying muscles, the names of the muscles often describe the arrangement of the muscle fibers (rectus abdominis), the location of the muscle (gluteus maximus), the shape of the muscle (piriformis), the number of origins (triceps), and even the origin and the insertion (sternocleidomastoid).

The accurate use of anatomical terms by medical personnel enables them to communicate with their colleagues both nationally and internationally. Without anatomical terms, one cannot discuss or record accurately the abnormal functions of joints, the actions of muscles, the alteration of position of organs, or the exact location of swellings or tumors.

### Skin

A general knowledge of the direction of the *lines of cleavage* greatly assists the surgeon in making incisions that will result in cosmetically acceptable scars. This is particularly important in the case of women, and in those areas of the body not normally covered by clothing. A salesman, for example, may lose his job if an operation leaves a hideous scar on his face.

The nail folds, hair follicles, and sebaceous glands are common sites for the entrance into the underlying tissues of pathogenic organisms such as *Staphylococcus aureus*. Infection occurring between the nail and the nail fold is called a *paro-nychia*. Infection of the hair follicle and sebaceous gland is responsible for the common *boil*. A *carbuncle* is a staphylococcal infection of the superficial fascia. It frequently occurs in the nape of the neck and usually starts as an infection of a hair follicle or a group of hair follicles.

A *sebaceous cyst* is due to obstruction of the mouth of a sebaceous duct and may be caused by damage from a comb or by infection. It occurs most frequently on the scalp.

A patient who is in a state of *shock* will be pale and exhibit gooseflesh due to overactivity of the sympathetic system, which causes vasoconstriction of the dermal arterioles and contraction of the arrector pili muscles.

The depth of a *burn* determines the method and rate of healing. A partial skin thickness burn will heal from the cells of the hair follicles, sebaceous glands, and sweat glands, as well as from the cells at the edge of the burn. A burn that extends deeper than the sweat glands will heal very slowly from the edges only, and there will be considerable contracture caused by fibrous tissue. To speed up healing and reduce the incidence of contracture, a deep burn should be grafted.

*Skin grafting* is of two main types, split-thickness grafting and full-thickness grafting. In a split-thickness graft the greater part of the epidermis, including the tips of the dermal papillae, are removed from the donor site and placed on the recipient site. This leaves at the donor site for repair purposes the epidermal cells on the sides of the dermal papillae and the cells of the hair follicles and sweat glands.

A full-thickness skin graft includes both the epidermis and dermis, and, to survive, requires the rapid establishment of a new circulation within it at the recipient site. The donor site is usually covered with a split-thickness graft. In certain circumstances the full-thickness graft is made in the form of a pedicle graft, in which a flap of full-thickness skin is turned and stitched in position at the recipient site, leaving the base of the flap with its blood supply intact at the donor site. Later, when the new blood supply to the graft has been established, the base of the graft is cut across.

## Fasciae

A knowledge of the arrangement of the deep fasciae will often explain the path taken by an infection when it spreads from its primary site. In the neck, for example, the various fascial planes explain how infection can extend from the region of the floor of the mouth to the larynx.

## Muscle

The determination of the *tone* of a muscle is an important clinical examination. If a muscle is found to be *flaccid*, then either the afferent or efferent neurons or both neurons involved in the reflex arc necessary for the production of muscle tone have been interrupted. For example, if the nerve trunk to a muscle is severed, both neurons will have been interrupted. If poliomyelitis has involved the motor anterior horn cells at a level in the spinal cord that innervates the muscle, the efferent motor neurons will not function. If, on the other hand, the muscle is found to be hypertonic, the possibility exists of a lesion involving higher motor neurons in the spinal cord or brain.

It is unnecessary to emphasize the importance of knowing the main attachments of all the major muscles of the body. Only with such knowledge is it possible to understand the normal and abnormal actions of individual muscles or muscle groups. How can one even attempt to analyze, for example, the abnormal gait of a patient without this information?

The general shape and form of muscles should also be noted, since a paralyzed muscle or one that is not used (such as occurs when a limb is immobilized in a cast) quickly atrophies and changes shape. In the case of the limbs it is always worth remembering that there is another muscle on the opposite side of the body for purposes of comparison.

## Bones

### FRACTURES OF BONES

Immediately following the fracture, the patient suffers severe local pain and is not able to use the injured part. Deformity may be visible if the bone fragments have been displaced relative to each other. The degree of deformity and the directions taken by the bony fragments will depend not only on the mechanism of injury but also on the pull of the muscles attached to the fragments. Ligamentous attachments also will influence the deformity. In certain situations, for example, the ileum, fractures result in no deformity, since the inner and outer surfaces of the bone are splinted by the extensive origins of muscles. In contrast, a fracture of the neck of the femur produces considerable displacement. The strong muscles of the thigh pull the distal fragment upward, so that the leg is shortened. The very strong lateral rotators rotate the distal fragment laterally.

Fracture of a bone is accompanied by a considerable hemorrhage of blood between the bone ends and into the surrounding soft tissue. The blood vessels and the fibroblasts and osteoblasts from the periosteum and endosteum take part in the repair process.

### RICKETS

In rickets, there is a defective mineralization of the cartilage matrix in growing bones. This produces a condition in which the cartilage cells continue to grow, producing excess cartilage and a widening of the epiphyseal plates. The poorly mineralized cartilaginous matrix and the osteoid matrix are soft, and they bend under the stress of bearing weight. The resulting deformities include enlarged costochondral junctions, bowing of the long bones of the lower limbs, and bossing of the frontal bones of the skull. Deformities of the pelvis may also occur.

### EPIPHYSEAL PLATE DISORDERS

The epiphyseal plate is the part of a growing bone concerned primarily with growth in length. Trauma, infection, diet, exercise, and endocrine disorders may disturb the growth of the hyaline cartilaginous plate, leading to deformity and loss of function. In the femur, for example, the proximal femoral epiphysis may slip due to mechanical stress or excessive loads. The length of the limbs can increase excessively due to increased vascularity in the region of the epiphyseal plate secondary to infection or in the presence of tumors. Shortening of a limb can follow trauma to the epiphyseal plate resulting form a diminished blood supply to the cartilage.

## Joints

The normal range of movement of all joints should be ascertained. When the bones of a joint are no longer in their normal anatomical relationship with one another, then the joint is said to be *dislocated*. Some of the joints are particularly susceptible to dislocation due to the lack of support by ligaments, the poor shape of the articular surfaces, or the absence of adequate muscular support. The shoulder joint, temporomandibular joint, and acromioclavicular joints are good examples. Dislocation of the hip is usually congenital, caused by inadequate development of the socket that normally holds the head of the femur firmly in position.

The presence of cartilaginous discs within joints, especially weight-bearing joints, as in the case of the knee, makes them particularly susceptible to injury in sports. During a rapid movement the disc loses its normal relationship to the bones and becomes crushed between the weight-bearing surfaces.

In certain diseases of the nervous system (e.g., *syringomyelia*), the sensation of pain in a joint is lost. This means that the warning sensations of pain felt when a joint moves beyond the normal range of movement are not experienced. This phenomenon results in the destruction of the joint.

A knowledge of the classification of joints is of great value since it is known that certain diseases affect only certain types of joints. For example, *gonococcal arthritis* affects large synovial joints such as the ankle, elbow, or wrist. Tuberculous arthritis also affects synovial joints and may start in the synovial membrane or in the bone.

It is also important to remember that more than one joint may receive the same nerve supply. For example, the hip and knee joints are both supplied by the *obturator nerve*. Thus, a patient with disease limited to one of these joints may experience pain in both.

## Bursae and Synovial Sheaths

Bursae and synovial sheaths are commonly the site of traumatic or infectious disease. For example, the extensor tendon sheaths of the hand may become inflamed following excessive or unaccustomed use; an inflammation of the prepatellar bursa may occur as the result of trauma from repeated kneeling on a hard surface.

## Blood Vessels

Diseases of blood vessels are very common. The surface anatomy of the main arteries, especially those of the limbs, should be learned in the appropriate sections of this book. The *collateral circulation* of most large arteries should be understood, and a distinction should be made between anatomical end arteries and functional end arteries.

All large arteries that cross over a joint are liable to be kinked during movements of the joint. However, the distal flow of blood is not interrupted, since there is usually an adequate anastomosis between branches of the artery that arise both proximal and distal to the joint. The alternative blood channels, which dilate under these circumstances, form the collateral circulation. A knowledge of the existence and position of such a circulation may be of vital importance should it be necessary to tie off a large artery that has been damaged by trauma or disease.

Coronary arteries are functional end arteries, and should they become blocked by disease (coronary arterial occlusion is common), the cardiac muscle normally supplied by that artery will receive insufficient blood and will undergo necrosis. Blockage of a large coronary artery will result in death of the patient.

## Lymphatic System

The lymphatic system is often de-emphasized by anatomists on the grounds that it is difficult to see on a cadaver. However, it is of vital importance to a practicing physician, and the lymphatic drainage of all major organs of the body, including the skin, should be known.

A patient may complain of a swelling produced by the enlargement of a lymph node. A physician must know the areas of the body that drain lymph to a particular node if he is going to be able to find the primary site of the disease. Quite often the patient ignores the primary disease, which may be a small, painless cancer of the skin.

Conversely, the patient may complain of a painful ulcer of the tongue, for example, and the physician must know the lymphatic drainage of the tongue to be able to determine whether or not the disease has spread beyond the limits of the tongue.

## Nervous System

The area of skin supplied by a single spinal nerve, and therefore a single segment of the spinal cord, is called a *dermatome*. On the trunk, adjacent dermatomes overlap considerably, so that to produce a region of complete anesthesia at least three contiguous spinal nerves have to be sectioned. Dermatomal charts for the anterior and posterior surfaces of the body are shown in Figures 1-31 and 1-32.

In the limbs the arrangement of the dermatomes is more complicated, and this is due to the embryological changes that take place as the limbs grow out from the body wall.

A physician should have a working knowledge of the segmental (dermatomal) innervation of skin, since with the help of a pin or a piece of cotton he can determine whether or not the sensory function of a particular spinal nerve or segment of the spinal cord is functioning normally.

Skeletal muscle also receives a segmental innervation. Most of these muscles are innervated by two, three, or four spinal nerves and therefore by the same number of segments of the spinal cord. To paralyze a muscle completely it would thus be necessary to section several spinal nerves or destroy several segments of the spinal cord.

It is an impossible task to learn the segmental innervation of all the muscles of the body. Nevertheless, the segmental innervation of the following muscles should be known, since it is possible to test them by eliciting simple muscle reflexes in the patient (Fig. 1-33).

*Biceps brachii tendon reflex* C**5** and 6 (flexion of the elbow joint by tapping the biceps tendon).

*Triceps tendon reflex* C6, **7**, and 8 (extension of the elbow joint by tapping the triceps tendon).

*Brachioradialis tendon reflex* C5, **6**, and 7 (supination of the radioulnar joints by tapping the insertion of the brachioradialis tendon).

*Abdominal superficial reflexes* (contraction of underlying abdominal muscles by stroking the skin). Upper abdominal skin T6–7; middle abdominal skin T8–9; lower abdominal skin T10–12.

*Patellar tendon reflex* (knee jerk) L2, **3**, and **4** (extension of knee joint on tapping the patellar tendon).

*Achilles tendon reflex* (ankle jerk) S**1** and S2 (plantar flexion of ankle joint on tapping the Achilles tendon).

### Autonomic Nervous System

Many drugs and surgical procedures are available that can modify the activity of the autonomic nervous system. For example, drugs can be administered to lower the blood pressure by affecting the normal function of sympathetic ganglia and causing vasodilatation of peripheral blood vessels. In patients with severe arterial disease affecting the main arteries of the lower limb, the limb can sometimes be saved by sectioning the sympathetic innervation to the blood vessels. This produces a vasodilatation and enables an adequate amount of blood to flow through the collateral circulation and so bypass the obstruction.

## Effect of Age on Structure

The fact that the structure and function of the human body change with age may seem obvious, but it is often overlooked. A few examples of such changes will be given here:

1. In the infant the bones of the skull are more resilient than in the adult, and for this reason fractures of the skull are much more common in the adult than in the young child.
2. The liver is relatively much larger in the child than in the adult. In the infant the lower margin of the liver extends inferiorly to a lower level

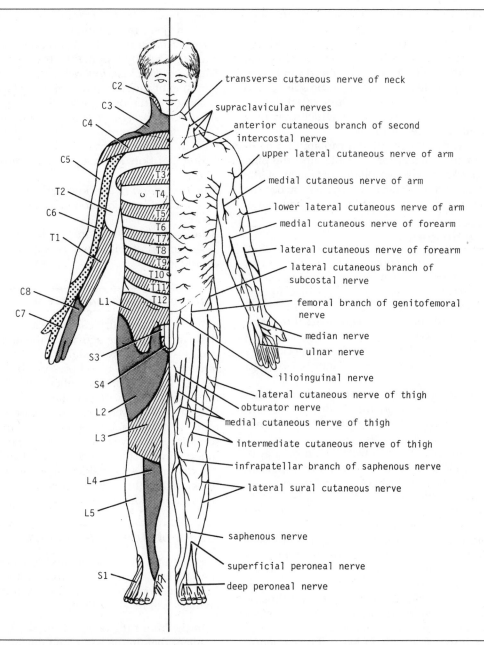

**Fig. 1-31. Dermatomes and distribution of cutaneous nerves on the anterior aspect of the body.**

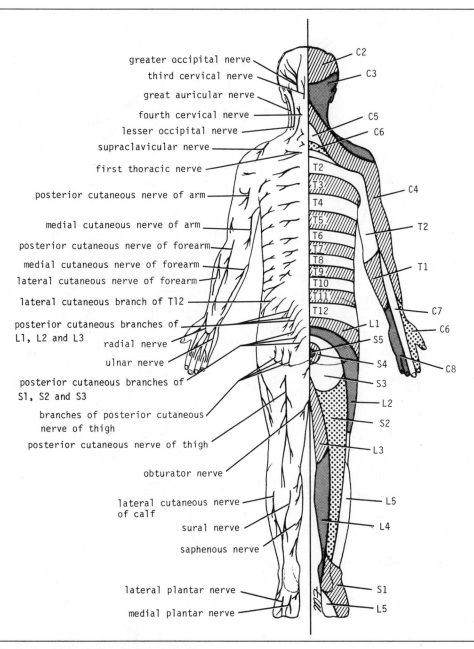

**Fig. 1-32. Dermatomes and distribution of cutaneous nerves on the posterior aspect of the body.**

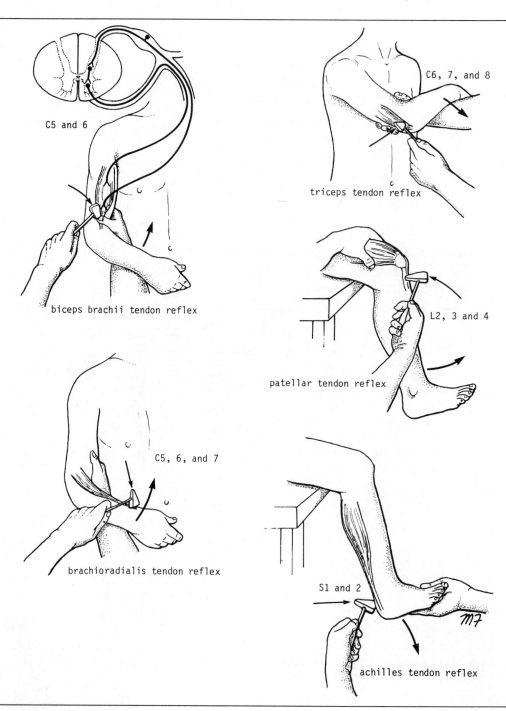

C5 and 6

biceps brachii tendon reflex

C6, 7, and 8

triceps tendon reflex

L2, 3 and 4

patellar tendon reflex

C5, 6, and 7

brachioradialis tendon reflex

S1 and 2

achilles tendon reflex

**Fig. 1-33. Some important tendon reflexes used in medical practice.**

than in the adult. This is an important consideration when making a diagnosis of hepatic enlargement.

3. The urinary bladder in the child cannot be accommodated entirely in the pelvis due to the small size of the pelvic cavity and is found in the lower part of the abdominal cavity. As the child grows, the pelvis enlarges and the bladder sinks down to become a true pelvic organ.

4. At birth all bone marrow is of the red variety. With advancing age the red marrow recedes up the bones of the limbs so that in the adult it is largely confined to the bones of the head, thorax, and abdomen.

5. Lymphatic tissues reach their maximum degree of development at puberty and thereafter atrophy, so that the volume of lymphatic tissue in the old is considerably reduced.

## Radiographic Anatomy

As a physician you will be frequently called upon to study normal and abnormal anatomy as seen on radiographs. Familiarity with normal radiographic

**Fig. 1-34. Postero-anterior radiograph of thorax.**

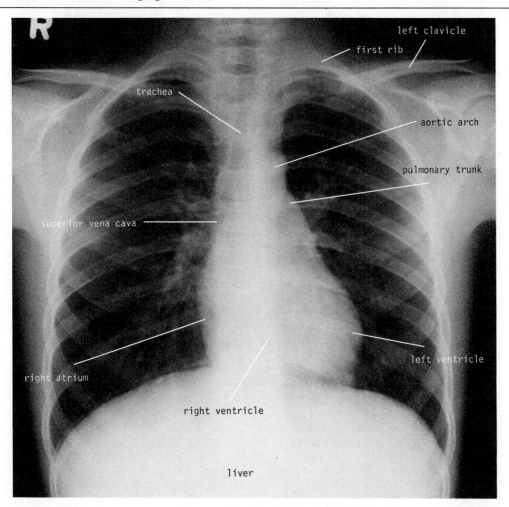

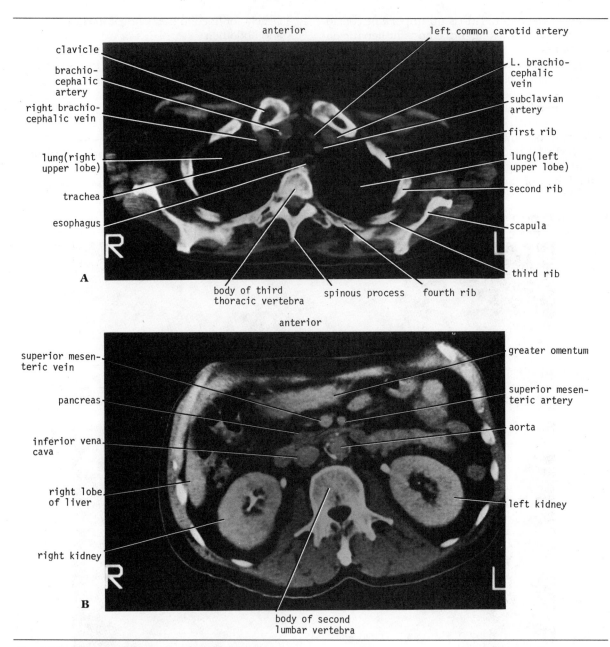

**Fig. 1-35. CT scans. (A) Upper thorax at level of third thoracic vertebra; (B) upper abdomen at level of second lumbar vertebra. All CT scans are viewed from below. Thus, the right side of the body appears on the left of the illustration.**

anatomy permits one to recognize abnormalities quickly, such as fractures or tumors.

The most common form of radiographic anatomy is studied on a radiograph (roentgenograph, X-ray film), which provides a two-dimensional image of the interior of the body (Fig. 1-34). To produce such a radiograph, a single barrage of X-rays is passed through the body and exposes the film. Tissues of differing densities show up as images of differing densities on the radiograph (or fluorescent screen). A tissue that is relatively dense absorbs (stops) more X-rays than tissues that are less dense. A very dense tissue is said to be radiopaque, but a less dense tissue is said to be radiolucent. Bone is very dense and fat is moderately dense; other soft tissues are the least dense.

Unfortunately, the examination of an ordinary radiograph shows the images of the different organs superimposed onto a flat sheet of film. This overlap of organs and tissues often makes it difficult to visualize them. This problem is overcome to some extent by taking films at right angles to one another or by making stereoscopic films.

In 1971, CT (computed tomography) scanning or CAT (computerized axial tomography) scanning was introduced so that tissue slices could be studied and tissues having minor differences in density could be recognized. CT scanning relies on the same physics as conventional X-rays but combines it with computer technology. A source of X-rays moves in an arc around the part of the body being studied and sends out a beam of X-rays. The beams of X-rays, having passed through the region of the body, are collected by a special X-ray detector. Here, the X-rays are converted into electronic impulses that produce readings of the density of the tissue in a one centimeter slice of the body. From these readings the computer is able to assemble a picture of the body called a CT scan, which can be viewed on a fluorescent screen and then photographed for later examination (Fig. 1-35). The procedure is safe and quick, lasts only a few seconds for each slice, and for most patients requires no sedation.

## CLINICAL PROBLEMS

*Answers on page 955*

1. The surgical notes of a patient state that she had a right infraumbilical paramedian incision through the skin of the anterior abdominal wall. Where exactly was this incision made?

2. A physician's letter states that a patient has a small, mobile tumor on the dorsum of the foot, just proximal to the base of the toes and lying superficial to the bones and extensor tendons, but deep to the superficial fascia. Examine your own foot and try to visualize where this tumor is located.

3. Following an attack of pericapsulitis of the shoulder joint, a patient finds that movements of the joint are restricted. On examination, abduction is limited to 30 degrees, there is no lateral rotation, and extention is limited to 10 degrees. Flexion is free. Demonstrate this disability on your own shoulder joint.

4. As the result of osteoarthritis, a 60-year-old patient has very restricted movements in the right hip joint. On examination, the patient is found to hold the joint partially flexed, abducted, and laterally rotated. The range of movement is limited in all directions, particularly in abduc-tion and internal rotation. Demonstrate this disability on your own hip joint.

5. A patient with a history of poliomyelitis affecting the anterior horn cells of the lower thoracic and lumbar segments of the spinal cord has a severe left lateral flexion deformity of the vertebral column. Explain this condition in anatomical and physiological terms.

6. Following an injury to the right elbow joint, it is found that a patient is unable to pronate his forearm. Demonstrate this on your own forearm.

7. After suffering a severe strain to the left ankle while playing tennis, a patient experiences considerable pain when she tries to move the foot so that the sole faces medially. What is the correct anatomical term for the movement of the foot that produces the pain?

8. Following a severe nerve injury on the anterior aspect of the forearm just proximal to the wrist, a patient is found to be unable to abduct her fingers. Demonstrate on your own hand what the patient is unable to do with her fingers.

9. A film actress is worried about a new scar on

her anterior abdominal wall. She says that from the age of 16, following an appendectomy, she has had a small oblique scar in the lower right side of her abdomen. This scar is small and hardly noticeable. Recently, she had her gallbladder removed through a right vertical supraumbilical paramedian incision, and the scar is wide and heaped-up. Can you explain why the scars are so different?

10. A patient has a large abscess in the neck that requires drainage through a surgical incision. In this situation would you use a vertical or a horizontal incision to obtain the best cosmetic results?

11. (a) A workman received a severe burn on his forearm, measuring about 4 inches (10 cm) square. In depth, it extended into the superficial part of the dermis. From which sites would the epidermal cells regenerate? (b) Another patient has a burn that penetrates as far as the superficial fascia. From where would the epidermal cells regenerate in this patient?

12. At which sites on the surface of intact skin are pathogenic organisms likely to enter the dermis or subcutaneous tissue?

13. A diagnosis has been made that a tumor of the vertebral column is pressing on the lumbar segments of a patient's spinal cord. He has a loss of sensation in the skin over the anterior surface of the thigh and is unable to extend his knee joint. On examination, it is found that the muscles of the front of the thigh have atrophied and have no tone and that the knee jerk is absent. Explain these findings in anatomical and physiological terms.

14. A 65-year-old patient has had a cerebral hemorrhage that has destroyed the upper motor neurons on one side of the brain. On examination of his right leg, the muscles are found to be hypertonic. Explain this in anatomical and physiological terms.

15. A housewife recently took up employment in a factory. She is a machinist, and for 6 hours a day has to move a lever repeatedly, which requires that she extend and flex her wrist joint. At the end of the second week of her employment, she began to experience pain over the posterior surface of her wrist and noticed a swelling in the area. What is the anatomical explanation for her discomfort?

16. A 40-year-old man decided to lay adhesive tiles on the floor of his large kitchen and family room. This involved many hours of work on his knees on a hard surface. After 5 days he noticed a tender swelling on the anterior surface of his knee joint. What is the anatomical explanation for his discomfort?

17. A young baseball player, on returning to his team after being hospitalized for a severe injury to his left knee, stated that the medial cartilaginous disc had been badly torn. He added that the cartilage had regenerated, the tear had healed, and his knee was as good as before the accident. Comment on this statement.

18. A young woman had a severe infection under the lateral edge of the nail of her right index finger. On examination, a series of red lines were seen to extend up the back of the hand and around to the front of the forearm and arm, up to the armpit. Palpation of the armpit revealed a number of enlarged, tender lymph nodes. What were the red lines, and why were the lymph nodes enlarged?

19. At postmortem, a branch of a patient's left coronary artery was found to be blocked. A large area of cardiac muscle was found to be necrosed. What is the anatomical explanation for the necrosis of the muscle?

20. A 19-year-old boy was suspected of having leukemia. It was decided to confirm the diagnosis by performing a bone marrow biopsy. The resident obtained a marrow specimen from the lower end of the tibia, but was surprised to learn that the specimen was useless for diagnostic purposes. What comments would you make on this?

# NATIONAL BOARD TYPE QUESTIONS

*Answers on Page 983*

**Select the best response.**

1. A patient who is standing in the anatomical position is:
   (a) Facing laterally
   (b) Has the palms of the hands directed medially
   (c) Has the ankles several inches apart
   (d) Is standing on his toes
   (e) Has the upper limbs by the sides of the trunk

2. A patient is performing the movement of flexion of the hip joint when she:
   (a) Moves the lower limb away from the midline in the coronal plane
   (b) Moves the lower limb posteriorly in the paramedian plane
   (c) Moves the lower limb anteriorly in the paramedian plane
   (d) Rotates the lower limb so that the anterior surface faces medially
   (e) Moves the lower limb toward the median sagittal plane

3. Inversion of the foot is the movement so that the sole faces:
   (a) Downward and posteriorly
   (b) Medially
   (c) Laterally
   (d) Downward
   (e) Downward and laterally

4. The lines of cleavage or Langer's lines are:
   (a) Finger prints
   (b) Skin creases over joints
   (c) Lines representing the interface between the superficial and deep layers of fascia
   (d) The direction of the rows of elastic fibers in the dermis
   (e) The direction of the rows of collagen fibers in the dermis

**Match each structure listed on the left with a structure or occurrence listed on the right with which it is most closely associated. Each lettered answer may be used more than once.**

5. Superficial fascia    (a) Divides up interior of limbs into compartments

6. Deep fascia    (b) Adipose tissue
      (c) Tendon spindles

7. Skeletal muscle    (d) None of the above

**For each joint listed on the left, indicate which type of movement it is associated with on the right.**

8. Sternoclavicular joint    (a) Flexion
9. Superior radioulnar joint    (b) Gliding
      (c) Both (a) and (b)
10. Ankle joint    (d) Neither (a) nor (b)

**For each joint listed on the left, give the most appropriate classification out of the list on the right.**

11. Joints between vertebral bodies    (a) Synovial joint
12. Inferior tibiofibular joint    (b) Cartilaginous
13. Sutures between bones of vault of skull    (c) Fibrous
14. Wrist joint    (d) None of the above

**For each type of synovial joint listed on the left, give an appropriate example from the list of joints on the right.**

15. Hinge joint    (a) Metacarpophalangeal joint of index finger
16. Condyloid joint    (b) Shoulder joint
17. Ball-and-socket joint    (c) Wrist joint
      (d) Carpometacarpal joint of the thumb
18. Saddle joint    (e) None of the above

**For questions 19 through 25, answer:**
   (a) IF (1), (2), AND (3) ONLY ARE CORRECT
   (b) IF (1) AND (3) ONLY ARE CORRECT
   (c) IF (2) AND (4) ONLY ARE CORRECT
   (d) IF (4) ONLY IS CORRECT
   (e) IF ALL ARE CORRECT

19. (1) A prime mover is the chief muscle responsible for a particular movement.
    (2) A fixator is a muscle that contracts isometrically to stabilize the origin of the prime mover.
    (3) A synergist is a muscle that prevents unwanted movements in an intermediate joint so that a prime mover can cross that joint and act primarily on a distal joint.

(4) An antagonist is a muscle that opposes the action of a fixator.

20. (1) The stability of a joint depends on the shape of the articular surfaces, the strength of the ligaments, and the tone of the muscles around the joint.
    (2) The capsule and ligaments of a joint are devoid of a sensory nerve supply.
    (3) Occasionally, the cavity of a bursa communicates with the cavity of a synovial joint.
    (4) The articular surfaces of all synovial joints are covered with fibrocartilage.

21. (1) Arterioles are vessels larger than 1 mm in diameter but smaller than 2 mm in diameter.
    (2) Venules, unlike arteries, possess small valves.
    (3) A portal system begins as an arteriovenous anastomosis and ends as a capillary network.
    (4) Anatomical end arteries are vessels whose terminal branches do not anastomose with branches or arteries supplying adjacent areas.

22. (1) Lymph capillaries are a network of fine vessels that do not communicate with the tissue spaces.
    (2) Large lymph vessels possess numerous valves.
    (3) The right lymphatic duct drains the right side of the head and neck, the right upper limb, and the right side of the thorax.
    (4) Before lymph enters the bloodstream it has to pass through at least one lymph node.

23. (1) A synapse is a site where two neurons come into close proximity but not into anatomical continuity.
    (2) Each spinal nerve is connected to the spinal cord by an anterior root and a posterior root.
    (3) There are eight cervical spinal nerves and only seven cervical vertebrae.
    (4) Preganglionic nerve fibers pass to spinal nerves in the gray rami communicantes.

24. (1) Motor nerve fibers to skeletal muscle always leave the spinal cord in an anterior root.
    (2) The anterior ramus of a spinal nerve does not supply the muscles of the limbs.
    (3) A dermatome is an area of skin supplied by one segment of the spinal cord.
    (4) The biceps brachii tendon reflex tests the integrity of C7–8 segments of the spinal cord.

25. (1) Long bones of the limbs are developed by membranous ossification.
    (2) A sesamoid bone is a small nodule of bone found in a tendon.
    (3) Growth that takes place in an epiphyseal plate is largely responsible for increasing the diameter of a long bone.
    (4) The epiphysis is the center of ossification found at the end of a long bone.

# 2. The Thorax: Part I
# The Thoracic Wall

The thorax (or chest) is the region of the body between the neck and the abdomen. It is flattened in front and behind but rounded at the sides. The framework of the walls of the thorax, which is referred to as the *thoracic cage*, is formed by the vertebral column behind, the ribs and intercostal spaces on either side, and the sternum and costal cartilages in front. Superiorly the thorax communicates with the neck through the *thoracic inlet*,* and inferiorly it is separated from the abdomen by the diaphragm. The thoracic cage protects the lungs and heart and affords attachment for the muscles of the thorax, upper extremity, abdomen, and back.

The cavity of the thorax may be divided into a median partition, called the *mediastinum*, and the laterally placed pleurae and lungs. The lungs are covered by a thin membrane called the *visceral pleura*, which passes from each lung at its root (i.e., where the main air passages and blood vessels enter) to the inner surface of the chest wall, where it is called the *parietal pleura*. In this manner two membranous sacs called the *pleural cavities* are formed, one on each side of the thorax, between the lungs and the thoracic walls.

## SURFACE ANATOMY

As physicians you will be examining the chest to detect evidence of disease. Your examination will consist of inspection, palpation, percussion, and auscultation.

*Inspection* shows the configuration of the chest, the range of respiratory movement, and any inequalities on the two sides. The type and rate of respiration will also be noted.

*Palpation* will enable the physician to confirm the impressions gained by inspection, especially of the respiratory movements of the chest wall. Abnormal protuberances or recession of part of the chest wall will be noted. Abnormal pulsations will also be felt and tender areas detected.

*Percussion* is a sharp tapping of the chest wall with the fingers. This produces vibrations that ex-

---

*Clinicians often loosely refer to this opening as the thoracic outlet, since important vessels and nerves emerge from the thorax here to enter the neck and upper limb.

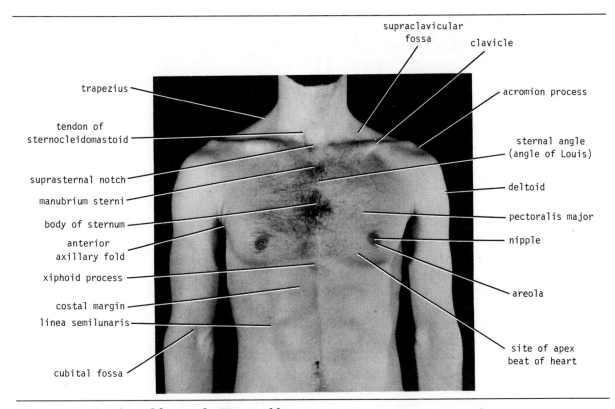

**Fig. 2-1. Anterior view of thorax of a 27-year-old male.**

tend through the tissues of the thorax. Air-containing organs such as the lungs produce a resonant note; on the other hand, a more solid viscus such as the heart produces a dull note. With practice, it is possible to distinguish the lungs from the heart or the liver by percussion.

*Auscultation* enables the physician to listen to the breath sounds as the air enters and leaves the respiratory passages. Should the alveoli or bronchi be diseased and filled with fluid, the nature of the breath sounds will be altered. The rate and rhythm of the heart can be confirmed by auscultation, and the various sounds produced by the heart and its valves during the different phases of the cardiac cycle can be heard. It may be possible to detect friction sounds produced by the rubbing together of diseased layers of pleura or pericardium.

To make these examinations, a physician must be familiar with the normal structure of the thorax

and must have a mental image of the normal position of the lungs and heart in relation to identifiable surface landmarks. Furthermore, it is essential that a physician be able to relate his abnormal findings to easily identifiable bony landmarks, so that he can accurately record and communicate them to his colleagues.

Since the thoracic wall actively participates in the movements of respiration, many bony landmarks change their levels with each phase of respiration. In practice, to simplify matters, the levels given are those usually found at about midway between full inspiration and full expiration.

## Examination of the Anterior Surface of the Thorax

**SUPRASTERNAL NOTCH.** This is the superior margin of the manubrium sterni and is easily felt between the prominent medial ends of the clavicles in the midline (Figs. 2-1 and 2-2). It lies opposite

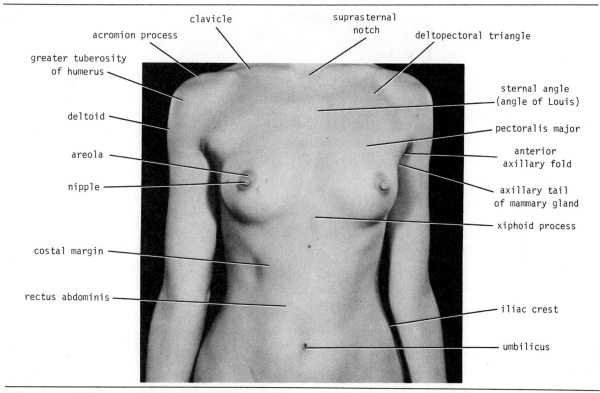

greater tuberosity
of humerus

acromion process

clavicle

suprasternal
notch

deltopectoral triangle

deltoid

areola

nipple

costal margin

rectus abdominis

sternal angle
(angle of Louis)

pectoralis major

anterior
axillary fold

axillary tail
of mammary gland

xiphoid process

iliac crest

umbilicus

**Fig. 2-2. Anterior view of thorax and abdomen of a 29-year-old female.**

the lower border of the body of the second thoracic vertebra (Fig. 2-10).

**STERNAL ANGLE (ANGLE OF LOUIS).** This is the angle made between the manubrium and body of the sternum (Figs. 2-1 and 2-2); at this level the *second costal cartilage* joins the lateral margin of the sternum. The sternal angle lies opposite the intervertebral disc between the fourth and fifth thoracic vertebrae (Fig. 2-10).

**XIPHISTERNAL JOINT.** This is the joint between the xiphoid process of the sternum and the body of the sternum (Fig. 2-4). It lies opposite the body of the ninth thoracic vertebra (Fig. 2-10).

**SUBCOSTAL ANGLE.** This is situated at the inferior end of the sternum, between the sternal attachments of the seventh costal cartilages (Fig. 2-4).

**COSTAL MARGIN.** This is the lower boundary of the thorax and is formed by the cartilages of the seventh, eighth, ninth, and tenth ribs and the ends

of the eleventh and twelfth cartilages (Figs. 2-1 and 2-2). The lowest part of the costal margin is formed by the tenth rib and lies at the level of the third lumbar vertebra.

**CLAVICLE.** This bone is subcutaneous throughout its entire length and can be easily palpated (Figs. 2-1 and 2-2). It articulates at its lateral extremity with the acromion process of the scapula.

**RIBS.** The first rib lies deep to the clavicle and cannot be palpated. The lateral surfaces of the remaining ribs can be felt by pressing the fingers upward into the axilla and drawing them downward over the lateral surface of the chest wall. The twelfth rib, if short, may be difficult to palpate. To identify a particular rib, always first identify the second costal cartilage at the sternal angle and then count the cartilages and ribs downward from this point.

**NIPPLE.** In the male it usually lies in the fourth intercostal space about 4 inches (10 cm) from the midline. In the female its position is not constant.

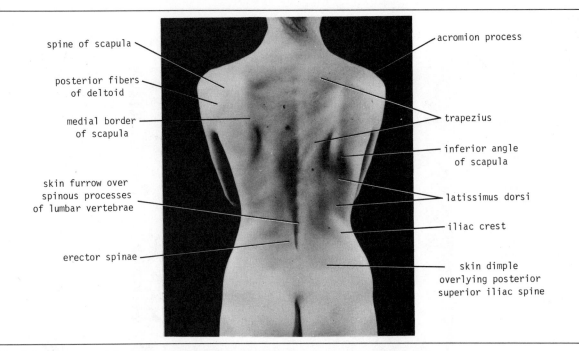

spine of scapula

posterior fibers
of deltoid

medial border
of scapula

skin furrow over
spinous processes
of lumbar vertebrae

erector spinae

acromion process

trapezius

inferior angle
of scapula

latissimus dorsi

iliac crest

skin dimple
overlying posterior
superior iliac spine

**Fig. 2-3. Posterior view of thorax of a 29-year-old female.**

**APEX BEAT.** The apex of the heart is formed by the lower portion of the left ventricle. The apex beat is due to the apex of the heart being thrust forward against the thoracic wall as the heart contracts. The apex beat can usually be felt by placing the flat of the hand on the chest wall over the heart. After the area of cardiac pulsation has been determined, the apex beat is accurately localized by placing two fingers over the intercostal spaces, and moving them until the point of maximum pulsation is found. The apex beat is normally found in the fifth left intercostal space 3½ inches (9 cm) from the midline. Should you have difficulty in finding the apex beat, have the patient lean forward in the sitting position.

In the female with pendulous breasts, the examining fingers should gently raise the left breast from below as the intercostal spaces are palpated.

**AXILLARY FOLDS.** The *anterior fold* is formed by the lower border of the pectoralis major muscle (Figs. 2-1 and 2-2). This may be made to stand out by asking the patient to press his hand hard against his hip. The *posterior fold* is formed by the tendon

of latissimus dorsi muscle as it passes round the lower border of the teres major muscle.

## Examination of the Posterior Surface of the Thorax

**SPINOUS PROCESSES OF THORACIC VERTEBRAE.** These can all be palpated in the midline posteriorly (Fig. 2-5). The index finger should be placed on the skin in the midline on the posterior surface of the neck and drawn downward in the nuchal groove. The first spinous process to be felt is that of the seventh cervical vertebra (*vertebra prominens*). Below this level are the overlapping spines of the thoracic vertebrae. The cervical spines 1–6 are covered by a large ligament, the ligamentum nuchae. It should be noted that the tip of a spinous process of a thoracic vertebra lies posterior to the body of the next vertebra below.

**SCAPULA (SHOULDER BLADE).** This bone is flat and triangular in shape and is located on the upper part of the posterior surface of the thorax. The *superior angle* lies opposite the spine of the second thoracic vertebra (Figs. 2-3 and 2-5). The *spine of* *the scapula* is subcutaneous, and the root of the spine lies on a level with the spine of the third thoracic vertebra (Figs. 2-3 and 2-5). The *inferior angle* lies on a level with the spine of the seventh thoracic vertebra (Figs. 2-3 and 2-5).

## Lines of Orientation

**MIDSTERNAL LINE.** This lies in the median plane over the sternum (Fig. 2-4).

**MIDCLAVICULAR LINE.** This runs vertically downward from the midpoint of the clavicle (Fig. 2-4).

**ANTERIOR AXILLARY LINE.** This runs vertically downward from the anterior axillary fold (Fig. 2-4).

**POSTERIOR AXILLARY LINE.** This runs vertically downward from the posterior axillary fold.

**MIDAXILLARY LINE.** This runs vertically downward from a point situated midway between the anterior and posterior axillary folds.

**SCAPULAR LINE.** This runs vertically downward on the posterior wall of the thorax (Fig. 2-5), passing through the inferior angle of the scapula (arms at the sides).

## Surface Markings of the Trachea, Lungs, and Pleura

### TRACHEA

The trachea extends from the lower border of the cricoid cartilage (opposite the body of the sixth cervical vertebra) in the neck to the level of the sternal angle in the thorax (Fig. 2-6). It commences in the midline and ends just to the right of the midline by dividing into the right and left principal bronchi. At the root of the neck it may be palpated in the midline in the suprasternal notch.

### LUNGS

The *apex of the lung* projects into the neck. It can be mapped out on the anterior surface of the body by drawing a curved line, convex upward, from the sternoclavicular joint to a point 1 inch (2.5 cm) above the junction of the medial and intermediate thirds of the clavicle (Fig. 2-6).

The *anterior border of the right lung* begins behind the sternoclavicular joint and runs downward

almost reaching the midline behind the sternal angle. It then continues downward until it reaches the xiphisternal joint (Fig. 2-6). The *anterior border of the left lung* has a similar course, but at the level of the fourth costal cartilage it deviates laterally and extends for a variable distance beyond the lateral margin of the sternum to form the *cardiac notch* (Fig. 2-6). This notch is produced by the heart displacing the lung to the left. The anterior border then turns sharply downward to the level of the xiphisternal joint.

The *lower border of the lung* in midinspiration follows a curving line, which crosses the sixth rib in the midclavicular line and the eighth rib in the midaxillary line, and reaches the tenth rib adjacent to the vertebral column posteriorly (Figs. 2-6, 2-7, and 2-8). It is important to understand that the level of the inferior border of the lung changes during inspiration and expiration.

The *posterior border of the lung* extends downward from the spinous process of the seventh cervical vertebra to the level of the tenth thoracic vertebra and lies about 1½ inches (4 cm) from the midline (Fig. 2-7).

The *oblique fissure* of the lung can be indicated on the surface by a line drawn from the root of the spine of the scapula obliquely downward, laterally and anteriorly, following the course of the sixth rib to the sixth costochondral junction. In the left lung the upper lobe lies above and anterior to this line; the lower lobe lies below and posterior to it (Figs. 2-6 and 2-7).

In the right lung there is an additional fissure, the *horizontal fissure*, which may be represented by a line drawn horizontally along the fourth costal cartilage to meet the oblique fissure in the midaxillary line (Figs. 2-6 and 2-8). Above the horizontal fissure lies the upper lobe and below it, the middle lobe; below and posterior to the oblique fissure lies the lower lobe.

### PLEURA

The boundaries of the pleural sac can be marked out as lines on the surface of the body. The lines, which will indicate the limits of the parietal pleura where it lies close to the body surface, are referred to as the *lines of pleural reflection*.

The *cervical pleura* bulges upward into the neck and has a surface marking identical to that of the

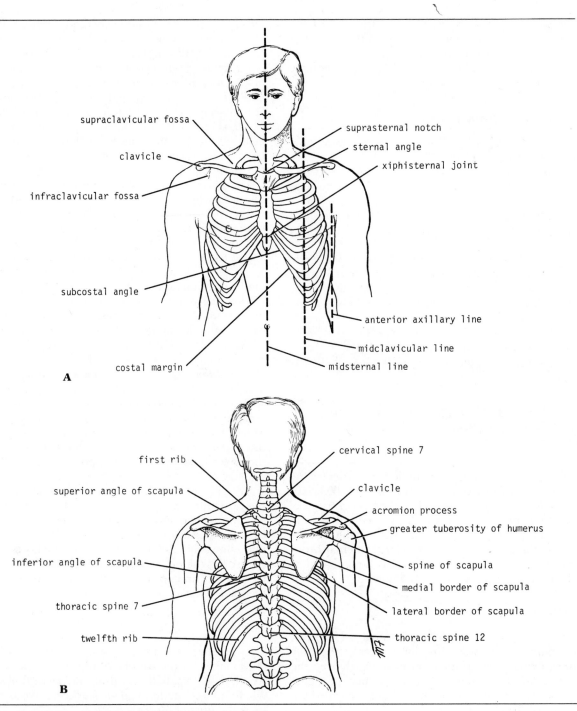

**Fig. 2-4. Surface landmarks of (A) anterior and (B) posterior thoracic walls.**

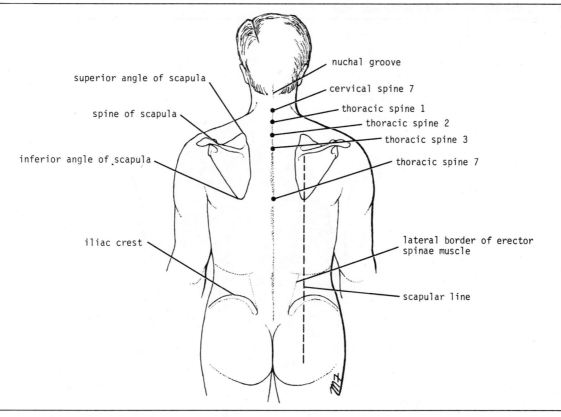

**Fig. 2-5. Surface landmarks of posterior thoracic wall.**

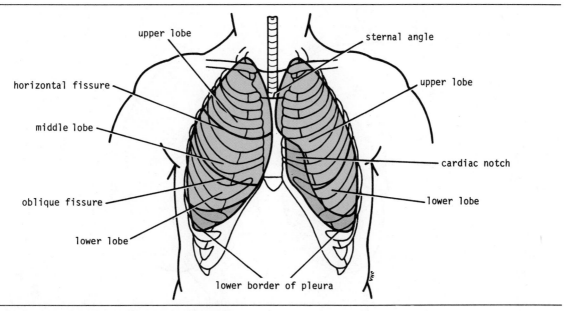

**Fig. 2-6. Surface markings of lungs and parietal pleura on anterior thoracic wall.**

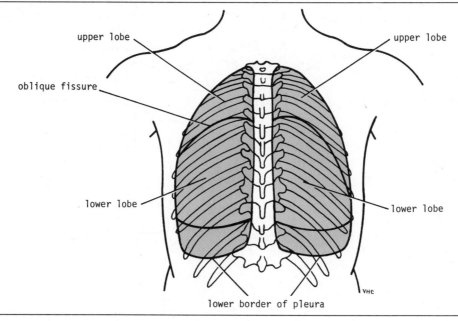

**Fig. 2-7. Surface markings of lungs and parietal pleura on posterior thoracic wall.**

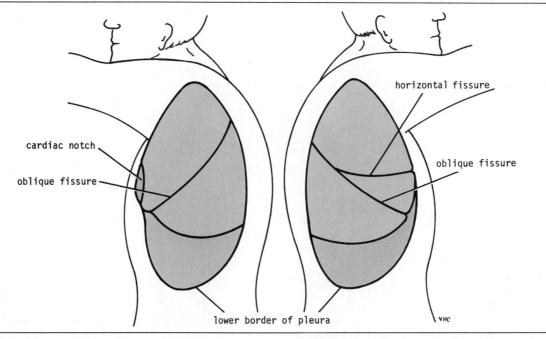

**Fig. 2-8. Surface markings of lungs and parietal pleura on lateral thoracic walls.**

apex of the lung. A curved line may be drawn, convex upward, from the sternoclavicular joint to a point 1 inch (2.5 cm) above the junction of the medial and intermediate thirds of the clavicle (Fig. 2-6).

The *anterior border of the right pleura* runs down behind the sternoclavicular joint almost reaching the midline behind the sternal angle. It then continues downward until it reaches the xiphisternal joint. The *anterior border of the left pleura* has a similar course, but at the level of the fourth costal cartilage it deviates laterally and extends to the lateral margin of the sternum to form the *cardiac notch.* (Note that the pleural cardiac notch is not as large as the cardiac notch of the lung). It then turns sharply downward to the xiphisternal joint (Fig. 2-6).

The *lower border of the pleura* on both sides follows a curved line, which crosses the eighth rib in the midclavicular line and the tenth rib in the midaxillary line, and reaches the twelfth rib adjacent to the vertebral column, i.e., at the lateral border of the erector spinae muscle (Figs. 2-6, 2-7, and 2-8). Note that the lower margins of the lungs cross the sixth, eighth, and tenth ribs at the midclavicular lines, the midaxillary lines, and the sides of the

vertebral column, respectively, and the lower margins of the pleura cross, at the same points respectively, the eighth, tenth, and twelfth ribs. The distance between the two borders corresponds to the *costodiaphragmatic recess.* (See p. 86.)

## Surface Markings of the Heart

For practical purposes the heart may be considered to have both an *apex* and *four borders.*

The *apex,* formed by the left ventricle, corresponds to the apex beat and is found in the fifth left intercostal space 3½ inches (9 cm) from the midline (Fig. 2-9).

The *superior border,* formed by the roots of the great blood vessels, extends from a point on the second left costal cartilage (remember sternal angle) ½ inch (1.3 cm) from the edge of the sternum to a point on the third right costal cartilage ½ inch (1.3 cm) from the edge of the sternum (Fig. 2-9).

The *right border,* formed by the right atrium, extends from a point on the third right costal cartilage ½ inch (1.3 cm) from the edge of the sternum downward to a point on the sixth right costal cartilage ½ inch (1.3 cm) from the edge of the sternum (Fig. 2-9).

The *left border,* formed by the left ventricle, extends from a point on the second left costal carti-

**Fig. 2-9. Surface markings of heart.**

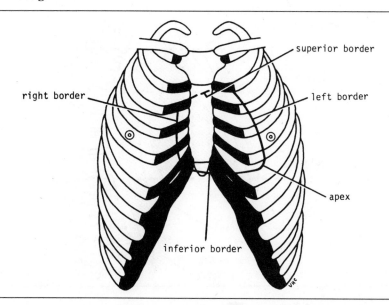

lage ½ inch (1.3 cm) from the edge of the sternum to the apex beat of the heart (Fig. 2-9).

The *inferior border,* formed by the right ventricle and the apical part of the left ventricle, extends from the sixth right costal cartilage ½ inch (1.3 cm) from the sternum to the apex beat (Fig 2-9).

## Surface Markings of the Vessels

The *arch of the aorta* and the roots of the *brachiocephalic* and *left common carotid arteries* lie behind the manubrium sterni (Fig. 2-10).

The *superior vena cava* and the terminal parts of the *right* and *left brachiocephalic veins* also lie behind the manubrium sterni.

The *internal thoracic vessels* run vertically downward posterior to the costal cartilages, ½ inch (1.13 cm) lateral to the edge of the sternum (Fig. 2-17), as far as the sixth intercostal space.

The *intercostal vessels and nerve* ("vein, artery, nerve"—V.A.N.—is the order from above downward) are situated immediately below their corresponding ribs (Fig. 2-16).

## Surface Anatomy of the Mammary Gland

The mammary gland lies in the superficial fascia covering the anterior chest wall (Fig. 2-2). In the child and in men it is rudimentary. In the female after puberty it enlarges and assumes its hemispherical shape. In the young adult female it overlies the second to the sixth ribs and their costal cartilages and extends from the lateral margin of the sternum to the midaxillary line. Its upper lateral edge extends around the lower border of the pectoralis major and enters the axilla. In middle-aged multiparous women the breasts may be large and pendulous. In older women past the menopause the adipose tissue of the breast may become reduced in amount and the hemispherical shape lost; the breasts then become smaller and the overlying skin is wrinkled.

The structure of the mammary gland is described fully on page 412.

**Fig. 2-10. Lateral view of thorax, showing relationship of surface markings to vertebral levels.**

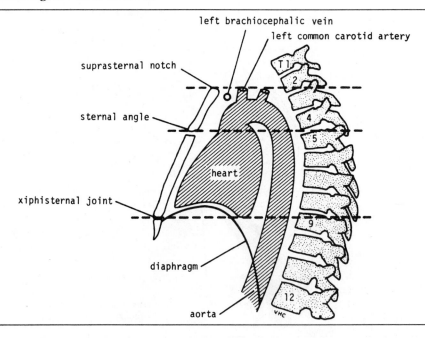

left brachiocephalic vein

left common carotid artery

suprasternal notch

sternal angle

T 1

2

4

5

heart

xiphisternal joint

diaphragm

9

12

VHC

aorta

## Openings of the Thorax

The thoracic cavity communicates with the root of the neck through an opening called the *thoracic inlet* (clinicians often call it the thoracic outlet). The opening is bounded posteriorly by the first thoracic vertebra, laterally by the medial borders of the first ribs and their costal cartilages, and anteriorly by the superior border of the manubrium sterni. The opening is obliquely placed facing upward and forward. Through this small opening pass the esophagus and trachea and many vessels and nerves. Because of the obliquity of the opening, the apices of the lung and pleurae project upward into the neck.

The thoracic cavity communicates with the abdomen through a large opening called the *thoracic outlet.* The opening is bounded posteriorly by the twelfth thoracic vertebra, laterally by the curving costal margin, and anteriorly by the xiphisternal joint. Through this large opening, which is closed by the diaphragm, pass the esophagus and many large vessels and nerves, all of which pierce the diaphragm.

## STRUCTURE OF THE THORACIC WALL

The thoracic wall is covered on the outside by skin and by muscles attaching the shoulder girdle to the trunk. It is lined with parietal pleura.

The thoracic wall is formed posteriorly by the thoracic part of the vertebral column; anteriorly, by the sternum and costal cartilages; laterally, by the ribs and intercostal spaces; superiorly, by the suprapleural membrane; and inferiorly, by the diaphragm, which separates the thoracic cavity from the abdominal cavity.

## Thoracic Part of the Vertebral Column

The thoracic part of the vertebral column is concave forward and is made up of twelve vertebrae, together with their intervertebral discs (Fig. 2-11). Thoracic vertebrae (Fig. 2-12) have the following characteristics:

1. The body is medium-sized and heart-shaped.
2. The spines are long and inclined downward.
3. Costal facets are present on the sides of the bodies where the heads of the ribs articulate, and on the transverse processes for articulation with the tubercles of the ribs. (T11 and 12 have no facets on the transverse processes.)
4. The articular processes lie on the arc of a circle whose center is located near the center of the vertebral body. This allows rotary movements to take place between adjacent vertebrae. (See also *vertebral column*, p. 931.)

## Sternum

The sternum is a flat bone that may be divided into three parts: (1) manubrium sterni, (2) body of the sternum, and (3) xiphoid process.

The *manubrium* is the upper part of the sternum, and it articulates with the clavicles and the first and upper part of the second costal cartilages on each side (Fig. 2-11). It lies opposite the third and fourth thoracic vertebrae.

The *body of the sternum* articulates above with the manubrium by means of a fibrocartilaginous joint, the *manubriosternal joint.* Below, it articulates with the xiphoid process. On each side are notches for articulation with the lower part of the second costal cartilage and the third to the seventh costal cartilages (Fig. 2-11). The second to the seventh costal cartilages articulate with the sternum at synovial joints.

The *xiphoid process* (Fig. 2-11) is the lowest and smallest part of the sternum. It is a thin plate of hyaline cartilage that becomes ossified at its proximal end in adult life.

The *sternal angle* (angle of Louis), formed by the articulation of the manubrium with the body of the sternum, can be recognized by the presence of a transverse ridge on the anterior aspect of the sternum (Fig. 2-10). The transverse ridge lies at the level of the second costal cartilage, the point from which all costal cartilages and ribs are counted. The sternal angle lies opposite the intervertebral disc between the fourth and fifth thoracic vertebrae.

The *xiphisternal joint* lies opposite the body of the ninth thoracic vertebra (Fig. 2-10).

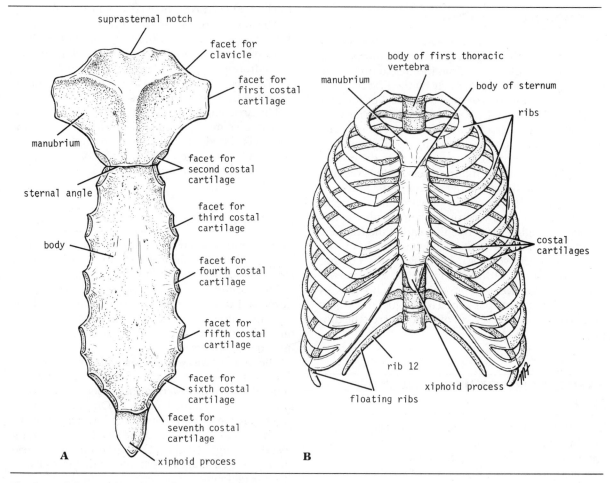

**Fig. 2-11. (A) Anterior view of sternum; (B) sternum, ribs, and costal cartilages forming thoracic skeleton.**

## Costal Cartilages

Costal cartilages are bars of hyaline cartilage connecting the upper seven ribs to the lateral edge of the sternum, and the eighth, ninth, and tenth ribs to the cartilage immediately above. The cartilages of the eleventh and twelfth ribs end in the abdominal musculature (Fig. 2-11).

The costal cartilages contribute significantly to the elasticity and mobility of the thoracic walls. In old age, the costal cartilages tend to lose some of their flexibility as the result of superficial calcification.

## Ribs

There are twelve pairs of ribs, all of which are attached posteriorly to the thoracic vertebrae (Figs. 2-11, 2-12, 2-13, and 2-14). The upper seven pairs are attached anteriorly to the sternum by their costal cartilages. The eighth, ninth, and tenth pairs of ribs are attached anteriorly to each other and to the seventh rib by means of their costal cartilages and small synovial joints. The eleventh and twelfth pairs have no anterior attachment and are referred to as *floating ribs*.

A *typical rib* is a long, twisted, flat bone having

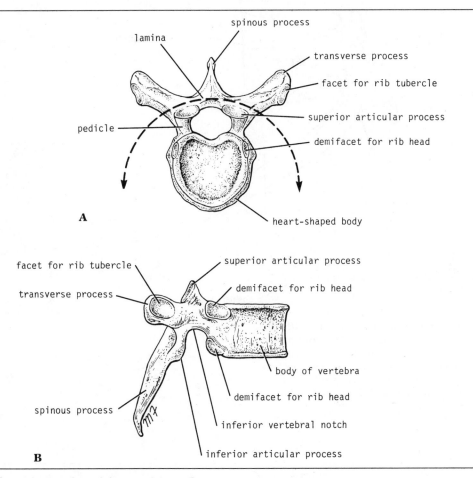

**A**

- spinous process
- lamina
- transverse process
- facet for rib tubercle
- pedicle
- superior articular process
- demifacet for rib head
- heart-shaped body

**B**

- facet for rib tubercle
- transverse process
- spinous process
- superior articular process
- demifacet for rib head
- body of vertebra
- demifacet for rib head
- inferior vertebral notch
- inferior articular process

**Fig. 2-12. Thoracic vertebra. (A) Superior surface. (B) Lateral surface.**

a rounded, smooth superior border and a sharp, thin inferior border (Figs. 2-13 and 2-14). The inferior border overhangs and forms the *costal groove*, which accommodates the intercostal vessels and nerve.

A rib has a *head*, *neck*, *tubercle*, *shaft*, and *angle* (Figs. 2-13 and 2-14). The *head* has two facets for articulation with the numerically corresponding vertebral body and that of the vertebra immediately above (Fig. 2-13). The *neck* is a constricted portion situated between the head and the tubercle.

The *tubercle* is a prominence on the outer surface of the rib at the junction of the neck with the shaft.

It has a facet for articulation with the transverse process of the numerically corresponding vertebra (Fig. 2-13). The *shaft or body* is thin and flattened and twisted on its long axis. Its inferior border has the costal groove. The *angle* is where the shaft of the rib bends sharply forward. The anterior end of each rib is attached to the corresponding costal cartilage.

The first rib is *atypical*. It is important because of its close relationship to the nerves of the brachial plexus and the main vessels to the arm, namely, the subclavian vessels (Fig. 2-15). This rib is flattened from above downward. It has a tubercle on the inner border, known as the *scalene tubercle*, for the insertion of the scalenus anterior muscle. Anterior to the tubercle the subclavian vein crosses the rib; posterior to the tubercle is the *subclavian groove*,

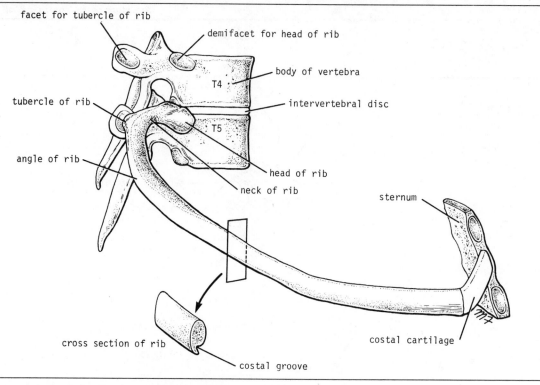

**Fig. 2-13. Fifth right rib as it articulates with vertebral column posteriorly and with sternum anteriorly. Note that rib head articulates with vertebral body of its own number and that of vertebra immediately above. Note also presence of costal groove along inferior border of the rib.**

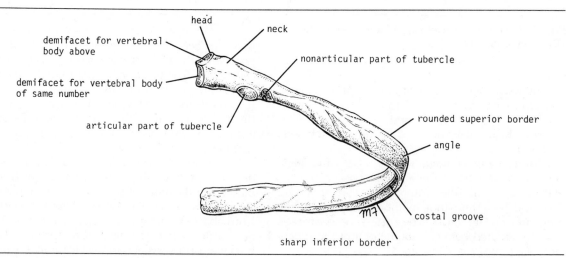

**Fig. 2-14. Fifth right rib, as seen from posterior aspect.**

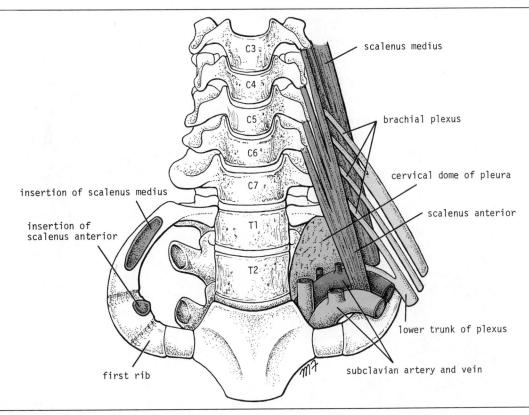

**Fig. 2-15. Thoracic inlet, showing cervical dome of pleura on left side of body and its relationship to inner border of first rib. Note also the presence of brachial plexus and subclavian vessels.**

where the subclavian artery and the lower trunk of the brachial plexus cross the rib and lie in contact with the bone.

## Intercostal Spaces

The spaces between the ribs are called intercostal spaces. Each space contains three muscles, comparable to those of the anterior abdominal wall, and a neurovascular bundle (Fig. 2-16).

## Intercostal Muscles

The *external intercostal muscle* forms the most superficial layer. Its fibers are directed downward and forward from the inferior border of the rib above to the superior border of the rib below (Fig. 2-16). The muscle extends forward from the rib tubercle behind to the costochondral junction in front, where the muscle is replaced by an aponeurosis, the *anterior (external) intercostal membrane* (Fig. 2-17).

The *internal intercostal muscle* forms the intermediate layer. Its fibers are directed downward and backward from the subcostal groove of the rib above to the upper border of the rib below (Fig. 2-16). The muscle extends backward from the sternum in front to the angles of the ribs behind, where the muscle is replaced by an aponeurosis, the *posterior (internal) intercostal membrane* (Fig. 2-17).

The *transversus thoracis muscle* forms the deepest layer and corresponds to the transversus abdominis muscle in the anterior abdominal wall. It

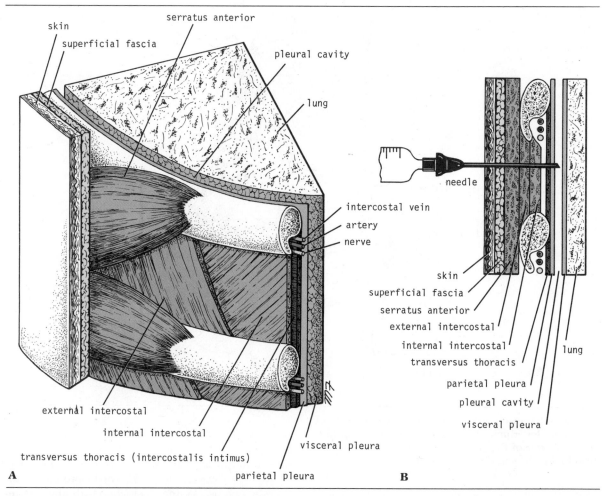

**Fig. 2-16. (A) Section through an intercostal space. (B) Structures penetrated by a needle when it passes from skin surface to pleural cavity. Needle has been introduced into seventh intercostal space in midaxillary line.**

is an incomplete muscle layer and crosses more than one intercostal space within the ribs. It is related internally to fascia (endothoracic fascia) and parietal pleura and externally to the intercostal nerves and vessels. The transversus thoracis muscle may be divided into three portions (Fig. 2-17), which are more or less separate from one another: (1) the subcostalis, (2) the intercostalis intimus, and (3) the sternocostalis. It is not necessary to

learn the details of the attachments of these muscle fibers.

## Action of Intercostal Muscles

When the intercostal muscles contract, they all tend to pull the ribs nearer to one another. If the first rib is fixed by the contraction of the muscles in the root of the neck, namely, the scaleni muscles, the intercostal muscles will raise the second to the twelfth ribs toward the first. If, on the other hand, the twelfth rib is fixed by the quadratus lumborum muscle and the oblique muscles of the abdomen, the first to the eleventh ribs will be lowered by the

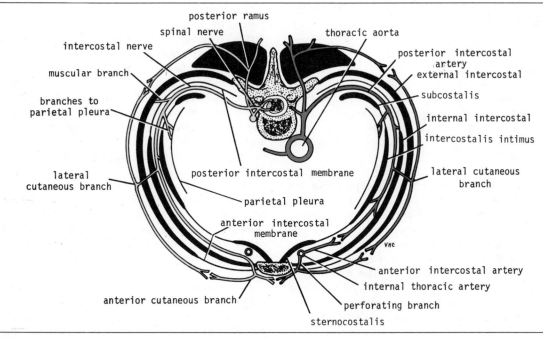

**Fig. 2-17. Cross section of thorax, showing distribution of a typical intercostal nerve and a posterior and an anterior intercostal artery.**

contraction of the intercostal muscles. In addition, the tone of the intercostal muscles during the different phases of respiration serves to strengthen the tissues of the intercostal spaces, thus preventing the sucking in or the blowing out of the tissues with changes in intrathoracic pressure. For further details concerning the action of these muscles, see mechanics of respiration on page 95.

### *Nerve Supply of Intercostal Muscles*

The intercostal muscles are supplied by the corresponding intercostal nerves.

The intercostal nerves and blood vessels (the neurovascular bundle), as in the abdominal wall, run between the middle and innermost layers of muscles (Figs. 2-16 and 2-17). They are arranged in the following order from above downward: intercostal vein, intercostal artery, and intercostal nerve (i.e., V.A.N.).

## Intercostal Arteries and Veins

Each intercostal space possesses a large single *posterior intercostal artery* and two small *anterior intercostal arteries*.

The *posterior intercostal arteries* of the first two spaces are branches from the superior intercostal artery, a branch of the costocervical trunk of the subclavian artery. The posterior intercostal arteries of the lower nine spaces are branches of the thoracic aorta (Fig. 2-17).

The *anterior intercostal arteries* of the first six spaces are branches of the internal thoracic artery (Fig. 2-17). The anterior intercostal arteries of the lower spaces are branches of the musculophrenic artery, one of the terminal branches of the internal thoracic artery.

Each intercostal artery gives off branches to the muscles, skin, and parietal pleura. In the region of the breast in the female, the branches to the superficial structures are particularly large.

The corresponding *posterior intercostal veins* drain backward into the azygos or hemiazygos veins, and the *anterior intercostal veins* drain forward into the internal thoracic and musculo-phrenic veins.

## Intercostal Nerves

The intercostal nerves are the anterior rami of the first eleven thoracic spinal nerves. The anterior ramus of the twelfth thoracic nerve lies in the abdomen and runs forward in the abdominal wall as the *subcostal nerve.*

Each intercostal nerve enters an intercostal space between the parietal pleura and the posterior intercostal membrane (Figs. 2-16 and 2-17). It then runs forward inferiorly to the intercostal vessels in the subcostal groove of the corresponding rib, between the transversus thoracis and internal intercostal muscle. The first six nerves are distributed within their intercostal spaces. The seventh to ninth intercostal nerves leave the anterior ends of their intercostal spaces by passing deep to the costal cartilages, to enter the anterior abdominal wall. In the case of the tenth and eleventh nerves, since the corresponding ribs are floating, these nerves pass directly into the abdominal wall.

### BRANCHES

The branches of the intercostal nerves (Fig. 2-17) are as follows:

1. *Rami communicantes* connect the intercostal nerve to a ganglion of the sympathetic trunk (see Fig. 1-26). The gray ramus joins the nerve medial to the point at which the white ramus leaves it.
2. A *collateral branch*, which runs forward inferiorly to the main nerve on the upper border of the rib below.
3. A *lateral cutaneous branch*, which reaches the skin near the midaxillary line. It divides into an anterior and a posterior branch.
4. An *anterior cutaneous branch*, which reaches the skin near the midline. It divides into a medial and a lateral branch.
5. Numerous *muscular branches* are given off by the main nerve and its collateral branch.
6. *Pleural* and *peritoneal* (7–11 intercostal nerves only) *sensory branches.*

The *first intercostal nerve* is joined to the brachial plexus by a large branch that is equivalent to the lateral cutaneous branch of typical intercostal nerves. The remainder of the first intercostal nerve is small, and there is no anterior cutaneous branch.

The *second intercostal nerve* is joined to the medial cutaneous nerve of the arm by a branch, called the *intercostobrachial nerve*, that is equivalent to the lateral cutaneous branch of other nerves. The second intercostal nerve therefore supplies the skin of the armpit and the upper medial side of the arm.

With the exceptions noted, the first six intercostal nerves therefore supply (1) the skin and the parietal pleura covering the outer and inner surfaces of each intercostal space, respectively, and (2) the intercostal muscles of each intercostal space and the levator costarum and serratus posterior muscles.

In addition, the seventh to the eleventh intercostal nerves supply (1) the skin and the parietal peritoneum covering the outer and inner surfaces of the abdominal wall, respectively, and (2) the anterior abdominal muscles, which include the external oblique, internal oblique, transversus abdominis, and rectus abdominis muscles.

## Suprapleural Membrane

Superiorly, the thorax opens into the root of the neck by a narrow aperture known as the *thoracic inlet.* As stated previously, it is bounded by the superior border of the manubrium sterni, by the medial borders of the first ribs, and by the body of the first thoracic vertebra. The thoracic inlet transmits structures that pass between the thorax and the neck (esophagus, trachea, blood vessels, etc.) and for the most part lie close to the midline. On either side of these structures the inlet is closed by a dense fascial layer called the *suprapleural membrane* (Fig. 2-18). This tent-shaped fibrous sheet is attached laterally to the medial border of the first rib and costal cartilage. It is attached at its apex to the tip of the transverse process of the seventh cervical vertebra and medially to the fascia investing the structures passing from the thorax into the neck. It protects the underlying cervical pleura and resists the changes in intrathoracic pressure occurring during respiratory movements.

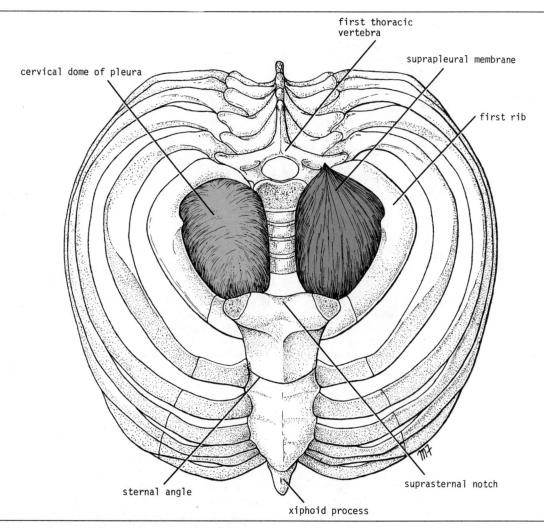

first thoracic
vertebra

suprapleural membrane

cervical dome of pleura

first rib

sternal angle

suprasternal notch

xiphoid process

**Fig. 2-18. Thoracic cage viewed from above, showing thoracic inlet. Note the cervical dome of parietal pleura on the right side and the suprapleural membrane on the left side. Note that the fibrous suprapleural membrane is attached superiorly to the transverse process of the seventh cervical vertebra.**

## Diaphragm

Inferiorly, the thorax opens into the abdomen by a wide aperture known as the *thoracic outlet*. It is bounded by the xiphisternal joint, the costal margin, and the body of the twelfth thoracic vertebra. It is closed by a muscular and tendinous septum, the diaphragm, which is pierced by the structures that pass between the thorax and the abdomen.

The diaphragm is the primary muscle of respiration. It is dome-shaped and consists of a peripheral muscular part, which arises from the margins

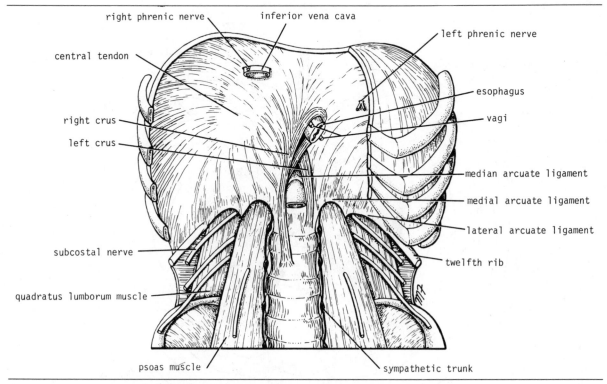

**Fig. 2-19. Diaphragm as seen from below. The anterior portion of the right side has been removed. Note sternal, costal, and vertebral origins of muscle and important structures that pass through it.**

of the thoracic outlet, and a centrally placed tendon (Fig. 2-19). The origin of the diaphragm may be divided into three parts:

1. A *sternal part* consisting of small right and left slips arising from the posterior surface of the xiphoid process (Fig. 2-19).
2. A *costal part* consisting of six slips that arise from the deep surfaces of the lower six ribs and their costal cartilages (Fig. 2-19).
3. A *vertebral part* arising by means of vertical columns or *crura* and from the arcuate ligaments.

The *right crus* arises from the sides of the bodies of the first three lumbar vertebrae; and the intervertebral discs; the *left crus* arises from the sides of the bodies of the first two lumbar vertebrae and the intervertebral disc (Fig. 2-19).

Lateral to the crura the diaphragm arises from

the *medial and lateral arcuate ligaments* (Fig. 2-19). The medial ligament is the thickened upper margin of the fascia covering the anterior surface of the psoas muscle (Fig. 2-19), and the lateral ligament is the thickened upper margin of the fascia covering the anterior surface of the quadratus lumborum muscle. The medial arcuate ligament extends from the side of the body of the second lumbar vertebra to the tip of the transverse process of the first lumbar vertebra. The lateral arcuate ligament extends from the tip of the transverse process of the first lumbar vertebra to the lower border of the twelfth rib.

The fibrous medial borders of the two crura are connected by a *median arcuate ligament*, which crosses over the anterior surface of the aorta (Fig. 2-19).

The diaphragm is inserted into a *central tendon*, which is trefoil in shape. The superior surface of the tendon is partially fused with the inferior surface of the fibrous pericardium. Some of the muscle fibers of the right crus pass up to the left and surround the esophageal orifice in a slinglike loop. These fibers appear to act as a sphincter and possibly assist in the prevention of regurgitation of the stomach contents into the thoracic part of the esophagus (Fig. 2-19).

## Shape of the Diaphragm

As seen from in front, the diaphragm curves up into right and left domes. The right dome reaches as high as the upper border of the fifth rib, and the left dome may reach the lower border of the fifth rib. The central tendon lies at the level of the xiphisternal joint. The levels of the diaphragm will vary with the phase of respiration, posture, and the degree of distention of the abdominal viscera. The diaphragm is lower when a person is sitting or standing; it is higher in the supine position and after a large meal.

When seen from the side, the diaphragm has the appearance of an inverted J, the long limb extending up from the vertebral column and the short limb extending forward to the xiphoid process.

## Action

1. *Muscle of inspiration.* On contraction, the diaphragm pulls down its central tendon and increases the vertical diameter of the thorax. The diaphragm is the most important muscle used in inspiration.
2. *Muscle of abdominal straining.* Its contraction aids that of the muscles of the anterior abdominal wall in raising the intra-abdominal pressure to evacuate the pelvic contents (micturition, defecation, parturition). This mechanism is further aided by the person taking a deep breath and closing the glottis. The diaphragm is unable to rise due to the air trapped in the respiratory tract. Now and again, air is allowed to escape, producing a grunting sound.
3. *Weight-lifting muscle.* By taking a deep breath and fixing the diaphragm as described above, it is possible to raise the intra-abdominal pressure to such an extent that it will help support the vertebral column and prevent flexion. This greatly assists the postvertebral muscles in the lifting of heavy weights. Needless to say, it is important to have adequate sphincteric control of

the bladder and anal canal under these circumstances.

4. *Thoraco-abdominal pump.* The descent of the diaphragm decreases the intrathoracic pressure and at the same time increases the intra-abdominal pressure. This pressure change compresses the blood in the inferior vena cava and forces it upward into the right atrium of the heart. Lymph within the abdominal lymph vessels is also compressed and its passage upward within the thoracic duct is aided by the negative intrathoracic pressure. The presence of valves within the thoracic duct prevents backflow.

## Nerve Supply

The motor nerve supply is from the phrenic nerve (C3, 4, 5) only. The sensory nerve supply to the parietal pleura and peritoneum covering the central surfaces of the diaphragm is from the phrenic nerve. The sensory supply to the periphery of the diaphragm is from the lower five intercostal nerves.

## Openings in the Diaphragm

The diaphragm has three main openings:

1. The *aortic opening*, which lies anterior to the body of the twelfth thoracic vertebra between the crura (Fig. 2-19). It transmits the aorta, the thoracic duct, and the azygos vein.
2. The *esophageal opening*, which lies at the level of the tenth thoracic vertebra in a sling of muscle fibers derived from the right crus (Fig. 2-19). It transmits the esophagus, the right and left vagus nerves, the esophageal branches of the left gastric vessels, and the lymphatics from the lower one-third of the esophagus.
3. The *caval opening*, which lies at the level of the eighth thoracic vertebra in the central tendon (Fig. 2-19). It transmits the inferior vena cava and terminal branches of the right phrenic nerve.

In addition to these structures, the greater, lesser, and lowest splanchnic nerves pierce the crura, the sympathetic trunk passes posterior to the medial arcuate ligament on each side, and the superior epigastric vessels pass between the sternal and costal origins of the diaphragm on each side

(Fig. 2-19). The left phrenic nerve pierces the left dome to supply the peritoneum on its undersurface, and the neurovascular bundles of the seventh to the eleventh intercostal spaces pass into the anterior abdominal wall between the muscular slips of the costal origin of the diaphragm.

## Internal Thoracic Artery

The internal thoracic artery supplies the anterior wall of the body from the clavicle to the umbilicus. It is a branch of the first part of the subclavian artery in the neck. It descends vertically on the pleura behind the costal cartilages, a fingerbreadth lateral to the sternum, and ends in the sixth intercostal space by dividing into the superior epigastric and musculophrenic arteries (Fig. 2-17).

### BRANCHES

The branches of the internal thoracic artery are as follows:

1. Two *anterior intercostal arteries* for the upper six intercostal spaces.
2. *Perforating arteries*, which accompany the terminal branches of the corresponding intercostal nerves.
3. The *pericardiacophrenic artery*, which accompanies the phrenic nerve and supplies the pericardium.
4. *Mediastinal arteries* to the contents of the anterior mediastinum, e.g., the thymus gland.
5. The *superior epigastric artery*, which enters the rectus sheath and supplies the rectus muscle as far as the umbilicus.
6. The *musculophrenic artery*, which runs around the costal margin of the diaphragm and supplies the lower intercostal spaces and the diaphragm.

## Internal Thoracic Vein

The internal thoracic vein begins as venae comitantes of the internal thoracic artery. The venae eventually join to form a single vessel, which drains into the brachiocephalic vein on each side.

## Levatores Costarum

There are twelve pairs of muscles. Each levator costae is triangular in shape and arises by its apex from the tip of the transverse process and is inserted into the rib below.

### Action

Each levator costae raises the rib below and is therefore an inspiratory muscle.

### Nerve Supply

Posterior rami of thoracic spinal nerves.

## Serratus Posterior Superior Muscle

The *serratus posterior superior* is a thin, flat muscle that arises from the lower cervical and upper thoracic spines. Its fibers pass downward and laterally and are inserted into the upper ribs.

### Action

It elevates the ribs and is therefore an inspiratory muscle.

### Nerve Supply

Intercostal nerves.

## Serratus Posterior Inferior Muscle

The *serratus posterior inferior* is a thin, flat muscle that arises from the upper lumbar and lower thoracic spines. Its fibers pass upward and laterally and are inserted into the lower ribs.

### Action

It depresses the ribs and is therefore an expiratory muscle.

### Nerve Supply

Intercostal nerves.

A summary of the muscles of the thorax, their nerve supply, and their actions is given in Table 2-1.

Table 2-1. Muscles of the Thorax

| Name of muscle | Origin | Insertion | Nerve supply | Action |
|---|---|---|---|---|
| External intercostal muscle (11) (Fibers pass downward and forward) | Inferior border of rib | Superior border of rib below | Intercostal nerves | With first rib fixed they raise ribs during inspiration and thus increase anteroposterior and transverse diameters of thorax. With last rib fixed by abdominal muscles they lower ribs during expiration |
| Internal intercostal muscle (11) (Fibers pass downward and backward) | Inferior border of rib | Superior border of rib below | Intercostal nerves | |
| Transversus thoracis (Incomplete layer— subcostalis, intercostalis intimus, and sternocostalis) | Adjacent ribs | Adjacent ribs | Intercostal nerves | Assist external and internal intercostal muscles |
| Diaphragm (Most important muscle of respiration) | Xiphoid process; lower six costal cartilages, first three lumbar vertebrae | Central tendon | Phrenic nerve | Very important muscle of inspiration; increases vertical diameter of thorax by pulling central tendon downward, assists in raising lower ribs. Also used in abdominal straining and weight-lifting |
| Levatores costarum (12) | Tip of transverse process of C7 and T1–11 vertebrae | Rib below | Posterior rami of thoracic spinal nerves | Raise ribs and therefore inspiratory muscles |
| Serratus posterior superior | Lower cervical and upper thoracic spines | Upper ribs | Intercostal nerves | Raises ribs and therefore inspiratory muscle |
| Serratus posterior inferior | Upper lumbar and lower thoracic spines | Lower ribs | Intercostal nerves | Depresses ribs and therefore expiratory muscle |

# CLINICAL NOTES

When one is examining the chest from in front, the *sternal angle* is a very important landmark. Its position can easily be felt and often seen by the presence of a transverse ridge. The finger moved to the right or to the left will pass directly onto the second costal cartilage and then the second rib. All other ribs may be counted from this point. The *twelfth rib* can usually be felt from behind, but in some obese persons this may prove difficult.

The *cutaneous innervation* of the anterior chest wall above the level of the sternal angle is derived from the *supraclavicular nerves* (C3 and 4). Below this level the anterior and lateral cutaneous branches of the intercostal nerves supply oblique bands of skin in regular sequence. The skin on the posterior surface of the chest wall is supplied by the posterior rami of the spinal nerves. The arrangement of the dermatomes is shown in Chapter 1, Figures 1-31 and 1-32.

The *lymphatic drainage* of the skin of the anterior chest wall passes to the anterior axillary lymph nodes; that from the posterior chest wall passes to the posterior axillary nodes (Fig. 2-20). The lymphatic drainage of the intercostal spaces passes forward to the internal thoracic nodes, situated along the internal thoracic artery, and posteriorly to the posterior intercostal nodes and the para-aortic nodes in the posterior mediastinum. The lymphatic drainage of the breast is described on page 412.

The *shape* of the thorax may be distorted by congenital anomalies of the vertebral column or by the ribs. Destructive disease of the vertebral column producing lateral flexion or scoliosis results in marked distortion of the thoracic cage.

*Traumatic injury* to the thorax is common, especially as a result of automobile accidents. In children the ribs are highly elastic, and fractures of ribs are therefore rare in this age group. In the adult the ribs tend to break at their weakest part in the region of their angles. The first two ribs are protected by the clavicle and the pectoralis major muscle, and the last two ribs are mobile and are rarely injured.

In *severe crush injuries* a number of ribs may break. If limited to one side, the fractures may occur near the rib angles and also anteriorly, near the costochondral junctions. This causes *flail chest*. If the fractures occur on either side of the sternum,

the sternum may be flail. In either case, the stability of the chest wall is lost, and the flail segment is sucked in during inspiration and driven out during expiration, thus producing paradoxical respiratory movements. One of the dangers of a fractured rib is that it may damage the underlying lung or upper abdominal organs, such as the liver or spleen.

A *cervical rib*, i.e., a rib arising from the anterior tubercle of the transverse process of the seventh cervical vertebra, occurs in about 0.5 percent of persons (Fig. 2-21). It may have a free anterior end, may be connected to the first rib by a fibrous band, or may articulate with the first rib. The importance of a cervical rib is that it may cause pressure on the lower trunk of the brachial plexus in some subjects, producing pain down the medial side of the forearm and hand and wasting of the small muscles of the hand. It may also exert pressure on the overlying subclavian artery and interfere with the circulation of the upper limb.

*Rib excision* is commonly performed by thoracic surgeons wishing to gain entrance to the thoracic cavity. A longitudinal incision is made through the periosteum on the outer surface of the rib and a segment of the rib is removed. A second longitudinal incision is then made through the bed of the rib, which is the inner covering of periosteum. Following the operation, the rib regenerates from the osteogenetic layer of the periosteum.

In situations requiring a smaller exposure, a *thoracotomy* may be performed simply by making an incision through an intercostal space.

Remember that costal cartilages sometimes become ossified, especially in old age, and this alters their usual radiographic appearance.

Since the *sternum* possesses red hematopoietic marrow throughout life, it is a common site for

**Fig. 2-21. Thoracic inlet as seen from above. Note presence of cervical ribs (solid black) on both sides. On right side of thorax, rib is almost complete and articulates anteriorly with first rib. On left side of thorax, rib is rudimentary but is continued forward as fibrous band that is attached to first costal cartilage. Note that cervical rib may exert pressure on lower trunk of brachial plexus and may kink subclavian artery.**

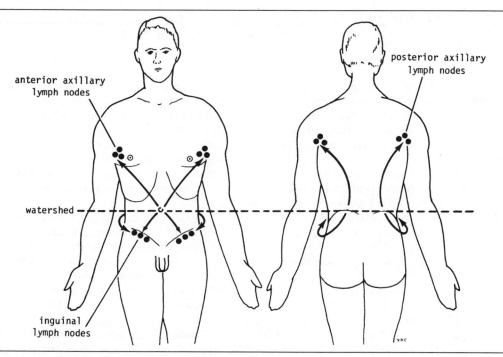

**Fig. 2-20. Lymphatic drainage of skin of thorax and abdomen. Note that levels of umbilicus anteriorly and iliac crests posteriorly may be regarded as watersheds for lymph flow.**

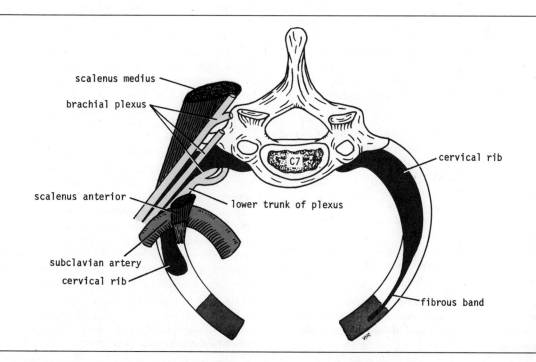

marrow biopsy. Under a local anesthetic, a wide-bore needle is introduced into the marrow cavity through the anterior surface of the bone. The sternum may also be split at operation to allow the surgeon to gain easy access to the heart, great vessels, and thymus gland.

It is important to realize that the *lower ribs* and *costal margin* overlap and protect the upper abdominal contents. Nevertheless, severe injuries to the lower part of the thoracic wall may well be associated with unsuspected injury to the liver, spleen, or kidney.

It is hardly necessary to emphasize the importance of knowing the surface markings of the pleural reflections and the lobes of the lungs. When listening to the breath sounds of the respiratory tract, it should be possible to have a mental image of the structures that lie beneath the stethoscope.

The *cervical dome of the pleura* and the *apex of the lungs* extend up into the neck so that at their highest point they lie about 1 inch (2.5 cm) above the clavicle (Figs. 2-6, 2-15, and 2-18). Consequently they are vulnerable to stab wounds in the root of the neck or to damage by an anesthetist's needle when a nerve block of the lower trunk of the brachial plexus is being performed.

Remember also that the *lower limit of the pleural reflection*, as seen from the back, may be damaged during nephrectomy operations. The pleura crosses the twelfth rib and may be damaged during removal of the kidney through an incision in the loin.

Certain anatomical and physiological changes take place in the thorax with advancing years:

1. The *rib cage* becomes more rigid and loses its elasticity as the result of calcification and even ossification of the costal cartilages.
2. The stooped posture (*kyphosis*), so often seen in the old due to degeneration of the intervertebral discs, decreases the chest capacity.
3. *Disuse atrophy* of the thoracic and abdominal muscles may result in poor respiratory movements.
4. *Degeneration of the elastic tissue* in the lungs and bronchi results in impairment of the movement of expiration.

These changes, when severe, diminish the efficiency of the respiratory movements and impair the ability of the individual to withstand respiratory disease.

*Hiccup* is the involuntary spasmodic contraction of the diaphragm accompanied by the approximation of the vocal folds and closure of the glottis of the larynx. It is a common condition in normal individuals and occurs after eating or drinking as a result of gastric irritation of the vagus nerve endings. It may, however, be a symptom of disease such as pleurisy, peritonitis, pericarditis, or uremia.

The surface markings of the *heart* and the position of the apex beat may enable a physician to determine whether the heart has shifted its position in relation to the chest wall or whether the heart is enlarged by disease. The apex beat can often be seen and almost always can be felt. The position of the margins of the heart can be determined by percussion.

The *arch of the aorta* lies behind the manubrium sterni. A gross dilatation of the aorta (*aneurysm*) may show itself as a pulsatile swelling in the suprasternal notch.

*Coarctation of the aorta*, a congenital anomaly, may produce a dilatation of the vessels taking part in a collateral circulation to bypass the narrowing. As a result, the intercostal arteries undergo extreme dilatation and erode the lower borders of the ribs, producing characteristic notching that is seen on radiographic examination.

It may be necessary to pass a hypodermic needle through an intercostal space to withdraw a sample of pleural fluid or to drain fluid (pus or blood) away from the pleural cavity (Fig. 2-16). Or it may be necessary to anesthetize an intercostal nerve and produce a nerve block. In both cases it is essential that the physician know the structures that his needle penetrates. For example, in the midaxillary line over the eighth intercostal space, the needle pierces the following structures from without inward: (1) skin, (2) superficial fascia, (3) digitation of serratus anterior muscle, (4) external intercostal muscle, (5) internal intercostal muscle, (6) transversus thoracis muscle, and (7) parietal pleura. In order to avoid damage to the intercostal vessels and nerve, the needle should be inserted close to the upper border of the rib. To block an intercostal nerve, the anesthetic is infiltrated around the nerve trunk as it lies in the subcostal groove.

The *diaphragm is developed* from three main sources in the embryo: (1) the septum transversum (the fused myotomes of segments C3, 4, and 5); (2)

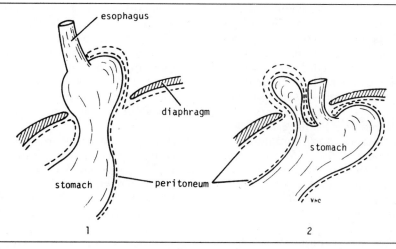

**Fig. 2-22. (1) Sliding esophageal hernia and (2) para-esophageal hernia.**

the dorsal mesentery; and (3) the pleuroperitoneal membranes from the body wall. Normally, these fuse together. Occasionally, fusion is incomplete and *congenital herniae* may occur through the following structures:

1. The pleuroperitoneal canal. This is more common on the left side and is due to failure of fusion of the septum transversum with the pleuroperitoneal membrane.
2. The opening between the xiphoid and costal origins of the diaphragm.
3. The esophageal hiatus.

*Acquired herniae* may occur in middle-aged persons with weak musculature around the esophageal opening in the diaphragm. These herniae may be either (1) sliding or (2) para-esophageal (Fig. 2-22).

A single dome of the diaphragm may be *paralyzed* by crushing or sectioning of the phrenic nerve in the neck. This may be necessary in the treatment of certain forms of lung tuberculosis, when the physician wishes to rest the lower lobe of the lung on one side. Occasionally, the contribution from the fifth cervical spinal nerve joins the phrenic nerve late as a branch from the nerve to the subclavius muscle. This is known as the *accessory phrenic nerve*. To obtain complete paralysis under these circumstances, the nerve to the subclavius muscle must also be sectioned.

## CLINICAL PROBLEMS

*Answers on page 956*

1. On examination, a patient is found to have a soft swelling over the fifth left intercostal space in the midaxillary line. Indicate on yourself the position of the swelling.
2. A patient is found to be tender over the spine of the thoracic vertebra that is on a level with the inferior angle of the scapula. To which thoracic vertebra does the spine belong?
3. On percussion of the right border of the heart, the margin is found to lie 2 inches (5 cm) to the right of the edge of the sternum. Which chamber of the heart is likely to be enlarged?
4. A soldier who had received a shrapnel wound in the neck 2 years previously noticed that when he blew his nose or sneezed, the skin above the right clavicle bulged upward. Explain this upward bulging of the skin in anatomical terms.
5. A young resident obtained a sample of pleural fluid from a patient's pleural cavity. He inserted the needle close to the lower border of the

eighth rib in the anterior axillary line. The next morning he was surprised to hear that the patient had complained of altered skin sensation extending from the point where the needle was inserted downward and forward to the midline of the abdominal wall above the umbilicus. Can you explain this altered sensation in anatomical terms?

6. A patient complained of a swelling in the skin on the back of the chest. On examination, a hard lump was found in the skin in the right scapula line opposite the seventh thoracic vertebra. It was a malignant tumor. Where would you examine for enlargement of the lymph nodes draining this region?

7. A 65-year-old man and a 10-year-old boy were both involved in a severe automobile accident. In both cases the thorax had been badly crushed. On X-ray examination, the man was seen to have five fractured ribs, but the boy had no fractures. Comment on these findings.

8. On examination of a routine postero-anterior chest radiograph of an 85-year-old woman, it

was noticed that many of the costal cartilages showed scattered radiopaque areas. What is the likely explanation for these opacities?

9. On examination of a postero-anterior chest radiograph of a young woman, it was seen that the left dome of the diaphragm was higher than the right dome and reached to the upper border of the fourth rib. Comment on this finding.

10. A patient suffering from right-sided pleurisy complained of pain over the tenth right intercostal space that extended downward and forward across the costal margin as far as the umbilicus. Explain in anatomical terms why the pain was felt over such an extensive area.

11. Certain anatomical and physiological changes take place in the thorax with advancing years and, when severe, impair the ability of the individual to withstand respiratory disease. Describe these changes.

12. Describe the main anatomical structures involved in the production of hiccup.

## NATIONAL BOARD TYPE QUESTIONS

*Answers on page 983*

**In each of the following questions, answer:**

    (a) If (1) is correct only

    (b) If (2) is correct only

    (c) If both (1) and (2) are correct, and

    (d) If neither (1) nor (2) is correct

1. (1) The trachea bifurcates opposite the manubriosternal joint (angle of Louis) in the midrespiratory position.

   (2) The arch of the aorta lies behind the body of the sternum.

2. (1) The apex beat of the heart can normally be felt in the fifth left intercostal space about 3½ inches (9 cm) from the midline.

   (2) The lower margin of the right lung on full inspiration could extend down in the midclavicular line to the eighth costal cartilage.

3. (1) All intercostal nerves are derived from anterior rami of thoracic spinal nerves.

   (2) The parietal pleura is sensitive to the sensations of pain and touch.

**Match the statement on the left with the best response on the right.**

4. The thoracic duct passes through the _____ opening of the diaphragm.

5. The superior epigastric artery passes through the _____ opening of the diaphragm.

6. The right phrenic nerve passes through the _____ opening of the diaphragm.

7. The left vagus nerve passes through the _____ opening of the diaphragm.

(a) Aortic

(b) Esophageal

(c) Caval

(d) None of the above

**In the following questions, answer:**
- (a) IF (1), (2), AND (3) ONLY ARE CORRECT
- (b) IF (1) AND (3) ONLY ARE CORRECT
- (c) IF (2) AND (4) ONLY ARE CORRECT
- (d) IF (4) ONLY IS CORRECT, OR
- (e) IF ALL ARE CORRECT

8. Which of the following statements regarding structures in the intercostal spaces are correct?
   - (1) The intercostal blood vessels and nerves are positioned in the order of vein, nerve, and artery from superior to inferior in a subcostal groove.
   - (2) The anterior intercostal arteries of the upper six intercostal spaces are branches of the internal thoracic artery.
   - (3) The intercostal nerves travel forward in an intercostal space between the external and internal intercostal muscles.
   - (4) The lower five intercostal nerves supply sensory innervation to the skin of the lateral thoracic and anterior abdominal walls.

9. Which of the following statements concerning the diaphragm are true?
   - (1) The right crus provides a muscular sling around the esophagus and possibly prevents regurgitation of stomach contents into the esophagus.
   - (2) On contraction, the diaphragm raises the intra-abdominal pressure and assists in the return of the venous blood to the right atrium of the heart.
   - (3) On contraction, the central tendon descends, reducing the intrathoracic pressure.
   - (4) The level of the diaphragm will be higher in the recumbent position than in the standing position.

**Select the best response:**

10. The sixth thoracic vertebra articulates by means of synovial joints with all the following structures **except:**
    - (a) The head of the sixth rib
    - (b) The body of the fifth thoracic vertebra
    - (c) The tubercle of the sixth rib
    - (d) The inferior articular process of the fifth thoracic vertebra
    - (e) The superior articular process of the seventh thoracic vertebra

11. Which of the following costal cartilages do **not** articulate directly with the body of the sternum?
    - (a) Second
    - (b) Fourth
    - (c) Fifth
    - (d) Eighth
    - (e) Third

12. Which of the following statements is **incorrect** concerning the intercostal nerves?
    - (a) They provide motor innervation to the peripheral parts of the diaphragm.
    - (b) They provide motor innervation to the intercostal muscles.
    - (c) They provide sensory innervation to the costal parietal pleura.
    - (d) They contain sympathetic fibers to innervate the vascular smooth muscle.
    - (e) The seventh to the eleventh intercostal nerves provide sensory innervation to the parietal peritoneum.

13. With a patient in the standing position, fluid in the left pleural cavity tends to gravitate down to the:
    - (a) Oblique fissure
    - (b) Cardiac notch
    - (c) Costomediastinal recess
    - (d) Horizontal fissure
    - (e) Costodiaphragmatic recess

14. In order to pass a needle into the pleural cavity in the midaxillary line, the following structures will have to be pierced **except** the:
    - (a) Internal intercostal muscle
    - (b) Levator costarum
    - (c) External intercostal muscle
    - (d) Parietal pleura
    - (e) Transversus thoracis

15. The following statements concerning the thoracic inlet are true **except:**
    - (a) The manubrium sterni forms the anterior border.
    - (b) On each side, the lower trunk of the brachial plexus and the subclavian artery emerge through the inlet and pass laterally over the upper surface of the first rib.
    - (c) The body of the seventh cervical vertebra forms the posterior boundary.
    - (d) The first ribs form the lateral boundaries.
    - (e) The esophagus and trachea pass through the inlet.

# 3. The Thorax: Part II
# The Thoracic Cavity

The thoracic cavity may be divided into a median partition, called the mediastinum, and the laterally placed pleurae and lungs (Fig. 3-3).

## MEDIASTINUM

The mediastinum, though thick, is movable and extends superiorly to the thoracic inlet and the root of the neck and inferiorly to the diaphragm. It extends anteriorly to the sternum and posteriorly to the twelve thoracic vertebrae of the vertebral column. It contains the remains of the thymus, the heart and large blood vessels, the trachea and esophagus, the thoracic duct and lymph nodes, the vagus and phrenic nerves, and the sympathetic trunks.

For purposes of description, the mediastinum is divided into *superior* and *inferior mediastina* by an imaginary plane passing from the sternal angle anteriorly to the lower border of the body of the fourth thoracic vertebra posteriorly (Fig. 3-1). The inferior mediastinum is further subdivided into the *middle mediastinum*, which consists of the pericardium and heart; the *anterior mediastinum*, which

is a space between the pericardium and the sternum; and the *posterior mediastinum*, which lies between the pericardium and the vertebral column.

For purposes of orientation, it is convenient to remember that the major mediastinal structures are arranged in the following order from anterior to posterior.

### Superior Mediastinum

(1) Thymus, (2) large veins, (3) large arteries, (4) trachea, (5) esophagus and thoracic duct, (6) sympathetic trunks.

The superior mediastinum is bounded in front by the manubrium sterni and behind by the first four thoracic vertebrae (Fig. 3-1).

### Inferior Mediastinum

(1) Thymus, (2) heart within the pericardium with the phrenic nerves on each side, (3) esophagus and thoracic duct, (4) descending aorta, (5) sympathetic trunks.

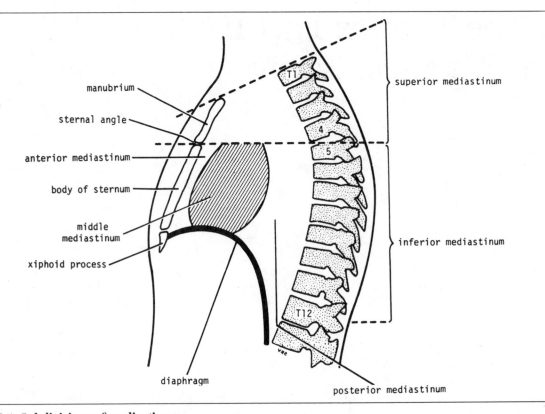

**Fig. 3-1. Subdivisions of mediastinum.**

The inferior mediastinum is bounded in front by the body of the sternum and behind by the lower eight thoracic vertebrae (Fig. 3-1).

## PLEURAE

Before discussing the pleurae, it might be helpful to look at the illustrations of the development of the lungs in Figure 3-2.

Each pleura has two parts: (1) a *parietal layer,* which lines the thoracic wall, covers the thoracic surface of the diaphragm and the lateral aspect of the mediastinum, and extends into the root of the neck to line the undersurface of the suprapleural membrane at the thoracic inlet; and (2) a *visceral layer,* which completely covers the outer surfaces of the lungs and extends into the depths of the interlobar fissures (Figs. 3-2, 3-3, 3-5, and 3-30).

The two layers become continuous with one another by means of a cuff of pleura that surrounds the structures entering and leaving the lung at the lung root (Figs. 3-2, 3-3, and 3-5). To allow for movement of the lung root during respiration, the pleural cuff hangs down as a loose fold called the *pulmonary ligament* (Fig. 3-5).

The parietal and visceral layers of pleura are separated from one another by a slitlike space, the *pleural cavity* (Figs. 3-2 and 3-3). This normally contains a small amount of tissue fluid, the *pleural fluid,* which covers the surfaces of the pleura as a thin film and permits the two layers to move on each other with the minimum of friction.

For purposes of description, it is customary to divide the parietal pleura according to the region in which it lies or the surface that it covers. The *cervical pleura* extends up into the neck, lining the undersurface of the suprapleural membrane (Fig. 3-4). It reaches a level about 1 to 1½ inches (2.5–4 cm) above the medial third of the clavicle.

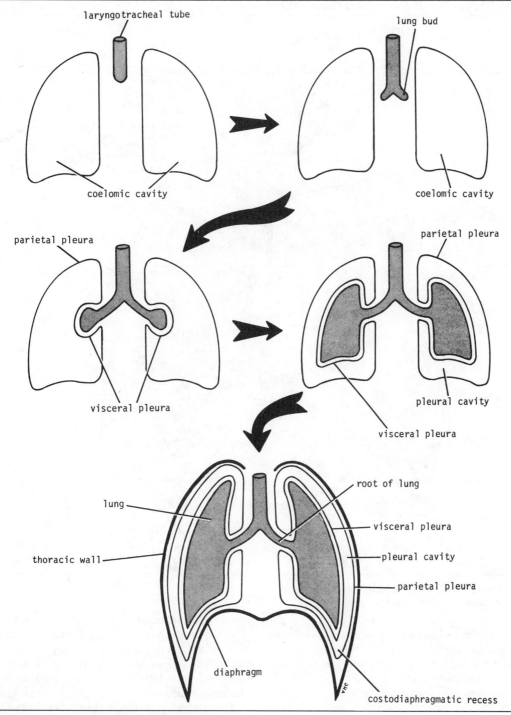

**Fig. 3-2. Formation of lungs.** Note that each lung bud invaginates wall of coelomic cavity and then grows to fill greater part of cavity. Note also that lung is covered with visceral pleura and thoracic wall lined with parietal pleura. Original coelomic cavity is reduced to a slitlike space called the pleural cavity, as the result of growth of lung.

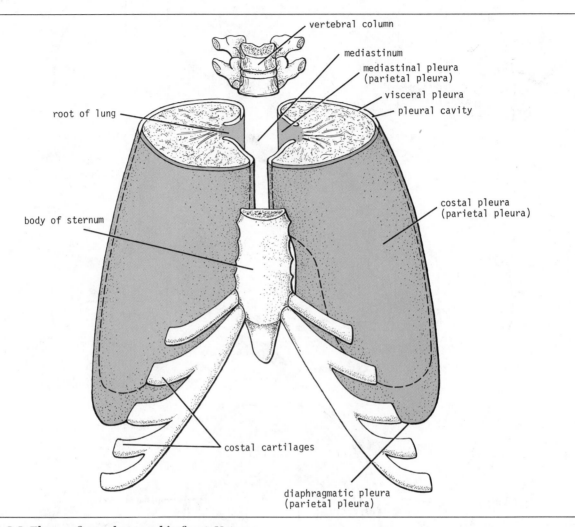

**Fig. 3-3. Pleurae from above and in front. Note position of mediastinum and root of each lung.**

The *costal pleura* lines the inner surfaces of the ribs, the costal cartilages, the intercostal spaces, the sides of the vertebral bodies, and the back of the sternum (Fig. 3-3).

The *diaphragmatic pleura* covers the thoracic surface of the diaphragm (Figs. 3-3 and 3-5). In quiet respiration the costal and diaphragmatic pleurae are in apposition to each other below the lower border of the lung. In deep inspiration the margins of the base of the lung descend, and the costal and diaphragmatic pleurae separate. This lower area of the pleural cavity into which the lung expands on inspiration is referred to as the *costo-diaphragmatic recess* (Figs. 3-2 and 3-5). The recess is 2 inches (5 cm) deep in the scapular line posteriorly; 3 to 3½ inches (8–9 cm) in the midaxillary line; and 1 to 1½ inches (2.5–4 cm) in the midclavicular line.

The *mediastinal pleura* covers and forms the lateral boundary of the mediastinum (Figs. 3-3 and 3-5). At the root of the lung it is reflected as a cuff around the vessels and bronchi and here becomes continuous with the visceral pleura. It is thus seen that each lung lies free except at its root, where it is attached to the blood vessels and bronchi. During

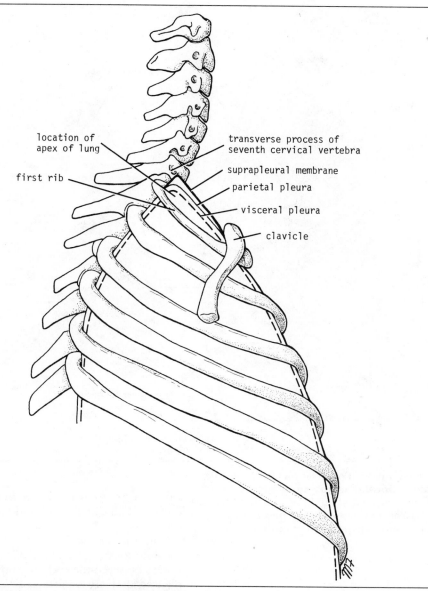

location of
apex of lung

transverse process of
seventh cervical vertebra

suprapleural membrane

parietal pleura

first rib

visceral pleura

clavicle

**Fig. 3-4. Lateral view of upper opening of thoracic cage, showing how apex of lung projects superiorly into the root of the neck. Note that lung apex is covered with visceral and parietal layers of pleura and is protected by the suprapleural membrane.**

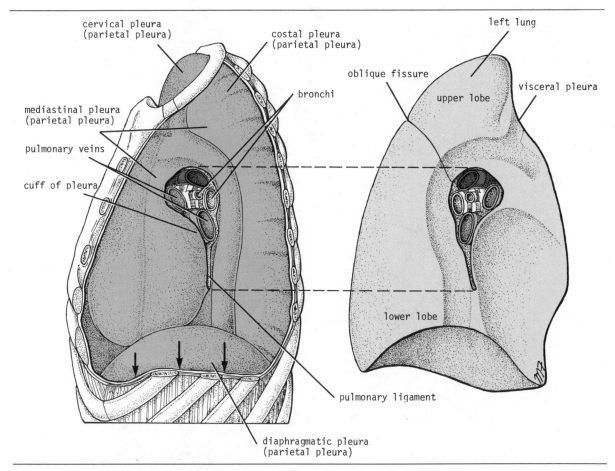

**Fig. 3-5. Different areas of parietal pleura. Note cuff of pleura (dotted lines) that surrounds structures entering and leaving root of left lung. It is here that parietal and visceral layers of pleura become continuous. Arrows indicate position of costodiaphragmatic recess.**

full inspiration the lungs expand and fill the pleural cavities. However, during quiet inspiration there are three sites where the lungs do not fully occupy the pleural cavities. These are the right and left costodiaphragmatic recesses and the right and left costomediastinal recesses.

The *costodiaphragmatic recesses* are slitlike spaces between the costal and diaphragmatic parietal pleurae that are separated only by a capillary layer of pleural fluid. During inspiration the lower margins of the lungs descend into the recesses.

During expiration the lower margins of the lungs ascend so that the costal and diaphragmatic pleurae come together again.

The *costomediastinal recesses* are situated along the anterior margins of the pleura. They are slitlike spaces between the costal and mediastinal parietal pleurae that are separated by a capillary layer of pleural fluid. During inspiration and expiration the anterior borders of the lungs slide in and out of the recesses.

The surface marking of the lungs and pleurae have already been described on page 57.

## Nerve Supply of the Pleura

The parietal pleura is supplied as follows: The costal pleura is segmentally supplied by the intercostal nerves; the mediastinal pleura is supplied by the phrenic nerve; and the diaphragmatic pleura is supplied over the domes by the phrenic nerve and around the periphery by the lower five intercostal nerves. The visceral pleura covering the lungs receives an autonomic vasomotor supply, but is insensitive to common sensations such as pain and touch.

## Trachea

The trachea is a tube about 5 inches (13 cm) long and 1 inch (2.5 cm) in diameter (Fig. 3-6). It has a fibroelastic wall in which are embedded a series of U-shaped bars of hyaline cartilage that keep the lumen patent. The trachea commences in the neck below the cricoid cartilage of the larynx at the level of the body of the sixth cervical vertebra. It ends below in the thorax at the level of the sternal angle (lower border of the fourth thoracic vertebra) by dividing into the right and left principal (main) bronchi. In deep inspiration the bifurcation descends to the level of the sixth thoracic vertebra.

The relations of the trachea in the neck are described on page 736.

The relations of the trachea in the superior mediastinum of the thorax are as follows:

### Anteriorly

The sternum, the thymus, the left brachiocephalic vein, the origins of the brachiocephalic and left common carotid arteries, the arch of the aorta (Figs. 3-6, 3-14, and 3-29).

### Posteriorly

The esophagus and the left recurrent laryngeal nerve (Fig. 3-29).

### Right Side

The azygos vein, the right vagus nerve, and the pleura (Figs. 3-24 and 3-29).

### Left Side

The arch of the aorta, the left common carotid and left subclavian arteries, the left vagus and left phrenic nerves, and the pleura (Figs. 3-25 and 3-29).

## Principal Bronchi

The right principal (main) bronchus is wider, shorter, and more vertical than the left (Fig. 3-6). It is about 1 inch (2.5 cm) long. Before entering the hilum of the right lung, the principal bronchus gives off the *superior lobar bronchus*. On entering the hilum it divides into a *middle and inferior lobe bronchus*.

The left principal (main) bronchus is narrower, longer, and more horizontal than the right and is about 2 inches (5 cm) long. It passes to the left below the arch of the aorta and *in front of the esophagus*. On entering the hilum of the left lung, it divides into a *superior and an inferior lobar bronchus*.

## LUNGS

During life the two lungs are soft and spongy. They are very elastic and should the thoracic cavity be opened they immediately shrink to one-third or less in volume. In the child they are pink in color, but with age they become dark and mottled due to the inhalation of dust particles that become trapped in the phagocytes of the lung. This is especially well seen in city dwellers and coal miners. The lungs are situated so that one lies on each side of the mediastinum. They are therefore separated from each other by the heart and great vessels and other structures in the mediastinum. Each lung is conical in shape and is covered with visceral pleura. It is suspended free in its own pleural cavity, being attached to the mediastinum only by its root (Fig. 3-2).

Each lung has a blunt *apex*, which projects upward into the neck for about 1 inch (2.5 cm) above the clavicle, a convex *costal surface*, which corresponds to the chest wall, and a concave *mediastinal surface*, which is molded to the pericardium and other mediastinal structures (Figs. 3-7, and 3-8). At about the middle of this surface, the *hilum* is located, a depression in which the bronchi, vessels, and nerves enter the lung to form the *root*.

The *anterior border* is thin and overlaps the heart; it is here on the left lung that the *cardiac notch* is found. The *posterior border* is thick and lies beside the vertebral column.

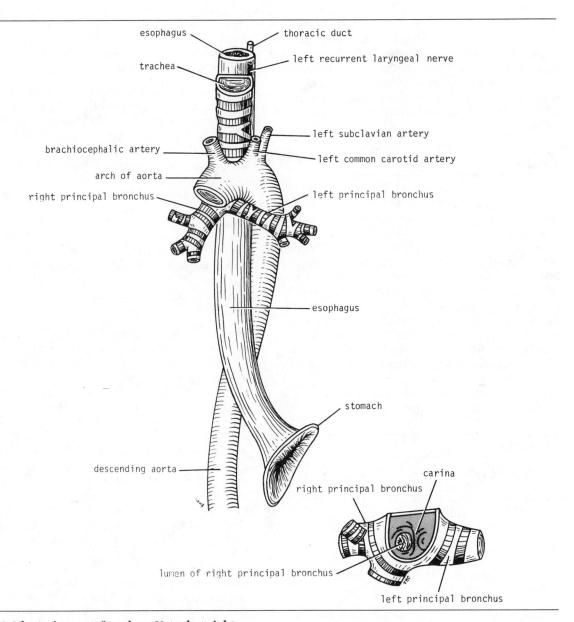

esophagus

thoracic duct

trachea

left recurrent laryngeal nerve

brachiocephalic artery

left subclavian artery

left common carotid artery

arch of aorta

right principal bronchus

left principal bronchus

esophagus

stomach

carina

right principal bronchus

descending aorta

lumen of right principal bronchus

left principal bronchus

**Fig. 3-6. Thoracic part of trachea. Note that right principal bronchus is the wider and more direct continuation of trachea, as compared with the left. Bifurcation of trachea viewed from above is also shown.**

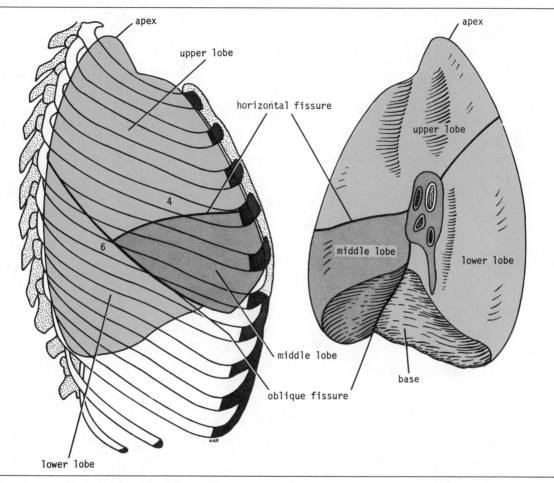

**Fig. 3-7. Lateral and medial surfaces of right lung.**

## Lobes and Fissures

The right lung is slightly larger than the left and is divided by the oblique and horizontal fissures into three lobes, the *upper*, *middle*, and *lower lobes* (Figs. 3-7 and 3-8). The oblique fissures runs from the inferior border upward and backward across the medial and costal surfaces until it cuts the posterior border about 2½ inches (6.25 cm) below the apex. The *horizontal fissure* runs horizontally across the costal surface at the level of the fourth costal cartilage to meet the oblique fissure in the midaxillary line. The middle lobe is thus a small triangular lobe bounded by the horizontal and oblique fissures.

The left lung is divided by a similar oblique fissure into two lobes, the *upper* and the *lower lobes* (Fig. 3-8). There is no horizontal fissure in the left lung.

## Bronchopulmonary Segments

Each lobar (secondary) bronchus, which passes to a lobe of the lung, gives off branches called *segmental* (tertiary) *bronchi* (Fig. 3-9). Each segmental bronchus passes to a structurally and functionally independent unit of a lung lobe called a *bronchopulmonary segment* (Fig. 3-10). A bronchopulmonary segment of lung tissue is pyramidal in shape, having its apex toward the root of the lung

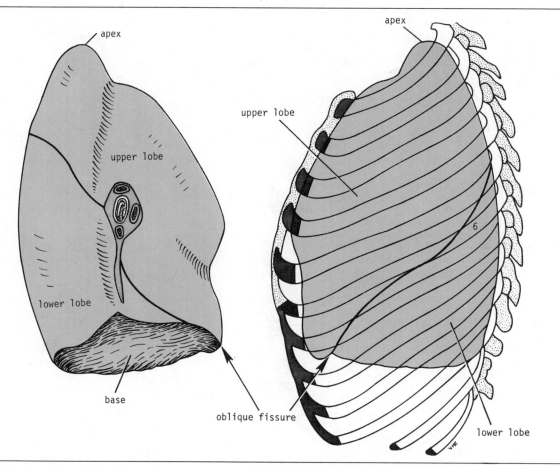

**Fig. 3-8. Lateral and medial surfaces of left lung.**

and its base toward the lung surface. Each bronchopulmonary segment is surrounded by connective tissue, and in addition to its own bronchus it receives an artery, a vein, lymph vessels, and autonomic nerves. The main bronchopulmonary segments (Fig. 3-10) are as follows:

## RIGHT LUNG

| | |
|---|---|
| *Superior lobe* | (1) Apical, (2) posterior, (3) anterior. |
| *Middle lobe* | (4) Lateral, (5) medial. |
| *Inferior lobe* | (6) Superior (apical), (7) medial basal, (8) anterior basal, (9) lateral basal, (10) posterior basal. |

## LEFT LUNG

| | |
|---|---|
| *Superior lobe* | (1) Apical, (2) posterior, (3) anterior, (4) superior lingular, (5) inferior lingular. |
| *Inferior lobe* | (6) Superior (apical), (7) medial basal, (8) anterior basal, (9) lateral basal, (10) posterior basal. |

While the general arrangement of the bronchopulmonary segments is of clinical importance, it is unnecessary to memorize the details unless one is intending to specialize in pulmonary medicine or surgery.

The *root of the lung* is formed of structures that are entering or leaving the lung. It is made up of the bronchi, pulmonary artery and veins, lymph

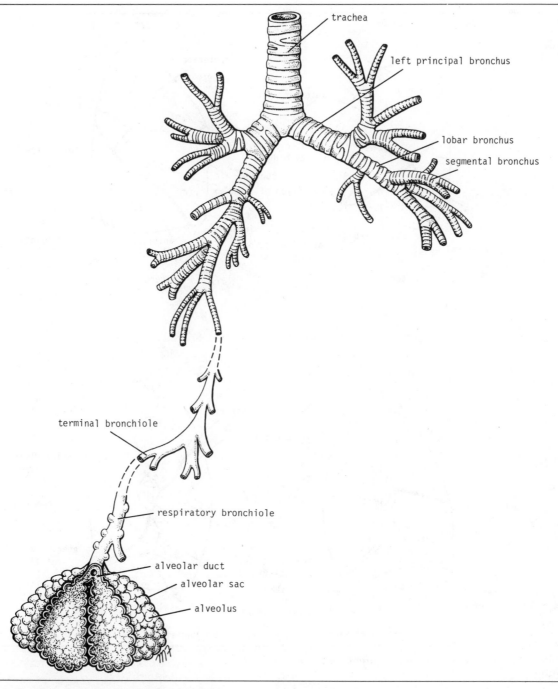

trachea

left principal bronchus

lobar bronchus

segmental bronchus

terminal bronchiole

respiratory bronchiole

alveolar duct

alveolar sac

alveolus

**Fig. 3-9. Trachea, bronchi, bronchioles, alveolar ducts, alveolar sacs, and alveoli. Note the path taken by inspired air from the trachea to the alveoli.**

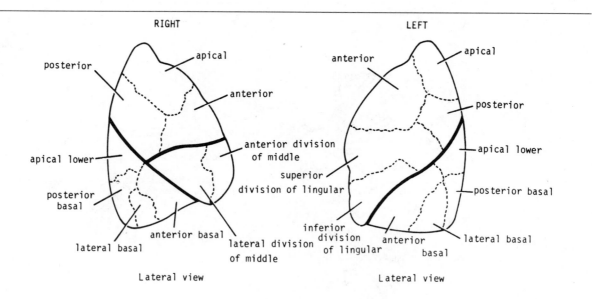

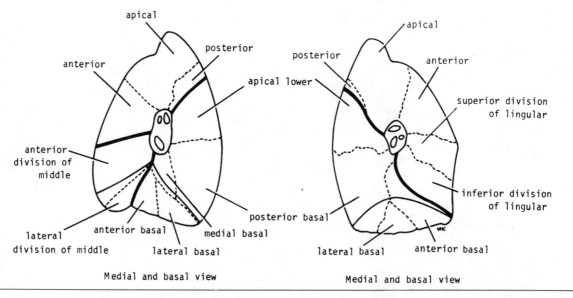

**Fig. 3-10. Bronchopulmonary segments of each lung.**

vessels, bronchial vessels, and nerves. It is surrounded by a tubular sheath of pleura, which joins the mediastinal parietal pleura to the visceral pleural covering the lungs (Figs. 3-5, 3-24, and 3-25).

## Blood Supply of Lungs

The bronchi, the connective tissue of the lung, and the visceral pleura receive their blood supply from the bronchial arteries, which are branches of the descending aorta. The bronchial veins (which communicate with the pulmonary veins) drain into the azygos and hemiazygos veins.

The alveoli receive deoxygenated blood from the terminal branches of the pulmonary arteries. The oxygenated blood leaving the alveolar capillaries drains into the tributaries of the pulmonary veins, which follow the intersegmental connective tissue septa to the lung root. Two pulmonary veins leave each lung root (Figs. 3-24 and 3-25).

## Lymph Drainage of Lungs

The lymph vessels originate in superficial and deep plexuses (Fig. 3-11). The *superficial plexus* lies beneath the visceral pleura; the *deep plexus* travels along the bronchi and pulmonary vessels toward the root of the lung. The vessels from the deep plexus drain into *pulmonary nodes*, which are located within the lung close to the hilum. All lymph from the pulmonary nodes and from the superficial plexus drain into the *bronchopulmonary nodes* in the hilum. The lymph then drains into the *bronchomediastinal lymph trunks*. The two trunks ascend on either side of the trachea and drain into the brachiocephalic vein or the thoracic or right lymphatic ducts.

## Nerve Supply of Lungs

At the root of each lung is a *pulmonary plexus* composed of efferent and afferent autonomic nerve fibers. The plexus is formed from branches of the sympathetic trunk and receives parasympathetic fibers from the vagus nerve.

The sympathetic efferent fibers produce bronchodilatation and vasoconstriction. The parasym-pathetic efferent fibers produce bronchoconstriction, vasodilatation, and increased glandular secretion.

Afferent impulses derived from the bronchial mucous membrane and from stretch receptors in the alveolar walls pass to the central nervous system in both sympathetic and parasympathetic nerves.

# The Mechanics of Respiration

Respiration consists of two phases, inspiration and expiration. They are accomplished by the alternate increase and decrease of the capacity of the thoracic cavity. The rate varies between 16 and 20 per minute in normal resting subjects, and is faster in children and slower in the old.

## INSPIRATION

### Quiet Inspiration

Compare the thoracic cavity to a box with a single entrance at the top, which is a tube, the trachea (Fig. 3-12). The capacity of the box can be increased by elongating all its diameters, and this will result in air under atmospheric pressure entering the box through the tube.

Consider now the three diameters of the thoracic cavity and how they may be increased (Fig. 3-12).

### Vertical Diameter

Theoretically, the roof could be raised and the floor lowered. The roof is formed by the suprapleural membrane and is fixed. On the other hand, the floor is formed by the mobile diaphragm. When the diaphragm contracts, the domes become flattened and the level of the diaphragm is lowered (Fig. 3-12).

### Anteroposterior Diameter

If the downward-sloping ribs were raised at their sternal ends, the anteroposterior diameter of the thoracic cavity would be increased and the lower end of the sternum would be thrust forward (Fig. 3-12). This can be brought about by fixing the first rib by the contraction of the scaleni muscles of the

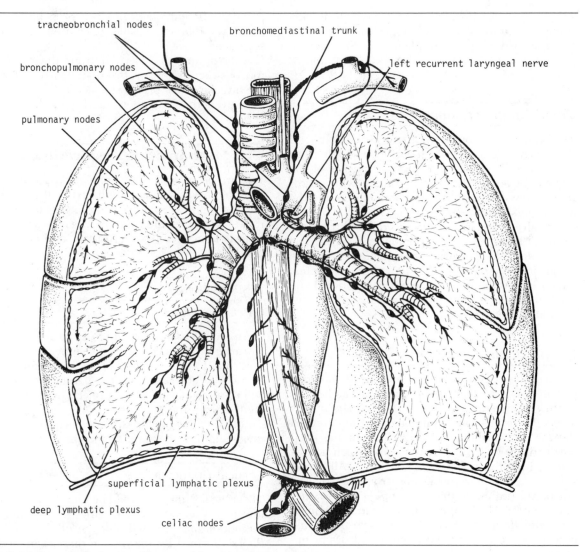

tracheobronchial nodes

bronchomediastinal trunk

left recurrent laryngeal nerve

bronchopulmonary nodes

pulmonary nodes

superficial lymphatic plexus

deep lymphatic plexus

celiac nodes

**Fig. 3-11. Lymph drainage of lung and lower end of esophagus.**

**Fig. 3-12. The different ways in which capacity of thoracic cavity is increased during inspiration.**

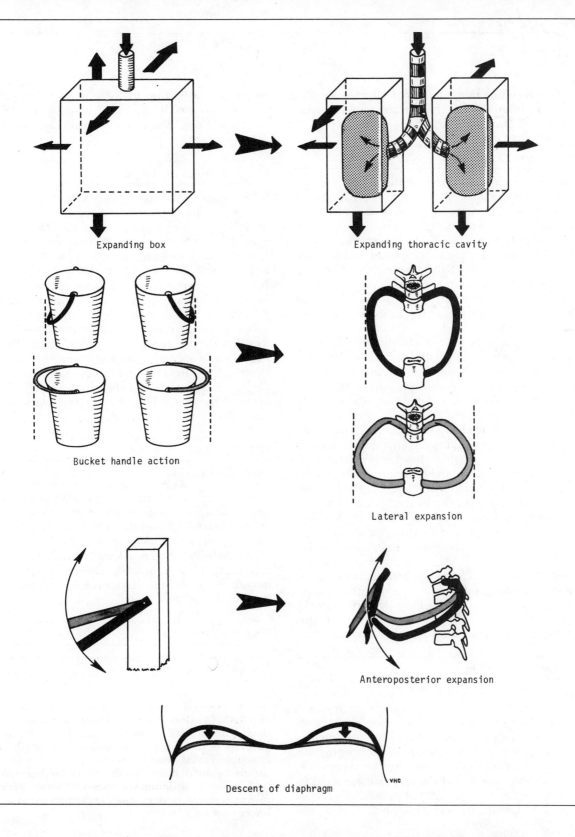

Expanding box

Expanding thoracic cavity

Bucket handle action

Lateral expansion

Anteroposterior expansion

Descent of diaphragm

neck and contracting the intercostal muscles (Fig. 3-13). By this means all the ribs are drawn together and raised toward the first rib.

## Transverse Diameter

The ribs articulate in front with the sternum via their costal cartilages and behind with the vertebral column. Since the ribs curve downward as well as forward around the chest wall, they resemble bucket handles (Fig. 3-12). It therefore follows that if the ribs are raised (like bucket handles), the transverse diameter of the thoracic cavity will be increased. As described previously, this can be accomplished by fixing the first rib and raising the other ribs to it by contracting the intercostal muscles (Fig. 3-13).

An additional factor, which must not be overlooked, is the effect of the descent of the diaphragm on the abdominal viscera, and the tone of the muscles of the anterior abdominal wall. As the diaphragm descends on inspiration, the intra-abdominal pressure rises. This rise in pressure is accommodated by the reciprocal relaxation of the abdominal wall musculature. However, a point is reached when no further abdominal relaxation is possible, and the liver and other upper abdominal viscera act as a platform that resists further diaphragmatic descent. On further contraction the diaphragm will now have its central tendon supported from below, and its shortening muscle fibers will assist the intercostal muscles in raising the lower ribs (Fig. 3-13).

Apart from the diaphragm and the intercostals, other less important muscles also contract on inspiration and assist in elevating the ribs, namely, the *levatores costarum muscles* and the *serratus posterior superior muscles.*

## Forced Inspiration

In deep forced inspiration there is a maximum increase in the capacity of the thoracic cavity. Every muscle that can raise the ribs is brought into action, including the scalenus anterior and medius and the sternocleidomastoid. In respiratory distress the action of all the muscles already engaged becomes more violent and the scapulae are fixed by the trapezius, levator scapulae, and rhomboid muscles, enabling the serratus anterior and pector-

alis minor to pull up the ribs. If the upper limbs can be supported by grasping a chair back or table, the sternal origin of the pectoralis major muscles can also assist the process.

### Lung Changes on Inspiration

In inspiration, the root of the lung descends and the level of the bifurcation of the trachea may be lowered by as much as two vertebrae. The bronchi elongate and dilate and the alveolar capillaries dilate, thus assisting the pulmonary circulation. Air is drawn into the bronchial tree as the result of the positive atmospheric pressure exerted through the upper part of the respiratory tract, and the negative pressure on the outer surface of the lungs brought about by the increased capacity of the thoracic cavity. With the expansion of the lungs, the elastic tissue in the bronchial walls and connective tissue is stretched. As the diaphragm descends, the costodiaphragmatic recess of the pleural cavity opens up, and the expanding sharp lower edges of the lungs descend to a lower level.

## EXPIRATION

### Quiet Expiration

Quiet expiration is largely a passive phenomenon and is brought about by the elastic recoil of the lungs, the relaxation of the intercostal muscles and diaphragm, and an increase in tone of the muscles of the anterior abdominal wall, which forces the relaxing diaphragm upward. The *serratus posterior inferior muscles* play a minor role in pulling down the lower ribs.

### Forced Expiration

Forced expiration is an active process brought about by the forcible contraction of the muscula-

**Fig. 3-13. (A) How intercostal muscles raise ribs during inspiration. Note that scaleni muscles fix first rib, or in forced inspiration raise first rib. (B) How intercostal muscles can be used in forced expiration provided that twelfth rib is fixed or made to descend by abdominal muscles. (C) How liver provides platform to enable diaphragm to raise lower ribs.**

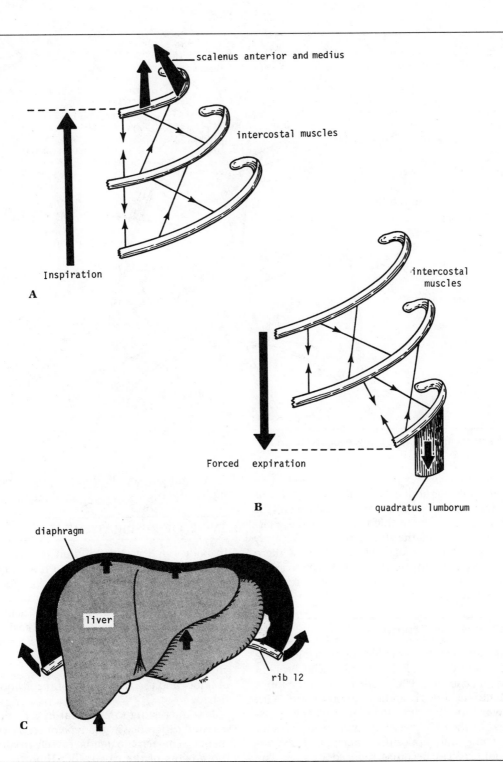

scalenus anterior and medius

intercostal muscles

Inspiration

**A**

intercostal muscles

Forced  expiration

**B**

quadratus lumborum

diaphragm

liver

rib 12

**C**

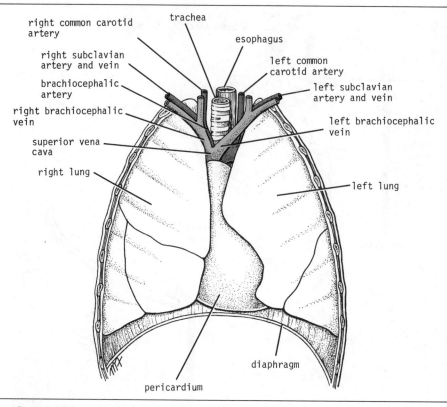

right common carotid artery

trachea

esophagus

right subclavian artery and vein

left common carotid artery

brachiocephalic artery

left subclavian artery and vein

right brachiocephalic vein

left brachiocephalic vein

superior vena cava

right lung

left lung

pericardium

diaphragm

**Fig. 3-14. Pericardium and lungs exposed from in front.**

ture of the anterior abdominal wall. The quadratus lumborum also contracts and pulls down the twelfth ribs. It is conceivable that under these circumstances some of the intercostal muscles may contract, pull the ribs together, and depress them to the lowered twelfth rib (Fig. 3-13). The serratus posterior inferior and the latissimus dorsi muscles may also play a minor role.

*Lung Changes on Expiration*

In expiration, the roots of the lungs ascend along with the bifurcation of the trachea. The bronchi shorten and contract. The elastic tissue of the lungs recoils, and the lungs become reduced in size. With the upward movement of the diaphragm, increasing areas of the diaphragmatic and costal parietal pleura come into apposition, and the costodiaphragmatic recess becomes reduced in size. The lower margins of the lungs shrink and rise to a higher level.

TYPES OF RESPIRATION

In babies and young children the ribs are nearly horizontal. Thus, they have to rely mainly on the descent of the diaphragm to increase their thoracic capacity on inspiration. Since this is accompanied by a marked inward and outward excursion of the anterior abdominal wall, which is easily seen, respiration at this age is referred to as the *abdominal type of respiration.*

After the second year the ribs become more oblique, and the adult form of respiration is established.

It is interesting to note that in the adult there is a sexual difference in the type of respiratory movements. The female tends to rely mainly on the movements of the ribs rather than the descent of the diaphragm on inspiration. This is referred to as the *thoracic type of respiration*. The male uses both the thoracic and abdominal forms of respiration, but mainly the abdominal form.

## Pericardium

The pericardium is a sac that encloses the heart and the roots of the great vessels. It lies within the middle mediastinum (Figs. 3-1, 3-14, 3-15, and 3-16). It is located posterior to the body of the sternum and the second to the sixth costal cartilages.

The *fibrous pericardium* is the fibrous part of the sac and serves to limit the movement of the heart. It is firmly attached below to the central tendon of the diaphragm. It blends with the outer coats of the great blood vessels passing through it (Fig. 3-15), namely, the aorta, the pulmonary trunk, the superior and inferior venae cavae, and the pulmonary veins (Fig. 3-16). It is attached in front to the sternum by the *sternopericardial ligaments*.

The *serous pericardium* is divided into parietal and visceral layers (Fig. 3-15). The parietal layer lines the fibrous pericardium and is reflected around the roots of the great vessels to become continuous with the visceral layer of serous pericardium (Fig. 3-16).

The visceral layer is closely applied to the heart and is often called the *epicardium*. The slitlike space between the parietal and visceral layers is referred to as the *pericardial cavity* (Fig. 3-15). Normally, the cavity contains a small amount of tissue fluid, which acts as a lubricant to facilitate movements of the heart.

On the posterior surface of the heart, the reflection of the serous pericardium around the large veins forms a recess called the *oblique sinus* (Fig. 3-16). Also on the posterior surface of the heart is the *transverse sinus*, which is a short passage that lies between the reflection of serous pericardium around the aorta and pulmonary trunk and the reflection around the large veins (Fig. 3-16).

## Heart

The heart is a hollow muscular organ that is somewhat pyramidal in shape and lies within the pericardium in the mediastinum (Fig. 3-17). It is connected at its base to the great blood vessels but otherwise lies free within the pericardium.

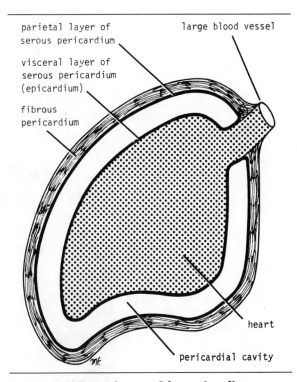

Fig. 3-15. Different layers of the pericardium.

The heart has three surfaces: sternocostal (anterior), diaphragmatic (inferior), and a base (posterior). It also has an apex, which is directed downward, forward, and to the left.

The *sternocostal surface* is formed mainly by the right atrium and the right ventricle, which are separated from each other by the vertical atrioventricular groove (Fig. 3-17). The right border is formed by the right atrium and the left border, by the left ventricle and part of the left auricle. The right ventricle is separated from the left ventricle by the anterior interventricular groove.

The *diaphragmatic surface* of the heart is formed mainly by the right and left ventricles separated by the posterior interventricular groove. The inferior surface of the right atrium, into which the inferior vena cava opens, also forms part of this surface.

The *base of the heart*, or posterior surface, is formed mainly by the left atrium, into which open the four pulmonary veins (Fig. 3-18). The right

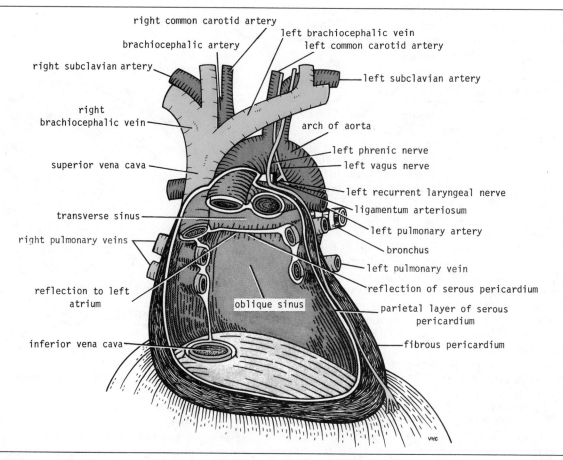

**Fig. 3-16. Great blood vessels and interior of pericardium.**

atrium also contributes to a lesser extent to this surface.

The *apex of the heart*, formed by the left ventricle, is directed downward, forward, and to the left (Fig. 3-17). It lies at the level of the fifth left intercostal space, 3½ inches (9 cm) from the midline. In the region of the apex, the apex beat can usually be seen and palpated in the living subject.

Note that the base of the heart is called the base, since the heart is pyramid-shaped and the base lies opposite the apex. The heart does not rest on its base; it rests on its diaphragmatic (inferior) surface.

## Chambers of the Heart

The heart is divided by vertical septa into four chambers, the right and left atria and the right and left ventricles. The right atrium lies anterior to the left atrium and the right ventricle lies anterior to the left ventricle.

The walls of the heart are composed of cardiac muscle, the *myocardium*, covered externally with serous pericardium, called the *epicardium*, and lined internally with a layer of endothelium, the *endocardium*.

### RIGHT ATRIUM

The right atrium consists of a main cavity and an auricle (Figs. 3-17 and 3-19). Externally at the junc-

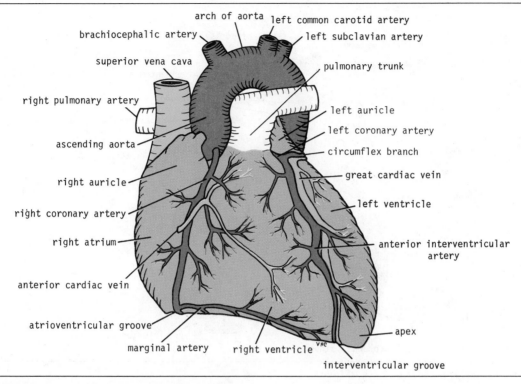

**Fig. 3-17. Anterior surface of heart and great blood vessels. Note course of coronary arteries and cardiac veins.**

tion of the two parts is a vertical groove, the *sulcus terminalis,* which on the inside forms a ridge, the *crista terminalis.* The main part of the atrium that lies posterior to the ridge is smooth-walled and is derived embryologically from the sinus venosus. The part of the atrium in front of the ridge is trabeculated by bundles of muscle fibers, the *musculi pectinati,* which run from the crista terminalis to the auricle. This anterior part is derived embryologically from the primitive atrium.

### Openings into the Right Atrium

The *superior vena cava* opens into the upper part of the right atrium; it is devoid of any valve. It returns the blood to the heart from the upper half of the body. The *inferior vena cava* (larger than that of the superior vena cava) opens into the lower part of the right atrium; it is guarded by a rudimentary, nonfunctioning valve. It returns the blood to the heart from the lower half of the body.

The *coronary sinus* opens into the right atrium between the inferior vena cava and the atrioventricular orifice; it is guarded by a rudimentary, nonfunctioning valve.

The *right atrioventricular orifice* lies anterior to the inferior vena caval opening and is guarded by the tricuspid valve.

There are also many small orifices of small veins that drain the wall of the heart and open directly into the right atrium.

### Fetal Remnants in the Right Atrium

In addition to the rudimentary valve of the inferior vena cava, there are the *fossa ovalis* and *anulus ovalis.* These latter structures lie on the *atrial septum* that separates the right atrium from the left atrium (Fig. 3-19). The fossa ovalis is a shallow depression, which is the site of the *foramen ovale* in the fetus (Fig. 3-21). The anulus ovalis forms the

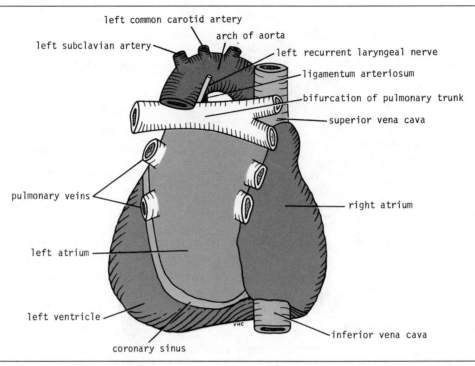

left common carotid artery

left subclavian artery

arch of aorta

left recurrent laryngeal nerve

ligamentum arteriosum

bifurcation of pulmonary trunk

superior vena cava

pulmonary veins

right atrium

left atrium

left ventricle

inferior vena cava

coronary sinus

**Fig. 3-18. Posterior surface or base of heart.**

upper margin of the fossa. The floor of the fossa represents the persistent septum primum of the heart of the embryo, and the anulus is formed from the lower edge of the septum secundum (Fig. 3-21).

## RIGHT VENTRICLE

The right ventricle communicates with the right atrium through the atrioventricular orifice, and with the pulmonary trunk through the pulmonary orifice (Fig. 3-19). As the cavity approaches the pulmonary orifice it becomes funnel-shaped, at which point it is referred to as the *infundibulum*.

The walls of the right ventricle are much thicker than those of the right atrium and show a number of internal projecting ridges formed of muscle bundles. The projecting ridges give the ventricular wall a sponge-like appearance and are known as *trabeculae carneae*. The trabeculae carneae are composed of three types. The first type comprises the *papillary muscles*, which project inward, being attached

by their bases to the ventricular wall; their apices are connected by fibrous chords (the *chordae tendineae*) to the cusps of the tricuspid valve (Fig. 3-19). The second type are attached at their ends to the ventricular wall, being free in the middle. One of these, the *moderator band*, crosses the ventricular cavity from the septal to the anterior wall. It conveys the right branch of the atrioventricular bundle, which is part of the conducting system of the heart. The third type is simply comprised of prominent ridges.

The *tricuspid valve* guards the atrioventricular orifice (Figs. 3-19 and 3-20). It consists of three cusps formed by a fold of endocardium with some fibrous tissue enclosed. The cusps are *anterior, septal,* and *inferior*. The anterior cusp lies anteriorly, the septal cusp lies against the ventricular septum, and the inferior cusp lies inferiorly. The bases of the cusps are attached to the fibrous ring of the skeleton of the heart (see page 107) while their free edges and ventricular surfaces are attached to the *chordae tendineae*. The chordae tendineae connect the cusps to the *papillary muscles*. When the ven-

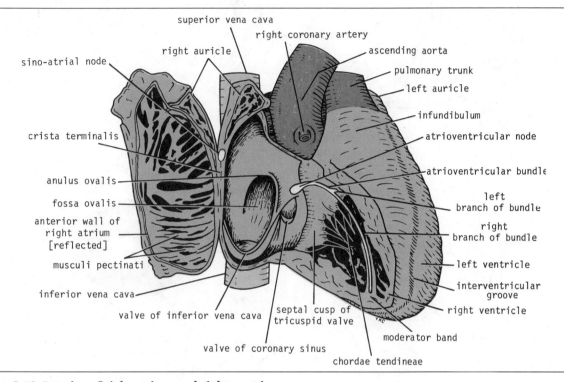

**Fig. 3-19. Interior of right atrium and right ventricle. Note positions of sino-atrial node and atrioventricular node and bundle.**

tricle contracts, the papillary muscles contract and prevent the cusps from being forced into the atrium and turning inside out as the intraventricular pressure rises. To assist in this process, the chordae tendineae of one papillary muscle are connected to the adjacent parts of two cusps.

The *pulmonary valve* guards the pulmonary orifice (Fig. 3-20A). It consists of three semilunar cusps formed by folds of endocardium with some fibrous tissue enclosed. The curved lower margin of each cusp is attached to the arterial wall. The open mouths of the cusps are directed upward into the pulmonary trunk. At the root of the pulmonary trunk are three dilatations called the *sinuses*, and one is situated external to each cusp (see *aortic valve*).

The three semilunar cusps are arranged with one posterior (left cusp) and two anterior (anterior and right cusps).* During ventricular systole, the cusps of the valve are pressed against the wall of the pulmonary trunk by the outrushing blood. During diastole, blood flows back toward the heart and enters the sinuses; the valve cusps fill, come into apposition in the center of the lumen, and close the pulmonary orifice.

## LEFT ATRIUM

The left atrium consists of a main cavity and an auricle. The left atrium is situated behind the right atrium and forms the greater part of the base or posterior surface of the heart (Fig. 3-18). Behind it lies the oblique sinus of the serous pericardium, and the fibrous pericardium separates it from the esophagus (Figs. 3-16 and 3-30).

*The cusps of the pulmonary and aortic valves are named according to their position in the fetus before the heart has rotated to the left. This unfortunately causes a great deal of unnecessary confusion.

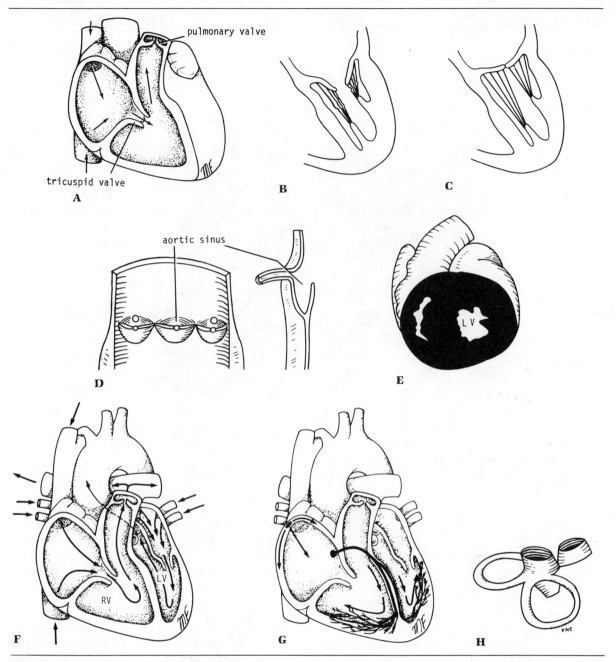

Fig. 3-20. (A) Position of tricuspid and pulmonary valves. (B) Mitral cusps with valve open and (C) mitral cusps with valve closed. (D) Semilunar cusps of aortic valve. (E) Cross section of ventricles of heart. (F) Path taken by blood through heart. (G) Path taken by cardiac impulse from sino-atrial node to Purkinje network. (H) Fibrous skeleton of heart.

The interior of the left atrium is smooth, but the auricle possesses muscular ridges as on the right side.

## Openings into the Left Atrium

The four *pulmonary veins*, two from each lung, open through the posterior wall (Fig. 3-18) and are devoid of valves. The left *atrioventricular* orifice is guarded by the mitral valve.

## LEFT VENTRICLE

The left ventricle communicates with the left atrium through the atrioventricular orifice and with the aorta through the aortic orifice. The walls of the left ventricle (Fig. 3-20) are three times thicker than those of the right ventricle. (The left intraventricular blood pressure is six times higher than that inside the right ventricle.) In cross section, the left ventricule is circular; the right is crescentic because of the bulging of the ventricular septum into the cavity of the right ventricle (Fig. 3-20). There are well-developed trabeculae carneae, two large papillary muscles, but there is no moderator band. The part of the ventricle below the aortic orifice is called the *aortic vestibule*.

The *mitral valve* guards the atrioventricular orifice (Fig. 3-20). It consists of two cusps, one anterior and one posterior, which have a similar structure to those of the tricuspid valve. The anterior cusp is the larger and intervenes between the atrioventricular and the aortic orifices. The attachment of the chordae tendineae to the cusps and the papillary muscles is similar to the tricuspid valve.

The *aortic valve* guards the aortic orifice and is precisely similar in structure to the pulmonary valve (Fig. 3-20). One cusp is situated on the anterior wall (right cusp) and two are located on the posterior wall (left and posterior cusps). Behind each cusp the aortic wall bulges to form an *aortic sinus*. The anterior aortic sinus gives origin to the right coronary artery, and the left posterior sinus gives origin to the left coronary artery.

## Structure of the Heart

The walls of the heart are composed of a thick layer of cardiac muscle, the *myocardium*, covered externally by the *epicardium* and lined internally by the *endocardium*. The atrial portion of the heart has relatively thin walls and is divided by the *atrial (interatrial) septum* into the right and left atria. The septum runs from the anterior wall of the heart backward and to the right. The ventricular portion of the heart has thick walls and is divided by the *ventricular (interventricular) septum* into right and left ventricles. The septum is placed obliquely, with one surface facing forward and to the right and the other backward and to the left. Its position is indicated on the surface of the heart by the anterior and posterior interventricular grooves. The lower part of the septum is thick and formed of muscle. The smaller upper part of the septum is thin and membranous and attached to the fibrous skeleton.

The so-called *skeleton of the heart* (Fig. 3-20) consists of fibrous rings that surround the atrioventricular, pulmonary, and aortic orifices and are continuous with the membranous upper part of the ventricular septum. The fibrous rings around the atrioventricular orifices separate the muscular walls of the atria from those of the ventricles, but provide attachment for the muscle fibers. The fibrous rings support the bases of the valve cusps and prevent the valves from stretching and becoming incompetent.

## Conducting System of the Heart

The normal human heart contracts rhythmically at about 70 beats per minute in the resting adult. The rhythmic contractile process originates spontaneously in the conducting system and the impulse travels to different regions of the heart, so the atria contract first and together, to be followed later by the contractions of both ventricles together. The slight delay in the passage of the impulse from the atria to the ventricles allows time for the atria to empty their blood into the ventricles before the ventricles contract.

The conducting system of the heart consists of specialized cardiac muscle present in the *sino-atrial node*, the *atrioventricular node*, the *atrioventricular bundle* and its right and left terminal branches, and the subendocardial plexus of *Purkinje fibers*.[†] The *sino-atrial node* is the site where the contraction of the heart muscle is initiated and

---

[†]The specialized cardiac muscle fibers that form the conducting system of the heart are known as Purkinje fibers.

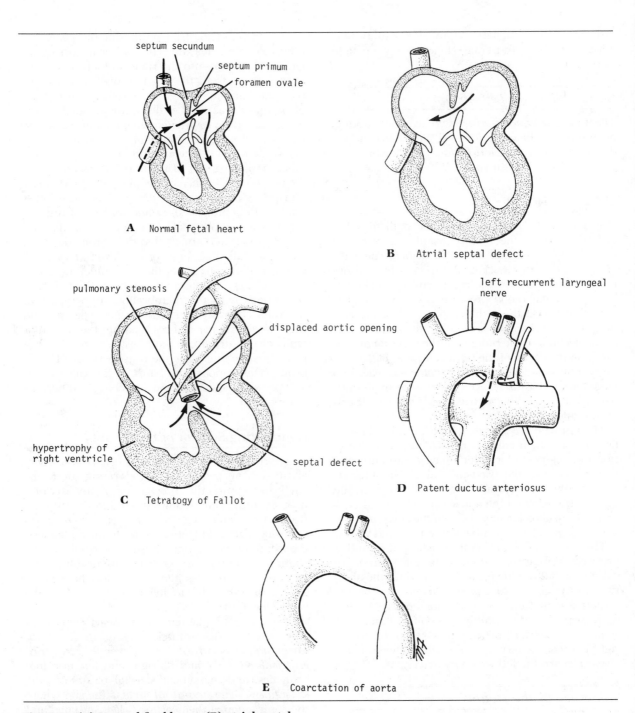

**Fig. 3-21. (A) Normal fetal heart, (B) atrial septal defect, (C) tetralogy of Fallot, (D) patent ductus arteriosus (note close relationship to left recurrent laryngeal nerve), and (E) coarctation of aorta.**

is often called the *pacemaker*. It is small, but forms the full thickness of the myocardium of the right atrium (Figs. 3-19 and 3-20). It is situated at the upper part of the sulcus terminalis just to the right of the opening of the superior vena cava into the right atrium. Once initiated, the cardiac impulse spreads through the atrial myocardium to reach the atrioventricular node.

The *atrioventricular node* is situated in the lower part of the atrial septum just above the attachment of the septal cusp of the tricuspid valve (Figs. 3-19 and 3-20). From it, the cardiac impulse is conducted to the ventricles by the *atrioventricular bundle*. It is important to realize that the atrioventricular bundle is the only muscular connection between the myocardium of the atria and the myocardium of the ventricles.

The *atrioventricular bundle* (Figs. 3-19 and 3-20) descends behind the septal cusp of the tricuspid valve to reach the inferior border of the membranous part of the ventricular septum. At the upper border of the muscular part of the septum it divides into two branches, one for each ventricle. The right branch passes down on the right side of the ventricular septum to reach the moderator band, by means of which it crosses to the anterior wall of the right ventricle. Here, it becomes continuous with the fibers of the Purkinje plexus (Fig. 3-20).

The left branch of the bundle pierces the septum and passes down on its left side beneath the endocardium. It usually divides into two branches, which eventually become continuous with the fibers of the Purkinje plexus of the left ventricle.

It is thus seen that the conducting system of the heart is responsible not only for generating rhythmical cardiac impulses, but also for conducting these impulses rapidly throughout the myocardium of the heart, so that the different chambers contract in a coordinated and efficient manner.

The activities of the conducting system can be influenced by the autonomic nerve supply to the heart. The parasympathetic nerves slow the rhythm and diminish the rate of conduction of the impulse; the sympathetic nerves have the opposite effect.

## Arterial Supply of the Heart

The arterial supply of the heart is provided by the right and left coronary arteries, which arise from the aorta immediately above the aortic valve (Fig. 3-22).

The *right coronary artery* arises from the anterior aortic sinus and runs forward between the pulmonary trunk and the right auricle (Fig. 3-17). It descends in the atrioventricular groove, giving branches to the right atrium and right ventricle. At the inferior border of the heart it continues posteriorly along the atrioventricular groove to anastomose with the left coronary artery (Fig. 3-22). It gives off a *marginal branch*, which supplies the right ventricle, and a *posterior interventricular branch*, which supplies both ventricles (Fig. 3-22). The posterior interventricular branch anastomoses with the anterior interventricular branch of the left coronary artery in the posterior interventricular groove.

The *left coronary artery*, which is larger than the right coronary artery, arises from the left posterior aortic sinus and passes forward between the pulmonary trunk and the left auricle (Fig. 3-17). It then enters the atrioventricular groove and divides into an anterior interventricular branch and a circumflex branch. The *anterior interventricular branch* runs downward to the apex of the heart in the anterior interventricular groove. It then passes around the apex to anastomose with the posterior interventricular branch of the right coronary artery. The anterior interventricular branch supplies the right and left ventricles and the ventricular septum.

The *circumflex branch* follows the atrioventricular groove, winds around the left margin of the heart, and ends by anastomosing with the right coronary artery (Fig. 3-22). The circumflex branch supplies the left atrium and the left ventricle.

The foregoing description of the coronary arteries and their branches should be memorized, but it must be understood that variations are common. The commonest variations affect the blood supply to the diaphragmatic surface of both ventricles. Here the origin, size, and distribution of the posterior interventricular artery are variable. In the case of "right dominance" the posterior interventricular artery is a large branch of the right coronary artery, whereas in the case of "left dominance" the posterior interventricular artery is a branch of the left coronary artery.

Although anastomoses between the terminal branches of the coronary arteries do occur, they are

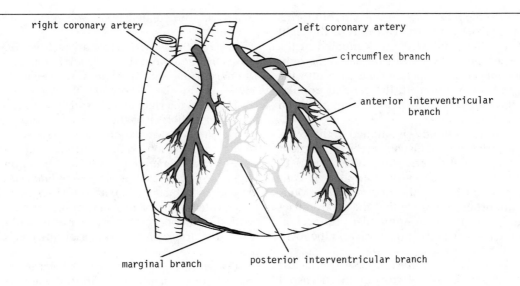

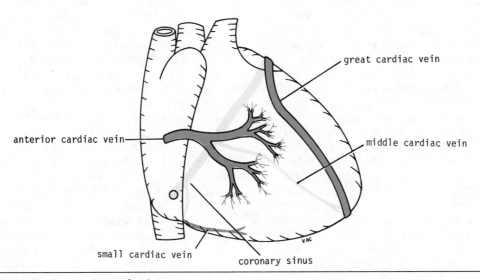

**Fig. 3-22. Coronary arteries and veins.**

not large enough to provide an adequate supply of blood to the cardiac muscle should one of the larger branches become blocked by disease.

## BLOOD SUPPLY OF CONDUCTING SYSTEM

The sino-atrial node is supplied by the right and left coronary arteries. The atrioventricular node and the atrioventricular bundle are supplied by the right coronary artery. The right terminal branch of the atrioventricular bundle is supplied by the right coronary artery; the left terminal branch is supplied by the right and left coronary arteries.

### Venous Drainage of the Heart

Most of the blood from the heart wall drains into the right atrium by means of the *coronary sinus* (Fig. 3-22). This lies in the posterior part of the atrioventricular groove and is a continuation of the *great cardiac vein*. It opens into the right atrium to the left of the inferior vena cava. The *small cardiac vein* and the *middle cardiac vein* are tributaries of the coronary sinus. The remainder of the blood is returned to the right atrium by the *anterior cardiac vein* (Fig. 3-22) and also by small veins that open directly into the heart chambers.

### Nerve Supply of the Heart

The heart is innervated by sympathetic and parasympathetic fibers of the autonomic nervous system via the *cardiac plexuses*. The sympathetic supply arises from the cervical and upper thoracic portions of the sympathetic trunks, and the parasympathetic supply comes from the vagus nerves.

Efferent postganglionic fibers pass to the sino-atrial and atrioventricular nodes and are also distributed to the remainder of the heart as nerve plexuses around the coronary arteries.

Afferent fibers running with the vagus nerves take part in cardiovascular reflexes. Afferent fibers running with the sympathetic nerves carry nervous impulses that normally do not reach consciousness. However, should the blood supply to the myocardium become impaired, pain impulses reach consciousness via this pathway.

### Action of the Heart

The heart is a muscular pump. The series of changes that take place within it as it fills with blood and empties is referred to as the *cardiac cycle*. The normal heart beats about 70 times per minute in the resting adult and about 130 times a minute in the newborn child.

Blood is continuously returning to the heart, and during ventricular systole (contraction), when the atrioventricular valves are closed, the blood is temporarily accommodated in the large veins and atria. Once ventricular diastole (relaxation) occurs, the atrioventricular valves open, and blood passively flows from the atria to the ventricles (Fig. 3-20). When the ventricles are nearly full, atrial systole occurs and forces the remainder of the blood in the atria into the ventricles. The sino-atrial node initiates the wave of contraction in the atria, which commences around the openings of the large veins and "milks" the blood toward the ventricles. By this means there is no reflux of blood into the veins.

The cardiac impulse, having reached the atrioventricular node, is conducted to the papillary muscles by the atrioventricular bundle and its branches (Fig. 3-20). The papillary muscles now begin to contract and take up the slack of the chordae tendineae. Meanwhile, the ventricles start contracting and the atrioventricular valves close. The spread of the cardiac impulse along the atrioventricular bundle (Fig. 3-20) and its terminal branches, including the Purkinje fibers, ensures that myocardial contraction occurs at almost the same time throughout the ventricles.

Once the intraventricular blood pressure exceeds that present in the large arteries (aorta and pulmonary trunk), the semilunar valve cusps are pushed aside, and the blood is ejected from the heart. On the conclusion of ventricular systole, blood begins to move back toward the ventricles and immediately fills the pockets of the semilunar valves. The cusps now float into apposition and completely close the aortic and pulmonary orifices.

## SURFACE ANATOMY OF THE HEART VALVES

The surface projection of the heart has already been described on page 61. The surface markings

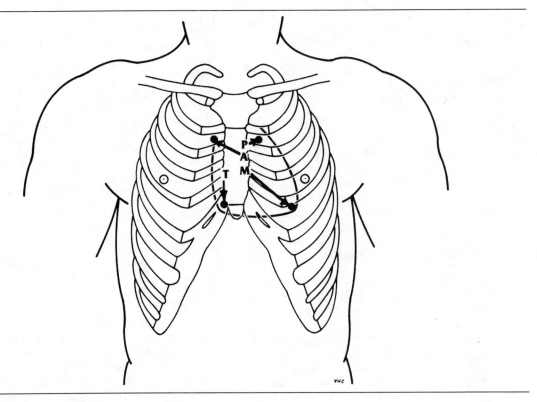

**Fig. 3-23. Position of heart valves. P = pulmonary valve, A = aortic valve, M = mitral valve, T = tricuspid valve. Arrows indicate positions where valves may be heard with least interference.**

of the heart valves (Fig. 3-23) are of academic value only; the clinician is more interested in listening to the valves in action.

The *tricuspid valve* lies behind the right half of the sternum opposite the fourth intercostal space.

The *mitral valve* lies behind the left half of the sternum opposite the fourth costal cartilage.

The *pulmonary valve* lies behind the medial end of the third left costal cartilage and the adjoining part of the sternum.

The *aortic valve* lies behind the left half of the sternum opposite the third intercostal space.

## AUSCULTATION OF THE HEART VALVES

On listening to the heart with a stethoscope, one can hear two sounds: lūb-dŭp. The first sound is produced by the contraction of the ventricles and the closure of the tricuspid and mitral valves. The second sound is produced by the sharp closure of the aortic and pulmonary valves. It is important for a physician to know where to place his stethoscope on the chest wall so that he will be able to hear sounds produced at each valve with the minimum of distraction or interference.

The *tricuspid valve* is best heard over the right half of the lower end of the body of the sternum (Fig. 3-23).

The *mitral valve* is best heard over the apex beat (Fig. 3-23).

The *pulmonary valve* is heard with least interference over the medial end of the second left intercostal space (Fig. 3-23).

The *aortic valve* is best heard over the medial end of the second right intercostal space (Fig. 3-23).

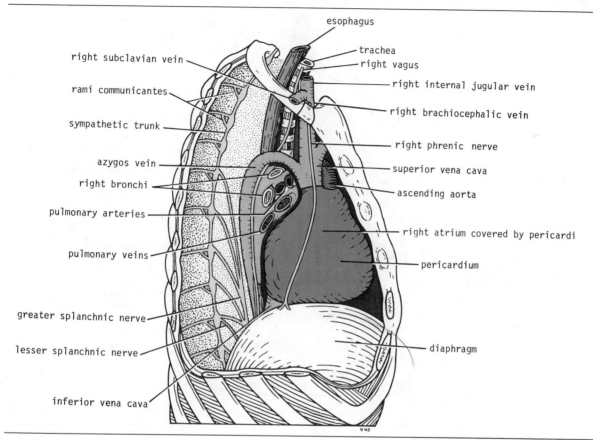

right subclavian vein

rami communicantes

sympathetic trunk

azygos vein

right bronchi

pulmonary arteries

pulmonary veins

greater splanchnic nerve

lesser splanchnic nerve

inferior vena cava

esophagus

trachea

right vagus

right internal jugular vein

right brachiocephalic vein

right phrenic nerve

superior vena cava

ascending aorta

right atrium covered by pericardi

pericardium

diaphragm

**Fig. 3-24. Right side of mediastinum.**

## CONGENITAL ANOMALIES OF THE HEART

Of the many congenital anomalies that may occur in the heart and large arteries, four common varieties are illustrated in Figure 3-21.

## Large Veins of the Thorax

### Brachiocephalic Veins

The *right brachiocephalic vein* is formed at the root of the neck by the union of the right subclavian and the right internal jugular veins (Figs. 3-24 and 3-26). The *left brachiocephalic vein* has a similar origin (Figs. 3-14 and 3-16). It passes obliquely downward and to the right behind the manubrium sterni and in front of the large branches of the aortic arch. It joins the right brachiocephalic vein to form the superior vena cava (Fig. 3-26).

### Superior Vena Cava

The superior vena cava contains all the venous blood from the head and neck and both upper limbs and is formed by the union of the two brachiocephalic veins (Figs. 3-16 and 3-26). It passes downward to end in the right atrium of the heart (Fig. 3-19). The vena azygos joins the posterior aspect of the superior vena cava just before it enters the pericardium (Figs. 3-24 and 3-26).

### Inferior Vena Cava

The inferior vena cava pierces the central tendon of the diaphragm opposite the eighth thoracic verte-

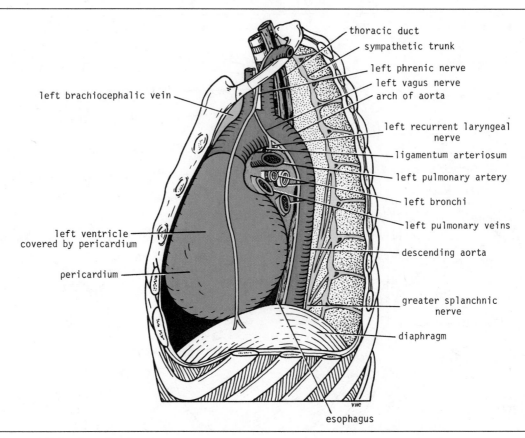

left brachiocephalic vein

thoracic duct
sympathetic trunk
left phrenic nerve
left vagus nerve
arch of aorta
left recurrent laryngeal nerve
ligamentum arteriosum
left pulmonary artery
left bronchi
left pulmonary veins
descending aorta

left ventricle covered by pericardium

pericardium

greater splanchnic nerve

diaphragm

esophagus

**Fig. 3-25. Left side of mediastinum.**

bra and almost immediately enters the lowest part of the right atrium (Figs. 3-19, 3-24, and 3-26).

### *Pulmonary Veins*

Two pulmonary veins leave each lung carrying oxygenated blood to the left atrium of the heart (Figs. 3-18, 3-24, 3-25, and 3-30).

## **Large Arteries of the Thorax**

### *Aorta*

The aorta in the thorax may be divided into three parts: the ascending aorta, the arch of the aorta, and the descending aorta.

### ASCENDING AORTA

The ascending aorta commences at the base of the left ventricle and runs upward and forward to come to lie behind the right half of the sternal angle, where it becomes continuous with the arch of the aorta (Fig. 3-17). Together with the pulmonary trunk, it is enclosed in a sheath of serous pericardium (Fig. 3-16).

### *Branches*

The *right coronary artery* arises from the anterior aortic sinus, and the *left coronary artery* arises from the left posterior aortic sinus (Figs. 3-17 and 3-22). The further course of these important arteries is described on page 109.

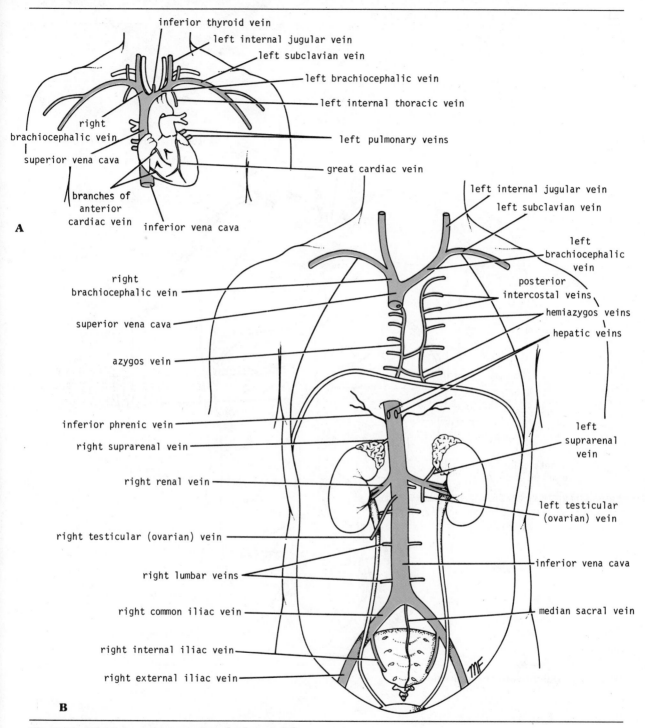

**Fig. 3-26. (A) Major veins entering the heart. (B) Major veins draining into the superior vena cava and the inferior vena cava.**

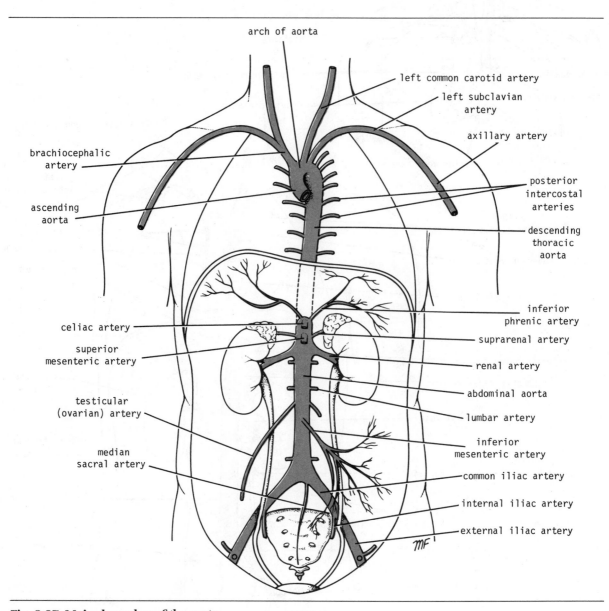

**Fig. 3-27. Major branches of the aorta.**

## ARCH OF THE AORTA

The arch of the aorta is a continuation of the ascending aorta (Fig. 3-17). It lies behind the manubrium sterni and runs upward, backward, and to the left in front of the trachea. It then passes downward to the left of the trachea, and at the level of the sternal angle becomes continuous with the descending aorta.

### Branches

The *brachiocephalic artery* arises from the convex surface of the aortic arch (Figs. 3-17 and 3-27). It passes upward and to the right of the trachea and divides into the right subclavian and common carotid arteries behind the right sternoclavicular joint.

The *left common carotid artery* arises from the aortic arch on the left side of the brachiocephalic artery (Figs. 3-17 and 3-27). It runs upward and to the left of the trachea and enters the neck behind the left sternoclavicular joint.

The *left subclavian artery* arises from the aortic arch behind the left common carotid artery (Figs. 3-17, 3-18, and 3-27). It runs upward along the left side of the trachea and the esophagus to enter the root of the neck (Fig. 3-25).

## DESCENDING AORTA

The descending aorta begins as a continuation of the arch of the aorta on the left side of the lower border of the body of the fourth thoracic vertebra (i.e., opposite the sternal angle). It extends downward in the posterior mediastinum (Figs. 3-25 and 3-27) to the level of the twelfth thoracic vertebra, where it passes through the aortic opening of the diaphragm in the midline and becomes continuous with the abdominal aorta.

### Branches

*Posterior intercostal arteries* are given off to the lower nine intercostal spaces on each side (Fig. 3-27). *Subcostal arteries* are given off on each side and run along the lower border of the twelfth rib.

*Pericardial*, *esophageal*, and *bronchial arteries* are small branches that are distributed to these organs.

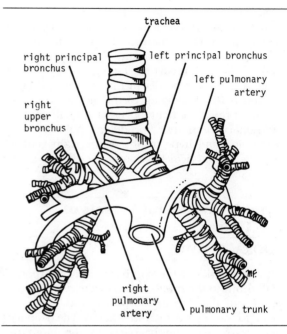

**Fig. 3-28. Relationship of pulmonary arteries to bronchial tree.**

## PULMONARY TRUNK

The pulmonary trunk conveys deoxygenated blood from the right ventricle of the heart to the lungs. It leaves the upper part of the right ventricle and runs upward, backward, and to the left (Fig. 3-17). It is about 2 inches (5 cm) long and terminates in the concavity of the aortic arch by dividing into right and left pulmonary arteries (Fig. 3-28). Together with the ascending aorta, it is enclosed in a sheath of serous pericardium (Fig. 3-16).

### Branches

The *right pulmonary artery* runs to the right behind the ascending aorta and superior vena cava to enter the root of the right lung (Figs. 3-17, 3-24, and 3-28).

The *left pulmonary artery* runs to the left in front of the descending aorta to enter the root of the left lung (Figs. 3-17, 3-25, and 3-28).

The *ligamentum arteriosum* is a fibrous band that connects the bifurcation of the pulmonary trunk to the lower concave surface of the aortic arch (Figs. 3-16 and 3-18). The ligamentum arteriosum is the remains of the ductus arteriosus, which in the fetus conducts blood from the pulmonary trunk to the aorta, thus bypassing the lungs. The left recurrent laryngeal nerve hooks around the lower border of this structure (Figs. 3-16 and 3-18). Following birth, the ductus closes. Should it remain patent, aortic blood will enter the pulmonary circulation, producing pulmonary hypertension and hypertrophy of the right ventricle (Fig. 3-21). Surgical ligation of the ductus is then necessary.

# Lymph Nodes and Vessels of the Thorax

## THORACIC WALL

The lymphatics of the skin of the anterior thoracic wall drain to the *anterior axillary nodes*. The lymphatics of the skin of the posterior thoracic wall drain to the *posterior axillary nodes*. The deep lymphatics of the anterior parts of the intercostal spaces drain forward to the *internal thoracic nodes* along the internal thoracic blood vessels. From here, the lymph passes to the thoracic duct on the left side and the bronchomediastinal trunk on the right side. The deep lymphatics of the posterior parts of the intercostal spaces drain backward to the *posterior intercostal nodes* lying near the heads of the ribs. From here, the lymph enters the thoracic duct.

## LUNGS

The lymph passes to the root of the lung by means of a superficial and a deep lymphatic plexus of vessels (Fig. 3-11). At the lung roots the lymph passes through the *bronchopulmonary nodes* and emerges in the bronchomediastinal trunk. The bronchomediastinal trunks drain into the brachiocephalic vein or into the thoracic or right lymphatic ducts.

## MEDIASTINUM

In addition to the nodes draining the lungs, other nodes are found scattered through the mediastinum. They drain lymph from mediastinal structures and empty into the bronchomediastinal trunks and thoracic duct. Disease and enlargement of these nodes may exert pressure on important neighboring mediastinal structures.

## THORACIC DUCT

The thoracic duct begins below in the abdomen as a dilated sac, the *cisterna chyli*. It ascends through the aortic opening in the diaphragm, on the right side of the descending aorta. It gradually crosses the median plane behind the esophagus and reaches the left border of the esophagus (Fig. 3-29B) at the level of the lower border of the body of the fourth thoracic vertebra (sternal angle). It then runs upward along the left edge of the esophagus to enter the root of the neck (Fig. 3-29A). Here, it bends laterally behind the carotid sheath and in front of the vertebral vessels. It turns downward in front of the left phrenic nerve and crosses the subclavian artery to enter the beginning of the left brachiocephalic vein.

At the root of the neck the thoracic duct receives the *left jugular, subclavian*, and *bronchomediastinal lymph trunks*, although they may drain directly into the adjacent large veins.

The thoracic duct thus conveys to the blood all lymph from the lower limbs, pelvic cavity, abdominal cavity, left side of the thorax, and left side of the head, neck, and left arm. (See also p. 25.)

## RIGHT LYMPHATIC DUCT

The right jugular, subclavian, and bronchomediastinal trunks, which drain the right side of the head and neck, the right upper limb, and the right side of the thorax, respectively, may join to form the right lymphatic duct. This common duct, if present, is about ½ inch (1.3 cm) long and opens into the beginning of the right brachiocephalic vein. Alternatively, the trunks open independently into the great veins at the root of the neck.

# Nerves of the Thorax

## VAGUS NERVES

The *right vagus nerve* descends in the thorax, first lying posterolateral to the brachiocephalic artery (Fig. 3-29), then lateral to the trachea and medial to the terminal part of the azygos vein (Fig. 3-24). It passes *behind* the root of the right lung and as-

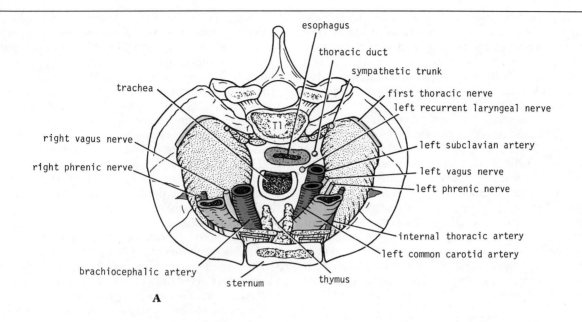

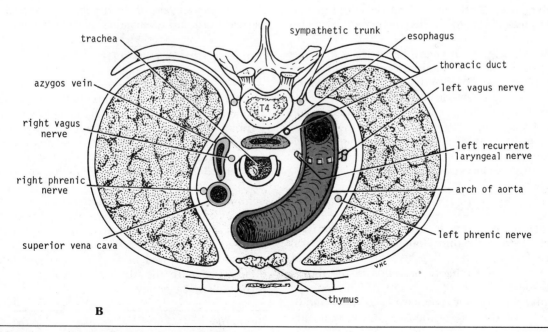

**Fig. 3-29. Cross sections of thorax. (A) At inlet and
(B) at fourth thoracic vertebra.**

sists in the formation of the *pulmonary plexus.* On leaving the plexus, the vagus passes onto the posterior surface of the esophagus and takes part in the formation of the *esophageal plexus.* It then passes through the esophageal opening of the diaphragm behind the esophagus to reach the posterior surface of the stomach.

The *left vagus nerve* descends in the thorax between the left common carotid and left subclavian arteries (Figs. 3-25 and 3-29). It then crosses the left side of the aortic arch and is itself crossed by the left phrenic nerve. The vagus then turns backwards *behind* the root of the left lung and assists in the formation of the *pulmonary plexus.* On leaving the plexus, the vagus passes onto the anterior surface of the esophagus and takes part in the formation of the *esophageal plexus.* It then passes through the esophageal opening in the diaphragm in front of the esophagus to reach the anterior surface of the stomach.

### Branches

Both vagi supply the lungs and esophagus. The right vagus gives off cardiac branches, and the left vagus gives origin to the left recurrent laryngeal nerve.[†]

The *left recurrent laryngeal nerve* arises from the left vagus trunk as the nerve crosses the arch of the aorta (Figs. 3-16, 3-18, and 3-25). It hooks around the ligamentum arteriosum and ascends in the groove between the trachea and the esophagus on the left side (Fig. 3-29). It supplies all the muscles acting on the left vocal cord (except the cricothyroid muscle, a tensor of the cord, which is supplied by the external laryngeal branch of the vagus).

### PHRENIC NERVES

The phrenic nerves arise from the neck from the anterior rami of the third, fourth, and fifth cervical nerves. (See p. 734.)

The *right phrenic nerve* descends in the thorax along the right side of the right brachiocephalic vein and the superior vena cava (Figs. 3-24 and 3-29). It passes *in front of* the root of the right lung

and runs along the right side of the pericardium, which separates the nerve from the right atrium. It then descends on the right side of the inferior vena cava to the diaphragm. Its terminal branches pass through the caval opening in the diaphragm to supply the central part of the peritoneum on its under-aspect.

The *left phrenic nerve* descends in the thorax along the left side of the left subclavian artery. It crosses the left side of the aortic arch (Fig. 3-25) and here crosses the left side of the left vagus nerve. It passes *in front of* the root of the left lung and then descends over the left surface of the pericardium, which separates the nerve from the left ventricle. On reaching the diaphragm, the terminal branches pierce the muscle and supply the central part of the peritoneum on its under-aspect.

The phrenic nerves possess efferent and afferent fibers. The efferent fibers are the *sole nerve supply* to the muscle of the diaphragm.

The afferent fibers carry sensation to the central nervous system from (1) the peritoneum covering the central region of the undersurface of the diaphragm; (2) the pleura covering the central region of the upper surface of the diaphragm; and (3) the pericardium and mediastinal parietal pleura.

## THORACIC PART OF THE SYMPATHETIC TRUNK

The thoracic part of the sympathetic trunk is continuous above with the cervical and below with the lumbar parts of the sympathetic trunk. It is the most laterally placed structure in the mediastinum and runs downward on the heads of the ribs (Figs. 3-24 and 3-25). It leaves the thorax on the side of the body of the twelfth thoracic vertebra by passing behind the medial arcuate ligament.

The sympathetic trunk has twelve (often only eleven) segmentally arranged ganglia, each with a *white* and *gray ramus communicans* passing to the corresponding spinal nerve. The first ganglion is often fused with the inferior cervical ganglion to form the *stellate ganglion.*

### Branches

1. Gray rami communicantes go to all the thoracic spinal nerves. The postganglionic fibers are distributed through the branches of the spinal

---

[†]The right recurrent laryngeal nerve arises from the right vagus in the neck and hooks around the subclavian artery and ascends between the trachea and esophagus.

nerves to the blood vessels, sweat glands, and arrector pili muscles of the skin.

2. The first five ganglia give postganglionic fibers to the heart, aorta, lungs, and esophagus.
3. The lower eight ganglia mainly give preganglionic fibers, which are grouped together to form the splanchnic nerves (Figs. 3-24 and 3-25) and supply the abdominal viscera. They enter the abdomen by piercing the crura of the diaphragm. The *greater splanchnic nerve* arises from ganglia 5–9, the *lesser spanchnic nerve* arises from ganglia 10 and 11, and the *lowest splanchnic nerve* arises from ganglion 12. For details of the distribution of these nerves in the abdomen, see page 264.

## Esophagus

The esophagus is a tubular structure about 10 inches (25 cm) long, which is continuous above with the laryngeal part of the pharynx opposite the sixth cervical vertebra. It passes through the diaphragm at the level of the tenth thoracic vertebra to join the stomach (Fig. 3-6).

In the neck, it lies in front of the vertebral column. Laterally, it is related to the lobes of the thyroid gland and anteriorly, it is in contact with the trachea and the recurrent laryngeal nerves. (See p. 737).

In the thorax, it passes downward and to the left through the superior and then the posterior mediastina. At the level of the sternal angle the aortic arch pushes the esophagus over to the midline (Fig. 3-29).

The relations of the thoracic part of the esophagus from above downward are as follows:

### Anteriorly

The trachea and the left recurrent laryngeal nerve; the left principal bronchus, which constricts it; and the pericardium, which separates the esophagus from the left atrium (Figs. 3-29 and 3-30).

### Posteriorly

The bodies of the thoracic vertebrae, the thoracic duct, the azygos veins, the right posterior intercostal arteries, and, at its lower end, the descending thoracic aorta (Figs. 3-29 and 3-30).

### Right Side

The mediastinal pleura and the terminal part of the azygos vein (Fig. 3-24).

### Left Side

The left subclavian artery, the aortic arch, the thoracic duct, and the mediastinal pleura (Fig. 3-25).

Inferiorly to the level of the roots of the lungs, the vagus nerves leave the pulmonary plexus and join with sympathetic nerves to form the *esophageal plexus.* The left vagus lies anterior to the esophagus and the right vagus, posterior. At the opening in the diaphragm the esophagus is accompanied by the two vagi, branches of the left gastric blood vessels, and lymphatic vessels. Fibers from the right crus of the diaphragm pass around the esophagus in the form of a sling.

In the abdomen the esophagus descends for about ½ inch (1.3 cm) and then enters the stomach. It is related to the left lobe of the liver anteriorly and to the left crus of the diaphragm posteriorly.

### Blood Supply

The upper third of the esophagus is supplied by the inferior thyroid artery; the middle third by branches from the descending thoracic aorta; and the lower third by branches from the left gastric artery. The veins from the upper third drain into the inferior thyroid veins; from the middle third into the azygos veins; and from the lower third into the left gastric vein, a tributary of the portal vein.

### Lymphatic Drainage

Lymph vessels from the upper third of the esophagus drain into the deep cervical nodes; from the middle third, into the superior and posterior mediastinal nodes; and from the lower third, into nodes along the left gastric blood vessels and the celiac nodes (Fig. 3-11).

### Nerve Supply

The esophagus is supplied by parasympathetic and sympathetic efferent and afferent fibers via the

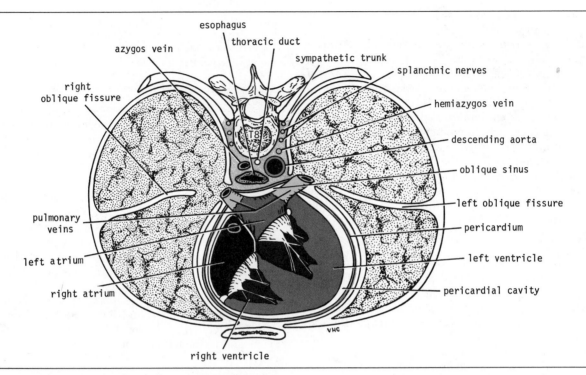

esophagus
thoracic duct
azygos vein
sympathetic trunk
splanchnic nerves
right
oblique fissure
hemiazygos vein
descending aorta
oblique sinus
left oblique fissure
pericardium
pulmonary
veins
left atrium
left ventricle
right atrium
pericardial cavity
right ventricle

**Fig. 3-30. Cross section of thorax at eighth thoracic vertebra.**

vagi and sympathetic trunks. In the lower part of its thoracic course, the esophagus is surrounded by the esophageal nerve plexus.

## Thymus

The thymus is a flattened, bilobed structure (Fig. 3-29) lying between the sternum and the pericardium in the anterior mediastinum. In the newborn infant it reaches its largest size relative to the size of the body, at which time it may extend up through the superior mediastinum in front of the great vessels into the root of the neck. It continues to grow until puberty, but thereafter undergoes involution. It has a pink, lobulated appearance and is an important source of T-lymphocytes.

### Blood Supply

The blood supply of the thymus is from the inferior thyroid and internal thoracic arteries.

### Cross-Sectional Anatomy of the Thorax

In order to assist in the interpretation of CT scans of the thorax, study the labeled cross sections of the thorax shown in Figure 3-31. The sections have been photographed on their *inferior surfaces*. (See Figs. 3-40 and 3-41 for CT scans.)

## RADIOGRAPHIC APPEARANCES OF THE THORAX

Only the more important features seen in standard postero-anterior and oblique lateral radiography of the chest will be discussed.

### Postero-Anterior Radiography

A postero-anterior radiograph is taken with the anterior wall of the patient's chest touching the cassette holder and with the X-rays traversing the thorax from the posterior to the anterior aspect

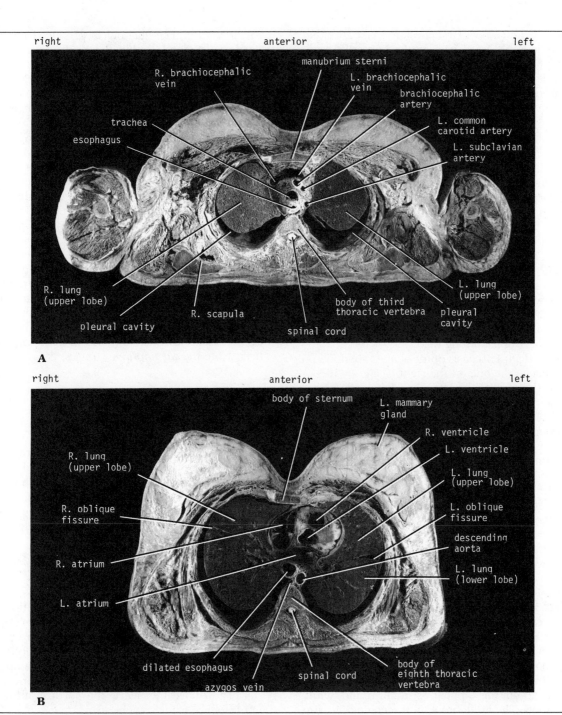

**A**

right  anterior  left

R. brachiocephalic vein

manubrium sterni

L. brachiocephalic vein

brachiocephalic artery

trachea

L. common carotid artery

esophagus

L. subclavian artery

R. lung (upper lobe)

L. lung (upper lobe)

R. scapula

body of third thoracic vertebra

pleural cavity

pleural cavity

spinal cord

**B**

right  anterior  left

body of sternum

L. mammary gland

R. ventricle

L. ventricle

R. lung (upper lobe)

L. lung (upper lobe)

R. oblique fissure

L. oblique fissure

R. atrium

descending aorta

L. atrium

L. lung (lower lobe)

dilated esophagus

spinal cord

body of eighth thoracic vertebra

azygos vein

Fig. 3-31. Cross sections of thorax viewed from below. (A) At level of body of third thoracic vertebra;
(B) at level of eighth thoracic vertebra.

(Figs. 3-32 and 3-33). First check to make sure that the radiograph is a true postero-anterior radiograph and is not slightly oblique. Look at the sternal ends of both clavicles; they should be equidistant from the vertebral spines.

Now examine the following in a systematic order.

1. *Superficial soft tissues.* The nipples in both sexes and the breasts in the female may be seen superimposed on the lung fields. The pectoralis major may also cast a soft shadow.
2. *Bones.* The thoracic vertebrae are imperfectly seen. The costotransverse joints and each rib should be examined in order from above downward and compared with the fellows of the opposite side (Fig. 3-32). The costal cartilages are not usually seen, but should they be calcified, they will be visible. The clavicles are clearly seen crossing the upper part of each lung field. The medial borders of the scapulae may overlap the periphery of each lung field.
3. *Diaphragm.* This casts dome-shaped shadows on each side; the one on the right is slightly higher than the one on the left. Note the costophrenic angle, where the diaphragm meets the thoracic wall (Fig. 3-32). Beneath the right dome is the homogeneous, dense shadow of the liver, and beneath the left dome a gas bubble may be seen in the fundus of the stomach.
4. *Trachea.* The radiotranslucent, air-filled shadow of the trachea is seen in the midline of the neck as a dark area (Fig. 3-32). This is superimposed on the lower cervical and upper thoracic vertebrae.
5. *Lungs.* Looking first at the lung roots, one sees relatively dense shadows due to the presence of the blood-filled pulmonary and bronchial vessels, the large bronchi, and the lymph nodes (Fig. 3-32). The lung fields, by virtue of the air they contain, readily permit the passage of X-rays. For this reason the lungs are more translucent on full inspiration than on expiration. The pulmonary blood vessels are seen as a series of shadows radiating from the lung root. When seen end on, they appear as small, round, white shadows. The large bronchi, if seen end on, also cast similar round shadows. *The smaller bronchi are not seen.*

6. *Mediastinum.* The shadow is produced by the various structures within the mediastinum, superimposed one upon the other (Figs. 3-32 and 3-33). Note the outline of the heart and great vessels. The transverse diameter of the heart should not exceed half the width of the thoracic cage. Remember that on deep inspiration, when the diaphragm descends, the vertical length of the heart increases and the transverse diameter is narrowed. In infants the heart is always wider and more globular in shape than in adults.

The right border of the mediastinal shadow from above downward consists of the right brachiocephalic vein, the superior vena cava, the right atrium, and sometimes the inferior vena cava (Figs. 3-32 and 3-33). The left border consists of a prominence, the *aortic knuckle,* caused by the aortic arch; below this are the left margin of the pulmonary trunk, the left auricle, and the left ventricle (Figs. 3-32 and 3-33). The inferior border of the mediastinal shadow (lower border of the heart) blends with the diaphragm and liver. Note the *cardiophrenic angles.*

## Right Oblique Radiograph

A right oblique radiograph is obtained by rotating the patient so that his right anterior chest wall is touching the cassette holder and the X-rays traverse the thorax from posterior to anterior in an oblique direction (Figs. 3-34 and 3-35). The heart shadow is largely made up by the right ventricle. A small part of the posterior border is formed by the right atrium. For further details of structures seen on this view, see Figures 3-34 and 3-35.

## Left Oblique Radiograph

A left oblique radiograph is obtained by rotation of the patient so that his left anterior chest wall is touching the cassette holder and the X-rays traverse the thorax from posterior to anterior in an oblique direction. The heart shadow is largely made up of the right ventricle anteriorly and the left ventricle posteriorly. Above the heart, the aortic arch and the pulmonary trunk may be seen.

An example of a left-lateral radiograph of the chest is shown in Figures 3-36 and 3-37.

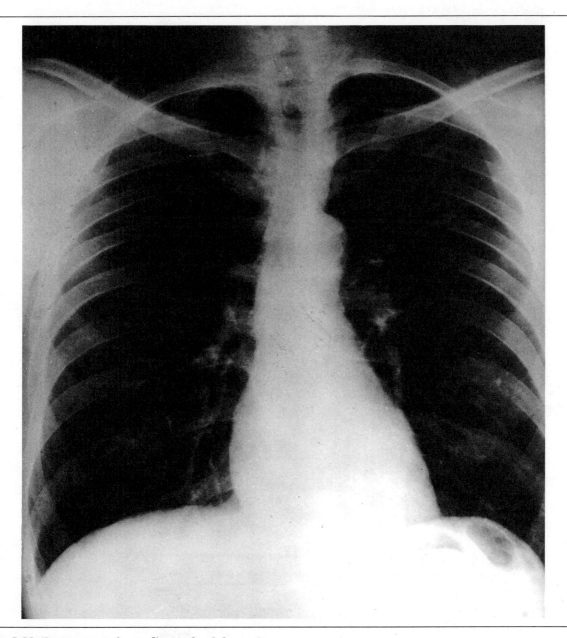

**Fig. 3-32. Postero-anterior radiograph of chest of a
normal adult male.**

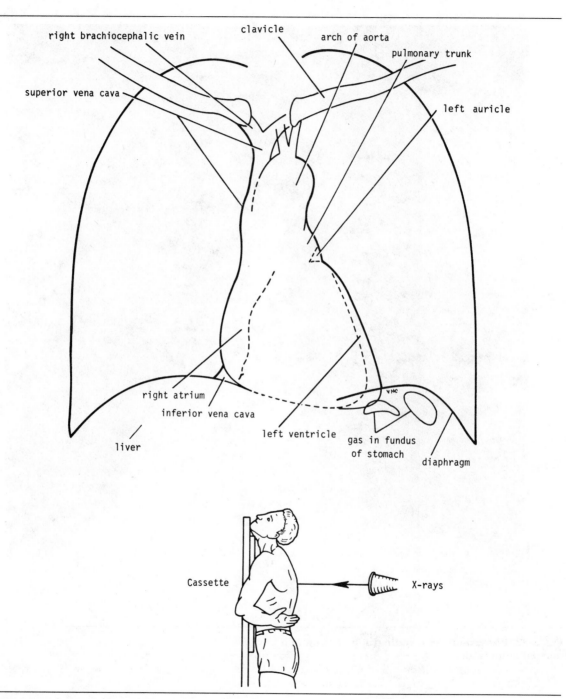

**Fig. 3-33. Main features observable in postero-anterior radiograph of the chest in Figure 3-32. Note position of patient in relation to X-ray source and cassette holder.**

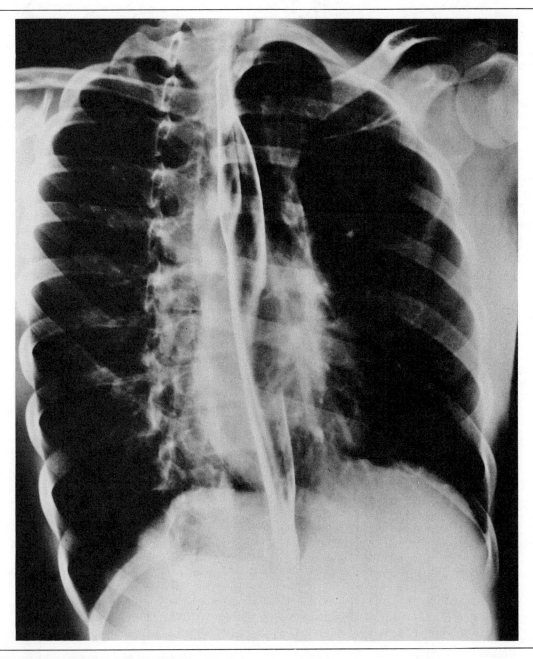

**Fig. 3-34. Right-oblique radiograph of chest of a normal adult male following a barium swallow.**

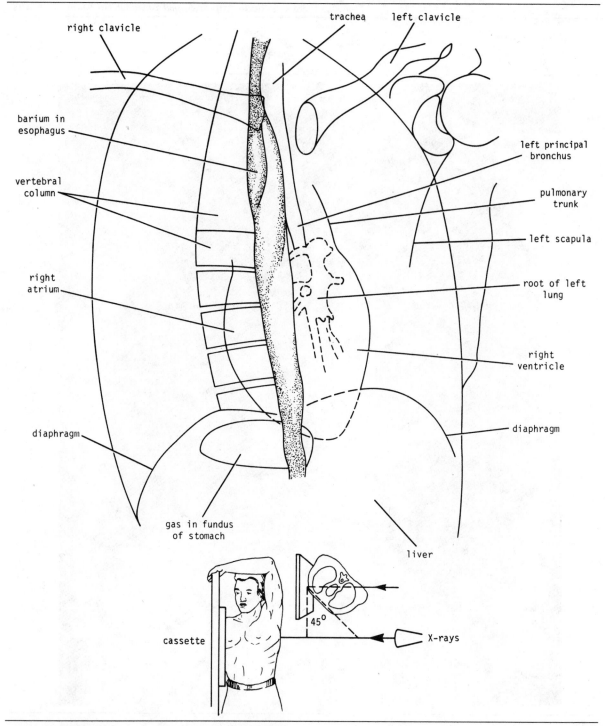

**Fig. 3-35. Main features observable in right-oblique radiograph of chest in Figure 3-34. Note position of patient in relation to X-ray source and cassette holder.**

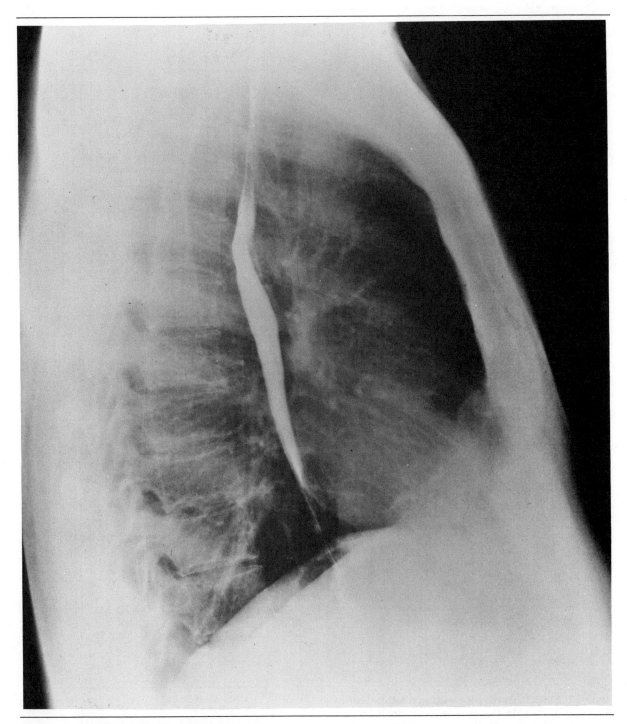

**Fig. 3-36. Left-lateral radiograph of chest of a normal adult male following a barium swallow.**

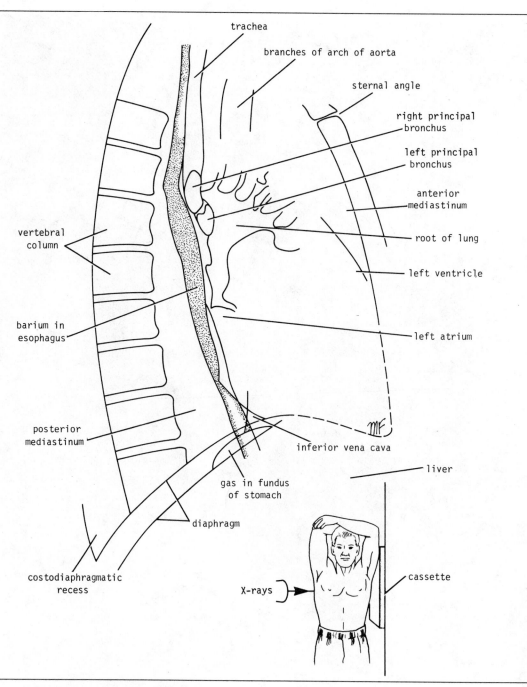

**Fig. 3-37. Main features observable in left-lateral radiograph of chest in Figure 3-36. Note position of patient in relation to X-ray source and cassette holder.**

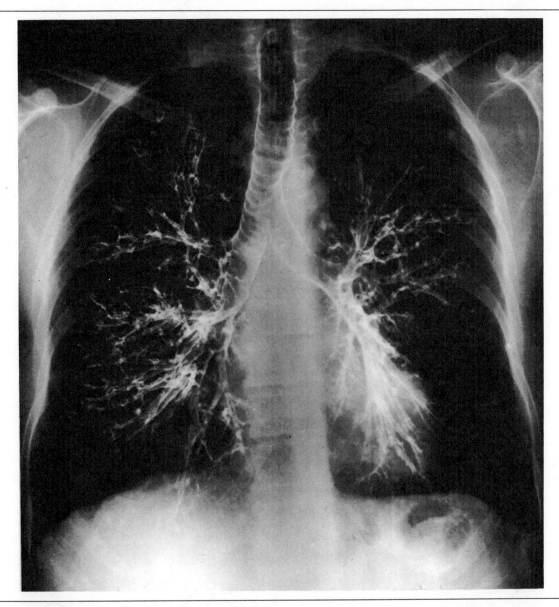

**Fig. 3-38. Postero-anterior bronchogram of chest.**

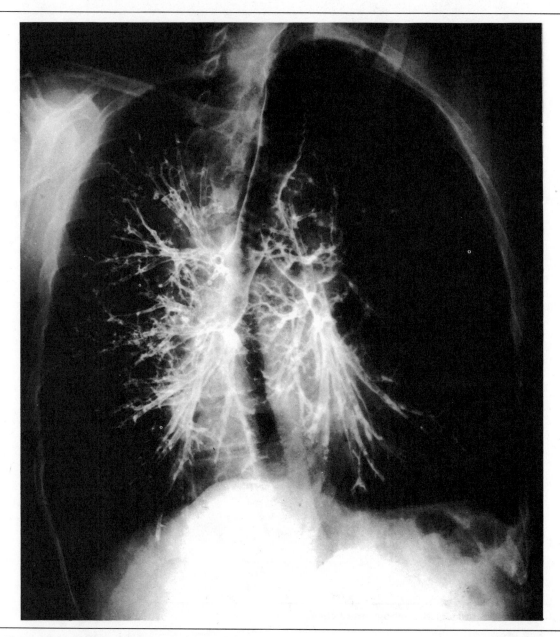

Fig. 3-39. Right-oblique bronchogram of chest.

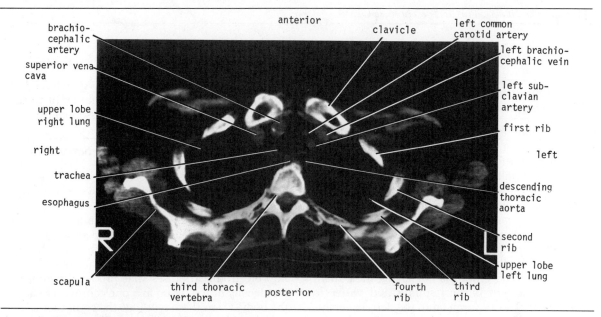

**Fig. 3-40. CT scan of upper part of thorax at the level of the third thoracic vertebra. The section is viewed from below.**

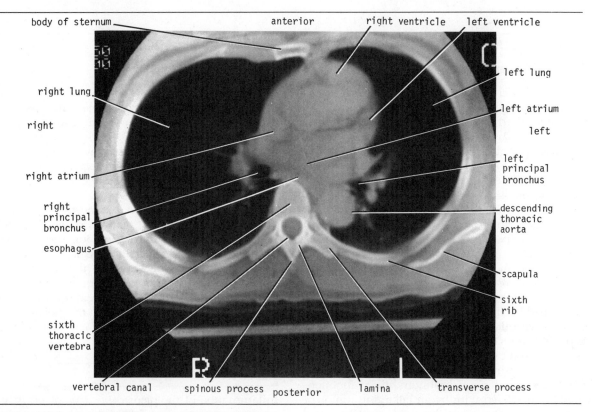

**Fig. 3-41. CT scan of middle part of thorax, at level of sixth thoracic vertebra. The section is viewed from below.**

## Bronchography and Contrast Visualization of the Esophagus

Bronchography is a special study of the bronchial tree by means of the introduction of iodized oil or other contrast medium into a particular bronchus or bronchi, usually under fluoroscopic control. The contrast media are nonirritating and sufficiently radiopaque to allow good visualization of the bronchi (Figs. 3-38 and 3-39). After the radiographic examination is completed, the patient is asked to cough and expectorate the contrast medium.

Contrast visualization of the esophagus (Figs. 3-34 and 3-36) is accomplished by giving the patient a creamy paste of barium sulfate and water to swallow. The aortic arch and the left bronchus cause a smooth indentation on the anterior border of the barium-filled esophagus. This procedure can also be used to outline the posterior border of the left atrium in a right oblique view. An enlarged left atrium will cause a smooth indentation of the anterior border of the barium-filled esophagus.

## CT (Computed Tomography) Scanning of the Thorax

CT scanning relies on the same physics as conventional X-rays but combines it with computer technology. A source of X-rays moves in an arc around the thorax and sends out a beam of X-rays. The beams of X-rays, having passed through the thoracic wall and the thoracic viscera, are converted into electronic impulses that produce readings of the density of the tissue in a one centimeter slice of the body. From these readings the computer is able to assemble a picture of the thorax called a CT scan, which can be viewed on a fluorescent screen and then photographed (Figs. 3-40 and 3-41).

# CLINICAL NOTES

## Mediastinum

In the cadaver, the mediastinum, as the result of the hardening effect of the preserving fluids, is an inflexible, fixed structure. In the living, it is very mobile; the lungs, heart, and large arteries are in rhythmic pulsation and the esophagus distends as each bolus of food passes through it.

If air should enter the pleural cavity (a condition called *pneumothorax*), the lung on that side would immediately collapse and the mediastinum would be displaced to the opposite side. This condition would reveal itself by the patient's being breathless and in a state of shock, and, on examination, the trachea and the heart would be found to be displaced to the opposite side.

The structures that make up the mediastinum are embedded in loose connective tissue that is continuous with that of the root of the neck. Thus, it is possible for a deep infection of the neck to spread readily into the thorax, producing a *mediastinitis*.

Because many vital structures are crowded together within the mediastinum, their functions may be interfered with by an enlarging tumor or organ. A tumor of the left lung may rapidly spread to involve the mediastinal lymph nodes, which on enlargement may compress the left recurrent laryngeal nerve, producing paralysis of the left vocal fold. An expanding cyst or tumor may partially occlude the superior vena cava, causing severe congestion of the veins of the upper part of the body. Other pressure effects may be seen on the sympathetic trunks, phrenic nerves, and sometimes on the trachea, main bronchi, and esophagus.

*Mediastinoscopy* is a diagnostic procedure whereby specimens of tracheobronchial lymph nodes may be obtained without opening the pleural cavities. A small incision is made in the midline in the neck just above the suprasternal notch, and the superior mediastinum is explored down to the region of the bifurcation of the trachea. The procedure may be used to determine the diagnosis and degree of spread of carcinoma of the bronchus.

## Pleurae

Inflammation of the pleura (*pleuritis* or *pleurisy*), secondary to inflammation of the lung (i.e., *pneumonia*), results in the pleural surfaces becoming coated with inflammatory exudate, causing the

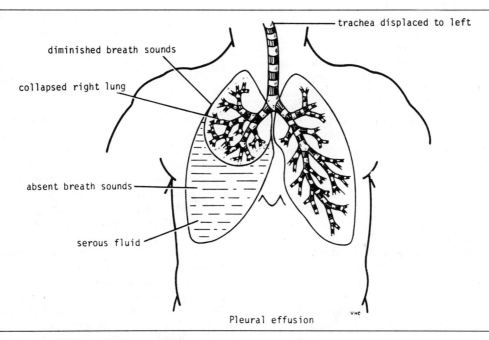

Fig. 3-42. Case of right-sided pleural effusion. Mediastinum is displaced to left, right lung is compressed, and bronchi are narrowed. Auscultation would reveal only faint breath sounds over compressed lung and absent breath sounds over fluid in pleural cavity.

surfaces to be roughened. This roughening produces friction and a *pleural rub* may be heard with the stethoscope on inspiration and expiration. Often the exudate becomes invaded by fibroblasts, which lay down collagen and bind the visceral pleura to the parietal pleura, forming *pleural adhesions*.

As the result of disease or injury, air may enter the pleural cavity from the lungs or through the chest wall (pneumothorax). In the treatment of tuberculosis, air may purposely be injected into the pleural cavity to collapse and rest the lung. This is known as *artificial pneumothorax*. A *spontaneous pneumothorax* is a condition in which air enters the pleural cavity suddenly without its cause being immediately apparent. After investigation, it is usually found that air has entered from a diseased lung and a bullus (bleb) has ruptured.

Stab wounds of the thoracic wall may pierce the parietal pleura so that the pleural cavity is open to the outside air. This condition is called *open pneumothorax*. Each time the patient inspires it is possible to hear air under atmospheric pressure being sucked into the pleural cavity. Sometimes the clothing and the layers of the thoracic wall combine to form a valve so that air enters on inspiration but cannot exit through the wound. In these circumstances, the air pressure builds up on the wounded side and pushes the mediastinum toward the opposite side. In this situation, you have a collapsed lung on the injured side and the opposite lung will be compressed by the deflected mediastinum. This dangerous condition is called a *positive-pressure pneumothorax*.

Air in the pleural cavity associated with serous fluid is known as *hydropneumothorax*; associated with pus, as *pyopneumothorax*; and associated with blood, as *hemopneumothorax*. A collection of pus (without air) in the pleural cavity is called an *empyema*. The presence of serous fluid in the pleural cavity is referred to as a *pleural effusion* (Fig. 3-42). Fluid (serous, blood, or pus) may be drained from the pleural cavity through a wide-bore needle, as described on page 78.

## Trachea and Bronchi

The trachea is a fibroelastic tube kept patent under normal conditions by U-shaped bars of cartilage. In the neck, a unilateral or bilateral enlargement of the thyroid gland may cause gross displacement or compression of the trachea. A dilatation of the aortic arch (*aneurysm*) may compress the trachea. With each cardiac systole the pulsating aneurysm may tug at the trachea and left bronchus, a clinical sign that may be felt by palpating the trachea in the suprasternal notch.

The mucosa lining the trachea is innervated by the recurrent laryngeal nerve, and in the region of its bifurcation, by the pulmonary plexus. A *tracheitis* or *bronchitis* gives rise to a raw, burning sensation felt deep to the sternum, rather than actual pain. Many thoracic and abdominal viscera, when diseased, give rise to discomfort that is felt in the midline. (See p. 279.) It would appear that organs possessing a sensory innervation that is not under normal conditions directly relayed to consciousness display this phenomenon. It is interesting to note that the afferent fibers from these organs traveling to the central nervous system accompany autonomic nerves.

Inhalation of *foreign bodies* into the lower respiratory tract is common, especially in children. Pins, screws, nuts, bolts, peanuts, and parts of chicken bones and toys have all found their way into the bronchi. Parts of teeth may be inhaled while a patient is under anesthesia during a difficult dental extraction. Since the right bronchus is the wider and more direct continuation of the trachea (Fig. 3-9), foreign bodies tend to enter the right rather than the left bronchus. From there, they usually pass into the middle or lower lobe bronchi.

*Bronchoscopy* enables a physician to examine the interior of the trachea, its bifurcation, called the *carina*, and the main bronchi. With experience it is possible to examine the interior of the lobar bronchi and the beginning of the first segmental bronchi. By means of this instrument, it is also possible to obtain biopsy specimens of mucous membrane and remove inhaled foreign bodies (even an open safety pin).

Lodgement of a foreign body in the larynx or edema of the mucous membrane of the larynx secondary to infection or trauma may require immediate relief to prevent asphyxiation. One of the methods commonly used to relieve complete obstruction is tracheotomy (see p. 898).

## Lungs

A physician must always remember that the apex of the lung projects up into the neck (1 inch [2.5 cm] above the clavicle) and may be damaged in stab wounds or bullet wounds in this area.

Although the lungs are well protected by the bony thoracic cage, a splinter from a fractured rib may nevertheless penetrate the lung and air may escape into the pleural cavity, causing a pneumothorax and collapse of the lung. It may also find its way into the lung connective tissue. From there, the air moves under the visceral pleura until it reaches the lung root. It then passes into the mediastinum and up to the neck. Here, it may distend the subcutaneous tissue, a condition known as *subcutaneous emphysema*.

In physical examination of the patient, it is well to remember that the upper lobes of the lungs are most easily examined from in front of the chest and the lower lobes from the back. In the axillae, areas of all lobes may be examined.

Lung tissue and the visceral pleura are devoid of pain-sensitive nerve endings, so that pain in the chest is always the result of conditions affecting the surrounding structures. In tuberculosis or pneumonia, for example, pain may never be experienced.

Once lung disease crosses the visceral pleura and the pleural cavity to involve the parietal pleura, pain becomes a prominent feature. Lobar pneumonia with pleurisy, for example, produces a severe tearing pain, accentuated by inspiring deeply or coughing. Since the lower part of the costal parietal pleura receives its sensory innervation from the lower five intercostal nerves, which also innervate the skin of the anterior abdominal wall, pleurisy in this area commonly produces pain that is referred to the abdomen. This has sometimes resulted in a mistaken diagnosis of an acute abdominal lesion.

In a similar manner, pleurisy of the central part of the diaphragmatic pleura, which receives sensory innervation from the phrenic nerve (C3, 4, and 5), may lead to referred pain over the shoulder, since the skin of this region is supplied by the supraclavicular nerves (C3 and 4).

Surgical access to the lung or mediastinum

is commonly undertaken through an intercostal space. Special rib retractors are used that allow the ribs to be widely separated. The costal cartilages are sufficiently elastic to permit considerable bending. Good exposure of the lungs is obtained by this method.

A localized chronic lesion such as that of tuberculosis or a benign neoplasm may require surgical removal. If it is restricted to a bronchopulmonary segment, it is possible carefully to dissect out a particular segment and remove it, leaving the surrounding lung intact (*segmental resection*). Segmental resection requires that the radiologist and thoracic surgeon have a sound knowledge of the bronchopulmonary segments and that they cooperate fully to localize the lesion accurately before operation.

*Bronchogenic carcinoma* accounts for about one-third of all cancer deaths in men. It commences in the majority of patients in the mucous membrane lining the larger bronchi and is therefore situated close to the hilus of the lung. The neoplasm rapidly spreads to the tracheobronchial and bronchomediastinal nodes and may involve the recurrent laryngeal nerves. Lymphatic spread via the bronchomediastinal trunks may result in early involvement in the lower deep cervical nodes just above the level of the clavicle. Hematogenous spread to bones and the brain commonly occurs.

The following are some diseases that decrease respiratory efficiency.

## Constriction of the Bronchi as in Bronchial Asthma

One of the problems associated with bronchial asthma is the spasm of the smooth muscle in the wall of the bronchioles. This particularly reduces the diameter of the bronchioles during expiration, usually causing the asthmatic patient to experience great difficulty in expiring, although inspiration is accomplished normally. The lungs consequently become greatly distended and the thoracic cage becomes permanently enlarged, forming the so-called *barrel chest*. In addition, the air flow through the bronchioles is further impeded by the presence of excess mucus, which the patient is unable to clear because an effective cough cannot be produced.

## Loss of Lung Elasticity

Many diseases of the lungs, such as *emphysema* and *pulmonary fibrosis*, destroy the elasticity of the lungs, and thus the lungs are unable to recoil adequately, causing incomplete expiration. The respiratory muscles in these patients have to assist in expiration, which no longer is a passive phenomenon.

## Loss of Lung Distensibility

Diseases such as *silicosis*, *asbestosis*, *cancer*, and *pneumonia* interfere with the process of expanding the lung in inspiration. There then occurs a decrease in the compliance of the lungs and the chest wall, and a greater effort has to be undertaken by the inspiratory muscles to inflate the lungs.

## Postural Drainage

Excessive accumulation of bronchial secretions in a lobe or segment of a lung may seriously interfere with the normal flow of air into the alveoli. Furthermore, the stagnation of such secretions is often quickly followed by infection. To aid in the normal drainage of a bronchial segment, a physiotherapist will often alter the position of the patient so that gravity will assist in the process of drainage. It is clear that a sound knowledge of the bronchial tree is necessary to determine the optimum position of the patient for good postural drainage.

# Pericardium

In inflammation of the serous pericardium, called *pericarditis*, there may be an excessive accumulation of pericardial fluid. This may compress the thin-walled atria and interfere with the filling of the heart during diastole. This compression of the heart is called *cardiac tamponade*.

Roughening of the visceral and parietal layers of serous pericardium by inflammatory exudate in acute pericarditis will produce *pericardial friction rub*, which can be felt on palpation and heard through a stethoscope.

Pericardial fluid may be aspirated from the pericardial cavity should excessive amounts accumulate in pericarditis. This process is called paracentesis. The needle can be introduced to the left of the xiphoid process in an upward and backward direction at an angle of. 45 degrees to the skin. When paracentesis is performed at this site, the pleura and lung are not damaged, due to the presence of the cardiac notch in this area.

# Heart

## Atrial Septal Defects

After birth the foramen ovale becomes completely closed as the result of the fusion of the septum primum with the septum secundum. In 25 percent of hearts, a small opening persists, but this is usually of such a minor nature that it has no clinical significance. Occasionally the opening is much larger and results in oxygenated blood from the left atrium passing over into the right atrium.

## Ventricular Septal Defects

The ventricular septum is formed in a complicated manner and is only complete when the membranous part fuses with the muscular part. Ventricular septal defects are less frequent than atrial septal defects. They are found in the membranous part of the septum and may measure from 1 to 2 cm in diameter. Blood under high pressure passes through the defect from left to right, causing enlargement of the right ventricle. Large defects are serious and can shorten life if surgery is not performed.

## Myocardial Ischemia

Inadequate blood supply to the myocardium (*myocardial ischemia*) results in the patient's experiencing severe pain over the middle of the sternum, often spreading to one or both arms, the root of the neck, and even the jaw. The pain is assumed to be caused by the accumulation of metabolites and by oxygen deficiency, which stimulate the sensory nerve endings in the myocardium. The afferent nerve fibers ascend to the central nervous system through the cardiac branches of the sympathetic trunk and enter the spinal cord via the posterior roots of the upper four thoracic nerves.

It is interesting to note that cardiac pain is not felt in the heart, but is referred to the skin areas supplied by the corresponding spinal nerves. The skin areas supplied by the upper four intercostal nerves and by the intercostal brachial nerve (T2) are therefore affected. A certain amount of spread of nervous information must occur within the central nervous system, for the pain is sometimes felt in the neck and jaw.

It must be emphasized that although the coronary arteries anastomose with each other at the arteriolar level, they are essentially *functional end arteries*. (See p. 21.) A sudden block of one of the large branches of either coronary artery will inevitably lead to necrosis of the cardiac muscle (*myocardial infarction*) in that vascular area and often to death of the patient.

The conducting system of the heart receives its blood supply from the coronary arteries. In the majority of persons, the atrioventricular bundle, for example, is supplied by the posterior interventricular branch of the right coronary artery. Occlusion of the artery supplying the atrioventricular bundle is a serious condition resulting in *heart block*. The ventricles no longer receive the cardiac impulse from the atria, and they start to beat independently of the atria, at a slower rate.

## Valvular Disease of the Heart

The inflammatory response may cause the edges of the valve cusps to stick together. Later, fibrous thickening occurs, followed by loss of flexibility and shrinkage. Narrowing (stenosis) and valvular incompetence (regurgitation) result, and the heart ceases to function as an efficient pump. In rheumatic disease of the mitral valve, for example, not only do the cusps undergo fibrosis and shrink, but the chordae tendineae shorten, preventing closure of the cusps during ventricular systole.

## Valvular Heart Murmurs

Apart from the sounds of the valves closing, lŭb-dŭp, the blood passes through the normal heart silently. Should the valve orifices become narrowed or the valve cusps distorted and shrunken by disease, however, a rippling effect is set up that leads to turbulence and vibrations that are heard as heart murmurs.

## Esophagus

The esophagus has three anatomical and physiological constrictions. The first is where the pharynx joins the upper end; the second is where the aortic arch and the left bronchus cross its anterior surface; and the third occurs where the esophagus passes through the diaphragm into the stomach. These constrictions are of considerable clinical importance since they are sites where swallowed foreign bodies may lodge or through which it may be difficult to pass an *esophagoscope*. Since a slight delay in the passage of food or fluid occurs at these levels, strictures develop here following the drinking of caustic fluids. Those constrictions are also the common sites of carcinoma of the esophagus. It is useful to remember that their respective distances from the upper incisor teeth are 6 inches (15 cm), 10 inches (25 cm), and 16 inches (41 cm) (Fig. 3-43).

At the lower third of the esophagus there is an important *portal-systemic venous anastomosis.* (For other portal-systemic anastomoses, see p. 238.) Here, the esophageal tributaries of the azygos veins (systemic veins) anastomose with the esophageal tributaries of the left gastric vein (which drains into the portal vein). Should the portal vein become obstructed, as, for example, in *cirrhosis of the liver*, *portal hypertension* develops, resulting in the dilatation and varicosity of the portal systemic anastomoses. Varicosed esophageal veins may rupture during the passage of food, causing *hematemesis* (vomiting of blood), which may prove fatal.

The lymphatic drainage of the lower third of the esophagus descends through the esophageal opening in the diaphragm and ends in the celiac nodes around the celiac artery (Fig. 3-11). A malignant tumor of this area of the esophagus would therefore tend to spread below the diaphragm along this route. Consequently, surgical removal of the lesion would include not only the removal of the primary lesion, but also the celiac lymph nodes and all regions that drain into these nodes, namely, the stomach, the upper half of the duodenum, the spleen, and the omenta. Restoration of continuity of the gut is accomplished by performing an esophagojejunostomy.

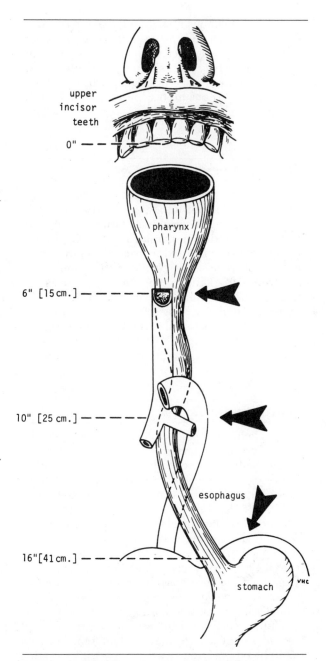

**Fig. 3-43. Levels of normal constrictions of esophagus. It is at these levels that (1) swallowed foreign bodies tend to become arrested, (2) strictures may develop following the swallowing of corrosive fluids, and (3) cancer commonly develops.**

The close relationship between the anterior wall of the esophagus and the posterior wall of the left atrium has already been emphasized. A "barium swallow" may help a physician to assess the size of the left atrium in cases of left-sided heart failure, in which the left atrium becomes distended due to back pressure of venous blood.

## Phrenic Nerves

The phrenic nerve may be paralyzed due to pressure from malignant tumors in the mediastinum. Surgical crushing or sectioning of the phrenic nerve in the neck producing paralysis of the diaphragm on one side may be used as part of the treatment of lung tuberculosis, especially of the lower lobes. The immobile dome of the diaphragm rests the lung. Additional immobility and rest to the lower lobe of the lung can be accomplished by introducing air into the peritoneal cavity, a procedure called *pneumoperitoneum*, which further raises the flaccid dome of the diaphragm up into the thorax.

## Sympathetic Trunk

Preganglionic sympathectomy of the second and third thoracic ganglia may be performed to increase the blood flow to the fingers for such conditions as *Raynaud's disease.* The sympathectomy causes vasodilatation of the arterioles in the upper limb.

Removal of the upper four or five thoracic ganglia and the stellate ganglion is sometimes performed to interrupt the afferent pain fibers passing from the heart to the central nervous system via the sympathetic nerves. This relieves the severe anginal pain in selected cases of *angina pectoris.*

Bilateral thoracolumbar sympthectomy was sometimes performed in cases of severe *essential hypertension.* The resulting vasodilatation caused a fall in blood pressure.

A high *spinal anesthetic* may block the preganglionic sympathetic fibers passing out from the lower thoracic segments of the spinal cord. This will produce temporary vasodilatation below this level, with a consequent fall in blood pressure.

## Patent Ductus Arteriosus

The ductus arteriosus represents the distal portion of the sixth left aortic arch and connects the left pulmonary artery to the descending aorta. During fetal life blood passes through it from the pulmonary artery to the aorta, thus bypassing the lungs. After birth, it normally constricts, later closes, and becomes the *ligamentum arteriosum.*

Failure of the ductus arteriosus to close may occur as an isolated congenital abnormality or may be present in association with congenital heart disease. A persistent patent ductus arteriosus results in high-pressure aortic blood passing into the pulmonary artery, which raises the pressure in the pulmonary circulation. A patent ductus arteriosus is life-threatening and should be ligated and divided surgically.

## Coarctation of the Aorta

Coarctation of the aorta is a congenital narrowing of the aorta just proximal to, opposite, or distal to the site of attachment of the ligamentum arteriosum. This condition is believed to result from an unusual quantity of ductus arteriosus muscle tissue being present in the wall of the aorta. When the ductus arteriosus contracts, the ductal muscle in the aortic wall also contracts and the aortic lumen becomes narrowed. Later, when fibrosis takes place, the aortic wall also is involved and permanent narrowing occurs.

Clinically, the cardinal sign of aortic coarctation is absent or diminished pulses in the femoral arteries of both lower limbs. To compensate for the diminished volume of blood reaching the lower part of the body, an enormous collateral circulation develops with dilation of the internal thoracic, subclavian, and posterior intercostal arteries. The condition should be treated surgically.

## Azygos and Hemi-Azygos Veins

In obstruction of the superior or inferior venae cavae, the azygos veins provide an alternative pathway for the return of venous blood to the right atrium of the heart. This is possible since these veins and their tributaries connect the superior and inferior venae cavae.

# CLINICAL PROBLEMS

*Answers on page 957*

1. A 55-year-old man states that his wife has recently noticed an alteration in his voice. He has lost 40 pounds in weight and has a persistent cough with blood-stained sputum. He smokes fifty cigarettes a day. On examination, the left vocal fold is immobile and lies in the adducted position. A postero-anterior chest radiograph reveals a large mass in the upper lobe of the left lung with an increase in width of the mediastinal shadow on the left side. Explain in anatomical terms the alteration of the voice, and, if possible, relate the voice changes to the other findings.

2. A 50-year-old patient with a history of syphilis in her youth has a swelling that protrudes from beneath the upper margin of the sternum in the midline of the neck. The swelling expands with each systole of the heart. On examination, the trachea is found to be displaced to the right in the neck, and there is a distinct tugging sensation felt on palpation of the trachea. What anatomical structure lying within the superior mediastinum is likely to have an expansile swelling that tugs at the trachea?

3. A 35-year-old woman says that she has difficulty in breathing and sleeping at night. She falls asleep, only to wake up with a choking sensation. She finds that she has to sleep propped up in bed on pillows with her neck flexed to the right. On examination, she is found to have an enlarged thyroid gland and congested veins in the root of the neck. Explain on anatomical grounds (a) why she should have difficulty in breathing when asleep and (b) why the veins in the neck are congested.

4. A 15-year-old boy was rescued from a lake after falling through thin ice. The next day a severe cold developed, and three days later his general condition deteriorated. He became more febrile and started to cough up blood-stained sputum. At first, he had no chest pain, but later, when he coughed, he experienced a severe pain over the right fifth intercostal space in the midclavicular line. The diagnosis of lobar pneumonia was made. Explain the following: (a) Why did he not experience chest pain early in the disease? (b) What is the pain due to and why is it

worse on coughing? (c) Which lobe of the lung is likely to be diseased?

5. Following a broken engagement, a 33-year-old woman attempted to commit suicide by swallowing a solution of caustic soda. Her family physician managed to start treatment early and she recovered. Five years later, now happily married, she was admitted to the hospital complaining of difficulty in swallowing. She stated that she felt that solid foods tended to "stick here"—and she pointed to the region of the sternal angle. Where is the obstruction likely to be situated?

6. While changing the diaper on her 2-year-old boy, a mother failed to find one of the small safety pins. Two days later the child developed a cough and became febrile. What is likely to have occurred to the safety pin, and where, anatomically, would you expect to find it?

7. A young man aged 25 was found on routine chest X-ray to have a localized tuberculous lesion in the right upper lobe of the right lung. It failed to respond to medication. Is it possible to remove a local area of the lung surgically?

8. A 45-year-old man has a history of rheumatic fever at age 10. Following the illness, stenosis of the mitral valve developed. He was recently admitted to the hospital with the signs and symptoms of right-sided heart failure. Based on your anatomical knowledge, by what method would you assess the size of the left atrium?

9. A colleague reports that a patient has altered breath sounds in the upper lobe of the left lung. To verify this, where would you place your stethoscope on the chest wall?

10. A patient is suspected of suffering from bronchiectasis (permanent dilatation of the bronchi) in the lower lobe of the right lung. (a) Would you expect to be able to see a small bronchus on a postero-anterior radiograph of the thorax? (b) What special methods are available to examine the lumen of a small bronchus?

11. A 36-year-old patient with a known history of emphysema (dilatation of alveoli and destruction of alveolar walls with tendency to form cystic spaces) suddenly experiences a severe pain in the chest, is breathless, and is obviously

in a state of shock. On examination, the trachea is found displaced to the right in the suprasternal notch, and the apex beat of the heart can be felt in the fifth left intercostal space just lateral to the sternum. Assuming the patient has had a spontaneous pneumothorax, explain the following: (a) Why are the trachea and apex beat displaced to the right? (b) Why is the left lung collapsed? (c) What is the air pressure in the left pleural cavity?

12. A wife was told that her husband was suffering from cancer of the lower end of the esophagus. She was informed that to save his life the surgeon would have to remove the lower part of the esophagus, the stomach, the spleen, and the upper part of the duodenum. She could not understand why such a drastic operation was required to remove such a small tumor. Can you explain why in anatomical terms?

13. A blue-nosed, 50-year-old man, who is a chronic alcoholic patient, was told by his physician that he had cirrhosis of the liver with portal hypertension. Recently, he informed his physician that he had vomited a cupful of blood. Using your anatomical knowledge, can you explain his last medical episode?

14. A mother, on looking at her newborn baby lying on its back in a crib, was astonished and horrified to see its anterior abdominal wall bulging in and out with each respiration. Can you explain this in anatomical terms?

15. A chronic asthmatic patient recently had a severe attack in a restaurant. It was noticed that he stood up and grabbed the tabletop and chair back to help him inspire. Is this likely? If so, explain how he was able to assist his inspiratory efforts by this means.

16. A physician treating a patient with acute pericarditis decided it would be advantageous to aspirate some of the fluid in the pericardial cavity. Where should he insert the needle and why?

17. A 55-year-old man states that he experiences pain down the inner side of this left arm when he walks upstairs or when he becomes excited. He emphatically denies that he has heart disease since, as he says, "I have no pain in my heart." Is he correct? Can you explain the pathway taken by the pain fibers from the heart to the central nervous system?

18. Three years ago a 58-year-old man had a venous graft operation for disease of his right coronary artery. After 3 years of freedom from the pain, he now complains of severe angina pectoris on exertion. His surgeons are against performing a further graft operation. Is it possible to interrupt the pain fibers from the heart?

19. A patient has severe aortic incompetence. Where would you place your stethoscope on the chest wall to hear the aortic valve with the least distortion?

20. A patient has disease of the mitral valve. Where would you place your stethoscope on the chest wall to hear the mitral valve?

21. Can you explain why an obese patient is more likely to develop postoperative pulmonary complications than a thin patient?

22. A doctor was driving across a road intersection when another car passing through a red light hit his car broadside at high speed. Within a few minutes, an ambulance was on the scene and the attendants managed to free the doctor from the wreckage of his car. He was found to be deeply cyanotic and all respiratory movements had ceased. His heart was still beating but his pulse was rapid and weak. Given that the doctor's spinal cord had been severed at the level of the second cervical segment, explain in anatomical terms why all respiratory movements had ceased.

# NATIONAL BOARD TYPE QUESTIONS
*Answers on page 983*

**In each of the following questions, answer:**

    (a) IF (1), (2), AND (3) ONLY ARE CORRECT
    (b) IF (1) AND (3) ONLY ARE CORRECT
    (c) IF (2) AND (4) ONLY ARE CORRECT
    (d) IF (4) ONLY IS CORRECT
    (e) IF ALL ARE CORRECT

1. Which of the following statements are **true** with regard to the trachea?
    (1) It lies posterior to the esophagus in the superior mediastinum.
    (2) In deep inspiration the bifurcation may descend as far as the level of the sixth tho-

racic vertebra.

(3) Its left principal bronchus is more vertical than the right principal bronchus.

(4) The arch of the aorta lies on its anterior and left sides in the superior mediastinum.

2. Which of the following statements are **true** with regard to the root of the right lung?

(1) The right phrenic nerve passes anterior to the lung root.

(2) The azygos vein arches forward over the superior margin of the lung root.

(3) The right pulmonary artery lies anterior to the principal bronchus in the lung root.

(4) The right vagus nerve passes posterior to the lung root.

3. Which of the following statements are **true** regarding the right lung?

(1) It possesses a horizontal and an oblique fissure.

(2) Its covering of visceral pleura is sensitive to pain and temperature sensations.

(3) The lymph from the substance of the lung reaches the hilus by the superficial and deep lymphatic plexuses.

(4) The pulmonary ligament anchors the right lung to the central tendon of the diaphragm.

4. The sternocostal (anterior) surface of the heart is formed by the:

(1) Right atrium

(2) Right ventricle

(3) Left ventricle

(4) Left atrium

5. As seen in a postero-anterior (PA) radiograph of the thorax, the left margin of the heart shadow includes which of the following structures?

(1) Left auricle

(2) Pulmonary trunk

(3) Arch of aorta

(4) Left ventricle

**In each of the following questions, answer:**

(a) IF (1) IS CORRECT ONLY

(b) IF (2) IS CORRECT ONLY

(c) IF BOTH (1) AND (2) ARE CORRECT, AND

(d) IF NEITHER (1) NOR (2) IS CORRECT

6. (1) Stab wounds of the neck immediately above the medial one-third of the clavicle may perforate the pleural cavity.

(2) The base of the heart is formed mainly by the left atrium.

7. (1) The second heart sound (dŭp) is produced by the closure of the mitral and tricuspid valves.

(2) The lymph nodes at the hilus of the lung are drained superiorly into the broncho-mediastinal trunk.

8. (1) The right coronary artery gives off the anterior interventricular branch, which supplies the right and left ventricles.

(2) The anterior cardiac vein opens into the left atrium.

9. (1) The lymphatic drainage of the lower one-third of the esophagus empties into the celiac nodes.

(2) Both the sino-atrial and atrioventricular nodes lie in the atrial portion of the heart.

**Select the best response:**

10. All of the following statements concerning the esophagus are correct **except:**

(a) It receives an arterial blood supply from both the descending thoracic aorta and the left gastric artery.

(b) It is constricted by the presence of the left principal bronchus.

(c) It crosses from right to left posterior to the descending aorta.

(d) It pierces the diaphragm with the left vagus on its anterior surface and the right vagus on its posterior surface.

(e) It joins the stomach about 16 inches (41 cm) from the incisor teeth.

11. All of the following statements concerning the mediastinum are correct **except:**

(a) The mediastinum forms a partition between the two pleural cavities.

(b) The mediastinal pleura demarcates the lateral boundaries of the mediastinum.

(c) The heart occupies the middle mediastinum.

(d) Should air enter the left pleural cavity, the structures forming the mediastinum will be deflected over to the right.

(e) The anterior boundary of the mediastinum extends to a lower level than the posterior boundary.

12. All of the following statements regarding the conducting system of the heart are true **except:**
    (a) The impulse for cardiac contraction spontaneously begins in the sino-atrial node.
    (b) The atrioventricular bundle is the sole pathway for conduction of the waves of contraction between the atria and the ventricles.
    (c) The sino-atrial node is frequently supplied by the right and left coronary arteries.
    (d) The sympathetic nerves to the heart slow the rate of discharge from the sino-atrial node.
    (e) The atrioventricular bundle descends behind the septal cusp of the tricuspid valve.

13. All of the following statements regarding the mechanics of inspiration are true **except:**
    (a) The diaphragm is the most important muscle of inspiration.
    (b) The suprapleural membrane can be raised.
    (c) The sternum moves anteriorly.
    (d) The ribs are raised superiorly.
    (e) The tone of the muscles of the anterior abdominal wall is diminished.

14. Which of the following statements concerning the lungs is **correct**?
    (a) There are no lymph nodes within the lungs.
    (b) The right lung is in direct contact with the arch of the aorta and the descending thoracic aorta.
    (c) Inhaled foreign bodies most frequently enter the right lung.
    (d) The structure of the lungs receives its blood supply form the pulmonary arteries.
    (e) The costodiaphragmatic recesses are lined with visceral pleura.

15. Which of the following statements concerning the blood supply to the heart is **incorrect?**
    (a) The coronary arteries are branches of the ascending aorta.
    (b) The right coronary artery supplies both the right atrium and the right ventricle.
    (c) The circumflex branch of the left coronary artery descends in the anterior interventricular groove and passes around the apex of the heart.
    (d) Arrhythmias (abnormal heart beats) can occur following the occlusion of a coronary artery.
    (e) Coronary arteries can be classified as functional end arteries.

16. Which of the following statements is **incorrect** concerning bronchopulmonary segments?
    (a) The veins are intersegmental.
    (b) The segments are separated by connective tissue septa.
    (c) The arteries are intrasegmental.
    (d) Each segment is supplied by a secondary bronchus.
    (e) Each pyramid-shaped segment has its base pointing toward the lung surface.

**Match the structures on the left with the regions of the heart on the right. Each lettered region may be selected once or more than once.**

17. Coronary sinus (opening)
18. Moderator band
19. Anulus ovalis
20. Right pulmonary veins (openings)

    (a) Left atrium
    (b) Right ventricle
    (c) Right atrium
    (d) Left ventricle
    (e) Right auricle

# 4. The Abdomen: Part I
# The Abdominal Wall

The abdomen may be defined as the region of the trunk that lies between the diaphragm above and the inlet of the pelvis below.

## SURFACE ANATOMY

### Surface Landmarks

#### Xiphoid Process

This is the thin cartilaginous lower part of the sternum. It is easily palpated in the depression where the costal margins meet in the upper part of the anterior abdominal wall (Figs. 4-1 and 4-2). The *xiphisternal junction* is identified by feeling the lower edge of the body of the sternum, and it lies opposite the body of the ninth thoracic vertebra.

#### Costal Margin

This is the curved lower margin of the thoracic wall and is formed in front by the cartilages of the seventh, eighth, ninth, and tenth ribs (Figs. 4-1 and 4-2) and behind by the cartilages of the eleventh and twelfth ribs. The costal margin reaches its low-est level at the tenth costal cartilage, which lies opposite the body of the third lumbar vertebra. The twelfth rib may be short and difficult to palpate.

#### Iliac Crest

This may be felt along its entire length and ends in front at the *anterior superior iliac spine* (Figs. 4-1 and 4-2) and behind at the *posterior superior iliac spine* (Fig. 4-3). Its highest point lies opposite the body of the fourth lumbar vertebra.

About 2 inches (5 cm) posterior to the anterior superior iliac spine, the outer margin projects to form the *tubercle of the crest* (Fig. 4-2). The tubercle lies at the level of the body of the fifth lumbar vertebra.

#### Inguinal Ligament

This is the rolled-under inferior margin of the aponeurosis of the external oblique muscle (Figs. 4-1, 4-8, and 4-9). It is attached laterally to the anterior superior iliac spine and curves downward and medially, to be attached to the pubic tubercle. The *pu-*

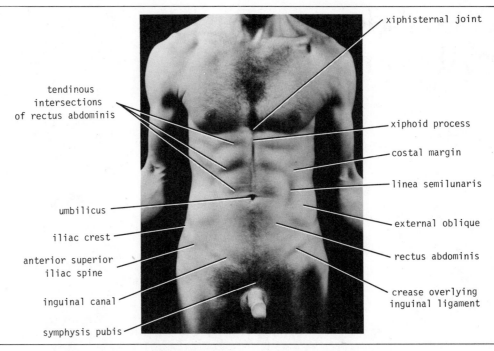

tendinous
intersections
of rectus abdominis

umbilicus

iliac crest

anterior superior
iliac spine

inguinal canal

symphysis pubis

xiphisternal joint

xiphoid process

costal margin

linea semilunaris

external oblique

rectus abdominis

crease overlying
inguinal ligament

**Fig. 4-1. Anterior abdominal wall of a 27-year-old male.**

bic tubercle may be identified as a small protuberance along the superior surface of the pubis (Figs. 4-2, 4-10, and 4-26).

## Symphysis Pubis

This is the cartilaginous joint that lies in the midline between the bodies of the pubic bones (Fig. 4-1). It is felt as a solid structure beneath the skin in the midline at the lower extremity of the anterior abdominal wall. The *pubic crest* is the name given to the ridge on the superior surface of the pubic bones medial to the pubic tubercle (Fig. 4-26).

## Superficial Inguinal Ring

This is a triangular aperture in the aponeurosis of the external oblique muscle situated above and medial to the pubic tubercle (Figs. 4-2, 4-8, 4-10, and 4-16). In the adult male, the margins of the ring can be felt by invaginating the skin of the upper part of the scrotum with the tip of the little finger.

The soft tubular *spermatic cord* can be felt emerging from the ring and descending over or medial to the pubic tubercle into the scrotum (Fig. 4-16). Palpate the spermatic cord in the upper part of the scrotum between the finger and thumb and note the presence of a firm cordlike structure in its posterior part called the *vas deferens* (Figs. 4-18 and 4-21).

In the female the superficial inguinal ring is smaller and difficult to palpate; it transmits the *round ligament of the uterus*.

## Scrotum

This is a pouch of skin containing the testes, the epididymides, and the lower ends of the spermatic cords. The testis on each side is a firm ovoid body surrounded on its lateral, anterior, and medial surfaces by the two layers of the *tunica vaginalis* (Fig. 4-18). The testis should therefore lie free and not tethered to the skin or subcutaneous tissue. Posterior to the testis is an elongated structure, the *epididymis* (Fig. 4-18). It has an enlarged upper end called the *head*, a *body*, and a narrow lower end,

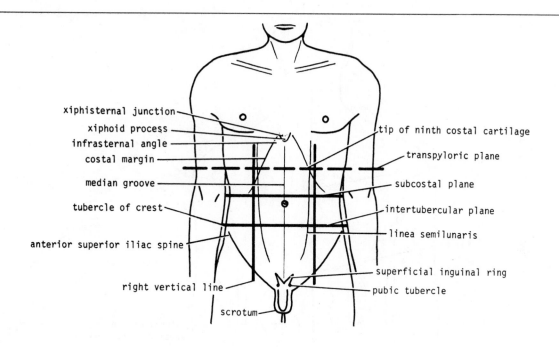

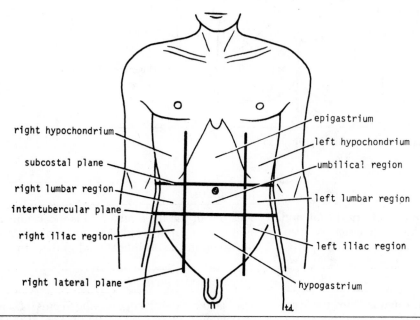

**Fig. 4-2. Surface landmarks and regions of anterior abdominal wall.**

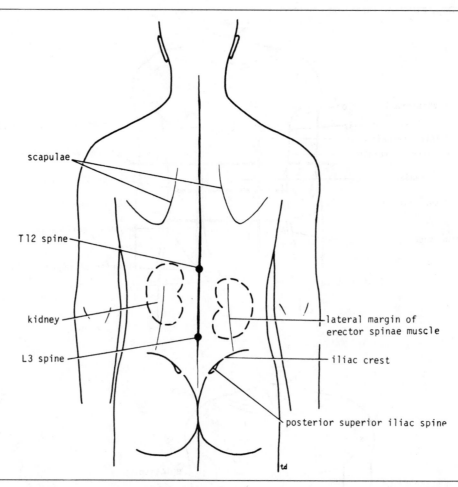

**Fig. 4-3. Surface landmarks of posterior abdominal wall.**

the *tail*. The vas deferens emerges from the tail and ascends medial to the epididymis to enter the spermatic cord.

## Linea Alba

This is a fibrous band that extends from the symphysis pubis to the xiphoid process and lies in the midline (Fig. 4-10). It is formed by the fusion of the aponeuroses of the muscles of the anterior abdominal wall and is represented on the surface by a slight median groove (Figs. 4-1 and 4-2).

## Umbilicus

This lies in the linea alba and is inconstant in position. It is a puckered scar and is the site of attachment of the umbilical cord in the fetus.

## Linea Semilunaris

This is the lateral edge of the rectus abdominis muscle and crosses the costal margin at the tip of the ninth costal cartilage (Figs. 4-1 and 4-2). To accentuate the semilunar lines, the patient is asked to lie on his back and raise his shoulders off the couch without using his arms. To accomplish this, he contracts his rectus abdominis muscles so that their lateral edges stand out.

## Abdominal Regions

For clinical purposes it is customary to divide the abdomen into nine regions by two vertical and two horizontal lines (Fig. 4-2).

Each *vertical line* passes through the midpoint between the anterior superior iliac spine and the symphysis pubis. The upper horizontal line, sometimes referred to as the *subcostal plane*, joins the lowest point of the costal margin on each side. This is the inferior margin of the tenth costal cartilage and lies opposite the third lumbar vertebra.

The lowest horizontal line, often referred to as the *intertubercular plane*, joins the tubercles on the iliac crests. This plane lies at the level of the body of the fifth lumbar vertebra.

The regions thus marked out (Fig. 4-2) are:

*In the upper abdomen:* the right hypochondrium, epigastrium, and left hypochondrium.
*In the middle abdomen:* the right lumbar, umbilical, and left lumbar.
*In the lower abdomen:* the right iliac region, hypogastrium, and left iliac region.

### Abdominal Quadrants

Many physicians simply divide the abdomen into quadrants by using a vertical and a horizontal line that intersect at the umbilicus. The quadrants are named upper right, upper left, lower right, and lower left.

### Transpyloric Plane

This plane is in common use clinically (Fig. 4-2). It passes through the tips of the ninth costal cartilages on the two sides, i.e., the point where the lateral margin of the rectus abdominis (*linea semilunaris*) crosses the costal margin. To identify these points clearly, ask the supine patient to sit up without using his arms. To accomplish this, he contracts the rectus abdominis muscles on both sides, and the lateral margins of these muscles then stand out. This plane passes through the pylorus, the duodenojejunal junction, the neck of the pancreas, and the hili of the kidneys.

### Intercristal Plane

This plane passes across the highest points on the iliac crests and lies on the level of the body of the fourth lumbar vertebra.

## Abdominal Viscera

It must be emphasized that the positions of the majority of the abdominal viscera show individual variations as well as variations in the same person at different times. Posture and respiration have a profound influence on the position of viscera.

The following organs are more or less fixed, and their surface markings are of clinical value.

### *Liver*

The liver lies under cover of the lower ribs, and most of its bulk lies in the right hypochondrium and epigastrium (Fig. 4-4). In infants, until about the end of the third year, the lower margin of the liver extends 1 or 2 fingerbreadths below the costal margin (Fig. 4-4). In the adult who is obese or has a well-developed right rectus abdominis muscle, the liver is impalpable. In a thin adult the lower edge of the liver may be felt a fingerbreadth below the costal margin. It is most easily felt when the patient inspires deeply and the diaphragm contracts and pushes down the liver.

### *Gallbladder*

The fundus of the gallbladder lies opposite the tip of the right ninth costal cartilage, i.e., where the lateral edge of the right rectus abdominis muscle crosses the costal margin (Fig. 4-4).

### *Spleen*

The spleen is situated in the left hypochondrium and lies under cover of the ninth, tenth, and eleventh ribs (Fig. 4-4). Its long axis corresponds to that of the tenth rib, and in the adult it does not normally project forward in front of the midaxillary line. In infants the lower pole of the spleen may just be felt (Fig. 4-4).

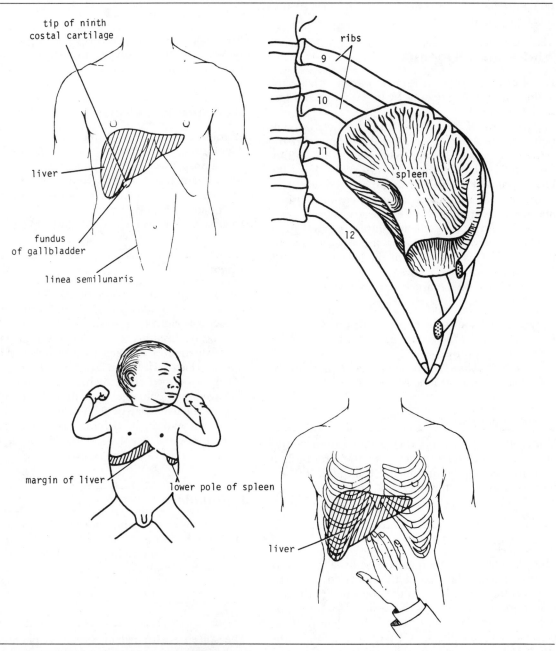

**Fig. 4-4. Surface markings of fundus of gallbladder, spleen, and liver. Note that in a young child, lower margin of normal liver and lower pole of normal spleen may be palpated. In a thin adult, lower margin of normal liver may just be felt at the end of deep inspiration.**

## Pancreas

The pancreas lies across the transpyloric plane. The head lies below and to the right, the neck lies on the plane, and the body and tail lie above and to the left.

## Kidneys

The right kidney lies at a slightly lower level than the left kidney (due to the bulk of the right lobe of the liver), and the lower pole may be palpated in the right lumbar region at the end of deep inspiration in a person with poorly developed abdominal musculature. Each kidney moves about 1 inch (2.5 cm) in a vertical direction during full respiratory movement of the diaphragm. The normal left kidney is impalpable.

On the anterior abdominal wall the hilus of each kidney lies on the transpyloric plane, about 3 fingerbreadths from the midline. On the back, the kidneys extend from the twelfth thoracic spine to the third lumbar spine, and the hili are opposite the first lumbar vertebra (Fig. 4-3).

## Cecum

The cecum is situated in the right iliac region. It is often distended with gas and gives a resonant sound when percussed. It can be palpated through the anterior abdominal wall.

## Appendix

The appendix lies in the right iliac region. The base of the appendix is situated one-third of the way up the line, joining the anterior superior iliac spine to the umbilicus (McBurney's point). The position of the free end of the appendix is very variable.

## Ascending Colon

The ascending colon extends upward from the cecum on the lateral side of the right vertical line and disappears under the right costal margin. It can be palpated through the anterior abdominal wall.

## Transverse Colon

The transverse colon extends across the abdomen, occupying the umbilical and hypogastric regions. It arches downward with its concavity directed upward. Because it has a mesentery, its position is variable.

## Descending Colon

The descending colon extends downward from the left costal margin on the lateral side of the left vertical line. In the left iliac region it curves medially and downward to become continuous with the sigmoid colon. The descending colon has a smaller diameter than the ascending colon and can be palpated through the anterior abdominal wall.

## Aorta

The aorta lies in the midline of the abdomen and bifurcates below into the right and left common iliac arteries opposite the fourth lumbar vertebra, i.e., on the intercristal plane. It may be palpated through the upper part of the anterior abdominal wall just to the left of the midline.

## External Iliac Artery

The pulsations of this artery may be felt as it passes under the inguinal ligament to become continuous with the femoral artery. It may be located at a point halfway between the anterior superior iliac spine and the symphysis pubis.

## Urinary Bladder and Pregnant Uterus

The full bladder and pregnant uterus may be palpated through the lower part of the anterior abdominal wall in the hypogastrium (See p. 300.)

## STRUCTURE OF THE ABDOMINAL WALL

Superiorly, the abdominal wall is formed by the *diaphragm*, which separates the abdominal cavity from the thoracic cavity. For a description of the diaphragm, see page 71.

Inferiorly, the abdominal cavity is continuous

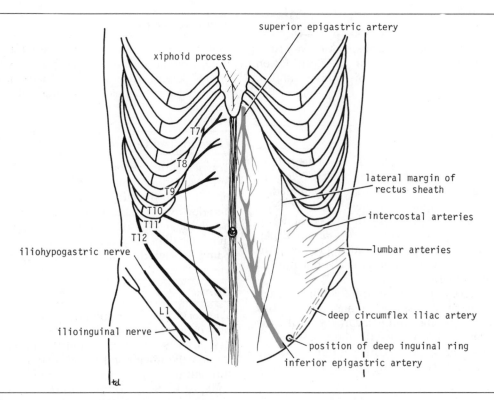

**Fig. 4-5. On left, segmental innervation of anterior abdominal wall. On right, arterial supply to anterior abdominal wall.**

with the pelvic cavity through the pelvic inlet. (See p. 305.)

Anteriorly, the abdominal wall is formed above by the lower part of the thoracic cage and below by the rectus abdominis muscles, the external oblique, the internal oblique, and the transversus abdominis muscles and fasciae.

Posteriorly, the abdominal wall is formed in the midline by the five lumbar vertebrae and their intervertebral discs; laterally, by the twelfth ribs, the upper part of the bony pelvis, the psoas muscles, the quadratus lumborum muscles, and the aponeuroses of origin of the transversus abdominis muscles. The iliacus muscles lie in the upper part of the bony pelvis.

Laterally, the abdominal wall is formed above by the lower part of the thoracic wall, including the lungs and pleura, and below, by the external oblique, internal oblique, and transversus abdominis muscles.

The abdominal wall is lined by a fascial envelope and the parietal peritoneum.

## Anterior and Lateral Abdominal Walls

### Skin

The natural lines of cleavage in the skin are constant and run almost horizontally around the trunk. This is important clinically, since an incision along a cleavage line will heal as a narrow scar, whereas one that crosses the lines will heal as a wide or heaped-up scar. (For details, see p. 5.)

### Nerve Supply

The cutaneous nerve supply to the anterior abdominal wall is derived from the anterior rami of the

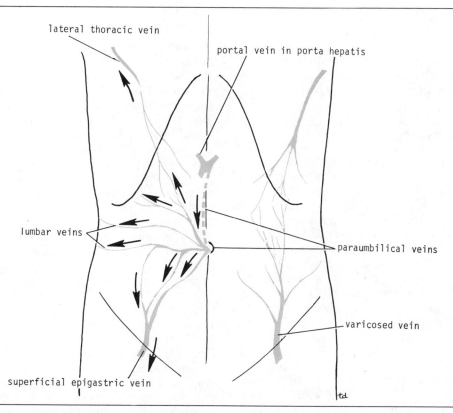

lateral thoracic vein

portal vein in porta hepatis

lumbar veins

paraumbilical veins

varicosed vein

superficial epigastric vein

**Fig. 4-6. Superficial veins of anterior abdominal wall. On the left are shown anastomoses between systemic veins and portal vein via paraumbilical veins. Arrows indicate direction taken by venous blood when there is obstruction of portal vein. On the right is shown an enlarged anastomosis between lateral thoracic vein and superficial epigastric vein. This occurs if there is obstruction to either superior or inferior vena cava.**

lower six thoracic and the first lumbar nerves (Fig. 4-5). The thoracic nerves are represented by the lower five intercostal and the subcostal nerves, and the lumbar nerve is represented by the iliohypogastric and ilioinguinal nerves. The dermatome of T7 is situated in the epigastrium just over the xiphoid process; that of T10 includes the umbilicus; and that of L1 lies just above the inguinal ligament and the symphysis pubis. For the dermatomes of the anterior abdominal wall, see page 44.

## Blood Supply

Cutaneous arteries, which are branches of the superior and inferior epigastric arteries, supply the area near the midline, and branches from the intercostal, lumbar, and deep circumflex iliac arteries supply the flanks (Fig. 4-5).

The venous blood is collected into a network of veins that radiates out from the umbilicus (Fig. 4-6). The network is drained above into the axillary vein via the lateral thoracic vein and below into the femoral vein via the superficial epigastric and great saphenous veins. A few small veins, the *paraumbilical veins*, connect the network through the umbilicus and along the ligamentum teres to the portal vein. They form an important portal-systemic venous anastomosis.

### Lymph Drainage

The cutaneous lymph vessels above the level of the umbilicus drain upward into the anterior axillary lymph nodes. The vessels below this level drain downward into the superficial inguinal nodes.

## Superficial Fascia

The superficial fascia may be divided into a superficial fatty layer and a deep membranous layer (Fig. 4-7). The fatty layer is continuous with the superficial fat over the rest of the body and may be extremely thick (3 inches [8 cm] or more in obese patients). The membranous layer fades out over the thoracic wall above and along the midaxillary line laterally. Inferiorly, it passes onto the front of the thigh, where it fuses with the deep fascia one fingerbreadth below the inguinal ligament (Fig. 4-7). In the midline it is not attached to the pubis, but forms a tubular sheath for the penis (or clitoris). Below these structures it is continued over the perineum and forms a saclike investment for the scrotum (or labia majora). In the perineum it widens out and is attached on each side to the margins of the pubic arch. Posteriorly, it fuses with the perineal body and the posterior margin of the perineal membrane (Fig. 4-7).

Clinicians often refer to the fatty layer of superficial fascia as the *fascia of Camper*; the membranous layer where it is situated on the anterior abdominal wall, as *Scarpa's fascia*; and the membranous layer in the perineum, as *Colles' fascia*.

## Deep Fascia

In the anterior abdominal wall the deep fascia is merely a thin layer of areolar tissue covering the muscles.

## Muscles of the Anterior and Lateral Abdominal Walls

The muscles of the anterior and lateral abdominal walls include the external oblique, the internal oblique, the transversus, the rectus abdominis, and the pyramidalis muscles.

### EXTERNAL OBLIQUE

The external oblique muscle is a broad, thin, muscular sheet that arises from the outer surfaces of the lower eight ribs and fans out to be inserted into the xiphoid process, the linea alba, the pubic crest, the pubic tubercle, and the anterior half of the iliac crest (Fig. 4-8). The majority of the fibers are inserted by means of a broad aponeurosis. Note that the most posterior fibers passing down to the iliac crest form a posterior free border.

A triangular-shaped defect in the external oblique aponeurosis lies immediately above and medial to the pubic tubercle. This is known as the *superficial inguinal ring* (Figs. 4-8, 4-10, and 4-16). The spermatic cord (or round ligament of the uterus) passes through this opening and carries the *external spermatic fascia* (or the external covering of the round ligament of the uterus) from the margins of the ring (Figs. 4-20 and 4-21).

Between the anterior superior iliac spine and the pubic tubercle, the lower border of the aponeurosis is folded backward on itself, forming the *inguinal ligament* (Figs. 4-8 and 4-9). From the medial end of the ligament the *lacunar ligament* extends backward and upward to the pectineal line on the superior ramus of the pubis (Fig. 4-9). Its sharp, free crescentic edge forms the medial margin of the *femoral ring*. (See p. 588.) On reaching the pectineal line, the lacunar ligament becomes continuous with a thickening of the periosteum called the *pectineal ligament* (Fig. 4-9).

The lateral part of the posterior edge of the inguinal ligament gives origin to part of the internal oblique and transversus abdominis muscles. To the inferior rounded border of the inguinal ligament is attached the deep fascia of the thigh, the *fascia lata* (Fig. 4-7).

### INTERNAL OBLIQUE

The internal oblique muscle is also a broad, thin, muscular sheet that lies deep to the external oblique; the majority of its fibers run at right angles to those of the external oblique (Fig. 4-8). It arises from the lumbar fascia, the anterior two-thirds of the iliac crest, and the lateral two-thirds of the inguinal ligament. The muscle fibers radiate as they pass upward and forward. The muscle is

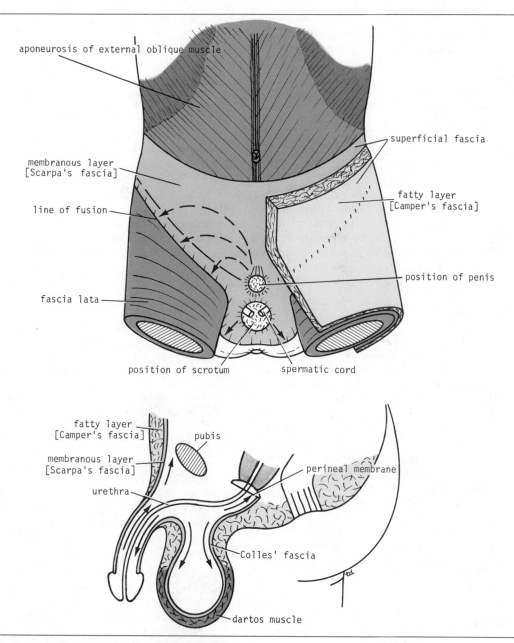

**Fig. 4-7.** Arrangement of fatty layer and membranous layer of superficial fascia in lower part of anterior abdominal wall. Note line of fusion between membranous layer and deep fascia of thigh (fascia lata). In lower diagram, note attachment of membranous layer to posterior margin of perineal membrane. Arrows indicate paths taken by urine in cases of ruptured urethra.

inserted into the lower borders of the lower three ribs and their costal cartilages, the xiphoid process, the linea alba, and the symphysis pubis. The internal oblique has a lower free border that arches over the spermatic cord (or round ligament of the uterus) and then descends behind it to be attached to the pubic crest and the pectineal line. Near their insertion, the lowest tendinous fibers are joined by similar fibers from the transversus abdominis to form the *conjoint tendon* (Figs. 4-12 and 4-16). The conjoint tendon is attached medially to the linea alba, but it has a lateral free border.

As the spermatic cord (or round ligament of the uterus) passes under the lower border of the internal oblique, it carries with it some of the muscle fibers that are called the *cremaster muscle* (Figs. 4-12 and 4-16). The *cremasteric fasica* is the term used to describe the cremaster muscle and its fascia.

## TRANSVERSUS

The transversus muscle is a thin sheet of muscle that lies deep to the internal oblique, and its fibers run horizontally forward (Fig. 4-8). It arises from the deep surface of the lower six costal cartilages (interdigitating with the diaphragm), the lumbar fascia, the anterior two-thirds of the iliac crest, and the lateral third of the inguinal ligament. It is inserted into the xiphoid process, the linea alba, and the symphysis pubis. The lowest tendinous fibers join similar fibers from the internal oblique to form the conjoint tendon, which is fixed to the pubic crest and the pectineal line (Figs. 4-12 and 4-16).

Note that the posterior border of the external oblique muscle is free, while the posterior borders of the internal oblique and transversus muscles are attached to the lumbar vertebrae by the lumbar fascia (Figs. 4-8 and 4-15).

## RECTUS ABDOMINIS

The rectus abdominis is a long strap muscle that extends along the whole length of the anterior abdominal wall. It is broader above and lies close to the midline, being separated from its fellow by the linea alba.

The rectus abdominis muscle arises by two heads, from the front of the symphysis pubis and

from the pubic crest (Figs. 4-9 and 4-13). It is inserted into the fifth, sixth, and seventh costal cartilages and the xiphoid process (Fig. 4-10). When it contracts, its lateral margin forms a curved ridge that can be palpated and often seen and is termed the *linea semilunaris* (Figs. 4-1, 4-2, and 4-10). This extends from the tip of the ninth costal cartilage to the pubic tubercle.

The anterior surface of the muscle is crossed by three *tendinous intersections:* at the tip of the xiphoid, at the umbilicus, and halfway between the two (Fig. 4-10). These intersections are strongly attached to the anterior wall of the rectus sheath. (See below.)

The rectus abdominis is enclosed between the aponeuroses of the external oblique, the internal oblique, and the transversus, which form the *rectus sheath.*

## PYRAMIDALIS

The pyramidalis muscle is often absent. It arises by its base from the anterior surface of the pubis and is inserted into the linea alba (Fig. 4-10). It lies in front of the lower part of the rectus abdominis.

## RECTUS SHEATH

The rectus sheath is a long sheath that encloses the rectus abdominis muscle and pyramidalis muscle (if present) and contains the anterior rami of the lower six thoracic nerves and the superior and inferior epigastric vessels and lymphatics. It is formed largely by the aponeuroses of the three lateral abdominal muscles (Figs. 4-8, 4-10, and 4-11).

For ease of description it will be considered at four levels (Fig. 4-13).

1. Above the costal margin the anterior wall is formed by the aponeurosis of the external oblique. The posterior wall is formed by the thoracic wall, i.e., the fifth, sixth, and seventh costal cartilages and the intercostal spaces.
2. Between the costal margin and the level of the anterior superior iliac spine, the aponeurosis of the internal oblique splits to enclose the rectus muscle; the external oblique aponeurosis is directed in front of the muscle, and the transversus aponeurosis is directed behind the muscle.
3. Between the level of the anterior superior iliac

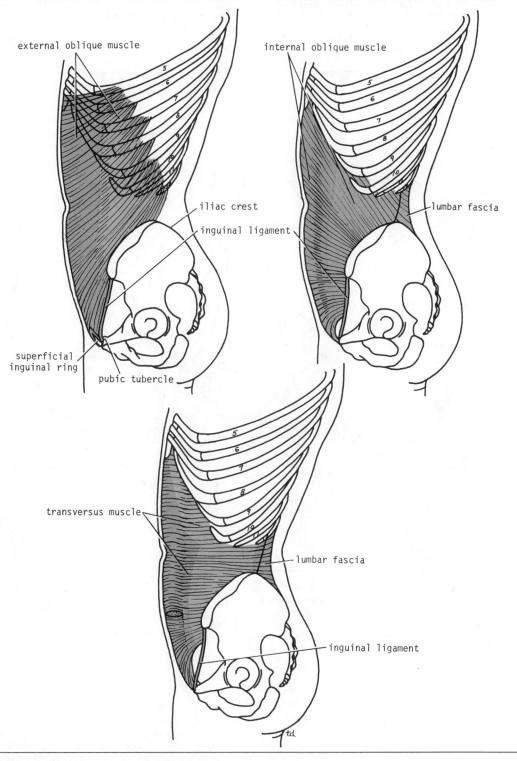

**Fig. 4-8. External oblique, internal oblique, and transversus muscles of anterior and lateral abdominal walls.**

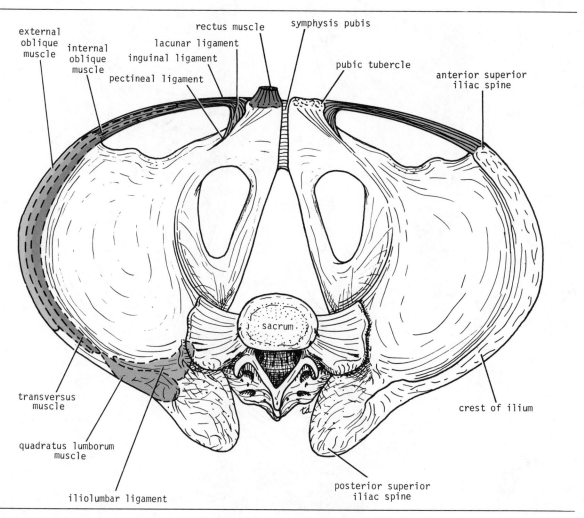

**Fig. 4-9. Bony pelvis viewed from above. Note attachments of inguinal, lacunar, and pectineal ligaments.**

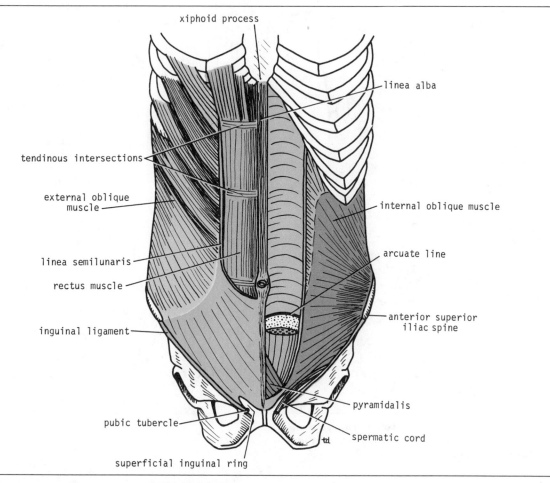

**Fig. 4-10. Anterior view of rectus abdominis muscle and rectus sheath. On left, anterior wall of sheath has been partly removed, revealing rectus muscle with its tendinous intersections. On right, posterior wall of rectus sheath is shown. Edge of arcuate line is shown at level of anterior superior iliac spine.**

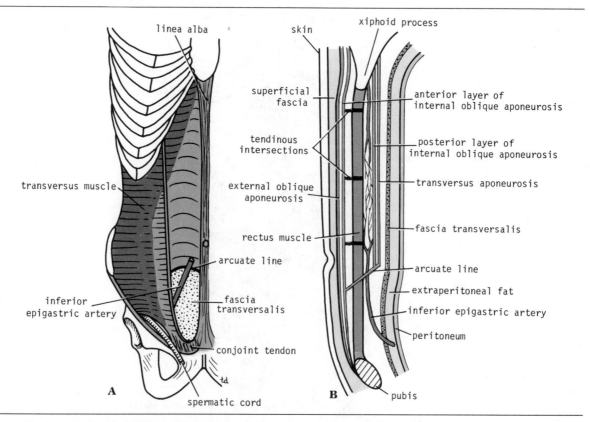

**Fig. 4-11. Rectus sheath (A) in anterior view and (B) in sagittal section. Note arrangement of aponeuroses forming rectus sheath.**

spine and the pubis, the aponeuroses of all three muscles form the anterior wall. The posterior wall is absent, and the rectus muscle lies in contact with the fascia transversalis.

4. In front of the pubis the origin of the rectus muscle and the pyramidalis (if present) is covered anteriorly by the aponeuroses of all three muscles. The posterior wall is formed by the body of the pubis.

It should be noted that where the aponeuroses forming the posterior wall pass in front of the rectus at the level of the anterior superior iliac spine, the posterior wall has a free, curved lower border

called the *arcuate line* (Figs. 4-10 and 4-11). At this site the inferior epigastric vessels enter the rectus sheath and pass upward to anastomose with the superior epigastric vessels.

The rectus sheath is separated from its fellow on the opposite side by the *linea alba* (Figs. 4-10, 4-12, and 4-13). This extends from the xiphoid process down to the symphysis pubis and is formed by the fusion of the aponeuroses of the lateral muscles of the two sides. Wider above the umbilicus, it narrows down below the umbilicus to be attached to the symphysis pubis.

The posterior wall of the rectus sheath is not attached to the rectus abdominis muscle. The anterior wall is firmly attached to it by the muscle's tendinous intersections (Figs. 4-10 and 4-11).

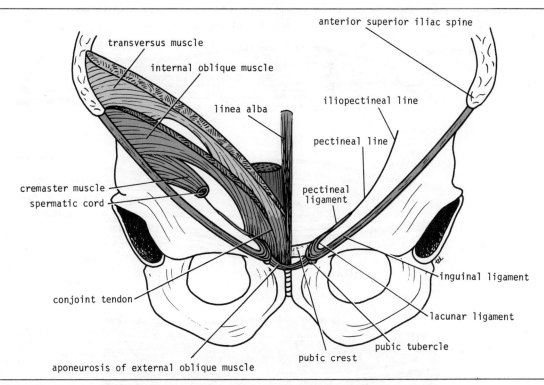

**Fig. 4-12. Anterior view of pelvis, showing attachment of conjoint tendon to pubic crest and adjoining part of pectineal line.**

## Function of the Anterior and Lateral Abdominal Wall Muscles

The oblique muscles laterally flex and rotate the trunk (Fig. 4-14). The rectus abdominis flexes the trunk and stabilizes the pelvis, and the pyramidalis keeps the linea alba taut during the process.

The muscles of the anterior and lateral abdominal walls assist the diaphragm during inspiration by relaxing as the diaphragm descends, so that the abdominal viscera may be accommodated.

The muscles assist in the act of forced expiration that occurs during coughing and sneezing by pulling down the ribs and sternum. Their tone plays a very important part in supporting and protecting the abdominal viscera. By contracting simultaneously with the diaphragm, with the glottis closed, they increase the intra-abdominal pressure and help in micturition, defecation, vomiting, and parturition.

## Nerve Supply

The oblique and transversus abdominis muscles are supplied by the lower six thoracic nerves and the iliohypogastric and ilioinguinal nerves (L1). The rectus muscle is supplied by the lower six thoracic nerves (Figs. 4-5 and 4-15). The pyramidalis is supplied by the twelfth thoracic nerve.

A summary of the muscles of the anterior and lateral abdominal walls, their nerve supply, and their action is given in Table 4-1.

## Nerves of the Anterior and Lateral Abdominal Walls

The nerves of the anterior and lateral abdominal walls are the anterior rami of the lower six thoracic and the first lumbar nerves (Figs. 4-5 and 4-15). They pass forward in the interval between the in-

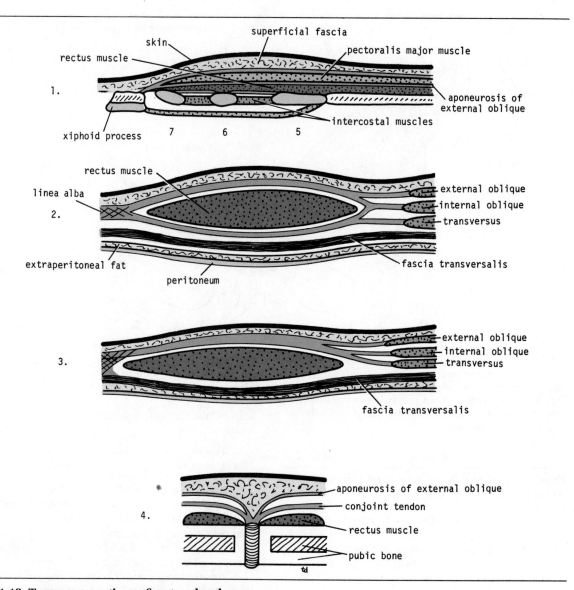

**Fig. 4-13. Transverse sections of rectus sheath seen at four levels. (1) Above costal margin. (2) Between costal margin and level of anterior superior iliac spine. (3) Below level of anterior superior iliac spine and above pubis. (4) At level of pubis.**

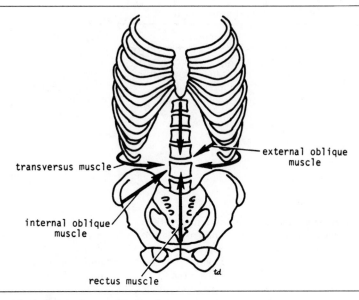

transversus muscle

external oblique muscle

internal oblique muscle

rectus muscle

td

**Fig. 4-14. Action of muscles of anterior and lateral abdominal walls. Arrows indicate line of pull of different muscles.**

ternal oblique and transversus muscles. They supply the skin of the anterior abdominal wall, the muscles (see above), and the parietal peritoneum. (Compare with the intercostal nerves, which run forward in the intercostal spaces between the internal intercostal and transversis thoracis muscles. See p. 70.) The lower six thoracic nerves then pierce the posterior wall of the rectus sheath to supply the rectus muscle and the pyramidalis (T12 only). They terminate by piercing the anterior wall of the sheath and supplying the skin.

The first lumbar nerve has a similar course, but it does not enter the rectus sheath (Figs. 4-5 and 4-15). It is represented by the iliohypogastric nerve, which pierces the external oblique aponeurosis above the superficial inguinal ring, and by the ilioinguinal nerve, which emerges through the ring. They end by supplying the skin just above the inguinal ligament and the symphysis pubis.

## Arteries of the Anterior and Lateral Abdominal Walls

The *superior epigastric artery*, one of the terminal branches of the internal thoracic artery, enters the upper part of the rectus sheath between the sternal and costal origins of the diaphragm (Fig. 4-5). It descends behind the rectus muscle, supplying the upper central part of the anterior abdominal wall, and anastomoses with the inferior epigastric artery.

The *inferior epigastric artery* is a branch of the external iliac artery just above the inguinal ligament. It runs upward and medially along the medial side of the deep inguinal ring (Figs. 4-5, 4-16, and 4-20). It pierces the fascia transversalis to enter the rectus sheath anterior to the arcuate line (Fig. 4-11). It ascends behind the rectus muscle, supplying the lower central part of the anterior abdominal wall, and anastomoses with the superior epigastric artery.

The *deep circumflex iliac artery* is a branch of the external iliac artery just above the inguinal ligament (Fig. 4-5). It runs upward and laterally toward the anterior superior iliac spine and then continues along the iliac crest. It supplies the lower lateral part of the abdominal wall.

The lower two *posterior intercostal arteries*, branches of the descending thoracic aorta, and the four *lumbar arteries*, branches of the abdominal aorta, pass forward between the muscle layers and supply the lateral part of the abdominal wall (Fig. 4-5).

Table 4-1. Muscles of Anterior and Lateral Abdominal Walls

| Name of muscle | Origin | Insertion | Nerve supply | Action |
| --- | --- | --- | --- | --- |
| External oblique | Lower eight ribs | Xiphoid process, linea alba, pubic crest, pubic tubercle, iliac crest | Lower six thoracic nerves and iliohypogastric and ilioinguinal nerves (L1) | Supports abdominal contents; compresses abdominal contents; assists in flexing and rotation of trunk. Assists in forced expiration, micturition, defecation, partuition, and vomiting |
| Internal oblique | Lumbar fascia, iliac crest, lateral two-thirds of inguinal ligament | Lower three ribs and costal cartilages, xiphoid process, linea alba, symphysis pubis | Lower six thoracic nerves, iliohypogastric and ilioinguinal nerves (L1) | As above |
| Transversus | Lower six costal cartilages, lumbar fascia, iliac crest, lateral third of inguinal ligament | Xiphoid process, linea alba, symphysis pubis | Lower six thoracic nerves, iliohypogastric and ilioinguinal nerves (L1) | Compresses abdominal contents |
| Rectus abdominis | Symphysis pubis and pubic crest | Fifth, sixth, and seventh costal cartilages and xiphoid process | Lower six thoracic nerves | Compresses abdominal contents and flexes vertebral column; accessory muscle of expiration |
| Pyramidalis (if present) | Anterior surface of pubis | Linea alba | Twelfth thoracic nerve | Tenses the linea alba |

## Veins of the Anterior and Lateral Abdominal Walls

The superficial veins have been described on page 153. The superior epigastric, inferior epigastric, and deep circumflex iliac veins follow the arteries of the same name and drain into the internal thoracic and external iliac veins. The posterior intercostal veins drain into the azygos veins, and the lumbar veins drain into the inferior vena cava.

## Lymph Drainage of the Anterior and Lateral Abdominal Walls

The cutaneous lymph vessels above the level of the umbilicus drain upward into the anterior axillary lymph nodes. The vessels below this level drain downward into the superficial inguinal nodes. The deep lymph vessels follow the arteries and drain into the internal thoracic, external iliac, posterior mediastinal, and para-aortic (lumbar) nodes.

## Inguinal Canal

The inguinal canal is an oblique passage through the lower part of the anterior abdominal wall and is present in both sexes. It allows structures to pass to and from the testis to the abdomen in the male. In the female it permits the passage of the round ligament of the uterus from the uterus to the labium majus. In addition, it transmits the ilioinguinal nerve in both sexes (Fig. 4-16).

The canal is about 1½ inches (4 cm) long in the adult and extends from the deep inguinal ring, a hole in the fascia transversalis (see p. 180), downward and medially to the superficial inguinal ring, a hole in the aponeurosis of the external oblique muscle (Figs. 4-16 and 4-20). It lies parallel to and immediately above the inguinal ligament. In the newborn child, the deep ring lies almost directly posterior to the superficial ring, so that the canal is considerably shorter at this age. Later, as the result of growth, the deep ring moves laterally.

The *deep inguinal ring,** an oval opening in the fascia transversalis, lies about ½ inch (1.3 cm) above the inguinal ligament midway between the anterior superior iliac spine and the symphysis pubis (Figs. 4-16 and 4-20). Related to it medially are the inferior epigastric vessels, which pass upward from the external iliac vessels. The margins of the ring give origin to the *internal spermatic fascia* (or the internal covering of the round ligament of the uterus).

The *superficial inguinal ring** is a triangular-shaped defect in the aponeurosis of the external oblique muscle and the base is formed by the pubic crest (Figs. 4-16, 4-20, and 4-21). The margins of the ring, sometimes called the *crura,* give origin to the *external spermatic fascia.*

The *anterior wall of the canal* is formed along its entire length by the aponeurosis of the external oblique muscle. It is reinforced in its lateral third by the fibers of origin of the internal oblique (Figs.

4-16 and 4-20). This wall is therefore strongest where it lies opposite the weakest part of the posterior wall, namely, the deep inguinal ring.

The *posterior wall of the canal* is formed along its entire length by the fascia transversalis. It is reinforced in its medial third by the conjoint tendon, the common tendon of insertion of the internal oblique and transversus, which is attached to the pubic crest and pectineal line (Figs. 4-16 and 4-20). This wall is therefore strongest where it lies opposite the weakest part of the anterior wall, namely, the superficial inguinal ring.

The *inferior wall* or *floor of the canal* is formed by the rolled-under inferior edge of the aponeurosis of the external oblique muscle, namely, the inguinal ligament and, at its medial end, the lacunar ligament (Fig. 4-12).

The *superior wall* or *roof of the canal* is formed by the arching lowest fibers of the internal oblique and transversus abdominis muscles (Fig. 4-12).

## MECHANICS OF THE INGUINAL CANAL

The presence of the inguinal canal in the lower part of the anterior abdominal wall in both sexes constitutes a potential weakness. It is interesting to consider how the design of this canal attempts to lessen this weakness.

1. Except in the newborn infant, the canal is an oblique passage with the weakest areas, namely, the superficial and deep rings, lying some distance apart.
2. The anterior wall of the canal is reinforced by the fibers of the internal oblique muscle immediately in front of the deep ring.
3. The posterior wall of the canal is reinforced by the strong conjoint tendon immediately behind the superficial ring.
4. On coughing and straining, as in micturition, defecation, and parturition, the arching lowest fibers of the internal oblique and transversus abdominis muscles contract, flattening out the arched roof so that it is lowered toward the floor. The roof may actually compress the contents of the canal against the floor so that the canal is virtually closed (Fig. 4-17).
5. When great straining efforts may be necessary, as in defecation and parturition, the person naturally tends to assume the squatting position;

---

*A common frustration for medical students is the inability to observe these rings as openings. One must remember that the internal spermatic fascia is attached to the margins of the deep inguinal ring and the external spermatic fasica is attached to the margins of the superficial inguinal ring, so that the edges of the rings cannot be observed externally. Compare this arrangement to the openings for the fingers seen inside a glove with the absence of openings for the fingers when the glove is viewed from the outside.

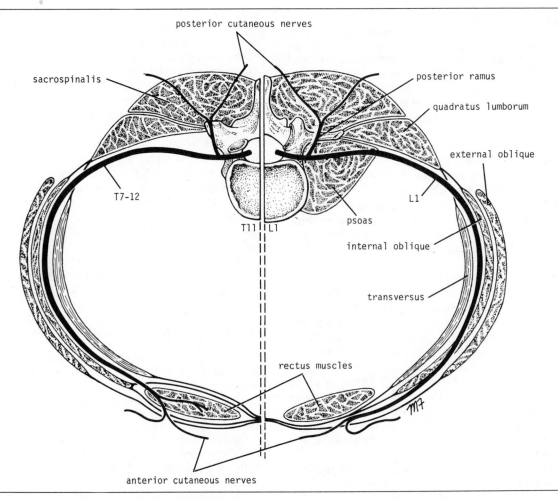

posterior cutaneous nerves

sacrospinalis

posterior ramus

quadratus lumborum

external oblique

T7-12

L1

psoas

T11  L1

internal oblique

transversus

rectus muscles

anterior cutaneous nerves

**Fig. 4-15. Cross section of abdomen, showing courses of lower thoracic and first lumbar nerves.**

**Fig. 4-16. Inguinal canal, showing arrangement of (1) external oblique muscle, (2) internal oblique muscle, (3) transversus muscle, (4) fascia transversalis. Note that anterior wall of canal is formed by external oblique and internal oblique, and posterior wall is formed by fascia transversalis and conjoint tendon. Deep inguinal ring lies lateral to inferior epigastric artery.**

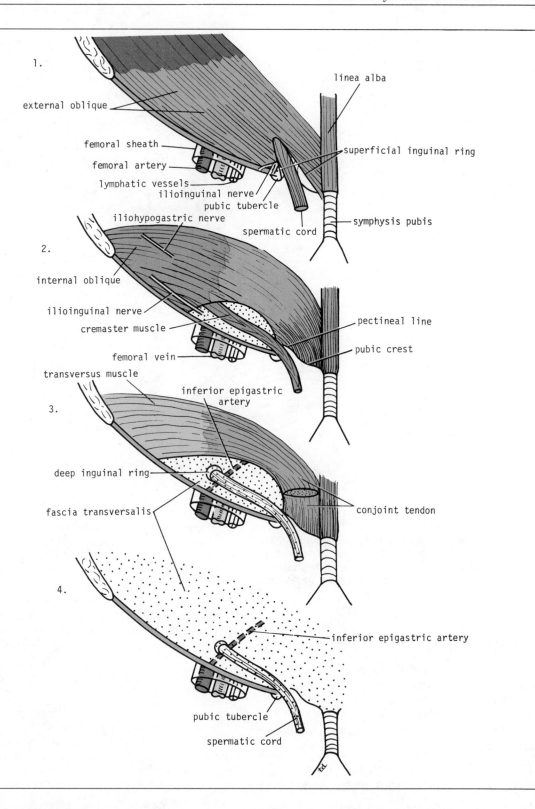

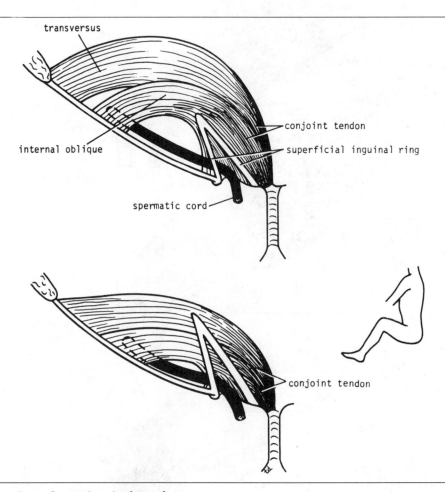

**Fig. 4-17. Action of muscles on inguinal canal. Note that canal is "obliterated" when muscles contract. Note also that anterior surface of thigh protects inguinal region when one assumes squatting position.**

the hip joints are flexed, and the anterior surfaces of the thighs are brought up against the anterior abdominal wall. By this means the lower part of the anterior abdominal wall is protected by the thighs (Fig. 4-17).

## *Spermatic Cord*

The spermatic cord is a collection of structures that traverse the inguinal canal and pass to and from the testis (Fig. 4-18). It is covered with three concentric layers of fascia derived from the layers of the anterior abdominal wall (Figs. 4-20 and 4-21). It begins at the deep inguinal ring lateral to the inferior epigastric artery and ends at the testis.

## STRUCTURES OF THE SPERMATIC CORD

### Vas Deferens

This is a cordlike structure (Figs 4-18 and 4-21), which can be palpated between finger and thumb in the upper part of the scrotum. It is a thick-walled muscular duct, which transports spermatozoa from the epididymis to the urethra.

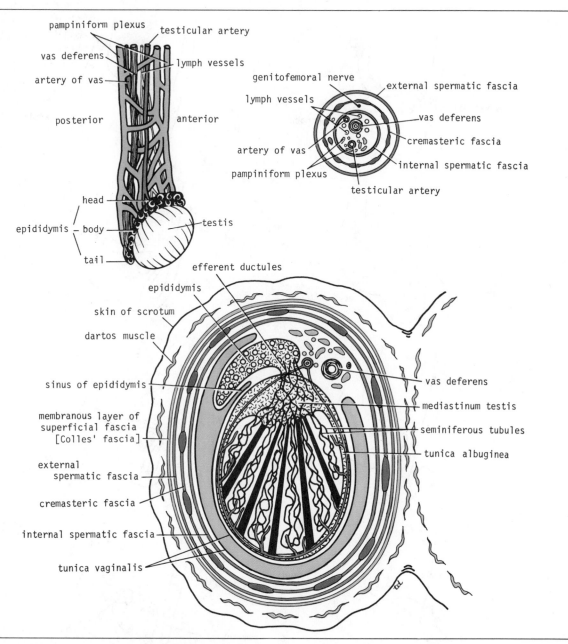

**Fig. 4-18. Testis and epididymis, spermatic cord, and scrotum. Lower diagram shows testis and epididymis cut across in horizontal section.**

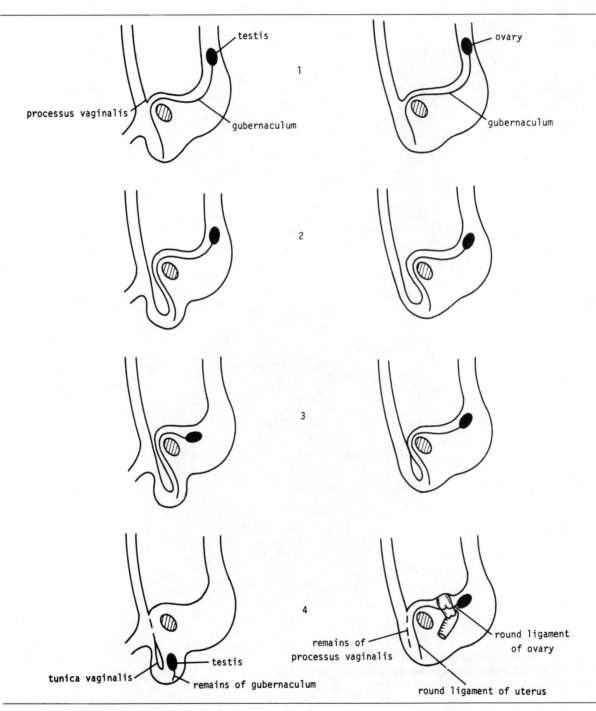

**Fig. 4-19. Origin, development, and fate of processus vaginalis in the two sexes. Note descent of testis into scrotum and descent of ovary into pelvis.**

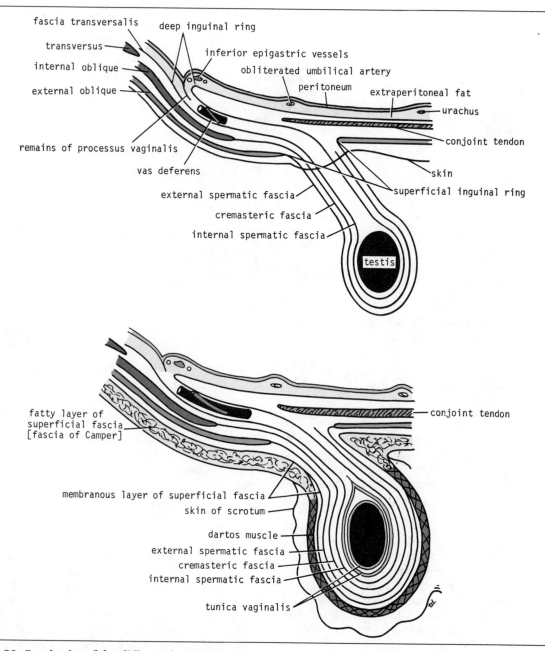

**Fig. 4-20. Continuity of the different layers of anterior abdominal wall with coverings of spermatic cord. In lower diagram, skin and superficial fascia of abdominal wall and scrotum have been included, and tunica vaginalis is also shown.**

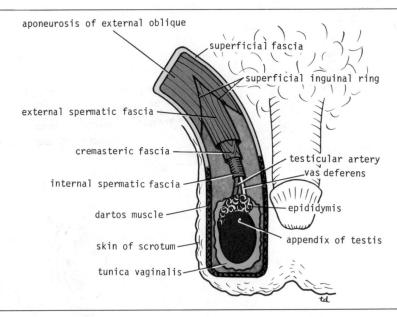

aponeurosis of external oblique

superficial fascia

superficial inguinal ring

external spermatic fascia

cremasteric fascia

testicular artery

vas deferens

internal spermatic fascia

dartos muscle

epididymis

skin of scrotum

appendix of testis

tunica vaginalis

td

**Fig. 4-21. Scrotum dissected from in front. Note spermatic cord and its coverings.**

## Testicular Artery

A branch of the abdominal aorta (at the level of the second lumbar vertebra), the testicular artery is long and slender and descends on the posterior abdominal wall. It traverses the inguinal canal and supplies the testis and the epididymis (Fig. 4-18).

## Testicular Veins

An extensive venous plexus, the *pampiniform plexus*, leaves the posterior border of the testis (Fig. 4-18). As the plexus ascends, it becomes reduced in size, so that at about the level of the deep inguinal ring, a single testicular vein is formed. This runs up on the posterior abdominal wall and drains into the left renal vein on the left side, and into the inferior vena cava on the right side.

## Lymph Vessels

The testicular lymph vessels ascend through the inguinal canal and pass up over the posterior abdominal wall to reach the lumbar (para-aortic) lymph nodes on the side of the aorta at the level of the first lumbar vertebra (Fig. 4-22).

## Autonomic Nerves

Sympathetic fibers run with the testicular artery from the renal or aortic sympathetic plexuses. Afferent sensory nerves accompany the efferent sympathetic fibers.

## Processus Vaginalis

The remains of the processus vaginalis are present within the cord.

In addition to the structures described are (1) the small *cremasteric artery,* a branch of the inferior epigastric artery, which supplies the cremasteric fascia (see the next section); (2) the small *artery to the vas deferens,* a branch of the inferior vesical artery; and (3) the *genital branch of the genitofemoral nerve,* which supplies the cremaster muscle (Fig. 4-18).

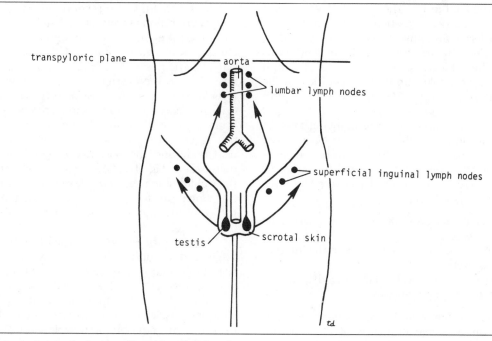

**Fig. 4-22. Lymphatic drainage of testis and skin of scrotum.**

## COVERINGS OF THE SPERMATIC CORD (THE SPERMATIC FASCIAE) AND THE DEVELOPMENT OF THE INGUINAL CANAL

To understand the coverings of the spermatic cord, one must first consider the development of the inguinal canal.

Prior to the descent of the testis and the ovary from their site of origin high upon the posterior abdominal wall (L1), a peritoneal diverticulum called the *processus vaginalis* is formed (Fig. 4-19). The processus vaginalis passes through the layers of the lower part of the anterior abdominal wall and, as it does so, acquires a tubular covering from each layer. It traverses the fascia transversalis at the deep inguinal ring and acquires a tubular convering, the *internal spermatic fascia* (Fig. 4-20). As it passes through the lower part of the internal oblique muscle, it takes with it some of its lowest fibers, which form the *cremaster muscle*. The mus-

cle fibers are embedded in fascia, and thus the second tubular sheath is known as the *cremasteric fascia* (Fig. 4-20). The processus vaginalis passes under the arching fibers of the transversus abdominis muscle and therefore does not acquire a covering from this abdominal layer. On reaching the aponeurosis of the external oblique, it evaginates this to form the superficial inguinal ring and acquires a third tubular fascial coat, the *external spermatic fascia* (Figs. 4-20 and 4-21). It is in this manner that the inguinal canal is formed in both sexes. (In the female the term "spermatic" fascia should be replaced by *the covering of the round ligament of the uterus.*)

Meanwhile, a band of mesenchyme, extending from the lower pole of the developing gonad through the inguinal canal to the labioscrotal swelling, has condensed to form the *gubernaculum* (Fig. 4-19).

In the male the testis descends through the pelvis and inguinal canal during the seventh and eighth months of fetal life. The normal stimulus for the descent of the testis is testosterone, which is secreted by the fetal testes. The testis follows the gubernaculum and descends behind the peritoneum on the posterior abdominal wall. The testis then passes behind the processus vaginalis and pulls down its duct, blood vessels, nerves, and lymph vessels. The testis takes up its final position in the developing scrotum by the end of the eighth month.

Since the testis and its accompanying vessels, ducts, and so on follow the course previously taken by the processus vaginalis, they acquire the same three coverings as they pass down the inguinal canal. Thus, the spermatic cord is covered by three concentric layers of fascia:

1. The internal spermatic fascia, derived from the fascia transversalis.
2. The cremasteric fascia, derived from the internal oblique muscle.
3. The extenal spermatic fascia, derived from the external oblique aponeurosis.

In the female the ovary descends into the pelvis following the gubernaculum (Fig. 4-19). The gubernaculum becomes attached to the side of the developing uterus, and the gonad descends no further. That part of the gubernaculum extending from the uterus into the developing labium majus persists as the *round ligament of the uterus.* Thus, in the female the only structures that pass through the inguinal canal from the abdominal cavity are the round ligament of the uterus and a few lymph vessels. The lymph vessels convey a small amount of lymph from the body of the uterus to the superficial inguinal nodes.

## Scrotum, Testes, and Epididymides

### SCROTUM

The scrotum may be considered as an outpouching of the lower part of the anterior abdominal wall. It contains the testes, the epididymides, and the lower ends of the spermatic cords (Figs. 4-18 and 4-20).

The *skin* of the scrotum is thin, wrinkled, and pigmented and forms a single pouch. A slightly raised ridge in the midline indicates the line of fusion of the two lateral labioscrotal swellings. (In the female the swellings remain separate and form the labia majora.)

The *superficial fascia* is continuous with the fatty and membranous layers of the anterior abdominal wall; the fat is, however, replaced by smooth muscle called the *dartos muscle.* This is innervated by sympathetic nerve fibers and is responsible for the wrinkling of the overlying skin. The membranous layer of the superficial fascia (often referred to as Colles' fascia) is continuous in front with the membranous layer of the anterior abdominal wall (Scarpa's fascia), and behind it is attached to the perineal body and the posterior edge of the perineal membrane (Fig. 4-7). At the sides it is attached to the ischiopubic rami. Both layers of superficial fascia contribute to a median partition that crosses the scrotum and separates the testes from each other.

The *spermatic fasciae* lie beneath the superficial fascia and are derived from the layers of the anterior abdominal wall on each side, as previously explained. (See p. 173.) The external spermatic fascia is derived from the aponeurosis of the external oblique muscle; the cremasteric fascia is derived from the internal oblique muscle; and, finally, the internal spermatic fascia is derived from the fascia transversalis. The cremaster muscle is supplied by the genital branch of the genitofemoral nerve.

The cremaster muscle can be made to contract by stroking the skin on the medial aspect of the thigh. This is called the *cremasteric reflex.* The afferent fibers of this reflex arc travel in the femoral branch of the genitofemoral nerve (L1 and 2) and the efferent motor nerve fibers travel in the genital branch of the genitofemoral nerve. The function of the cremaster muscle is to raise the testis and the scrotum upward for warmth and for protection against injury. For testicular temperature and fertility, see below.

The *tunica vaginalis* (Figs. 4-18, 4-20, and 4-21) lies within the spermatic fasciae and covers the anterior, medial, and lateral surfaces of each testis. It is the lower expanded part of the processus vaginalis, and normally, just before birth, it becomes shut off from the upper part of the processus and the peritoneal cavity. The tunica vaginalis is thus a closed sac, invaginated from behind by the testis.

### TESTIS

The *testis* is a mobile organ lying within the scrotum (Figs. 4-18 and 4-21). The left testis usually lies at a lower level than the right. The upper pole of the gland is tilted slightly forward. Each testis is surrounded by a tough fibrous capsule, the *tunica albuginea.*

Extending from the inner surface of the capsule is a series of fibrous septa that divide the interior of the organ into *lobules.* Lying within each lobule are one to three coiled *seminiferous tubules.* The tubules open into a network of channels called the *rete testis.* Small *efferent ductules* connect the rete testis to the upper end of the epididymis (Fig. 4-18).

Normal spermatogenesis can occur only if the testes are at a temperature lower than that of the abdomen. When they are located in the scrotum, they are at a temperature about 3°C lower than the abdominal temperature. The control of testicular temperature in the scrotum is not fully understood, but the surface area of the scrotal skin can be changed reflexly by the contraction of the dartos and cremaster muscles. It is now recognized that the testicular veins in the spermatic cord that form the pampiniform plexus—together with the branches of the testicular arteries, which lie close to the veins—probably assist in stabilizing the temperature of the testes by a countercurrent heat exchange mechanism. By this means, the hot blood arriving in the artery from the abdomen loses heat to the blood ascending to the abdomen within the veins.

## EPIDIDYMIS

The *epididymis* is a firm structure lying posterior to the testis, with the vas deferens on its medial side (Fig. 4-18). It has an expanded upper end, the *head,* a *body,* and a pointed *tail* inferiorly. Laterally, there is a distinct groove between the testis and the epididymis, which is lined with the inner visceral layer of the tunica vaginalis and is called the *sinus of the epididymis* (Fig. 4-18).

The epididymis is a much coiled tube nearly 20 feet (6 m) long, embedded in connective tissue. The tube emerges from the tail of the epididymis as the *vas deferens,* which enters the spermatic cord.

The long length of the duct of the epididymis provides storage space for the spermatozoa and allows them to mature. One of the main functions of the epididymis is the absorption of fluid. Another

function may be the addition of substances to the seminal fluid to nourish the maturing sperm.

## LYMPHATIC DRAINAGE OF THE SCROTUM AND CONTENTS

The lymphatic drainage of the wall of the scrotum, i.e., from the skin and fasciae, including the tunica vaginalis, is into the superficial inguinal lymph nodes (Fig. 4-22).

The lymphatic drainage of the testis and epididymis (Fig. 4-22) ascends in the spermatic cord and ends in the lumbar (para-aortic) lymph nodes at the level of the first lumbar vertebra (i.e., on the transpyloric plane). This is to be expected, since the testis during development has migrated from high up on the posterior abdominal wall, down through the inguinal canal, and into the scrotum, dragging its blood supply and lymph vessels after it.

### Labia Majora

The labia majora are prominent, hair-bearing folds of skin formed by the enlargement of the genital swellings in the fetus. (In the male the genital swellings fuse in the midline to form the scrotum). Within the labia is a large amount of adipose tissue and the terminal strands of the round ligaments of the uterus.

## Posterior Abdominal Wall

The posterior abdominal wall is formed in the midline by the five lumbar vertebrae and their intervertebral discs; laterally, by the twelfth ribs, the upper part of the bony pelvis (Fig. 4-23), the psoas muscles, the quadratus lumborum muscles, and the aponeuroses of origin of the transversus abdominis muscles. The iliacus muscles lie in the upper part of the bony pelvis.

### Lumbar Vertebrae

The *body* of each vertebra (Fig. 4-24) is massive and kidney-shaped, and it has to bear the greater part of the body weight. The *pedicles* are strong and directed backward. The *laminae* are thick and enclose a small triangular *vertebral foramen.* The *transverse processes* are long, with sharp ends. The *spinous process* is short and flat and projects

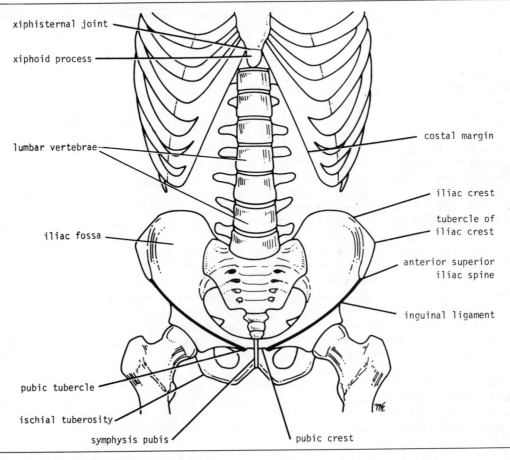

Fig. 4-23. Costal margin and bones of the abdomen.

Labels in figure: xiphisternal joint, xiphoid process, lumbar vertebrae, iliac fossa, pubic tubercle, ischial tuberosity, symphysis pubis, pubic crest, costal margin, iliac crest, tubercle of iliac crest, anterior superior iliac spine, inguinal ligament

**Fig. 4-23. Costal margin and bones of the abdomen.**

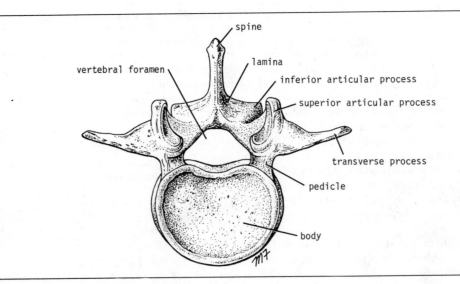

Labels in figure: spine, vertebral foramen, lamina, inferior articular process, superior articular process, transverse process, pedicle, body

**Fig. 4-24. Fifth lumbar vertebra.**

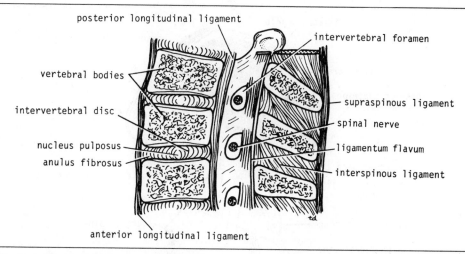

posterior longitudinal ligament

intervertebral foramen

vertebral bodies

intervertebral disc

nucleus pulposus

anulus fibrosus

supraspinous ligament

spinal nerve

ligamentum flavum

interspinous ligament

anterior longitudinal ligament

**Fig. 4-25. Sagittal section of lumbar part of vertebral column, showing intervertebral discs and ligaments.**

straight back. The *superior articular processes* face medially, and the *inferior articular processes* face laterally. The fifth lumbar vertebra articulates with the base of the sacrum at the *lumbosacral joint.*

The *intervertebral discs* (Fig. 4-25) in the lumbar region are thicker than in other regions of the vertebral column. They are wedge-shaped and are responsible for the normal lordosis found in the lumbar region. For a full description of the structure of intervertebral discs, see page 926.

## Twelfth Pair of Ribs

The ribs are described on page 64. It should be noted that the head has a single facet for articulation with the body of the twelfth thoracic vertebra. The anterior end is pointed and has a small costal cartilage, which is embedded in the musculature of the anterolateral abdominal wall. In many persons it is so short that it fails to protrude beyond the lateral border of the erector spinae muscle on the back.

## Ilium

The ilium, together with the ischium and pubis, forms the hip bone (Fig. 4-26); they meet one another at the acetabulum. The medial surface of the ilium is divided into two parts by the *arcuate line.* Above this line is a concave surface called the iliac fossa; below this line is a flattened surface that is continuous with the medial surfaces of the pubis and ischium. It should be noted that the arcuate line of the ilium forms the posterior part of the *iliopectineal line*, and the *pectineal line* forms the anterior part of the iliopectineal line. The iliopectineal line runs forward and demarcates the false from the true pelvis. For further details on the structure of the hip bone, see page 563.

## Muscles of the Posterior Abdominal Wall

### PSOAS MAJOR

The psoas muscle[†] arises from the roots of the transverse processes, the sides of the vertebral bodies, and the intervertebral discs, from the twelfth thoracic to the fifth lumbar vertebrae (Fig. 4-27). The fibers run downward and laterally and leave the abdomen to enter the thigh by passing behind the inguinal ligament. The muscle is inserted into the lesser trochanter of the femur. The psoas is enclosed in a fibrous sheath that is derived from the lumbar fascia. The sheath is thickened above to form the *medial arcuate ligament.*

[†]The psoas minor is a small muscle with a long tendon that lies anterior to the psoas major. It is unimportant and is absent in 40 percent of subjects.

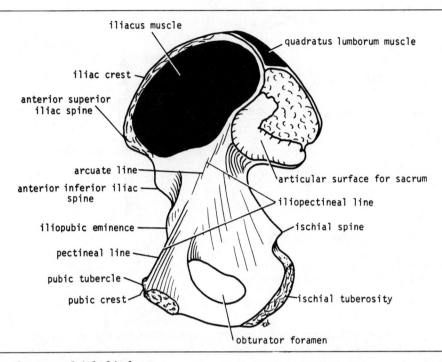

**Fig. 4-26. Internal aspect of right hip bone.**

### Nerve Supply

This muscle is supplied by the lumbar plexus.

### Action

The psoas flexes the thigh at the hip joint on the trunk; or if the thigh is fixed, it flexes the trunk on the thigh, as in sitting up from a lying position.

## QUADRATUS LUMBORUM

The quadratus lumborum is a flat, quadrilateral-shaped muscle that lies alongside the vertebral column. It arises below from the iliolumbar ligament, the adjoining part of the iliac crest, and the tips of the transverse processes of the lower lumbar vertebrae (Fig. 4-27). The fibers run upward and medially and are inserted into the lower border of the twelfth rib and the transverse processes of the upper four lumbar vertebrae. The anterior surface of the muscle is covered by lumbar fascia, which is thickened above to form the *lateral arcuate ligament* and below, to form the *iliolumbar ligament*.

### Nerve Supply

This muscle is supplied by the lumbar plexus.

### Action

It fixes or depresses the twelfth rib during respiration (see p. 100) and laterally flexes the vertebral column to the same side.

## TRANSVERSUS ABDOMINIS

The transversus abdominis muscle is fully described on page 156.

## ILIACUS

The iliacus muscle is fan-shaped and arises from the upper part of the iliac fossa (Figs. 4-26 and

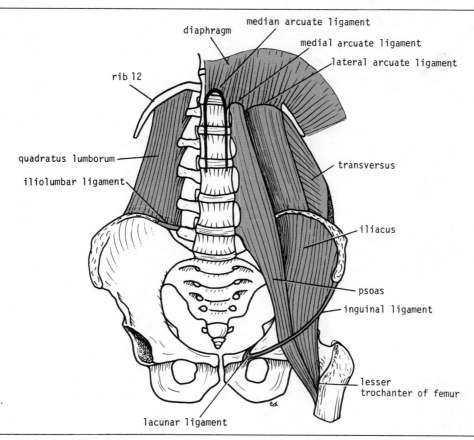

**Fig. 4-27. Muscles and bones forming posterior abdominal wall.**

4-27). Its fibers join the lateral side of the psoas tendon to be inserted into the lesser trochanter of the femur. The combined muscles are often referred to as the *iliopsoas*.

## Nerve Supply

This muscle is supplied by the femoral nerve, a branch of the lumbar plexus.

## Action

The iliopsoas flexes the thigh on the trunk at the hip joint; or if the thigh is fixed, it flexes the trunk on the thigh.

The posterior part of the *diaphragm* (Fig. 4-27) also forms part of the posterior abdominal wall. It is described on page 71. A summary of the muscles of the posterior abdominal wall, their nerve supply, and their action is given in Table 4-2.

## Fascial Lining of the Abdominal Walls

The abdominal walls are lined by one continuous layer of connective tissue that lies between the parietal peritoneum and the muscles (Fig. 4-28). It is continuous below with a similar fascial layer lining the pelvic walls. It is customary to name the fascia according to the structure it overlies. For example, the *diaphragmatic fascia* covers the undersurface of the diaphragm; the *transversalis fascia* lines the

Table 4-2. Muscles of the Posterior Abdominal Wall

| Name of muscle | Origin | Insertion | Nerve supply | Action |
|---|---|---|---|---|
| Psoas | Transverse processes, bodies, and intervertebral discs of twelfth thoracic and five lumbar vertebrae | With iliacus into lesser trochanter of femur | Lumbar plexus | Flexes thigh on trunk; if thigh is fixed it flexes trunk on thigh as in sitting up from lying position |
| Quadratus lumborum | Iliolumbar ligament, iliac crest, tips of transverse processes of lower lumbar vertebrae | Twelfth rib | Lumbar plexus | Fixes twelfth rib during inspiration; depresses twelfth rib during forced expiration; laterally flexes vertebral column same side |
| Iliacus | Iliac fossa | With psoas into lesser trochanter of femur | Femoral nerve | Flexes thigh on trunk; if thigh is fixed it flexes the trunk on the thigh as in sitting up from lying position |

*transversus abdominis;* the *psoas fascia* covers the psoas muscle; the *quadratus lumborum fascia* covers the quadratus lumborum; and the *iliaca fascia* covers the iliacus muscle.

It is interesting to note that the abdominal blood and lymph vessels lie within this fascial lining, whereas the principal nerves lie outside the fascia. This fact is important in the understanding of the femoral sheath (Fig. 4-28). This is simply a downward prolongation of the fascial lining around the femoral vessels and lymphatics, for about 1½ inches (4 cm) into the thigh, behind the inguinal ligament. Since the femoral nerve lies outside the fascial envelope, it has no sheath. (See p. 590.)

In certain areas of the abdominal wall, the fascial lining performs particularly important functions. Inferior to the level of the anterior superior iliac spines, the posterior wall of the rectus sheath is devoid of muscular aponeuroses (Figs. 4-11 and 4-13) and is formed by the fascia transversalis and peritoneum only. (See p. 160.)

At the midpoint between the anterior superior iliac spine and the symphysis pubis, the spermatic cord pierces the fascia transversalis to form the deep inguinal ring (Fig. 4-16). From the margins of the ring, the fascia is continued over the cord as a tubular sheath, the internal spermatic fascia (Fig. 4-20).

## Peritoneal Lining of the Abdominal Walls

The walls of the abdomen are lined with parietal peritoneum. This is a thin serous membrane consisting of a layer of mesothelium resting on connective tissue. It is continuous below with the parietal peritoneum lining the pelvis (Fig. 4-28). For further details, see pages 343 and 353.

### Nerve Supply

The central part of the diaphragmatic peritoneum is supplied by the phrenic nerves and the peripheral part, by the lower intercostal nerves. The peritoneum lining the anterior, lateral, and posterior abdominal walls is supplied segmentally by intercostal and lumbar nerves, which also supply the overlying muscles and skin.

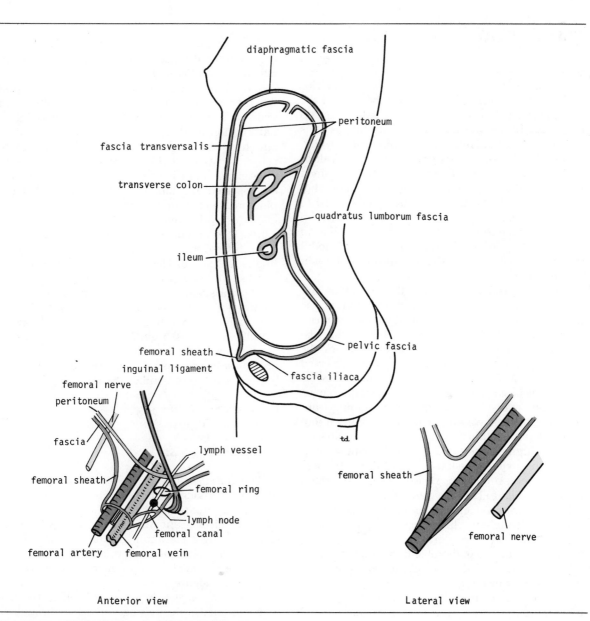

diaphragmatic fascia

peritoneum

fascia transversalis

transverse colon

quadratus lumborum fascia

ileum

pelvic fascia

femoral sheath

inguinal ligament

fascia iliaca

femoral nerve

peritoneum

fascia

lymph vessel

femoral sheath

femoral ring

lymph node

femoral canal

femoral artery    femoral vein

femoral sheath

femoral nerve

Anterior view

Lateral view

**Fig. 4-28. Sagittal section of abdomen, showing arrangement of fascial and peritoneal linings of walls. Femoral sheath with its contained vessels is also shown. Note that femoral nerve is devoid of a fascial sheath.**

## CLINICAL NOTES

### Skin

The *lymphatic drainage* of the skin of the anterior and lateral abdominal walls above the level of the umbilicus is upward to the anterior axillary group of nodes; below the level of the umbilicus it is downward to the superficial inguinal nodes (Fig. 4-29). The skin of the posterior abdominal wall or back above the level of the iliac crests is drained upward to the posterior axillary group of nodes; below the level of the iliac crests it is downward to the superficial inguinal nodes (Fig. 4-29). Clinically, it is therefore possible to find a swelling in the groin (an enlarged superficial inguinal node) due to an infection or malignant tumor of the skin of the buttock.

The *superficial veins* around the umbilicus and the paraumbilical veins connecting them to the portal vein may become grossly distended in cases of portal vein obstruction (Fig. 4-29). The distended subcutaneous veins radiate out from the umbilicus, producing the clinical picture referred to as *caput Medusae*. If there is obstruction to the superior vena cava or inferior vena cava, the venous blood causes distention of the veins running from the anterior chest wall to the thigh. The lateral thoracic vein, a tributary of the axillary vein, anastomoses with the superficial epigastric vein, a tributary of the great saphenous vein of the leg. In these circumstances a tortuous varicose vein may extend from the axilla to the lower abdomen (Fig. 4-6).

The *nerves* of the anterior and lateral abdominal walls supply the skin, the muscles, and the parietal peritoneum. They are derived from the anterior rami of the lower six thoracic nerves and the first lumbar nerves. The skin of the back is supplied by the posterior rami of the spinal nerves. It is important to remember that the seventh to the eleventh thoracic anterior rami are intercostal nerves and also supply the skin, intercostal muscles, and parietal pleura of the thoracic wall.

Inflammation of the parietal peritoneum will cause not only pain in the overlying skin, but also a reflex increase in tone of the abdominal musculature in the same area. For example, a localized peritonitis in the right iliac region will cause pain in that region. Palpation of the abdominal wall in the right iliac region will detect a reflex rigidity of the abdominal muscles, as compared with a softness of the abdominal muscles elsewhere.

Sometimes it is difficult for a physician to decide whether the muscles of the anterior abdominal wall of a patient are rigid due to underlying inflammation of the parietal peritoneum, or if the patient is voluntarily contracting the muscles because he resents being examined or because the physician's hand is cold. This problem is usually easily solved by asking the patient, who is lying supine on the examination table, to rest his arms by his sides and draw up his knees to flex the hip joints. It is practically impossible for a patient to keep the abdominal musculature tensed when the thighs are flexed. Needless to say, the examiner's hand should be warm.

A pleurisy involving the lower costal parietal pleura will cause pain in the overlying skin that may radiate down into the abdomen. Although it is unlikely to cause rigidity of the abdominal muscles, it may cause confusion in making a diagnosis unless these anatomical facts are remembered.

It is useful to remember the following:

Dermatomes over:
| | |
|---|---|
| The xiphoid process | T7 |
| The umbilicus | T10 |
| The pubis | L1 |

The *umbilicus* is a consolidated scar representing the site of attachment of the umbilical cord in the fetus; it is situated in the linea alba. In the adult it often receives scant attention in the bath and is consequently a common site for infection. It possesses a number of embryological remains that may give rise to clinical problems. These are summarized in Figure 4-30.

A *patent urachus* may reveal itself in the newborn infant by the passage of urine through the umbilicus if there is a congenital urethral obstruction. More often, it remains undiscovered until old age, when enlargement of the prostate may obstruct the urethra (Fig. 4-30).

Persistence of the *vitello-intestinal duct* may result in an umbilical fecal fistula (Fig. 4-30). If the duct remains as a fibrous band, a loop of bowel may become wrapped around it, causing intestinal obstruction (Fig. 4-30).

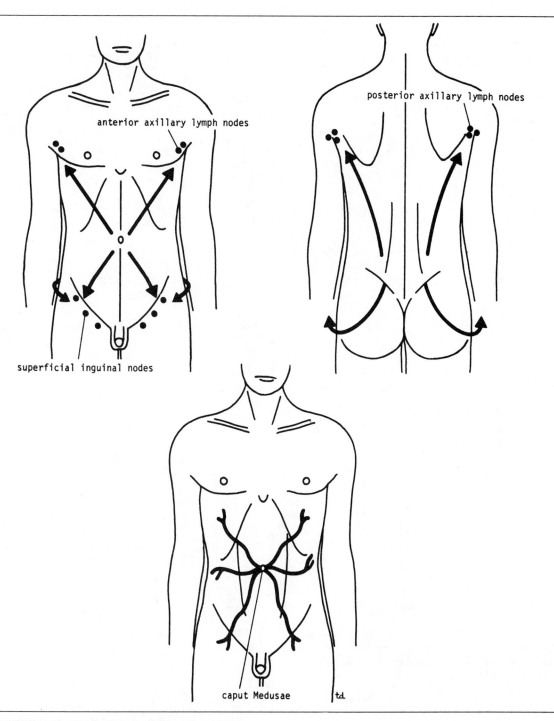

**Fig. 4-29.** Lymphatic drainage of skin of anterior and posterior abdominal walls. Lower diagram shows example of caput Medusae in case of portal obstruction due to cirrhosis of liver.

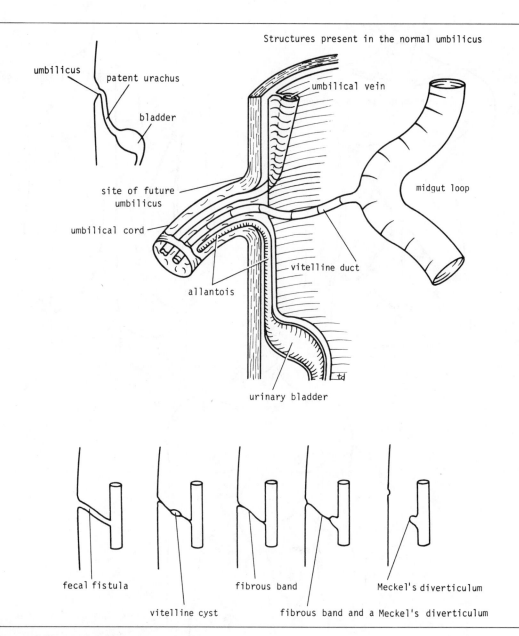

**Fig. 4-30. Umbilicus, showing some common congenital defects.**

A *Meckel's diverticulum* is present in 2 percent of subjects (Fig. 4-30). It occurs about 2 feet from the ileocolic junction and is about 2 inches long. It may become ulcerated or cause intestinal obstruction.

## Superficial Fascia

The *membranous layer of the superficial fascia* is important clinically, since beneath it there is a potential closed space that does not open into the thigh, but is continuous with the superficial perineal pouch via the penis and scrotum. Rupture of the penile urethra may be followed by extravasation of urine into the scrotum, perineum, and penis and then up into the lower part of the anterior abdominal wall deep to the membranous layer of fascia. The urine is excluded from the thigh because of the attachment of the fascia to the fascia lata (Fig. 4-7).

When closing abdominal wounds it is usual for a surgeon to put in a continuous suture uniting the divided membranous layer of superficial fascia. This strengthens the healing wound, prevents stretching of the skin scar, and makes for a more cosmetically acceptable result.

## Muscles

Remember that the abdominal muscles contract and relax with respiration, and the abdominal wall conforms to the volume of the abdominal viscera. There is an *abdominothoracic rhythm*. Normally, during inspiration, when the sternum moves forward and the chest expands, the anterior abdominal wall also moves forward. If, when the chest expands, the anterior abdominal wall remains stationary or contracts inward, it is highly probable that the parietal peritoneum is inflamed and has caused a reflex contraction of the abdominal muscles.

The shape of the anterior abdominal wall is dependent on the tone of its muscles. A middle-aged woman with poor abdominal muscles who has had multiple pregnancies is often incapable of supporting her abdominal viscera. The lower part of the anterior abdominal wall protrudes forward, a condition known as *visceroptosis*. This should not be confused with an abdominal tumor such as an ovarian cyst, or with the excessive accumulation of fat in the fatty layer of the superficial fascia.

## Surgical Incisions

The length and direction of surgical incisions through the anterior abdominal wall to expose the underlying viscera are largely governed by the position and direction of the nerves of the abdominal wall, the direction of the muscle fibers, and the arrangement of the aponeuroses forming the rectus sheath. Ideally, the incision should be made in the direction of the lines of cleavage in the skin so that a hairline scar is produced. The surgeon usually has to compromise, placing the safety of the patient first and the cosmetic result second.

Incisions that necessitate the division of one of the main segmental nerves lying within the abdominal wall will result in paralysis of part of the anterolateral abdominal musculature and a segment of the rectus abdominis. The consequent weakness of the abdominal musculature will cause an unsightly bulging forward of the abdominal wall and visceroptosis; extreme cases may require a surgical belt for support.

If the incision can be made in the line of the muscle fibers or aponeurotic fibers as each layer is traversed, on closing the incision the fibers fall back into position and function normally.

Incisions through the rectus sheath are widely used, provided that the rectus abdominis muscle and its nerve supply are kept intact. On closure of the incisions, the anterior and posterior walls of the sheath are sutured separately, and the rectus muscle springs back into position between the suture lines. The result is a very strong repair, with the minimum interference with function.

The following incisions are commonly used:

1. *Paramedian incision.* This may be supraumbilical, for exposure of the upper part of the abdominal cavity, or infraumbilical, for the lower abdomen and pelvis. In extensive operations in which a large exposure is required, the incision can run the full length of the rectus sheath. The anterior wall of the rectus sheath is exposed and incised about 1 inch (2.5 cm) from the midline. The medial edge of the incision is dissected medially, freeing the anterior wall of the sheath from the tendinous intersections of the rectus

muscle. The rectus abdominis muscle is retracted laterally with its nerve supply intact, and the posterior wall of the sheath is exposed. The posterior wall is then incised, together with the fascia transversalis and the peritoneum. The wound is closed in layers.

2. *Pararectus incision.* The anterior wall of the rectus sheath is incised medially and parallel to the lateral margin of the rectus muscle. The rectus is freed and retracted medially, exposing the segmental nerves entering its posterior surface. If the opening into the abdominal cavity is to be small, these nerves may be retracted upward and downward. The posterior wall of the sheath is then incised, as in the paramedian incision. The great disadvantage of this incision is that the opening is small, and any longitudinal extension requires that one or more segmental nerves to the rectus abdominis be divided, with resultant postoperative rectus muscle weakness.

3. *Midline incision.* This incision is made through the linea alba. The fascia transversalis, the extraperitoneal connective tissue, and the peritoneum are then incised. It is easier to perform above the umbilicus because the linea alba is wider in that region. It is a rapid method of gaining entrance to the abdomen and has the obvious advantage that it does not damage muscles or their nerve and blood supply. It has the additional advantage that it may be converted into a T-shaped incision for greater exposure. The anterior and posterior walls of the rectus sheath are then cut across transversely, and the rectus muscle is retracted laterally.

4. *Transrectus incision.* The technique in the making and closing of this incision is the same as that used in the paramedian incision, except that the rectus abdominis muscle is incised longitudinally and not retracted laterally from the midline. This incision has the great disadvantage of sectioning the nerve supply to that part of the muscle that lies medial to the muscle incision.

5. *Transverse incision.* This may be made above or below the umbilicus and can be small or so large that it extends from flank to flank. It may be made through the rectus sheath and the rectus abdominis muscles and through the oblique and transversus abdominis muscles laterally. It is rare to damage more than one segmental nerve, so that there is minimal postoperative abdomi-

nal weakness. The incision gives good exposure and is well tolerated by the patient. Closure of the wound is made in layers. It is unnecessary to suture the cut ends of the rectus muscles, provided that the sheaths are carefully repaired.

6. *Muscle splitting*, or *McBurney's incision.* This is chiefly used for cecostomy and appendectomy. It gives a limited exposure only, and should there be any doubt about the diagnosis, an infraumbilical right paramedian incision should be used instead.

An oblique skin incision is made in the right iliac region about 2 inches (5 cm) above and medial to the anterior superior iliac spine. The external and internal oblique and transversus muscles are incised or split in the line of their fibers and retracted to expose the fascia transversalis and the peritoneum. The latter are now incised and the abdominal cavity is opened. The incision is closed in layers, with no postoperative weakness.

7. *Abdominothoracic incision.* This is used to expose the lower end of the esophagus, as, for example, in esophagogastric resection for carcinoma of this region. An upper oblique or paramedian abdominal incision is extended upward and laterally into the seventh, eighth, or ninth intercostal space, the costal arch is transected, and the diaphragm is incised. Wide exposure of the upper abdomen and thorax is then obtained by the use of a rib-spreading retractor.

On completion of the operation, the diaphragm is repaired with nonabsorbable sutures, the costal margin is reconstructed, and the abdominal and thoracic wounds are closed.

## Abdominal Herniae

A hernia is the protrusion of part of the abdominal contents beyond the normal confines of the abdominal wall (Fig. 4-31). It consists of three parts: the sac, the contents of the sac, and the coverings of the sac. The *hernial sac* is a diverticulum of peritoneum and has a neck and a body. The *hernial contents* may consist of any structure found within the abdominal cavity and may vary from a small piece of omentum to a large viscus such as the kidney. The *hernial coverings* are formed from the layers of the abdominal wall through which the hernial sac passes.

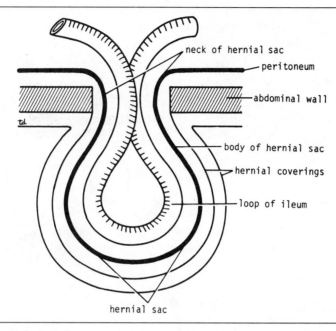

Fig. 4-31. Different parts of a hernia.

Abdominal herniae are of the following common types:

1. Inguinal, which may be (a) indirect or (b) direct
2. Femoral
3. Umbilical, which may be (a) congenital or (b) acquired
4. Epigastric
5. Divarication of the recti abdominis

### Indirect Inguinal Hernia

Indirect inguinal hernia is the most common form of hernia. The hernial sac is believed to be the remains of the processus vaginalis. It therefore enters the inguinal canal through the deep inguinal ring *lateral to the inferior epigastric vessels* (Fig. 4-32). It may extend part of the way along the canal or the full length, as far as the superficial inguinal ring. If the processus vaginalis has undergone no obliteration, then the hernia will be complete and will extend through the superficial inguinal ring down into the scrotum or labium majus. Under these circumstances the neck of the hernial sac will lie at the deep inguinal ring lateral to the inferior epigastric vessels, and the body of the sac will reside in the inguinal canal and scrotum (or base of labium majus).

An indirect inguinal hernia is about twenty times more common in males than in females, and nearly one-third are bilateral. It is more common on the right (normally, the right processus vaginalis becomes obliterated after the left; the right testis descends later than the left). It is most common in children and the young adult.

### Direct Inguinal Hernia

Direct inguinal hernia comprises about 15 percent of all inguinal hernias. The sac of a direct hernia bulges directly anteriorly through the posterior wall of the inguinal canal *medial to the inferior epigastric vessels* (Fig. 4-32). Because of the presence of the strong conjoint tendon, this hernia is usually nothing more than a generalized bulge, and therefore the neck of the hernial sac is wide.

Direct inguinal hernias are rare in women, and the majority are bilateral. It is a disease of old men with weak abdominal muscles.

*An inguinal hernia may be distinguished from a*

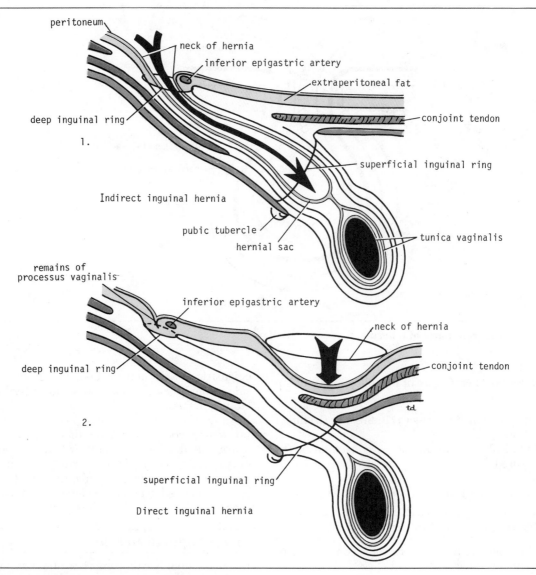

**Fig. 4-32. (1) Indirect inguinal hernia and (2) direct inguinal hernia. Note that neck of indirect inguinal hernia lies lateral to inferior epigastric artery, and neck of direct inguinal hernia lies medial to inferior epigastric artery.**

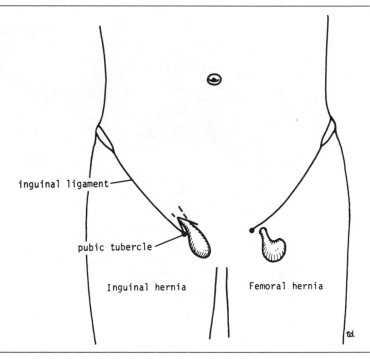

inguinal ligament

pubic tubercle

Inguinal hernia

Femoral hernia

**Fig. 4-33. Relation of inguinal and femoral hernial sacs to pubic tubercle.**

*femoral hernia by the fact that the sac, as it emerges through the superficial inguinal ring, lies above and medial to the pubic tubercle, while that of a femoral hernia lies below and lateral to the tubercle* (Fig. 4-33).

## Femoral Hernia

The femoral sheath is fully described on page 588. It is a protrusion of the fascial envelope lining the abdominal walls and surrounds the femoral vessels and lymphatics for about 1 inch (2.5 cm) below the inguinal ligament (Fig. 4-34). The *femoral canal*, the compartment for the lymphatics, occupies the medial part of the sheath. It is about ½ inch (1.3 cm) long, and its upper opening is referred to as the *femoral ring*. The *femoral septum*, which is a condensation of extraperitoneal tissue, closes the femoral ring.

A femoral hernia is more common in women than in men (possibly due to a wider pelvis and femoral canal). The hernial sac passes down the femoral canal, pushing the femoral septum before it. On escaping through the lower end, it expands to form a swelling in the upper part of the thigh (Fig. 4-34). With further expansion the hernial sac may turn upward to cross the anterior surface of the inguinal ligament.

The neck of the sac always lies below and lateral to the *pubic tubercle* (Fig. 4-33), and this serves to distinguish it from an inguinal hernia. The neck of the sac is narrow and lies at the femoral ring. The ring is related anteriorly to the inguinal ligament, posteriorly to the pectineal ligament and the pubis, medially to the sharp free edge of the lacunar ligament, and laterally to the femoral vein. Because of the presence of these anatomical structures, the neck of the sac is unable to expand. Once an abdominal viscus has passed through the neck into the body of the sac, it may be difficult to push it up and return it to the abdominal cavity (*irreducible hernia*). Furthermore, after straining or coughing, a piece of bowel may be forced through the neck and its blood vessels may be compressed by the femoral ring, seriously impairing its blood supply (*strangulated hernia*). A femoral hernia is a

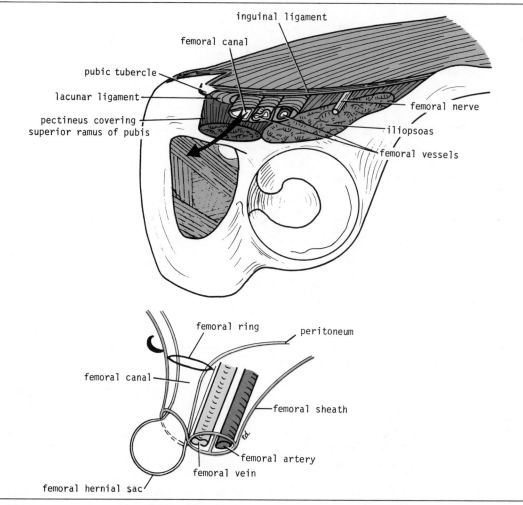

**Fig. 4-34. Femoral sheath as seen from below. Arrow emerging from femoral canal indicates path taken by femoral hernial sac. Note relations of femoral ring.**

dangerous disease and should always be treated surgically.

### Umbilical Herniae

*Congenital umbilical hernia*, or exomphalos (omphalocele), is a failure of part of the midgut to return to the abdominal cavity from the extraembryonic coelom during fetal life. For a diagram of the hernial sac and its relationship to the umbilical cord, see Figure 4-35.

An *acquired infantile umbilical hernia* is a small hernia that sometimes occurs in children and is due to a weakness in the scar of the umbilicus in

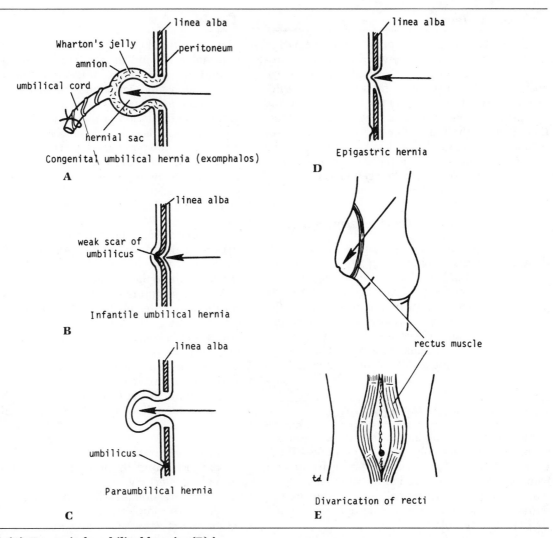

**Fig. 4-35. (A) Congenital umbilical hernia, (B) infantile umbilical hernia, (C) paraumbilical hernia, (D) epigastric hernia, and (E) divarication of recti abdominis.**

the linea alba (Fig. 4-35). The majority become smaller and disappear without treatment as the abdominal cavity enlarges.

An *acquired umbilical hernia of adults* is more correctly referred to as a *paraumbilical hernia*. The hernial sac does not protrude through the umbilical scar, but through the linea alba in the region of the umbilicus (Fig. 4-35). Paraumbilical herniae gradually increase in size and hang downward. The neck of the sac may be narrow, but the body of the sac often contains coils of small and large intenstine and omentum. Paraumbilical herniae are much more common in women than in men.

## *Epigastric Hernia*

Epigastric hernia occurs through the widest part of the linea alba, anywhere between the xiphoid process and the umbilicus. It is usually small in size and starts off as a small protrusion of extraperitoneal fat between the fibers of the linea alba. During the following months or years the fat is forced farther through the linea alba and eventually drags behind it a small peritoneal sac. The body of the sac often contains a small piece of greater omentum. It is common in middle-aged laborers.

## *Divarication of the Recti Abdominis*

Divarication of the recti abdominis occurs in elderly multiparous women with weak abdominal muscles (Fig. 4-35). In this condition, the aponeuroses forming the rectus sheath become excessively stretched. When the patient coughs or strains the recti separate widely, and a large hernial sac, containing abdominal viscera, bulges forward between the medial margins of the recti. This may be corrected by wearing a suitable abdominal belt.

## Paracentesis of the Abdomen

Paracentesis of the abdomen may be necessary to withdraw excessive collections of peritoneal fluid. Under a local anesthetic, a wide-bore trocar and cannula are inserted through the anterior abdominal wall. The underlying coils of intestine are not damaged, since they are mobile and are pushed away by the cannula.

If the cannula is inserted in the midline (Fig. 4-36), it will pass through the following anatomical structures: (1) skin, (2) superficial fascia, (3) deep fascia (very thin), (4) linea alba (virtually bloodless), (5) fascia transversalis and extraperitoneal fat, and (6) parietal peritoneum.

If the cannula is inserted in the flank (Fig. 4-36) lateral to the inferior epigastric artery and above the deep circumflex artery, it will pass through the following: (1) skin, (2) superficial fascia, (3) deep fascia (very thin), (4) aponeurosis or muscle of external oblique, (5) internal oblique muscle, (6) transversus muscle, (7) fascia transversalis and extraperitoneal fat, and (8) parietal peritoneum.

## Testis

The testis develops high up on the posterior abdominal wall, and in late fetal life it "descends" behind the peritoneum, dragging its blood supply, nerve supply, and lymphatic drainage after it.

A *variocele* is a condition in which there is elongation and dilatation of the veins of the pampiniform plexus. It is a common disorder, found in adolescents and young adults. The great majority occur on the left side. This is thought to be due to the fact that the right testicular vein joins the low-pressure inferior vena cava, whereas the left vein joins the left renal vein, in which the venous pressure is higher. Rarely, malignant disease of the left kidney may extend along the renal vein and block the exit of the testicular vein. A rapidly developing left-sided variocele should therefore always lead one to examine the left kidney.

A malignant tumor of the testis spreads upward via the lymph vessels to the lumbar (para-aortic) lymph nodes at the level of the first lumbar vertebra. It is only later, when the tumor spreads locally to involve the tissues and skin of the scrotum, that the superficial inguinal lymph nodes are involved.

The process of the descent of the testis is shown diagrammatically in Figure 4-19. The testis may be subject to the following congenital anomalies:

1. *Anterior inversion,* in which the epididymis lies anteriorly and the testis and the tunica vaginalis posteriorly.
2. *Polar inversion,* in which the testis and epididymis are completely inverted.
3. *Imperfect descent (cryptorchidism)*
   (a) *Incomplete descent* (Fig. 4-37), in which the testis, although traveling down its normal path, fails to reach the floor of the scrotum. It may be found within the abdomen, within the inguinal canal, at the superficial inguinal ring, or high up in the scrotum.
   (b) *Maldescent* (Fig. 4-38), in which the testis travels down an abnormal path and fails to reach the scrotum. It may be found in the superficial fascia of the anterior abdominal wall above the inguinal ligament, in front of the pubis, in the perineum, or in the thigh.

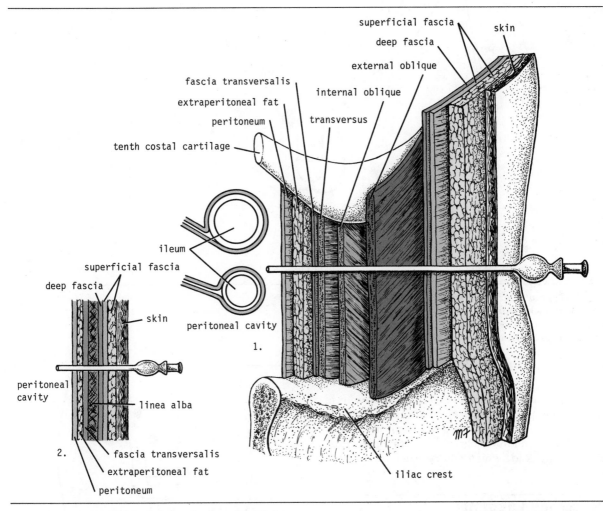

**Fig. 4-36. Paracentesis of abdominal cavity (1) in flanks and (2) in midline.**

It is necessary for the testes to leave the abdominal cavity since the temperature there retards the normal process of spermatogenesis. If an incompletely descended testis is brought down into the scrotum by surgery before puberty, it will develop and function normally. A maldescended testis, although often developing normally, is very susceptible to traumatic injury and for this reason should be placed in the scrotum. Many authorities believe that there is a greater incidence of tumor formation in testes that have not descended into the scrotum.

The *appendix of the testis* and the *appendix of the epididymis* are embryological remnants found at the upper poles of these organs that may become cystic. The appendix of the testis is derived from the paramesonephric ducts, and the appendix of the epididymis is a remnant of the mesonephric tubules.

## Vasectomy

Bilateral vasectomy is a simple operation performed to produce infertility. Under local anesthesia, a small incision is made in the upper part of the scrotal wall and the vas deferens is divided

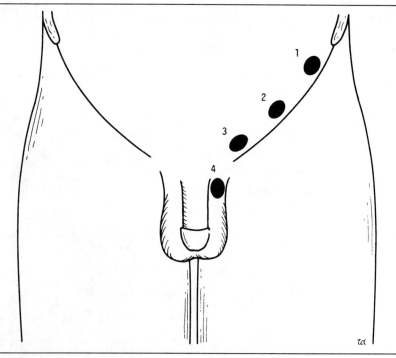

Fig. 4-37. Four degrees of incomplete descent of testis. (1) In abdominal cavity close to deep inguinal ring. (2) In inguinal canal. (3) At superficial inguinal ring. (4) In upper part of scrotum.

between ligatures. Spermatozoa may be present in the first few postoperative ejaculations, but that is simply an emptying process. Now only the secretions of the seminal vesicles and prostate constitute the seminal fluid, which can be ejaculated as before.

## Processus Vaginalis

The formation of the processus vaginalis and its passage through the lower part of the anterior abdominal wall with the formation of the inguinal canal in both sexes has been described. (See p. 173.) Normally, the upper part becomes obliterated just prior to birth, and the lower part remains as the tunica vaginalis.

The processus is subject to the following common congenital anomalies:

1. It may persist partially or in its entirety as a *preformed hernial sac* for an indirect inguinal hernia (Fig. 4-39).
2. It may become very much narrowed, but its lumen remains in communication with the ab-

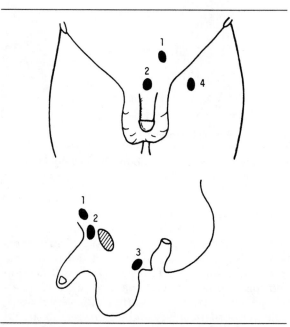

Fig. 4-38. Four types of maldescent of testis. (1) In superficial fascia of anterior abdominal wall, above superficial inguinal ring. (2) At root of penis. (3) In the perineum. (4) In the thigh.

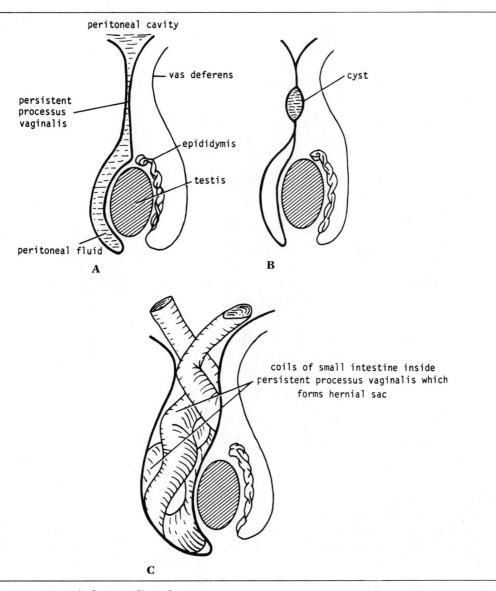

**Fig. 4-39. Common congenital anomalies of processus vaginalis. (A) Congenital hydrocele. (B) Encysted hydrocele of cord. (C) Preformed hernial sac for indirect inguinal hernia.**

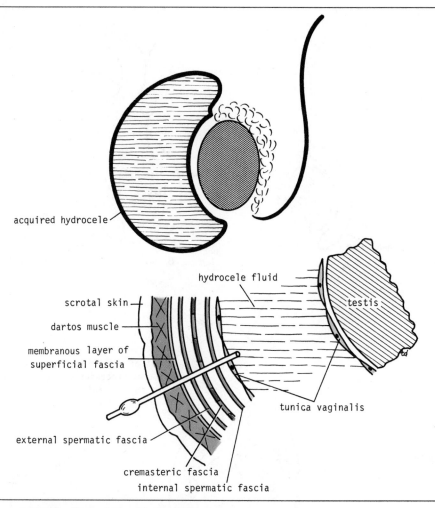

**Fig. 4-40. Tunica vaginalis distended with fluid (hydrocele). Lower diagram shows various anatomical layers traversed by a trocar and cannula when a hydrocele is tapped.**

dominal cavity. Peritoneal fluid will accumulate in it, forming a *congenital hydrocele* (Fig. 4-39).

3. The upper and lower ends of the processus may become obliterated, leaving a small intermediate cystic area referred to as an *encysted hydrocele of the cord* (Fig. 4-39).

The tunica vaginalis is closely related to the front and sides of the testis. It is therefore not surprising to find that inflammation of the testis may cause an accumulation of fluid within the tunica vaginal-

is. This is referred to, simply, as a *hydrocele* (Fig. 4-40). The great majority of hydroceles are idiopathic.

## Tapping a Hydrocele

To remove excess fluid from the tunica vaginalis, a fine trocar and cannula are inserted through the scrotal skin (Fig. 4-40). The following anatomical structures are traversed by the cannula: (1) skin, (2) dartos muscle and membranous layer of fascia (Colles' fascia), (3) external spermatic fascia, (4) cremasteric fascia, (5) internal spermatic fascia, and (6) parietal layer of the tunica vaginalis.

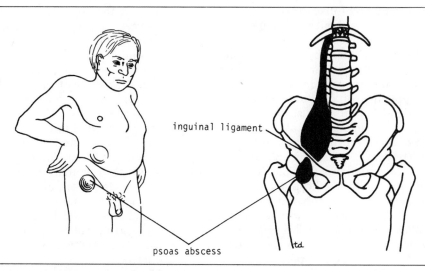

inguinal ligament

psoas abscess

**Fig. 4-41. Case of advanced tuberculous disease of thoracolumbar region of vertebral column. A psoas abscess is present, and there are swellings in right groin above and below right inguinal ligament.**

## Psoas Fascia

The psoas fascia covers the anterior surface of the psoas muscle and may influence the direction taken by a tuberculous abscess. Tuberculous disease of the thoracolumbar region of the vertebral column results in the destruction of the vertebral bodies, with possible extension of pus laterally under the psoas fascia (Fig. 4-41). From there, the pus tracks downward, following the course of the psoas muscle, and appears as a swelling in the upper part of the thigh below the inguinal ligament. It may be mistaken for a femoral hernia.

## CLINICAL PROBLEMS

*Answers on page 959*

1. Following an automobile accident a 35-year-old man was admitted to the hospital with a fracture of the tenth rib on the left side. He was noted to be pale, restless, and sweating, and his blood pressure was low. Should an examination of the abdomen have been made in this case? Which organ was most likely to have been damaged?

2. A fat, flatulent 40-year-old woman complained of pain over the right shoulder and back. The pain was made worse by eating fatty foods. Her condition was diagnosed as inflammation of the gallbladder (cholecystitis). Where would you palpate the gallbladder? Why should a painful lesion under the diaphragm give rise to pain over the shoulder?

3. A patient returning from the tropics was suspected of having an amebic abscess of the liver. He had a history of amebic dysentery 3 months previously. He experienced a dull, aching pain in the right hypochondrium that was occasionally felt over the right shoulder. Where exactly is the right hypochondrium? Why should he have pain over the right shoulder?

4. An 8-year-old boy was admitted to the hospital with a pyrexia, furred tongue, and pain in the right iliac region. On examination the muscles of the anterior abdominal wall in the lower right quadrant were noted to be contracted and rigid. A diagnosis of appendicitis with perforation of the wall of the appendix was made. Why were the muscles of the anterior abdomi-

nal wall rigid in the right iliac region?

5. In 1930, a film actress about to embark on a world tour was told that she had to be vaccinated. She insisted that her physician vaccinate her on the edge of the umbilicus, so that her admirers would not notice the scar. One week later she complained of painful swellings in both armpits and in both groins. Can you explain these swellings in anatomical terms?

6. A workman engaged on the erection of a skyscraper in New York City lost his balance and fell astride a horizontal girder on the floor below. He was admitted to the hospital suffering from severe shock. He was unable to micturate normally and passed only a few drops of blood-stained urine. On examination, he was found to have extensive swelling of his perineum, scrotum, and penis. The lower part of the anterior abdominal wall was also swollen, but his thighs were normal. Can you explain the distribution of the swelling in anatomical terms?

7. Following a sudden severe blow on the anterior abdominal wall from the hind leg of a horse, a patient complained of pain and swelling below the umbilicus. On examination, extensive bruising of the skin was observed over the lower part of the right rectus muscle. On gentle palpation, a deep swelling confined to the right rectus sheath was felt. Given that the deep swelling was due to a collection of blood (hematoma), which blood vessel was likely to have been ruptured?

8. A soldier suffered a severe shrapnel wound in the right lumbar region. Recently, in addition to diminished skin sensation over the right lumbar region and umbilicus, he has noticed a bulging forward of the right side of his anterior abdominal wall. Can you explain these defects in anatomical terms?

9. A 35-year-old man with a long history of duodenal ulcer suddenly complained of severe epigastric pain. He was admitted to the hospital in a state of shock, and it was noticed that his anterior abdominal wall did not move on respiration. On examination, the abdominal musculature was noted to be rigid in the epigastrium. A perforated duodenal ulcer was diagnosed. Can you explain why the anterior abdominal wall failed to move on respiration?

Why were the upper abdominal muscles rigid?

10. A 9-week-old boy was admitted to hospital with a swelling in the right groin that extended down into the upper part of the scrotum. When he cried, the swelling enlarged. On careful palpation it was possible to reduce the size of the swelling, and this procedure was accompanied by a gurgling noise. The swelling was situated above and medial to the pubic tubercle on the right side. In anatomical terms, explain what was wrong with the infant.

11. A 45-year-old woman noticed a painful swelling in her left groin after moving some heavy furniture. On examination, a small tender swelling was noted in the left groin, situated below and lateral to the pubic tubercle. What is your diagnosis? Give your reasons in anatomical terms.

12. A 35-year-old man was digging up a tree root in his garden when he experienced a dragging sensation in his right groin. On examination, the patient was found to have a swelling above the medial end of the right inguinal ligament, which expanded on coughing. On palpation, it was possible to reduce the size of the swelling. By exerting pressure just above and medial to the midpoint of the inguinal ligament, one could prevent the reappearance of the swelling when the patient coughed. What is your diagnosis? Why is it possible to prevent the reappearance of the swelling by exerting pressure over the area indicated above?

13. A 75-year-old man with chronic bronchitis noticed that a bulge was developing in his left groin. On examination, an elongated swelling was seen above the medial end of the left inguinal ligament. When the patient coughed, the swelling enlarged, but did not descend into the scrotum. The patient had weak abdominal muscles. What is the diagnosis? Why did the swelling enlarge when the patient coughed?

14. A 5-year-old girl has a small, painless swelling of the umbilicus. The mother notices that the swelling enlarges when the child strains while sitting on the toilet. What is the diagnosis? In anatomical terms, explain the defect in the anterior abdominal wall.

15. A laborer on a building site noticed that a small, tender swelling was developing in the

epigastrium. He first experienced the tenderness months ago, and since then a swelling has appeared and is slowly increasing in size. On examination, a small, soft swelling about the size of a pea was seen in the midline just below the xiphoid process. It was impossible to reduce the size of the swelling by digital pressure. What is the diagnosis? Explain in anatomical terms how the swelling developed.

16. A 65-year-old woman has an excessive accumulation of fluid within the peritoneal cavity (ascites) due to a malignant tumor of the right ovary. Her physician has decided she will be more comfortable if some of the fluid is drained off. Where would you insert the trocar and cannula? Name the anatomical structures that would be traversed by the cannula.

17. A 6-year-old boy had no testicle present in the right side of his scrotum. On careful palpation, a deep, firm ovoid structure could be felt above the medial part of the inguinal ligament. What is the diagnosis? Is surgical treatment necessary?

18. An 18-year-old boy, at a medical examination for admission to the army, was found to have no testis in the left side of the scrotum. Nothing abnormal could be palpated in the inguinal cana, but a small, firm ovoid structure could be felt in front of the upper part of the left thigh. What is the diagnosis? Is surgical treatment required?

19. In a 25-year-old man, a swelling developed above the medial end of the right inguinal ligament. It was associated with a dull, aching pain, but it did not expand on coughing. On palpation, the swelling appeared to fluctuate, and on grasping the right testis through the scrotal wall and gently pulling it inferiorly, the swelling moved medially along the inguinal canal. What was the swelling? Why did it move with the testis?

20. A senior medical student taking a surgical examination was asked to look at a 55-year-old man. On examination of the abdomen, he found a hard, fixed mass in the midline, about 4 inches (10 cm) in diameter, lying on the transpyloric plane. The patient told the student that he had recently lost 20 pounds and had a poor appetite. The student told the examiner

that the patient was suffering from carcinoma of the stomach and should have an immediate gastrectomy. The examiner then asked the student if he had examined the patient's scrotum, and the student admitted he had not. On examination, the scrotum was found to contain a large, hard mass on the right side that was not tethered to the skin. The inguinal lymph nodes on the right side were normal. (a) Can you explain the connection between the abdominal swelling and the scrotal swelling? (b) Why were the inguinal lymph nodes normal?

21. A chronic alcoholic patient was told by his physician that he had cirrhosis of the liver. On examination of the anterior abdominal wall, the right lobe of the liver was felt to extend 4 fingerbreadths below the costal margin. A number of superficial veins around the umbilicus were seen to be enlarged. What is the cause of the congested superficial veins?

22. On studying a patient's records, it was found that she had had a right infra-umbilical paramedian incision. What are the advantages and disadvantages of this incision as compared with a pararectus incision and with a transrectus incision?

23. A McBurney's incision is often made for appendectomies. What are the advantages and disadvantages of this incision?

24. A patient suffering from tuberculosis of the left epididymis was found to have an ulcer on the posterior wall of the scrotum. Which group of lymph nodes would you examine for *local spread* of the disease?

25. A resident was asked to examine the vas deferens of a patient. Where would you examine the vas deferens, and what does a normal vas feel like?

26. In a patient with a history of tuberculosis, an angular kyphosis of the lumbar vertebral column suddenly developed. On examination, a swelling was found in the groin, just below the right inguinal ligament. On deep palpation of the anterior abdominal wall above the right inguinal ligament, a further swelling could be felt. Digital pressure on the first swelling caused expansion of the second swelling and vice versa. What is the diagnosis? Explain the swelling in anatomical terms.

27. In a patient with a history of venereal disease, a large fluctuant swelling developed in front of the testis. From your knowledge of anatomy, and given that there is fluid present in the scrotum, where in the scrotum is the fluid likely to collect?

28. A 50-year-old man is found on examination to have a small cystic swelling above his left testis. What anatomical structure is likely to be involved? Can you explain the presence of this structure embryologically?

29. A 75-year-old man complained of a "weeping belly button." Being questioned, he stated that he had difficulty in micturition. His physician informed him that he had an enlarged prostate. Can you explain the "weeping" anatomically and embryologically?

30. In a young man of 25 years, an enlargement of the pampiniform plexus on the left side suddenly developed. The physician examined the scrotum and then the left kidney. Can you explain why he should examine the left kidney?

## NATIONAL BOARD TYPE QUESTIONS

*Answers on page 984*

**In each of the following questions, answer:**
  (a) IF (1) ONLY IS CORRECT
  (b) IF (2) ONLY IS CORRECT
  (c) IF BOTH (1) AND (2) ARE CORRECT, AND
  (d) IF NEITHER (1) OR (2) IS CORRECT

1. (1) The fundus of the gallbladder lies opposite the tip of the right ninth costal cartilage where the linea semilunaris crosses the costal margin.
   (2) The abdominal part of the aorta bifurcates at the level of the intertubercular line.

2. (1) The spleen is situated in the left hypochondrium and its lower pole can easily be palpated in a normal thin adult.
   (2) The lower border of the liver can be felt three fingerbreadths below the costal margin in a normal thin adult.

3. (1) The costal margin is formed in front by the cartilages of the seventh, eighth, ninth, and tenth ribs.
   (2) The lower pole of the left kidney cannot be palpated through the anterior abdominal wall in the left lumbar region at the end of deep inspiration.

4. (1) The transpyloric plane can be identified by asking the supine patient to sit up without using his arms. Where the lateral border of the rectus muscle crosses the costal margin signifies the level of the plane.
   (2) The intercristal plane passes across the highest points on the iliac crests and lies on the level of the body of the third lumbar vertebra.

**In the following questions, answer:**
  (a) IF (1), (2), AND (3) ONLY ARE CORRECT
  (b) IF (1) AND (3) ONLY ARE CORRECT
  (c) IF (2) AND (4) ONLY ARE CORRECT
  (d) IF (4) ONLY IS CORRECT, OR
  (e) IF ALL ARE CORRECT

5. Which of the following statements are **true** concerning the rectus abdominis muscle?
   (1) It arises by three heads from the front of the symphysis pubis, the pubic crest, and the pubic tubercle.
   (2) It is inserted into the eighth, ninth, and tenth costal cartilages.
   (3) The tendinous intersections are attached to the posterior wall of the rectus sheath.
   (4) It is innervated by the lower six intercostal nerves.

6. Which of the following statements concerning the function of the anterior and lateral wall muscles of the abdomen are **true**?
   (1) They laterally flex and rotate the vertebral column.
   (2) They assist the diaphragm during inspiration by relaxing.
   (3) They support and protect the abdominal viscera.
   (4) They assist in increasing the intra-abdominal pressure during micturition and defecation.

7. The rectus sheath contains the following important structures:
   (1) Terminal branches of lower intercostal nerves

(2) Inferior epigastric artery

(3) Lymph vessels

(4) Ilioinguinal nerve

8. The walls of the inguinal canal are formed by many structures that include:

(1) Conjoint tendon

(2) Transversus abdominis

(3) Fascia transversalis

(4) Lacunar ligament

9. The weakest parts of the inguinal canal include:

(1) The roof

(2) The deep inguinal ring

(3) The floor

(4) The superficial inguinal ring

10. In the female, the inguinal canal contains the following structures:

(1) Ilioinguinal nerve

(2) Remnant of the processus vaginalis

(3) Round ligament of the uterus

(4) Lymph vessels from the fundus of the uterus

11. Which of the following statements concerning the spermatic cord are correct?

(1) It is covered by the five layers of spermatic fascia.

(2) It extends from the deep inguinal ring to the scrotum.

(3) It does not contain the testicular artery.

(4) It contains the pampiniform plexus.

**Select the best response:**

12. All the following structures are present in the inguinal canal in the male **except** the:

(a) Internal spermatic fascia

(b) Genital branch of the genitofemoral nerve

(c) Testicular vessels

(d) Deep circumflex iliac artery

(e) Ilioinguinal nerve

13. All the following statements concerning the conjoint tendon are true **except:**

(a) It forms part of the posterior wall of the inguinal canal.

(b) It is formed by the fusion of the aponeuroses of the transversus abdominis and internal oblique muscles.

(c) It reinforces the superficial inguinal ring.

(d) It is continuous with the inguinal ligament.

(e) It may bulge forward in a direct inguinal hernia.

14. All the following statements concerning an indirect inguinal hernia are true **except:**

(a) It is the most common form of abdominal hernia.

(b) The neck of the hernial sac lies medial to the inferior epigastric artery.

(c) The sac is the remains of the processus vaginalis.

(d) The hernial sac can extend into the scrotum.

(e) At the superficail inguinal ring, the hernial sac will lie above and medial to the pubic tubercle.

15. In order to pass a needle into the cavity of the tunica vaginalis in the scrotum, the following structures have to be pierced **except:**

(a) Skin

(b) Dartos muscle and Colles' fascia

(c) Tunica albuginea

(d) Internal spermatic fascia

(e) Cremasteric fascia

16. The following statements are true about muscles forming the posterior abdominal wall **except:**

(a) The psoas major muscle has a fascial sheath that extends down into the thigh as far as the lesser trochanter of the femur.

(b) The quadratus lumborum is covered anteriorly by fascia that forms the lateral arcuate ligament.

(c) The ilacus muscle is innervated by the femoral nerve.

(d) The transversus abdominis muscle does form part of the posterior abdominal wall.

(e) The diaphragm does not contribute to the musculature on the posterior abdominal wall.

17. Below the level of the anterior superior iliac spine, which structure forms the posterior wall of the rectus sheath?

(a) Transversus abdominis

(b) External oblique

(c) Internal oblique

(d) Conjoint tendon

(e) Transversalis fascia.

**Match the structure on the left with an appropriate structure listed on the right:**

18. External spermatic fascia
19. Round ligament of the uterus
20. Cremasteric fascia
21. Internal spermatic fascia
22. Deep inguinal ring

(a) Internal oblique

(b) Fascia transversalis

(c) Gubernaculum

(d) External oblique

(e) None of the above

**Match the structure on the left with the appropriate group of lymph nodes that drain the area listed on the right:**

23. Testis
24. Skin of anterior abdominal wall below the level of the umbilicus
25. Epididymis
26. Skin of scrotum

(a) Anterior axillary lymph nodes
(b) Para-aortic or lumbar lymph nodes
(c) Superficial inguinal nodes
(d) External iliac nodes
(e) None of the above

# 5. The Abdomen: Part II
# The Abdominal Cavity

## GENERAL ARRANGEMENT OF THE ABDOMINAL VISCERA

### Liver

The liver is a very large organ that occupies the upper part of the abdominal cavity (Figs. 5-1, 5-2). It lies almost entirely under cover of the ribs and costal cartilages in the right hypochondrium and epigastric regions.

### Gallbladder

The gallbladder is a pear-shaped sac that is adherent to the undersurface of the right lobe of the liver; its blind end, or fundus, projects below the inferior border of the liver (Figs. 5-1, 5-2).

### Esophagus

The esophagus is a tubular structure that joins the pharynx to the stomach. The esophagus pierces the diaphragm slightly to the left of the midline and after a short course of about ½ inch (1.25 cm) en-

ters the stomach on its right side. It is deeply placed, lying behind the left lobe of the liver (Fig. 5-1).

### Stomach

The stomach is a dilated part of the alimentary canal between the esophagus and the small intestine (Figs. 5-1, 5-2). It occupies the left hypochondriac, epigastric, and umbilical regions and much of it lies under cover of the ribs. Its long axis passes downward and forward to the right and then backward and slightly upward.

### Small Intestine

The small intestine is divided into three regions: duodenum, jejunum, and ileum. The *duodenum* is the first part of the small intestine and most of it is deeply placed on the posterior abdominal wall. It is situated in the epigastric and umbilical regions. It is a C-shaped tube that extends from the stomach around the head of the pancreas to join the jejunum (Fig. 5-1). About halfway down its

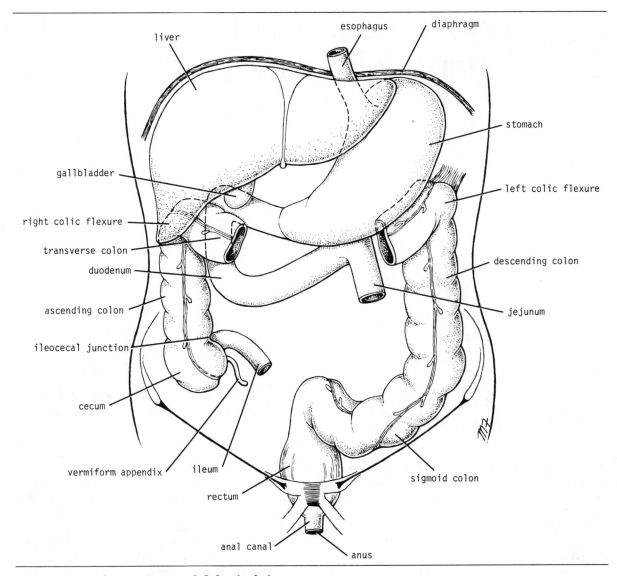

**Fig. 5-1. General arrangement of abdominal viscera.**

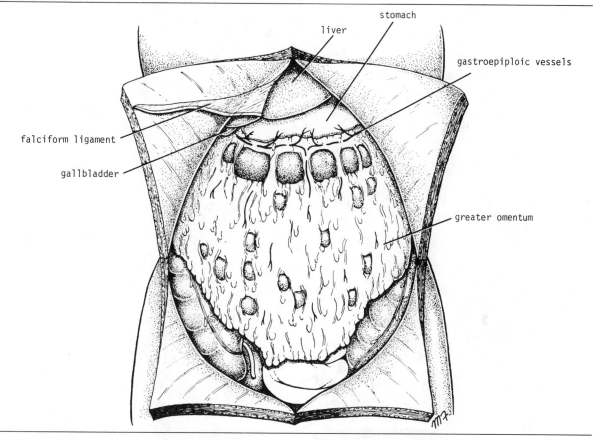

**Fig. 5-2. Abdominal organs in situ. Note that greater omentum hangs down in front of small and large intestines.**

length it receives the bile and the pancreatic ducts.

The *jejunum and ileum* measure about 20 feet (6 m) long, the upper two-fifths of this length being the jejunum. The jejunum begins at the duodenojejunal junction, and the ilium ends at the ileocecal junction (Fig. 5-1). The coils of jejunum occupy the upper left part of the abdominal cavity, while the ileum tends to occupy the lower right part of the abdominal cavity and the pelvic cavity (Fig. 5-3).

## Large Intestine

The large intestine is divided into the cecum, the vermiform appendix, the ascending colon, the transverse colon, the descending colon, the sigmoid (pelvic) colon, the rectum, and the anal canal (Fig. 5-1). The large intestine arches around and encloses the coils of the small intestine (Fig. 5-3). It tends to be more fixed than the small intestine.

The *cecum* is a blind-ended sac that projects downward in the right iliac region below the ileocecal junction (Figs. 5-1, 5-3). The *vermiform appendix* is a worm-shaped tube that arises from its medial side (Fig. 5-1).

The *ascending colon* extends upward from the cecum to the inferior surface of the right lobe of the liver occupying the right iliac and right lumbar regions (Figs. 5-1, 5-3). On reaching the liver it bends to the left, forming the right colic (hepatic) flexure.

The *transverse colon* crosses the abdomen in the umbilical region from the right colic flexure to the

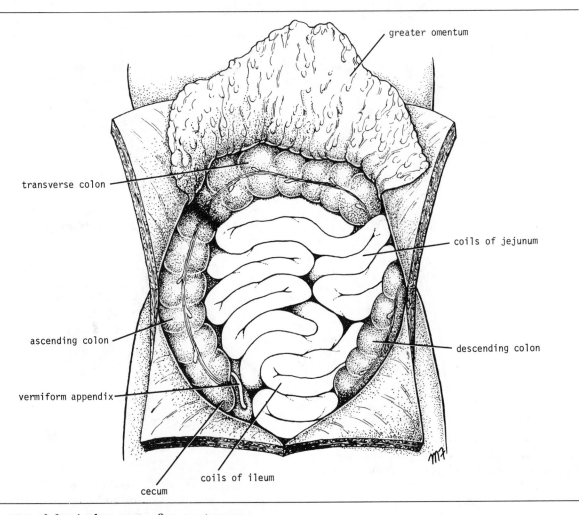

greater omentum

transverse colon

coils of jejunum

ascending colon

descending colon

vermiform appendix

coils of ileum

cecum

**Fig. 5-3. Abdominal contents after greater omentum has been reflected upward. Coils of small intestine occupy central part of abdominal cavity, while ascending, transverse, and descending parts of colon are located at periphery.**

left colic flexure (Figs. 5-1, 5-3). It forms a wide U-shaped curve. In the erect position, the lower part of the U may extend down into the pelvis. The transverse colon, on reaching the region of the spleen, bends downward, forming the left colic (splenic) flexure to become the descending colon.

The *descending colon* extends from the left colic flexure to the pelvic inlet below (Fig. 5-1, 5-3). It occupies the left lumbar and iliac regions.

The *sigmoid colon* begins at the pelvic inlet,

where it is a continuation of descending colon (Fig. 5-1). It hangs down into the pelvic cavity in the form of a loop. It joins the rectum in front of the sacrum.

The *rectum* occupies the posterior part of the pelvic cavity (Fig. 5-1). It is continuous above with the sigmoid colon and descends in front of the sacrum to leave the pelvis by piercing the pelvic floor. Here it becomes continuous with the anal canal in the perineum.

## Pancreas

The pancreas is a soft, lobulated organ that stretches obliquely across the posterior abdominal wall in the epigastric region (Fig. 5-1). It is situated behind the stomach, and it extends from the duodenum to the spleen.

## Spleen

The spleen is a soft, friable mass of lymphatic tissue that occupies the left upper part of the abdomen between the stomach and the diaphragm (Fig. 5-1). It lies in the left hypochondriac region along the long axis of the tenth left rib.

## Kidneys

The kidneys are two reddish-brown organs situated high up on the posterior abdominal wall, one on each side of the vertebral column (Fig. 5-1). The left kidney lies slightly higher than the right. Each kidney gives rise to a *ureter* that runs vertically downward on the psoas muscle.

## Suprarenal Glands

The suprarenal glands are two yellowish organs that lie on the upper poles of the kidneys (Fig. 5-12) on the posterior abdominal wall.

## THE PERITONEUM

The peritoneum is a thin serous membrane lining the walls of the abdominal and pelvic cavities and clothing the abdominal and pelvic viscera. It resembles the pleura, though it is arranged in a more complex manner. The *parietal peritoneum* lines the walls of the abdominal and pelvic cavities, and the *visceral peritoneum* covers the organs. The peritoneum secretes a small amount of serous fluid, which lubricates the surfaces of the peritoneum and facilitates free movement between the viscera. Between the parietal peritoneum and the fascial lining of the abdominal and pelvic walls is a layer of connective tissue called the *extraperitoneal tissue.* It varies in amount in different regions and in the area of the kidneys contains a large amount of fat. The visceral peritoneum is closely bound to the underlying viscus by only a small amount of connective tissue.

The potential space between the parietal and visceral layers of peritoneum is called the *peritoneal cavity.* In the male this is a closed cavity, but in the female there is a communication with the exterior through the uterine tubes, the uterus, and the vagina.

The peritoneal cavity may be divided into two parts, the greater sac and the lesser sac (Figs. 5-4, 5-5 and 5-6). The *greater sac* is the main compartment of the peritoneal cavity and extends across the whole breadth of the abdomen, and from the diaphragm to the pelvis. The *lesser sac* is the smaller compartment and lies behind the stomach; as a diverticulum from the greater sac, it opens through an oval window called the *opening of the lesser sac*, or the *epiploic foramen* (Figs. 5-5 and 5-7).

An organ is said to be *retroperitoneal* if it lies behind the peritoneal cavity. This means that the organ is covered only in the front by peritoneum. Examples of retroperitoneal organs are the pancreas, duodenum, ascending colon, and descending colon. The kidneys, ureters, inferior vena cava, and aorta are further examples.

The following specialized areas of the peritoneum should be understood (Fig. 5-4).

1. A *mesentery*, which is a two-layered fold of peritoneum that attaches part of the intestines to the posterior abdominal wall, and includes the *mesentery of the small intestine*, the *transverse mesocolon*, and the *sigmoid mesocolon.* A mesentery permits the intestine to be very mobile within the abdominal cavity.

2. An *omentum*, which is a two-layered fold of peritoneum that attaches the stomach to another viscus. The *greater omentum* is attached to the greater curvature of the stomach and hangs down like an apron in the space between the coils of small intestine and the anterior abdominal wall (Fig. 5-2). It is folded back on itself and is attached to the inferior border of the transverse colon. The lesser omentum slings the lesser curvature of the stomach to the undersurface of the liver. The *gastrosplenic omentum* (ligament) connects the stomach to the spleen.

3. The *peritoneal ligaments*, which are two-layered folds of peritoneum that attach the lesser mobile solid viscera to the abdominal walls. (They do not possess the dense fibrous tissue seen in ligaments associated with bones.) The liver, for example, is attached by the *falciform ligament* to

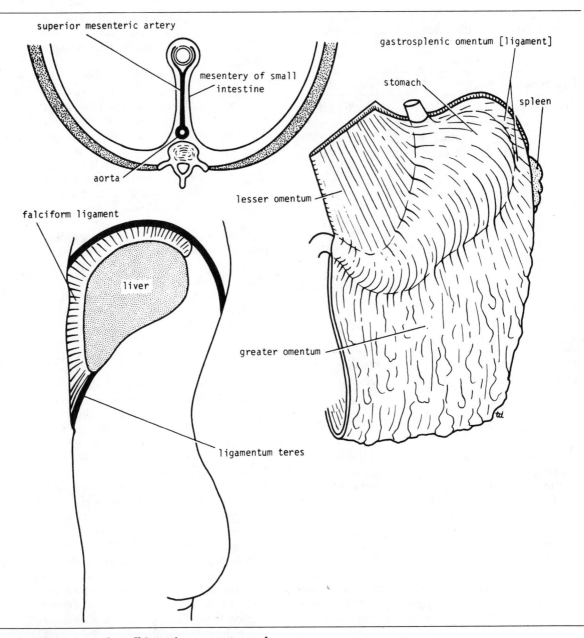

**Fig. 5-4. Mesentery of small intestine, omenta, and falciform ligament. Note that right edge of greater omentum has been cut to show layers of peritoneum.**

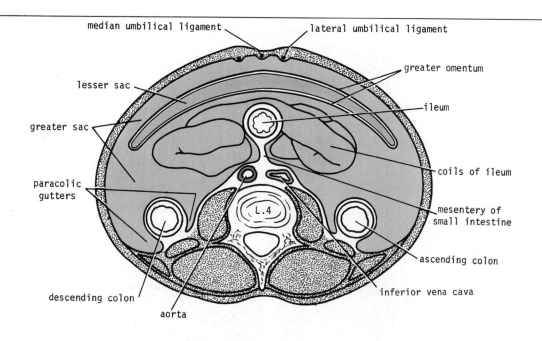

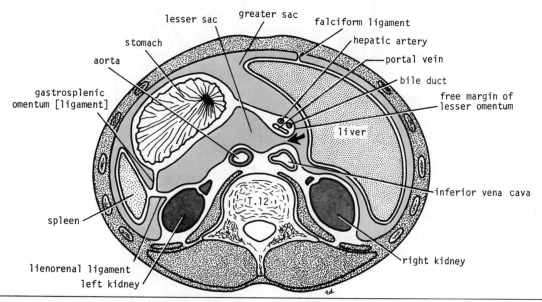

**Fig. 5-5. Transverse sections of abdomen, showing arrangement of peritoneum. Arrow in lower diagram indicates position of opening of lesser sac.**

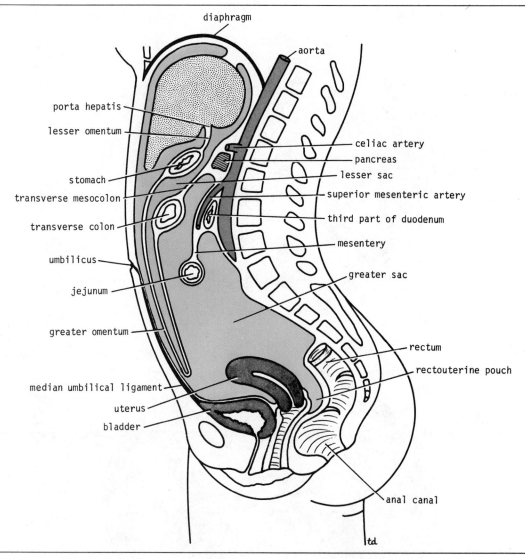

**Fig. 5-6. Sagittal section of female abdomen, showing arrangement of peritoneum.**

the anterior abdominal wall and to the undersurface of the diaphragm.

The mesenteries, omenta, and peritoneal ligaments allow blood vessels, lymphatics, and nerves to reach the various viscera.

To understand the attachments of the peritoneal ligaments, mesenteries, and so on, it is advisable to trace the peritoneum around the abdominal cavity, first in a transverse direction and then in a vertical direction.

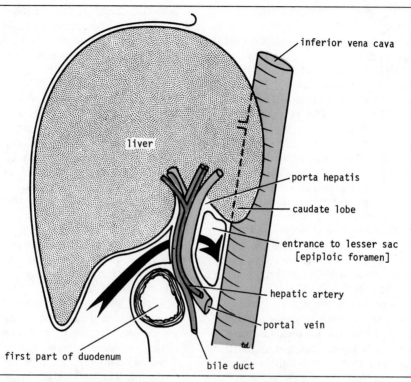

inferior vena cava

liver

porta hepatis

caudate lobe

entrance to lesser sac
[epiploic foramen]

hepatic artery

portal vein

first part of duodenum

bile duct

**Fig. 5-7. Sagittal section through entrance into lesser sac, showing important structures that form boundaries to opening.**

## Transverse Section of the Abdomen Through the Fourth Lumbar Vertebra

The parietal peritoneum lining the anterior abdominal wall below the umbilicus is smooth, apart from the low ridges produced by the *median umbilical ligament* (the urachus, the remains of the fetal allantois, which passes from the apex of the bladder to the umbilicus) and the *lateral umbilical ligaments* (the obliterated umbilical arteries, which pass from the internal iliac arteries to the umbilicus) (Fig. 5-5).

The parietal peritoneum passes onto the posterior abdominal wall and becomes continuous with the visceral peritoneum covering the sides and anterior surfaces of the ascending colon and descending colon (Fig. 5-5). In the region of the aorta and inferior vena cava, the parietal peritoneum becomes continuous with the mesentery of the small

intestine. Note the *right* and *left paracolic grooves*, or *gutters*, which lie lateral and medial to the ascending colon and the descending colon, respectively (Fig. 5-5). Note also that the peritoneum forms a continuous layer that may be traced around the abdominal cavity without interruption.

## Transverse Section of Abdomen Through the Twelfth Thoracic Vertebra

The parietal peritoneum lining the anterior abdominal wall forms a sickle-shaped fold called the *falciform ligament* (Figs. 5-2 and 5-4). This connects the anterior surface of the liver to the anterior abdominal wall above the umbilicus and to the diaphragm. It lies slightly to the right of the midline (Fig. 5-5). In the free border of the ligament, where the two layers of peritoneum are continuous with each other, lies the *ligamentum teres* (Figs. 5-2 and 5-4). This is the obliterated umbilical vein of the fetus, which passes upward to enter the groove between the quadrate lobe and the left lobe of the liver.

If the parietal peritoneum is followed around the abdominal wall on the left side, it reaches the lateral margin of the left kidney (Fig. 5-5). Here, it becomes continuous with the visceral peritoneum covering the lateral margin and part of the anterior surface of the left kidney. The peritoneum then leaves the kidney and passes to the hilus of the spleen as the posterior layer of the *lienorenal ligament*. The visceral peritoneum covers the spleen and, on reaching the hilus again, is reflected onto the greater curvature of the stomach as the anterior layer of the *gastrosplenic omentum* (ligament). The visceral peritoneum covers the anterior surface of the stomach and leaves the lesser curvature to form the anterior layer of the *lesser omentum* (Fig. 5-5). On the right, the lesser omentum has a free border, and here the peritoneum folds around the *bile duct*, *hepatic artery*, and *portal vein*. The free border of the lesser omentum forms the anterior margin of the opening into the lesser sac (Figs. 5-5 and 5-7).

The peritoneum forms the posterior layer of the *lesser omentum* and becomes continuous with the visceral layer of peritoneum covering the posterior wall of the stomach. Note that the peritoneum here forms the anterior wall of the lesser sac (Fig. 5-5). At the greater curvature of the stomach, the peritoneum leaves the stomach, forming the posterior layer of the *gastrosplenic omentum* (ligament) and reaches the hilus of the spleen. Here, it is reflected backward to the posterior abdominal wall, forming the anterior layer of the *lienorenal ligament*. The peritoneum now covers the anterior surface of the pancreas, the aorta, and the inferior vena cava, forming the posterior wall of the lesser sac (Fig. 5-5). The peritoneum passes onto the anterior surface of the right kidney and sweeps around the lateral abdominal wall to reach the anterior abdominal wall. Once again, note that the peritoneum forms a continuous layer around the abdomen (Fig. 5-5).

## Sagittal Section of the Abdomen and Pelvis

The parietal peritoneum lining the anterior abdominal wall may be traced upward to the left of the falciform ligament to reach the undersurface of the diaphragm (Fig. 5-6). Here, it is reflected onto the upper surface of the liver as the anterior layer of the *left triangular ligament*. The visceral peritoneum then covers the anterior and inferior surfaces of the liver until it reaches the *porta hepatis*. Here, the peritoneum passes to the lesser curvature of the stomach as the anterior layer of the *lesser omentum*. Having covered the anterior surface of the stomach, the peritoneum leaves the greater curvature forming the anterior layer of the *greater omentum* (Figs. 5-2 and 5-6).

The greater omentum hangs down as a fold in front of the coils of intestine and contains within it the lower part of the lesser sac. Having reached the lowest limit of the greater omentum, the peritoneum folds upward and forms the posterior layer of the greater omentum. On reaching the inferior border of the transverse colon (Figs. 5-3 and 5-6), the peritoneum covers its posterior surface and then leaves the colon to form the posterior layer of the *transverse mesocolon*. The peritoneum then passes to the anterior border of the pancreas and runs downward anterior to the third part of the duodenum (Fig. 5-6).

The peritoneum now leaves the posterior abdominal wall as the anterior layer of the *mesentery of the small intestine*. The visceral peritoneum covers the jejunum and then forms the posterior layer of the mesentery. On returning to the posterior abdominal wall, the peritoneum runs downward into the pelvis and covers the anterior surface of the upper part of the rectum (Fig. 5-6). From here, it is reflected onto the posterior surface of the upper part of the vagina, forming the important *rectouterine pouch (pouch of Douglas)*. In the male the peritoneum is reflected onto the upper part of the posterior surface of the bladder and the seminal vesicles, forming the *rectovesical pouch*.

The peritoneum passes over the upper surface of the uterus in the female and is reflected from its anterior surface onto the upper surface of the bladder (Fig. 5-6). In both sexes the peritoneum passes from the bladder onto the anterior abdominal wall.

## Peritoneal Pouches or Fossae

### LESSER SAC

The lesser sac is an extensive peritoneal pouch situated behind the lesser omentum and stomach and

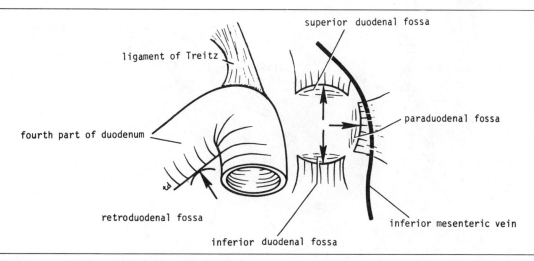

superior duodenal fossa

ligament of Treitz

paraduodenal fossa

fourth part of duodenum

retroduodenal fossa

inferior mesenteric vein

inferior duodenal fossa

**Fig. 5-8. Peritoneal fossae that may be present in region of duodenojejunal junction. Note presence of inferior mesenteric vein in peritoneal fold, forming paraduodenal fossa.**

lying in front of structures situated on the posterior abdominal wall (Figs. 5-5, 5-6, and 5-13). It projects upward as far as the diaphragm and downward between the layers of the greater omentum. The lower part of the lesser sac is often obliterated by the adherence of the anterior layers of the greater omentum to the posterior layers. Its left margin is formed by the spleen (Fig. 5-13) and the gastrosplenic omentum and lienorenal ligaments; below it is formed by the left free border of the greater omentum (Fig. 5-5). The right margin of the sac opens into the greater sac, i.e., the main part of the peritoneal cavity, through the *opening of the lesser sac,* or *epiploic foramen.* Below the opening the right margin is formed by the right free border of the greater omentum.

The *opening into the lesser sac* (epiploic foramen) has the following boundaries (Fig. 5-7).

## Anteriorly

The free border of the lesser omentum, containing the *bile duct,* the *hepatic artery,* and the *portal vein* (Fig. 5-13). Note that the bile duct lies to the right and in front; the hepatic artery lies to the left and in front; and the portal vein lies posteriorly (Figs. 5-5 and 5-13).

## Posteriorly

The inferior vena cava.

## Superiorly

The caudate process of the caudate lobe of the liver.

## Inferiorly

The first part of the duodenum.

## DUODENAL FOSSAE

In the region of the duodenojejunal junction there may be four small pocket-like pouches of peritoneum (Fig. 5-8). The mouths of the *superior duodenal, inferior duodenal,* and *paraduodenal fossae* face each other, but the mouth of the *retroduodenal fossa* does not face the others. The inferior mesenteric vein often runs in the free margin of the paraduodenal fossa. Commonly, only one or two of these fossae are present.

## CECAL FOSSAE

The presence of folds of peritoneum in the vicinity of the cecum creates three peritoneal fossae: the *superior ileocecal,* the *inferior ileocecal,* and the *retrocecal fossae* (Fig. 5-9).

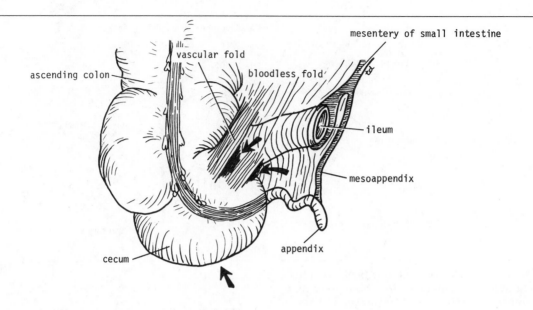

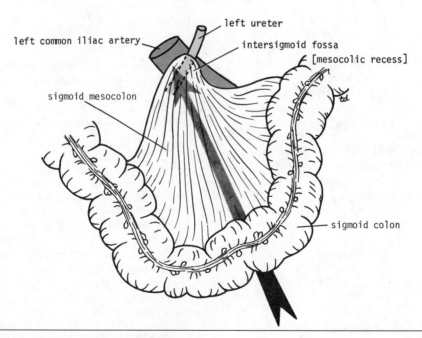

**Fig. 5-9. Peritoneal fossae (arrows) in region of
cecum, and fossa related to sigmoid mesocolon.**

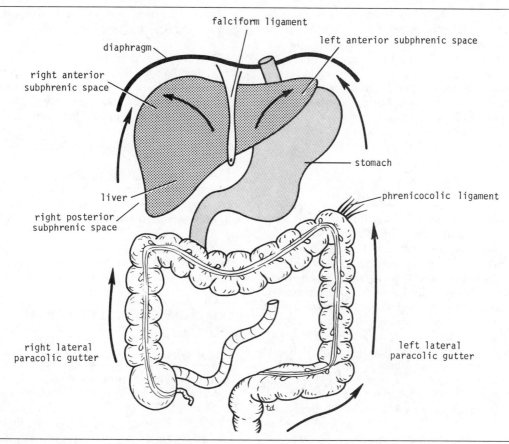

**Fig. 5-10. Normal direction of flow of peritoneal fluid from different parts of peritoneal cavity to subphrenic spaces.**

## INTERSIGMOID FOSSA

The intersigmoid fossa is situated at the apex of the inverted, V-shaped root of the sigmoid mesocolon (Fig. 5-9). Its mouth opens downward and lies in front of the left ureter.

The important pelvic peritoneal fossae are described on pages 343 and 353.

## SUBPHRENIC SPACES

The existence of the intraperitoneal subphrenic spaces is due to the complicated arrangement of the peritoneum in the region of the liver. The *right* and *left anterior subphrenic spaces* lie between the diaphragm and the liver, one on each side of the falciform ligament (Fig. 5-10). The *right posterior subphrenic space* lies between the right lobe of the liver, the right kidney, and the right colic flexure (Fig. 5-65). The *right extraperitoneal space* lies between the layers of the coronary ligament and is therefore situated between the liver and the diaphragm. (See p. 244.)

## PARACOLIC GUTTERS

The arrangement of the ascending colon and descending colon, the attachments of the transverse mesocolon and the mesentery of the small intestine to the posterior abdominal wall, result in the formation of four important paracolic gutters (Figs. 5-5 and 5-10). These gutters lie on the lateral and medial sides of the ascending and descending colons, respectively. It is interesting to note that the right medial paracolic gutter is closed off from the pelvic cavity inferiorly by the mesentery of the small intestine, while the others are in free communication with the pelvic cavity. The right lateral paracolic gutter is in communication with the right posterior subphrenic space; but the left lateral gutter is separated from the area around the spleen by the *phrenicocolic ligament*, a fold of peritoneum that passes from the left colic flexure to the diaphragm.

### Nerve Supply of the Peritoneum

The *parietal peritoneum* is sensitive to pain, temperature, touch, and pressure. The parietal peritoneum lining the anterior and lateral abdominal walls is supplied by the lower six thoracic and first lumbar nerves, i.e., the same nerves that innervate the overlying muscles and skin. The central part of the diaphragmatic peritoneum is supplied by the phrenic nerves; peripherally, the diaphragmatic peritoneum is supplied by the lower six thoracic nerves. The parietal peritoneum in the pelvis is mainly supplied by the obturator nerve.

The *visceral peritoneum* is sensitive to stretch and is supplied by autonomic afferent nerves. Overdistention of a viscus will lead to the sensation of pain. The mesenteries of the small and large intestines are sensitive to mechanical stretching.

### Function of the Peritoneum

The peritoneal fluid, which is pale yellow in color and somewhat viscid, contains leukocytes. It is secreted by the peritoneum and ensures that the mobile viscera glide easily upon one another. As the result of the movements of the diaphragm and the abdominal muscles, together with the peristaltic movements of the intestinal tract, the peritoneal fluid is not static. Experimental evidence has shown that particulate matter introduced into the lower part of the peritoneal cavity reaches the subphrenic peritoneal spaces very rapidly, whatever the position of the body. It seems that there is a continuous intraperitoneal movement of fluid toward the diaphragm (Fig. 5-10), and there it is quickly absorbed into the subperitoneal lymphatic capillaries.

This can be explained on the basis that the area

of peritoneum is very extensive in the region of the diaphragm and the respiratory movements of the diaphragm aid lymph flow in the lymph vessels.

The peritoneal coverings of the intestine tend to stick together in the presence of infection. The greater omentum, which is kept constantly on the move by the peristalsis of the neighboring intestinal tract, may adhere to other peritoneal surfaces around a focus of infection. In this manner, many of the intraperitoneal infections are sealed off and remain localized.

The peritoneal folds play an important part in suspending the various organs within the peritoneal cavity and serve as a means of conveying the blood vessels, lymphatics, and nerves to these organs.

Large amounts of fat are stored in the peritoneal ligaments and mesenteries, and especially large amounts may be found in the greater omentum.

## THE GASTROINTESTINAL TRACT
### Abdominal Part of the Esophagus

The esophagus is a muscular collapsible tube about 10 inches (25 cm) long that joins the pharynx to the stomach. The greater part of the esophagus lies within the thorax. (See p. 121.) The esophagus enters the abdomen through an opening in the right crus of the diaphragm (Fig. 5-12). After a course of about ½ inch (1.25 cm), it enters the stomach on its right side. It is covered on its anterior and lateral surfaces by peritoneum. The esophagus is related anteriorly to the posterior surface of the left lobe of the liver and posteriorly to the left crus of the diaphragm. The left and right vagi lie on its anterior and posterior surfaces, respectively.

The blood supply, nerve supply, and lymphatic drainage of the lower part of the esophagus is the same as that for the stomach. (For details, see p. 220.)

The function of the esophagus is to conduct food from the pharynx into the stomach. Wavelike contractions of the muscular coat, called *peristalsis*, propel the food onward. Although no anatomical sphincter exists at the lower end of the esophagus, there is no doubt that the circular coat of smooth muscle in this region serves physiologically as a sphincter. As the food descends through the esoph-

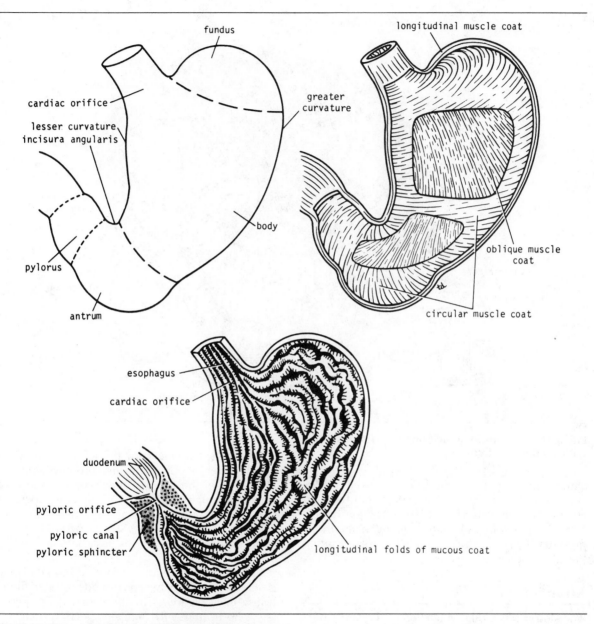

**Fig. 5-11. Stomach, showing different parts, muscular coats, and mucosal lining. Note increased thickness of circular muscle forming pyloric sphincter.**

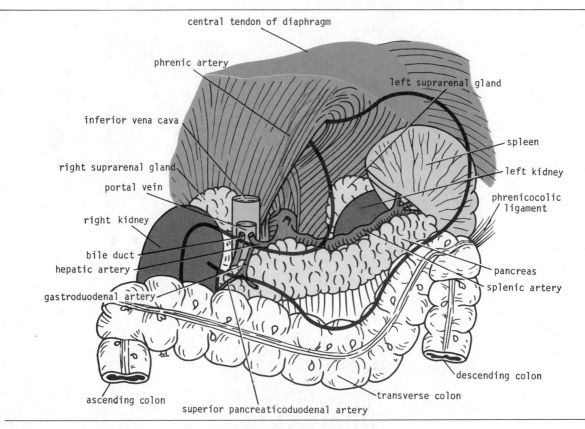

central tendon of diaphragm

phrenic artery

left suprarenal gland

inferior vena cava

spleen

right suprarenal gland

left kidney

portal vein

phrenicocolic
ligament

right kidney

bile duct

hepatic artery

pancreas

gastroduodenal artery

splenic artery

descending colon

ascending colon

transverse colon

superior pancreaticoduodenal artery

**Fig. 5-12. Structures situated on posterior abdominal wall behind stomach.**

agus, relaxation of the muscle at the lower end occurs ahead of the peristaltic wave, so that the food
enters the stomach. The tonic contraction of this
so-called physiological *gastroesophageal sphincter*
prevents the stomach contents from regurgitating
into the esophagus.

## Stomach

The stomach is a dilated portion of the alimentary
canal and has three main functions: (1) storage of
food; in the adult it has a capacity of about 1,500
ml; (2) mixes the food with gastric secretions to
form a semifluid *chyme*; and (3) controls the rate
of delivery of the chyme to the small intestine so
that efficient digestion and absorption can take
place.

The stomach is situated in the upper part of the
abdomen, extending from the left hypochondriac
region into the epigastric and umbilical regions.
Much of the stomach lies under cover of the lower
ribs. It is roughly J-shaped and has two openings,
the *cardiac* and *pyloric orifices*, two curvatures
known as the *greater* and *lesser curvatures*, and two
surfaces, an *anterior* and a *posterior surface* (Fig.
5-11).

The stomach is relatively fixed at both ends, but
is very mobile in between. It tends to be high and
transversely arranged in the short, obese person
(steer-horn stomach) and elongated vertically in
the tall, thin person (J-shaped stomach). Its shape
undergoes considerable variation in the same person and depends on the volume of its contents, the
position of the body, and the phase of respiration.

For purposes of description it is usual to divide
the stomach (Fig. 5-11) into the following parts:
The *fundus* is dome-shaped and projects upward
and to the left of the cardiac orifice. It is usually

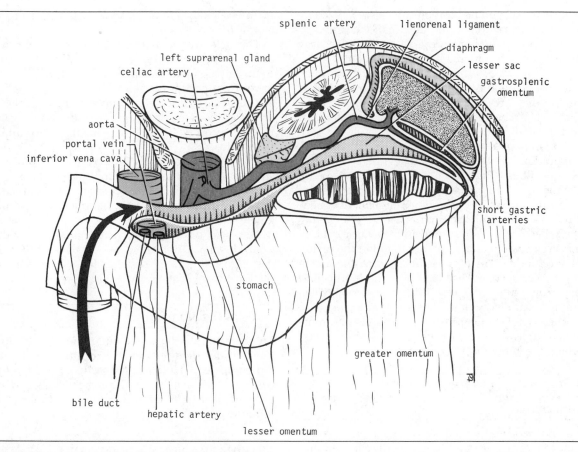

splenic artery

lienorenal ligament

left suprarenal gland

diaphragm

celiac artery

lesser sac

gastrosplenic
omentum

aorta

portal vein

inferior vena cava

short gastric
arteries

stomach

greater omentum

bile duct

hepatic artery

lesser omentum

**Fig. 5-13. Lesser sac, showing arrangement of peritoneum in formation of lesser omentum, gastrosplenic omentum, and lienorenal ligament. Arrow indicates position of opening of lesser sac.**

full of gas. The *body* extends from the level of the cardiac orifice to the level of the *incisura angularis*, a constant notch in the lower part of the lesser curvature. The *pyloric antrum* extends from the incisura angularis to the proximal limit of the pylorus. The *pylorus* is the most tubular part of the stomach. Its thick muscular wall forms the *pyloric sphincter*. The cavity of the pylorus is called the *pyloric canal*.

The *lesser curvature* forms the right border of the stomach and extends from the cardiac orifice to the pylorus (Fig. 5-11). The lesser omentum extends from the lesser curvature to the liver. The *greater curvature* is much longer than the lesser curvature and extends from the left of the cardiac orifice,

over the dome of the fundus, and then sweeps around and to the right to the inferior part of the pylorus (Fig. 5-11). The gastrosplenic omentum (ligament) extends from the upper part of the greater curvature to the spleen, and the greater omentum extends from the lower part of the greater curvature to the transverse colon (Fig. 5-13).

The *cardiac orifice* is where the abdominal part of the esophagus enters the stomach (Fig. 5-11). Although no anatomical sphincter can be demonstrated here, there is little doubt that a physiological mechanism exists that prevents regurgitation of stomach contents into the esophagus. See page 218.

The *pyloric orifice* is formed by the *pyloric canal*, which is about 1 inch (2.5 cm) long. The circular muscle coat of the stomach is much thicker here and forms the anatomical and physiological *pyloric sphincter* (Fig. 5-11). The pylorus lies on the transpyloric plane, and its position can be recognized by a slight constriction on the surface of the stomach. The pyloric sphincter controls the rate of discharge of the stomach contents into the duodenum.

The *mucous membrane* of the stomach is thick and vascular and is thrown into numerous folds, or *rugae*, that are mainly longitudinal in direction (Fig. 5-11). The folds flatten out when the stomach is distended.

The *muscular wall* of the *stomach* contains (1) longitudinal fibers, (2) circular fibers, and (3) oblique fibers (Fig. 5-11). The longitudinal fibers are the most superficial and are most concentrated along the curvatures. The inner circular fibers encircle the body of the stomach and are greatly thickened at the pylorus to form the pyloric sphincter. Very few circular fibers are found in the region of the fundus. The oblique fibers form the innermost muscle coat. They loop over the fundus and pass down along the anterior and posterior walls, running parallel with the lesser curvature.

The *peritoneum* completely surrounds the stomach and leaves its curvatures as double layers, which are known as the *omenta*.

## RELATIONS OF THE STOMACH

### Anteriorly

The anterior abdominal wall, the left costal margin, the left pleura and lung, the diaphragm, and the left lobe of the liver (Figs. 5-2 and 5-6).

### Posteriorly

The lesser sac, the diaphragm, the spleen, the left suprarenal gland, the upper part of the left kidney, the splenic artery, the pancreas, the transverse mesocolon, and the transverse colon (Figs. 5-6, 5-12, and 5-13).

## BLOOD SUPPLY OF THE STOMACH

### Arterial Supply to the Stomach

These are derived from the branches of the celiac artery (Fig. 5-14).

The *left gastric artery* arises from the celiac artery. It passes upward and to the left to reach the esophagus and then descends along the lesser curvature of the stomach. It supplies the lower third

of the esophagus and the upper right part of the stomach.

The *right gastric artery* arises from the hepatic artery at the upper border of the pylorus and runs to the left along the lesser curvature. It supplies the lower right part of the stomach.

The *short gastric arteries* arise from the splenic artery at the hilus of the spleen and pass forward in the gastrosplenic omentum (ligament) to supply the fundus.

The *left gastroepiploic artery* arises from the splenic artery at the hilus of the spleen and passes forward in the gastrosplenic omentum (ligament) to supply the stomach along the upper part of the greater curvature.

The *right gastroepiploic artery* arises from the gastroduodenal branch of the hepatic artery. It passes to the left and supplies the stomach along the lower part of the greater curvature.

### Veins of the Stomach

These drain into the portal circulation (Fig. 5-29). The *left* and *right gastric veins* drain directly into the portal vein. The *short gastric veins* and the *left gastroepiploic veins* join the splenic vein. The *right gastroepiploic vein* joins the superior mesenteric vein.

## LYMPHATIC DRAINAGE OF THE STOMACH

The *lymph vessels* of the stomach (Fig. 5-15) follow the arteries and are arranged in four main groups:

1. Those that drain into lymph nodes along the left gastric vessels. The efferents from these nodes pass to the celiac nodes, which surround the origin of the celiac artery.
2. Those that drain into lymph nodes along the right gastric vessels. The efferents from these nodes pass to nodes along the hepatic artery and then to the celiac nodes.
3. Those that drain into nodes along the short gastric arteries and the left gastroepiploic artery and then drain into lymph nodes at the hilus of the spleen. From there, they pass to pancreaticosplenic nodes along the splenic artery, which in turn drain into the celiac nodes.
4. Those that drain into the right gastroepiploic

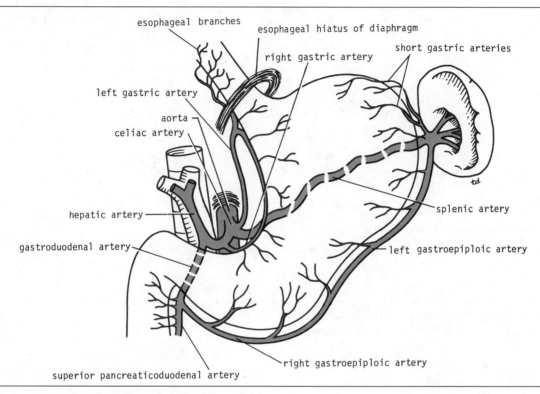

esophageal branches

esophageal hiatus of diaphragm

short gastric arteries

right gastric artery

left gastric artery

aorta

celiac artery

hepatic artery

gastroduodenal artery

splenic artery

left gastroepiploic artery

superior pancreaticoduodenal artery

right gastroepiploic artery

**Fig. 5-14. Arteries that supply stomach. Note that all the arteries are derived from branches of celiac artery.**

nodes, which lie along the lower part of the greater curvature of the stomach. The efferent lymph vessels join nodes along the gastroduodenal artery, which, in turn, drain to the celiac nodes.

It is thus seen that all the gastric lymph vessels drain ultimately into the celiac lymph nodes (Fig. 5-15).

## NERVE SUPPLY OF THE STOMACH

The *nerves of the stomach* are derived from the celiac sympathetic plexus and from the right and left vagus nerves (Fig. 5-16).

The *anterior vagal trunk*, which is formed in the thorax mainly from the left vagus nerve, enters the abdomen on the anterior surface of the esophagus.

The trunk, which may be single or multiple, then divides into branches that supply the anterior surface of the stomach. A large hepatic branch passes up to the liver, and from this a pyloric branch passes down to the pylorus (Fig. 5-16).

The *posterior vagal trunk*, which is formed in the thorax mainly from the right vagus nerve, enters the abdomen on the posterior surface of the esophagus. The trunk then divides into branches that supply mainly the posterior surface of the stomach. A large branch passes to the celiac and superior mesenteric plexuses and is distributed to the intestine as far as the splenic flexure and to the pancreas (Fig. 5-16).

The sympathetic innervation of the stomach carries a proportion of pain-transmitting nerve fibers, while the parasympathetic vagal fibers are secretomotor to the gastric glands and motor to the muscular wall of the stomach. The pyloric sphincter receives motor fibers from the sympathetic system and inhibitory fibers form the vagi.

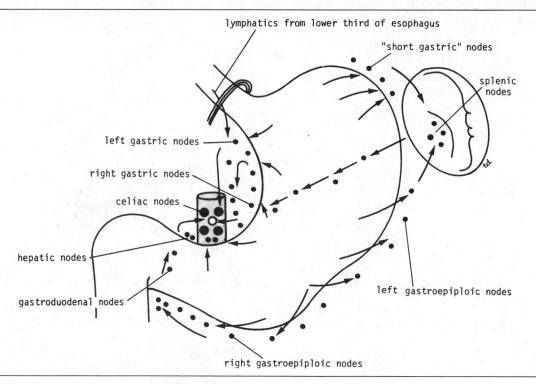

Fig. 5-15. Lymphatic drainage of stomach. Note that all the lymph eventually passes through celiac lymph nodes.

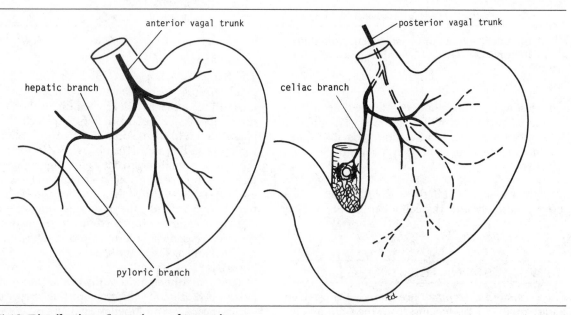

Fig. 5-16. Distribution of anterior and posterior vagal trunks within abdomen. Note: The celiac branch of the posterior vagal trunk is distributed with the sympathetic nerves as far down the intestinal tract as the left colic flexure.

## Celiac Artery

The celiac artery or trunk is very short and arises from the commencement of the abdominal aorta at the level of the twelfth thoracic vertebra. It is the artery that supplies the foregut. It is surrounded by the celiac plexus and lies behind the lesser sac of peritoneum. It has three terminal branches, the left gastric, splenic, and hepatic arteries.

### LEFT GASTRIC ARTERY

The small left gastric artery runs to the cardiac end of the stomach, gives off a few esophageal branches, then turns to the right along the lesser curvature of the stomach. It anastomoses with the right gastric artery.

### SPLENIC ARTERY

The large splenic artery runs to the left in a wavy course along the upper border of the pancreas and behind the stomach. On reaching the left kidney the artery enters the lienorenal ligament and runs to the hilus of the spleen.

#### Branches

1. *Pancreatic branches.*
2. The *left gastro-epiploic artery* arises near the hilus of the spleen and reaches the greater curvature of the stomach in the gastrosplenic omentum. It passes to the right along the greater curvature of the stomach between the layers of the greater omentum. It anastomoses with the right gastroepiploic artery.
3. The *short gastric arteries*, five or six in number, arise from the end of the splenic artery and reach the fundus of the stomach in the gastrosplenic omentum. They anastomose with the left gastric artery and the left gastroepiploic artery.

### HEPATIC ARTERY

The medium-sized hepatic artery* runs forward and to the right and then ascends between the lay-

*For purposes of description the hepatic artery is sometimes divided into the *common hepatic artery*, which extends from its origin to the gastroduodenal branch, and the *hepatic artery proper*, which is the remainder of the artery.

ers of the lesser omentum. It lies in front of the opening into the lesser sac and is placed to the left of the bile duct and in front of the portal vein. At the porta hepatis it divides into right and left branches to supply the corresponding lobes of the liver.

#### Branches

1. The *right gastric artery* arises from the hepatic artery at the upper border of the pylorus and runs to the left in the lesser omentum along the lesser curvature of the stomach. It anastomoses with the left gastric artery.
2. The *gastroduodenal artery* is a large branch that descends behind the first part of the duodenum. It divides into the *right gastroepiploic artery* that runs along the greater curvature of the stomach between the layers of the greater omentum, and the *superior pancreaticoduodenal artery* that descends between the second part of the duodenum and the head of the pancreas.
3. The *right and left hepatic arteries* that enter the porta hepatis. The right hepatic artery usually gives off the *cystic artery*, which runs to the neck of the gallbladder.

## Small Intestine

The small intestine, the longest part of the alimentary canal, is divided into three regions: the duodenum, the jejunum, and the ileum. The primary function of the small intestine is digestion and the absorption of the products of digestion.

### Duodenum

The duodenum is a C-shaped tube about 10 inches (25 cm) long that joins the stomach to the jejunum. It is very important because it receives the openings of the bile and the pancreatic ducts. The duodenum curves around the head of the pancreas (Fig. 5-17). The first inch (2.5 cm) of the duodenum resembles the stomach in that it is covered on its anterior and posterior surfaces with peritoneum and has the lesser omentum attached to its upper border and the greater omentum attached to its lower border; the lesser sac lies behind this short segment. The remainder of the duodenum is

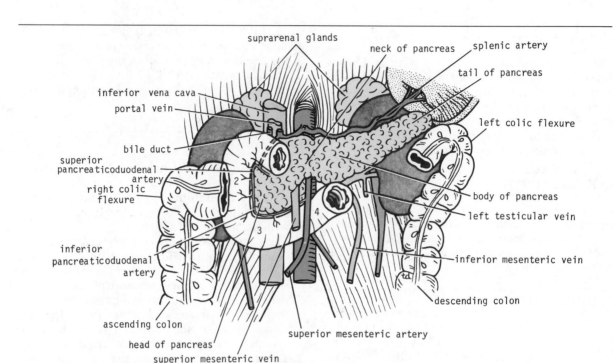

inferior vena cava
portal vein

bile duct

superior
pancreaticoduodenal
artery
right colic
flexure

inferior
pancreaticoduodenal
artery

ascending colon
head of pancreas
superior mesenteric vein

suprarenal glands

neck of pancreas    splenic artery

tail of pancreas

left colic flexure

body of pancreas

left testicular vein

inferior mesenteric vein

descending colon

superior mesenteric artery

**Fig. 5-17. Pancreas and anterior relations of kidneys.**

retroperitoneal, being only partially covered by peritoneum.

The duodenum is situated in the epigastric and umbilical regions and for purposes of description is divided into four parts:

## FIRST PART OF THE DUODENUM

The first part of the duodenum is 2 inches (5 cm) long (Figs. 5-17 and 5-18). It begins at the pylorus and runs upward and backward on the right side of the first lumbar vertebra. It thus lies on the transpyloric plane.

### *Relations*

### Anteriorly

The quadrate lobe of the liver and the gallbladder (Fig. 5-35).

### Posteriorly

The lesser sac (first inch only), the gastroduodenal artery, the bile duct and portal vein, and the inferior vena cava (Fig. 5-18).

### Superiorly

The entrance into the lesser sac (the epiploic foramen) (Figs. 5-7 and 5-13).

### Inferiorly

The head of the pancreas (Fig. 5-17).

## SECOND PART OF THE DUODENUM

The second part of the duodenum is 3 inches (8 cm) long. It runs vertically downward in front of the hilus of the right kidney on the right side of the second and third lumbar vertebrae (Figs. 5-17 and 5-18).

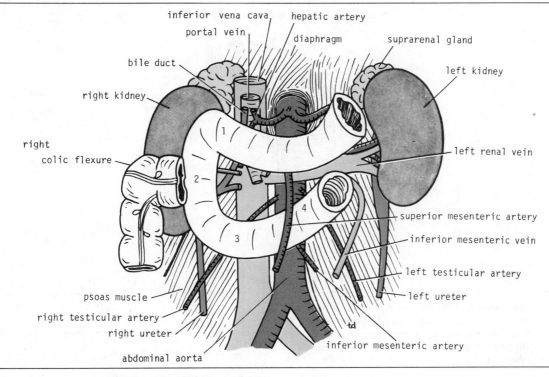

**Fig. 5-18. Posterior relations of duodenum and pancreas.**

## Relations

### Anteriorly

The fundus of the gallbladder and the right lobe of the liver, the transverse colon, and the coils of the small intestine (Fig. 5-36).

### Posteriorly

The hilus of the right kidney and the right ureter (Fig. 5-18).

### Laterally

The ascending colon, the right colic flexure, and the right lobe of the liver (Fig. 5-18).

### Medially

The head of the pancreas. About halfway down its posteromedial aspect, the bile duct and main pancreatic duct pierce the duodenal wall. The accessory pancreatic duct, if present, opens into the duodenum a little higher up (Figs. 5-18 and 5-19).

## THIRD PART OF THE DUODENUM

The third part of the duodenum is 3 inches (8 cm) long. It runs horizontally to the left on the subcostal plane, following the lower margin of the head of the pancreas (Figs. 5-17 and 5-18).

## Relations

### Anteriorly

The root of the mesentery of the small intestine, the superior mesenteric vessels contained within it, and coils of jejunum (Figs. 5-17 and 5-18).

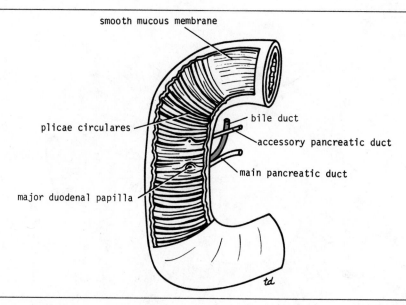

smooth mucous membrane

plicae circulares

bile duct

accessory pancreatic duct

main pancreatic duct

major duodenal papilla

td

**Fig. 5-19. Entrance of bile duct and main and accessory pancreatic ducts into second part of duodenum. Note smooth lining of first part of duodenum, plicae circulares of second part, and major duodenal papilla.**

## Posteriorly

The right ureter, the right psoas muscle, the inferior vena cava, and the aorta (Fig. 5-18).

## Superiorly

The head of the pancreas (Fig. 5-17).

## Inferiorly

Coils of jejunum.

## FOURTH PART OF THE DUODENUM

The fourth part of the duodenum is 2 inches (5 cm) long (Figs. 5-17 and 5-18). It runs upward and to the left and then turns forward at the *duodenojejunal junction*. Here, a well-marked peritoneal fold, the *ligament of Treitz*, ascends to the right crus of the diaphragm and holds the junction in position (Fig. 5-8). Note the position of the duodenal fossae. (See p. 213.)

## Relations

### Anteriorly

The beginning of the root of the mesentery and coils of jejunum (Fig. 5-20).

### Posteriorly

The left margin of the aorta and the medial border of the left psoas muscle (Fig. 5-18).

The *mucous membrane* of the duodenum is thick. In the first part of the duodenum it is smooth (Fig. 5-19). In the remainder of the duodenum it is thrown into numerous circular folds called the *plicae circulares*. At the site where the bile duct and the main pancreatic duct pierce the medial wall of the second part, there is a small, rounded elevation called the *major duodenal papilla* (Fig. 5-19). The accessory pancreatic duct, if present, opens into the duodenum on a smaller papilla about ¾ inch (1.9 cm) above the major duodenal papilla.

The *arterial supply* of the upper half of the duodenum is the superior pancreaticoduodenal artery, a branch of the gastroduodenal artery (Figs. 5-14 and 5-17). The lower half of the duodenum is supplied by the inferior pancreaticoduodenal artery, a branch of the superior mesenteric artery. The cor-

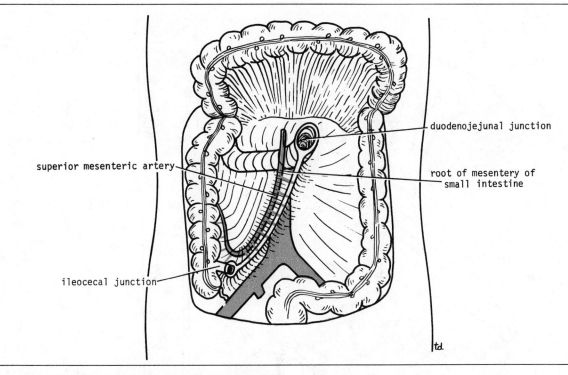

duodenojejunal junction

root of mesentery of small intestine

superior mesenteric artery

ileocecal junction

td

**Fig. 5-20. Attachment of root of mesentery of small intestine to posterior abdominal wall. Note that it extends from duodenojejunal junction on left of aorta, downward and to right to ileocecal junction. Superior mesenteric artery lies in root of mesentery.**

responding *veins of the duodenum* drain into the portal circulation, the superior vein joining the portal vein directly and the inferior vein joining the superior mesenteric vein (Fig. 5-29). The vessels lie between the concavity of the duodenum and the head of the pancreas.

The *lymph vessels* of the duodenum follow the arteries and drain upward via pancreaticoduodenal nodes to the gastroduodenal nodes and thence to the celiac nodes; and downward, via pancreaticoduodenal nodes to the superior mesenteric nodes around the origin of the superior mesenteric artery.

The *nerves of the duodenum* are derived from sympathetic and parasympathetic (vagus) nerves from the celiac and superior mesenteric plexuses.

## Jejunum and Ileum

The jejunum and ileum measure about 20 feet (6 m) long, the upper two-fifths of this length being the jejunum. Each has distinctive features, but there is a gradual change from one to the other. The jejunum begins at the duodenojejunal junction, and the ileum ends at the ileocecal junction.

The coils of jejunum and ileum are attached to the posterior abdominal wall by a fan-shaped fold of peritoneum known as the *mesentery of the small intestine* (Fig. 5-20). The long free edge of the fold encloses the mobile intestine. The short root of the fold is continuous with the parietal peritoneum on the posterior abdominal wall along a line that extends downward and to the right from the left side of the second lumbar vertebra to the region of the right sacroiliac joint. The root of the mesentery permits the entrance and exit of the branches of the superior mesenteric artery and vein, lymph vessels, and nerves into the space between the two layers of peritoneum forming the mesentery.

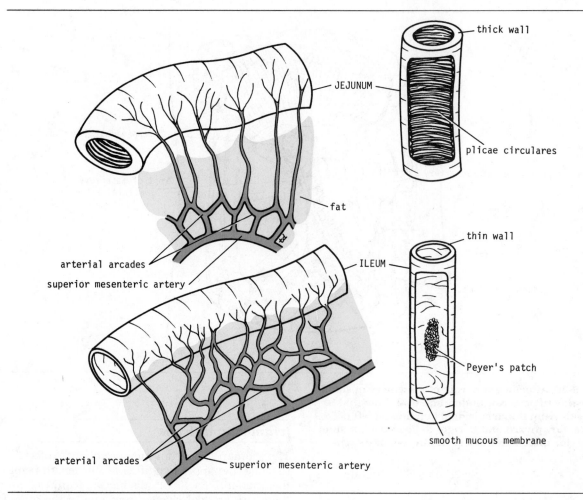

**Fig. 5-21. Some external and internal differences between jejunum and ileum.**

In the living the jejunum may be distinguished from the ileum by the following features:

1. The jejunum lies coiled in the upper part of the peritoneal cavity below the left side of the transverse mesocolon; the ileum is in the lower part of the cavity and in the pelvis (Fig. 5-3).
2. The jejunum is wider-bored, thicker-walled, and redder than the ileum. The jejunal wall feels thicker because the permanent infoldings of the mucous membrane, the plicae circulares, are larger, more numerous, and closely set in the jejunum; whereas in the upper part of the ileum they are smaller and more widely separated and in the lower part, they are absent (Fig. 5-21).
3. The jejunal mesentery is attached to the posterior abdominal wall above and to the left of the aorta, whereas the ileal mesentery is attached below and to the right of the aorta.
4. The jejunal mesenteric vessels form only one or two arcades, with long and infrequent branches passing to the intestinal wall. The ileum receives numerous short terminal vessels that arise from a series of three or four or even more arcades (Fig. 5-21).
5. At the jejunal end of the mesentery, the fat is deposited near the root and is scanty near the

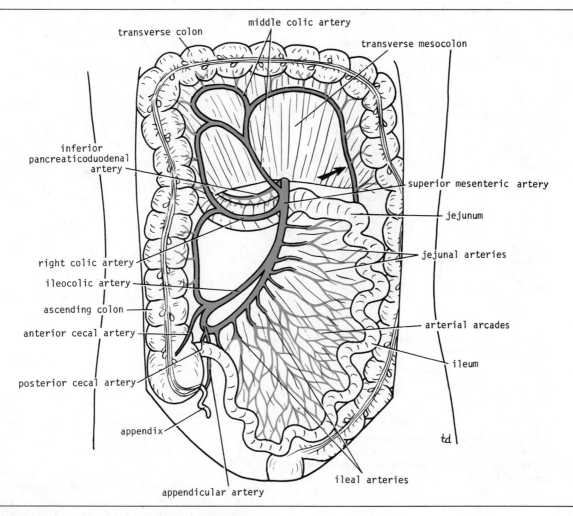

**Fig. 5-22. Superior mesenteric artery and its branches. Note that this artery supplies gut from halfway down second part of duodenum to distal third of transverse colon (arrow).**

intestinal wall. At the ileal end of the mesentery the fat is deposited throughout, so that it extends from the root to the intestinal wall (Fig. 5-21).

6. Aggregations of lymphoid tissue (Peyer's patches) are present in the mucous membrane of the lower ileum along the antimesenteric border (Fig. 5-21). In the living these may be visible through the wall of the ileum from the outside.

The *arterial supply of the jejunum and ileum* is from branches of the superior mesenteric artery (Fig. 5-22). The intestinal branches arise from the left side of the artery and run in the mesentery to reach the gut. They anastomose with one another to form a series of arcades. The lowest part of the ileum is also supplied by the ileocolic artery.

The *veins of the jejunum and ileum* correspond to the branches of the superior mesenteric artery and drain into the superior mesenteric vein (Fig. 5-29).

The *lymph vessels of the jejunum and ileum* pass through a large number of mesenteric nodes and finally reach the superior mesenteric nodes, which are situated around the origin of the superior mesenteric artery.

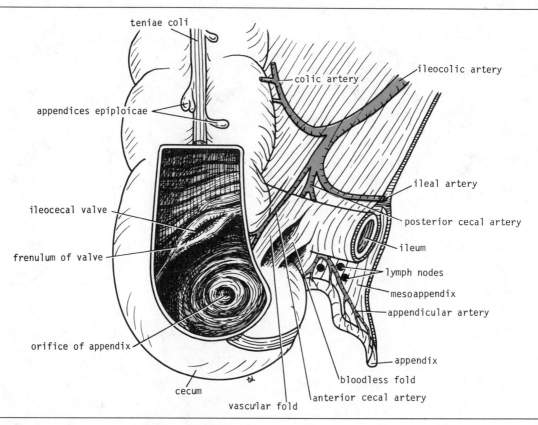

teniae coli

appendices epiploicae

colic artery

ileocolic artery

ileal artery

posterior cecal artery

ileocecal valve

frenulum of valve

ileum

lymph nodes

mesoappendix

appendicular artery

orifice of appendix

appendix

cecum

bloodless fold

anterior cecal artery

vascular fold

**Fig. 5-23. Cecum and vermiform appendix. Note that appendicular artery is a branch of posterior cecal artery. Edge of mesoappendix has been cut to show peritoneal layers.**

The *nerves of the jejunum and ileum* are derived from the sympathetic and parasympathetic (vagus) nerves from the superior mesenteric plexus.

## Large Intestine

The large intestine is divided into the cecum, the vermiform appendix, the ascending colon, the transverse colon, the descending colon, and the sigmoid colon. The rectum and anal canal will be considered in the sections on the pelvis and perineum. The primary function of the large intestine is the absorption of water and electrolytes and the storage of undigested material until it can be expelled from the body as feces.

## Cecum

The cecum is that part of the large intestine that lies below the level of the junction of the ileum with the large intestine (Figs. 5-22 and 5-23). It is situated in the right iliac fossa, is about 2½ inches (6 cm) long, and is completely covered with peritoneum. It possesses a considerable amount of mobility, although it does not have a mesentery. The presence of peritoneal folds in the vicinity of the cecum (Fig. 5-23) creates the superior ileocecal, the inferior ileocecal, and the retrocecal fossae. (See p. 213.)

As in the colon, the longitudinal muscle is restricted to three flat bands, the *teniae coli*, which converge on the base of the appendix and provide for it a complete longitudinal muscle coat (Fig 5-23). The cecum is often distended with gas and can then be palpated through the anterior abdominal wall in the living subject.

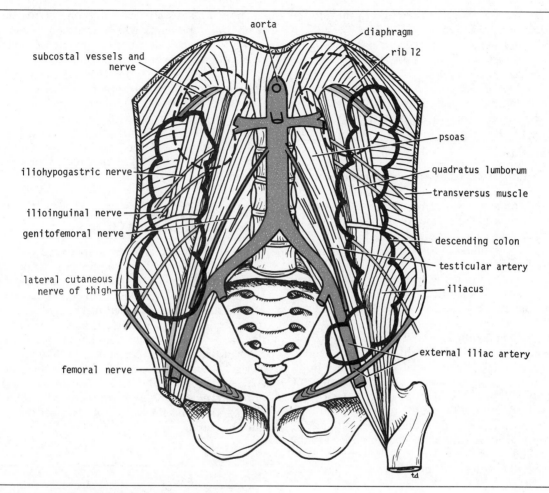

**Fig. 5-24. Posterior abdominal wall, showing posterior relations of kidneys and colon.**

The ileum enters the cecum at the junction of the cecum with the colon. The opening is provided with two folds, or lips, which form the so called ileocecal valve (see below). The vermiform appendix communicates with the cavity of the cecum through an opening that is located below and behind the ileocecal opening.

## Relations

### Anteriorly

Coils of small intestine, sometimes part of the greater omentum, and the anterior abdominal wall in the right iliac region.

### Posteriorly

The psoas and the iliacus muscles, the femoral nerve, and the lateral cutaneous nerve of the thigh (Fig. 5-24). The vermiform appendix is commonly found behind the cecum.

### Medially

The vermiform appendix arises from the cecum on its medial side (Fig. 5-23).

The *arterial supply* to the cecum is by way of the anterior and posterior cecal arteries, which are branches of the ileocolic artery, which is a branch of the superior mesenteric artery (Fig. 5-23). The *veins* correspond to the arteries and drain into the superior mesenteric vein.

The *lymph vessels* of the cecum pass through a number of mesenteric nodes and finally reach the superior mesenteric nodes.

The *nerves* of the cecum are derived from the sympathetic and parasympathetic (vagus) nerves from the superior mesenteric plexus.

## Appendix

The vermiform appendix is an organ containing a large amount of lymphoid tissue. It varies in length from 3 to 5 inches (8–13 cm). The base is attached to the posteromedial surface of the cecum about 1 inch (2.5 cm) below the ileocecal junction (Fig. 5-23). The remainder of the appendix is free. It has a complete peritoneal covering, which is attached to the lower layer of the mesentery of the small intestine to form a short mesentery of its own, the *mesoappendix*. The size of the mesoappendix is variable, and sometimes as much as the distal one-third of the appendix is devoid of a mesentery.

The appendix lies in the right iliac region, and in relation to anterior abdominal wall its base is situated one-third of the way up the line joining the right anterior superior iliac spine to the umbilicus (McBurney's point). Inside the abdomen the base of the appendix is easily found by identifying the teniae coli of the cecum and following them to the base of the appendix, where they converge to form a complete longitudinal muscular coat (Figs. 5-22 and 5-23).

The tip of the appendix is subject to a considerable range of movement and may be found in the following positions: (1) hanging down into the pelvis against the right pelvic wall; (2) coiled up behind the cecum in the retrocecal fossa; (3) projecting upward along the lateral side of the cecum; (4) in front of or behind the terminal part of the ileum. The first and second positions are the commonest sites.

The common congenital anomalies of the appendix are shown in Figure 5-32.

The *arterial supply of the appendix* is by means of the appendicular artery, a branch of the posterior cecal artery (Fig. 5-23). It is often double and passes to the tip of the appendix in the mesoappendix. The *appendicular vein* joins the posterior cecal vein.

The *lymph vessels* drain into one or two nodes lying in the mesoappendix. From there, the lymph passes through a number of mesenteric nodes to reach the superior mesenteric nodes.

The *nerves of the appendix* are derived from sympathetic and parasympathetic (vagus) nerves from the superior mesenteric plexus. Afferent nerve fibers concerned with the conduction of visceral pain from the appendix are believed to accompany the sympathetic nerves and enter the spinal cord at the level of the tenth thoracic segment.

## Ileocecal (Ileocolic) Valve

The ileocecal valve is a vestigial structure situated at the end of the ileum at the site of junction of the cecum with the colon (Fig. 5-23). It consists of two horizontal folds of mucous membrane that project around the orifice of the ileum. The folds meet medially and laterally in single ridges, the *frenula* of the valve. The circular and longitudinal muscle coats of the ileum are continued into the lips of the valve. The valve only partially controls the reflux of colic contents into the ileum. The circular muscle of the lower end of the ileum serves as a sphincter and controls the flow of contents from the ileum into the colon.

## Ascending Colon

The ascending colon is about 5 inches (13 cm) long and lies in the right iliac region (Fig. 5-25). It extends upward from the cecum to the inferior surface of the right lobe of the liver, where it turns sharply to the left, forming the *right colic flexure*, and becomes continuous with the transverse colon. The peritoneum covers the front and the sides of the ascending colon and binds it to the posterior abdominal wall.

### Relations

#### Anteriorly

Coils of small intestine, the greater omentum, and the anterior abdominal wall (Figs. 5-2 and 5-3).

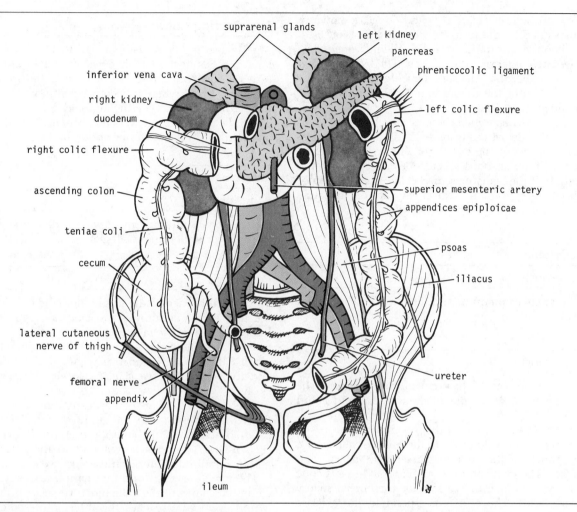

**Fig. 5-25. Abdominal cavity, showing terminal part of ileum, cecum, appendix, ascending colon, right colic flexure, left colic flexure, and descending colon. Note teniae coli and appendices epiploicae.**

## Posteriorly

The iliacus, the iliac crest, the quadratus lumborum, the origin of the transversus abdominis muscle, and the lower pole of the right kidney. The iliohypogastric and the ilioinguinal nerves cross behind it (Fig. 5-24).

The *arterial supply* to the ascending colon is from the ileocolic and right colic branches of the superior mesenteric artery (Fig. 5-22). The *veins* correspond to the arteries and drain into the superior mesenteric vein.

The *lymph vessels* drain into lymph nodes lying along the course of the colic blood vessels and ultimately reach the superior mesenteric nodes.

The *nerves of the ascending colon* are derived from sympathetic and parasympathetic (vagus) nerves from the superior mesenteric plexus.

### Transverse Colon

The transverse colon is about 15 inches (38 cm) long and extends across the abdomen, occupying the umbilical and hypogastric regions. It begins at the right colic (hepatic) flexure below the right lobe of the liver (Fig. 5-12) and hangs downward, suspended by the transverse mesocolon (Fig. 5-6). It then ascends to the *left colic* (splenic) *flexure* immediately inferior to the spleen. The left colic flexure is higher than the right colic flexure and is suspended from the diaphragm by the *phrenicocolic ligament* (Fig. 5-25).

The *transverse mesocolon* is attached to the superior border of the transverse colon, and the posterior layers of the greater omentum are attached to the inferior border (Fig. 5-6). Because of the length of the transverse mesocolon, the position of the transverse colon is extremely variable and may reach the pelvis.

## Relations

### Anteriorly

The greater omentum and the anterior abdominal wall (Fig. 5-6).

### Posteriorly

The second part of the duodenum, the head of the pancreas, and the coils of the jejunum and ileum (Fig. 5-25).

The *arterial supply* of the proximal two-thirds of the transverse colon is from the middle colic artery, a branch of the superior mesenteric artery (Fig. 5-22). The distal third is supplied by the left colic artery, a branch of the inferior mesenteric artery (Fig. 5-26). The *veins* correspond to the arteries and drain into the superior and inferior mesenteric veins.

The *lymph vessels* drain into lymph nodes lying along the course of the colic blood vessels. Lymph from the proximal two-thirds of the transverse colon will drain into the superior mesenteric nodes, while that from the region of the distal third will drain into the inferior mesenteric nodes.

The *nerves* supplying the transverse colon are derived from sympathetic and vagus nerves and from the pelvic parasympathetic nerves. The sympathetic fibers pass from the superior and inferior mesenteric plexuses. The fibers of the vagi supply only the proximal two-thirds of the transverse colon; the distal third is supplied by the pelvic parasympathetic nerves.

## Descending Colon

The descending colon is about 10 inches (25 cm) long and lies in the left iliac region (Fig. 5-25). It extends downward from the left colic flexure, to the pelvic brim, where it becomes continuous with the sigmoid colon. (For the sigmoid colon, see p. 331.) The peritoneum covers the front and the sides and binds it to the posterior abdominal wall.

## Relations

### Anteriorly

Coils of small intestine, the greater omentum, and the anterior abdominal wall (Figs. 5-2 and 5-3).

### Posteriorly

The lateral border of the left kidney, the origin of the transversus abdominis muscle, the quadratus lumborum, the iliac crest, the iliacus, and the left psoas. The iliohypogastric and the ilioinguinal nerves, the lateral cutaneous nerve of the thigh, and the femoral nerve (Fig. 5-24) also lie posteriorly.

The *arterial supply* to the descending colon is from the left colic and the sigmoid (inferior left colic) branches of the inferior mesenteric artery (Fig. 5-26). The *veins* correspond to the arteries and drain into the inferior mesenteric vein.

The *lymph vessels* drain into lymph nodes lying along the course of the colic blood vessels, and the lymph eventually reaches the inferior mesenteric nodes around the origin of the inferior mesenteric artery.

The *nerves* supplying the descending colon are derived from sympathetic fibers from the inferior mesenteric plexus and from pelvic parasympathetic nerves.

# Blood Supply of the Gastrointestinal Tract

The arterial supply to the gut and its relationship to the development of the different parts of the gut are illustrated diagrammatically in Figure 5-27. The celiac artery is the artery of the foregut and supplies the gastrointestinal tract from the lower

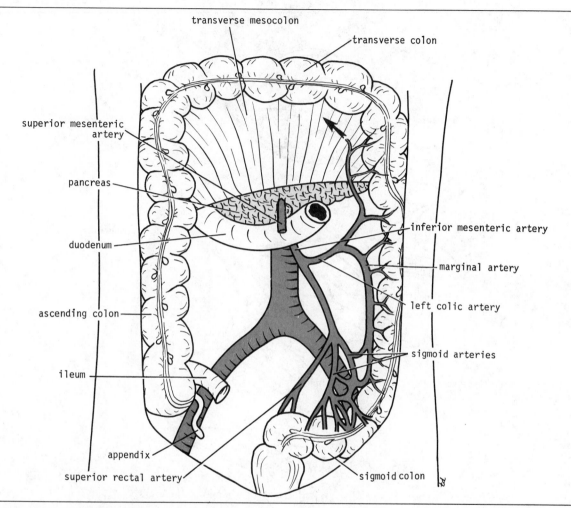

**Fig. 5-26. Inferior mesenteric artery and its branches. Note that this artery supplies large bowel from distal third of transverse colon to half-way down anal canal. It anastomoses with middle colic branch of superior mesenteric artery (arrow).**

one-third of the esophagus down as far as the middle of the second part of the duodenum. The superior mesenteric artery is the artery of the midgut and supplies the gastrointestal tract from the middle of the second part of the duodenum as far as the distal one-third of the transverse colon. The inferior mesenteric artery is the artery of the hindgut and supplies the large intestine from the distal one-third of the transverse colon to halfway down the anal canal.

## Celiac Artery

The celiac artery and its branches are described on page 223.

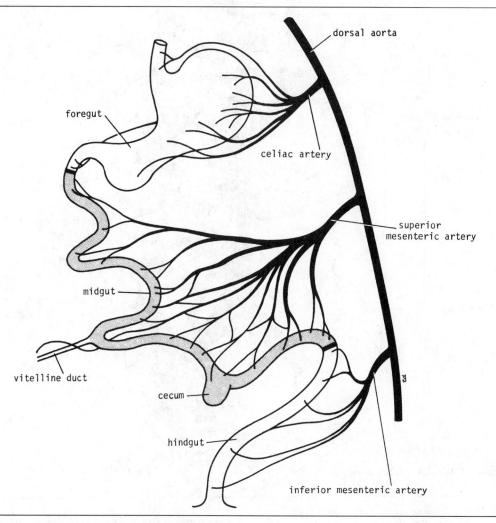

**Fig. 5-27. Arterial supply of developing gastrointestinal tract in fetus. Celiac artery supplies foregut; superior mesenteric artery, midgut (shaded); and inferior mesenteric artery, hindgut.**

## Superior Mesenteric Artery

The superior mesenteric artery supplies the distal part of the duodenum, the jejunum, ileum, cecum, appendix, ascending colon, and most of the transverse colon. It arises from the front of the abdominal aorta just below the celiac artery (Fig. 5-22). It runs downward and to the right behind the neck of the pancreas and in front of the third part of the duodenum. It continues downward to the right between the layers of the mesentery of the small intestine and ends by anastomosing with the ileal branch of its own ileocolic branch.

## Branches

1. The *inferior pancreaticoduodenal artery* passes to the right as a single or double branch along the upper border of the third part of the duodenum and the head of the pancreas. It supplies the pancreas and the adjoining part of the duodenum.
2. The *middle colic artery* runs forward in the transverse mesocolon to supply the transverse colon. It divides into right and left branches.
3. The *right colic artery* is often a branch of the ileocolic artery. It passes to the right to supply the ascending colon. It divides into ascending and descending branches.
4. The *ileocolic artery* passes downward and to the right. It gives rise to a *superior branch* that anastomoses with the right colic artery and an *inferior branch* that anastomoses with the end of the superior mesenteric artery. The inferior branch gives rise to the *anterior* and *posterior cecal arteries;* the *appendicular artery* is a branch of the posterior cecal artery.
5. *Jejunal and ileal branches.* These branches are twelve to fifteen in number and arise from the left side of the superior mesenteric artery. Each artery divides into two, which unite with adjacent branches to form a series of arcades. Branches from the arcades divide and unite to form a second, third, and fourth series of arcades. Fewer arcades supply the jejunum as compared with the ileum. From the terminal arcades small straight vessels supply the intestine.

## Inferior Mesenteric Artery

The inferior mesenteric artery supplies the distal third of the transverse colon, the left colic flexure, descending colon, sigmoid colon, rectum, and upper half of the anal canal. It arises from the abdominal aorta about 1½ inches (3.8 cm) above its bifurcation (Fig. 5-26). The artery runs downward and to the left and crosses the left common iliac artery. Here, it changes its name and becomes the superior rectal artery.

## Branches

1. The *left colic artery* runs upward and to the left and supplies the distal third of the transverse colon, the left colic (splenic) flexure, and the upper part of the descending colon. It divides into ascending and descending branches.
2. *Sigmoid arteries.* These are two or three in number and supply the descending and sigmoid colon.
3. The *superior rectal artery* is a continuation of the inferior mesenteric artery as it crosses the left common iliac artery. It descends into the pelvis behind the rectum. The artery supplies the rectum and upper half of the anal canal and anastomoses with the middle rectal and inferior rectal arteries.

## MARGINAL ARTERY

The anastomosis of the colic arteries around the concave margin of the large intestine forms a single arterial trunk called the *marginal artery.* This begins at the ileocecal junction, where it anastomoses with the ileal branches of the superior mesenteric artery, and it ends where it anastomoses less freely with the superior rectal artery (Fig. 5-26).

## Veins

The venous blood from the greater part of the gastrointestinal tract and its accessory organs drains to the liver by the portal venous system.

The proximal tributaries drain directly into the portal vein, but the veins forming the distal tributaries correspond to the branches of the celiac artery and the superior and inferior mesenteric arteries.

## Portal Vein

This important vein is about 2 inches (5 cm) long and is formed behind the neck of the pancreas by the union of the superior mesenteric and the splenic veins (Fig. 5-28). It runs upward and to the right, posterior to the first part of the duodenum, and enters the lesser omentum (Figs. 5-7 and 5-13). It then ascends in front of the opening into the lesser sac to the porta hepatis, where it divides into right and left terminal branches.

The portal circulation begins as a capillary plexus in the organs it drains and ends by emptying its blood into sinusoids within the liver. The portal vein drains blood from the gastrointestinal tract

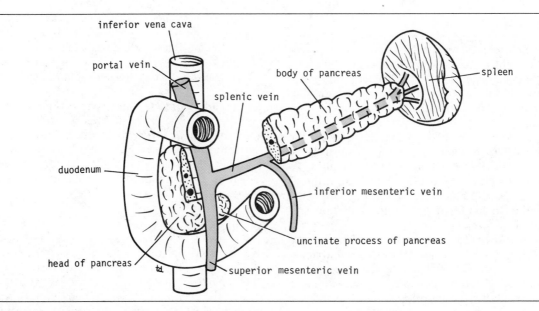

**Fig. 5-28. Formation of portal vein behind neck of pancreas.**

from the lower end of the esophagus to halfway down the anal canal; from the pancreas, gallbladder, and bile ducts; and from the spleen (Fig. 5-29).

For the relations of the portal vein in the lesser omentum, see page 212.

## PORTAL-SYSTEMIC ANASTOMOSES

Under normal conditions the portal venous blood traverses the liver and drains into the inferior vena cava of the systemic venous circulation by way of the hepatic veins. This is the direct route. However, other, smaller communications exist between the portal and systemic systems, which become important should the direct route become blocked (Fig. 5-30).

These communications are as follows:

1. At the lower third of the esophagus, the esophageal branches of the left gastric vein (portal tributary) anastomose with the esophageal veins draining the middle third of the esophagus into the azygos veins (systemic tributary).
2. Halfway down the anal canal, the superior rectal veins (portal tributary) draining the upper half of the anal canal anastomose with the middle

and inferior rectal veins (systemic tributaries), which are tributaries of the internal iliac and internal pudendal veins, respectively.
3. The paraumbilical veins connect the left branch of the portal vein with the superficial veins of the anterior abdominal wall (systemic tributaries). The paraumbilical veins travel in the falciform ligament and accompany the ligamentum teres.
4. The veins of the ascending colon, descending colon, duodenum, pancreas, and liver (portal tributary) anastomose with the renal, lumbar, and phrenic veins (systemic tributaries).

## Differences Between the Small and Large Intestine

### External Differences (Fig. 5-31)

1. The small intestine (with the exception of the duodenum) is mobile, while the ascending and descending parts of the colon are fixed.
2. The caliber of the full small intestine is normally smaller than that of the filled large intestine.
3. The small intestine (with the exception of the duodenum) has a mesentery that passes downward across the midline into the right iliac fossa.
4. The longitudinal muscle of the small intestine forms a continuous layer around the gut. In the

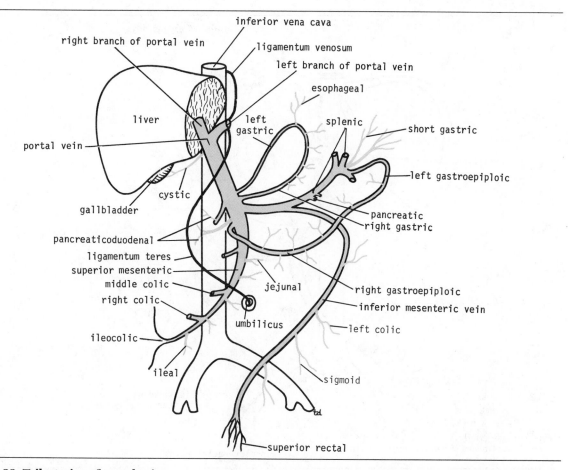

**Fig. 5-29. Tributaries of portal vein.**

large intestine (with the exception of the appendix) the longitudinal muscle is collected into three bands, the teniae coli.

5. The small intestine has no fatty tags attached to its wall. The large intestine has fatty tags, called the *appendices epiploicae.*

6. The wall of the small intestine is smooth, whereas that of the large intestine is sacculated.

*Internal Differences (Fig. 5-31)*

1. The mucous membrane of the small intestine has permanent folds, called *plicae circulares,* which are absent in the large intestine.

2. The mucous membrane of the small intestine has villi, which are absent in the large intestine.

3. Aggregations of lymphoid tissue called Peyer's patches are found in the mucous membrane of the small intestine; these are absent in the large intestine.

## Congenital Anomalies of the Gastrointestinal Tract

Some of the more common congenital anomalies of the gastrointestinal tract are shown in Figure 5-32.

## LIVER

The liver is the largest gland in the body and has a wide variety of functions. Three of its basic functions are (1) the production and secretion of bile,

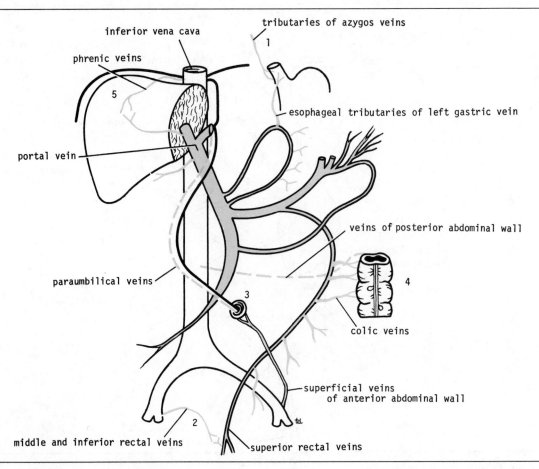

inferior vena cava

tributaries of azygos veins

phrenic veins

1

5

esophageal tributaries of left gastric vein

portal vein

veins of posterior abdominal wall

paraumbilical veins

3

4

colic veins

superficial veins
of anterior abdominal wall

2

middle and inferior rectal veins

superior rectal veins

**Fig. 5-30. Important portal-systemic anastomoses. (See text.)**

which is passed into the intestinal tract, (2) the involvement in many metabolic activities related to carbohydrate, fat, and protein metabolism, and (3) the filtration of the blood removing bacteria and other foreign particles that have gained entrance to the blood from the lumen of the intestine. The liver is soft and pliable and occupies the right hypochondrium, extending into the epigastrium (Fig. 5-2). The greater part of the liver is situated under cover of the ribs and costal cartilages and is in contact with the diaphragm, which separates it from the pleura, lungs, pericardium, and heart. The convex upper surface of the liver is molded to the undersurface of the domes of the diaphragm. The *pos-*

*tero-inferior*, or *visceral surface* is molded to adjacent viscera and is therefore irregualr in shape; it lies in contact with the abdominal part of the esophagus, the stomach, the duodenum, the right colic flexure, the right kidney and suprarenal gland, and the gallbladder.

For purposes of description it is customary to divide the liver into a large *right lobe* and a small *left lobe* by the attachment of the peritoneum of the falciform ligament (Fig. 5-33). The right lobe is further divided into a *quadrate lobe* and a *caudate lobe* by the presence of the gallbladder, the fissure for the ligamentum teres, the inferior vena cava, and the fissure for the ligamentum venosum. Experiments have shown that, in fact, the quadrate and caudate lobes are a functional part of the left lobe of the liver. Thus, the right and left branches

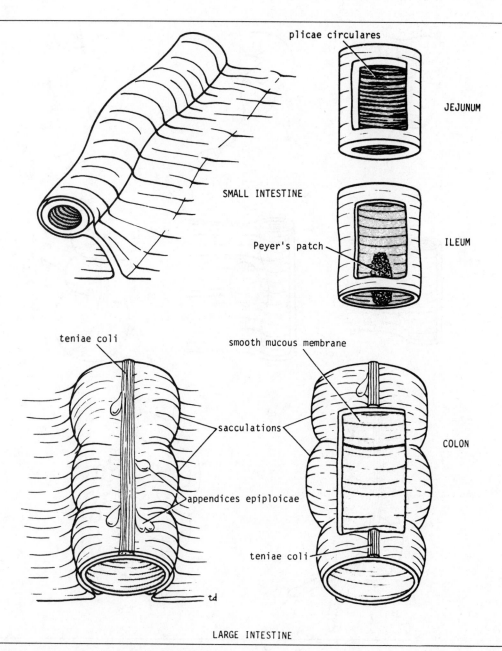

plicae circulares

JEJUNUM

SMALL INTESTINE

Peyer's patch

ILEUM

teniae coli

smooth mucous membrane

sacculations

COLON

appendices epiploicae

teniae coli

td

LARGE INTESTINE

**Fig. 5-31. Some external and internal differences between small and large intestine.**

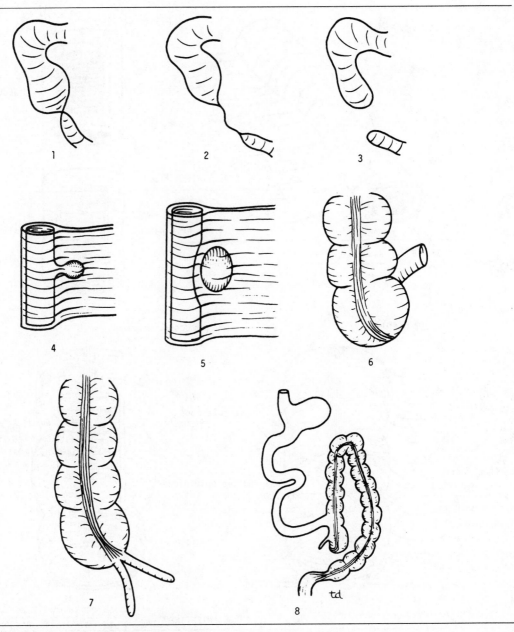

Fig. 5-32. Some common congenital anomalies of intestinal tract. (1–3) Congenital atresias of small intestine; (4) diverticulum of duodenum or jejunum; (5) mesenteric cyst of small intestine; (6) absence of vermiform appendix; (7) double appendix; (8) malrotation of gut, with appendix lying in left iliac fossa. For Meckel's diverticulum, see Figure 4-30.

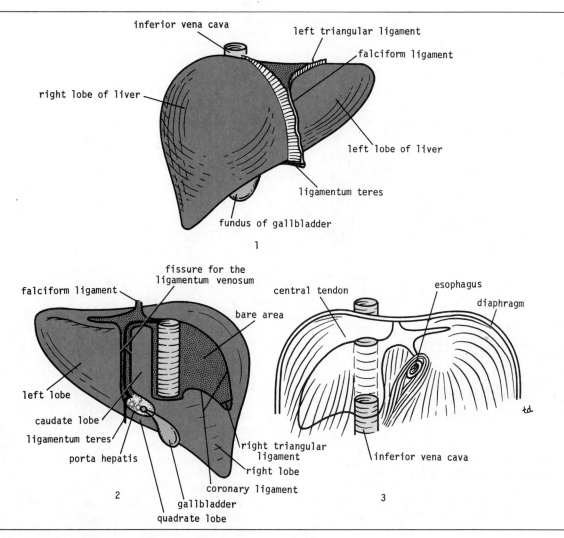

**Fig. 5-33. Liver as seen from in front (1) and from behind (2). On posterior surface of liver, note peritoneal reflections onto diaphragm (3).**

of the hepatic artery and portal vein, and the right and left hepatic ducts, are distributed to the right lobe and the left lobe (plus quadrate plus caudate lobes), respectively. There is apparently very little overlap between the two sides.

The *porta hepatis*, or hilus of the liver, is found on the postero-inferior surface (Fig. 5-33). The upper part of the free edge of the lesser omentum is attached to its margins. In it lie the right and left hepatic ducts, the right and left branches of the hepatic artery, the portal vein, and sympathetic and parasympathetic nerve fibers (Fig. 5-34). A few hepatic lymph nodes lie here; they drain the liver and gallbladder and send their efferent vessels to the celiac lymph nodes.

The liver, surrounded by a fibrous capsule, is made up of *liver lobules.* The *central vein* of each lobule is a tributary of the hepatic veins. In the spaces between the lobules are the *portal canals.* These contain branches of the hepatic artery, portal vein, and a tributary of a bile duct (portal triad). The arterial and venous blood passes be-

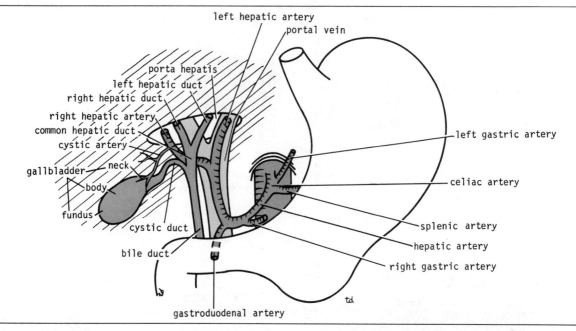

**Fig. 5-34. Structures entering and leaving porta hepatis.**

tween the liver cells by means of *sinusoids* and drains into the central vein.

## Peritoneal Attachments and Ligaments of the Liver

The *falciform ligament*, which is a two-layered fold of the peritoneum, ascends from the umbilicus to the liver (Figs. 5-4 and 5-33). Within the falciform ligament lies the ligamentum teres, the remains of the umbilical vein (left umbilical vein). The falciform ligament passes on to the anterior and then to the superior surfaces of the liver and then splits into two layers. The right layer forms the upper layer of the *coronary ligament*; the left layer forms the upper layer of the *left triangular ligament* (Fig. 5-33). The right extremity of the coronary ligament is known as the *right triangular ligament* of the liver. It should be noted that the peritoneal layers forming the coronary ligament are widely separated, leaving an area of liver devoid of peritoneum. Such an area is referred to as a *"bare" area of the liver* (Fig. 5-33).

The *ligamentum teres* passes into a fissure on the visceral surface of the liver and joins the left branch of the portal vein in the porta hepatis (Figs. 5-29 and 5-33). The *ligamentum venosum*, a fibrous band that is the remains of the *ductus venosus*, is attached to the left branch of the portal vein and ascends in a fissure on the visceral surface of the liver to be attached above to the inferior vena cava (Figs. 5-29 and 5-33). In the fetus, oxygenated blood is brought to the liver in the umbilical vein (ligamentum teres). The greater proportion of the blood bypasses the liver in the ductus venosus (ligamentum venosum) and joins the inferior vena cava. At birth, the umbilical vein and ductus venosus close and become fibrous cords.

The *lesser omentum* arises from the edges of the porta hepatis and the fissure for the ligamentum venosum and passes down to the lesser curvature of the stomach (Fig. 5-35).

## Blood and Lymph Vessels and Nerve Supply of the Liver

### BLOOD VESSELS

The *blood vessels* (Fig. 5-34) conveying blood to the liver are the hepatic artery (30 percent) and portal

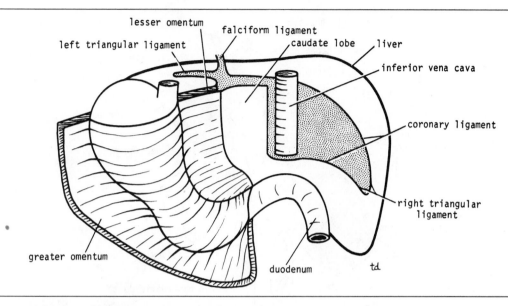

left triangular ligament

lesser omentum

falciform ligament

caudate lobe

liver

inferior vena cava

coronary ligament

right triangular
ligament

greater omentum

duodenum

td

**Fig. 5-35. Attachment of lesser omentum to stomach and posterior surface of liver.**

vein (70 percent). The hepatic artery brings oxygenated blood to the liver, while the portal vein brings venous blood rich in the products of digestion, which have been absorbed from the gastrointestinal tract. The arterial and venous blood is conducted to the central vein of each liver lobule by the liver sinusoids. The central veins drain into the right and left *hepatic veins,* and these leave the posterior surface of the liver and open directly into the inferior vena cava.

## LYMPH VESSELS

The liver produces a large amount of lymph—about one-third to one-half of all body lymph. The *lymph vessels* leave the liver and enter a number of lymph nodes in the porta hepatis. The efferent vessels pass to the celiac nodes. A small number of vessels pass from the bare area of the liver through the diaphragm to the posterior mediastinal lymph nodes.

## NERVE SUPPLY

The *nerve supply* of the liver is derived from the sympathetic and parasympathetic nerves by way of

the celiac plexus. The anterior vagal trunk gives rise to a large hepatic branch, which passes directly to the liver.

## EXTRAHEPATIC BILIARY APPARATUS

Bile is secreted by the liver cells, stored, and concentrated in the gallbladder; later it is delivered to the intestinal tract. The extrahepatic biliary apparatus consists of the *right* and *left hepatic ducts,* the *common hepatic duct,* the *bile duct,* the *gallbladder,* and the *cystic duct.*

The smallest interlobular tributaries of the bile ducts are situated in the portal canals of the liver; they receive the bile canaliculi. The interlobular ducts join one another to form progressively larger ducts and, eventually, at the porta hepatis, from the right and left hepatic ducts. The right hepatic duct drains the right lobe of the liver and the left duct drains the left lobe, caudate lobe, and quadrate lobe.

## Hepatic Ducts

The *right* and *left hepatic ducts* emerge from the liver at the porta hepatis (Fig. 5-34). After a short

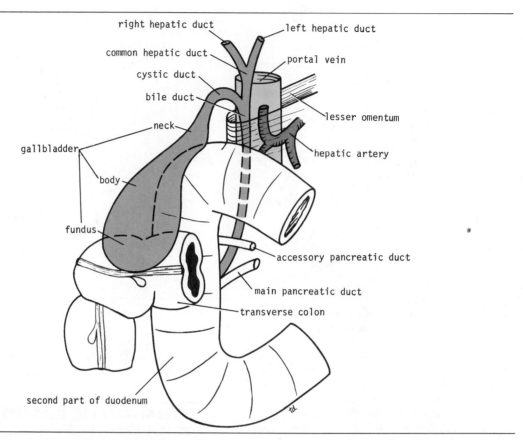

**Fig. 5-36. Different parts of extrahepatic biliary apparatus. Note relation of gallbladder to transverse colon and duodenum.**

course the hepatic ducts unite to form the common hepatic duct (Fig. 5-36).

The *common hepatic duct* is about 1.5 inches (4 cm) long and descends within the free edge of the lesser omentum. It is joined on the right side by the cystic duct from the gallbladder to form the bile duct (Fig. 5-36).

## Bile Duct

The bile duct (common bile duct) is about 3 inches (8 cm) long. In the first part of its course it lies in the right free edge of the lesser omentum in front of the opening into the lesser sac. Here it lies in front of the right margin of the portal vein and on the right of the hepatic artery (Fig. 5-13). In the

second part of its course it is situated behind the first part of the duodenum (Fig. 5-7) to the right of the gastroduodenal artery (Fig. 5-12). In the third part of its course it lies in a groove on the posterior surface of the head of the pancreas (Fig. 5-36). Here, the bile duct comes into contact with the main pancreatic duct.

The bile duct ends below by piercing the medial wall of the second part of the duodenum about halfway down its length (Fig. 5-37). It is usually joined by the main pancreatic duct, and together they open into a small ampulla in the duodenal wall, called the *ampulla of Vater*. The ampulla itself opens into the lumen of the duodenum by means of a small papilla, the *major duodenal papilla* (Fig. 5-37). The terminal parts of both ducts and the ampulla are surrounded by circular muscle fibers, known as the *sphincter of Oddi* (Fig. 5-37). Occasionally, the bile and pancreatic ducts open

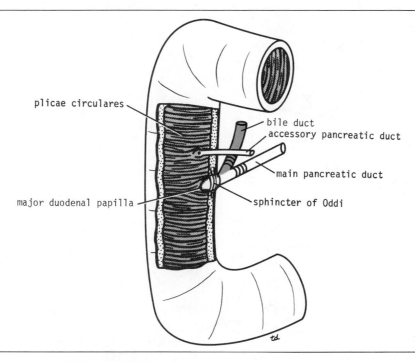

plicae circulares

bile duct
accessory pancreatic duct

main pancreatic duct

major duodenal papilla

sphincter of Oddi

**Fig. 5-37. Terminal parts of bile and pancreatic ducts as they enter second part of duodenum. Note sphincter of Oddi and smooth muscle around ends of bile duct and main pancreatic duct.**

separately into the duodenum. The common variations of this arrangement are shown diagrammatically in Figure 5-38.

## Gallbladder

The *gallbladder* is a pear-shaped sac lying on the visceral surface of the liver (Figs. 5-33 and 5-36). For descriptive purposes it is divided into the *fundus*, *body*, and *neck*. The *fundus* is rounded and usually projects below the inferior margin of the liver, where it comes in contact with the anterior abdominal wall at the level of the tip of the ninth right costal cartilage. The *body* lies in contact with the visceral surface of the liver and is directed upward, backward, and to the left. The *neck* becomes continuous with the cystic duct, which turns into the lesser omentum to join the right side of the common hepatic duct, to form the bile duct (Fig. 5-36).

The peritoneum completely surrounds the fundus of the gallbladder and binds the body and neck to the visceral surface of the liver.

### RELATIONS OF GALLBLADDER

#### Anteriorly

The anterior abdominal wall and the visceral surface of the liver (Fig. 5-2).

#### Posteriorly

The transverse colon and the first and second parts of the duodenum (Fig. 5-36).

### FUNCTION OF GALLBLADDER

The gallbladder serves as a reservoir for bile, with a capacity of about 50 ml. It has the ability to concentrate the bile, and to aid this process the mucous membrane is thrown into permanent folds that unite with each other, giving the surface a honeycombed appearance. The columnar cells lin-

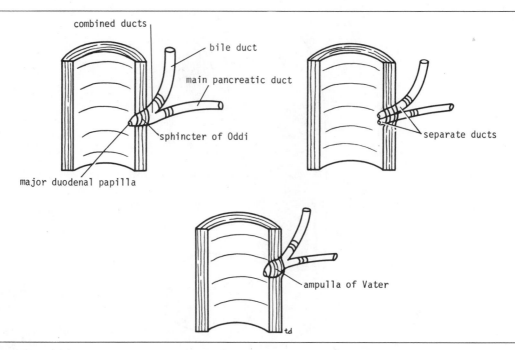

**Fig. 5-38. Three common variations of terminations of bile and main pancreatic ducts as they enter second part of duodenum.**

ing the surface also have numerous microvilli on their free surface.

Bile is delivered to the duodenum as the result of contraction and partial emptying of the gallbladder. This mechanism is initiated by the entrance of fatty foods into the duodenum. The fat causes the release of the hormone *cholecystokinin* from the mucous membrane of the duodenum; the hormone then enters the blood, causing the gallbladder to contract. At the same time the smooth muscle around the distal end of the bile duct and the ampulla is relaxed, thus allowing the passage of concentrated bile into the duodenum. The bile salts in the bile are important in emulsifying the fat in the intestine and assisting in its digestion and absorption.

The *arterial supply* of the gallbladder is from the *cystic artery*, a branch of the right hepatic artery (Fig. 5-34). The *cystic vein* drains directly into the portal vein. A number of very small arteries and veins also run between the liver and gallbladder.

The *lymph vessels* pass to a *cystic lymph node* situated near the neck of the gallbladder. From here, the lymph vessels pass by way of the hepatic nodes along the course of the hepatic artery to the celiac nodes.

The *nerves* to the gallbladder are derived from the celiac plexus.

## Cystic Duct

The *cystic duct* is about 1.5 inches (4 cm) long and connects the neck of the gallbladder to the common hepatic duct to form the bile duct (Fig. 5-36). It is usually somewhat S-shaped and descends for a variable distance in the right free edge of the lesser omentum.

The mucous membrane of the cystic duct is raised to form a spiral fold that is continuous with a similar fold in the neck of the gallbladder. The fold is commonly known as the "spiral valve." The function of the spiral valve is believed to be strengthening of the wall and assisting in keeping the lumen open.

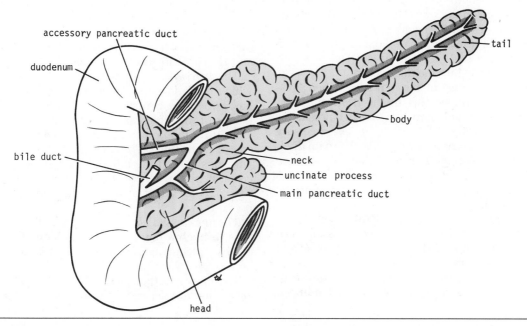

**Fig. 5-39. Different parts of pancreas dissected to reveal duct system.**

# PANCREAS

The pancreas is both an exocrine and an endocrine gland. The exocrine portion of the gland produces a secretion that contains enzymes that are capable of hydrolyzing proteins, fats, and carbohydrates. The endocrine portion of the gland, the *islets of Langerhans*, produces the hormones *insulin* and *glucagon* that play a key role in carbohydrate metabolism.

The pancreas is a soft, lobulated organ that lies on the posterior abdominal wall behind the peritoneum. It crosses the transpyloric plane. For purposes of description it is divided into a *head, neck, body*, and *tail* (Fig. 5-39).

The *head* of the pancreas is disc-shaped and lies within the concavity of the duodenum (Figs. 5-17 and 5-39). A part of the head extends to the left behind the superior mesenteric vessels and is called the *uncinate process.*

The *neck* is the constricted portion of the pancreas and connects the head to the body. It lies in front of the beginning of the portal vein and the origin of the superior mesenteric artery from the aorta (Fig. 5-17).

The *body* runs upward and to the left across the midline (Fig. 5-12). It is somewhat triangular in cross section.

The *tail* passes forward in the lienorenal ligament and comes in contact with the hilus of the spleen (Fig. 5-12).

## Relations

### Anteriorly

From right to left: the transverse colon and the attachment of the transverse mesocolon, the lesser sac, and the stomach (Figs. 5-6 and 5-12).

### Posteriorly

From right to left: the bile duct, the portal and splenic veins, the inferior vena cava, the aorta, the origin of the superior mesenteric artery, the left psoas muscle, the left suprarenal gland, the left kidney, and the hilus of the spleen (Figs. 5-12 and 5-18).

The *main duct of the pancreas* begins in the tail

and runs the length of the gland, receiving numerous tributaries on the way (Fig. 5-39). It pierces the posteromedial wall of the second part of the duodenum at about its middle and opens into the ampulla of Vater with the bile duct (Fig. 5-37) or drains separately into the duodenum.

The *accessory duct* of the pancreas, when present, drains the upper part of the head and then opens into the duodenum a short distance above the main duct (Figs. 5-37 and 5-39). The accessory duct frequently communicates with the main duct.

The *arterial supply* is from the splenic and the superior and inferior pancreaticoduodenal arteries (Fig. 5-17). The corresponding *veins* drain into the portal system.

The *lymph vessels* drain into lymph nodes situated along the arteries that supply the gland. The efferent vessels ultimately drain into the celiac and superior mesenteric lymph nodes.

The *nerve supply* to the pancreas is from sympathetic and parasympathetic (vagal) nerve fibers.

# SPLEEN

The spleen is reddish in color and lies in the left hypochondrium. Its long axis lies along the shaft of the tenth rib, and its lower pole extends forward only as far as the midaxillary line and cannot be palpated on clinical examination (Fig. 5-40). The spleen is the largest single mass of lymphoid tissue in the body and is generally ovoid in shape. It has a *notched anterior border*.

The spleen is surrounded by peritoneum (Fig. 5-5 and 5-40), which passes from it at the hilus as the gastrosplenic omentum (ligament) to the greater curvature of the stomach (carrying the short gastric and left gastroepiploic vessels). The peritoneum also passes to the left kidney as the lienorenal ligament (carrying the splenic vessels and the tail of the pancreas).

## *Relations*

### Anteriorly

The stomach, the tail of the pancreas, and the left colic flexure. The left kidney lies along its medial border (Figs. 5-12 and 5-13).

### Posteriorly

The diaphragm, the left pleura (left costodiaphragmatic recess), the left lung, and the ninth, tenth, and eleventh ribs (Figs. 5-13 and 5-40).

The *arterial supply* is from the splenic artery, which is the largest branch of the celiac artery. It has a tortuous course as it runs along the upper border of the pancreas. The splenic artery then divides up into about six branches, which enter the spleen at the hilus. The splenic vein leaves the hilus and runs behind the tail and the body of the pancreas. Behind the neck of the pancreas the splenic vein joins the superior mesenteric vein to form the portal vein.

The *lymph vessels* emerge from the hilus and pass through a few lymph nodes along the course of the splenic artery and drain into the celiac nodes.

The *nerves* to the spleen accompany the splenic artery and are derived from the celiac plexus.

# URINARY TRACT

## Kidneys

The two kidneys function to excrete most of the waste products of metabolism. They play a major role in controlling the water and electrolyte balance within the body and maintaining the acid-base balance of the blood. The waste products leave the kidneys as *urine*, which passes down the *ureters* to the *urinary bladder*, located within the pelvis. The urine leaves the body in the *urethra*.

The kidneys are reddish-brown in color and lie behind the peritoneum high up on the posterior abdominal wall, largely under cover of the costal margin (Fig. 5-41). The right kidney lies slightly lower than the left kidney, due to the bulk of the right lobe of the liver. With contraction of the diaphragm during respiration, both kidneys move downward in a vertical direction by as much as 1 inch (2.5 cm). On the medial border of each kidney is a vertical slit, which is bounded by thick lips of renal substance and is called the *hilus* (Fig. 5-42). The hilus transmits, from the front backward, the renal vein, two branches of the renal artery, the ureter, and the third branch of the renal artery

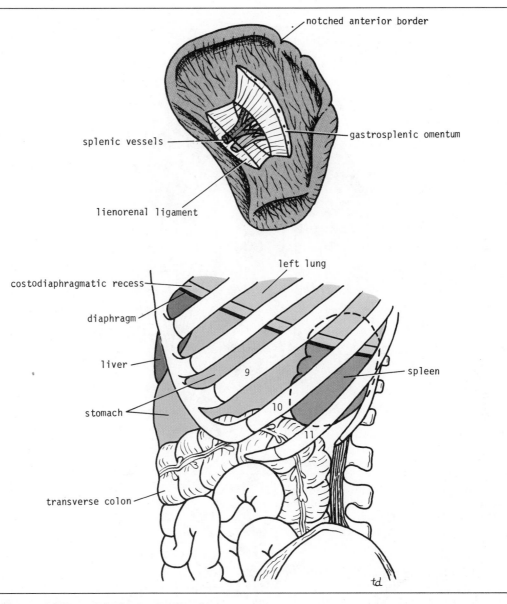

**Fig. 5-40. Spleen, with its notched anterior border, and its relation to adjacent structures.**

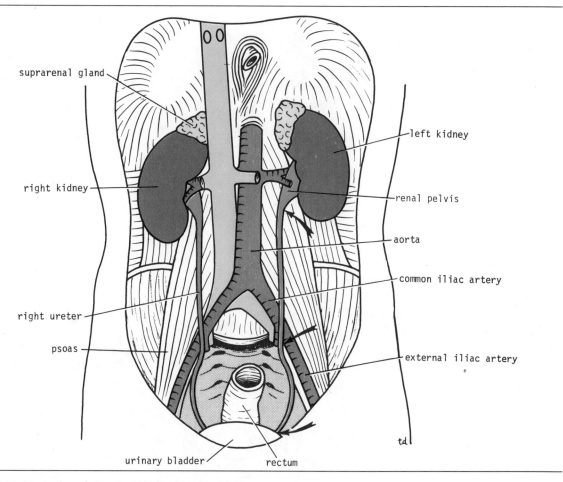

suprarenal gland

left kidney

right kidney

renal pelvis

aorta

common iliac artery

right ureter

psoas

external iliac artery

td

urinary bladder          rectum

**Fig. 5-41. Posterior abdominal wall, showing kidneys and ureters in situ. Arrows indicate three sites where ureter is narrowed.**

(V.A.U.A.). Lymph vessels and sympathetic fibers also pass through the hilus.

The kidneys are surrounded by a *fibrous capsule,* which is closely applied to the cortex. Outside the fibrous capsule is a covering of fat known as the *perirenal fat* (Fig. 5-42). The *renal fascia* surrounds the perirenal fat and encloses the kidneys and the suprarenal glands. The renal fascia is a condensation of areolar tissue, which is continuous laterally with the fascia transversalis. Behind the renal fascia there is usually a large amount of fat called *pararenal fat.*

## Relations

### RIGHT KIDNEY

#### Anteriorly

The suprarenal gland, the liver, the second part of the duodenum, and the right colic flexure (Figs. 5-12 and 5-43).

**Fig. 5-42. (A) Right kidney, anterior surface. (B) Right kidney, coronal section; shows cortex, medulla, pyramids, renal papillae, and calyces. (C) Section of kidney showing positions of nephrons and arrangement of blood vessels within kidney.**

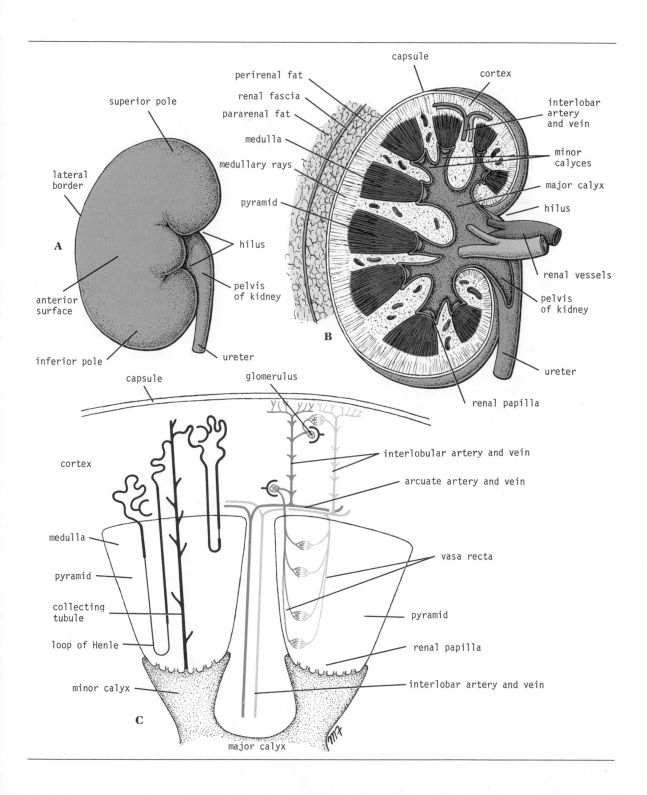

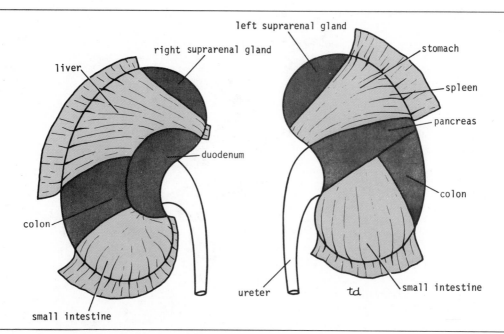

**Fig. 5-43. Anterior relations of both kidneys. Visceral peritoneum covering kidneys has been left in position. Shaded areas indicate where kidney is in direct contact with adjacent viscera.**

## Posteriorly

The diaphragm, the costodiaphragmatic recess of the pleura, the twelfth rib, and the psoas, quadratus lumborum, and transversus abdominis muscles. The subcostal (T12), iliohypogastric, and ilioinguinal nerves (L1) run downward and laterally (Fig. 5-24).

## LEFT KIDNEY

### Anteriorly

The suprarenal gland, the spleen, the stomach, the pancreas, the left colic flexure, and coils of jejunum (Figs. 5-12 and 5-43).

## Posteriorly

The diaphragm, the costodiaphragmatic recess of the pleura, the eleventh (the left kidney is higher)

and twelfth ribs, and the psoas, quadratus lumborum, and transversus abdominis muscles. The subcostal (T12), iliohypogastric, and ilioinguinal nerves (L1) run downward and laterally (Fig. 5-24).

Note that many of the structures are directly in contact with the kidneys, while others are separated by visceral layers of peritoneum. For details, see Figure 5-43.

Within the kidney, the upper expanded end of the ureter, the renal pelvis (pelvis of the ureter) divides into *two* or *three major calyces*, each of which divides into *two* or *three minor calyces* (Fig. 5-42). Each minor calyx is indented by the apex of a *medullary pyramid* called the *renal papilla*. A coronal section through the kidney showing the relationships of the cortex, medulla, renal papillae, and minor calyces appears in Figure 5-42.

The *arterial supply* to the kidney is by way of the renal artery, a branch of the aorta. Each renal artery usually divides into five *segmental arteries* that enter the hilus of the kidney, four in front and one behind the renal pelvis. They are distributed to different segments or areas of the kidney. *Lobar arteries* arise from each segmental artery, one for each renal pyramid. Before entering the renal substance, each lobar artery gives off two or three *interlobar arteries* (Fig. 5-42). The interlobar arteries run toward the cortex on each side of the renal pyramid. At the junction of the cortex and the medulla, the interlobar arteries give off the *arcuate arteries*, which arch over the bases of the pyramids (Fig. 5-42). The arcuate arteries give off a number of *interlobular arteries* that ascend in the cortex. The *afferent glomerular arterioles* arise as branches of the interlobular arteries.

The *renal vein* emerges from the hilus in front of the renal artery. The renal vein drains into the inferior vena cava.

The *lymph vessels* follow the renal artery to lateral aortic lymph nodes around the origin of the renal artery.

The *nerves* to the kidney originate in the renal sympathetic plexus and are distributed along the branches of the renal vessels. The afferent fibers that travel through the renal plexus enter the spinal cord in the tenth, eleventh, and twelfth thoracic nerves.

Some of the more *common forms of congenital anomalies of the kidney* are shown in Figure 5-44.

# Ureter

The ureters convey the urine from the kidneys to the urinary bladder. The urine is propelled along the ureter by peristaltic contractions of the muscle coat, assisted by the filtration pressure of the glomeruli.

The ureter is 10 inches (25 cm) long and resembles the esophagus (also 10 inches long) in having three constrictions along its course: (1) where the renal pelvis joins the ureter, (2) where it is kinked as it crosses the pelvic brim, and (3) where it pierces the bladder wall (Fig. 5-41).

The *renal pelvis (pelvis of the ureter)* is the funnel-shaped expanded upper end of the ureter. It lies within the hilus of the kidney and receives the major calyces (Fig. 5-42). The ureter emerges from the hilus of the kidney and runs vertically downward behind the parietal peritoneum (adherent to it) on the psoas muscle, which separates it from the tips of the transverse processes of the lumbar vertebrae. It enters the pelvis by crossing the bifurcation of the common iliac artery in front of the sacroiliac joint (Fig. 5-41). The ureter then runs down the lateral wall of the pelvis to the region of the ischial spine and turns forward to enter the lateral angle of the bladder. The pelvic course of the ureter is described in detail on pages 340 and 344.

## Relations

### RIGHT URETER

#### Anteriorly

The duodenum, the terminal part of the ileum, the right colic and ileocolic vessels, the right testicular or ovarian vessels, and the root of the mesentery of the small intestine (Fig. 5-18).

#### Posteriorly

The right psoas muscle, which separates it from the lumbar transverse processes, and the bifurcation of the right common iliac artery (Fig. 5-41).

### LEFT URETER

#### Anteriorly

The sigmoid colon and sigmoid mesocolon, the left colic vessels, and the left testicular or ovarian vessels (Figs. 5-9 and 5-18).

#### Posteriorly

The left psoas muscle, which separates it from the lumbar transverse processes, and the bifurcation of the left common iliac artery (Fig. 5-41).

The inferior mesenteric vein lies along the medial side of the left ureter (Fig. 5-18).

The *arterial supply* to the ureter is from: (1) the renal artery, (2) the testicular or ovarian artery, and below, in the pelvis, from (3) the superior vesical artery. The *venous blood* drains into corresponding veins.

The *lymph vessels* drain into the lateral aortic nodes and the iliac nodes.

The *nerves* of the ureter are derived from the renal, testicular (or ovarian), and hypogastric plexuses (in the pelvis). Afferent fibers travel with the sympathetic nerves and enter the spinal cord in the first and second lumbar segments.

The common congenital anomalies of the ureter are shown in Figure 5-45.

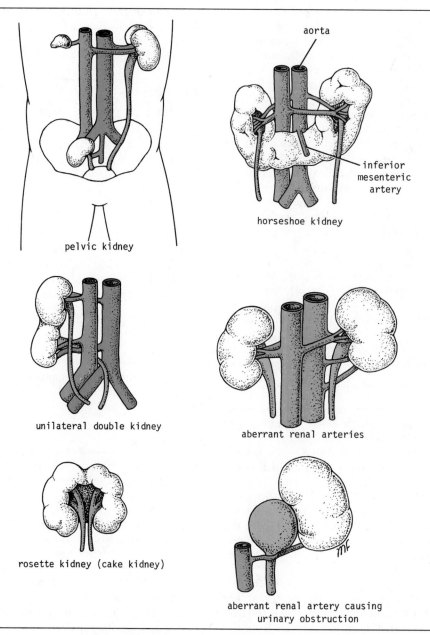

**Fig. 5-44. Some common congenital anomalies of kidney.**

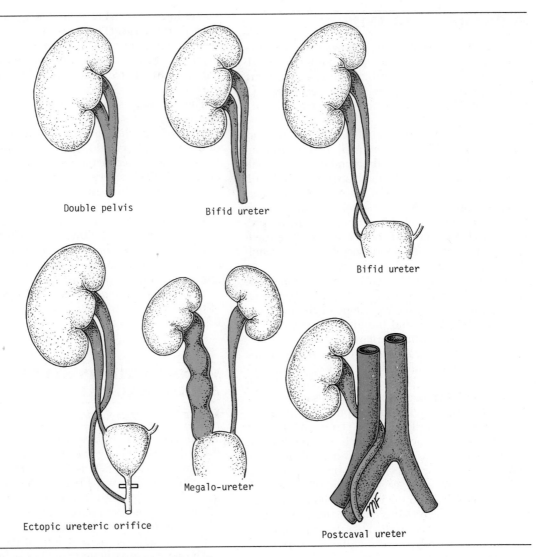

Double pelvis

Bifid ureter

Bifid ureter

Ectopic ureteric orifice

Megalo-ureter

Postcaval ureter

**Fig. 5-45. Some common congenital anomalies of ureter.**

## SUPRARENAL GLANDS

The suprarenal glands are yellowish retroperitoneal organs that lie on the upper poles of the kidneys. They are surrounded by renal fascia (but are separated from the kidneys by the perirenal fat). Each gland has a yellow-colored *cortex* and a dark brown *medulla.*

The cortex of the suprarenal glands secretes hormones that include (1) *mineral corticoids* concerned with the control of fluid and electrolyte balance, (2) *glucocorticoids*, which are concerned with the control of the metabolism of carbohydrates, fats, and proteins, and (3) small amounts of *sex hormones*, which probably play a role in the prepubertal development of the sex organs. The medulla of the suprarenal glands secretes the catecholamines *epinephrine* and *norepinephrine.*

The *right suprarenal gland* is pyramidal in shape and caps the upper pole of the right kidney (Fig. 5-12). It lies behind the right lobe of the liver and extends medially behind the inferior vena cava. It rests posteriorly on the diaphragm.

The *left suprarenal gland* is crescentic in shape and extends along the medial border of the left kidney from the upper pole to the hilus (Fig. 5-12). It lies behind the pancreas, the lesser sac, and the stomach and rests posteriorly on the diaphragm.

The *arteries* supplying each gland are three in number: (1) suprarenal branch of the inferior phrenic artery, (2) suprarenal branch of the aorta, and (3) suprarenal branch of the renal artery.

A single *vein* emerges from the hilus of each gland and drains into the inferior vena cava on the right and into the renal vein on the left.

*Lymph vessels* pass to the lateral aortic nodes.

The nerves are predominantly preganglionic sympathetic fibers and are derived from the splanchnic nerves; the majority of the nerves end in the medulla of the gland.

## ARTERIES ON THE POSTERIOR ABDOMINAL WALL

### Aorta

The aorta enters the abdomen through the aortic opening of the diaphragm in front of the twelfth thoracic vertebra. It descends anteriorly on the bodies of the lumbar vertebrae, and in front of the fourth lumbar vertebra it divides into the two common iliac arteries (Fig. 5-46). On its right side lie the inferior vena cava, the cisterna chyli, and the beginning of the azygos vein. On its left side the left sympathetic trunk is closely related to its margin.

The aorta gives off the branches shown in Diagram 5-1.

The *common iliac arteries* arise at the bifurcation of the aorta and run downward and laterally along the medial border of the psoas muscle (Figs. 4-41 and 5-46). Each artery ends in front of the sacroiliac joint by dividing into the external and internal iliac arteries.

The *external iliac artery* runs along the medial border of the psoas, following the pelvic brim (Fig. 5-41). It gives off the *inferior epigastric* and *deep circumflex iliac* branches before it passes under the inguinal ligament to become the *femoral artery.*

The *internal iliac artery* enters the pelvis in front of the sacroiliac joint (Fig. 5-46). Its further course is described on page 320.

## VEINS ON THE POSTERIOR ABDOMINAL WALL

### Inferior Vena Cava

The inferior vena cava conveys most of the blood from the body below the diaphragm to the right atrium of the heart. It is formed by the union of the common iliac veins behind the right common iliac artery (Fig. 5-46). It ascends on the right side of the aorta, pierces the central tendon of the diaphragm at the level of the eighth thoracic vertebra, and drains into the right atrium of the heart.

The right sympathetic trunk lies behind its right margin and the right ureter lies ½ inch (1.3 cm) from its right border. The entrance into the lesser sac separates the inferior vena cava from the portal vein (Fig. 5-7).

The inferior vena cava receives the tributaries shown in Diagram 5-2.

If one remembers that the venous blood from the abdominal portion of the gastrointestinal tract drains to the liver by means of the tributaries of the portal vein, and that the left suprarenal and testicular or ovarian veins drain first into the left renal vein, then it is apparent that the tributaries of the

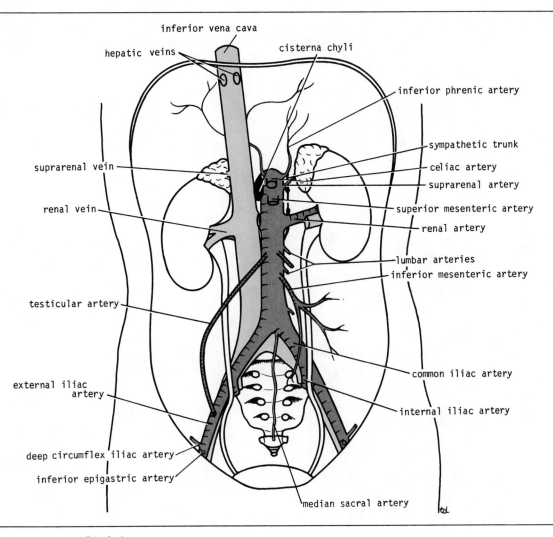

**Fig. 5-46. Aorta and inferior vena cava.**

inferior vena cava correspond rather closely to the branches of the abdominal portion of the aorta.

## Inferior Mesenteric Vein

The inferior mesenteric vein is a tributary of the portal circulation. It begins halfway down the anal canal as the superior rectal vein (Figs. 5-17, 5-28, and 5-29). It passes up the posterior abdominal wall on the left side of the inferior mesenteric artery and the duodenojejunal junction and joins the splenic vein behind the pancreas. It receives tributaries that correspond to the branches of the artery.

## Splenic Vein

The splenic vein is a tributary of the portal circulation. It begins at the hilus of the spleen by the union of several veins and is then joined by the short gastric and the left gastroepiploic veins (Figs. 5-28 and 5-29). It passes to the right within the lienorenal ligament and runs behind the pancreas below the splenic artery. It joins the superior mesenteric vein behind the neck of the pancreas to

Diagram 5-1. Branches of Abdominal Aorta.

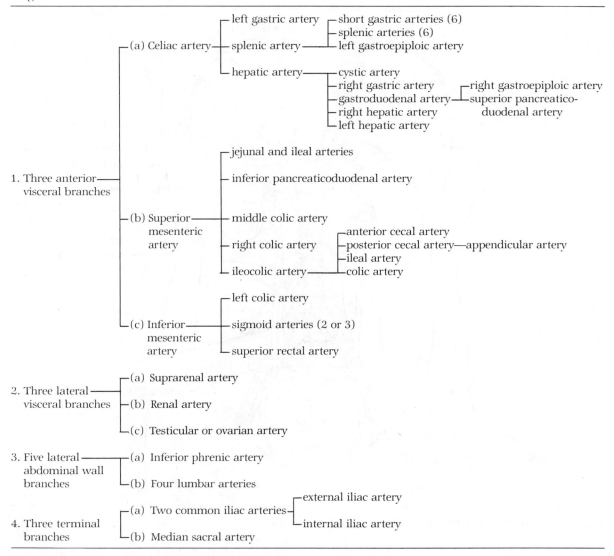

form the portal vein. It is joined by veins from the pancreas and the inferior mesenteric vein.

## Superior Mesenteric Vein

The superior mesenteric vein is a tributary of the portal circulation (Figs. 5-17, 5-28, and 5-29). It begins at the ileocecal junction and runs upward on the posterior abdominal wall within the root of the mesentery of the small intestine and on the right side of the superior mesenteric artery. It passes in front of the third part of the duodenum and behind the neck of the pancreas, where it joins the splenic vein to form the portal vein. It receives tributaries that correspond to the branches of the superior mesenteric artery and also receives the inferior pancreaticoduodenal vein and the right gastroepiploic vein (Fig. 5-29).

Diagram 5-2. Tributaries of Inferior Vena Cava.

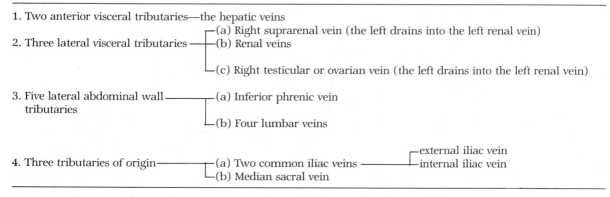

1. Two anterior visceral tributaries—the hepatic veins

2. Three lateral visceral tributaries
   ┌(a) Right suprarenal vein (the left drains into the left renal vein)
   ├(b) Renal veins
   └(c) Right testicular or ovarian vein (the left drains into the left renal vein)

3. Five lateral abdominal wall tributaries
   ┌(a) Inferior phrenic vein
   └(b) Four lumbar veins

4. Three tributaries of origin
   ┌(a) Two common iliac veins ──── ┌external iliac vein
   │                                 └internal iliac vein
   └(b) Median sacral vein

## Portal Vein

The portal vein is described on page 237.

## LYMPHATICS ON THE POSTERIOR ABDOMINAL WALL

### Lymph Nodes

The lymph nodes are closely related to the aorta and form a preaortic and a right and left lateral aortic (para-aortic or lumbar) chain (Fig. 5-47).

The *preaortic lymph nodes* lie around the origins of the celiac, superior mesenteric, and inferior mesenteric arteries and are referred to as the *celiac*, *superior mesenteric*, and *inferior mesenteric lymph nodes*, respectively. They drain the lymph from the gastrointestinal tract, extending from the lower one-third of the esophagus to halfway down the anal canal; and from the spleen, pancreas, gallbladder, and greater part of the liver. The efferent lymph vessels form the large *intestinal trunk* (see below).

The *lateral aortic (para-aortic or lumbar) lymph nodes* drain lymph from the kidneys and suprarenals; from the testes in the male and from the ovaries, the uterine tubes, and the fundus of the uterus in the female; from the deep lymph vessels of the abdominal walls and from the common iliac nodes. The efferent lymph vessels form the *right* and *left lumbar trunks* (see below).

## Lymph Vessels

The *thoracic duct* commences in the abdomen as an elongated lymph sac, the *cisterna chyli*. This lies just below the diaphragm in front of the first two lumbar vertebrae and on the right side of the aorta (Fig. 5-47).

The cisterna chyli receives: (1) the intestinal trunk, (2) the right and left lumbar trunks, and (3) some small lymph vessels that descend from the lower part of the thorax.

## NERVES ON THE POSTERIOR ABDOMINAL WALL

### Lumbar Plexus

The lumbar plexus is formed in the psoas muscle from the anterior rami of the upper four lumbar nerves (Fig. 5-48). The anterior rami receive gray rami communicantes from the sympathetic trunk, and the upper two give off white rami communicantes to the sympathetic trunk. The branches of the plexus emerge from the lateral and medial borders of the muscle and from its anterior surface.

The iliohypogastric, ilioinguinal, lateral cutaneous nerve of the thigh, and femoral nerves emerge from the lateral border of the psoas in that order from above downward (Fig. 5-24). The *iliohypogastric* and *ilioinguinal nerves* (L1) enter the lateral and anterior abdominal walls. (See p. 163.) The iliohypogastric nerve supplies the skin of the lower part of the anterior abdominal wall and the ilioinguinal nerve passes through the inguinal

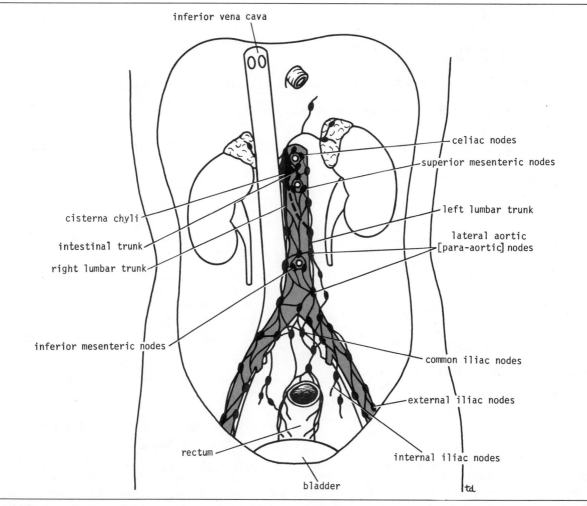

**Fig. 5-47. Lymphatic vessels and nodes on posterior abdominal wall.**

canal to supply the skin of the groin and the scrotum or labium majus. The *lateral cutaneous nerve of the thigh* crosses the iliac fossa in front of the iliacus muscle and enters the thigh behind the lateral end of the inguinal ligament. (See p. 577.) It supplies the skin over the lateral surface of the thigh. The *femoral nerve* (L2, 3, and 4) runs downward and laterally between the psoas and the iliacus muscles and enters the thigh behind the inguinal ligament and lateral to the femoral sheath. (See p. 590.) In the abdomen it supplies the iliacus muscle.

The obturator nerve and the fourth lumbar root of the lumbosacral trunk emerge from the medial border of the psoas at the brim of the pelvis. The *obturator nerve* (L2, 3, and 4) crosses the pelvic brim in front of the sacroiliac joint and behind the common iliac vessels. It leaves the pelvis by passing through the obturator foramen into the thigh. (For a description of its course in the pelvis, see p. 320, and in the thigh, see p. 596.) The *fourth lumbar root of the lumbosacral trunk* takes part in the formation of the sacral plexus. (See p. 314.) It descends anterior to the ala of the sacrum and joins the first sacral nerve.

The *genitofemoral nerve* (L1 and 2) emerges on

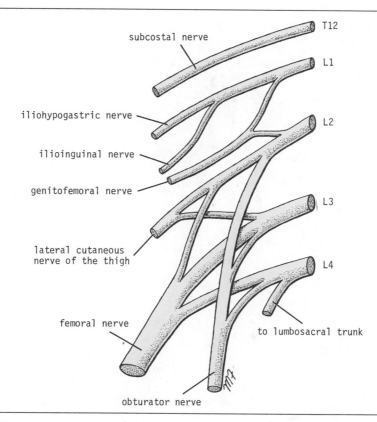

**Fig. 5-48. Lumbar plexus of nerves.**

the anterior surface of the psoas. It runs downward in front of the muscle and divides into (1) a *genital branch*, which enters the spermatic cord and supplies the cremaster muscle, and (2) a *femoral branch*, which supplies a small area of the skin of the thigh. (See p. 577.) It is the nervous pathway involved in the *cremasteric reflex*, in which stimulation of the skin of the thigh in the male results in reflex contraction of the cremaster muscle and the drawing upward of the testis within the scrotum.

## Abdominal Part of the Sympathetic Trunk

The abdominal part of the sympathetic trunk is continuous above with the thoracic and below with the pelvic parts of the sympathetic trunk. It runs downward along the medial border of the psoas muscle on the bodies of the lumbar vertebrae (Fig. 5-49). It enters the abdomen from behind the medial arcuate ligament and gains entrance to the pelvis below by passing behind the common iliac vessels. The *right sympathetic trunk* lies behind the right border of the inferior vena cava; the *left sympathetic trunk* lies close to the left border of the aorta.

The sympathetic trunk possesses four segmentally arranged ganglia, the first and second often being fused together. The upper two ganglia receive a white ramus communicans from the first and second lumbar nerves.

### Branches

1. Gray rami communicantes to the lumbar spinal nerves. The postganglionic fibers are distributed through the branches of the spinal nerves to the

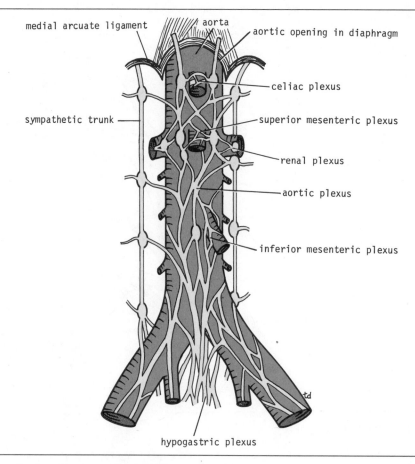

medial arcuate ligament

aorta

aortic opening in diaphragm

celiac plexus

superior mesenteric plexus

sympathetic trunk

renal plexus

aortic plexus

inferior mesenteric plexus

hypogastric plexus

**Fig. 5-49. Aorta and related sympathetic plexuses.**

blood vessels, sweat glands, and arrector pili muscles of the skin. (See Fig. 1-4.)

2. Fibers pass medially to the sympathetic plexuses on the abdominal aorta and its branches. (These plexuses also receive fibers from splanchnic nerves and the vagus.)

3. Fibers pass downward and medially in front of the common iliac vessels into the pelvis, where, together with branches from sympathetic nerves in front of the aorta, they form a large bundle of fibers called the *hypogastric plexus* (Fig. 5-49).

## Aortic Plexuses

Preganglionic and postganglionic sympathetic fibers, preganglionic parasympathetic fibers, and visceral afferent fibers form a plexus of nerves, the *aortic plexus*, around the abdominal part of the aorta (Fig. 5-49). Regional concentrations of this plexus around the origins of the celiac, renal, superior mesenteric, and inferior mesenteric arteries form the *celiac plexus*, *renal plexus*, *superior mesenteric plexus*, and *inferior mesenteric plexus*, respectively.

The celiac plexus consists mainly of two *celiac ganglia* connected together by a large network of fibers that surrounds the origin of the celiac artery. The ganglia receive the greater and lesser splanchnic nerves (preganglionic sympathetic fibers). Postganglionic branches accompany the branches of the celiac artery and follow them to their distribution. Parasympathetic vagal fibers also accompany the branches of the artery.

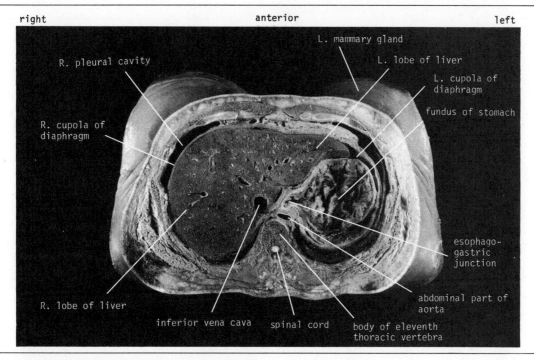

**Fig. 5-50. Cross section of abdomen at level of body of eleventh thoracic vertebra, viewed from below.**

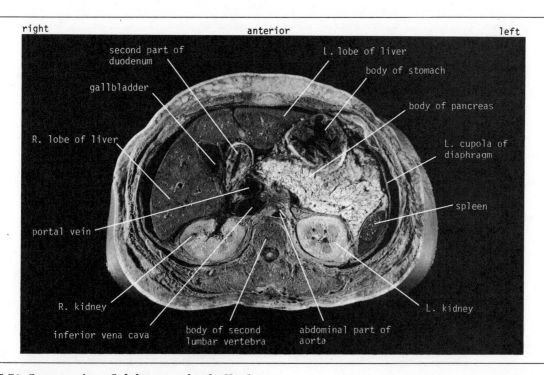

**Fig. 5-51. Cross section of abdomen at level of body of second lumbar vertebra, viewed from below.**

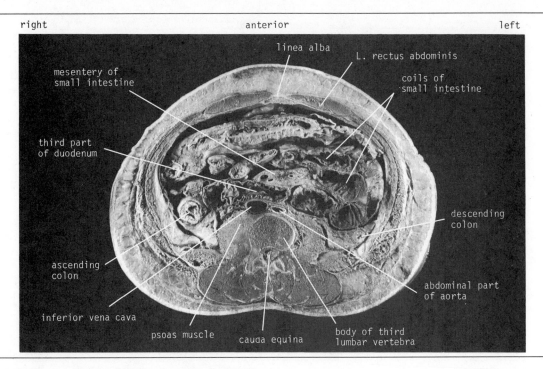

right           anterior           left

linea alba

L. rectus abdominis

mesentery of
small intestine

coils of
small intestine

third part
of duodenum

descending
colon

ascending
colon

abdominal part
of aorta

inferior vena cava

psoas muscle

cauda equina

body of third
lumbar vertebra

**Fig. 5-52. Cross section of abdomen at level of body
of third lumbar vertebra, viewed from below.**

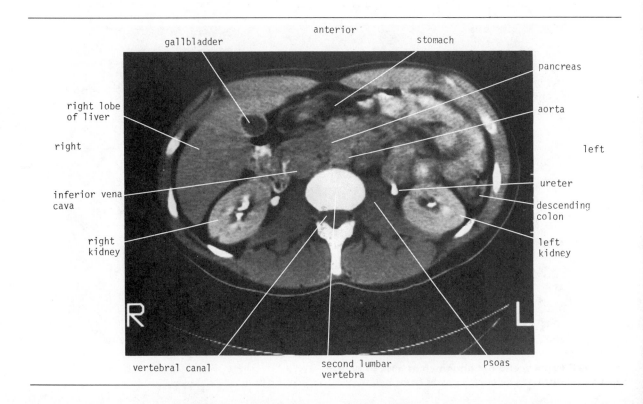

anterior

gallbladder

stomach

pancreas

right lobe
of liver

aorta

right

left

inferior vena
cava

ureter

descending
colon

right
kidney

left
kidney

R

L

vertebral canal

second lumbar
vertebra

psoas

The renal and superior mesenteric plexuses are smaller than the celiac plexus. They are distributed along the branches of the corresponding arteries. The inferior mesenteric plexus is similar, but receives parasympathetic fibers from the sacral parasympathetic.

## Cross-Sectional Anatomy of the Abdomen

In order to assist you in the interpretation of CT scans of the abdomen, you should study the labeled cross sections of the abdomen shown in Figures 5-50, 5-51, and 5-52. The sections have been photographed on their *inferior surfaces*. See Figure 5-53 for CT scan.

# RADIOGRAPHIC APPEARANCES OF THE ABDOMEN

Only the more important features seen in a standard anteroposterior radiograph of the abdomen, with the patient in the supine position, will be described (Figs. 5-54 and 5-55).

Examine the following in a systematic order.

1. *Bones.* In the upper part of the radiograph the lower ribs will be seen. Running down the middle of the radiograph are the lower thoracic and lumbar vertebrae and the sacrum and coccyx. On either side are the sacroiliac joints, the pelvic bones, and the hip joints.
2. *Diaphragm.* This casts dome-shaped shadows on each side; the one on the right is slightly higher than the one on the left (not shown in Fig. 5-54).
3. *Psoas muscle.* On either side of the vertebral column the lateral borders of the psoas muscle cast a shadow that passes downward and laterally from the twelfth thoracic vertebra.
4. *Liver.* This forms a homogeneous opacity in the upper part of the abdomen.
5. *Spleen.* This may cast a soft shadow, which can

be seen in the left ninth and tenth intercostal spaces (not shown in Fig. 5-54).
6. *Kidneys.* These are usually visible because the perirenal fat surrounding the kidneys produces a transradiant line.
7. *Stomach and intestines.* Gas may be seen in the fundus of the stomach and in the intestines. Fecal material may also be seen in the colon.
8. *Urinary bladder.* If this contains sufficient urine, it will cast a shadow in the pelvis.

## Radiographic Appearances of the Stomach

The stomach may be demonstrated radiologically (Figs. 5-56 and 5-57) by the administration of a watery suspension of barium sulfate (barium meal). With the patient in the erect position, the first few mouthfuls pass into the stomach and form a triangular shadow with the apex downward. The gas bubble in the fundus will show above the fluid level at the top of the barium shadow. As the stomach is filled, the greater and lesser curvatures will be outlined, and the body and pyloric portions will be recognized. The pylorus will be seen to move downward and come to lie at the level of the third lumbar vertebra.

Fluoroscopic examination of the stomach as it is filled with the barium emulsion will reveal peristaltic waves of contraction of the stomach wall, which commence near the middle of the body and pass to the pylorus. The respiratory movements of the diaphragm will cause displacement of the fundus.

## Radiographic Appearances of the Duodenum

A barium meal passes into the first part of the duodenum and forms a triangular homogeneous shadow, the *duodenal cap*, which has its base toward the pylorus (Fig. 5-58). Under the influence of peristalsis, the barium quickly leaves the duodenal cap and passes rapidly through the remaining portions of the duodenum. The outline of the barium shadow in the first part of the duodenum is smooth, due to the absence of mucosal folds. In the remainder of the duodenum, the presence of plicae circulares breaks up the barium emulsion, giving it a floccular appearance.

**Fig. 5-53. CT scan of abdomen at the level of the second lumbar vertebra following intravenous pyelography. The radiopaque material can be seen in the renal pelvis and the ureters. The section is viewed from below.**

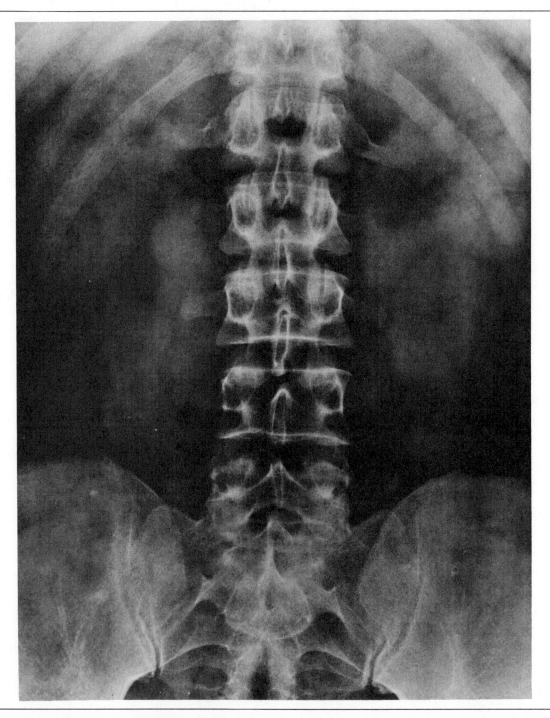

**Fig. 5-54. Anteroposterior radiograph of abdomen.**

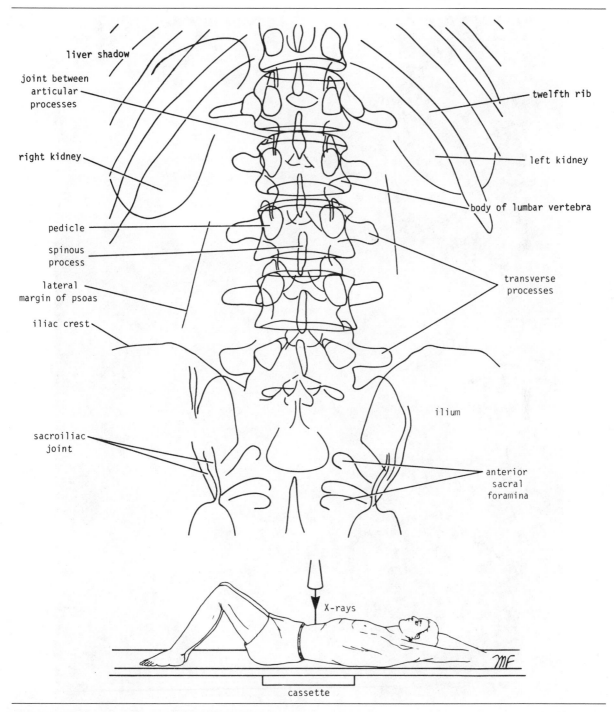

**Fig. 5-55. Diagrammatic representation of main features seen in anteroposterior radiograph in Figure 5-54.**

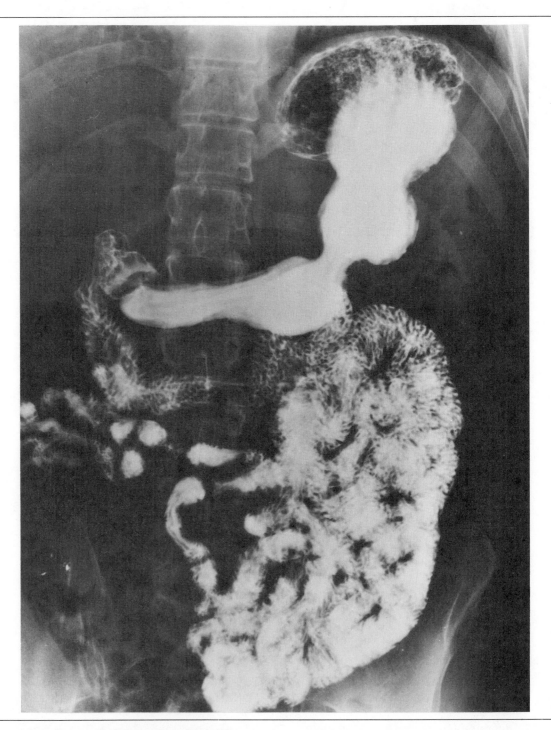

**Fig. 5-56. Radiograph of stomach and small intestine following ingestion of barium meal.**

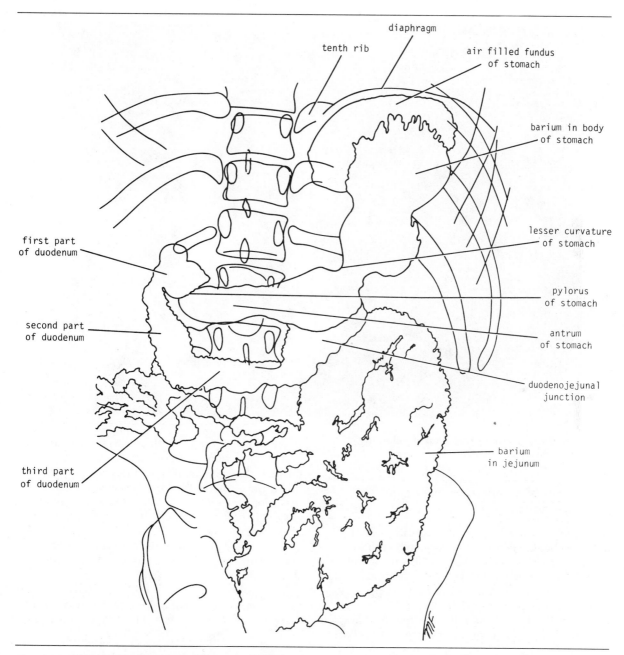

**Fig. 5-57. Diagrammatic representation of main features seen in radiograph in Figure 5-56.**

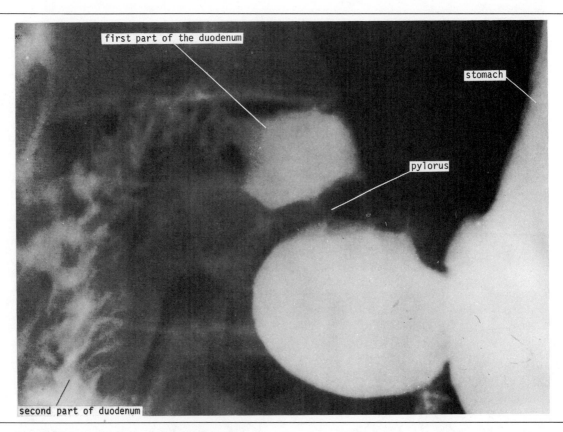

first part of the duodenum

stomach

pylorus

second part of duodenum

**Fig. 5-58. Radiograph of duodenum following ingestion of barium meal.**

## Radiographic Appearances of the Jejunum and Ileum

A barium meal enters the jejunum in a few minutes, reaches the ileocecal junction in 30 minutes to 2 hours, and the greater part has left the small intestine in 6 hours. In the jejunum and upper part of the ileum, the mucosal folds and the peristaltic activity scatter the barium shadow (Figs. 5-56 and 5-59). In the last part of the ileum, the barium meal tends to form a continuous mass of barium.

## Radiographic Appearances of the Large Intestine

The large intestine may be demonstrated by the administration of a barium enema or a barium meal. The former is more satisfactory.

The bowel may be outlined by the administration of 2 to 3 pints (1 L) of barium sulfate emulsion through the anal canal. When the large intestine is filled, the entire outline may be seen in an anteroposterior projection (Fig. 5-60). Oblique and lateral views of the colic flexures may be necessary. The characteristic sacculations are well seen when the bowel is filled, and, after the enema has been evacuated, the mucosal pattern is clearly demonstrated.

The appendix frequently fills with barium after an enema. The radiographic appearances of the sigmoid colon and rectum are described on page 357.

## Radiographic Appearances of the Extrahepatic Biliary Apparatus

The bile passages normally are not visible on a radiograph. Their lumina may be outlined by the administration of various iodine-containing com-

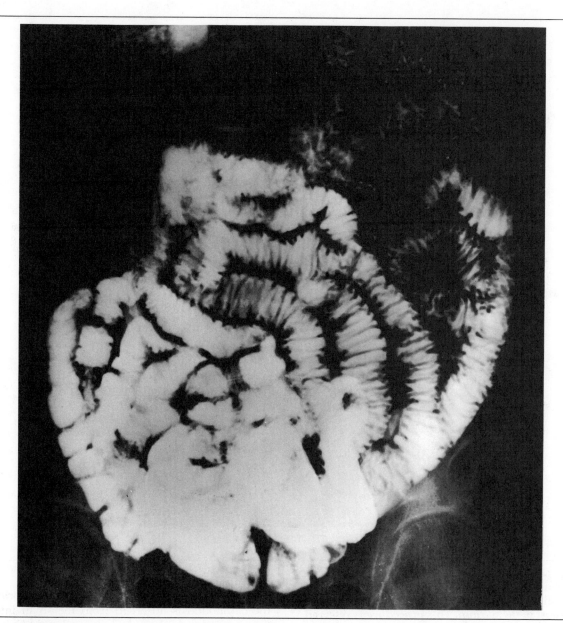

**Fig. 5-59. Radiograph of small intestine following barium meal.**

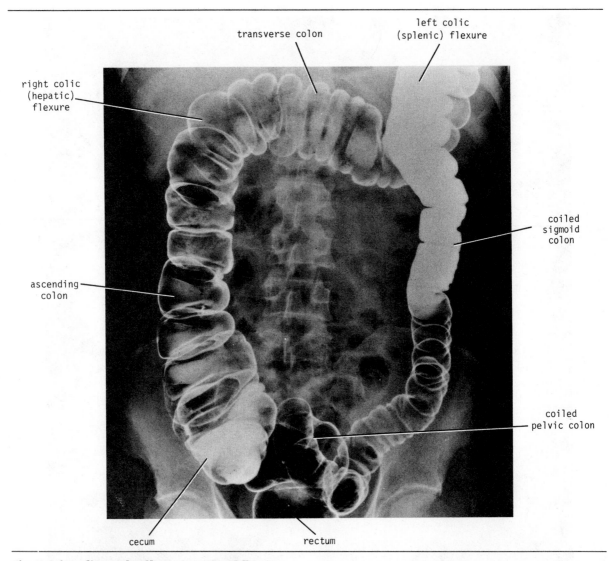

**Fig. 5-60. Radiograph of large intestine following barium enema. Air has been introduced into the intestine through the enema tube after evacuation of most of the barium. This procedure is referred to as a contrast enema.**

pounds orally or by injection. When taken orally, the compound is absorbed from the small intestine, carried to the liver, and excreted with the bile. On reaching the gallbladder, it is concentrated with the bile. The concentrated iodine compound, mixed with the bile, is now radiopaque and will reveal the gallbladder as a pear-shaped opacity in the angle between the right twelfth rib and the vertebral column (Figs. 5-61 and 5-62). If the patient is given a fatty meal, the gallbladder contracts, and the cystic and bile ducts will become visible as the

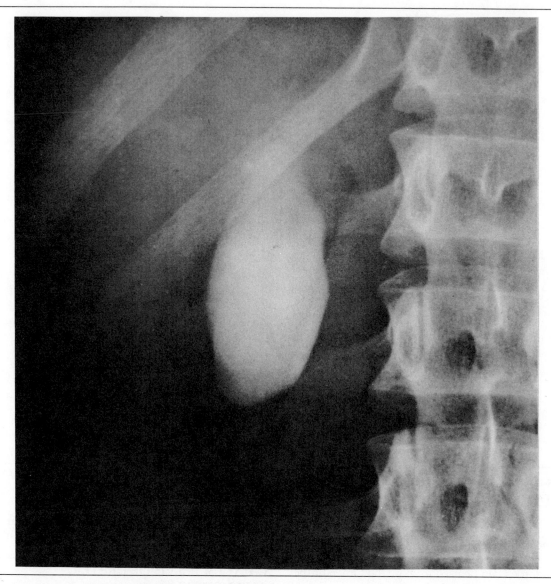

**Fig. 5-61. Radiograph of gallbladder after administration of iodine-containing compound.**

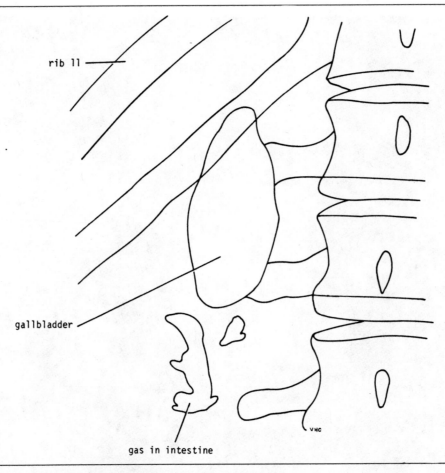

rib 11

gallbladder

gas in intestine

VHC

**Fig. 5-62. Diagrammatic representation of main features seen in radiograph in Figure 5-61.**

opaque medium passes down to the second part of the duodenum. A sonogram of the upper part of the abdomen can be used to show the lumen of the gallbladder (Fig. 5-73).

### Radiographic Appearances of the Urinary Tract

#### KIDNEYS

The kidneys are usually visible on a standard anteroposterior radiograph of the abdomen because the perirenal fat surrounding the kidneys produces a transradiant line. They can be made to show up very clearly by the introduction of air under the renal fascia.

### CALYCES, PELVIS OF URETER, AND URETER

These structures are not normally visible on a standard radiograph. The lumen can be demonstrated by the use of radiopaque compounds in *intravenous pyelography* or *retrograde pyelography*.

With *intravenous pyelography*, an iodine-containing compound is injected into a subcutaneous arm vein. It is excreted and concentrated by the kidneys, thus rendering the calyces and ureter opaque to X-rays (Figs. 5-63 and 5-64). When

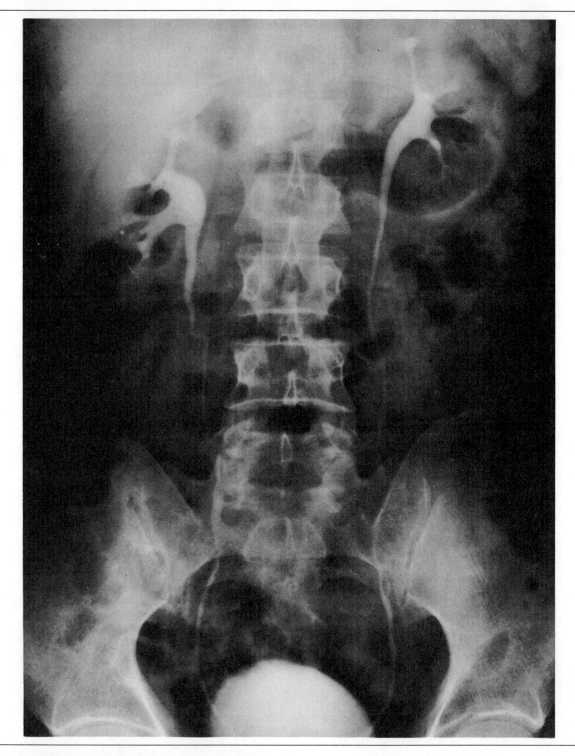

Fig. 5-63. Radiograph of ureter and renal pelvis after intravenous injection of iodine-containing compound, which is excreted by kidney. Major and minor calyces are also shown.

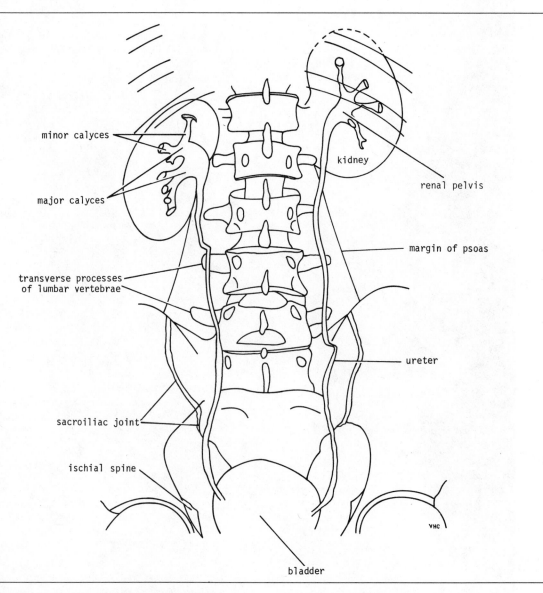

minor calyces

major calyces

kidney

renal pelvis

margin of psoas

transverse processes
of lumbar vertebrae

ureter

sacroiliac joint

ischial spine

bladder

**Fig. 5-64. Diagrammatic representation of main
features seen in radiograph in Figure 5-63.**

enough of the opaque medium has been excreted, the bladder will also be revealed. The ureters are seen superimposed on the transverse processes of the lumbar vertebrae. They cross the sacroiliac joints and enter the pelvis. In the vicinity of the ischial spines, they turn medially to enter the bladder. The three normal constrictions of the ureters (at the junction of the renal pelvis with the ureter, at the pelvic brim, and where the ureter enters the bladder) may be recognized.

With *retrograde pyelography*, a cystoscope is passed through the urethra into the bladder, and a ureteric catheter is inserted into the ureter. A solution of sodium iodide is then injected along the catheter into the ureter. When the minor calyces become filled with the radiopaque medium, the detailed anatomy of the minor and major calyces and the pelvis of the ureter can be clearly seen. Each minor calyx has a cup-shaped appearance caused by the renal papilla projecting into it.

## CLINICAL NOTES

### Peritoneum

The peritoneal cavity is the largest cavity in the body, since the surface area of the parietal and visceral layers of peritoneum is enormous. The living peritoneum, unlike that seen in the cadaver, possesses a certain degree of mobility on the extraperitoneal fat and can be stretched without tearing. The peritoneal fluid not only lubricates the surfaces of the mobile viscera, but also contains leukocytes and antibodies, which have remarkable powers of resisting infection.

As has been described previously, the *peritoneal fluid* circulates around the peritoneal cavity and quickly finds its way into the lymphatics of the diaphragm. While it is probable that peritoneal fluid can be absorbed at other sites in the peritoneal cavity, it is generally accepted that absorption from under the diaphragm is the most rapid route. In order to delay the absorption of toxins from intraperitoneal infections, it is common nursing practice to sit a patient up in bed with the back at an angle of 45 degrees. In this position the infected peritoneal fluid gravitates downward into the pelvic cavity, where the rate of toxin absorption is slow (Fig. 5-65).

The *peritoneal cavity* is divided into an upper part within the abdomen and a lower part in the pelvis. The abdominal part is further subdivided by the many peritoneal reflections into important fossae and spaces, which, in turn, are continued into the paracolic gutters. The attachment of the transverse mesocolon and the mesentery of the small intestine to the posterior abdominal wall provides natural peritoneal barriers that may hinder the movement of infected peritoneal fluid from the upper part to the lower part of the peritoneal cavity.

With the patient in the supine position, it is interesting to note that the right subphrenic peritoneal space and the pelvic cavity are the lowest areas of the peritoneal cavity, and the region of the pelvic brim may be regarded as a watershed (Fig. 5-65).

Collection of infected peritoneal fluid in one of the *subphrenic spaces* is often accompanied by infection of the adjoining pleural cavity. It is not uncommon to find a localized empyema in a patient with a subphrenic abscess. It is believed that the infection spreads from the peritoneum to the pleura via the diaphragmatic lymphatics. A patient with a subphrenic abscess often complains of pain over the shoulder. The skin of the shoulder is supplied by the supraclavicular nerves (C3 and 4), which have the same segmental origin as the phrenic nerve, which supplies the peritoneum in the center of the undersurface of the diaphragm.

*Abdominal pain* arising from the parietal peritoneum is of the somatic type and can be *precisely localized* to the site of origin. It is usually severe in nature. The visceral peritoneum, including the mesentery, is sensitive to stretch and is innervated by autonomic nerves. Overdistention of a viscus will lead to the sensation of pain. The gastrointestinal tract itself arises embryologically as a midline structure and receives a bilateral nerve supply. Pain arising from an abdominal viscus is dull and *poorly localized* and is referred to the midline. The sites of origin of visceral pain are shown in Figure 5-66.

*Infection* may gain entrance to the peritoneal cavity through a number of routes: (1) from the interior of the gastrointestinal tract; (2) through

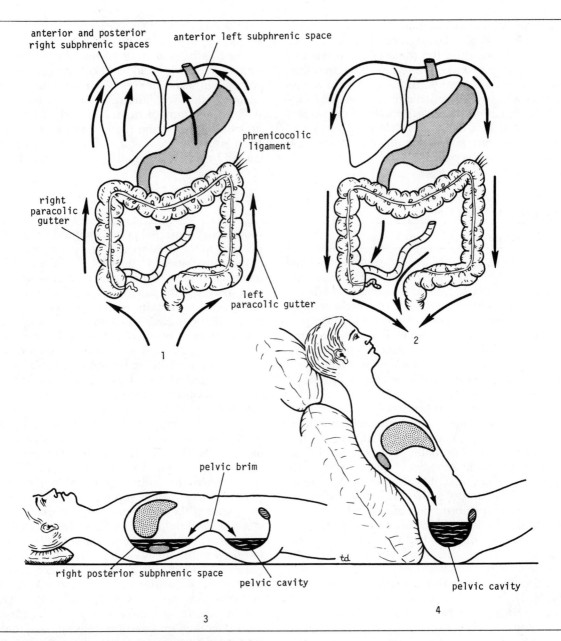

**Fig. 5-65. Direction of flow of peritoneal fluid. (1) Normal flow upward to subphrenic spaces. (2) Flow of inflammatory exudate in peritonitis. (3) The two sites where inflammatory exudate tends to collect when patient is nursed in the supine position. (4) Accumulation of inflammatory exudate in pelvis when patient is nursed in inclined position.**

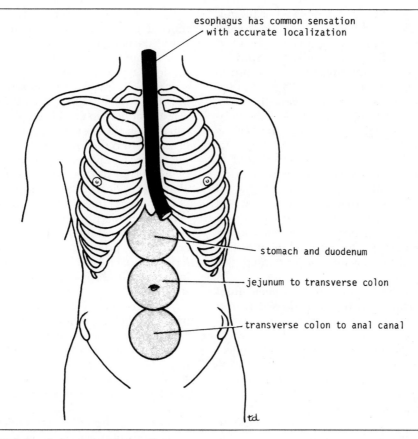

esophagus has common sensation
with accurate localization

stomach and duodenum

jejunum to transverse colon

transverse colon to anal canal

**Fig. 5-66. Sites of origin of visceral pain in alimentary tract.**

the abdominal walls; (3) via the uterine tubes in the female (gonococcal peritonitis in the adult and pneumococcal peritonitis in the child occur through this route); and (4) by the bloodstream, as in septicemia.

An inflamed parietal peritoneum is extremely sensitive to stretching. This is made use of clinically in diagnosing peritonitis. Pressure is applied to the abdominal wall with a single finger over the site of inflammation. The pressure is then removed by suddenly withdrawing the finger. The abdominal wall rebounds, resulting in extreme local pain, which is known as *rebound tenderness.*

It should always be remembered that the peritoneum of the pelvis can be palpated through the rectum and the vagina. (See pp. 395 and 399.) An inflamed appendix may be hanging down in the pelvis, irritating the parietal peritoneum. In these circumstances, a rectal or vaginal examination with a gloved finger may detect extreme tenderness of the peritoneum on the right side.

Since the peritoneum is a semipermeable membrane, it allows rapid bidirectional transfer of substances across itself. Because its surface area is enormous, this transfer property has been made use of in *peritoneal dialysis* for acute renal insufficiency. A watery solution, the dialysate, is introduced through a catheter into the peritoneal cavity. The products of metabolism, such as urea, diffuse through the peritoneal lining cells from the blood vessels into the dialysate and are removed from the patient.

The *greater omentum* is often referred to by the surgeons as the "abdominal policeman." The lower and the right and left margins are free, and it moves about the peritoneal cavity in response to the peristaltic movements of the neighboring gut. In the first 2 years of life, it is poorly developed and thus is less protective in a young child. Later, however, in an acutely inflamed appendix, for example, the inflammatory exudate causes the omentum to adhere to the appendix and wrap itself around the infected organ (Fig. 5-67). By this means the infection is often localized to a small area of the peritoneal cavity, thus saving the patient from a serious diffuse peritonitis.

The greater omentum has also been found to plug the neck of a hernial sac and prevent the entrance of coils of small intestine.

The greater omentum may undergo torsion and if this is extensive, the blood supply to a part of it may be cut off, causing necrosis. Surgeons sometimes use the omentum to buttress an intestinal anastomosis or in the closure of a perforated gastric or duodenal ulcer.

The excessive accumulation of peritoneal fluid within the peritoneal cavity is called *ascites*. In a thin patient, as much as 1,500 ml has to accumulate before ascites can be recognized clinically. In obese subjects a far greater amount has to collect before it may be detected. The withdrawal of peritoneal fluid from the peritoneal cavity is described on page 192.

Occasionally, a loop of intestine enters a peritoneal fossa (for example, the lesser sac or the duodenal fossae) and becomes strangulated at the edges of the fossa. It is well to remember that very important structures form the boundaries of the entrance into the lesser sac, and that the inferior mesenteric vein often lies in the anterior wall of the paraduodenal fossa.

## Stomach

The mucous membrane of the body of the stomach and, to a lesser extent, that of the fundus, produces acid and pepsin. The secretion of the antrum and pyloric canal is mucous and is weakly alkaline (Fig. 5-68). The secretion of acid and pepsin is controlled by two mechanisms: (1) nervous and (2) hormonal. The vagus nerves are responsible for the nervous control, and the hormone *gastrin*, produced by the antral mucosa, is responsible for the hormonal control. In the surgical treatment of chronic gastric and duodenal ulcers, attempts are made to reduce the amount of acid secretion by

section of the vagus nerves (vagotomy) and by removing the gastrin-bearing area of mucosa, the antrum (partial gastrectomy).

A chronic gastric ulcer occurs in the alkaline-producing mucosa, usually on or close to the lesser curvature. It invades the muscular coats and will, in time, involve the peritoneum, so that the stomach will adhere to neighboring structures. An ulcer situated on the posterior wall of the stomach may perforate into the lesser sac or become adherent to the pancreas. Erosion of the pancreas will produce pain referred to the back. The splenic artery runs along the upper border of the pancreas and erosion of this artery may produce fatal hemorrhage. A penetrating ulcer of the anterior stomach wall may result in the escape of stomach contents into the greater sac, producing diffuse peritonitis. The anterior stomach wall may, however, adhere to the liver, and the chronic ulcer may penetrate the liver substance.

The sensation of pain in the stomach is caused by the stretching or spasmodic contraction of the smooth muscle in its walls and is referred to the epigastrium. It is believed that the pain-transmitting fibers leave the stomach in company with the sympathetic nerves. They pass through the celiac ganglia and reach the spinal cord via the greater splanchnic nerves.

Since the lymphatic vessels of the mucous membrane and submucosa of the stomach are in continuity, it is possible for cancer cells to travel to different parts of the stomach some distance away from the primary site. Cancer cells also often pass through or bypass the local lymph nodes and are held up in the regional nodes. For these reasons, malignant disease of the stomach is treated by total gastrectomy, which includes the removal of the lower end of the esophagus and the first part of the duodenum; the spleen and the gastrosplenic and lienorenal ligaments and their associated lymph nodes; the splenic vessels; the tail and body of the pancreas and their associated nodes; the nodes along the lesser curvature of the stomach; and the nodes along the greater curvature, along with the greater omentum. This radical operation is a desperate attempt to remove the stomach en bloc and, with it, its lymphatic field. The continuity of the gut is restored by anastomosing the esophagus with the jejunum.

*Gastroscopy* is the viewing of the mucous mem-

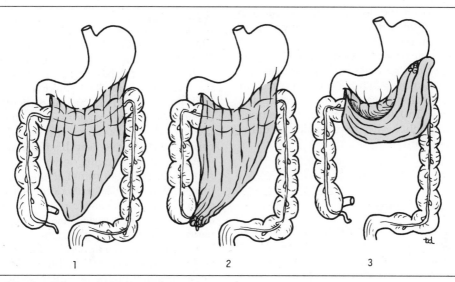

Fig. 5-67. (1) Normal greater omentum; (2) greater omentum wrapped around an inflamed appendix; (3) greater omentum adherent to base of gastric ulcer. One important function of greater omentum is to attempt to limit spread of intraperitoneal infections.

brane of the stomach through an illuminated tube fitted with a lens system. The patient is anesthetized and the gastroscope is passed into the stomach, which is then inflated with air. By using a flexible fiberoptic instrument, direct visualization of different parts of the gastric mucous membrane is possible. It is also possible to perform a mucosal biopsy through a gastroscope.

## Duodenum

As the stomach empties its contents into the duodenum, the acid chyme is squirted against the anterolateral wall of the first part of the duodenum. This is thought to be an important factor in the production of a duodenal ulcer at this site. An ulcer of the anterior wall of the first inch of the duodenum may perforate into the upper part of the greater sac, above the transverse colon. The transverse colon directs the escaping fluid into the right lateral paracolic gutter and thus down to the right iliac fossa. The differential diagnosis between a perforated duodenal ulcer and a perforated appendix may be very difficult.

An ulcer of the posterior wall of the first part of the duodenum may penetrate the wall and erode the relatively large gastroduodenal artery, causing a very severe hemorrhage.

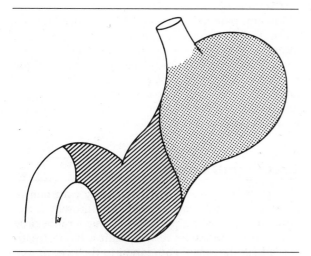

Fig. 5-68. Areas of stomach that produce acid and pepsin (stipple) and alkali and gastrin (hatched).

The importance of the duodenal fossae and the occurrence of herniae of the intestine have already been alluded to on page 282.

The relation to the duodenum of the gallbladder, the transverse colon, and the right kidney should be remembered. Cases have been reported in which a large gallstone has ulcerated through the gallbladder wall into the duodenum. Operations on the colon and right kidney have resulted in damage to the duodenum.

# Jejunum and Ileum

A physician should be able to distinguish between the large and small intestine. He may be called upon to examine a case of postoperative burst abdomen, where coils of gut are lying free in the bed. The macroscopic differences are described on page 238.

The line of attachment of the mesentery of the small intestine to the abdominal wall should be remembered. A tumor or cyst of the mesentery, when palpated through the anterior abdominal wall, will be more mobile in a direction at right angles to the line of attachment than along the line of attachment.

Pain fibers from the jejunum and ileum traverse the superior mesenteric sympathetic plexus and pass to the spinal cord via the splanchnic nerves. Referred pain from this segment of the gastrointestinal tract is felt in the dermatomes supplied by the ninth, tenth, and eleventh thoracic nerves. Strangulation of a coil of small intestine in an inguinal hernia first gives rise to pain in the region of the umbilicus. Only later, when the parietal peritoneum of the hernial sac becomes inflamed, does the pain become more intense and localized to the inguinal region.

It will be appreciated that the superior mesenteric artery and vein supply an extensive territory of the gut, from halfway down the second part of the duodenum to the left colic flexure. The lodgement of an *embolus* in the artery or a *thrombosis* of the vein will result in death of this segment of the bowel, with fatal intestinal obstruction. If the diagnosis can be made early enough, surgical relief of the vessel obstruction can be carried out.

# Appendix

The inconstancy of the position of the appendix should be borne in mind when one is attempting to diagnose an appendicitis. A retrocecal appendix, for example, may lie behind a cecum distended with gas, and thus it may be difficult to elicit tenderness on palpation in the right iliac region. Irritation of the psoas muscle, on the other hand, may cause the patient to keep his right hip joint flexed.

An appendix hanging down in the pelvis may result in absent abdominal tenderness in the right iliac region, but deep tenderness may be experienced in the hypogastric region. Rectal or vaginal examination may reveal tenderness of the peritoneum in the pelvis on the right side.

The artery to the appendix does not anastomose with any other artery. The blind end of the appendix is supplied by the terminal branches of the appendicular artery. Inflammatory edema of the appendicular wall compresses the blood supply to the appendix, and often leads to thrombosis of the appendicular artery. These conditions commonly result in necrosis or gangrene of the appendicular wall, with perforation.

Perforation of the appendix or transmigration of bacteria through the inflamed appendicular wall results in infection of the peritoneum of the greater sac. The part that the greater omentum may play in arresting the spread of the peritoneal infection has already been described on page 282.

Visceral pain in the appendix is produced by distention of its lumen or spasm of its muscle. The afferent pain fibers enter the spinal cord at the level of the tenth thoracic segment, and a *vague referred pain* is felt in the region of the umbilicus. Later, the pain shifts to where the inflamed appendix irritates the parietal peritoneum. Here, the *pain* is *precise*, *severe*, and *localized*.

# Cecum and Colon

*Cancer of the large bowel* is relatively common in persons over 50 years of age. The growth is restricted to the bowel wall for a considerable time before it spreads via the lymphatics. Bloodstream spread via the portal circulation to the liver occurs late. If a diagnosis is made early and a partial col-

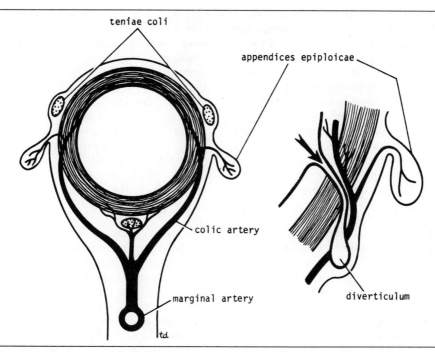

teniae coli

appendices epiploicae

colic artery

marginal artery

td

diverticulum

**Fig. 5-69. (1) Blood supply to colon and (2) formation of diverticulum. Note passage of mucosal diverticulum through circular muscle coat along course of artery.**

ectomy is performed, accompanied by removal of the lymph vessels and lymph nodes draining the area, then a cure can be anticipated.

*Diverticulosis* of the colon is a common clinical condition. It consists of a herniation of the lining mucosa through the circular muscle between the teniae coli. It occurs at points where the circular muscle is weakest, i.e., where the blood vessels pierce the muscle (Fig. 5-69). The common site for herniation is shown in Figure 5-69.

Because of the anatomical mobility of the cecum and transverse colon, they may be brought to the surface through a small opening in the anterior abdominal wall. If the cecum or transverse colon is then opened, the bowel contents may be allowed to drain by this route. These procedures are referred to as *cecostomy* or *transverse colostomy*, respectively, and are used to relieve large-bowel obstructions.

The *congenital anomaly* of undescended cecum

or a failure of rotation of the gut so that the cecum lies in the left fossa may give rise to confusion in diagnosis. The pain of appendicitis, for example, although initially starting in the umbilical region, may shift not to the right iliac fossa but to the right hypochondrium or to the left iliac fossa.

## Liver

The liver is the largest gland in the body and is concerned with the metabolism of the products of carbohydrate and protein digestion, which are conveyed to it through the portal vein. The carbohydrate is stored in the liver as glycogen, and the amino acids are synthesized into proteins or deaminized with the formation of urea. The liver synthesizes heparin, an anticoagulant substance, and has an important detoxicating function. It produces bile pigments from the hemoglobin of worn-out red blood corpuscles and secretes bile salts; these together are conveyed to the duodenum by the biliary ducts.

The liver is held in position in the upper part of

the abdominal cavity by the attachment of the hepatic veins to the inferior vena cava. The peritoneal ligaments and the tone of the abdominal muscles play a minor role in its support. This fact is important surgically, since even if the peritoneal ligaments are cut, the liver can be only slightly rotated.

The liver is a soft, friable structure enclosed in a fibrous capsule. Its close relationship to the lower ribs must be emphasized. Fractures of the lower ribs or penetrating wounds of the thorax or upper abdomen are common causes of liver injury. Blunt traumatic injuries from automobile accidents are also common, and severe hemorrhage accompanies tears of this organ.

Because anatomical research has shown that the bile ducts, hepatic arteries, and portal vein are distributed in a segmental manner, appropriate ligation of these structures allows the surgeon to remove large portions of the liver in patients with severe traumatic lacerations of the liver or with a liver tumor. Even large, localized carcinomatous metastatic tumors have been successfully removed.

The majority of amebic abscesses in the liver are solitary and are located in the upper part of the right lobe. Diaphragmatic irritation may cause referred pain over the shoulder, since the nervous impulses ascend in the phrenic nerves (C3, 4, and 5); and the supraclavicular nerves (C3 and 4) supply the skin in this area.

*Liver biopsy* is a common diagnostic procedure. With the patient holding his or her breath in full expiration, to reduce the size of the costodiaphragmatic recess and to reduce the likelihood of damage to the lung, a needle is inserted through the right eighth or ninth intercostal space in the midaxillary line. The needle passes through the diaphragm into the liver and a small biopsy of liver tissue is removed for microscopic examination.

## Portal Vein

The *portal vein* conveys about 70 percent of the blood to the liver. The remaining 30 percent is oxygenated blood, which passes to the liver via the hepatic artery. The wide angle of union of the splenic vein with the superior mesenteric vein to form the portal vein has been shown to lead to streaming of the blood flow in the portal vein. The right lobe of the liver receives blood mainly from the intestine, whereas the left lobe plus the quad-

rate lobe and the caudate lobe receive blood from the stomach and the spleen. This distribution of blood may explain the distribution of secondary malignant deposits in the liver.

*Portal hypertension* is a common clinical condition, and for this reason the list of portal-systemic anastomoses should be remembered. (See p. 238.) Enlargement of the portal-systemic connections is frequently accompanied by congestive enlargement of the spleen. *Portacaval shunts* for the treatment of portal hypertension may involve the anastomosis of the portal vein, since it lies within the lesser omentum, to the anterior wall of the inferior vena cava behind the entrance into the lesser sac. The splenic vein may be anastomosed to the left renal vein, after removing the spleen.

The important *subphrenic spaces* and their relationship to the liver have been described on page 215. It should be realized that under normal conditions these are potential spaces only, and the peritoneal surfaces are in contact. An abnormal accumulation of gas or fluid is necessary for the separation of the peritoneal surfaces. The anterior surface of the liver is normally dull on percussion. Perforation of a gastric ulcer is often accompanied by a loss of liver dullness, due to the accumulation of gas over the anterior surface of the liver and in the subphrenic spaces.

## Gallbladder and Biliary Ducts

The liver excretes bile at a constant rate of about 40 ml per hour. When digestion is not taking place, the sphincter of Oddi remains closed, and bile accumulates in the gallbladder. The gallbladder has the following functions: It (1) concentrates bile, (2) stores bile, (3) selectively absorbs bile salts, keeping the bile acid, (4) excretes cholesterol, and (5) secretes mucus.

A number of congenital anomalies occur in the biliary system, as well as a number of variations in the blood supply to the gallbladder (Figs. 5-70, 5-71, and 5-72). The medical student should be aware of their existence, though the details need not be committed to memory.

The gallbladder is supplied by nerves from both sides of the body via the celiac plexus. Spasm of the smooth muscle of the wall of the gallbladder in attempting to expel a gallstone gives rise to referred pain in the epigastrium. Afferent fibers from the

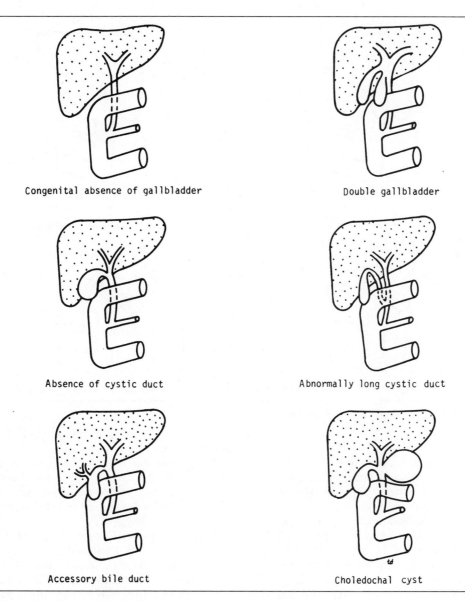

Congenital absence of gallbladder

Double gallbladder

Absence of cystic duct

Abnormally long cystic duct

Accessory bile duct

Choledochal cyst

**Fig. 5-70. Some common congenital anomalies of gallbladder.**

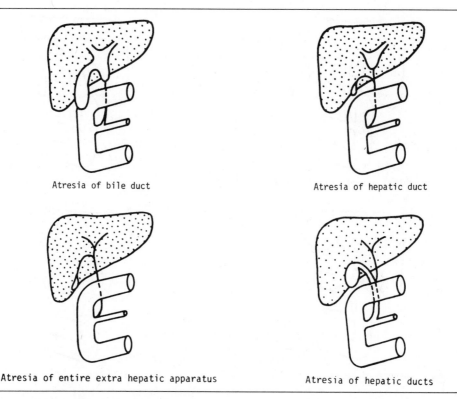

Atresia of bile duct

Atresia of hepatic duct

Atresia of entire extra hepatic apparatus

Atresia of hepatic ducts

**Fig. 5-71. Some common congenital anomalies of biliary ducts.**

gallbladder and stomach are believed to enter the same segment of the spinal cord, so that gallbladder disease may produce symptoms similar to those of gastric ulceration.

Inflammation of the gallbladder may cause irritation of the subdiaphragmatic parietal peritoneum, which is supplied in part by the phrenic nerve (C3, 4, and 5). This may give rise to referred pain over the shoulder, since the skin in this area is supplied by the supraclavicular nerves (C3 and 4).

Obstruction of the biliary ducts with a gallstone or by compression by a tumor of the pancreas results in the backup of bile in the ducts and the development of *jaundice*. The impaction of a stone in the ampulla of Vater may result in the passage of infected bile into the pancreatic duct, producing *pancreatitis*. The anatomical arrangement of the terminal part of the bile duct and the main pancreatic duct is subject to considerable variation. The type of duct system present will determine

whether or not infected bile is likely to enter the pancreatic duct.

Unlike the appendix, which has a single arterial supply, the gallbladder very rarely becomes gangrenous. In addition to the cystic artery, the gallbladder also receives small vessels from the visceral surface of the liver.

Gallstones present in the gallbladder have been known to ulcerate through the wall into the transverse colon or into the duodenum. In the former case they are passed naturally per rectum, but in the latter case they may be held up at the ileocecal junction, producing intestinal obstruction.

Sonograms can now be used to demonstrate the gallbladder (Fig. 5-73).

## Pancreas

Anatomically, the pancreas is deeply placed within the abdomen and is well protected. However, blows on the abdomen and automobile accidents may cause tears in its structure.

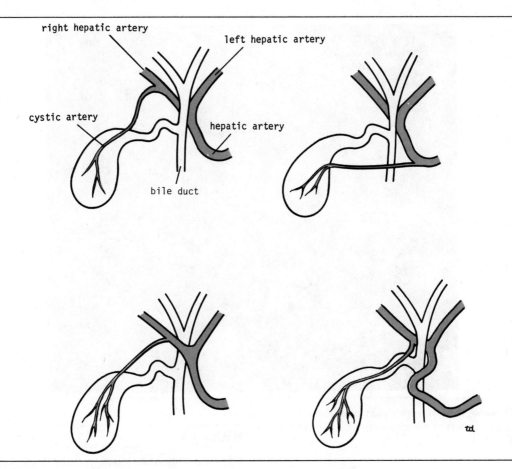

**Fig. 5-72. Some common variations of blood supply to gallbladder.**

The close relation of the pancreas to the posterior wall of the stomach may result in the organ's involvement in a deep, penetrating ulcer of the stomach. The lesser sac is the immediate anterior relation of the pancreas, and inflammation or damage to the pancreas may result in effusion of peritoneal fluid into this space. *Pseudocysts of the pancreas* are due to cystic accumulations of fluid in the lesser sac.

Because of the intimate relation of the head of the pancreas to the bile duct, cancer of the head of the pancreas often causes obstructive jaundice.

The presence of the tail of the pancreas in the lienorenal ligament sometimes results in its damage during splenectomy.

During development, the ventral bud of the pancreas may become tethered to the posterior abdominal wall and fail to fuse correctly with the dorsal bud. Such a rare condition is known as *anular pancreas* and may cause duodenal obstruction.

## Spleen

A pathologically enlarged spleen extends downward and medially. The left colic flexure and the phrenicocolic ligament prevent a direct downward enlargement of the organ. As the enlarged spleen projects below the left costal margin, its notched anterior border can be recognized by palpation through the anterior abdominal wall.

The tail of the pancreas lies in the lienorenal ligament and in contact with the hilus of the spleen.

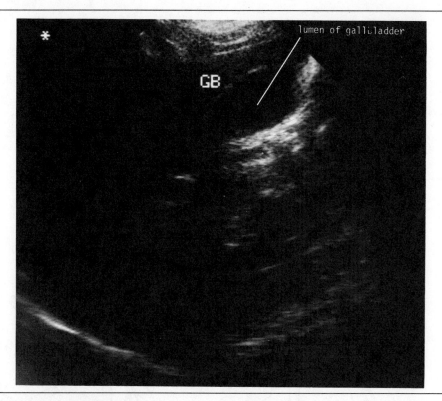

**Fig. 5-73. Sonogram of upper part of abdomen showing the lumen of the gallbladder. (Courtesy of Dr. M. C. Hill.)**

Damage to the pancreas may occur during splenectomy operations.

The spleen is situated at the beginning of the splenic vein, and in cases of portal hypertension it often enlarges from venous congestion.

Although anatomically the spleen gives the appearance of being well protected, automobile accidents of the crushing or runover type commonly produce laceration of the spleen. Penetrating wounds of the lower left thorax may also damage the spleen.

*Accessory spleens* may be present in the area of the hilus or in the splenic ligaments. Sometimes they are present in the transverse mesocolon or mesentery of the small intestine. They exist in about 10 percent of subjects, and if forgotten during a splenectomy for the treatment of acholuric jaundice, for example, they may enlarge and result in a return of the symptoms.

## Kidneys

The kidney is maintained in its normal position by intra-abdominal pressure and by its connections with the perirenal fat and renal fascia. Each kidney moves slightly with respiration. The right kidney lies at a slightly lower level than the left kidney, and the lower pole may be palpated in the right lumbar region at the end of deep inspiration in a person with poorly developed abdominal musculature. Should the amount of perirenal fat be reduced, the mobility of the kidney may become excessive and produce symptoms of renal colic, due to kinking of the ureter. Excessive mobility of the kidney leaves the suprarenal gland undisturbed, since the latter occupies a separate compartment in the renal fascia.

Malignant tumors of the kidney have a strong tendency to spread along the renal vein. The left renal vein receives the left testicular vein in the male, and this may rarely become blocked, producing left-sided *variocele*. (See p. 192.)

The kidney develops a pelvic organ and only later ascends into the abdomen to take up its final position. Rarely is the ascent arrested. For a consideration of horseshoe kidney and other congenital anomalies, see Figure 5-44.

Supernumerary renal arteries are relatively common. They represent persistent fetal renal arteries, which grow in sequence from the aorta to supply the kidney as it ascends from the pelvis. Their occurrence is clinically important, since a supernumerary artery may cross the pelviureteral junction and obstruct the outflow of urine, producing dilation of the calyces and pelvis, a condition known as *hydronephrosis* (Fig. 5-44).

The afferent nerves from the kidney enter the spinal cord at segments T10 and T12. Most common diseases of the kidney such as nephritis are unaccompanied by pain. If pain does occur, it is usually a dull ache in the lumbar region.

## Ureter

There are three sites of anatomical narrowing of the ureter where urinary calculi may be arrested, namely, the pelviureteral junction, the pelvic brim, and where the ureter enters the bladder. The majority of calculi cast a shadow on radiographic examination, and these are usually seen along the normal line of the ureter. It should be remembered that the ureter runs down in front of the tips of the transverse processes of the lumbar vertebrae, crosses the region of the sacroiliac joint, swings out to the ischial spine, and then turns medially to the bladder.

The renal pelvis and the ureter send their afferent nerves into the spinal cord at segments T11 and T12 and L1 and L2. In ureteral colic, strong peristaltic waves of contraction pass down the ureter in an attempt to pass the stone onward. The spasm of the smooth muscle causes an agonizing colicky pain, which is referred to the skin areas that are supplied by these segments of the spinal cord.

The pain extends from the loin to the groin. When a calculus enters the low part of the ureter, the pain is felt at a lower level and is often referred to the testis in the male and the labium majus in the female. The cremaster muscle commonly contracts reflexly due to nervous impulses passing down the genital branch of the genitofemoral nerve. A calculus in the lower pelvic part of the ureter may produce pain that is referred to the tip of the penis in the male. Sometimes ureteral pain is referred along the femoral branch of the genitofemoral nerve, so that pain is experienced in the front of the thigh.

The physician should be aware of the common congenital anomalies of the ureter, which are summarized in Figure 5-45. It should be noted that in ectopic ureter, the lower end of the ureter does not open into the bladder, but in the male drains into the seminal vesicle or prostatic urethra, and in the female may open into the vagina. In all cases the opening is below the bladder sphincter, and the patient is incontinent.

## Suprarenal Glands

The suprarenal glands, together with the kidneys, are enclosed within the renal fascia; the suprarenal glands, however, lie in a separate compartment, which allows the two organs to be separated easily at operation.

Because of their position on the posterior abdominal wall, few tumors of the suprarenal glands can be palpated. The injection of oxygen into the retroperitoneal tissues is a safe and easy procedure, which allows the glands to be visualized radiographically. This procedure may be combined with intranveous pyelography (see p. 276) and aortography for clearer visualization. CT scans can also be used.

At birth, the suprarenal glands are relatively very large due to the presence of the fetal cortex; later, when this part of the cortex involutes, the gland becomes reduced in size. While the process of involution is proceeding, the cortex is friable and it is very susceptible to damage and severe hemorrhage.

*Suprarenal cortical hyperplasia* is the commonest cause of Cushing's syndrome, the clinical manifestations of which include moon-shaped face, truncal obesity, hirsutism, and hypertension; if the syndrome occurs later in life, it may result from an adenoma or carcinoma of the cortex. Adrenocortical insufficiency (Addison's disease), which is characterized clinically by increased pigmentation, muscular weakness, weight loss, and hypotension, may be caused by tuberculous destruction or bilateral atrophy of both cortices.

*Pheochromocytoma*, a tumor of the medulla, produces a paroxysmal or sustained hypertension. The symptoms and signs result from the production of a large amount of catecholamines, which are then poured into the bloodstream.

*Lumbar sympathectomy* is performed mainly to produce a vasodilation of the arteries of the lower limb in patients with vasospastic disorders. The preganglionic sympathetic fibers that supply the vessels of the lower limb leave the spinal cord from segments T11 to L2. They synapse in the lumbar and sacral ganglia of the sympathetic trunks. The postganglionic fibers join the lumbar and sacral nerves and are distributed to the vessels of the limb as branches of these nerves. Additional postganglionic fibers pass directly from the lumbar ganglia to the common and external iliac arteries, but they follow the latter artery only down as far as the inguinal ligament. In the male a bilateral lumbar sympathectomy may be followed by loss of ejaculatory power, but erection is not impaired.

## Aorta

Localized or diffuse dilatations of the abdominal part of the aorta (aneurysms) usually occur below the origin of the renal arteries. The majority result from atherosclerosis and occur most commonly in elderly men. Large aneurysms should be surgically excised and replaced with a prosthetic graft.

Gradual occlusion of the bifurcation of the abdominal aorta, produced by atherosclerosis, results in the characteristic clinical symptoms of pain in the legs on walking (claudication) and impotence, the latter caused by lack of blood in the internal iliac arteries. In otherwise healthy individuals, surgical treatment by thromboendarterectomy or a bypass graft should be considered.

The bifurcation of the abdominal aorta where the lumen suddenly narrows may be a lodging site for an embolus discharged from the heart. Severe ischemica of the lower limbs results.

## Inferior Vena Cava

The inferior vena cava is commonly compressed during the later stages of pregnancy by the enlarged uterus. This produces edema of the ankles and feet and temporary varicose veins. Malignant retroperitoneal tumors may also compress the inferior vena cava.

## CLINICAL PROBLEMS

*Answers on page 961*

1. A 45-year-old man was admitted to the emergency room complaining of severe pain in the right iliac region. He had repeatedly vomited, and his temperature and pulse rate were elevated. His history indicated that he had been suffering from acute appendicitis and that the pain had suddenly increased. On examination, the muscles of the lower part of the anterior abdominal wall in the right iliac region showed rigidity. The diagnosis of peritonitis following perforation of the appendix was made. Can you explain the rigidity of the abdominal muscles in anatomical terms? What anatomical structure helps the body to localize inflammatory lesions of the peritoneum?

2. A 50-year-old man was admitted to the hospital following perforation of a gastric ulcer. He complained of severe epigastric pain, but also stated that he had discomfort over his right shoulder. Explain in anatomical terms the right shoulder discomfort.

3. A 65-year-old man who had a history of a chronic duodenal ulcer over a 20-year period was admitted to the hospital unconscious and exhibiting all the signs of a severe internal hemorrhage. Which blood vessel was likely to have been eroded?

4. At operation for treatment of a chronic gastric ulcer, it was found that the posterior wall of the patient's stomach was stuck down to the posterior abdominal wall. Which structures lie behind the stomach and were likely to be involved in the disease process? Name a large artery that runs behind the stomach and may become eroded by a chronic ulcer.

5. A medical student was surprised to see a patient who had a perforated peptic ulcer propped up in bed at an angle of about 45 de-

grees. In anatomical terms, explain the reason for the patient's being placed in this position.

6. A collection of infected material is present in the peritoneal cavity in the region of the anterior surface of the second part of the duodenum. Indicate in anatomical terms the pathways along which the material may spread to reach (a) the undersurface of the diaphragm and (b) the pelvic cavity.

7. Name the sites in the abdominal part of the peritoneal cavity where herniation of loops of small intestine may occur.

8. The diagnosis in a 50-year-old woman was carcinoma of the greater curvature of the stomach. What is the lymphatic drainage of the greater curvature of the stomach? Describe briefly the extent of the operation necessary to treat cancer of the stomach, and give your anatomical reasons.

9. Following a barium meal in a normal person, the outline of the barium shadow in the first part of the duodenum was smooth. In the remainder of the duodenum, the barium was broken up, giving a floccular appearance. In anatomical terms, can you explain these differences in appearance?

10. In the treatment of chronic gastric and duodenal ulcers, the surgeon attempts to reduce the amount of acid produced by the mucous membrane of the stomach. In anatomical and physiological terms, explain how this might be achieved.

11. The differential diagnosis between a perforated duodenal ulcer and a perforated appendix may prove difficult. Can you explain this in anatomical terms?

12. A 40-year-old obese woman known to have a history of gallstones was admitted to the hospital with intestinal obstruction. Knowing that large gallstones occasionally ulcerate through the wall of the gallbladder into adjacent viscera, where would you expect that such a gallstone is likely to cause obstruction?

13. A 5-year-old child, recovering from an extensive abdominal operation, persistently cried for her parents. After 4 days it was decided to reduce the amount of sedation, whereupon she had attacks of screaming, was extremely restless, and could not be pacified. The attending physician was suddenly called to the bedside and was told by the nurse that the abdominal wound had burst open and coils of gut were lying between the sheets. The physician was asked whether the gut was small or large intestine. How would you recognize a piece of small or large intestine? How would you distinguish between a piece of jejunum or ileum?

14. A surgeon performing a difficult appendectomy was confronted by a large number of adhesions in the peritoneal cavity in the right iliac region. He was unable to find the base of the appendix, although he could recognize the cecum. What anatomical structure on the cecum would enable him, without fail, to find the base of the appendix?

15. Gangrene of an inflamed appendix is common, but gangrene of an inflamed gallbladder is rare. What is the blood supply of these two organs and how do they differ?

16. The pain of acute appendicitis is felt first in the umbilical region and later in the right iliac region. Can you explain this in anatomical terms?

17. Diverticulosis of the colon is a common disease. Is there an anatomical explanation for the development of this condition?

18. Diseases of the gallbladder and the stomach produce pain or discomfort in the epigastrium. Trace the pain fibers from the gallbladder and stomach, and explain in anatomical terms why the pain sensation in these two organs is referred to the epigastrium.

19. The anatomical arrangement of the terminal part of the bile duct and the main pancreatic duct is subject to considerable variation. Which variations are likely to be associated with a pancreatitis should a gallstone become impacted at the lower end of the bile duct?

20. Following a splenectomy, it was noticed that pancreatic juice was exuding through the patient's abdominal wound. Is the pancreas likely to be damaged during a splenectomy? Which part of the pancreas?

21. On examination of the abdomen of a 35-year-old woman, a large swelling was found to extend downward and medially below the left costal margin. On percussion, a continuous band of dullness was noted to extend upward from the left of the umbilicus to the left axillary region. On palpation, a notch was felt along the anterior border of the swelling. Name the ana-

tomical structure producing the swelling, and state your reasons.

22. A patient with thrombocytopenic purpura was advised to have a splenectomy to stop the episodes of bleeding from the gums and gastrointestinal tract. The operation was successful. Eighteen months later the clinical features returned. Can you explain in anatomical terms the recurrence of the bleeding after the condition had apparently been cured by splenectomy?

23. An explorer in the Amazon jungle was found alive after having lost contact with the outside world for 6 months. On physical examination, he was found to be in an emaciated condition. On palpation of the abdomen, a rounded, smooth swelling appeared in the right loin at the end of inspiration. On expiration, the swelling moved upward and could no longer be felt. What anatomical structure could produce such a swelling?

24. An intravenous pyelogram revealed that a patient's left kidney was in its normal position, but the right kidney was situated in front of the right sacroiliac joint. Can you explain this on embryological grounds?

25. An examination of a patient revealed that she had a horseshoe kidney. What anatomical structure prevents a horseshoe kidney from ascending to a level above the umbilicus?

26. An intravenous pyelogram revealed that the calyces and pelvis of a patient's right kidney were grossly dilated (a condition known as *hydronephrosis*). What embryological anomaly may be responsible for this condition?

27. A 55-year-old woman was found rolling on her kitchen floor, crying out from agonizing pain in her abdomen. The pain came in waves and extended from the right loin to the groin and to the front of the right thigh. An anteroposterior radiograph of the abdomen revealed a calculus in the right ureter. (a) What causes the pain when a ureteral calculus is present? (b) Why is the pain felt in such an extensive area? (c) Where does one look for the course of the ureter in a radiograph? (d) Where along the ureter is a calculus likely to be held up?

28. Which congenital anomaly of the ureter is likely to present as a case of urinary incontinence?

29. The marginal artery is important to surgeons operating on the colon. What is the so-called marginal artery?

30. A 58-year-old man complained of pain in the right leg when walking. He found he was able to walk a hundred yards, and then the intensity of the pain made him rest. After 5 minutes' rest he could progress for another hundred yards before resting again. This syndrome is known as *intermittent claudication*. In this case it was due to occlusive disease of the arteries of the right leg. The patient was advised by his vascular surgeon to have a right lumbar sympathectomy. In anatomical terms, explain why such a sympathectomy may improve the condition.

31. A 65-year-old woman was admitted to the hospital with progressive jaundice of 3 months' duration and weight loss. She had not experienced any colicky pain. On examination, a soft swelling could be felt in the abdomen in the region of the tip of the right ninth costal cartilage. A diagnosis of cancer of the head of the pancreas was made. What anatomical structure is responsible for the swelling?

32. A 50-year-old woman with a history of flatulent dyspepsia suddenly experienced an excruciating colicky pain across the upper part of the abdomen. On examination after the attack, some rigidity and tenderness was noted in the right hypochondrium. Two days later the patient became jaundiced, and it was noticed that the degree of jaundice varied from day to day. The diagnosis of biliary colic was made. (a) Why should a person passing a gallstone experience pain? (b) Why is the pain experienced in the area described above? (c) Why does the jaundice vary in intensity?

33. A patient with advanced renal failure is waiting for a suitable kidney donor for a transplantation operation. His blood urea nitrogen level is high and he is developing some of the signs of uremia. What method(s) is (are) available to keep this patient alive until a suitable donor can be found?

34. A 56-year-old man visited his physician complaining that he experiences severe pain in both legs when taking long walks. He has noticed recently that the cramplike pain occurs after walking only a hundred yards. On questioning, he said that the pain quickly disappears on rest

only to return after he walks the same distance. After questioning the patient further on his general health, the physician asked him about his sexual activity. The patient replied that his sex life had been entirely satisfactory until 6 weeks ago when he experienced difficulty with erection. Using your knowledge of anatomy, explain what is wrong with this patient.

35. A 23-year-old woman, who was 8 months pregnant, told her obstetrician that she had recently noticed that her feet and ankles were swollen at the end of the day. She said that the swelling was worse if she had been standing for long periods. She also noticed that the veins around her ankles were becoming very prominent. Assuming that this patient's general health is normal, can you explain the changes in her legs?

# NATIONAL BOARD TYPE QUESTIONS

*Answers on page 984*

**In each of the following questions, answer:**
- (a) If (1) ONLY IS CORRECT
- (b) If (2) ONLY IS CORRECT
- (c) If BOTH (1) AND (2) ARE CORRECT, AND
- (d) If NEITHER (1) NOR (2) IS CORRECT

1. Which of the following statements is (are) correct?
   - (1) The right lateral paracolic gutter communicates superiorly with the right subphrenic spaces and inferiorly with the pelvic cavity.
   - (2) Referred pain resulting from acute appendicitis is felt initially in the region of the umbilicus.

2. Which of the following statements is (are) correct?
   - (1) The body of the gallbladder lies between the caudate and the right lobes of the liver.
   - (2) The ureters enter the pelvis at the point where the common iliac arteries bifurcate into internal and external iliac arteries.

3. Which of the following statements is (are) correct?
   - (1) The phrenicocolic ligament extends from the diaphragm to the right colic flexure.
   - (2) The tail of the pancreas extends into the gastrosplenic omentum and thus lies at the hilus of the spleen.

4. Which of the following statements is (are) correct?
   - (1) The portal vein is formed by the union of the inferior mesenteric vein and the splenic vein.
   - (2) The fundus of the gallbladder lies opposite the tip of the right seventh costal cartilage.

5. Which of the following statements is (are) correct?
   - (1) The left kidney is related posteriorly to the eleventh and twelfth ribs.
   - (2) Sympathetic innervation to the transverse colon is by nerves derived from both the superior and inferior mesenteric plexuses.

**In the following questions, answer:**
- (a) If (1), (2), AND (3) ONLY ARE CORRECT
- (b) If (1) AND (3) ONLY ARE CORRECT
- (c) If (2) AND (4) ONLY ARE CORRECT
- (d) If (4) ONLY IS CORRECT, OR
- (e) If ALL ARE CORRECT

6. Which of the following statements is (are) correct?
   - (1) The splenic artery runs along the upper border of the pancreas and lies behind the stomach.
   - (2) The lesser curvature of the stomach receives blood from the right and left gastroepiploic arteries.
   - (3) The lymphatic drainage from the lower one-third of the esophagus passes into the celiac nodes.
   - (4) The parasympathetic innervation of the sigmoid colon is from the vagus nerves.

7. Which of the following statements is (are) correct?
   - (1) The lesser omentum forms the anterior boundary into the lesser sac.
   - (2) The thoracic duct leaves the abdominal cavity by passing through the aortic opening.
   - (3) The bile duct (common bile duct) descends behind the first part of the duodenum.
   - (4) The quadrate lobe of the liver drains into the left hepatic duct.

8. Which of the following statements is (are) correct?
   (1) The preaortic lymph nodes drain lymph from the suprarenals and kidneys.
   (2) The branches of the superior mesenteric artery serving the jejunum form more arcades than those serving the ileum.
   (3) The coronary ligament of the liver lies anterior to the abdominal portion of the esophagus.
   (4) The attachment of the hepatic veins to the inferior vena cava is one of the most important supports of the liver.

9. Which of the following statements is (are) correct?
   (1) The circular muscle of the lower end of the ileum serves as a sphincter at the junction of the ileum with the cecum.
   (2) The innervation of the parietal peritoneum at the periphery of the undersurface of the diaphragm is from the lower thoracic spinal nerves.
   (3) The pyloric sphincter closes as the result of the activity of the sympathetic part of the autonomic nervous system.
   (4) The anterior vagal trunk in front of the abdominal part of the esophagus is derived mainly from the left vagus nerve.

10. Which of the following statements is (are) correct?
    (1) The plicae circulares are more prominent at the distal end of the ileum than in the jejunum.
    (2) Peyer's patches are present in the mucous membrane of the lower ileum along the antimesenteric border.
    (3) The spleen is easily recognized by the presence of notches on its posterior border.
    (4) The cystic duct joins the common hepatic duct to form the bile duct.

11. Which of the following statements concerning the pancreas is (are) correct?
    (1) The main pancreatic duct opens into the third part of the duodenum.
    (2) The pancreas receives part of its arterial supply from the splenic artery.
    (3) The uncinate process projects from the body of the pancreas.
    (4) The bile duct (common bile duct) lies posterior to the head of the pancreas.

12. The hilus of the right kidney contains the following important structures:
    (1) Branches of the renal artery
    (2) Renal pelvis
    (3) Sympathetic nerve fibers
    (4) Tributaries of the renal vein

13. Which of the following statements is (are) **true** concerning the abdominal part of the aorta?
    (1) It enters the abdomen in front of the twelfth thoracic vertebra.
    (2) From its anterior surface arise the celiac, superior mesenteric, and inferior mesenteric arteries.
    (3) It bifurcates into the two common iliac arteries in front of the fourth lumbar vertebra.
    (4) It lies on the right side of the inferior vena cava.

**Select the best response:**

14. All of the following structures lie in front of the right kidney **except** the:
    (a) Right lobe of liver
    (b) Second part of the duodenum
    (c) Right colic flexure
    (d) Coils of intestine
    (e) Head of the pancreas

15. The following statements concerning the abdominal part of the sympathetic trunk are **not true except** that:
    (a) It enters the abdomen behind the lateral arcuate ligament.
    (b) The trunk possesses six segmentally arranged ganglia.
    (c) All the ganglia receive white rami communicantes.
    (d) Gray rami communicantes are given off to the lumbar spinal nerves.
    (e) The splanchnic nerves from the thorax join the trunks below the diaphragm.

16. The following statements concerning the lumbar plexus are true **except:**
    (a) The plexus lies within the psoas muscle.
    (b) The plexus is formed from the posterior rami of the upper four lumbar nerves.
    (c) The femoral nerve emerges from the lateral border of the psoas muscle.
    (d) The obturator nerve emerges from the medial border of the psoas muscle.

(e) The iliohypogastric nerve emerges from the lateral border of the psoas muscle.

17. The following veins form important portal-systemic anastomoses **except:**
    (a) The esophageal branches of the left gastric vein and tributaries of the azygos veins.
    (b) The superior rectal vein and the inferior vena cava.
    (c) The paraumbilical veins and the superficial veins of the anterior abdominal wall.
    (d) The veins of the ascending and descending parts of the colon with the lumbar veins.
    (e) The veins from the bare areas of the liver with the phrenic veins.

18. Which of the following roots of the lumbar plexus contribute to the sacral plexus?
    (a) L3
    (b) L2
    (c) L4
    (d) L1
    (e) All of the above

19. The following statements concerning the ureters are true **except:**
    (a) Both have three anatomical sites that are constricted.
    (b) Both receive their blood supply from the testicular and ovarian arteries.
    (c) Both are separated from the transverse processes of the lumbar vertebrae by the psoas muscles.
    (d) Both pass anterior to the testicular or ovarian vessels.
    (e) Both lie anterior to the sacroiliac joints.

20. Concerning the inferior mesenteric artery, all of the following statements are true **except:**
    (a) Its colic branch supplies the descending colon.
    (b) It gives off the inferior pancreaticoduodenal artery.
    (c) It supplies the sigmoid colon.
    (d) Its branches contribute to the marginal artery.

    (e) It arises from the aorta immediately below the third part of the duodenum.

21. The following statements concerning the left suprarenal gland are incorrect **except:**
    (a) It extends behind the inferior vena cava.
    (b) It is separated from the left kidney by the pararenal fat.
    (c) Its vein drains into the left renal vein.
    (d) It is usually located on the upper pole and lateral border of the left kidney.
    (e) The medulla is innervated by postganglionic sympathetic nerve fibers.

22. Which of the following structures is **not** present within the lesser omentum?
    (a) Portal vein
    (b) Bile duct
    (c) Inferior vena cava
    (d) Hepatic artery
    (e) Lymph nodes

**Match the area of the stomach on the left with the appropriate arterial supply on the right:**

23. Fundus
24. Right half of greater curvature
25. Left half of greater curvature

(a) Left gastroepiploic artery
(b) Left gastric artery
(c) Short gastric arteries
(d) Right gastric artery
(e) None of the above

**Match the arterial branches on the left with their origin on the right:**

26. Gastroduodenal artery
27. Middle colic artery
28. Left gastroepiploic artery
29. Ileocolic artery
30. Superior rectal artery

(a) Inferior mesenteric
(b) Hepatic
(c) Splenic
(d) Superior mesenteric
(e) None of the above

# 6. The Pelvis: Part I
# The Pelvic Walls

The pelvis* is the region of the trunk that lies below the abdomen. Although the abdominal and pelvic cavities are continuous, the two regions are described separately.

## SURFACE ANATOMY
### Surface Landmarks

#### Iliac Crest

This can be felt through the skin along its entire length (Figs. 6-1, 6-2, and 6-3).

#### Anterior Superior Iliac Spine

This is situated at the anterior end of the iliac crest and lies at the upper lateral end of the fold of the groin (Figs. 6-1, 6-2, and 6-3).

*The term pelvis is loosely used to describe the region where the trunk and lower limbs meet. The word "pelvis" means a basin and is more correctly applied to the skeleton of the region, i.e., the pelvic girdle or bony pelvis.

#### Posterior Superior Iliac Spine

This is situated at the posterior end of the iliac crest (Fig. 6-3). It lies at the bottom of a small skin dimple and on a level with the second sacral spine, which coincides with the lower limit of the subarachnoid space.

#### Pubic Tubercle

This can be felt on the upper border of the pubis (Figs. 6-1, 6-2, and 6-3). Attached to it is the medial end of the inguinal ligament. The tubercle can be palpated easily in the male by inverting the scrotum from below with the examining finger. In the female the pubic tubercle can be palpated through the lateral margin of the labium majus.

#### Symphysis Pubis

This is the cartilaginous joint that lies in the midline between the bodies of the pubic bones (Figs. 6-3 and 6-5). It can be palpated as a solid structure through the fat that is present in this region.

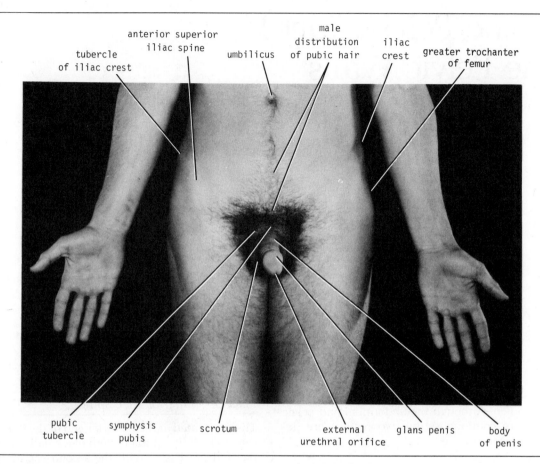

Fig. 6-1. Anterior view of pelvis of a 27-year-old male.

## Pubic Crest

This is the ridge of bone on the superior surface of the pubic bone, medial to the pubic tubercle (Figs. 6-3 and 6-5).

## Spinous Processes of Sacrum

These processes (Fig. 6-3) are fused with each other in the midline to form the *median sacral crest.* The crest can be felt beneath the skin in the uppermost part of the natal cleft between the buttocks.

## Sacral Hiatus

This is situated on the posterior aspect of the lower end of the sacrum, and it is here that the extradural space terminates (Fig. 6-3). The hiatus lies about 2 inches (5 cm) above the tip of the coccyx and beneath the skin of the natal cleft.

## Coccyx

The inferior surface and tip of the coccyx (Fig. 6-3) can be palpated in the natal cleft about 1 inch (2.5 cm) behind the anus. The anterior surface of the coccyx may be palpated with a gloved finger in the anal canal.

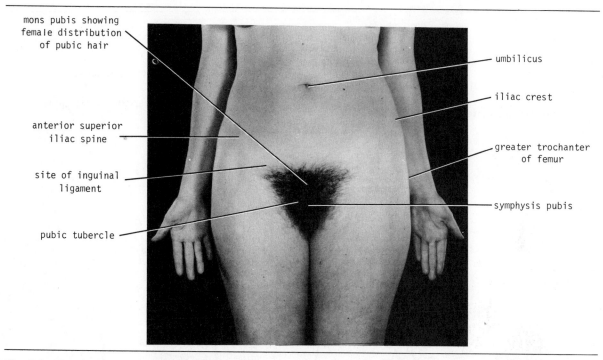

mons pubis showing female distribution of pubic hair

umbilicus

iliac crest

anterior superior iliac spine

greater trochanter of femur

site of inguinal ligament

symphysis pubis

pubic tubercle

**Fig. 6-2. Anterior view of pelvis of a 29-year-old female.**

## Viscera

### URINARY BLADDER

In the young child the bladder is an abdomino-pelvic organ, and when filled, it can be palpated through the anterior abdominal wall above the symphysis pubis. With growth, the pelvic cavity enlarges and the bladder becomes a pelvic organ (Fig. 6-4). However, even in the adult, when the bladder fills, its superior wall rises out of the pelvis and may be palpated through the anterior abdominal wall.

### UTERUS

Toward the end of the second month of pregnancy, the fundus of the uterus may be palpated through the lower part of the anterior abdominal wall. With the progressive enlargement of the uterus, the fundus rises above the level of the umbilicus and reaches the region of the xiphoid process by the ninth month of pregnancy (Fig. 6-4). Later, when the presenting part of the fetus, usually the head, descends into the pelvis, the fundus of the uterus descends also.

### Rectal and Vaginal Examinations as a Means of Palpating the Pelvic Viscera

Bimanual recto-abdominal and vaginal-abdominal examinations are extremely valuable methods of palpating the pelvic viscera; they are described in detail on pages 395–399.

## THE PELVIS

The bony pelvis is composed of four bones: the two innominate bones, or hip bones, which form the lateral and anterior walls, and the sacrum and the coccyx, which are part of the vertebral column, form the back wall (Fig. 6-5). The pelvis is divided into two parts by the *pelvic brim*, which is formed by the *sacral promontory* behind, the *iliopectineal lines* laterally, and the *symphysis pubis* anteriorly.

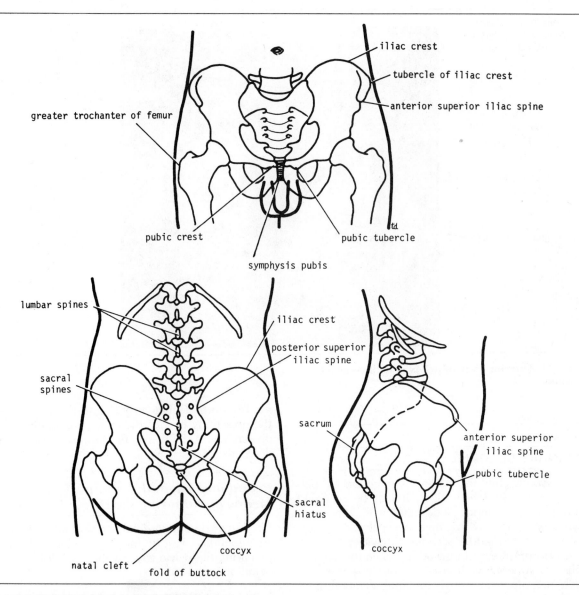

**Fig. 6-3. Relationship between different parts of pelvis and body surface.**

**Fig. 6-4. (A) Surface anatomy of empty bladder and full bladder and (B) height of fundus of uterus at various months of pregnancy.**

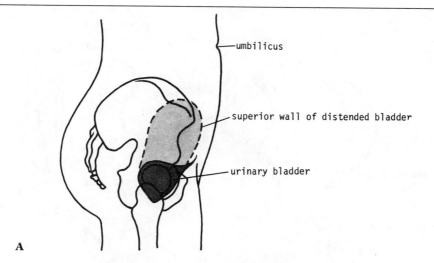

A

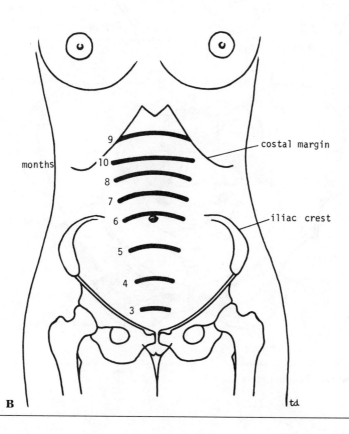

B

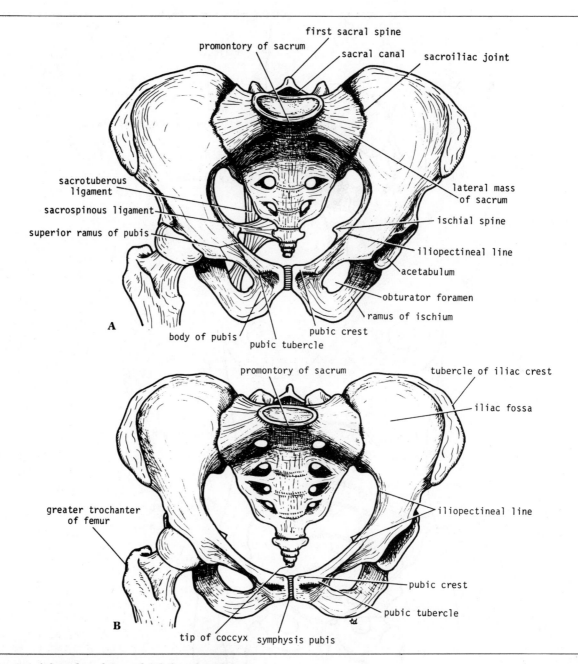

**Fig. 6-5. (A) Male pelvis and (B) female pelvis, as seen on anterior view.**

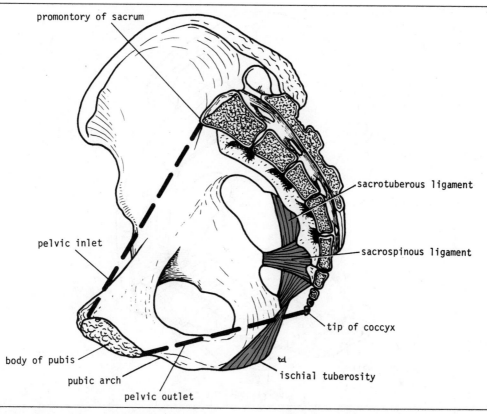

promontory of sacrum

pelvic inlet

body of pubis

pubic arch

pelvic outlet

sacrotuberous ligament

sacrospinous ligament

tip of coccyx

ischial tuberosity

**Fig. 6-6. Right half of pelvis, showing the pelvic inlet, pelvic outlet, and sacrotuberous and sacrospinous ligaments.**

Above the brim is the *false pelvis*, or *greater pelvis*, which forms part of the abdominal cavity. Below the brim is the *true pelvis*, or *lesser pelvis*.

It is important that the student, at the outset, understand the correct orientation of the bony pelvis relative to the trunk, with the individual standing in the anatomical position. The front of the symphysis pubis and the anterior superior iliac spines should lie in the same vertical plane. This means that the pelvic surface of the symphysis pubis faces upward and backward and the anterior surface of the sacrum is directed forward and downward.

## The False Pelvis

The false pelvis is of little clinical importance. It is bounded behind by the lumbar vertebrae, laterally by the iliac fossae and the iliacus muscles, and in front by the lower part of the anterior abdominal wall. The false pelvis flares out at its upper end and should be considered as part of the abdominal cavity. It supports the abdominal contents and after the third month of pregnancy helps to support the gravid uterus. During the early stages of labor it helps to guide the fetus into the true pelvis.

## The True Pelvis

The true pelvis is a bowl-shaped structure that contains and protects the lower parts of the intestinal and urinary tracts and the internal organs of reproduction. Knowledge of the shape and dimensions of the female pelvis is of great importance for obstetrics, since it is the bony canal through which the child passes during birth.

The true pelvis has an inlet, an outlet, and a cavity. *The pelvic inlet*, or *pelvic brim* (Fig. 6-6), is

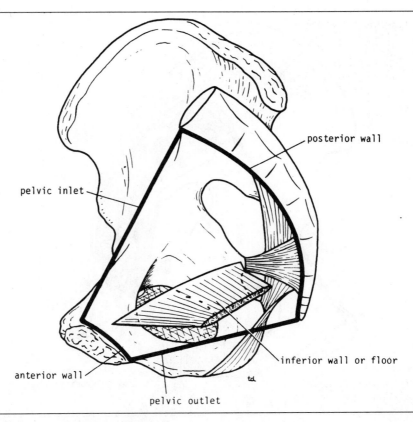

**Fig. 6-7. Right half of pelvis, showing pelvic walls.**

bounded posteriorly by the sacral promontory, laterally by the iliopectineal lines, and anteriorly by the symphysis pubis (Fig. 6-5).

The *pelvic outlet* (Fig. 6-6) is bounded posteriorly by the coccyx, laterally by the ischial tuberosities, and anteriorly by the pubic arch (Figs. 6-6 and 6-12). The pelvic outlet does not present a smooth outline but has three wide notches. Anteriorly, there is the *pubic arch* between the ischiopubic rami, and, laterally, there are the sciatic notches. The sciatic notches are divided by the *sacrotuberous* and *sacrospinous ligaments* (Figs. 6-5 and 6-6) into the *greater and lesser sciatic foramina* (see page 312). From a practical standpoint, since the sacrotuberous ligaments are strong and relatively inflexible, they should be considered to form part of the perimeter of the pelvic outlet. Thus, the outlet is diamond-shaped, with the ischiopubic rami and the symphysis pubis forming the boundaries

in front, and the sacrotuberous ligaments and the coccyx forming the boundaries behind.

The *pelvic cavity* lies between the inlet and the outlet. It is a short, curved canal, with a shallow anterior wall and a much deeper posterior wall (Fig. 6-6).

## Structure of the Pelvic Walls

The walls of the pelvis are formed by bones and ligaments that are partly lined with muscles covered with fascia and parietal peritoneum. The pelvis has anterior, posterior, and lateral walls and it also has an inferior wall or floor (Fig. 6-7).

### ANTERIOR PELVIC WALL

The anterior pelvic wall is the shallowest wall and is formed by the posterior surfaces of the bodies of the pubic bones, the pubic rami, and the symphysis pubis (Fig. 6-8).

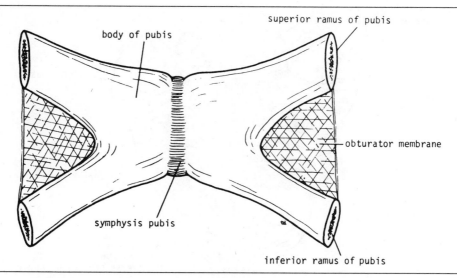

**Fig. 6-8. Anterior wall of pelvis (posterior view).**

## POSTERIOR PELVIC WALL

The posterior pelvic wall is extensive and is formed by the sacrum and coccyx (Fig. 6-9) and by the piriformis muscles (Fig. 6-11) and their covering of parietal pelvic fascia.

### Sacrum

The sacrum consists of five rudimentary vertebrae fused together to form a wedge-shaped bone, which is concave anteriorly (Fig. 6-9). The upper border or base of the bone articulates with the fifth lumbar vertebra. The narrow inferior border articulates with the coccyx. Laterally, the sacrum articulates with the two innominate (hip) bones to form the *sacroiliac joints* (Fig. 6-5). The anterior and upper margins of the first sacral vertebra bulge forward as the posterior margin of the pelvic inlet, which is known as the *sacral promontory*.

The vertebral foramina are present and, together, form the sacral canal. The laminae of the fifth sacral vertebra, and sometimes those of the fourth also, fail to meet in the midline, forming the *sacral hiatus* (Fig. 6-9). The *sacral canal* contains part of the cauda equina, filum terminale, and meninges, down as far as the lower border of the second sacral vertebra. The lower part of the sacral canal contains the lower sacral and coccygeal nerve roots, the filum terminale, and fibro-fatty material (Fig. 6-10).

The anterior and posterior surfaces of the sacrum possess on each side four foramina for the passage of the anterior and posterior rami of the upper four sacral nerves (Fig. 6-9).

The sacrum is usually wider in proportion to its length in the female than in the male. The sacrum is tilted forward so that it forms an angle with the fifth lumbar vertebra, called the *lumbosacral angle*.

### Coccyx

The coccyx consists of four rudimentary vertebrae fused together to form a small triangular bone, which articulates at its base with the lower end of the sacrum (Fig. 6-9).

The coccygeal vertebrae consist of bodies only, but the first vertebra possesses a rudimentary *transverse process* and *cornua*. The cornua are the remains of the pedicles and superior articular processes and project upward to articulate with the sacral cornua (Fig. 6-9).

### Piriformis Muscle

The piriformis muscle arises from the front of the lateral masses of the sacrum and leaves the pelvis to enter the gluteal region by passing laterally

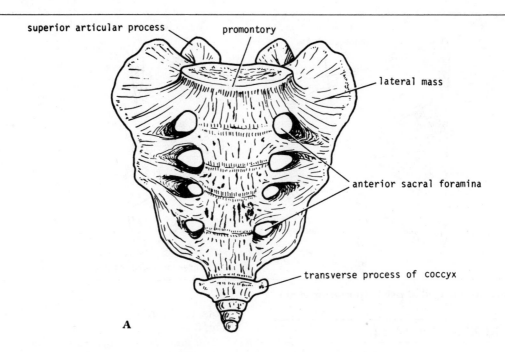

superior articular process

promontory

lateral mass

anterior sacral foramina

transverse process of coccyx

**A**

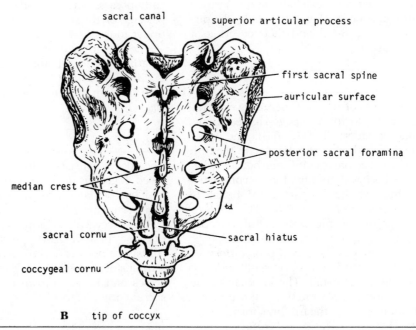

sacral canal

superior articular process

first sacral spine

auricular surface

posterior sacral foramina

median crest

sacral cornu

sacral hiatus

coccygeal cornu

**B**    tip of coccyx

**Fig. 6-9. Sacrum. (A) Anterior view and (B) posterior view.**

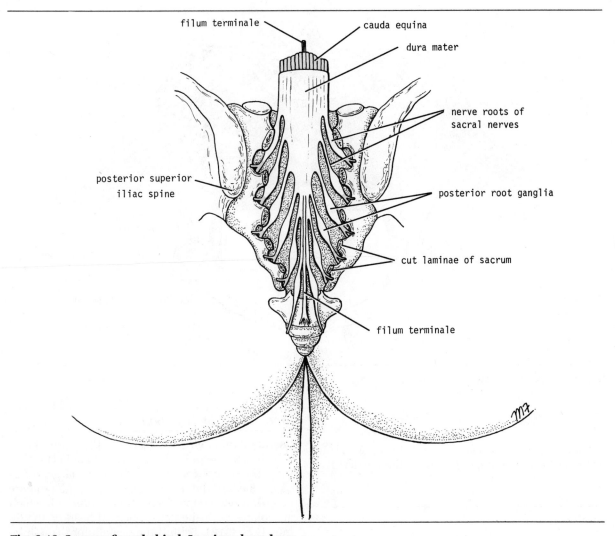

**Fig. 6-10. Sacrum from behind. Laminae have been removed to show sacral nerve roots lying within sacral canal. Note that in the adult, the spinal cord ends below at the level of the lower border of the first lumbar vertebra.**

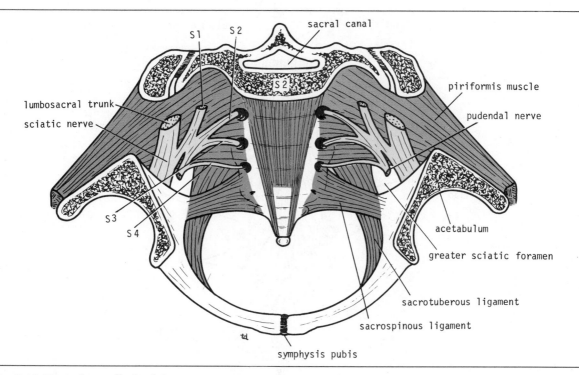

**Fig. 6-11. Posterior wall of pelvis.**

through the greater sciatic foramen (Fig. 6-11). It is inserted into the upper border of the greater trochanter of the femur.

## Action

It is a lateral rotator of the femur at the hip joint.

## Nerve Supply

It receives branches from the sacral plexus.

## LATERAL PELVIC WALL

The lateral pelvic wall is formed by part of the innominate bone below the pelvic inlet, the obturator membrane, the sacrotuberous and sacrospinous ligaments, and the obturator internus muscle and its covering fascia.

### Innominate, or Hip, Bone

Each innominate bone consists of the ilium, the ischium, and the pubis (Fig. 6-12). At puberty these three bones fuse together to form one large, irregular bone. On the outer surface of the innominate bone is a deep depression, the *acetabulum*, which articulates with the hemispherical head of the femur (Fig. 6-12). Behind the acetabulum is a large notch, the *greater sciatic notch*, which is separated from the *lesser sciatic notch* by the *spine of the ischium* (Fig. 6-12).

The upper flattened part of the bone is formed by the *ilium* and has the *iliac crest* running between the *anterior* and *posterior superior iliac spines*. Below these spines are the corresponding *in-*

**Fig. 6-12. Right innominate bone. (A) Medial surface and (B) lateral surface. Note the lines of fusion between the three bones—the ilium, the ischium, and the pubis.**

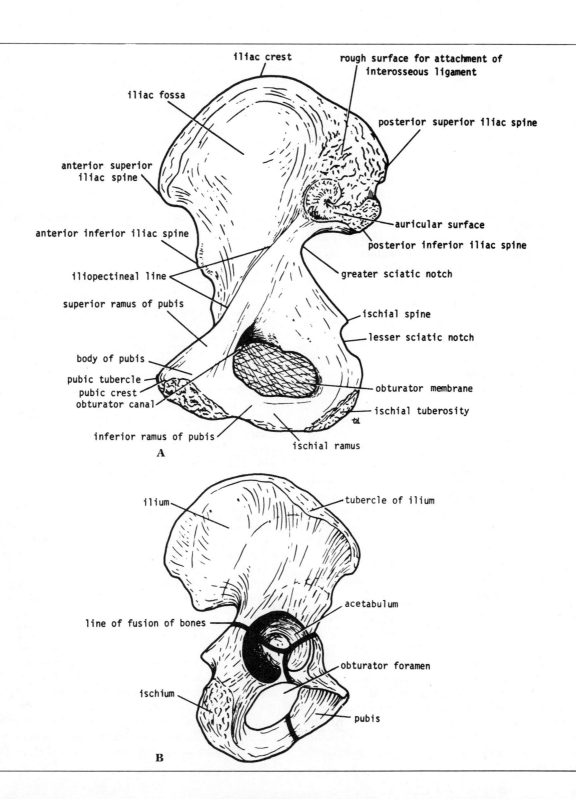

iliac crest

rough surface for attachment of
interosseous ligament

iliac fossa

posterior superior iliac spine

anterior superior
iliac spine

auricular surface

anterior inferior iliac spine

posterior inferior iliac spine

iliopectineal line

greater sciatic notch

superior ramus of pubis

ischial spine

lesser sciatic notch

body of pubis

pubic tubercle

pubic crest

obturator canal

obturator membrane

ischial tuberosity

inferior ramus of pubis

ischial ramus

A

ilium

tubercle of ilium

acetabulum

line of fusion of bones

obturator foramen

ischium

pubis

B

*ferior iliac spines.* On the inner surface of the ilium is the large *auricular surface* for articulation with the sacrum. The *iliopectineal line* runs downward and forward around the inner surface of the ilium and serves to divide the false from the true pelvis.

The *ischium* is the inferior and posterior part of the innominate bone and possesses an *ischial spine* and an *ischial tuberosity* (Fig. 6-12).

The *pubis* is the anterior part of the innominate bone and has a *body* and *superior and inferior pubic rami.* The body of the pubis bears the *pubic crest* and the *pubic tubercle* and articulates with the pubic bone of the opposite side at the *symphysis pubis.* In the lower part of the innominate bone is a large opening, the *obturator foramen*, which is bounded by the parts of the ischium and pubis (Fig. 6-12).

## Obturator Membrane

The obturator membrane is a fibrous sheet that almost completely closes the obturator foramen, leaving a small gap, the *obturator canal*, for the passage of the obturator nerve and vessels as they leave the pelvis to enter the thigh (Fig. 6-12).

## Sacrotuberous Ligament

The sacrotuberous ligament is strong and extends from the lateral part of the sacrum and coccyx and the posterior inferior iliac spine to the ischial tuberosity (Figs. 6-6 and 6-11).

## Sacrospinous Ligament

The sacrospinous ligament is strong and triangular in shape. It is attached by its base to the lateral part of the sacrum and coccyx and by its apex to the spine of the ischium (Figs. 6-6 and 6-11).

The sacrotuberous and sacrospinous ligaments prevent the lower end of the sacrum and the coccyx from being rotated upward at the sacroiliac joint by the weight of the body (Fig. 6-19). The two ligaments also convert the greater and lesser sciatic notches into foramina, the *greater* and *lesser sciatic foramina.*

## Obturator Internus Muscle

The obturator internus muscle arises from the pelvic surface of the obturator membrane and the adjoining part of the hip bone (Fig. 6-13). The muscle fibers converge to a tendon, which leaves the pelvis through the lesser sciatic foramen and is inserted into the greater trochanter of the femur.

## Action

The obturator internus is a lateral rotator of the femur at the hip joint.

## Nerve Supply

The muscle is supplied by the nerve to obturator internus, a branch from the sacral plexus.

## INFERIOR PELVIC WALL, OR PELVIC FLOOR

The floor of the pelvis supports the pelvic viscera and is formed by the pelvic diaphragm.

The pelvic floor stretches across the pelvis and divides it into the main pelvic cavity above, which contains the pelvic viscera, and the perineum below. The perineum is considered in detail in Chapter 8.

## PELVIC DIAPHRAGM

The pelvic diaphragm is formed by the important levatores ani muscles and the small coccygeus muscles and their covering fasciae (Fig. 6-14). It is incomplete anteriorly, to allow for the passage of the urethra and in the female, the vagina also.

## Levator Ani Muscle

The levator ani muscle is a wide thin sheet that has a linear origin from the back of the body of the pubis, a tendinous arch formed by a thickening of the pelvic fascia covering the obturator internus, and the spine of the ischium (Fig. 6-14). From this extensive origin, groups of fibers sweep downward and medially to their insertion (Fig. 6-15), as follows:

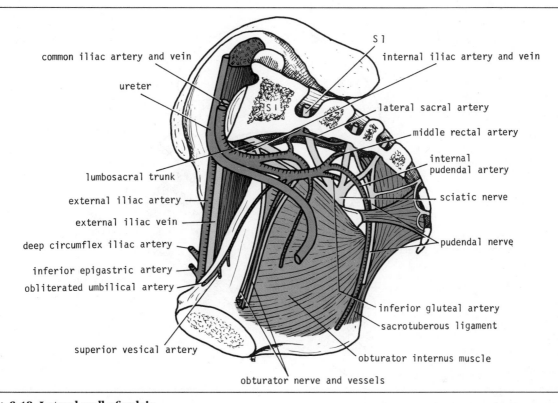

common iliac artery and vein

ureter

lumbosacral trunk

external iliac artery

external iliac vein

deep circumflex iliac artery

inferior epigastric artery

obliterated umbilical artery

superior vesical artery

obturator nerve and vessels

S 1

internal iliac artery and vein

lateral sacral artery

middle rectal artery

internal pudendal artery

sciatic nerve

pudendal nerve

inferior gluteal artery

sacrotuberous ligament

obturator internus muscle

**Fig. 6-13. Lateral wall of pelvis.**

1. *Anterior fibers.* The *levator prostatae* or *sphincter vaginae* forms a sling around the prostate or vagina and is inserted into the *perineal body.*
2. *Intermediate fibers.* The *puborectalis* forms a sling around the junction of the rectum and anal canal. The *pubococcygeus* passes posteriorly to be inserted into a median raphe, the *anococcygeal body*, between the tip of the coccyx and the anal canal; it may also be attached to the coccyx.
3. *Posterior fibers.* The *iliococcygeus* is inserted into the anococcygeal body and the coccyx.

## Action

The levatores ani muscles of the two sides form an efficient muscular sling that supports and maintains the pelvic viscera in position. They resist the rise in intrapelvic pressure during the straining and expulsive efforts of the abdominal muscles (as occurs in coughing). Further, they have an important

sphincter action on the anorectal junction, and in the female they serve also as a sphincter of the vagina.

## Nerve Supply

This is derived from the perineal branch of the fourth sacral nerve and from the perineal branch of the pudendal nerve (or the inferior rectal nerve).

## Coccygeus Muscle

The coccygeus muscle is a small muscle that arises from the spine of the ischium and is inserted into the lower end of the sacrum and upper part of the coccyx (Figs. 6-14 and 6-15).

## Action

The two muscles assist the levatores ani in supporting the pelvic viscera. They can also flex the coccyx.

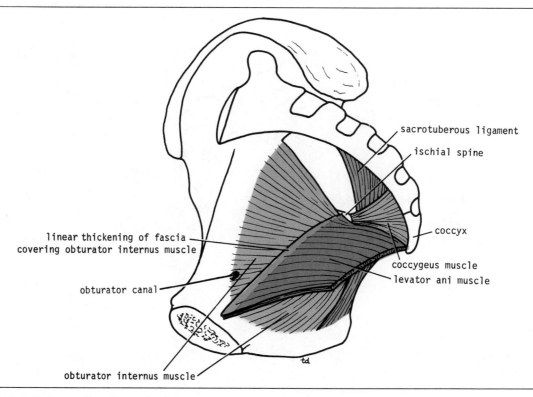

**Fig. 6-14. Inferior wall or floor of pelvis.**

### Nerve Supply

The coccygeus muscle is supplied by a branch of the fourth and fifth sacral nerves.

A summary of the muscles of the pelvic walls and floor, their nerve supply, and their action is given in Table 6-1.

## PELVIC FASCIA

The pelvic fascia may be divided into the parietal and visceral layers.

### Parietal Pelvic Fascia

The parietal pelvic fascia lines the walls of the pelvis and may be named according to the muscle it overlies. For example, over the obturator internus muscle it is dense and strong and is known as the obturator internus fascia (Fig. 6-16). Above the pelvic inlet it is continuous with the fascia lining the abdominal walls.

### Visceral Pelvic Fascia

The visceral pelvic fascia is a layer of loose connective tissue that covers all the pelvic viscera. It is continuous with the fascia on the upper surface of the levator ani muscle (Fig. 6-16).

## PERITONEUM

The parietal peritoneum lines the pelvic walls and is reflected onto the pelvic viscera, where it becomes continuous with the visceral peritoneum (Fig. 6-16).

## Nerves of the Pelvic Walls

### Sacral Plexus

The sacral plexus is formed from the anterior rami of the fourth and fifth lumbar nerves and the an-

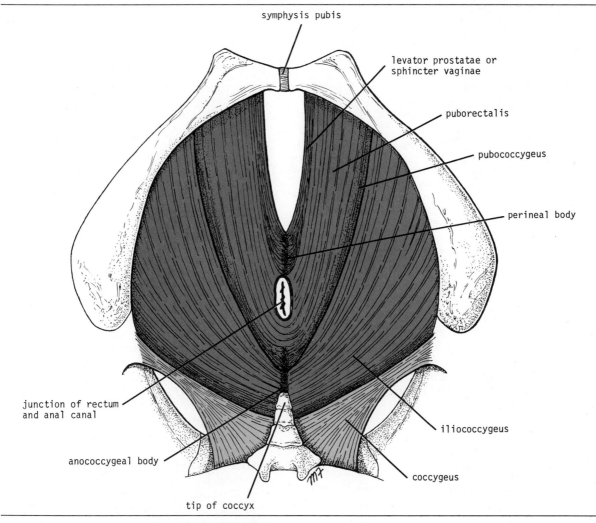

symphysis pubis

levator prostatae or
sphincter vaginae

puborectalis

pubococcygeus

perineal body

iliococcygeus

coccygeus

junction of rectum
and anal canal

anococcygeal body

tip of coccyx

**Fig. 6-15. Levator ani muscle (dark brown) and coccygeus muscle (light brown) seen on their inferior aspects. Note that the levator ani is made up of a number of different muscle groups. The levator ani and coccygeus muscles with their fascial coverings form a continuous muscular floor to the pelvis known as the pelvic diaphragm.**

Table 6-1. Muscles of the Pelvic Walls and Floor

| Name of muscle | Origin | Insertion | Nerve supply | Action |
|---|---|---|---|---|
| Piriformis | Front of sacrum | Greater trochanter of femur | Sacral plexus | Lateral rotator of femur at hip joint |
| Obturator internus | Obturator membrane and adjoining part of hip bone | Great trochanter of femur | Nerve to obturator internus from sacral plexus | Lateral rotator of femur at hip joint |
| Levator ani | Body of pubis, fascia of obturator internus, spine of ischium | Perineal body, anococcygeal body, walls of prostate, vagina, rectum, and anal canal | Fourth sacral nerve, pudendal nerve | Supports pelvic viscera; sphincter to anorectal junction and vagina |
| Coccygeus | Spine of ischium | Lower end of sacrum; coccyx | Fourth and fifth sacral nerve | Assists levator ani to support pelvic viscera; flexes coccyx |

terior rami of the first, second, third, and fourth sacral nerves (Fig. 6-17). Note that the contribution from the fourth lumbar nerve joins the fifth lumbar nerve to form the *lumbosacral trunk*. The lumbosacral trunk passes down into the pelvis and joins the sacral nerves as they emerge from the anterior sacral foramina.

## Relations

### Anteriorly

The parietal pelvic fascia, which separates the plexus from the internal iliac vessels and their branches, and the rectum (Fig. 6-13).

### Posteriorly

The piriformis muscle (Fig. 6-18).

## Branches

1. Branches to the lower limb that leave the pelvis through the greater sciatic foramen (Fig. 6-13):
   (a) The *sciatic nerve* (L4 and 5; S1, 2, and 3) is the largest branch of the plexus and the largest nerve in the body (Fig. 6-11).
   (b) The *superior gluteal nerve*, which supplies the gluteus medius and minimus and the tensor fasciae latae muscles.
   (c) The *inferior gluteal nerve*, which supplies the gluteus maximus muscle.
   (d) The *nerve to the quadratus femoris muscle*, which also supplies the inferior gamellus muscle.
   (e) The *nerve to the obturator internus muscle*, which also supplies the superior gamellus muscle.
   (f) The *posterior cutaneous nerve of the thigh*, which supplies the skin of the buttock and the back of the thigh.
2. Branches to the pelvic muscles, pelvic viscera, and perineum.
   (a) The *pudendal nerve* (S2, 3, and 4), which leaves the pelvis through the greater sciatic foramen and enters the perineum through the lesser sciatic foramen (Fig. 6-13).
   (b) The *nerves to the piriformis muscle*.
   (c) The *pelvic splanchnic nerves*. These constitute the sacral part of the parasympathetic system and arise from the second, third, and fourth sacral nerves. They are distributed to the pelvic viscera.
3. The *perforating cutaneous nerve*, which supplies the skin of the lower medial part of the buttock.

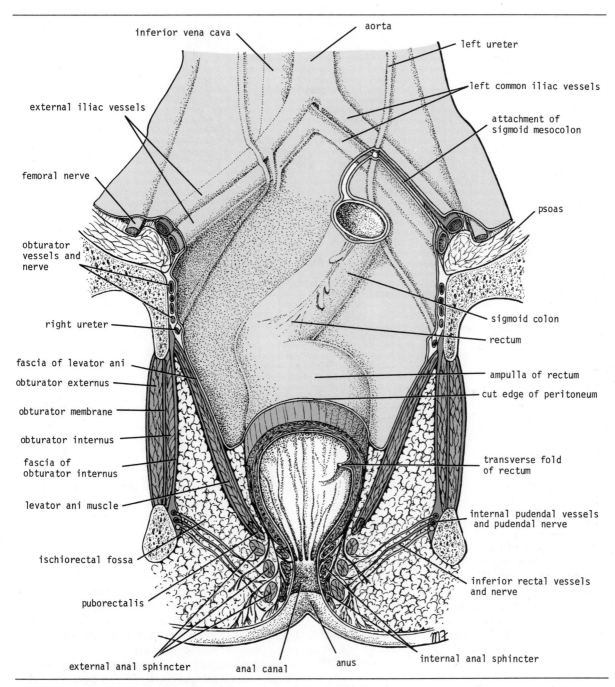

**Fig. 6-16. Coronal section through pelvis.**

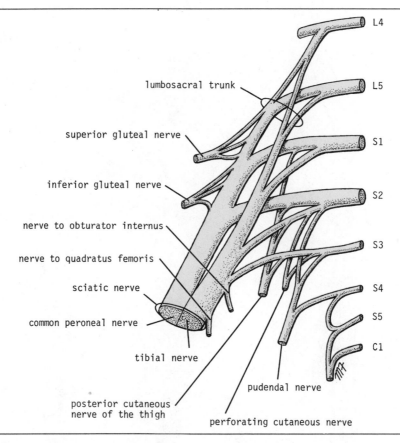

lumbosacral trunk

superior gluteal nerve

inferior gluteal nerve

nerve to obturator internus

nerve to quadratus femoris

sciatic nerve

common peroneal nerve

tibial nerve

posterior cutaneous
nerve of the thigh

pudendal nerve

perforating cutaneous nerve

L4

L5

S1

S2

S3

S4

S5

C1

**Fig. 6-17. Sacral plexus.**

## Pelvic Part of the Sympathetic Trunk

The pelvic part of the sympathetic trunk is continuous above, behind the common iliac vessels, with the abdominal part (Fig. 6-18). It runs down behind the rectum on the front of the sacrum, medial to the anterior sacral foramina. The sympathetic trunk has four or five segmentally arranged ganglia. Below, the two trunks converge and finally unite in front of the coccyx.

## Branches

1. Gray rami communicantes to the sacral and coccygeal spinal nerves.

2. Fibers that join the pelvic plexuses. Note that no white rami communicantes pass to this part of the sympathetic trunk.

## Superior Hypogastric Plexus (Presacral Nerve)

The superior hypogastric plexus lies in the retroperitoneal tissue in front of the promontory of the sacrum and between the common iliac arteries (Fig. 6-18). It is formed from (1) the aortic sympathetic plexus and (2) branches of the lumbar sympathetic ganglia. As the superior hypogastric plexus enters the pelvis, it divides into *right and left inferior hypogastric plexuses* (pelvic plexuses). Each inferior hypogastric plexus runs down on the medial side of the internal iliac artery and its branches and lateral to the rectum. It is joined on

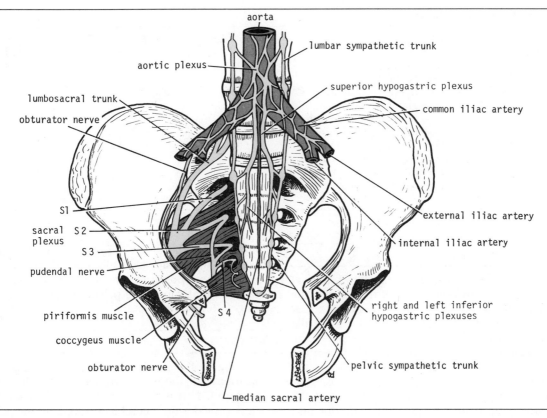

aorta

lumbar sympathetic trunk

aortic plexus

superior hypogastric plexus

lumbosacral trunk

common iliac artery

obturator nerve

S1

sacral plexus    S2

S3

pudendal nerve

piriformis muscle

coccygeus muscle

obturator nerve

S4

external iliac artery

internal iliac artery

right and left inferior hypogastric plexuses

pelvic sympathetic trunk

median sacral artery

**Fig. 6-18. Posterior pelvic wall, showing sacral plexus, superior hypogastric plexus, and right and left inferior hypogastric plexuses. Pelvic parts of sympathetic trunks are also shown.**

each side by parasympathetic nerve fibers called the *pelvic splanchnic nerve.* The right and left inferior hypogastric plexuses thus contain both sympathetic and parasympathetic nerve fibers, which are distributed to the pelvic viscera along the branches of the internal iliac artery.

### Pelvic Splanchnic Nerves

The pelvic splanchnic nerves constitute the sacral part of the parasympathetic system and contain preganglionic fibers from the second, third, and fourth sacral nerves. Branches pass to the right and left inferior hypogastric plexuses and are distributed to the pelvic viscera. Some of the parasympathetic fibers ascend from the inferior hypogastric plexuses to the superior hypogastric plexus and

thence via the aortic plexus to the inferior mesenteric plexus. The fibers are then distributed along the branches of the inferior mesenteric artery to supply the large bowel from the left colic flexure to the upper half of the anal canal.

The preganglionic fibers of the pelvic splanchnic nerves synapse with postganglionic neurones that are located either in the inferior hypogastric plexuses or in the walls of the viscera.

### Branches of the Lumbar Plexus

#### LUMBOSACRAL TRUNK

Part of the anterior ramus of the fourth lumbar nerve emerges on the medial border of the psoas muscle and joins the anterior ramus of the fifth lumbar nerve to form the lumbosacral trunk (Figs. 6-17 and 6-18). This trunk now enters the pelvis by passing down in front of the sacroiliac joint and joins the sacral plexus.

Diagram 6-1. Branches of Internal Iliac Artery.

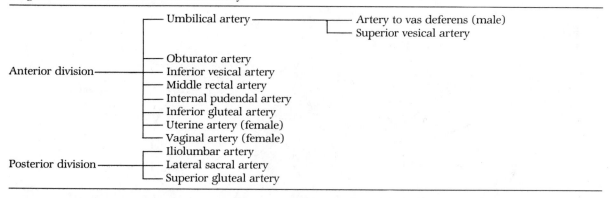

## OBTURATOR NERVE

This branch of the lumbar plexus (L2, 3, and 4) emerges from the medial border of the psoas in the abdomen and accompanies the lumbosacral trunk to the pelvic inlet (Fig. 6-18). Here, it crosses the front of the sacroiliac joint and runs forward on the lateral wall of the pelvis in the angle between the internal and external iliac vessels (Fig. 6-13). On reaching the obturator canal (i.e., the upper part of the obturator foramen, which is devoid of the obturator membrane), it splits into anterior and posterior divisions. The distribution of the obturator nerve in the thigh is considered on page 596.

### Branches

Fibers are given off that supply the parietal peritoneum on the lateral wall of the pelvis.

## Arteries of the Pelvic Walls

Each *common iliac artery* ends at the pelvic inlet in front of the sacroiliac joint by dividing into the external and internal iliac arteries (Figs. 6-13 and 6-18).

The *external iliac artery* runs along the medial border of the psoas muscle, following the pelvic brim (Fig. 6-13), and gives off the *inferior epigastric* and *deep circumflex iliac* branches. It leaves the false pelvis by passing under the inguinal ligament, to become the *femoral artery.*

The *internal iliac artery* enters the pelvis in front of the sacroiliac joint, and at this point it is crossed

anteriorly by the ureter (Fig. 6-13). At the upper border of the greater sciatic foramen, it divides into anterior and posterior divisions, which give off branches that supply the pelvic viscera, perineum, buttock, and sacral canal.

## BRANCHES

The origin of the terminal branches is subject to variation, but the usual arrangement is shown in Diagram 6-1. It is thus seen that the internal iliac artery provides the main blood supply to the pelvic viscera. The distribution of the visceral branches is discussed with the individual viscera in Chapter 7.

The *median sacral artery* arises from the aorta at the point where it bifurcates into the two common iliac arteries (Fig. 6-18). It enters the pelvis in front of the sacral promontory and runs downward on the anterior surface of the sacrum, where it anastomoses with the lateral sacral arteries.

It is thus seen that in the male, three arteries enter the pelvis: (1) the internal iliac arteries, (2) the median sacral artery, and (3) the superior rectal artery. In the female, the paired ovarian arteries also enter the pelvis.

## Veins of the Pelvic Walls

The *external iliac vein* begins behind the inguinal ligament as a continuation of the femoral vein. It runs along the medial side of the corresponding artery and joins the internal iliac vein to form the *common iliac vein* (Fig. 6-13). It receives the *inferior epigastric* and *deep circumflex iliac veins.*

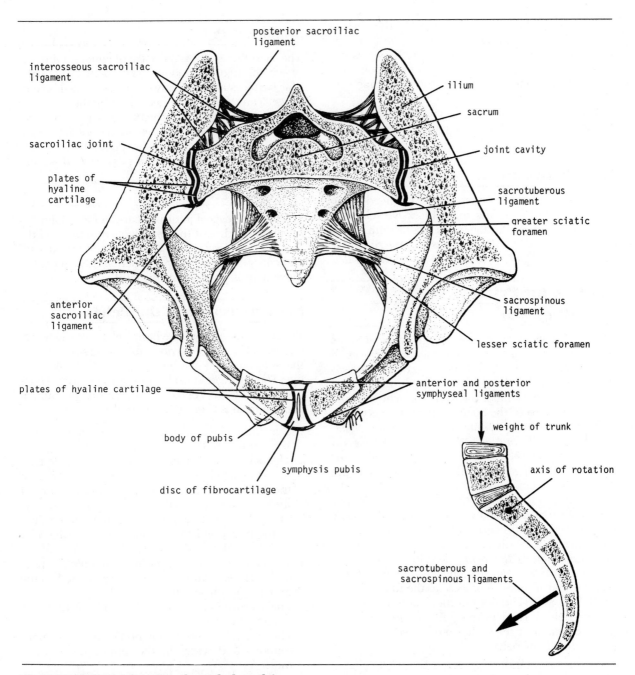

**Fig. 6-19. Horizontal section through the pelvis, showing the sacroiliac joints and the symphysis pubis. The lower diagram shows the function of the sacrotuberous and sacrospinous ligaments in resisting the rotation force exerted on the sacrum by the weight of the trunk.**

The *internal iliac vein* begins by the joining together of tributaries that correspond to the branches of the internal iliac artery. It passes upward in front of the sacroiliac joint and joins the external iliac vein to form the common iliac vein (Fig. 6-13).

The *median sacral veins* accompany the corresponding artery and end by joining the left common iliac vein.

## Lymphatics on the Pelvic Walls

The lymph nodes and vessels are arranged in an obvious chain along the main blood vessels. The nodes are named after the blood vessels with which they are associated. Thus, there are *external iliac nodes*, *internal iliac nodes*, and *common iliac nodes*.

## Joints of the Pelvis

### Sacroiliac Joints

The sacroiliac joints are very strong synovial joints and are formed between the auricular surfaces of the sacrum and the iliac bones (Fig. 6-19). The sacrum carries the weight of the trunk, and, apart from the interlocking of the irregular articular surfaces, the shape of the bones contributes very little to the stability of the joints. The very strong *posterior* and *interosseous sacroiliac ligaments* suspend the sacrum between the two iliac bones. The *anterior sacroiliac ligament* is thin and situated on the anterior aspect of the joint.

The weight of the trunk tends to thrust the upper end of the sacrum downward and rotate the lower end of the bone upward (Fig. 6-19). This rotatory movement is prevented by the strong *sacrotuberous* and *sacrospinous ligaments* described previously. The *iliolumbar ligament* connects the tip of the fifth lumbar transverse process to the iliac crest. A small but limited amount of movement is possible at these joints. Their primary function is to transmit the weight of the body from the vertebral column to the bony pelvis.

### Nerve Supply

The sacroiliac joint is supplied by branches from the sacral plexus and by posterior rami of the first two sacral nerves.

### Symphysis Pubis

The symphysis pubis is a cartilaginous joint between the two pubic bones (Fig. 6-19). The articular surfaces are covered by a layer of hyaline cartilage and are connected together by a fibrocartilaginous disc. The disc has a small cavity in the midline. The joint is surrounded by ligaments that extend from one pubic bone to the other. Almost no movement is possible at this joint.

### Sacrococcygeal Joint

The sacrococcygeal joint is a cartilaginous joint between the bodies of the last sacral vertebra and the first coccygeal vertebra. The cornua of the sacrum and coccyx are joined by ligaments. A great deal of movement is possible at this joint.

## Sex Differences of the Pelvis

The sex differences of the bony pelvis are easily recognized. The more obvious differences are due to the adaptation of the female pelvis for childbearing. The stronger muscles in the male are responsible for the thicker bones and more prominent bony markings (Figs. 6-5 and 6-20).

1. The false pelvis is shallow in the female and deep in the male.
2. The pelvic inlet is transversely oval in the female but is heart-shaped in the male. This is due to the indentation produced by the promontory of the sacrum in the male.
3. The pelvic cavity is roomier in the female than in the male, and the distance between the inlet and the outlet is much shorter.
4. The pelvic outlet is larger in the female than the male. The ischial tuberosities are everted in the female and turned in in the male.
5. The sacrum is shorter, wider, and flatter in the female than the male.
6. The subpubic angle, or pubic arch, is more rounded and wider in the female than in the male.

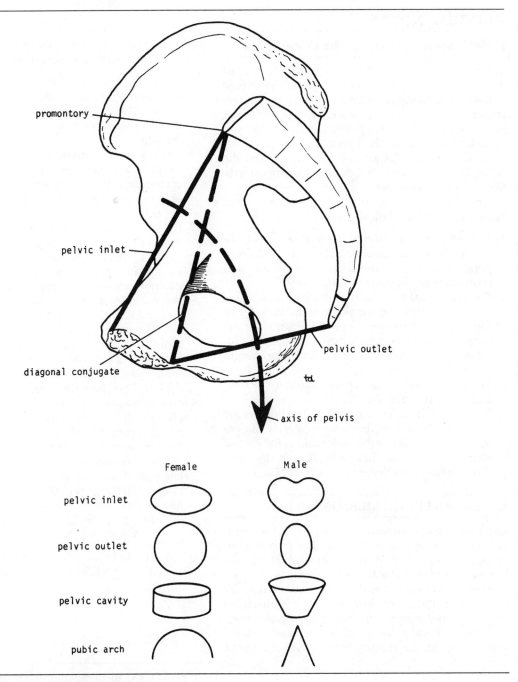

**Fig. 6-20. Pelvic inlet, pelvic outlet, diagonal conjugate, and axis of pelvis. Lower diagrams illustrate some of the main differences between female and male pelvis.**

# CLINICAL NOTES

## Pelvic Measurements in Obstetrics

The capacity and shape of the female pelvis are of fundamental importance in obstetrics. The female pelvis is well adapted for the process of parturition (childbirth). The pelvis is shallower and the bones are smoother than in the male. The size of the pelvic inlet is similar in the two sexes, but in the female the cavity is larger and cylindrical, and the pelvic outlet is wider in both the anteroposterior and transverse diameters.

Four terms relating to areas of the pelvis are commonly used in clinical practice:

1. The *pelvic inlet or brim* of the true pelvis (Fig. 6-20) is bounded anteriorly by the symphysis pubis, laterally by the iliopectineal lines, and posteriorly by the sacral promontory.
2. The *pelvic outlet* of the true pelvis (Fig. 6-20) is bounded in front by the pubic arch, laterally by the ischial tuberosities, and posteriorly by the coccyx. It must be remembered that the sacrotuberous ligaments also form part of the margin of the outlet.
3. The *pelvic cavity* is the space between the inlet and the outlet (Fig. 6-20).
4. The *axis of the pelvis* is an imaginary line joining the central points of the anteroposterior diameters from the inlet to the outlet and is the curved course taken by the baby's head as it descends through the pelvis during parturition (Fig. 6-20).

### EXTERNAL PELVIC MEASUREMENTS

External pelvic measurements are no longer considered to be of practical value. The only way to obtain accurate factual information about the dimensions and shape of the pelvic cavity is by *X-ray pelvimetry*. In view of the danger of radiation to the fetus, especially to the gonads, X-ray pelvimetry is seldom used during pregnancy. However, it is used to obtain precise pelvic measurements after delivery to help in the management of a subsequent pregnancy.

### INTERNAL PELVIC ASSESSMENTS

Internal pelvic assessments are made by vaginal examination during the later weeks of pregnancy, when the pelvic tissues are softer and more yielding than in the newly pregnant condition.

1. *Pubic arch.* Spread the fingers under the pubic arch and examine its shape. Is it broad or angular? The examiner's four fingers should be able to rest comfortably in the angle below the symphysis.
2. *Lateral walls.* Palpate the lateral walls and determine whether they are concave, straight, or converging. The prominence of the ischial spines and the position of the sacrospinous ligaments are noted.
3. *Posterior wall.* The sacrum is palpated to determine whether it is straight or well curved. Finally, if the patient has relaxed the perineum sufficiently, an attempt is made to palpate the promontory of the sacrum. The second finger of the examining hand is placed on the promontory, and the index finger of the free hand, outside the vagina, is placed at the point on the examining hand where it makes contact with the lower border of the symphysis. The fingers are then withdrawn and the distance measured (Fig. 6-21B). This gives the measurement of the *diagonal conjugate*, which is normally about 5 inches (13 cm). The anteroposterior diameter from the sacrococcygeal joint to the lower border of the symphysis is then estimated.
4. *Ischial tuberosities.* The distance between the ischial tuberosities may be estimated by using the closed fist (Fig. 6-21D). It measures about 4 inches (10 cm), but is difficult to measure exactly.

Needless to say, considerable clinical experience is required to be able to assess the shape and size of the pelvis by vaginal examination.

**Fig. 6-21. (A) Birth canal. Interrupted line indicates axis of canal. (B) Procedure used in measuring diagonal conjugate. (C) Different types of pelvic inlet, according to Caldwell and Moloy. (D) Estimation of width of pelvic outlet by means of closed fist.**

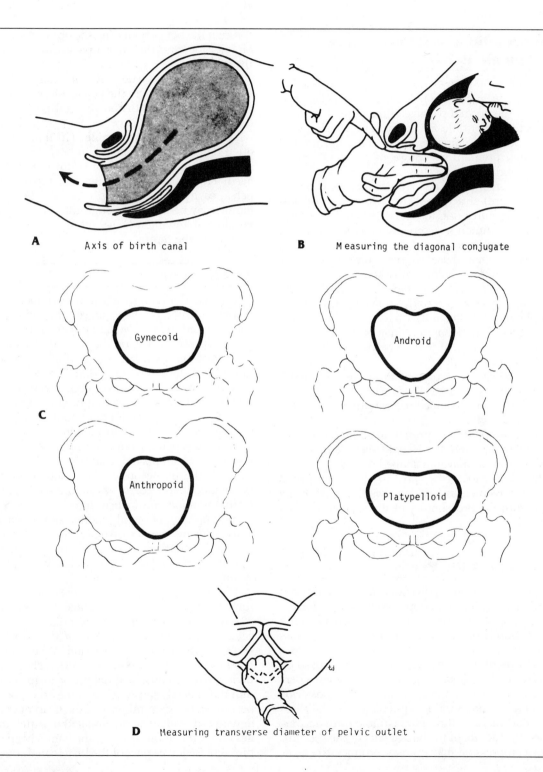

**A**    Axis of birth canal

**B**    Measuring the diagonal conjugate

**C**    Gynecoid    Android

Anthropoid    Platypelloid

**D**    Measuring transverse diameter of pelvic outlet

## Abnormalities and Varieties of the Female Pelvis

Deformities of the pelvis may be responsible for *dystocia* (difficult labor). A contracted pelvis may obstruct the normal passage of the fetus. It may be indirectly responsible for dystocia by causing conditions such as malpresentation or malposition of the fetus, premature rupture of the fetal membranes, and uterine inertia.

The cause of pelvic deformities may be congenital (very rare) or acquired from disease, poor posture, or fractures due to injury. Pelvic deformities are more common in women who have grown up in a poor environment and are undernourished. It is probable that these women suffered in their youth from minor degrees of rickets.

In 1933, Caldwell and Moloy classified pelves into four groups: gynecoid, android, anthropoid, and platypelloid (Fig. 6-21).

The *gynecoid* type, present in about 41 percent of women, is the typical female pelvis that has been previously described.

The *android* type, present in about 33 percent (white females) and 16 percent (black females), is the male or funnel-shaped pelvis with a contracted outlet.

The *anthropoid* type, present in about 24 percent (white females) and 41 percent (black females), is long, narrow, and oval in shape.

The *platypelloid* type, present in only about 2 percent of women, is a wide pelvis flattened at the brim, with the promontory of the sacrum pushed forward.

## Fractures of the Pelvis

Fractures of the false pelvis due to direct trauma occasionally occur. The upper part of the ilium is seldom displaced because of the attachment of the iliacus muscle on the inside and the gluteal muscles on the outside.

Fractures of the true pelvis are usually caused by a severe crushing injury, such as occurs when a person is run over by an automobile. Anteroposterior compression may produce dislocation of the symphysis, or fractures through the pubic rami accompanied by fracture of the lateral part of the sacrum. Other compression injuries may produce fractures through the bones bounding the obturator foramen on one side, and the lateral part of the sacrum on the other side.

Fractures of the true pelvis are commonly associated with damage to the pelvic viscera. The pelvic veins that lie beneath the parietal peritoneum, if damaged, are a source of severe hemorrhage, which may be life-threatening. The male urethra is often damaged; the bladder in both sexes is occasionally damaged, but the rectum is rarely damaged. Fragments of bone may be thrust into the pelvic cavity, tearing the viscera. The head of the femur may be driven through the floor of the acetabulum following a heavy fall on the greater trochanter of the femur.

## Pelvic Floor

The pelvic diaphragm is a gutter-shaped sheet of muscle, formed by the levatores ani and coccygeus muscles and their covering fasciae. From their origin, the muscle fibers on the two sides slope downward and backward to the midline, producing a gutter that slopes downward and forward.

A rise in the intra-abdominal pressure, caused by the contraction of the diaphragm and the muscles of the anterior and lateral abdominal walls, is counteracted by the contraction of the muscles forming the pelvic floor. By this means the pelvic viscera are supported and do not "drop out" through the pelvic outlet. Contraction of the puborectalis fibers greatly assists the anal sphincters in maintaining continence under these conditions by pulling the anorectal junction upward and forward. During the act of defecation, however, the levator ani continues to support the pelvic viscera, but the puborectalis fibers relax with the anal sphincters.

The female pelvic floor serves an important function during the second stage of labor (Fig. 6-22). At the pelvic inlet the widest diameter is transverse, so that the longest axis of the baby's head takes up the transverse position. When the head reaches the pelvic floor, the gutter shape of the floor tends to cause the baby's head to rotate, so that its long axis comes to lie in the anteroposterior position. The occipital part of the head now moves downward and forward along the gutter until it lies under the pubic arch. As the baby's head passes through the lower part of the birth canal, the small

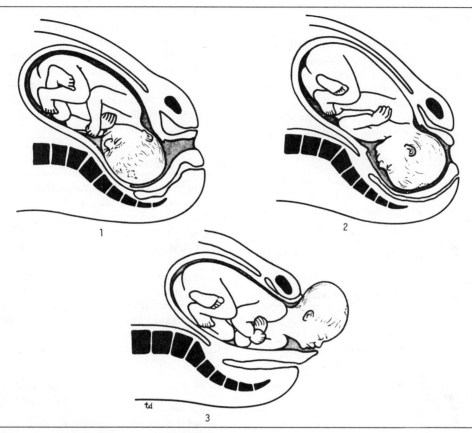

1

2

td

3

**Fig. 6-22. Stages in rotation of baby's head during second stage of labor. Shape of pelvic floor plays an important part in this process.**

gap that exists in the anterior part of the pelvic diaphragm becomes enormously enlarged, so that the head may slip through into the perineum. Once the baby has passed through the perineum, the levatores ani muscles recoil and take up their previous position.

Injury to the pelvic floor during a difficult childbirth can result in the loss of support for the pelvic viscera leading to *uterine* and *vaginal prolapse*, herniation of the bladder (*cystocele*), and alteration in the position of the bladder neck and urethra, leading to *stress incontinence*. In the latter condition, the patient dribbles urine whenever the intra-abdominal pressure is raised, as in coughing. *Prolapse of the rectum* may also occur.

## Sacral Plexus

### Pressure from Fetal Head

During the later stages of pregnancy, when the fetal head has descended into the pelvis, the mother often complains of discomfort or aching pain extending down one or other of the lower limbs. This is due to the pressure from the fetal head. The discomfort is often relieved by changing position, such as lying on the side in bed.

### Invasion By a Malignant Tumor

The nerves of the sacral plexus may become invaded by malignant tumors extending from neighboring viscera. A carcinoma of the rectum, for example, can cause severe intractable pain down the lower limbs.

## Obturator Nerve

The obturator nerve lies on the lateral wall of the pelvis and supplies the parietal peritoneum. An inflamed vermiform appendix hanging down into the pelvic cavity could cause irritation of the obturator nerve endings, leading to referred pain down the inner side of the right thigh. Inflammation of the ovaries can produce similar symptoms.

## Sacrum

The first sacral vertebra may be partly or completely separated from the second sacral vertebra. Occasionally, on radiography of the vertebral column, examples are seen in which the fifth lumbar vertebra has fused with the first sacral vertebra.

## Caudal Anesthesia (Analgesia)

Anesthetic solutions may be injected into the sacral canal through the sacral hiatus. The solutions then act on the spinal roots of the second, third, fourth, and fifth sacral and coccygeal segments of the cord as they emerge from the dura mater. The roots of higher spinal segments may also be blocked by this method. The needle must be confined to the lower part of the sacral canal, since the meninges extend down as far as the lower border of the second sacral vertebra. Caudal anesthesia is used in obstetrics to block pain fibers from the cervix of the uterus and to anesthetize the perineum.

## Pelvic Joints

During pregnancy, the symphysis pubis and the ligaments of the sacroiliac and sacrococcygeal joints undergo temporary changes that permit freer movement of the joints. This makes possible an enlargement of the pelvic cavity and thus facilitates parturition.

The ligaments of the pelvic joints are softened during pregnancy by the hormones *estrogen* and *progesterone*, produced by the ovary and the placenta. An additional hormone, called *relaxin*, produced by the these organs, may also have a relaxing effect on the pelvic ligaments.

## CLINICAL PROBLEMS

*Answers on page 964*

1. A patient suffering from a severe form of anemia was found to be unresponsive to treatment. It was decided to perform a marrow biopsy. Where in the pelvic region is it safe to obtain a marrow biopsy? What structures are penetrated by the needle as it passes into the red marrow?
2. A medical student, during a clinical examination, was asked to place a finger on a patient's back to indicate the position of the lower level of the subarachnoid space within the vertebral canal. Where would you place your finger?
3. A young actress, finding herself pregnant, told her friends that she hated the idea of going through the pain of childbirth, but she nonetheless detested the thought of having a general anesthetic. Is there a specialized local analgesic technique that will provide painless labor? List the anatomical structures that you will pierce with the needle in order to induce the analgesia.
4. A 65-year-old man with a history of prostatic enlargement complained that he could not micturate. The last time that he passed urine had been 6 hours previously. He was found lying on his bed in great distress, clutching his anterior abdominal wall with both hands and pleading for something to be done quickly. On examination, a large ovoid swelling could be palpated in the hypogastrium, reaching up to the region of the umbilicus. The swelling was dull on percussion, and pressure on it exacerbated the symptoms. Explain in anatomical terms the nature of the swelling.
5. A high school girl, rejected by her parents, went to see a physician because she had noticed that she had a large abdominal swelling. On questioning, it became apparent that she was mentally retarded, and she volunteered the fact that she had recently missed several menstrual periods. On examination, a large ovoid swelling was found rising out of the pelvis. The upper end of the swelling reached the level of the umbilicus. The breasts were seen to be enlarged

and the superficial veins congested. What is the diagnosis?

6. An enthusiastic young surgeon, while operating on a patient to drain an abscess in the region of the anorectal junction, decided to divide the puborectalis muscle to obtain better access. Three days later he was surprised to learn that the patient was incontinent. Can you explain this on anatomical grounds?

7. A pregnant woman visited an antenatal clinic. A vaginal examination revealed that the sacral promontory could be easily palpated and that the diagonal conjugate measured less than 4 inches. On the basis of your anatomical knowledge, do you think that this patient would have a normal labor?

8. A man who had already drunk five pints of beer decided to have one more for the road. On leaving the bar, he staggered out into the street and was run over by a passing bus. On admission to the hospital he was unconscious and showed signs indicating a severe degree of shock. He had extensive bruising of the lower part of the anterior abdominal wall, and the body of the pubis was prominent on the right side. A drop of blood-stained fluid could be expressed from the external meatus of the penis. In anatomical terms, can you explain the blood-stained discharge from the penis?

9. A young woman visited her obstetrician and was told that she was pregnant. The process of childbirth was carefully but simply explained to her, and she was instructed to exercise her pelvic floor twelve times a day. Using your knowledge of anatomy of the region, explain how you would instruct the patient to carry out these exercises. Why was this advice given to the patient by the obstetrician?

10. A heavily built, middle-aged man, running down a flight of stone steps, misjudged the position of one of the steps and fell suddenly onto his buttocks. Following the fall he complained of severe bruising of the area of the natal cleft and persistent pain in this area. Which bone is he likely to have damaged?

11. A patient was injured in a plane crash. X-ray of the pelvis revealed a fracture of the ilium and iliac crest on the right side. From your knowledge of anatomy, would you expect much displacement of the bone fragments? Do you think that the bone fragments need splinting?

12. A middle-aged Indian woman recently admitted to this country had a history of osteomalacia during a previous pregnancy. *Osteomalacia* is a disease of adult life and is associated with gross deficiency of calcium and vitamin D in the diet. She asked her physician whether or not she would be able to have another child. What effects, if any, do you think osteomalacia has on the anatomical shape of the pelvis?

# NATIONAL BOARD TYPE QUESTIONS

*Answers on page 984*

**In each of the following questions, answer:**
   (a) If (1) only is correct
   (b) If (2) only is correct
   (c) If (1) and (2) are correct, and
   (d) If neither (1) nor (2) is correct

1. (1) The obturator internus muscle leaves the pelvis through the greater sciatic foramen.
   (2) The sacrospinous ligaments prevent the superior end of the sacrum from being rotated upward at the sacroiliac joints.

2. (1) In the female pelvis, the ischial tuberosities are usually everted.
   (2) The pelvic brim is bounded posteriorly by the sacral promontary, laterally by the iliopectineal line, and anteriorly by the symphysis pubis.

3. (1) The sacral part of the sympathetic trunk descends over the pelvic brim anterior to the common iliac vessels.
   (2) The sacral plexus receives contributions from the third and fourth lumbar spinal nerves.

4. (1) The obturator nerve supplies the parietal peritoneum of the pelvis and may be stimulated by an inflamed ovary.
   (2) The puborectalis, a part of the levator ani muscle, forms an important part of the sphincteric mechanism of the anal canal.

5. (1) The pelvic diaphragm is formed by the levatores ani and coccygeus muscles and their covering fasciae.
   (2) The sacroiliac joint, which transfers the

weight of the trunk to the innominate bone, is a cartilaginous joint.

**Match the nerve on the left with the segmental origin on the right:**

6. Sciatic nerve                (a) L2, 3, and 4
7. Pudendal nerve               (b) L4, 5, S1, 2, and 3
8. Pelvic splanchnic            (c) S2, 3, and 4
   nerve                        (d) S1 and 2
9. Obturator nerve              (e) L3, 4, S1, and 2

**Match the artery on the left with its origin on the right:**

10. Superior rectal             (a) Superior mesenteric
    artery                          artery
11. Ovarian artery              (b) Abdominal part of
12. Uterine artery                  aorta
13. Middle rectal artery        (c) Renal artery
14. Superior gluteal            (d) Internal iliac artery
    artery                      (e) None of the above

**In the following questions, answer:**

    (a) If (1), (2), AND (3) ONLY ARE CORRECT
    (b) If (1) AND (3) ONLY ARE CORRECT
    (c) If (2) AND (4) ONLY ARE CORRECT
    (d) If (4) ONLY IS CORRECT, OR
    (e) If ALL ARE CORRECT

15. (1) The sacrum is shorter, wider, and flatter in the female than in the male.
    (2) The female sex hormones have no effect on the ligaments of the pelvis during pregnancy.
    (3) The inferior hypogastric plexus contains both sympathetic and parasympathetic nerves.
    (4) The pelvic part of the sympathetic trunk possesses both white and gray rami communicantes.

16. (1) The levator ani muscle is innervated by the perineal branch of the fourth sacral nerve and from the perineal branch of the pudendal nerve.
    (2) The iliococcygeus muscle arises from a thickening of the obturator internus fascia.
    (3) The obturator nerve leaves the pelvis through the obturator canal.

(4) The ilium, ischium, and pubis are three bones that fuse together to form the innominate bone at the twenty-fifth year of life.

17. (1) The lymph nodes in the pelvis are concentrated around the median sacral artery.
    (2) The platypelloid type of pelvis occurs in about 50 percent of women.
    (3) External pelvic measurements have great practical importance in determining whether there is likely to be a disproportion between the size of the fetal head and the size of the pelvic inlet.
    (4) The pelvic outlet is formed by the symphysis pubis anteriorly, the ischial tuberosities laterally, the sacrotuberous ligaments laterally, and the coccyx posteriorly.

18. (1) In the pelvis, the fascia is divided into parietal and visceral layers.
    (2) The false pelvis helps to guide the fetus into the true pelvis during labor.
    (3) In the young child, the urinary bladder is an abdominopelvic organ.
    (4) The internal iliac arteries are the main blood supply to the pelvic viscera.

19. (1) The veins of the pelvic walls are very variable and extensive; they may be a serious source of hemorrhage in a fracture of the pelvis.
    (2) Very little movement is possible at the sacrococcygeal joint.
    (3) The superior hypogastric plexus is formed from the aortic sympathetic plexus and branches of the lumbar sympathetic ganglia.
    (4) The sciatic nerve leaves the pelvis through the lesser sciatic foramen.

20. (1) When the patient is in the standing position, the anterior superior iliac spines lie vertically above the anterior surface of the symphysis pubis.
    (2) The piriformis leaves the pelvis through the greater sciatic foramen.
    (3) The anterior sacral foramina permit the passage of the anterior rami of the upper four sacral nerves.
    (4) The external iliac vein begins behind the inguinal ligament as a continuation of the femoral vein.

# 7. The Pelvis: Part II
# The Pelvic Cavity

The pelvic cavity, or cavity of the true pelvis, may be defined as the area situated between the pelvic inlet and the pelvic outlet. It is customary to subdivide it by the pelvic diaphragm into the main pelvic cavity above and the perineum below (Fig. 7-1). This chapter is concerned with the contents of the main pelvic cavity. A detailed description of the perineum will be given in Chapter 8.

## CONTENTS OF THE PELVIC CAVITY

### Sigmoid Colon

The sigmoid colon is about 10 to 15 inches (25–38 cm) long and begins as a continuation of the descending colon in front of the left external iliac artery. Below, it becomes continuous with the rectum in front of the third sacral vertebra. It hangs down into the pelvic cavity in the form of a loop.

The pelvic colon is attached to the posterior pelvic wall by the fan-shaped *pelvic mesocolon*. The root of the mesocolon resembles an inverted V. One limb of the V is attached along the medial side of the left external iliac artery, and the other limb runs from the bifurcation of the left common iliac artery, downward in front of the sacrum, as far as the third sacral vertebra. At the apex of the V is a small peritoneal recess, called the *recess of the pelvic mesocolon*. Lying beneath the floor of the recess is the left ureter (Fig. 5-9). The clinical significance of the recess is described on page 282.

### Relations

#### Anteriorly

In the male, urinary bladder; in the female, the posterior surface of the uterus and upper part of the vagina.

#### Posteriorly

The rectum and the sacrum. The sigmoid colon is also related to the more dependent coils of the terminal part of the ileum.

The *arterial supply* to the sigmoid colon is from the sigmoid branches of the inferior mesenteric ar-

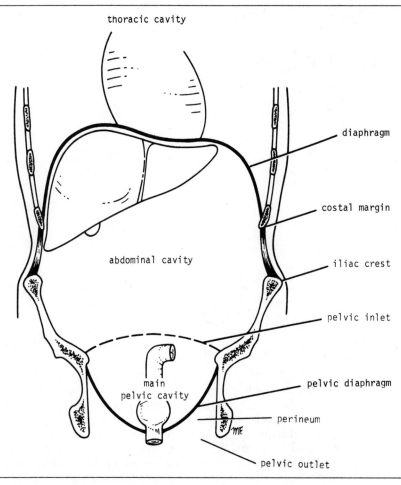

thoracic cavity

diaphragm

costal margin

iliac crest

pelvic inlet

abdominal cavity

main pelvic cavity

pelvic diaphragm

perineum

pelvic outlet

**Fig. 7-1. Coronal section through thorax, abdomen, and pelvis, showing thoracic, abdominal, and pelvic cavities, and perineum.**

tery. The *veins* of the sigmoid colon correspond to the arteries and are tributaries of the portal venous system.

The *lymph vessels* from the sigmoid colon drain into nodes along the course of the sigmoid arteries; from these nodes the lymph travels to the inferior mesenteric nodes.

The *nerve supply* to the sigmoid colon is derived from the inferior hypogastric plexuses. The nerves ascend in the root of the sigmoid mesocolon.

## Rectum

The rectum is about 5 inches (13 cm) long and begins in front of the third sacral vertebra as a continuation of the sigmoid colon. It passes downward, following the curve of the sacrum and coccyx, and ends 1 inch (2.5 cm) in front of the tip of the coccyx by piercing the pelvic diaphragm and becoming continuous with the anal canal. The lower part of the rectum that lies immediately above the pelvic diaphragm is dilated to form the *rectal ampulla.*

When examined from in front, the rectum is seen to deviate to the left, but quickly returns to the me-

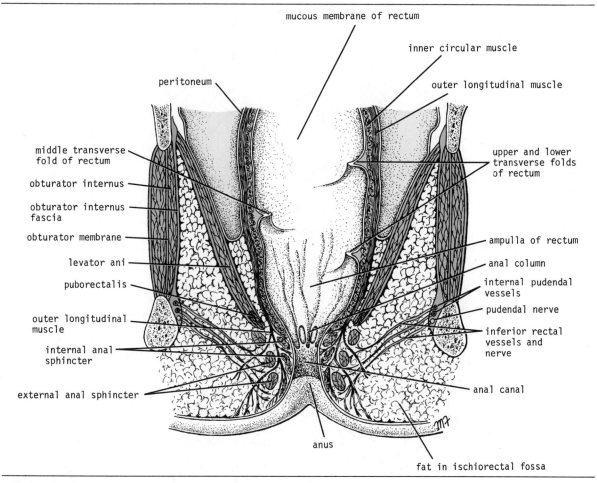

mucous membrane of rectum

inner circular muscle

outer longitudinal muscle

peritoneum

middle transverse
fold of rectum

obturator internus

obturator internus
fascia

obturator membrane

levator ani

puborectalis

outer longitudinal
muscle

internal anal
sphincter

external anal sphincter

anus

upper and lower
transverse folds
of rectum

ampulla of rectum

anal column

internal pudendal
vessels

pudendal nerve

inferior rectal
vessels and
nerve

anal canal

fat in ischiorectal fossa

**Fig. 7-2. Coronal section through pelvis, showing rectum and pelvic floor.**

dian plane (Fig. 7-2). When seen on lateral view, the rectum follows the anterior concavity of the sacrum before bending downward and backward at its junction with the anal canal (Fig. 7-3).

The puborectalis portion of the levator ani muscles forms a sling (see page 378) at the junction of the rectum with the anal canal and is responsible for pulling this part of the bowel forward, producing the anorectal angle.

The *peritoneum* covers the anterior and lateral surfaces of the first third of the rectum and only the anterior surface of the middle third, leaving the

lower third devoid of peritoneum (Figs. 7-3 and 7-9).

The *muscular coat* of the rectum is arranged in the usual outer longitudinal and inner circular layers of smooth muscle. The three teniae coli of the sigmoid colon, however, come together, so that the longitudinal fibers form a broad band on the anterior and posterior surfaces of the rectum.

The *mucous membrane* of the rectum, together with the circular muscle layers, forms three permanent folds called the *transverse folds of the rectum* (Fig. 7-2). These folds are semicircular; two are placed on the left rectal wall and one on the right wall.

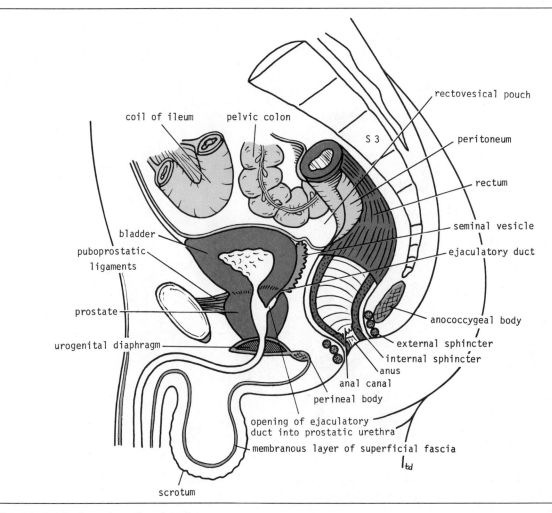

**Fig. 7-3. Sagittal section of male pelvis.**

## Relations

### Posteriorly

The rectum is in contact with the sacrum and coccyx, the piriformis, coccygeus, and levatores ani muscles, the sacral plexus, and the sympathetic trunks (Fig. 6-18).

### Anteriorly

*In the male* the upper two-thirds of the rectum, which is covered by peritoneum, is related to the sigmoid colon and coils of ileum that occupy the rectovesical pouch. The lower third of the rectum, which is devoid of peritoneum, is related to the posterior surface of the bladder, to the termination of the vas deferens and the seminal vesicles on each side, and to the prostate. These structures are embedded in visceral pelvic fascia (Fig. 7-3).

*In the female* the upper two-thirds of the rectum, which is covered by peritoneum, is related to the sigmoid colon and coils of ileum that occupy the rectouterine pouch (pouch of Douglas). The lower third of the rectum, which is devoid of peritoneum, is related to the posterior surface of the vagina (Fig. 7-9).

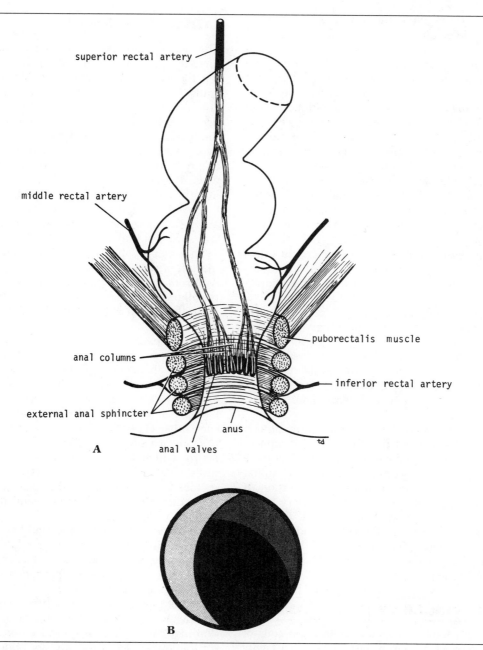

Fig. 7-4. (A) Blood supply to rectum and (B) transverse folds of rectum as seen through sigmoidoscope.

The *arterial supply* to the rectum is from the superior, middle, and inferior rectal arteries (Fig. 7-4).

The *superior rectal artery*, a continuation of the inferior mesenteric artery, is the chief artery supplying the mucous membrane. It enters the pelvis by descending in the root of the sigmoid mesocolon and divides into right and left branches. These at first lie behind the rectum and then pierce the muscular coat and supply the mucous membrane. They anastomose with one another and with the middle and inferior rectal arteries.

The *middle rectal artery* is a small branch of the internal iliac artery. It runs forward and medially to the rectum, to be distributed mainly to the muscular coat.

The *inferior rectal artery* is a branch of the internal pudendal artery in the perineum. It anastomoses with the middle rectal artery at the anorectal junction.

The *veins of the rectum* correspond to the arteries. The *superior rectal vein* is a tributary of the portal circulation and drains into the inferior mesenteric vein. The *middle* and *inferior rectal veins* drain into the internal iliac and internal pudendal veins, respectively. The union between the rectal veins forms an important portal-systemic anastomosis. (See Chap. 5.)

The *lymph vessels* of the rectum drain into a group of nodes embedded in the perirectal connective tissue just outside the muscular coat. These are called the *pararectal nodes*. Lymph vessels draining the upper and middle members of this group of nodes accompany the superior rectal artery to the inferior mesenteric nodes. Lymph vessels from the lower part of the rectum follow the middle rectal artery to the internal iliac nodes.

The *nerve supply* to the rectum is from the inferior hypogastric plexuses.

## PELVIC VISCERA IN THE MALE

The rectum, sigmoid colon, and terminal coils of ileum occupy the posterior part of the pelvic cavity in both sexes, as described above. The contents of the anterior part of the pelvic cavity in the male are described in the following sections.

## Urinary Bladder

The urinary bladder is situated immediately behind the pubic bones (Fig. 7-3). It is a receptacle for the storage of urine, with strong muscular walls. Its shape and relations vary according to the amount of urine that it contains. The empty bladder in the adult lies entirely within the pelvis; as the bladder fills, its superior wall rises up into the hypogastric region (Fig. 7-5). In the young child the empty bladder projects above the pelvic inlet; later, when the pelvic cavity enlarges, the bladder sinks into the pelvis to take up the adult position.

The empty bladder is pyramidal in shape, having an apex, a base, and a superior and two inferolateral surfaces; it also has a neck.

The *apex* of the bladder lies behind the upper margin of the symphysis pubis (Figs. 7-3 and 7-5). A fibrous cord, known as the *urachus* (remains of the allantois), passes upward in the extraperitoneal fat to the umbilicus, forming the *median umbilical ligament*.

The *base*, or *posterior surface*, of the bladder is triangular in shape. The superolateral angles are joined by the ureters, and the inferior angle gives rise to the urethra (Fig. 7-5). The two vasa deferentia lie side by side on the posterior surface of the bladder and separate the seminal vesicles from each other (Fig. 7-5). The upper part of the posterior surface of the bladder is covered by peritoneum, which forms the anterior wall of the rectovesical pouch. The lower part of the posterior surface is separated from the rectum by the vasa deferentia, the seminal vesicles, and the rectovesical fascia (Fig. 7-3).

The *superior surface* of the bladder is completely covered with peritoneum and is related to coils of ileum or sigmoid colon (Fig. 7-3). Along the lateral margins of this surface, the peritoneum is reflected onto the lateral pelvic walls.

As the bladder fills, the superior surface of the bladder enlarges and bulges upward into the abdominal cavity. The peritoneal covering is peeled off the lower part of the anterior abdominal wall, so that the bladder comes into direct contact with the anterior abdominal wall.

The *inferolateral surfaces* are related in front to the *retropubic pad of fat* and the pubic bones. More posteriorly, they lie in contact with the obturator

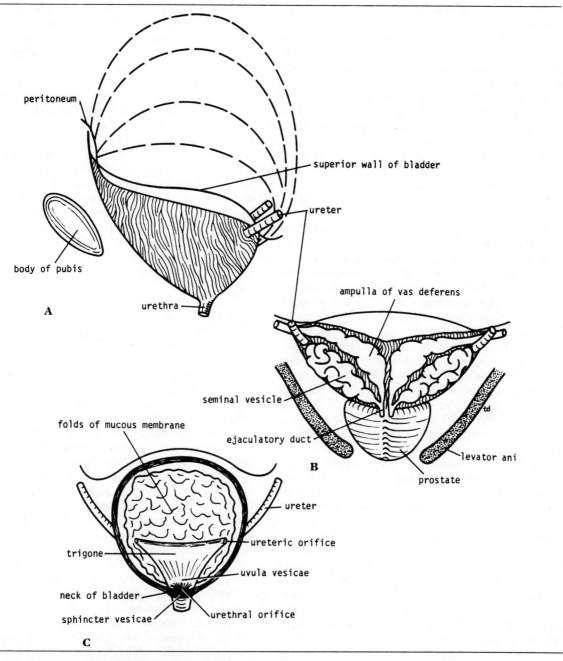

**Fig. 7-5. (A) Lateral view of bladder. Note that superior wall rises as viscus fills with urine. (B) Posterior view of bladder, prostate, vasa deferentia, and seminal vesicles. (C) Interior of bladder.**

internus muscle above and the levator ani muscle below (Fig. 7-5).

The *neck* of the bladder lies inferiorly and rests on the upper surface of the prostate (Fig. 7-5). Here, the smooth muscle fibers of the bladder wall are continuous with those of the prostate. The neck of the bladder is held in position by the *puboprostatic ligaments* in the male and the *pubovesical ligaments* in the female. These ligaments are thickenings of the pelvic fascia.

When the bladder fills, it loses its pyramidal shape and becomes ovoid. The posterior surface and neck remain more or less unchanged in position, but the superior surface rises into the abdomen as described above.

The *mucous membrane* of the greater part of the empty bladder is thrown into folds, but these disappear when the bladder is full. The area of mucous membrane covering the internal surface of the base of the bladder is referred to as the *trigone*. Here, the mucous membrane is always smooth, even when the viscus is empty (Fig. 7-5) because the mucous membrane over the trigone is firmly adherent to the underlying muscular coat.

The superior angles of the trigone correspond to the openings of the ureters and its inferior angle, to the internal urethral orifice (Fig. 7-5). The ureters pierce the bladder wall obliquely, and this provides a valvelike action, which prevents a reverse flow of urine toward the kidneys as the bladder fills.

The trigone is limited above by a muscular ridge, which runs from the opening of one ureter to that of the other and is known as the *interureteric ridge*. The *uvula vesicae* is a small elevation situated immediately behind the urethral orifice that is produced by the underlying median lobe of the prostate.

The *muscular coat of the bladder* is composed of smooth muscle and is arranged as three layers of interlacing bundles, known as the *detrusor muscle*. At the neck of the bladder, the circular component of the muscle coat is thickened to form the *sphincter vesicae* (Fig. 7-5).

The *arterial supply* to the bladder is from the superior and inferior vesical arteries, branches of the internal iliac arteries. The *veins* form the *vesical venous plexus*, which communicates below with the prostatic plexus; it is drained into the internal iliac vein.

The *lymph* vessels from the bladder drain into the internal and external iliac nodes.

The *nerve supply to the bladder* is from the inferior hypogastric plexuses. The sympathetic postganglionic fibers originate in the first and second lumbar ganglia and descend to the bladder via the hypogastric plexuses. The parasympathetic preganglionic fibers arise as the pelvic splanchnic nerves from the second, third, and fourth sacral nerves; they pass through the inferior hypogastric plexuses to reach the bladder wall, where they synapse with postganglionic neurons. The majority of afferent sensory fibers arising in the bladder reach the central nervous system via the pelvic splanchnic nerves. Some afferent fibers travel with the sympathetic nerves via the hypogastric plexuses and enter the first and second lumbar segments of the spinal cord.

### Micturition

Micturition is a reflex action, which in the toilet-trained individual is controlled by higher centers in the brain. The reflex is initiated by the stretching of the bladder muscle as the organ fills with urine. The afferent impulses pass up the pelvic splanchnic nerves and enter the second, third, and fourth sacral segments of the spinal cord (Fig. 7-6). Efferent impulses leave the cord from the same segments and pass via the parasympathetic preganglionic nerve fibers through the pelvic splanchnic nerves and the inferior hypogastirc plexuses to the bladder wall, where they synapse with postganglionic neurons. By means of this nervous pathway, the smooth muscle of the bladder wall (the detrusor muscle) is made to contract, and the sphincter vesicae is made to relax. Efferent impulses also pass to the urethral sphincter via the pudendal nerve (S2, 3, and 4), and this undergoes relaxation. Once urine enters the urethra, additional afferent impulses pass to the spinal cord from the urethra and reinforce the reflex action.

In young children, micturition is a simple reflex act and takes place whenever the bladder becomes distended. In the adult, this simple stretch reflex is inhibited by the activity of the cerebral cortex until the time and place for micturition are favorable. The inhibitory fibers pass downward with the corticospinal tracts to the second, third, and fourth sacral segments of the cord. The contraction of the

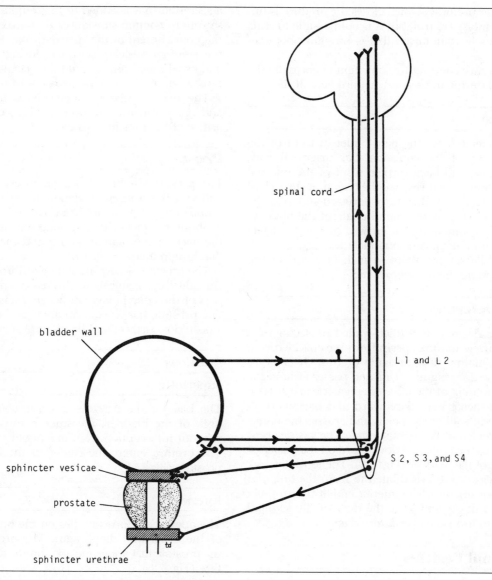

spinal cord

bladder wall

L 1 and L 2

sphincter vesicae

prostate

S 2, S 3, and S 4

sphincter urethrae

**Fig. 7-6. Nervous control of bladder. Sympathetic fibers have been omitted for the sake of simplification.**

sphincter urethrae, which closes the urethra, is under voluntary control, but it is not possible to relax this muscle voluntarily; this is brought about reflexly.

Voluntary control of micturition is normally developed during the second or third year of life.

## Ureter

The ureter crosses the pelvic inlet in front of the bifurcation of the common iliac artery. It runs downward and backward in front of the internal iliac artery until it reaches the region of the ischial spine (Fig. 7-7). It then turns forward and medially and enters the upper lateral angle of the bladder. Near its termination it is crossed by the vas deferens. The ureter passes obliquely through the wall of the bladder for about ¾ inch (1.9 cm) before opening into the bladder cavity.

## Vas Deferens

The vas deferens is a thick-walled muscular tube that conveys mature sperm from the epididymis to the ejaculatory duct and the urethra. It emerges from the deep inguinal ring and passes around the lateral margin of the inferior epigastric artery (Fig. 7-7). It then passes downward and backward on the lateral wall of the pelvis and crosses the ureter in the region of the ischial spine. The vas deferens then runs medially and downward on the posterior surface of the bladder (Fig. 7-5). The terminal part of the vas deferens is dilated to form the *ampulla of the vas deferens.* The inferior end of the ampulla narrows down and joins the duct of the seminal vesicle to form the *ejaculatory duct.*

## Seminal Vesicles

The seminal vesicles are two lobulated sacs about 2 inches (5 cm) long, lying on the posterior surface of the bladder (Fig. 7-5). Their upper ends are widely separated, and their lower extremities are close together. On the medial side of each vesicle lies the terminal part of the vas deferens. Posteriorly, the seminal vesicles are related to the rectum (Fig. 7-3). Inferiorly, each seminal vesicle narrows and joins the vas deferens of the same side to form the *ejaculatory duct.*

The function of the seminal vesicles is to produce a secretion that is added to the seminal fluid. The secretions contain substances that are essential for the nourishment of the spermatozoa. The walls of the seminal vesicles contract during ejaculation and expel their contents into the ejaculatory ducts, thus washing the spermatozoa out of the urethra.

The two ejaculatory ducts pierce the prostate and open into the prostatic urethra, close to the orifice of the prostatic utricle.

## Prostate

The prostate is shaped like an inverted pyramid and is a fibromuscular glandular organ that surrounds the prostatic urethra (Figs. 7-3 and 7-8). It is about 1¼ inches (3 cm) long and lies between the neck of the bladder above and the urogenital diaphragm below (Fig. 7-8).

The prostate is surrounded by a fibrous capsule. Outside the capsule is a fibrous sheath, which is part of the visceral layer of pelvic fascia (Fig. 7-8). The prostate has a base, an apex, an anterior and a posterior surface, and two lateral surfaces.

### Relations

#### Superiorly

The base of the prostate is continuous with the neck of the bladder, the smooth muscle passing without interruption from one organ to the other. The urethra enters the center of the base of the prostate (Fig. 7-3).

#### Inferiorly

The apex of the prostate lies on the upper surface of the urogenital diaphragm. The urethra leaves the prostate just above the apex on the anterior surface (Fig. 7-8).

#### Anteriorly

The anterior surface of the prostate is related to the symphysis pubis, separated from it by the *extraperitoneal fat* in the retropubic space (*cave of Retzius*). The fibrous sheath of the prostate is connected to the posterior aspect of the pubic bones by the *puboprostatic ligaments.* These ligaments lie one on either side of the midline and are condensations of pelvic fascia (Fig. 7-3).

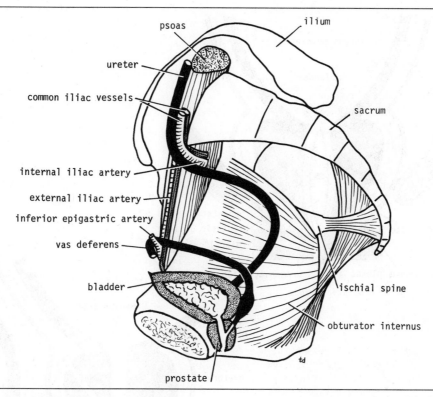

psoas

ureter

common iliac vessels

internal iliac artery

external iliac artery

inferior epigastric artery

vas deferens

bladder

prostate

ilium

sacrum

ischial spine

obturator internus

td

**Fig. 7-7. Right half of pelvis, showing relations of ureter and vas deferens.**

## Posteriorly

The posterior surface of the prostate (Figs. 7-3 and 7-8) is closely related to the anterior surface of the rectal ampulla and is separated from it by the *rectovesical septum (fascia of Denonvillier).* This septum is formed in fetal life by the fusion of the walls of the lower end of the rectovesical pouch of peritoneum, which originally extended down to the perineal body.

## Laterally

The lateral surfaces of the prostate are embraced by the anterior fibers of the levator ani as they run posteriorly from the pubis (Fig. 7-8).

The *ejaculatory ducts* pierce the upper part of the posterior surface of the prostate, to open into the prostatic urethra at the lateral margins of the orifice of the *prostatic utricle* (Fig. 7-8).

The prostate is incompletely divided into five lobes (Fig. 7-8). The *anterior lobe,* or *isthmus,* lies in front of the urethra and is devoid of glandular tissue. The *median,* or *middle, lobe* is the wedge of gland situated between the urethra and the ejaculatory ducts. Its upper surface is related to the trigone of the bladder; it is rich in glands. The *posterior lobe* is situated behind the urethra and below the ejaculatory ducts and also contains glandular tissue. The *right* and *left lateral lobes* lie on either side of the urethra and are separated from one another by a shallow vertical groove on the posterior surface of the prostate. The lateral lobes contain many glands.

The function of the prostate is the production of a thin, milky fluid containing citric acid and acid phosphatase. It is added to the seminal fluid at the time of ejaculation. The smooth muscle in the capsule and stroma contract, and the secretion from the many glands is squeezed into the prostatic ure-

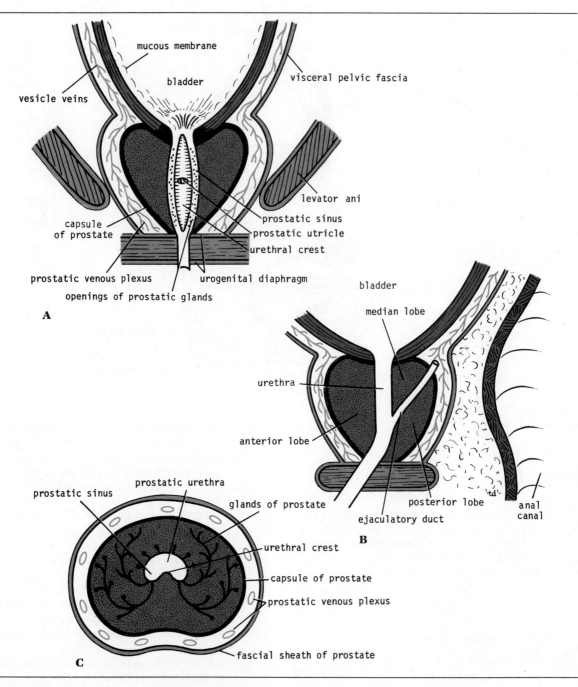

**Fig. 7-8. Prostate in (A) coronal section, (B) sagittal section, and (C) horizontal section.**

thra. The prostatic secretion is alkaline and helps to neutralize the acidity in the vagina.

The *arterial supply to the prostate* is from branches of the inferior vesical and middle rectal arteries. The *veins* form the *prostatic venous plexus*, which is situated between the capsule of the prostate and the fibrous sheath (Fig. 7-8). The prostatic plexus receives the deep dorsal vein of the penis and numerous vesical veins, and drains into the internal iliac veins.

The *lymph vessels* from the prostate drain into the internal iliac nodes.

The *nerve supply* to the prostate is from the inferior hypogastric plexuses.

## Prostatic Urethra

The prostatic urethra is about 1¼ inches (3 cm) long and begins at the neck of the bladder. It traverses the prostate and emerges on its anterior aspect a little above the apex, where it becomes continuous with the membranous part of the urethra (Fig. 7-8).

The *prostatic urethra is the widest and most dilatable part of the entire urethra.* On the posterior wall is a longitudinal elevation, called the *urethral crest* (Fig. 7-8). The grooves along the sides of the crest are called the *prostatic sinuses* and receive the numerous openings of the prostatic glands. At the summit of the crest, a small diverticulum, the *prostatic utricle* (remains of fused paramesonephric ducts), opens into the urethra. At the lateral margins of the utricle, the ejaculatory ducts open into the urethra (Fig. 7-8).

## Visceral Pelvic Fascia

The visceral pelvic fascia is a layer of connective tissue that covers and supports the pelvic viscera. It fills in the spaces between the viscera and supports the blood vessels, lymphatic vessels, and nerves that pass to the viscera. It is condensed to form the fascial sheath of the prostate and the puboprostatic ligaments (Figs. 7-3 and 7-8). The visceral fascia is continuous below with the fascia covering the upper surface of the levator ani and coccygeus muscles, and with the parietal pelvic fascia on the walls of the pelvis.

## Peritoneum

The peritoneum is best understood by tracing it around the pelvis in a sagittal plane (Fig. 7-3).

The peritoneum passes down from the anterior abdominal wall onto the upper surface of the urinary bladder. It then runs down on the posterior surface of the bladder for a short distance, until it reaches the upper ends of the seminal vesicles. Here, it sweeps backward to reach the anterior aspect of the rectum, forming the shallow *rectovesical pouch*. The peritoneum then passes up on the front of the middle third of the rectum and the front and lateral surfaces of the upper third of the rectum. It then becomes continuous with the parietal peritoneum on the posterior abdominal wall. It is thus seen that the most dependent part of the abdomino–pelvic peritoneal cavity, with the patient in the erect position, is the rectovesical pouch (Fig. 7-3).

The peritoneum covering the superior surface of the bladder passes laterally to the lateral pelvic walls and does not cover the lateral surfaces of this viscus. It is important to remember that as the bladder fills, the superior wall rises up into the abdomen and peels off the peritoneum from the anterior abdominal wall.

## PELVIC VISCERA IN THE FEMALE

The rectum, sigmoid colon, and terminal coils of ileum occupy the posterior part of the pelvic cavity, as described previously. The contents of the anterior part of the pelvic cavity in the female are described in the following sections.

## Urinary Bladder

As in the male, the urinary bladder is situated immediately behind the pubic bones (Fig 7-9). Because of the absence of the prostate, the bladder lies at a lower level than in the male pelvis, and the neck rests directly on the upper surface of the urogenital diaphragm. The close relation of the bladder to the uterus and the vagina is of considerable clinical importance (Fig. 7-9).

The *apex* of the bladder lies behind the symphysis pubis (Fig. 7-9). The *base*, or *posterior sur-*

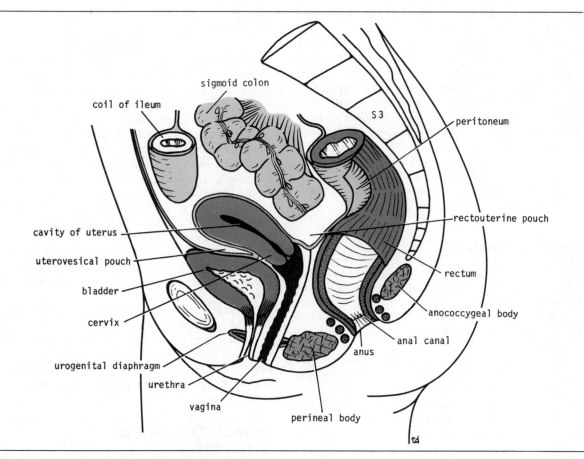

**Fig. 7-9. Sagittal section of female pelvis.**

*face*, is separated by the vagina from the rectum. The *superior* surface is related to the uterovesical pouch of peritoneum and to the body of the uterus. The *inferolateral surfaces* are related in front to the *retropubic pad of fat* and the pubic bones. More posteriorly, they lie in contact with the obturator internus muscle above and the levator ani muscle below. The *neck* of the bladder rests on the upper surface of the urogenital diaphragm.

The general shape and structure of the bladder, its blood supply, lymphatic drainage, and nerve supply, and the process of micturition are identical to the male's.

## Ureter

The ureter crosses the pelvic inlet in front of the bifurcation of the common iliac artery (Fig. 7-10). It runs downward and backward in front of the internal iliac artery and behind the ovary, until it reaches the region of the ischial spine. It then turns forward and medially beneath the base of the broad ligament, where it is crossed by the uterine artery (Figs. 7-10 and 7-11). The ureter then runs forward, lateral to the lateral fornix of the vagina, to enter the bladder.

## Ovary

Each ovary is an almond-shaped organ, measuring 1½ by ¾ inches (4 × 2 cm), and is attached to the

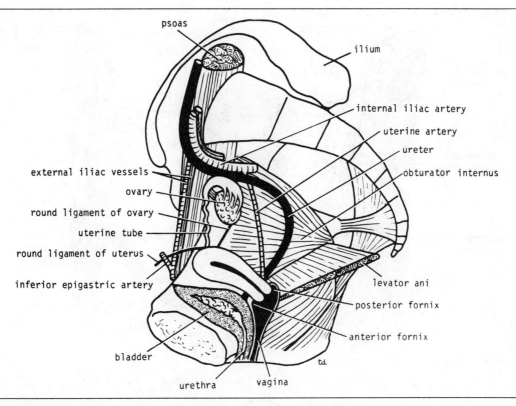

psoas

ilium

internal iliac artery

uterine artery

ureter

external iliac vessels

obturator internus

ovary

round ligament of ovary

uterine tube

round ligament of uterus

inferior epigastric artery

levator ani

posterior fornix

anterior fornix

bladder

urethra          vagina

**Fig. 7-10. Right half of pelvis, showing ovary, uterine tube, and vagina.**

back of the broad ligament by the *mesovarium* (Fig. 7-11). Usually, the ovary lies with its long axis vertical, but it shares in any movement of the broad ligament and uterus.

That part of the broad ligament extending between the attachment of the mesovarium and the lateral wall of the pelvis is sometimes called the *suspensory ligament of the ovary* (Fig. 7-11).

The *round ligament of the ovary* is the remains of the upper part of the gubernaculum. (The round ligament of the uterus is the remains of the lower part of the gubernaculum.) It extends from the upper end of the lateral wall of the uterus to the medial margin of the ovary (Figs. 7-10 and 7-11).

The ovary usually lies against the lateral wall of the pelvis in a depression called the *ovarian fossa*. The fossa is bounded by the external iliac vessels above and by the internal iliac vessels and the ureter behind (Fig. 7-10). The obturator nerve crosses

the floor of the fossa. The position of the ovary is, however, extremely variable, and it is often found hanging down in the rectouterine pouch (pouch of Douglas). During pregnancy the enlarging uterus pulls the ovary up into the abdominal cavity. After parturition, when the broad ligament is lax, the ovary takes up a variable position in the pelvis.

The ovaries are surrounded by a thin fibrous capsule, the *tunica albuginea*. This capsule is covered externally by a single layer of cuboid cells called the *germinal epithelium*. The term germinal epithelium is a misnomer, because the layer does not give rise to ova. Oogonia develop before birth from primordial germ cells. The germinal epithelium is merely a modified area of peritoneum and is continuous with the squamous mesothelial cells of the general peritoneum at the hilus of the ovary, where the mesovarium is attached.

Before puberty the ovary is smooth, but after puberty the ovary becomes progressively scarred as successive corpora lutea degenerate. After the

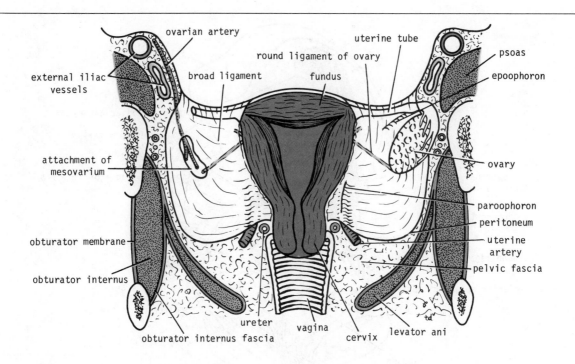

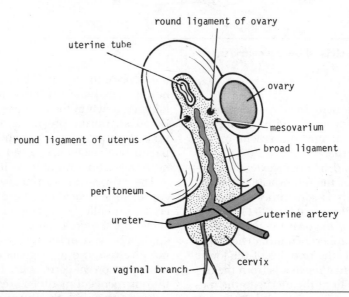

**Fig. 7-11. (Above) Coronal section of pelvis, showing uterus, broad ligaments, and right ovary on posterior view. The left ovary and part of the left uterine tube have been removed. (Below) Uterus on lateral view. Note structures that lie within broad ligament.**

menopause the ovary becomes shrunken and its surface is pitted with scars.

The *arterial supply* to the ovary is the *ovarian artery*, which arises from the aorta at the level of the first lumbar vertebra. The *ovarian vein* drains into the inferior vena cava on the right side and into the left renal vein on the left.

The *lymph vessels* of the ovary follow the ovarian artery and drain into the para-aortic nodes at the level of the first lumbar vertebra.

The *nerve supply* to the ovary is derived from the aortic plexus and accompanies the ovarian artery.

The blood supply, lymphatic drainage, and nerve supply of the ovary pass over the pelvic inlet and cross the external iliac vessels (Fig. 7-11). They then enter the lateral end of the broad ligament, the part known as the suspensory ligament of the ovary. The vessels and nerves finally enter the hilus of the ovary via the mesovarium. (Compare the blood supply and the lymphatic drainage of the ovary with that of the testis.)

## Uterine Tube

There are two uterine tubes; each is about 4 inches (10 cm) long and lies in the upper border of the broad ligament (Figs. 7-10 and 7-11). Each connects the peritoneal cavity in the region of the ovary with the cavity of the uterus. For purposes of description it is divided into four parts: the infundibulum, the ampulla, the isthmus, and the intramural part.

The *infundibulum* is the funnel-shaped lateral extremity, which projects beyond the broad ligament and overlies the ovary. The free edge of the funnel is broken up into a number of finger-like processes, known as *fimbriae*, which are draped over the ovary (Figs. 7-11 and 7-12).

The *ampulla* is the widest part of the tube (Fig. 7-12); it is in this portion that fertilization of the ovum takes place.

The *isthmus* is the narrowest part of the tube and lies just lateral to the uterus (Fig. 7-12).

The *intramural part* is the segment that pierces the uterine wall (Fig. 7-12).

The *arterial supply* to the uterine tube is from the uterine and ovarian arteries (Fig. 7-12). The *veins* correspond to the arteries.

The *lymphatic vessels* follow the corresponding arteries and drain into the internal iliac and para-aortic nodes.

The *nerve supply* is from the inferior hypogastric plexuses.

### Function of the Uterine Tube

The uterine tube receives the ovum from the ovary and provides a site where fertilization can occur. The tube also provides a conduit along which the spermatozoa travel to reach the ovum.

## Uterus

The uterus is a hollow, pear-shaped organ with thick muscular walls. In the young nulliparous adult it measures 3 inches (8 cm) long, 2 inches (5 cm) wide, and 1 inch (2.5 cm) thick. For purposes of description it is divided into the fundus, body, and cervix (Fig. 7-12).

The *fundus* is the part of the uterus that lies above the entrance of the uterine tubes.

The *body* is the part of the uterus that lies below the entrance of the uterine tubes. It narrows below, where it becomes continuous with the *cervix*. The cervix pierces the anterior wall of the vagina and is divided into the *supravaginal* and *vaginal parts of the cervix.*

The *cavity* of the uterine body is triangular in coronal section, but it is merely a cleft in the sagittal plane (Fig. 7-12). The cavity of the cervix, the *cervical canal*, is spindle-shaped and communicates with the cavity of the body through the *internal os*, and with that of the vagina, through the *external os*. In a nullipara, the external os is circular. In a parous woman, the vaginal part of the cervix is larger, and the external os is opened out transversely, so that it possesses an anterior lip and a posterior lip (Fig. 7-12).

Normally, in the majority of women, the long axis of the uterus is bent forward on the long axis of the vagina, forming an angle of 90 degrees. This position is referred to as *anteversion of the uterus* (Fig. 7-12). Furthermore, the long axis of the body of the uterus is bent forward at the level of the internal os with the long axis of the cervix, forming an angle of about 170 degrees. This position is termed *anteflexion of the uterus* (Fig. 7-12). Thus, in the erect position, with the bladder empty, the uterus lies in an almost horizontal plane.

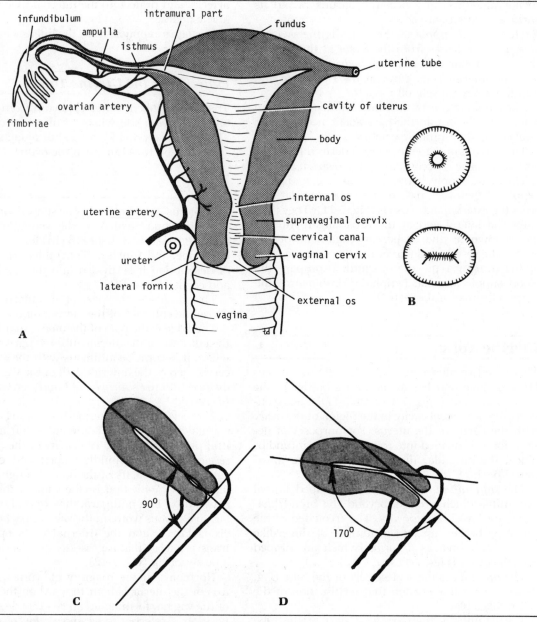

**Fig. 7-12. (A)** Different parts of uterine tube and uterus. **(B)** External os of cervix: (above) nulliparous; (below) parous. **(C)** Anteverted position of uterus. **(D)** Anteverted and anteflexed position of uterus.

In some women, the fundus and body of the uterus are bent backward on the vagina, so that they lie in the rectouterine pouch (pouch of Douglas). In this situation the uterus is said to be *retroverted*. If the body of the uterus is, in addition, bent backward on the cervix, it is said to be *retroflexed.*

## Structure

The uterus is covered with peritoneum except anteriorly, below the level of the internal os, where the peritoneum is reflected onto the bladder. Laterally, there is also a space between the attachment of the layers of the broad ligament.

The *muscular wall*, or *myometrium*, is thick and made up of smooth muscle supported by connective tissue.

The *mucous membrane* lining the body of the uterus is known as the *endometrium*. It is continuous above with the mucous membrane lining the uterine tubes, and below with the mucous membrane lining the cervix. The endometrium is applied directly to the muscle, there being no submucosa. From puberty to the menopause, the endometrium undergoes extensive changes during the menstrual cycle in response to the ovarian hormones.

## Relations

### Anteriorly

The body of the uterus is related anteriorly to the uterovesical pouch and the superior surface of the bladder (Fig. 7-9). The supravaginal cervix is related to the superior surface of the bladder. The vaginal cervix is related to the anterior fornix of the vagina.

### Posteriorly

The body of the uterus is related posteriorly to the rectouterine pouch (pouch of Douglas) with coils of ileum or sigmoid colon within it (Fig. 7-9).

### Laterally

The body of the uterus is related laterally to the broad ligament and the uterine artery and vein (Fig. 7-11). The supravaginal cervix is related to the

ureter as it passes forward to enter the bladder. The vaginal cervix is related to the lateral fornix of the vagina. The uterine tubes enter the superolateral angles of the uterus, and the round ligaments of the ovary and of the uterus are attached to the uterine wall just below this level.

The *arterial supply* to the uterus is mainly from the uterine artery, a branch of the internal iliac artery. It reaches the uterus by running medially in the base of the broad ligament (Fig. 7-11). It crosses above the ureter at right angles and reaches the cervix at the level of the internal os (Fig. 7-12). The artery then ascends along the lateral margin of the uterus within the broad ligament and ends by anastomosing with the ovarian artery, which also assists in supplying the uterus. The uterine artery gives off a small descending branch that supplies the cervix and the vagina.

The *uterine vein* follows the artery and drains into the internal iliac vein.

The *lymphatic vessels* from the fundus of the uterus accompany the ovarian artery and drain into the para-aortic nodes at the level of the first lumbar vertebra. The vessels from the body and cervix drain into the internal and external iliac lymph nodes. A few lymph vessels follow the round ligament of the uterus through the inguinal canal and drain into the superficial inguinal lymph nodes.

The *nerve supply* to the uterus is from branches of the inferior hypogastric plexuses.

## Supports of the Uterus

The uterus is supported mainly by (1) the tone of the levatores ani muscles and (2) the condensations of pelvic fascia, which form three important ligaments.

### THE LEVATORES ANI MUSCLES AND THE PERINEAL BODY

The origin and the insertion of the levatores ani muscles have been described in Chapter 6. They form a broad muscular sheet stretching across the pelvic cavity, and together with the pelvic fascia on their upper surface, they effectively support the pelvic viscera and resist the intra-abdominal pressure transmitted downward through the pelvis. The medial edges of the anterior parts of the leva-

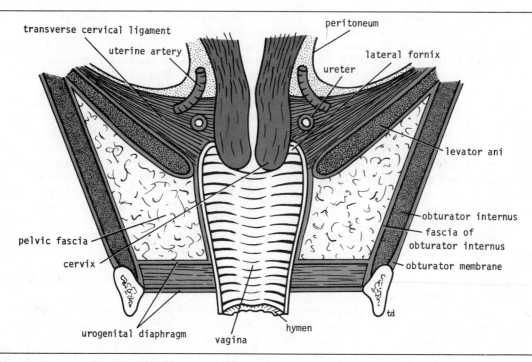

**Fig. 7-13. Coronal section of pelvis, showing relation of levatores ani muscles and transverse cervical ligaments to uterus and vagina.**

tores ani muscles are attached to the cervix of the uterus by the pelvic fascia (Fig. 7-13).

Some of the fibers of levatore ani are inserted into a fibromuscular structure called the *perineal body* (Fig. 7-9). This structure is important in maintaining the integrity of the pelvic floor, and should it be damaged during childbirth, prolapse of the pelvic viscera may occur. The perineal body lies in the perineum between the vagina and the anal canal. It is slung up to the pelvic walls by the levatores ani, and thus supports the vagina and indirectly supports the uterus.

## THE TRANSVERSE CERVICAL, PUBOCERVICAL, AND SACROCERVICAL LIGAMENTS

These three ligaments are condensations of pelvic fascia on the upper surface of the levatores ani muscles. They are attached to the cervix and the vault of the vagina and play an important part in

supporting the uterus and keeping the cervix in its correct position (Figs. 7-13 and 7-14).

### Transverse Cervical Ligaments

Transverse cervical ligaments are fibromuscular condensations of pelvic fascia that pass to the cervix and the upper end of the vagina from the lateral walls of the pelvis.

### Pubocervical Ligaments

The pubocervical ligaments consist of two firm bands of connective tissue that pass to the cervix from the posterior surface of the pubis. They are positioned on either side of the neck of the bladder, to which they give some support (*pubovesical ligaments*).

### Sacrocervical Ligaments

The sacrocervical ligaments consist of two firm fibromuscular bands of pelvic fascia that pass to the cervix and the upper end of the vagina from the

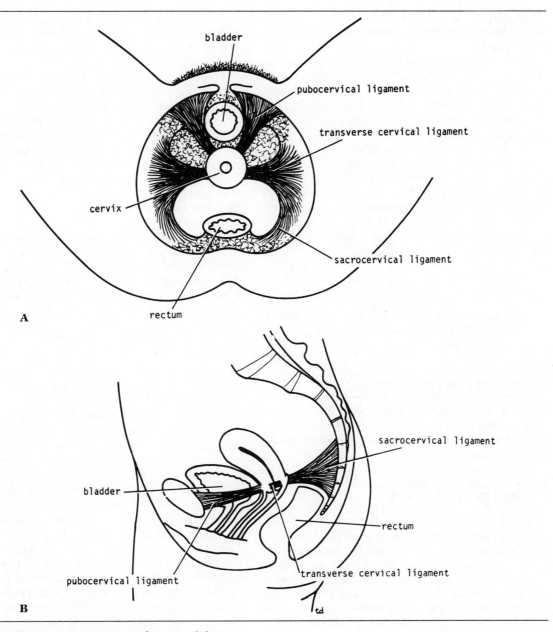

bladder

pubocervical ligament

transverse cervical ligament

cervix

sacrocervical ligament

rectum

**A**

sacrocervical ligament

bladder

rectum

pubocervical ligament

transverse cervical ligament

**B**

td

**Fig. 7-14. Ligamentous supports of uterus. (A) As seen from below, and (B) lateral view.**

lower end of the sacrum. They form two ridges, one on either side of the rectouterine pouch (pouch of Douglas).

The broad ligaments and the round ligaments of the uterus are lax structures, and the uterus may be pulled up or pushed down for a considerable distance before they become taut. Clinically, they are considered to play a very minor role in supporting the uterus.

The *round ligament of the uterus*, which represents the remains of the lower half of the gubernaculum, extends between the superolateral angle of the uterus, through the deep inguinal ring and inguinal canal, to the subcutaneous tissue of the labium majus (Fig. 7-10). It helps to keep the uterus anteverted and anteflexed, but it is considerably stretched during pregnancy.

## UTERUS IN THE CHILD

The uterus remains small until puberty, when it enlarges greatly in response to the estrogens secreted by the ovaries.

## UTERUS AFTER THE MENOPAUSE

After the menopause, the uterus atrophies and becomes smaller and less vascular. These changes occur because the ovaries no longer produce estrogens and progesterone.

## UTERUS IN PREGNANCY

During pregnancy, the uterus becomes greatly enlarged as the result of the increasing production of estrogens and progesterone, first by the corpus luteum of the ovary and later by the placenta. At first it remains as a pelvic organ but by the third month, the fundus rises out of the pelvis, and by the ninth month it has reached the xiphoid process. The increase in size is largely due to hypertrophy of the smooth muscle fibers of the myometrium, although some hyperplasia does take place.

## ROLE OF THE UTERUS IN LABOR

Labor, or parturition, is the series of processes by which the baby, the fetal membranes, and the pla-

centa are expelled from the genital tract of the mother. Normally this process takes place at the end of the tenth lunar month, at which time the pregnancy is said to be at *term.*

The cause of the onset of labor is not definitely known. By the end of pregnancy, the contractility of the uterus has been fully developed in response to estrogen, and it is particularly sensitive to the actions of oxytocin at this time. It is possible that the onset of labor is triggered by the sudden withdrawal of progesterone. Once the presenting part (usually the fetal head) starts to stretch the cervix, it is thought that a nervous reflex mechanism is initiated that increases the force of the contractions of the uterine body.

It is interesting to note that the uterine muscular activity is largely independent of the extrinsic innervation. In women in labor, spinal anesthesia does not interfere with the normal uterine contractions. Severe emotional disturbance, however, may cause premature parturition.

# Vagina

The vagina is the female organ for copulation; it serves as the excretory duct for the uterus and forms part of the birth canal.

The vagina extends upward and backward from the vulva (Fig. 7-9). It measures about 3 inches (8 cm) long and has anterior and posterior walls, which are normally in apposition. At its upper end the anterior wall is pierced by the cervix, which projects downward and backward into the vagina. It is important to remember that the upper half of the vagina lies above the pelvic floor and the lower half lies within the perineum (Figs. 7-9 and 7-13). The area of the vaginal lumen, which surrounds the cervix, is divided for purposes of description into four regions, or *fornices:* anterior, posterior, right lateral, and left lateral. A thin mucosal fold called the *hymen* surrounds the entrance to the vaginal orifice. After childbirth the hymen usually consists only of tags.

## Relations

### Anteriorly

The vagina is closely related to the bladder above and to the urethra below (Fig. 7-9).

## Posteriorly

The upper third of the vagina is related to the rectouterine pouch (pouch of Douglas) and its middle third, to the ampulla of the rectum. The lower third is related to the perineal body, which separates it from the anal canal (Fig. 7-9).

## Laterally

In its upper part, the vagina is related to the ureter; its middle part is related to the anterior fibers of the levator ani, as they run backward to reach the perineal body and hook around the anorectal junction (Figs. 7-11 and 7-13). Contraction of the fibers of levator ani compress the walls of the vagina together. In its lower part, the vagina is related to the urogenital diaphragm (see Chap. 8) and the bulb of the vestibule.

The *arterial supply* to the vagina is from the vaginal artery, a branch of the internal iliac artery; it is also supplied by a vaginal branch of the uterine artery. The *vaginal veins* form a plexus around the vagina that drains into the internal iliac vein.

The *lymphatic vessels* from the upper third of the vagina drain to the external and internal iliac nodes; from the middle third, to the internal iliac nodes; and from the lower third, to the superfical inguinal nodes.

The *nerve supply* to the vagina is from the inferior hypogastric plexuses.

### SUPPORTS OF THE VAGINA

The upper part of the vagina is supported by the levatores ani muscles and the transverse cervical, pubocervical, and sacrocervical ligaments. These structures are attached to the vaginal wall by pelvic fascia (Figs. 7-13 and 7-14).

The middle part of the vagina is supported by the urogenital diaphragm. (See Chap. 8.)

The lower part of the vagina, especially the posterior wall, is supported by the perineal body (Fig. 7-9).

## Visceral Pelvic Fascia

The visceral pelvic fascia is a layer of connective tissue, which, as in the male, covers and supports the pelvic viscera. It is condensed to form the pubocervical, transverse cervical, and sacrocervical ligaments of the uterus (Fig. 7-14). Clinically, the pelvic fascia in the region of the uterus is often referred to as the *parametrium*. The visceral fascia is continuous below with the fascia covering the upper surface of the levatores ani and coccygeus muscles, and on the walls of the pelvis, with the parietal pelvic fascia.

## Peritoneum

The peritoneum in the female, as in the male, is best understood by tracing it around the pelvis in a sagittal plane (Fig. 7-9).

The peritoneum passes down from the anterior abdominal wall onto the upper surface of the urinary bladder. It then runs directly onto the anterior surface of the uterus, at the level of the internal os. The peritoneum now passes upward over the anterior surface of the body and fundus of the uterus and then downward over the posterior surface. It continues downward and covers the upper part of the posterior surface of the vagina, where it forms the anterior wall of the rectouterine pouch (pouch of Douglas). The peritoneum is then reflected onto the front of the rectum, as in the male.

In the female the most dependent part of the abdomino-pelvic peritoneal cavity in the erect position is the rectouterine pouch.

## BROAD LIGAMENTS

The *broad ligaments* are two-layered folds of peritoneum that extend across the pelvic cavity from the lateral margins of the uterus to the lateral pelvic walls (Fig. 7-11). Superiorly, the two layers are continuous and form the upper free edge. Inferiorly, at the base of the ligament, the layers separate to cover the pelvic floor. The ovary is attached to the posterior layer by the *mesovarium*. That part of the broad ligament that lies lateral to the attachment of the mesovarium is sometimes referred to as the *suspensory ligament of the ovary*. The part of the broad ligament between the uterine tube and the mesovarium is called the *mesosalpinx*.

At the base of the broad ligament, the uterine artery crosses the ureter (Figs. 7-11 and 7-13).

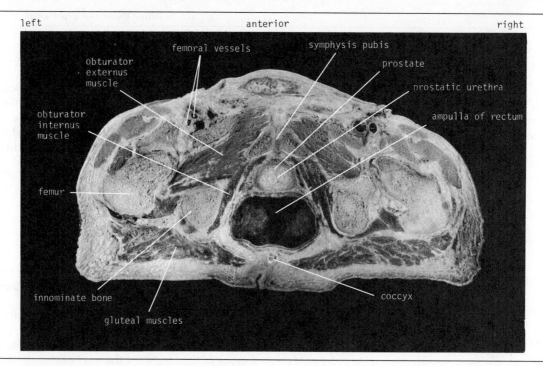

Fig. 7-15. Cross section of the male pelvis as seen from above.

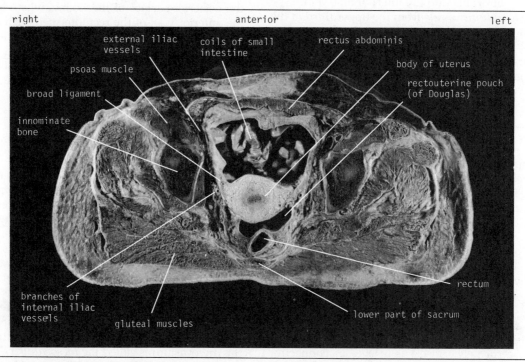

Fig. 7-16. Cross section of the female pelvis as seen from below.

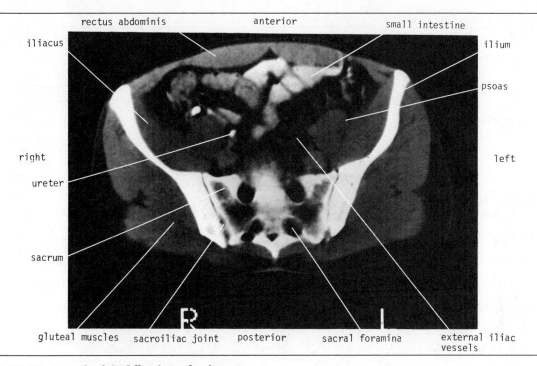

rectus abdominis    anterior    small intestine

iliacus    ilium

psoas

right    left

ureter

sacrum

gluteal muscles    sacroiliac joint    posterior    sacral foramina    external iliac vessels

Fig. 7-17. CT scan of pelvis following a barium meal and intravenous pyelography. Note the presence of the radiopaque material in the small intestine and the right ureter. The section is viewed from below.

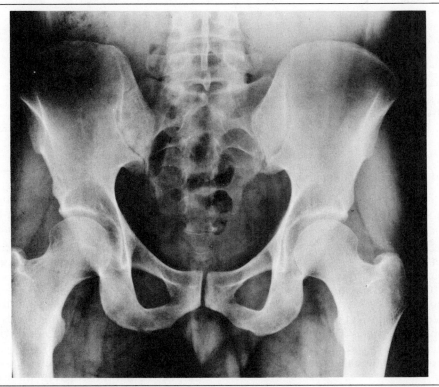

Fig. 7-18. Anteroposterior radiograph of the male pelvis.

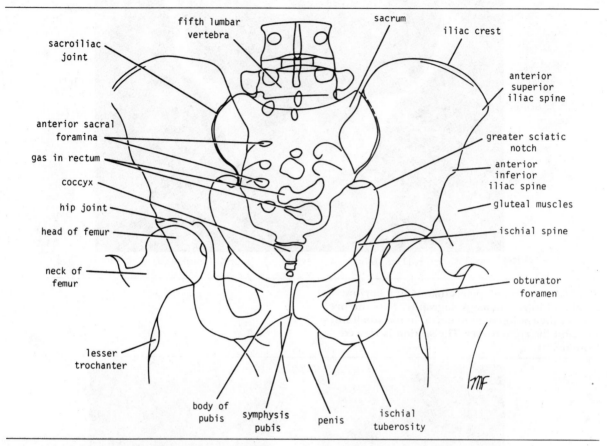

**Fig. 7-19. Diagrammatic representation of radiograph of pelvis seen in Figure 7-18.**

Each broad ligament contains the following:

1. The uterine tube in its upper free border.
2. The round ligament of the ovary and the round ligament of the uterus.
3. The uterine and ovarian blood vessels, lymphatics, and nerves.
4. The epoophoron. This is a vestigial structure that lies in the broad ligament above the attachment of the mesovarium. It represents the remains of the mesonephros (Fig. 7-11).
5. The paroophoron. This is also a vestigial structure that lies in the broad ligament just lateral to the uterus. It is a mesonephric remnant (Fig. 7-11).

## Cross-Sectional Anatomy of the Pelvis

In order to assist you in the interpretation of CT scans of the pelvis, you should study the labeled cross sections of the pelvis shown in Figures 7-15 and 7-16. (See Fig. 7-17 for CT scan.)

## Radiographic Appearances of the Bony Pelvis

A routine anteroposterior view of the pelvis is taken with the patient in the supine position and with the cassette underneath the tabletop. Unfortunately, a somewhat distorted view of the lower part of the sacrum and coccyx is obtained, and these bones may be partially obscured by the symphysis pubis.

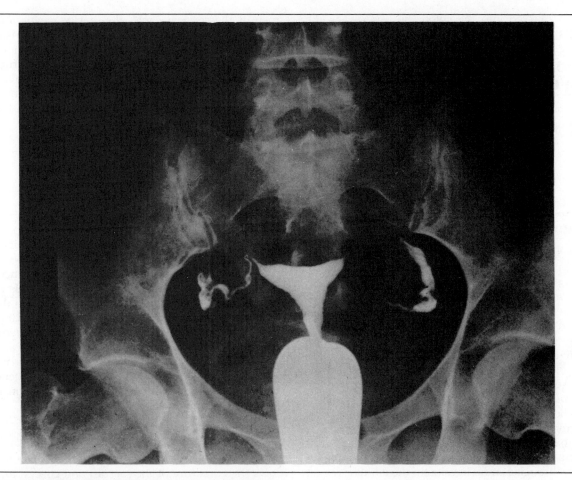

**Fig. 7-20. Anteroposterior radiograph of female pelvis following injection of radiopaque compound into uterine cavity (hysterosalpingogram).**

A better view of the sacrum and coccyx may be obtained by slightly tilting the X-ray tube.

An anteroposterior radiograph should be systematically examined (Figs. 7-18 and 7-19). The lower lumbar vertebrae, sacrum, and coccyx may be looked at first, followed by the sacroiliac joints, the different parts of the innominate bones, and finally the hip joints and the upper ends of the femurs. Gas and fecal material may be seen in the large bowel, and soft-tissue shadows of the skin and subcutaneous tissues may also be visualized.

To demonstrate the sacrum and sacroiliac joints more clearly, lateral and oblique views of the pelvis are often taken.

## Radiographic Appearances of the Sigmoid Colon and Rectum

### Barium Enema

The pelvic colon and rectum can be demonstrated by the administration of 2 to 3 pints (1 L) of barium sulfate emulsion slowly through the anus. The appearances of the pelvic colon are similar to those seen in the more proximal parts of the colon, but a distended sigmoid colon usually shows no sacculations. The rectum is seen to have a wider caliber than the colon.

A *contrast enema* is sometimes very useful for examining the mucous membrane of the sigmoid colon. The barium enema is partly evacuated and air is injected into the colon. By this means the walls of the colon become outlined (see Fig. 5-60).

## Radiographic Appearances of the Female Genital Tract

The instillation of viscous iodine preparations through the external os of the uterus allows the lu-men of the cervical canal, the uterine cavity, and the different parts of the uterine tubes to be visualized (Fig. 7-20). This procedure is known as *hysterosalpingography*. The patency of these structures will be demonstrated by the entrance into the peritoneal cavity of some of the opaque medium.

A sonogram of the female pelvis shows the uterus and the vagina (Fig. 7-23).

## CLINICAL NOTES

### Sigmoid Colon (Pelvic Colon)

The sigmoid colon shows great variation in length and may measure as much as 36 inches (91 cm). In the young child, since the pelvis is of small size, this segment of the colon may lie mainly in the abdomen. Because of its extreme mobility, the pelvic colon sometimes rotates around its mesentery. This may correct itself spontaneously, or the rotation may continue until the blood supply of the gut is cut off completely. The rotation commonly occurs in a counterclockwise direction, and is referred to as *volvulus*.

The sigmoid colon is often selected as a site for performing a *colostomy* in patients with carcinoma of the rectum. Its mobility allows the surgeon to bring out a loop of colon, with its blood supply intact, through a small incision in the left iliac region of the anterior abdominal wall. Its mobility also makes it suitable for implantation of the ureters following surgical removal of the bladder.

Diverticula of the mucous membrane along the course of the arteries supplying the sigmoid colon is a common clinical condition and has been described on page 285. In patients with diverticulitis or ulcerative colitis, the sigmoid colon may become adherent to the bladder, rectum, ileum, or ureter and produce an internal fistula. Since the sigmoid colon lies only a short distance from the anus (6½ inches, or 17 cm), it is possible to examine the mucous membrane under direct vision for pathological conditions. A tube fitted with lenses and illuminated internally is introduced through the anus and carefully passed up through the anal canal and rectum to the sigmoid colon. This examination, called *sigmoidoscopy*, can be carried out without an anesthetic in an outpatient clinic. Biopsy specimens of the mucous membrane may be obtained through this instrument.

The sigmoid colon is a common site for cancer of the large bowel. Since the lymphatic vessels of this segment of the colon drain ultimately into the inferior mesenteric nodes, it follows that an extensive resection of the gut and its associated lymphatic field is necessary to extirpate the growth and its local lymphatic metastases. The colon is removed from the left colic flexure to the distal end of the sigmoid colon, and the transverse colon is anastomosed to the rectum.

### Rectum

The anteroposterior flexure of the rectum, as it follows the curvature of the sacrum and coccyx, and the three lateral flexures, must be remembered when one is passing a sigmoidoscope, to avoid causing the patient unnecessary discomfort.

The three crescentic transverse mucosal folds of the rectum, two on the left rectal wall and one on the right, must also be borne in mind when passing an instrument into the rectum. It is thought that these folds serve to support the weight of the feces and to prevent excessive distention of the rectal ampulla.

The chief arterial supply to the rectum is from the superior rectal artery, a continuation of the inferior mesenteric artery. In front of the third sacral vertebra, the artery divides into right and left branches. Halfway down the rectum, the right branch divides into an anterior and posterior branch. The tributaries of the superior rectal vein are arranged in a similar manner, so that it is not surprising to find that *internal hemorrhoids* are arranged in three groups (see Chap. 8): two on the right side of the lower rectum and anal canal, and one on the left.

*Partial* and *complete prolapses of the rectum* through the anus are relatively common clinical conditions. In partial prolapse, the rectal mucous membrane and submucous coat protrude for a short distance outside the anus. In complete prolapse, the whole thickness of the rectal wall protrudes through the anus. In both conditions, many causative factors may be involved. However, damage to the levatores ani muscles as the result of childbirth and poor muscle tone in the aged are important contributing factors. A complete rectal prolapse may be regarded as a sliding hernia through the pelvic diaphragm.

Cancer (carcinoma) of the rectum is a very common clinical finding. Fortunately, it remains localized to the rectal wall for a considerable time. At first, it tends to spread locally in the lymphatics around the circumference of the bowel. Later, it spreads upward and laterally along the lymph vessels, following the superior rectal and middle rectal arteries. Venous spread occurs late, and since the superior rectal vein is a tributary of the portal vein, the liver is a common site for secondary deposits.

Once the malignant tumor has extended beyond the confines of the rectal wall, a knowledge of the anatomical relations of the rectum will enable a physician to assess the structures and organs likely to be involved. In both sexes, a posterior penetration will involve the sacral plexus and may cause severe intractable pain down the leg in the distribution of the sciatic nerve. A lateral penetration may involve the ureter. An anterior penetration in the male may involve the prostate, seminal vesicles, or bladder; in the female, the vagina and uterus may be invaded.

It is clear from the anatomy of the rectum and its lymphatic drainage that a wide resection of the rectum with its lymphatic field offers the best chance of cure. When the tumor has spread to contiguous organs and is of a low grade of malignancy, some form of pelvic evisceration may be justifiable.

It is most important for a medical student to remember that the interior of the lower part of the rectum can be examined by a gloved index finger introduced through the anal canal. The anal canal is about 1½ inches (4 cm) long, so that the pulp of the index finger can easily feel the mucous membrane lining the lower end of the rectum. The great majority of cancers of the rectum can be diagnosed by this means. This examination can be extended in both sexes by placing the other hand on the lower part of the anterior abdominal wall. With the bladder empty, the anterior rectal wall can be examined bimanually. In the female, the placing of one finger in the vagina and another in the rectum may enable the physician to make a thorough examination of the lower part of the anterior rectal wall.

*Proctoscopy*, the introduction of an internally illuminated tubular instrument through the anus, enables the physician to examine the greater part of the rectal mucosa under direct vision. If the rectopelvic junction cannot be seen, a sigmoidoscope should be used.

## "Pelvic Appendix"

If an inflamed appendix is hanging down into the pelvis there may be no abdominal tenderness in the right iliac region, but deep tenderness may be experienced in the hypogastric region. Rectal examination (or vaginal examination in the female) may reveal tenderness of the peritoneum in the pelvis on the right side. Should such an inflamed appendix perforate, a localized pelvic peritonitis may result.

## Urinary Bladder

The full bladder in the adult projects up into the abdomen and may be palpated in the hypogastrium. In a patient with a full bladder, a severe blow on the lower part of the anterior abdominal wall may therefore result in intraperitoneal rupture of the bladder.

As the bladder fills, the superior wall rises out of the pelvis and peels the peritoneum off the posterior surface of the anterior abdominal wall. In cases of acute retention of urine, when catheterization has failed, it is possible to pass a needle into the bladder through the anterior abdominal wall above the symphysis pubis without entering the peritoneal cavity. This is a simple method of draining off the urine in an emergency; but if the bladder is allowed to refill, leakage may occur into the extraperitoneal space through the puncture hole.

*Bimanual palpation* of the empty bladder with

or without a general anesthetic is an important method of examining the bladder. In the male, one hand is placed on the anterior abdominal wall above the symphysis pubis, and the gloved index finger of the other hand is inserted into the rectum. From your knowledge of anatomy, you can see that the bladder wall can be palpated between the examining fingers. In the female, an abdominovaginal examination can be similarly made. In the child, the bladder is in a higher position than in the adult, due to the relatively smaller size of the pelvis. For this reason, when making a low abdominal incision, the surgeon must make sure that the child's bladder is empty.

*Cystoscopy.* The mucous membrane of the bladder, the two ureteric orifices, and the urethral meatus can easily be observed by means of a cystoscope. With the bladder distended with fluid, an illuminated tube fitted with lenses is introduced into the bladder through the urethra. Over the trigone the mucous membrane is pink and smooth. If the bladder is partially emptied, the mucous membrane over the trigone remains smooth, but is thrown into folds elsewhere. The ureteric orifices are slitlike and eject a drop of urine at intervals of about 1 minute. The interureteric ridge and the uvula vesicae may be recognized easily.

The bladder is normally supported by the visceral pelvic fascia, which in certain areas is condensed to form ligaments. However, the most important support for the bladder is the tone of the levatores ani muscles. In the female, a difficult labor, especially one in which forceps are used, excessively stretches the supports of the bladder neck, and the normal angle between the urethra and the posterior wall of the bladder is lost. This injury causes *stress incontinence*, a condition of *partial urinary incontinence* occurring when the patient coughs or strains or laughs excessively.

## SPINAL CORD INJURIES

Following injuries to the spinal cord, the nervous control of micturition is disrupted.

The *normal bladder* is innervated as follows:

*Sympathetic outflow* is from the first and second lumbar segments of the spinal cord.
*Parasympathetic outflow* is from the second, third, and fourth sacral segments of the spinal cord.

*Sensory nerve fibers* enter the spinal cord at the above segments.

The *atonic bladder* occurs during the phase of spinal shock immediately following the injury and may last for a few days to several weeks. The bladder wall muscle is relaxed, the sphincter vesicae tightly contracted, and the sphincter urethrae relaxed. The bladder becomes greatly distended and finally overflows. Depending on the level of the cord injury, the patient either will or will not be aware that the bladder is full.

The *automatic reflex bladder* (Fig. 7-21) occurs after the patient has recovered from spinal shock, provided that the cord lesion lies above the level of the parasympathetic outflow (S2, 3, and 4). It is the type of bladder normally found in infancy. The bladder fills and empties reflexly. Stretch receptors in the bladder wall are stimulated as the bladder fills, and the afferent impulses pass to the spinal cord (segments S2, 3, and 4). Efferent impulses pass down to the bladder muscle, which contracts; the sphincter vesicae and the urethral sphincter both relax. This simple reflex occurs every 1 to 4 hours.

The *autonomous bladder* (Fig. 7-21) is the condition that occurs if the sacral segments of the spinal cord are destroyed. It should be remembered that the sacral segments of the spinal cord are situated in the upper part of the lumbar region of the vertebral column (see p. 936). The bladder is without any external reflex control. The bladder wall is flaccid, and the capacity of the bladder is greatly increased. It merely fills to capacity and overflows; continual dribbling is the result. The bladder may be partially emptied by manual compression of the lower part of the anterior abdominal wall, but infection of the urine and backpressure effects on the ureters and kidneys are inevitable.

## Ureter

In the female the close relation of the ureter to the cervix and vagina is an important fact clinically. Disease of the lower end of the ureter may be diagnosed by digital palpation through the lateral fornix of the vagina. When performing a hysterectomy, the surgeon must always remember the relation of the uterine artery to the ureter beneath the

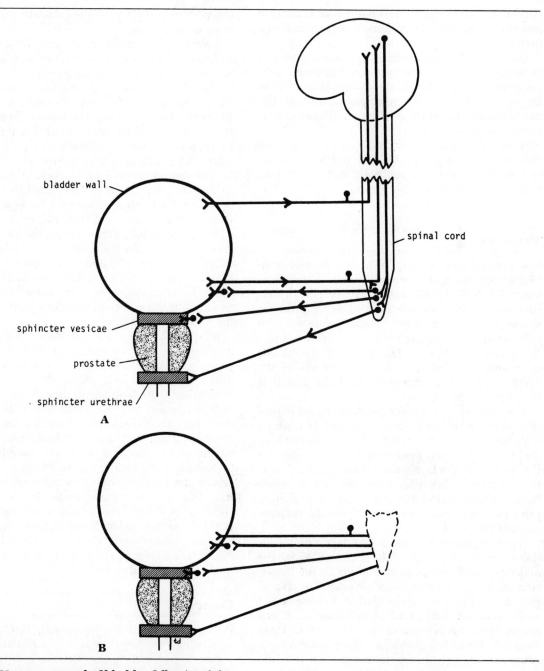

**Fig. 7-21.** Nervous control of bladder following (A) section of spinal cord in high thoracic region and (B) destruction of sacral segments of cord. Sympathetic fibers have been omitted for the sake of simplification.

base of the broad ligament. A badly placed ligature around the uterine artery could easily include the ureter.

*Ureteric calculi* are discussed on page 291. It will be remembered that the ureter is narrowed anatomically where it bends down into the pelvis at the pelvic brim and where it passes through the bladder wall. It is at these sites that urinary calculi may be arrested.

When a calculus enters the lower pelvic part of the ureter, the pain is often referred to the testis and tip of the penis in the male and the labium majus in the female.

# Prostate

The normal prostate has been seen to have a fibrous capsule, and external to this, a prostatic sheath of visceral pelvic fascia. Between the fibrous capsule and the prostatic sheath lies the prostatic venous plexus. Microscopically, the prostate is seen to be made up of numerous glands embedded in a fibromuscular stroma. The ducts of the glands open into the prostatic sinus on either side of the urethral crest. The arrangement of the glands is shown in Figure 7-8.

The urethra and ejaculatory ducts make it possible to divide up the prostate into five lobes, which have been described on page 341. The median, or middle, lobe, which lies between the urethra and the ejaculatory ducts, contains more glandular material than the other lobes. Clinically, it is therefore not surprising to find that the median lobe is often principally affected in the cases of benign enlargement of the prostate.

It is now generally believed that the normal glandular activity of the prostate is controlled by the androgens and estrogens circulating in the bloodstream. The secretions of the prostate are poured into the urethra during ejaculation and are added to the seminal fluid. Acid phosphatase is an important enzyme present in the secretion in large amounts. When the glandular cells producing this enzyme cannot discharge their secretion into the ducts, as in carcinoma of the prostate, the serum acid phosphatase level of the blood rises.

The prostate can be examined clinically by palpation by performing a rectal examination (see page 395). The examiner's gloved finger can feel the posterior surface of the prostate through the anterior rectal wall.

Benign enlargement of the prostate is common in men over 50 years of age. The cause is possibly an imbalance in the hormonal control of the gland. The median lobe of the gland enlarges upward and encroaches within the sphincter vesicae, located at the neck of the bladder. The leakage of urine into the prostatic urethra causes an intense reflex desire to micturate. The enlargement of the median and lateral lobes of the gland produces elongation and lateral compression and distortion of the urethra, so that the patient experiences difficulty in passing urine, and the stream is weak. Back-pressure effects on the ureters and both kidneys are a common complication. The enlargement of the uvula vesicae (due to the enlarged median lobe) results in the formation of a pouch of stagnant urine behind the urethral orifice within the bladder (Fig. 7-22). The stagnant urine frequently becomes infected, and the inflamed bladder (*cystitis*) adds to the patient's symptoms.

In all operations on the prostate, the surgeon regards the prostatic venous plexus with respect. The veins have very thin walls, are valveless, and are drained by several large trunks directly into the internal iliac veins. Damage to these veins may result in a severe hemorrhage. Batson has shown that there are many connections between the prostatic venous plexus and the vertebral veins. During coughing and sneezing or abdominal straining, it is possible for prostatic venous blood to flow in a reverse direction and enter the vertebral veins. This may well be the explanation for the frequent occurrence of skeletal metastases in the lower vertebral column and pelvic bones of patients with carcinoma of the prostate. Cancer cells could enter the skull via this route by floating up the valveless prostatic and vertebral veins.

# Ovary

Before puberty, the ovary is smooth, but it becomes increasingly puckered by repeated ovulation. After the menopause the ovary becomes shriveled and shrinks in size.

The ovary is kept in position by the broad ligament and the mesovarium. Following pregnancy the broad ligament is lax, and the ovaries may prolapse into the rectouterine pouch (pouch of Doug-

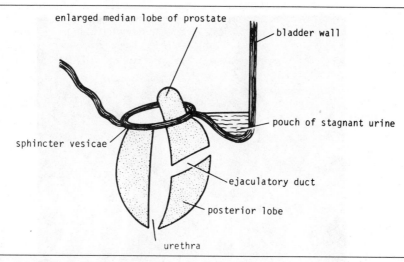

enlarged median lobe of prostate

bladder wall

pouch of stagnant urine

sphincter vesicae

ejaculatory duct

posterior lobe

urethra

**Fig. 7-22. Sagittal section of a prostate that had undergone benign enlargement of median lobe. Note presence of bladder pouch filled with stagnant urine behind prostate.**

las). In these circumstances, the ovary may be tender and cause discomfort on sexual intercourse (dyspareunia). An ovary situated in the rectouterine pouch may be palpated through the posterior fornix of the vagina.

## Uterine Tube

The uterine tube lies in the upper free border of the broad ligament and is a direct route of communication from the vulva through the vagina and uterine cavity to the peritoneal cavity. It is the route along which gonorrhea spreads to involve the pelvic peritoneum.

Acute inflammation of the uterine tube and ovary (salpingo-oophoritis) produces lower abdominal pain, which is usually bilateral.

Implantation of a fertilized ovum may occur outside the uterine cavity in the wall of the uterine tube. This is a variety of *ectopic pregnancy.* There being no decidua formation in the tube, the eroding action of the trophoblast quickly destroys the wall of the tube. Tubal abortion or rupture of the tube, with the effusion of a large quantity of blood into the peritoneal cavity, is the common result.

*Ligation and division of the uterine tubes* is a method of obtaining permanent birth control and is usually restricted to those women who already have children. The ova that are discharged from the ovarian follicles degenerate in the tube proximal to the obstruction. If, later, the woman wishes to have an additional child, restoration of the continuity of the uterine tubes can be attempted, and, in about 20 percent of women, fertilization occurs.

## Uterus

A great deal of useful clinical information can be obtained about the state of the uterus, uterine tubes, and ovaries from a bimanual examination. The examination is easiest in parous women who are able to relax while the examination is in progress. In patients in whom it causes distress, the examination may be performed under an anesthetic. With the bladder empty, the vaginal portion of the cervix is first palpated with the index finger of the right hand. The external os is circular in the nulliparous woman, but has anterior and posterior lips in the multiparous woman. The cervix normally has the consistency of the end of the nose, but in the pregnant uterus it is soft and vascular and has the consistency of the lips. The left hand is then placed gently on the anterior abdominal wall in the hypogastrium, and the fundus and body of the uterus may be palpated between the abdominal and vaginal fingers situated in the anterior fornix. The size, shape, and mobility of the uterus can then be ascertained.

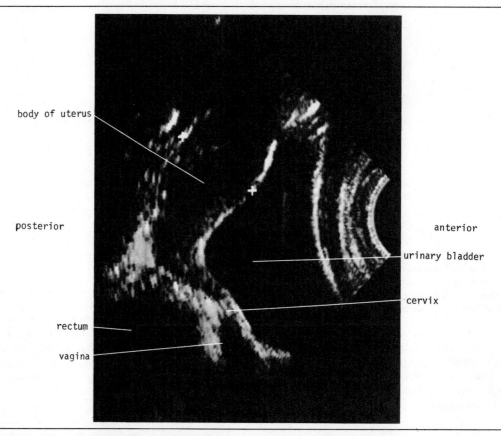

body of uterus

posterior

anterior

urinary bladder

cervix

rectum

vagina

**Fig. 7-23. Sonogram of female pelvis, showing the uterus and the vagina. (Courtesy of Dr. M. C. Hill.)**

In the majority of women, the uterus is anteverted and anteflexed. A retroverted, retroflexed uterus can be palpated through the posterior vaginal fornix.

The great importance of the tone of the levatores ani muscles in supporting the uterus has already been emphasized. The importance of the transverse cervical, pubocervical, and sacrocervical ligaments in positioning the cervix within the pelvic cavity has been considered. Damage to these structures during childbirth, or general poor body muscular tone, may result in downward displacement of the uterus, called *uterine prolapse*. It most commonly reveals itself after the menopause, when the visceral pelvic fascia tends to atrophy along with the pelvic organs. In advanced cases, the cervix descends the length of the vagina and may protrude through the orifice.

Because of the attachment of the cervix to the vaginal vault, it follows that prolapse of the uterus is always accompanied by some prolapse of the vagina.

A sonogram of the female pelvis can be used to visualize the uterus and the developing fetus, and the vagina (Fig. 7-23).

## Vagina

The anatomical relations of the vagina are of great clinical importance. Many pathological conditions occurring in the female pelvis may be diagnosed by a simple vaginal examination.

The following structures may be palpated through the vaginal walls from above downward.

## Anteriorly

(1) The bladder and (2) the urethra.

## Posteriorly

(1) Loops of ileum and sigmoid colon in the rectouterine peritoneal pouch (pouch of Douglas), (2) the rectal ampulla, and (3) the perineal body.

## Laterally

(1) The ureters, (2) the pelvic fascia and the anterior fibers of the levatores ani muscles, and (3) the urogenital diaphragm.

The vaginal vault is supported by the same structures that support the uterine cervix. Prolpase of the uterus is necessarily associated with some degree of sagging of the vaginal walls. However, if the supports of the bladder, urethra, or anterior rectal wall are damaged in childbirth, prolapse of the vaginal walls occurs with the uterus remaining in its correct position.

Sagging of the bladder results in the bulging of the anterior wall of the vagina, a condition known as a *cystocele*. When the ampulla of the rectum sags against the posterior vaginal wall, the bulge is called a *rectocele*.

The closeness of the peritoneal cavity to the posterior vaginal fornix enables the physician to drain a pelvic abscess through the vagina without performing a major operation. Unfortunately, the closeness of the peritoneal cavity to the vagina may be a disadvantage. Many a woman in the past died as the result of an amateur abortionist's thrusting his nonsterile instruments through the posterior vaginal fornix instead of through the external os of the cervix. Death from peritonitis has been the coroner's verdict.

## CLINICAL PROBLEMS

*Answers on page 966*

1. A 45-year-old woman visited her physician complaining of colicky pain in the lower abdomen. She had noticed that her bowel movements had recently become irregular, and she often had a strong desire to defecate, but only succeeded in passing blood-stained mucus. On vaginal examination, the genital tract was found to be normal, but a firm, mobile swelling could be palpated through the posterior fornix, situated in the rectouterine pouch. Digital rectal examination revealed nothing abnormal. Using your knowledge of anatomy, name the structure in the rectouterine pouch that is likely to be diseased. What additional examination would you perform?

2. A 30-year-old man involved in a barroom brawl was admitted to the hospital in a state of severe shock. He was found to have a blood-stained tear on the seat of his trousers and lacerations of the anal margin. During the fight he was knocked down and fell in the sitting position on the leg of an upturned chair. While he was under observation, the signs and symptoms of peritonitis developed. What parts of the lower bowel are related to the peritoneal cavity?

3. An embarrassed 58-year-old man visited the outpatient clinic complaining that he had something in his "back passage" that he could not remove. Digital rectal examination revealed a hard object located about 3 inches (8 cm) above the anus that filled the distended lumen of the rectum. Examination with a protoscope showed the base of a glass container. What anatomical structures are present in the rectum that might resist the patient's efforts to pass the foreign object?

4. A middle-aged man had been treating himself for hemorrhoids for the past 3 years. He was using a well-advertised brand of suppository. He had noticed that his feces were often slightly blood-stained. For the past 12 months, he had noticed that when he had his bowels open, he always felt that there was more to come. Sometimes he went to the toilet several times a day, but was only able to pass flatus and blood-stained mucus. Recently, pain had developed down the outside of his right leg. What anatomical structure is closely related to the rectum that, when involved by disease, would give the patient pain down the leg?

5. An inebriated 40-year-old man was involved in a fight over a woman. The woman's husband

gave the man a severe blow in his hypogastrium, whereupon he doubled up with pain and collapsed on the floor. Several hours later, the man was admitted to the hospital in a state of shock and complaining of pain in the lower abdominal region. On questioning, he said he had been unable to pass urine since the fight. Even though the patient had consumed a considerable volume of liquor, there was no dullness on percussion of the anterior abdominal wall above the symphysis pubis. Rectal examination revealed a bulging backward of the rectovesical pouch. Which pelvic viscus is likely to have been damaged by the blow?

6. A medical student, tired of the pressure of work and examinations, decided to go rock climbing. After a successful climb, he was returning to base when he missed his footing and fell 100 feet down a ravine. When he was admitted to the hospital, radiographic and neurological examination revealed a severe fracture dislocation of the midthoracic region of the vertebral column and extensive local damage to the spinal cord. Assuming that the patient had a complete transection of the spinal cord, (a) would he ever again be able to tell if his urinary bladder was full? (b) Would he have an automatic bladder or an autonomous bladder?

7. A 35-year-old man went to his physician complaining of pain on micturition. He said the pain was worse toward the end of the act and was sometimes referred to the end of his penis. He found that the pain was aggravated by jolting movements and relieved by lying down. Occasionally, he passed a few drops of blood at the end of micturition. Which pelvic organ is likely to be diseased? Why is the pain referred to the penis? Why is the pain relieved by lying down?

8. A 74-year-old man was admitted to the hospital as an emergency case. He had a past history of having difficulty with micturition and had not passed urine for 8 hours. On examination of the abdomen, it was found that the bladder extended above the level of the umbilicus. Digital rectal examination revealed a large, firm, fixed swelling anterior to the lower part of the rectum. Catheterization was found to be impossible. The surgeon decided to relieve the condition by passing a suprapubic tube into the bladder. What anatomical structure lies anterior to the lower part of the rectum, which, if diseased, is likely to interfere with micturition? Will the surgeon's suprapubic tube cross the peritoneal cavity before entering the bladder?

9. A 65-year-old man with a history of prostatic disease was found on radiography of his skeleton to have extensive carcinomatous metastases in his skull and lumbar vertebrae. His serum acid phosphatase level was abnormally high. Using your knowledge of anatomy, can you suggest a possible route taken by the cancer cells as they migrated from the prostate to (1) the lumbar vertebrae and (2) the skull?

10. An 88-year-old man had a history of prostatic disease. His latest symptoms included difficulty in starting to micturate, a poor urinary stream, and difficulty in stopping the flow of urine. Which lobe or lobes of the prostate are related to the sphincter vesicae? The enlargement of which lobe is likely to interfere with the sphincter's function?

11. A 25-year-old woman with a medical history of tuberculosis of the left lung visited her physician complaining of frequency of micturition and the passage of blood-stained urine (*hematuria*). Using your knowledge of anatomy, can you describe a clinical method of examining the lower part of the ureters?

12. A 23-year-old woman was admitted to the hospital as an emergency case. She complained of severe spasmodic pain in the right iliac fossa. Just prior to admission, the pain had suddenly intensified and the patient had collapsed. On physical examination, the patient was seen to be pale, in a state of shock, and with the signs and symptoms of internal hemorrhage. There was extreme tenderness in the right iliac fossa and some rigidity of the abdominal muscles. A vaginal examination revealed a softening of the cervix and a "doughlike" sensation through the posterior fornix, suggestive of fluid in the rectouterine pouch. On questioning about her menstrual history, she disclosed that she had missed her last period. The attending physician made a diagnosis of a ruptured ectopic pregnancy. Using your anatomical knowledge, can you explain the relation of the uterine tube to the peritoneal cavity?

13. The postnatal vaginal examination of a young

woman revealed that the fundus and body of the uterus could be palpated in the rectouterine pouch (pouch of Douglas). What is the normal position of the uterus in the majority of women?

14. A middle-aged woman was found on vaginal examination to have an advanced carcinoma of the cervix. What is the lymphatic drainage of the uterine cervix?

15. A multiparous 57-year-old woman visited her physician complaining of a "bearing-down" feeling in the pelvis and of low backache, both of which were worse when she was tired. On vaginal examination, the external os of the cervix was found to be located just within the vaginal orifice. A diagnosis of uterine prolapse was made. What are the main supports of the uterus?

16. During vaginal examination of a multiparous woman, she was asked to strain downward. The anterior wall of the vagina was found to sag downward. A diagnosis of prolapsed anterior vaginal wall was made. What structures lie anterior to the vagina and will sag downward with the vaginal wall?

---

# NATIONAL BOARD TYPE QUESTIONS

*Answers on page 984*

**In each of the following questions, answer:**

   (a) If (1) ONLY IS CORRECT
   (b) If (2) ONLY IS CORRECT
   (c) If BOTH (1) AND (2) ARE CORRECT, AND
   (d) If NEITHER (1) NOR (2) IS CORRECT

1. Which of the following statements is (are) correct?
   (1) The sigmoid colon becomes continuous with the rectum at the level of the sacral promontory.
   (2) The uterine artery crosses the ureter, just lateral to the lateral fornix of the vagina.

2. Which of the following statements is (are) correct concerning the rectum?
   (1) The rectal ampulla is the lower part of the rectum that is dilated and lies immediately above the pelvic diaphragm.
   (2) There are usually two transverse folds of mucous membrane on the left rectal wall and one on the right wall.

3. Which of the following statements concerning the bladder is (are) correct?
   (1) The base of the bladder lies inferiorly and in the male rests on the upper surface of the prostate.
   (2) The mucous membrane of the empty bladder is thrown into folds throughout the interior of the bladder.

4. Which of the following statements concerning the prostate is (are) correct?
   (1) The prostatic urethra is the widest and most dilatable part of the entire urethra.
   (2) The median or middle lobe of the prostate is situated behind the urethra and below the ejaculatory ducts.

**Select the best response:**

5. The following statements concerning the uterus are correct **except:**
   (a) The fundus is part of the uterus above the openings of the uterine tubes.
   (b) The long axis of the uterus is usually bent anteriorly on the long axis of the vagina (anteversion).
   (c) The nerve supply of the uterus is from the inferior hypogastric plexuses.
   (d) The anterior surface of the cervix is completely covered with peritoneum.
   (e) The uterine veins drain into the internal iliac veins.

6. Concerning the broad ligament of the uterus, all the following statements are true **except:**
   (a) It extends from the lateral margins of the uterus to the side wall of the pelvis.
   (b) The uterine tube lies within its upper free border.
   (c) The ovary lies between its anterior and posterior layers.
   (d) The suspensory ligament of the ovary extends from the ovary to the lateral pelvic wall.
   (e) The ovarian and uterine arteries anastomose between its layers of peritoneum.

7. Concerning the vas deferens, all of the following statements are true **except:**

  (a) It emerges from the deep inguinal ring and passes around the lateral margin of the inferior epigastric artery.

  (b) It crosses the ureter in the region of the ischial spine.

  (c) The terminal part is dilated to form the ampulla.

  (d) It lies on the posterior surface of the prostate but is separated from it by the peritoneum.

  (e) It joins the duct of the seminal vesicle to form the ejaculatory duct.

8. Concerning the pelvic part of the ureter, the following statements are true **except:**

  (a) It enters the pelvis in front of the bifurcation of the common iliac artery.

  (b) The ureter enters the bladder by passing directly through its wall, there being no valvular mechanism at its entrance.

  (c) It has a close relationship to the ischial spine before it turns medially toward the bladder.

  (d) The blood supply of the distal part of the ureter is from the superior vesical artery.

  (e) It enters the bladder at the upper lateral angle of the trigone.

**In the following questions, answer:**

  (a) IF (1), (2), AND (3) ONLY ARE CORRECT

  (b) IF (1) AND (3) ONLY ARE CORRECT

  (c) IF (2) AND (4) ONLY ARE CORRECT

  (d) IF (4) ONLY IS CORRECT, OR

  (e) IF ALL ARE CORRECT

9. Which of the following statements is (are) correct concerning the seminal vesicle?

  (1) The seminal vesicles are two lobulated sacs that store spermatozoa.

  (2) The seminal vesicles are related posteriorly to the rectum and can be palpated through the rectal wall.

  (3) The upper ends of the seminal vesicles are not covered with peritoneum.

  (4) The seminal vesicles are related anteriorly to the bladder and there is no peritoneum separating these structures.

10. Which of the following statements concerning the ovary is (are) correct?

  (1) The lymphatic drainage is into the para-aortic (lumbar) lymph nodes at the level of the first lumbar vertebra.

  (2) The round ligament of the ovary extends from the ovary to the upper end of the lateral wall of the uterus.

  (3) The ovarian fossa is bounded above by the external iliac vessels and behind by the internal iliac vessels.

  (4) The obturator nerve usually lies lateral to the ovary.

11. Which of the following statements concerning the nerve supply to the urinary bladder is (are) correct?

  (1) The sympathetic postganglionic fibers originate in the first and second lumbar ganglia.

  (2) The parasympathetic preganglionic fibers synapse with postganglionic neurons in the inferior hypogastric plexuses.

  (3) The afferent sensory fibers arising in the bladder reach the spinal cord via the pelvic splanchnic nerves and also travel with the sympathetic nerves.

  (4) The parasympathetic preganglionic fibers arise from the second, third, and fourth sacral segments of the spinal cord.

12. Which of the following statements is (are) correct concerning the vagina?

  (1) The area of the vaginal lumen around the cervix is divided into four fornices.

  (2) The upper part of the vagina is supported by the levator ani muscles and the transverse cervical ligaments.

  (3) The perineal body lies posterior to and supports the lower part of the vagina.

  (4) The upper part of the vagina is not covered with peritoneum.

13. Which of the following statements is (are) correct concerning the visceral layer of pelvic fascia in the female?

  (1) It covers the obturator internus muscle.

  (2) In the region of the uterus, it is called the parametrium.

  (3) It is continuous above with the fascia transversalis.

  (4) It is condensed to form the pubocervical, transverse cervical, and sacrocervical ligaments of the uterus.

**Match the structures listed on the left with the most likely route of lymphatic drainage listed on the right:**

14. Cervix of uterus

15. Prostate gland

16. Posterior fornix of vagina

(a) Internal iliac lymph nodes
(b) Internal and external iliac lymph nodes
(c) Superficial inguinal lymph nodes
(d) Para-aortic (lumbar) lymph nodes
(e) None of the above

**Match the structures listed on the left with the appropriate main venous drainage listed on the right:**

17. Left ovary
18. Prostate
19. Urinary bladder
20. Mucous membrane of the rectum

(a) External iliac vein
(b) Internal iliac vein
(c) Inferior vena cava
(d) Common iliac vein
(e) None of the above

# 8. The Perineum

The pelvic cavity, or cavity of the true pelvis, has been defined as the area situated between the pelvic inlet and the pelvic outlet. The pelvic diaphragm subdivides the cavity into the main pelvic cavity above and the perineum below. This chapter is concerned with the perineum.

## SURFACE ANATOMY

The perineum when seen from below with the thighs abducted (Figs. 8-1 and 8-2) is diamond-shaped and is bounded anteriorly by the *symphysis pubis*, posteriorly by the tip of the *coccyx*, and laterally by the *ischial tuberosities*.

## SYMPHYSIS PUBIS

This is the cartilaginous joint that lies in the midline between the bodies of the pubic bones (Fig. 8-1). It is felt as a solid structure beneath the skin in the midline at the lower extremity of the anterior abdominal wall.

## COCCYX

The inferior surface and tip of the coccyx can be palpated in the natal cleft about 1 inch (2.5 cm) behind the anus (Fig. 8-1).

## ISCHIAL TUBEROSITY

This can be palpated in the lower part of the buttock (Fig. 8-1). In the standing position, the tuberosity is covered by the gluteus maximus. In the sitting position, the ischial tuberosity emerges from beneath the lower border of the gluteus maximus and supports the weight of the body.

It is customary to divide the perineum into two triangles by joining the ischial tuberosities by an imaginary line (Fig. 8-2). The posterior triangle, which contains the anus, is called the *anal triangle*; the anterior triangle, which contains the urogenital orifices, is called the *urogenital triangle*.

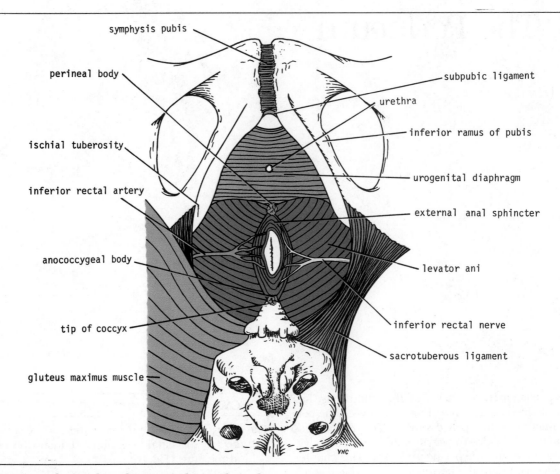

symphysis pubis

perineal body

subpubic ligament

urethra

ischial tuberosity

inferior ramus of pubis

urogenital diaphragm

inferior rectal artery

external anal sphincter

anococcygeal body

levator ani

tip of coccyx

inferior rectal nerve

sacrotuberous ligament

gluteus maximus muscle

YHC

**Fig. 8-1. Anal triangle and urogenital triangle in the male as seen from below.**

## Anal Triangle

### ANUS

The anus is the lower opening of the anal canal and it lies in the midline. In the living the anal margin is reddish-brown in color and is puckered by the contraction of the *external anal sphincter*. Around the anal margin are a number of coarse hairs.

## Male Urogenital Triangle

This region contains the penis and the scrotum.

## PENIS

This consists of a root, a body, and a glans (Figs. 8-3 and 8-12). The *root of the penis* consists of three masses of erectile tissue, called the *bulb of the penis* and the *right* and *left crura of the penis*. The bulb may be felt on deep palpation in the midline of the perineum, posterior to the scrotum.

The *body of the penis* is the free portion of the penis that is suspended from the symphysis pubis. Note that the dorsal surface (anterior surface of the flaccid organ) usually possesses a *superficial dorsal vein* in the midline (Fig. 8-12).

The *glans penis* forms the extremity of the body of the penis (Figs. 8-3 and 8-12). At the summit of the glans is the *external urethral meatus*. Extending from the lower margin of the external meatus is a

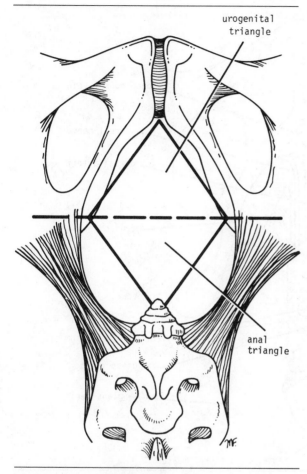

urogenital
triangle

anal
triangle

**Fig. 8-2. Diamond-shaped perineum divided by broken line into urogenital triangle and anal triangle.**

fold connecting the glans to the prepuce, called the *frenulum*. The edge of the base of the glans is called the *corona* (Fig. 8-3). The *prepuce* is formed by a fold of skin attached to the neck of the penis. The prepuce covers the glans for a variable extent and it should be possible to retract it over the glans.

## SCROTUM

This is a sac of skin and fascia (Fig. 8-9) containing the testes and the epididymides. The skin of the scrotum is rugose and is covered with sparse hairs. The bilateral origin of the scrotum is indicated by

the presence of a dark line in the midline called the *scrotal raphe.*

## TESTES

The testes should be palpated. They are oval in shape and have a firm consistency.

## EPIDIDYMIDES

Each epididymis can be palpated posterior to the testis. The epididymis is a long, narrow, firm structure having the cordlike *vas deferens* on its medial side. The epididymis has an expanded upper end or *head*, a *body*, and a pointed *tail* inferiorly (see Fig. 4-18).

# Female Urogenital Triangle

### VULVA

This is the name applied to the female external genitalia (Fig. 8-4).

### Mons Pubis

This is the rounded, hair-bearing elevation of skin found anterior to the pubis (Fig. 8-4). The pubic hair in the female has an abrupt horizontal superior margin, whereas in the male it extends upward to the umbilicus.

### Labia Majora

These are prominent, hair-bearing folds of skin extending posteriorly from the mons pubis to unite posteriorly in the midline (Fig. 8-4).

### Labia Minora

These are two smaller, hairless folds of soft skin that lie between the labia majora (Fig. 8-4). Their posterior ends are united to form a sharp fold, the *fourchette.* Anteriorly they split to enclose the clitoris, forming an anterior *prepuce* and a posterior *frenulum* (Fig. 8-4).

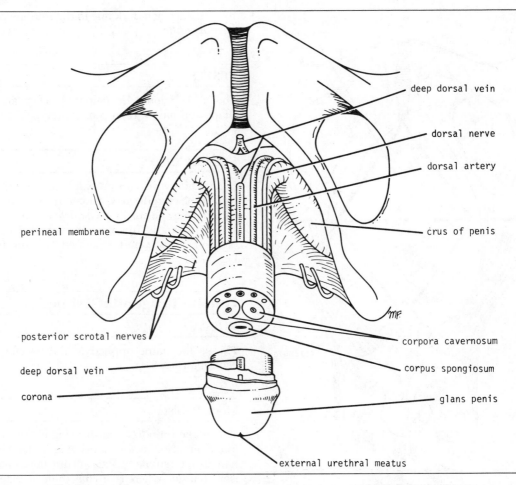

**Fig. 8-3. Root and body of penis.**

## Vestibule

This is a smooth triangular area bounded laterally by the labia minora, with the clitoris at its apex and the fourchette at its base (Fig. 8-4).

## Vaginal Orifice

This is protected in the virgin by a thin mucosal fold called the *hymen*, which is perforated at its center (Fig. 8-4). At the first coitus the hymen tears, usually posteriorly or posterolaterally, and after childbirth only a few tags of the hymen remain (Fig. 8-4).

## Orifices of the Ducts of the Greater Vestibular Glands

These are small orifices, one on each side, in the groove between the hymen and the posterior part of the labium minus (Fig. 8-4).

## Clitoris

This is situated at the apex of the vestibule anteriorly (Fig. 8-4). The *glans of the clitoris* is partly hidden by the *prepuce*.

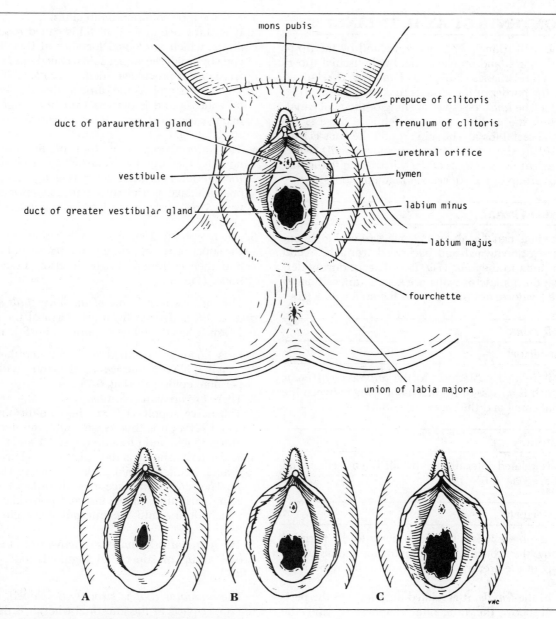

**Fig. 8-4. Vulva. Note different appearances of hymen in (A) virgin, (B) woman who has had sexual intercourse, and (C) multiparous woman.**

## CONTENTS OF ANAL TRIANGLE

The anal triangle is bounded behind by the tip of the coccyx, and on each side, by the ischial tuberosity and the sacrotuberous ligament, overlapped by the border of the gluteus maximus muscle (Fig. 8-1). The *anus*, or lower opening of the anal canal, lies in the midline, and on each side there is the ischiorectal fossa. The skin around the anus is supplied by the inferior rectal (hemorrhoidal) nerve. The lymphatic vessels of the skin drain into the medial group of the superficial inguinal nodes.

## Anal Canal

The anal canal is about 1½ inches (4 cm) long and passes downward and backward from the rectal ampulla to the anus (Fig. 8-5). Except during defecation, its lateral walls are kept in apposition by the levatores ani muscles and the anal sphincters.

### Relations

#### Posteriorly

It is related posteriorly to the *anococcygeal body*, which is a mass of fibrous tissue lying between the anal canal and the coccyx (Fig. 8-5).

#### Laterally

It is related laterally to the fat-filled ischiorectal fossae (Fig. 8-6).

#### Anteriorly

In the *male* it is related anteriorly to the perineal body, the urogenital diaphragm, the membranous part of the urethra, and the bulb of the penis (Fig. 8-5).

In the *female*, it is related anteriorly to the perineal body, the urogenital diaphragm, and the lower part of the vagina (Fig. 8-5).

### Structure of the Anal Canal

The *mucous membrane of the upper half of the anal canal* is derived from hindgut entoderm (Fig. 8-7). It has the following important anatomical features:

1. It is lined by columnar epithelium.
2. It is thrown into vertical folds called *anal columns*, which are joined together at their lower ends by small semilunar folds, called *anal valves* (remains of proctodeal membrane) (Fig. 8-6).
3. The nerve supply is the same as that for the rectal mucosa and is derived from the autonomic hypogastric plexuses. It is sensitive only to stretch (Fig. 8-7).
4. The arterial supply is that of the hindgut, namely, the superior rectal artery, a branch of the inferior mesenteric artery (Fig. 8-7). The venous drainage is mainly by the superior rectal vein, a tributary of the inferior mesenteric vein (Fig. 8-6).
5. The lymphatic drainage is mainly upward along the superior rectal artery to the pararectal nodes and then eventually to the inferior mesenteric nodes (Fig. 8-7).

The *mucous membrane of the lower half of the anal canal* is derived from ectoderm of the proctodeum. It has the following important features:

1. It is lined by stratified squamous epithelium, which gradually merges at the anus with the perianal epidermis (Fig. 8-7).
2. There are *no* anal columns.
3. The nerve supply is from the somatic inferior rectal nerve; it is thus sensitive to pain, temperature, touch, and pressure (Figs. 8-1 and 8-7).
4. The arterial supply is the inferior rectal artery, a branch of the internal pudendal artery (Fig. 8-1). The venous drainage is by the inferior rectal vein, a tributary of the internal pudendal vein, which drains into the internal iliac vein (Fig. 8-7).
5. The lymphatic drainage is downward to the medial group of superficial inguinal nodes (Fig. 8-7).

The *muscular coat* is strongly developed. As in the upper parts of the intestinal tract, it is divided into an outer longitudinal and an inner circular layer of smooth muscle (Fig. 8-6). The circular coat is thickened at the upper end of the anal canal to form the *involuntary internal sphincter*. The internal sphincter is enclosed by a sheath of striped muscle that forms the voluntary external sphincter (Figs. 8-6 and 8-7).

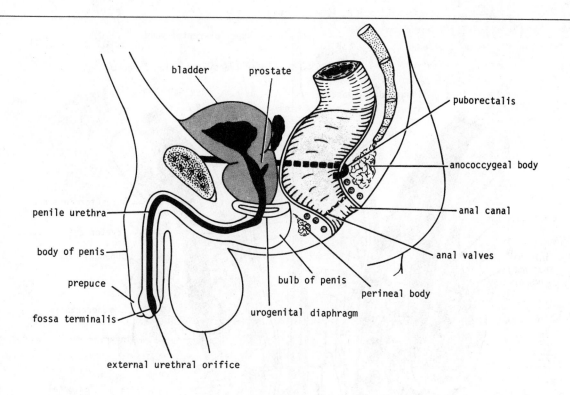

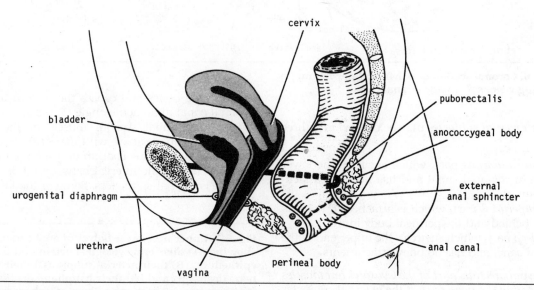

**Fig. 8-5. Sagittal sections of male and female pelvis.**

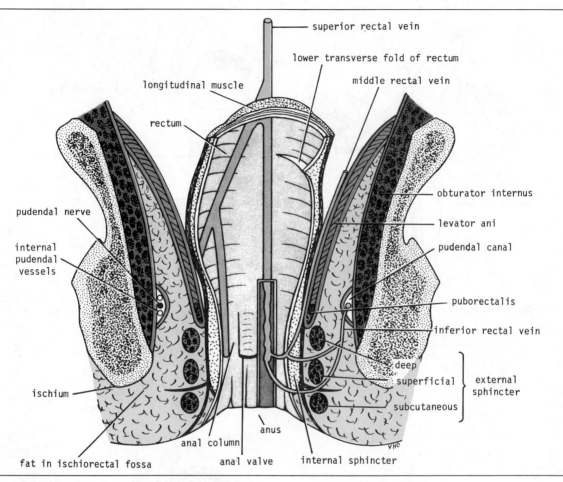

superior rectal vein

lower transverse fold of rectum

middle rectal vein

longitudinal muscle

rectum

obturator internus

levator ani

pudendal canal

pudendal nerve

internal pudendal vessels

puborectalis

inferior rectal vein

deep

superficial

external sphincter

subcutaneous

ischium

anus

fat in ischiorectal fossa

anal column

anal valve

internal sphincter

**Fig. 8-6. Coronal section of pelvis and perineum, showing venous drainage of anal canal.**

The *external sphincter* may be divided into three parts:

1. A *subcutaneous* part, which encircles the lower end of the anal canal and has no bony attachments.
2. A *superficial* part, which is attached to the coccyx behind and the perineal body in front.
3. A *deep* part, which encircles the upper end of the anal canal and has no bony attachments.

The *puborectalis part of the levatores ani muscles* blends with the deep part of the external sphincter (Figs. 8-6 and 8-7). The puborectalis fibers of the two sides form a sling, which is attached in front

to the pubic bones and causes the rectum to join the anal canal at an acute angle (Fig. 8-7).

The longitudinal smooth muscle of the anal canal is continuous above with that of the rectum. It forms a continuous coat around the anal canal and descends in the interval between the internal and external anal sphincters. Some of the longitu-

**Fig. 8-7. Upper and lower halves of anal canal, showing (A) their embryological origin and lining epithelium, (B) their arterial supply, (C) their venous drainage, and (D) their lymphatic drainage. (E) Arrangement of muscle fibers of puborectalis muscle and different parts of external anal sphincter.**

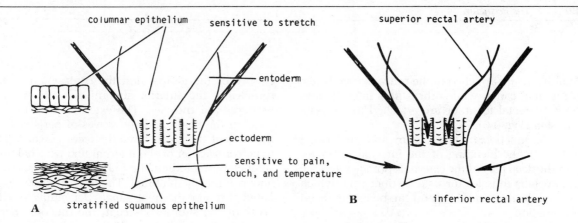

A

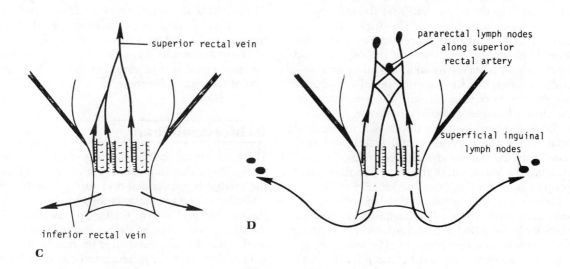

C

D

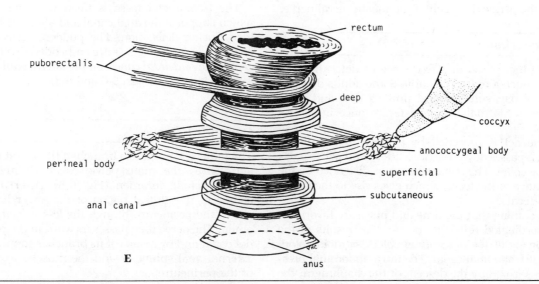

E

dinal fibers are attached to the mucous membrane of the anal canal, while others pass laterally into the ischiorectal fossa, or are attached to the perianal skin (Fig. 8-6).

At the junction of the rectum and anal canal (Fig. 8-7), the blending of the internal sphincter with the deep part of the external sphincter and the puborectalis muscles forms a distinct ring, which is called the *anorectal ring* and can be felt on rectal examination.

The *arterial supply* to the anal canal is from the superior and inferior rectal arteries; the superior artery supplies the upper half and the inferior artery, the lower half (Fig. 8-7). The veins correspond to the arteries. It should be remembered that the union between the tributaries of the superior and inferior rectal veins forms an important portal-systemic anastomosis; the superior rectal vein drains into the portal circulation, and the inferior rectal vein drains into the systemic circulation.

The *lymph vessels* of the upper half of the anal canal follow the superior rectal artery and ultimately join the inferior mesenteric nodes. The vessels from the lower half of the canal join the medial group of superficial inguinal nodes (Fig. 8-7).

The *nerve supply* to the mucous membrane has been described above. The involuntary internal sphincter is supplied by sympathetic fibers from the *inferior hypogastric* plexuses. The voluntary external sphincter is supplied by the inferior rectal nerve, a branch of the pudendal nerve (Fig. 8-1), and the perineal branch of the fourth sacral nerve.

## Defecation

The time, place, and frequency of defecation are very much a matter of habit. Some adults defecate once a day, some several times a day, and some perfectly normal people defecate once in several days.

The act is preceded by a wave of peristalsis, which passes down the descending and pelvic parts of the colon. The rectum becomes distended by the entrance of the feces, which gives rise to the desire to defecate.

Assuming that the time and place are favorable, a coordinated reflex act occurs that results in the emptying of the descending colon, sigmoid colon, rectum, and anal canal. The intra-abdominal pressure is raised by the descent of the diaphragm, the closure of the glottis, and the contraction of the muscles of the anterior abdominal walls and the levatores ani muscles. The external pressure applied to the colon and the waves of peristalsis in the wall of the colon force the feces onward. The tonic contraction of the internal and external anal sphincters, including the puborectalis muscles, is now voluntarily inhibited. The feces are now evacuated through the anal canal. Depending on the laxity of the submucous coat, the mucous membrane of the lower part of the anal canal is extruded through the anus ahead of the fecal mass At the end of the act, the mucosa is returned to the anal canal by the tone of the longitudinal fibers of the anal walls and the contraction and upward pull of the puborectalis muscle. The empty lumen of the anal canal is now closed by the tonic contraction of the anal sphincters.

## Ischiorectal Fossa

The ischiorectal fossa is a wedge-shaped space on each side of the anal canal (Fig. 8-6). The base of the wedge is superficial and formed by the skin. The edge of the wedge is formed by the junction of the medial and lateral walls. The medial wall is formed by the sloping levator ani muscle and the anal canal. The lateral wall is formed by the lower part of the obturator internus muscle, covered with pelvic fascia.

The ischiorectal fossa is filled with dense fat, which supports the anal canal and allows it to distend during defecation. The pudendal nerve and internal pudendal vessels are embedded in a fascial canal, the *pudendal canal*, on the lateral wall of the ischiorectal fossa (Figs. 8-6 and 8-8).

## Pudendal Nerve

The pudendal nerve is a branch of the sacral plexus and leaves the main pelvic cavity through the greater sciatic foramen (Fig. 8-8). After a brief course in the gluteal region of the lower limb, it enters the perineum through the lesser sciatic foramen. The nerve then passes forward in the pudendal canal and by means of its branches supplies the external anal sphincter and the muscles and skin of the perineum.

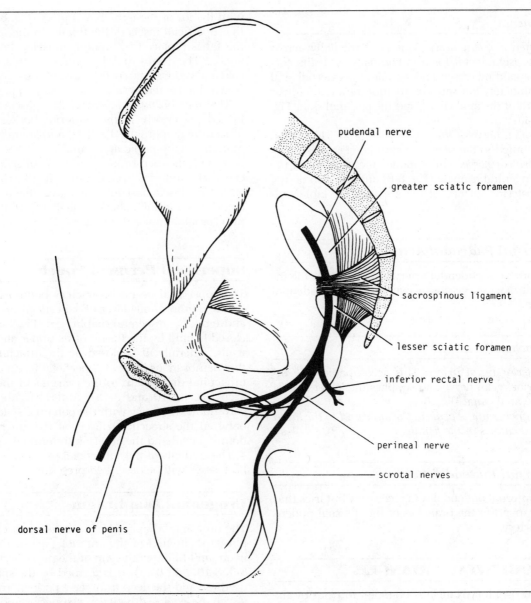

**Fig. 8-8. Course and branches of pudendal nerve in male.**

### Branches

1. *Inferior rectal nerve.* This runs medially across the ischiorectal fossa in company with the corresponding vessels and supplies the external anal sphincter, the mucous membrane of the lower half of the anal canal, and the perianal skin (Fig. 8-1).
2. *Dorsal nerve of the penis (or clitoris).* This is distributed to the penis (or clitoris) (Fig. 8-8).
3. *Perineal nerve.* This supplies the muscles in the urogenital triangle (Fig. 8-8) and the skin on the posterior surface of the scrotum (or labia majora).

### Internal Pudendal Artery

The internal pudendal artery is a branch of the internal iliac artery and accompanies the pudendal nerve.

### Branches

1. *Inferior rectal artery.* This accompanies the inferior rectal nerve and supplies the lower half of the anal canal (Fig. 8-1).
2. *Branches to the penis in the male and to the labia and clitoris in the female.*

### Internal Pudendal Vein

The internal pudendal vein receives tributaries that correspond to the branches of the internal pudendal artery.

## UROGENITAL TRIANGLE

The urogenital triangle is bounded in front by the pubic arch and laterally by the ischial tuberosities (Fig. 8-1).

## Superficial Fascia

The superficial fascia of the urogenital triangle may be divided into a fatty layer and a membranous layer.

The *fatty layer* (fascia of Camper) is continuous with the fat of the ischiorectal fossa (Fig. 8-9) and the superficial fascia of the thighs. In the scrotum, the fat is replaced by smooth muscle, the *dartos muscle.* The dartos muscle contracts in response to cold and reduces the surface area of the scrotal skin (see testicular temperature and fertility, p. 193).

The *membranous layer* (Colles' fascia) is attached posteriorly to the posterior border of the urogenital diaphragm (Fig. 8-9) and laterally to the margins of the pubic arch; anteriorly, it is continuous with the membranous layer of superficial fascia of the anterior abdominal wall (Scarpa's fascia). The fascia is continued over the penis (or clitoris) as a tubular sheath (Fig. 8-12). In the scrotum (or labia majora) it forms a distinct layer (Fig. 8-9).

## Superficial Perineal Pouch

The superficial perineal pouch is bounded below by the membranous layer of superficial fascia and above by the urogenital diaphragm (Fig. 8-9). It is closed behind by the fusion of its upper and lower walls. Laterally, it is closed by the attachment of the membranous layer of superficial fascia and the urogenital diaphragm to the margins of the pubic arch (Figs. 8-10 and 8-13). Anteriorly, the space communicates freely with the potential space lying between the superficial fascia of the anterior abdominal wall and the anterior abdominal muscles.

The contents of the superficial perineal pouch in both sexes will be described presently.

## Urogenital Diaphragm

The urogenital diaphragm is a musculofascial diaphragm, situated in the anterior part of the perineum and filling in the gap of the pubic arch (Figs. 8-9, 8-10, and 8-13). It is formed by the sphincter urethrae and the deep transverse perineal muscles, which are enclosed between a superior and an inferior layer of fascia of the urogenital diaphragm. The inferior layer of fascia is often referred to as the *perineal membrane.*

Anteriorly, the two layers of fascia fuse, leaving a small gap beneath the symphysis pubis. Posteriorly, the two layers of fascia fuse with each other, and with the membranous layer of the superficial fascia and the perineal body (Fig. 8-9). Laterally, the layers of fascia are attached to the pubic arch.

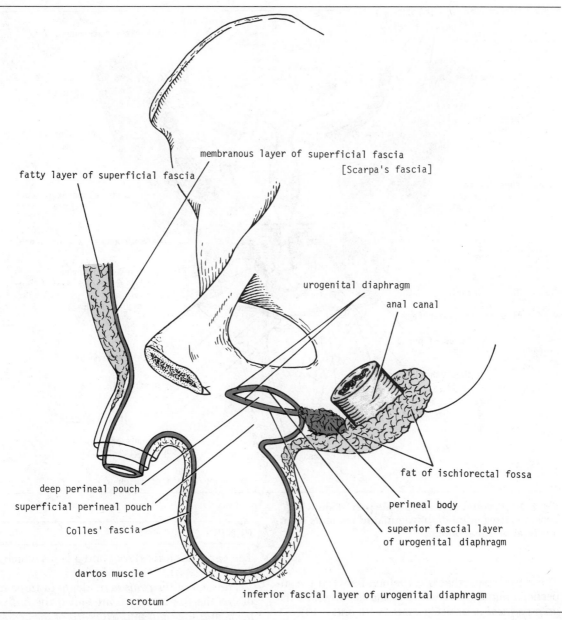

fatty layer of superficial fascia

membranous layer of superficial fascia
[Scarpa's fascia]

urogenital diaphragm

anal canal

fat of ischiorectal fossa

deep perineal pouch

superficial perineal pouch

Colles' fascia

perineal body

superior fascial layer
of urogenital diaphragm

dartos muscle

scrotum

inferior fascial layer of urogenital diaphragm

**Fig. 8-9. Arrangement of superficial fascia in urogenital triangle. Note superficial and deep perineal pouches.**

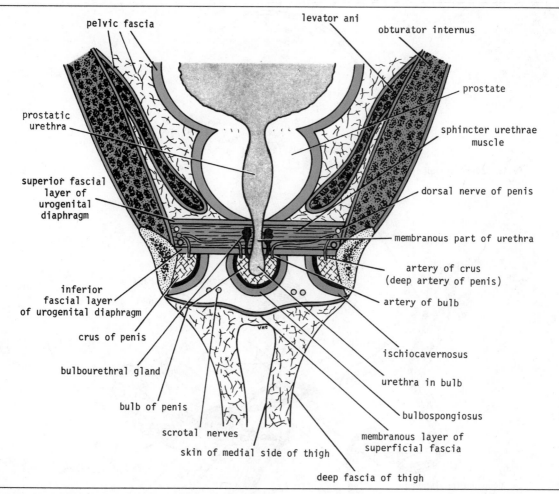

**Fig. 8-10. Coronal section of male pelvis, showing prostate, urogenital diaphragm, and contents of superficial perineal pouch.**

The closed space that is contained between the superficial and deep layers of fascia is known as the *deep perineal pouch* (Figs. 8-9, 8-10, and 8-13).

The contents of the deep perineal pouch in both sexes will be described in subsequent sections.

## Contents of the Male Urogenital Triangle

In the male the triangle contains the penis and scrotum.

## PENIS

The penis has a fixed *root* and a *body*, which hangs free (Figs. 8-3).

The *root of the penis* is made up of three masses of erectile tissue, which are called the *bulb of the penis* and the *right* and *left crura of the penis* (Figs. 8-3 and 8-11). The bulb is situated in the midline and is attached to the undersurface of the urogenital diaphragm. It is traversed by the urethra and is covered on its outer surface by the *bulbospongiosus muscles.* Each crus is attached to the side of the pubic arch and is covered on its outer surface by the *ischiocavernosus muscle.* The bulb is continued forward into the body of the penis and forms the

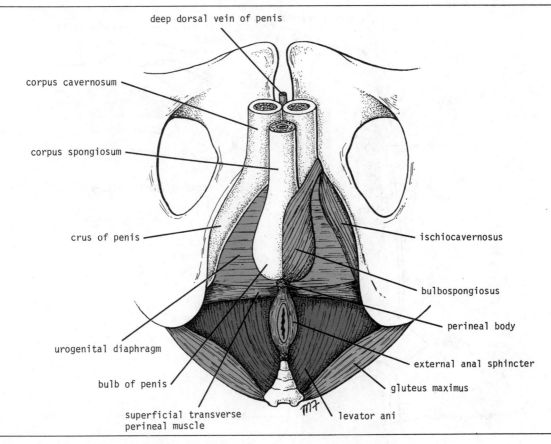

deep dorsal vein of penis

corpus cavernosum

corpus spongiosum

crus of penis

urogenital diaphragm

bulb of penis

superficial transverse
perineal muscle

ischiocavernosus

bulbospongiosus

perineal body

external anal sphincter

gluteus maximus

levator ani

**Fig. 8-11. Root of penis and perineal muscles.**

*corpus spongiosum* (Fig. 8-12). The two crura converge anteriorly and come to lie side by side in the dorsal part of the body of the penis, forming the *corpora cavernosa* (Fig. 8-3 and 8-12).

The *body of the penis* is essentially composed of three cylinders of erectile tissue enclosed in a tubular sheath of fascia. The erectile tissue is made up of two dorsally placed corpora cavernosa and a single corpus spongiosum applied to their ventral surface (Figs. 8-3 and 8-12). At its distal extremity, the corpus spongiosum expands to form the *glans penis*, which covers the distal ends of the corpora cavernosa. On the tip of the glans penis is the slit-like orifice of the urethra, called the *external urethral meatus*.

The *prepuce* is a hoodlike fold of skin that covers the glans (Fig. 8-12). It is connected to the glans

just below the urethral orifice by a fold called the *frenulum*.

The body of the penis is supported by two condensations of deep fascia that extend downward from the linea alba and symphysis pubis to be attached to the fascia of the penis.

## Lymphatic Drainage of the Penis

The skin of the penis is drained into the medial group of superficial inguinal nodes. The deep structures of the penis are drained into the internal iliac nodes.

## SCROTUM

The scrotum is a pouch of skin that contains the testes, the epididymides, and the lower ends of the spermatic cords. Since the structure of the scro-

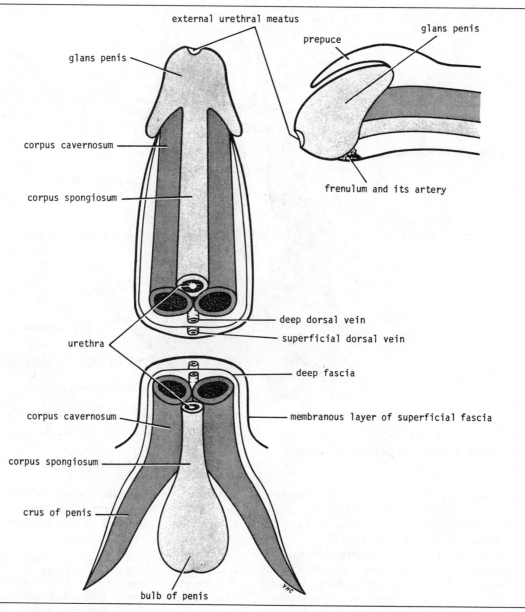

**Fig. 8-12. Body and root of penis.**

tum, the descent of the testes, and the formation of the inguinal canal are interrelated, these are fully described in Chapter 4.

## Lymphatic Drainage of the Scrotum and Contents

The wall of the scrotum is drained into the medial group of superficial inguinal lymph nodes. The lymphatic drainage of the testis and epididymis ascends in the spermatic cord and ends in the lumbar (paraaortic) lymph nodes at the level of the first lumbar vertebra. This is to be expected, since the testis during development has migrated from high up on the posterior abdominal wall, down through the inguinal canal, and into the scrotum, dragging its blood supply and lymph vessels after it.

## Contents of the Superficial Perineal Pouch in the Male

The superficial perineal pouch contains structures forming the root of the penis, together with the muscles that cover them, namely, the bulbospongiosus muscles and the ischiocavernosus muscles (Fig. 8-11). The *bulbospongiosus muscles*, situated one on each side of the midline (Fig. 8-11), cover the bulb of the penis and the posterior portion of the corpus spongiosum. Their function is to compress the penile part of the urethra and empty it of residual urine or semen. The anterior fibers also compress the deep dorsal vein of the penis, thus impeding the venous drainage of the erectile tissue and thereby assisting in the process of erection of the penis.

## Ischiocavernosus Muscles

The ischiocavernosus muscles cover the crus penis on each side (Fig. 8-11). The action of each muscle is to compress the crus penis and assist in the process of erection of the penis.

## Superficial Transverse Perineal Muscles

The superficial transverse perineal muscles lie in the posterior part of the superficial perineal pouch (Fig. 8-11). Each muscle arises from the ischial ramus and is inserted into the perineal body. The

function of these muscles is to fix the perineal body in the center of the perineum.

## Nerve Supply

All the muscles of the superficial perineal pouch are supplied by the perineal branch of the pudendal nerve.

## Perineal Body

This is a small mass of fibrous tissue, which is attached to the center of the posterior margin of the urogenital diaphragm (Figs. 8-9 and 8-11). It serves as a point of attachment for the following muscles: (1) external anal sphincter, (2) bulbospongiosus muscle, and (3) the superficial transverse perineal muscles.

## Perineal Branch of Pudendal Nerve

The perineal branch of the pudendal nerve on each side terminates in the superficial perineal pouch by supplying the muscles and skin (Fig. 8-8).

## Contents of the Deep Perineal Pouch in the Male

The deep perineal pouch contains (1) the membranous part of the urethra, (2) the sphincter urethrae, (3) the bulbourethral glands, (4) the deep transverse perineal muscles, (5) the internal pudendal vessels and their branches, and (6) the dorsal nerves of the penis.

## Membranous Part of Urethra

This is about ½ inch (1.3 cm) long and is continuous above with the prostatic urethra and below with the penile urethra. It is the shortest and least dilatable part of the urethra (Fig. 8-10).

## Sphincter Urethrae Muscle

The sphincter urethrae muscle arises from the pubic arch and passes medially to surround the urethra (Fig. 8-10).

## Nerve Supply

The perineal branch of the pudendal nerve.

## Action

It is the means by which micturition can be voluntarily stopped.

### Bulbourethral Glands

The bulbourethral glands are two small glands that lie beneath the sphincter urethrae muscle (Fig. 8-10). Their ducts pierce the perineal membrane (inferior fascial layer of the urogenital diaphragm) and enter the penile portion of the urethra. The secretion is poured into the urethra as the result of erotic stimulation.

### Deep Transverse Perineal Muscles

The deep transverse perineal muscles lie posterior to the sphincter urethrae muscle. Each muscle arises from the ischial ramus and passes medially to be inserted into the perineal body. These muscles are clinically unimportant.

### Internal Pudendal Artery

The internal pudendal artery (Fig. 8-10) on each side enters the deep perineal pouch and passes forward, giving rise to (1) the *artery to the bulb of the penis*, (2) the *arteries to the crura of the penis* (deep artery of penis), and (3) the *dorsal artery of the penis*, which supplies the skin and fascia of the penis.

### Dorsal Nerve of the Penis

The dorsal nerve of the penis on each side passes forward through the deep perineal pouch and supplies the skin of the penis (Fig. 8-10).

## Erection of the Penis

Erection in the male is gradually built up as a consequence of various sexual stimuli. Pleasurable sight, sound, smell, and other psychic stimuli, fortified later by direct touch sensory stimuli from the general body skin and genital skin, result in a bombardment of the central nervous system by afferent stimuli. Efferent nervous impulses pass down the spinal cord to the parasympathetic outflow in the second, third, and fourth sacral segments. The parasympathetic preganglionic fibers enter the inferior hypogastric plexuses and synapse on the postganglionic neurons. The postganglionic fibers join the internal pudendal arteries and are distributed along their branches, which enter the erectile tissue at the root of the penis. Vasodilatation of the arteries now occurs, producing a great increase in blood flow through the blood spaces of the erectile tissue. The corpora cavernosa and the corpus spongiosum become engorged with blood and expand, compressing their draining veins against the surrounding fascia. By this means, the outflow of blood from the erectile tissue is retarded, so that the internal pressure is further accentuated and maintained. The penis thus increases in length and diameter and assumes the erect position.

Once the climax of sexual excitement is reached and ejaculation takes place, or the excitement passes off or is inhibited, the arteries supplying the erectile tissue undergo vasoconstriction. The penis then returns to its flaccid state.

## Ejaculation

During the increasing sexual excitement that occurs during sex play the external urinary meatus of the glans penis becomes moist, due to the secretions of the bulbourethral glands.

Friction on the glans penis, reinforced by other afferent nervous impulses, results in a discharge along the sympathetic nerve fibers to the smooth muscle of the duct of the epididymis and the vas deferens on each side, the seminal vesicles, and the prostate. The smooth muscle contracts, and the spermatozoa, together with the secretions of the seminal vesicles and prostate, are discharged into the prostatic urethra. The fluid now joins the secretions of the bulbourethral glands and penile urethral glands and is then ejected from the penile urethra as a result of the rhythmic contractions of the bulbospongiosus muscles, which compress the urethra. Meanwhile, the sphincter of the bladder contracts and prevents a reflux of the spermatozoa into the bladder. The spermatozoa and the secretions of the several accessory glands constitute the *seminal fluid*, or *semen*.

At the climax of male sexual excitement, a mass discharge of nervous impulses takes place in the central nervous system. Impulses pass down the spinal cord to the sympathetic outflow (T1–L2). The nervous impulses that pass to the genital organs are thought to leave the cord at the first and second lumbar segments in the preganglionic sympathetic fibers. Many of these fibers synapse with postganglionic neurons in the first and second lumbar ganglia. Other fibers may synapse in ganglia in the lower lumbar or pelvic parts of the sympathetic trunks. The postganglionic fibers are then distributed to the vas deferens, the seminal vesicles, and the prostate via the inferior hypogastric plexuses.

## MALE URETHRA

The male urethra is about 8 inches (20 cm) long and extends from the neck of the bladder to the external meatus on the glans penis (Fig. 8-5). It is divided into three parts: (1) prostatic, (2) membranous, and (3) penile.

The *prostatic urethra* is described on page 343. It lies within the prostate and is the widest and most dilatable portion of the urethra (Fig. 8-10).

The *membranous urethra* lies within the urogenital diaphragm surrounded by the sphincter urethrae muscle. It is the least dilatable portion of the urethra (Fig. 8-10).

The *penile urethra* is enclosed in the bulb and the corpus spongiosum of the penis (Figs. 8-5 and 8-10 through 8-12). The external meatus is the narrowest part of the entire urethra. The part of the urethra that lies within the glans penis is dilated to form the *fossa terminalis* (navicular fossa) (Fig. 8-5).

## Contents of the Female Urogenital Triangle

In the female the triangle contains the external genitalia and the orifices of the urethra and the vagina.

## CLITORIS

The clitoris, which corresponds to the penis in the male, is situated at the apex of the vestibule anteriorly. It has a structure similar to the penis. The *glans* of the clitoris is partly hidden by the *prepuce*.

The *root of the clitoris* is made up of three masses of erectile tissue, which are called the bulb of the vestibule and the right and left crura of the clitoris (Figs. 8-13 and 8-14).

The *bulb of the vestibule* corresponds to the bulb of the penis, but because of the presence of the vagina, it is divided into two halves (Fig. 8-14). It is attached to the undersurface of the urogenital diaphragm and is covered by the *bulbospongiosus muscles*. Anteriorly, the two halves unite to form the *glans clitoris* (Fig. 8-14).

The *crura of the clitoris* correspond to the crura of the penis and become the corpora cavernosa anteriorly. Each remains separate and is covered by an *ischiocavernosus muscle* (Fig. 8-14).

## Contents of the Superficial Perineal Pouch in the Female

The superficial perineal pouch contains structures forming the root of the clitoris and the muscles that cover them, namely, the bulbospongiosus muscles and the ischiocavernosus muscles (Figs. 8-13 and 8-14).

### Bulbospongiosus Muscle

The bulbospongiosus muscle surrounds the orifice of the vagina and covers the vestibular bulbs. Its fibers extend forward to gain attachment to the corpora cavernosa of the clitoris. The bulbospongiosus muscle reduces the size of the vaginal orifice and compresses the deep dorsal vein of the clitoris, thereby assisting in the mechanism of erection in the clitoris.

### Ischiocavernosus Muscle

The ischiocavernosus muscle on each side covers the crus clitoridis. Contraction of this muscle assists in causing the erection of the clitoris.

### Superficial Transverse Perineal Muscles

The superficial transverse perineal muscles are identical in structure and function to those of the male.

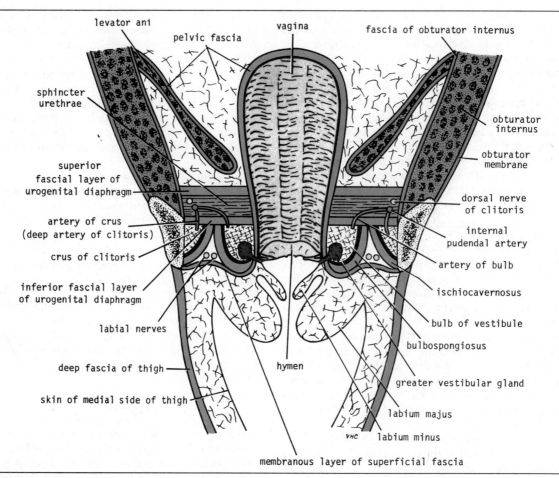

**Fig. 8-13. Coronal section of female pelvis, showing vagina, urogenital diaphragm, and contents of superficial perineal pouch.**

## Nerve Supply

All the muscles of the superficial perineal pouch are supplied by the perineal branch of the pudendal nerve.

### Perineal Body

The perineal body is larger than that of the male and it is clinically important. It is a wedge-shaped mass of fibrous tissue situated between the lower end of the vagina and the anal canal (Figs. 8-5 and 8-14). It is the point of attachment of many peri-neal muscles (as in the male), including the levatores ani muscles; the latter assist the perineal body in supporting the posterior wall of the vagina.

### Perineal Branch of Pudendal Nerve

The perineal branch of the pudendal nerve on each side terminates in the superficial perineal pouch by supplying the muscles and skin (Fig. 8-8).

## Contents of the Deep Perineal Pouch in the Female

The deep perineal pouch (Fig. 8-13) contains (1) part of the urethra; (2) part of the vagina; (3) the sphincter urethrae, which is pierced by the urethra

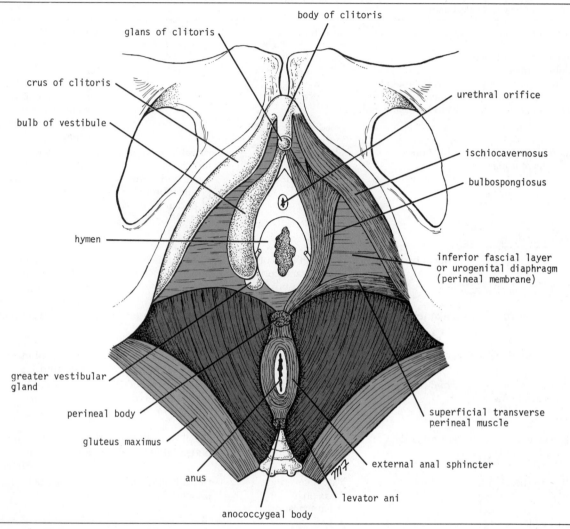

**Fig. 8-14. Root and body of clitoris and perineal muscles.**

and the vagina; (4) the deep transverse perineal muscles; (5) the internal pudendal vessels and their branches; and (6) the dorsal nerves of the clitoris.

The urethra is described on page 393, and the vagina on page 352.

The *sphincter urethrae* and the *deep transverse perineal muscles* are described on page 387. The *internal pudendal vessels* and the *dorsal nerves of the clitoris* have an arrangement similar to the corresponding structures found in the male.

A summary of the muscles of the perineum,

their nerve supply, and their action is given in Table 8-1.

## Erection of the Clitoris

Sexual excitement produces engorgement of the erectile tissue within the clitoris in exactly the same manner as in the male.

## Orgasm in the Female

As in the male, vision, hearing, smell, touch, and other psychic stimuli gradually build up the inten-

Table 8-1. Muscles of Perineum

| Name of muscle | Origin | Insertion | Nerve supply | Action |
|---|---|---|---|---|
| External anal sphincter | | | Inferior rectal nerve | Together with puborectalis muscle forms voluntary sphincter of anal canal |
| Subcutaneous part | Encircles anal canal | | | |
| Superficial part | Perineal body | Coccyx | | |
| Deep part | Encircles anal canal | | | |
| Puborectalis (part of levator ani) | Pubic bones | Sling around junction of rectum and anal canal | Perineal branch of fourth sacral nerve and from perineal branch of pudendal nerve | Together with external anal sphincter forms voluntary sphincter for anal canal |
| MALE UROGENITAL MUSCLES | | | | |
| Bulbospongiosus | Perineal body | Fascia of bulb of penis and corpus spongiosum and cavernosum | Perineal branch of pudendal nerve | Compress urethra and assists in erection of penis |
| Ischiocavernosus | Ischial tuberosity | Fascia covering corpus cavernosum | Perineal branch of pudendal nerve | Assists in erection of penis |
| Sphincter urethrae | Pubic arch | Surrounds urethra | Perineal branch of pudendal nerve | Voluntary sphincter of urethra |
| Deep transverse perineal muscle | Ischial ramus | Perineal body | Perineal branch of pudendal nerve | Fixes perineal body |
| FEMALE UROGENITAL MUSCLES | | | | |
| Bulbospongiosus | Perineal body | Fascia of corpus cavernosum | Perineal branch of pudendal nerve | Sphincter of vagina and assists in erection of clitoris |
| Ischiocavernosus | Ischial tuberosity | Fascia covering corpus cavernosum | Perineal branch of pudendal nerve | Causes erection of clitoris |
| Superficial transverse perineal muscle | Same as in male | | | |
| Sphincter urethrae | Same as in male | | | |
| Deep transverse perineal muscle | Same as in male | | | |

sity of sexual excitement. During this process the vaginal walls become moist, due to transudation of fluid through the congested mucous membrane. In addition, the greater vestibular glands at the vaginal orifice secrete a lubricating mucus.

The upper part of the vagina, which resides in the pelvic cavity, is supplied by the hypogastric plexuses and is sensitive to stretch only. The region of the vaginal orifice, the labia minora, and the clitoris are extremely sensitive to touch and are supplied by the ilioinguinal nerves and the dorsal nerves of the clitoris.

Appropriate sexual stimulation of these sensitive areas, reinforced by afferent nervous impulses from the breasts and other regions, results in a climax of pleasurable sensory impulses reaching the central nervous system. Impulses then pass down the spinal cord to the sympathetic outflow (T1–L2).

The nervous impulses that pass to the genital organs are thought to leave the cord at the first and second lumbar segments in preganglionic sympathetic fibers. Many of these fibers synapse with postganglionic neurons in the first and second lumbar ganglia; other fibers may synapse in ganglia in the lower lumbar or pelvic parts of the sympathetic trunks. The postganglionic fibers are then distributed to the smooth muscle of the vaginal wall, which rhythmically contracts. In addition, nervous impulses travel in the pudendal nerve (S2, 3, and 4), to reach the bulbospongiosus and ischiocavernosus muscles, which also undergo rhythmic contraction. In many women a single orgasm brings about sexual contentment, but other women require a series of orgasms to feel replete.

## FEMALE URETHRA

The female urethra is about 1½ inches (3.8 cm) long. It extends from the neck of the bladder to the vestibule, where it opens about 1 inch (2.5 cm) below the clitoris (Figs. 8-5 and 8-14). It traverses the sphincter urethrae and lies immediately in front of the vagina.

The *paraurethral glands*, which correspond to the prostate in the male, open into the vestibule by small ducts on either side of the urethral orifice (Fig. 8-4).

The *greater vestibular glands* are a pair of small mucus-secreting glands that lie under cover of the posterior parts of the bulb of the vestibule and the labia majora (Figs. 8-13 and 8-14). Each drains its secretion into the vestibule by a small duct, which opens into the groove between the hymen and the posterior part of the labium minus (Fig. 8-4). These glands secrete a lubricating mucus during sexual intercourse.

## VAGINA

The vagina serves as the excretory duct of the uterus, the female organ for copulation, and part of the birth canal. It extends upward and backward from the vulva (Fig. 8-5) and measures about 3 inches (8 cm) long. The vaginal orifice possesses a thin mucosal fold called the *hymen*, which is perforated at its center. The upper half of the vagina lies above the pelvic floor and the lower half lies within the perineum (Fig. 8-13).

The *arterial supply* to the vagina is from the vaginal artery, a branch of the internal iliac artery; it is also supplied by the vaginal branch of the uterine artery. The *vaginal veins* drain into the internal iliac veins.

The *lymphatic vessels* from the upper third of the vagina drain to the external and internal iliac nodes; from the middle third, to the internal iliac nodes; and from the lower third, to the superficial inguinal nodes.

The *supports of the vagina* are discussed on page 353.

The *nerve supply* to the vagina is from the inferior hypogastric plexuses.

## Lymphatic Drainage of the Vulva

The skin of the vulva is drained into the medial group of superficial inguinal nodes.

## CLINICAL NOTES

### Anal Canal

In the submucosa of the anal canal is a plexus of veins that is principally drained upward by the superior rectal vein. The small tributaries of the middle and inferior rectal veins communicate with each other and with the superior rectal vein through this plexus. The rectal venous plexus therefore forms an important portal systemic anastomosis, since the superior rectal vein drains into the portal vein, and the middle and inferior rectal veins drain into the systemic system.

*Internal hemorrhoids* (piles) are varicosities of the tributaries of the superior rectal (hemorrhoidal) vein and are covered by mucous membrane (Fig. 8-15). The tributaries of the vein, which lie in the anal columns at the 3, 7, and 11 o'clock positions when the patient is viewed in the lithotomy

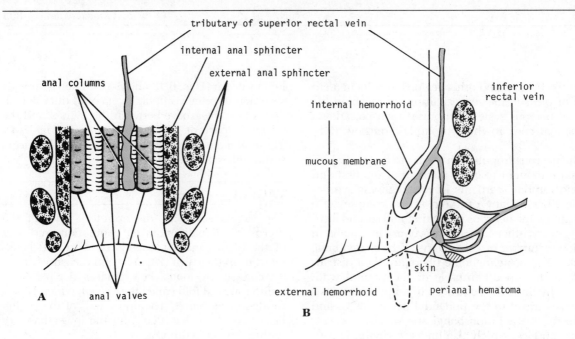

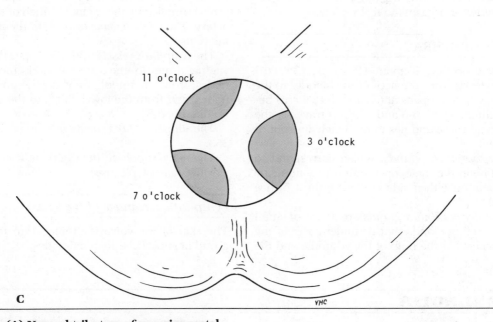

Fig. 8-15. (A) Normal tributary of superior rectal
vein within anal column. (B) Varicosed tributary
of superior rectal vein forming internal hemorrhoid (dotted lines indicate degrees of severity of
condition). (C) Positions of three internal hemorrhoids as seen through proctoscope with patient in
lithotomy position.

position,* are particularly liable to become varicosed. Anatomically, a hemorrhoid is therefore a fold of mucous membrane and submucosa containing a varicosed tributary of the superior rectal vein and a terminal branch of the superior rectal artery. To begin with, hemorrhoids are contained within the anal canal (first degree). As they enlarge, they are extruded from the canal on defecation, but return at the end of the act (second degree). With further elongation, they prolapse on defecation and remain outside the anus (third degree).

The causes of internal hemorrhoids are many. They frequently occur in members of the same family, which suggests a congenital weakness of the vein walls. Varicose veins of the legs and hemorrhoids often go together. The superior rectal vein is the most dependent part of the portal circulation and is valveless. The weight of the column of venous blood is thus greatest in the veins in the upper half of the anal canal. Here, the loose connective tissue of the submucosa gives little support to the walls of the veins. Moreover, the venous return is interrupted by the contraction of the muscular coat of the rectal wall during defecation. Chronic constipation, associated with prolonged straining at stool, is a common predisposing factor. Pregnancy hemorrhoids are common, due to pressure on the superior rectal veins by the gravid uterus. Portal hypertension due to cirrhosis of the liver may also cause hemorrhoids. The possibility that cancerous tumors of the rectum are blocking the superior rectal vein must never be overlooked.

*External hemorrhoids* are varicosities of the tributaries of the inferior rectal (hemorrhoidal) vein as they run laterally from the anal margin. They are covered by skin (Fig. 8-15) and are commonly associated with well-established internal hemorrhoids. A more important clinical condition is rupture of the tributaries of the inferior rectal vein as the result of coughing or straining, with the appearance of a small clot of blood in the subcutaneous tissue near the anus (Fig. 8-15). This small blue swelling is called a *perianal hematoma* (wrongly called thrombosed external pile).

The lower ends of the anal columns are connected by small folds called *anal valves*. In persons suffering from chronic constipation, the anal valves may be torn down to the anus as the result of the edge of the fecal mass's catching on the fold of mucous membrane. The elongated ulcer so formed, known as an *anal fissure* (Fig. 8-16), is extremely painful. It is interesting to note that this condition occurs most commonly in the midline posteriorly or anteriorly, and this may be due to the lack of support provided by the superficial part of the external sphincter in these areas. (The superficial part of the external sphincter does not encircle the anal canal, but sweeps past its lateral sides.)

The site of the anal fissure in the sensitive lower half of anal canal, which is innervated by the inferior rectal nerve, results in reflex spasm of the external anal sphincter, aggravating the condition. Because of the intense pain, anal fissures may have to be examined under local anesthesia.

*Perianal abscesses* are produced by fecal trauma to the anal mucosa (Fig. 8-16). Infection may gain entrance to the submucosa through a small mucosal lesion, or the abscess may complicate an anal fissure. The abscess may be localized to the submucosa (submucous abscess), may occur beneath the perianal skin (subcutaneous abscess), or may occupy the ischiorectal fossa (ischiorectal abscess). Sometimes an abscess may be found in the space between the ampulla of the rectum and the upper surface of the levator ani (pelvirectal abscess). Anatomically, these abscesses are closely related to the different parts of the external sphincter and levator ani muscles, as seen in Figure 8-16).

*Anal fistulae* develop as the result of spread or inadequate treatment of anal abscesses. The fistula opens at one end at the lumen of the anal canal or lower rectum and at the other end on the skin surface close to the anus (Fig. 8-16). If the abscess opens onto only one surface, it is known as a *sinus*, not a fistula. The high-level fistulae are rare and run from the rectum to the perianal skin. They are located above the anorectal ring, and, as a result, fecal material constantly soils the clothes. The low-level fistulae occur below the level of the anorectal ring, as shown in Figure 8-16).

The most important part of the sphincteric mechanism of the anal canal is the *anorectal ring*. It consists of the deep part of the external sphincter, the internal sphincter, and the puborectalis part of the levator ani. Surgical operations on the

---

*The patient is in the supine position with both hip joints flexed and abducted; the feet are held in position by stirrups. The position is commonly used for pelvic examinations in the female.

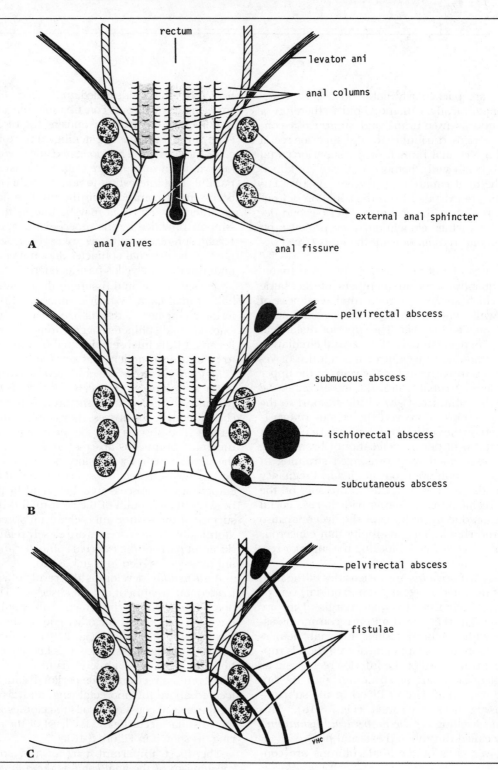

**Fig. 8-16. (A) Tearing downward of anal valve to form anal fissure. (B) Common locations of perianal abscesses. (C) Common positions of perianal fistulae.**

anal canal that results in damage to the anorectal ring will produce fecal incontinence.

## *Rectal Examination*

The following structures can be palpated by the gloved index finger inserted into the anal canal and rectum in the normal patient.

### Anteriorly

*Male:*

1. *Opposite the terminal phalanx* are the contents of the rectovesical pouch, the posterior surface of the bladder, the seminal vesicles, and the vasa deferentia (Fig. 8-17).
2. *Opposite the middle phalanx* are the rectoprostatic fascia and the prostate.
3. *Opposite the proximal phalanx* are the perineal body, the urogenital diaphragm, and the bulb of the penis.

### Anteriorly

*Female:*

1. *Opposite the terminal phalanx* are the rectouterine pouch, the vagina, and the cervix.
2. *Opposite the middle phalanx* are the urogenital diaphragm and the vagina.
3. *Opposite the proximal phalanx* are the perineal body and the lower part of the vagina.

### Posteriorly

The sacrum, coccyx, and anococcygeal body.

### Laterally

The ischiorectal fossae and ischial spines.

## Ischiorectal Fossa

The ischiorectal fossae are filled with fat that is poorly vascularized. The close proximity to the anal canal makes them particularly vulnerable to infection. Infection commonly tracks laterally from the anal mucosa through the external anal sphincter. Infection of the perianal hair follicles or sweat glands may also be the cause of infection in the fossae. Rarely, a perirectal abscess bursts down-ward through the levator ani muscle. An ischiorectal abscess may involve the opposite fossa by the spread of infection across the midline behind the anal canal.

The *pudendal nerve*, as it runs forward in the fascial pudendal canal on the lateral wall of the ischiorectal fossa, may be blocked by an anesthetic to produce analgesia of the perineum in forceps delivery.

## Lymphatic Drainage of Anal Canal

It is important to remember the lymphatic drainage of the anal canal. The upper half of the mucous membrane is drained upward to lymph nodes along the course of the superior rectal artery. The lower half of the mucous membrane is drained downward to the medial group of superficial inguinal nodes. Many a patient has thought himself to have an inguinal hernia, and the physician has found a cancer of the lower half of the anal canal, with secondary deposits in the inguinal lymph nodes.

## Male Urogenital Triangle

### Circumcision

This is the operation of removing the greater part of the prepuce, or foreskin. In many newborn males, the prepuce cannot be retracted over the glans. This may result in the infection of the secretions beneath the prepuce leading to inflammation, swelling, and fibrosis of the prepuce. Repeated inflammation leads to constriction of the orifice of the prepuce (*phimosis*) with obstruction to urination. It is now generally believed that chronic inflammation of the prepuce predisposes to carcinoma of the glans penis. For these reasons prophylactic circumcision is commonly practiced and in Jews it is a religious rite.

### Rupture of the Urethra

This may complicate a severe blow on the perineum. The common site of rupture is within the bulb of the penis, just below the perineal membrane. The urine extravasates into the superficial perineal pouch and then passes forward over the scrotum beneath the membranous layer of the su-

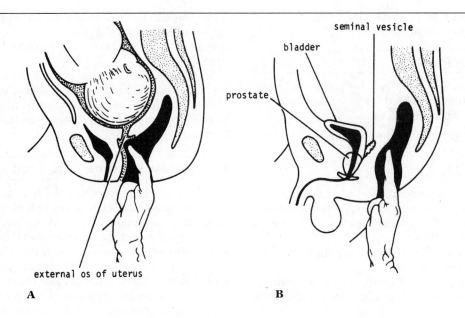

A

external os of uterus

B

bladder

seminal vesicle

prostate

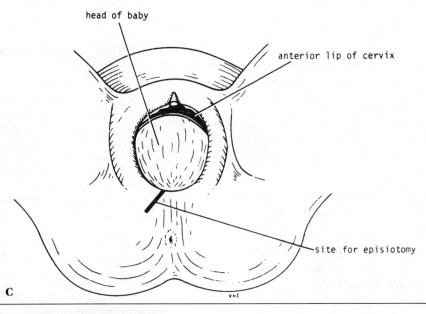

C

head of baby

anterior lip of cervix

site for episiotomy

**Fig. 8-17. (A) Rectal examination in pregnant woman, showing how it is possible to palpate cervix through anterior rectal wall. (B) Rectal examination in male, showing how it is possible to palpate prostate and seminal vesicles through anterior rectal wall. (C) Position of episiotomy incision in a woman during second stage of labor. Baby's head is presenting at vaginal orifice.**

perficial fascia, as described in Chapter 4. If the membranous part of the urethra is ruptured, urine escapes into the deep perineal pouch and may extravasate upward around the prostate and bladder, or downward into the superficial perineal pouch.

## Catheterization

The following anatomical facts should be remembered before passing a catheter or other instrument along the male urethra.

1. The external orifice at the glans penis is the narrowest part of the entire urethra.
2. Within the glans the urethra dilates to form the *fossa terminalis.*
3. Near the posterior end of the fossa, a fold of mucous membrane projects into the lumen from the roof.
4. The membranous part of the urethra is narrow and fixed.
5. The prostatic part of the urethra is the widest and most dilatable part of the urethra.
6. By holding the penis upward, the S-shaped curve to the urethra is converted into a J-shaped curve.

If the point of the catheter will pass through the external orifice and is then directed toward the urethral floor until it has passed the mucosal fold (see 3 above), it should easily pass along a normal urethra into the bladder.

The most dependent part of the male urethra is that which lies within the bulb. Here, it is subject to chronic inflammation and stricture formation.

The many glands that open into the urethra, including those of the prostate, the bulbourethral glands, and many small penile urethral glands, are commonly the site of chronic gonococcal infection.

Clinical notes on the anatomy of the scrotum have been presented in Chapter 4.

# Female Urogenital Triangle

In the region of the vulva, the presence of numerous glands and ducts opening onto the surface makes this area prone to infection. The sebaceous glands of the labia majora, the ducts of the greater vestibular glands, the vagina (with its indirect communication with the peritoneal cavity), the urethra, and the paraurethral glands can all become infected. The vagina itself has no glands and is lined with stratified squamous epithelium. Provided that the pH of its interior is kept low, it is capable of resisting infection to a remarkable degree.

An important sign in the diagnosis of pregnancy is the appearance of a bluish discoloration of the vulva and vagina due to venous congestion. It appears at approximately the eighth to twelfth week and increases as the pregnancy progresses.

## *Vaginal Examination*

Digital examination of the vagina may provide the physician with much valuable information concerning the health of the vaginal walls, the uterus, and the surrounding structures (Fig. 8-17). The anatomical relations of the vagina must therefore be known and are considered in detail in Chapter 7.

The *perineal body* is a wedge of fibromuscular tissue that lies between the lower part of the vagina and the anal canal. It is held in position by the insertion of the perineal muscles and by the attachment of the levator ani muscles. In the female it is a much larger structure than in the male, and it serves to support the posterior wall of the vagina. Damage by laceration during childbirth may be followed by permanent weakness of the pelvic floor.

Very few women escape some injury to the birth canal during delivery. In the majority this is little more than an abrasion of the posterior vaginal wall. Spontaneous delivery of the child with the patient unattended may result in a severe tear of the lower third of the posterior wall of the vagina, the perineal body, and overlying skin. In severe tears the lacerations may extend backward into the anal canal and damage the external sphincter. In these cases it is imperative that an accurate repair of the walls of the anal canal, vagina, and perineal body be undertaken as soon as possible.

In the management of childbirth, when it is obvious to the obstetrician that the perineum will tear before the baby's head emerges through the vaginal orifice, a planned surgical incision is made through the perineal skin in a posterolateral direction to avoid the anal sphincters. This procedure is known as an *episiotomy* (Fig. 8-17). Breech deliveries and forceps deliveries are usually preceded by an episiotomy.

## *Pudendal Nerve Block*

During the second stage of labor, when the presenting part of the fetus, usually the head, is descending through the vulva, the area may be anesthetized by performing a pudendal nerve block.

The local anesthetic is injected around the pudendal nerve as it lies within the pudendal canal on the medial side of the ischial tuberosity. This procedure relieves the pain caused by stretching of the soft tissues of the perineum.

## CLINICAL PROBLEMS

*Answers on page 967*

1. A 53-year-old man stated that for the past 4 years he had frequently passed blood-stained stools. The blood was bright red in color but small in amount, and he did not think it was serious enough to consult his physician. Recently, he had noticed that his "bowel" protruded from his anus after defecation, and this caused him considerable discomfort. He had also started to have intense perianal irritation, especially at night. On physical examination, the perianal skin was noted to be red, moist, and excoriated. On digital examination of the anal canal, no tumor was palpated. On proctoscopic examination, the mucous membrane at the level of the anal valves was found to bulge downward when the patient strained. The mucous membrane over the swellings was pink in color, but large, congested veins could be seen beneath the surface. With the patient in the lithotomy position, the swellings were arranged at 3, 7, and 11 o'-clock. From your knowledge of anatomy, what is the diagnosis?

2. A 42-year-old woman visited her physician because of an agonizing pain in the "rectum," which occurred on defecation. She had first noticed the pain a week before when she tried to defecate. The pain lasted for about an hour, then passed off, only to return with the next bowel movement. On questioning the patient, the physician learned that the woman suffered from chronic constipation. She admitted that the stools were sometimes streaked with blood. On examination, the anus was found to be tightly closed. An attempt to examine the anal canal digitally failed because of the severe pain it caused the patient. By gently everting the anal margin, however, the lower edge of a linear tear in the posterior wall of the anal canal could be seen. A diagnosis of anal fissure was made. In anatomical terms, explain the possible causes of anal fissure. Why is this condition so painful, when, for example, carcinomatous ulcers of the upper half of the anal canal are painless? Why was the anus tightly closed?

3. An enthusiastic young surgeon, while operating on a perianal fistula, found that it was necessary to extend his dissection of the fistulous tract superiorly. Unfortunately, his exposure of the area was poor, and troublesome bleeding kept hiding important structures from view. Ten days after the operation, he was surprised to learn that the patient was suffering from fecal incontinence. What anatomical structure must be preserved in this region in order to prevent fecal incontinence?

4. A 51-year-old woman with a history of a chronic anal fissure visited her physician complaining of a painful swelling in the region of the anus. Physical examination revealed a hot, red, tender swelling lateral to the right anal margin. A diagnosis of ischiorectal abscess was made. Why is an ischiorectal abscess a common complication of anal fissure? Are there any important structures in the ischiorectal fossa that might be damaged by a surgical incision of an abscess in this region?

5. A 16-year-old boy was taking part in a bicycle race when, on approaching a steep incline, he stood up on his pedals to increase his speed. Unfortunately, his right foot slipped off the pedal and he fell violently, his perineum hitting the bar of the bicycle. Several hours later he was admitted to the hospital unable to micturate. He had extensive swelling of his penis and scrotum. The diagnosis of ruptured urethra was made. Which part of the urethra was likely to have been damaged? What is the cause of the extensive swelling of the penis and scrotum?

6. An inexperienced but learned student was asked to catheterize a male patient who was suffering

from postoperative retention of urine following an appendectomy. The patient's urinary tract was otherwise normal. Remembering his anatomy, the student knew that once he got the end of the catheter through the external urethral meatus, everything should go smoothly, and the catheter should pass without difficulty into the bladder. For reasons he could not understand, the point of the catheter refused to go beyond the first ¾ of an inch (1.9 cm) of the urethra. After six attempts, the patient was jumping up and down on the bed and threatening to call the fire brigade. The embarrassed student, red-faced and wishing the floor would swallow him up, unobtrusively left the ward and sought advice from the resident. What anatomical structure had the student forgotten?

7. A 35-year-old woman with a history of chronic gonorrhea visited her physician complaining of a swelling in the genital region. On examination, a tense cystic swelling was found beneath the posterior two-thirds of the left labium majus and minus. Name the secretory organs that lie under cover of the posterior portions of the labia.

8. A 65-year-old woman went to her physician complaining of irritation, discharge, and bleeding of the genital region. On physical examination, a hard-based ulcer was found on the medial aspect of the right labium majus. A diagnosis of squamous cell carcinoma of the skin was made. What is the lymphatic drainage of this region, and which group of lymph nodes would you therefore examine for evidence of metastases?

---

# NATIONAL BOARD TYPE QUESTIONS

*Answers on page 984*

### In each of the following questions, answer:

(a) IF (1) IS CORRECT ONLY
(b) IF (2) IS CORRECT ONLY
(c) IF BOTH (1) AND (2) ARE CORRECT, AND
(d) IF NEITHER (1) NOR (2) IS CORRECT

1. Which of the following statements is (are) true:
   (1) The anal valves mark the boundary between the upper and lower halves of the anal canal.
   (2) The mucous membrane of the upper and lower halves of the anal canal is sensitive to touch and pain.

2. Which of the following statements is (are) true concerning the male urethra?
   (1) The narrowest part of the entire male urethra is the external urethral meatus.
   (2) The widest part of the male urethra is the membranous urethra.

3. Which of the following statements is (are) true concerning the vulva?
   (1) The ducts of the greater vestibular glands open into the vulva between the labia minora and labia majora.
   (2) The lymph vessels of the skin of the vulva are drained into the medial group of superficial inguinal nodes.

### In the following questions, answer:

(a) IF (1), (2), AND (3) ONLY ARE CORRECT
(b) IF (1) AND (3) ONLY ARE CORRECT
(c) IF (2) AND (4) ARE ONLY CORRECT
(d) IF (4) ONLY IS CORRECT, OR
(e) IF ALL ARE CORRECT

4. (1) The anorectal ring is formed by the subcutaneous, superficial, and deep fibers of the external anal sphincter.
   (2) The urogenital diaphragm is attached laterally to the inferior ramus of the pubis and the ischial ramus.
   (3) The bulbourethral glands are situated in the superficial perineal pouch.
   (4) The crura of the clitoris anteriorly form the corpora cavernosa of the body of the clitoris.

5. Which of the following structures may be palpated by a vaginal examination?
   (1) Sigmoid colon
   (2) Ureters
   (3) Perineal body
   (4) Ischial spines

6. Concerning the ischiorectal fossa:
   (1) The pudendal nerve lies in the lateral wall.
   (2) The medial wall is formed in part by the levator ani muscle.
   (3) The floor is formed by superficial fascia and skin.
   (4) The lateral wall is formed by the obturator internus muscle and its fascia.

7. Concerning the penis:
   (1) Its root is formed in the midline by the bulb of the penis, which continues anteriorly as the corpus spongiosum.
   (2) Its roots laterally are formed by the crura, which continue anteriorly as the corpora cavernosa.
   (3) The penile urethra lies within the corpus spongiosum.
   (4) The glans penis is a distal expansion of the fused corpora cavernosa.

8. Concerning the female urethra:
   (1) It lies immediately anterior to the vagina.
   (2) Its external orifice lies about 2 inches (5 cm) from the clitoris.
   (3) It is about 1½ inches (3.8 cm) long.
   (4) It does not pierce the deep perineal pouch.

**Select the best response:**

9. The urogenital diaphragm is formed by all of the following structures **except** the:
   (a) Deep transverse perineal muscle
   (b) Perineal membrane
   (c) Sphincter urethrae muscle
   (d) Colles' fascia (membranous layer of superficial fascia)
   (e) Parietal pelvic fascia covering the upper surface of the sphincter urethrae muscle

10. Which of the following structures **cannot** be palpated on rectal examination in the male?
    (a) Bulb of the penis
    (b) Urogenital diaphragm
    (c) Anorectal ring
    (d) The anterior surface of the sacrum
    (e) Ureter

11. Which statement is **not true** concerning the anal canal?
    (a) It is about 1½ inches (3.75 cm) long.
    (b) It pierces the urogenital diaphragm.
    (c) It is related laterally to the external anal sphincter.
    (d) It is the site of an important portal-systemic anastomosis.
    (e) The mucous membrane of the lower half receives its arterial supply from the inferior rectal artery.

12. Which of the following features **is true** concerning the subcutaneous part of the external anal sphincter?
    (a) It encircles the anal canal.
    (b) It is attached to the anococcygeal body.
    (c) It is composed of smooth muscle fibers.
    (d) It causes the rectum to join the anal canal at an acute angle.
    (e) It is innervated by the middle rectal nerve.

13. The following facts concerning defecation are true **except:**
    (a) The act is often preceded by the entrance of feces into the rectum, which gives rise to the desire to defecate.
    (b) The muscles of the anterior abdominal wall contract.
    (c) The external anal sphincters and the puborectalis muscle relax.
    (d) The internal sphincter contracts and causes the evacuation of the feces.
    (e) The mucous membrane of the lower part of the anal canal is extruded through the anus ahead of the fecal mass.

14. The process of ejaculation depends on the following processes **except:**
    (a) The sphincter of the bladder contracts.
    (b) The sympathetic preganglionic nerve fibers arising from the first and second lumbar segments of the spinal cord must be intact.
    (c) The smooth muscle of the epididymis, vas deferens, seminal vesicles, and prostate contracts.
    (d) The bulbourethral glands and urethral glands are active.
    (e) The bulbospongiosus muscles relax.

15. Which of the following structures does not receive innervation from the branches of the pudendal nerve?
    (a) Labia minora
    (b) Urethral sphincter
    (c) The posterior fornix of the vagina
    (d) Ischiocavernosus muscles
    (e) Skin of the penis or clitoris

# 9. The Upper Limb

The upper limb may be regarded as a multijointed lever that is freely movable on the trunk at the shoulder joint. At the distal end of the upper limb is the important prehensile organ, the hand. Much of the importance of the hand is dependent on the pincer-like action of the thumb, which enables one to grasp objects between the thumb and index finger.

The upper limb may be divided into the shoulder (junction of the trunk with the arm), arm, elbow, forearm, wrist, and hand.

## SURFACE ANATOMY

The following information should be verified on the living body. Remember that much of the information that you obtain on physical examination of a patient depends on your adequate knowledge of surface anatomy.

### The Mammary Glands

The mammary glands in the female after puberty (Fig. 9-1) are usually hemispherical in shape, are slightly pendulous, and extend from the second to the sixth rib and from the lateral margin of the sternum to the midaxillary line. The greater part of a mammary gland lies in the superficial fascia and can be moved freely in all directions. A small part of the breast, known as the *axillary tail* (Fig. 9-1), extends upward and laterally, pierces the deep fascia at the lower border of the pectoralis major muscle, and comes into close relationship with the axillary vessels.

In the living subject, the breast is soft because the fat contained within it is fluid. On careful palpation with the open hand, the breast has a firm overall lobulated consistency, produced by its glandular tissue.

The *nipple* projects from the lower half of the breast (Fig. 9-1), but its position in relation to the chest wall varies greatly and depends on the degree of development of the gland. The base of the nipple is surrounded by a circular area of pigmented skin called the *areola* (Fig. 9-1). Pink in color in the young girl, the areola becomes darker in color in the second month of the first pregnancy and never regains its former tint. Tiny tubercles on the areola

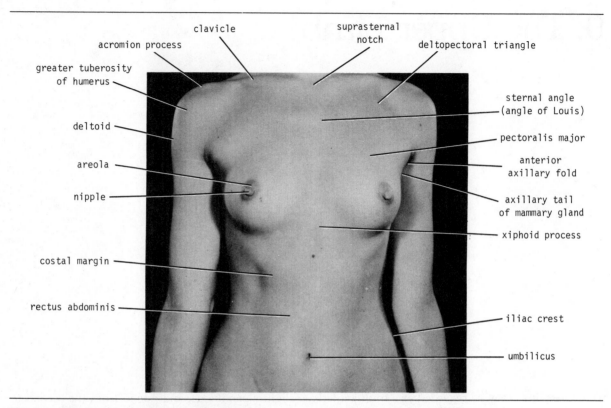

greater tuberosity
of humerus

acromion process

clavicle

suprasternal
notch

deltopectoral triangle

sternal angle
(angle of Louis)

deltoid

pectoralis major

areola

anterior
axillary fold

nipple

axillary tail
of mammary gland

xiphoid process

costal margin

rectus abdominis

iliac crest

umbilicus

**Fig. 9-1. Anterior view of thorax and abdomen in a 29-year-old female.**

are produced by the underlying *areolar glands.*

In the immature female and in the male, the mammary glands are rudimentary. The nipples are small and usually lie over the fourth intercostal space.

## The Clavicles

The clavicle is situated at the root of the neck and can be palpated throughout its length (Figs. 9-1, 9-2, and 9-3). The positions of the *sternoclavicular* and *acromioclavicular joints* can be easily identified. Note that the medial end of the clavicle projects above the margin of the manubrium sterni.

## The Deltopectoral Triangle

This is a small triangular depression situated below the outer third of the clavicle; it is bounded by the pectoralis major and deltoid muscles (Figs. 9-1, 9-2, and 9-3).

## The Scapulae

The tip of the *coracoid process* of the scapula (Fig. 9-3) can be felt on deep palpation in the lateral part of the deltopectoral triangle; it is covered by the anterior fibers of the deltoid.

The *acromion process of the scapula* forms the lateral extremity of the spine of the scapula. It is subcutaneous and easily located (Figs. 9-1 and 9-2).

Immediately below the lateral edge of the acromion process, the smooth rounded curve of the shoulder is produced by the *deltoid muscle,* which covers the *greater tuberosity of the humerus* (Fig. 9-2).

The *crest of the spine of the scapula* can be palpated and traced medially to the medial border of

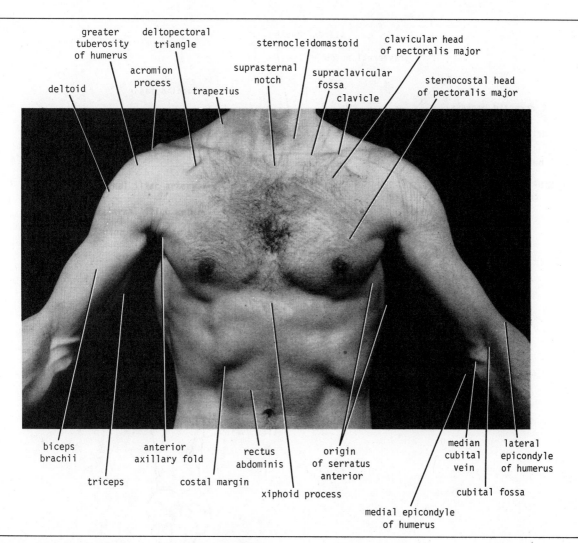

**Fig. 9-2. The pectoral region in a 27-year-old male.**

the scapula, which it joins at the level of the third thoracic spine (Fig. 9-5).

The *inferior angle of the scapula* can be palpated opposite the seventh thoracic spine (Figs. 9-4 and 9-5).

## Axillary Folds and Axilla

The *anterior axillary fold*, which is formed by the lower margin of the pectoralis major muscle, can be palpated between the finger and thumb (Figs. 9-1, 9-2, and 9-3).

The *posterior axillary fold*, which is formed by the tendon of latissimus dorsi winding around the lower border of the teres major muscle, can be similarly palpated between finger and thumb (Fig. 9-5).

The *axilla* should be examined with the forearm supported and the pectoral muscles relaxed. With the arm by the side, the inferior part of the *head of the humerus* can be easily palpated through the floor of the axilla. The pulsations of the *axillary artery* may be felt high up in the axilla, and around the artery may be palpated the *cords of the brachial plexus.* The medial wall of the axilla is

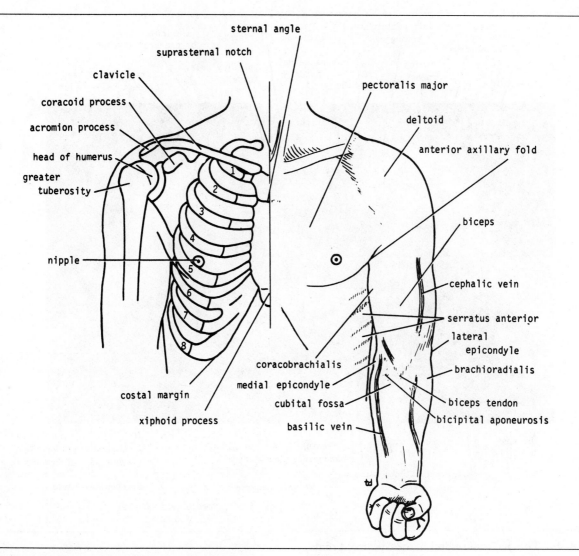

**Fig. 9-3. Surface anatomy of chest, shoulder, and upper limb as seen anteriorly.**

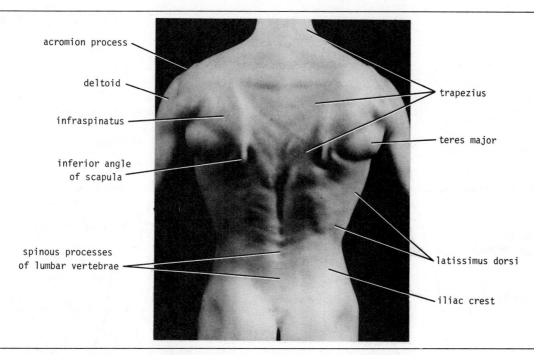

acromion process

deltoid

infraspinatus

inferior angle
of scapula

spinous processes
of lumbar vertebrae

trapezius

teres major

latissimus dorsi

iliac crest

**Fig. 9-4. The back in a 27-year-old male.**

formed by the *upper ribs* covered by the *serratus anterior muscle*, the serrations of which can be seen and felt in a muscular subject (Fig. 9-2). The lateral wall is formed by the *coracobrachialis* and *biceps brachii muscles* and the bicipital groove of the humerus.

## Elbow Region

The *medial and lateral epicondyles* of the humerus (Figs. 9-2 and 9-5) and the olecranon process of the ulna can be palpated (Fig. 9-5). When the elbow joint is extended, these bony points lie on the same straight line; when the elbow is flexed, these three points form the boundaries of an equilateral triangle.

The *head of the radius* can be palpated in a depression on the posterolateral aspect of the extended elbow, distal to the lateral epicondyle. The head of the radius can be felt to rotate during pronation and supination of the forearm.

The boundaries of the *cubital fossa* (Figs. 9-3

and 9-6) can be seen and felt; the brachioradialis muscle forms the lateral boundary and the pronator teres forms the medial boundary. The *tendon of the biceps muscle* can be palpated as it passes downward into the fossa, and the *bicipital aponeurosis* can be felt as it leaves the tendon to join the deep fascia on the medial side of the forearm (Figs. 9-3 and 9-6). The tendon and aponeurosis are most easily felt if the elbow joint is flexed against resistance.

The *ulnar nerve* can be palpated where it lies behind the medial epicondyle of the humerus. It feels like a rounded cord, and when it is compressed, a "pins-and-needles" sensation is felt along the medial part of the hand.

## The Brachial Artery

This artery may be felt to pulsate as it passes down the arm, overlapped by the medial border of the biceps muscle. In the cubital fossa, it lies beneath the bicipital aponeurosis, and at a level just below the head of the radius, it divides into its terminal branches.

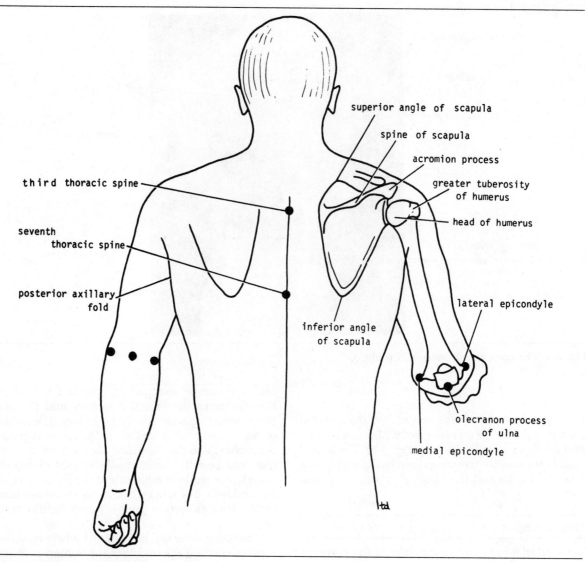

superior angle of scapula

spine of scapula

acromion process

greater tuberosity
of humerus

third thoracic spine

head of humerus

seventh
thoracic spine

posterior axillary
fold

lateral epicondyle

inferior angle
of scapula

olecranon process
of ulna

medial epicondyle

**Fig. 9-5. Surface anatomy of scapula, shoulder, and
elbow regions as seen posteriorly.**

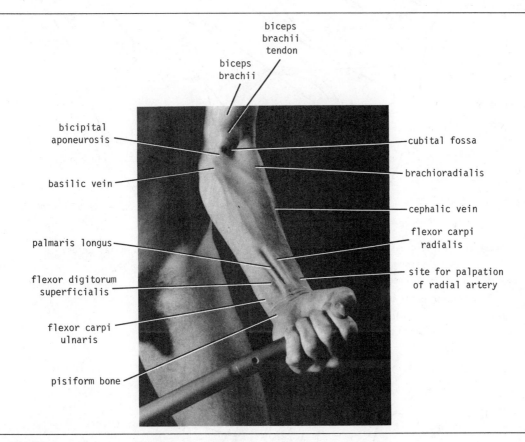

biceps
brachii
tendon

biceps
brachii

bicipital
aponeurosis

basilic vein

palmaris longus

flexor digitorum
superficialis

flexor carpi
ulnaris

pisiform bone

cubital fossa

brachioradialis

cephalic vein

flexor carpi
radialis

site for palpation
of radial artery

**Fig. 9-6. The cubital fossa and anterior surface of forearm in a 27-year-old male.**

## The Ulna Bone

The *posterior border of the ulna bone* is subcutaneous and can be palpated along its entire length.

## The Wrist and Hand

At the wrist, the *styloid processes of the radius* (Fig. 9-7) and *ulna* can be palpated. The styloid process of the radius lies about ¾ inch (1.9 cm) distal to that of the ulna.

The *dorsal tubercle of the radius* is palpable on the posterior surface of the lower end of the radius (Fig. 9-7).

The *pisiform bone* can be felt on the medial side of the anterior aspect of the wrist between the two transverse creases (Figs. 9-6 and 9-7). The *hook of the hamate bone* can be felt on deep palpation of the hypothenar eminence, a fingerbreadth distal and lateral to the pisiform bone.

The *transverse creases* seen in front of the wrist are important landmarks (Fig. 9-7). The proximal transverse crease lies at the level of the wrist joint. The distal transverse crease corresponds to the proximal border of the flexor retinaculum.

The following important structures lie in front of the wrist region and should be palpated. The pulsations of the *radial artery* may be felt anterior to the distal third of the radius (Figs. 9-6 and 9-7). The tendon medial to this is the tendon of the *flexor carpi radialis*. Medial to this, the tendon of *palmaris longus* may be present overlying the *median nerve*. Medial to this, a rounded group of tendons will be seen when the wrist and fingers are flexed and extended; this is the *flexor digitorum su-*

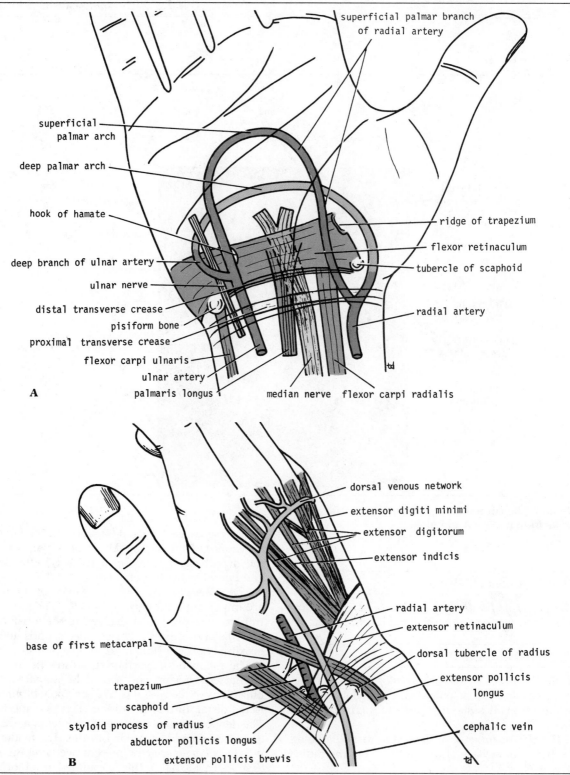

**Fig. 9-7. Surface anatomy of wrist region.**    *Fig. 7. Surface anatomy of wrist region.*

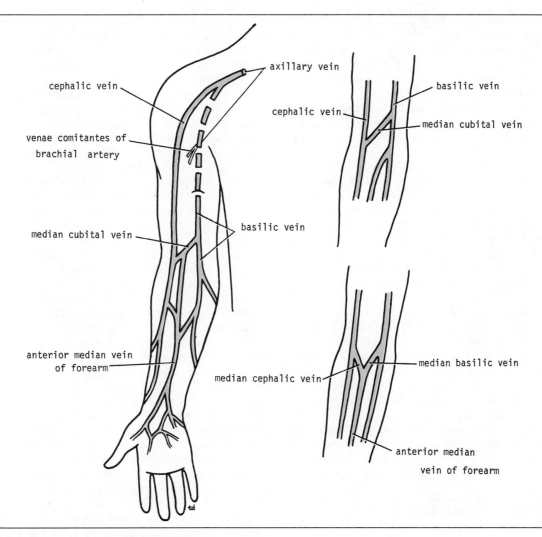

**Fig. 9-8. Superficial veins of upper limb. Note common variations seen in region of elbow.**

*perficialis muscle.* The tendon of the *flexor carpi ulnaris* lies most medially as it goes to its insertion on the pisiform bone (Figs. 9-6 and 9-7). By careful palpation, the pulsations of the *ulnar artery* can be felt lateral to this tendon. The *ulnar nerve* lies immediately medial to the ulnar artery in this region (Fig. 9-7).

On the lateral aspect of the wrist, distal to the styloid process of the radius, lies the "*anatomical snuffbox*" (Fig. 9-7). It is bounded medially by the tendon of *extensor pollicis longus* and laterally by the tendons of *abductor pollicis longus* and *extensor pollicis brevis.* In its floor are the *styloid process of the radius*, the *scaphoid* and *trapezium*, and the base of the *first metacarpal bone* (Fig. 9-7). The *radial artery* may be palpated within the snuffbox as the artery winds around the lateral margin of the wrist to reach the dorsum of the hand (Fig. 9-7).

The *superficial palmar arterial arch* (Fig. 9-7) is located in the central part of the palm and lies on a line drawn across the palm at the level of the distal border of the fully extended thumb. The *deep palmar arterial arch* (Fig. 9-7) lies a fingerbreadth proximal to this.

The *metacarpal phalangeal joints* lie approximately at the level of the distal transverse palmar crease. The *interphalangeal joints* lie at the level of the middle and distal finger creases.

On the dorsum of the hand, the tendons of the *extensor digitorum*, the *extensor indicis*, and the *extensor digiti minimi* can be seen passing to the bases of the fingers (Fig. 9-7).

The *venous network of superficial veins* can be identified on the dorsum of the hand (Fig. 9-7). These drain upward into a lateral *cephalic vein* and a medial *basilic vein*. The cephalic vein crosses the anatomical snuffbox and winds around onto the anterior aspect of the forearm. It then ascends into the arm and runs along the lateral border of the biceps (Fig. 9-8). It ends by piercing the deep fascia in the deltopectoral triangle and enters the axillary vein.

The basilic vein can be traced from the dorsum of the hand around the medial side of the forearm; it reaches the anterior aspect of the forearm just below the elbow (Fig. 9-8). It pierces the deep fascia at about the middle of the arm. The *median cubital vein* (or median cephalic and median basilic veins) links the cephalic and basilic veins in the cubital fossa (Fig. 9-8).

To identify these veins easily, apply firm pressure around the upper arm and repeatedly clench and relax the fist. By this means the veins become distended with blood.

## THE PECTORAL REGION AND THE AXILLA

### Mammary Glands

The mammary glands are specialized accessory glands of the skin that are capable of secreting milk. They are present in both sexes. In the male and the immature female, the mammary glands are similar. The nipples are small, and each is surrounded by a colored area of skin called the areola. The breast tissue consists of little more than a system of ducts embedded in connective tissue, and they do not extend beyond the margin of the areola.

At puberty in the female, the mammary glands gradually enlarge and assume their hemispherical shape under the influence of the ovarian hormones

(Figs. 9-1 and 9-9). The ducts elongate, but the increased size of the glands is due to the deposition of fat. The base of the breast extends from the second to the sixth rib and from the lateral margin of the sternum to the midaxillary line. The greater part of the gland lies in the superficial fascia. A small part, called the *axillary tail* (Figs. 9-1 and 9-9), extends upward and laterally, pierces the deep fascia at the lower border of the pectoralis major muscle, and comes into close relationship with the axillary vessels.

Each mammary gland consists of fifteen to twenty *lobes*, which radiate out from the nipple. The main duct from each lobe opens separately on the summit of the nipple and possesses a dilated *ampulla* just prior to its termination. The base of the nipple is surrounded by the *areola* (Figs. 9-1 and 9-9). Tiny tubercles on the areola are produced by the underlying *areolar glands.*

The lobes of the gland are separated by fibrous septa. The septa in the upper part of the gland are well developed and extend from the skin to the deep fascia; they serve as *suspensory ligaments* (Fig. 9-9). The mammary glands are separated from the deep fascia covering the underlying muscles by an area of loose connective tissue known as the *retromammary space* (Fig. 9-9).

In the young woman the breasts tend to protrude forward from a circular base; in the older woman they tend to be pendulous. They reach their maximum size during lactation.

The *arterial supply* of the mammary gland is from perforating branches of the internal thoracic artery and the intercostal arteries. The axillary artery also supplies the gland via its lateral thoracic and thoracoacromial branches. The *veins* correspond to the arteries.

The *lymphatic drainage* of the mammary gland is of considerable clinical importance because of the frequent development of cancer in the gland and the subsequent dissemination of the malignant cells along the lymph vessels to the lymph nodes.

The lymphatic capillaries of the breast form an anastomosing network, which is continuous across the midline with that of the opposite side, and below with that of the abdominal wall. The efferent lymphatic vessels from this network accompany the arteries supplying the gland. The lateral part of the gland is drained into the anterior axillary or pectoral nodes (Fig. 9-10). The medial part of the

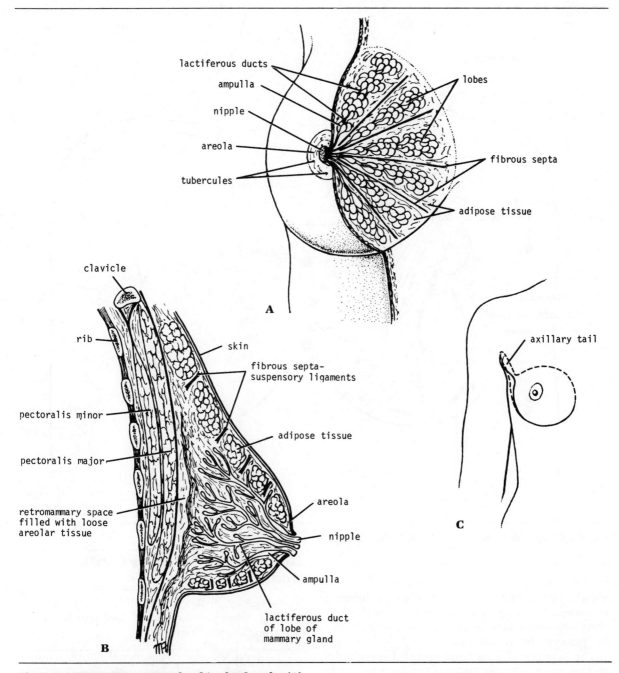

**Fig. 9-9. Mature mammary gland in the female. (A) Anterior view with skin partially removed to show internal structure. (B) Sagittal section. (C) The axillary tail, which pierces the deep fascia and extends into the axilla.**

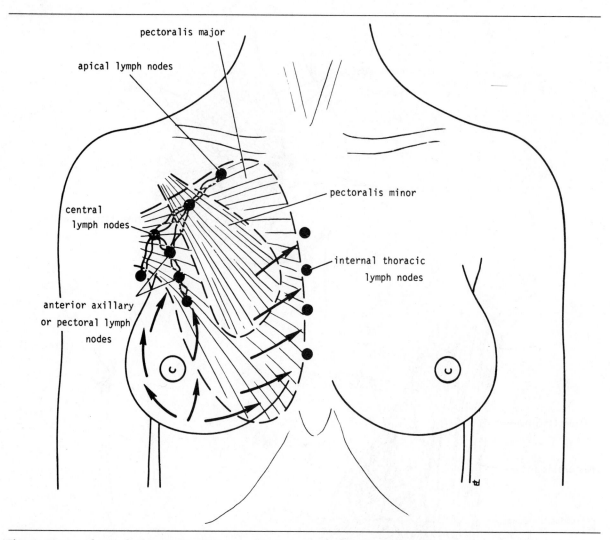

**Fig. 9-10. Lymphatic drainage of mammary gland.**

gland is drained into nodes lying along the internal thoracic artery. A few lymph vessels follow the posterior intercostal arteries and drain posteriorly into the posterior intercostal nodes.

The superficial lymphatic plexus beneath the areola (*subareola plexus*), and the deep plexus in the fascia covering the pectoralis major (*submammary plexus*), are no longer regarded as especially important in the drainage of lymph from the mammary gland.

## Bones of the Shoulder Girdle and Arm

The shoulder girdle consists of the clavicle and the scapula, which articulate with one another at the acromioclavicular joint.

### CLAVICLE

The clavicle lies horizontally and articulates with the sternum and first costal cartilage medially and with the acromion process of the scapula laterally (Fig. 9-11). The clavicle acts as a strut, which holds

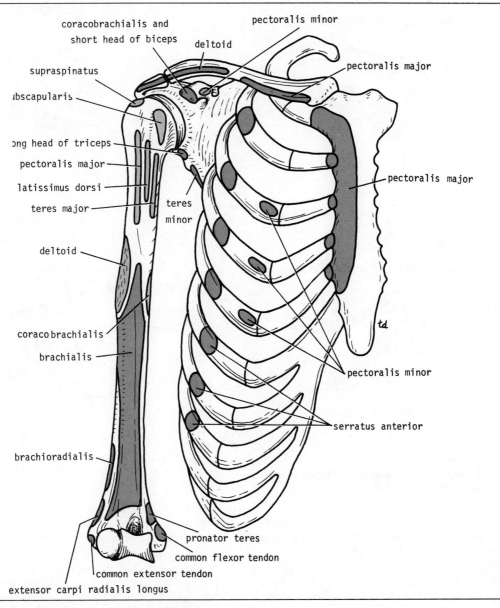

coracobrachialis and
short head of biceps

pectoralis minor

deltoid

supraspinatus

pectoralis major

ibscapularis

ong head of triceps

pectoralis major

latissimus dorsi

teres major

teres
minor

pectoralis major

deltoid

coracobrachialis

brachialis

td

pectoralis minor

serratus anterior

brachioradialis

pronator teres

common flexor tendon

common extensor tendon

extensor carpi radialis longus

**Fig. 9-11. Muscle attachments to bones of thorax,
clavicle, scapula, and humerus.**

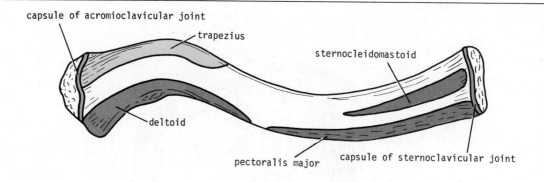

Superior surface

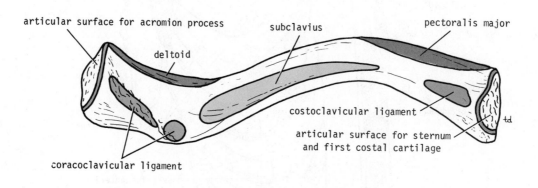

Inferior surface

**Fig. 9-12. Important muscular and ligamentous attachments to right clavicle.**

the arm away from the trunk. It also serves to transmit forces from the upper limb to the axial skeleton, and it provides attachment for muscles.

The clavicle is subcutaneous throughout its length; its medial two-thirds are convex forward and its lateral third, concave forward. The important muscles and ligaments that are attached to the clavicle are shown in Figure 9-12).

## SCAPULA

The scapula is a flat triangular bone (Fig. 9-5), which lies on the posterior thoracic wall between the second and the seventh ribs. On its posterior surface the *spine of the scapula* projects backward

(Fig. 9-13). The lateral end of the spine is free and forms the *acromion*, which articulates with the clavicle. The superolateral angle of the scapula forms the pear-shaped *glenoid cavity*, or *fossa*, which articulates with the head of the humerus. The *coracoid process* projects upward and forward above the glenoid cavity and provides attachment for muscles and ligaments. Medial to the base of the coracoid process is the *suprascapular notch* (Fig. 9-13).

The anterior surface of the scapula is concave and forms the shallow *subscapular fossa*. The posterior surface of the scapula is divided by the spine into the *supraspinous fossa* above and an *infraspinous fossa* below (Fig. 9-13). The *inferior angle* of the scapula can be palpated easily in the living

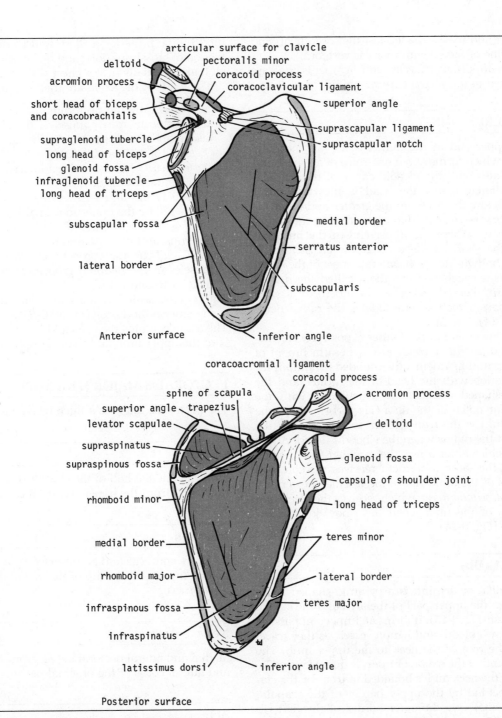

**Fig. 9-13. Important muscular and ligamentous attachments to right scapula.**

subject and marks the level of the seventh rib and the spine of the seventh thoracic vertebra.

The important muscles and ligaments that are attached to the scapula are shown in Figure 9-13.

## HUMERUS

The upper end of the humerus has a *head* (Fig. 9-14), which forms about one-third of a sphere and articulates with the glenoid cavity of the scapula. Immediately below the head is the *anatomical neck*. Below the neck are the *greater* and *lesser tuberosities*, separated from each other by the *bicipital groove*. Where the upper end of the humerus joins the shaft, there is a narrow *surgical neck*. About halfway down the lateral aspect of the shaft, there is a roughened elevation, called the *deltoid tuberosity*. Behind and below the tuberosity is a *spiral groove*, which accommodates the radial nerve (Fig. 9-14).

The lower end of the humerus possesses the *medial* and *lateral epicondyles* for the attachment of muscles and ligaments; the rounded *capitulum* for articulation with the head of the radius; and the pulley-shaped *trochlea* for articulation with the trochlear notch of the ulna (Fig. 9-14). Above the capitulum is the *radial fossa*, which receives the head of the radius when the elbow is flexed. Above the trochlea anteriorly is the *coronoid fossa*, which during the same movement receives the coronoid process of the ulna. Above the trochlea posteriorly is the *olecranon fossa*, which receives the olecranon process of the ulna when the elbow joint is extended (Fig. 9-14).

## The Axilla

The axilla, or armpit, is a pyramid-shaped space between the upper part of the arm and the side of the chest (Fig. 9-15). It forms an important passage for nerves, blood, and lymph vessels as they travel from the root of the neck to the upper limb. The upper end of the axilla, or *apex*, is directed into the root of the neck and is bounded in front by the clavicle, behind by the upper border of the scapula, and medially by the outer border of the first rib (Fig. 9-15). The lower end, or *base*, is bounded in front by the anterior axillary fold (formed by the lower border of the pectoralis major muscle), be-

hind by the posterior axillary fold (formed by the tendon of latissimus dorsi and the teres major muscle), and medially by the chest wall (Fig. 9-15).

The walls of the axilla are made up as follows:

*Anterior wall*, by the pectoralis major, subclavius, and pectoralis minor muscles, the clavipectoral fascia, and the suspensory ligament of the axilla (Figs. 9-16, 9-17, and 9-19).

*Posterior wall*, by the subscapularis, latissimus dorsi, and teres major muscles from above down (Figs. 9-17 through 9-19).

*Medial wall*, by the upper four or five ribs and the intercostal spaces covered by the serratus anterior muscle (Figs. 9-18 and 9-19).

*Lateral wall*, by the coracobrachialis and biceps muscles in the bicipital groove of the humerus (Figs. 9-18 and 9-19).

The *base* is formed by the skin stretching between the anterior and posterior walls (Fig. 9-19).

The axilla contains the principal vessels and nerves to the upper limb and many lymph nodes.

## PECTORALIS MAJOR (FIG. 9-16)

The pectoralis major is a thick triangular muscle.

### Origin

From the medial half of the clavicle, from the sternum, and from the upper six costal cartilages.

### Insertion

Its fibers converge and are inserted by a bilaminar tendon into the lateral lip of the bicipital groove of the humerus.

### Nerve Supply

Medial and lateral pectoral nerves from the medial and lateral cords of the brachial plexus.

### Action

It adducts the arm and rotates it medially; the clavicular fibers also flex the arm.

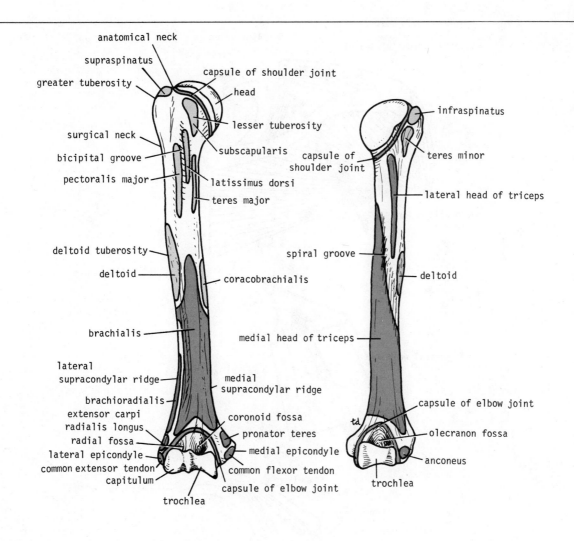

**Fig. 9-14. Important muscular and ligamentous attachments to right humerus.**

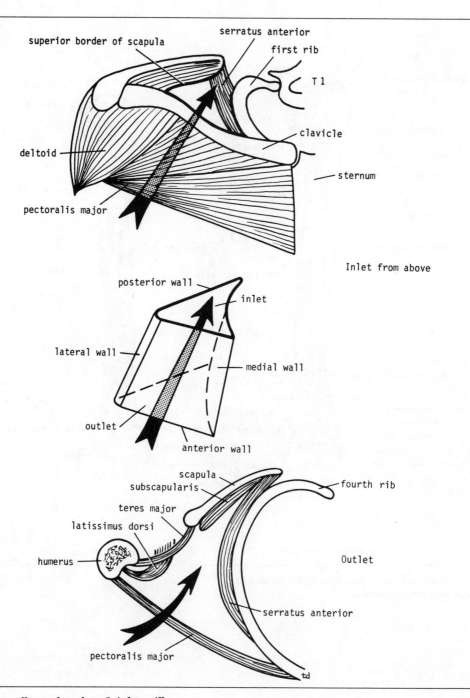

**Fig. 9-15. Inlet, walls, and outlet of right axilla.**

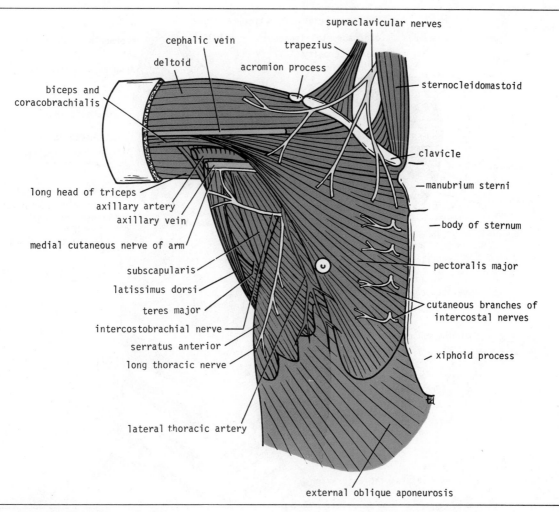

**Fig. 9-16. Pectoral region and axilla.**

## SUBCLAVIUS (FIG. 9-18)

### Origin

From the first costal cartilage.

### Insertion

Its fibers pass upward and laterally and are inserted into the groove on the inferior surface of the clavicle.

### Nerve Supply

The nerve to the subclavius from the upper trunk of the brachial plexus.

### Action

It depresses the clavicle and steadies this bone during movements of the shoulder girdle.

## PECTORALIS MINOR (FIG. 9-17)

The pectoralis minor is a thin triangular muscle.

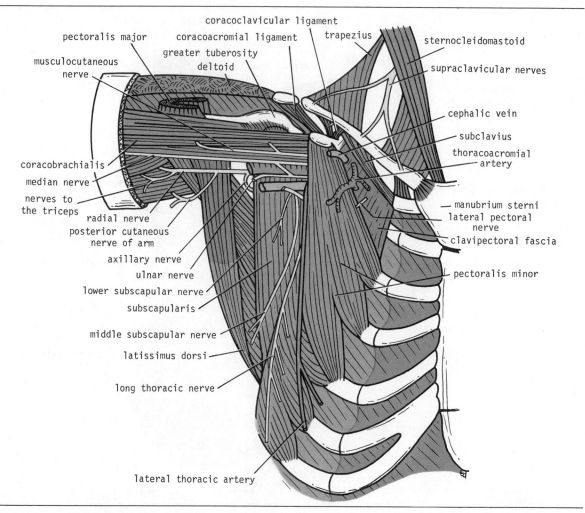

**Fig. 9-17. Pectoral region and axilla; pectoralis major muscle has been removed to display underlying structures.**

## Origin

From the third, fourth, and fifth ribs.

## Insertion

Its fibers converge to be inserted into the medial border of the coracoid process.

## Nerve Supply

From the medial pectoral nerve, a branch of the medial cord of the brachial plexus.

## Action

It pulls the shoulder downward and forward; if the shoulder is fixed, it elevates the ribs of origin.

## CLAVIPECTORAL FASCIA

The clavipectoral fascia is a strong sheet of connective tissue, which is split above to enclose the sub-

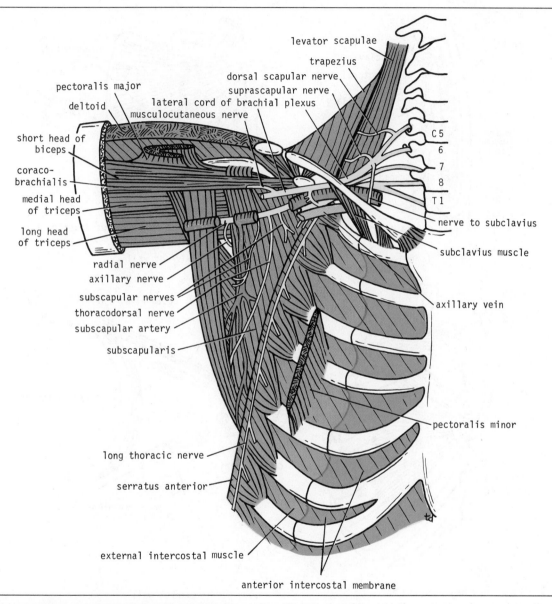

**Fig. 9-18. Pectoral region and axilla; pectoralis major and minor muscles and clavipectoral fascia have been removed to display underlying structures.**

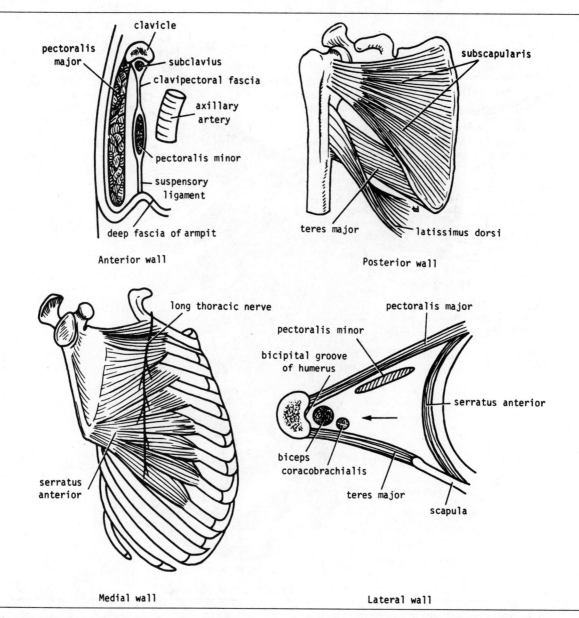

**Fig. 9-19. Various structures that form walls of axilla. The lateral wall is indicated by arrow.**

clavius muscle and is attached to the clavicle (Figs. 9-17 and 9-19). Below, it splits to enclose the pectoralis minor muscle and then continues downward as the *suspensory ligament of the axilla* and joins the fascial floor of the armpit.

### Function

It protects the contents of the axilla by filling in the interval between the clavicle and the pectoralis minor muscle. The suspensory ligament of the axilla is responsible for raising the skin of the armpit when the clavicle is elevated.

The clavipectoral fascia is pierced by (1) the cephalic vein in order that it may enter the axillary vein; (2) the thoracoacromial artery, a branch of the axillary artery; (3) lymphatic vessels from the infraclavicular nodes; and (4) the lateral pectoral nerve as it passes to the pectoralis major muscle (Fig. 9-17).

## SUBSCAPULARIS (FIGS. 9-18 AND 9-19)

### Origin

From the subscapular fossa on the anterior surface of the scapula.

### Insertion

Its fibers converge and are inserted on the lesser tuberosity of the humerus.

### Nerve Supply

The upper and lower subscapular nerves, branches of the posterior cord of the brachial plexus.

### Action

Medially rotates the arm and acts with the other short muscles around the shoulder joint in helping to stabilize this joint.

## LATISSIMUS DORSI (FIGS. 9-17 AND 9-26)

The latissimus dorsi is a large, flat, triangular muscle that extends over the lumbar region and the lower part of the thorax.

### Origin

From the posterior part of the iliac crest, the lumbar fascia, and the spines of the lower six thoracic vertebrae (deep to the trapezius), from the lower three or four ribs, and sometimes by a few fibers from the inferior angle of the scapula.

### Insertion

Its tendon wraps around the lower border of the teres major muscle and is inserted into the floor of the bicipital groove of the humerus.

### Nerve Supply

The thoracodorsal nerve, a branch of the posterior cord of the brachial plexus.

### Action

It extends, adducts, and medially rotates the arm.

## TERES MAJOR (FIGS. 9-16 AND 9-19)

### Origin

From the lower third of the posterior surface of the lateral border of the scapula.

### Insertion

Into the medial lip of the bicipital groove of the humerus.

### Nerve Supply

Lower subscapular nerve from the posterior cord of the brachial plexus.

### Action

It medially rotates and adducts the arm.

## SERRATUS ANTERIOR (FIGS. 9-16 AND 9-19)

The serratus anterior is a large, thin muscle that covers the lateral chest wall.

## Origin

From the outer surfaces of the upper eight ribs.

## Insertion

Into the anterior surface of the medial border of the scapula. A great part of this muscle is inserted in the region of the inferior angle.

## Nerve Supply

The long thoracic nerve, which arises from roots C5, 6, and 7 of the brachial plexus.

## Action

It draws the scapula forward around the thoracic wall and, because of the greater pull exerted on the inferior angle, rotates it so that the inferior angle passes laterally and forward and the glenoid cavity is raised upward and forward; in this action the muscle is assisted by the trapezius. This rotation of the scapula takes place when the arm is raised from the horizontal abducted position upward to a vertical position above the head. This muscle is also used when the arm is pushed forward in the horizontal position as in a forward punch.

The *biceps brachii* and the *coracobrachialis* muscles are described on pages 453 and 456.

## Contents of the Axilla

The axilla contains the axillary artery and its branches, which supply blood to the upper limb; the axillary vein and its tributaries, which drain blood from the upper limb; and lymph vessels and lymph nodes, which drain lymph from the upper limb and the mammary gland and from the skin of the trunk, down as far as the level of the umbilicus. Lying among these structures in the axilla is an important nerve plexus, the brachial plexus, which innervates the upper limb.

### AXILLARY ARTERY

The axillary artery (Figs. 9-16, 9-17, and 9-18) begins at the lateral border of the first rib as a continuation of the subclavian (Fig. 9-20), and ends at the lower border of the teres major muscle, where it continues as the brachial artery. Throughout its course, the artery is closely related to the cords of the brachial plexus and their branches and is enclosed with them in a connective tissue sheath, called the *axillary sheath*. If this sheath is traced upward into the root of the neck, it is seen to be continuous with the prevertebral fascia.

The pectoralis minor muscle crosses in front of the axillary artery and, for purposes of description, is said to divide it into three parts (Figs. 9-17, 9-18, and 9-20).

### *First Part of the Axillary Artery*

The first part of the axillary artery extends from the lateral border of the first rib to the upper border of the pectoralis minor (Fig. 9-20).

### *Relations*

#### Anteriorly

The pectoralis major and the covering fasciae and skin. The cephalic vein crosses the artery (Figs. 9-17 and 9-18).

#### Posteriorly

The long thoracic nerve (nerve to the serratus anterior) (Fig. 9-18).

#### Laterally

The three cords of the brachial plexus (Fig. 9-18).

#### Medially

The axillary vein (Fig. 9-18).

### *Second Part of the Axillary Artery*

The second part of the axillary artery lies behind the pectoralis minor muscle (Fig. 9-20).

### *Relations*

#### Anteriorly

The pectoralis minor, the pectoralis major, and the covering fasciae and skin (Figs. 9-17 and 9-20).

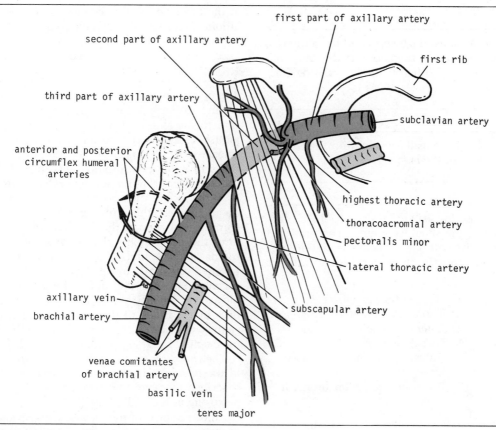

first part of axillary artery

second part of axillary artery

first rib

third part of axillary artery

subclavian artery

anterior and posterior
circumflex humeral
arteries

highest thoracic artery

thoracoacromial artery

pectoralis minor

lateral thoracic artery

axillary vein

brachial artery

subscapular artery

venae comitantes
of brachial artery

basilic vein

teres major

**Fig. 9-20. Different parts of axillary artery and its border of teres major muscle.** Note formation of axillary vein at lower border of teres major muscle.

## Posteriorly

The posterior cord of the brachial plexus and the subscapularis muscle (Fig. 9-18).

## Laterally

The lateral cord of the brachial plexus (Figs. 9-17 and 9-18).

## Medially

The medial cord of the brachial plexus and the axillary vein (Figs. 9-18 and 9-22).

## Third Part of the Axillary Artery

The third part of the axillary artery extends from the lower border of the pectoralis minor to the lower border of the teres major (Fig. 9-20).

## Relations

### Anteriorly

The pectoralis major for a short distance; lower down the artery is crossed by the medial root of the median nerve (Fig. 9-17).

### Posteriorly

The subscapularis, the latissimus dorsi, and the teres major. The axillary and radial nerves also lie behind the artery (Fig. 9-18).

## Laterally

The coracobrachialis, the biceps, and the humerus. The lateral root of the median and the musculocutaneous nerves also lie on the lateral side (Fig. 9-17).

## Medially

The ulnar nerve, the axillary vein, and the medial cutaneous nerve of the arm (Fig. 9-17).

### *Branches*

The branches of the axillary artery supply the thoracic wall and the shoulder region. The first part of the artery gives off one branch, the second part two branches, and the third part three branches (Fig. 9-20).

| | |
|---|---|
| *First Part* | (1) Highest thoracic artery |
| *Second Part* | (1) Thoracoacromial artery |
| | (2) Lateral thoracic artery |
| *Third Part* | (1) Subscapular artery |
| | (2) Anterior circumflex humeral artery |
| | (3) Posterior circumflex humeral artery |

The *highest thoracic artery* is small and runs along the upper border of the pectoralis minor. The *thoracoacromial artery* pierces the clavipectoral fascia and immediately divides into terminal branches. The *lateral thoracic artery* runs along the lower border of the pectoralis minor (Fig. 9-20). The subscapular artery runs along the lower border of the subscapularis muscle. The *anterior and posterior circumflex humeral arteries* wind around the front and the back of the surgical neck of the humerus, respectively (Fig. 9-20).

## AXILLARY VEIN

The axillary vein (Fig. 9-16) is formed in the region of the lower border of the teres major muscle by the union of the venae comitantes of the brachial artery and the basilic vein (Fig. 9-20). It runs upward on the medial side of the axillary artery and ends at the lateral border of the first rib by becoming the subclavian vein.

The vein receives tributaries, which correspond to the branches of the axillary artery, and, in addition, it receives the cephalic vein.

## BRACHIAL PLEXUS

The nerves entering the upper limb provide the following important functions: (1) sensory innervation to the skin and deep structures, such as the joints; (2) motor innervation to the muscles; (3) influence over the diameters of the blood vessels by the sympathetic vasomotor nerves; and (4) sympathetic secretomotor supply to the sweat glands.

At the root of the upper limb, the nerves that are about to enter the limb come together to form a complicated plexus, the *brachial plexus.* This allows the nerve fibers derived from different segments of the spinal cord to be arranged and distributed efficiently in different nerve trunks to the various parts of the upper limb. The brachial plexus is formed in the posterior triangle of the neck by the union of the anterior rami of the fifth, sixth, seventh, and eighth cervical and the first thoracic spinal nerves (Fig. 9-21 and 9-22).

The *roots, trunks,* and *divisions* of the plexus lie in the posterior triangle of the neck and are fully described on page 718. The cords of the brachial plexus lie in the axilla (Fig. 9-22).

All three cords of the plexus lie above and lateral to the first part of the axillary artery (Figs. 9-18 and 9-23). The medial cord crosses behind the artery to reach the medial side of the second part of the artery (Fig. 9-23). The posterior cord lies behind the second part of the artery, and the lateral cord lies on the lateral side of the second part of the artery (Fig. 9-23). Thus the cords of the plexus have the relationship to the second part of the axillary artery that is indicated by their names.

The majority of the branches of the cords that form the main nerve trunks of the upper limb continue this relationship to the artery in its third part (Fig. 9-23).

The *branches* of the different parts of the brachial plexus (Fig. 9-21 and 9-24) are as follows:

| | |
|---|---|
| *Roots* | Dorsal scapular nerve (C5) |
| | Long thoracic nerve (C5, 6, and 7) |
| *Upper* | Nerve to subclavius (C5 and 6) |
| *Trunk* | Suprascapular nerve (supplies the supraspinatus and infraspinatus muscles) |

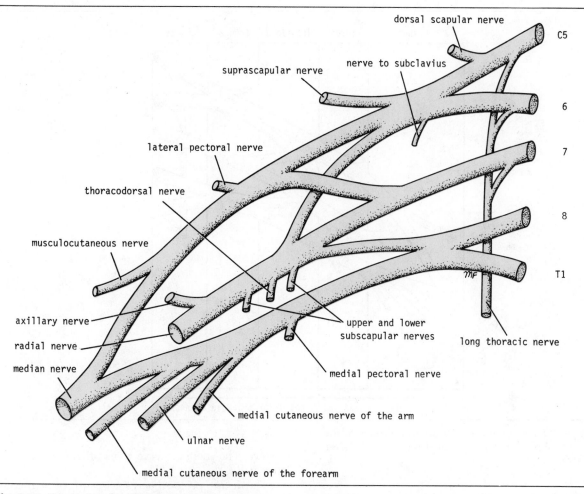

**Fig. 9-21. Roots, trunks, divisions, cords, and terminal branches of brachial plexus.**

| *Lateral Cord* | Lateral pectoral nerve |
| | Musculocutaneous nerve |
| | Lateral root of median nerve |
| *Medial Cord* | Medial pectoral nerve |
| | Medial cutaneous nerve of arm and medial cutaneous nerve of forearm |
| | Ulnar nerve |
| | Medial root of median nerve |
| *Posterior Cord* | Upper and lower subscapular nerves |
| | Thoracodorsal nerve |
| | Axillary nerve |
| | Radial nerve |

## Branches of the Brachial Plexus Found in the Axilla

The *nerve to the subclavius* (C5 and 6), having descended in front of the trunks of the brachial plexus and the subclavian artery in the neck, supplies the subclavius muscle (Figs. 9-18, 9-21, and 9-23). It is important clinically, since it may give a contribution (C5) to the phrenic nerve; this branch, when present, is referred to as the *accessory phrenic nerve.*

The *long thoracic nerve* (C5, 6, and 7) arises from the roots of the brachial plexus in the neck and enters the axilla by passing down over the lateral border of the first rib behind the axillary vessels and brachial plexus (Figs. 9-18 and 9-21). It descends

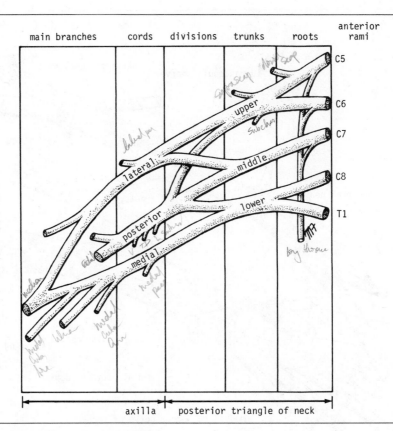

**Fig. 9-22. The formation of the main parts of the brachial plexus. Note the location of the different parts.**

over the lateral surface of the serratus anterior muscle, which it supplies.

The *lateral pectoral nerve* arises from the lateral cord of the brachial plexus, pierces the clavipectoral fascia, and supplies the pectoralis major muscle (Figs. 9-17 and 9-23).

The *musculocutaneous nerve* arises from the lateral cord of the brachial plexus, supplies the coracobrachialis muscle, and leaves the axilla by piercing that muscle (Figs. 9-17 and 9-23). A summary diagram of the complete distribution of the musculocutaneous nerve is given on page 533.

The *lateral root of the median nerve* is the direct continuation of the lateral cord of the brachial plexus (Figs. 9-17 and 9-23). It is joined by the medial root to form the median nerve trunk, and this passes downward on the lateral side of the axillary

artery. The median nerve gives off no branches in the axilla.

The *medial pectoral nerve* arises from the medial cord of the brachial plexus, supplies and pierces the pectoralis minor muscle, and supplies the pectoralis major muscle (Fig. 9-21).

The *medial cutaneous nerve of the arm* (T1) arises from the medial cord of the brachial plexus (Figs. 9-17 and 9-23) and is joined by the intercostal brachial nerve (lateral cutaneous branch of the second intercostal nerve). It supplies the skin on the medial side of the arm.

The *medial cutaneous nerve of the forearm* arises from the medial cord of the brachial plexus and descends in front of the axillary artery (Fig. 9-23).

The *ulnar nerve* (C8 and T1) arises from the medial cord of the brachial plexus and descends in the interval between the axillary artery and vein (Figs. 9-17 and 9-23). The ulnar nerve gives off no branches in the axilla. A summary diagram of the

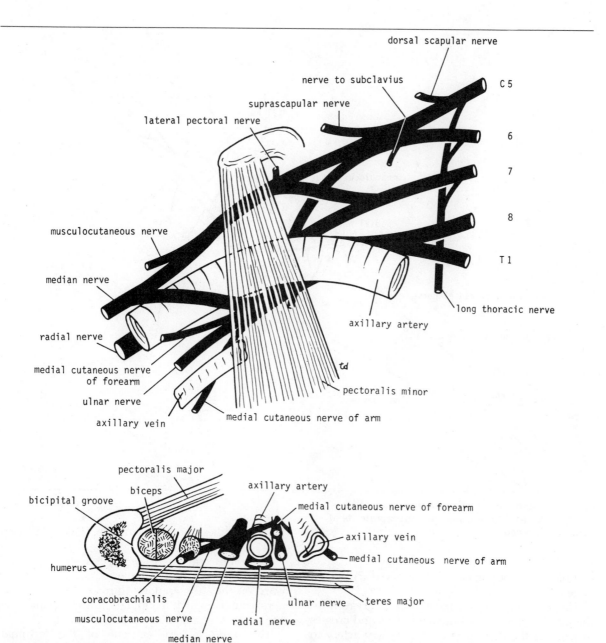

**Fig. 9-23. Relations of brachial plexus and its branches to axillary artery and vein. Lower diagram is section through the axilla at level of teres major muscle.**

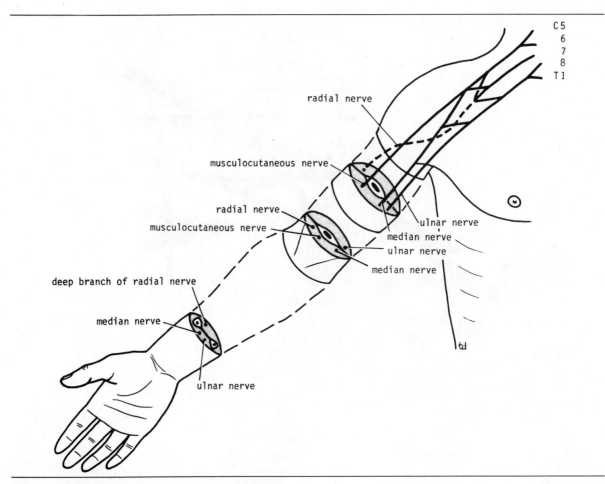

**Fig. 9-24. Distribution of main branches of brachial plexus to different fascial compartments of arm and forearm.**

complete distribution of the ulnar nerve is given on page 535.

The *medial root of the median nerve* arises from the medial cord of the brachial plexus and crosses in front of the third part of the axillary artery to join the lateral root of the median nerve (Figs. 9-17 and 9-23). A summary diagram of the complete distribution of the median nerve is given on page 533.

The *upper and lower subscapular nerves* arise from the posterior cord of the brachial plexus and supply the upper and lower parts of the subscapularis muscle. In addition, the lower subscapular

nerve supplies the teres muscle (Figs. 9-18 and 9-21).

The *thoracodorsal nerve* arises from the posterior cord of the brachial plexus and runs downward on the subscapularis to reach the latissimus dorsi muscle, which it supplies. It accompanies the subscapular vessels (Figs. 9-18 and 9-21).

The *axillary nerve* is one of the terminal branches of the posterior cord of the brachial plexus (Figs. 9-18 and 9-21). At the lower border of the subscapularis muscle, it turns backward and passes through the quadrilatereal space (see p. 439) in company with the posterior circumflex humeral artery. Having given off a branch to the shoulder joint, it divides into anterior and posterior branches. (See p. 440.) A summary diagram of the

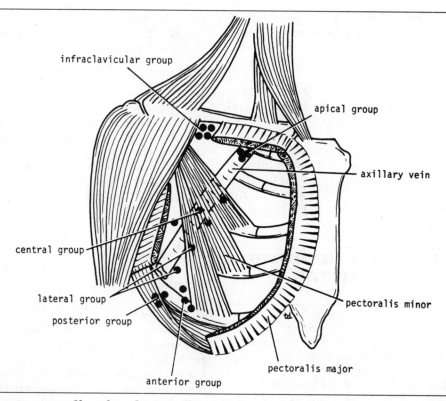

infraclavicular group

apical group

axillary vein

central group

lateral group

posterior group

pectoralis minor

anterior group

pectoralis major

**Fig. 9-25. Different groups of lymph nodes in axilla.**

complete distribution of the axillary nerve is given on page 529.

The *radial nerve* is the direct continuation of the posterior cord of the brachial plexus and lies behind the axillary artery. It is the largest branch of the brachial plexus (Figs. 9-18, 9-21, and 9-23). Before leaving the axilla, it gives off branches to the long and medial heads of the triceps muscle and the posterior cutaneous nerve of the arm (Fig. 9-17). The latter branch is distributed to the skin on the middle of the back of the arm. A summary diagram of the complete distribution of the radial nerve is given on page 530.

## LYMPH NODES OF THE AXILLA

The axillary lymph nodes (20 to 30 in number) drain lymphatic vessels from the lateral part of the breast, the superficial lymphatic vessels from the thoracoabdominal walls above the level of the umbilicus, and the vessels from the upper limb.

The lymph nodes are arranged in six groups (Fig. 9-25).

1. *Anterior (pectoral) group.* Lying along the lower border of the pectoralis minor behind the pectoralis major, these nodes receive lymph vessels from the lateral part of the breast and superficial vessels from the anterolateral abdominal wall above the level of the umbilicus.

2. *Posterior (subscapular) group.* Lying in front of the subscapularis muscle in association with the subscapular vessels, these nodes receive superficial lymph vessels from the back, down as far as the level of the iliac crests.

3. *Lateral group.* Lying along the medial side of the axillary vein, these nodes receive most of the lymph vessels of the upper limb (except those superficial vessels draining the lateral side—see infraclavicular nodes below).

4. *Central group.* Lying in the center of the axilla in the axillary fat, these nodes receive lymph from the above three groups.

5. *Infraclavicular (deltopectoral) group.* Lying on

the clavipectoral fascia in the deltopectoral triangle, these nodes receive superficial lymph vessels from the lateral side of the hand, forearm, and arm; the lymph vessels accompany the cephalic vein.

6. *Apical group.* Lying at the apex of the axilla at the lateral border of the first rib, these nodes receive the efferent lymph vessels from all the other axillary nodes.

The apical nodes drain into the *subclavian lymph trunk*. On the left side, this trunk drains into the thoracic duct and on the right side, into the right lymphatic trunk. Alternatively, the lymph trunks may drain directly into one of the large veins at the root of the neck.

## THE SUPERFICIAL PART OF THE BACK AND THE SCAPULAR REGION

### Skin

The *sensory nerve supply* to the skin of the back is from the posterior rami of the spinal nerves (Fig. 1-32). The first and eighth cervical nerves do not supply the skin, and the posterior rami of the upper three lumbar nerves run downward to supply the skin over the buttock.

The *blood supply* to the skin is from the posterior branches of the posterior intercostal arteries and the lumbar arteries. The *veins* correspond to the arteries and drain into the azygos veins and the inferior vena cava.

The *lymphatic drainage* of the skin and superficial fascia of the back above the level of the iliac crests is upward into the posterior or subscapular group of axillary lymph nodes.

### Muscles

#### TRAPEZIUS (FIG. 9-27)

The trapezius is a large, flat triangular muscle that extends over the back of the neck and thorax.

#### Origin

From the medial third of the superior nuchal line of the occipital bone (Fig. 9-26), the external occipital protuberance, and the ligamentum nuchae; from the spine of the seventh cervical vertebra and the spines and supraspinous ligaments of all the thoracic vertebrae.

#### Insertion

The upper fibers are directed downward and laterally into the lateral third of the clavicle; the middle fibers are directed horizontally into the acromion process and the upper border of the spine of the scapula; the lowest fibers are directed upward and laterally and are inserted on the medial end of the spine of the scapula.

#### Nerve Supply

Motor fibers from the spinal part of the accessory nerve and sensory fibers from the third and fourth cervical nerves.

#### Action

The trapezius muscle suspends the shoulder girdle from the skull and the vertebral column. The upper fibers elevate the scapula. The middle fibers pull the scapula medially. The lower fibers pull the medial border of the scapula downward so that the glenoid cavity faces upward and forward.

Knowing that the scapula rotates around the point of attachment of the coracoid process to the clavicle by the coracoclavicular ligament, it is easy to understand that the superior and inferior fibers of the trapezius assist the serratus anterior muscle in rotating the scapula when the arm is raised above the head (Fig. 9-33).

#### LATISSIMUS DORSI (FIG. 9-27)

The latissimus dorsi is a large, flat, triangular muscle that extends over the lumbar region and the lower part of the thorax. The latissimus dorsi muscle is considered on page 425.

#### LEVATOR SCAPULAE (FIGS. 9-27 AND 9-28)

#### Origin

From the transverse processes of the upper four cervical vertebrae.

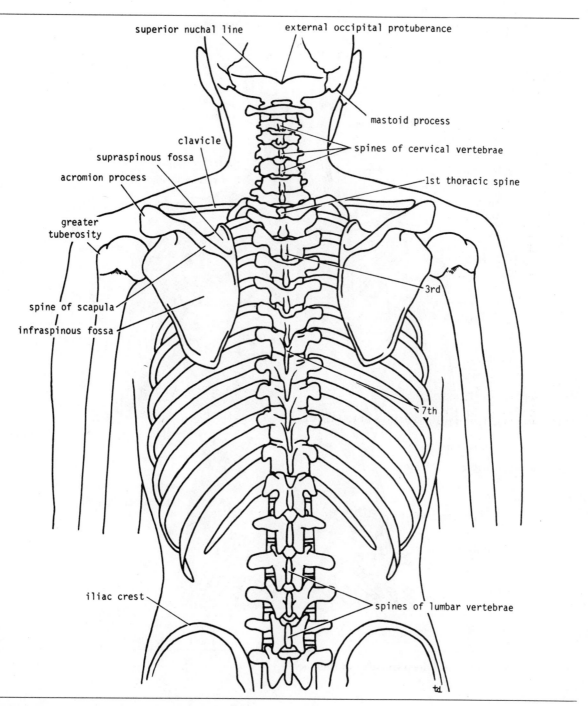

**Fig. 9-26. Bones of the back.**

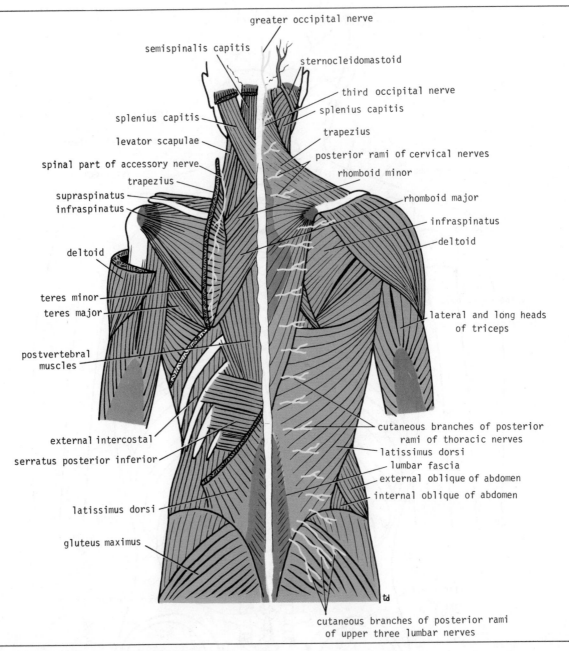

greater occipital nerve

semispinalis capitis

sternocleidomastoid

third occipital nerve

splenius capitis

splenius capitis

levator scapulae

trapezius

spinal part of accessory nerve

posterior rami of cervical nerves

trapezius

rhomboid minor

supraspinatus

rhomboid major

infraspinatus

infraspinatus

deltoid

deltoid

teres minor

teres major

lateral and long heads
of triceps

postvertebral
muscles

external intercostal

cutaneous branches of posterior
rami of thoracic nerves

serratus posterior inferior

latissimus dorsi

lumbar fascia

external oblique of abdomen

internal oblique of abdomen

latissimus dorsi

gluteus maximus

cutaneous branches of posterior rami
of upper three lumbar nerves

**Fig. 9-27. Superficial and deep muscles of the back.**

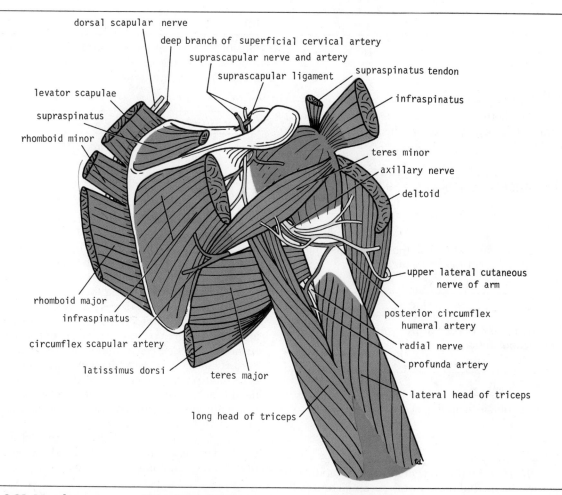

dorsal scapular nerve

deep branch of superficial cervical artery

suprascapular nerve and artery

suprascapular ligament

supraspinatus tendon

levator scapulae

infraspinatus

supraspinatus

rhomboid minor

teres minor

axillary nerve

deltoid

rhomboid major

infraspinatus

circumflex scapular artery

latissimus dorsi

teres major

long head of triceps

upper lateral cutaneous nerve of arm

posterior circumflex humeral artery

radial nerve

profunda artery

lateral head of triceps

**Fig. 9-28. Muscles, nerves, and blood vessels of scapular region. Note close relationship of axillary nerve to shoulder joint.**

### Insertion

Into the medial border of the scapula opposite the supraspinous fossa.

### Nerve Supply

From the third and fourth cervical nerves and from the dorsal scapular nerve (C5).

### Action

It raises the medial border of the scapula. When it acts in conjunctiuon with the middle fibers of the trapezius and with the rhomboids, it pulls the scapula medially and upward, i.e., braces the shoulder backward.

### RHOMBOID MINOR (FIGS. 9-27 AND 9-28)

### Origin

From the lower part of the ligamentum nuchae and the spines of the seventh cervical and first thoracic vertebrae.

### Insertion

Into the medial border of the scapula opposite the root of the spine.

## Nerve Supply

From the dorsal scapular nerve (C5).

## Action

With the rhomboid major and levator scapulae, it elevates the medial border of the scapula and pulls it medially.

## RHOMBOID MAJOR (FIGS. 9-27 AND 9-28)

### Origin

From the second to the fifth thoracic spines and the corresponding supraspinous ligaments.

### Insertion

Into the medial border of the scapula opposite the infraspinous fossa.

### Nerve Supply

From the dorsal scapular nerve (C5).

### Action

With the rhomboid minor and levator scapulae, it elevates the medial border of the scapula and pulls it medially.

## DELTOID (FIGS. 9-16, 9-27, AND 9-28)

The deltoid muscle is thick and triangular in shape and covers the shoulder joint. It forms the rounded contour of the shoulder.

### Origin

*Anterior fibers* arise from the lateral third of the anterior border of the clavicle. *Middle fibers* arise from the lateral border of the acromion process. *Posterior fibers* arise from the lower border of the spine of the scapula.

## Insertion

Its fibers converge to be inserted into the deltoid tuberosity, on the middle of the lateral surface of the shaft of the humerus.

## Nerve Supply

From the axillary nerve (C5 and 6).

## Action

With the help of the supraspinatus muscle, the deltoid abducts the upper limb at the shoulder joint. The main effort is undertaken by the strong multipennate middle (acromial) fibers; the weaker anterior and posterior fibers serve as stays and prevent the arm from swaying forward or backward. Note that for every 3 degrees of abduction of the arm, a 2-degree abduction occurs in the shoulder joint and 1 degree occurs by rotation of the scapula. At about 120 degrees of abduction the greater tuberosity of the humerus hits the lateral edge of the acromion process. Elevation of the arm above the head is accomplished by rotating the scapula. This is brought about by the contraction of the trapezius and serratus anterior muscles.

In addition, the anterior fibers of the deltoid can flex and medially rotate the arm, and the posterior fibers can extend and laterally rotate the arm.

## SUPRASPINATUS (FIGS. 9-27 AND 9-28)

### Origin

From the supraspinous fossa of the scapula.

### Insertion

Into the upper facet of the greater tuberosity of the humerus and into the capsule of the shoulder joint.

### Nerve Supply

Suprascapular nerve.

## Action

It assists the deltoid muscle in the movement of abduction of the arm at the shoulder joint by fixing the head of the humerus against the glenoid cavity.

## INFRASPINATUS (FIGS. 9-27 AND 9-28)

### Origin

From the infraspinous fossa of the scapula.

### Insertion

Into the middle facet of the greater tuberosity of the humerus and into the capsule of the shoulder joint.

### Nerve Supply

Suprascapular nerve.

### Action

It laterally rotates the arm and stabilizes the shoulder joint.

## TERES MINOR (FIGS. 9-27 AND 9-28)

### Origin

From the upper two-thirds of the posterior surface of the lateral border of the scapula.

### Insertion

Into the lower facet of the greater tuberosity of the humerus and into the capsule of the shoulder joint.

### Nerve Supply

A branch of the axillary nerve.

### Action

It laterally rotates the arm and stabilizes the shoulder joint.

## TERES MAJOR (FIGS. 9-19, 9-27, AND 9-28)

The teres major is considered on page 425.

## SUBSCAPULARIS (FIGS. 9-18 AND 9-19)

The subscapularis muscle is considered on page 425.

## Rotator Cuff

The four muscles, the supraspinatus, the infraspinatus, the teres minor, and the subscapularis form what is termed the *rotator cuff.* The tone of these muscles assists in holding the head of the humerus in the glenoid cavity of the scapula during movements at the shoulder joint. Therefore, they assist in stabilizing the shoulder joint. Note that the cuff lies on the anterior, superior, and posterior aspects of the joint. The cuff is deficient inferiorly and this is a site of potential weakness.

## Anatomical Spaces

Two potential intermuscular spaces, the quadrilateral and triangular spaces, are found in the shoulder region.

### Quadrilateral Space

The quadrilateral space is bounded above by the subscapularis in front and the teres minor behind, and between these two muscles is the capsule of the shoulder joint. It is bounded below by the teres major muscle. The space is bounded medially by the long head of the triceps and laterally by the surgical neck of the humerus.

The axillary nerve and the posterior circumflex humeral vessels pass backward through this space (Fig. 9-28).

### Triangular Space

The triangular space is bounded above by the teres minor, below by the teres major, and laterally by the long head of the triceps muscle (Fig. 9-28).

The circumflex scapular artery, a branch of the subscapular artery, passes backward through this space to enter the infraspinous fossa.

## Spinal Part of the Accessory Nerve

The spinal part of the accessory nerve runs downward in the posterior triangle of the neck on the levator scapulae muscle. It is accompanied by branches from the anterior rami of the third and fourth cervical nerves. The accessory nerve runs beneath the anterior border of the trapezius muscle (Fig. 9-27) at the junction of its middle and lower thirds, and together with the cervical nerves, supplies the trapezius muscle.

## Suprascapular Nerve

The suprascapular nerve arises from the upper trunk of the brachial plexus (C5 and 6) in the posterior triangle in the neck. It runs downward and laterally and passes beneath the *suprascapular ligament*, which bridges the suprascapular notch, to reach the supraspinous fossa (Fig. 9-28). It supplies the supraspinatus and infraspinatus muscles and the shoulder joint.

## Axillary Nerve

The axillary nerve arises from the posterior cord of the brachial plexus (C5 and 6) in the axilla. (See p. 432.) It passes backward around the lower border of the subscapularis muscle and enters the quadrilateral space with the posterior circumflex humeral artery (Fig. 9-28). As the nerve passes through the space, it comes into close relationship with the inferior aspect of the capsule of the shoulder joint and with the medial side of the surgical neck of the humerus. It terminates by dividing into anterior and posterior branches (Fig. 9-28).

### Branches

The axillary nerve has the following branches:

1. An *articular branch* to the shoulder joint.
2. An *anterior terminal branch*, which winds around the surgical neck of the humerus beneath the deltoid muscle; it supplies the deltoid and the skin that covers its lower part.
3. A *posterior terminal branch*, which gives off a *branch to the teres minor muscle* and a few branches to the deltoid, then emerges from the posterior border of the deltoid as the *superior* or *upper lateral cutaneous nerve of the arm* (Fig. 9-28).

It is thus seen that the axillary nerve supplies the shoulder joint, two muscles, and the skin covering the lower half of the deltoid muscle.

## The Arterial Anastomosis Around the Shoulder Joint

The extreme mobility of the shoulder joint may result in kinking of the axillary artery and a temporary occlusion of its lumen. To compensate for this, an important arterial anastomosis exists between the branches of the first part of the subclavian artery and the third part of the axillary artery, thus ensuring that an adequate blood flow takes place into the upper limb irrespective of the position of the arm (Fig. 9-29).

### Branches from the Subclavian Artery

1. The *suprascapular artery*, which is distributed to both the supraspinous and infraspinous fossae of the scapula.
2. The *superficial cervical artery*, which gives off a deep branch that runs down the medial border of the scapula in company with the dorsal scapular nerve.

### Branches from the Axillary Artery

1. The *subscapular artery* and its circumflex scapular branch supply the subscapular and infraspinous fossae of the scapula, respectively.
2. The *anterior circumflex humeral artery*.
3. The *posterior circumflex humeral artery*.

Both the circumflex arteries form an anastomosing circle around the surgical neck of the humerus (Fig. 9-29).

## Sternoclavicular Joint (Fig. 9-30)

### Articulation

This occurs between the sternal end of the clavicle, the manubrium sterni, and the first costal cartilage.

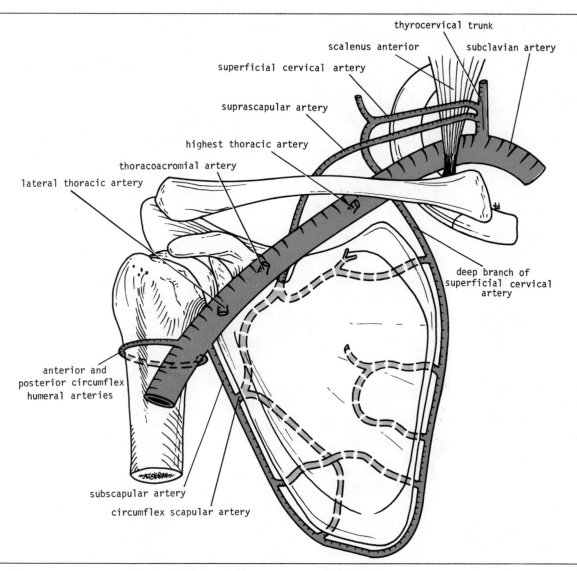

**Fig. 9-29. Arteries that take part in anastomosis around shoulder joint.**

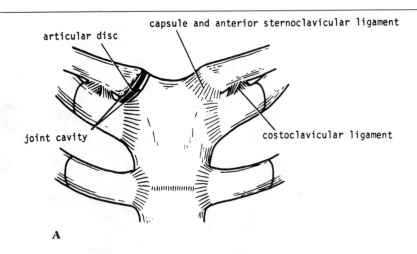

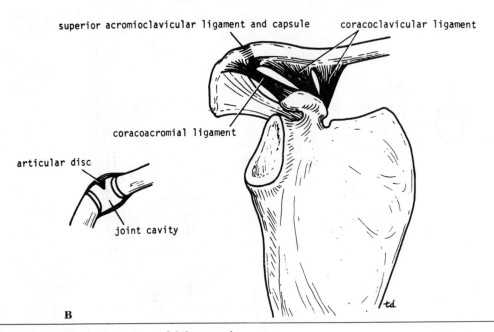

Fig. 9-30. (A) Sternoclavicular joint and (B) acromioclavicular joint.

## Type

It is a synovial, double plane joint.

## Capsule

This surrounds the joint and is attached to the margins of the articular surfaces.

## Ligaments

The capsule is reinforced in front of and behind the joint by the strong *sternoclavicular ligaments*.

The *articular disc* of the sternoclavicular joint is a flat fibrocartilaginous disc, which lies within the joint and divides the joint's interior into two compartments (Fig. 9-30). Its circumference is attached to the interior of the capsule, but it is also strongly attached to the superior margin of the articular surface of the clavicle above to the first costal cartilage below.

## Accessory Ligament

The *costoclavicular ligament* is a strong ligament, which runs from the junction of the first rib with the first costal cartilage to the inferior surface of the sternal end of the clavicle (Fig. 9-30).

## Synovial Membrane

This lines the capsule and is attached to the margins of the cartilage covering the articular surfaces.

## Nerve Supply

The supraclavicular nerve and the nerve to the subclavius muscle.

## Movements

Forward and backward movement of the clavicle takes place in the medial compartment. Elevation and depression of the clavicle takes place in the lateral compartment.

## Muscles Producing Movement

The forward movement of the clavicle is produced by the serratus anterior muscle. The backward movement is produced by the trapezius and rhomboid muscles. Elevation of the clavicle is produced by the trapezius, sternocleidomastoid, levator scapulae, and rhomboid muscles. Depression of the clavicle is produced by the pectoralis minor and the subclavius muscles.

### *Important Relations*

### Anteriorly

The skin and some fibers of the sternocleidomastoid and pectoralis major muscles.

### Posteriorly

The sternohyoid muscle; on the right, the brachiocephalic artery; on the left, the left brachiocephalic vein and the left common carotid artery.

## Acromioclavicular Joint (Fig. 9-30)

### Articulation

This occurs between the acromion process of the scapula and the lateral end of the clavicle.

### Type

This is a synovial plane joint.

### Capsule

This surrounds the joint and is attached to the margins of the articular surfaces.

### Ligaments

*Superior* and *inferior acromioclavicular ligaments* reinforce the capsule; from the capsule, a wedge-shaped *fibrocartilaginous disc* projects into the joint cavity from above (Fig. 9-30).

### Accessory Ligament

The very strong *coracoclavicular ligament* extends from the coracoid process to the undersurface of the clavicle (Fig. 9-30). It is largely responsible for suspending the weight of the scapula and the upper limb from the clavicle.

## Synovial Membrane

This lines the capsule and is attached to the margins of the cartilage covering the articular surfaces.

## Nerve Supply

The suprascapular nerve.

## Movements

A gliding movement takes place when the scapula rotates, or when the clavicle is elevated or depressed.

### *Important Relations*

### Anteriorly

The deltoid muscle.

### Posteriorly

The trapezius muscle.

### Superiorly

The skin.

# Shoulder Joint (Fig. 9-31)

## Articulation

This occurs between the rounded head of the humerus and the shallow, pear-shaped glenoid cavity of the scapula. The articular surfaces are covered by hyaline articular cartilage, and the glenoid cavity is deepened by the presence of a fibrocartilaginous rim called the *glenoid labrum* (Figs. 9-31 and 9-32).

## Type

It is a synovial ball-and-socket joint.

## Capsule

This surrounds the joint and is attached medially to the margin of the glenoid cavity outside the labrum; laterally, it is attached to the anatomical neck of the humerus, extending downward onto the medial side of the shaft for a short distance (Fig. 9-32).

The capsule is thin and lax, allowing a wide range of movement. It is strengthened by fibrous slips from the tendons of the subscapularis, supraspinatus, infraspinatus, and teres minor muscles (the rotator cuff muscles).

## Ligaments

The *glenohumeral ligaments* are three weak bands of fibrous tissue that strengthen the front of the capsule.

The *transverse humeral ligament* strengthens the capsule and bridges the gap between the two tuberosities (Fig. 9-31).

The *coracohumeral ligament* strengthens the capsule above and stretches from the root of the coracoid process to the greater tuberosity of the humerus (Fig. 9-31).

## Accessory Ligaments

The *coracoacromial ligament* extends between the coracoid process and the acromion. Its function is to protect the superior aspect of the joint (Fig. 9-31).

## Synovial Membrane

This lines the capsule and is attached to the margins of the cartilage covering the articular surfaces (Figs. 9-31 and 9-32). It surrounds the tendon of the biceps and extends beyond the transverse humeral ligament for a short distance as a tubular sheath around the tendon of the long head of the biceps brachii. It protrudes forward through the anterior wall of the capsule to form a bursa, which lies beneath the subscapularis muscle (Fig. 9-31).

## Nerve Supply

The axillary and suprascapular nerves.

## Movements

The shoulder joint has a wide range of movement, and the stability of the joint has been sacrificed to permit this. (Compare with the hip joint, which is

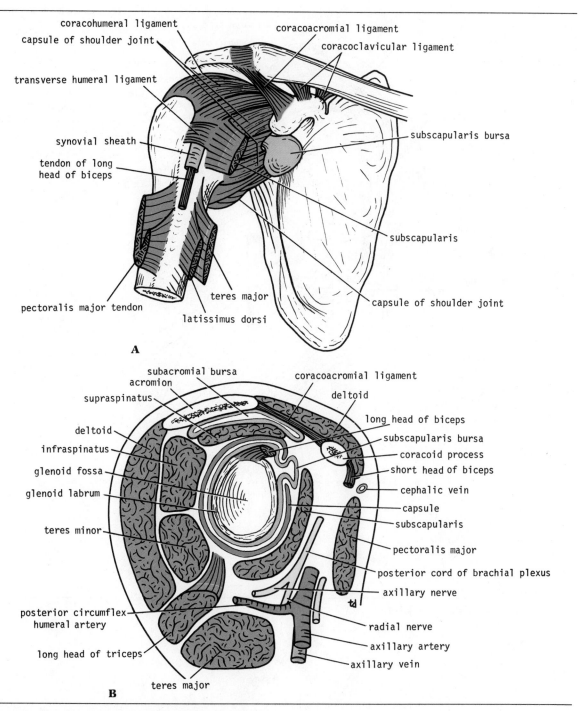

**Fig. 9-31. Shoulder joint, showing its relations. (A)
Anterior view and (B) sagittal section.**

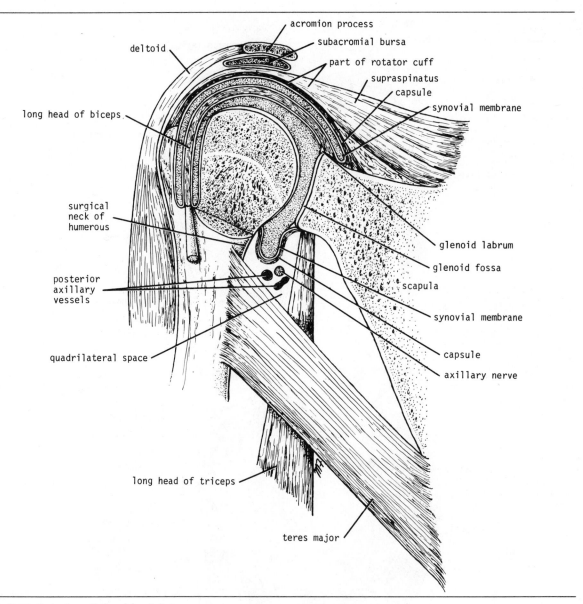

**Fig. 9-32. Interior of shoulder joint.**

stable, but limited in its movements.) The strength of the joint depends on the tone of the short rotator cuff muscles that cross in front, above, and behind the joint, namely, the subscapularis, supraspinatus, infraspinatus, and teres minor. When the joint is abducted, the lower surface of the head of the humerus is supported by the long head of the triceps, which bows downward because of its length and gives little actual support to the humerus. In addition, the inferior part of the capsule is the weakest area.

The following movements are possible:

*Flexion* is performed by the anterior fibers of the deltoid, pectoralis major, biceps, and coracobrachialis muscles.

*Extension* is performed by the posterior fibers of the deltoid, latissimus dorsi, and teres major muscles.

*Abduction* is performed by the middle fibers of the deltoid, assisted by the supraspinatus muscle (Fig. 9-33).

*Adduction* is performed by the pectoralis major, latissimus dorsi, teres major, and teres minor muscles.

*Lateral rotation* is performed by the infraspinatus, the teres minor, and the posterior fibers of the deltoid muscles.

*Medial rotation* is performed by the subscapularis, latissimus dorsi, and teres major muscles.

*Circumduction* is a combination of these movements.

## *Important Relations*

### Anteriorly

The subscapularis muscle and the axillary vessels and brachial plexus.

### Posteriorly

The infraspinatus and teres minor muscles.

### Superiorly

The supraspinatus muscle, subacromial bursa, coracoacromial ligament, and deltoid muscle.

### Inferiorly

The long head of the triceps muscle, the axillary nerve, and the posterior circumflex humeral vessels.

The tendon of the long head of the biceps muscle passes through the joint and emerges beneath the transverse ligament.

## *Muscles—Their Nerve Supply and Action*

Students wishing to review the muscles discussed so far should study Tables 9-1, 9-2, and 9-3.

## THE UPPER ARM

### Skin

The *sensory nerve supply* (Fig. 9-34) to the skin over the point of the shoulder to halfway down the deltoid muscle is from the *supraclavicular nerves* (C3 and 4). The skin over the lower half of the deltoid is supplied by the *upper lateral cutaneous nerve of the arm*, a branch of the axillary nerve (C5 and 6). The skin over the lateral surface of the arm below the deltoid is supplied by the *lower lateral cutaneous nerve of the arm*, a branch of the radial nerve (C5 and 6). The skin of the armpit and the medial side of the arm is supplied by the *medial cutaneous nerve of the arm* (T1) and the *intercostobrachial nerves* (T2). The skin of the back of the arm (Fig. 9-34) is supplied by the *posterior cutaneous nerve of the arm*, a branch of the radial nerve (C8).

The *superficial veins* of the arm (Fig. 9-8) lie in the superficial fascia.

The *cephalic vein* ascends in the superficial fascia on the lateral side of the biceps and, on reaching the infraclavicular fossa, pierces the clavipectoral fascia to drain into the axillary vein.

The *basilic vein* ascends in the superficial fascia on the medial side of the biceps (Fig. 9-8). Halfway up the arm, it pierces the deep fascia and at the lower border of the teres major joins the venae comitantes of the brachial artery to form the axillary vein.

The *superficial lymphatic vessels* draining the superficial tissues of the upper arm pass upward to the axilla (Fig. 9-35). Those from the lateral side of the arm follow the cephalic vein to the infraclavicular group of nodes; those from the medial side

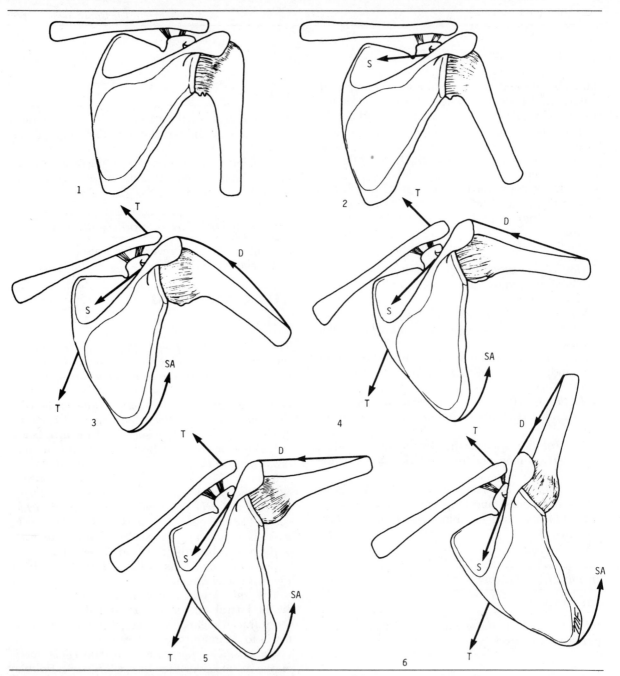

**Fig. 9-33. Movements of abduction of shoulder joint and rotation of scapula and the muscles producing these movements. Note that for every 3 degrees of abduction of the arm, a 2-degree abduction occurs in the shoulder joint, and 1 degree occurs by rotation of the scapula. At about 120 degrees of abduction the greater tuberosity of the humerus hits the lateral edge of the acromion process. Elevation of the arm above the head is accomplished by rotating the scapula. S = supraspinatus, D = deltoid, T = trapezius, and SA = serratus anterior.**

Table 9-1. Muscles Connecting the Upper Limb to the Thoracic Wall

| Name of muscle | Origin | Insertion | Nerve supply | Action |
|---|---|---|---|---|
| Pectoralis major | Clavicle, sternum, and upper six costal cartilages | Lateral lip of bicipital groove of humerus | Medial and lateral pectoral nerves from brachial plexus | Adducts arm and rotates it medially; clavicular fibers also flex arm |
| Pectoralis minor | Third, fourth, and fifth ribs | Coracoid process of scapula | Medial pectoral nerve from brachial plexus | Depresses point of shoulder; if the scapula is fixed, it elevates the ribs of origin |
| Subclavius | First costal cartilage | Clavicle | Nerve to subclavius from upper trunk of brachial plexus | Depresses the clavicle and steadies this bone during movements of the shoulder girdle |
| Serratus anterior | Upper eight ribs | Medial border and inferior angle of scapula | Long thoracic nerve | Draws the scapula forward around the thoracic wall; rotates scapula |

Table 9-2. Muscles Connecting the Upper Limb to the Vertebral Column

| Name of muscle | Origin | Insertion | Nerve supply | Action |
|---|---|---|---|---|
| Trapezius | Occipital bone, ligamentum nuchae, spine of seventh cervical vertebra, spines of all thoracic vertebrae | Upper fibers into lateral third of clavicle; middle and lower fibers into acromion process and spine of scapula | Accessory nerve and C3 and 4 | Upper fibers elevate the scapula; middle fibers pull scapula medially; lower fibers pull medial border of scapula downward |
| Latissimus dorsi | Iliac crest, lumbar fascia, spines of lower six thoracic vertebrae, lower three or four ribs, and inferior angle of scapula | Floor of bicipital groove of humerus | Thoracodorsal nerve | Extends, adducts, and medially rotates the arm |
| Levator scapulae | Transverse processes of first four cervical vertebrae | Medial border of scapula | C3 and 4 and dorsal scapular nerve | Raises medial border of scapula |
| Rhomboid minor | Ligamentum nuchae and spines of seventh cervical and first thoracic vertebrae | Medial border of scapula | Dorsal scapular nerve | Raises medial border of scapula upward and medially |
| Rhomboid major | Second to fifth thoracic spines | Medial border of scapula | Dorsal scapular nerve | Raises medial border of scapula upward and medially |

Table 9-3. Muscles Connecting the Scapula to the Humerus

| Name of muscle | Origin | Insertion | Nerve supply | Action |
|---|---|---|---|---|
| Deltoid | Lateral third of clavicle, acromion process, spine of scapula | Middle of lateral surface of shaft of humerus | Axillary nerve | Abducts arm; anterior fibers flex and medially rotate arm; posterior fibers extend and laterally rotate arm |
| Supraspinatus | Supraspinous fossa of scapula | Greater tuberosity of humerus; capsule of shoulder joint | Suprascapular nerve | Abducts arm and stabilizes shoulder joint |
| Infraspinatus | Infraspinous fossa of scapula | Greater tuberosity of humerus; capsule of shoulder joint | Suprascapular nerve | Laterally rotates arm and stabilizes shoulder joint |
| Teres major | Lower third lateral border of scapula | Medial lip of bicipital groove of humerus | Lower subscapular nerve | Medially rotates and adducts arm and stabilizes shoulder joint |
| Teres minor | Upper two-thirds lateral border of scapula | Greater tuberosity of humerus; capsule of shoulder joint | Axillary nerve | Laterally rotates arm and stabilizes shoulder joint |
| Subscapularis | Subscapular fossa | Lesser tuberosity of humerus | Upper and lower subscapular nerves | Medially rotates the arm and stabilizes the shoulder joint |

follow the basilic vein to the lateral group of axillary nodes.

The *deep lymphatic vessels* draining the muscles and deep structures of the arm drain into the lateral group of axillary nodes.

## Fascial Compartments of the Upper Arm

The upper arm is enclosed in a sheath of deep fascia (Fig. 9-36). Two fascial septa, one on the medial and one on the lateral side, extend from this sheath and are attached to the medial and lateral supracondylar ridges of the humerus, respectively. By this means the upper arm is divided into an anterior and a posterior fascial compartment, each having its muscles, nerves, and arteries.

## Contents of the Anterior Fascial Compartment of the Upper Arm

### Muscles

Biceps brachii, coracobrachialis, and brachialis.

### Blood Supply

Brachial artery (Fig. 9-37).

### Nerve Supply to the Muscles

Musculocutaneous nerve.

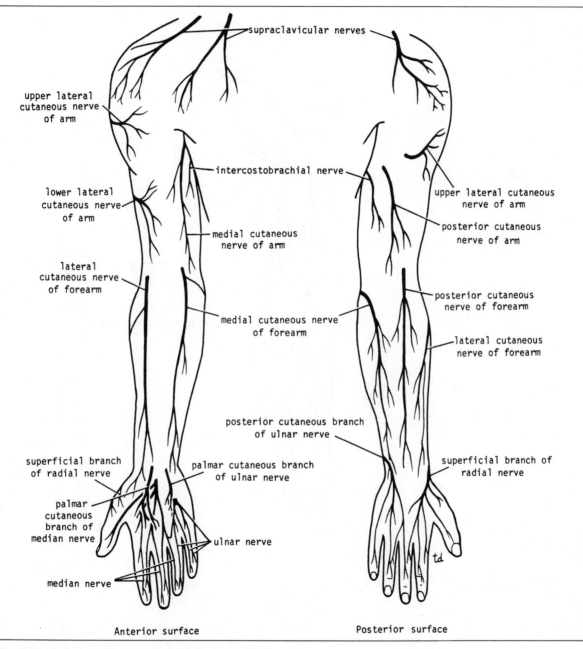

**Fig. 9-34. Cutaneous innervation of the upper limb.**

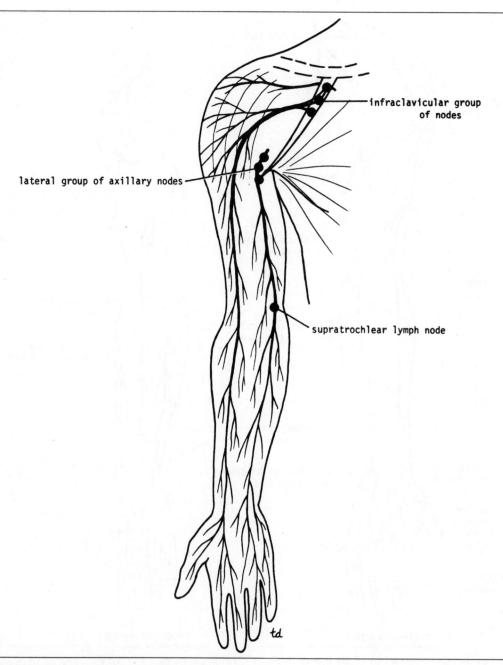

**Fig. 9-35. Superficial lymphatics of upper limb.**
**Note positions of lymph nodes.**

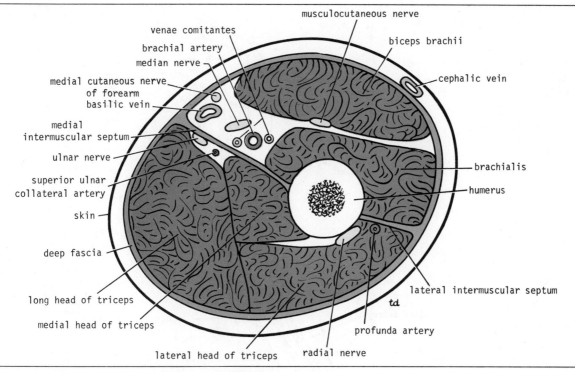

**Fig. 9-36. Cross section of upper arm just below level of insertion of deltoid muscle. Note division of arm by humerus and medial and lateral intermuscular septa into anterior and posterior compartments.**

## Structures Passing Through the Compartment

Musculocutaneous, median, and ulnar nerves; brachial artery and basilic vein. The radial nerve is present in the lower part of the compartment.

## BICEPS BRACHII (FIG. 9-38)

### Origin

The *long head* from the supraglenoid tubercle of the scapula; the *short head* from the tip of the coracoid process of the scapula.

The tendon of the long head crosses the humeral head within the capsule of the shoulder joint and emerges from the joint in the bicipital groove of the humerus. It is joined in the middle of the upper arm by the short head.

### Insertion

Into the posterior part of the tuberosity of the radius, and by an aponeurotic band called the *bicipital aponeurosis*, into the deep fascia on the medial aspect of the forearm. The aponeurosis protects underlying structures present in the cubital fossa.

### Nerve Supply

Musculocutaneous nerve.

### Action

The biceps is a strong supinator of the forearm. Corkscrews and the threads of screws are designed to make use of this powerful supinator action in twisting the corkscrew into the cork or driving the screw into wood with a screwdriver. The biceps also is a powerful flexor of the elbow joint and a weak flexor of the shoulder joint.

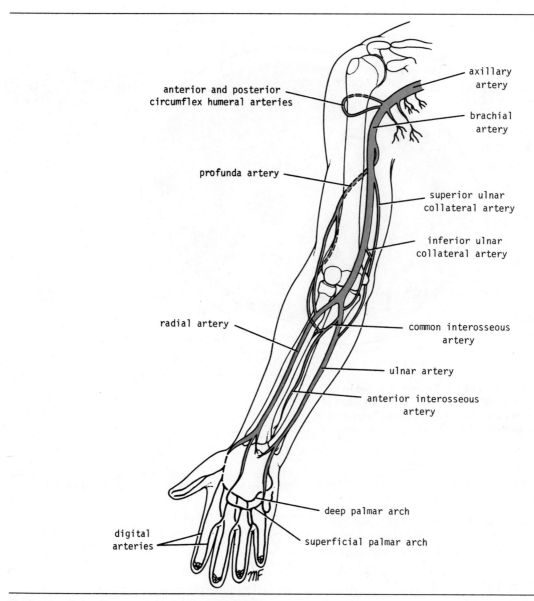

anterior and posterior
circumflex humeral arteries

axillary
artery

brachial
artery

profunda artery

superior ulnar
collateral artery

inferior ulnar
collateral artery

radial artery

common interosseous
artery

ulnar artery

anterior interosseous
artery

deep palmar arch

digital
arteries

superficial palmar arch

**Fig. 9-37. The main arteries of the upper limb.**

**Fig. 9-38. Anterior view of upper arm. Middle portion of biceps brachii has been removed to show musculocutaneous nerve lying in front of brachialis.**

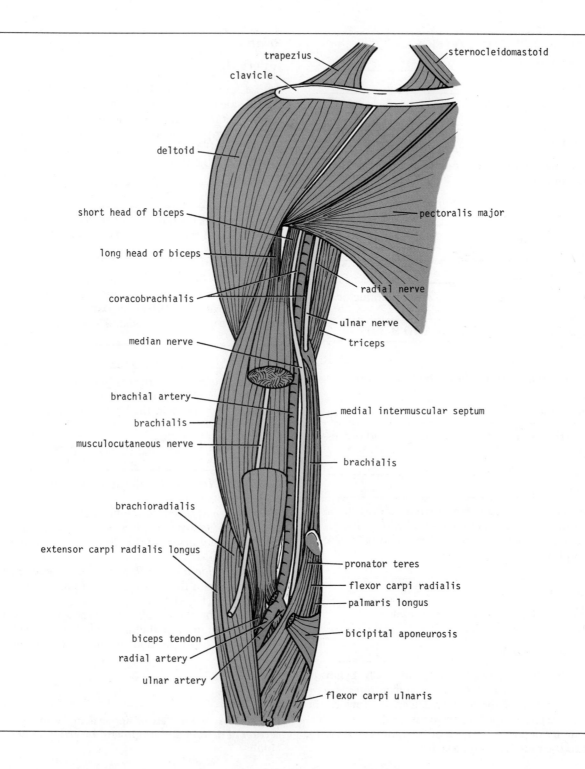

## CORACOBRACHIALIS (FIGS. 9-38 AND 9-39)

### Origin

From the tip of the coracoid process.

### Insertion

Into the middle of the medial side of the shaft of the humerus.

### Nerve Supply

Musculocutaneous nerve.

### Action

It flexes the arm and is also a weak adductor.

## BRACHIALIS (FIGS. 9-38 AND 9-39)

### Origin

From the front of the lower half of the humerus.

### Insertion

Into the anterior surface of the coronoid process of the ulna.

### Nerve Supply

Musculocutaneous nerve. A small part of the muscle that arises behind the deltoid tuberosity, and is therefore located in the posterior compartment, is supplied by the radial nerve.

### Action

It is a strong flexor of the elbow joint.

## BRACHIAL ARTERY

The brachial artery (Figs. 9-37 and 9-38) begins at the lower border of the teres major muscle as a continuation of the axillary artery. It provides the main arterial supply to the arm. (Fig. 9-37). It terminates opposite the neck of the radius by dividing into the radial and ulnar arteries.

### Relations

### Anteriorly

The vessel is superficial and is overlapped from the lateral side by the coracobrachialis and biceps. The medial cutaneous nerve of the forearm lies in front of the upper part; the median nerve crosses its middle part; and the bicipital aponeurosis crosses its lower part (Fig. 9-38).

### Posteriorly

The artery lies on the triceps, the coracobrachialis insertion, and the brachialis (Fig. 9-38).

### Medially

The ulnar nerve and the basilic vein in the upper part of the arm; in the lower part of the arm, the median nerve lies on its medial side (Fig. 9-38).

### Laterally

The median nerve and the coracobrachialis and biceps muscles above; the tendon of the biceps lies lateral to the artery in the lower part of its course (Fig. 9-38).

### Branches

1. *Muscular branches* to the anterior compartment of the upper arm.
2. The *nutrient artery* to the humerus.
3. The *profunda artery* arises near the beginning of the brachial artery and follows the radial nerve (Fig. 9-40).
4. The *superior ulnar collateral artery* arises near the middle of the upper arm and follows the ulnar nerve (Fig. 9-40).
5. The *inferior ulnar collateral artery* arises near the termination of the artery and takes part in the anastomosis around the elbow joint (Fig. 9-40).

**Fig. 9-39. Anterior view of upper arm, showing insertion of deltoid and origin and insertion of brachialis.**

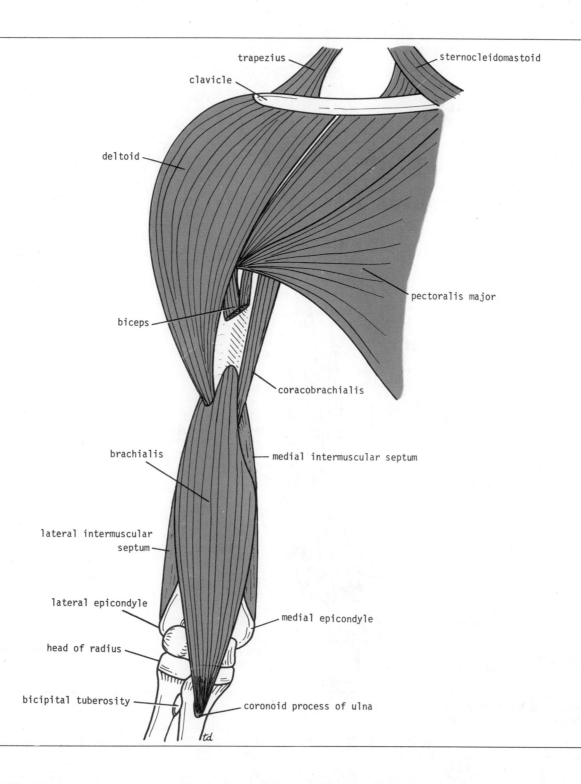

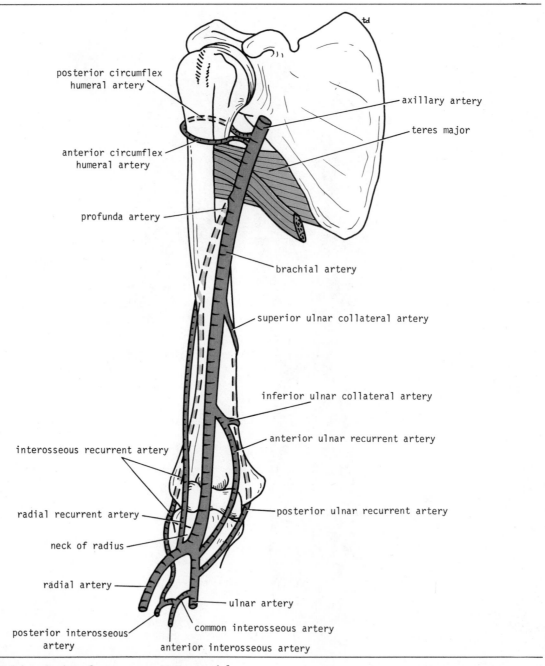

posterior circumflex
humeral artery

anterior circumflex
humeral artery

profunda artery

interosseous recurrent artery

radial recurrent artery

neck of radius

radial artery

posterior interosseous
artery

anterior interosseous artery

axillary artery

teres major

brachial artery

superior ulnar collateral artery

inferior ulnar collateral artery

anterior ulnar recurrent artery

posterior ulnar recurrent artery

ulnar artery

common interosseous artery

**Fig. 9-40. Main arteries of upper arm. Note arterial anastomosis around elbow joint.**

## MUSCULOCUTANEOUS NERVE

The origin of the musculocutaneous nerve from the lateral cord of the brachial plexus (C5, 6, and 7) in the axilla is described on page 430. It runs downward and laterally, pierces the coracobrachialis muscle (Fig. 9-18) and then passes downward between the biceps and brachialis muscles (Fig. 9-38). It appears at the lateral margin of the biceps tendon and pierces the deep fascia just above the elbow. It runs down the lateral aspect of the forearm as the *lateral cutaneous nerve of the forearm* (Fig. 9-34).

### Branches (Fig. 9-79)

1. *Muscular branches* to the biceps, coracobrachialis, and brachialis.
2. *Cutaneous branches.* The *lateral cutaneous nerve of the forearm* supplies the skin of the front and lateral aspect of the forearm down as far as the root of the thumb.
3. *Articular branches* to the elbow joint.

## MEDIAN NERVE

The origin of the median nerve from the medial and lateral cords of the brachial plexus in the axilla is described on page 430. It runs downward on the lateral side of the brachial artery (Fig. 9-38). Halfway down the upper arm, it crosses the brachial artery and continues downward on its medial side.

The nerve, like the artery, is therefore superficial, but at the elbow it is crossed by the bicipital aponeurosis. The further course of this nerve is described on page 476.

The median nerve has no branches in the upper arm (Fig. 9-79), except for a small vasomotor nerve to the brachial artery.

## ULNAR NERVE

The origin of the ulnar nerve from the medial cord of the brachial plexus in the axilla is described on page 430. It runs downward on the medial side of the brachial artery as far as the middle of the arm (Fig. 9-38). Here, at the insertion of the coracobrachialis, the nerve pierces the medial fascial septum, accompanied by the superior ulnar collateral artery, and enters the posterior compartment of the arm. (See p. 463.)

The ulnar nerve has no branches in the anterior compartment of the upper arm (Fig. 7-81).

## RADIAL NERVE

On leaving the axilla, the radial nerve immediately enters the posterior compartment of the arm and only enters the anterior compartment just above the lateral epicondyle.

### Contents of the Posterior Fascial Compartment of the Upper Arm

#### Muscle

The three heads of the triceps muscle.

#### Nerve Supply to the Muscle

Radial nerve.

#### Blood Supply

Profunda brachii and ulnar collateral arteries.

#### Structures Passing Through the Compartment

Radial nerve and ulnar nerve.

## TRICEPS (FIG. 9-41)

The triceps is a large muscle that forms the greater part of the substance of the back of the arm.

#### Origin

*Long head* from the infraglenoid tubercle of the scapula; *lateral head* from the upper half of the posterior surface of the shaft of the humerus above the spiral groove; *medial head* from the posterior surface of the lower half of the shaft of the humerus below the spiral groove.

#### Insertion

The common tendon is inserted into the upper surface of the olecranon process of the ulna.

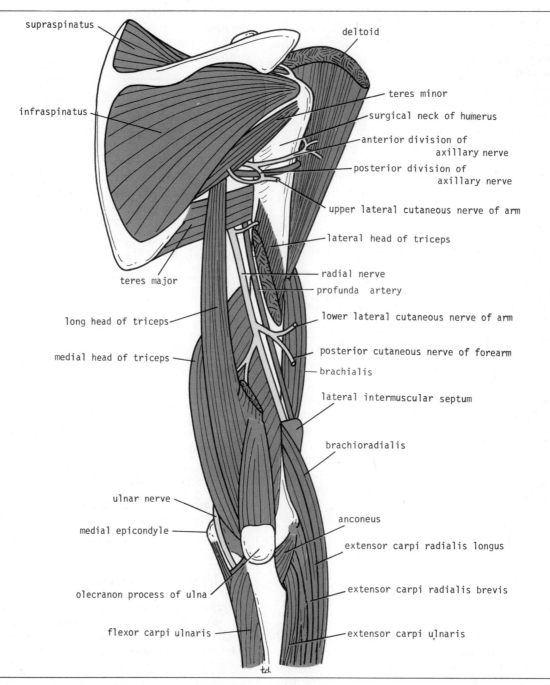

**Fig. 9-41. Posterior view of upper arm. Lateral head of triceps has been divided to display radial nerve and profunda artery in spiral groove of humerus.**

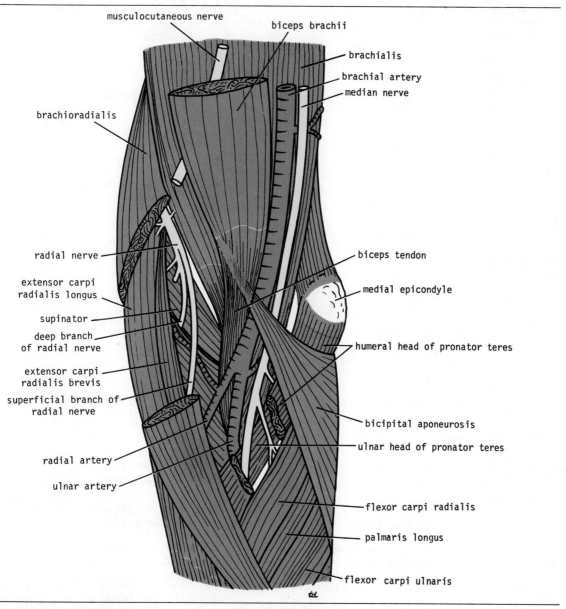

musculocutaneous nerve

biceps brachii

brachialis

brachial artery

median nerve

brachioradialis

biceps tendon

radial nerve

medial epicondyle

extensor carpi
radialis longus

supinator

humeral head of pronator teres

deep branch
of radial nerve

extensor carpi
radialis brevis

bicipital aponeurosis

superficial branch of
radial nerve

ulnar head of pronator teres

radial artery

flexor carpi radialis

ulnar artery

palmaris longus

flexor carpi ulnaris

**Fig. 9-42. Right cubital fossa.**

Table 9-4. Muscles of the Arm

| Name of muscle | Origin | Insertion | Nerve supply | Action |
|---|---|---|---|---|
| ANTERIOR COMPARTMENT | | | | |
| Biceps brachii | | | | |
| Long head | Supraglenoid tubercle of scapula | Tuberosity of radius | Musculocutaneous nerve | Supinator of forearm and flexor of elbow joint; weak flexor of shoulder joint |
| Short head | Coracoid process of scapula | | | |
| Coracobrachialis | Coracoid process of scapula | Medial aspect of shaft of humerus | Musculocutaneous nerve | Flexes arm and also weak adductor |
| Brachialis | Front of lower half of humerus | Coronoid process of ulna | Musculocutaneous nerve | Flexor of elbow joint |
| POSTERIOR COMPARTMENT | | | | |
| Triceps | | | | |
| Long head | Infraglenoid tubercle of scapula | | | |
| Lateral head | Upper half of posterior surface of shaft of humerus | Olecranon process of ulna | Radial nerve | Extensor of the elbow joint |
| Medial head | Lower half of posterior surface of shaft of humerus | | | |

## Nerve Supply

Radial nerve.

## Action

This muscle is a strong extensor of the elbow joint.

## RADIAL NERVE

The origin of the radial nerve from the posterior cord of the brachial plexus in the axilla is described on page 433. The nerve winds around the back of the arm, first between the long and medial heads of triceps, then in the spiral groove on the back of the humerus, between the lateral and medial heads of the triceps (Fig. 9-41). It pierces the lateral fascial septum above the elbow and continues down-ward into the cubital fossa, between the brachialis and the brachioradialis muscles (Fig. 9-42). In the spiral groove the nerve is accompanied by the profunda vessels, and it lies directly in contact with the shaft of the humerus (Fig. 9-41).

### Branches (Fig. 9-77)

*Branches in the axilla.* Branches are given to the long and medial heads of the triceps, and the *posterior cutaneous nerve of the arm* is given off.

*Branches in the spiral groove* (Fig. 9-41). Branches are given to the lateral and medial heads of triceps and to the anconeus. The *lower lateral cutaneous nerve of the arm* supplies the skin over the lateral and anterior aspects of the lower part of the arm. The *posterior cutaneous nerve of the forearm* runs down the middle of the back of the forearm as far as the wrist.

*Branches in the anterior compartment of the arm.* After the nerve has pierced the lateral fascial septum, it gives branches to the brachialis, brachioradialis, and the extensor carpi radialis longus muscles (Fig. 9-42). It also gives *articular branches* to the elbow joint.

### ULNAR NERVE

Having pierced the medial fascial septum halfway down the upper arm (see p. 459), the ulnar nerve descends behind the septum, covered posteriorly by the medial head of the triceps. The nerve is accompanied by the superior ulnar collateral vessels. At the elbow, it lies *behind the medial epicondyle of the humerus* (Fig. 9-41) on the medial ligament of the elbow joint. It continues downward to enter the forearm between the two heads of origin of the flexor carpi ulnaris. (See p. 477.)

#### Branches (Fig. 9-81)

The ulnar nerve has an articular branch to the elbow joint.

### PROFUNDA BRACHII ARTERY

The profunda brachii artery arises from the brachial artery near its origin (Fig. 9-40). It accompanies the radial nerve through the spiral groove, supplies the triceps muscle, and takes part in the anastomosis around the elbow joint.

### SUPERIOR AND INFERIOR ULNAR COLLATERAL ARTERIES

The superior and inferior ulnar collateral arteries arise from the brachial artery and take part in the anastomosis around the elbow joint.

#### Muscles—Their Nerve Supply and Action

Students wishing to review the muscles of the arm should study Table 9-4.

## THE CUBITAL FOSSA

The cubital fossa is a depression that lies in front of the elbow and is triangular in shape (Figs. 9-6 and 9-42). It has the following boundaries:

### Laterally

The brachioradialis muscle.

### Medially

The pronator teres muscle.

The *base* of the triangle is formed by an imaginary line drawn between the two epicondyles of the humerus.

The *floor* of the fossa is formed by the supinator muscle laterally and the brachialis muscle medially.

The *roof* is formed by skin and fascia and is reinforced by the bicipital aponeurosis.

The cubital fossa (Fig. 9-42) contains the following structures, enumerated from the medial to the lateral side: the median nerve; the bifurcation of the brachial artery into the ulnar and radial arteries; the tendon of the biceps muscle; and the radial nerve and its deep branch.

The *supratrochlear lymph node* lies in the superficial fascia over the upper part of the fossa, above the trochlea (Fig. 9-35). It receives afferent lymphatic vessels from the third, fourth, and fifth fingers, the medial part of the hand, and the medial side of the forearm. The efferent lymphatic vessels pass up to the axilla and enter the lateral axillary group of nodes (Fig. 9-35).

## Bones of the Forearm

The forearm contains two bones, the radius and the ulna.

### RADIUS

The radius is the lateral bone of the forearm (Fig. 9-43). Its upper end articulates with the humerus at the elbow joint and with the ulna at the superior radioulnar joint. Its lower end articulates with the scaphoid and lunate bones of the carpus at the wrist joint, and with the ulna at the inferior radioulnar joint.

At the *upper end* of the radius is the small circular *head* (Fig. 9-43). The upper surface of the head is concave and articulates with the convex capitulum of the humerus. The circumference of the head articulates with the radial notch of the ulna. Below the head, the bone is constricted to

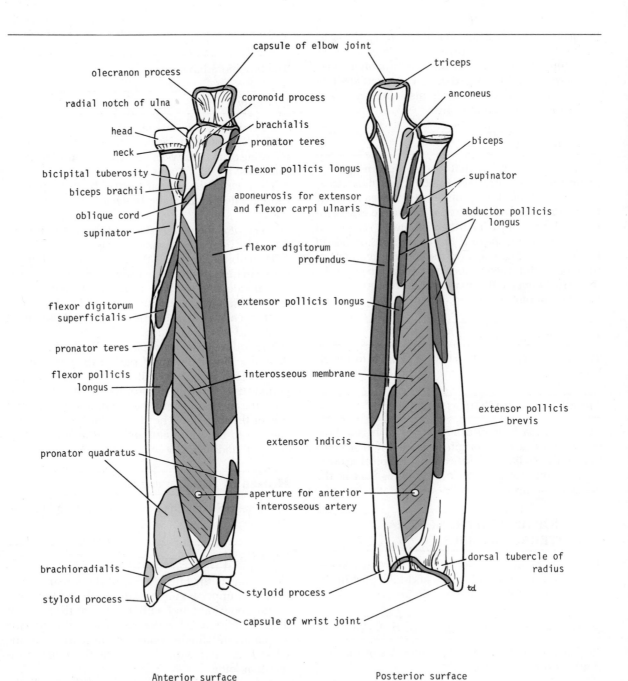

capsule of elbow joint

olecranon process

coronoid process

radial notch of ulna

brachialis

head

pronator teres

neck

bicipital tuberosity

flexor pollicis longus

biceps brachii

aponeurosis for extensor
and flexor carpi ulnaris

oblique cord

supinator

flexor digitorum
profundus

flexor digitorum
superficialis

extensor pollicis longus

pronator teres

flexor pollicis
longus

interosseous membrane

pronator quadratus

extensor indicis

aperture for anterior
interosseous artery

brachioradialis

styloid process

styloid process

capsule of wrist joint

triceps

anconeus

biceps

supinator

abductor pollicis
longus

extensor pollicis
brevis

dorsal tubercle of
radius

td

Anterior surface

Posterior surface

**Fig. 9-43. Important muscular and ligamentous attachments to radius and ulna.**

form the *neck*. Below the neck is the *bicipital tuberosity*.

The shaft of the radius, in contradistinction to that of the ulna, is wider below than above (Fig. 9-43). It has a sharp *interosseous border* medially for the attachment of the interosseous membrane. It has an *oblique line*, or ridge, on its anterior surface, which extends downward and laterally from the bicipital tuberosity to the *pronator tubercle* on its lateral surface.

At the *lower end* of the radius is the *styloid process*; this projects distally from its lateral margin (Fig. 9-43). On the medial surface is the *ulnar notch*, which articulates with the round head of the ulna. The inferior articular surface is divided in two by a ridge; the lateral area articulates with the scaphoid bone and the medial area with the lunate bone. On the posterior aspect of the lower end is a small tubercle, the *dorsal tubercle*, which is grooved on its medial aspect by the tendon of the extensor pollicis longus (Fig. 9-43).

The important muscles and ligaments that are attached to the radius are shown in Figure 9-43.

## ULNA

The ulna is the medial bone of the forearm (Fig. 9-43). Its upper end articulates with the humerus at the elbow joint, and with the head of the radius at the superior radioulnar joint. Its lower end articulates with the radius at the inferior radioulnar joint, but it is excluded from the wrist joint by the articular disc.

The *upper end* of the ulna is large and is known as the *olecranon process* (Fig. 9-43). It has a notch on its anterior surface, the *trochlear notch*, which articulates with the trochlea of the humerus. Below the trochlear notch is the triangular *coronoid process*, which has on its lateral surface the *radial notch* for articulation with the head of the radius.

The *shaft* of the ulna tapers from above down (Fig. 9-43). It has three surfaces and three borders. The *lateral*, or *interosseous*, border is sharp and gives attachment to the interosseous membrane. The *posterior border* is rounded and subcutaneous and can be easily palpated throughout its length. The *anterior border* is also rounded, but it is covered by muscles. Below the radial notch is a depression, the *supinator fossa*, which gives clearance for the movement of the bicipital tuberosity of the ra-

dius. The posterior border of the fossa is sharp and is known as the *supinator crest*; it gives origin to the supinator muscles.

At the *lower end* of the ulna is the small rounded *head*, which has projecting from its medial aspect the *styloid process* (Fig. 9-43).

The important muscles and ligaments that are attached to the ulna are shown in Figure 9-43.

## Bones of the Carpus

There are eight carpal bones, made up of two rows of four (Figs. 9-44 and 9-45). The *proximal row* consists of (from lateral to medial) the *scaphoid*, *lunate*, *triquetral*, and *pisiform* bones. The *distal row* consists of (from lateral to medial) the *trapezium*, *trapezoid*, *capitate*, and *hamate* bones. Together, the bones of the carpus present on their anterior surface a concavity, to the lateral and medial edges of which is attached the flexor retinaculum. In this manner, an osteofascial tunnel is formed for the passage of the median nerve and the flexor tendons of the fingers.

The carpus is cartilaginous at birth. The capitate begins to ossify during the first year, and the others begin to ossify at intervals thereafter, until the twelfth year, when all the bones are ossified.

While a detailed knowledge of the bones of the hand is unnecessary for a medical student, the position, shape, and size of the scaphoid bone should be studied, since it is commonly fractured. The ridge and groove of the trapezium, the hook of the hamate, and the pea-like pisiform bone should be examined.

## The Metacarpals and Phalanges

There are five metacarpal bones, each of which has a *base*, a *shaft*, and a *head* (Figs. 9-44 and 9-45).

The first metacarpal bone of the thumb is the shortest and most mobile. It does not lie in the same plane as the others, but occupies a more anterior position. It is also rotated medially through a right angle, so that its extensor surface is directed laterally and not backward.

The bases of the metacarpal bones articulate with the distal row of the carpal bones; the heads, which form the knuckles, articulate with the proximal phalanges (Figs. 9-44 and 9-45). The shaft of each metacarpal bone is slightly concave forward

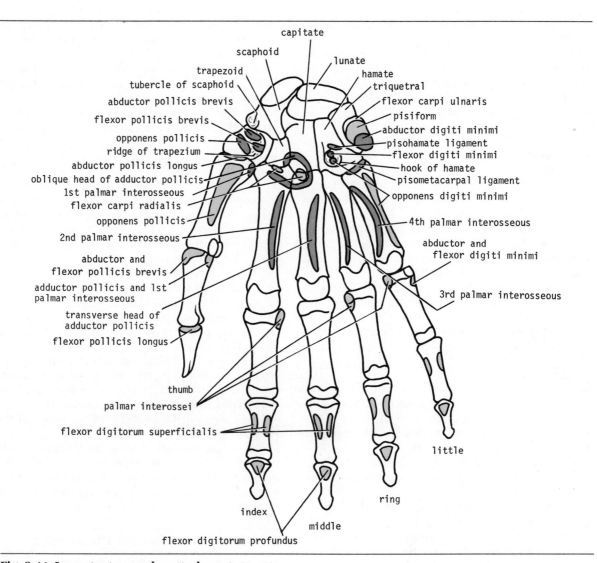

**Fig. 9-44. Important muscular attachments to anterior surfaces of bones of the hand.**

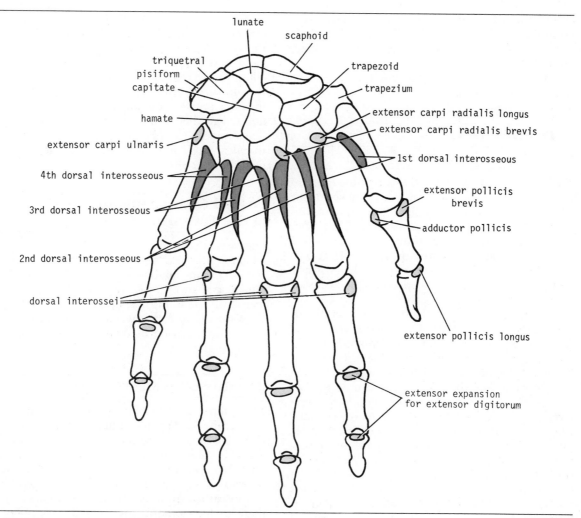

**Fig. 9-45. Important muscular attachments to posterior surfaces of bones of the hand.**

and is triangular in transverse section. Its surfaces are posterior, lateral, and medial.

There are three phalanges for each of the fingers, but only two for the thumb.

The important muscles that are attached to the bones of the hand and fingers are shown in Figures 9-44 and 9-45.

# THE FOREARM

## Skin

The *sensory nerve supply* to the skin of the forearm is from the anterior and posterior branches of the lateral cutaneous nerve of the forearm, a continuation of the musculocutaneous nerve, and from the anterior and posterior branches of the medial cutaneous nerve of the forearm (Fig. 9-34). A narrow strip of skin down the middle of the posterior surface of the forearm is supplied by the posterior cutaneous nerve of the forearm.

The *superficial veins* of the forearm lie in the superficial fascia (Fig. 9-8). The *cephalic vein* arises from the lateral side of the dorsal venous arch on the back of the hand and winds around the lateral border of the forearm; it then ascends into the cubital fossa and up the front of the arm on the lateral side of the biceps. Its termination in the axillary vein is described on page 428. As the cephalic vein passes up the upper limb, it receives a variable number of tributaries from the lateral and posterior surfaces of the limb (Fig. 9-8). The *median cubital vein*, a branch of the cephalic vein in the cubital fossa, runs upward and medially and joins the basilic vein. In the cubital fossa the median cubital vein crosses in front of the brachial artery and the median nerve, but it is separated from them by the bicipital aponeurosis.

The *basilic vein* arises from the medial side of the dorsal venous arch on the back of the hand and winds around the medial border of the forearm; it then ascends into the cubital fossa and up the front of the arm on the medial side of the biceps (Fig. 9-8). Its termination by joining the venae comitantes of the brachial artery to form the axillary vein is described on page 428. It receives the median cubital vein and a variable number of tributaries from the medial and posterior surfaces of the upper limb.

The *superficial lymphatic vessels* from the thumb and lateral fingers and the lateral areas of the hand

and forearm follow the cephalic vein to the infraclavicular group of nodes (Fig. 9-35). Those from the medial fingers and the medial areas of the hand and forearm follow the basilic vein to the cubital fossa. Here, some of the vessels drain into the *supratrochlear lymph node*, while others bypass the node and accompany the basilic vein to the axilla, where they drain into the lateral group of axillary nodes. The efferent vessels from the supratrochlear node also drain into the lateral axillary nodes (Fig. 9-35).

# Fascial Compartments of the Forearm

The forearm is enclosed in a sheath of deep fascia, which is attached to the periosteum of the posterior subcutaneous border of the ulna (Fig. 9-46). This fascial sheath, together with the interosseous membrane and fibrous intermuscular septa, divides up the forearm into a number of compartments, each having its own muscles, nerves, and blood supply.

## INTEROSSEOUS MEMBRANE

The interosseous membrane is a thin but strong membrane uniting the radius and ulna; it is attached to their interosseous borders (Figs. 9-43 and 9-46). Its fibers run obliquely downward and medially, so that a force applied to the lower end of the radius (e.g., falling on the outstretched hand) is transmitted from the radius to the ulna and from there to the humerus and scapula. Its fibers are taut when the forearm is in the midprone position, i.e., the position of function. The interosseous membrane provides attachment for neighboring muscles. Its lower part is pierced by the anterior interosseous vessels.

The *oblique cord* is a narrow, ligamentous structure, which extends from the radius, below the tuberosity, to the apex of the coronoid process of the ulna (Fig. 9-43). Its function is unknown.

## FLEXOR AND EXTENSOR RETINACULA

The flexor and extensor retinacula are specialized bands of deep fascia seen in the region of the wrist and hand.

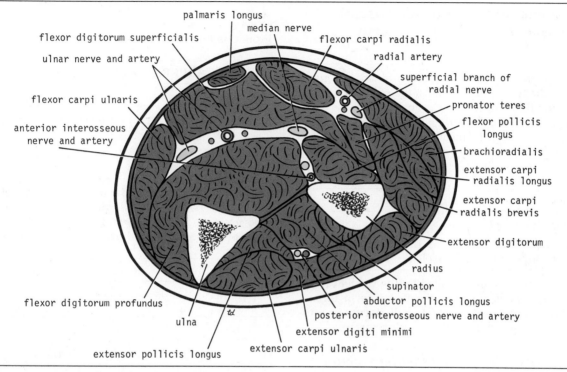

Fig. 9-46. Cross section of forearm at level of insertion of pronator teres muscle.

## Flexor Retinaculum

The flexor retinaculum is a thickening of deep fascia, which stretches across the front of the wrist and converts the concave anterior surface of the carpus into an osteofascial tunnel, called the *carpal tunnel,* for the passage of the median nerve and the flexor tendons of the thumb and fingers (Fig. 9-52). It is attached medially to the pisiform bone and the hook of the hamate, and laterally to the tubercle of the scaphoid and the trapezium bones. The attachment to the trapezium consists of superficial and deep parts and forms a synovial-lined tunnel for passage of the tendon of the flexor carpi radialis.

The upper border of the retinaculum corresponds to the distal transverse crease in front of the wrist and is continuous with the deep fascia of the forearm. The lower border is attached to the palmar aponeurosis (Fig. 9-53).

The relations of the flexor retinaculum are described on page 486.

## Extensor Retinaculum

The extensor retinaculum is a thickening of deep fascia that stretches across the back of the wrist (Fig. 9-61). It converts the grooves on the posterior surface of the distal ends of the radius and ulna into six separate tunnels for the passage of the long extensor tendons. Each tunnel is lined with a synovial sheath, which extends above and below the retinaculum on the tendons. The tunnels are separated from one another by fibrous septa that pass from the deep surface of the retinaculum to the bones.

The retinaculum is attached medially to the pisiform bone and the hook of the hamate, and laterally to the distal end of the radius. The upper and lower borders of the retinaculum are continuous with the deep fascia of the forearm and hand, respectively.

The contents of the tunnels beneath the extensor retinaculum are described on page 487.

# Contents of the Anterior Fascial Compartment of the Forearm

## Muscles

A *superficial group*, consisting of the pronator teres, the flexor carpi radialis, the palmaris longus, and the flexor carpi ulnaris; an *intermediate group* consisting of the flexor digitorum superficialis; and a *deep group* consisting of the flexor pollicis longus, the flexor digitorum profundus, and the pronator quadratus.

## Blood Supply to the Muscles

Ulnar and radial arteries.

## Nerve Supply to the Muscles

All the muscles are supplied by the median nerve and its branches, except the flexor carpi ulnaris and the medial part of the flexor digitorum profundus, which are supplied by the ulnar nerve.

## Muscles of the Fascial Compartments of the Forearm: Superficial Group

The superficial group of muscles possesses a common tendon of origin, which is attached to the medial epicondyle of the humerus.

## PRONATOR TERES (FIGS. 9-47 AND 9-48)

### Origin

A *humeral head*, which arises from the common tendon attached to the medial epicondyle of the humerus, and an *ulnar head*, which springs from the medial border of the coronoid process of the ulna.

### Insertion

The two heads unite to be inserted into the pronator tuberosity on the lateral aspect of the shaft of the radius.

### Nerve Supply

Median nerve.

### Action

Pronation and flexion of the forearm.

## FLEXOR CARPI RADIALIS (FIG. 9-47)

### Origin

From the common tendon attached to the medial epicondyle of the humerus.

### Insertion

The tendon runs through a synovial-lined tunnel in the lateral part of the flexor retinaculum in a groove on the trapezium and is inserted into the bases of the second and third metacarpal bones.

### Nerve Supply

Median nerve.

### Action

Flexes and abducts the hand at the wrist joint.

## PALMARIS LONGUS (FIG. 9-47)

The palmaris longus muscle is often absent.

### Origin

From the common tendon attached to the medial epicondyle of the humerus.

### Insertion

Into the flexor retinaculum and palmar aponeurosis.

### Nerve Supply

Median nerve.

**Fig. 9-47. Anterior view of forearm. Middle portion of brachioradialis muscle has been removed to display superficial branch of radial nerve and the radial artery.**

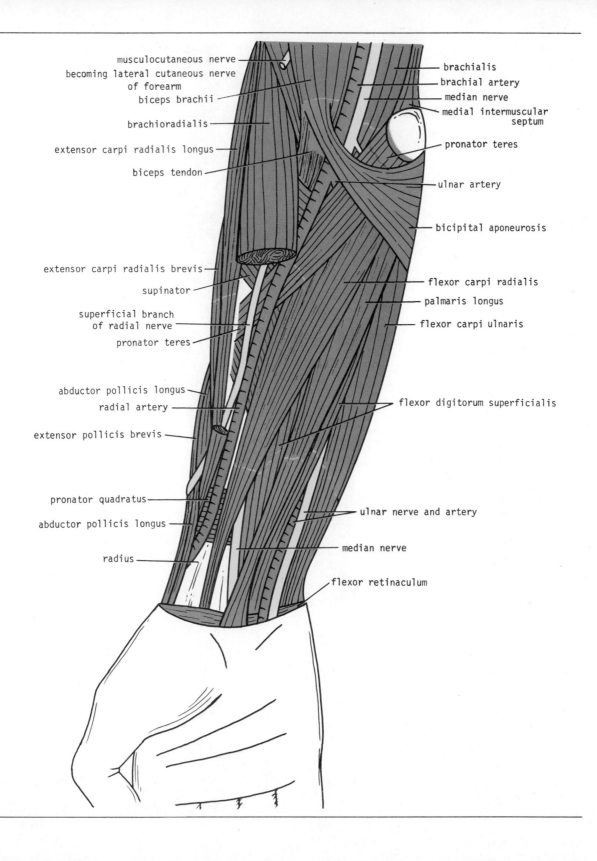

musculocutaneous nerve
becoming lateral cutaneous nerve
of forearm
biceps brachii
brachioradialis
extensor carpi radialis longus
biceps tendon

extensor carpi radialis brevis
supinator
superficial branch
of radial nerve
pronator teres

abductor pollicis longus
radial artery
extensor pollicis brevis

pronator quadratus
abductor pollicis longus
radius

brachialis
brachial artery
median nerve
medial intermuscular
septum
pronator teres
ulnar artery

bicipital aponeurosis

flexor carpi radialis
palmaris longus
flexor carpi ulnaris

flexor digitorum superficialis

ulnar nerve and artery

median nerve

flexor retinaculum

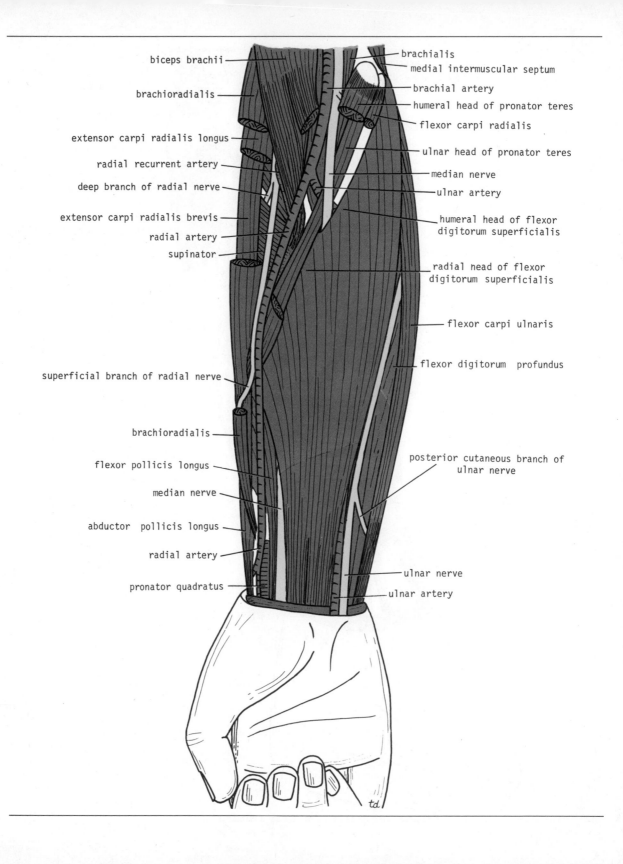

biceps brachii

brachioradialis

extensor carpi radialis longus

radial recurrent artery

deep branch of radial nerve

extensor carpi radialis brevis

radial artery

supinator

superficial branch of radial nerve

brachioradialis

flexor pollicis longus

median nerve

abductor pollicis longus

radial artery

pronator quadratus

brachialis

medial intermuscular septum

brachial artery

humeral head of pronator teres

flexor carpi radialis

ulnar head of pronator teres

median nerve

ulnar artery

humeral head of flexor digitorum superficialis

radial head of flexor digitorum superficialis

flexor carpi ulnaris

flexor digitorum profundus

posterior cutaneous branch of ulnar nerve

ulnar nerve

ulnar artery

## Action

Flexes the hand.

## FLEXOR CARPI ULNARIS (FIG. 9-47)

### Origin

A *humeral head*, which arises from the common tendon attached to the medial epicondyle of the humerus, and an *ulnar head*, which springs from the medial aspect of the olecranon process of the ulna and the posterior border of the ulna.

### Insertion

The two heads unite to form a long tendon, which is inserted into the pisiform bone, and by two ligaments, called the pisohamate and pisometacarpal ligaments into the hook of the hamate, and the base of the fifth metacarpal bone, respectively.

### Nerve Supply

Ulnar nerve.

### Action

Flexes and adducts the hand at the wrist joint.

## Muscles of the Anterior Fascial Compartments of the Forearm: Intermediate Muscle

### FLEXOR DIGITORUM SUPERFICIALIS (FIG. 9-48)

### Origin

A *humeroulnar head* from the common tendon, attached to the medial epicondyle of the humerus and the medial margin of the coronoid process of the ulna; it is also attached to the medial ligament of the elbow joint. A *radial head*, arising from the

Fig. 9-48. Anterior view of forearm. Most of superficial muscles have been removed to display flexor digitorum superficialis, median nerve, superficial branch of radial nerve, and radial artery. Note that ulnar head of pronator teres separates median nerve from ulnar artery.

oblique line on the anterior surface of the shaft of the radius.

### Insertion

The two heads unite to form the muscle belly, and in the lower part of the forearm this gives rise to four tendons; these enter the hand by passing behind the flexor retinaculum. Here, the tendons for the middle and ring fingers lie anterior to those for the index and little fingers (Fig. 9-56).

On reaching the proximal phalanges, each tendon divides into two slips; these then unite, and the tendon finally divides again into two slips, which are inserted into the sides of the middle phalanx (Fig. 9-55). The corresponding tendon of the flexor digitorum profundus passes through the division of each superficialis tendon and is inserted into the base of the distal phalanx.

### Nerve Supply

Median nerve.

### Action

Flexes the middle phalanx of the fingers and also assists in flexing the proximal phalanx and the hand.

Since the profundus tendons pierce the superficialis tendons, the latter serve as pulleys to the profundus muscle and enhance efficiency (Fig. 9-55) of this muscle.

## Muscles of the Anterior Fascial Compartments of the Forearm: Deep Group

### FLEXOR POLLICIS LONGUS (FIG. 9-49)

### Origin

From the middle of the anterior surface of the shaft of the radius and from the adjoining part of the interosseous membrane.

### Insertion

The tendon passes behind the flexor retinaculum and is inserted into the base of the distal phalanx of the thumb.

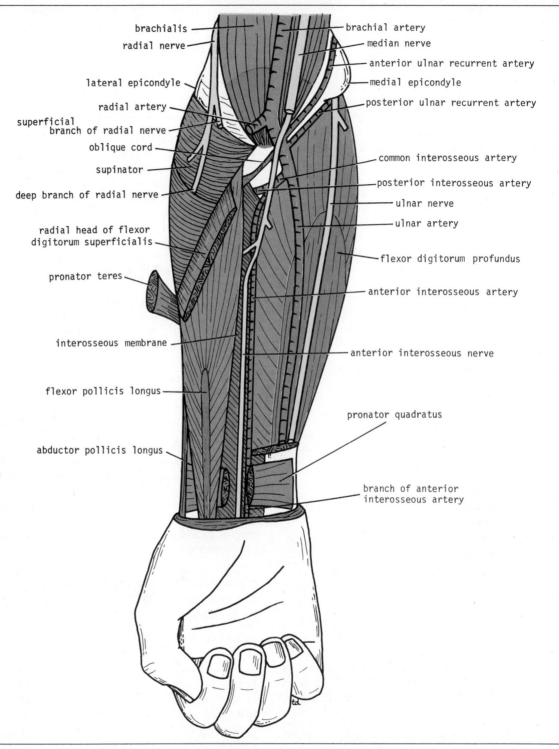

brachialis

radial nerve

lateral epicondyle

radial artery

superficial
branch of radial nerve

oblique cord

supinator

deep branch of radial nerve

radial head of flexor
digitorum superficialis

pronator teres

interosseous membrane

flexor pollicis longus

abductor pollicis longus

brachial artery

median nerve

anterior ulnar recurrent artery

medial epicondyle

posterior ulnar recurrent artery

common interosseous artery

posterior interosseous artery

ulnar nerve

ulnar artery

flexor digitorum profundus

anterior interosseous artery

anterior interosseous nerve

pronator quadratus

branch of anterior
interosseous artery

**Fig. 9-49. Anterior view of forearm, showing deep structures.**

Nerve Supply

The anterior interosseous branch of the median nerve.

Action

Flexes the distal phalanx of the thumb.

## FLEXOR DIGITORUM PROFUNDUS (FIG. 9-49)

### Origin

From the upper three-fourths of the anterior and medial surfaces of the shaft of the ulna and the adjoining part of the interosseous membrane.

### Insertion

Above the wrist, the muscle divides into four tendons, which pass down behind the flexor digitorum superficialis and the flexor retinaculum (Fig. 9-52). Each tendon passes through a division in the corresponding tendon of the superficialis muscle and is inserted into the base of the distal phalanx (Fig. 9-55).

### Nerve Supply

The ulnar nerve supplies the medial half of the muscle (going to little and ring fingers); the anterior interosseous branch of the median nerve supplies the lateral half (going to index and middle fingers).

### Action

Flexes the distal phalanx of the fingers and then assists in the flexion of the middle and proximal phalanges. It also assists in flexing the wrist.

## PRONATOR QUADRATUS (FIG. 9-49)

### Origin

From the lower quarter of the anterior surface of the shaft of the ulna.

Insertion

Into the lower quarter of the anterior surface of the shaft of the radius.

Nerve Supply

The anterior interosseous branch of the median nerve.

Action

Pronates the forearm at the superior and inferior radioulnar joints.

## *Arteries of the Anterior Fascial Compartment of the Forearm*

### ULNAR ARTERY

The ulnar artery, one of the terminal branches of the brachial artery, arises in the cubital fossa at the level of the neck of the radius (Figs. 9-37 and 9-48). It passes downward and medially, deep to the pronator teres, flexor carpi radialis, palmaris longus, and flexor digitorum superficialis muscles. It then runs downward under the flexor carpi ulnaris on the lateral side of the ulnar nerve. The artery is related posteriorly first to the brachialis muscle and then to the flexor digitorum profundus (Fig. 9-49). At the wrist, it is quite superficial and lies with the ulnar nerve between the tendon of the flexor carpi ulnaris muscle and the tendons of the flexor digitorum superficialis muscle (Fig. 9-48).

The ulnar artery enters the palm on the lateral side of the ulnar nerve. It crosses in front of the flexor retinaculum and lies lateral to the pisiform bone (Fig. 9-56). It ends by joining the superficial palmar branch of the radial artery to form the *superficial palmar arch.* (See p. 501.)

### *Branches*

1. *Muscular branches* to neighboring muscles.
2. *Recurrent branches* that take part in the arterial anastomosis around the elbow joint (Fig. 9-49).
3. *Branches that take part in the arterial anastomosis around the wrist joint.*
4. The *common interosseous artery,* which arises from the upper part of the ulnar artery and after

a brief course divides into the anterior and posterior interosseous arteries (Fig. 9-49).

## ANTERIOR INTEROSSEOUS ARTERY

The anterior interosseous is a small artery that arises from the common interosseous artery and passes downward, with the anterior interosseous nerve, on the front of the interosseous membrane (Fig. 9-49). At the upper border of the pronator quadratus, it pierces the interosseous membrane and descends behind it to take part in the arterial anastomosis around the wrist joint.

### Branches

1. *Muscular branches* to neighboring muscles.
2. *Nutrient branches* to the radius and ulna bones.
3. *Median artery* to the median nerve.

## POSTERIOR INTEROSSEOUS ARTERY

The posterior interosseous is a small artery that arises from the common interosseous artery and passes backward over the upper border of the interosseous membrane between the radius and ulna, to enter the posterior compartment of the forearm (Fig. 9-49). Its further course is described on page 483.

## RADIAL ARTERY

The radial artery, one of the terminal branches of the brachial artery, arises in the cubital fossa at the level of the neck of the radius (Figs. 9-47 and 9-48). It passes downward and laterally, deep to the brachioradialis muscle. In the lower half of the forearm, it emerges on the medial side of the tendon of the brachioradialis and lies on the lateral side of the tendon of the flexor carpi radialis. It is covered only by skin and fascia.

As the artery descends (Fig. 9-48), it rests on the following structures, from above downward: the tendon of the biceps, the supinator, the pronator teres, the radial head of the flexor digitorum superficialis, the flexor pollicis longus, the pronator quadratus, and finally the radius, where its pulsations may be felt in the living subject. In the middle third of the forearm, the superficial branch of the radial nerve lies lateral to it.

The radial artery leaves the forearm by winding around the lateral aspect of the wrist to reach the posterior surface of the hand. (See p. 506.)

### Branches

1. *Muscular branches* to neighboring muscles.
2. *Recurrent branch*, which takes part in the arterial anastomosis around the elbow joint (Fig. 9-48).
3. *Superficial palmar branch*, which arises just above the wrist (Fig. 9-48) and enters the palm of the hand by passing either superficially or through the muscles of the thenar eminence. It frequently joins the ulnar artery to form the *superficial palmar arch.*

## Nerves of the Anterior Fascial Compartment of the Forearm

### MEDIAN NERVE

The median nerve leaves the cubital fossa by passing between the two heads of the pronator teres (Fig. 9-48). It runs behind the humeral head of the pronator teres and is separated from the ulnar artery by the ulnar head of that muscle. It continues downward behind the flexor digitorum superficialis and is attached to its deep surface by connective tissue. It rests posteriorly on the flexor digitorum profundus. At the wrist, the median nerve emerges from the lateral border of the flexor digitorum superficialis muscle and lies behind the tendon of the palmaris longus (Figs. 9-47 and 9-48). It enters the palm by passing *behind* the flexor retinaculum. (See pp. 487 and 501.)

### Branches (Fig. 9-79)

1. *Muscular branches* in the cubital fossa to the pronator teres, the flexor carpi radialis, the palmaris longus, and the flexor digitorum superficialis.
2. *Articular branches* to the elbow joint.
3. *Anterior interosseous nerve.* This arises from the median nerve as it emerges from between the two heads of the pronator teres muscle (Fig. 9-49).
4. *Palmar cutaneous branch.* This arises in the lower part of the forearm and is distributed to

the skin over the lateral part of the palm (Fig. 9-34).

## ANTERIOR INTEROSSEOUS NERVE

The anterior interosseous nerve arises from the median nerve as it emerges from between the two heads of the pronator teres. It passes downward on the anterior surface of the interosseous membrane, between the flexor pollicis longus and the flexor digitorum profundus, and is accompanied by the anterior interosseous artery (Fig. 9-49). It ends on the anterior surface of the carpus.

### Branches

1. *Muscular branches* to the flexor pollicis longus, the pronator quadratus, and the lateral half of the flexor digitorum profundus.
2. *Articular branches* to the wrist and inferior radioulnar joints. It also supplies the joints of the carpus.

## ULNAR NERVE

The ulnar nerve (Fig. 9-49) passes from behind the medial epicondyle of the humerus, crosses the medial ligament of the elbow joint, and enters the front of the forearm by passing between the two heads of the flexor carpi ulnaris. It then runs down the forearm between the flexor carpi ulnaris and the flexor digitorum profundus muscles. In the lower two-thirds of the forearm, the ulnar artery lies on the lateral side of the ulnar nerve (Fig. 9-49). At the wrist, the ulnar nerve becomes superficial and lies between the tendons of the flexor carpi ulnaris and flexor digitorum superficialis muscles (Fig. 9-47). The ulnar nerve enters the palm of the hand by passing *in front* of the flexor retinaculum and lateral to the pisiform bone; here, it has the ulnar artery lateral to it. (See p. 486.)

### Branches (Fig. 9-81)

1. *Muscular branches* to the flexor carpi ulnaris and to the medial half of the flexor digitorum profundus.
2. *Articular branches* to the elbow joint.
3. *Palmar cutaneous branch.* This is a small branch that arises in the middle of the forearm (Fig. 9-

34) and supplies the skin over the hypothenar eminence.
4. *Dorsal, or posterior cutaneous branch.* This is a large branch that arises in the distal third of the forearm. It passes medially between the tendon of the flexor carpi ulnaris and the ulna, and is distributed on the posterior surface of the hand and fingers.

## Contents of the Lateral Fascial Compartment of the Forearm*

### Muscles

Brachioradialis and extensor carpi radialis longus.

### Blood Supply

Radial and brachial arteries.

### Nerve Supply to the Muscles

Radial nerve.

### Muscles of the Lateral Fascial Compartment of the Forearm

#### BRACHIORADIALIS (FIG. 9-47)

##### Origin

From the upper two-thirds of the lateral supracondylar ridge of the humerus and from the adjoining lateral intermuscular septum.

##### Insertion

Into the base of the styloid process of the radius.

##### Nerve Supply

Radial nerve.

##### Action

This muscle flexes the forearm at the elbow joint; it also assists in rotating the forearm to the mid-

*This may be regarded as part of the posterior fascial compartment.

prone position, or restoring the forearm to the mid-prone position from the full prone position.

## EXTENSOR CARPI RADIALIS LONGUS (FIGS. 9-47 AND 9-48)

### Origin

From the lower third of the lateral supracondylar ridge of the humerus and from the adjoining lateral intermuscular septum.

### Insertion

The long tendon passes under the extensor retinaculum and is inserted into the posterior surface of the base of the second metacarpal bone.

### Nerve Supply

Radial nerve.

### Action

It extends and abducts the hand at the wrist joint.

## Arteries of the Lateral Compartment of the Forearm

The arterial supply is derived from branches of the radial and brachial arteries.

## Nerve of the Lateral Compartment of the Forearm

### RADIAL NERVE

The radial nerve pierces the lateral intermuscular septum in the lower part of the arm and passes forward into the cubital fossa (Fig. 9-42). It then passes downward in front of the septum and the lateral epicondyle of the humerus, lying between the brachialis on the medial side, and the brachioradialis and extensor carpi radialis longus on the lateral side (Fig. 9-48). At the level of the lateral epicondyle it divides into superficial and deep branches (Figs. 9-48 and 9-49).

### Branches (Fig. 9-77)

1. *Muscular branches* to the brachioradialis, to the extensor carpi radialis longus, and a small branch to the lateral part of the brachialis muscle.
2. *Articular branches* to the elbow joint.
3. *Deep branch of the radial nerve.* This winds around the neck of the radius, between the superficial and deep layers of the supinator (Fig. 9-49), and enters the posterior compartment of the forearm (Fig. 9-51).
4. *Superficial branch of the radial nerve.*

### Superficial Branch of the Radial Nerve

The superficial branch of the radial nerve is the direct continuation of the nerve after its main stem has given off its deep branch in front of the lateral epicondyle of the humerus (Fig. 9-48). It runs down under cover of the brachioradialis muscle and lies on the supinator and the pronator teres muscles. It is situated close to the lateral side of the radial artery in the middle third of the forearm. In the lower part of the forearm, it leaves the artery and passes backward under the tendon of the brachioradialis (Fig. 9-48). It reaches the posterior surface of the wrist, where it divides into terminal branches that supply the skin on the lateral two-thirds of the posterior surface of the hand (Fig. 9-34) and the posterior surface of the lateral two and one-half fingers over the proximal phalanx. The area of skin supplied by the nerve on the dorsum of the hand is variable.

## Contents of the Posterior Fascial Compartment of the Forearm

### Muscles

*Superficial group:* extensor carpi radialis brevis, extensor digitorum, extensor digiti minimi, extensor carpi ulnaris, anconeus. These muscles possess a common tendon of origin, which is attached to the lateral epicondyle of the humerus.

*Deep group:* supinator, abductor pollicis longus, extensor pollicis brevis, extensor pollicis longus, extensor indicis.

## Blood Supply

Posterior and anterior interosseous arteries.

## Nerve Supply to the Muscles

Deep branch of the radial nerve.

## *Muscles of the Posterior Fascial Compartments of the Forearm: Superficial Group*

### EXTENSOR CARPI RADIALIS BREVIS (FIG. 9-50)

#### Origin

From the common tendon attached to the lateral epicondyle of the humerus.

#### Insertion

The tendon passes under the extensor retinaculum and is inserted into the posterior surface of the base of the third metacarpal bone.

#### Nerve Supply

Deep branch of the radial nerve.

#### Action

It extends and abducts the hand at the wrist joint.

### EXTENSOR DIGITORUM (FIG. 9-50)

#### Origin

From the common tendon attached to the lateral epicondyle of the humerus.

#### Insertion

The muscle divides into four tendons, which pass under the extensor retinaculum and then diverge to the fingers. The tendons are connected to one another on the posterior surface of the hand. The first and fourth tendons are also connected to the extensor indicis and the extensor digiti minimi, respectively.

On the posterior surface of each finger, the extensor tendon becomes incorporated into a fascial expansion, called the *extensor expansion* (Fig. 9-61). Near the proximal interphalangeal joint, the extensor expansion splits into three parts: a central part, which is inserted into the base of the middle phalanx; and two lateral parts, which converge to be inserted into the base of the distal phalanx (Fig. 9-55).

#### Nerve Supply

Deep branch of the radial nerve.

#### Action

It extends the metacarpophalangeal joints and, through the extensor expansion, assists the lumbrical muscles and interossei to extend the proximal and distal interphalangeal joints. It also assists in extending the hand.

Because of the presence of connections between the tendons, complete extension of one finger at the metacarpophalangeal joint is impossible so long as the remaining fingers are kept flexed. The index finger has a greater freedom of movement, since its tendon is not connected to the tendons of the other fingers.

### EXTENSOR DIGITI MINIMI (FIG. 9-50)

#### Origin

From the common tendon attached to the lateral epicondyle of the humerus.

#### Insertion

The tendon passes under the extensor retinaculum and divides into two slips, which are inserted into the extensor expansion for the little finger (Fig. 9-61). It is joined by the small fourth tendon of the extensor digitorum.

#### Nerve Supply

Deep branch of the radial nerve.

#### Action

It extends the metacarpophalangeal joint of the little finger.

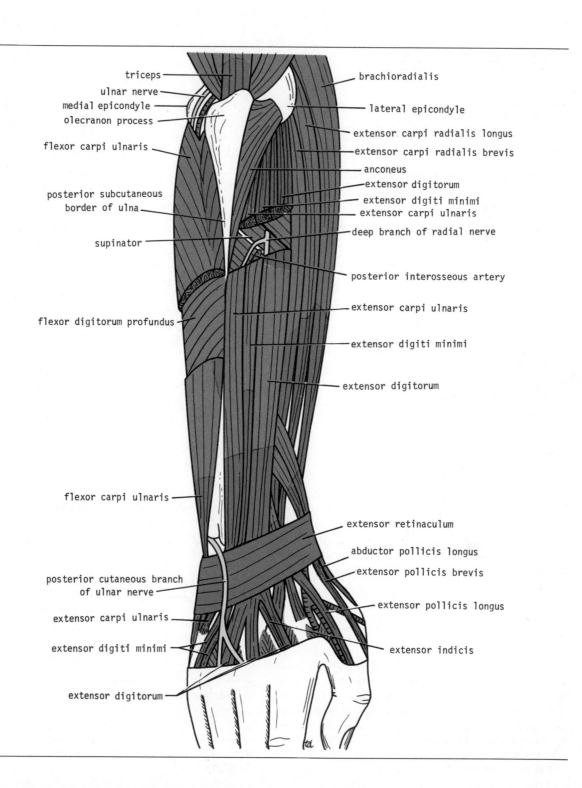

triceps

ulnar nerve

medial epicondyle

olecranon process

flexor carpi ulnaris

posterior subcutaneous border of ulna

supinator

flexor digitorum profundus

flexor carpi ulnaris

posterior cutaneous branch of ulnar nerve

extensor carpi ulnaris

extensor digiti minimi

extensor digitorum

brachioradialis

lateral epicondyle

extensor carpi radialis longus

extensor carpi radialis brevis

anconeus

extensor digitorum

extensor digiti minimi

extensor carpi ulnaris

deep branch of radial nerve

posterior interosseous artery

extensor carpi ulnaris

extensor digiti minimi

extensor digitorum

extensor retinaculum

abductor pollicis longus

extensor pollicis brevis

extensor pollicis longus

extensor indicis

## EXTENSOR CARPI ULNARIS (FIG. 9-50)

### Origin

From the common tendon attached to the lateral epicondyle of the humerus.

### Insertion

The tendon passes under the extensor retinaculum and is inserted into the posterior surface of the base of the fifth metacarpal bone.

### Nerve Supply

Deep branch of the radial nerve.

### Action

It extends and adducts the hand at the wrist joint.

## ANCONEUS (FIG. 9-50)

The anconeus is a small triangular muscle, which should be considered as part of the triceps muscle. It does not belong to the posterior fascial compartment of the forearm, but for convenience it will be described here.

### Origin

From the posterior aspect of the lateral epicondyle of the humerus.

### Insertion

Into the lateral surface of the olecranon process of the ulna.

### Nerve Supply

Radial nerve.

**Fig. 9-50. Posterior view of forearm. Parts of extensor digitorum, extensor digiti minimi, and extensor carpi ulnaris have been removed to show deep branch of radial nerve and posterior interosseous artery.**

### Action

It assists the triceps to extend the elbow joint.

## Muscles of the Posterior Fascial Compartment of the Forearm: Deep Group

### SUPINATOR (FIG. 9-49)

### Origin

It arises from the lateral epicondyle of the humerus, the lateral ligament of the elbow joint, the anular ligament of the superior radioulnar joint, and the supinator crest and fossa of the ulna.

### Insertion

Its fibers are arranged in two planes, between which the deep branch of the radial nerve lies. The two planes of muscle fibers wind around the posterior and lateral surface of the neck of the radius and are inserted into the posterior, lateral, and anterior aspects of the neck and shaft of the radius, down as far as the oblique line.

### Nerve Supply

Deep branch of the radial nerve.

### Action

It assists in supination of the forearm at the superior and inferior radioulnar joints. (The biceps brachii muscle is the chief supinator.)

## ABDUCTOR POLLICIS LONGUS (FIGS. 9-50 AND 9-51)

### Origin

From the middle of the posterior surface of the shaft of the ulna and radius and the intervening interosseous membrane.

### Insertion

The tendon passes under the extensor retinaculum and is inserted into the posterior surface of the base of the first metacarpal bone.

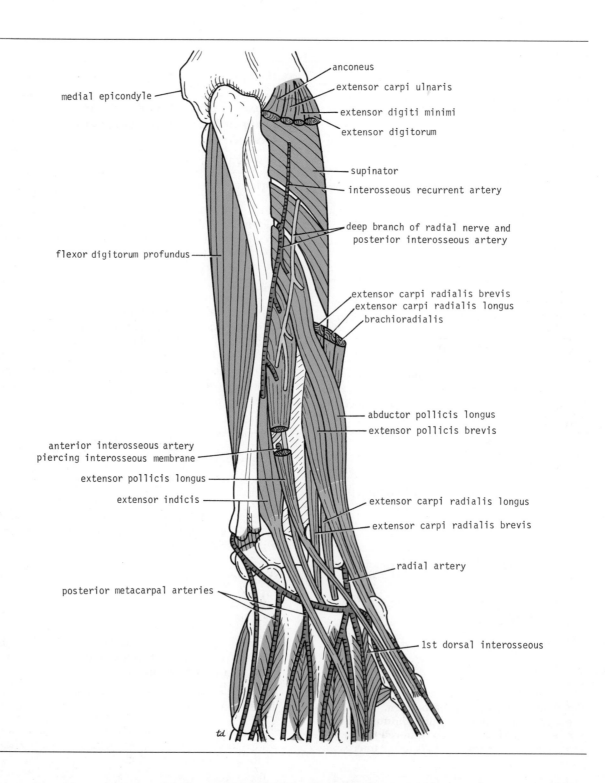

anconeus

extensor carpi ulnaris

extensor digiti minimi

extensor digitorum

medial epicondyle

supinator

interosseous recurrent artery

deep branch of radial nerve and
posterior interosseous artery

flexor digitorum profundus

extensor carpi radialis brevis
extensor carpi radialis longus
brachioradialis

abductor pollicis longus
extensor pollicis brevis

anterior interosseous artery
piercing interosseous membrane

extensor pollicis longus

extensor indicis

extensor carpi radialis longus

extensor carpi radialis brevis

radial artery

posterior metacarpal arteries

1st dorsal interosseous

td

Nerve Supply

Deep branch of the radial nerve.

Action

Abducts and extends the thumb at the carpometacarpal joint.

## EXTENSOR POLLICIS BREVIS (FIG. 9-51)

### Origin

From the posterior surface of the radius and the adjacent part of the interosseous membrane.

### Insertion

The tendon passes under the extensor retinaculum and is inserted into the posterior surface of the base of the proximal phalanx of the thumb.

### Nerve Supply

Deep branch of the radial nerve.

### Action

Extends the metacarpophalangeal joint of the thumb.

## EXTENSOR POLLICIS LONGUS (FIGS. 9-50 AND 9-51)

### Origin

From the posterior surface of the ulna and the adjacent part of the interosseous membrane.

### Insertion

The tendon passes beneath the extensor retinaculum and uses the medial side of the dorsal tubercle of the radius as a pulley. It is inserted into the posterior surface of the base of the distal phalanx of the thumb.

**Fig. 9-51. Posterior view of forearm. Superficial muscles have been removed to display deep structures.**

Nerve Supply

Deep branch of the radial nerve.

Action

Extends the distal phalanx of the thumb.

## "ANATOMICAL SNUFFBOX"

The anatomical snuffbox is a term commonly used to describe a triangular depression on the lateral side of the wrist that is bounded medially by the tendon of extensor pollicis longus and laterally by the tendons of abductor pollicis longus and extensor pollicis brevis (Fig. 9-50). Its clinical importance lies in the fact that the scaphoid bone is most easily palpated here and that the pulsations of the radial artery can be felt here (Fig. 9-7).

## EXTENSOR INDICIS (FIG. 9-51)

### Origin

From the posterior surface of the ulna and the adjacent part of the interosseous membrane.

### Insertion

The tendon passes beneath the extensor retinaculum in company with the tendons of the extensor digitorum. It is inserted into the extensor expansion of the index finger.

### Nerve Supply

Deep branch of the radial nerve.

### Action

It extends the metacarpophalangeal joint of the index finger.

## *Arteries of the Posterior Fascial Compartment of the Forearm*

### POSTERIOR INTEROSSEOUS ARTERY

The posterior interosseous artery arises from the common interosseous artery, a branch of the ulnar

Table 9-5. Muscles of the Forearm

| Name of muscle | Origin | Insertion | Nerve supply | Action |
|---|---|---|---|---|
| MUSCLES OF ANTERIOR FASCIAL COMPARTMENT | | | | |
| Pronator teres | | | | |
| Humeral head | Medial epicondyle of humerus | Lateral aspect of shaft of radius | Median nerve | Pronation and flexion of forearm |
| Ulnar head | Medial border of coronoid process of ulna | | | |
| Flexor carpi radialis | Medial epicondyle of humerus | Bases of second and third metacarpal bones | Median nerve | Flexes and abducts hand at wrist joint |
| Palmaris longus | Medial epicondyle of humerus | Flexor retinaculum and palmar aponeurosis | Median nerve | Flexes hand |
| Flexor carpi ulnaris | | | | |
| Humeral head | Medial epicondyle of humerus | Pisiform bone, hook of the hamate, base of fifth metacarpal bone | Ulnar nerve | Flexes and adducts the hand at the wrist joint |
| Ulnar head | Medial aspect of olecranon process and posterior border of ulna | | | |
| Flexor digitorum superficialis | | | | |
| Humeroulnar head | Medial epicondyle of humerus; medial border of coronoid process of ulna | Middle phalanx of medial four fingers | Median nerve | Flexes middle phalanx of fingers and assists in flexing proximal phalanx and hand |
| Radial head | Oblique line on anterior surface of shaft of radius | | | |
| Flexor pollicis longus | Anterior surface of shaft of radius | Distal phalanx of thumb | Anterior interosseous branch of median nerve | Flexes distal phalanx of thumb |
| Flexor digitorum profundus | Anteromedial surface of shaft of ulna | Distal phalanges of medial four fingers | Ulnar and median nerves | Flexes distal phalanx of the fingers; then assists in flexion of middle and proximal phalanges and the wrist |
| Pronator quadratus | Anterior surface of shaft of ulna | Anterior surface of shaft of radius | Anterior interosseous branch of median nerve | Pronates forearm |

Table 9-6. Muscles of the Forearm

| Name of muscle | Origin | Insertion | Nerve supply | Action |
| --- | --- | --- | --- | --- |
| MUSCLES OF THE LATERAL FASCIAL COMPARTMENT OF THE FOREARM | | | | |
| Brachioradialis | Lateral supracondylar ridge of humerus | Base of styloid process of radius | Radial nerve | Flexes forearm at elbow joint; rotates forearm to the midprone position |
| Extensor carpi radialis longus | Lateral supracondylar ridge of humerus | Posterior surface of base of second metacarpal bone | Radial nerve | Extends and abducts hand at wrist joint |

Table 9-7. Muscles of the Forearm

| Name of muscle | Origin | Insertion | Nerve supply | Action |
| --- | --- | --- | --- | --- |
| MUSCLES OF THE POSTERIOR FASCIAL COMPARTMENT | | | | |
| Extensor carpi radialis brevis | Lateral epicondyle of humerus | Posterior surface of base of third metacarpal bone | Deep branch of radial nerve | Extends and abducts the hand at the wrist joint |
| Extensor digitorum | Lateral epicondyle of humerus | Middle and distal phalanges of the medial four fingers | Deep branch of radial nerve | Extends fingers and hand (see text for details) |
| Extensor digiti minimi | Lateral epicondyle of humerus | Extensor expansion of little finger | Deep branch of radial nerve | Extends metacarpal phalangeal joint of little finger |
| Extensor carpi ulnaris | Lateral epicondyle of humerus | Base of fifth metacarpal bone | Deep branch of radial nerve | Extends and adducts hand at the wrist joint |
| Anconeus | Lateral epicondyle of humerus | Lateral surface of olecranon process of ulna | Radial nerve | Extends elbow joint |
| Supinator | Lateral epicondyle of humerus, anular ligament of superior radioulnar joint, and ulna | Neck and shaft of radius | Deep branch of radial nerve | Supination of forearm |
| Abductor pollicis longus | Posterior surface of shafts of radius and ulna | Base of first metacarpal bone | Deep branch of radial nerve | Abducts and extends thumb |
| Extensor pollicis brevis | Posterior surface of shaft of radius | Base of proximal phalanx of thumb | Deep branch of radial nerve | Extends metacarpophalangeal joints of thumb |
| Extensor pollicis longus | Posterior surface of shaft of ulna | Base of distal phalanx of thumb | Deep branch of radial nerve | Extends distal phalanx of thumb |
| Extensor indicis | Posterior surface of shaft of ulna | Extensor expansion of index finger | Deep branch of radial nerve | Extends metacarpophalangeal joint of index finger |

artery (Figs. 9-49 and 9-51). It passes backward above the upper margin of the interosseous membrane between the radius and ulna. It then passes downward between the supinator and abductor pollicis longus and reaches the interval between the superficial and deep groups of muscles (Fig. 9-51). It ends by anastomosing with the anterior interosseous artery and taking part in the arterial anastomosis around the wrist joint.

## Branches

1. *Muscular branches* to neighboring muscles.
2. *Recurrent branch*, which takes part in the arterial anastomosis around the elbow joint (Fig. 9-51).

## ANTERIOR INTEROSSEOUS ARTERY

The anterior interosseous artery, as described previously (p. 476), arises from the common interosseous branch of the ulnar artery (Fig. 9-49). It descends in front of the interosseous membrane and pierces the membrane in the lower third of the forearm, to enter the posterior compartment (Fig. 9-51). It ends by anastomosing with the posterior interosseous artery and taking part in the arterial anastomosis around the wrist joint.

## Branches

It has *muscular branches* to neighboring muscles.

## Nerve of the Posterior Fascial Compartment of the Forearm

### DEEP BRANCH OF THE RADIAL NERVE

The deep branch arises from the radial nerve in front of the lateral epicondyle of the humerus in the cubital fossa (Fig. 9-49). It pierces the supinator, and winds around the lateral aspect of the neck of the radius in the substance of the muscle, to reach the posterior compartment of the forearm. The nerve emerges from the supinator (Fig. 9-51) and descends in the interval between the superficial and deep groups of muscles in company with the posterior interosseous artery. It eventually reaches the posterior surface of the interosseous membrane and runs with the anterior interosseous

artery. It terminates on the back of the carpus in an enlargement from which branches pass to the carpal joints.

## Branches

1. *Muscular branches* to the extensor carpi radialis brevis and the supinator, the extensor digitorum, the extensor digiti minimi, the extensor carpi ulnaris, the abductor pollicis longus, the extensor pollicis brevis, the extensor pollicis longus, and the extensor indicis.
2. *Articular branches* to the wrist and carpal joints.

## Muscles—Their Nerve Supply and Action

Students wishing to review the muscles of the forearm should study Tables 9-5, 9-6, and 9-7.

# THE REGION OF THE WRIST

Before learning the anatomy of the hand, it is essential that a student have a sound knowledge of the arrangement of the tendons, arteries, and nerves in the region of the wrist joint. From the clinical standpoint, the wrist is a common site for injury.

In the drawing of a transverse section through the wrist shown in Figure 9-52, identify the structures from medial to lateral. At the same time, examine your own wrist and identify as many of the structures as possible.

## Anterior Aspect of the Wrist

### Structures that Pass Anterior to the Flexor Retinaculum from Medial to Lateral

1. *Flexor carpi ulnaris tendon*, ending on the pisiform bone. (This tendon does not actually cross the flexor retinaculum but is included for the sake of completeness.)
2. The *ulnar nerve*, lying lateral to the pisiform bone.
3. The *ulnar artery*.
4. The *palmar cutaneous branch of the ulnar nerve*.
5. The *palmaris longus tendon*, passing to its insertion into the flexor retinaculum and the palmar aponeurosis.
6. The *palmar cutaneous branch of the median nerve*.

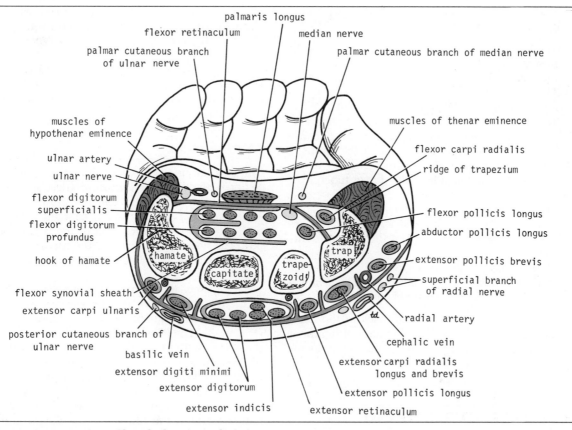

**Fig. 9-52. Cross section of hand, showing relationship of tendons, nerves, and arteries to flexor and extensor retinacula.**

## Structures that Pass Beneath the Flexor Retinaculum from Medial to Lateral (Fig. 9-52)

1. The *flexor digitorum superficialis tendons* and, posterior to these, the *tendons of flexor digitorum profundus;* both groups of tendons share a common synovial sheath.
2. The *median nerve.*
3. The *flexor pollicis longus tendon,* surrounded by a synovial sheath.
4. The *flexor carpi radialis tendon* going through the split in the flexor retinaculum. The tendon is surrounded by a synovial sheath.

## Structures that Pass Posterior to the Extensor Retinaculum from Medial to Lateral (Fig. 9-52)

1. The *dorsal (posterior) cutaneous branch of the ulnar nerve.*
2. The *basilic vein.*
3. The *cephalic vein.*
4. The *superficial branch of the radial nerve.*

## Structures that Pass Beneath the Extensor Retinaculum

Beneath the extensor retinaculum, fibrous septa pass to the underlying radius and ulna and form six compartments that contain the tendons of the extensor muscles. Each compartment is provided with a synovial sheath, which extends above and below the retinaculum.

From medial to lateral, these compartments (Fig. 9-52) are the following:

1. The *extensor carpi ulnaris tendon*, which occupies the most medial compartment and grooves the posterior aspect of the head of the ulna.
2. The *extensor digiti minimi tendon* is situated posterior to the inferior radioulnar joint.
3. The *extensor digitorum and extensor indicis tendons* share a common synovial sheath and are situated on the posterior surface of the radius.
4. The *extensor pollicis longus tendon* winds around the medial side of the dorsal tubercle of the radius.
5. The *extensor carpi radialis longus* and *brevis tendons* share a common synovial sheath and are situated on the lateral part of the posterior surface of the radius.
6. The *abductor pollicis longus* and the *extensor pollicis brevis tendons* have separate synovial sheaths, but share a common compartment.

The *radial artery* reaches the back of the hand by passing between the lateral collateral ligament of the wrist joint and the tendons of the abductor pollicis longus and extensor pollicis brevis (Fig. 9-51).

## THE PALM OF THE HAND
### Skin

The skin of the palm of the hand is thick and hairless. It is bound down to the underlying deep fascia by numerous fibrous bands. The skin shows many flexure creases at the sites of skin movement, which are not necessarily placed at the site of joints. Sweat glands are present in large numbers.

The *palmaris brevis* (Fig. 9-53) is a small muscle that arises from the flexor retinaculum and palmar aponeurosis and is inserted into the skin of the palm. It is supplied by the superficial branch of the ulnar nerve. Its function is to corrugate the skin at the base of the hypothenar eminence and so improve the grip of the palm in holding a rounded object.

The *sensory nerve* supply to the skin of the palm (Figs. 9-34 and 9-53) is derived from the *palmar cutaneous branch of the median nerve*, which crosses in front of the flexor retinaculum and supplies the lateral part of the palm, and the *palmar cutaneous branch of the ulnar nerve*; the latter nerve also crosses in front of the flexor retinaculum (Fig. 9-52) and supplies the medial part of the palm.

The skin over the base of the thenar eminence is supplied by the *lateral cutaneous nerve of the forearm* or the *superficial branch of the radial nerve* (Fig. 9-34).

### Deep Fascia

The deep fascia of the wrist and palm is thickened to form the *flexor retinaculum* (described on p. 469) and the *palmar aponeurosis*.

The *palmar aponeurosis* is triangular in shape and occupies the central area of the palm (Fig. 9-53). The deep fascia covering the medial hypothenar muscles and the lateral thenar muscles is thin and weak.

The apex of the palmar aponeurosis is attached to the distal border of the flexor retinaculum and receives the insertion of the palmaris longus tendon (Fig. 9-53). The base of the aponeurosis divides at the bases of the fingers into four slips. Each slip divides into two bands, one passing superficially to the skin and the other passing deeply to the root of the finger; here, each deep band divides into two, which diverge around the flexor tendons and finally fuse with the fibrous flexor sheath and the deep transverse ligaments.

The medial and lateral borders of the palmar aponeurosis are continuous with the thinner deep fascia covering the hypothenar and thenar muscles. From each of these borders, fibrous septa pass posteriorly into the palm and take part in the formation of the palmar fascial spaces. (See p. 502.)

The function of the palmar aponeurosis is to give firm attachment to the overlying skin and so improve the grip, and to protect the underlying tendons.

### The Carpal Tunnel

The carpus is deeply concave on its anterior surface and forms a bony gutter. The gutter is converted into a tunnel by the flexor retinaculum (Fig. 9-52).

The long flexor tendons to the fingers and thumb pass through the tunnel and are accompanied by the median nerve. The four separate tendons of the flexor digitorum superficialis muscle are arranged

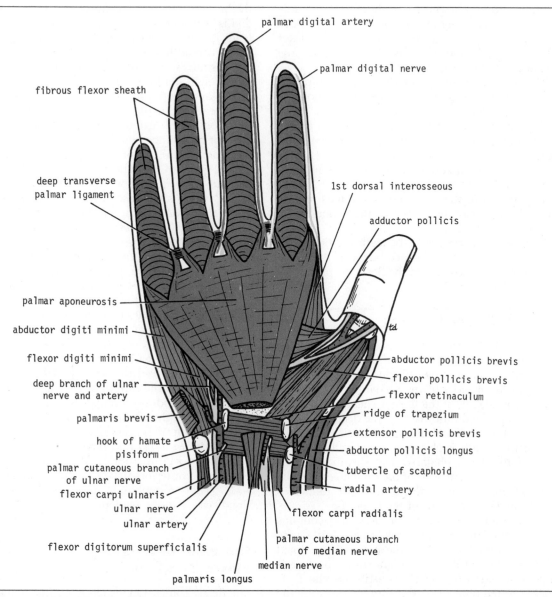

**Fig. 9-53. Anterior view of palm of hand. Palmar
aponeurosis has been left in position.**

in anterior and posterior rows, those to the middle and ring fingers lying in front of those to the index and little fingers. At the lower border of the flexor retinaculum, the four tendons diverge and become arranged on the same plane (Fig. 9-56).

The tendons of the flexor digitorum profundus muscle are on the same plane and lie behind the superficialis tendons.

All eight tendons of the flexor digitorum superficialis and profundus invaginate a common synovial sheath from the lateral side (Fig. 9-52). This allows the arterial supply to the tendons to enter them from the lateral side.

The tendon of the flexor pollicis longus muscle runs through the lateral part of the tunnel in its own synovial sheath.

The median nerve passes beneath the flexor retinaculum in a *restricted* space between the flexor digitorum superficialis and the flexor carpi radialis muscles (Fig. 9-52).

## Fibrous Flexor Sheaths

The anterior surface of each finger, from the head of the metacarpal to the base of the distal phalanx, is provided with a strong fibrous sheath, which is attached to the sides of the phalanges (Fig. 9-54). The proximal ends of the fibrous sheaths of the fingers (except the thumb) receive the deeper parts of the four slips of the palmar aponeurosis. The distal end of the sheath is closed and is attached to the base of the distal phalanx. The sheath, together with the anterior surfaces of the phalanges and the interphalangeal joints, forms a blind tunnel in which the flexor tendons of the finger lie.

In the thumb, the osteofibrous tunnel contains the tendon of the flexor pollicis longus. In the case of the four medial fingers, the tunnel is occupied by the tendons of the flexor digitorum superficialis and profundus (Fig. 9-54). The fibrous sheath is thick over the phalanges, but thin and lax over the joints.

## Synovial Flexor Sheaths

The crowded long flexor tendons emerge from the carpal tunnel and diverge as they pass down into the hand.

The flexor pollicis longus tendon enters the osteofibrous tunnel of the thumb and is inserted into the base of the distal phalanx (Fig. 9-53). The tendon is surrounded by a synovial sheath, which extends into the forearm for a distance equal to about a fingerbreadth proximal to the flexor retinaculum, and distally, it extends to the insertion.

The eight tendons of the flexor digitorum superficialis and profundus invaginate a common synovial sheath from the lateral side (Fig. 9-52). This common sheath extends proximally into the forearm for a distance equal to about a fingerbreadth proximal to the flexor retinaculum. Distally, the medial part of the sheath continues downward without interruption on the tendons of the little finger as far as the base of the distal phalanx (Fig. 9-54) The remainder of the sheath ends blindly approximately at the level of the proximal transverse crease of the palm.

The distal ends of the flexor tendons of the index, middle, and ring fingers have *digital synovial sheaths*, which commence at the level of the distal transverse crease of the palm and end at the bases of the distal phalanges (Fig. 9-54). Thus, for a short length, the tendons for these fingers are devoid of a synovial covering.

The synovial sheath of the flexor pollicis longus (sometimes referred to as the *radial bursa*) communicates with the common synovial sheath of the superficialis and profundus tendons (sometimes referred to as the *ulnar bursa*) at the level of the wrist in about 50 percent of subjects.

The *vincula longa* and *breva* are small vascular folds of synovial membrane that connect the tendons to the anterior surface of the phalanges (Fig. 9-55). They resemble a mesentery and convey blood vessels to the tendons.

## Insertion of the Long Flexor Tendons

The flexor pollicis longus tendon is inserted simply onto the anterior surface of the base of the distal phalanx of the thumb (Fig. 9-54).

Each tendon of the flexor digitorum superficialis enters the fibrous flexor sheath; opposite the proximal phalanx it divides into two halves, which pass around the profundus tendon and meet on its deep or posterior surface, where partial decussation of the fibers takes place (Fig. 9-55). The superficialis tendon, having united again, divides almost at once

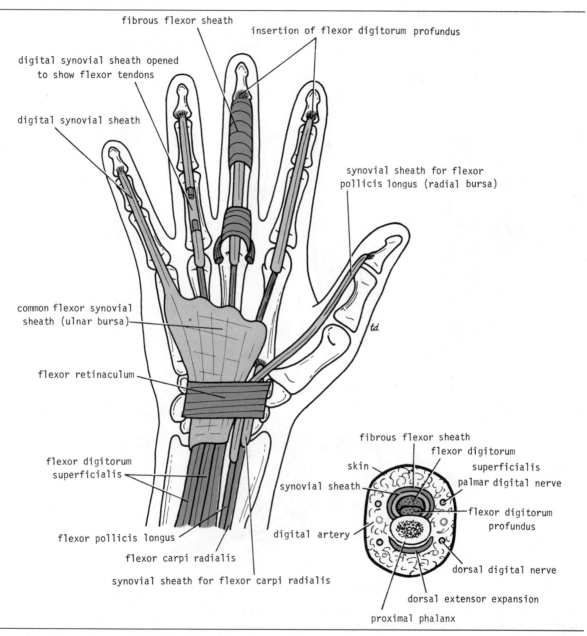

fibrous flexor sheath

insertion of flexor digitorum profundus

digital synovial sheath opened
to show flexor tendons

digital synovial sheath

synovial sheath for flexor
pollicis longus (radial bursa)

common flexor synovial
sheath (ulnar bursa)

td

flexor retinaculum

fibrous flexor sheath

flexor digitorum
superficialis

skin

palmar digital nerve

synovial sheath

flexor digitorum
profundus

flexor digitorum
superficialis

digital artery

flexor pollicis longus

flexor carpi radialis

dorsal digital nerve

synovial sheath for flexor carpi radialis

dorsal extensor expansion

proximal phalanx

**Fig. 9-54. Anterior view of palm of hand, showing
flexor synovial sheaths. Cross section of a finger is
also shown.**

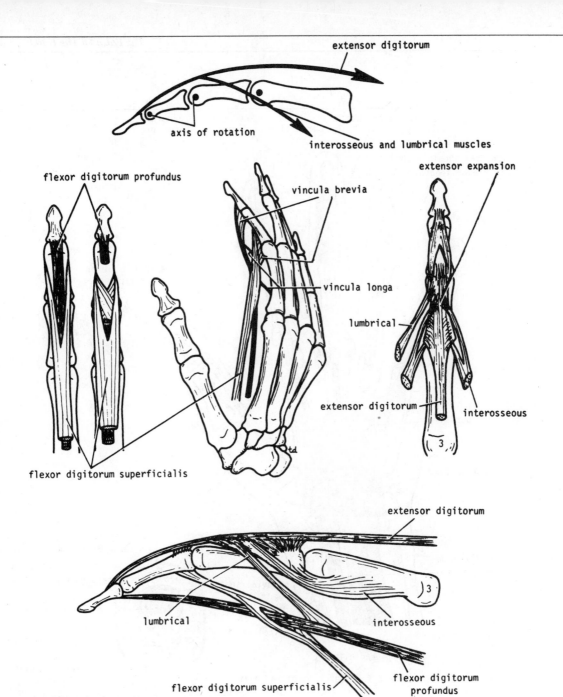

**Fig. 9-55.** Insertions of long flexor and extensor tendons in the fingers. Insertions of lumbrical and interossei muscles are also shown. Uppermost figure illustrates action of lumbricals and interossei muscles in flexing metacarpophalangeal joints and extending the interphalangeal joints.

into two further slips, which are attached to the borders of the middle phalanx.

Each tendon of the flexor digitorum profundus, having passed through the division of the superficialis tendon, continues downward, to be inserted into the anterior surface of the base of the distal phalanx (Fig. 9-55).

# Small Muscles of the Hand

## LUMBRICAL MUSCLES (FIG. 9-56)

### Origin

The lumbrical muscles are four in number and arise from the tendons of the flexor digitorum profundus in the palm.

### Insertion

Each muscle is inserted into the lateral side of the corresponding extensor expansion (Fig. 9-55).

### Nerve Supply

The first and second lumbricals, i.e., the lateral two lumbricals, are supplied by the median nerve; the third and fourth lumbricals are supplied by the deep branch of the ulnar nerve.

### Action

Assisted by the interossei, they flex the metacarpophalangeal joints and extend the interphalangeal joints (Fig. 9-55).

## The Interossei

There are eight interossei, consisting of four dorsal and four palmar muscles.[†] They occupy the spaces between the metacarpal bones. The dorsal muscles arise by two heads and are larger than the palmar muscles, which have only one head.

[†]Some authors describe only three palmar interossei and state that the first palmar interosseous is in reality a second head to the flexor pollicis brevis; others believe that it is part of the adductor pollicis muscle.

## PALMAR INTEROSSEI (FIGS. 9-57 TO 9-59)

### Origin

The first arises from the medial side of the base of the first metacarpal bone. The second, third, and fourth arise from the anterior surfaces of the second, fourth, and fifth metacarpal bones, respectively.

### Insertion

The first is inserted into the medial side of the base of the proximal phalanx of the thumb. The second is inserted into the medial side of the base of the proximal phalanx of the index finger. The third and fourth are inserted into the lateral side of the corresponding bones of the ring finger and little finger, respectively. In addition, all the interossei are inserted into the extensor expansion of the digit on which they act.

### Nerve Supply

Deep branch of the ulnar nerve.

### Action

They adduct the fingers toward the center of the third finger at the metacarpophalangeal joints, flex the metacarpophalangeal joints, and extend the interphalangeal joints (Fig. 9-59).

## DORSAL INTEROSSEI (FIGS. 9-59 TO 9-61)

### Origin

The four dorsal interossei arise from the contiguous sides of the first and second, second and third, third and fourth, and fourth and fifth metacarpal bones, respectively.

### Insertion

The first dorsal interosseous (Fig. 9-58) is inserted into the lateral side of the base of the proximal phalanx of the index finger; the second, into the lateral side of the base of the proximal phalanx of

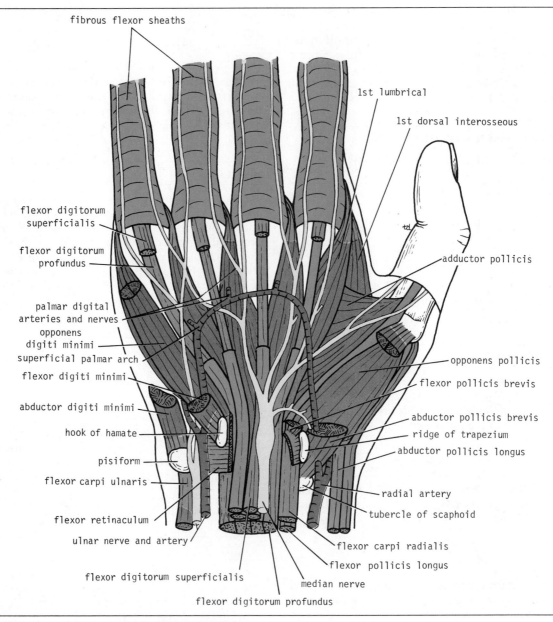

**Fig. 9-56. Anterior view of palm of hand. Palmar aponeurosis and greater part of flexor retinaculum have been removed to display superficial palmar arch, median nerve, and long flexor tendons. Segments of tendons of flexor digitorum superficialis have been removed to show underlying tendons of flexor digitorum profundus.**

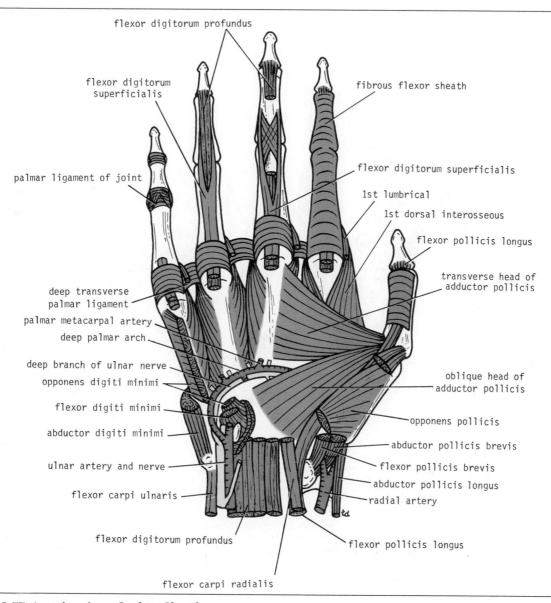

flexor digitorum profundus

flexor digitorum superficialis

fibrous flexor sheath

flexor digitorum superficialis

1st lumbrical

palmar ligament of joint

1st dorsal interosseous

flexor pollicis longus

transverse head of adductor pollicis

deep transverse palmar ligament

palmar metacarpal artery

deep palmar arch

deep branch of ulnar nerve

opponens digiti minimi

oblique head of adductor pollicis

flexor digiti minimi

opponens pollicis

abductor digiti minimi

abductor pollicis brevis

flexor pollicis brevis

ulnar artery and nerve

abductor pollicis longus

radial artery

flexor carpi ulnaris

flexor digitorum profundus

flexor pollicis longus

flexor carpi radialis

**Fig. 9-57. Anterior view of palm of hand. Long
flexor tendons have been removed from palm, but
their method of insertion into the fingers is shown.**

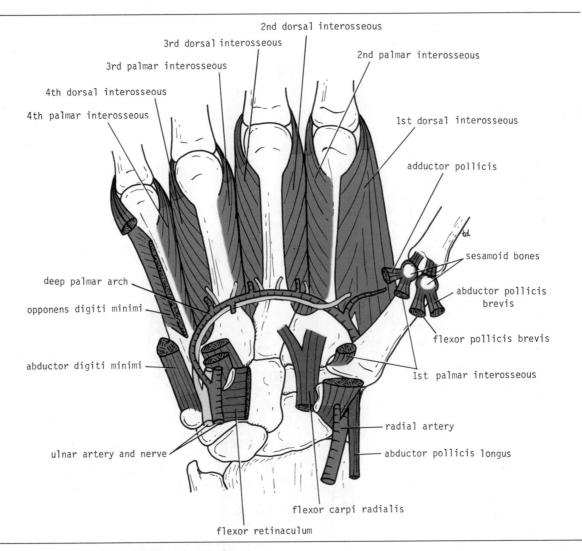

**Fig. 9-58. Anterior view of palm of hand, showing deep palmar arch and deep terminal branch of ulnar nerve; interossei are also shown.**

the middle finger (Fig. 9-59); the third, on the medial side of the same bone; and the fourth, on the medial side of the base of the proximal phalanx of the ring finger. In addition, all the interossei are inserted into the extensor expansion of the digit on which they act (Fig. 9-55).

## Nerve Supply

Deep branch of the ulnar nerve.

## Action

They abduct the fingers away from the center of the third finger at the metacarpophalangeal joints, flex the metacarpophalangeal joints, and extend the interphalangeal joints (Fig. 9-59).

## Short Muscles of the Thumb

The short muscles of the thumb are the abductor pollicis brevis, the flexor pollicis brevis, the opponens pollicis, and the adductor pollicis. The first three of these muscles form the *thenar eminence*.

## ABDUCTOR POLLICIS BREVIS (FIGS. 9-53 AND 9-56)

### Origin

From the scaphoid, the trapezium, and the flexor retinaculum.

### Insertion

Into the lateral aspect of the base of the proximal phalanx of the thumb with the flexor pollicis brevis.

### Nerve Supply

Median nerve.

### Action

Abduction of the thumb at the carpometacarpal joint and the metacarpophalangeal joint. Abduction of the thumb may be defined as a movement forward of the thumb in the anteroposterior plane.

## FLEXOR POLLICIS BREVIS (FIGS. 9-53 AND 9-56)

### Origin

From the anterior surface of the flexor retinaculum.

### Insertion

Into the lateral aspect of the base of the proximal phalanx of the thumb with the abductor pollicis brevis. A small sesamoid bone is usually present in the combined tendon.

### Nerve Supply

Median nerve.

### Action

Flexes the metacarpophalangeal joint of the thumb.

## OPPONENS POLLICIS (FIGS. 9-56 AND 9-57)

### Origin

From the anterior surface of the flexor retinaculum.

### Insertion

Into the whole length of the lateral border of the shaft of the first metacarpal bone.

### Nerve Supply

Median nerve.

### Action

Pulls the thumb medially and forward across the palm, so that the palmar surface of the tip of the thumb may come into contact with the palmar surface of the tips of the other fingers. It is a very important muscle and enables the thumb to form one claw in the pincer-like action used in the picking up of objects.

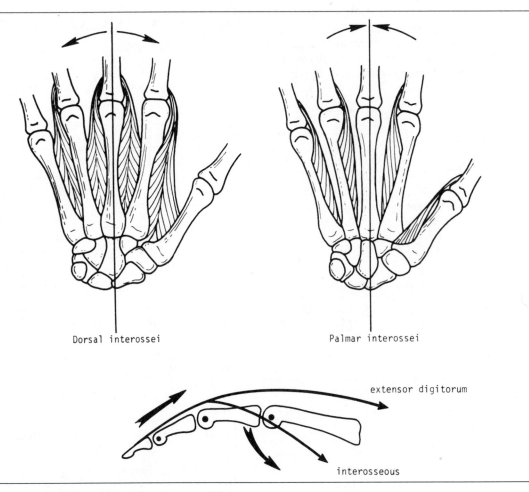

Dorsal interossei                    Palmar interossei

extensor digitorum

interosseous

**Fig. 9-59. Origins and insertions of palmar and dorsal interossei muscles; actions of these muscles are also shown.**

This complex movement involves a flexion of the carpometacarpal and metacarpophalangeal joints and a small amount of abduction and medial rotation of the metacarpal bone at the carpometacarpal joint.

## ADDUCTOR POLLICIS (FIG. 9-57)

### Origin

The *oblique head* arises from the anterior surfaces of the bases of the second and third metacarpals and the adjoining carpal bones. The *transverse head* arises from the anterior surface of the shaft of the third metacarpal bone.

### Insertion

The fibers from the two heads converge and are inserted with the first palmar interosseus muscle by a common tendon into the medial side of the base of the proximal phalanx of the thumb. A small sesamoid bone is usually present in the combined tendon.

### Nerve Supply

A deep branch of the ulnar nerve.

Action

Adduction of the thumb at the carpometacarpal and at the metacarpophalangeal joint.

Adduction of the thumb may be defined as a movement backward of the abducted thumb in the anteroposterior plane. It restores the thumb to its anatomical position, which is flush with the palm. The adductor pollicis is the muscle which, in association with the flexor pollicis longus and the opponens pollicis muscles, is largely responsible for the power of the pincers grip of the thumb.

## Short Muscles of the Little Finger

The short muscles of the little finger are the abductor digiti minimi, the flexor digiti minimi brevis, and the opponens digiti minimi, all of which together form the *hypothenar eminence.*

### ABDUCTOR DIGITI MINIMI (FIGS. 9-53 AND 9-56)

Origin

From the pisiform bone.

Insertion

Into the medial side of the base of the proximal phalanx of the little finger.

Nerve Supply

Deep branch of the ulnar nerve.

Action

Abducts the little finger at the metacarpophalangeal joint.

### FLEXOR DIGITI MINIMI (FIGS. 9-53 AND 9-56)

Origin

From the anterior surface of the flexor retinaculum.

Insertion

Into the medial side of the base of the proximal phalanx of the little finger.

Nerve Supply

Deep branch of the ulnar nerve.

Action

Flexes the little finger at the metacarpophalangeal joint.

### OPPONENS DIGITI MINIMI (FIGS. 9-56 AND 9-57)

Origin

From the anterior surface of the flexor retinaculum.

Insertion

Into the whole length of the medial border of the fifth metacarpal bone.

Nerve Supply

Deep branch of the ulnar nerve.

Action

This muscle is capable of rotating only the fifth metacarpal bone to a slight degree. However, it assists the flexor digiti minimi in flexing the carpometacarpal joint of the little finger, thereby pulling the fifth metacarpal bone forward and cupping the palm.

### *Small Muscles of the Hand—Their Nerve Supply and Action*

Students wishing to review the small muscles of the hand should study Table 9-8.

Table 9-8. Small Muscles of the Hand

| Name of muscle | Origin | Insertion | Nerve supply | Action |
|---|---|---|---|---|
| Palmaris brevis | Flexor retinaculum, palmar aponeurosis | Skin of palm | Superficial branch of ulnar nerve | Corrugates skin to improve grip of palm |
| Lumbricals (4) | Tendons of flexor digitorum profundus | Extensor expansion of medial four fingers | First and second, i.e., lateral two, median nerve; third and fourth deep branch of ulnar nerve | Flex metacarpophalangeal joints and extend interphalangeal joints of fingers except thumb |
| Interossei (8) Palmar (4) | First arises from base of first metacarpal; remaining three from anterior surface of shafts of 2, 3, and 4 metacarpals | Proximal phalanges of thumb, index, ring, and little fingers and dorsal extensor expansion of each finger (Fig. 9-59) | Deep branch of ulnar nerve | Palmar interossei adduct fingers toward center of third finger |
| Dorsal (4) | Contiguous sides of shafts of metacarpal bones | Proximal phalanges of index, middle, and ring fingers and dorsal extensor expansion (Fig. 9-59) | Deep branch of ulnar nerve | Dorsal interossei abduct fingers from center of third finger; both palmar and dorsal flex the metacarpophalangeal joints and extend the interphalangeal joints |
| SHORT MUSCLES OF THUMB | | | | |
| Abductor pollicis brevis | Scaphoid, trapezium, flexor retinaculum | Base of proximal phalanx of thumb | Median nerve | Abduction of thumb |
| Flexor pollicis brevis | Flexor retinaculum | Base of proximal phalanx of thumb | Median nerve | Flexes metacarpophalangeal joint of thumb |
| Opponens pollicis | Flexor retinaculum | Shaft of metacarpal bone of thumb | Median nerve | Pulls thumb medially and forward across palm |
| Adductor pollicis | Oblique head; second and third metacarpal bones; transverse head; third metacarpal bone | Base of proximal phalanx of thumb | Deep branch of ulnar nerve | Adduction of thumb |

Table 9-8. (continued)

| Name of muscle | Origin | Insertion | Nerve supply | Action |
|---|---|---|---|---|
| SHORT MUSCLES OF LITTLE FINGER | | | | |
| Abductor digiti minimi | Pisiform bone | Base of proximal phalanx of little finger | Deep branch of ulnar nerve | Abducts little finger |
| Flexor digiti minimi | Flexor retinaculum | Base of proximal phalanx of little finger | Deep branch of ulnar nerve | Flexes little finger |
| Opponens digiti minimi | Flexor retinaculum | Medial border of fifth metacarpal bone | Deep branch of ulnar nerve | Pulls fifth metacarpal forward as in cupping the palm |

## Arteries of the Palm of the Hand

### ULNAR ARTERY

The ulnar artery enters the hand anterior to the flexor retinaculum on the lateral side of the ulnar nerve and the pisiform bone (Fig. 9-56). The artery gives off a deep branch and then continues into the palm as the superficial palmar arch.

The *superficial palmar arch* is a direct continuation of the ulnar artery (Fig. 9-56). On entering the palm, it curves laterally behind the palmar aponeurosis and in front of the long flexor tendons. The arch is completed on the lateral side by one of the branches of the radial artery, either the superficial palmar branch, the radialis indicis, or the princeps pollicis. The curve of the arch lies across the palm, level with the distal border of the fully extended thumb.

Four *digital arteries* arise from the convexity of the arch and pass to the fingers (Fig. 9-56). The most medial artery supplies the medial side of the little finger, and the remaining three subdivide into two and supply the contiguous sides of the little, ring, middle, and index fingers, respectively.

The *deep branch of the ulnar artery* arises in front of the flexor retinaculum, passes between the abductor digiti minimi and the flexor digiti minimi, and joins the radial artery to complete the deep palmar arch (Figs. 9-57 and 9-58).

### RADIAL ARTERY

The radial artery leaves the dorsum of the hand by turning forward between the proximal ends of the first and second metacarpal bones and the two heads of the first dorsal interosseous muscle. (See p. 506.) On entering the palm, it curves medially between the oblique and transverse heads of the adductor pollicis and continues as the deep palmar arch (Figs. 9-57 and 9-58).

The *deep palmar arch* is a direct continuation of the radial artery (Fig. 9-58). It curves medially beneath the long flexor tendons and is in contact posteriorly with the metacarpal bones and the interosseous muscles. The arch is completed on the medial side by the deep branch of the ulnar artery. The curve of the arch lies across the upper part of the palm at a level with the proximal border of the extended thumb. The deep branch of the ulnar nerve lies within its concavity (Fig. 9-58).

The deep palmar arch sends branches superiorly, which take part in the anastomosis around the wrist joint, and inferiorly, to join the digital branches of the superficial palmar arch.

### Branches of the Radial Artery in the Palm

Immediately on entering the palm, the radial artery gives off: (1) the *arteria radialis indicis*, which supplies the lateral side of the index finger, and (2) the *arteria princeps pollicis*, which divides into two and supplies the lateral and medial sides of the thumb.

## Nerves of the Palm of the Hand

### MEDIAN NERVE

The median nerve enters the palm *beneath* the flexor retinaculum (Fig. 9-56) and immediately divides into lateral and medial branches. The *lateral branch* gives off: (1) a *muscular branch*, which supplies the abductor pollicis brevis, the flexor pollicis brevis, the opponens pollicis, and the first lumbrical muscles; and (2) *cutaneous branches* to both sides of the anterior surface of the thumb and the lateral side of the index finger.

The *medial branch* divides into two, which give off: (1) a *muscular branch* to the second lumbrical muscle, and (2) *cutaneous branches* to the adjacent sides of the index and middle fingers and the adjacent sides of the middle and ring fingers.

*In summary:* In the palm, the median nerve supplies the three muscles of the thenar eminence and the first two lumbrical muscles and gives cutaneous branches to the palmar aspect of the lateral three and one-half fingers. Note that the digital branches not only supply the whole palmar aspect of the fingers, but supply the distal half of the dorsal aspect of each finger as well. Note also that the *palmar cutaneous branch* of the median nerve given off in the front of the forearm (Fig. 9-53) crosses *anterior* to the flexor retinaculum and supplies the skin over the lateral part of the palm (Fig. 9-34).

## ULNAR NERVE

The ulnar nerve enters the palm *anterior* to the flexor retinaculum alongside the lateral border of the pisiform bone (Figs. 9-53 and 9-56). As it crosses the retinaculum, it divides into a superficial and a deep terminal branch.

### Superficial Branch of the Ulnar Nerve

The superficial branch of the ulnar nerve descends into the palm, lying in the subcutaneous tissue between the pisiform bone and the hook of the hamate (Figs. 9-53 and 9-56). The ulnar artery is on its lateral side. It gives off the following branches: (1) a *muscular branch* to the palmaris brevis and (2) *cutaneous branches* to the palmar aspect of the medial side of the little finger and the adjacent sides of the little and ring fingers (Fig. 9-56).

### Deep Branch of the Ulnar Nerve

The deep branch of the ulnar nerve runs backward between the abductor digiti minimi and the flexor

digiti minimi (Fig. 9-57). It pierces the opponens digiti minimi, winds around the lower border of the hook of the hamate, and passes laterally within the concavity of the deep palmar arch. The nerve lies behind the long flexor tendons and in front of the metacarpal bones and interosseous muscles. It gives off *muscular branches* to the abductor digiti minimi, the flexor digiti minimi, and the opponens digiti minimi. It supplies all the palmar and dorsal interossei, the third and fourth lumbrical muscles, and both heads of the adductor pollicis muscle.

*In summary:* In the palm, the ulnar nerve supplies the three muscles of the hypothenar eminence, the palmaris brevis, the third and fourth lumbrical muscles, all the interossei muscles, and the adductor pollicis muscle. It gives cutaneous branches to the palmar aspect of the medial one and one-half fingers. Note that the digital branches supply not only the whole palmar aspect of the fingers, but also the distal half of the dorsal aspect of each finger. Note also that the palmar cutaneous branch of the ulnar nerve given off in the front of the forearm (Fig. 9-52) crosses *anterior* to the flexor retinaculum (Fig. 9-52) and supplies the skin over the medial part of the palm (Fig. 9-34).

## Fascial Spaces of the Palm

Normally, the fascial spaces of the palm are potential spaces filled with loose connective tissue. Their boundaries are important clinically, since they may limit the spread of infection in the palm.

The triangle-shaped palmar aponeurosis fans out from the lower border of the flexor retinaculum (Fig. 9-53). From its medial border a fibrous septum passes backward and is attached to the anterior border of the fifth metacarpal bone (Fig. 9-60). Medial to this septum is a fascial compartment containing the three hypothenar muscles; this compartment is unimportant clinically. From the lateral border of the palmar aponeurosis, a second fibrous septum passes obliquely backward to the anterior border of the third metacarpal bone (Fig. 9-60). Usually, the septum passes between the long flexor tendons of the index and middle fingers. This second septum divides the palm up into the *thenar space*, which lies lateral to the septum (and must not be confused with the fascial compartment containing the thenar muscles), and the *midpalmar space*, which lies medial to the septum

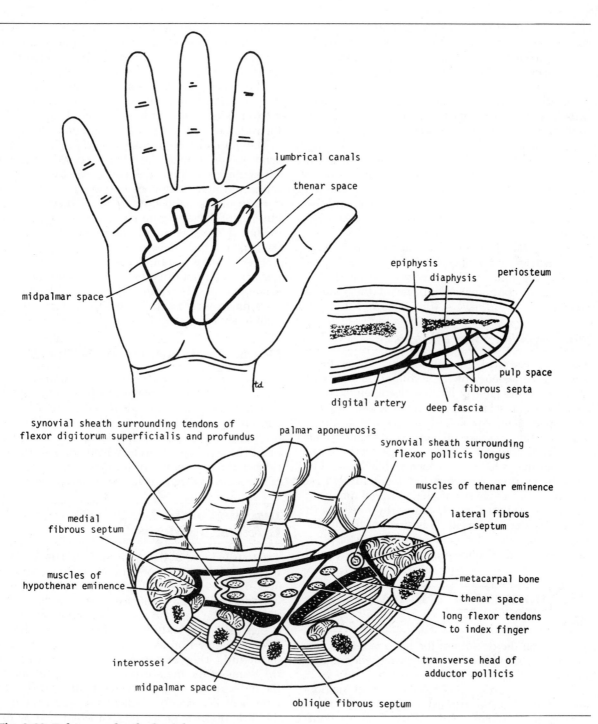

**Fig. 9-60. Palmar and pulp fascial spaces.**

(Fig. 9-60). Proximally, the thenar and midpalmar spaces are closed off from the forearm by the walls of the carpal tunnel. Distally, the two spaces are continuous with the appropriate lumbrical canals (Fig. 9-60).

*The thenar space* contains the first lumbrical muscle and lies posterior to the long flexor tendons to the index finger and in front of the adductor pollicis muscle (Fig. 9-60).

*The midpalmar space* contains the second, third, and fourth lumbrical muscles and lies posterior to the long flexor tendons to the middle, ring, and little fingers. It lies in front of the interossei and the third, fourth, and fifth metacarpal bones (Fig. 9-60)

*The lumbrical canal* is a potential space surrounding the tendon of each lumbrical muscle and is normally filled with connective tissue. Proximally, it is continuous with one of the palmar spaces.

## Pulp Space of the Fingers

The deep fascia of the pulp of each finger fuses with the periosteum of the terminal phalanx just distal to the insertion of the long flexor tendons and closes off a fascial compartment, known as the pulp space (Fig. 9-60). Each pulp space is subdivided by the presence of numerous septa, which pass from the deep fascia to the periosteum. Through the pulp space, which is filled with fat, runs the terminal branch of the digital artery that supplies the diaphysis of the terminal phalanx. Thrombosis of this vessel, caused by infection of the pulp space, will result in necrosis of the diaphysis of this bone. The epiphysis of the distal phalanx receives its blood supply proximal to the pulp space.

## THE DORSUM OF THE HAND

### Skin

The skin on the dorsum of the hand is thin, hairy, and freely mobile on the underlying tendons and bones.

The *sensory nerve supply* to the skin on the dorsum of the hand is derived from the superficial branch of the radial nerve and the posterior cutaneous branch of the ulnar nerve.

The *superficial branch of the radial nerve* winds

around the radius deep to the brachioradialis tendon, descends over the extensor retinaculum, and supplies the lateral two-thirds of the dorsum of the hand (Fig. 9-34). It divides into a number of dorsal digital nerves, which supply the thumb, the index finger, and the lateral side of the middle finger. The area of skin on the back of the hand supplied by the radial nerve is subject to variation.

The *posterior cutaneous branch of the ulnar nerve* winds around the ulna deep to the flexor carpi ulnaris tendon, descends over the extensor retinaculum, and supplies the medial third of the dorsum of the hand (Fig. 9-34). It divides into a number of dorsal digital nerves, which supply the medial side of the middle finger and the sides of the ring and little fingers.

The dorsal digital branches of the radial and ulnar nerves do not extend very far beyond the proximal phalanx. The remainder of the dorsum of each finger receives its nerve supply from palmar digital nerves.

## Dorsal Venous Arch (or Network)

The dorsal venous arch lies in the subcutaneous tissue proximal to the metacarpophalangeal joints and drains on the lateral side into the cephalic vein, and on the medial side, into the basilic vein (Fig. 9-7). The greater part of the blood from the whole hand drains into the arch, which receives digital veins and freely communicates with the deep veins of the palm through the interosseous spaces.

## Insertion of the Long Extensor Tendons

The four tendons of the extensor digitorum emerge from under the extensor retinaculum and fan out over the dorsum of the hand (Fig. 9-61). The tendons are embedded in the deep fascia, and together they form the roof of a *subfascial space*, which occupies the whole width of the dorsum of the hand. Strong oblique fibrous bands connect the tendons to the little, ring, and middle fingers, proximal to the heads of the metacarpal bones. The tendon to the index finger is joined on its medial side by the tendon of the extensor indicis, and the tendon to the little finger is joined on its medial side by the two tendons of the extensor digiti minimi (Fig. 9-61).

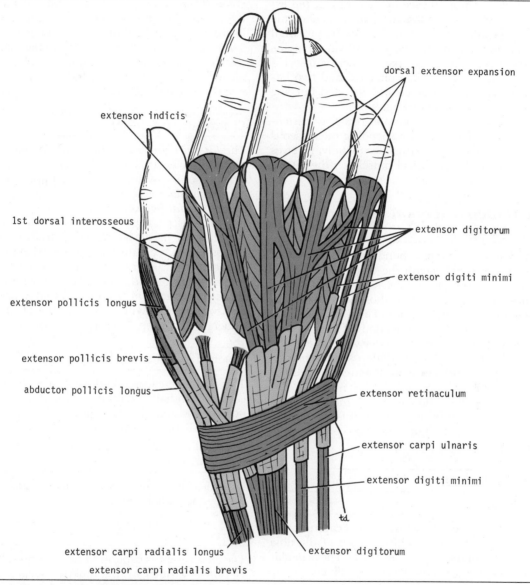

**Fig. 9-61. Dorsal surface of hand, showing long extensor tendons and their synovial sheaths.**

On the posterior surface of each finger, the extensor tendon joins the fascial expansion called the *extensor expansion* (Fig. 9-61). Near the proximal interphalangeal joint, the extensor expansion splits into three parts: a central part, which is inserted into the base of the middle phalanx, and two lateral parts, which converge to be inserted into the base of the distal phalanx (Fig. 9-55).

The dorsal extensor expansion receives the tendon of insertion of the corresponding interosseous muscle on each side, and farther distally, receives the tendon of the lumbrical muscle on the lateral side (Fig. 9-55).

## The Radial Artery on the Dorsum of the Hand

The radial artery winds around the lateral margin of the wrist joint, beneath the tendons of the abductor pollicis longus and extensor pollicis brevis, and lies on the lateral ligament of the joint (Fig. 9-51). On reaching the dorsum of the hand, the artery descends beneath the tendon of the extensor pollicis longus to reach the interval between the two heads of the first dorsal interosseous muscle; here, the artery turns forward to enter the palm of the hand. (See p. 501.)

*Branches of the radial artery on the dorsum of the hand* take part in the anastomosis around the wrist joint. Dorsal digital arteries pass to the thumb and index finger (Fig. 9-51).

## ELBOW JOINT

### Articulation

This occurs between the trochlea and capitulum of the humerus and the trochlear notch of the ulna and the head of the radius (Fig. 9-62). The articular surfaces are covered with hyaline cartilage.

### Type

It is a synovial hinge joint.

### Capsule

*Anteriorly*, it is attached above to the humerus along the upper margins of the coronoid and radial fossae and to the front of the medial and lateral epicondyles. Below, it is attached to the margin of the coronoid process of the ulna and to the anular ligament, which surrounds the head of the radius.

*Posteriorly*, it is attached above to the margins of the olecranon fossa of the humerus. Below, it is attached to the upper margin and sides of the olecranon process of the ulna and to the anular ligament.

### Ligaments (Fig. 9-62).

The *lateral ligament* is triangular in shape and is attached by its apex to the lateral epicondyle of the humerus, and by its base to the upper margin of the anular ligament.

The *medial ligament* is also triangular in shape and consists principally of three strong bands: (1) The anterior band, which passes from the medial epicondyle of the humerus to the medial margin of the coronoid process. (2) The posterior band, which passes from the medial epicondyle of the humerus to the medial side of the olecranon. (3) The transverse band, which passes between the ulnar attachments of the two preceding bands.

### Synovial Membrane

This lines the capsule and covers the floors of the coronoid, radial, and olecranon fossae; it is continuous below with the synovial membrane of the superior radioulnar joint.

### Nerve Supply

Branches from the median, ulnar, musculocutaneous, and radial nerves.

### Movements

The elbow joint is capable of flexion and extension. Flexion is limited by the anterior surfaces of the forearm and arm coming into contact. Extension is checked by the tension of the anterior ligament and the brachialis muscle. *Flexion* is performed by the brachialis, biceps brachii, brachioradialis, and pronator teres muscles. *Extension* is performed by the triceps and anconeus muscles.

It should be noted that the long axis of the extended forearm lies at an angle to the long axis of the arm. This angle, which opens laterally, is called the *carrying angle* and is about 170 degrees in the

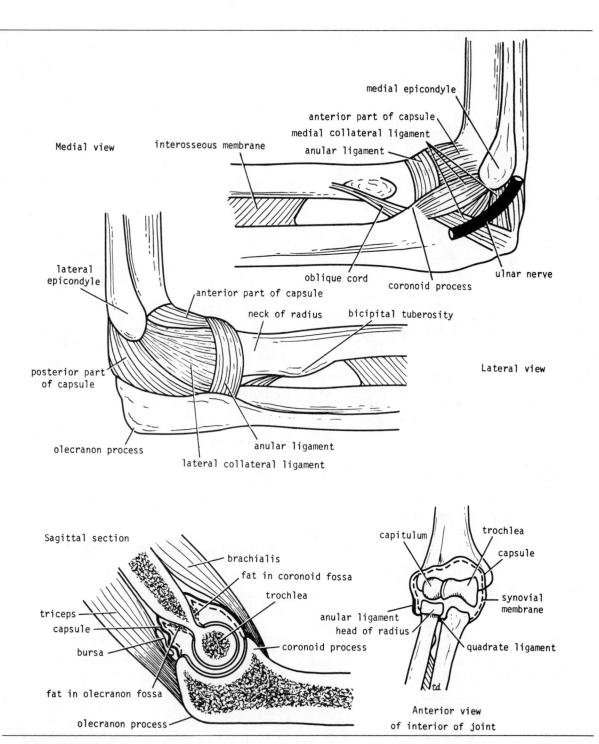

**Fig. 9-62. Right elbow joint.**

male and 167 degrees in the female. The angle disappears when the elbow joint is fully flexed.

### Important Relations

#### Anteriorly

The brachialis, the tendon of the biceps, the median nerve, and the brachial artery.

#### Posteriorly

The triceps muscle, a small bursa intervening.

#### Medially

*The ulnar nerve passes behind the medial epicondyle and crosses the medial ligament of the joint.*

#### Laterally

The common extensor tendon and the supinator.

The elbow joint is a very stable hinge joint because of the wrench shape of the trochlear notch of the ulna, which fits around the pulley-shaped trochlea of the humerus. The joint is also strengthened by strong medial and lateral collateral ligaments.

## SUPERIOR RADIOULNAR JOINT

### Articulation

Between the circumference of the head of the radius and the anular ligament and the radial notch on the ulna (Figs. 9-62 and 9-63).

### Type

It is a synovial pivot joint.

### Capsule

The capsule encloses the joint and is continuous with that of the elbow joint.

### Ligaments

The *anular ligament* is attached to the anterior and posterior margins of the radial notch on the ulna

and forms a collar around the head of the radius (Fig. 9-63). It is continuous above with the capsule of the elbow joint. It is not attached to the radius.

The *small quadrate ligament* extends between the neck of the radius and the ulna, just below the radial notch.

### Synovial Membrane

This is continuous above with that of the elbow joint. Below, it is attached to the inferior margin of the articular surface of the radius and the lower margin of the radial notch of the ulna.

### Nerve Supply

Branches of the median, ulnar, musculocutaneous, and radial nerves.

### Movements

Pronation and supination of the forearm (see below).

### Important Relations

#### Anteriorly

Supinator muscle and the radial nerve.

#### Posteriorly

Supinator muscle and the common extensor tendon.

## INFERIOR RADIOULNAR JOINT

### Articulation

Between the rounded head of the ulna and the ulnar notch on the radius (Fig. 9-63).

### Type

It is a synovial pivot joint.

### Capsule

The capsule encloses the joint, but is deficient superiorly.

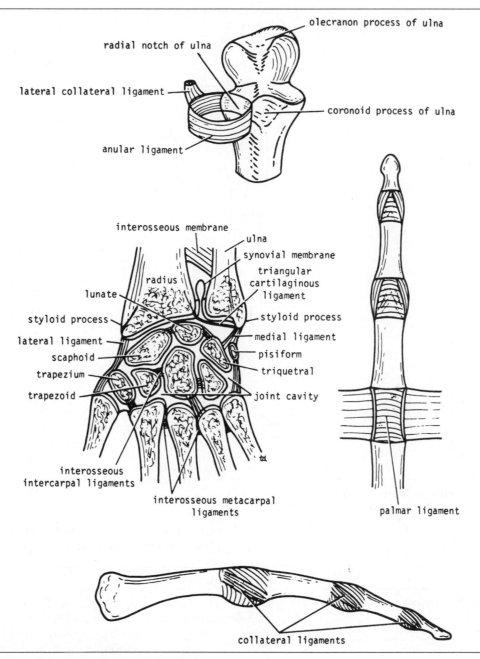

**Fig. 9-63. Ligaments of superior and inferior radioulnar joints, wrist joint, carpal joints, and joints of the fingers.**

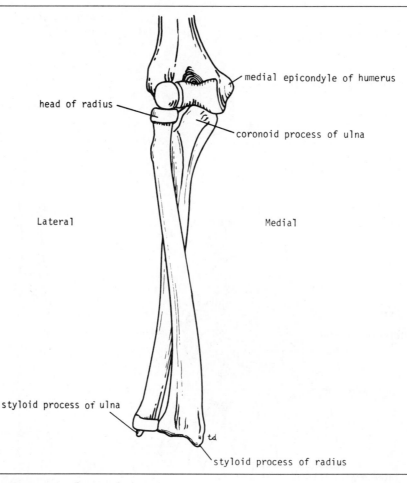

head of radius

medial epicondyle of humerus

coronoid process of ulna

Lateral

Medial

styloid process of ulna

td

styloid process of radius

**Fig. 9-64. Relative positions of radius and ulna when forearm is fully pronated.**

ferior radioulnar joint from the wrist and strongly unites the radius to the ulna.

## Ligaments

Weak *anterior* and *posterior* ligaments strengthen the capsule.

## Articular Disc

This is triangular in shape and composed of fibrocartilage. It is attached by its apex to the lateral side of the base of the styloid process of the ulna and by its base to the lower border of the ulnar notch of the radius (Fig. 9-63). It shuts off the in-

## Synovial Membrane

This lines the capsule passing from the edge of one articular surface to that of the other. A pouch of synovial membrane projects upward in front of the interosseous membrane for a variable distance beyond the joint.

## Nerve Supply

Anterior interosseous nerve and deep branch of radial nerve.

## Movements

The movements of pronation and supination of the forearm involve a rotary movement around a vertical axis at the superior and inferior radioulnar joints. The axis passes through the head of the radius above and the attachment of the apex of the triangular articular disc below.

In the movement of pronation, the head of the radius rotates within the anular ligament, while the distal end of the radius with the hand moves bodily forward, the ulnar notch of the radius moving around the circumference of the head of the ulna (Fig. 9-64). In addition, the distal end of the ulna moves laterally, so that the hand remains in line with the upper limb and is not displaced medially. This movement of the ulna is important when using an instrument such as a screwdriver, since it prevents side-to-side movement of the hand during the repetitive movements of supination and pronation.

The movement of pronation results in the hand's rotating medially in such a manner that the palm comes to face posteriorly, and the thumb lies on the medial side. The movement of supination is a reversal of this process, so that the hand returns to the anatomical position, and the palm faces anteriorly.

*Pronation* is performed by the pronator teres and the pronator quadratus.

*Supination* is performed by the biceps brachii and the supinator. Supination is the more powerful of the two movements because of the strength of the biceps muscle. Since supination is the more powerful movement, screw threads and the spiral of corkscrews are made so that the screw and corkscrews are driven inward by the movement of supination in right-handed people.

### *Important Relations*

#### Anteriorly

The tendons of flexor digitorum profundus.

#### Posteriorly

The tendon of extensor digiti minimi.

# RADIOCARPAL, OR WRIST, JOINT

## Articulation

Between the lower end of the radius and the articular disc above, and the scaphoid, lunate, and triquetral bones below (Fig. 9-63). The proximal articular surface forms an ellipsoid concave surface, which is adapted to the distal ellipsoid convex surface.

## Type

It is a synovial ellipsoid joint.

## Capsule

The capsule encloses the joint and is attached above to the lower ends of the radius and ulna, and below to the proximal row of carpal bones.

## Ligaments

*Anterior* and *posterior ligaments* strengthen the capsule.

The *medial ligament* is attached to the styloid process of the ulna and to the triquetral bone (Fig. 9-63).

The *lateral ligament* is attached to the styloid process of the radius and to the scaphoid bone (Fig. 9-63).

## Synovial Membrane

This lines the capsule and is attached to the margins of the articular surfaces. The joint cavity does not communicate with that of the inferior radioulnar joint, nor with the joint cavities of the intercarpal joints.

## Nerve Supply

Anterior interosseous nerve and the deep branch of the radial nerve.

## Movements

The following movements are possible: flexion, extension, abduction, adduction, and circumduction.

Rotation is *not* possible, since the articular surfaces are ellipsoid in shape. The lack of rotation is compensated for by the movements of pronation and supination of the forearm.

*Flexion* is performed by the flexor carpi radialis, the flexor carpi ulnaris, and the palmaris longus. These muscles are assisted by the flexor digitorum superficialis, the flexor digitorum profundus, and the flexor pollicis longus.

*Extension* is performed by the extensor carpi radialis longus, the extensor carpi radialis brevis, and the extensor carpi ulnaris. These muscles are assisted by the extensor digitorum, the extensor indicis, the extensor digiti minimi, and the extensor pollicis longus.

*Abduction* is performed by the flexor carpi radialis and the extensor carpi radialis longus and brevis. These muscles are assisted by the abductor pollicis longus and extensor pollicis longus and brevis.

*Adduction* is performed by the flexor and extensor carpi ulnaris.

### *Important Relations*

#### Anteriorly

The tendons of the flexor digitorum profundus and superficialis, the flexor pollicis longus, the flexor carpi radialis, the flexor carpi ulnaris, and the median and ulnar nerves.

#### Posteriorly

The tendons of the extensor carpi ulnaris, the extensor digiti minimi, the extensor digitorum, the extensor indicis, the extensor carpi radialis longus and brevis, the extensor pollicis longus and brevis, and the abductor pollicis longus.

#### Medially

The posterior cutaneous branch of the ulnar nerve.

#### Laterally

The radial artery.

## INTERCARPAL JOINTS

### Articulation

Between the individual bones of the proximal row of the carpus; between the individual bones of the distal row of the carpus; and finally, the *midcarpal joint*, between the proximal and distal rows of carpal bones (Fig. 9-63).

### Type

They are synovial plane joints.

### Capsule

The capsule surrounds each joint.

### Ligaments

The bones are united by strong *anterior, posterior,* and *interosseous ligaments.*

### Synovial Membrane

This lines the capsule and is attached to the margins of the articular surfaces. The joint cavity of the midcarpal joint extends not only between the two rows of carpal bones, but also upward between the individual bones forming the proximal row and downward between the bones of the distal row.

### Nerve Supply

Anterior interosseous nerve, deep branch of the radial nerve, and the deep branch of the ulnar nerve.

### Movements

A small amount of gliding movement is possible.

## CARPOMETACARPAL AND INTERMETACARPAL JOINTS

The carpometacarpal and intermetacarpal joints are synovial plane joints possessing anterior, posterior, and interosseous ligaments. They have a common joint cavity. A small amount of gliding movement is possible (Fig. 9-63).

## Carpometacarpal Joint of the Thumb

### Articulation

Between the trapezium and the saddle-shaped base of the first metacarpal bone (Fig. 9-63).

### Type

Synovial saddle-shaped joint.

### Capsule

The capsule surrounds the joint.

### Synovial Membrane

This lines the capsule and forms a separate joint cavity.

### Movements

The following movements are possible: flexion, extension, abduction, adduction, and a certain amount of rotation.

*Flexion.* Flexor pollicis brevis and opponens pollicis.

*Extension.* Extensor pollicis longus and brevis.

*Abduction.* Abductor pollicis longus and brevis.

*Adduction.* Adductor pollicis.

*Rotation (Opposition).* The thumb is rotated medially by the opponens pollicis.

## METACARPOPHALANGEAL JOINTS

### Articulations

Between the heads of the metacarpal bones and the bases of the proximal phalanges (Fig. 9-63).

### Type

Synovial ellipsoid joints.

### Capsule

The capsule surrounds the joint.

### Ligaments

The *palmar ligaments* are very strong and contain some fibrocartilage. They are firmly attached to the phalanx, but less so to the metacarpal bone (Fig. 9-63). The palmar ligaments of the second, third, fourth, and fifth joints are united by the *deep transverse metacarpal ligaments*, which hold the heads of the metacarpal bones together. The tendons of the lumbricals pass in front of the ligaments, and the tendons of the interossei pass behind the ligaments. The *collateral ligaments* are cordlike bands present on each side of the joints (Fig. 9-63). Each passes downward and forward from the head of the metacarpal bone to the base of the phalanx. The collateral ligaments are taut when the joint is in flexion and lax when the joint is in extension.

### Synovial Membrane

This lines the capsule and is attached to the margins of the articular surfaces.

### Movements

The following movements are possible: flexion, extension, abduction, and adduction.

*Flexion.* The lumbricals and the interossei, assisted by the flexor digitorum superficialis and profundus.

*Extension.* Extensor digitorum, extensor indicis, and extensor digiti minimi.

*Abduction.* Movement away from the midline of the third finger is performed by the dorsal interossei.

*Adduction.* Movement toward the midline of the third finger is performed by the palmar interossei. In the case of the metacarpophalangeal joint of the thumb, *flexion* is performed by the flexor pollicis longus and brevis, *extension* by the extensor pollicis longus and brevis. The movements of abduction and adduction are performed at the carpometacarpal joint.

## INTERPHALANGEAL JOINTS

Interphalangeal joints are synovial hinge joints that have a structure similar to that of the metacarpophalangeal joints (Fig. 9-63).

## THE HAND AS A FUNCTIONAL UNIT

The upper limb is a multijointed lever freely movable on the trunk at the shoulder joint. At the distal end of the upper limb is the important prehensile organ—the hand. Much of the importance of the hand is dependent on the pincers action of the thumb, which enables one to grasp objects between the thumb and index finger. The extreme mobility of the first metacarpal bone makes the thumb functionally as important as all the remaining fingers combined.

In order to comprehend fully the important positioning and movements of the hand described in this section, the reader is strongly advised to closely observe the movements in his own hand.

### Position of the Hand

For the hand to be able to perform delicate movements, such as those used in the holding of small instruments in watch repairing, the forearm is placed in the semiprone position, and the wrist joint is partially extended. It is interesting to note that the forearm bones are most stable in the midprone position, when the interosseous membrane is taut; in other positions of the forearm bones, the interosseous membrane is lax. With the wrist partially extended, the long flexor and extensor tendons of the fingers are working to their best mechanical advantage; at the same time, the flexors and extensors of the carpus can exert a balanced fixator action on the wrist joint, ensuring a stable base for the movements of the fingers.

The *position of rest* is the posture adopted by the hand when the fingers are at rest and the hand is relaxed (Fig. 9-65). The forearm is in the semiprone position; the wrist joint is slightly extended; the second, third, fourth, and fifth fingers are partially flexed, although the index finger is not flexed as much as the others; and the plane of the thumb-

nail lies at a right angle to the plane of the other fingernails.

The *position of function* is the posture adopted by the hand when it is about to grasp an object between the thumb and index finger (Fig. 9-65). The forearm is in the semiprone position, the wrist joint is partially extended (more so than in the position of rest), the fingers are partially flexed, the index finger being flexed as much as the others. The metacarpal bone of the thumb is rotated in such a manner that the plane of the thumbnail lies parallel with that of the index finger, and the pulp of the thumb and index finger are in contact.

The following movements are described with the hand in the anatomical position.

### Movements of the Thumb

*Flexion* is the movement of the thumb across the palm in such a manner as to maintain the plane of the thumbnail at right angles to the plane of the other fingernails (Fig. 9-65). The movement takes place between the trapezium and the first metacarpal bone, at the metacarpophalangeal and interphalangeal joints. The muscles producing the movement are the flexor pollicis longus and brevis and the opponens pollicis.

*Extension* is the movement of the thumb in a lateral or coronal plane away from the palm in such a manner as to maintain the plane of the thumbnail at right angles to the plane of the other fingernails (Figs. 9-65 and 9-66A). The movement takes place between the trapezium and the first metacarpal bone, at the metacarpophalangeal and interphalangeal joints. The muscles producing the movement are the extensor pollicis longus and brevis.

*Abduction* is the movement of the thumb in an anteroposterior plane away from the palm, the plane of the thumbnail being kept at right angles to the plane of the other nails (Figs. 9-65 and 9-67A). The movement takes place mainly between the trapezium and the first metacarpal bone; a small amount of movement takes place at the metacarpophalangeal joint. The muscles producing the movement are the abductor pollicis longus and brevis.

*Adduction* is the movement of the thumb in an anteroposterior plane toward the palm, the plane of the thumbnail being kept at right angles to the

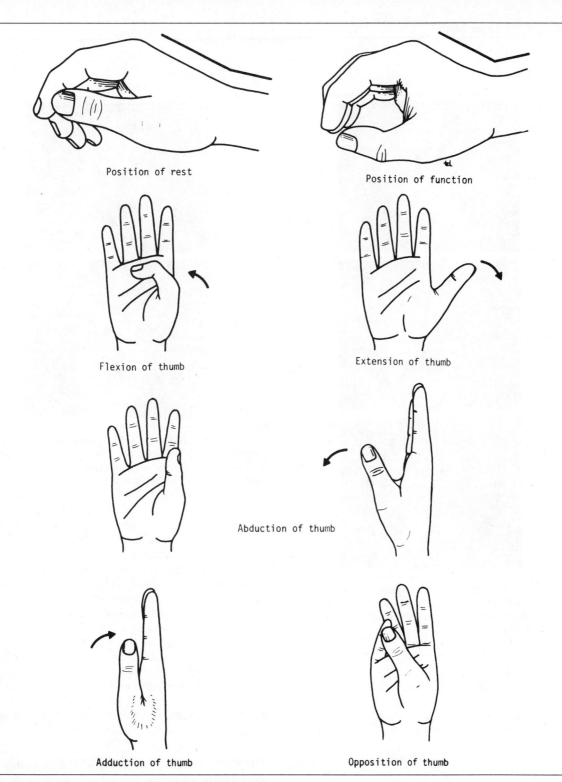

Position of rest

Position of function

Flexion of thumb

Extension of thumb

Abduction of thumb

Adduction of thumb

Opposition of thumb

**Fig. 9-65. Various positions of hand and movements of thumb.**

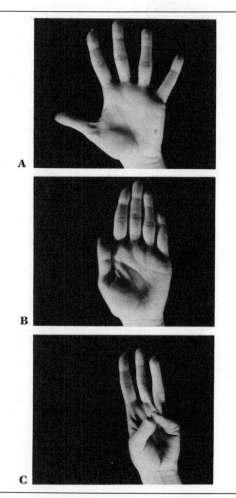

**Fig. 9-66. Left hand (A) with fingers abducted and thumb extended, (B) with fingers adducted and thumb adducted, and (C) with thumb in position of opposition.**

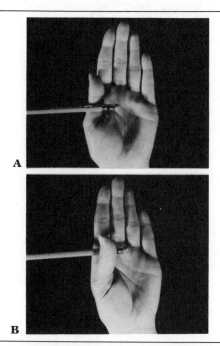

**Fig. 9-67. Left hand (A) with thumb about to move pencil away from palm to demonstrate abduction, and (B) with thumb about to move pencil in direction of palm to demonstrate adduction.**

plane of the other fingernails (Fig. 9-65 and 9-67B). The movement takes place between the trapezium and the first metacarpal bone. The muscle producing the movement is the adductor pollicis.

*Opposition* is the movement of the thumb across the palm in such a manner that the anterior surface of the tip comes into contact with the anterior surface of the tip of any of the other fingers (Figs. 9-65 and 9-66C). The movement is accomplished by the medial rotation of the first metacarpal bone and the attached phalanges on the trapezium. The plane of the thumbnail comes to lie parallel with the plane of the nail of the opposed finger. The muscle producing the movement is the opponens pollicis.

## Movements of the Index, Middle, Ring, and Little Fingers

*Flexion* is the movement forward of the finger in an anteroposterior plane. The movement takes place at the interphalangeal and metacarpophalangeal joints. The distal phalanx is flexed by the flexor digitorum profundus, the middle phalanx by the flexor digitorum superficialis, and the proximal phalanx by the lumbricals and the interossei.

*Extension* is the movement backward of the finger in an anteroposterior plane. The movements take place at the interphalangeal and metacarpophalangeal joints. The distal phalanx is extended by the lumbricals and interossei, the middle phalanx by the lumbricals and interossei, and the proximal phalanx by the extensor digitorum (in addition, by the extensor indicis for the index finger and the extensor digiti minimi for the little finger).

*Abduction* is the movement of the fingers (including the middle finger) away from the imaginary midline of the middle finger (Fig. 9-59 and 9-66A). The movement takes place at the metacarpophalangeal joint. The muscles producing the movement are the dorsal interossei; the abductor digiti minimi abducts the little finger.

*Adduction* is the movement of the fingers toward the midline of the middle finger (Fig. 9-66B). The movement takes place at the metacarpophalangeal joint. The muscles producing the movement are the palmar interossei.

It should be noted that abduction and adduction of the fingers is possible only in the extended position. In the flexed position of the finger, the articular surface of the base of the proximal phalanx lies in contact with the flattened anterior surface of the head of the metacarpal bone. The two bones are held in close contact by the collateral ligaments, which are taut in this position. In the extended position of the metacarpophalangeal joint, the base of the phalanx is in contact with the rounded part of the metacarpal head, and the collateral ligaments are slack.

## Cupping the Hand

In this position, the palm of the hand is formed into a deep concavity. To achieve this, the thumb is abducted and placed in a partially opposed position; it is also slightly flexed. This has the effect of drawing the thenar eminence forward.

The fourth and fifth metacarpal bones are flexed and slightly rotated at the carpometacarpal joints. This has the effect of drawing the hypothenar eminence forward. The palmaris brevis muscle contracts and pulls the skin over the hypothenar eminence medially; it also puckers the skin, which improves the gripping ability of the palm.

The index, middle, ring, and little fingers are partially flexed; the fingers are also rotated slightly at the metacarpophalangeal joints to increase the general concavity of the cupped hand.

## Making a Fist

Making a fist is accomplished by flexing the metacarpophalangeal joints and the interphalangeal joints of the fingers and thumb. It is performed by the contraction of the long flexor muscles of the fingers and thumb. For this movement to be carried out efficiently, there must be a synergic contraction of the extensor carpi radialis longus and brevis and the extensor carpi ulnaris muscles in order to extend the wrist joint. (Try to make a "strong fist" with the wrist joint flexed—it is very difficult.)

## RADIOGRAPHIC APPEARANCES OF THE UPPER LIMB

Radiological examination of the upper limb concentrates mainly on the bony structures, since the muscles, tendons, and nerves blend into a homogeneous mass. Blood vessels may be visualized by using special contrast media. The radiographic appearances of the upper limb of the adult as seen on routine X-ray examination will be described in this section. The practicing radiologist must be cognizant of the age changes that take place in the body and how these will influence the radiographic appearances. For example, knowing the times at which the primary and secondary centers of ossification appear in the different bones and the dates at which they fuse is fundamental, since without this information an epiphyseal line could be mistaken for a fracture. It is useful to remember that a person has two upper limbs and that the normal side may serve as a baseline for comparison with the potentially abnormal side.

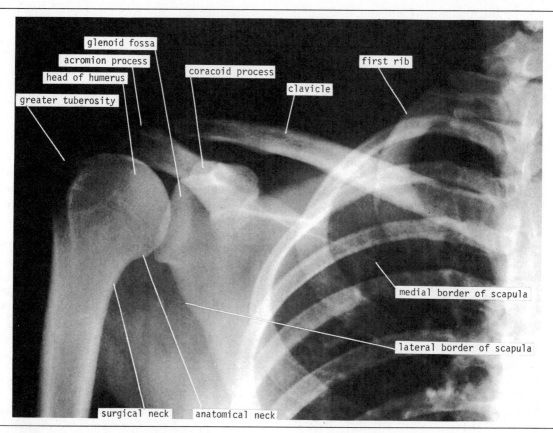

**Fig. 9-68. Anteroposterior radiograph of shoulder region in adult.**

## Radiographic Appearances of the Shoulder Region

The views of the shoulder region commonly used are (1) the anteroposterior and (2) the inferosuperior.

The *anteroposterior view* is taken with the film cassette placed posterior to the shoulder and the X-ray tube positioned in front of the shoulder. This view (Fig. 9-68) shows the outer two-thirds of the clavicle, separated from the acromion process of the scapula by a gap that represents the acromio-clavicular joint; the presence of the fibrocartilaginous disc within the joint explains the gap. The

acromion process is seen above the head of the humerus and continuous with the spine of the scapula. The coracoid process of the scapula is seen projecting upward and forward. The glenoid fossa is not seen in complete profile and is overlapped to a variable extent by the head of the humerus. It should be possible, however, to delineate the margins of the glenoid fossa. The greater part of the scapula is projected behind the upper part of the thoracic cage, and there is consequent loss of detail. The superior and inferior angles of the scapula are shown, as well as its superior, lateral, and medial borders.

**Fig. 9-69. Anteroposterior radiograph of elbow region in adult.**

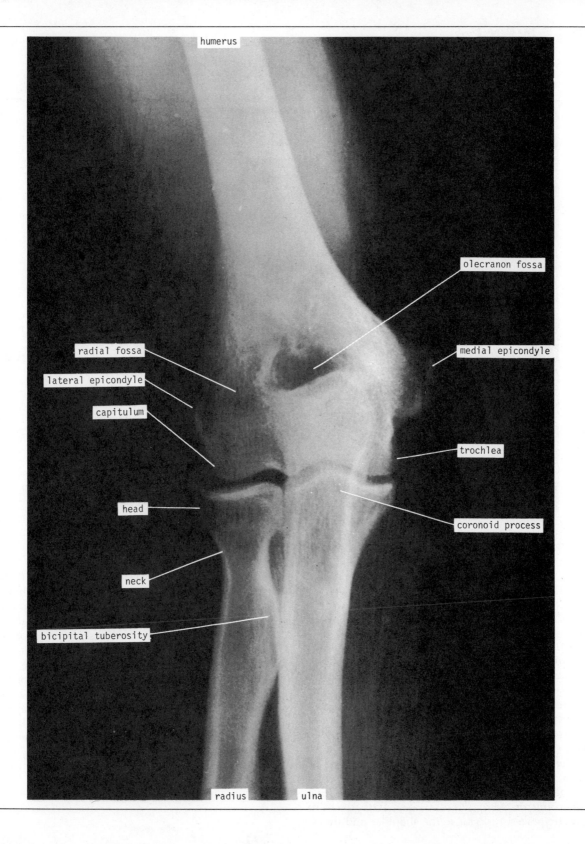

humerus

olecranon fossa

radial fossa

medial epicondyle

lateral epicondyle

capitulum

trochlea

head

coronoid process

neck

bicipital tuberosity

radius          ulna

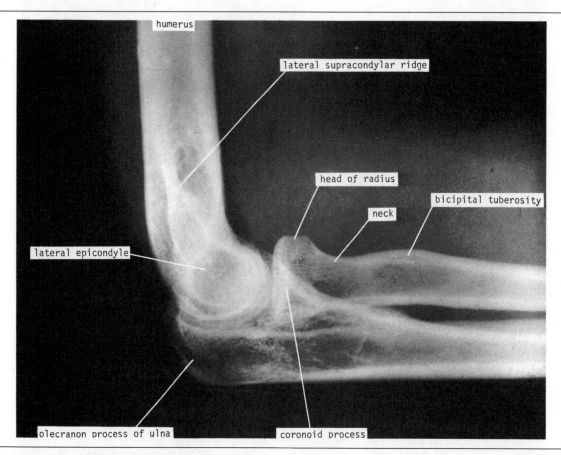

humerus

lateral supracondylar ridge

head of radius

bicipital tuberosity

neck

lateral epicondyle

olecranon process of ulna

coronoid process

**Fig. 9-70. Lateral radiograph of elbow region in adult.**

The upper third of the humerus is well visualized. The smooth, rounded head projects upward and medially, and the greater tuberosity projects laterally. The junction between the upper part of the head of the humerus and the anatomical neck shows as a notch. The lesser tuberosity is seen "face on"; although superimposed on the bone, it can usually be recognized. The bicipital groove cannot be seen. The surgical neck of the humerus is well seen.

The subacromial bursa and the tendons of the short muscles around the shoulder joint cannot normally be identified.

The *inferosuperior view* is taken with the film cassette placed superior to the shoulder and the X-ray tube positioned between the elbow and flank and directed upward through the axilla. The shoulder joint is abducted and externally rotated, and the forearm and hand are supported in a comfortable position. The coracoid process is seen projecting anteriorly; the acromion process and the spine of the scapula are seen posteriorly. The clavicle and the glenoid fossa are well seen. The lateral border of the scapula is seen as a shadow running posteromedially from the glenoid fossa, and since it is superimposed on other parts of the scapula, it may cause confusion. The rounded head of the humerus and the lesser tuberosity are well delineated. The surgical neck of the humerus is clearly seen.

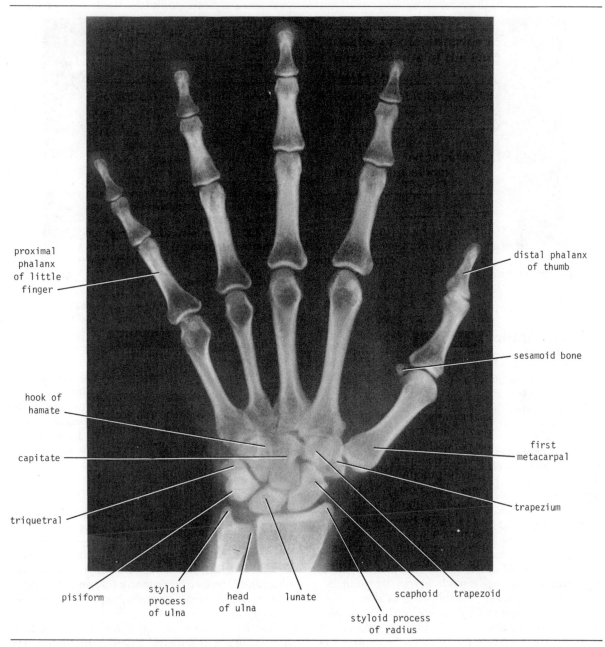

proximal
phalanx
of little
finger

distal phalanx
of thumb

sesamoid bone

hook of
hamate

first
metacarpal

capitate

triquetral

trapezium

pisiform

styloid
process
of ulna

head
of ulna

lunate

scaphoid

trapezoid

styloid process
of radius

**Fig. 9-71. Postero-anterior radiograph of adult
wrist and hand.**

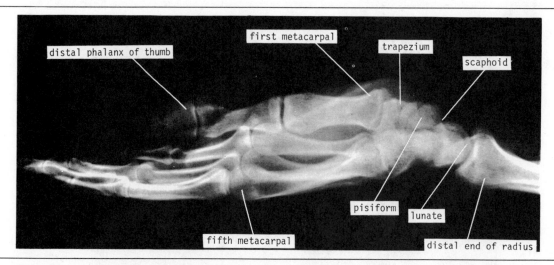

**Fig. 9-72. Lateral radiograph of adult wrist and hand.**

## Radiographic Appearances of the Elbow Region

The views of the elbow region commonly used are (1) anteroposterior and (2) lateral. The lower end of the humerus, the upper ends of the radius and ulna, and the elbow and superior radioulnar joints are visualized.

The *anteroposterior view* is taken with the arm immobilized, the elbow joint extended, and the radioulnar joints in the supine position. The film cassette is placed behind the elbow joint, and the X-ray tube is directed into the cubital fossa.

The lateral and medial epicondyles of the humerus are clearly seen (Fig. 9-69). The olecranon and coronoid fossae of the humerus, being superimposed, produce an area of transradiancy. A parallel translucent gap extends across the joint between the rounded capitulum and the upper surface of the head of the radius, and the trochlea and coronoid process. It is due to the presence of the articular cartilage covering the articular surfaces. The head, neck, and bicipital tuberosity of the radius are clearly seen. The olecranon and coronoid processes of the ulna are also seen, and the superior radioulnar joint may be visualized.

The *lateral view* is taken with the elbow joint flexed to 90 degrees. The shoulder joint is abducted to a right angle, and the arm is placed at the same level as the shoulder. The film cassette is placed against the medial epicondyle, and the X-ray tube is directed along an imaginary line connecting the two epicondyles. The medial and lateral supracondylar ridges and the medial and lateral epicondyles of the humerus are superimposed, but the latter may be recognized by tracing the long axis of the radius superiorly (Fig. 9-70). The olecranon and coronoid processes of the ulna may be seen. The greater part of the head of the radius may be visualized, although the posterior half is partially obscured by the coronoid process of the ulna.

## Radiographic Appearances of the Wrist and Hand

The views commonly used are (1) posteroanterior and (2) lateral. The lower ends of the radius and ulna, the inferior radioulnar joint, and the carpal and proximal ends of the metacarpal bones are visualized.

The *postero-anterior view* is taken with the forearm pronated and the fingers partially flexed. The film cassette is placed against the palm of the hand, and the X-ray tube is directed onto the dorsal surface of the hand. The lower ends of the radius and ulna, with their styloid processes, can be seen, and the radial styloid process is seen to extend farther distally than that of the ulna (Fig. 9-71). The

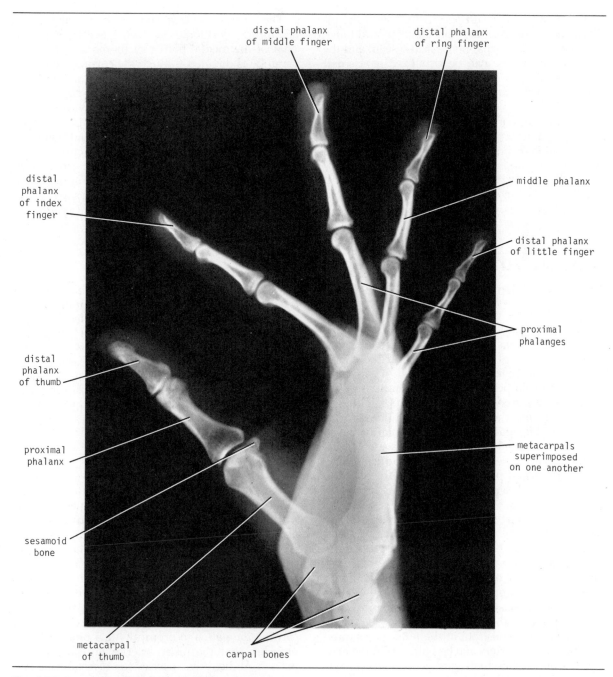

**Fig. 9-73. Lateral radiograph of adult wrist and hand with fingers at different degrees of flexion.**

proximal row of carpal bones is seen with the pisiform bone superimposed on the triquetral bone. The distal row of carpal bones is also seen, and the hook of the hamate can be visualized as a small oval area of increased density. The joint spaces of the carpal, wrist, and inferior radioulnar joints can be seen.

The different parts of the metacarpal bones and phalanges may also be seen. The sesamoid bones of the abductor pollicis brevis and flexor pollicis brevis tendons and the tendons of the adductor pollicis and the first palmar interosseous muscle can usually be recognized. The sesamoid bones overlap the first metacarpophalangeal joint.

The *lateral view* is taken with the forearm in the semiprone position. The film cassette is placed against the medial border of the hand and the X-ray tube is directed through the carpus (Figs. 9-72 and 9-73). The articulation of the radius with the lunate is well shown on this view. The concave distal surface of the lunate articulating with the capitate is also seen. The pisiform bone is visualized anteriorly and may overlap the scaphoid bone.

Owing to the great frequency of fractures of the scaphoid bone, and the difficulty often experienced in visualizing its midsection radiographically, a number of oblique views of the carpus are sometimes necessary.

## CLINICAL NOTES

### Arteries of the Upper Limb

The arteries of the upper limb may be damaged by penetrating wounds or may require ligation in amputation operations. Because of the existence of an adequate collateral circulation around the shoulder, elbow, and wrist joints, ligation of the main arteries of the upper limb is not followed by tissue necrosis or gangrene, provided, of course, that the arteries forming the collateral circulation are not diseased and the patient's general circulation is satisfactory. Nevertheless, it may take days or weeks for the collateral vessels to open up sufficiently to provide the distal part of the limb with the same volume of blood as previously supplied by the main artery.

A clinician must know where the arteries of the upper limb can be palpated or compressed in an emergency. The subclavian artery, as it crosses the first rib to become the axillary artery, may be palpated in the root of the posterior triangle of the neck. The artery may be compressed here against the first rib and so stop a catastrophic hemorrhage. The third part of the axillary artery may be felt in the axilla as it lies in front of the teres major muscle. The brachial artery can be palpated in the arm as it lies on the brachialis and is overlapped from the lateral side by the biceps brachii.

The radial artery lies superficially in front of the distal end of the radius, between the tendons of the brachioradialis and flexor carpi radialis; it is here that the clinician takes the radial pulse. If the pulse cannot be felt, try feeling for the radial artery on the other wrist; occasionally a congenitally abnormal radial artery may be difficult to feel. The radial artery may be less easily felt as it crosses the anatomical snuffbox.

The ulnar artery may be palpated as it crosses anterior to the flexor retinaculum in company with the ulnar nerve. The artery lies lateral to the pisiform bone, separated from it by the ulnar nerve. The artery is commonly damaged here in laceration wounds in front of the wrist.

### Allen Test

This test is used to determine the patency of the ulnar and radial arteries. With the patient's hands resting in his lap, compress the radial arteries against the anterior surface of each radius, and ask the patient to tightly clench his fists. The clenching of the fists closes off the superficial and deep palmar arterial arches. When the patient is asked to open his hands, the skin of the palms is at first white in color, and then normally the blood quickly flows into the arches through the ulnar arteries, causing the palms to promptly turn pink. This establishes that the ulnar arteries are patent. The patency of the radial arteries can be established by repeating the test, but this time compress the ulnar arteries as they lie lateral to the pisiform bones.

*The arteries of the upper limb are innervated by sympathetic nerves.* The preganglionic fibers originate from cell bodies in the second to eighth thoracic segments of the spinal cord. They ascend in the sympathetic trunk and synapse in the middle cervical, inferior cervical, first thoracic, or stellate ganglia. The postganglionic fibers join the nerves that form the brachial plexus and are distributed to the arteries within the branches of the plexus. For example, the digital arteries of the fingers are supplied by postganglionic sympathetic fibers that run in the digital nerves. Vasospastic diseases involving digital arterioles, such as *Raynaud's disease*, may require a cervicodorsal preganglionic sympathectomy to prevent necrosis of the fingers. The operation is followed by arterial vasodilation, with consequent increased blood flow to the upper limb.

## Veins of the Upper Limb

The veins of the upper limb may be divided into two groups, the superficial and the deep. The deep veins comprise the venae comitantes, which accompany all the large arteries, usually in pairs, and the axillary vein. *Spontaneous thrombosis* of the axillary vein occasionally occurs following excessive and unaccustomed movements of the arm at the shoulder joint. The superficial veins are clinically much more important and are used for venipuncture, transfusion, and cardiac catheterization. It is important that every physician, in an emergency, know where he can give blood or obtain blood from the arm. Unfortunately, when a patient is in a state of shock, the superficial veins are not always visible. The cephalic vein lies fairly constantly in the superficial fascia, immediately posterior to the styloid process of the radius. In the cubital fossa, the median cubital vein is separated from the underlying brachial artery by the bicipital aponeurosis. This is important, since it protects the artery from the mistaken introduction into its lumen of irritating drugs, which should have been injected into the vein. The cephalic vein, in the deltopectoral triangle, frequently communicates with the external jugular vein by a small vein that crosses in front of the clavicle. Fracture of the clavicle may result in rupture of this communicating vein, with the formation of a large hematoma.

## Lymphatics of the Upper Limb

Infection of the lymphatic vessels (*lymphangitis*) of the arm is a common occurrence. Red streaks along the course of the lymphatic vessels is characteristic of the condition. The lymphatics from the thumb and index finger and the lateral part of the hand follow the cephalic vein to the infraclavicular group of axillary nodes; those from the middle, ring, and little fingers, and from the medial part of the hand, follow the basilic vein to the supratrochlear node, which lies in the superficial fascia just above the medial epicondyle of the humerus, and thence to the lateral group of axillary nodes. Once the infection reaches the lymph nodes, they become enlarged and tender, a condition known as *lymphadenitis*. Most of the lymph vessels from the fingers and palm pass to the dorsum of the hand before passing up into the forearm. This explains the frequency of inflammatory edema, or even abscess formation, which may occur on the dorsum of the hand following infection of the fingers or palm.

## Mammary Glands

The mammary gland is one of the common sites of cancer in women. It is also the site of different types of benign tumors and may be subject to acute inflammation and abscess formation. For these reasons, the clinician must be familiar with the development, structure, and lymphatic drainage of this organ.

### Breast Examination

With the patient undressed to the waist and sitting upright, the mammary glands are first inspected for symmetry. Some degree of asymmetry is common and is the result of unequal breast development. Any swelling should be noted. A swelling may be due to an underlying tumor, cyst, or abscess formation. The nipples should be carefully examined for evidence of retraction. A carcinoma within the breast substance can cause retraction of the nipple by pulling on the lactiferous ducts. The patient is then asked to lie down so that the breasts can be palpated against the underlying thoracic wall. Finally the patient is asked to sit up again and raise both arms above her head. With this maneuver a carcinoma tethered to the skin, the suspensory ligaments, or the lactiferous ducts will produce dimpling of the skin or retraction of the nipple.

*Supernumerary nipples* occasionally occur along a line extending from the axilla to the groin; they may or may not be associated with breast tissue. This minor congenital anomaly may result in a mistaken diagnosis of warts or moles. A long-standing *retracted nipple* is a congenital deformity due to a failure in the complete development of the nipple. A retracted nipple of recent occurrence is usually due to an underlying carcinoma pulling on the lactiferous ducts.

The interior of the mammary gland is divided into fifteen to twenty compartments, which radiate out from the nipple by fibrous septa that extend from the deep surface of the skin. Each compartment contains a lobe of the gland. Normally, the skin feels completely mobile over the breast substance. However, should the fibrous septa become involved in a scirrhous carcinoma, or in a disease such as a breast abscess, which results in the production of contracting fibrous tissue, the septa will be pulled upon, causing dimpling of the skin. The fibrous septa are sometimes referred to as the *suspensory ligaments* of the mammary gland.

An acute infection of the mammary gland may occur during lactation. Pathogenic bacteria gain entrance to the breast tissue through a crack in the nipple. Because of the presence of the fibrous septa, the infection remains localized to one compartment or lobe to begin with. Should an abscess occur, it should be drained through a radial incision to avoid spreading of the infection into neighboring compartments; a radial incision will also minimize the damage to the radially arranged ducts.

The importance of knowing the lymphatic drainage of the breast in relation to the spread of cancer from that organ cannot be overemphasized. The lymphatic vessels from the medial half of the breast pierce the second, third, and fourth intercostal spaces and enter the thorax to drain into the lymph nodes alongside the internal thoracic artery. The lymphatic vessels from the lateral half of the breast drain into the anterior or pectoral group of axillary nodes. It follows, therefore, that a cancer occurring in the lateral half of the breast will tend to spread to the axillary nodes. Thoracic metastases are difficult or impossible to treat, but the lymph nodes of the axilla may be removed surgically.

Fortunately, about 60 percent of carcinomas of the breast occur in the upper lateral quadrant. The lymphatic spread of cancer to the opposite breast, to the abdominal cavity, or into lymph nodes in the root of the neck is due to obstruction of the normal lymphatic pathways by malignant cells or destruction of lymphatic vessels by surgery or radiotherapy. The cancer cells are swept along the lymphatic vessels and follow the lymph stream. The entrance of cancer cells into the blood vessels accounts for the metastases in distant bones.

In patients with localized cancer of the breast, most surgeons do a simple mastectomy followed by radiotherapy to the axillary lymph nodes. In patients with localized cancer of the breast with early metastases in the axillary lymph nodes, most authorities agree that radical mastectomy offers the best chance of cure. In those patients in whom the disease has already spread beyond these areas (e.g., into the thorax), simple mastectomy, followed by radiotherapy or hormone therapy, is the treatment of choice.

Radical mastectomy is designed to remove the primary tumor and the lymphatic vessels and nodes that drain the area. This means that the breast and the associated structures containing the lymphatic vessels and nodes must be removed en bloc. The excised mass is therefore made up of the following: (1) a large area of skin overlying the tumor and including the nipple; (2) all the breast tissue; (3) the pectoralis major and associated fascia through which the lymphatic vessels pass to the internal thoracic nodes; (4) the pectoralis minor and associated fascia related to the lymphatic vessels passing to the axilla; (5) all the fat, fascia, and lymph nodes in the axilla; (6) the fascia covering the upper part of the rectus sheath, the serratus anterior, the subscapularis, and the latissimus dorsi muscles. The axillary blood vessels, the brachial plexus, and the nerves to the serratus anterior and the latissimus dorsi are preserved. Some degree of postoperative edema of the arm is likely to follow such a radical removal of the lymphatic vessels draining the upper limb.

## Dermatomes and Cutaneous Nerves

It may be necessary for a physician to test the integrity of the spinal cord segments of C3 through T1. The diagrams in Figures 1-31 and 1-32 show the arrangement of the dermatomes of the upper limb. It is seen that the dermatomes for the upper cervical segments C3 to C6 are located along the lateral margin of the upper limb; the C7 dermatome is situated on the middle finger, and the dermatomes for C8, T1, and T2 are along the medial margin of the limb. It must be remembered that the nerve fibers from a particular segment of the spinal cord, although they exit from the cord in a spinal nerve of the same segment, pass to the skin in two or more different cutaneous nerves.

The skin over the point of the shoulder and halfway down the lateral surface of the deltoid muscle is supplied by the supraclavicular nerves (C3 and C4). Pain may be referred to this region as the result of inflammatory lesions involving the diaphragmatic pleura or peritoneum. The afferent stimuli reach the spinal cord via the phrenic nerves (C3, 4, and 5). Pleurisy, peritonitis, subphrenic abscess, or gallbladder disease may therefore be responsible for shoulder pain.

## Tendon Reflexes

The skeletal muscle receives a segmental innervation. Most muscles are innervated by several spinal nerves and therefore by several segments of the spinal cord. A physician should know the segmental innervation of the following muscles, since it is possible to test them by eliciting simple muscle reflexes in the patient.

*Biceps brachii tendon reflex* **C5** and 6 (flexion of the elbow joint by tapping the biceps tendon).
*Triceps tendon reflex* C6, **7,** and 8 (extension of the elbow joint by tapping the triceps tendon).
*Brachioradialis tendon reflex* C5, **6,** 7 (supination of the radioulnar joints by tapping the insertion of the brachioradialis tendon).

## Nerves of the Upper Limb

The roots, trunks, and divisions of the brachial plexus reside in the lower part of the posterior triangle of the neck, while the cords and most of the branches of the plexus lie in the axilla. Complete lesions involving all the roots of the plexus are rare. Incomplete injuries are common and are usually caused by traction or pressure; individual nerves may be divided by stab wounds.

### Upper Lesions of the Brachial Plexus (Erb-Duchenne Palsy)

Upper lesions of the brachial plexus are injuries resulting from excessive displacement of the head to the opposite side and depression of the shoulder on the same side. This causes excessive traction or even tearing of C5 and C6 roots of the plexus. It occurs in infants during a difficult delivery, or in adults following a blow or fall on the shoulder. The suprascapular nerve, the nerve to the subclavius, and the musculocutaneous and axillary nerves all possess nerve fibers derived from C5 and C6 roots and will therefore be functionless. The following muscles will consequently be paralyzed: (1) the supraspinatus and infraspinatus, (2) the subclavius, (3) the biceps brachii and the greater part of the brachialis and the coracobrachialis, and (4) the deltoid and the teres minor. Thus, the limb will hang limply by the side, medially rotated by the unopposed sternocostal part of the pectoralis major; the forearm will be pronated due to loss of the action of the biceps. The position of the upper limb in this condition has been likened to that of a porter or waiter hinting for a tip (Fig. 9-74). In addition, there will be a loss of sensation down the lateral side of the arm.

### Lower Lesions of the Brachial Plexus (Klumpke Palsy)

Lower lesions of the brachial plexus are usually traction injuries caused by excessive abduction of the arm, as occurs in the case of a person falling from a height clutching at an object to save himself. The first thoracic nerve is usually torn. The nerve fibers from this segment run in the ulnar and median nerves to supply *all the small muscles of the hand.* The hand has a clawed appearance due to hyperextension of the metacarpophalangeal joints and flexion of the interphalangeal joints. The extensor digitorum is unopposed by the lumbricals and interossei and extends the metacarpophalangeal joints; the flexor digitorum superficialis and profundus are unopposed by the lumbricals and interossei and flex the middle and terminal phalanges, respectively. There will be, in addition, a loss of sensation along the medial side of the arm. If the eighth cervical nerve is also damaged, the extent of

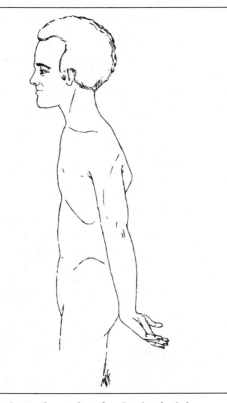

**Fig. 9-74. Erb-Duchenne's palsy (waiter's tip).**

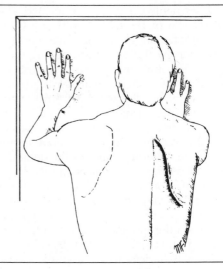

**Fig. 9-75. Winging of right scapula.**

anesthesia will be greater and will involve the medial side of the forearm, hand, and medial two fingers.

Lower lesions of the brachial plexus may also be produced by the presence of a cervical rib or malignant metastases from the lungs in the lower deep cervical lymph nodes.

## Axillary Sheath

The axillary sheath, formed of deep fascia derived from the prevertebral layer of deep fascia in the neck (see p. 713), encloses the axillary vessels and the brachial plexus. A brachial plexus nerve block can easily be obtained by closing the distal part of the sheath with finger pressure, inserting a syringe needle into the proximal part of the sheath, and then injecting a local anesthetic. The anesthetic solution is massaged along the sheath, producing a nerve block. The position of the sheath can be verified by feeling the pulsations of the third part of the axillary artery.

## LONG THORACIC NERVE

The long thoracic nerve, which arises from C5, 6, and 7 and supplies the serratus anterior muscle, may be injured by blows or pressure on the posterior triangle or during the surgical procedure of radical mastectomy. Paralysis of the serratus anterior results in the inability to rotate the scapula during the movement of abduction of the arm above a right angle. The patient therefore experiences difficulty in raising the arm above the head. The vertebral border and inferior angle of the scapula will no longer be kept closely applied to the chest wall and will protrude posteriorly, a condition known as *"winged scapula"* (Fig. 9-75).

## AXILLARY NERVE

The axillary nerve (Fig. 9-76), which arises from the posterior cord of the brachial plexus (C5 and 6), may be injured by the pressure of a badly adjusted crutch pressing upward into the armpit. The passage of the axillary nerve backward from the axilla through the quadrilateral space makes it particularly vulnerable here to downward displacement of the humeral head in shoulder dislocations

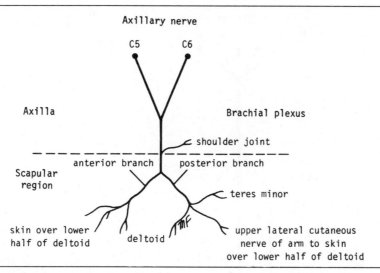

**Fig. 9-76. Summary diagram of main branches of axillary nerve.**

or fracture of the surgical neck of the humerus. Paralysis of the deltoid and teres minor muscles results. The cutaneous branches of the axillary nerve, including the upper lateral cutaneous nerve of the arm, are functionless, and consequently there is a loss of skin sensation over the *lower half* of the deltoid muscle. The paralyzed deltoid wastes rapidly, and the underlying greater tuberosity can be readily palpated. Since the supraspinatus is the only other abductor of the shoulder, this movement is much impaired. Paralysis of the teres minor is not recognizable clinically.

## RADIAL NERVE

The radial nerve (Fig. 9-77), which arises from the posterior cord of the brachial plexus, characteristically gives off its branches some distance proximal to the part to be innervated.

*In the axilla* it gives off three branches: (1) the posterior cutaneous nerve of the arm, which supplies the skin on the back of the arm down to the elbow; (2) the nerve to the long head of the triceps; and (3) the nerve to the medial head of the triceps.

*In the spiral groove* of the humerus it gives off four branches: (1) the lower lateral cutaneous nerve of the arm, which supplies the lateral surface of the arm down to the elbow; (2) the posterior cutaneous nerve of the forearm, which supplies the skin down the middle of the back of the forearm as far as the wrist; (3) the nerve to the lateral head of the triceps; and (4) the nerve to the medial head of the triceps and the anconeus.

*In the anterior compartment of the arm* above the lateral epicondyle it gives off three branches: (1) the nerve to a small part of the brachialis; (2) the nerve to the brachioradialis; and (3) the nerve to the extensor carpi radialis longus.

*In the cubital fossa* it gives off the deep branch of the radial nerve and continues as the superficial radial nerve. The deep branch supplies the extensor carpi radialis brevis and the supinator in the cubital fossa, and all the extensor muscles in the posterior compartment of the forearm. The superficial radial nerve is sensory and supplies the skin over the lateral part of the dorsum of the hand and the dorsal surface of the lateral three and one-half fingers proximal to the nail beds. (The ulnar nerve supplies the medial part of the dorsum of the hand and the dorsal surface of the medial one and one-half fingers; the exact cutaneous areas innervated by the radial and ulnar nerves on the hand are subject to variation.)

The radial nerve is commonly damaged in the axilla and in the spiral groove.

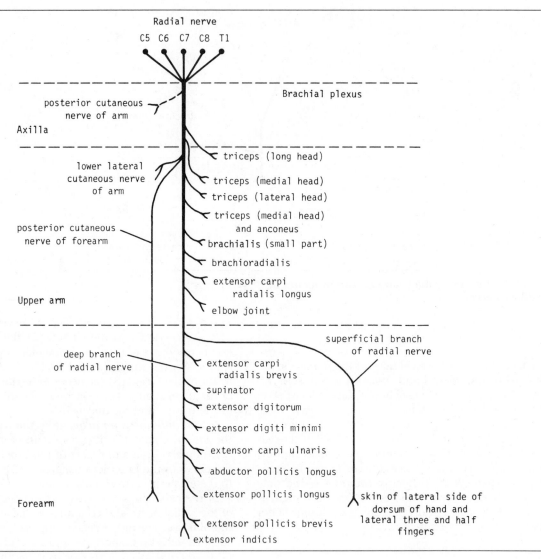

**Fig. 9-77. Summary diagram of main branches of radial nerve.**

## Injuries to Radial Nerve in Axilla

In the axilla the nerve may be injured by the pressure of the upper end of a badly fitting crutch pressing up into the armpit, or by a drunkard falling asleep with his arm over the back of a chair. It may also be badly damaged in the axilla by fractures and dislocations of the upper end of the humerus. When the humerus is displaced downward in dislocations of the shoulder, the radial nerve, which is wrapped around the back of the shaft of the bone, is pulled downward, stretching the nerve in the axilla excessively.

The clinical findings in injury to the radial nerve in the axilla are as follows:

### Motor

The triceps, the anconeus, and the long extensors of the wrist are paralyzed. The patient is unable to extend the elbow joint, the wrist joint, and the fingers. Wristdrop, or flexion of the wrist (Fig. 9-78), occurs as the result of the action of the unopposed flexor muscles of the wrist. Wristdrop is very disabling, since one is unable to flex the fingers strongly for the purpose of firmly gripping an object with the wrist fully flexed. (Try it on yourself.) If the wrist and proximal phalanges are passively extended by holding them in position with the opposite hand, the middle and distal phalanges of the fingers can be extended by the action of the lumbricals and interossei, which are inserted into the extensor expansions.

The brachioradialis and supinator muscles are also paralyzed, but supination is still performed well by the biceps brachii.

### Sensory

There is a small loss of skin sensation down the posterior surface of the lower part of the arm and down a narrow strip on the back of the forearm. There is also a variable area of sensory loss on the lateral part of the dorsum of the hand and base of the thumb. The area of total anesthesia is relatively small, due to the overlap of sensory innervation by adjacent nerves.

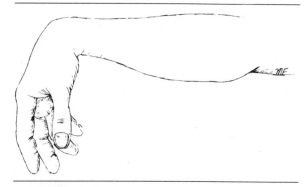

**Fig. 9-78. Wristdrop.**

### Trophic Changes

These are slight.

## Injuries to Radial Nerve in Spiral Groove

In the spiral groove of the humerus, the radial nerve may be injured at the time of fracture of the shaft of the humerus, or subsequently involved during the formation of the callus. The pressure of the back of the arm on the edge of the operating table in an unconscious patient has also been known to injure the nerve at this site. The prolonged application of a tourniquet to the arm in a person with a slender triceps muscle will often be followed by temporary radial palsy.

The clinical findings in injury to the radial nerve in the spiral groove are as follows:

The injury to the radial nerve occurs most commonly in the distal part of the groove, beyond the origin of the nerves to the triceps and the anconeus and beyond the origin of the cutaneous nerves.

### Motor

The patient is unable to extend the wrist and the fingers, and there is wristdrop (see above).

### Sensory

A variable small area of anesthesia is present over the root of the thumb.

## Trophic Changes

These are very slight or absent.

### Injuries to Deep Branch of Radial Nerve

The deep branch of the radial nerve, as it lies in the supinator muscle, may be accidentally damaged by the surgeon when trying to expose the head of the radius. In this instance, the nerve supply to the supinator and the extensor carpi radialis longus will be undamaged, and because the latter muscle is powerful, it will keep the wrist joint extended, and wristdrop will not occur. There will be no sensory loss, since this is a motor nerve.

### Injuries to Superficial Radial Nerve

Division of the superficial radial nerve, which is sensory, as in a stab wound, will result in a variable small area of anesthesia over the root of the thumb.

## MUSCULOCUTANEOUS NERVE

The musculocutaneous nerve (Fig. 9-79) is rarely injured because of its protected position beneath the biceps brachii muscle. If it is injured high up in the arm, the biceps and coracobrachialis are paralyzed and the brachialis muscle is weakened (the latter muscle is also supplied by the radial nerve). Flexion of the forearm at the elbow joint is then produced by the remainder of the brachialis muscle and the flexors of the forearm. When the forearm is in the prone position, the extensor carpi radialis longus and the brachioradialis muscles assist in flexion of the forearm. There is also sensory loss along the lateral side of the forearm. Wounds or cuts of the forearm can sever the lateral cutaneous nerve of the forearm, resulting in sensory loss along the lateral side of the forearm.

## MEDIAN NERVE

The median nerve (Fig. 9-79), which arises from the medial and lateral cords of the brachial plexus, gives off no cutaneous or motor branches in the axilla or in the arm. In the upper third of the front of the forearm, by unnamed branches or by its anterior interosseous branch, it supplies all the muscles of the front of the forearm except the flexor carpi ulnaris and the medial half of the flexor digitorum profundus, which are supplied by the ulnar nerve. In the lower third of the forearm, it gives rise to a palmar cutaneous branch, which crosses in front of the flexor retinaculum and supplies the skin on the lateral half of the palm. In the palm the median nerve supplies the muscles of the thenar eminence and the first two lumbricals and gives sensory innervation to the skin of the palmar aspect of the lateral three and one-half fingers, including the nail beds on the dorsum.

From the clinical standpoint, the median nerve is injured occasionally in the elbow region in supracondylar fractures of the humerus. It is most commonly injured by stab wounds or broken glass just proximal to the flexor retinaculum; here, it lies in the interval between the tendons of the flexor carpi radialis and flexor digitorum superficialis, overlapped by the palmaris longus.

The clinical findings in injury to the medial nerve are as follows:

### Injuries to the Median Nerve at the Elbow

#### Motor

The pronator muscles of the forearm and the long flexor muscles of the wrist and fingers, with the exception of the flexor carpi ulnaris and the medial half of the flexor digitorum profundus, will be paralyzed. As a result, the forearm is kept in the supine position; wrist flexion is weak and is accompanied by adduction. The latter deviation is due to the paralysis of the flexor carpi radialis and the strength of the flexor carpi ulnaris and the medial half of the flexor digitorum profundus. No flexion is possible at the interphalangeal joints of the index and middle fingers, although weak flexion of the metacarpophalangeal joints of these fingers is attempted by the interossei. When the patient tries to make a fist, the index and to a lesser extent the middle finger tend to remain straight, while the ring and little fingers flex (Fig. 9-80). The latter two fingers are, however, weakened by the loss of the flexor digitorum superficialis.

Flexion of the terminal phalanx of the thumb is lost due to paralysis of the flexor pollicis longus. The muscles of the thenar eminence are paralyzed and wasted, so that the eminence is flattened. The

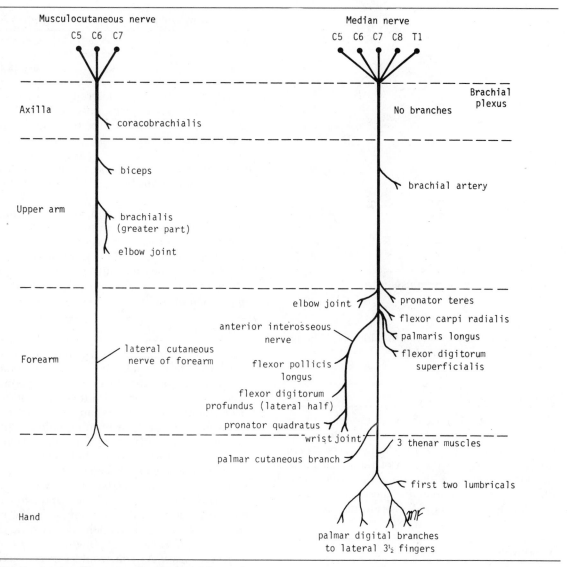

**Fig. 9-79. Summary diagram of main branches of musculocutaneous and median nerves.**

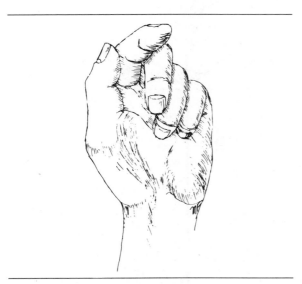

**Fig. 9-80. Median nerve palsy.**

thumb is laterally rotated and adducted. The hand looks flattened and "apelike."

## Sensory

There is loss of skin sensation of the lateral half or less of the palm of the hand and the palmar aspect of the lateral three and one-half fingers. There is also sensory loss of the skin of the distal parts of the dorsal surfaces of the lateral three and one-half fingers. The area of total anesthesia is considerably less, due to the overlap of adjacent nerves.

## Vasomotor Changes

The skin areas involved in sensory loss are warmer and drier than normal. This is due to the arteriolar dilatation and absence of sweating resulting from loss of sympathetic control.

## Trophic Changes

In long-standing cases, changes are found in the hand and fingers. The skin is dry and scaly, the nails crack easily, and there is atrophy of the pulp of the fingers.

## Injuries to the Median Nerve at the Wrist

### Motor

The muscles of the thenar eminence are paralyzed and wasted so that the eminence becomes flattened. The thumb is laterally rotated and adducted. The hand looks flattened and "apelike." Opposition movement of the thumb is impossible. The first two lumbricals are paralyzed, which can be recognized clinically when the patient is asked to make a fist slowly, and the index and middle fingers tend to lag behind the ring and little fingers.

*Sensory, vasomotor,* and *trophic changes* are identical to those found in the elbow lesions.

Perhaps the most serious disability of all in median nerve injuries is the loss of ability to oppose the thumb to the other fingers and the loss of sensation over the lateral fingers. The delicate pincerlike action of the hand is no longer possible.

## Carpal Tunnel Syndrome

The carpal tunnel, formed by the concave anterior surface of the carpal bones and closed by the flexor retinaculum, is tightly packed with the long flexor tendons of the fingers, with their surrounding synovial sheaths, and the median nerve. Clinically, the syndrome consists of a burning pain or "pins and needles" along the distribution of the median nerve to the lateral three and one-half fingers and weakness of the thenar muscles. It is produced by compression of the median nerve within the tunnel. The exact cause of the compression is difficult to determine, but thickening of the synovial sheaths of the flexor tendons or arthritic changes in the carpal bones are thought to be responsible in many cases. As you would expect, there is no paresthesia over the thenar eminence, since this area of skin is supplied by the palmar cutaneous branch of the median nerve, which passes superficially to the flexor retinaculum. The condition is dramatically relieved by decompressing the tunnel by making a longitudinal incision through the flexor retinaculum.

## Ulnar Nerve

The ulnar nerve (Fig. 9-81), which arises from the medial cord of the brachial plexus (C8 and T1),

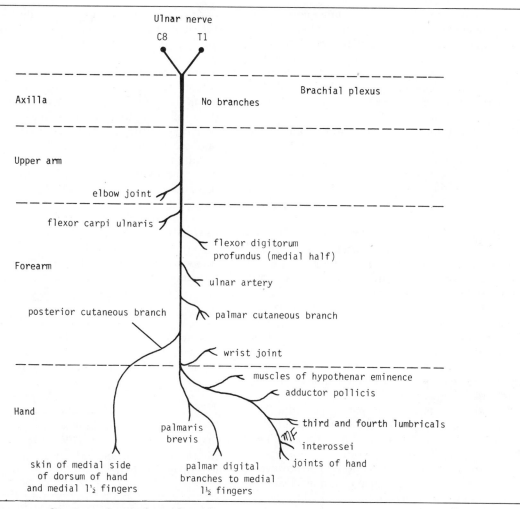

Ulnar nerve

C8     T1

Brachial plexus

Axilla          No branches

Upper arm

elbow joint

flexor carpi ulnaris

flexor digitorum
profundus (medial half)

Forearm          ulnar artery

posterior cutaneous branch     palmar cutaneous branch

wrist joint

muscles of hypothenar eminence

adductor pollicis

Hand

palmaris
brevis

third and fourth lumbricals

interossei

joints of hand

skin of medial side
of dorsum of hand
and medial 1½ fingers

palmar digital
branches to medial
1½ fingers

**Fig. 9-81. Summary diagram of main branches of
ulnar nerve.**

gives off no cutaneous or motor branches in the axilla or in the arm. As it enters the forearm from behind the medial epicondyle, it supplies the flexor carpi ulnaris and the medial half of the flexor digitorum profundus. In the distal third of the forearm, it gives off its palmar and posterior cutaneous branches. The palmar cutaneous branch supplies the skin over the hypothenar eminence; the posterior branch supplies the skin over the medial third of the dorsum of the hand and the medial one and one-half fingers. Not uncommonly, the posterior

branch supplies two and one-half instead of one and one-half fingers. It does not supply the skin over the distal part of the dorsum of these fingers.

Having entered the palm by passing in *front of the flexor retinaculum*, the *superficial branch* of the ulnar nerve supplies the skin of the palmar surface of the medial one and one-half fingers, including their nail beds; it also supplies the palmaris brevis muscle. The *deep branch* supplies all the small muscles of the hand except the muscles of the thenar eminence and the first two lumbricals, which are supplied by the median nerve.

The ulnar nerve is most commonly injured at the elbow, where it lies behind the medial epicondyle, and at the wrist, where it lies with the ulnar artery in front of the flexor retinaculum. The injuries at the elbow are usually associated with fractures of the medial epicondyle. The superficial position of the nerve at the wrist makes it very vulnerable to damage from cuts and stab wounds.

The clinical findings in injury to the ulnar nerve are as follows:

## Injuries to the Ulnar Nerve at the Elbow

### Motor

The flexor carpi ulnaris and the medial half of the flexor digitorum profundus muscles are paralyzed. The paralysis of the flexor carpi ulnaris can be observed by asking the patient to make a tightly clenched fist. Normally, the synergistic action of the flexor carpi ulnaris tendon can be observed as it passes to the pisiform bone; the tightening of the tendon will be absent if the muscle is paralyzed. The profundus tendons to the ring and little fingers will be functionless, and the terminal phalanges of these fingers are therefore not capable of being markedly flexed. Flexion of the wrist joint will result in abduction, owing to paralysis of the flexor carpi ulnaris. The medial border of the front of the forearm will show flattening, due to the wasting of the underlying ulnaris and profundus muscles.

The small muscles of the hand will be paralyzed, except the muscles of the thenar eminence and the first two lumbricals, which are supplied by the median nerve. The patient is unable to adduct and abduct the fingers and consequently is unable to grip a piece of paper placed between the fingers. Remember that the extensor digitorum can abduct the fingers to a small extent, but only when the metacarpophalangeal joints are hyperextended.

It is impossible to adduct the thumb, because the adductor pollicis muscle is paralyzed. If the patient is asked to grip a piece of paper between the thumb and the index finger, he does so by strongly contracting his flexor pollicis longus and flexing the terminal phalanx (Froment's sign).

The metacarpophalangeal joints become hyperextended due to the paralysis of the lumbrical and interosseous muscles, which normally flex these joints. Since the first and second lumbricals are not paralyzed (they are supplied by the median nerve), the hyperextension of the metacarpophalangeal joints is most prominent in the fourth and fifth fingers. The interphalangeal joints are flexed, due again to the paralysis of the lumbrical and interosseous muscles, which normally extend these joints through the extensor expansion. The flexion deformity at the interphalangeal joints of the fourth and fifth fingers is obvious, because the first and second lumbrical muscles of the index and middle fingers are not paralyzed. In long-standing cases the hand assumes the characteristic "claw" deformity (main en griffe). Wasting of the paralyzed muscles results in flattening of the hypothenar eminence and loss of the convex curve to the medial border of the hand. Examination of the dorsum of the hand will show hollowing between the metacarpal bones, due to wasting of the dorsal interosseous muscles (Fig. 9-82).

### Sensory

Loss of skin sensation will be observed over the anterior and posterior surfaces of the medial third of the hand and the medial one and one-half fingers.

### Vasomotor Changes

The skin areas involved in sensory loss are warmer and drier than normal. This is due to the arteriolar dilatation and absence of sweating resulting from loss of sympathetic control.

## Injuries to the Ulnar Nerve at the Wrist

### Motor

The small muscles of the hand will be paralyzed and show wasting, except for the muscles of the thenar eminence and the first two lumbricals, as described above. The clawhand is much more obvious in wrist lesions, since the flexor digitorum profundus muscle is not paralyzed, and there is marked flexion of the terminal phalanges.

### Sensory

The main ulnar nerve and its palmar cutaneous branch are usually severed; the posterior cutaneous branch, which arises from the ulnar nerve trunk

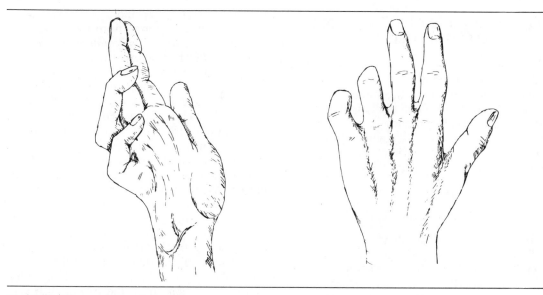

**Fig. 9-82. Ulnar nerve palsy.**

about 2½ inches (6 cm) above the pisiform bone, is usually unaffected. The sensory loss will therefore be confined to the palmar surface of the medial third of the hand and the medial one and one-half fingers and to the dorsal aspects of the middle and distal phalanges of the same fingers.

The *vasomotor* and *trophic changes* are the same as those described for injuries at the elbow. It is important to remember that with ulnar nerve injuries, the higher the lesion the less obvious is the clawing deformity of the hand.

Unlike median nerve injuries, lesions of the ulnar nerve leave a relatively efficient hand. The sensation over the lateral part of the hand is intact, and the pincer-like action of the thumb and index finger is reasonably good, although there is some weakness due to loss of the adductor pollicis.

## Muscles of the Upper Limb

### PECTORALIS MAJOR

Very occasionally parts of the pectoralis major muscle may be absent. The sternocostal origin is the most commonly missing part, and this causes weakness in adduction and medial rotation of the shoulder joint.

### PALMARIS LONGUS

This may be absent on one or both sides of the forearm in about 10 percent of persons. Others show variation in form such as centrally or distally placed muscle belly in the place of a proximal one. Since the muscle is relatively weak, its absence produces no disability.

### BICEPS BRACHII

The tendon of the long head of biceps is intracapsular but extrasynovial; it is attached to the supraglenoid tubercle within the shoulder joint. Advanced osteoarthritic changes in the joint may lead to erosion and fraying of the tendon by osteophytic outgrowths, and rupture of the tendon can occur.

### TENNIS ELBOW

This is believed to be due to a partial tearing or degeneration of the origin of the superficial extensor muscles from the lateral epicondyle of the humerus. It is a condition characterized by pain and tenderness over the lateral epicondyle of the humerus, with pain radiating down the lateral side of the forearm; it is common in tennis players, violinists, and housewives.

## ABDUCTOR·POLLICIS LONGUS TENDON AND EXTENSOR POLLICIS BREVIS TENDON

As the result of repeated friction between these tendons and the styloid process of the radius, they sometimes become edematous and swell. Later, fibrosis of the synovial sheath produces a condition known as *stenosing tenosynovitis* in which movement of the tendons becomes restricted. Advanced cases require surgical incision along the constricting sheath.

## EXTENSOR POLLICIS LONGUS TENDON

Rupture of this tendon can occur following fracture of the distal third of the radius. Roughening of the dorsal tubercle of the radius by the fracture line can cause excessive friction on the tendon, which may then rupture. Rheumatoid arthritis can also cause rupture of this tendon.

## MALLET FINGER

Avulsion of the insertion of one of the extensor tendons into the distal phalanges can occur if the distal phalanx is forcibly flexed when the extensor tendon is taut. The last 20 degrees of active extension is lost, resulting in a condition known as mallet finger.

## TRIGGER FINGER

In this condition there is a palpable and even audible snapping when a patient is asked to flex and extend his fingers. It is due to the presence of a localized swelling of one of the long flexor tendons that catches on a narrowing of the fibrous flexor sheath anterior to the metacarpophalangeal joint. It may take place either in flexion or in extension. A similar condition occurring in the thumb is called trigger thumb. The situation may be relieved surgically by incising the fibrous flexor sheath.

# Bones of the Upper Limb

## CLAVICLE

The clavicle is a strut that holds the arm laterally so that it may move freely on the trunk. Unfortu-

nately, because of its position, it is exposed to trauma and transmits forces from the upper limb to the trunk. *It is the most commonly fractured bone in the body.* The fracture usually occurs as the result of a fall on the shoulder or outstretched hand. The force is transmitted along the clavicle, which breaks at its weakest point, the junction of the middle and outer thirds. Following the fracture, the lateral fragment is depressed by the weight of the arm, and it is pulled medially and forward by the strong adductor muscles of the shoulder joint, especially the pectoralis major. The medial end is tilted upward by the sternocleidomastoid muscle.

The clavicle commences its ossification in membrane early in intrauterine life. A secondary center appears at the sternal end in the twentieth year, and this fuses with the shaft by about the twenty-fifth year. The epiphyseal plate that remains within the shaft of the bone until the twenty-fifth year must not be confused with a fracture line.

The close relationship of the supraclavicular nerves to the clavicle may result in their involvement in callus formation following fracture of the bone. This may be the cause of persistent pain over the side of the neck.

## SCAPULA

Fractures of the scapula are usually the result of severe trauma, such as occurs in runover accident victims or in occupants of automobiles involved in crashes. Injuries are usually associated with fractured ribs. Fortunately, most fractures of the scapula require little treatment because the muscles on the anterior and posterior surfaces adequately splint the fragments.

The position of the scapula on the posterior wall of the thorax is maintained by the tone and balance of the muscles attached to it. If one of these muscles is paralyzed, this balance is upset, as in dropped shoulder, which occurs with paralysis of the trapezius, or winged scapula (Fig. 9-75), which occurs with paralysis of the serratus anterior. Such imbalance can be detected by careful physical examination.

## HUMERUS

### Fractures of the Proximal End of the Humerus

The greater tuberosity may be fractured by direct trauma or avulsed by violent contractions of the supraspinatus muscle. The surgical neck of the humerus may be fractured by a direct blow on the lateral aspect of the shoulder or in an indirect manner by the person falling on the outstretched hand. In children, injury to the proximal epiphyseal cartilage may be followed by retardation of growth in the length of the humerus.

### Fractures of the Shaft of the Humerus

These fractures are common, with the displacement of the fragments dependent on the relation of the site of the fracture to the insertion of the deltoid. When the fracture line is proximal to the deltoid insertion, the proximal fragment is adducted by the pectoralis major, latissimus dorsi, and teres major muscles; the distal fragment is pulled proximally by the deltoid, biceps, and triceps. When the fracture is distal to the deltoid insertion, the proximal fragment is abducted by the deltoid, and the distal fragment is pulled proximally by the biceps and triceps. The radial nerve may be damaged in shaft fractures.

### Fractures of the Distal End of the Humerus

Supracondylar fractures are common in children and occur when the child falls on the outstretched hand with the elbow joint partially flexed. The median nerve and the brachial artery are occasionally injured in this type of fracture.

The medial epicondyle can be avulsed by the medial collateral ligament of the elbow joint if the forearm is forcibly abducted. The ulnar nerve may be injured at the time of a fracture, may become involved in the callus formation, or may undergo irritation on the irregular bony surface after the bone fragments are reunited.

## RADIUS AND ULNA

Fractures of the head of the radius can occur from falls on the outstretched hand. As the force is trans-mitted along the radius, the head of the radius is driven sharply against the capitulum, splitting or splintering the head.

Fractures of the neck of the radius occur in young children from falls on the outstretched hand.

Fractures of the shafts of the radius and ulna may or may not occur together. Displacement of the fragments is usually considerable and depends on the pull of the attached muscles. The proximal fragment of the radius is supinated by the supinator and the biceps brachii muscles. The distal fragment of the radius is pronated and pulled medially by the pronator quadratus muscle. The strength of the brachioradialis and extensor carpi radialis longus and brevis shortens and angulates the forearm. In fracture of the ulna, the ulna angulates posteriorly. In order to restore the normal movements of pronation and supination, the normal anatomical relationship of the radius, ulna, and interosseous membrane must be regained.

*Colles' fracture* is a fracture of the distal end of the radius resulting from a fall on the outstretched hand. The force drives the distal fragment posteriorly and superiorly, and the distal articular surface is inclined posteriorly. This posterior displacement produces a posterior bump, sometimes referred to as the "dinner-fork deformity," since the forearm and wrist resemble the shape of a dinner fork. Failure to restore the distal articular surface to its normal position will severely limit the range of flexion of the wrist joint.

*Smith's fracture* is a fracture of the distal end of the radius and occurs from a fall on the back of the hand. It is a reversed Colles' fracture because the distal fragment is displaced anteriorly.

## BONES OF THE HAND

Fracture of the *scaphoid* bone is common in young adults and, unless treated effectively, the fragments will not unite and permanent weakness of the wrist will result, with the subsequent development of osteoarthritis. The fracture line usually goes through the narrowest part of the bone, which because of its location is bathed in synovial fluid. The blood vessels to the scaphoid enter its proximal and distal ends, although the blood supply is occasionally confined to its distal end. If the latter occurs, a fracture deprives the proximal fragment of its arterial supply, and this fragment undergoes avascular necrosis. Deep tenderness in the anatomical snuffbox following a fall on the outstretched hand in a young adult makes one suspicious of a fractured scaphoid.

Dislocations of the *lunate* bone occasionally occur in young adults who fall on the outstretched hand in a way that causes hyperextension of the wrist joint. Involvement of the median nerve is common.

Fractures of the *metacarpal bones* can occur as the result of direct violence, in which case the fracture line is transverse and the fragments are displaced anteriorly. "Indirect violence," for example, striking an opponent with a clenched fist, commonly produces an oblique fracture of the necks of the fourth and fifth metacarpals. The distal fragment is commonly displaced proximally, thus shortening the finger posteriorly.

*Bennett's fracture* is a fracture of the base of the metacarpal of the thumb caused when violence is applied along the long axis of the thumb or the thumb is forcefully abducted. The fracture is oblique and enters the carpometacarpal joint of the thumb, causing joint instability.

Fractures of the phalanges are common and usually follow direct injury.

## Joints of the Upper Limb

### STERNOCLAVICULAR JOINT

The very strong costoclavicular ligament firmly holds the medial end of the clavicle to the first costal cartilage. Violent forces directed along the long axis of the clavicle usually result in fracture of that bone, but dislocation of the sternoclavicular joint takes place occasionally. The clavicle is then displaced forward and downward. Should the costoclavicular ligament rupture completely, it is difficult to maintain the normal position of the clavicle once reduction has been accomplished.

### ACROMIOCLAVICULAR JOINT

The plane of the articular surfaces of the acromioclavicular joint passes downward and medially, so that there is a tendency for the lateral end of the clavicle to ride up over the upper surface of the acromion. The strength of the joint is dependent on

the very strong coracoclavicular ligament, which binds the coracoid process to the undersurface of the lateral part of the clavicle. It should be remembered that the greater part of the weight of the upper limb is transmitted to the clavicle through this ligament, and that rotary movements of the scapula occur at this important ligament. A severe blow on the point of the shoulder, as is incurred during blocking or tackling in football or any severe fall, may result in the acromion's being thrust beneath the lateral end of the clavicle, tearing the coracoclavicular ligament. This condition is often referred to as "*shoulder separation*." The displaced outer end of the clavicle is easily palpable. As in the case of the sternoclavicular joint, the dislocation is easily reduced, but withdrawal of support results in immediate redislocation.

### SHOULDER JOINT

The shallowness of the glenoid fossa of the scapula and the lack of support provided by weak ligaments make this joint a very unstable structure. Its strength is almost entirely dependent on the tone of the short muscles that bind the upper end of the humerus to the scapula, namely, the subscapularis in front, the supraspinatus above, and the infraspinatus and teres minor behind. The tendons of these muscles together are referred to clinically as the *rotator cuff*. The least supported part of the joint lies in the inferior location, where it is unprotected by muscles.

*The shoulder joint is the most commonly dislocated large joint.* Sudden violence applied to the humerus with the joint fully abducted tilts the humeral head downward onto the inferior weak part of the capsule, which tears, and the humeral head comes to lie inferior to the glenoid fossa. During this movement the acromion has acted as a fulcrum. The strong flexors and adductors of the shoulder joint now usually pull the humeral head forward and upward into the subcoracoid position. Posterior dislocations are rare and are usually due to direct violence to the front of the joint. On inspection of the patient, the rounded appearance of the shoulder is seen to be lost, since the greater tuberosity of the humerus is no longer bulging laterally beneath the deltoid muscle. The subglenoid displacement of the head of the humerus into the quadrilateral space may cause damage to the axillary nerve, as indicated by paralysis of the deltoid muscle and loss of skin sensation over the lower half of the deltoid.

The *rotator cuff*, consisting of the tendons of the subscapularis, supraspinatus, infraspinatus, and teres minor muscles, which are fused to the underlying capsule of the shoulder joint, plays a very important role in stabilizing the shoulder joint. Lesions of the cuff are a common cause of pain in the shoulder region. During the movement of abduction of the shoulder joint, the supraspinatus tendon is exposed to friction against the acromion process. Under normal conditions the amount of friction is reduced to a minimum by the large subacromial bursa, which extends laterally beneath the deltoid. Degenerative changes in the bursa are followed by degenerative changes in the underlying supraspinatus tendon, and these may extend into the other tendons of the rotator cuff. Clinically, the condition is known as *subacromial bursitis*, *supraspinatus tendinitis*, or *pericapsulitis*. It is characterized by the presence of a spasm of pain in the middle range of abduction.

In advanced cases the necrotic supraspinatus tendon may become calcified or may rupture. Rupture of the tendon seriously interferes with the normal abduction movement of the shoulder joint. It will be remembered that the main function of the supraspinatus muscle is to hold the head of the humerus in the glenoid fossa at the commencement of abduction. The patient with a ruptured supraspinatus tendon is unable to initiate abduction of the arm. However, if the arm is passively assisted for the first 15 degrees of abduction, the deltoid can then take over and complete the movement to a right angle.

The synovial membrane, capsule, and ligaments of the shoulder joint are innervated by the axillary nerve and the suprascapular nerve. The joint is sensitive to pain, pressure, excessive traction, and distention. The muscles surrounding the joint undergo reflex spasm in response to pain originating in the joint, which in turn serves to immobilize the joint and thus reduce the pain.

Injury to the shoulder joint is followed by pain, limitation of movement, and muscle atrophy due to disuse. It is important to appreciate that pain in the shoulder region may be caused by disease elsewhere and that the shoulder joint may be normal; for example, diseases of the spinal cord and verte-bral column, and the pressure of a cervical rib (see p. 76) can all cause shoulder pain. Irritation of the diaphragmatic pleura or peritoneum can produce referred pain via the phrenic and supraclavicular nerves.

## ELBOW JOINT

The elbow joint is stable because of the wrench-shaped articular surface of the olecranon and the pulley-shaped trochlea of the humerus; it also has very strong medial and lateral ligaments. Posterior dislocations of the joint are common in children because the parts of the bones that stabilize the joint are incompletely developed in childhood. Avulsion of the epiphysis of the medial epicondyle is also common in childhood, because then the medial ligament is much stronger than the bond of union between the epiphysis and the diaphysis.

When examining the elbow joint, the physician must remember the normal relations of the bony points. In extension, the medial and lateral epicondyles and the top of the olecranon process are in a straight line; in flexion, the bony points form the boundaries of an equilateral triangle.

The anterior and posterior walls of the capsule are weak, and when the joint is distended with fluid, the posterior aspect of the joint becomes swollen. Aspiration of joint fluid may easily be performed through the back of the joint on either side of the olecranon process.

The close relationship of the ulnar nerve to the medial side of the joint often results in its becoming damaged in dislocations of the joint or in fracture dislocations in this region. The nerve lesion may occur at the time of injury or weeks, months, or years later. The nerve may be involved in scar tissue formation, or may become stretched due to lateral deviation of the forearm in a badly reduced supracondylar fracture of the humerus. During movements of the elbow joint, the continued friction between the medial epicondyle and the stretched ulnar nerve eventually results in ulnar palsy.

In examining lateral radiographs of the elbow region, it is important to remember that the lower end of the humerus is normally angulated forward 45 degrees on the shaft; and when examining a patient, the physician should see that the medial epicondyle, in the anatomical position, is directed medially and posteriorly and faces in the same direction as the head of the humerus.

A small subcutaneous *bursa* is present over the olecranon process of the ulna, and repeated minor trauma, such as occurs in miners, or students, often produces a *chronic bursitis.*

## RADIOULNAR JOINTS

The superior radioulnar joint communicates with the elbow joint, whereas the inferior radioulnar joint does not communicate with the wrist joint. In practical terms this means that infection of the elbow joint invariably involves the superior radioulnar joint. The strength of the superior radioulnar joint depends on the integrity of the strong anular ligament. Rupture of this ligament occurs in cases of anterior dislocation of the head of the radius on the capitulum of the humerus. In young children, in whom the head of the radius is still small and undeveloped, a sudden jerk on the arm may pull the radial head down through the anular ligament.

## WRIST JOINT

The wrist joint is essentially a synovial joint between the lower end of the radius and the proximal row of carpal bones. The head of the ulna is separated from the carpal bones by the strong triangular fibrocartilaginous ligament, which separates the wrist joint from the inferior radioulnar joint. The joint is stabilized by the very strong medial and lateral ligaments.

Because the styloid process of the radius is longer than that of the ulna, abduction of the wrist joint is less extensive than adduction. In flexion-extension movements, the hand can be flexed about 80 degrees, but extended to only about 45 degrees. The range of flexion is increased by movement at the midcarpal joint.

### *Falls on the Outstretched Hand*

In falls on the outstretched hand, forces are transmitted from the scaphoid to the lower end of the radius, from the radius across the interosseous membrane to the ulna, and from the ulna to the humerus; thence, through the glenoid fossa of the scapula to the coracoclavicular ligament and the clavicle and, finally, to the sternum. If the forces are excessive, different parts of the upper limb give way under the strain. The area affected seems to be related to age. In a young child, for example, there may be a posterior displacement of the distal radial epiphysis; in the teenager the clavicle might fracture; in the young adult the scaphoid is commonly fractured; and in the elderly the distal end of the radius is fractured about 1 inch (2.5 cm) proximal to the wrist joint (Colles' fracture).

### *Volkmann's Ischemic Contracture*

Volkmann's ischemic contracture is a contracture of the muscles of the forearm following fractures of the lower end of the humerus or fractures of the radius and ulna. In this syndrome a localized segment of the brachial artery goes into spasm, reducing the arterial flow to the flexor and the extensor muscles, so that they undergo ischemic necrosis. The flexor muscles are larger than the extensor muscles and they are therefore the ones mainly affected. The muscles are replaced by fibrous tissue, which contracts, producing the deformity. The arterial spasm is usually caused by an over-tight plaster cast, but in some cases the fracture itself may be responsible. The deformity can only be explained by understanding the anatomy of the region. Three types of deformity exist:

1. The long flexor muscles of the carpus and fingers are more contracted than the extensor muscles, and the wrist joint is flexed; the fingers are extended. If the wrist joint is extended passively, the fingers become flexed.
2. The long extensor muscles to the fingers, which are inserted into the extensor expansion that is attached to the proximal phalanx, are greatly contracted; the metacarpophalangeal joints and the wrist joint are extended, and the interphalangeal joints of the fingers are flexed.
3. Both the flexor and the extensor muscles of the forearm are contracted. The wrist joint is flexed, the metacarpophalangeal joints are extended, and the interphalangeal joints are flexed.

### *Palmar Aponeurosis*

The palmar aponeurosis may be regarded as the degenerated distal part of the tendon of the palmaris longus muscle. The tendon of the muscle is attached to the apex of the triangle-shaped aponeurosis. The base of the aponeurosis divides into four slips, one for each finger. The deeper part of each slip divides into two parts, which are inserted into the bases of the proximal phalanges and the fibrous flexor sheaths. The palmar aponeurosis is firmly attached to the skin of the palm and assists the latter in gripping an object; also, by virtue of its toughness, it protects the underlying tendons and synovial sheaths.

*Dupuytren's contracture*, which is a localized thickening and contracture of the palmar aponeurosis, commonly starts near the root of the ring finger and draws that finger into the palm, flexing it at the metacarpophalangeal joint. Later, the condition involves the little finger in the same manner. In long-standing cases, the pull on the fibrous sheaths of these fingers results in the flexion of the proximal interphalangeal joints. The distal interphalangeal joints are not involved and are actually extended by the pressure of the fingers against the palm.

### Synovial Sheaths of the Flexor Tendons

The arrangement of the synovial sheaths of the flexor tendons has been described previously (p. 490). They are essentially lubricating mechanisms, which permit the long flexor tendons to move smoothly, with the minimum of friction, beneath the flexor retinaculum. The vincula longa and brevia serve as mesenteries, which permit blood vessels to enter the tendons. The synovial sheath of the flexor pollicis longus (sometimes referred to as the *radial bursa*) communicates with the common synovial sheath of the superficialis and profundus tendons (sometimes referred to as the *ulnar bursa*) at the level of the wrist in about 50 percent of persons.

*Tenosynovitis* is an infection of a synovial sheath. It most commonly results from the introduction of bacteria into a sheath through a small penetrating wound, such as that made by the point of a needle. Rarely, the sheath may become infected by extension of a pulp-space infection.

Infection of a digital sheath results in the distention of the sheath with pus; the finger is held semiflexed and is very swollen. Any attempt to extend the finger is accompanied by extreme pain, since the distended sheath is stretched. As the inflammatory process continues, the pressure within the sheath rises, and the sheath may rupture at its proximal end. Anatomically, the digital sheath of the index finger is related to the thenar space, while that of the ring finger is related to the midpalmar space. The sheath for the middle finger is related to both the thenar and midpalmar spaces. These relationships explain how infection can extend from the digital synovial sheaths and involve the palmar fascial spaces.

In the case of infection of the digital sheaths of the little finger and thumb, the ulnar and radial bursae are quickly involved. Should such an infection be neglected, pus may burst through the proximal ends of these bursae and enter the fascial space of the forearm between the flexor digitorum profundus anteriorly and the pronator quadratus and the interosseous membrane posteriorly. This fascial space in the forearm is commonly referred to clinically as the *space of Parona*.

### Fascial Spaces of the Palm

The fascial spaces of the palm have been fully described previously. (See p. 502.) They are clinically important because they may become infected and distended with pus as the result of the spread of infection in acute suppurative tenosynovitis; rarely, they may become infected following penetrating wounds.

### Pulp Space of the Fingers

The pulp space of the fingers is a closed fascial compartment situated in front of the terminal phalanx of each finger. Infection of such a space is very common and serious, occurring most often in the thumb and index finger. Bacteria are usually introduced into the space by pinpricks or sewing needles. Since each space is subdivided into numerous smaller compartments by fibrous septa, it is easily understood that the accumulation of inflammatory exudate within these compartments will press upon and, in extreme cases, occlude the vessels traversing the compartments. Sometimes the tension is so great that soft-tissue necrosis occurs and the diaphysis of the terminal phalanx is destroyed. The proximally located epiphysis of this bone is saved, since it receives its arterial supply just proximal to the pulp space.

The close relationship of the proximal end of the pulp space to the digital synovial sheath accounts for the involvement of the sheath in the infectious process in cases in which the pulp-space infection has been neglected.

## Diseases of the Hand

From the clinical standpoint the hand is one of the most important organs of the body. Without a normally functioning hand the patient's livelihood is often in jeopardy. To the medical student who doubts this statement, I would suggest that he place his right (or left) hand in his pocket for 24 hours, and he will be astonished at the number of times he would like to use it if he could.

From the purely mechanical point of view, the hand may be regarded as a pincer-like mechanism between the thumb and fingers, situated at the end of a multijointed lever. *The most important part of the hand is the thumb*, and it is the physician's responsibility to preserve the thumb, or as much of it as possible, so that the pincer-like mechanism can be maintained. The pincer-like action of the thumb largely depends on its unique ability to be drawn across the palm and opposed to the other fingers. Unfortunately, this movement alone, although important, is insufficient for the mechanism to work effectively. The opposing skin surfaces must have tactile sensation—and this explains why median nerve palsy is so much more disabling than ulnar nerve palsy.

Should the hand require immobilization for the treatment of disease of any part of the upper limb, it should be immobilized (if possible) in the *position of function*. This means that if there is a loss of movement at the wrist joint, or at the joints of the hand or fingers, the patient will at least have a hand that is in a position of mechanical advantage, and one that can serve a useful purpose.

Physicians should also remember that when the fingers (excluding the thumb) are normally flexed into the palm, they point to the tubercle of the scaphoid; fingers requiring immobilization in flexion, on a splint or within a plaster cast, should therefore always be placed in this position.

*Always refer to the patient's fingers by name:* thumb, index, middle, ring, and little finger. Numbering the fingers is confusing (is the thumb a finger?) and has led to such disastrous results as amputating the wrong finger.

## CLINICAL PROBLEMS
*Answers on page 968*

1. A middle-aged man with a history of chronic duodenal ulcer was seen in the emergency room in a state of severe shock. He was pale, restless, and sweating, and his blood pressure was 80/60 mm Hg. The resident made a diagnosis of internal hemorrhage, probably due to erosion of the gastroduodenal artery or one of its branches, and decided to set up a blood transfusion immediately. Based on your knowledge of anatomy, into which superficial vein of the upper limb would you perform the transfusion: (1) in the elbow region or (2) in the forearm? If the veins were too collapsed to be recognized, where, in an emergency, could you cut down on a superficial vein in the upper limb?

2. A young woman, 18 years of age, visited her physician complaining of severe pain and redness around the base of the nail of the right index finger. She stated that three days previously she had trimmed the cuticle (eponychium) of her nail with scissors, and the following day the pain commenced. On physical examination, the skin folds around the root of the nail were found to be swollen, red, and extremely tender. The index finger was swollen and the back of the hand was slightly edematous. Red streaks could be seen coursing up the front of the forearm. On examination of the infraclavicular fossa, some small, tender nodules could be palpated. The patient's temperature was 101°F. From your knowledge of anatomy, explain the red streaks present on the front of the forearm and the tender nodules in the infraclavicular fossa.

3. A medical student, while rock-climbing, lost his footing on the edge of a crevasse. As he started to fall, he grabbed at a bush with his

outstretched hand. Fortunately, his companions were able to get to him just as the bush was becoming uprooted. On returning to the base of the mountain, he was examined by a physician and was found to have damaged the first thoracic spinal nerve. Using your knowledge of anatomy, where would you test for loss of skin sensation in such a nerve injury?

4. A 16-year-old girl, while demonstrating to her sisters her proficiency at standing on her hands, suddenly went off balance and put all her body weight on her right outstretched hand. A distinctive cracking noise was heard, and she felt a sudden pain in her right shoulder region. On examination, the smooth contour of her right clavicle was found to be absent. The shoulder and the lateral end of the clavicle were depressed, and the medial part of the clavicle was elevated. The clavicle was obviously fractured, and the edges of the bony fragments could be palpated. From your knowledge of anatomy, state: (a) Which part of the clavicle is most commonly fractured? (b) Why are the bony fragments displaced? (c) Why do the mechanical forces fracture the clavicle instead of dislocating the sternoclavicular or the acromioclavicular joints? (d) What structures are liable to be damaged in a simple fracture of the clavicle?

5. A 45-year-old woman visited her physician concerning a hard, painless lump in the left breast, which she had noticed while bathing the previous day. She stated that she had breast-fed her four children and had not seen any blood-stained discharge from her nipples. On examination in a good light, with her arms at her sides, the left nipple was seen to be higher than the right. A small dimple of skin was noted over the upper lateral quadrant of the left breast. On palpating the breast with the flat hand, the physician felt a hard lump, about 2 inches (5 cm) in diameter, in the upper left quadrant of the left breast. The right breast was normal. On examination of the left axilla, three small, hard discrete nodules could be palpated posterior to the lower border of the pectoralis major. The right axilla was normal. A diagnosis of scirrhous carcinoma of the left breast was made, with secondary deposits in the axilla. (a) Why was the left nipple higher than the right nipple? (b) What is the cause of dimpling of the skin? (c) What is the lymphatic drainage of the breast?

6. In a 25-year-old woman, an abscess developed in the lower medial quadrant of the right breast. She was breast-feeding her infant, which had been born 3 weeks previously. She had always breast-fed her children, and had experienced no difficulties with her three older babies. On physical examination, a hot, red, tender swelling was found in the lower medial quadrant of the right breast. The swelling was fluctuant, indicating the presence of pus. Using your knowledge of anatomy, would you use a radial or a transverse incision to drain the abscess?

7. A 50-year-old woman, when told by her surgeon that she had cancer of the right breast and that she should be admitted to the hospital for an operation, was indignant and dismayed to learn that she must have her breast removed. The neoplasm was located above the nipple. She could not understand why she was being advised to have deep X-ray therapy after the operation. On anatomical grounds, explain why it may be necessary to perform a radical mastectomy for carcinoma of the breast, to be followed by radiotherapy.

8. A 35-year-old farmer, while cutting his wheat, fell into a combine harvester. His left arm was wrenched off from the trunk at the shoulder, and he was found unconscious and bleeding profusely a few moments later. Fortunately, a physician was staying on the farm for his vacation. On hearing about the tragic accident, he ran to the field and was able to stop the bleeding. Where would you apply pressure to the main arterial supply of the upper limb in a case such as this, where there is no limb stump around which to place a tourniquet?

9. Palpation of the radial artery at the wrist can provide the experienced physician with considerable insight into the state of the patient's circulatory system. The degree of hardness of the arterial wall can be appreciated by the examining finger; the pulse rate and quality of the rhythm can be determined. What are the relations of the radial artery at the site where the pulse is taken?

10. A 20-year-old girl, riding pillion on a motorcycle, was involved in an accident. The motorcy-

cle hit a bridge abutment. The driver was killed instantly, and the girl was thrown 15 feet and landed on the left shoulder and the left side of the head. After hospitalization for 3 weeks, it was noticed that she kept her left arm internally rotated by her side with the forearm pronated. An area of anesthesia was present along the lateral side of the upper arm. A diagnosis of damage to the upper part of the brachial plexus was made. In anatomical terms, explain the position adopted by the left arm in this patient. The precise nature of the nerve lesions and the muscles paralyzed should be stated.

11. A baby was delivered as a breech presentation. During the second stage of labor, the right arm was carried up above the head, severely stretching the lower part of the brachial plexus. Assume that 3 months later you were asked to examine the child. Describe exactly the position assumed by the fingers and thumb of the right hand and name the muscles that would show evidence of wasting. It is assumed that the first thoracic root of the brachial plexus had been severely damaged. Would you expect to find any sensory loss?

12. Following a left radical mastectomy, a 53-year-old woman found that she was unable to raise her left arm above the head to comb her hair. During the physical examination, she was asked to face a wall and push hard against it with both outstretched hands. It was noted that the inferior angle and medial border of the left scapula projected markedly posteriorly during this maneuver. Which nerve had been cut during the mastectomy operation? Explain in anatomical terms the inability of the patient to raise her left arm above her head. Why does she demonstrate "winging" of her left scapula?

13. During examination of a 35-year-old patient for left inguinal hernia, it was noticed that he held his right forearm pronated and the wrist joint and the fingers flexed. He stated that he experienced difficulty in gripping objects in the hand with the wrist always in the flexed position. On examination of the upper arm, an old scar was seen crossing the posterior surface. On being questioned, the patient said that he had sustained a fracture of his arm when he was 12 years old, and this had been followed by an operation some months later, which his physi-

cians had considered to be unsuccessful. Name the muscles that were paralyzed in this patient. Which nerve was involved in the fracture? Why was an operation necessary after the fracture? Why was the operation a failure? Why does the patient experience difficulty in holding objects in the hand?

14. A 58-year-old woman fell down the stairs and dislocated her left shoulder joint. Several days after the dislocation had been reduced, it was noted that the patient had signs and symptoms indicative of radial nerve damage in the axilla. Name the muscles that would be paralyzed and the area of the skin that would show sensory loss. Describe the deformity of the upper limb in a person with a long-standing radial nerve palsy following damage to the nerve in the axilla.

15. A 45-year-old woman was seen in the emergency room with a dislocation of the right shoulder joint. A senior medical student was asked to examine the patient and make a report on possible nerve injuries. He carefully examined the arm for neurological defects of the median, ulnar, radial, and musculocutaneous nerves, but found nothing abnormal. He then remembered the axillary nerve and its close relationship to the inferior surface of the shoulder joint. He also remembered that this nerve supplies the deltoid muscle, which abducts the shoulder joint. He asked the patient to try and abduct the dislocated shoulder; she could not, of course, because of the severe pain. The medical student reported to the resident that he was unable to test the axillary nerve for a neurological defect. What test would you have carried out, using your knowledge of anatomy?

16. A 6-year-old boy, running along a concrete path with a jam jar in his hand, slipped and fell. The glass from the broken jar pierced the skin on the front of his left wrist. On examination, a small clean wound was seen to be present on the front of the left wrist, at the level of the proximal transverse crease. The palmarus longus tendon had been transected. The thumb was laterally rotated and adducted, and the boy was unable to oppose his thumb to the other fingers. When he was asked to make a fist slowly, the flexion of the index and middle fingers lagged behind the ring and little fingers.

There was diminished skin sensation over the lateral half of the palm of the hand and the palmar aspect of the lateral three and one-half fingers. There was also sensory loss of the skin of the distal parts of the dorsal surface of the lateral three and one-half fingers. Using your anatomical knowledge, name the structure or structures that had been damaged.

17. A 56-year-old woman visited her physician complaining of severe "pins and needles" in the hand and lateral fingers. The condition was becoming progressively worse and was more severe at night. She said she had experienced difficulty in buttoning up her clothes when dressing. On physical examination, the patient pointed to the thumb, index, middle, and ring fingers as the areas where she felt the discomfort. No objective impairment of sensation could be detected in those areas. The muscles of the thenar eminence appeared to be functioning normally, and there was no loss of power or wasting of the thenar eminence. What anatomical structure was diseased in this patient?

18. A 30-year-old man visted his physician complaining of "pins and needles" and numbness over the front and back surfaces of the little and ring fingers of the right hand. The patient had also noticed a weakness of his fingers of the right hand, especially when performing the upstroke in writing. On questioning, it was found that at the age of 11 he had a fracture of the lower end of the right humerus in an automobile accident. On examination, his carrying angle on the right side was noted to be greater than on the left (*cubitus valgus*), and there was some flattening along the medial border of the right forearm. There was also flattening of the hypothenar eminence and a loss of curvature along the medial border of the hand. On the dorsum of the right hand there was evidence of "hollowing-out" between the metacarpal bones. The patient could draw his thumb across the palm, but if asked to pinch a piece of paper between his thumb and fingers, the terminal phalanx of the thumb had to be flexed to grip the paper; the movement of adduction of the thumb was extremely weak. He was unable to grip a piece of paper by adducting the index and middle fingers. He also had loss of touch over the medial one and one-half fingers, both anteriorly and posteriorly.

Using your anatomical knowledge, explain this patient's disability, remembering that in his youth he suffered from a supracondylar fracture of the humerus. Why was the medial border of the forearm flattened? What muscles were wasted? Why was the hypothenar eminence flattened? Why was there "hollowing'out" on the dorsum of the hand? Why could the patient not adduct his thumb or his index and middle fingers?

19. A young typist, running from her office, had a glass door swing back in her face. In order to protect herself, she held out her left hand, which smashed through the glass. The physician found her sitting on the floor and bleeding profusely from a superficial laceration in front of her left wrist. She had sensory loss over the palmar aspect of the medial one and one-half fingers, but normal sensation on the back of these fingers. She was unable to grip a piece of paper between her left index and middle fingers. All her long flexor tendons were intact. Which artery and nerve were cut in the accident?

20. A 50-year-old man fell off the back of a truck he was unloading and struck the ground on his left arm and side. On getting up, he experienced severe pain and limitation of movement of his left shoulder. On examination, the normal rounded contour of the left shoulder was seen to be lost, the long axis of the left arm passed upward and medially, and some fullness was noted below the lateral part of the clavicle. Very little passive movement of the left shoulder was permitted by the patient. The skin sensation was normal over the lower part of the left deltoid muscle. Using your knowledge of anatomy, explain what was wrong with the patient. Why had the shoulder lost its normal contour? Why was there fullness below the lateral end of the clavicle?

21. A 40-year-old man visited his physician complaining of pain of 3 weeks' duration in his right shoulder. On examination, the patient could actively abduct his right shoulder to 40 degrees; thereafter, he experienced severe pain that prevented further movement. If the arm was then passively raised above a right angle, it

could be held actively without pain in that position. If the patient attempted to lower the arm, he again experienced severe pain in the middle range of abduction. What is your diagnosis?

22. A father, seeing his little 5-year-old boy playing in the garden, ran up and picked him up by both hands and swung him around in a circle. The child's obvious enjoyment suddenly turned to tears, and he said his right elbow hurt. On examination, the child held his right elbow joint semiflexed and his forearm pronated. What is your diagnosis?

23. A 20-year-old man fell off his bicycle onto his outstretched left hand. He thought he had sprained his wrist, but when the pain in his wrist persisted for 2 weeks, he sought medical advice. On physical examination of the backs of both hands, with the fingers and thumbs fully extended, a small amount of swelling was seen to be present over the anatomical snuffbox of his left hand. On deep palpation, a localized area of tenderness could be felt over the scaphoid bone. A diagnosis of a fractured left scaphoid bone was made. For what anatomical reasons do fractures of this bone sometimes fail to unite?

24. An 8-year-old boy fell off a swing and sustained a supracondylar fracture of his left humerus. Following the reduction of the fracture, a plaster cast was carefully applied, and the child was sent home. A few hours later, the child complained of pain in the forearm, which persisted. Four hours later, the parents decided to return to the hospital, since the child's left hand looked dusky white, and the pain in the forearm was still present. On examination, there was found to be a complete loss of cutaneous sensation of the hand. The color of the skin was bluish-white, and after the lower part of the plaster cast was removed, the pulse of the radial and ulnar arteries could not be palpated. Every possible effort was made to restore the circulation of the forearm, without avail. What deformity would you expect this child to have 1 year later? Explain the deformity on anatomical grounds.

25. A 64-year-old man consulted his physician because he had noticed during the past 3 months a thickening of the skin at the base of his left ring finger. As he described it: "There appears to be a band of tissue that is pulling my ring finger into the palm." On examination of the palms of both hands, a localized thickening of the palmar aponeurosis could be felt at the base of the left ring and little fingers. The metacarpophalangeal joint of the ring finger could not be fully extended, either actively or passively. What was wrong anatomically with the left ring finger?

26. A 26-year-old woman, while pruning some rose bushes, caught the anterior surface of the right index finger on a large thorn. Three days later the entire finger became swollen, red, and painful. On examination, the whole finger was found to be swollen and was held in a semiflexed position; it was tender, especially along the line of the flexor tendons. Flexion of the finger was difficult, and extension was limited by extreme pain. A small black dot over the anterior surface of the middle phalangeal region showed the site of entry of the thorn. With your knowledge of anatomy and assuming that the thorn was contaminated with infecting organisms, what is your diagnosis? If the diseased finger is left untreated, where is the infection likely to spread?

27. A 9-year-old boy, while climbing over the ruins of an old building, slipped and caught his right hand on a rusty nail. A deep puncture wound occurred on the palm of the hand in front of the fourth metacarpal bone. Two days later, the wound became painful and tender and the back of the hand became swollen. With your knowledge of anatomy, and assuming that the nail was contaminated with infecting organisms, what is your diagnosis?

28. A little 6-year-old girl decided to use the sewing machine when her mother was out of the room. When she started the machine, the point of the needle entered the pulp of the thumb of her left hand. Thirty-six hours later, the child woke up in the night complaining of severe pain in her thumb. On examination, the skin over the anterior surface of the terminal phalanx of the left thumb was found to be swollen, red, and extremely tender to touch. Over the center of the pulp was a small area of yellow devitalized skin. From your knowledge of the anatomy of the region, and assuming that the

needle was contaminated with pathogenic organisms, what is your diagnosis? What is the great danger of infections in this anatomical location? What is the lymphatic drainage of this area?

29. A 25-year-old woman fell from her horse and sustained a fracture of the lower end of the right radius. The orthopedic surgeon instructed the resident to be sure to immobilize the wrist joint in the position of function. What is the position of function? What is the clinical importance of this position?

30. A 50-year-old woman slipped on a shiny floor and sustained a fracture of the fifth right metacarpal bone. When a plaster cast is applied with the little finger flexed, in which direction should the little finger be pointing?

31. During a neurological examination of a 54-year-old man, the physician tested the biceps and triceps reflexes by smartly striking the tendons of these muscles with a reflex hammer. The reflexes were normal on the right side but absent on the left side. What is the significance of this finding?

32. A 23-year-old maid was making a bed in a hotel bedroom. As she straightened the sheet, by running her right hand with her fingers extended over the surface of the sheet, she caught the end of her index finger in a fold. She experienced a sudden severe pain over the base of the terminal phalanx. Several hours later when the pain had diminished, she noted that the end of her right index finger was swollen and she could not extend the terminal phalangeal joint completely. What is your diagnosis?

33. A 26-year-old woman was driving to work when she was hit broadside by another car at a road intersection. A radiological examination of her left humerus showed an oblique fracture of the shaft just proximal to the deltoid tuberosity. Using your knowledge of anatomy, explain the possible displacement of the proximal and distal fragments that might have occurred. Which nerve might have been damaged in a fracture of this type?

## NATIONAL BOARD TYPE QUESTIONS

*Answers on page 984*

### In each of the following questions, answer:
(a) If (1) is correct only
(b) If (2) is correct only
(c) If both (1) and (2) are correct, and
(d) If neither (1) nor (2) is correct

1. Concerning the mammary gland:
   (1) In the female, the greater part of the mammary gland lies in the superficial fascia; the axillary tail, however, pierces the deep fascia and becomes closely related to the axillary blood vessels.
   (2) In the male, the duct system does not extend beyond the margin of the areola.

2. Concerning the brachial artery:
   (1) Its pulsations may be felt in the arm beneath the medial border of the biceps brachii.
   (2) In the cubital fossa, the artery lies superficial to the bicipital aponeurosis.

3. Concerning the region of the elbow joint:
   (1) When the elbow joint is flexed, the medial and lateral epicondyles of the humerus and the olecranon process of the ulna lie on the same straight line.
   (2) The ulnar nerve can be palpated as it lies in front of the medial epicondyle of the humerus.

4. Concerning the region of the wrist joint:
   (1) The styloid process of the radius lies about ¾ inches (1.9 cm) distal to that of the ulna.
   (2) The ulnar nerve lies immediately medial to the pisiform bone.

5. Concerning the superficial veins of the upper limb:
   (1) The cephalic vein reaches the front of the forearm by winding around the medial side of the wrist.
   (2) The basilic vein ends by piercing the clavipectoral fascia and draining into the axillary vein.

6. (1) The pulse of the radial artery can be felt in front of the distal end of the radius, lateral to the tendon of the flexor carpi radialis.

(2) The pulse of the ulnar artery can be felt where it passes superficial to the flexor retinaculum.

**In the following questions, answer:**
(a) IF ONLY (1), (2), AND (3) ARE CORRECT
(b) IF ONLY (1) AND (3) ARE CORRECT
(c) IF ONLY (2) AND (4) ARE CORRECT
(d) IF ALL ARE CORRECT

7. The glenoid cavity of the scapula is rotated upward by which of the following muscles?
   (1) Levator scapulae
   (2) Serratus anterior
   (3) Rhomboid minor
   (4) Trapezius

8. Structures that pass posterior to the flexor retinaculum of the wrist include:
   (1) Flexor digitorium superficialis tendons
   (2) Median nerve
   (3) Anterior interosseous nerve
   (4) Ulnar nerve

9. Muscles that insert into the proximal phalanx of the thumbs include:
   (1) Abductor pollicis brevis
   (2) First palmar interosseous
   (3) Adductor pollicis
   (4) Abductor pollicis longus

10. Adductors of the hand at the wrist joint include:
    (1) Brachioradialis
    (2) Flexor carpi ulnaris
    (3) Extensor carpi radialis brevis
    (4) Extensor carpi ulnaris

11. Which of the following bones form the proximal row of carpal bones?
    (1) Scaphoid
    (2) Lunate
    (3) Triquetral
    (4) Capitate

12. The muscles of the rotator cuff include:
    (1) Teres minor
    (2) Subscapularis
    (3) Supraspinatus
    (4) Infraspinatus

13. Muscles that can flex the forearm include:
    (1) Brachialis
    (2) Brachioradialis
    (3) Pronator teres
    (4) Anconeus

14. Muscles that can extend the middle and distal phalanges of the index finger include:
    (1) The first dorsal interosseous
    (2) The first lumbrical
    (3) The second palmar interosseous
    (4) Extensor carpi radialis longus

**Match the statement on the left with the correct nerve on the right:**

15. Hyperextension of the proximal phalanges of the little and ring fingers (i.e., claw hand) can result from damage to the _____ nerve.

16. Wrist drop can result from damage to the _____ nerve.

17. An inability to oppose the thumb to the little finger can result from damage to the _____ nerve.

(a) Ulnar

(b) Axillary

(c) Radial

(d) Median

**Match the sensory innervation of the skin of the hand and fingers on the left with the most appropriate nerve on the right:**

18. Nail bed of index finger

19. Medial side of palm

20. Dorsal surface of root of thumb

21. Medial side of palmar aspect of ring finger

(a) Median nerve
(b) Radial nerve
(c) Dorsal cutaneous branch of ulnar nerve
(d) Superficial branch of ulnar nerve
(e) Palmar cutaneous branch of ulnar nerve

**Match the nerves on the left with their origins from the brachial plexus on the right:**

22. Musculocutaneous nerve
23. Suprascapular nerve
24. Median nerve
25. Thoracodorsal nerve
26. Axillary nerve

(a) Posterior cord
(b) Lateral cord
(c) Both medial and lateral cords
(d) Upper trunk
(e) None of the above

**Multiple choice:**

27. The quadrilateral space is bounded by the following structures, **except** the:
    (a) Surgical neck of the humerus
    (b) Long head of triceps
    (c) Deltoid
    (d) Teres major
    (e) Teres minor

28. The lymph from the upper lateral quadrant of the mammary gland drains mainly into the:
    (a) Lateral axillary nodes
    (b) Internal thoracic nodes
    (c) Posterior axillary nodes
    (d) Anterior axillary nodes
    (e) Deltopectoral group of nodes

29. The main ligamentous union between the clavicle and the remainder of the upper limbs is the:
    (a) Coracoacromial ligament
    (b) Costoclavicular ligament
    (c) Sternoclavicular ligament
    (d) Coracoclavicular ligament
    (e) Intertubercular ligament

30. Which of the following statements is **not** true of the metacarpophalangeal joint of the middle finger?
    (a) It is a synovial joint.
    (b) Flexion of the joint can be brought about by the lumbrical and interossei muscles.
    (c) Flexion, extension, abduction, and adduction are possible at this joint.
    (d) The dorsal ligament of the joint is united to the dorsal ligament of adjacent metacarpophalangeal joints by the deep transverse metacarpal ligaments.
    (e) The collateral ligaments pass downward and forward from the head of the metacarpal bone to the base of the phalanx.

31. Following injury to a nerve at the wrist, the thumb is laterally rotated and adducted. The hand has a flattened appearance and is "apelike." Which of the following nerves is damaged?
    (a) Ulnar nerve
    (b) Anterior interosseous nerve
    (c) Deep radial nerve
    (d) Musculocutaneous nerve
    (e) Median nerve

32. A patient is found to have pus in the midpalmar space. The space is:
    (a) The space around the hypothenar muscles.
    (b) The synovial sheath for the flexor pollicis longus tendon.
    (c) The ulnar bursa.
    (d) The space around the muscles of the thenar eminence.
    (e) A space lying medial to a fibrous septum, attaching the palmar aponeurosis to the third metacarpal bone.

33. The radial nerve gives off the following branches in the posterior compartment of the arm, **except** the:
    (a) Lateral head of the triceps
    (b) Lower lateral cutaneous nerve of the arm
    (c) Medial head of the triceps
    (d) Brachioradialis
    (e) Anconeus

34. The medial collateral ligament of the elbow joint is closely related to the following structure:
    (a) Brachial artery
    (b) Radial nerve
    (c) Ulnar artery
    (d) Basilic vein
    (e) Ulnar nerve

35. All of the following statements concerning the brachial plexus are true **except:**
    (a) The roots C8 and T1 join to form the lower trunk.
    (b) The roots, trunks, and divisions are not located in the axilla.
    (c) The nerve that innervates the levator scapulae is a branch of the upper trunk.
    (d) The cords are named according to their position relative to the first part of the axillary artery.
    (e) No nerves originate as branches from the individual divisions of the brachial plexus.

# 10. The Lower Limb

The primary function of the lower limbs is to support the weight of the body and to provide a stable foundation in standing, walking, and running; they have become specialized for locomotion.

The lower limbs, although similar in structure in many respects to the upper limbs, have less freedom of movement. Whereas the pectoral girdle of the upper limb is united to the trunk by only a small joint, the sternoclavicular joint, the two hip bones articulate posteriorly with the trunk at the strong sacroiliac joints and anteriorly with each other at the symphysis pubis. The result is that the lower limbs are more stable and can bear the weight of the body during standing, walking, and running.

Each lower limb may be divided into the gluteal region, the thigh, the knee, the leg, the ankle, and the foot.

## SURFACE ANATOMY

The following information should be verified on the living body. An adequate physical examination of the lower limb of a patient requires a sound knowledge of the surface anatomy of the region.

### Gluteal Region

The *iliac crests* are easily palpable along their entire length (Figs. 10-1 and 10-2). Each crest ends in front at the *anterior superior iliac spine* (Fig. 10-3), and behind at the *posterior superior iliac spine* (Fig. 10-1); the latter lies beneath a skin dimple at the level of the second sacral vertebra and the middle of the sacroiliac joint. The *iliac tubercle* is a prominence felt on the outer surface of the iliac crest about 2 inches (5 cm) posterior to the anterior superior iliac spine (Fig. 10-2).

The *ischial tuberosity* can be palpated in the lower part of the buttock (Figs. 10-1 and 10-2). In the standing position, the tuberosity is covered by the gluteus maximus. In the sitting position, the ischial tuberosity emerges from beneath the lower border of the gluteus maximus and supports the weight of the body; in this position, the tuberosity is separated from the skin by only a bursa and a pad of fat.

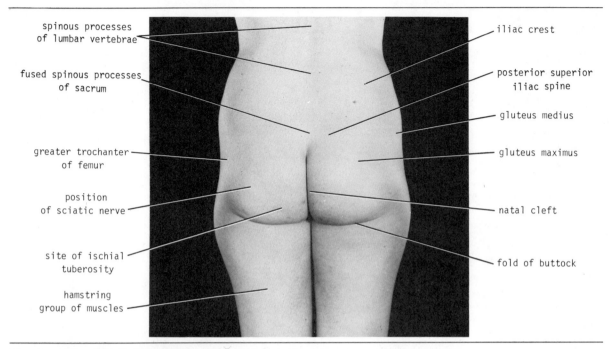

spinous processes
of lumbar vertebrae

fused spinous processes
of sacrum

greater trochanter
of femur

position
of sciatic nerve

site of ischial
tuberosity

hamstring
group of muscles

iliac crest

posterior superior
iliac spine

gluteus medius

gluteus maximus

natal cleft

fold of buttock

**Fig. 10-1. The gluteal region and posterior aspect of the thigh of a 25-year-old female.**

The *greater trochanter* of the femur can be felt on the lateral surface of the thigh (Figs. 10-1 and 10-2) and moves beneath the examining finger as the hip joint is flexed and extended. It is important to verify that in the normal hip joint, the upper border of the greater trochanter lies on a line connecting the anterior superior iliac spine to the ischial tuberosity (Fig. 10-2).

The tip of the *coccyx* may be palpated beneath the skin in the upper part of the natal cleft (Fig. 10-2).

The *fold of the nates*, or *buttocks*, is most prominent in the standing position; its lower border does not correspond to the lower border of the gluteus maximus muscle.

The *sciatic nerve* in the buttock lies under cover of the gluteus maximus muscle. As it curves laterally and downward, it is situated at first midway between the posterior superior iliac spine and the ischial tuberosity and lower down, midway between the tip of the greater trochanter and the ischial tuberosity (Figs. 10-1 and 10-2).

## Inguinal Region

The *inguinal ligament* may be felt along its length. It is attached laterally to the anterior superior iliac spine and medially to the pubic tubercle (Figs. 10-2 and 10-3).

The *symphysis pubis* is a cartilaginous joint that lies in the midline between the bodies of the pubic bones (Fig. 10-3). The *upper margin of the symphysis pubis* and the bodies of the pubic bones may be felt on palpation.

The *pubic tubercle* is easily palpated in the male by invaginating the scrotum with the examining finger, and the tubercle is felt on the upper border of the body of the pubis (Figs. 10-2 and 10-3). In both sexes, the tendon of the adductor longus forms the medial boundary of the upper part of the thigh, and if this is traced upward, it leads to the pubic tubercle.

## Femoral Triangle

The femoral triangle can be seen as a depression below the fold of the groin in the upper part of the thigh (Figs. 10-2 and 10-3). In a thin, muscular

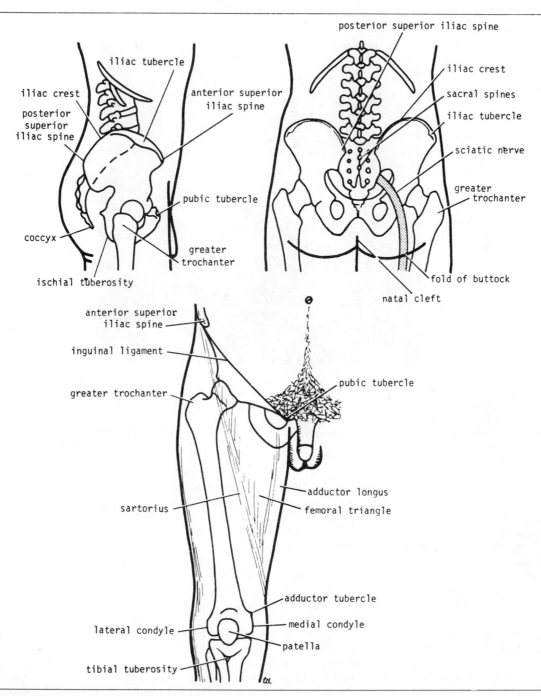

**Fig. 10-2. Surface markings in gluteal region and front of thigh.**

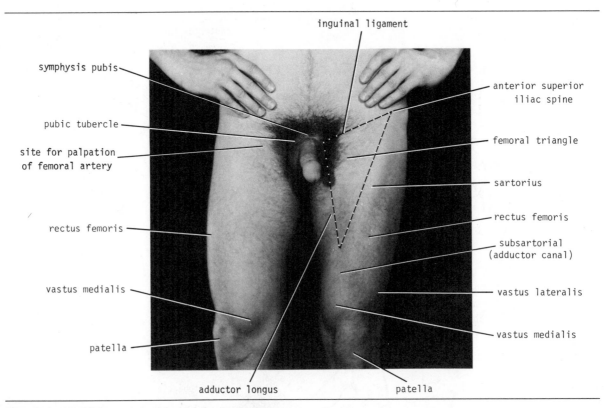

symphysis pubis

inguinal ligament

anterior superior
iliac spine

pubic tubercle

femoral triangle

site for palpation
of femoral artery

sartorius

rectus femoris

rectus femoris

subsartorial
(adductor canal)

vastus medialis

vastus lateralis

vastus medialis

patella

adductor longus

patella

**Fig. 10-3. Anterior aspect of the thigh of a 27-year-old male. The broken lines indicate the boundaries of the femoral triangle. The right leg is laterally rotated at the hip joint.**

subject, the boundaries of the triangle can be identified when the thigh is flexed, abducted, and laterally rotated. The *base* of the triangle is formed by the inguinal ligament, the *lateral border* by the sartorius muscle, and the *medial border* by the adductor longus muscle.

The horizontal group of *superficial inguinal lymph nodes* may be palpated in the superficial fascia just below and parallel to the inguinal ligament (Fig. 10-19).

The *femoral artery* enters the thigh behind the inguinal ligament (Fig. 10-21) at the midpoint of a line joining the symphysis pubis to the anterior superior iliac spine; its pulsations are easily felt (Fig. 10-3).

The *femoral vein* leaves the thigh by passing behind the inguinal ligament medial to the pulsating femoral artery (Fig. 10-21).

The lower opening of the *femoral canal* lies below and lateral to the pubic tubercle (Figs. 10-19 and 10-21).

The *femoral nerve* enters the thigh behind the midpoint of the inguinal ligament, i.e., lateral to the pulsating femoral artery (Fig. 10-21).

The *great saphenous vein* pierces the saphenous opening in the deep fascia (fascia lata) of the thigh and joins the femoral vein 1½ inches (4 cm) below and lateral to the pubic tubercle (Figs. 10-18 and 10-19).

## Adductor Canal

The *adductor* (*subsartorial*) *canal* lies in the middle third of the thigh (Fig. 10-3), immediately distal to the apex of the femoral triangle. It is an intermuscular cleft situated beneath the sartorius muscle and is bounded laterally by the vastus medialis muscle and posteriorly by the adductor longus and magnus muscles. It contains the femoral vessels and the saphenous nerve.

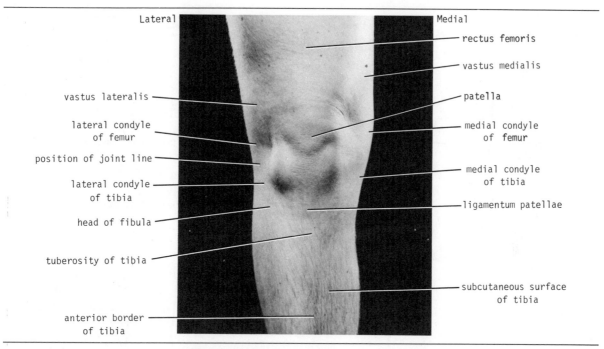

Lateral

rectus femoris

vastus medialis

patella

vastus lateralis

lateral condyle
of femur

medial condyle
of femur

position of joint line

lateral condyle
of tibia

medial condyle
of tibia

head of fibula

ligamentum patellae

tuberosity of tibia

subcutaneous surface
of tibia

anterior border
of tibia

Medial

**Fig. 10-4. Anterior aspect of the right knee of a 27-year-old male.**

## Knee Region

The *popliteal fossa* is a diamond-shaped depression situated behind the knee (Fig. 10-5). When the knee is flexed, the deep fascia, which roofs over the fossa, is relaxed and the boundaries are easily defined. Its upper part is bounded *laterally* by the tendon of the biceps femoris muscle and *medially* by the tendons of the semimembranosus and semitendinosus muscles. Its lower part is bounded on each side by one of the heads of the gastrocnemius muscle.

The *common peroneal nerve* can be palpated on the medial side of the tendon of the biceps femoris (Fig. 10-5), as the latter passes to its insertion on the head of the fibula. With the knee joint partially flexed, the nerve can be rolled beneath the finger.

The *popliteal artery* can be felt pulsating in the depths of the popliteal fossa, provided that the deep fascia is fully relaxed by passively flexing the knee joint.

In front of the knee, the *patella* and the *ligamentum patellae* can be readily palpated (Fig. 10-4).

On the sides of the knee, the *condyles* of the femur and tibia can be recognized, and the *joint line* can be identified. On the lateral aspect of the knee, the rounded *tendon of biceps* can be traced to the *head of the fibula* (Fig. 10-4). Just below the head of the fibula, the *common peroneal nerve* can be rolled beneath the examining finger as it passes around the lateral side of this bone (Fig. 10-5). On the medial aspect of the knee, the *adductor tubercle* can be palpated just above the medial condyle of the femur; the tendon of the hamstring part of the *adductor magnus* can be felt passing to it (Fig. 10-5).

## Tibia

The medial surface and anterior border of the *tibia* are subcutaneous and can be felt throughout their length (Fig. 10-4).

## Ankle Region and Foot

In the region of the ankle, the fibula is subcutaneous and may be followed downward to form the *lateral malleolus* (Figs. 10-5 and 10-6). The tip of the *medial malleolus* of the tibia lies about ½ inch

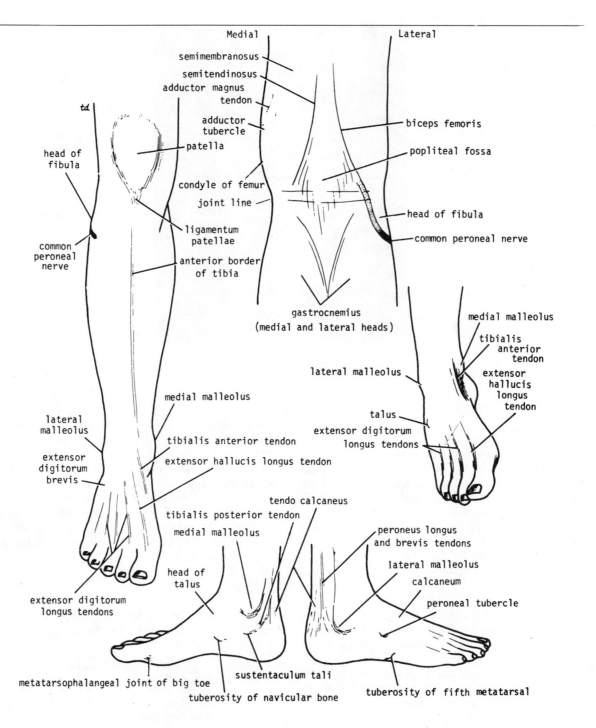

Medial                    Lateral

semimembranosus

semitendinosus

adductor magnus
tendon                                      biceps femoris

adductor
tubercle                                    popliteal fossa

patella

condyle of femur

joint line                                  head of fibula

ligamentum                                  common peroneal nerve
patellae

anterior border
of tibia

td

head of
fibula

common
peroneal
nerve

gastrocnemius
(medial and lateral heads)

medial malleolus

tibialis
anterior
tendon

lateral malleolus                           extensor
                                            hallucis
                                            longus
                                            tendon

medial malleolus                talus

lateral                         extensor digitorum
malleolus                       longus tendons

extensor
digitorum           tibialis anterior tendon
brevis
                    extensor hallucis longus tendon

                    tendo calcaneus

tibialis posterior tendon
medial malleolus                            peroneus longus
                                            and brevis tendons

head of                                     lateral malleolus
talus                                       calcaneum

extensor digitorum                          peroneal tubercle
longus tendons

metatarsophalangeal joint of big toe    sustentaculum tali

tuberosity of navicular bone            tuberosity of fifth metatarsal

Medial aspect of foot              Lateral aspect of foot

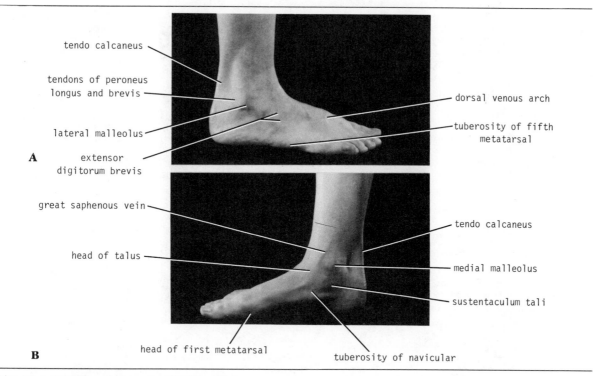

tendo calcaneus

tendons of peroneus longus and brevis

lateral malleolus

**A**      extensor digitorum brevis

dorsal venous arch

tuberosity of fifth metatarsal

great saphenous vein

head of talus

**B**      head of first metatarsal

tendo calcaneus

medial malleolus

sustentaculum tali

tuberosity of navicular

**Fig. 10-6. (A) Lateral aspect and (B) medial aspect of the right ankle of a 29-year-old female.**

(1.3 cm) proximal to the level of the tip of the lateral malleolus (Fig. 10-5 and 10-6).

In the interval behind the medial malleolus (Fig. 10-5) and the medial surface of the calcaneum lie the following structures, in the order named: (1) the *tendon of tibialis posterior*, (2) the *tendon of flexor digitorum longus*, (3) the *posterior tibial vessels*, (4) the *posterior tibial nerve*, and (5) the *tendon of flexor hallucis longus*. The pulsations of the posterior tibial artery may be felt halfway between the medial malleolus and the heel (Fig. 10-7). Behind the lateral malleolus are the *tendons of peroneus brevis* and *longus* (Figs. 10-6 and 10-7).

On the anterior surface of the ankle joint, the *tendon of tibialis anterior* can be seen when the foot is dorsiflexed and inverted (Figs. 10-5 and 10-

7). The *tendon of extensor hallucis longus* lies lateral to it and can be made to stand out by extending the big toe (Figs. 10-5 and 10-7). Lateral to the extensor hallucis longus lie the *tendons of extensor digitorum longus* and *peroneus tertius*. The pulsations of the *dorsalis pedis artery* may be felt between the tendons of extensor hallucis longus and extensor digitorum longus, midway between the two malleoli on the front of the ankle.

On the posterior surface of the ankle joint the prominence of the heel is formed by the *calcaneum*. Above the heel is the *tendo calcaneus* (Achilles tendon) (Fig. 10-7).

On the dorsum of the foot, the *head of the talus* can be palpated just in front of the malleoli (Fig. 10-6). The *tendons of extensor digitorum longus* and *extensor hallucis longus* can be made prominent by dorsiflexing the toes (Fig. 10-5).

The *dorsal venous arch* or *plexus* can be seen on the dorsal surface of the foot proximal to the toes (Figs. 10-6 and 10-18). The *great saphenous vein* leaves the medial part of the plexus and passes upward *in front of* the medial malleolus (Fig. 10-6)

**Fig. 10-5. Surface markings in popliteal fossa, front of leg, and foot.**

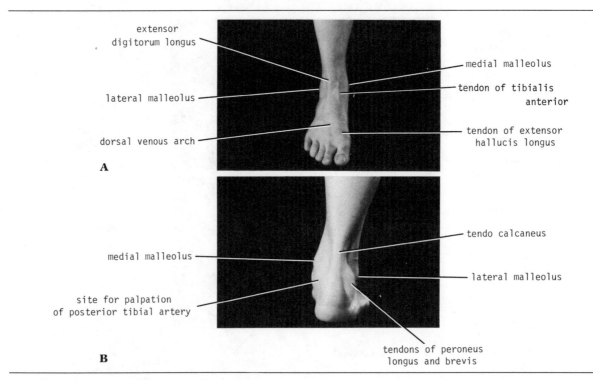

extensor digitorum longus

lateral malleolus

dorsal venous arch

**A**

medial malleolus

tendon of tibialis anterior

tendon of extensor hallucis longus

medial malleolus

site for palpation of posterior tibial artery

**B**

tendo calcaneus

lateral malleolus

tendons of peroneus longus and brevis

**Fig. 10-7. (A) Anterior aspect and (B) posterior aspect of the right foot and ankle of a 29-year-old female.**

The *small saphenous vein* drains the lateral part of the plexus and passes up *behind* the lateral malleolus (Fig. 10-18).

On the lateral aspect of the foot, the *peroneal tubercle* of the calcaneum can be palpated about 1 inch (2.5 cm) below and in front of the tip of the lateral malleolus (Fig. 10-5). Above the tubercle, the *tendon of peroneus brevis* passes forward to its insertion on the prominent tuberosity on the base of the *fifth metatarsal bone* (Fig. 10-6). Below the tubercle the *tendon of peroneus longus* passes forward to enter the groove on the under aspect of the cuboid bone.

On the medial aspect of the foot, the *sustentaculum tali* can be palpated about 1 inch (2.5 cm) below the tip of the medial malleolus (Fig. 10-6). The tendon of tibialis posterior lies immediately above the sustenaculum tali; the tendon of flexor digitorum longus crosses its medial surface; and

the tendon of flexor hallucis longus winds around its lower surface.

In front of the sustentaculum tali, the *tuberosity of the navicular bone* can be seen and palpated (Fig. 10-6). It receives the main part of the tendon of insertion of the tibialis posterior muscle.

## THE GLUTEAL REGION

The gluteal region, or buttock, is bounded superiorly by the iliac crest and inferiorly by the fold of the buttock. The region is largely made up of the gluteal muscles and a thick layer of superficial fascia.

### The Skin of the Buttock

The *cutaneous nerves* (Fig. 10-9) are derived from posterior and anterior rami of spinal nerves, as follows:

1. The upper medial quadrant is supplied by the posterior rami of the upper three lumbar nerves and the upper three sacral nerves.

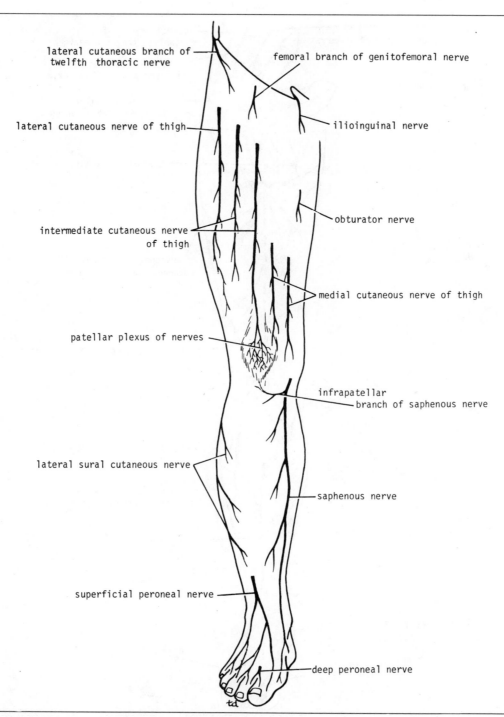

lateral cutaneous branch of twelfth thoracic nerve

femoral branch of genitofemoral nerve

lateral cutaneous nerve of thigh

ilioinguinal nerve

intermediate cutaneous nerve of thigh

obturator nerve

medial cutaneous nerve of thigh

patellar plexus of nerves

infrapatellar branch of saphenous nerve

lateral sural cutaneous nerve

saphenous nerve

superficial peroneal nerve

deep peroneal nerve

**Fig. 10-8. Cutaneous nerves of anterior surface of right lower limb.**

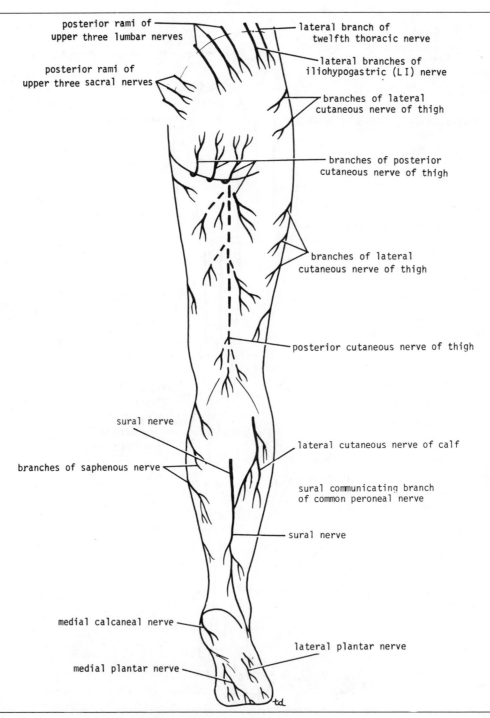

posterior rami of
upper three lumbar nerves

posterior rami of
upper three sacral nerves

lateral branch of
twelfth thoracic nerve

lateral branches of
iliohypogastric (L I) nerve

branches of lateral
cutaneous nerve of thigh

branches of posterior
cutaneous nerve of thigh

branches of lateral
cutaneous nerve of thigh

posterior cutaneous nerve of thigh

sural nerve

branches of saphenous nerve

lateral cutaneous nerve of calf

sural communicating branch
of common peroneal nerve

sural nerve

medial calcaneal nerve

lateral plantar nerve

medial plantar nerve

**Fig. 10-9. Cutaneous nerves of posterior surface of
right lower limb.**

2. The upper lateral quadrant is supplied by the lateral branches of the iliohypogastric (L1) and twelfth thoracic nerves (anterior rami).
3. The lower lateral quadrant is supplied by branches from the lateral cutaneous nerve of the thigh (L2 and 3, anterior rami).
4. The lower medial quadrant is supplied by branches from the posterior cutaneous nerve of the thigh (S1, 2, and 3, anterior rami).

The skin over the coccyx in the floor of the natal cleft is supplied by small branches of the lower sacral and coccygeal nerves.

The *lymphatic vessels* drain into the lateral group of the superficial inguinal nodes (Figs. 10-19 and 10-28).

## Fascia of the Buttock

The *superficial fascia* is thick, especially in women, and is impregnated with large quantities of fat. It is responsible for the prominence of the buttock.

The *deep fascia* is continuous below with the *deep fascia*, or *fascia lata*, of the thigh. In the gluteal region it splits to enclose the gluteus maximus muscle (Fig. 10-15). Above the gluteus maximus it continues as a single layer that covers the outer surface of the gluteus medius and is attached to the iliac crest. On the lateral surface of the thigh, the fascia is thickened to form a strong, wide band, the *iliotibial tract* (Figs. 10-15 and 10-21). This is attached above to the tubercle of the iliac crest and below to the lateral condyle of the tibia. The iliotibial tract forms a sheath for the tensor fasciae latae muscle and receives the greater part of the insertion of the gluteus maximus.

## Bones of the Gluteal Region

### HIP BONE

The ilium, together with the ischium and pubis, form the hip bone (Figs. 10-10 and 10-11). They meet one another at the acetabulum. The hip bones articulate with the sacrum at the sacroiliac joints and form the anterolateral walls of the pelvis; they also articulate with one another anteriorly at the symphysis pubis. The detailed structure of the internal aspect of the bony pelvis is considered on page 301

The important features found on the outer surface of the hip bone in the gluteal region are as follows:

The *ilium*, which is the upper flattened part of the bone, possesses the *iliac crest* (Fig. 10-11). This can be felt through the skin along its entire length; it ends in front at the *anterior superior iliac spine*, and behind at the *posterior superior iliac spine*. The *iliac tubercle* lies about 2 inches (5 cm) behind the anterior superior spine. Below the anterior superior iliac spine is a prominence, the *anterior inferior iliac spine*; a similar prominence, the *posterior inferior iliac spine*, is located below the posterior superior iliac spine. Above and behind the acetabulum, the ilium possesses a large notch, the *greater sciatic notch* (Figs. 10-10 and 10-11).

The outer surface of the ilium is undulating, being convex in front and concave behind. It is marked by three curved gluteal lines, the *posterior gluteal line*, the *middle gluteal line*, and the *inferior gluteal line* (Fig. 10-11).

The *ischium* is L-shaped, possessing an upper thicker part, the *body*, and a lower thinner part, the *ramus* (Figs. 10-10 and 10-11). The *ischial spine* projects from the posterior border of the ischium and intervenes between the *greater* and *lesser sciatic notches*. The *ischial tuberosity* forms the posterior aspect of the lower part of the body of the bone. The greater and lesser sciatic notches are converted into *greater* and *lesser sciatic foramina* by the presence of the sacrospinous and sacrotuberous ligaments. (See p. 312.)

The *pubis* may be divided into a *body*, a *superior ramus*, and an *inferior ramus* (Fig. 10-11). The bodies of the two pubic bones articulate with each other in the midline anteriorly at the *symphysis pubis*; the superior ramus joins the ilium and ischium at the acetabulum, and the inferior ramus joins the ischial ramus below the *obturator foramen*. The obturator foramen in life is filled in by the *obturator membrane*. (See p. 312.) The *pubic crest* forms the upper border of the body of the pubis, and it ends laterally as the *pubic tubercle* (Figs. 10-10 and 10-11).

On the outer surface of the hip bone there is a deep depression called the *acetabulum*; this articulates with the almost spherical head of the femur to form the hip joint (Figs. 10-11 and 10-12). The inferior margin of the acetabulum is deficient and is marked by the *acetabular notch* (Fig. 10-11). The articular surface of the acetabulum is limited to a

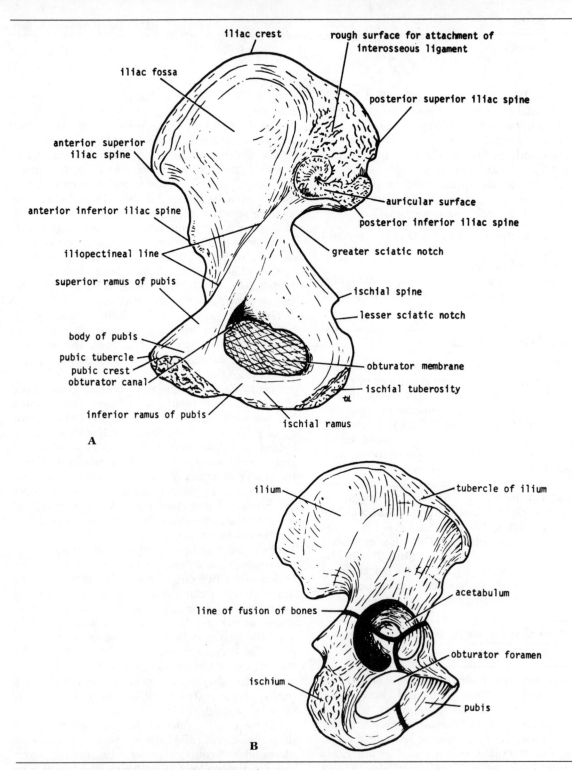

Fig. 10-10. (A) Medial surface and (B) lateral surface of right innominate bone. Note the lines of fusion between the three bones—the ilium, the ischium, and the pubis.

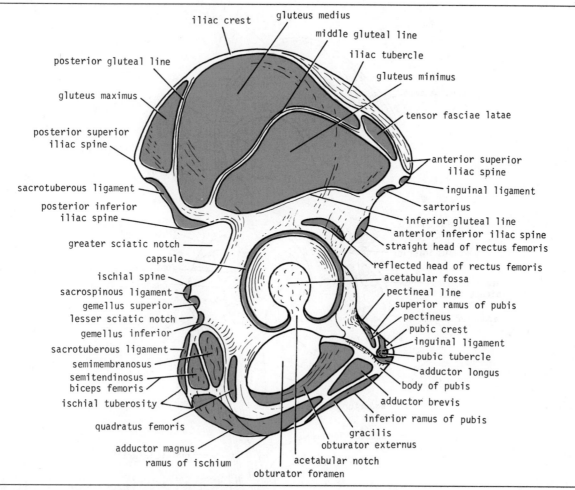

iliac crest
gluteus medius
middle gluteal line
iliac tubercle
posterior gluteal line
gluteus minimus
gluteus maximus
tensor fasciae latae
posterior superior
iliac spine
anterior superior
iliac spine
inguinal ligament
sacrotuberous ligament
sartorius
posterior inferior
iliac spine
inferior gluteal line
anterior inferior iliac spine
greater sciatic notch
straight head of rectus femoris
capsule
reflected head of rectus femoris
acetabular fossa
ischial spine
pectineal line
sacrospinous ligament
superior ramus of pubis
gemellus superior
pectineus
lesser sciatic notch
pubic crest
gemellus inferior
inguinal ligament
sacrotuberous ligament
pubic tubercle
semimembranosus
adductor longus
semitendinosus
body of pubis
biceps femoris
adductor brevis
ischial tuberosity
inferior ramus of pubis
quadratus femoris
gracilis
obturator externus
adductor magnus
acetabular notch
ramus of ischium
obturator foramen

**Fig. 10-11. Muscles and ligaments attached to external surface of right hip bone.**

horseshoe-shaped area and is covered with hyaline cartilage. The floor of the acetabulum is nonarticular and is called the *acetabular fossa* (Fig. 10-11).

It is important that the student understand the correct orientation of the bony pelvis relative to the trunk when the individual is standing in the anatomical position. The front of the symphysis pubis and the anterior superior iliac spines should lie in the same vertical plane. This means that the pelvic surface of the symphysis pubis faces upward and backward and the anterior surface of the sacrum is directed forward and downward.

The important muscles and ligaments that are attached to the outer surface of the hip bone are shown in Figure 10-11.

## FEMUR

The upper end of the femur has a head, neck, and greater and lesser trochanters (Figs. 10-13 and 10-14). The *head* forms about two-thirds of a sphere and articulates with the acetabulum of the hip bone to form the hip joint (Fig. 10-12). In the center of the head there is a small depression called the *fovea capitis,* for the attachment of the ligament of the head. Part of the blood supply to the head of the femur is conveyed along this ligament and enters the bone at the fovea.

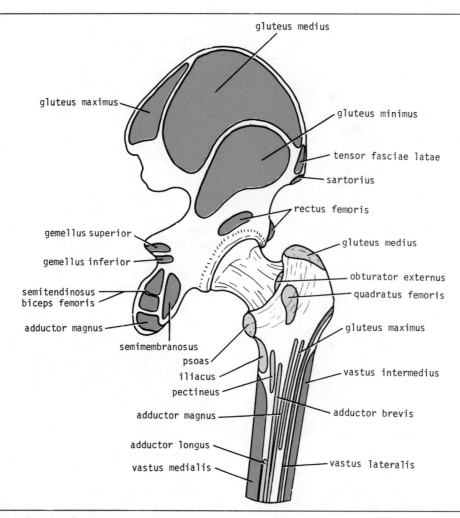

gluteus medius

gluteus maximus

gluteus minimus

tensor fasciae latae

sartorius

rectus femoris

gemellus superior

gemellus inferior

gluteus medius

obturator externus

quadratus femoris

semitendinosus
biceps femoris

adductor magnus

gluteus maximus

semimembranosus

psoas

iliacus

pectineus

vastus intermedius

adductor magnus

adductor brevis

adductor longus

vastus medialis

vastus lateralis

**Fig. 10-12. Muscles attached to external surface of right hip bone and posterior surface of femur.**

The *neck*, which connects the head to the shaft, passes downward, backward, and laterally and makes an angle of about 125 degrees (slightly less in the female) with the long axis of the shaft. The size of this angle should be remembered, since it can be altered by disease.

The *greater* and *lesser trochanters* are large eminences situated at the junction of the neck and the shaft (Figs. 10-13 and 10-14). Connecting the two trochanters are the *intertrochanteric line* anteriorly and a prominent *intertrochanteric crest* posteriorly, on which is the *quadrate tubercle* (Fig. 10-14).

The *shaft* of the femur shows a general forward convexity. It is smooth and rounded on its anterior surface, but has posteriorly a ridge, the *linea aspera* (Fig. 10-14). The margins of the linea aspera diverge above and below. The medial margin continues below as the *medial supracondylar ridge* to the *adductor tubercle* on the medial condyle (Fig. 10-14). The lateral margin becomes continuous below with the *lateral supracondylar ridge*. On the

**Fig. 10-13. Muscles and ligaments attached to anterior surface of right femur.**

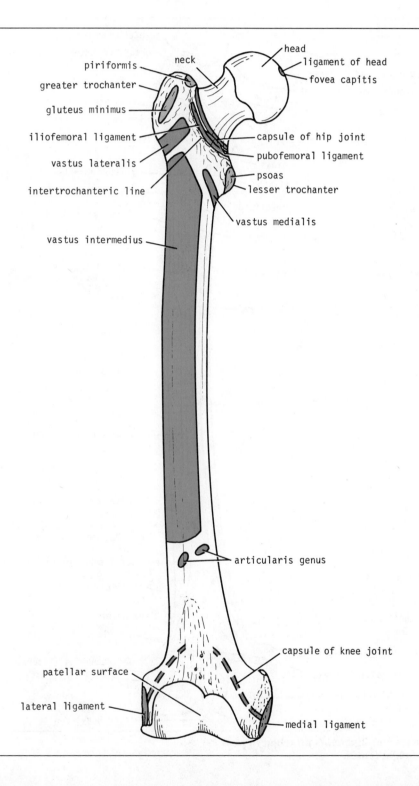

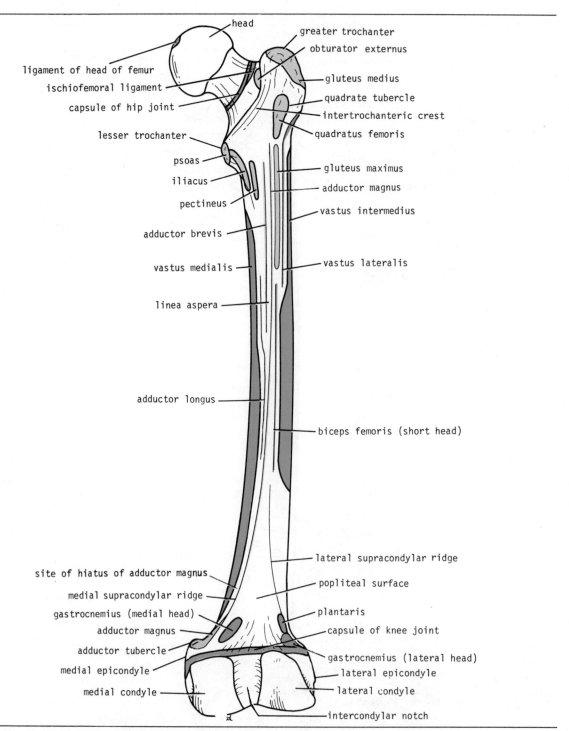

head

ligament of head of femur

ischiofemoral ligament

capsule of hip joint

lesser trochanter

psoas

iliacus

pectineus

adductor brevis

vastus medialis

linea aspera

adductor longus

site of hiatus of adductor magnus

medial supracondylar ridge

gastrocnemius (medial head)

adductor magnus

adductor tubercle

medial epicondyle

medial condyle

greater trochanter

obturator externus

gluteus medius

quadrate tubercle

intertrochanteric crest

quadratus femoris

gluteus maximus

adductor magnus

vastus intermedius

vastus lateralis

biceps femoris (short head)

lateral supracondylar ridge

popliteal surface

plantaris

capsule of knee joint

gastrocnemius (lateral head)

lateral epicondyle

lateral condyle

intercondylar notch

**Fig. 10-14. Muscles and ligaments attached to posterior surface of right femur.**

posterior surface of the shaft below the greater trochanter is the *gluteal tuberosity*, which is continuous below with the linea aspera. The shaft becomes broader toward its distal end and forms a flat triangular area on its posterior surface, called the *popliteal surface* (Fig. 10-14).

The lower end of the femur has *lateral* and *medial* condyles, separated posteriorly by the *intercondylar notch*. The anterior surfaces of the condyles are joined by an articular surface for the patella. The two condyles take part in the formation of the knee joint. Above the condyles are the *medial* and *lateral epicondyles* (Fig. 10-14). The adductor tubercle is continuous with the medial epicondyle.

The important muscles and ligaments that are attached to the femur are shown in Figures 10-13 and 10-14.

## Muscles of the Gluteal Region

### GLUTEUS MAXIMUS (FIG. 10-15)

The gluteus maximus is the largest muscle in the body. It lies superficial in the gluteal region and is largely responsible for the prominence of the buttock.

### Origin

From the outer surface of the ilium behind the posterior gluteal line; from the adjacent posterior surface of the sacrum and coccyx; and from the sacrotuberous ligament. (See p. 312.)

### Insertion

The fibers pass downward and laterally, and the majority are inserted into the iliotibial tract; some of the deeper fibers are inserted into the gluteal tuberosity of the femur.

### Nerve Supply

Inferior gluteal nerve.

### Action

It extends and laterally rotates the hip joint; through the iliotibial tract it helps to maintain the knee joint in extension. It is most commonly used

as an extensor of the trunk on the thigh, as, for example, when raising the trunk from the sitting or stooping positions.

Three bursae are usually associated with the gluteus maximus: (1) between the tendon of insertion and the greater trochanter; (2) between the tendon of insertion and the vastus lateralis; and (3) overlying the ischial tuberosity.

### GLUTEUS MEDIUS (FIGS. 10-15 AND 10-16)

The gluteus medius is a thick fan-shaped muscle and its posterior part is covered by the gluteus maximus.

### Origin

From the outer surface of the ilium between the iliac crest above, the posterior gluteal line behind, and the middle gluteal line below.

### Insertion

The fibers pass downward and laterally and are attached to the lateral surface of the greater trochanter.

### Nerve Supply

Superior gluteal nerve.

### Action

Acting with the gluteus minimus and the tensor fasciae latae, the thigh is powerfully abducted at the hip joint. Its most important action takes place in walking or running; the three muscles contract and steady the pelvis on the lower limb. When the foot of the opposite side is taken off the ground and thrust forward, the pelvis is held in position and does not tilt downward on the unsupported side. (See p. 687.) The anterior fibers also medially rotate the thigh.

### GLUTEUS MINIMUS (FIG. 10-16)

The gluteus minimus is fan-shaped and lies deep to the gluteus medius.

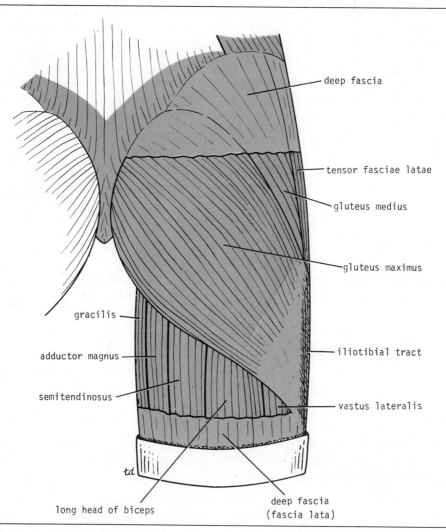

deep fascia

tensor fasciae latae

gluteus medius

gluteus maximus

gracilis

adductor magnus

semitendinosus

iliotibial tract

vastus lateralis

long head of biceps

deep fascia
(fascia lata)

td

**Fig. 10-15. Right gluteus maximus muscle.**

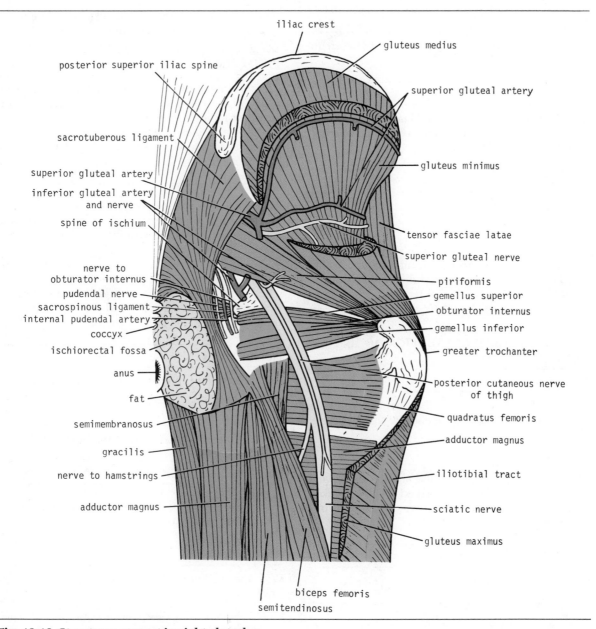

**Fig. 10-16. Structures present in right gluteal region; greater part of gluteus maximus and part of gluteus medius have been removed.**

### Origin

From the outer surface of the ilium between the middle and inferior gluteal lines.

### Insertion

The fibers pass downward and laterally and are attached to the anterior surface of the greater trochanter.

### Nerve Supply

Superior gluteal nerve.

### Action

Acting with the gluteus medius and the tensor fasciae latae, the thigh is powerfully abducted at the hip joint. (See p. 604.) The anterior fibers also medially rotate the thigh.

## TENSOR FASCIAE LATAE (FIGS. 10-16, 10-17, AND 10-21)

### Origin

From the outer edge of the iliac crest between the anterior superior iliac spine and the iliac tubercle.

### Insertion

The fibers run downward and slightly backward, enclosed within a sheath formed by the iliotibial tract of the fascia lata; the fibers are inserted into the iliotibial tract.

### Nerve Supply

Superior gluteal nerve.

### Action

It exerts traction on the iliotibial tract. It thus assists the gluteus maximus muscle in maintaining the knee in the extended position. As long as the iliotibial tract remains in front of the axis of flexion of the knee, it assists in keeping the knee extended. Very often, when one is standing upright, the upward pull of the iliotibial tract is the most impor-

tant factor in keeping the knee extended; the quadriceps muscles may be relaxed.

## PIRIFORMIS (FIG. 10-16)

The piriformis muscle lies partly within the pelvis at its origin. It emerges through the greater sciatic foramen to enter the gluteal region. Its position in the gluteal region serves to separate the superior gluteal vessels and nerves from the inferior gluteal vessels and nerves (Fig. 10-16).

### Origin

From the anterior surface of the second, third, and fourth sacral vertebrae within the pelvis.

### Insertion

The fibers pass downward and laterally through the greater sciatic foramen and are attached to the upper border of the greater trochanter.

### Nerve Supply

Anterior rami of the first and second sacral nerves.

### Action

Lateral rotator of the thigh at the hip joint.

## GEMELLUS SUPERIOR (FIG. 10-16)

### Origin

Spine of the ischium.

### Insertion

With the tendon of the obturator internus (see below).

### Nerve Supply

Nerve to the obturator internus from the sacral plexus.

### Action

Lateral rotator of the thigh at the hip joint.

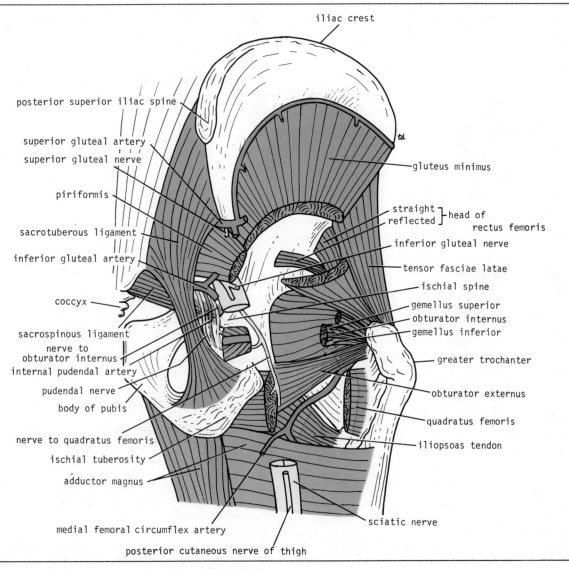

iliac crest

posterior superior iliac spine

superior gluteal artery
superior gluteal nerve

piriformis

sacrotuberous ligament

inferior gluteal artery

coccyx

sacrospinous ligament
nerve to
obturator internus
internal pudendal artery

pudendal nerve

body of pubis

nerve to quadratus femoris

ischial tuberosity

adductor magnus

medial femoral circumflex artery

posterior cutaneous nerve of thigh

td

gluteus minimus

straight ⎤ head of
reflected ⎦ rectus femoris

inferior gluteal nerve

tensor fasciae latae

ischial spine

gemellus superior
obturator internus
gemellus inferior

greater trochanter

obturator externus

quadratus femoris

iliopsoas tendon

sciatic nerve

**Fig. 10-17. Deep structures present in right gluteal region; gluteus maximus and gluteus medius muscles have been completely removed.**

## GEMELLUS INFERIOR (FIG. 10-16)

### Origin

Upper margin of the ischial tuberosity.

### Insertion

With the tendon of obturator internus (see below).

### Nerve Supply

Nerve to the quadratus femoris from the sacral plexus.

### Action

Lateral rotator of the thigh at the hip joint.

## OBTURATOR INTERNUS

The obturator internus lies partly within the pelvis at its origin. It emerges through the lesser sciatic foramen to enter the gluteal region.

### Origin

From the pelvic surface of the obturator membrane and the surrounding bones. (See p. 312.)

### Insertion

The tendon passes out of the pelvis through the lesser sciatic foramen and is joined by the superior and inferior gemelli. The common tendon is inserted into the upper border of the greater trochanter.

### Nerve Supply

Nerve to the obturator internus from the sacral plexus.

### Action

Lateral rotator of the thigh at the hip joint.

## QUADRATUS FEMORIS (FIGS. 10-16 AND 10-17)

### Origin

From the lateral border of the ischial tuberosity.

### Insertion

The fibers pass laterally to be inserted into the quadrate tubercle on the intertrochanteric crest of the femur.

### Nerve Supply

A branch from the sacral plexus.

### Action

Lateral rotator of the thigh at the hip joint.

*Muscles of the Gluteal Region—Their Nerve Supply and Action*

Students wishing to review these muscles should study Table 10-1.

## Nerves of the Lower Limb

The nerves entering the lower limb provide the following important functions: (1) sensory innervation to the skin and deep structures such as the joints, (2) motor innervation to the muscles, (3) sympathetic vasomotor nerves that influence the diameters of the blood vessels, and (4) sympathetic secretomotor supply to the sweat glands.

The nerves that innervate the lower limb originate from the lumbar plexus, situated in the abdomen (see p. 261), and the sacral plexus, situated in the pelvis (see p. 314). These plexuses permit nerve fibers derived from different segments of the spinal cord to be arranged and distributed efficiently in different nerve trunks to the various parts of the lower limb.

## Nerves of the Gluteal Region

### SCIATIC NERVE

The sciatic nerve, a branch of the sacral plexus (L4 and 5; S1, 2, and 3), emerges from the pelvis

Table 10-1. Muscles of the Lower Limb—Muscles of the Gluteal Region

| Name of muscle | Origin | Insertion | Nerve Supply | Action |
|---|---|---|---|---|
| Gluteus maximus | Outer surface of ilium, sacrum, coccyx, sacrotuberous ligament | Iliotibial tract and gluteal tuberosity of femur | Inferior gluteal nerve | Extends and laterally rotates hip joint; through iliotibial tract, it extends knee joint |
| Gluteus medius | Outer surface of ilium | Lateral surface of greater trochanter of femur | Superior gluteal nerve | Abducts thigh at hip joint; tilts pelvis when walking to permit opposite leg to clear ground |
| Gluteus minimus | Outer surface of ilium | Anterior surface of greater trochanter of femur | Superior gluteal nerve | Abducts thigh at hip joint; tilts pelvis when walking to permit opposite leg to clear ground |
| Tensor fasciae latae | Iliac crest | Iliotibial tract | Superior gluteal nerve | Assists gluteus maximus in extending the knee joint |
| Piriformis | Anterior surface of sacrum | Upper border of greater trochanter of femur | First and second sacral nerves | Lateral rotator of thigh at hip joint |
| Obturator internus | Inner surface of obturator membrane | Upper border of greater trochanter of femur | Sacral plexus | Lateral rotator of thigh at hip joint |
| Gemellus superior | Spine of ischium | Upper border of greater trochanter of femur | Sacral plexus | Lateral rotator of thigh at hip joint |
| Gemellus inferior | Ischial tuberosity | Upper border of greater trochanter of femur | Sacral plexus | Lateral rotator of thigh at hip joint |
| Quadratus femoris | Lateral border of ischial tuberosity | Quadrate tubercle of femur | Sacral plexus | Lateral rotator of thigh at hip joint |

through the lower part of the greater sciatic foramen (Figs. 10-16 and 10-17). It is the largest nerve in the body and consists of the tibial and common peroneal nerves bound together with fascia (Figs. 10-68 and 10-70). The nerve appears below the piriformis muscle and curves downward and laterally, lying successively on the root of the ischial spine, the superior gemellus, the obturator internus, the inferior gemellus, and the quadratus femoris, to reach the back of the adductor magnus muscle (Fig. 10-16). It is related posteriorly to the posterior cutaneous nerve of the thigh and the gluteus maximus. It leaves the buttock region by passing deep to the long head of the biceps femoris, to enter the back of the thigh. (See p. 600.)

Occasionally, the common peroneal nerve leaves the sciatic nerve high in the pelvis and appears in the gluteal region by passing above or through the piriformis muscle.

The sciatic nerve usually gives no branches in the gluteal region.

## POSTERIOR CUTANEOUS NERVE OF THE THIGH

The posterior cutaneous nerve of the thigh, a branch of the sacral plexus, enters the gluteal region through the lower part of the greater sciatic foramen below the piriformis muscle (Fig. 10-16). It passes downward on the posterior surface of the sciatic nerve deep to the gluteus maximus. It crosses superficial to the biceps femoris and runs down the back of the thigh beneath the deep fascia. In the popliteal fossa, it pierces the deep fascia and supplies the skin.

### Branches

1. *Gluteal branches* to the skin over the lower medial quadrant of the buttock (Fig. 10-9); these wind around the lower margin of the gluteus maximus.
2. *Perineal branch* to the skin of the back of the scrotum or labium majus. It arises in the gluteal region and passes medially across the origin of the hamstring muscles.
3. *Cutaneous branches* to the back of the thigh and the upper part of the leg (Fig. 10-9).

## SUPERIOR GLUTEAL NERVE

The superior gluteal nerve, a branch of the sacral plexus, leaves the pelvis through the upper part of the greater sciatic foramen above the piriformis (Fig. 10-16). It runs forward between the gluteus medius and minimus, supplies both, and ends by supplying the tensor fasciae latae.

## INFERIOR GLUTEAL NERVE

The inferior gluteal nerve, a branch of the sacral plexus, leaves the pelvis through the lower part of the greater sciatic foramen below the piriformis (Figs. 10-16 and 10-17). It lies close to the posterior cutaneous nerve of the thigh. It ends by supplying the gluteus maximus muscle.

## NERVE TO THE QUADRATUS FEMORIS

A branch of the sacral plexus, the nerve to the quadratus femoris leaves the pelvis through the lower part of the greater sciatic foramen (Fig. 10-

17). It crosses the root of the spine of the ischium and runs downward beneath the sciatic nerve, the gemelli, and the tendon of the obturator internus. It ends by supplying the quadratus femoris and the inferior gemellus.

## PUDENDAL NERVE AND THE NERVE TO THE OBTURATOR INTERNUS

These branches of the sacral plexus leave the pelvis through the lower part of the greater sciatic foramen, below the piriformis (Figs. 10-16 and 10-17). They cross the ischial spine with the internal pudendal artery and immediately reenter the pelvis through the lesser sciatic foramen; they then lie in the ischiorectal fossa. (See p. 380.) The pudendal nerve supplies structures in the perineum. The nerve to the obturator internus supplies the obturator internus muscle on its pelvic surface.

# Arteries of the Gluteal Region

## SUPERIOR GLUTEAL ARTERY

A branch from the internal iliac artery, the superior gluteal artery enters the gluteal region through the upper part of the greater sciatic foramen above the piriformis (Figs. 10-16 and 10-17). It immediately divides into branches that are distributed throughout the gluteal region.

## INFERIOR GLUTEAL ARTERY

A branch from the internal iliac artery, the inferior gluteal artery enters the gluteal region through the lower part of the greater sciatic foramen, below the piriformis (Figs. 10-16 and 10-17). It immediately divides into numerous branches that are distributed throughout the gluteal region.

### The Trochanteric Anastomosis

The trochanteric anastomosis provides the main blood supply to the head of the femur. The nutrient arteries pass along the femoral neck beneath the capsule (Fig. 10-30). The following arteries take part in the anastomisis: (1) the superior gluteal artery, (2) the inferior gluteal artery, (3) the medial femoral circumflex artery, and (4) the lateral femoral circumflex artery.

## The Cruciate Anastomosis

The cruciate anastomosis is situated at the level of the lesser trochanter of the femur and, together with the trochanteric anastomosis, provides a connection between the internal iliac and the femoral arteries. The following arteries take part in the anastomosis: (1) the inferior gluteal artery, (2) the medial femoral circumflex artery, (3) the lateral femoral circumflex artery, and (4) the first perforating artery, a branch of the profunda artery.

# THE FRONT AND MEDIAL ASPECTS OF THE THIGH

## Skin of the Thigh

### CUTANEOUS NERVES

The *lateral cutaneous nerve of the thigh*, a branch of the lumbar plexus (L2 and 3), enters the thigh behind the lateral end of the inguinal ligament (Fig. 10-8). Having divided into anterior and posterior branches, it supplies the skin of the lateral aspect of the thigh and knee. It also supplies the skin of the lower lateral quadrant of the buttock (Fig. 10-9).

The *femoral branch of the genitofemoral nerve*, a branch of the lumbar plexus (L1 and 2), enters the thigh behind the middle of the inguinal ligament and supplies a small area of skin (Fig. 10-8). The genital branch supplies the cremaster muscle. (See p. 262.)

The *ilioinguinal nerve*, a branch of the lumbar plexus (L1), enters the thigh through the superficial inguinal ring (Fig. 10-8). It is distributed to the skin of the root of the penis and adjacent part of the scrotum (or root of the clitoris and adjacent part of the labium majus in the female) and to a small skin area below the medial part of the inguinal ligament.

The *medial cutaneous nerve of the thigh*, a branch of the femoral nerve, supplies the medial aspect of the thigh and joins the patellar plexus (Fig. 10-8).

The *intermediate cutaneous nerve of the thigh*, a branch of the femoral nerve, divides into two branches that supply the anterior aspect of the thigh and joins the patellar plexus (Fig. 10-8).

The *obturator nerve*. Branches from the anterior division of the obturator nerve supply a variable area of skin on the medial aspect of the thigh (Fig. 10-8).

The *patellar plexus* lies in front of the knee and is formed from the terminal branches of the lateral, intermediate, and medial cutaneous nerves of the thigh and the infrapatellar branch of the saphenous nerve (Fig. 10-8).

## SUPERFICIAL VEINS

The superficial veins of the leg are the great and small saphenous veins and their tributaries (Fig. 10-18). They are of great clinical importance.

The *great saphenous vein* drains the medial end of the dorsal venous arch of the foot and passes upward *directly in front of* the medial malleolus (Fig. 10-18). It then ascends in company with the saphenous nerve in the superficial fascia over the medial side of the leg. The vein passes behind the knee and curves forward around the medial side of the thigh. It passes through the lower part of the saphenous opening in the deep fascia and joins the femoral vein about 1½ inches (4 cm) below and lateral to the pubic tubercle (Figs. 10-18 and 10-19).

The great saphenous vein possesses numerous valves; it is connected to the small saphenous vein by one or two branches that pass behind the knee. A number of *perforating veins* connect the great saphenous vein with the deep veins along the medial side of the calf (Fig. 10-18).

At the saphenous opening in the deep fascia, the great saphenous vein usually receives three tributaries that are variable in size and arrangement (Figs. 10-18 and 10-19): (1) the *superficial circumflex iliac vein*, (2) the *superficial epigastric vein*, and (3) the *superficial external pudendal vein*. These veins correspond with the three branches of the femoral artery found in this region.

An additional vein, known as the *accessory vein*, usually joins the main vein about the middle of the thigh or higher up at the saphenous opening.

The *small saphenous vein* is described on page 627.

## SUPERFICIAL INGUINAL LYMPH NODES

The superficial inguinal lymph nodes lie in the superficial fascia below the inguinal ligament, and

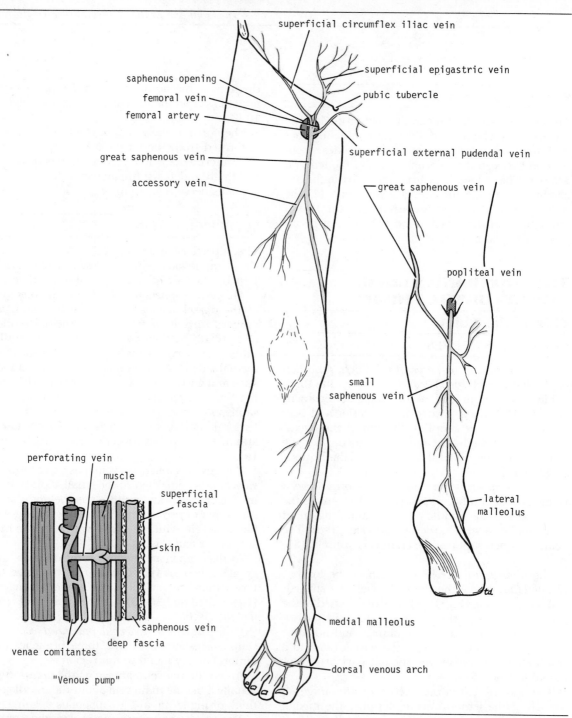

superficial circumflex iliac vein

superficial epigastric vein

saphenous opening

femoral vein

femoral artery

pubic tubercle

great saphenous vein

accessory vein

superficial external pudendal vein

great saphenous vein

popliteal vein

small saphenous vein

perforating vein

muscle

superficial fascia

skin

lateral malleolus

saphenous vein

venae comitantes

deep fascia

"Venous pump"

medial malleolus

dorsal venous arch

**Fig. 10-18. Superficial veins of right lower limb. Note importance of valved perforating veins in the "venous pump."**

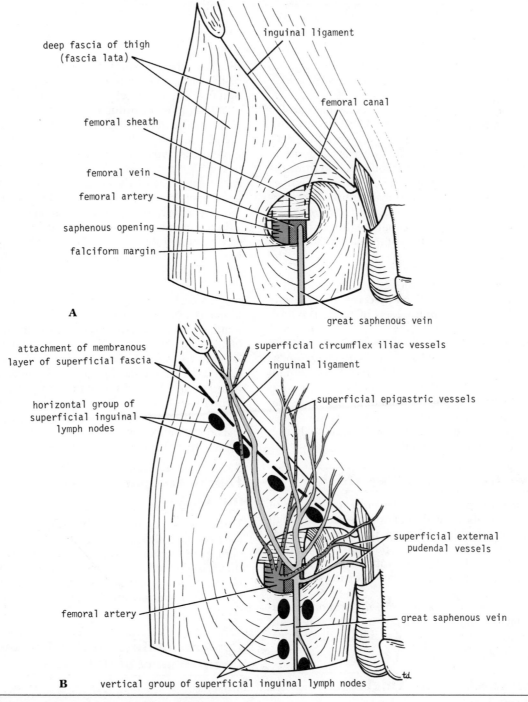

**Fig. 10-19. Superficial veins, arteries, and lymph nodes over right femoral triangle. Note saphenous opening in deep fascia and its relationship to femoral sheath. Note also line of attachment of membranous layer of superficial fascia to deep facia, about a fingerbreadth below inguinal ligament.**

for purposes of description, may be divided into a horizontal and a vertical group (Figs. 10-19 and 10-28).

The *horizontal group* lies just below and parallel to the inguinal ligament (Figs. 10-19 and 10-28). The medial members of the group receive superficial lymph vessels from the anterior abdominal wall below the level of the umbilicus, and from the perineum (Fig. 10-28). It is important to remember that the lymph vessels from the urethra, the external genitalia of both sexes (but not the testes), and the lower half of the anal canal are drained by this route. The lateral members of the group receive superficial lymph vessels from the back below the level of the iliac crests (Fig. 10-28).

The *vertical group* lies along the terminal part of the great saphenous vein and receives the majority of the superficial lymph vessels of the lower limb (Figs. 10-19 and 10-28).

The efferent lymph vessels from the superficial inguinal nodes pass through the saphenous opening in the deep fascia and join the deep inguinal nodes that lie along the medial side of the femoral vein (Fig. 10-22). The deep inguinal nodes are described on page 590; the efferent vessels from these nodes pass through the femoral canal to lymph nodes along the external iliac artery.

## Superficial Fascia of the Thigh

The *membranous layer of the superficial fascia* of the anterior abdominal wall extends into the thigh and is attached to the deep fascia (fascia lata) about a fingerbreadth below the inguinal ligament (Figs. 10-19 and 10-22). The importance of this fact in connection with extravasation of urine is fully described in Chapter 4.

The *fatty layer of the superficial fascia* on the anterior abdominal wall extends into the thigh and continues down over the lower limb without interruption (Fig. 10-22).

## Deep Fascia of the Thigh (Fascia Lata)

The deep fascia encloses the thigh like a trouser leg (Fig. 10-20), and at its upper end is attached to the pelvis and its associated ligaments. On its lateral aspect, it is thickened to form the *iliotibial tract* (Figs. 10-20 and 10-21), which is attached above to the iliac tubercle, and below to the lateral condyle of the tibia. The iliotibial tract receives the insertion of the tensor fasciae latae and the greater part of the gluteus maximus muscle. (See pp. 569 and 572.) In the gluteal region, the deep fascia forms sheaths, which enclose the tensor fasciae latae and the gluteus maximus muscles.

The *saphenous opening* is a gap in the deep fascia in the front of the thigh, which transmits the great saphenous vein, some small branches of the femoral artery, and lymph vessels (Fig. 10-19). It is situated about 1½ inches (4 cm) below and lateral to the pubic tubercle. The *falciform margin* is the lower lateral border of the opening, which lies anterior to the femoral vessels (Fig. 10-19). The medial border of the opening curves around behind the femoral vessels. It is thus seen that the deep fascia is attached to the whole length of the inguinal ligament above, and the falciform margin of the saphenous opening sweeps downward and laterally from the pubic tubercle. The border of the opening then curves upward and medially, and then laterally behind the femoral vessels, to be attached to the pectineal line of the superior ramus of the pubis.

The saphenous opening is filled with loose connective tissue called the *cribriform fascia*.

## Fascial Compartments of the Thigh

Three fascial septa pass from the inner aspect of the deep fascial sheath of the thigh to the linea aspera of the femur (Fig. 10-20). By this means, the thigh is divided into three compartments, each having muscles, nerves, and arteries. The compartments are anterior, medial, and posterior in position.

## Contents of the Anterior Fascial Compartment of the Thigh

### Muscles

Sartorius, iliacus, psoas, pectineus, and quadriceps femoris.

### Blood Supply

Femoral artery.

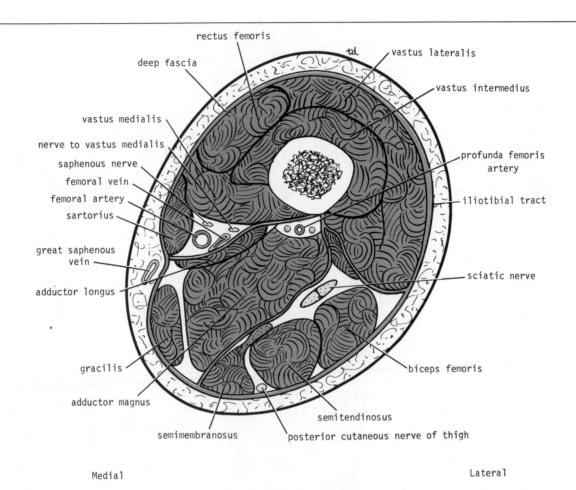

**Fig. 10-20. Transverse section through middle of right thigh as seen from above.**

### Nerve Supply

Femoral nerve.

### *Muscles of the Anterior Fascial Compartment of the Thigh*

#### SARTORIUS (FIG. 10-21)

The sartorius is a narrow strap-shaped muscle that covers the femoral artery in the middle one-third of the thigh.

### Origin

From the anterior superior iliac spine.

### Insertion

The muscle fibers run downward and medially and are attached to the upper part of the medial surface of the shaft of the tibia.

### Nerve Supply

Femoral nerve.

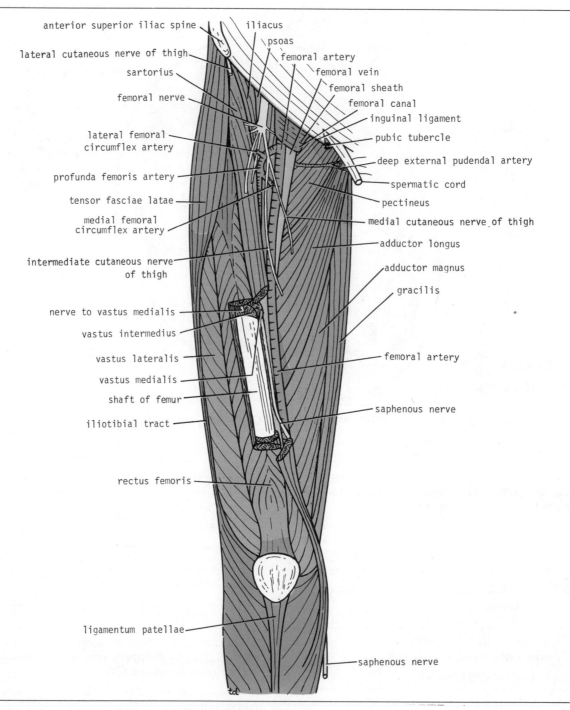

**Fig. 10-21. Femoral triangle and adductor (subsartorial) canal in right lower limb.**

## Action

Flexes, abducts, and laterally rotates the thigh at the hip joint. It flexes and medially rotates the leg at the knee joint.

## ILIACUS (FIGS. 10-21, 10-22, AND 10-23)

### Origin

This fan-shaped muscle arises from the iliac fossa within the abdomen. (See p. 178.)

### Insertion

The fibers converge and join the tendon of the psoas to form the iliopsoas muscle (see below).

### Nerve Supply

A branch of the femoral nerve within the abdomen.

### Action

The iliopsoas flexes the thigh on the trunk at the hip joint; or if the thigh is fixed, it flexes the trunk on the thigh; it also medially rotates the thigh.

## PSOAS (FIGS. 10-21, 10-22, AND 10-23)

The psoas is a long fusiform muscle that arises within the abdomen and descends into the thigh.

### Origin

From the roots of the transverse processes, the sides of the vertebral bodies, and the intervertebral discs, from the twelfth thoracic to the fifth lumbar vertebrae.

### Insertion

The fibers run downward and laterally and leave the abdomen to enter the thigh by passing behind the inguinal ligament. The iliopsoas tendon is attached to the lesser trochanter of the femur. A bursa intervenes between the tendon and the hip joint and may communicate with the joint.

### Nerve Supply

Branches from the lumbar plexus.

### Action

The iliopsoas flexes the thigh on the trunk at the hip joint; or if the thigh is fixed, it flexes the trunk on the thigh; it also medially rotates the thigh.*

The fascial sheath enclosing the muscle is described on pages 177 and 197.

## PECTINEUS (FIGS. 10-21 AND 10-22)

### Origin

From the superior ramus of the pubis.

### Insertion

The muscle fibers pass downward, backward, and laterally and are attached to the upper end of the linea aspera just below the lesser trochanter.

### Nerve Supply

Femoral nerve. (Occasionally, it receives a branch from the obturator nerve.)

### Action

Flexes and adducts the thigh at the hip joint.

The *quadriceps femoris* muscle consists of four parts: the rectus femoris, vastus lateralis, vastus medialis, and vastus intermedius, which have a common tendon of insertion into the upper lateral and medial borders of the patella and then, via the ligamentum patellae, into the tubercle of the tibia.

## RECTUS FEMORIS (FIG. 10-21)

### Origin

A *straight head* from the anterior inferior iliac spine and a *reflected head* from the ilium above the acetabulum.

*Electromyographic studies do not support this latter action.

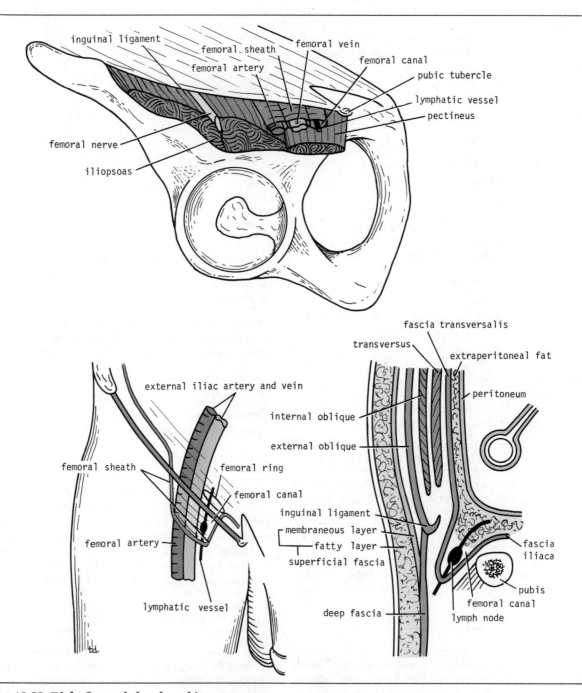

**Fig. 10-22. Right femoral sheath and its contents.**

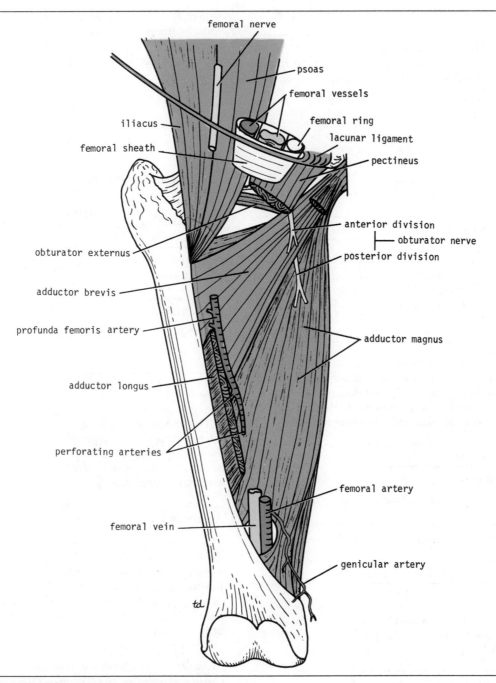

**Fig. 10-23. Relationship between obturator nerve and adductor muscles in right lower limb.**

## Insertion

The two heads unite in front of the hip joint, and the bipennate muscle is inserted into the quadriceps tendon and so into the patella.

## Nerve Supply

Femoral nerve.

## Action

See *Action of Quadriceps Femoris Muscle.*

## VASTUS LATERALIS (FIG. 10-21)

### Origin

From the intertrochanteric line, the base of the great trochanter, and the lateral lip of the linea aspera of the femur and the lateral deep fascial septum.

### Insertion

The fibers pass downward and forward to be inserted into the quadriceps tendon and so into the patella. Some of the tendinous fibers join the capsule of the knee joint and strengthen it.

### Nerve Supply

Femoral nerve.

### Action

See *Action of Quadriceps Femoris Muscle.*

## VASTUS MEDIALIS (FIG. 10-21)

### Origin

From the intertrochanteric line and the medial lip of the linea aspera of the femur and the medial deep fascial septum.

### Insertion

The fibers pass downward and forward, to be inserted into the quadriceps tendon and so into the patella. Some of the tendinous fibers join the cap-
sule of the knee joint and strengthen it. The lowest muscle fibers are almost horizontal in direction and prevent the patella from being pulled laterally during contraction of the quadriceps muscle.

## Nerve Supply

Femoral nerve.

## Action

See *Action of Quadriceps Femoris Muscle.*

## VASTUS INTERMEDIUS (FIG. 10-21)

### Origin

From the anterior and lateral surfaces of the shaft of the femur.

### Insertion

The fibers pass downward and join the deep aspect of the quadriceps tendon.

### Nerve Supply

Femoral nerve.

The *articularis genus* is a small part of the vastus intermedius that is inserted into the upper part of the synovial membrane of the knee joint. It serves to retract the synovial membrane superiorly during extension of the knee joint.

## *Action of Quadriceps Femoris Muscle*

The quadriceps femoris is a powerful extensor of the knee joint; the patella is a sesamoid bone in the tendon of insertion. The rectus femoris also flexes the hip joint. The tone of the quadriceps muscle plays an important role in strengthening the knee joint. The lower fibers of the vastus medialis stabilize the patella.

## *Muscles of the Anterior Fascial Compartment of the Thigh—Their Nerve Supply and Action*

Students wishing to review these muscles should study Table 10-2.

Table 10-2. Muscles of the Lower Limb—Muscles of the Anterior Fascial Compartment of the Thigh

| Name of muscle | Origin | Insertion | Nerve supply | Action |
|---|---|---|---|---|
| Sartorius | Anterior superior iliac spine | Upper medial surface shaft of tibia | Femoral nerve | Flexes, abducts, laterally rotates thigh at hip joint; flexes and medially rotates leg at knee joint |
| Iliacus | Iliac fossa within abdomen | With psoas into lesser trochanter of femur | Femoral nerve | Flexes thigh on trunk; if thigh is fixed, it flexes the trunk on the thigh as in sitting up from lying down |
| Psoas | Transverse processes, bodies, and intervertebral discs of the twelfth thoracic and five lumbar vertebrae | With iliacus into lesser trochanter of femur | Lumbar plexus | Flexes thigh on trunk; if thigh is fixed, it flexes the trunk on thigh as in sitting up from lying down |
| Pectineus | Superior ramus of pubis | Upper end of linea aspera of shaft of femur | Femoral nerve | Flexes and adducts thigh at hip joint |
| Quadriceps femoris | | | | |
| Rectus femoris | Straight head; anterior inferior iliac spine; reflected head; ilium above acetabulum | Quadriceps tendon into patella then via ligamentum patellae into tubercle of tibia | Femoral nerve | Extension of leg at knee joint; flexes thigh at hip joint |
| Vastus lateralis | Upper end and shaft of femur; lateral deep fascial septum | Quadriceps tendon into patella then via ligamentum patellae into tubercle of tibia | Femoral nerve | Extension of leg at knee joint |
| Vastus medialis | Upper end and shaft of femur; medial deep fascial septum | Quadriceps tendon into patella then via ligamentum patellae into tubercle of tibia | Femoral nerve | Extension of leg at knee joint; stabilizes patella |
| Vastus intermedius | Anterior and lateral surfaces of shaft of femur | Quadriceps tendon into patella then via ligamentum patellae into tubercle of tibia | Femoral nerve | Extension of leg at knee joint; articularis genus retracts synovial membrane |

## FEMORAL SHEATH

The femoral sheath (Figs. 10-19, 10-21, 10-22, and 10-23) is a downward protrusion into the thigh of the fascial envelope lining the abdominal walls. (See p. 179.) Its anterior wall is continuous above with the fascia transversalis, and its posterior wall with the fascia iliaca. The sheath surrounds the femoral vessels and lymphatics for about 1 inch (2.5 cm) below the inguinal ligament. The *femoral artery*, as it enters the thigh beneath the inguinal ligament, occupies the *lateral compartment* of the sheath. The *femoral vein*, which lies on its medial side and is separated from it by a fibrous septum, occupies the *intermediate compartment*. The *lymphatics*, which are separated from the vein by a fibrous septum, occupy the most *medial compartment* (Fig. 10-22).

The *femoral canal* is the term used to name the small medial compartment for the lymphatics (Fig. 10-22). It is about ½ inch (1.3 cm) long, and its upper opening is referred to as the *femoral ring*. The *femoral septum*, which is a condensation of extraperitoneal tissue, closes the ring. The femoral canal contains: (1) fatty connective tissue, (2) all the efferent lymph vessels from the deep inguinal lymph nodes, and (3) one of the deep inguinal lymph nodes.

The femoral sheath is adherent to the walls of the blood vessels and inferiorly blends with the tunica adventitia of these vessels. The part of the femoral sheath that forms the medially located femoral canal is not adherent to the walls of the small lymphatic vessels; it is this site that forms a potentially weak area in the abdomen. It is not difficult to imagine how a protrusion of peritoneum could be forced down the femoral canal, pushing the femoral septum before it. Such a condition is known as a *femoral hernia* and is described in detail on page 189.

The upper end of the canal, or femoral ring (Fig. 10-22), has the following important relations: anteriorly, the inguinal ligament; posteriorly, the superior ramus of the pubis; medially, the lacunar ligament; and laterally, the femoral vein.

The lower end of the canal is normally closed by the adherence of its medial wall to the tunica adventitia of the femoral vein. It lies close to the saphenous opening in the deep fascia of the thigh (Fig. 10-19).

## FEMORAL ARTERY

The femoral artery enters the thigh by passing behind the inguinal ligament, as a continuation of the external iliac artery (Figs. 10-21 and 10-24). Here, it lies midway between the anterior superior iliac spine and the symphysis pubis. The femoral artery is the main arterial supply to the lower limb. It descends almost vertically toward the adductor tubercle of the femur and ends at the opening in the adductor magnus muscle by entering the popliteal space as the popliteal artery (Fig. 10-23).

### Relations

#### Anteriorly

In the upper part of its course, it is superficial and is covered by skin and fascia. In the lower part of its course, it passes behind the sartorius muscle (Fig. 10-21). It is related to the anterior wall of the femoral sheath above and is crossed by the medial cutaneous nerve of the thigh and the saphenous nerve below.

#### Posteriorly

The artery lies on the psoas, which separates it from the hip joint, the pectineus, and the adductor longus (Fig. 10-21). The femoral vein intervenes between the artery and the adductor longus.

#### Medially

It is related to the femoral vein in the upper part of its course (Fig. 10-21).

#### Laterally

The femoral nerve and its branches (Fig. 10-21).

### Branches

1. The *superficial circumflex iliac artery* is a small branch that passes through the saphenous opening and runs up to the region of the anterior superior iliac spine (Fig. 10-19).
2. The *superficial epigastric artery* is a small branch that passes through the saphenous opening, crosses the inguinal ligament, and runs to the region of the umbilicus (Fig. 10-19).

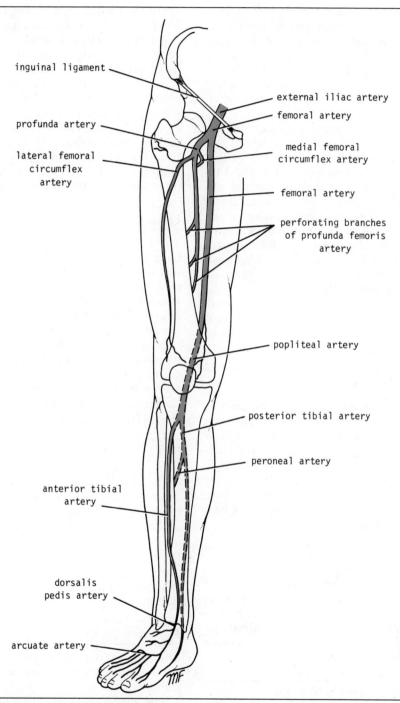

inguinal ligament

profunda artery

lateral femoral
circumflex
artery

external iliac artery

femoral artery

medial femoral
circumflex artery

femoral artery

perforating branches
of profunda femoris
artery

popliteal artery

posterior tibial artery

peroneal artery

anterior tibial
artery

dorsalis
pedis artery

arcuate artery

**Fig. 10-24. Major arteries of the lower limb.**

3. The *superficial external pudendal artery* (Fig. 10-19) is a small branch that passes through the saphenous opening and runs medially, to supply the skin of the scrotum (or labium majus).
4. The *deep external pudendal artery* (Fig. 10-21) runs medially in front of or behind the femoral vein and supplies the skin of the scrotum (or labium majus).
5. The *profunda femoris artery* is a large and important branch that arises from the lateral side of the femoral artery about 1½ inches (4 cm) below the inguinal ligament (Figs. 10-21 and 10-24). It passes medially behind the femoral vessels and enters the medial fascial compartment of the thigh by running behind the adductor longus (Figs. 10-23 and 10-25). It ends by becoming the *fourth perforating artery*. At its origin, it gives off the *medial* and *lateral femoral circumflex arteries*, and during its course it gives off *three perforating arteries* (Fig. 10-25).
6. The *descending genicular artery* is a small branch that arises from the femoral artery near its termination (Fig. 10-23). It assists in supplying the knee joint.

## FEMORAL VEIN

The femoral vein enters the thigh by passing through the opening in the adductor magnus as a continuation of the popliteal vein (Fig. 10-23). It ascends through the thigh, lying at first on the lateral side of the artery, then posterior to it, and finally on its medial side (Fig. 10-21). It leaves the thigh in the intermediate compartment of the femoral sheath and passes behind the inguinal ligament to become the external iliac vein.

### Tributaries

The tributaries of the femoral vein are the *great saphenous vein* and veins that correspond to the branches of the femoral artery (Fig. 10-19). The superficial circumflex iliac vein, the superficial epigastric vein, and the external pudendal veins drain into the great saphenous vein.

## DEEP INGUINAL LYMPH NODES

The deep inguinal lymph nodes are variable in number, but there are commonly three. They lie along the medial side of the terminal part of the femoral vein, and the most superior is usually located in the femoral canal (Fig. 10-22). They receive all the lymph from the superficial inguinal nodes via lymph vessels that pass through the cribriform fascia of the saphenous opening. They also receive lymph from the deep structures of the lower limb that have ascended in lymph vessels alongside the arteries, some having passed through the popliteal nodes. The efferent lymph vessels from the deep inguinal nodes ascend into the abdominal cavity through the femoral canal and drain into the external iliac nodes.

## FEMORAL NERVE

The femoral nerve is the largest branch of the lumbar plexus (L2, 3, and 4). It emerges from the lateral border of the psoas muscle within the abdomen (see p. 262), and passes downward in the interval between the psoas and iliacus. It lies behind the fascia iliaca and enters the thigh lateral to the femoral artery and the femoral sheath, behind the inguinal ligament (Figs. 10-21 and 10-22). About 1½ inches (4 cm) below the inguinal ligament, it terminates by dividing into anterior and posterior divisions. The femoral nerve supplies all the muscles of the anterior compartment of the thigh (Fig. 10-21).

### Branches (Fig. 10-67)

#### Anterior Division

This gives off two cutaneous and two muscular branches. The cutaneous branches are: (1) the *medial cutaneous nerve of the thigh* and (2) the *intermediate cutaneous nerves* that supply the skin of the medial and anterior surfaces of the thigh, respectively (Figs. 10-8 and 10-21). The muscular branches supply (1) the sartorius and (2) the pectineus.

#### Posterior Division

This gives off one cutaneous branch, the saphenous nerve, and muscular branches to the quadriceps muscle.

The *saphenous nerve* runs downward and medially and crosses the femoral artery from its lateral to its medial side (Fig. 10-21). It pierces the

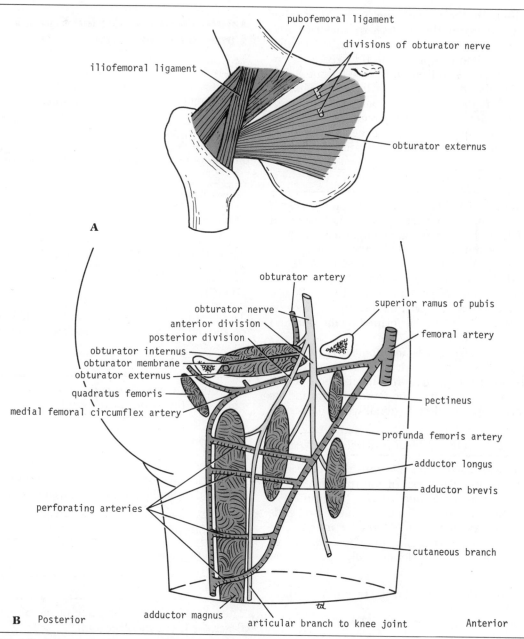

**Fig. 10-25. (A) Obturator externus muscle and (B) vertical section of medial compartment of thigh. Note courses taken by obturator nerve and its divisions, and profunda femoris artery and its branches. Note also anastomosis between perforating arteries and medial femoral circumflex artery.**

deep fascia on the medial side of the knee, after emerging between the tendons of sartorius and gracilis (Fig. 10-8). It then runs down the medial side of the leg in company with the great saphenous vein. It passes in front of the medial malleolus and along the medial border of the foot, where it terminates in the region of the ball of the big toe.

The *muscular branch* of the rectus femoris also supplies the hip joint; the branches to the three vasti muscles also supply the knee joint.

## FEMORAL TRIANGLE

The femoral triangle is a triangular depressed area situated in the upper part of the medial aspect of the thigh just below the inguinal ligament (Fig. 10-21). It is bounded *superiorly* by the inguinal ligament, *laterally* by the sartorius, and *medially* by the medial border of the adductor longus muscle. Its *floor* is gutter-shaped and formed from lateral to medial by the iliopsoas, the pectineus, and the adductor longus. Its *roof* is formed by the skin and fasciae of the thigh.

The femoral triangle is an important area, since it contains the terminal part of the femoral nerve and its branches, the femoral sheath, the femoral artery and its branches, the femoral vein and its tributaries, and the deep inguinal lymph nodes.

## ADDUCTOR (OR SUBSARTORIAL) CANAL

The adductor canal is an intermuscular cleft situated on the medial aspect of the middle third of the thigh (Figs. 10-20 and 10-21). It commences above at the apex of the femoral triangle and ends below at the opening in the adductor magnus. In cross section it is triangular in shape, having an anteromedial wall, a posterior wall, and a lateral wall.

The *anteromedial wall* is formed by a fibrous sheet deep to the sartorius.
The *posterior wall* is formed by the adductor longus and magnus.
The *lateral wall* is formed by the vastus medialis.

The adductor canal contains the terminal part of the femoral artery, the femoral vein, the deep lymph vessels, the saphenous nerve, the nerve to the vastus medialis, and the terminal part of the obturator nerve.

## Contents of the Medial Fascial Compartment of the Thigh

### Muscles

Gracilis, adductor longus, adductor brevis, adductor magnus, and obturator externus.

### Blood Supply

Profunda femoris artery and obturator artery.

### Nerve Supply

Obturator nerve.

### *Muscles of the Medial Fascial Compartment of the Thigh*

### GRACILIS (FIG. 10-21)

The gracilis muscle is long and straplike and lies on the medial side of the thigh and the knee.

### Origin

From the outer surface of the inferior ramus of the pubis and the ramus of the ischium.

### Insertion

The fibers pass downward along the medial side of the thigh, and the tendon is attached to the upper part of the medial surface of the shaft of the tibia.

### Nerve Supply

Obturator nerve.

### Action

The muscle adducts the thigh at the hip joint and flexes the leg at the knee joint.

### ADDUCTOR LONGUS (FIGS. 10-21 AND 10-23)

The adductor longus muscle is triangular in shape and is the most anterior of the three adductor muscles.

## Origin

From the front of the body of the pubis below and medial to the pubic tubercle.

## Insertion

The muscle fibers diverge as they pass downward and laterally and are attached to the medial lip of the linea aspera.

## Nerve Supply

Obturator nerve.

## Action

Adducts the thigh at the hip joint and assists in lateral rotation.

## ADDUCTOR BREVIS (FIGS. 10-23 AND 10-25)

The adductor brevis lies posterior to the pectineus and the adductor longus.

## Origin

From the outer surface of the inferior ramus of the pubis.

## Insertion

The muscle fibers diverge as they pass downward and laterally and are attached to the linea aspera.

## Nerve Supply

Obturator nerve.

## Action

Adducts the thigh at the hip joint and assists in lateral rotation.

## ADDUCTOR MAGNUS (FIGS. 10-23, 10-25, AND 10-27)

The adductor magnus is a very large triangular-shaped muscle consisting of adductor and hamstring portions.

## Origin

From the outer surface of the inferior ramus of the pubis; from the ramus of the ischium and the ischial tuberosity.

## Insertion

*Adductor portion.* The muscle fibers diverge as they pass downward and laterally to be attached to the posterior surface of the femur from the quadrate tubercle above, along the linea aspera to the medial supracondylar ridge below. *Hamstring portion.* The fibers that arise from the ischial tuberosity are inserted below on the adductor tubercle on the medial condyle of the femur.

There is a gap (*adductor hiatus*) in the attachment of this muscle to the medial supracondylar ridge, which permits the femoral vessels to pass from the adductor canal downward into the popliteal space.

## Nerve Supply

The adductor portion is supplied by the obturator nerve; the hamstring portion is supplied by the sciatic nerve.

## Action

The adductor portion adducts the thigh at the hip joint and also assists in lateral rotation. The hamstring portion extends the thigh at the hip joint.

## OBTURATOR EXTERNUS (FIG. 10-25)

The obturator externus is a deeply placed triangular-shaped muscle.

## Origin

From the outer surface of the obturator membrane and the adjacent margin of the pubic and ischial rami.

## Insertion

The muscle fibers converge as they pass laterally at first below and then behind the hip joint, to be in-

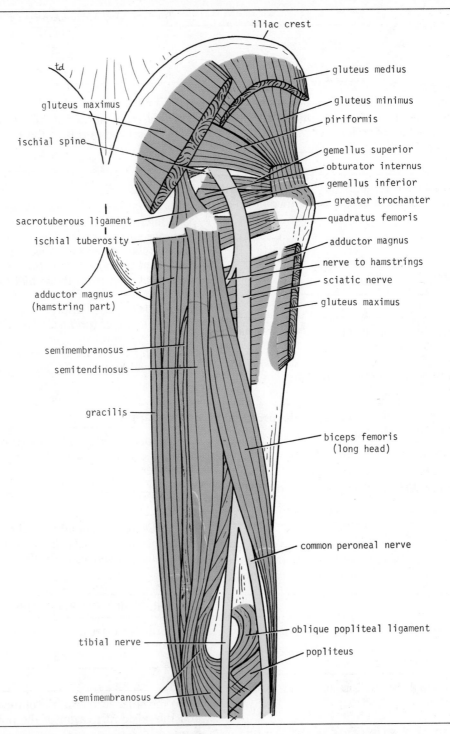

**Fig. 10-26. Structures present on posterior aspect of right thigh.**

Labels (clockwise from top): iliac crest, gluteus medius, gluteus minimus, piriformis, gemellus superior, obturator internus, gemellus inferior, greater trochanter, quadratus femoris, adductor magnus, nerve to hamstrings, sciatic nerve, gluteus maximus, biceps femoris (long head), common peroneal nerve, oblique popliteal ligament, popliteus, semimembranosus, tibial nerve, gracilis, semitendinosus, semimembranosus, adductor magnus (hamstring part), ischial tuberosity, sacrotuberous ligament, ischial spine, gluteus maximus, td

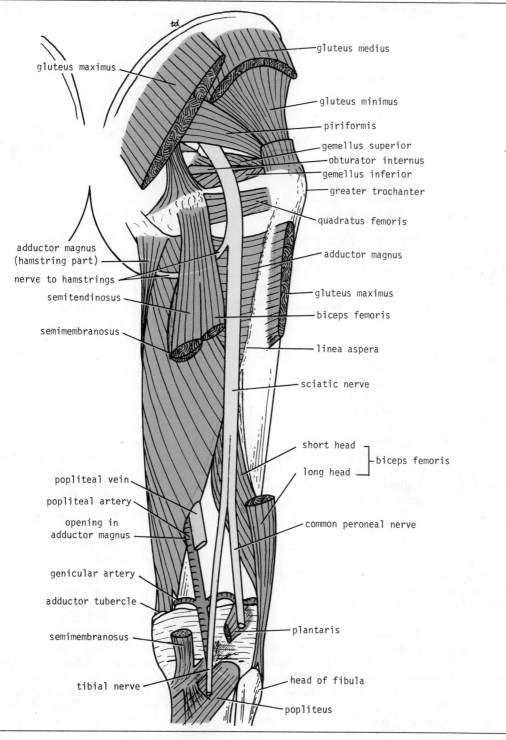

**Fig. 10-27. Deep structures present on posterior aspect of right thigh.**

serted onto the medial surface of the greater trochanter.

## Nerve Supply

Obturator nerve.

## Action

Laterally rotates the thigh at the hip joint.

## Muscles of the Medial Fascial Compartment of the Thigh—Their Nerve Supply and Action

Students wishing to review these muscles should study Table 10-3.

## PROFUNDA FEMORIS ARTERY

The profunda femoris is a large artery, which arises from the lateral side of the femoral artery in the femoral triangle, about 1½ inches (4 cm) below the inguinal ligament (Figs. 10-21 and 10-24). It leaves the anterior fascial compartment by passing down behind the adductor longus muscle (Fig. 10-23). It descends in the interval between the adductor longus and adductor brevis and then lies on the adductor magnus, where it ends as the fourth perforating artery (Fig. 10-25).

### Branches

1. *Medial femoral circumflex artery.* This passes backward between the muscles that form the floor of the femoral triangle and gives off muscular branches in the medial fascial compartment of the thigh (Fig. 10-25). It takes part in the formation of the cruciate anastomosis.
2. *Lateral femoral circumflex artery.* This passes laterally between the terminal branches of the femoral nerve (Fig. 10-21). It breaks up into branches that supply the muscles of the region and takes part in the formation of the cruciate anastomosis.
3. *Four perforating arteries.* Three of these arise as branches of the profunda femoris artery; the fourth performing artery is the terminal part of the profunda artery (Fig. 10-25). The perforating arteries run backward and laterally, piercing the various muscle layers as they go. They supply the muscles and terminate by anastomosing with one another and with the inferior gluteal artery and the circumflex femoral arteries above and the muscular branches of the popliteal artery below.

## PROFUNDA FEMORIS VEIN

The profunda femoris vein receives tributaries that correspond to the branches of the artery. It drains into the femoral vein.

## OBTURATOR ARTERY

The obturator artery is a branch of the internal iliac artery. (See p. 320.) It passes downward and forward on the lateral wall of the pelvis and accompanies the obturator nerve through the obturator canal (i.e., the upper part of the obturator foramen) (Fig. 10-25). On entering the medial fascial compartment of the thigh, it divides into medial and lateral branches, which pass around the margin of the outer surface of the obturator membrane. It gives off muscular branches and an articular branch to the hip joint.

## OBTURATOR VEIN

The obturator vein receives tributaries that correspond to the branches of the artery. It drains into the internal iliac vein.

## OBTURATOR NERVE

The obturator nerve arises from the lumbar plexus (L2, 3, and 4) and emerges on the medial border of the *psoas* muscle within the abdomen. (See p. 262.) It runs downward and forward on the lateral wall of the pelvis to reach the upper part of the obturator foramen (Fig. 6-13), where it divides into anterior and posterior divisions (Fig. 10-25).

Table 10-3. Muscles of the Lower Limb—Muscles of the Medial Fascial Compartment of the Thigh

| Name of muscle | Origin | Insertion | Nerve supply | Action |
|---|---|---|---|---|
| Gracilis | Inferior ramus of pubis; ramus of ischium | Upper part of shaft of tibia on medial surface | Obturator nerve | Adducts thigh at hip joint; flexes leg at knee joint |
| Adductor longus | Body of pubis, medial to pubic tubercle | Posterior surface of shaft of femur (medial lip linea aspera) | Obturator nerve | Adducts thigh at hip joint and assists in lateral rotation |
| Adductor brevis | Inferior ramus of pubis | Posterior surface of shaft of femur (linea aspera) | Obturator nerve | Adducts thigh at hip joint and assists in lateral rotation |
| Adductor magnus | Inferior ramus of pubis, ramus of ischium, ischial tuberosity | Posterior surface of shaft of femur; adductor tubercle of femur | Adductor portion: obturator nerve; hamstring portion: sciatic nerve | Adducts thigh at hip joint and assists in lateral rotation; hamstring portion extends thigh at hip joint |
| Obturator externus | Outer surface of obturator membrane and pubic and ischial rami | Medial surface greater trochanter | Obturator nerve | Laterally rotates thigh at hip joint |

## Branches (Fig. 10-71)

1. The *anterior division* passes downward in front of the obturator externus and the adductor brevis and behind the pectineus and adductor longus (Fig. 10-25). It gives muscular branches to the gracilis, adductor brevis, and adductor longus, and occasionally to the pectineus. It gives articular branches to the hip joint and terminates as a small nerve that supplies the femoral artery. It contributes a variable branch to the subsartorial plexus and supplies the skin on the medial side of the thigh.

2. The *posterior division* pierces the obturator externus and passes downward behind the adductor brevis and in front of the adductor magnus (Fig. 10-25). It terminates by descending through the opening in the adductor magnus to supply the knee joint. It gives muscular branches to the obturator externus, to the adductor part of the adductor magnus, and occasionally to the adductor brevis.

# THE BACK OF THE THIGH

## Skin

### CUTANEOUS NERVES

The *posterior cutaneous nerve of the thigh*, a branch of the sacral plexus, leaves the gluteal region by emerging from beneath the lower border of the gluteus maximus muscle (Fig. 10-9). It descends on the back of the thigh beneath the deep fascia and crosses superficial to the biceps femoris. In the popliteal fossa it pierces the deep fascia and supplies the skin. It gives off numerous cutaneous branches to the skin on the back of the thigh and the upper part of the leg (Fig. 10-9).

### SUPERFICIAL VEINS

Many small veins curve around the medial and lateral aspects of the thigh and ultimately drain into the great saphenous vein (Fig. 10-18). Superficial veins from the lower part of the back of the thigh join the small saphenous vein in the popliteal fossa.

## LYMPH VESSELS

Lymph from the skin and superficial fascia on the back of the thigh drains upward and forward into the vertical group of superficial inguinal lymph nodes (Fig. 10-28).

# Contents of the Posterior Fascial Compartment of the Thigh

## Muscles

Biceps femoris, semitendinosus, semimembranosus, and a small part of the adductor magnus (hamstring muscles).

## Blood Supply

Branches of the profunda femoris artery.

## Nerve Supply

Sciatic nerve.

## *Muscles of the Posterior Fascial Compartment of the Thigh*

### BICEPS FEMORIS (FIGS. 10-26 AND 10-27)

### Origin

The *long head* from the ischial tuberosity in common with the tendon of origin of the semitendinosus muscle. The *short head* from the linea aspera, the lateral supracondylar ridge, and the lateral intermuscular septum.

### Insertion

The two heads unite just above the knee joint, and the common tendon is inserted into the head of the fibula.

### Nerve Supply

The long head is supplied by the tibial part of the sciatic; the short head is supplied by the common peroneal part of the sciatic.

### Action

Flexes and laterally rotates the leg at the knee joint; the long head also extends the thigh at the hip joint.

### SEMITENDINOSUS (FIG. 10-26)

### Origin

From the ischial tuberosity in common with the long head of the biceps.

### Insertion

By a long tendon into the upper part of the medial surface of the shaft of the tibia.

### Nerve Supply

The tibial portion of the sciatic.

### Action

Flexes and medially rotates the leg at the knee joint; it also extends the thigh at the hip joint.

### SEMIMEMBRANOSUS (FIGS. 10-26 AND 10-27)

### Origin

From the ischial tuberosity.

### Insertion

Into a groove on the posteromedial surface of the medial condyle of the tibia. It sends a fibrous expansion upward and laterally, which reinforces the capsule on the back of the knee joint; the expansion is called the *oblique popliteal ligament.*

**Fig. 10-28. Lymphatic drainage of superficial tissues of right lower limb, and abdominal walls below level of umbilicus. Note arrangement of superficial and deep inguinal lymph nodes and their relationship to saphenous opening in deep fascia. Note also that all lymph from these nodes ultimately drains into external iliac nodes via femoral canal.**

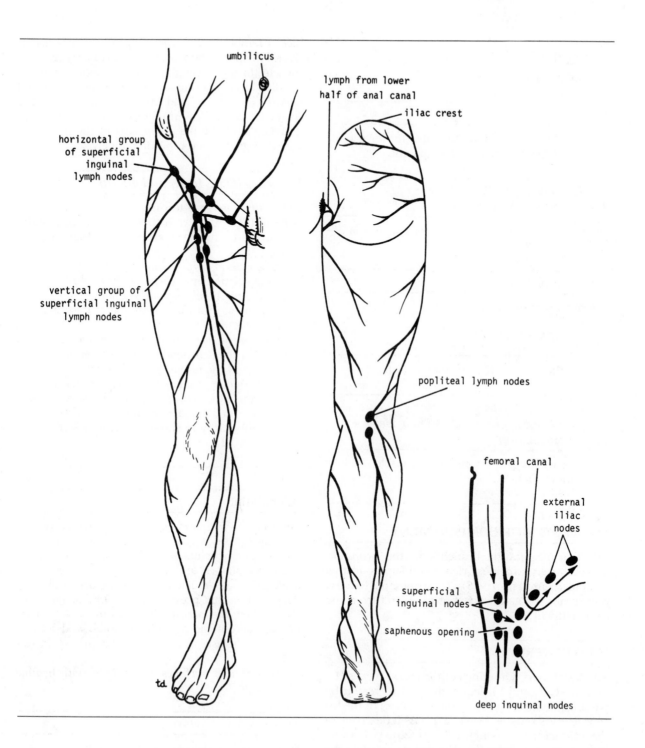

umbilicus

lymph from lower
half of anal canal

iliac crest

horizontal group
of superficial
inguinal
lymph nodes

vertical group of
superficial inguinal
lymph nodes

popliteal lymph nodes

femoral canal

external
iliac
nodes

superficial
inguinal nodes

saphenous opening

deep inguinal nodes

td

## Nerve Supply

The tibial portion of the sciatic.

## Action

Flexes and medially rotates the leg at the knee joint; it also extends the thigh at the hip joint.

## ADDUCTOR MAGNUS (HAMSTRING PORTION)

The hamstring portion of adductor magnus is described on page 593.

## Nerve Supply

The tibial portion of the sciatic.

## Action

Extends the thigh at the hip joint.

### Muscles of the Posterior Fascial Compartment of the Thigh—Their Nerve Supply and Action

Students wishing to review these muscles should study Table 10-4.

## VESSELS OF THE POSTERIOR COMPARTMENT

The four perforating branches of the profunda femoris artery provide a rich blood supply to this compartment (Fig. 10-25). The profunda femoris vein drains the greater part of the blood from the compartment.

## SCIATIC NERVE

The sciatic nerve, a branch of the sacral plexus (L4 and 5; S1, 2, and 3), leaves the gluteal region by passing downward deep to the long head of the biceps femoris muscle. (See p. 575.) As it descends in the midline of the thigh (Fig. 10-26), it is overlapped posteriorly by the adjacent margins of the biceps femoris and semimembranosus muscles. It lies on the posterior aspect of the adductor magnus muscle. In the lower third of the thigh it ends by dividing into the tibial and common peroneal nerves (Figs. 10-26 and 10-27). Occasionally, the sciatic nerve divides into its two terminal parts at a higher level—in the upper part of the thigh, the gluteal region, or even inside the pelvis.

### Branches (Figs. 10-68 and 10-70)

1. The *tibial nerve*, a terminal branch of the sciatic nerve (Figs. 10-26 and 10-27), enters the popliteal fossa. Its further course is described on page 616.
2. The *common peroneal nerve*, a terminal branch of the sciatic nerve (Figs. 10-26 and 10-27), enters the popliteal fossa on the lateral side of the tibial nerve. Its further course is described on page 617.
3. *Muscular branches* to the long head of the biceps femoris, the semitendinosus, the semimembranosus, and the ischial fibers (hamstring part) of the adductor magnus. These branches arise from the tibial component of the sciatic nerve and run medially a short distance below the ischial tuberosity to supply the muscles (Figs. 10-26 and 10-27).

## HIP JOINT

### Articulation

Between the hemispherical head of the femur and the cup-shaped acetabulum of the hip bone (Fig. 10-30). The articular surface of the acetabulum is horseshoe-shaped and is deficient inferiorly at the *acetabular notch*. The cavity of the acetabulum is deepened by the presence of a fibrocartilaginous rim, called the *acetabular labrum*. The labrum bridges across the acetabular notch and is here called the *transverse acetabular ligament* (Fig. 10-30).

The articular surfaces are covered with hyaline cartilage.

### Type

Synovial ball-and-socket.

Table 10-4. Muscles of the Lower Limb—Muscles of the Posterior Fascial Compartment of the Thigh

| Name of muscle | Origin | Insertion | Nerve supply | Action |
|---|---|---|---|---|
| Biceps femoris | Long head: ischial tuberosity; Short head: linea aspera, lateral supracondylar ridge of shaft of femur | Head of fibula | Long head: tibial portion of sciatic nerve; Short head: common peroneal portion of sciatic nerve | Flexes and laterally rotates leg at knee joint; long head also extends thigh at hip joint |
| Semitendinosus | Ischial tuberosity | Upper part medial surface of shaft of tibia | Tibial portion of sciatic nerve | Flexes and medially rotates leg at knee joint; extends thigh at hip joint |
| Semimembranosus | Ischial tuberosity | Medial condyle of tibia | Tibial portion of sciatic nerve | Flexes and medially rotates leg at knee joint; extends thigh at hip joint |
| Adductor magnus (hamstring portion) | Ischial tuberosity | Adductor tubercle of femur | Tibial portion of sciatic nerve | Extends thigh at hip joint |

## Capsule

This encloses the joint and is attached to the acetabular labrum medially (Fig. 10-30). Laterally, it is attached to the intertrochanteric line of the femur in front and halfway along the posterior aspect of the neck of the bone behind. At its attachment to the intertrochanteric line in front, some of its fibers, accompanied by blood vessels, are reflected upward along the neck as bands, called *retinacula*. These blood vessels supply the head and neck of the femur.

## Ligaments

The *iliofemoral ligament* is strong and Y-shaped (Fig. 10-29). Its base is attached to the anterior inferior iliac spine above; below, the two limbs of the Y are attached to the upper and lower parts of the intertrochanteric line of the femur. This very strong ligament prevents overextension during standing.

The *pubofemoral ligament* is triangular in shape (Fig. 10-29). The base of the ligament is attached to the superior ramus of the pubis, and the apex is attached below to the lower part of the intertrochanteric line. This ligament limits extension and abduction.

The *ischiofemoral ligament* is spiral in shape and is attached to the body of the ischium near the acetabular margin (Fig. 10-29). The fibers pass upward and laterally and are attached to the greater trochanter. This ligament limits extension.

The *transverse acetabular ligament* is formed by the acetabular labrum as it bridges the acetabular notch (Fig. 10-30). The ligament converts the notch into a tunnel through which the blood vessels and nerves enter the joint.

The *ligament of the head of the femur* is flat and triangular in shape (Fig. 10-30). It is attached by its apex to the pit on the head of the femur (fovea capitis), and by its base to the transverse ligament and the margins of the acetabular notch. It lies within the joint and is ensheathed by synovial membrane (Fig. 10-30).

## Synovial Membrane

This lines the capsule and is attached to the margins of the articular surfaces (Fig. 10-30). It covers the portion of the neck of the femur that lies within the joint capsule. It ensheathes the ligament of the head of the femur and covers the pad of fat contained in the acetabular fossa. A pouch of synovial membrane frequently protrudes through a gap in

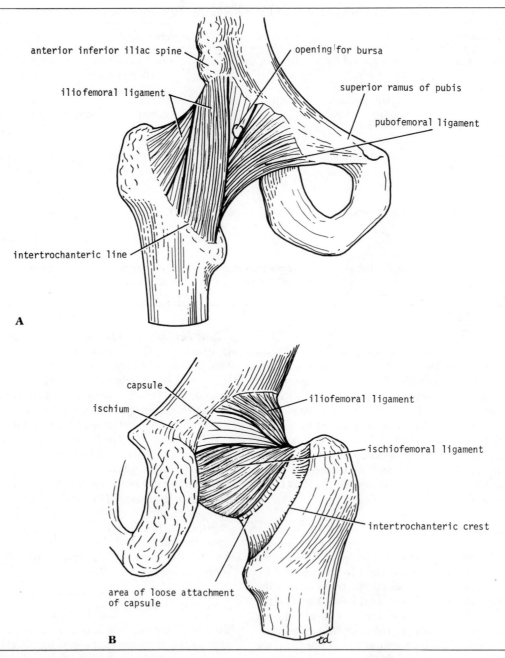

anterior inferior iliac spine

iliofemoral ligament

intertrochanteric line

opening for bursa

superior ramus of pubis

pubofemoral ligament

**A**

capsule

ischium

iliofemoral ligament

ischiofemoral ligament

intertrochanteric crest

area of loose attachment
of capsule

**B**

**Fig. 10-29. (A) Anterior aspect and (B) posterior aspect of right hip joint.**

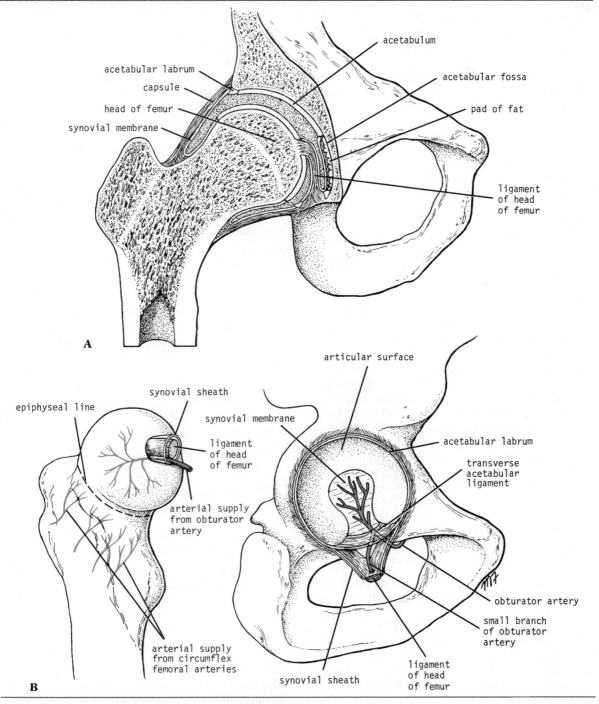

**Fig. 10-30. (A) Coronal section of the right hip joint and (B) articular surfaces of the right hip joint and arterial supply of head of femur.**

the anterior wall of the capsule, between the pubofemoral and iliofemoral ligaments, and forms the *psoas bursa* beneath the psoas tendon (Figs. 10-29 and 10-31).

## Nerve Supply

Femoral, obturator, and sciatic nerves and the nerve to the quadratus femoris.

## Movements

The hip joint has a wide range of movement, but less so than the shoulder joint. Some of the movement has been sacrificed in order to provide strength and stability. The strength of the joint depends largely on the shape of the bones taking part in the articulation and on the very strong ligaments. When the knee is flexed, flexion is limited by the anterior surface of the thigh coming into contact with the anterior abdominal wall. When the knee is extended, flexion is limited by the tension of the hamstring group of muscles. Extension, which is the movement of the flexed thigh backward to the anatomical position, is limited by the tension of the iliofemoral, pubofemoral, and ischiofemoral ligaments. Abduction is limited by the tension of the pubofemoral ligament, and adduction is limited by contact with the opposite limb and by the tension in the ligament of the head of the femur. Lateral rotation is limited by the tension in the iliofemoral and pubofemoral ligaments, and medial rotation is limited by the ischiofemoral ligament. The following movements take place:

*Flexion* is performed by the iliopsoas, rectus femoris, and sartorius, and also by the adductor muscles.

*Extension* (a backward movement of the flexed thigh) is performed by the gluteus maximus and the hamstring muscles.

*Abduction* is performed by the gluteus medius and minimus, assisted by the sartorius, tensor fasciae latae, and piriformis.

*Adduction* is performed by the adductor longus and brevis and the adductor fibers of the adductor magnus. These muscles are assisted by the pectineus and the gracilis.

*Lateral rotation* is performed by the piriformis, obturator internus and externus, superior and in-

ferior gemelli, and quadratus femoris, assisted by the gluteus maximus.

*Medial rotation* is performed by the anterior fibers of gluteus medius and gluteus minimus and the tensor fasciae latae.

*Circumduction* is a combination of the above movements.

It should be remembered that the extensor group of muscles is more powerful than the flexor group, and that the lateral rotators are more powerful than the medial rotators.

### *Important Relations*

#### Anteriorly

Iliopsoas, pectineus, and rectus femoris muscles. The iliopsoas and pectineus separate the femoral vessels and nerve from the joint (Fig. 10-31).

#### Posteriorly

The obturator internus, the gemelli, and the quadratus femoris muscles separate the joint from the sciatic nerve (Fig. 10-31).

#### Superiorly

Piriformis and gluteus minimus (Fig. 10-31).

#### Inferiorly

Obturator externus tendon (Fig. 10-31).

## Bones of the Leg

The leg is the part of the lower limb between the knee joint and the ankle joint.

### PATELLA

The patella (Fig. 10-32) is a sesamoid bone lying within the quadriceps tendon. It is triangular in shape, and its apex lies inferiorly; the apex is connected to the tuberosity of the tibia by the ligamentum patellae. The posterior surface articulates with the condyles of the femur. It is situated in an exposed position in front of the knee joint and can easily be palpated through the skin. It is separated

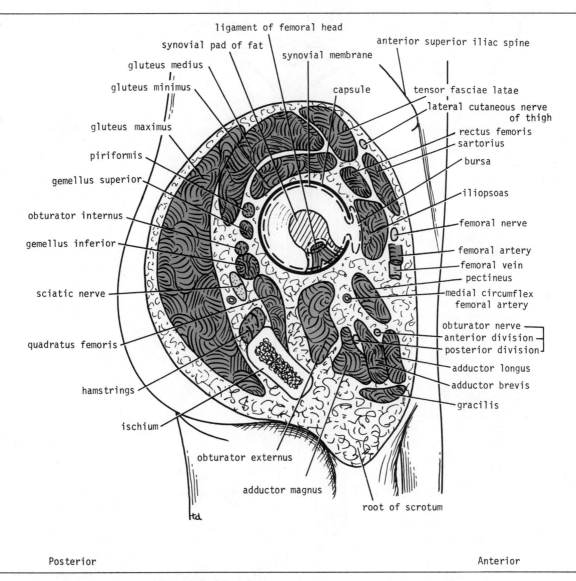

ligament of femoral head
synovial pad of fat
synovial membrane
capsule
anterior superior iliac spine
gluteus medius
gluteus minimus
tensor fasciae latae
gluteus maximus
lateral cutaneous nerve of thigh
rectus femoris
sartorius
piriformis
bursa
gemellus superior
iliopsoas
obturator internus
femoral nerve
gemellus inferior
femoral artery
femoral vein
pectineus
sciatic nerve
medial circumflex femoral artery
quadratus femoris
obturator nerve
anterior division
posterior division
adductor longus
hamstrings
adductor brevis
gracilis
ischium
obturator externus
adductor magnus
root of scrotum

Posterior

Anterior

**Fig. 10-31. Structures surrounding right hip joint.**

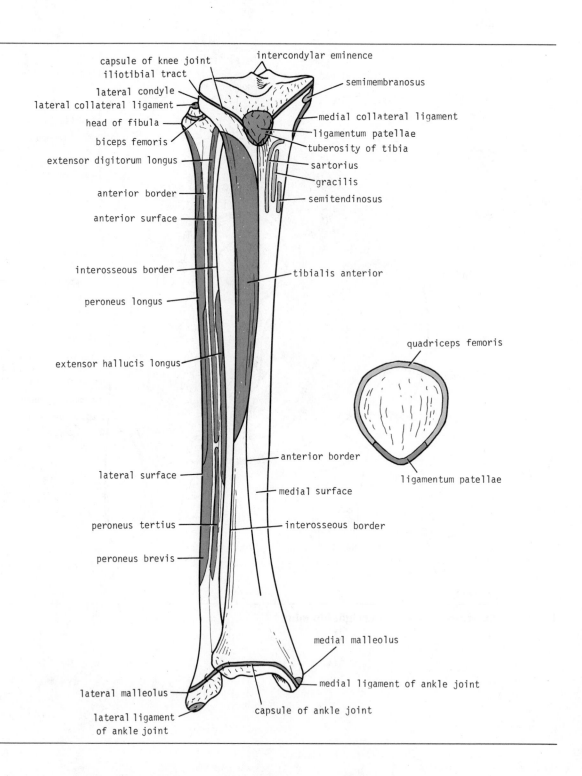

capsule of knee joint
iliotibial tract
lateral condyle
lateral collateral ligament
head of fibula
biceps femoris
extensor digitorum longus
anterior border
anterior surface
interosseous border
peroneus longus
extensor hallucis longus
lateral surface
peroneus tertius
peroneus brevis
lateral malleolus
lateral ligament
of ankle joint

intercondylar eminence
semimembranosus
medial collateral ligament
ligamentum patellae
tuberosity of tibia
sartorius
gracilis
semitendinosus
tibialis anterior

quadriceps femoris
ligamentum patellae

anterior border
medial surface
interosseous border

medial malleolus
medial ligament of ankle joint
capsule of ankle joint

from the skin by an important subcutaneous bursa (Fig. 10-54).

The upper, lateral, and medial margins give attachment to the different parts of the quadriceps femoris muscle. It is prevented from being displaced laterally during the action of the quadriceps muscle by the lower horizontal fibers of the vastus medialis and by the large size of the lateral condyle of the femur.

## TIBIA

The tibia is the large medial bone of the leg (Figs. 10-32 and 10-33). It articulates with the condyles of the femur and the head of the fibula above, and with the talus and the distal end of the fibula below. It has an expanded upper end, a smaller lower end, and a shaft.

At the *upper end* are the *lateral* and *medial condyles* (sometimes called medial and lateral *tibial plateaus*), which articulate with the lateral and medial condyles of the femur, the *lateral* and *medial semilunar cartilages* intervening. Separating the upper articular surfaces of the tibial condyles are *anterior* and *posterior intercondylar areas*; lying between these areas is the *intercondylar eminence* (Fig. 10-32).

The lateral condyle possesses on its lateral aspect a *circular articular facet for the head of the fibula.* The medial condyle shows a groove on its posterior aspect for the insertion of the semimembranosus muscle (Fig. 10-33).

The *shaft of the tibia* is triangular in cross section, presenting three borders and three surfaces. Its anterior and medial borders, with the medial surface between them, are subcutaneous. The anterior border is prominent and forms the shin. At the junction of the anterior border with the upper end of the tibia is the *tuberosity*, which receives the attachment of the ligamentum patellae. The anterior border becomes rounded below, where it becomes continuous with the medial malleolus. The lateral or interosseous border gives attachment to the interosseous membrane.

The posterior surface of the shaft shows an oblique line, the *soleal line* (Fig. 10-33). Below the soleal line a vertical ridge passes downward, dividing the posterior surface into medial and lateral areas.

The *lower end* of the tibia is slightly expanded and on its inferior aspect shows a saddle-shaped articular surface for the talus. The lower end is prolonged downward medially to form the *medial malleolus*. The lateral surface of the medial malleolus articulates with the talus. The lower end of the tibia shows a wide, rough depression on its lateral surface for articulation with the fibula.

The important muscles and ligaments that are attached to the tibia are shown in Figures 10-32 and 10-33.

## FIBULA

The fibula is the slender lateral bone of the leg (Figs. 10-32 and 10-33). It takes no part in the articulation at the knee joint, but below, it forms the lateral malleolus of the ankle joint. It takes no part in the transmission of body weight. The fibula has an expanded upper end, a shaft, and a lower end.

The *upper end*, or *head*, is surmounted by a *styloid process*. It possesses an *articular surface* for articulation with the lateral condyle of the tibia.

The *shaft of the fibula* is long and slender, and its shape is subject to considerable variation. Typically, it has four borders and four surfaces. The anterior surface is very narrow in its upper part, where the anterior and medial borders run close together or may become confluent. The medial or *interosseous* border gives attachment to the interosseous membrane.

The *lower end of the fibula* forms the triangular lateral malleolus, which is subcutaneous. On the medial surface of the lateral malleolus is a triangular *articular facet* for articulation with the lateral aspect of the talus. Below and behind the articular facet is a depression called the *malleolar fossa*.

The important muscles and ligaments that are attached to the fibula are shown in Figures 10-32 and 10-33.

## Bones of the Foot

The bones of the foot are the *tarsal bones*, the *metatarsals*, and the *phalanges*.

**Fig. 10-32. Muscles and ligaments attached to anterior surfaces of right tibia and fibula; attachments to patella are also shown.**

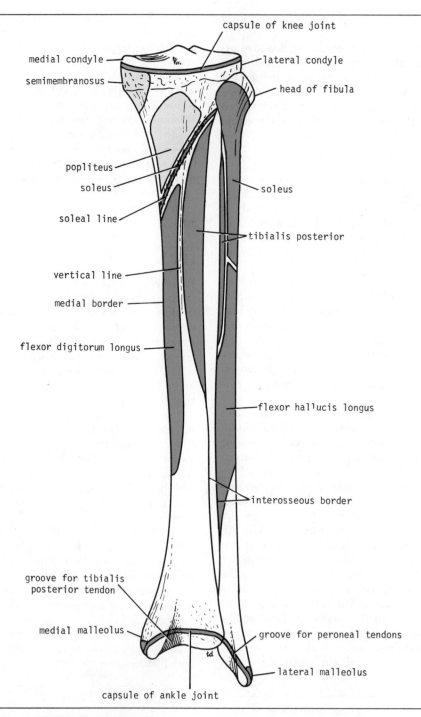

**Fig. 10-33. Muscles and ligaments attached to posterior surfaces of right tibia and fibula.**

## Tarsal Bones

The tarsal bones are the calcaneum, the talus, the navicular, the cuboid, and the three cuneiform bones. Only the talus articulates with the tibia and the fibula at the ankle joint.

## CALCANEUM

The calcaneum is the largest bone of the tarsus and forms the prominence of the heel (Figs. 10-34, 10-35, and 10-36). It articulates above with the talus and in front with the cuboid. It has six surfaces.

The *anterior surface* is small and forms the articular facet that articulates with the cuboid bone.

The *posterior surface* forms the prominence of the heel. It is marked by a *transverse ridge*, which gives attachment to the tendo calcaneus (Achilles tendon). The smooth area above the ridge is separated from the tendon by a bursa.

The *superior surface* is dominated by two articular facets for the talus: an anterior and a large posterior facet, separated by a roughened groove, the *sulcus calcanei*.

The *inferior surface* has an *anterior tubercle* in the midline and a large *medial* and a smaller *lateral* tubercle at the junction of the inferior and posterior surfaces.

The *medial surface* possesses a large shelf-like process, which projects medially from its upper border, termed the *sustentaculum tali*.

The *lateral surface* is almost flat. On its anterior part is a small elevation called the *peroneal tubercle*.

The important muscles and ligaments that are attached to the calcaneum are shown in Figures 10-34, 10-35, and 10-36.

## TALUS

The talus articulates above at the ankle joint with the tibia and fibula, below with the calcaneum, and in front with the navicular bone. It possesses a head, a neck, and a body (Figs. 10-34 and 10-35).

The *head* of the talus is directed distally and has an oval convex articular surface for articulation with the navicular bone. This articular surface is continued on its inferior surface, where it rests on the sustentaculum tali behind and the calcaneonavicular ligament in front.

The *neck* of the talus lies posterior to the head and is slightly narrowed. Its upper surface is roughened and gives attachment to ligaments, and its lower surface shows a deep groove, the *sulcus tali*. The sulcus tali and the sulcus calcanei in the articulated foot form a tunnel, the *sinus tarsi*, which is occupied by the *interosseous talocalcaneal ligament*.

The *body* of the talus is cuboidal in shape. Its superior surface articulates with the distal end of the tibia; it is convex from before backward and slightly concave from side to side. Its lateral surface presents a triangular *articular facet* for articulation with the lateral malleolus of the fibula. Its medial surface has a small, comma-shaped *articular facet* for articulation with the medial malleolus of the tibia. The posterior surface is marked by two small *tubercles*, separated by a groove for the flexor hallucis longus tendon.

Numerous important ligaments are attached to the talus, but no muscles are attached to this bone.

The remaining tarsal bones should be identified and the following important features noted.

## NAVICULAR BONE (FIGS. 10-34, 10-35, AND 10-36)

The *tuberosity* of the navicular bone can be seen and felt on the medial border of the foot 1 inch (2.5 cm) in front of and below the medial malleolus; it gives attachment to the main part of the tibialis posterior tendon.

## CUBOID BONE (FIGS. 10-34, 10-35, AND 10-36)

There is a deep *groove* on its inferior aspect, which lodges the tendon of the peroneus longus muscle.

## CUNEIFORM BONES (FIGS. 10-35 AND 10-36)

These are three small, wedge-shaped bones, which articulate proximally with the navicular bone and distally with the first three metatarsal bones. Their wedge shape contributes greatly to the formation and maintenance of the transverse arch of the foot. (See p. 668.)

The tarsal bones, unlike those of the carpus, start to ossify before birth. Centers of ossification for the

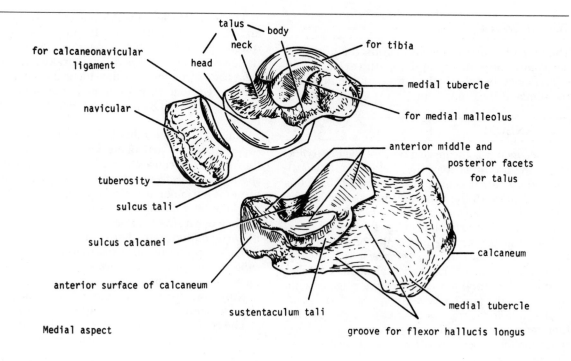

Medial aspect

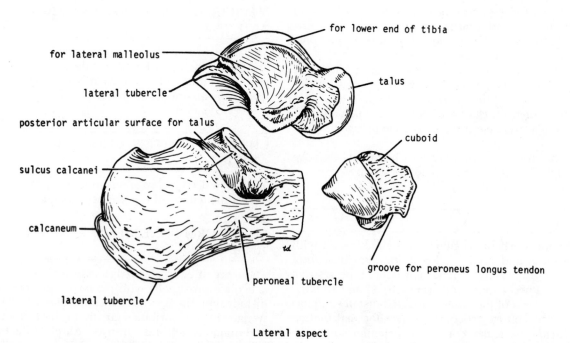

Lateral aspect

**Fig. 10-34. Calcaneum, talus, navicular, and cuboid bones.**

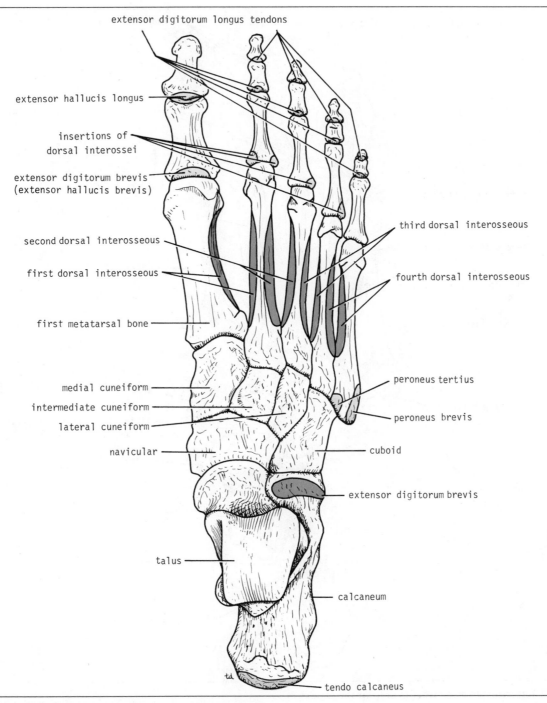

extensor digitorum longus tendons

extensor hallucis longus

insertions of
dorsal interossei

extensor digitorum brevis
(extensor hallucis brevis)

second dorsal interosseous

first dorsal interosseous

first metatarsal bone

medial cuneiform

intermediate cuneiform

lateral cuneiform

navicular

talus

third dorsal interosseous

fourth dorsal interosseous

peroneus tertius

peroneus brevis

cuboid

extensor digitorum brevis

calcaneum

tendo calcaneus

**Fig. 10-35. Muscle attachments on dorsal aspect of
bones of right foot.**

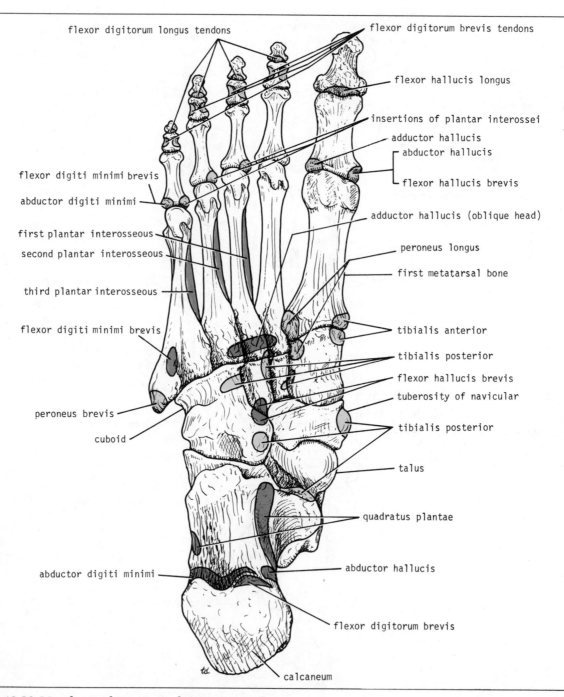

**Fig. 10-36. Muscle attachments on plantar aspect of bones of right foot.**

calcaneum and the talus, and often for the cuboid, are present at birth. By the fifth year, ossification is taking place in all the tarsal bones.

## Metatarsal Bones and Phalanges (Figs. 10-35 and 10-36)

The metatarsal bones and phalanges resemble the metacarpals and phalanges of the hand, and each possesses a *head*, distally, a *shaft*, and a *base*, proximally. The metatarsals are numbered from the medial to the lateral side.

The *first metatarsal* bone is large and strong and plays an important role in supporting the weight of the body. The head is grooved on its inferior aspect by the medial and lateral *sesamoid bones* in the tendons of the flexor hallucis brevis.

The *fifth metatarsal* has a prominent *tubercle* on its base, which can be easily palpated along the lateral border of the foot. The tubercle gives attachment to the peroneus brevis tendon.

## Popliteal Fossa

The popliteal fossa is a diamond-shaped intermuscular space situated at the back of the knee (Fig. 10-37). The fossa is most prominent when the knee joint is flexed. It contains the popliteal vessels, the small saphenous vein, the common peroneal and tibial nerves, the posterior cutaneous nerve of the thigh, the genicular branch of the obturator nerve, connective tissue, and lymph nodes.

### Boundaries

#### Laterally

The biceps femoris above and the lateral head of the gastrocnemius and plantaris below (Fig. 10-37).

#### Medially

The semimembranosus and semitendinosus above and the medial head of gastrocnemius below (Fig. 10-37).

The *anterior wall* or *floor* of the fossa is formed by the popliteal surface of the femur, the posterior ligament of the knee joint, and the popliteus muscle (Figs. 10-37 and 10-38).

The *roof* is formed by skin, superficial fascia, and the deep fascia of the thigh.

The *biceps femoris*, the *semimembranosus*, and the *semitendinosus* muscles are described in the section on the back of the thigh, on page 598. The *gastrocnemius* and *plantaris* are described in the section on the back of the leg, on pages 628 and 630.

## POPLITEUS MUSCLE (FIGS. 10-38 AND 10-45)

### Origin

From the lateral surface of the lateral condyle of the femur by a rounded tendon and by a few fibers from the lateral semilunar cartilage.

### Insertion

The fibers pass downward and medially and are attached to the posterior surface of the tibia, above the soleal line. The muscle arises within the capsule of the knee joint, and its tendon separates the lateral semilunar cartilage from the lateral ligament of the joint. It emerges through the lower part of the posterior surface of the capsule of the joint to pass to its insertion.

### Nerve Supply

Tibial nerve.

### Action

Medial rotation of the tibia on the femur; or, if the foot is on the ground, lateral rotation of the femur on the tibia. The latter action occurs at the commencement of flexion of the extended knee, and its rotatory action slackens the ligaments of the knee joint; this action is sometimes referred to as "unlocking the knee joint." Because of its attachment to the lateral semilunar cartilage, it also pulls the cartilage backward at the commencement of flexion of the knee.

## POPLITEAL ARTERY

The popliteal artery is deeply placed and enters the popliteal fossa through the opening in the adductor magnus, as a continuation of the femoral artery

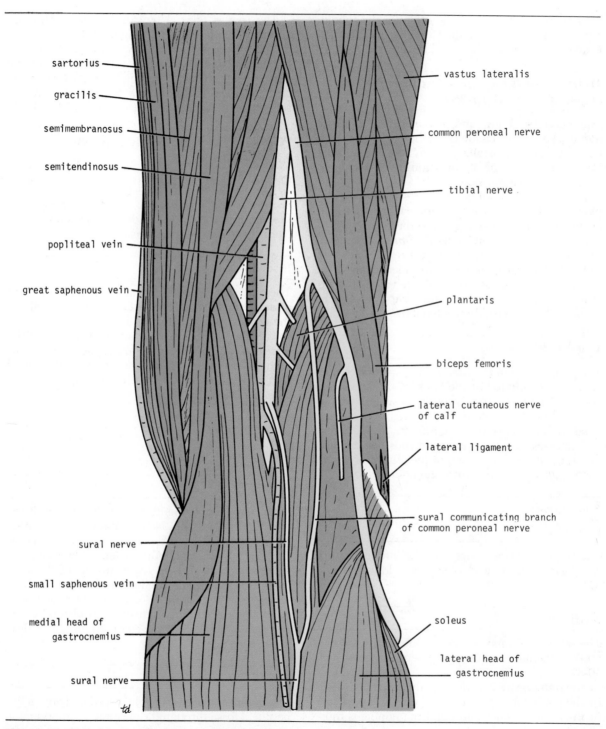

sartorius

gracilis

semimembranosus

semitendinosus

popliteal vein

great saphenous vein

vastus lateralis

common peroneal nerve

tibial nerve

plantaris

biceps femoris

lateral cutaneous nerve of calf

lateral ligament

sural communicating branch of common peroneal nerve

sural nerve

small saphenous vein

medial head of gastrocnemius

sural nerve

soleus

lateral head of gastrocnemius

**Fig. 10-37. Boundaries and contents of right popliteal fossa.**

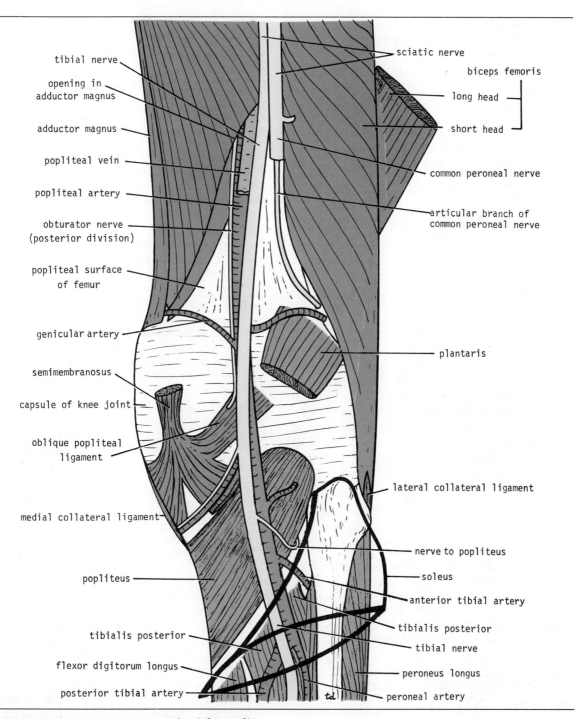

tibial nerve

opening in
adductor magnus

adductor magnus

popliteal vein

popliteal artery

obturator nerve
(posterior division)

popliteal surface
of femur

genicular artery

semimembranosus

capsule of knee joint

oblique popliteal
ligament

medial collateral ligament

popliteus

tibialis posterior

flexor digitorum longus

posterior tibial artery

sciatic nerve

biceps femoris

long head

short head

common peroneal nerve

articular branch of
common peroneal nerve

plantaris

lateral collateral ligament

nerve to popliteus

soleus

anterior tibial artery

tibialis posterior

tibial nerve

peroneus longus

peroneal artery

**Fig. 10-38. Deep structures present in right popliteal fossa. Origin of soleus muscle is shown in outline only.**

(Fig. 10-38). It ends at the level of the lower border of the popliteus muscle by dividing into anterior and posterior tibial arteries.

## Relations

### Anteriorly

The popliteal surface of the femur, the oblique popliteal ligament of the knee joint, and the popliteus muscle (Fig. 10-38).

### Posteriorly

The popliteal vein and the tibial nerve, fascia, and skin (Figs. 10-37 and 10-38).

## Branches

The popliteal artery has (1) *muscular branches* and (2) *articular branches* to the knee.

## POPLITEAL VEIN

The popliteal vein is formed by the junction of the venae comitantes of the anterior and posterior tibial arteries at the lower border of the popliteus muscle on the medial side of the popliteal artery. As it ascends through the fossa, it crosses behind the popliteal artery, so that it comes to lie on its lateral side (Figs. 10-37 and 10-38). It passes through the opening in the adductor magnus to become the femoral vein.

## Tributaries

The tributaries of the popliteal vein are as follows:

1. Veins that correspond to branches given off by the popliteal artery.
2. *Small saphenous vein.* This perforates the deep fascia and passes between the two heads of the gastrocnemius muscle to end in the popliteal vein. The origin of this vein is described on page 627.

## ARTERIAL ANASTOMOSIS AROUND THE KNEE JOINT

To compensate for the narrowing of the popliteal artery, which occurs during extreme flexion of the knee, there is around the knee joint a profuse anastomosis of small branches of the femoral artery with muscular and articular branches of the popliteal artery and with branches of the anterior and posterior tibial arteries.

## POPLITEAL LYMPH NODES

About six lymph nodes are embedded in the fatty connective tissue of the popliteal fossa (Fig. 10-28). They receive superficial lymph vessels from the lateral side of the foot and leg; these accompany the small saphenous vein into the popliteal fossa. They also receive lymph from the knee joint and from deep lymph vessels accompanying the anterior and posterior tibial arteries.

## TIBIAL NERVE

The larger terminal branch of the sciatic nerve (see p. 600), the tibial nerve arises in the lower third of the thigh. It runs downward through the popliteal fossa, lying first on the lateral side of the popliteal artery, then posterior to it, and finally, medial to it (Figs. 10-37 and 10-38). The popliteal vein lies between the nerve and the artery throughout its course. The nerve enters the posterior compartment of the leg by passing beneath the soleus muscle. Its further course is described on page 634.

## Branches (Fig. 10-70)

1. *Cutaneous.* The *sural nerve* descends between the two heads of the gastrocnemius muscle and is usually joined by the *sural communicating* branch of the common peroneal nerve (Fig. 10-37). Numerous small branches arise from the sural nerve to supply the skin of the calf and the back of the leg. The sural nerve accompanies the small saphenous vein behind the lateral malleolus and is distributed to the skin along the lateral border of the foot and the lateral side of the little toe.
2. *Muscular* branches supply both heads of the gastrocnemius and the plantaris, soleus, and popliteus (Figs. 10-37 and 10-38).
3. *Articular* branches supply the knee joint.

## COMMON PERONEAL NERVE

The smaller terminal branch of the sciatic nerve (see p. 600), the common peroneal nerve arises in the lower third of the thigh. It runs downward through the popliteal fossa, closely following the medial border of the biceps muscle (Fig. 10-37). It leaves the fossa by crossing superficially the lateral head of the gastrocnemius muscle. It then passes behind the head of the fibula, winds laterally around the neck of the bone, pierces the peroneus longus muscle, and divides into two terminal branches, (1) the superficial peroneal nerve and (2) the deep peroneal nerve (Fig. 10-42). As the nerve lies on the lateral aspect of the neck of the fibula, it is subcutaneous and can easily be rolled against the bone.

### Branches (Fig. 10-68)

1. *Cutaneous.*
    a. The *sural communicating branch* (Fig. 10-37) runs downward and joins the sural nerve.
    b. *The lateral cutaneous nerve of the calf* supplies the skin on the lateral side of the back of the leg (Figs. 10-9 and 10-37).
2. *Muscular* branch to the short head of the biceps femoris muscle, which arises high up in the popliteal fossa (Fig. 10-38).
3. *Articular* branches to the knee joint.

## POSTERIOR CUTANEOUS NERVE OF THE THIGH

The course of the posterior cutaneous nerve of the thigh through the gluteal region and the back of the thigh has been described on page 576. It terminates by supplying the skin over the popliteal fossa (Fig. 10-9).

## OBTURATOR NERVE

The course of the posterior division of the obturator nerve in the medial compartment of the thigh has been described on page 596. It leaves the subsartorial canal with the femoral artery by passing through the opening in the adductor magnus (Fig. 10-38). The nerve terminates by supplying the knee joint.

## FASCIAL COMPARTMENTS OF THE LEG

The deep fascia invests the leg and is continuous above with the deep fascia of the thigh. Below the tibial condyles it is attached to the anterior and medial borders of the tibia, where it is fused with the periosteum (Fig. 10-39). Two intermuscular septa pass from its deep aspect to be attached to the fibula. These, together with the interosseous membrane, divide the leg into three compartments: anterior, lateral, and posterior, each having its own muscles, blood supply, and nerve supply.

### INTEROSSEOUS MEMBRANE

The interosseous membrane is a thin but strong membrane connecting the interosseous borders of the tibia and fibula (Figs. 10-39 and 10-42). The majority of fibers run obliquely downward and laterally. A large opening exists in the upper part of the membrane to permit the anterior tibial vessels to enter the anterior fascial compartment of the leg. A small opening is present in the lower part of the membrane for the perforating branch of the peroneal artery to enter the anterior fascial compartment. The membrane is continuous below with the interosseous ligament of the inferior tibiofibular joint. The interosseous membrane binds the tibia and fibula together and provides attachment for neighboring muscles.

### RETINACULA OF THE ANKLE

In the region of the ankle joint, the deep fascia is thickened to form a series of retinacula, which serve to keep the long tendons in position and act as modified pulleys.

The *superior extensor retinaculum* is a thickened band of deep fascia that is attached to the distal ends of the anterior borders of the fibula and tibia (Fig. 10-40). Near its medial end, it splits to enclose the tendon of the tibialis anterior muscle.

The *inferior extensor retinaculum* is a Y-shaped band of deep fascia that is attached by its stem to the upper surface of the anterior part of the calcaneum (Figs. 10-40 and 10-41). The upper limb of the Y is attached to the medial malleolus, and the lower limb is continuous with the plantar fascia on

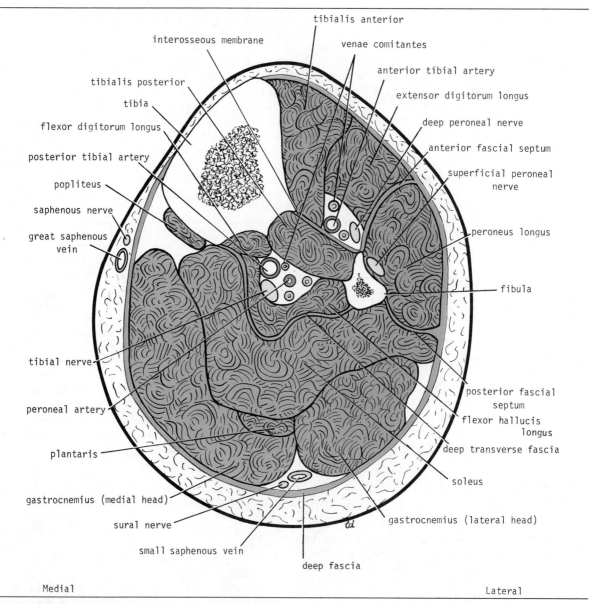

tibialis anterior

interosseous membrane

venae comitantes

anterior tibial artery

tibialis posterior

extensor digitorum longus

tibia

deep peroneal nerve

flexor digitorum longus

anterior fascial septum

posterior tibial artery

superficial peroneal nerve

popliteus

saphenous nerve

peroneus longus

great saphenous vein

fibula

tibial nerve

posterior fascial septum

flexor hallucis longus

peroneal artery

deep transverse fascia

plantaris

soleus

gastrocnemius (medial head)

gastrocnemius (lateral head)

sural nerve

small saphenous vein

deep fascia

Medial

Lateral

**Fig. 10-39. Transverse section through middle of right leg as seen from above.**

**Fig. 10-40. Structures present on anterior and lateral aspects of right leg and on dorsum of foot.**

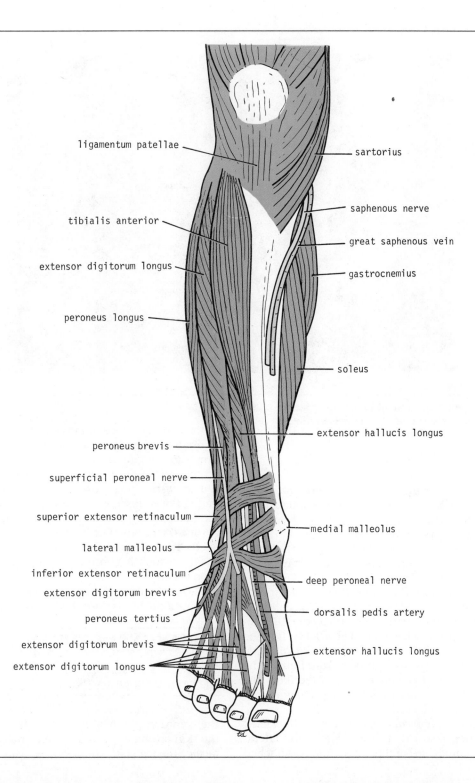

ligamentum patellae

sartorius

tibialis anterior

saphenous nerve

great saphenous vein

extensor digitorum longus

gastrocnemius

peroneus longus

soleus

extensor hallucis longus

peroneus brevis

superficial peroneal nerve

superior extensor retinaculum

medial malleolus

lateral malleolus

inferior extensor retinaculum

deep peroneal nerve

extensor digitorum brevis

peroneus tertius

dorsalis pedis artery

extensor digitorum brevis

extensor hallucis longus

extensor digitorum longus

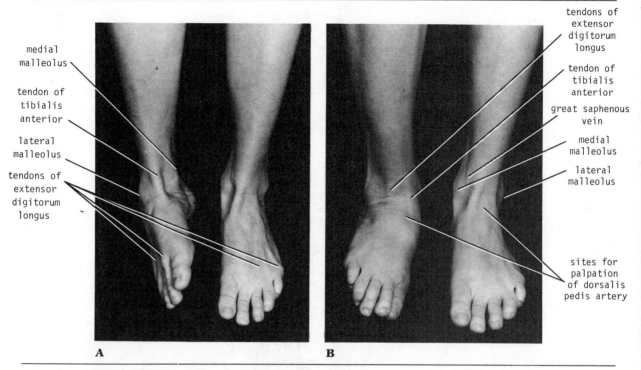

medial
malleolus

tendon of
tibialis
anterior

lateral
malleolus

tendons of
extensor
digitorum
longus

tendons of
extensor
digitorum
longus

tendon of
tibialis
anterior

great saphenous
vein

medial
malleolus

lateral
malleolus

sites for
palpation
of dorsalis
pedis artery

A                                    B

**Fig. 10-41. Anterior view of ankles and feet of a 29-year-old female showing (A) inversion of right foot and (B) eversion of right foot.**

the medial border of the foot. The tendons of the tibialis anterior, the extensor hallucis longus, the extensor digitorum longus, and the peroneus tertius split the upper limb of the retinaculum into superficial and deep layers. Fibrous bands separate the tendons into compartments, each of which is lined by a synovial sheath.

The *flexor retinaculum* is a thickened band of deep fascia that extends from the medial malleolus downward and backward to be attached to the medial surface of the calcaneum (Fig. 10-43). It binds the tendons of the deep muscles to the medial side of the ankle as they pass forward from behind the medial malleolus to enter the sole of the foot. The tendons lie in compartments, each of which is lined by a synovial sheath.

The *superior peroneal retinaculum* is a thickened band of deep fascia that extends from the lateral malleolus downward and backward to be attached

to the lateral surface of the calcaneum (Fig. 10-43). It binds the tendons of the peroneus longus and brevis to the lateral side of the ankle. The tendons are provided with a common synovial sheath.

The *inferior peroneal retinaculum* is a thickened band of deep fascia that is attached to the peroneal tubercle and to the calcaneum above and below the peroneal tendons (Fig. 10-43). The tendons of peroneus longus and brevis each possess a synovial sheath, which is continuous above with the common sheath.

The arrangement of the tendons beneath the different retinacula is described on page 634.

## THE FRONT OF THE LEG
### Skin

#### CUTANEOUS NERVES

The *lateral cutaneous nerve of the calf,* a branch of the common peroneal nerve (see p. 616), supplies the skin on the upper part of the anterolateral surface of the leg (Fig. 10-8).

The *superficial peroneal nerve*, a branch of the common peroneal nerve (see p. 617), supplies the skin of the lower part of the anterolateral surface of the leg (Fig. 10-8).

The *saphenous nerve*, a branch of the femoral nerve (see p. 590), supplies the skin on the antero-medial surface of the leg (Fig. 10-8).

## SUPERFICIAL VEINS

Numerous small veins curve around the medial aspect of the leg and ultimately drain into the great saphenous vein (Fig. 10-18).

## LYMPH VESSELS

The greater part of the lymph from the skin and superficial fascia on the front of the leg drains upward and medially in vessels that follow the great saphenous vein, to end in the vertical group of superficial inguinal lymph nodes (Fig. 10-28). A small amount of lymph from the upper lateral part of the front of the leg may pass via vessels that accompany the small saphenous vein and drain into the popliteal nodes (Fig. 10-28).

## Contents of the Anterior Fascial Compartment of the Leg

### Muscles

The tibialis anterior, extensor digitorum longus, peroneus tertius, and extensor hallucis longus.

### Blood Supply

Anterior tibial artery.

### Nerve Supply

Deep peroneal nerve.

### *Muscles of the Anterior Fascial Compartment of the Leg*

#### TIBIALIS ANTERIOR (FIG. 10-40)

#### Origin

From the upper half of the lateral surface of the tibia and from the interosseous membrane.

### Insertion

The tendon passes through both extensor retinacula and is attached to the medial cuneiform bone and the adjoining base of the first metatarsal bone.

### Nerve Supply

Deep peroneal nerve.

### Action

Extends (dorsiflexes)[†] the foot at the ankle joint and inverts the foot at the subtalar and transverse tarsal joints (see p. 663). It assists in holding up the medial longitudinal arch of the foot.

## EXTENSOR DIGITORUM LONGUS (FIG. 10-40)

### Origin

From the upper two-thirds of the anterior surface of the fibula and from the interosseous membrane.

### Insertion

The tendons pass behind the superior and through the inferior extensor retinacula. The four tendons then diverge and pass to the lateral four toes.

On the dorsal surface of each toe, the extensor tendon becomes incorporated into a fascial expansion, called the *extensor expansion*. The central part of the expansion is inserted into the base of the middle phalanx, and the two lateral parts converge to be inserted into the base of the distal phalanx. (Compare with the insertion of extensor digitorum in the hand.)

### Nerve Supply

Deep peroneal nerve.

### Action

Extends the toes and extends the foot at the ankle joint.

[†]Extension, or dorsiflexion, of the ankle is the movement of the foot away from the ground.

## PERONEUS TERTIUS (FIG. 10-40)

### Origin

This muscle is part of the extensor digitorum longus. It arises from the lower third of the anterior surface of the fibula and the interosseous membrane.

### Insertion

Its tendon follows the tendons of extensor digitorum longus behind the superior and through the inferior extensor retinacula and shares their synovial sheath. It is inserted into the medial side of the dorsal aspect of the base of the fifth metatarsal bone.

### Nerve Supply

Deep peroneal nerve.

### Action

Extends the foot at the ankle joint and everts the foot at the subtalar and transverse tarsal joints.

## EXTENSOR HALLUCIS LONGUS (FIG. 10-40)

### Origin

From the middle half of the anterior surface of the fibula and from the interosseous membrane.

### Insertion

The tendon passes behind the superior and through the inferior extensor retinacula. It is inserted into the base of the distal phalanx of the great toe.

### Nerve Supply

Deep peroneal nerve.

### Action

Extends the big toe and extends the foot at the ankle joint; it also assists in inversion of the foot at the subtalar and transverse tarsal joints.

## *Muscles of the Anterior Fascial Compartment of the Leg—Their Nerve Supply and Action*

Students wishing to review these muscles should study Table 10-5.

## *Artery of the Anterior Fascial Compartment of the Leg*

### ANTERIOR TIBIAL ARTERY

The anterior tibial artery is the smaller of the terminal branches of the popliteal artery. It arises at the level of the lower border of the popliteus muscle (see p. 613) and passes forward into the anterior compartment of the leg through an opening in the upper part of the interosseous membrane (Fig. 10-38). It descends on the anterior surface of the interosseous membrane, accompanied by the deep peroneal nerve (Fig. 10-42). In the upper part of its course it lies deep beneath the muscles of the compartment. In the lower part of its course it lies superficial in front of the lower end of the tibia (Figs. 10-40 and 10-42). Having passed behind the superior extensor retinaculum, it has the tendon of the extensor hallucis longus on its medial side and the deep peroneal nerve and the tendons of extensor digitorum longus on its lateral side. It is here that its pulsations can easily be felt in the living subject. In front of the ankle joint, the artery becomes the dorsalis pedis artery. (See p. 652.)

### *Branches*

The anterior tibial artery has the following branches:

1. *Muscular branches* to neighboring muscles.
2. *Anastomotic branches*, which anastomose with branches of other arteries around the knee and ankle joints.

*Venae comitantes* of the anterior tibial artery join those of the posterior tibial artery in the popliteal fossa, to form the popliteal vein.

Table 10-5. Muscles of the Lower Limb—Muscles of Anterior Fascial Compartment of the Leg

| Name of muscle | Origin | Insertion | Nerve supply | Action |
|---|---|---|---|---|
| Tibialis anterior | Lateral surface of shaft of tibia and interosseous membrane | Medial cuneiform and base of first metatarsal bone | Deep peroneal nerve | Extends* the foot at ankle joint; inverts foot at subtalar and transverse tarsal joints; holds up medial longitudinal arch of foot |
| Extensor digitorum longus | Anterior surface of shaft of fibula | Extensor expansion of lateral four toes | Deep peroneal nerve | Extends toes; extends foot at ankle joint |
| Peroneus tertius | Anterior surface of shaft of fibula | Base of fifth metatarsal bone | Deep peroneal nerve | Extends foot at ankle joint; everts foot at subtalar and transverse tarsal joints |
| Extensor hallucis longus | Anterior surface of shaft of fibula | Base of distal phalanx of great toe | Deep peroneal nerve | Extends big toe; extends foot at ankle joint; inverts foot at subtalar and transverse tarsal joints |
| Extensor digitorum brevis | Calcaneum | By four tendons into the proximal phalanx of big toe and long extensor tendons to second, third, and fourth toes | Deep peroneal nerve | Extends toes |

*Extension, or dorsiflexion, of the ankle is the movement of the foot away from the ground.

## Nerve of the Anterior Fascial Compartment of the Leg

### DEEP PERONEAL NERVE

The deep peroneal nerve is one of the terminal branches of the common peroneal nerve. (See p. 617.) It arises in the substance of the peroneus longus muscle on the lateral side of the neck of the fibula (Fig. 10-42). The nerve enters the anterior compartment by piercing the anterior fascial septum. It then descends deep to the extensor digitorum longus muscle, first lying lateral, then anterior, and finally lateral to the anterior tibial artery

(Fig. 10-42). The nerve passes behind the extensor retinacula. Its further course in the foot is described on page 652.

### Branches

The deep peroneal nerve has the following branches:

1. *Muscular branches* to the tibialis anterior, the extensor digitorum longus, the peroneus tertius, and the extensor hallucis longus.
2. *Articular branch* to the ankle joint.

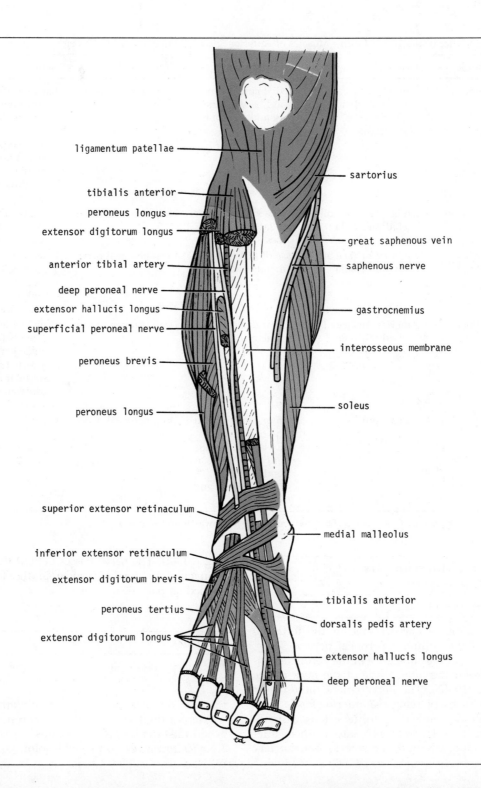

ligamentum patellae

tibialis anterior

peroneus longus

extensor digitorum longus

anterior tibial artery

deep peroneal nerve

extensor hallucis longus

superficial peroneal nerve

peroneus brevis

peroneus longus

superior extensor retinaculum

inferior extensor retinaculum

extensor digitorum brevis

peroneus tertius

extensor digitorum longus

sartorius

great saphenous vein

saphenous nerve

gastrocnemius

interosseous membrane

soleus

medial malleolus

tibialis anterior

dorsalis pedis artery

extensor hallucis longus

deep peroneal nerve

## Contents of the Lateral Fascial Compartment of the Leg

### Muscles

Peroneus longus and peroneus brevis.

### Blood Supply

Branches from the peroneal artery.

### Nerve Supply

Superficial peroneal nerve.

### *Muscles of the Lateral Fascial Compartment of the Leg*

### PERONEUS LONGUS (FIGS. 10-42 AND 10-43)

#### Origin

From the upper two-thirds of the lateral surface of the fibula.

#### Insertion

The tendon runs downward behind the lateral malleolus and is held in position by the superior peroneal retinaculum. The tendon then runs forward on the lateral surface of the calcaneum below the peroneal tubercle. Here, it is held in place by the inferior peroneal retinaculum. On reaching the lateral aspect of the cuboid, it winds around the lateral margin and enters a groove on its inferior aspect. It is inserted into the medial cuneiform and the base of the first metatarsal.

#### Nerve Supply

Superficial peroneal nerve.

#### Action

Plantar flexes the foot at the ankle joint and everts the foot at the subtalar and transverse tarsal joints.

**Fig. 10-42. Deep structures present on anterior and lateral aspects of right leg and on dorsum of foot.**

It plays an important part in holding up the lateral longitudinal arch in the foot and serves as a tie to the transverse arch of the foot.

### PERONEUS BREVIS (FIGS. 10-42 AND 10-43)

#### Origin

From the lower two-thirds of the lateral surface of the fibula.

#### Insertion

The tendon passes downward behind and directly in contact with the lateral malleolus and is held in position by the superior peroneal retinaculum. The tendon runs forward above the peroneal tubercle of the calcaneum and is held in place by the inferior peroneal retinaculum. It is inserted into the tubercle on the base of the fifth metatarsal bone.

#### Nerve Supply

Superficial peroneal nerve.

#### Action

Plantar flexes the foot at the ankle joint and everts the foot at the subtalar and transverse tarsal joints. It assists in holding up the lateral longitudinal arch of the foot.

### *Muscles of the Lateral Fascial Compartment of the Leg—Their Nerve Supply and Action*

Students wishing to review these muscles should study Table 10-6.

### *Artery of the Lateral Fascial Compartment of the Leg*

Numerous branches from the peroneal artery (see p. 634), which lies in the posterior compartment of the leg, pierce the posterior fascial septum and supply the peroneal muscles.

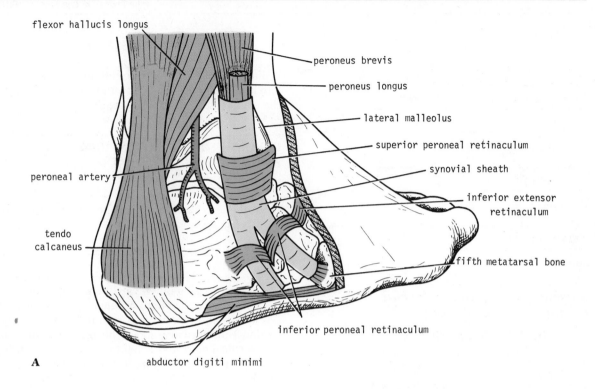

flexor hallucis longus

peroneus brevis

peroneus longus

lateral malleolus

superior peroneal retinaculum

synovial sheath

inferior extensor retinaculum

peroneal artery

tendo calcaneus

fifth metatarsal bone

inferior peroneal retinaculum

abductor digiti minimi

**A**

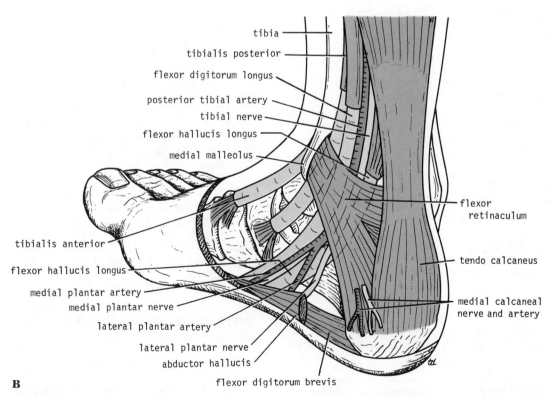

tibia

tibialis posterior

flexor digitorum longus

posterior tibial artery

tibial nerve

flexor hallucis longus

medial malleolus

flexor retinaculum

tibialis anterior

tendo calcaneus

flexor hallucis longus

medial plantar artery

medial plantar nerve

medial calcaneal nerve and artery

lateral plantar artery

lateral plantar nerve

abductor hallucis

flexor digitorum brevis

**B**

## Nerve of the Lateral Fascial Compartment of the Leg

### SUPERFICIAL PERONEAL NERVE

The superficial peroneal nerve is one of the terminal branches of the common peroneal nerve. (See p. 617.) It arises in the substance of the peroneus longus muscle on the lateral side of the neck of the fibula (Fig. 10-42). It descends first between the peroneus longus and brevis muscles and then between the peroneus brevis and the extensor digitorum longus. In the lower part of the leg, it pierces the deep fascia and becomes cutaneous (Fig. 10-40).

### Branches

The superficial peroneal nerve has the following branches:

1. *Muscular* branches to the peroneus longus and brevis (Fig. 10-42).
2. *Cutaneous.* Medial and lateral cutaneous branches are distributed to the skin on the lower part of the front of the leg and the dorsum of the foot. In addition, it supplies the dorsal surfaces of the skin of all the toes, except the adjacent sides of the first and second toes and the lateral side of the little toe. (See p. 650.)

## THE BACK OF THE LEG

## Skin

### CUTANEOUS NERVES

The *posterior cutaneous nerve of the thigh* descends on the back of the thigh. (See p. 576.) In the popliteal fossa, it pierces the deep fascia and supplies the skin over the popliteal fossa and the upper part of the back of the leg (Fig. 10-9).

The *lateral cutaneous nerve of the calf,* a branch

**Fig. 10-43. Structures passing behind (A) lateral malleolus and (B) medial malleolus. Synovial sheaths of tendons are shown in blue. Note positions of retinacula.**

of the common peroneal nerve (see p. 617), supplies the skin on the upper part of the posterolateral surface of the leg (Fig. 10-9).

The *sural nerve,* a branch of the tibial nerve (see p. 616), supplies the skin on the lower part of the posterolateral surface of the leg (Fig. 10-9).

The *saphenous nerve,* a branch of the femoral nerve (see p. 590), gives off branches that supply the skin on the posteromedial surface of the leg (Fig. 10-9).

### SUPERFICIAL VEINS

The *small saphenous vein* arises from the lateral part of the dorsal venous arch of the foot (Fig. 10-18). It ascends *behind* the lateral malleolus in company with the sural nerve. It follows the lateral border of the tendo calcaneus and then runs up the middle of the back of the leg. The vein pierces the deep fascia and passes between the two heads of the gastrocnemius muscle in the lower part of the popliteal fossa (Figs. 10-18 and 10-37); it ends in the popliteal vein. (See p. 616.) The small saphenous vein has numerous valves along its course.

### Tributaries

1. Numerous small veins from the back of the leg.
2. Communicating veins with the deep veins of the foot.
3. Important anastomotic branches that run upward and medially and join the great saphenous vein (Fig. 10-18).

The mode of termination of the small saphenous vein is subject to variation: (1) It may join the popliteal vein; (2) it may join the great saphenous vein; or (3) it may split in two, one division joining the popliteal and the other joining the great saphenous vein.

### LYMPH VESSELS

Lymph vessels from the skin and superficial fascia on the back of the leg drain upward and either pass forward around the medial side of the leg to end in the vertical group of superficial inguinal nodes, or drain into the popliteal nodes (Fig. 10-28).

Table 10-6. Muscles of the Lower Limb—Muscles of the Lateral Fascial Compartment of the Leg

| Name of muscle | Origin | Insertion | Nerve supply | Action |
|---|---|---|---|---|
| Peroneus longus | Lateral surface of shaft of fibula | Base of first metatarsal and the medial cuneiform | Superficial peroneal nerve | Plantar flexes foot at ankle joint; everts foot at subtalar and transverse tarsal joints; supports lateral longitudinal and transverse arches of foot |
| Peroneus brevis | Lateral surface of shaft of fibula | Base of fifth metatarsal bone | Superficial peroneal nerve | Plantar flexes foot at ankle joint; everts foot at subtalar and transverse tarsal joint; supports lateral longitudinal arch of foot |

## Contents of the Posterior Fascial Compartment of the Leg

The *deep transverse fascia* of the leg is a septum that divides the muscles of the posterior compartment into superficial and deep groups.

### Superficial Group of Muscles

Gastrocnemius, plantaris, and soleus.

### Deep Group of Muscles

Popliteus, flexor digitorum longus, flexor hallucis longus, and tibialis posterior.

### Blood Supply

Posterior tibial artery.

### Nerve Supply

Tibial nerve.

## Muscles of the Posterior Fascial Compartment of the Leg: Superficial Group

### GASTROCNEMIUS (FIG. 10-44)

The gastrocnemius is the most superficial of the calf muscles.

### Origin

*Lateral head*, from the lateral aspect of the lateral condyle of the femur; *medial head*, from the popliteal surface of the femur above the medial condyle.

### Insertion

The two large and powerful fleshy bellies join the posterior part of the common tendon called the tendo calcaneus, which is attached to the posterior surface of the calcaneum. A small bursa separates the tendon from the upper part of the posterior surface of the bone.

### Nerve Supply

Tibial nerve.

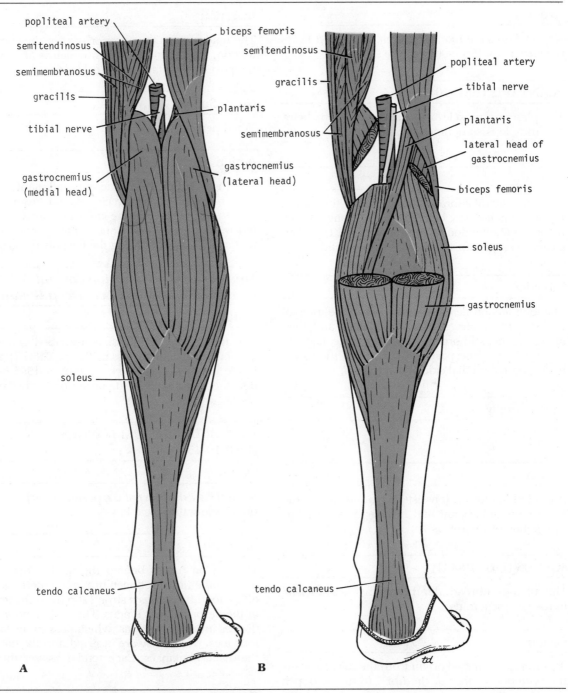

popliteal artery

semitendinosus

semimembranosus

gracilis

tibial nerve

gastrocnemius
(medial head)

soleus

tendo calcaneus

biceps femoris

plantaris

gastrocnemius
(lateral head)

semitendinosus

gracilis

semimembranosus

popliteal artery

tibial nerve

plantaris

lateral head of
gastrocnemius

biceps femoris

soleus

gastrocnemius

tendo calcaneus

A

B

**Fig. 10-44. Structures present on posterior aspect of right leg. In (B), part of gastrocnemius has been removed.**

## Action

Plantar flexes the foot at the ankle joint and flexes the knee joint. (See action of the soleus.)

## PLANTARIS (FIG. 10-44)

The plantaris muscle has a small fusiform belly. The muscle is sometimes double or it may be absent.

## Origin

From the lateral supracondylar ridge of the femur. It has a small fleshy belly and a very long narrow tendon. (The tendon is commonly used in reconstructive surgery of the tendons of the hand.)

## Insertion

The long tendon descends obliquely in the interval between the gastrocnemius and soleus and then along the medial border of the tendo calcaneus, to be attached to the posterior surface of the calcaneum, on the medial side of the tendon.

## Nerve Supply

Tibial nerve.

## Action

It is a feeble muscle. It assists in plantar flexing the foot at the ankle joint and flexing the knee joint. (See action of the soleus.)

## SOLEUS (FIG. 10-44)

The soleus is a broad, flat muscle that lies anterior to the gastrocnemius.

## Origin

An inverted V-shaped origin from the soleal line on the posterior surface of the tibia, from the upper quarter of the posterior surface of the shaft of the fibula, and from a fibrous arch between these bones.

## Insertion

The tendon joins the anterior part of the common tendon, the tendo calcaneus, which is attached to the posterior surface of the calcaneum.

## Nerve Supply

Tibial nerve.

## Action

Together with the gastrocnemius and plantaris, the three muscles act as powerful plantar flexors of the ankle joint. They provide the main forward propulsive force in walking and running by using the foot as a lever and raising the heel off the ground.

### *Muscles of the Posterior Fascial Compartment of the Leg: Deep Group*

## POPLITEUS

This flat triangular muscle is described in the section on the popliteal fossa. (See p. 613.) It arises inside the capsule of the knee joint and is inserted into the upper part of the posterior surface of the tibia.

## FLEXOR DIGITORUM LONGUS (FIG. 10-45)

## Origin

From the medial part of the posterior surface of the tibia, below the soleal line.

## Insertion

The tendon passes behind the medial malleolus, deep to the flexor retinaculum, and enters the sole of the foot. It receives a strong slip from the tendon of the flexor hallucis longus. The main tendon now divides into four tendons, which pass to the lateral four toes, where they are inserted into the bases of the distal phalanges. Each tendon passes through

**Fig. 10-45. Deep structures present on posterior aspect of right leg.**

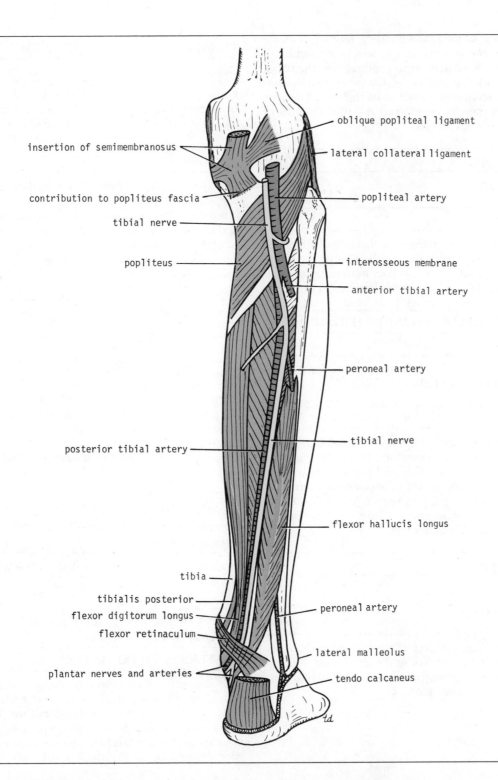

insertion of semimembranosus

oblique popliteal ligament

lateral collateral ligament

contribution to popliteus fascia

popliteal artery

tibial nerve

popliteus

interosseous membrane

anterior tibial artery

peroneal artery

posterior tibial artery

tibial nerve

flexor hallucis longus

tibia

tibialis posterior

flexor digitorum longus

peroneal artery

flexor retinaculum

plantar nerves and arteries

lateral malleolus

tendo calcaneus

an opening in the corresponding tendon of flexor digitorum brevis, the arrangement being similar to that in the flexor digitorum profundus in the hand. (See p. 493.) The quadratus plantae is inserted into the main tendon proximal to its division. The four lumbrical muscles arise from the four tendons of flexor digitorum longus.

## Nerve Supply

Tibial nerve.

## Action

Flexes the distal phalanges of the lateral four toes and assists in plantar flexing the foot at the ankle joint. It plays an important part in maintaining the medial and lateral longitudinal arches in the foot.

## FLEXOR HALLUCIS LONGUS (FIG. 10-45)

### Origin

From the lower two-thirds of the posterior surface of the shaft of the fibula.

### Insertion

The tendon passes behind the medial malleolus, deep to the flexor retinaculum. It grooves the posterior surface of the talus and passes forward on the sole of the foot beneath the sustentaculum tali. It gives off a strong slip to the tendon of flexor digitorum longus. It is inserted into the base of the distal phalanx of the big toe.

### Nerve Supply

Tibial nerve.

### Action

Flexes the distal phalanx of the big toe and assists in plantar flexing the foot at the ankle joint. It plays an important part in maintaining the medial longitudinal arch in the foot.

## TIBIALIS POSTERIOR (FIG. 10-45)

### Origin

From the lateral part of the posterior surface of the tibia, the interosseous membrane, and the upper half of the posterior surface of the fibula.

### Insertion

The tendon passes behind the medial malleolus deep to the flexor retinaculum. It runs forward into the sole of the foot above the sustentaculum tali and is inserted mainly into the tuberosity of the navicular bone. Small tendinous slips pass to the cuboid, the cuneiforms, and the bases of the second, third, and fourth metatarsals.

### Nerve Supply

Tibial nerve.

### Action

Plantar flexes the foot at the ankle joint and inverts the foot at the subtalar and transverse tarsal joints. It plays an important part in holding up the medial longitudinal arch in the foot. The small tendinous slips of insertion assist in holding the bones of the foot together.

### *Muscles of the Posterior Fascial Compartment of the Leg—Their Nerve Supply and Action*

Students wishing to review these muscles should study Table 10-7.

### *Artery of the Posterior Fascial Compartment of the Leg*

#### POSTERIOR TIBIAL ARTERY

The posterior tibial artery is one of the terminal branches of the popliteal artery. (See p. 613.) It begins at the level of the lower border of the popliteus muscle and passes downward deep to the gastrocnemius and soleus and the deep transverse fascia of the leg (Figs. 10-38, 10-39, and 10-45). It lies on

Table 10-7. Muscles of the Lower Limb—Muscles of the Posterior Fascial Compartment of the Leg

| Name of muscle | Origin | Insertion | Nerve supply | Action |
|---|---|---|---|---|
| SUPERFICIAL GROUP | | | | |
| Gastrocnemius | Lateral head from lateral condyle of femur and medial head from above medial condyle | Via tendo calcaneus into posterior surface of calcaneum | Tibial nerve | Plantar flexes foot at ankle joint; flexes knee joint |
| Plantaris | Lateral supracondylar ridge of femur | Posterior surface of calcaneum | Tibial nerve | Plantar flexes foot at ankle joint; flexes knee joint |
| Soleus | Shafts of tibia and fibula | Via tendo calcaneus into posterior surface of calcaneum | Tibial nerve | Together with gastrocnemius and plantaris is powerful plantar flexor of ankle joint; provides main propulsive force in walking and running |
| DEEP GROUP | | | | |
| Popliteus | Lateral surface of lateral condyle of femur | Posterior surface of shaft of tibia above soleal line | Tibial nerve | Flexes leg at knee joint; unlocks knee joint by lateral rotation of femur on tibia and slackens ligaments of joint |
| Flexor digitorum longus | Posterior surface of shaft of tibia | Bases of distal phalanges of lateral four toes | Tibial nerve | Flexes distal phalanges of lateral four toes; plantar flexes foot at ankle joint; supports medial and lateral longitudinal arches of foot |
| Flexor hallucis longus | Posterior surface of shaft of fibula | Base of distal phalanx of big toe | Tibial nerve | Flexes distal phalanx of big toe; plantar flexes foot at ankle joint; supports medial longitudinal arch of foot |
| Tibialis posterior | Posterior surface of shafts of tibia and fibula and interosseous membrane | Tuberosity of navicular bone and other neighboring bones | Tibial nerve | Plantar flexes foot at ankle joint; inverts foot at subtalar and transverse tarsal joints; supports medial longitudinal arch of foot |

the posterior surface of the tibialis posterior muscle above and on the posterior surface of the tibia below. The tibial nerve lies first on its medial side, then crosses posterior to it, and finally lies on its lateral side. In the lower part of the leg, the artery lies about 1 inch (2.5 cm) in front of the medial border of the tendo calcaneus and is here covered only by skin and fascia. The artery passes behind the medial malleolus deep to the flexor retinaculum and terminates by dividing into medial and lateral plantar arteries (Fig. 10-43).

## Branches

1. *Peroneal artery.* This is a large artery that arises close to the origin of the posterior tibial artery (Fig. 10-45). It descends behind the fibula, either within the substance of the flexor hallucis longus muscle or posterior to it. The peroneal artery gives off numerous *muscular branches* and a *nutrient artery to the fibula* and ends by taking part in the anastomosis around the ankle joint. A *perforating branch* pierces the interosseous membrane to reach the lower part of the front of the leg.
2. *Muscular branches* are distributed to muscles in the posterior compartment of the leg.
3. *Nutrient artery to the tibia.*
4. *Anastomotic branches*, which join other arteries around the ankle joint.
5. *Medial and lateral plantar arteries.* (See p. 648.)

*Venae comitantes* of the posterior tibial artery join those of the anterior tibial artery in the popliteal fossa, to form the popliteal vein.

## Nerve of the Posterior Fascial Compartment of the Leg

### TIBIAL NERVE

The tibial nerve is the larger terminal branch of the sciatic nerve (Fig. 10-70) in the lower third of the back of the thigh. (See p. 616.) It descends through the popliteal fossa and passes deep to the gastrocnemius and soleus muscles (Figs. 10-44 and 10-45). It lies on the posterior surface of the tibialis posterior, and lower down the leg, on the posterior surface of the tibia (Fig. 10-45). The nerve accompan-

ies the posterior tibial artery and lies at first on its medial side, then crosses posterior to it, and finally lies on its lateral side. The nerve, with the artery, passes behind the medial malleolus, between the tendons of the flexor digitorum longus and the flexor hallucis longus (Fig. 10-43). It is covered here by the flexor retinaculum and divides into the medial and lateral plantar nerves.

## Branches in the Leg (Below the Popliteal Fossa)

1. *Muscular branches* to the soleus, flexor digitorum longus, flexor hallucis longus, and tibialis posterior.
2. *Cutaneous.* The *medial calcaneal branch* supplies the skin over the medial surface of the heel (Fig. 10-43).
3. *Articular branch* to the ankle joint.
4. *Medial and lateral plantar nerves.* (See p. 648.)

## THE REGION OF THE ANKLE

Before learning the anatomy of the foot, it is essential that a student have a sound knowledge of the arrangement of the tendons, arteries, and nerves in the region of the ankle joint. From the clinical standpoint, the ankle is a common site for fractures, sprains, and dislocations.

In the drawing of a transverse section through the ankle joint shown in Figure 10-46, identify the structures from medial to lateral; at the same time, examine your own ankle and identify as many of the structures as possible.

## ANTERIOR ASPECT OF THE ANKLE (FIG. 10-46)

### Structures that Pass Anterior to the Extensor Retinacula from Medial to Lateral

1. Saphenous nerve and great saphenous vein (in front of the medial malleolus).
2. Superficial peroneal nerve (medial and lateral branches).

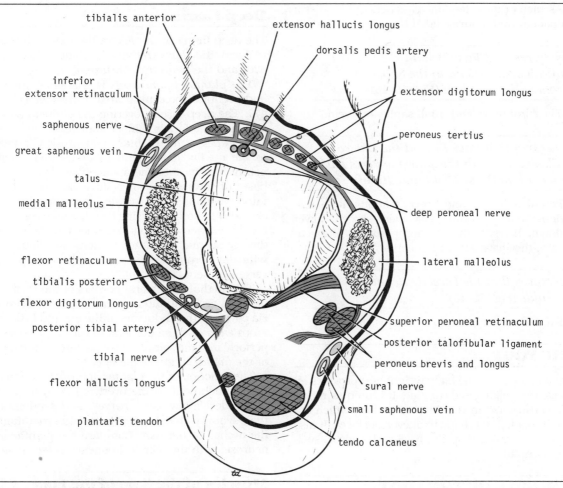

**Fig. 10-46. Relations of right ankle joint.**

*Structures that Pass Beneath or Through the Extensor Retinacula (Fig. 10-46) from Medial to Lateral*

1. Tibialis anterior tendon.
2. Extensor hallucis longus tendon.
3. Anterior tibial artery with venae comitantes.
4. Deep peroneal nerve.
5. Extensor digitorum longus tendons.
6. Peroneus tertius.

As each of the above tendons passes beneath or through the extensor retinacula, it is surrounded by a synovial sheath. The tendons of extensor digito-rum longus and the peroneus tertius share a common synovial sheath.

## POSTERIOR ASPECT OF THE ANKLE

*Structures that Pass Behind the Medial Malleolus Beneath the Flexor Retinaculum from Medial to Lateral (Figs. 10-43 and 10-46)*

1. Tibialis posterior tendon.
2. Flexor digitorum longus.
3. Posterior tibial artery with venae comitantes.
4. Tibial nerve.
5. Flexor hallucis longus.

As each of these tendons pass beneath the flexor retinaculum, it is surrounded by a synovial sheath.

*Structures that Pass Behind the Lateral Malleolus Superficial to the Superior Peroneal Retinaculum (Fig. 10-46)*

1. The sural nerve and small saphenous vein.

*Structures that Pass Behind the Lateral Malleolus Beneath the Superior Peroneal Retinaculum (Figs. 10-43 and 10-46)*

1. Peroneus longus and brevis tendons. Both tendons share a common synovial sheath; lower down, beneath the inferior peroneal retinaculum, they have separate sheaths.

*Structures that Lie Directly Behind the Ankle (Fig. 10-46)*

1. Fat and the large tendo calcaneus.

## THE FOOT

The foot supports the body weight and provides leverage for walking and running. It is unique in that it is constructed in the form of arches, which enable it to adapt its shape to uneven surfaces. It also serves as a resilient spring to absorb shocks, such as in jumping.

## THE SOLE OF THE FOOT

### Skin

The skin of the sole of the foot is thick and hairless. It is firmly bound down to the underlying deep fascia by numerous fibrous bands. The skin shows a few flexure creases at the sites of skin movement. Sweat glands are present in large numbers.

The *sensory nerve supply* to the skin of the sole of the foot is derived from the *medial calcaneal branch* of the tibial nerve, which innervates the medial side of the heel; branches from the *medial plantar nerve*, which innervate the medial two-thirds of the sole; and branches from the *lateral plantar nerve*, which innervate the lateral third of the sole (Figs. 10-9 and 10-47).

## Deep Fascia

The deep fascia of the sole of the foot is thickened to form the *flexor retinaculum* (described on p. 620) and the *plantar aponeurosis*.

The *plantar aponeurosis* is triangular in shape and occupies the central area of the sole (Fig. 10-47). The deep fascia covering the abductors of the big and little toes is thinner and weak.[‡]

The apex of the plantar aponeurosis is attached to the medial and lateral tubercles of the calcaneum. The base of the aponeurosis divides at the bases of the toes into five slips. Each slip divides into two bands, one passing superficially to the skin and the other passing deeply to the root of the toe; here, each deep band divides into two, which diverge around the flexor tendons and finally fuse with the fibrous flexor sheath and the deep transverse ligaments (Fig. 10-47).

The medial and lateral borders of the thick aponeurosis are continuous with the thinner deep fascia covering the abductors of the big and little toes. From each of these borders, fibrous septa pass superiorly into the sole and take part in the formation of the *fascial spaces of the sole*.

The function of the plantar aponeurosis is to give firm attachment to the overlying skin, to protect the underlying vessels, nerves, and tendons and their synovial sheaths, and to assist in maintaining the arches of the foot. Compare the plantar aponeurosis with the palmar aponeurosis (see p. 488).

## Muscles of the Sole of the Foot

For descriptive purposes these are conveniently divided into four layers.

| | |
|---|---|
| First Layer | Abductor hallucis |
| | Flexor digitorum brevis |
| | Abductor digiti minimi |
| Second Layer | Quadratus plantae (flexor digitorum accessorius) |
| | Lumbricals |
| | Flexor digitorum longus tendon |
| | Flexor hallucis longus tendon |

[‡]Many authorities describe the plantar aponeurosis as covering the whole sole and divide it into a thick central part and weak medial and lateral parts.

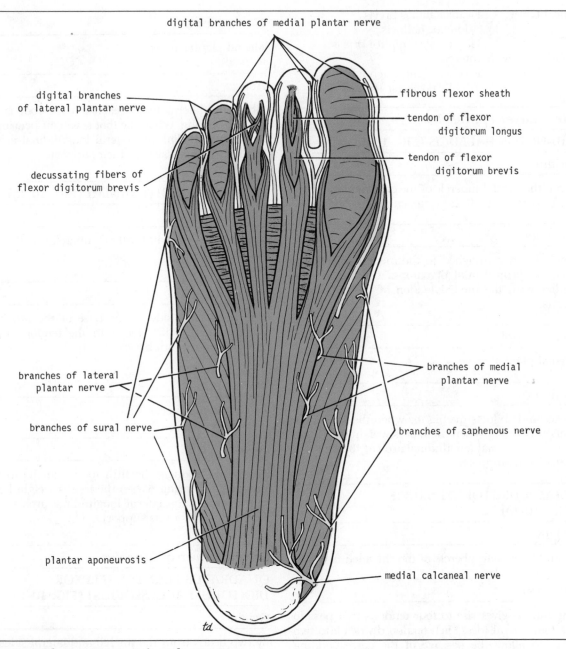

digital branches of medial plantar nerve

digital branches
of lateral plantar nerve

fibrous flexor sheath

tendon of flexor
digitorum longus

tendon of flexor
digitorum brevis

decussating fibers of
flexor digitorum brevis

branches of medial
plantar nerve

branches of lateral
plantar nerve

branches of saphenous nerve

branches of sural nerve

plantar aponeurosis

medial calcaneal nerve

**Fig. 10-47. Plantar aponeurosis and cutaneous
nerves of sole of right foot.**

| | |
|---|---|
| Third Layer | Flexor hallucis brevis |
| | Adductor hallucis |
| | Flexor digiti minimi brevis |
| Fourth Layer | Interossei |
| | Peroneus longus tendon |
| | Tibialis posterior tendon |

### *First Layer of Muscles*

### ABDUCTOR HALLUCIS (FIG. 10-48)

#### Origin

From the medial tubercle of the calcaneum and the lower edge of the flexor retinaculum.

#### Insertion

The tendon is attached to the medial side of the base of the proximal phalanx of the big toe, together with the medial tendon of flexor hallucis brevis.

#### Nerve Supply

Medial plantar nerve.

#### Action

Flexes and abducts the big toe when the foot is not weight-bearing. When the foot is weight-bearing it braces the medial longitudinal arch of the foot and assists in its support.

### FLEXOR DIGITORUM BREVIS (FIG. 10-48)

#### Origin

From the medial tubercle of the calcaneum.

#### Insertion

The muscle gives rise to four tendons that pass to the four lateral toes. Each tendon divides into two slips, to allow the passage of the corresponding tendon of the flexor digitorum longus. The two slips then unite and partially decussate; then the tendon divides again and is inserted into the borders of the middle phalanx. Compare with the insertion of the flexor digitorum superficialis in the fingers.

#### Nerve Supply

Medial plantar nerve.

#### Action

Flexes the lateral four toes when the foot is not weight-bearing. When the foot is weight-bearing, it braces the medial and lateral longitudinal arches of the foot and assists in their support.

### ABDUCTOR DIGITI MINIMI (FIG. 10-48)

#### Origin

From the medial and lateral tubercles of the calcaneum.

#### Insertion

Into the lateral side of the base of the proximal phalanx of the fifth toe with the tendon of the flexor digiti minimi brevis.

#### Nerve Supply

Lateral plantar nerve.

#### Action

Flexes and abducts the fifth toe when the foot is not weight-bearing. When the foot is weight-bearing, it braces the lateral longitudinal arch of the foot and assists in its support.

### *Second Layer of Muscles*

### QUADRATUS PLANTAE (FLEXOR DIGITORUM ACCESSORIUS) (FIG. 10-49)

#### Origin

By two heads from the medial and lateral sides of the calcaneum.

#### Insertion

Into the posterolateral margin of the tendon of flexor digitorum longus.

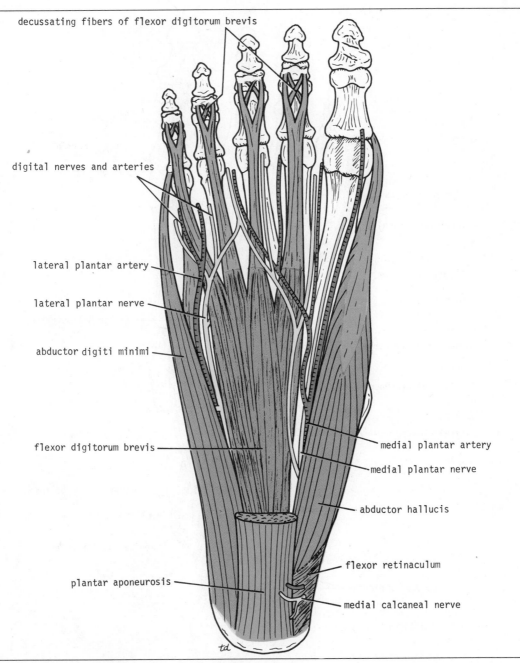

decussating fibers of flexor digitorum brevis

digital nerves and arteries

lateral plantar artery

lateral plantar nerve

abductor digiti minimi

flexor digitorum brevis

medial plantar artery

medial plantar nerve

abductor hallucis

flexor retinaculum

plantar aponeurosis

medial calcaneal nerve

**Fig. 10-48. Plantar muscles of right foot, first layer. Medial and lateral plantar arteries and nerves are also shown.**

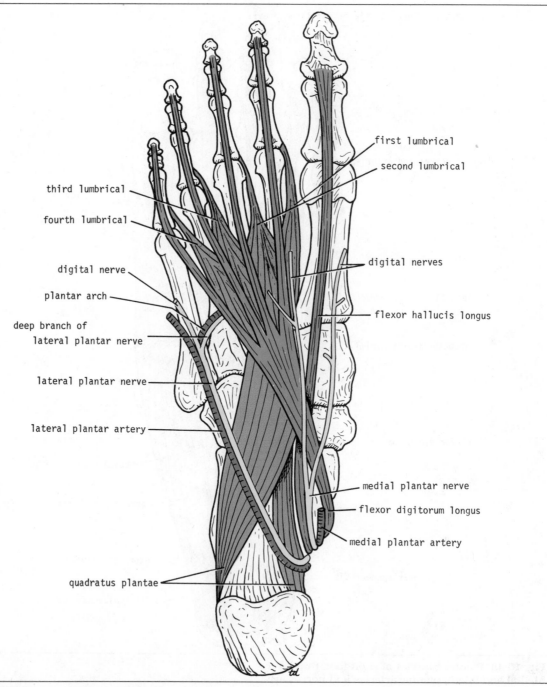

first lumbrical

second lumbrical

third lumbrical

fourth lumbrical

digital nerve

plantar arch

deep branch of
lateral plantar nerve

lateral plantar nerve

lateral plantar artery

quadratus plantae

digital nerves

flexor hallucis longus

medial plantar nerve

flexor digitorum longus

medial plantar artery

**Fig. 10-49. Plantar muscles of right foot, second layer. Medial and lateral plantar arteries and nerves are also shown.**

## Nerve Supply

Lateral plantar nerve.

## Action

It pulls the tendon of flexor digitorum longus directly posteriorly and thus aids this muscle in flexing the lateral four toes. It is also able to act alone and flex these toes, using the tendon of flexor digitorum longus when the belly of this muscle is relaxed.

## LUMBRICALS (FIG. 10-49)

### Origin

The four lumbricals arise from the tendons of the flexor digitorum longus.

### Insertion

Each is inserted into the medial side of the dorsal expansion of the corresponding tendon of extensor digitorum longus; they are also attached to the bases of the proximal phalanges of the lateral four toes.

### Nerve Supply

The first lumbrical (i.e., the most medial) is supplied by the medial plantar nerve and the remainder are supplied by the deep branch of the lateral plantar nerve.

### Action

To prevent the lateral four toes from buckling under during walking or running. This is accomplished by extending the toes at the interphalangeal joints when the flexor digitorum longus tendons are flexing the toes.

## FLEXOR DIGITORUM LONGUS TENDON (FIG. 10-49)

The flexor digitorum longus tendon enters the sole by passing behind the medial malleolus beneath the flexor retinaculum (Fig. 10-43). It passes forward across the medial surface of the sustentaculum tali and then crosses the tendon of flexor hallucis longus, from which it receives a strong slip. It is here that it receives on its lateral border the insertion of the quadratus plantae muscle. The tendon now divides into its four tendons of insertion, which pass forward, giving origin to the lumbrical muscles. The tendons then enter the fibrous sheaths of the lateral four toes (Fig. 10-47). Each tendon perforates the corresponding tendon of flexor digitorum brevis and passes on to be inserted into the base of the distal phalanx. It should be noted that the method of insertion is similar to that found for the flexor digitorum profundus in the hand. (See p. 490.)

## FLEXOR HALLUCIS LONGUS TENDON (FIG. 10-49)

The flexor hallucis longus tendon enters the sole by passing behind the medial malleolus beneath the flexor retinaculum. It runs forward below the sustentaculum tali and crosses deep to the flexor digitorum longus tendon, to which it gives a strong slip. It then enters the fibrous sheath of the big toe and is inserted into the base of the distal phalanx.

### *Fibrous Flexor Sheaths*

The inferior surface of each toe, from the head of the metatarsal bone to the base of the distal phalanx, is provided with a strong fibrous sheath, which is attached to the sides of the phalanges (Fig. 10-47). The arrangement is similar to that found in the fingers (see p. 490). The proximal ends of the fibrous sheaths of the toes receive the deeper parts of the five slips of the plantar aponeurosis. The distal end of the sheath is closed and is attached to the base of the distal phalanx. The sheath, together with the inferior surfaces of the phalanges and the interphalangeal joints, forms a blind tunnel in which lie the flexor tendons of the toe (Fig. 10-50).

In the big toe, the osteofibrous tunnel contains the tendon of the flexor hallucis longus (Fig. 10-50). In the case of the four lateral toes, the tunnel is occupied by the tendons of the flexor digitorum brevis and longus. The fibrous sheath is thick over the phalanges, but thin and lax over the joints.

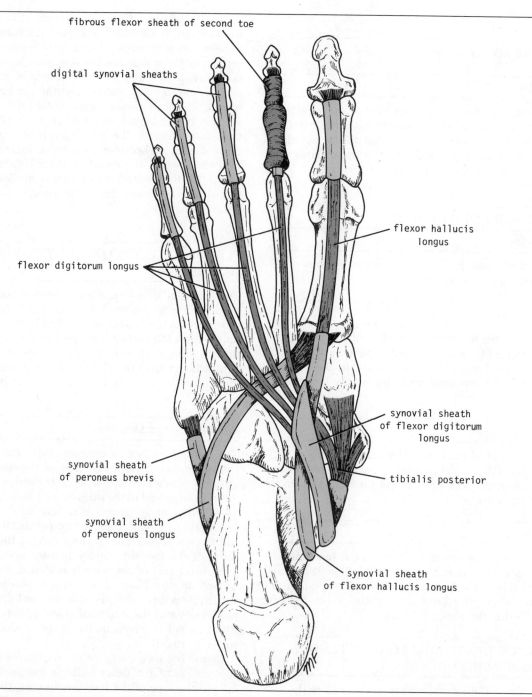

**Fig. 10-50. Synovial sheaths of tendons seen on sole of right foot.**

## Synovial Flexor Sheaths

The tendon of the flexor hallucis longus is surrounded by a synovial sheath, which extends upward behind the medial malleolus for a short distance above the flexor retinaculum (Fig. 10-43). Distally, the sheath extends as far as the base of the first metatarsal bone (Fig. 10-50). As the tendon enters the fibrous flexor sheath, it acquires a digital synovial sheath, which surrounds the tendon as far as its insertion (Fig. 10-50).

The tendon of the flexor digitorum longus is also surrounded by a synovial sheath, which extends upward behind the medial malleolus for a short distance above the flexor retinaculum (Fig. 10-43). Distally, the sheath extends as far as the navicular bone (Fig. 10-50). As each of the four tendons enters the fibrous flexor sheath of the lateral four toes, it acquires a digital synovial sheath, which surrounds the tendon as far as its insertion (Fig. 10-50).

## Third Layer of Muscles

### FLEXOR HALLUCIS BREVIS (FIG. 10-51)

#### Origin

From the cuboid and lateral cuneiform bones and from the tendinous extensions derived from the tibialis posterior insertion.

#### Insertion

The muscle splits into two bellies. The tendon of the medial belly joins that of the abductor hallucis and is attached to the medial side of the base of the proximal phalanx of the big toe. The tendon of the lateral belly joins that of the adductor hallucis and is attached to the lateral side of the base of the same bone. A *sesamoid bone* is present in each common tendon.

#### Nerve Supply

Medial plantar nerve.

#### Action

Flexion of the metatarsophalangeal joint of the big toe. It helps to support the medial longitudinal arch of the foot.

### ADDUCTOR HALLUCIS (FIG. 10-51)

#### Origin

*Oblique head* from the bases of the second, third, and fourth metatarsal bones; *transverse head* from the plantar ligaments of the metatarsophalangeal joints of the third, fourth, and fifth toes.

#### Insertion

The combined tendons join the lateral tendon of the flexor hallucis brevis and are inserted into the lateral side of the base of the proximal phalanx of the big toe.

#### Nerve Supply

Deep branch of the lateral plantar nerve.

#### Action

The oblique head assists the flexor hallucis brevis in the flexion of the metatarsophalangeal joint of the big toe. The transverse head holds together the heads of the metatarsal bones and assists in the stabilization of the forepart of the foot. It thus helps to support the transverse arch of the foot.

### FLEXOR DIGITI MINIMI BREVIS (FIG. 10-51)

#### Origin

From the base of the fifth metatarsal bone.

#### Insertion

Into the lateral side of the base of the proximal phalanx of the little toe.

#### Nerve Supply

Superficial branch of the lateral plantar nerve.

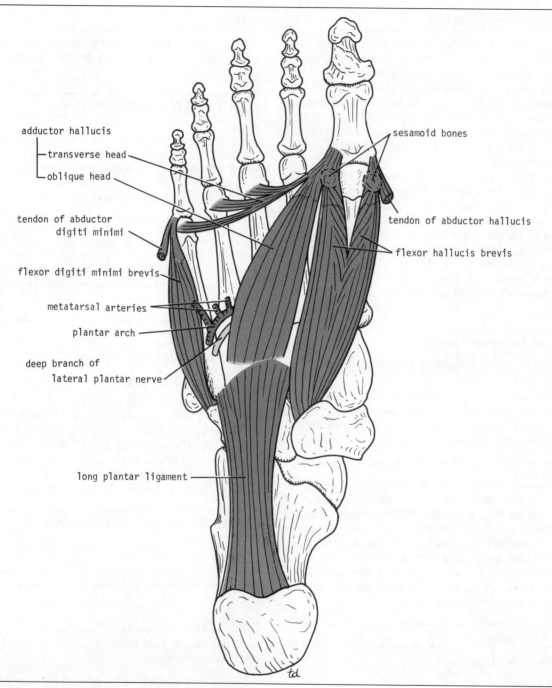

**Fig. 10-51. Plantar muscles of right foot, third layer. Deep branch of lateral plantar nerve and plantar arterial arch are also shown.**

## Action

Flexion of the metatarsophalangeal joint of the little toe.

### *Fourth Layer of Muscles*

### THE INTEROSSEI

There are seven interossei in the foot (eight in the hand), consisting of four dorsal and three plantar muscles. They occupy the spaces between the metatarsal bones. The dorsal muscles arise by two heads and are larger than the plantar muscles, which only have one head.

### *Plantar Interossei (Fig. 10-52)*

### Origin

The three muscles arise from the inferior surfaces of the third, fourth, and fifth metatarsal bones.

### Insertion

Into the medial side of the bases of the proximal phalanges and the dorsal expansion of the corresponding toes.

### Nerve Supply

Lateral plantar nerve.

### Action

Adduct the toes toward the center of the second toe at the metatarsophalangeal joints; flex the metatarsophalangeal joints and extend the interphalangeal joints.

### *Dorsal Interossei (Figs. 10-52 and 10-53)*

### Origin

The four muscles arise from the adjacent sides of the metatarsal bones.

### Insertion

The bases of the proximal phalanges; the first is inserted into the medial side of the second toe, and the three others are inserted into the lateral sides of the second, third, and fourth toes, respectively. They are also inserted into the dorsal expansion of the corresponding toes.

### Nerve Supply

Lateral plantar nerve.

### Action

Abduct the toes away from the center of the second toe at the metatarsophalangeal joints; flex the metatarsophalangeal joints and extend the interphalangeal joints. Since each muscle arises from two metatarsal bones, it serves to bind together the metatarsals and stabilize the forepart of the foot.

### PERONEUS LONGUS TENDON (FIG. 10-52)

The peroneus longus tendon enters the foot from behind the lateral malleolus. It passes beneath the superior and inferior peroneal retinacula and lies below the peroneal tubercle on the lateral side of the calcaneum. On reaching the cuboid bone, the tendon winds around its lateral margin and enters a groove on its inferior aspect. The tendon is held in place by a strong fibrous band, derived from the long plantar ligament. The tendon runs obliquely across the sole and is inserted into the base of the first metatarsal bone and the adjacent part of the medial cuneiform.

The tendon is surrounded by a synovial sheath as it passes beneath the peroneal retinacula. As the tendon winds around the lateral margin of the cuboid, it is thickened and contains a *sesamoid cartilage.* A second synovial sheath surrounds the tendon as it crosses the sole.

The action of the peroneus longus muscle is described on page 625.

### TIBIALIS POSTERIOR TENDON (FIG. 10-52)

The tibialis posterior tendon enters the foot from behind the medial malleolus. It passes beneath the flexor retinaculum and runs downward and forward above the sustentaculum tali, to be inserted mainly into the tuberosity of the navicular. Small

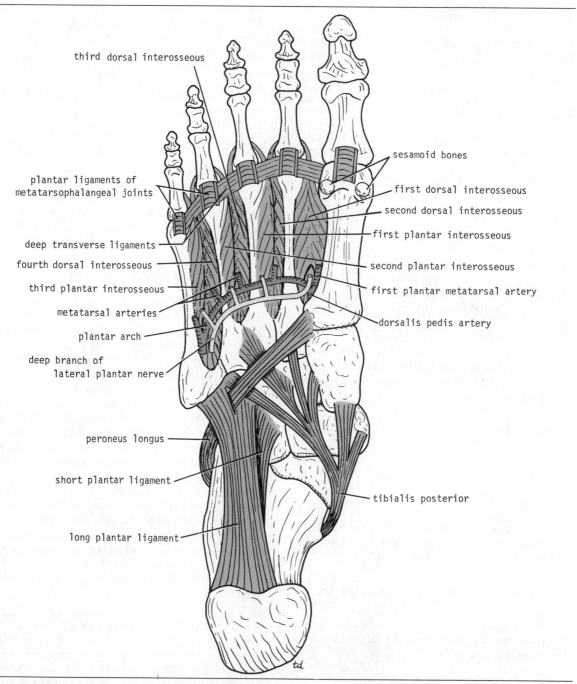

third dorsal interosseous

plantar ligaments of
metatarsophalangeal joints

deep transverse ligaments

fourth dorsal interosseous

third plantar interosseous

metatarsal arteries

plantar arch

deep branch of
lateral plantar nerve

peroneus longus

short plantar ligament

long plantar ligament

sesamoid bones

first dorsal interosseous

second dorsal interosseous

first plantar interosseous

second plantar interosseous

first plantar metatarsal artery

dorsalis pedis artery

tibialis posterior

td

**Fig. 10-52. Plantar muscles of right foot, fourth
layer. Deep branch of lateral plantar nerve and
plantar arterial arch are also shown. Note deep
transverse ligaments.**

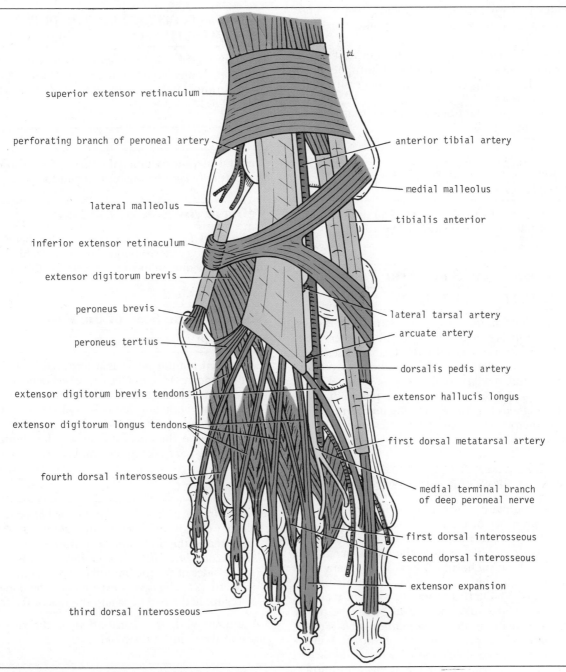

**Fig. 10-53. Structures present on dorsal aspect of right foot.**

superior extensor retinaculum

perforating branch of peroneal artery

lateral malleolus

inferior extensor retinaculum

extensor digitorum brevis

peroneus brevis

peroneus tertius

extensor digitorum brevis tendons

extensor digitorum longus tendons

fourth dorsal interosseous

third dorsal interosseous

anterior tibial artery

medial malleolus

tibialis anterior

lateral tarsal artery

arcuate artery

dorsalis pedis artery

extensor hallucis longus

first dorsal metatarsal artery

medial terminal branch
of deep peroneal nerve

first dorsal interosseous

second dorsal interosseous

extensor expansion

tendinous slips pass to the cuboid and the cuneiforms and to the bases of the second, third, and fourth metatarsals.

The tendon is surrounded by a synovial sheath as it passes beneath the flexor retinaculum; the sheath ends below, just proximal to the tuberosity of the navicular bone.

The action of the tibialis posterior muscle is described on page 632.

---

*Muscles of the Sole of the Foot—Their Nerve Supply and Action*

---

Students wishing to review the small muscles of the sole of the foot should study Table 10-8.

## Arteries of the Sole of the Foot

### MEDIAL PLANTAR ARTERY

The medial plantar artery is the smaller of the terminal branches of the posterior tibial artery. (See p. 634.) It arises beneath the flexor retinaculum and passes forward deep to the abductor hallucis muscle and medial to the medial plantar nerve (Fig. 10-43). It ends by supplying the medial side of the big toe (Fig. 10-48). During its course it gives off numerous muscular, cutaneous, and articular branches.

### LATERAL PLANTAR ARTERY

The lateral plantar artery is the larger of the terminal branches of the posterior tibial artery. (See p. 634.) It arises beneath the flexor retinaculum and passes forward deep to the abductor hallucis and the flexor digitorum brevis, in company with the lateral plantar nerve, which lies on its medial side (Figs. 10-43, 10-48, and 10-49). On reaching the base of the fifth metatarsal bone, the artery curves medially to form the *plantar arch* (Fig.10-51) and at the proximal end of the first intermetatarsal space, joins the dorsalis pedis artery (Fig. 10-52). During its course, it gives off numerous muscular, cutaneous, and articular branches. The plantar arch gives off plantar digital arteries to the adjacent sides of the lateral four toes and the lateral side of the little toe.

## DORSALIS PEDIS ARTERY (THE DORSAL ARTERY OF THE FOOT)

On entering the sole between the two heads of the first dorsal interosseous muscle, the dorsalis pedis artery immediately joins the lateral plantar artery (Fig. 10-52).

---

*Branches*

---

The first plantar metatarsal artery, which supplies the cleft between the big and second toes.

## Veins of the Sole of the Foot

*Medial* and *lateral plantar veins* accompany the corresponding arteries, and they unite behind the medial malleolus to form the posterior tibial venae comitantes.

## Nerves of the Sole of the Foot

### MEDIAL PLANTAR NERVE

The medial plantar nerve is a terminal branch of the tibial nerve. (See p. 634.) It arises beneath the flexor retinaculum (Fig. 10-43) and runs forward deep to the abductor hallucis, with the medial plantar artery on its medial side (Fig. 10-48). It comes to lie in the interval between the abductor hallucis and the flexor digitorum brevis.

---

*Branches*

---

1. *Muscular branches* to the abductor hallucis, the flexor digitorum brevis, the flexor hallucis brevis, and the first lumbrical muscle.
2. *Cutaneous branches. Plantar digital nerves* run to the sides of the medial three and one-half toes (Fig. 10-47). The nerves extend onto the dorsum and supply the nail beds and the tips of the toes.

Compare with the distribution of the median nerve in the palm of the hand.

### LATERAL PLANTAR NERVE

The lateral plantar nerve is a terminal branch of the tibial nerve. (See p. 634.) It arises beneath the flexor retinaculum (Fig. 10-43) and runs forward

Table 10-8. Muscles of the Lower Limb—Muscles of the Sole of the Foot

| Name of muscle | Origin | Insertion | Nerve supply | Action |
|---|---|---|---|---|
| **FIRST LAYER** | | | | |
| Abductor hallucis | Medial tuberosity of calcaneum and flexor retinaculum | Base of proximal phalanx of big toe | Medial plantar nerve | Flexes and abducts big toe; braces medial longitudinal arch |
| Flexor digitorum brevis | Medial tuberosity of calcaneum | Four tendons to four lateral toes— inserted into borders of middle phalanx; tendons perforated by those of flexor digitorum longus | Medial plantar nerve | Flexes lateral four toes; braces medial and lateral longitudinal arches |
| Abductor digiti minimi | Medial and lateral tuberosities of calcaneum | Base of proximal phalanx of fifth toe | Lateral plantar nerve | Flexes and abducts fifth toe; braces lateral longitudinal arch |
| **SECOND LAYER** | | | | |
| Quadratus plantae | Medial and lateral sides of calcaneum | Tendon of flexor digitorum longus | Lateral plantar nerve | Assists flexor digitorum longus in flexing lateral four toes |
| Lumbricals (4) | Tendons of flexor digitorum longus | Dorsal extensor expansion; bases of proximal phalanges of lateral four toes | First lumbrical medial plantar nerve; remainder lateral plantar nerve | Extends toes at interphalangeal joints |
| Flexor digitorum longus tendon | See Table 10-7 | | | |
| Flexor hallucis longus tendon | See Table 10-7 | | | |
| **THIRD LAYER** | | | | |
| Flexor hallucis brevis | Cuboid, lateral cuneiform, tibialis posterior insertion | Medial tendon into medial side of base of proximal phalanx of big toe; lateral tendon into lateral side of base of proximal phalanx of big toe | Medial plantar nerve | Flexes metatarso- phalangeal joint of big toe; supports medial longitudinal arch |
| Adductor hallucis | Oblique head bases of second, third, and fourth metatarsal bones; transverse head plantar ligaments | Lateral side base proximal phalanx of big toe | Deep branch lateral plantar nerve | Flexes metatarso- phalangeal joint of big toe; holds together metatarsal bones |

Table 10-8. (continued)

| Name of muscle | Origin | Insertion | Nerve supply | Action |
|---|---|---|---|---|
| Flexor digiti minimi brevis | Base of fifth metatarsal bone | Lateral side base proximal phalanx of little toe | Lateral plantar nerve | Flexes metatarso-phalangeal joint of little toe |
| FOURTH LAYER | | | | |
| Interossei | | | | |
| Dorsal (4) | Adjacent sides of metatarsal bones | Bases of proximal phalanges—first medial side second toe; remainder lateral sides second, third, and fourth toes—also dorsal extensor expansion | Lateral plantar nerve | Abduction of toes; flexes metatarso-phalangeal joints and extends interphalangeal joints |
| Plantar (3) | Inferior surfaces of third, fourth, and fifth metatarsal bones | Medial side of bases of proximal phalanges of lateral three toes | Lateral plantar nerve | Adduction of toes; flexes metatarso-phalangeal joints and extends interphalangeal joints |
| Peroneus longus tendon | See Table 10-6 | | | |
| Tibialis posterior tendon | See Table 10-7 | | | |

deep to the abductor hallucis and the flexor digitorum brevis, in company with the lateral plantar artery, which lies on its lateral side (Fig. 10-49). On reaching the base of the fifth metatarsal bone, it divides into superficial and deep branches (Fig. 10-49).

### Branches

1. *From the main trunk* to the quadratus plantae and abductor digiti minimi; cutaneous branches to the skin of the lateral part of the sole.
2. *From the superficial terminal branch* to the flexor digiti minimi and the interosseous muscles of the fourth intermetatarsal space. Plantar digital branches pass to the sides of the lateral one and one-half toes. The nerves extend onto the dorsum and supply the nail beds and tips of the toes.
3. *From the deep terminal branch* (Fig. 10-52). This

branch curves medially with the lateral plantar artery and supplies the adductor hallucis, the second, third, and fourth lumbricals, and all the interossei, except those in the fourth intermetatarsal space. (See superficial branch above.)

Compare with the distribution of the ulnar nerve in the palm of the hand.

## THE DORSUM OF THE FOOT

### Skin

The skin on the dorsum of the foot is thin, hairy, and freely mobile on the underlying tendons and bones.

The *sensory nerve supply* (Fig. 10-8) to the skin on the dorsum of the foot is derived from the superficial peroneal nerve, assisted by the deep peroneal, saphenous, and sural nerves.

The *superficial peroneal nerve* emerges from between the peroneus brevis and the extensor digitorum longus muscle in the lower part of the leg and pierces the deep fascia. (See p. 627.) It now divides into medial and lateral cutaneous branches. The *medial branch* supplies the skin on the dorsum of the foot, the medial side of the big toe, and the adjacent sides of the second and third toes. The *lateral branch* also supplies the skin on the dorsum of the foot and the adjacent sides of the third, fourth, and fifth toes.

The *deep peroneal nerve* supplies the skin of the adjacent sides of the big and second toes (Fig. 10-8).

The *saphenous nerve* passes onto the dorsum of the foot in front of the medial malleolus (Fig. 10-8). It supplies the skin along the medial side of the foot as far forward as the head of the first metatarsal bone.

The *sural nerve* (Fig. 10-9) enters the foot behind the lateral malleolus and supplies the skin along the lateral margin of the foot and the lateral side of the little toe.

The nail beds and the skin covering the dorsal surfaces of the terminal phalanges are supplied by the medial and lateral plantar nerves. (See p. 648.)

## Dorsal Venous Arch (or Network)

The dorsal venous arch lies in the subcutaneous tissue over the heads of the metatarsal bones and drains on the medial side into the great saphenous vein, and on the lateral side into the small saphenous vein (Fig. 10-18). The great saphenous vein leaves the dorsum of the foot by ascending into the leg in front of the medial malleolus. Its further course is described on page 577. The small saphenous vein ascends into the leg behind the lateral malleolus. Its course in the back of the leg is described on page 627. The greater part of the blood from the whole foot drains into the arch via digital veins and communicating veins from the sole, which pass through the interosseous spaces.

## EXTENSOR DIGITORUM BREVIS (FIG. 10-53)

### Origin

From the anterior part of the upper surface of the calcaneum and from the inferior extensor retinaculum.

### Insertion

The muscle gives rise to four tendons that pass forward and medially. The most medial tendon (sometimes called tendon of *extensor hallucis brevis*) is inserted into the base of the proximal phalanx of the big toe. The lateral three tendons join the long extensor tendons passing to the second, third, and fourth toes.

### Nerve Supply

Terminal part of the deep peroneal nerve.

### Action

Extends the first, second, third, and fourth toes at the interphalangeal and metatarsophalangeal joints. It is used particularly when the ankle joint is dorsiflexed and the extensor digitorum longus is unable to act.

## THE INSERTION OF THE LONG EXTENSOR TENDONS

The tendon of extensor digitorum longus passes beneath the superior extensor retinaculum and through the inferior extensor retinaculum, in company with the peroneus tertius muscle (Fig. 10-53). The tendon divides into four, which fan out over the dorsum of the foot and pass to the lateral four toes. Opposite the metatarsophalangeal joints of the second, third, and fourth toes, each tendon is joined on its lateral side by a tendon of extensor digitorum brevis (Fig. 10-53).

On the dorsal surface of each toe, the extensor tendon joins the fascial expansion called the *extensor expansion*. Near the proximal interphalangeal joint, the extensor expansion splits into three parts: a central part, which is inserted into the base of the middle phalanx, and two lateral parts, which converge to be inserted into the base of the distal phalanx (Fig. 10-53).

The dorsal expansion, as in the fingers, receives the tendons of insertion of the interosseous and lumbrical muscles.

## Synovial Sheath of the Tendon of Extensor Digitorum Longus

The extensor digitorum longus and peroneus tertius tendons are surrounded by a common synovial sheath as they pass beneath the extensor retinacula (Fig. 10-53). The sheath extends proximally for a short distance above the malleoli and distally to the level of the base of the fifth metatarsal bone.

## DORSALIS PEDIS ARTERY (THE DORSAL ARTERY OF THE FOOT)

The dorsalis pedis artery begins in front of the ankle joint as a continuation of the anterior tibial artery. (See p. 622.) It terminates by passing downward into the sole between the two heads of the first dorsal interosseous muscles, where it joins the lateral plantar artery and completes the plantar arch (Fig. 10-52). It is superficial in position and is crossed by the inferior extensor retinaculum and the first tendon of extensor digitorum brevis (Fig. 10-53). On its lateral side lie the terminal part of the deep peroneal nerve and the extensor digitorum longus tendons. On the medial side lies the tendon of extensor hallucis longus (Fig. 10-53). Its pulsations can easily be felt.

### Branches

1. *Lateral tarsal artery*, which crosses the dorsum of the foot just below the ankle joint (Fig. 10-53).
2. *Arcuate artery*, which runs laterally under the extensor tendons opposite the bases of the metatarsal bones (Fig. 10-53). It gives off metatarsal branches to the toes.
3. *First dorsal metatarsal artery*, which supplies both sides of the big toe (Fig. 10-53).

## DEEP PERONEAL NERVE

The deep peroneal nerve enters the dorsum of the foot by passing deep to the extensor retinacula on the lateral side of the dorsalis pedis artery. (See p. 623.) It divides into terminal, medial, and lateral branches. The medial branch supplies the skin of the adjacent sides of the big and second toes (Fig. 10-53). The lateral branch supplies the extensor digitorum brevis muscle. Both terminal branches give articular branches to the joints of the foot.

# Joints of the Lower Limb

The hip joint is fully described on page 600.

## KNEE JOINT

The knee joint is the largest and most complicated joint in the body. Basically, it consists of two condylar joints between the medial and lateral condyles of the femur and the corresponding condyles of the tibia, and a gliding joint, between the patella and the patellar surface of the femur. Note that the fibula is not directly involved in the joint.

### Articulation

Above are the rounded condyles of the femur; below are the condyles of the tibia and their semilunar cartilages (Fig. 10-54); in front is the articulation between the lower end of the femur and the patella.

The articular surfaces of the femur, tibia, and patella are covered with hyaline cartilage. Note that the articular surfaces of the medial and lateral condyles of the tibia are often referred to clinically as the medial and lateral *tibial plateaus*.

### Type

The joint between the femur and tibia is a synovial joint of the hinge variety, but some degree of rotatory movement is possible. The joint between the patella and femur is a synovial joint of the plane gliding variety.

### Capsule

This is attached to the margins of the articular surfaces and surrounds the sides and posterior aspect of the joint. On the front of the joint, the capsule is absent, permitting the synovial membrane to pouch upward beneath the quadriceps tendon, forming the *suprapatellar bursa* (Fig. 10-54). On each side of the patella, the capsule is strengthened by expansions from the tendons of vastus lateralis and medialis. Behind the joint the capsule is strengthened by an expansion of the semimembranous muscle called the *oblique popliteal ligament* (Fig. 10-54). There is an opening in the capsule behind the lateral tibial condyle, which

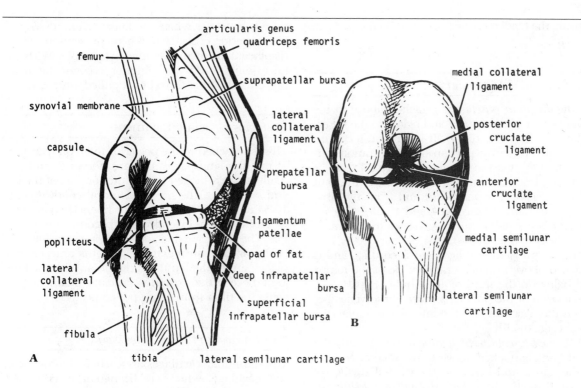

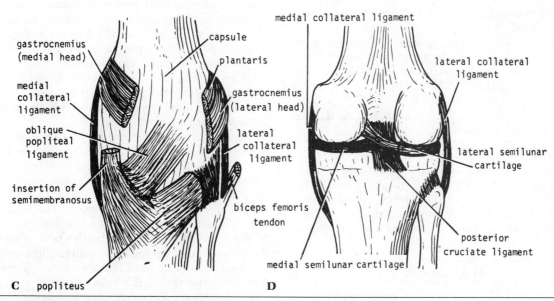

**Fig. 10-54. Right knee joint as seen from (A) lateral aspect, (B) anterior aspect, with joint flexed, and (C and D) posterior aspect.**

permits the tendon of the popliteus to emerge (Fig. 10-54).

## Extracapsular Ligaments

The *ligamentum patellae* is attached above to the lower border of the patella and below to the tubercle of the tibia (Fig. 10-54). It is, in fact, a continuation of the central portion of the common tendon of the quadriceps femoris muscle. It is separated from the synovial membrane of the joint by the *infrapatellar pad of fat* and from the tibia by a small bursa (Figs. 10-54 and 10-55). The *superficial infrapatellar bursa* separates the ligament from the skin (Fig. 10-54).

The *lateral collateral ligament* is cordlike and is attached above to the lateral condyle of the femur and below to the head of the fibula (Fig. 10-54). The tendon of the popliteus muscle intervenes between the ligament and the lateral semilunar cartilage (Fig. 10-55).

The *medial collateral ligament* is a broad, flat band and is attached above to the medial condyle of the femur and below to the medial surface of the shaft of the tibia (Fig. 10-54). *It is firmly attached to the edge of the medial semilunar cartilage* (Fig. 10-55).

The *oblique popliteal ligament* is a tendinous expansion derived from the semimembranosus muscle. It strengthens the posterior aspect of the capsule (Fig. 10-54).

## Intracapsular Ligaments

The *cruciate ligaments* are two very strong intracapsular ligaments, which cross each other within the joint cavity (Fig. 10-54). They are named anterior and posterior, according to their tibial attachments (Fig. 10-55). These important ligaments are the main bond between the femur and the tibia.

## Anterior Cruciate Ligament

This is attached to the anterior intercondylar area of the tibia and passes upward, backward, and laterally, to be attached to the posterior part of the medial surface of the lateral femoral condyle (Figs. 10-54 and 10-55). It is slack when the knee is flexed, but taut when the knee is fully extended. The anterior cruciate ligament prevents posterior displacement of the femur on the tibia. With the knee joint flexed, the anterior cruciate ligament prevents the tibia from being pulled anteriorly.

## Posterior Cruciate Ligament

This is attached to the posterior intercondylar area of the tibia and passes upward, forward, and medially, to be attached to the anterior part of the lateral surface of the medial femoral condyle (Figs. 10-54 and 10-55). The anterior fibers become slack when the knee is extended, but become taut in flexion. The posterior fibers are taut in extension. The posterior cruciate ligament prevents anterior displacement of the femur on the tibia. With the knee joint flexed, the posterior cruciate ligament prevents the tibia from being pulled posteriorly.

## Semilunar Cartilages (Menisci)

The semilunar cartilages are C-shaped lamellae of fibrocartilage, which are triangular in cross section. The peripheral border is thick and convex and attached to the capsule, and the inner border is thin and concave and forms a free edge (Figs. 10-54 and 10-55). The upper surfaces are concave and are in contact with the femoral condyles. The lower surfaces are flat and in contact with the tibial condyles. Their function is to deepen the articular surfaces of the tibial condyles to receive the convex femoral condyles.

The *medial semilunar cartilage* is nearly semicircular and is much broader behind than in front (Fig. 10-55). The anterior horn is attached to the anterior intercondylar area of the tibia and is connected to the lateral semilunar cartilage by a few fibers called the *transverse ligament*. The posterior horn is attached to the posterior intercondylar area of the tibia. The peripheral border is attached to the capsule and the medial collateral ligament of the joint, and because of this attachment the medial semilunar cartilage is relatively fixed.

The *lateral semilunar cartilage* is nearly circular and is uniformly wide throughout (Fig. 10-55). The anterior horn is attached to the anterior intercondylar area, immediately in front of the intercondylar eminence. The posterior horn is attached to

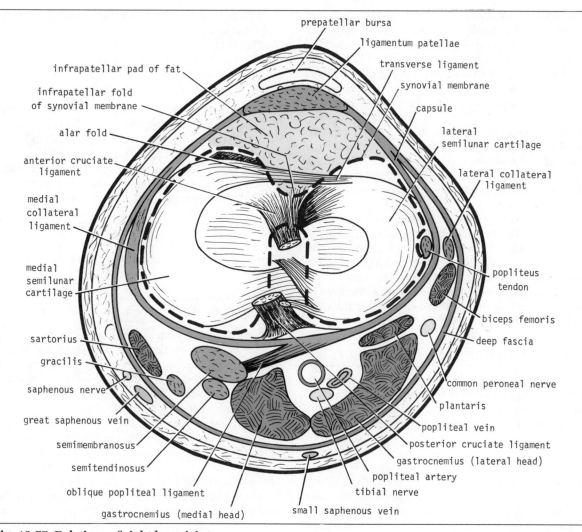

**Fig. 10-55. Relations of right knee joint.**

the posterior intercondylar area, immediately behind the eminence. A band of fibrous tissue commonly leaves the posterior horn and follows the posterior cruciate ligament to the medial condyle of the femur. The peripheral border of the cartilage is separated from the lateral collateral ligament by the tendon of the popliteus, a small part of the tendon being attached to the cartilage (Fig. 10-54). The result of this arrangement is that the lateral semilunar cartilage is less fixed in position than the medial semilunar cartilage.

## Synovial Membrane

The synovial membrane lines the capsule and is attached to the margins of the articular surfaces and to the peripheral edges of the semilunar cartilages (Figs. 10-54 and 10-55). On the front of the joint it forms a pouch, which extends up beneath the quadriceps femoris muscle for 3 fingerbreadths above the patella, forming the *suprapatellar bursa.* This is held in position by the attachment of a small portion of the vastus intermedius muscle, called the *articularis genus* muscle (Fig. 10-54).

At the back of the joint the synovial membrane

is prolonged downward on the deep surface of the tendon of the popliteus, forming the *popliteal bursa*. A bursa is interposed between the medial head of the gastrocnemius and the medial femoral condyle and the semimembranosus tendon; this is termed the *semimembranosus bursa*, and it frequently communicates with the synovial cavity of the joint.

The synovial membrane is reflected forward from the posterior part of the capsule around the front of the cruciate ligaments (Fig. 10-55). As a direct result of this, the cruciate ligaments lie behind the synovial cavity; i.e., they are intracapsular but extrasynovial.

The synovial membrane is reflected backward from the posterior surface of the ligamentum patellae (Fig. 10-55). This fold, the *infrapatellar fold*, converges to form a band, which is attached to the intercondylar fossa of the femur. The base of the synovial fold is filled by the *infrapatellar pad of fat*; the free borders of the fold are termed the *alar folds* (Fig. 10-55).

## BURSAE RELATED TO THE KNEE JOINT

There are numerous bursae related to the knee joint. Four are situated in front of the joint, and six are found behind the joint. They are found wherever skin, muscle, or tendon rubs against bone.

### Anterior Bursae

1. The *suprapatellar bursa* lies beneath the quadriceps muscle and communicates with the joint cavity (Fig. 10-54). It has been described above.
2. The *prepatellar bursa* lies in the subcutaneous tissue between the skin and the front of the lower half of the patellar and the upper part of the ligamentum patellae (Figs. 10-54 and 10-55).
3. The *superficial infrapatellar bursa* lies in the subcutaneous tissue between the skin and the front of the lower part of the ligamentum patellae (Fig. 10-54).
4. The *deep infrapatellar bursa* lies between the ligamentum patellae and the tibia (Fig. 10-54).

### Posterior Bursae

1. The *popliteal bursa* is found in association with the tendon of the popliteus and communicates with the joint cavity. It has been described on the preceding page.
2. The *semimembranosus bursa* is found related to the insertion of the semimembranosus muscle and frequently communicates with the joint cavity. It has been described previously.

The remaining four bursae are found related to (1) the tendon of insertion of the biceps femoris; (2) the tendons of the sartorius, gracilis, and semitendinosus muscles as they pass to their insertion on the tibia; (3) beneath the lateral head of origin of the gastrocnemius muscle; and (4) beneath the medial head of origin of the gastrocnemius muscle.

### Nerve Supply of the Knee Joint

Femoral, obturator, common peroneal, and tibial nerves.

### Movements of the Knee Joint

These are flexion, extension, and rotation.

*Flexion* is performed by the biceps femoris and the semitendinosus and the semimembranosus muscles, assisted by the gracilis, sartorius, and popliteus muscles. Flexion is limited by the contact of the back of the leg with the thigh.

*Extension* is performed by the quadriceps femoris and is limited to begin with by the anterior cruciate ligament becoming taut. Further extension of the knee is accompanied by medial *rotation* of the femur on the tibia, and the medial and lateral collateral ligaments and the oblique popliteal ligament become taut; the posterior fibers of the posterior cruciate ligament also are tightened. It is thus seen that as the knee joint assumes the position of full extension, or slight hyperextension, the medial rotation of the femur results in a twisting and tightening of all the major ligaments of the joint, and the knee becomes a mechanically rigid structure. The rotation of the femur in actual fact *"screws home"* the femur on the tibia, and the semilunar cartilages are compressed like rubber cushions between the femoral and tibial condyles. The hyperextended knee is said to be in the locked position.

During the initial stages of extension, the rounded femoral condyles roll forward, like a wheel on the ground, on the upper surfaces of the semilunar cartilages and the tibial condyles. Later, when the forward movement of the femur is held back by the posterior cruciate ligament, the rolling movement of the femoral condyles is converted into a spinning motion. As extension progresses, the flatter part of the curve of the femoral condyles moves inferiorly, and the semilunar cartilages have to adapt their shape to the changing contour of the femoral condyles. During the final stages of extension, when the femur undergoes medial rotation, the lateral femoral condyle moves forward, forcing the lateral semilunar cartilage to move forward with it.

Before flexion of the knee joint can occur, it is essential that the major ligaments be untwisted and slackened to permit movements between the joint surfaces to take place. This unlocking or untwisting process is accomplished by the popliteus muscle, which laterally rotates the femur on the tibia. As the lateral femoral condyle moves backward, the attachment of the popliteus to the lateral semilunar cartilage results in that structure being pulled backward also. Once again the semilunar cartilages have to adapt their shape to the changing contour of the femoral condyles.

When the knee joint is flexed to a right angle, a considerable range of rotation is possible. Medial rotation is performed by the sartorius, gracilis, and semitendinosus muscles, and lateral rotation is performed by the biceps femoris.

In the flexed position, the tibia can also be moved passively forward and backward on the femur to a limited extent; this is possible because the major ligaments, especially the cruciate ligaments, are slack in this position.

It is thus seen that the stability of the knee joint depends on the tone of the strong muscles acting on the joint and the strength of the ligaments. Of these factors, the tone of the muscles is the most important, and it is the job of the physiotherapist to build up the strength of these muscles, especially the quadriceps femoris, following injury to the knee joint.

## *Important Relations of the Knee Joint*

### Anteriorly

The prepatellar bursa and tendinous expansions from the vastus medialis and lateralis (Fig. 10-55).

### Posteriorly

The popliteal vessels; tibial and common peroneal nerves; lymph nodes and the muscles that form the boundaries of the popliteal fossa, namely, the semimembranosus, the semitendinosus, the biceps femoris, the two heads of the gastrocnemius, and the plantaris (Fig. 10-55).

### Medially

Sartorius, gracilis, and semitendinosus muscles (Fig. 10-55).

### Laterally

Biceps femoris (Fig. 10-55).

## Superior Tibiofibular Joint

### Articulation

Between the lateral condyle of the tibia and the head of the fibula (Fig. 10-54). The articular surfaces are flattened and covered by hyaline cartilage.

### Type

Synovial, plane, gliding.

### Capsule

Surrounds the joint and is attached to the margins of the articular surfaces.

### Ligaments

*Anterior and posterior ligaments* that strengthen the capsule. The *interosseous membrane*, which connects the shafts of the tibia and fibula together, also greatly strengthen the joint.

### Synovial Membrane

This lines the capsule and is attached to the margins of the articular surfaces.

### Nerve Supply

Common peroneal nerve.

## Movements

A small amount of gliding movement takes place during movements at the ankle joint.

# Inferior Tibiofibular Joint

## Articulation

Between the fibular notch at the lower end of the tibia and the lower end of the fibula (Figs. 10-56 and 10-57). The opposed bony surfaces are roughened.

## Type

Fibrous.

## Capsule

None.

## Ligaments

*The interosseous ligament* is a strong, thick band of fibrous tissue that binds the two bones together. The *interosseous membrane*, which connects the shafts of the tibia and fibula together, also greatly strengthen the joint.

The *anterior and posterior ligaments* are flat bands of fibrous tissue connecting the two bones together in front and behind the interosseous ligament.

The *inferior transverse ligament* runs from the medial surface of the upper part of the lateral malleolus to the posterior border of the lower end of the tibia.

## Nerve Supply

Deep peroneal and tibial nerves.

## Movements

A small amount of movement takes place during movements at the ankle joint.

# Ankle Joint

The ankle joint consists of a deep socket formed by the lower ends of the tibia and fibula, into which is fitted the upper part of the body of the talus. The talus is able to move on a transverse axis in a hinge-like manner. The shape of the bones and the strength of the ligaments and the surrounding tendons make this joint strong and stable.

## Articulation

Between the lower end of the tibia, the two malleoli, and the body of the talus (Figs. 10-56 and 10-57). The inferior transverse tibiofibular ligament deepens the socket into which the body of the talus fits snugly. The articular surfaces are covered with hyaline cartilage.

## Type

Synovial, hinge.

## Capsule

This encloses the joint and is attached to the bones near their articular margins.

## Ligaments

The *medial*, or *deltoid*, *ligament* is very strong and is attached by its apex to the tip of the medial malleolus (Fig. 10-56). Below, the deep fibers are attached to the nonarticular area on the medial surface of the body of the talus; the superficial fibers are attached to the medial side of the talus, the sustentaculum tali, the plantar calcaneonavicular ligament, and the tuberosity of the navicular bone.

The *lateral ligament* is weaker than the medial ligament and consists of three distinct bands.

1. The *anterior talofibular ligament* (Fig. 10-56) runs from the lateral malleolus to the lateral surface of the talus.
2. The *calcaneofibular ligament* (Fig. 10-56) runs from the tip of the lateral malleolus downward and backward to the lateral surface of the calcaneum.

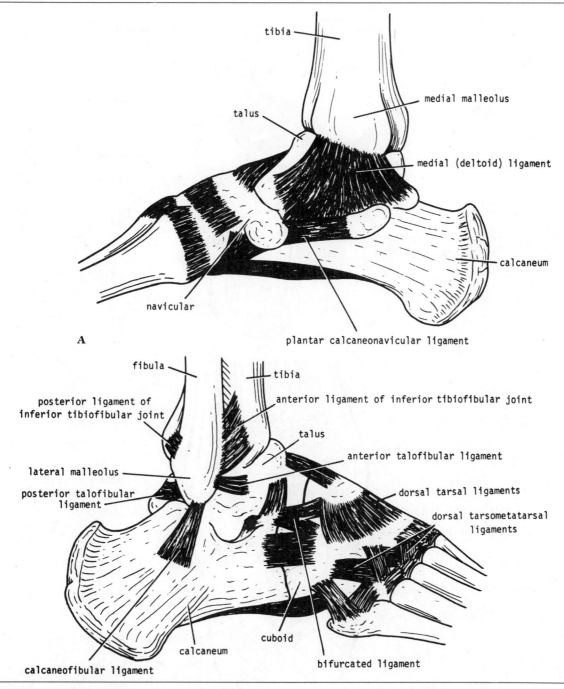

**Fig. 10-56. Right ankle joint as seen from (A) medial aspect and (B) lateral aspect.**

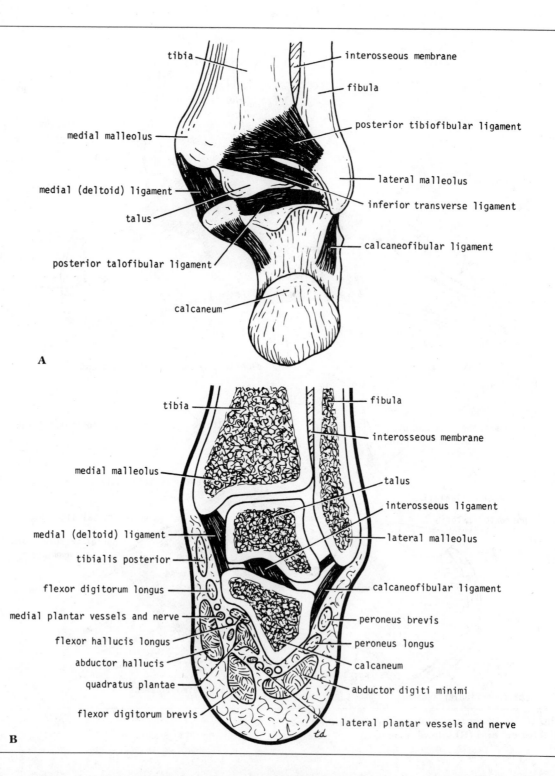

tibia

interosseous membrane

fibula

posterior tibiofibular ligament

medial malleolus

lateral malleolus

medial (deltoid) ligament

inferior transverse ligament

talus

calcaneofibular ligament

posterior talofibular ligament

calcaneum

A

tibia

fibula

interosseous membrane

medial malleolus

talus

interosseous ligament

medial (deltoid) ligament

lateral malleolus

tibialis posterior

flexor digitorum longus

calcaneofibular ligament

medial plantar vessels and nerve

peroneus brevis

flexor hallucis longus

peroneus longus

abductor hallucis

calcaneum

quadratus plantae

abductor digiti minimi

flexor digitorum brevis

lateral plantar vessels and nerve

td

B

3. The *posterior talofibular ligament* (Fig. 10-56) runs from the lateral malleolus to the posterior tubercle of the talus.

## Synovial Membrane

This lines the capsule and may pass upward for a short distance in front of the interosseous ligament of the inferior tibiofibular joint.

## Nerve Supply

Deep peroneal and tibial nerves.

## Movements

Dorsiflexion (toes pointing upward) and plantar flexion (toes pointing downward). The movements of inversion and eversion take place at the tarsal joints and *not at the ankle joint.*

*Dorsiflexion* is performed by the tibialis anterior, extensor hallucis longus, extensor digitorum longus, and peroneus tertius. It is limited by the tension of the tendo calcaneus, the posterior fibers of the medial ligament, and the calcaneofibular ligament.

*Plantar flexion* is performed by the gastrocnemius, soleus, plantaris, peroneus longus, peroneus brevis, tibialis posterior, flexor digitorum longus, and flexor hallucis longus. It is limited by the tension of the opposing muscles, the anterior fibers of the medial ligament, and the anterior talofibular ligament.

Note that during dorsiflexion of the ankle joint, the wider anterior part of the articular surface of the talus is forced between the medial and lateral malleoli, causing them to separate slightly and tighten the ligaments of the inferior tibiofibular joint. This arrangement greatly increases the stability of the ankle joint when the foot is in the initial position for major thrusting movements in walking, running, and jumping.

Note also that when the ankle joint is fully plantar flexed, the ligaments of the inferior tibiofibular joint are less taut and small amounts of rotation, abduction, and adduction are possible.

**Fig. 10-57. Right ankle joint as seen (A) from posterior aspect and (B) in coronal section.**

## Important Relations

### Anteriorly

The tibialis anterior, the extensor hallucis longus, the anterior tibial vessels, the deep peroneal nerve, the extensor digitorum longus, the peroneus tertius (Fig. 10-46).

### Posteriorly

The tendo calcaneus and plantaris (Fig. 10-46).

### Posterolaterally (behind the lateral malleolus)

The peroneus longus and brevis (Fig. 10-46).

### Posteromedially (behind the medial malleolus)

The tibialis posterior, the flexor digitorum longus, the posterior tibial vessels, the tibial nerve, and the flexor hallucis longus (Fig. 10-46).

## Tarsal Joints

### SUBTALAR JOINT

The subtalar joint is the posterior joint between the talus and the calcaneum.

### Articulation

Between the inferior surface of the body of the talus and the facet on the middle of the upper surface of the calcaneum (Fig. 10-34). The articular surfaces are covered with hyaline cartilage.

### Type

Synovial, of the plane variety.

### Capsule

Encloses the joint and is attached to the margins of the articular areas of the two bones.

## Ligaments

*Medial* and *lateral (talocalcaneal) ligaments* strengthen the capsule. The *interosseous (talocalcaneal) ligament* (Fig. 10-57) is very strong and is the main bond of union between the two bones. It is attached above to the sulcus tali and below to the sulcus calcanei.

## Synovial Membrane

This lines the capsule.

## Movements

The movements are described in the section on the calcaneocuboid joint.

## TALOCALCANEONAVICULAR JOINT

The talocalcaneonavicular joint is the anterior joint between the talus and the calcaneum and also involves the navicular bone (Fig. 10-34).

## Articulation

Between the rounded head of the talus, the upper surface of the sustentaculum tali, and the posterior concave surface of the navicular bone. The articular surfaces are covered with hyaline cartilage.

## Type

Synovial.

## Capsule

Incompletely encloses the joint.

## Ligaments

The *plantar calcaneonavicular ligament* is very strong and runs from the anterior margin of the sustentaculum tali to the inferior surface and tuberosity of the navicular bone. The superior surface of the ligament is covered with fibrocartilage and supports the head of the talus.

## Synovial Membrane

This lines the capsule.

## Movements

Gliding and rotatory movements are possible. The movements are described in the discussion of the calcaneocuboid joint that follows.

## CALCANEOCUBOID JOINT

### Articulation

Between the anterior end of the calcaneum and the posterior surface of the cuboid (Fig. 10-34). The articular surfaces are covered with hyaline cartilage.

### Type

Synovial, plane.

### Capsule

Encloses the joint.

### Ligaments

The *bifurcated ligament* is a strong ligament on the upper surface of the joint (Fig. 10-56). It is Y-shaped, and the stem is attached to the upper surface of the anterior part of the calcaneum. The lateral limb is attached to the upper surface of the cuboid, and the medial limb to the upper surface of the navicular bone.

The *long plantar ligament* is a strong ligament on the lower surface of the joint (Figs. 10-51 and 10-52). It is attached to the undersurface of the calcaneum behind, and to the undersurface of the cuboid and the bases of the third, fourth, and fifth metatarsal bones in front. It bridges over the groove for the peroneus longus tendon, converting it into a tunnel.

*The short plantar ligament* is a wide, strong ligament that is attached to the anterior tubercle on the undersurface of the calcaneum and to the adjoining part of the cuboid bone (Fig. 10-52).

## Synovial Membrane

This lines the capsule.

## Movements in the Subtalar, Talocalcaneonavicular, and Calcaneocuboid Joints

The talocalcaneonavicular and the calcaneocuboid joints are together referred to as the *midtarsal or transverse tarsal joints*

The important movements of inversion and eversion of the foot take place at the subtalar and transverse tarsal joints. *Inversion* is the movement of the foot so that the sole faces medially. *Eversion* is the opposite movement of the foot so that the sole faces in the lateral direction. The movement of inversion is more extensive than eversion.

*Inversion* is performed by the tibialis anterior, the extensor hallucis longus, and the medial tendons of extensor digitorum longus; the tibialis posterior also assists.

*Eversion* is performed by the peroneus longus, peroneus brevis, and peroneus tertius; the lateral tendons of the extensor digitorum longus also assist.

## CUNEONAVICULAR JOINT

The cuneonavicular joint is the *articulation* between the navicular bone and the three cuneiform bones. It is a synovial joint of the gliding variety. The *capsule* is strengthened by dorsal and plantar ligaments. The *joint cavity* is continuous with those of the intercuneiform and cuneocuboid joints, and also with the cuneometatarsal and intermetatarsal joints, between the bases of the second and third, and third and fourth, metatarsal bones.

## CUBOIDEONAVICULAR JOINT

The cuboideonavicular joint is usually a fibrous joint, with the two bones connected by dorsal, plantar, and interosseous ligaments.

## INTERCUNEIFORM AND CUNEOCUBOID JOINTS

The intercuneiform and cuneocuboid joints are synovial joints of the plane variety. Their joint cavities are continuous with that of the cuneonavicular joint. The bones are connected by dorsal, plantar, and interosseous ligaments.

## Tarsometatarsal and Intermetatarsal Joints

The tarsometatarsal and intermetatarsal joints are synovial joints of the plane variety. The bones are connected by dorsal, plantar, and interosseous ligaments. The tarsometatarsal joint of the big toe has a separate joint cavity.

## Metatarsophalangeal and Interphalangeal Joints

The metatarsophalangeal and interphalangeal joints closely resemble those of the hand. (See pp. 513 and 514.) The deep transverse ligaments connect the joints of the five toes.

The movements of abduction and adduction of the toes, performed by the interossei muscles, are minimal, and take place from the midline of the second digit and not the third, as in the hand.

## THE FOOT AS A FUNCTIONAL UNIT

### *The Foot as a Weight-Bearer and a Lever*

The foot has two important functions: (1) to support the body weight and (2) to serve as a lever to propel the body forward in walking and running. If the foot possessed a single strong bone, instead of a series of small bones, it could sustain the body weight and serve well as a rigid lever for forward propulsion (Fig. 10-58). However, with such an arrangement, the foot could not adapt itself to uneven surfaces, and the forward propulsive action would depend entirely on the activities of the gastrocnemius and soleus muscles. Because the lever is segmented with multiple joints, the foot is pliable and can adapt itself to uneven surfaces. Moreover, the long flexor muscles and the small muscles of the foot can exert their action on the bones of the forepart of the foot and toes, i.e., the takeoff point of the foot, and greatly assist the forward propulsive action of the gastrocnemius and soleus muscles (Fig. 10-58).

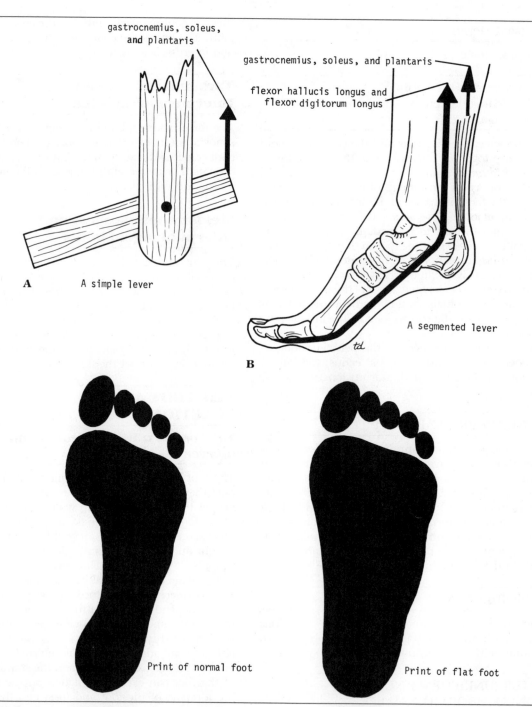

**Fig. 10-58. The foot as (A) a simple lever and (B) a segmented lever. Floor prints of normal foot and flat foot are also shown.**

# THE ARCHES OF THE FOOT

A segmented structure can hold up weight only if it is built in the form of an arch. In the foot there are three such arches, which are present at birth: the *medial longitudinal*, the *lateral longitudinal*, and the *transverse arches* (Fig. 10-59). In the young child, the foot appears to be flat because of the presence of a large amount of subcutaneous fat on the sole of the foot at this age.

If one examines the imprint of a wet foot on the floor made with the person in the standing position, it will be seen that the heel, the lateral margin of the foot, the pad under the metatarsal heads, and the pads of the distal phalanges are in contact with the ground (Fig. 10-58). The medial margin of the foot, from the heel to the first metatarsal head, is arched above the ground, due to the important medial longitudinal arch. The pressure exerted on the ground by the lateral margin of the foot is greatest at the heel and the fifth metatarsal head and least between these areas, due to the presence of the low-lying lateral longitudinal arch. The transverse arch involves the bases of the five metatarsals, and the cuboid and the cuneiform bones. This is, in fact, only half an arch, with its base on the lateral border of the foot and its summit on the foot's medial border. The foot has been likened to a half-dome, so that when the medial borders of the two feet are placed together, a complete dome is formed.

From this description, it can be understood that the body weight on standing is distributed through a foot via the the heel behind and six points of contact with the ground in front, namely, the two sesamoid bones under the head of the first metatarsal and the heads of the remaining four metatarsals.

## The Bones of the Arches

An examination of an articulated foot, or a lateral X-ray of the foot, will show the bones that form the arches.

## Medial Longitudinal Arch

This consists of the calcaneum, the talus, the navicular bone, the three cuneiform bones, and the first three metatarsal bones (Fig. 10-59).

## Lateral Longitudinal Arch

This consists of the calcaneum, the cuboid, and the fourth and fifth metatarsal bones (Fig. 10-59).

## Transverse Arch

This consists of the bases of the metatarsal bones and the cuboid and the three cuneiform bones (Fig. 10-59).

## Mechanisms of Arch Support

The examination of the design of any stone bridge will reveal the following engineering methods used for its support (Fig. 10-60).

1. *The shape of the stones.* The most effective way of supporting the arch is to make the stones wedge-shaped, with the thin edge of the wedge lying inferiorly. This applies particularly to the important stone that occupies the center of the arch and is referred to as the keystone.
2. *The inferior edges of the stones are tied together.* This is accomplished by interlocking the stones or binding their lower edges together with metal staples. This method effectively counteracts the tendency of the lower edges of the stones to separate when the arch is weight-bearing.
3. *The use of the tie beams.* If the span of the bridge is large and the foundations at either end are insecure, a tie beam connecting the ends will effectively prevent separation of the pillars and consequent sagging of the arch.
4. *A suspension bridge.* Here, the maintenance of the arch depends on multiple supports suspending the arch from a cable above the level of the bridge.

Using the bridge analogy, one can now examine the methods used to support the arches of the feet (Fig. 10-60).

## Maintenance of the Medial Longitudinal Arch

1. *Shape of the bones.* The sustentaculum tali holds up the talus; the concave proximal surface of the navicular bone receives the rounded head of the talus; the slight concavity of the proximal sur-

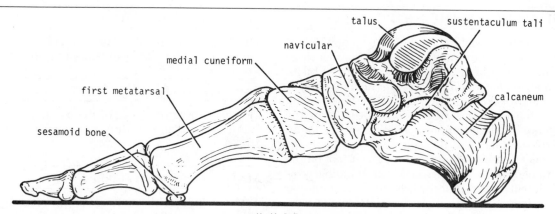

Medial longitudinal arch

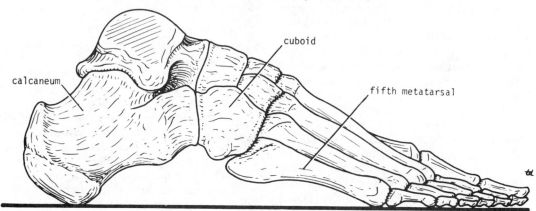

Lateral longitudinal arch

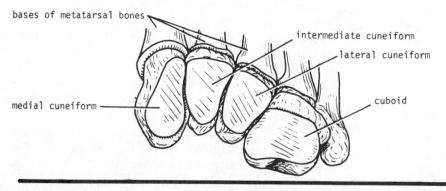

Transverse arch

**Fig. 10-59. Bones forming medial longitudinal, lateral longitudinal, and transverse arches of right foot.**

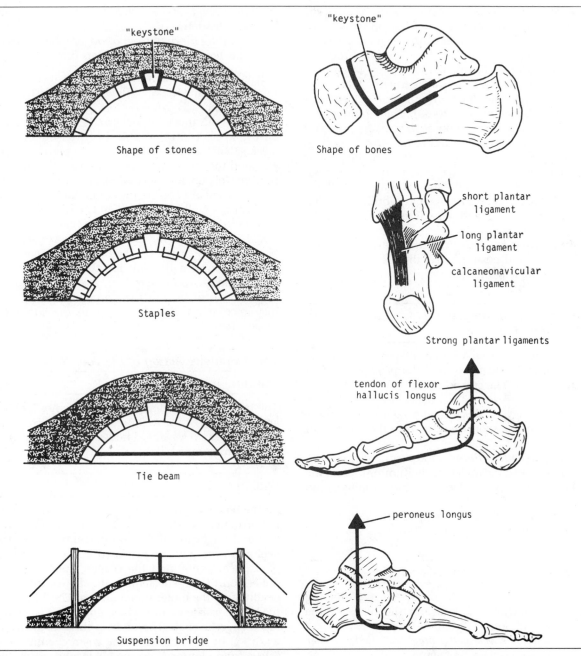

"keystone"

Shape of stones

"keystone"

Shape of bones

Staples

short plantar
ligament

long plantar
ligament

calcaneonavicular
ligament

Strong plantar ligaments

Tie beam

tendon of flexor
hallucis longus

peroneus longus

Suspension bridge

**Fig. 10-60. Different methods by which arches of foot may be supported.**

face of the medial cuneiform bone receives the navicular. The rounded head of the talus is the "keystone" in the center of the arch (Fig. 10-60).

2. *The inferior edges of the bones are tied together* by the plantar ligaments, which are larger and stronger than the dorsal ligaments. The most important ligament is the plantar calcaneonavicular ligament (Fig. 10-60). The tendinous extensions of the insertion of the tibialis posterior muscle play an important role in this respect.

3. *Tying the ends of the arch together* are the plantar aponeurosis, the medial part of the flexor digitorum brevis, the abductor hallucis, the flexor hallucis longus, the medial part of the flexor digitorum longus, and the flexor hallucis brevis (Fig. 10-60).

4. *Suspending the arch from above* are the tibialis anterior and posterior and the medial ligament of the ankle joint.

## Maintenance of the Lateral Longitudinal Arch

1. *Shape of the bones.* Minimal shaping of the distal end of the calcaneum and the proximal end of the cuboid. The cuboid is the "keystone."

2. *The inferior edges of the bones are tied together* by the long and short plantar ligaments and the origins of the short muscles from the forepart of the foot (Fig. 10-60).

3. *Tying the ends of the arch together* are the plantar aponeurosis, the abductor digiti minimi, and the lateral part of the flexor digitorum longus and brevis.

4. *Suspending the arch from above* are the peroneus longus and the brevis (Fig. 10-60).

## Maintenance of the Transverse Arch

1. *Shape of the bones.* The marked wedge-shaping of the cuneiform bones and the bases of the metatarsal bones (Fig. 10-59).

2. *The inferior edges of the bones are tied together* by the deep transverse ligaments, the very strong plantar ligaments, and the origins of the plantar muscles from the forepart of the foot; the dorsal interossei and the transverse head of the adductor hallucis are particularly important in this respect.

3. *Tying the ends of the arch together* is the peroneus longus tendon.

4. *Suspending the arch from above* are the peroneus longus tendon and the peroneus brevis.

It is clear that the arches of the feet are maintained by (1) the shape of the bones, (2) strong ligaments, and (3) muscle tone. Which of these factors is the most important? It has been demonstrated electromyographically by Basmajian and Stecko that the tibialis anterior, the peroneus longus, and the small muscles of the foot play no important role in the normal static support of the arches. They are commonly totally inactive. However, during walking and running all these muscles become very active. Standing immobile for long periods, especially if the person is overweight, will place excessive strain on the bones and ligaments of the feet and will result in fallen arches or flat feet. Athletes, route-marching soldiers, or nurses will be able to sustain their arches provided that they receive adequate training to develop their muscle tone.

## The Propulsive Action of the Foot

### Standing Immobile

The body weight is distributed via the heel behind and the heads of the metatarsal bones in front (including the two sesamoid bones under the head of the first metatarsal).

### Walking

As the body weight is thrown forward, the weight is born successively on the lateral margin of the foot and the heads of the metatarsal bones. As the heel rises, the toes are extended at the metatarsophalangeal joints, and the plantar aponeurosis is pulled on, thus shortening the tie beams and heightening the longitudinal arches. The "slack" in the long flexor tendons is taken up, thereby increasing their efficiency. The body is then thrown forward (1) by the actions of the gastrocnemius and soleus (and plantaris) on the ankle joint, using the foot as a lever and (2) the toes being strongly flexed by the long and short flexors of the foot, providing the final thrust forward. The lumbricals and interossei contract and keep the toes extended, so that they do not fold under because of the strong action of the flexor digitorum longus. In this action the long flexor tendons also assist in plantar flexing the ankle joint.

## Running

When a person runs, the weight is borne on the forepart of the foot, and the heel does not touch the ground. The forward thrust to the body is provided by mechanisms (1) and (2) described above.

# RADIOGRAPHIC APPEARANCES OF THE LOWER LIMB

Radiological examination of the lower limb concentrates mainly on the bony structures, since most of the muscles, tendons, and nerves blend into a homogeneous mass. Blood vessels may be visualized by using special contrast media. The radiographic appearances of the lower limb of the adult as seen on routine X-ray examination will be described in this section. As in the upper limb, the practicing radiologist must be cognizant of the age changes that take place in the body and how these will influence the radiographic appearances. For example, knowing the times at which the primary and secondary centers of ossification appear in the different bones, and the dates at which they fuse, is essential, since without this information an epiphyseal line could be mistaken for a fracture. It is useful to remember that a person has two lower limbs, and that the normal side may serve as a baseline for comparison with the potentially abnormal side.

## *Radiographic Appearances of the Hip Region*

The views commonly used are (1) the anteroposterior and (2) the lateral view.

*The anteroposterior view* is taken with the patient in the supine position. The film cassette is placed behind the hip, and the X-ray tube is positioned in front of the hip, centered over a point 1 inch (2.5 cm) below the midpoint of the inguinal ligament. The subject is asked to medially rotate his hip joint slightly so that his toes touch; this is important, so that the full length of the neck of the femur is visualized and it is not foreshortened. It may be desirable to view the whole pelvis so that

the two hips may be compared. In this case, the entire pelvis must be symmetrical, and the X-ray tube must be centered over a point about 1 inch (2.5 cm) above the symphysis pubis.

First, examine the relevant features seen in the pelvis (Fig. 10-61). The sacrum and sacroiliac joints should be recognized. The iliopectineal line and the symphysis pubis are well shown. The boundaries of the obturator foramen and the ischial tuberosity can be identified. The superior shelving margin of the acetabulum can be seen. The articulating surfaces of the hip joint are seen to be parallel and separated by a narrow space occupied by radiotranslucent articular cartilage. The head, the neck, the greater and lesser trochanters, and the intertrochanteric crest of the femur can all be visualized. The axial relationships of the hip joint should be studied. The inferior margin of the neck of the femur should form a smooth continuous curve with the superior margin of the obturator foramen (*Shenton's line*). The angle formed by the long axis of the neck of the femur with the long axis of the shaft of the femur measures between 120 and 130 degrees.

*The lateral view* is taken with the patient in the supine position, and the X-ray tube is directed either from the medial or lateral aspect of the thigh; a horizontal X-ray beam is employed. The film cassette is placed perpendicular to the tabletop.

First, identify as many of the relevant parts of the pelvis as possible. The obturator foramen, the ischial spine and tuberosity, the pubic ramus, and the body of the pubis may all be recognized. The acetabular rims and the head and the whole neck of the femur are demonstrated. The greater and lesser trochanters and the proximal part of the shaft are visualized.

## *Radiographic Appearances of the Knee Region*

The views commonly used are (1) anteroposterior and (2) lateral.

*The anteroposterior view* is taken with the patient supine and the film cassette placed behind the knee. The X-ray tube is placed in front of the knee and centered over a point about ½ inch (1.3 cm) below the apex of the patella.

The lower part of the shaft of the femur, the lat-

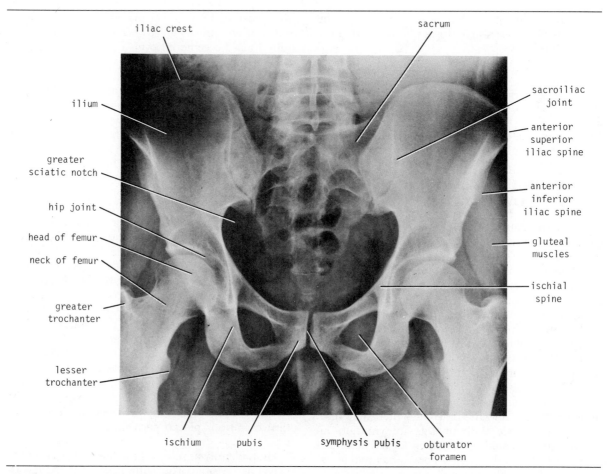

iliac crest

sacrum

sacroiliac joint

ilium

anterior superior iliac spine

greater sciatic notch

anterior inferior iliac spine

hip joint

head of femur

gluteal muscles

neck of femur

greater trochanter

ischial spine

lesser trochanter

ischium     pubis     symphysis pubis     obturator foramen

**Fig. 10-61. Anteroposterior radiograph of adult pelvis.**

eral and medial epincondyles, and the adductor tubercle are easily visualized (Fig. 10-62). The patella is seen superimposed in front of the lateral and medial femoral condyles. The *fabella*, a sesamoid bone in the lateral head of the gastrocnemius, is sometimes seen superimposed on the lateral femoral condyle. The parallel joint surfaces, separated by a wide space occupied by the articular cartilage and the semilunar cartilages, which cast no shadow, are easily recognized. The intercondylar notch of the femur and the intercondylar eminence of the tibia are well shown. The medial and lateral condyles of the tibia are seen. The head of the fibula partly overlaps the lateral condyle of the tibia. The neck of the fibula and the upper parts of the

shafts of the fibula and tibia are usually clearly seen.

*The lateral view* is taken with the knee joint partially flexed. The film cassette is placed against the lateral aspect of the joint, and the X-ray tube is centered on the medial side of the joint line. The patient reclines on his side on the table.

The lower part of the shaft of the femur is seen, and the lateral and medial femoral condyles are partly superimposed on each other (Fig. 10-63). The patella is clearly visualized in front of the femoral condyles. The intercondylar eminence of the tibia projects upward into the intercondylar notch of the femur, and its summit is overlapped by the femoral condyles. The lateral and medial tibial condyles are superimposed, and the tibial tuberosity is seen on the anterior surface of the bone.

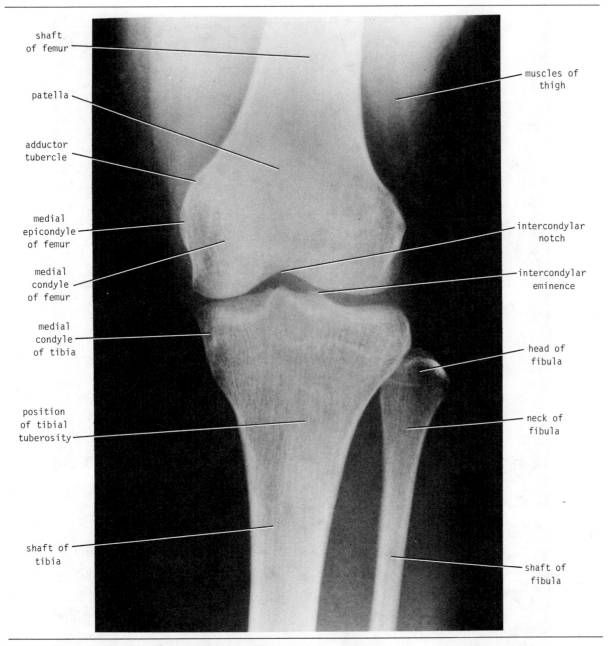

shaft
of femur

patella

adductor
tubercle

medial
epicondyle
of femur

medial
condyle
of femur

medial
condyle
of tibia

position
of tibial
tuberosity

shaft of
tibia

muscles of
thigh

intercondylar
notch

intercondylar
eminence

head of
fibula

neck of
fibula

shaft of
fibula

**Fig. 10-62. Anteroposterior radiograph of adult
knee.**

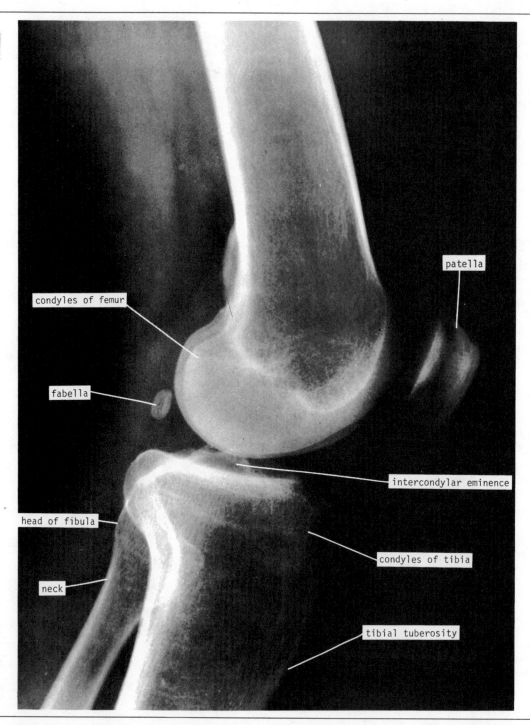

**Fig. 10-63. Lateral radiograph of adult knee.**

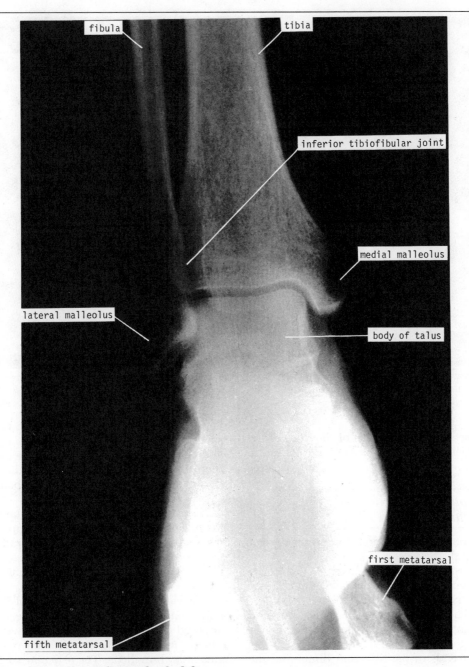

**Fig. 10-64. Anteroposterior radiograph of adult ankle.**

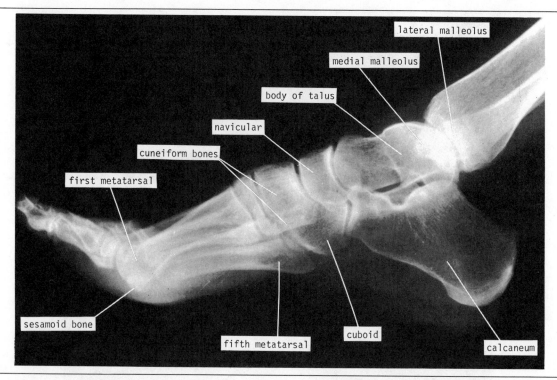

**Fig. 10-65. Lateral radiograph of adult ankle.**

The head, neck, and upper part of the shaft of the fibula are seen, the fibula overlapping the tibia to some extent.

## Radiographic Appearances of the Ankle Region

The views commonly used are: (1) anteroposterior and (2) lateral.

*The anteroposterior view* is taken with the patient in the supine position, the ankle joint is dorsiflexed to a right angle, and the big toe is pointed slightly medially. The film cassette is placed behind the ankle joint, and the X-ray tube is centered over the front of the joint.

The lower ends of the tibia and fibula and the inferior tibiofibular joint are well shown (Fig. 10-64). The medial and lateral malleoli and the articular surfaces of the tibia and the body of the talus are easily seen. The lateral malleolus usually partly overlaps the lateral aspect of the talus. The articu-

lar surfaces of the lower end of the tibia and the superior surface of the talus are seen to be parallel and separated by a narrow space occupied by the articular cartilage, which is radiotranslucent. Other than the talus, the tarsal bones are not clearly visualized.

*The lateral view* is taken with the lateral malleolus against the film cassette. It is important that the sagittal plane of the leg be parallel with the plane of the film. The X-ray tube is centered over a point about ¾ inch (1.9 cm) proximal to the tip of the lateral malleolus.

This view shows the lower ends of the tibia and fibula; the lateral and medial malleoli are superimposed (Fig. 10-65). It should, however, be possible to make out the anterior and posterior margins of both malleoli. The articular surfaces of the ankle joint are clearly visualized. The talus and calcaneum are seen in profile, and the subtalar and transverse tarsal joints can be identified. The cuneiform bones and the cuboid are overlapped and not clearly seen.

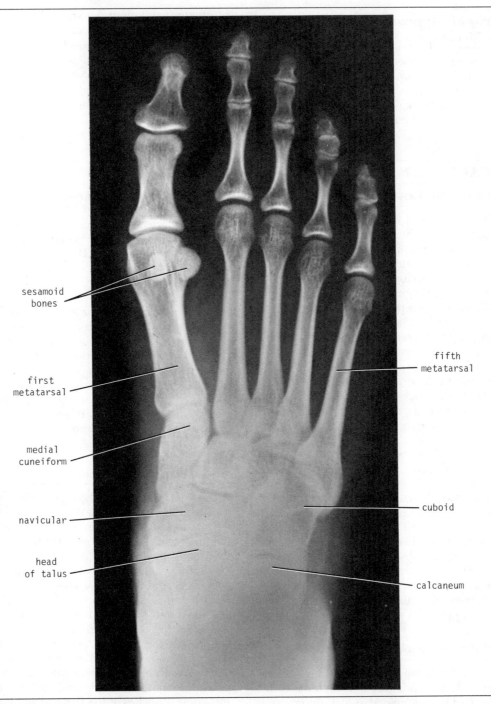

sesamoid
bones

fifth
metatarsal

first
metatarsal

medial
cuneiform

navicular

cuboid

head
of talus

calcaneum

**Fig. 10-66. Anteroposterior radiograph of adult
foot.**

### Radiographic Appearances of the Tarsus, Metatarsus, and Phalanges

The views commonly used are: (1) anteroposterior, (2) lateral, and (3) oblique.

The particular view used will depend on which bone is needed to be visualized to best advantage. The oblique view of the metatarsal bones is often of greater value than the lateral view, since, in the latter, the bones are superimposed. In the anteroposterior view, the film cassette is placed in contact with the sole. The tarsal bones, the metatarsals, and the phalanges are seen (Fig. 10-66). The two sesamoid bones of the big toe overlap the head of the first metatarsal bone.

## CLINICAL NOTES

### Arteries of the Lower Limb

Every physician should know the precise position of the main arteries within the lower limb, for he may be called upon to arrest a severe hemorrhage or palpate different parts of the arterial tree in patients with arterial occlusion.

The *femoral artery* enters the thigh behind the inguinal ligament, at a point midway between the anterior superior iliac spine and the symphysis pubis. The artery is easily palpated here, since it may be pressed backward against the pectineus and the superior ramus of the pubis.

The *dorsalis pedis artery* lies between the tendons of extensor hallucis longus and extensor digitorum longus on the dorsal surface of the foot. It is covered only by skin and fascia and is easily palpated.

The *posterior tibial artery* passes behind the medial malleolus, beneath the flexor retinaculum, and lies between the tendons of flexor digitorum longus and flexor hallucis longus. Gentle palpation, without too great pressure, enables one to detect the artery without much difficulty.

It should be remembered that the dorsalis pedis artery is sometimes absent and is replaced by a large perforating branch of the peroneal artery. In the same manner, the peroneal artery may be larger than normal and replace the posterior tibial artery in the lower part of the leg.

### Femoral Artery Catheterization

A long, fine catheter may be inserted into the femoral artery as it descends through the femoral triangle. The catheter is guided under fluoroscopic view along the external and common iliac arteries into the aorta. The catheter can then be passed into the inferior mesenteric, superior mesenteric, celiac arteries, or renal arteries. Contrast medium may then be injected into the artery under examination and a permanent record obtained by taking a radiograph. Pressure records can also be obtained by guiding the catheter through the aortic valve into the left ventricle.

The *popliteal artery* is deeply placed in the popliteal fossa, and when it is examined with the knee extended, the deep fascia roofing over the fossa is made taut, and the pulse is difficult to detect. With the knee flexed and the patient kneeling or in the prone position, the fascia and neighboring muscles are relaxed, and the pulse is easily felt. Atherosclerosis is common in the femoral artery at the hiatus in the adductor magnus. In severe cases there will be absence of a popliteal pulse.

### Collateral Circulation

If the arterial supply to the leg is occluded, necrosis or gangrene will follow unless there is an adequate bypass to the obstruction, that is, a collateral circulation. Sudden occlusion of the femoral artery by ligature or embolism, for example, is usually followed by gangrene. However, gradual occlusion such as occurs in atherosclerosis is less likely to be followed by necrosis, since the collateral blood vessels have time to dilate fully. The collateral circulation for the proximal part of the femoral artery is through the cruciate and trochanteric anastomoses; for the femoral artery in the adductor canal, it is through the perforating branches of the profunda femoris artery and the articular and muscular branches of the femoral and popliteal arteries.

*Arterial occlusive disease of the leg* is common in

men. Ischemia of the muscles produces a cramp-like pain with exercise. If the femoral artery is obstructed, there is an inadequate supply of blood to the calf muscles; the patient is forced to stop walking after a limited distance because of the intensity of the pain. With rest, the oxygen depletion is corrected and the pain disappears. However, on the resumption of walking, the pain recurs. This condition is known as *intermittent claudication.*

## Sympathetic Innervation of the Arteries

Sympathetic innervation of the arteries to the leg is derived from the lower three thoracic and upper two or three lumbar segments of the spinal cord. The preganglionic fibers pass to the lower thoracic and upper lumbar ganglia via white rami. The fibers synapse in the lumbar and sacral ganglia, and the postganglionic fibers reach the blood vessels via branches of the lumbar and sacral plexuses. The femoral artery receives its sympathetic fibers from the femoral and obturator nerves. The more distal arteries receive their postganglionic fibers via the common peroneal and tibial nerves.

*Lumbar sympathectomy* may be advocated as a form of treatment in occlusive arterial disease of the lower limb in order to increase the blood flow through the collateral circulation. Preganglionic sympathectomy is performed by removing the upper three lumbar ganglia and the intervening parts of the sympathetic trunk.

## Veins of the Lower Limb

The veins of the lower limb may be divided into three groups: (1) superficial, (2) deep, and (3) perforating. The *superficial veins* consist of the great and small saphenous veins and their tributaries; these are situated beneath the skin in the superficial fascia. The constant position of the great saphenous vein in front of the medial malleolus should be remembered for patients requiring emergency blood transfusion. The *deep veins* are the venae comitantes to the anterior and posterior tibial, popliteal, and femoral arteries and their branches. The *perforating veins* are communicating vessels that run between the superficial and deep veins. A number of these veins are found, par-

ticularly in the region of the ankle and the medial side of the lower part of the leg. They possess valves that are arranged to prevent the flow of blood from the deep to the superficial veins.

## Venous Pump of the Lower Limb

Within the closed fascial compartments of the lower limb, the thin-walled, valved venae comitantes are subjected to intermittent pressure both at rest and during exercise. The pulsations of the adjacent arteries help to move the blood up the limb. However, the contractions of the large muscles within the compartments during exercise compress these deeply placed veins and force the blood up the limb.

The superficial saphenous veins, except near their termination, lie within the superficial fascia and are not subject to these compression forces. The valves in the perforating veins prevent the high-pressure venous blood from being forced outward into the low-pressure superficial veins. Moreover, as the muscles within the closed fascial compartments relax, venous blood is sucked from the superficial into the deep veins.

## Varicose Veins

A vein is said to be varicose when its diameter is greater than normal and it is elongated and tortuous. Varicosity of the esophageal and rectal veins is described elsewhere (pp. 139 and 393). This condition commonly occurs in the superficial veins of the lower limb, and although not a fatal disease, it is responsible for considerable discomfort, pain, and suffering in innumerable persons.

Varicose veins have many causes, including hereditary weakness of the vein walls and incompetent valves; elevated intra-abdominal pressure as the result of multiple pregnancies or abdominal tumors; and thrombophlebitis of the deep veins, which results in the superficial veins becoming the main venous pathway for the lower limb. It is easy to understand how this condition could be produced by the incompetence of a valve in a perforating vein. Every time the patient exercised, high-pressure venous blood would escape from the deep veins into the superficial veins and produce a varicosity, which might be localized to begin with, but would become more extensive later.

The successful operative treatment of varicose veins depends on the ligation and division of all the main tributaries of the great or small saphenous veins, to prevent a collateral venous circulation from developing; and the ligation and division of all the perforating veins responsible for the leakage of high-pressure blood from the deep to the superficial veins. It is now common practice to remove or strip the superficial veins in addition. Needless to say, it is imperative to ascertain that the deep veins are patent before operative measures are taken.

### *Coronary Bypass Surgery*

In patients with occlusive coronary disease caused by atherosclerosis, the diseased arterial segment may be bypassed by inserting a graft consisting of a portion of the great saphenous vein. The venous segment is reversed so that its valves do not obstruct the arterial flow. Following removal of the great saphenous vein at the donor site, the superficial venous blood ascends the lower limb by passing through perforating veins and entering the deep veins.

The great saphenous vein can also be used to bypass obstructions of the brachial or femoral arteries.

## Lymphatics of the Lower Limb

The arrangement of the superficial and deep inguinal lymph nodes is fully described on page 577. It is important to remember that they not only drain all the lymph from the lower limb, but also drain lymph from the skin and superficial fascia of the anterior and posterior abdominal walls below the level of the umbilicus; lymph from the external genitalia and the mucous membrane of the lower half of the anal canal also drains into these nodes. Remember the large distances the lymph has had to travel in some instances before it reaches the inguinal nodes. For example, a patient may present with an enlarged, painful inguinal lymph node that is due to lymphatic spread of pathogenic organisms, which entered the body through a small scratch on the undersurface of the big toe.

## Muscles of the Lower Limb

### GLUTEUS MAXIMUS

The gluteus maximus is a very large, thick muscle with coarse fasciculi that can be easily separated without damage. The great thickness of this muscle makes it ideal for intramuscular injections. To avoid injury to the underlying sciatic nerve, the injection should be given well forward on the upper outer quadrant of the buttock.

### GLUTEUS MEDIUS AND MINIMUS

The gluteus medius and minimus muscles may be paralyzed when poliomyelitis involves the lower lumbar and sacral segments of the spinal cord. They are supplied by the superior gluteal nerve (L4 and 5 and S1). Paralysis of these muscles seriously interferes with the ability of the patient to tilt the pelvis when walking.

### QUADRICEPS FEMORIS

The quadriceps femoris is a most important extensor muscle for the knee joint. Its tone greatly strengthens the joint; therefore this muscle mass must be carefully examined when disease of the knee joint is suspected. Both thighs should be examined and the size, consistency, and strength of the quadriceps muscles should be tested. Reduction in size due to muscle atrophy may be tested by measuring the circumference of each thigh a fixed distance above the superior border of the patella.

The vastus medialis muscle extends farther distally than the vastus lateralis. Remember that the vastus medialis is the first part of the quadriceps muscle to atrophy in knee joint disease and the last to recover.

The rectus femoris muscle may rupture in sudden violent extension movements of the knee joint. The muscle belly retracts proximally, leaving a gap which may be palpable on the anterior surface of the thigh. In complete rupture of the muscle surgical repair is indicated.

### ADDUCTOR MUSCLES

In patients with cerebral palsy who have marked spasticity of the adductor group of muscles, it is common practice to perform a tenotomy of the adductor longus tendon and to divide the anterior division of the obturator nerve. In addition, in some severe cases the posterior division of the obturator nerve is crushed. This operation overcomes the spasm of the adductor group of muscles and permits slow recovery of the muscles supplied by the posterior division of the obturator nerve.

## GASTROCNEMIUS AND SOLEUS

Rupture of the tendo calcaneus is common in middle-aged men and frequently occurs in tennis players. The rupture occurs at its narrowest part, about 2 inches (5 cm) above its insertion. There is a sudden sharp pain with immediate disability. The gastrocnemius and soleus muscles retract proximally, leaving a palpable gap in the tendon. The tendon should be sutured as soon as possible and the leg immobilized in plaster with the ankle joint plantarflexed and the knee joint flexed.

## PLANTARIS

Rupture of the plantaris tendon is rare, although tearing of the fibers of the soleus or partial tearing of the tendo calcaneus is frequently diagnosed as such a rupture.

The plantaris muscle, which is often missing, can be used for tendon autografts in repairing severed flexor tendons to the fingers; the tendon of the palmaris longus muscle can also be used for this purpose.

## PERONEUS LONGUS AND BREVIS

Tenosynovitis (inflammation of the synovial sheaths) can affect the tendon sheaths of the peroneus longus and brevis muscles as they pass posterior to the lateral malleolus. Treatment consists of immobilization, heat, and physiotherapy. The tendons of peroneus longus and brevis can be dislocated forward from behind the lateral malleolus. For this condition to occur the superior peroneal retinaculum must be torn. It usually occurs in older children and is caused by trauma.

## TENDON REFLEXES

Skeletal muscles receive a segmental innervation. Most muscles are innervated by two, three, or four spinal nerves and therefore by the same number of segments of the spinal cord. The segmental innervation of the following muscles in the lower limb should be known, since it is possible to test them by eliciting simple muscle reflexes in the patient.

*Patellar tendon reflex* (knee jerk) L2, **3,** and **4** (extension of the knee joint on tapping the patellar tendon).

*Achilles tendon reflex* (ankle jerk) **S1** and S2 (plantar flexion of ankle joint on tapping the Achilles tendon).

## ANTERIOR COMPARTMENT OF THE LEG SYNDROME

The anterior compartment syndrome is produced by an increase in the intracompartmental pressure that results from an increased production of tissue fluid. Marked exertion may produce this condition. The deep, aching pain in the anterior compartment of the leg that is characteristic of this syndrome can become very severe. Dorsiflexion of the foot at the ankle joint increases the severity of the pain. As the pressure rises, the venous return is diminished, thus producing a further rise in pressure. In severe cases the arterial supply is eventually cut off by compression, and the dorsalis pedis arterial pulse disappears. The tibialis anterior, the extensor digitorum longus, and the extensor hallucis longus muscles are paralyzed. Loss of sensation is limited to the area supplied by the deep peroneal nerve, i.e., the skin cleft between the first and second toes. The surgeon can open up the anterior compartment of the leg by making a longitudinal incision through the deep fascia and thus decompress the area and prevent anoxic necrosis of the muscles.

# Nerves of the Lower Limb

## FEMORAL NERVE

The femoral nerve (L2, 3, and 4) enters the thigh from behind the inguinal ligament, at a point midway between the anterior superior iliac spine and the pubic tubercle; it lies about a fingerbreadth lateral to the femoral pulse. About 2 inches (5 cm) below the inguinal ligament, the nerve splits up into its terminal branches (Fig. 10-67).

The femoral nerve may be injured in stab or gunshot wounds, but a complete division of the nerve is rare. The following clinical features are present when the nerve is completely divided:

## Motor

The quadriceps femoris muscle is paralyzed, and the knee cannot be extended. In walking, this is compensated for to some extent by the use of the adductor muscles.

## Sensory

There is loss of sensation over the medial side of the lower part of the leg and along the medial border of the foot as far as the ball of the big toe; this area is normally supplied by the saphenous nerve.

## SCIATIC NERVE

The sciatic nerve (L4 and 5; S1, 2, and 3) curves laterally and downward through the gluteal region, situated at first midway between the posterior superior iliac spine and the ischial tuberosity, and lower down, midway between the tip of the greater trochanter and the ischial tuberosity. The nerve then passes downward in the midline on the posterior aspect of the thigh and divides into the common peroneal and tibial nerves, at a variable site above the popliteal fossa (Figs. 10-68 and 10-70).

The nerve is sometimes injured by penetrating wounds, fractures of the pelvis, or dislocations of the hip joint. It is most frequently injured by badly placed intramuscular injections in the gluteal region. To avoid this injury, nurses should be instructed to give injections into the gluteus maximus or the gluteus medius well forward on the upper outer quadrant of the buttock. The majority of the nerve lesions are incomplete, and in 90 percent of injuries, the common peroneal part of the nerve is the most affected. The following clinical features are present:

## Motor

The hamstring muscles are paralyzed, but weak flexion of the knee is possible, due to the action of the sartorius (femoral nerve) and gracilis (obturator nerve). All the muscles below the knee are paralyzed, and the weight of the foot causes it to assume the plantar-flexed position, or *footdrop*.

## Sensory

There is loss of sensation below the knee, except for a narrow area down the medial side of the lower part of the leg and along the medial border of the foot as far as the ball of the big toe, which is supplied by the saphenous nerve (femoral nerve).

The result of operative repair of a sciatic nerve injury is poor. It is rare for active movement to return to the small muscles of the foot, and sensory recovery is rarely complete. Loss of sensation in the sole of the foot makes the development of trophic ulcers inevitable.

## SCIATICA

This term is used to describe the condition in which patients have pain along the sensory distribution of the sciatic nerve. Thus the pain is experienced in the posterior aspect of the thigh, the posterior and lateral sides of the leg, and the lateral part of the foot. Sciatica can be caused by prolapse of an intervertebral disc (see p. 947), with pressure on one or more roots of the lower lumbar and sacral spinal nerves, pressure on the sacral plexus or sciatic nerve by an intrapelvic tumor, or inflammation of the sciatic nerve or its terminal branches.

## COMMON PERONEAL NERVE

The common peroneal nerve (Fig. 10-68) is in a very exposed position as it leaves the popliteal fossa and winds around the neck of the fibula to enter the peroneus longus muscle.

It is commonly injured in fractures of the neck of the fibula and by pressure from plaster casts or splints. The following clinical features are present:

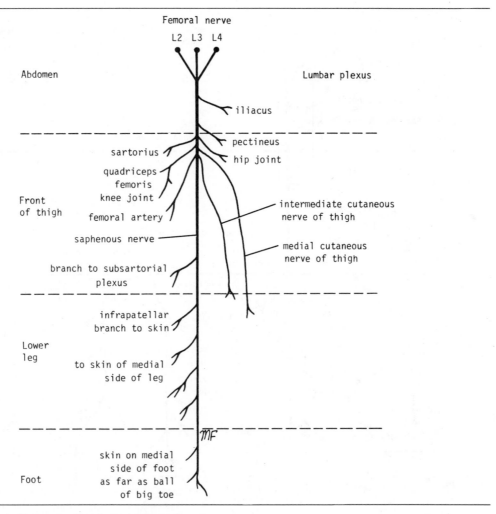

Femoral nerve

L2  L3  L4

Abdomen

Lumbar plexus

iliacus

pectineus

sartorius

hip joint

quadriceps
femoris

knee joint

Front
of thigh

intermediate cutaneous
nerve of thigh

femoral artery

saphenous nerve

medial cutaneous
nerve of thigh

branch to subsartorial
plexus

infrapatellar
branch to skin

Lower
leg

to skin of medial
side of leg

skin on medial
side of foot
as far as ball
of big toe

Foot

**Fig. 10-67. Summary diagram of main branches of femoral nerve.**

## Motor

The muscles of the anterior and lateral compartments of the leg are paralyzed; namely, the tibialis anterior, the extensor digitorum longus and brevis, the peroneus tertius, the extensor hallucis longus (supplied by the deep peroneal nerve), and the peroneus longus and brevis (supplied by the superficial peroneal nerve). As a result, the opposing muscles, the plantar flexors of the ankle joint and the invertors of the subtalar and transverse tarsal

joints, cause the foot to be plantar-flexed (footdrop) and inverted, an attitude referred to as *equinovarus* (Fig. 10-69).

## Sensory

There is loss of sensation down the anterior and lateral sides of the leg and dorsum of the foot and toes, including the medial side of the big toe. The lateral border of the foot and the lateral side of the little toe are virtually unaffected (sural nerve, mainly formed from tibial nerve). The medial bor-

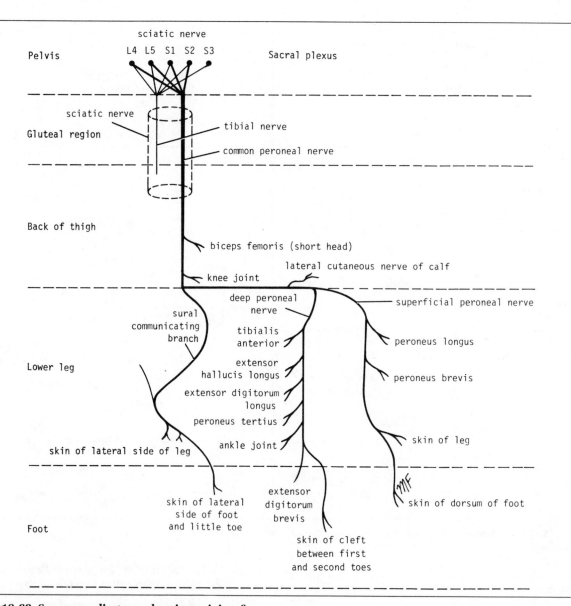

**Fig. 10-68. Summary diagram showing origin of sciatic nerve and main branches of common peroneal nerve.**

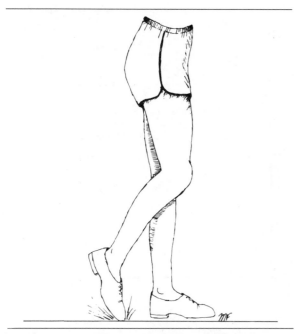

**Fig. 10-69. Footdrop. With this condition the individual catches his toes on the ground as he walks.**

der of the foot as far as the ball of the big toe is completely unaffected (saphenous nerve).

When the injury occurs distal to the site of origin of the lateral cutaneous nerve of the calf, the loss of sensibility is confined to the area of the foot and toes.

## TIBIAL NERVE

The tibial nerve (Fig. 10-70), leaves the popliteal fossa by passing deep to the gastrocnemius and soleus muscles. Because of its deep and protected position, it is rarely injured. Complete division results in the following clinical features:

### Motor

All the muscles in the posterior compartment of the leg and the sole of the foot are paralyzed. The opposing muscles dorsiflex the foot at the ankle joint and evert the foot at the subtalar and transverse tarsal joints, an attitude referred to as *calcaneovalgus*.

### Sensory

There is loss of sensation on the sole of the foot.

## OBTURATOR NERVE

The obturator nerve (L2, 3, and 4) enters the thigh as anterior and posterior divisions thorugh the upper part of the obturator foramen. The anterior division descends in front of the obturator externus and the adductor brevis, deep to the floor of the femoral triangle. The posterior division descends behind the adductor brevis and in front of the adductor magnus (Fig. 10-71).

It is very rarely injured in penetrating wounds, in anterior dislocations of the hip joint, or in abdominal herniae through the obturator foramen. It may be damaged during parturition. The following clinical features occur:

### Motor

All the adductor muscles are paralyzed except the hamstring part of adductor magnus, which is supplied by the sciatic nerve. The sensory loss is minimal on the medial aspect of the thigh.

## Bones

### FEMUR

*The head of the femur,* i.e., that part that is not intraacetabular, can be palpated on the anterior aspect of the thigh just inferior to the inguinal ligament and just lateral to the pulsating femoral artery. Tenderness over the head of the femur usually indicates the presence of arthritis of the hip joint.

The *neck of the femur* is inclined at an angle with the shaft; the angle is about 160 degrees in the young child and about 125 degrees in the adult. An increase in this angle is referred to as *coxa valga,* and it occurs, for example, in cases of congenital dislocation of the hip. In this condition, adduction of the hip joint is limited. A decrease in this angle is referred to as *coxa vara,* and it occurs in fractures of the neck of the femur and in slipping of the femoral epiphysis. In this condition, abduction of the hip joint is limited. Shenton's line is a useful means of assessing the angle of the femoral neck on a radiograph of the hip region. (See p. 669.)

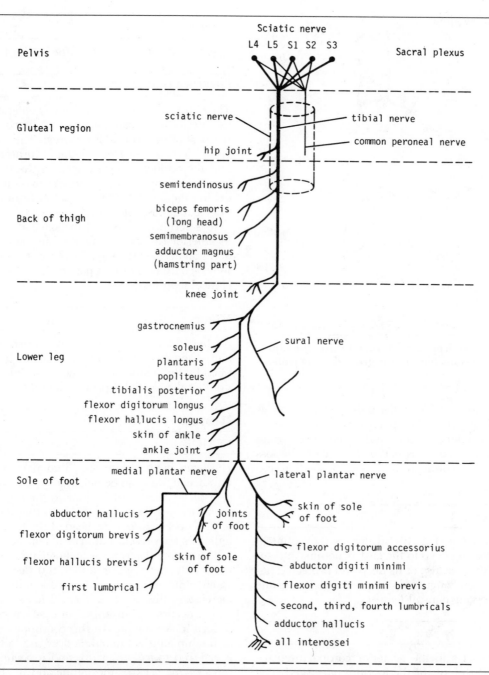

**Fig. 10-70. Summary diagram, showing origin of sciatic nerve and main branches of tibial nerve.**

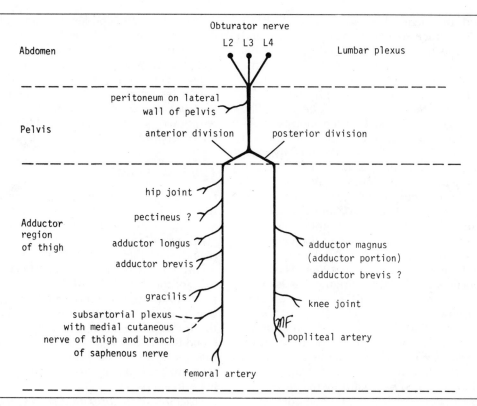

**Fig. 10-71. Summary diagram of main branches of obturator nerve.**

An anatomical knowledge of the *blood supply to the femoral head* explains why avascular necrosis of the head may occur following fractures of the neck of the femur. The epiphysis of the head is supplied by a small branch of the obturator artery, which passes to the head along the ligament of the femoral head. The upper part of the neck of the femur receives a profuse blood supply from the medial femoral circumflex artery. These branches pierce the capsule and ascend the neck deep to the synovial membrane. As long as the epiphysial cartilage remains, there is no communication between the two sources of blood. In the adult, after the epiphysial cartilage disappears, an anastomosis between the two sources of blood supply is established. It is not difficult to understand that fractures of the femoral neck will interfere with or completely interrupt the blood supply from the root of the femoral neck to the femoral head. The

scant blood flow along the small artery that accompanies the round ligament may be insufficient to sustain the viability of the femoral head, and ischemic necrosis gradually takes place.

*Fractures of the neck of the femur* are common and are of two types, (1) subcapital and (2) trochanteric. The subcapital fracture occurs in the elderly and is usually produced by a minor trip or stumble. Subcapital femoral neck fractures are particularly common in women following the menopause. This sexual predisposition is due to a thinning of the cortical and trabecular bone caused by estrogen deficiency. Avascular necrosis of the head is a common complication. If the fragments are not impacted, there is considerable displacement. The strong muscles of the thigh pull the distal fragment upward, so that the leg is shortened. The gluteus maximus, the piriformis, the obturator internus, the gamelli, and the quadratus femoris rotate the distal fragment laterally.

Trochanteric fractures commonly occur in the

young and middle-aged as the result of direct trauma. The fracture line is extracapsular, and both fragments have a profuse blood supply. If the bone fragments are not impacted, the pull of the strong muscles will produce shortening and lateral rotation of the leg, as previously explained.

*In fractures of the upper third of the femur,* the proximal fragment is flexed by the iliopsoas, abducted by the gluteus medius and minimus, and laterally rotated by the gluteus maximus, the piriformis, the obturator internus, the gemelli, and the quadratus femoris. The lower fragment is adducted by the adductor muscles, pulled upward by the hamstrings and quadriceps, and laterally rotated by the adductors and the weight of the foot.

*In fractures of the middle third of the femur,* the distal fragment is pulled upward by the hamstrings and the quadriceps, so that there is considerable shortening. The distal fragment is also rotated backward by the pull of the two heads of the gastrocnemius.

*In fractures of the distal third of the shaft of the femur,* the same displacement of the distal fragment occurs as seen in fractures of the middle third of the shaft. However, the distal fragment is smaller and is rotated backward by the gastrocnemius muscle to a greater degree and may exert pressure on the popliteal artery and interfere with the blood flow through the leg and foot.

From these accounts, it is clear that a knowledge of the different actions of the muscles of the leg is necessary if one is to understand the displacement of the fragments of a fractured femur. Considerable traction on the distal fragment is usually required to overcome the powerful muscles and restore the limb to its correct length. Manipulation of the bone is necessary to bring the proximal and distal fragments into correct alignment.

## PATELLA

The *patella* is a sesamoid bone lying within the quadriceps tendon. The importance of the lower horizontal fibers of the vastus medialis and the large size of the lateral condyle of the femur in preventing lateral displacement of the patella has been emphasized. Cases of congenital recurrent dislocation of the knee are due to underdevelopment of the lateral femoral condyle.

Should the patella be fractured as the result of

direct violence, it is broken into a number of small fragments. Since the bone lies within the quadriceps femoris tendon, little separation of the fragments takes place. Fracture of the patella due to indirect violence is caused by the sudden contraction of the quadriceps snapping the patella across the front of the femoral condyles. The knee is in the semiflexed position, and the fracture line is transverse. Separation of the fragments usually occurs.

## TIBIA AND FIBULA

*Fractures of the tibia and fibula* are common. Fortunately, if only one bone is fractured, the other acts as a splint and there is minimal displacement. Fractures of the shaft of the tibia are often compound, since the entire length of the medial surface is covered only by skin and superficial fascia. Fractures of the distal third of the shaft of the tibia are prone to delayed union or nonunion. This may be due to the fact that the nutrient artery is torn at the fracture line, with a consequent reduction in blood flow to the distal fragment; it is also possible that the splintlike action of the intact fibula prevents the proximal and distal fragments from coming into apposition.

*Fractures of the proximal end of the tibia.* Fractures of the tibial condyles are common in the middle-aged and elderly; they usually result from direct violence to the side of the knee joint, as when a person is hit by an automobile. The tibial condyle may show a split fracture or be broken up, or the fracture line may pass between both condyles in the region of the intercondylar eminence. As the result of forced abduction of the knee joint, the medial collateral ligament can also be ruptured.

## BONES OF THE FOOT

*Fractures of the bones of the foot.* Fractures of the talus occur at the neck or body of the talus. Neck fractures occur during violent dorsiflexion of the ankle joint when the neck is driven against the anterior edge of the distal end of the tibia. The body of the talus can be fractured by jumping from a height, although the two malleoli prevent displacement of the fragments.

Compression fractures of the calcaneum result from falls from a height. The weight of the body drives the talus downward into the calcaneum, crushing it in such a way that it loses vertical height and becomes wider laterally. The posterior portion of the calcaneum above the insertion of the tendo calcaneus can be fractured by the posterior displacement of the talus. The sustentaculum tali can be fractured by forced inversion of the foot.

*Fractures of the metatarsal bones.* The base of the fifth metatarsal can be fractured during forced inversion of the foot, at which time the tendon of insertion of the peroneus brevis muscle pulls off the base of the metatarsal.

Stress fracture of a metatarsal bone is common in soldiers after long marches; it can also occur in nurses and hikers. It occurs most frequently in the distal third of the second, third, or fourth metatarsal bone.

## Joints

### HIP JOINT

The stability of the hip joint depends on the ball-and-socket arrangement of the articular surfaces and the very strong ligaments. In *congenital dislocation of the hip*, the upper lip of the acetabulum fails to develop adequately, and the head of the femur, having no stable platform under which it can lodge, rides up out of the acetabulum onto the gluteal surface of the ilium.

Traumatic dislocation of the hip is rare because of its strength. However, should it occur, it usually does so when the joint is flexed and adducted. The head of the femur is displaced posteriorly out of the acetabulum, and it comes to rest on the gluteal surface of the ilium (posterior dislocation). The close relation of the sciatic nerve to the posterior surface of the joint makes it prone to injury in posterior dislocations.

The stability of the hip joint when a person stands on one leg with the foot of the opposite leg raised above the ground depends on three factors:

1. The gluteus medius and minimus must be functioning normally.
2. The head of the femur must be located normally within the acetabulum.
3. The neck of the femur must be intact and must have a normal angle with the shaft of the femur.

If any one of these factors is defective, then the pelvis will sink downward on the opposite, unsupported side. The patient is then said to exhibit a positive *Trendelenburg's sign* (Fig. 10-72).

Normally, when walking, a person alternately contracts the gluteus medius and minimus, first on one side and then on the other. By this means he is able to raise the pelvis first on one side and then on the other, to allow the leg to be flexed at the hip joint and moved forward; i.e., the leg is raised clear of the ground before it is thrust forward in taking the forward step. A patient with a right-sided congenital dislocation of the hip, when asked to stand on his right leg and raise the opposite leg clear of the ground, will exhibit a positive Trendelenburg's sign, and the unsupported side of the pelvis will sink below the horizontal. If the patient is asked to walk, he will show the characteristic "dipping" gait. In patients with bilateral congenital dislocation of the hip, the gait is typically "waddling" in nature.

A patient with an inflamed hip joint will place the femur in the position that gives him minimum discomfort, i.e., the position in which the joint cavity has the greatest capacity to contain the increased amount of synovial fluid secreted. The hip joint is partially flexed, abducted, and externally rotated.

*Osteoarthritis*, the most common disease of the hip joint in the adult, causes pain, stiffness, and deformity. The pain may be in the hip joint itself or referred to the knee (the obturator nerve supplies both joints). The stiffness is due to the pain and reflex spasm of the surrounding muscles. The deformity is flexion, adduction, and external rotation and is produced initially by muscle spasm and later by muscle contracture.

### KNEE JOINT

The strength of the *knee joint* depends on the strength of the ligaments that bind the femur to the tibia and on the tone of the muscles acting on the joint. The most important muscle group is the quadriceps femoris, and, provided this is well developed, it is capable of stabilizing the knee in the presence of torn ligaments.

The *synovial membrane* of the knee joint is very extensive, and if the articular surfaces, semilunar cartilages, or ligaments of the joint are damaged, the large synovial cavity becomes distended with fluid. The wide communication between the su-

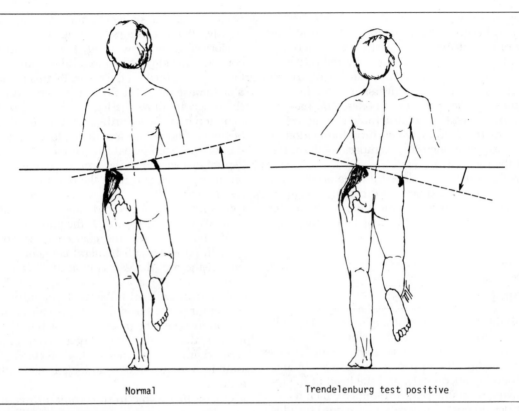

Normal                                  Trendelenburg test positive

**Fig. 10-72. The Trendelenburg test.**

prapatellar bursa and the joint cavity results in this structure's becoming distended also. The swelling of the knee extends some 3 or 4 fingerbreadths above the patella and laterally and medially beneath the aponeuroses of insertion of the vastus lateralis and medialis, respectively.

Injuries of the *semilunar cartilages* are very common. The medial cartilage is damaged much more frequently than the lateral, and this is probably due to its strong attachment to the medial collateral ligament of the knee joint, which restricts its mobility. The injury occurs when the femur is rotated on the tibia, or the tibia rotated on the femur, with the knee joint partially flexed and taking the weight of the body. The tibia is usually abducted on the femur, and the medial cartilage is pulled into an abnormal position between the femoral and tibial condyles (Fig. 10-73). A sudden movement between the condyles results in the cartilage being subjected to a severe grinding force, and it splits

along its length. When the torn part of the cartilage becomes wedged between the articular surfaces, further movement is impossible, and the joint is said to "lock."

Injury to the lateral semilunar cartilage is less common, probably due to the fact that it is not attached to the lateral collateral ligament of the knee joint and is consequently more mobile. The poplit-

**Fig. 10-73. (A) Mechanism involved in damage to the medial semilunar cartilage of the knee joint in a football game. Note that the right knee joint is semiflexed and that there is medial rotation of the femur on the tibia. The impact causes forced abduction of the tibia on the femur, and the medial semilunar cartilage is pulled into an abnormal position. The cartilage is then ground between the femur and the tibia. (B) Testing for integrity of the anterior cruciate ligament. (C) Testing for integrity of the posterior cruciate ligament.**

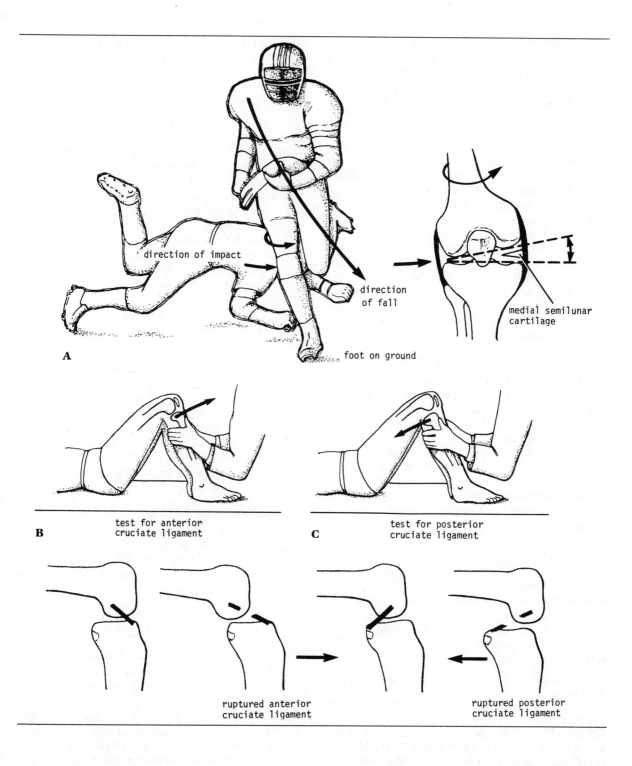

direction of impact

direction
of fall

foot on ground

medial semilunar
cartilage

**A**

**B**    test for anterior
cruciate ligament

**C**    test for posterior
cruciate ligament

ruptured anterior
cruciate ligament

ruptured posterior
cruciate ligament

eus muscle sends a few of its fibers into the lateral cartilage, and these may pull the cartilage into a more favorable position during sudden movements of the knee joint.

Forced abduction of the tibia on the femur may result in partial tearing of the *medial collateral ligament*, which may occur at its femoral or tibial attachments. It is useful to remember that tears of the semilunar cartilages result in localized tenderness on the joint line, whereas sprains of the medial collateral ligament result in tenderness over the femoral or tibial attachments of the ligament.

Tears or rupture of the *cruciate ligaments* occur when severe force is applied to the knee joint. The injury is always accompanied by damage to other knee structures; the collateral ligaments are commonly torn, or the capsule may be damaged. The joint cavity quickly fills with blood (*hemarthrosis*), so that the joint is swollen. Examination of patients with a ruptured anterior cruciate ligament shows that the tibia can be pulled excessively forward on the femur; with rupture of the posterior cruciate ligament, the tibia can be made to move excessively backward on the femur (Fig. 10-73). Since the stability of the knee joint depends largely on the tone of the quadriceps femoris muscle and the integrity of the collateral ligaments, operative repair of isolated torn cruciate ligaments is not often attempted. The knee is immobilized in slight flexion in a plaster cast, and active physiotherapy on the quadriceps femoris muscle is begun at once. Should, however, the capsule of the joint and the collateral ligaments be torn in addition, early operative repair is essential.

## Pneumoarthrography

Air can be injected into the synovial cavity of the knee joint so that soft tissues can be studied. This technique is based on the fact that air is less radiopaque than structures such as the medial and lateral semilunar cartilages, so their outline can be visualized on a radiograph.

## Arthroscopy

This procedure involves the introduction of a lighted instrument into the synovial cavity of the knee joint through a small incision. This technique permits the direct visualization of structures, such as the cruciate ligaments and the semilunar cartilages, for diagnostic purposes.

## ANKLE JOINT

The *ankle joint* is a hinge joint possessing great stability. The deep mortise formed by the lower end of the tibia and the medial and lateral malleoli securely holds the talus in position.

*Sprains of the ankle* are usually caused by excessive inversion of the foot. The anterior talofibular ligament and the calcaneofibular ligament are partially torn, giving rise to great pain and local swelling. A similar but less common injury may occur to the medial or deltoid ligament as the result of excessive eversion. The great strength of the medial ligament usually results in the ligament pulling off the tip of the medial malleolus.

*Fracture dislocations of the ankle joint* are common. Forced external rotation and overeversion of the foot is the most frequent cause of fracture dislocation. The talus is externally rotated forcibly against the lateral malleolus of the fibula. The torsion effect on the lateral malleolus causes it to fracture spirally. If the force continues, the talus moves laterally, and the medial ligament of the ankle joint becomes taut and pulls off the tip of the medial malleolus. If the talus is forced to move still farther, its rotary movement results in its violent contact with the posterior inferior margin of the tibia, which shears off. Other less common types of fracture dislocation are due to forced overeversion (without rotation), in which the talus presses the lateral malleolus laterally and causes it to fracture transversely. Overinversion (without rotation), in which the talus presses against the medial malleolus, will produce a vertical fracture through the base of the medial malleolus.

## METATARSOPHALANGEAL JOINT OF BIG TOE

*Hallux valgus*, which is a lateral deviation of the great toe at the metatarsophalangeal joint, is a common condition. Its incidence is greater in women than men and is associated with badly fitting shoes. It is often accompanied by the presence of a short first metatarsal bone. Once the deformity is established, it is progressively worsened by the pull of the flexor hallucis longus and the extensor hallucis longus muscles. Later, osteoarthritic changes occur in the metatarsophalangeal joint, which then becomes stiff and painful; the condition is then known as *hallux rigidus.*

## Bursae

A number of bursae are found in the lower limb where skin, tendons, ligaments, or muscles repeatedly rub against bony points or ridges. Unaccustomed exercise or direct trauma may result in their becoming inflamed and distended with excessive amounts of fluid. The following bursae are prone to inflammation: the bursa over the ischial tuberosity (weaver's bottom); the prepatellar and superficial infrapatellar bursae (housemaid's or clergyman's knee); the bursa between the tendo calcaneus and the upper part of the calcaneum (long-distance runner's ankle).

Two important bursae communciate with the knee joint, and they may become distended if there is an accumulation of excessive amounts of synovial fluid within the joint. The suprapatellar bursa extends proximally about 3 fingerbreadths above the patella beneath the quadriceps femoris muscle. The bursa, which is associated with the insertion of the semimembranosus muscle, may enlarge in patients with osteoarthritis of the knee joint.

The anatomical bursae described should not be confused with *adventitious bursae,* which develop in response to abnormal and excessive friction. For example, a subcutaneous bursa sometimes develops over the tendo calcaneus in response to badly fitting shoes. A *bunion* is an adventitial bursa located over the medial side of the head of the first metatarsal bone.

## Femoral Sheath and Femoral Hernia

The *femoral sheath* is a protrusion of the fascial envelope lining the abdominal walls and surrounds the femoral vessels and lymphatics for about 1 inch (2.5 cm) below the inguinal ligament (Fig. 4-34). The *femoral canal,* the compartment for the lymphatics, occupies the medial part of the sheath. It is about ½ inch (1.3 cm) long, and its upper opening is referred to as the *femoral ring.* The *femoral septum,* which is a condensation of extraperitoneal tissue, closes the femoral ring.

A *femoral hernia* is more common in women than in men (possibly due to their wider pelvis and femoral canal). The hernial sac passes down the femoral canal, pushing the femoral septum before it. On escaping through the lower end, it lies beneath the saphenous opening in the deep fascia of the thigh. With further expansion, the hernial sac may turn forward and upward to form a swelling in the upper part of the thigh (Fig. 4-34).

The neck of the sac always lies below and lateral to the *pubic tubercle* (Fig. 4-33). This serves to distinguish it from an inguinal hernia, which lies above and medial to the pubic tubercle. The neck of the sac is narrow and lies at the femoral ring. The ring is related anteriorly to the inguinal ligament; posteriorly, to the pectineal ligament and the superior ramus of the pubis; medially, to the sharp free edge of the lacunar ligament; and laterally, to the femoral vein. Because of these anatomical structures, the neck of the sac is unable to expand. Once an abdominal viscus has passed through the neck into the body of the sac, it may be difficult to push it up and return it to the abdominal cavity (*irreducible hernia*). Furthermore, after the patient strains or coughs, a piece of bowel may be forced through the neck, and its blood vessels may be compressed by the femoral ring, seriously impairing its blood supply (*strangulated hernia*). A femoral hernia is a dangerous condition and should always be treated surgically.

When considering the differential diagnosis of a femoral hernia, it is important to consider diseases that may involve other anatomical structures close to the inguinal ligament. For example:

1. *Inguinal canal.* The swelling of an inguinal hernia lies above the medial end of the inguinal ligament. Should the hernial sac emerge through the superficial inguinal ring to start its descent into the scrotum, the swelling will lie above and medial to the pubic tubercle. The sac of a femoral hernia lies below and lateral to the pubic tubercle.
2. *Superficial inguinal lymph nodes.* Usually, there is more than one lymph node enlarged. In patients with inflammation of the nodes (*lymphadenitis*), carefully examine the entire area of the body that drains its lymph into these nodes. A small, unnoticed skin abrasion may be found.

Never forget the mucous membrane of the lower half of the anal canal—it may have an undiscovered carcinoma.

3. *Great saphenous vein.* A localized dilatation of the terminal part of the great saphenous vein, a *saphenous varix,* may cause confusion, especially since a hernia and a varix increase in size when the patient is asked to cough. (Elevated intraabdominal pressure drives the blood downward.) The presence of varicose veins elsewhere in the leg should help in the diagnosis.

4. *Psoas sheath.* Tuberculous infection of a lumbar vertebra may result in the extravasation of pus down the psoas sheath into the thigh. The presence of a swelling above and below the inguinal ligament, together with clinical signs and symptoms referred to the vertebral column, should make the diagnosis obvious.

5. *Femoral artery.* An expansile swelling lying along the course of the femoral artery that fluctuates in time with the pulse rate should make the diagnosis of *aneurysm of the femoral artery* certain.

## Arches of the Foot

The *arches of the foot* have been fully described on page 665. Of the three arches, the medial longitudinal is the largest and clinically the most important. The shape of the bones, the strong ligaments, especially those on the plantar surface of the foot, and the tone of muscles all play an important role in supporting the arches. It has been shown that in the active foot the tone of muscles is a very important factor in arch support. When the muscles are fatigued by excessive exercise (a long route march by an army recruit), by standing for long periods (waitress or nurse), by overweight, or by illness, the muscular support gives way, the ligaments are stretched, and pain is produced.

*Pes planus* (flat foot) is a condition in which the medial longitudinal arch is depressed or collapsed. As a result, the forefoot is displaced laterally and everted. The head of the talus is no longer supported, and the body weight forces it downward and medially between the calcaneum and the navicular bone. When the deformity has existed for some time, the plantar, calcaneonavicular, and medial ligaments of the ankle joint become permanently stretched, and the bones change shape. The muscles and tendons are also permanently stretched. There are both congenital and acquired causes of flat foot.

*Pes cavus* (clawfoot) is a condition in which the medial longitudinal arch is unduly high. Most cases are due to muscle imbalance, in many instances resulting from poliomyelitis.

## Plantar Aponeurosis

Plantar fasciitis, which occurs in individuals who do a great deal of standing or walking, produces pain and tenderness of the sole of the foot. It is believed to be due to trauma or a pulling on the plantar aponeurosis. Repeated attacks of this condition induce ossification in the posterior attachment of the aponeurosis, forming a calcaneal spur.

## CLINICAL PROBLEMS                                    *Answers on page 972*

1. Following a major abdominal operation, a patient was given a course of antibiotics by intramuscular injection. The nurse was instructed to give the injections into the buttock. During convalescence, the patient started to experience numbness and a tingling sensation down the anterior and lateral sides of the right leg and dorsum of the foot. He also stated that his right leg felt heavy and that his right foot tended to catch on steps and on the edges of the carpet. On examination, there was evidence of impaired skin sensation on the anterior and lateral sides of the right leg and dorsum of the right foot. The patient tended to hold his foot plantar flexed and slightly inverted. On comparing the relative strengths of the plantar and dorsiflexors of the ankle joint on both legs, it was found that dorsiflexion of the right ankle was weaker than normal. The evertor muscles were also weaker than normal. Using your knowledge of anatomy, explain the patient's signs and symptoms. What precautions should a nurse take when giving intramuscular injections into the buttock?

2. A 5-year-old girl of Italian extraction was brought to the orthopedic clinic because of her dipping gait on walking. The history revealed that she started to walk late and that she always had a slight limp on the "left" side. On observing her walk, it was seen that her pelvis sagged down on the left side. When the patient was asked to stand on her left leg and raise the opposite leg off the ground, the pelvis on the unsupported side was lifted above the horizontal. When she was asked to stand on her right leg and raise the left leg off the ground, the pelvis on the unsupported side sagged down below the horizontal. With the patient standing upright on both legs, the top of the greater trochanter on the right side was situated above a line drawn between the anterior superior iliac spine and ischial tuberosity. On the left side the greater trochanter was normally placed on this line. Radiological examination of the right hip joint showed that Shenton's line was not intact. Using your anatomical knowledge, make a diagnosis. What is the anatomical basis for the Trendelenburg's sign?

3. A 45-year-old man visited his physician complaining of a lump in the groin. He had first noticed it 3 months earlier, and it had gradually become larger. The lump caused him neither pain nor discomfort. On examination, a small discrete hard swelling was found about 2 inches (5 cm) below and lateral to the pubic tubercle on the front of the right thigh. Two smaller hard swellings were also found immediately below the other swelling. The patient's skin on the front and the back of the body was carefully examined from the level of the umbilicus to the sole of the foot. The external genitalia were also carefully scrutinized. Examination of the anal canal revealed nothing abnormal. The only abnormal lesion discovered was a small pigmented tumor beneath the nail of the right second toe. What is the possible connection between the small painless mole on the second toe and the painless swellings in the groin? Why did the physician examine such a wide area of the patient's body?

4. A 45-year-old woman was seen in the emergency room complaining of abdominal pain and repeated vomiting. On questioning, the patient stated that the pain was severe and colicky in nature and most intense in the region of the umbilicus. On examination, the patient showed obvious signs of dehydration, namely, dry skin, dry tongue, and sunken eyes. The abdomen showed no distention, but excessively loud bowel peristaltic sounds (*borborygmi*) could be heard with the stethoscope. A small, tender, tense swelling was found in the front of the left thigh. When the patient was asked to cough, there was no expansion of the swelling. The swelling was located below and lateral to the left pubic tubercle. The patient said she had first noticed the swelling about 3 years ago and that 2 days ago, after coughing, it had increased in size and become tender. Given that the patient has acute intestinal obstruction and using your knowledge of anatomy, what is your diagnosis?

5. A 47-year-old woman visited her physician complaining of a dull, aching pain in the lower part of both legs. She stated that this had been getting progressively worse since the birth of her sixth child and was particularly bad at the end of a long day of standing at the ironing board. She also noticed that the skin along the medial side of the legs had started to become irritated. On examination, the patient was seen to have widespread varicose veins, extending from the lower thigh down to the dorsum of the foot in both legs. The skin showed marked discoloration over the medial malleoli and was dry and scaly. This patient had severe varicose veins of both legs, which required surgical treatment. Using your anatomical knowledge, state which veins of the legs were involved. What simple test must be performed before a radical venous operation is carried out? Which important tributaries have to be ligated at operation? Suppose that, on placing your hand over the varicose veins, you noticed that they expanded when the patient coughed. Why would this occur? Describe the so-called venous pump of the leg.

6. On a routine anteroposterior X-ray examination of a patient's right hip joint, the long axis of the neck of the femur was found to be at an angle of 160 degrees with the long axis of the femoral shaft. Is this angle normal in a 5-year-old child? In a 35-year-old man? What is the clinical condition called in which the angle is smaller than

normal? Which movement of the hip joint is limited by this condition?

7. Fractures of the neck of the femur in the adult commonly result in avascular necrosis of part of the femoral head. Can you explain this on anatomical grounds? Trochanteric fractures are never accompanied by avascular necrosis. Why?

8. A 25-year-old man was admitted to the hospital following an automobile accident. Apart from other superficial injuries, he was found to have a fracture of the middle third of the right femur. On examination, the right leg showed 2 inches (5 cm) of shortening. A lateral radiograph showed overlap of the fragments, with the distal fragment rotated backward. Using your knowledge of anatomy, can you explain the shortening of the right leg? Why was the distal fragment rotated posteriorly? Will a large or a small amount of force be necessary to restore the leg to its original length?

9. A 65-year-old man was admitted to the hospital suffering from severe intermittent claudication of the left leg. He could walk about 50 yards (46 m) before the cramplike pain in his calf muscles forced him to rest. On examination, his femoral pulses were found to be normal in both legs. The popliteal, dorsalis pedis, and posterior tibial arteries were present in the right leg, but completely absent in the left leg. Arteriography revealed a normal right femoral artery, but the artery on the left was completely blocked at the level of the adductor tubercle. Other blocks were seen at different levels lower down the arterial tree. In view of the extensive arteriosclerosis involving many areas of the main arterial supply, it was decided that a venous grafting operation would not be performed. Instead, an attempt would be made to increase the blood flow through the collateral circulation. (a) How is it possible to increase the blood flow through the collateral circulation? (b) How does blood reach the distal part of the leg in the presence of a block in the femoral artery at the opening in the adductor magnus?

10. What is the sympathetic innervation of the digital arteries of the foot?

11. On the death of her husband, a 50-year-old woman decided to earn her living by becoming an office cleaner. Part of her work involved the scrubbing of a flight of stone steps. After 3 weeks she noticed a painful swelling in front of the lower part of the left knee. On the basis of your knowledge of anatomy, what is the diagnosis?

12. A medical student, while playing football, collided with another player and fell to the ground. As he fell, the right knee, which was taking the weight of his body, was partially flexed, the femur was rotated medially, and the leg abducted on the thigh. A sudden pain was felt in the right knee joint, and he was unable to extend it. When examined by a physician an hour later, the student was still unable to extend the knee, which was by now greatly swollen. Severe local tenderness was felt along the medial side of the joint line. What is the diagnosis? Why was the knee locked? Why was the swelling of the knee so extensive over the front of the joint?

13. A 30-year-old man suspected of having a neurological disorder was asked to cross his knees and relax while his ligamentum patellae was tapped briskly with a small hammer. The process was repeated on the opposite leg. The patient was then asked to kneel on a chair, and while the examiner gently held his foot, the tendo calcaneus was tapped briskly with a hammer. This was repeated on the opposite side. Finally, the patient was asked to lie flat on his back, and a sharp, pointed instrument was drawn steadily along the lateral border of the sole of the foot (*Babinski test*). What are the anatomical reasons for performing (1) the knee jerk, (2) the ankle jerk, and (3) the Babinski test?

14. Following a severe automobile accident, a 25-year-old woman was found to have an unstable right knee joint. On examination under an anesthetic, it was possible to pull the tibia forward excessively on the femur. The patient's knee joint was fixed in plaster for 3 months, and vigorous physiotherapy was carried out on her quadriceps femoris muscle. What structure was damaged in the knee joint? Why was it necessary to have physiotherapy on the quadriceps femoris muscle?

15. An 18-year-old girl was running across a plowed field when she stumbled and overinverted her left foot. One hour later she was ex-

amined by her physician. The lateral side of the ankle was tender and swollen. Movements of the ankle were restricted, especially when the foot was inverted. Careful, gentle palpation demonstrated a small area of great tenderness below and in front of the lateral malleolus. X-ray examination of the ankle joint was negative. What is the diagnosis? In which joints do eversion and inversion of the foot take place?

16. A 25-year-old man was running across open country when he caught his left foot in a rabbit hole. As he fell, the left foot was violently rotated laterally and overeverted. On attempting to stand, he found he could place no weight on his left foot. Half an hour later he was examined by a physician. The left ankle was considerably swollen, especially on the lateral side. The left heel appeared to be unduly prominent. Anteroposterior and lateral radiographs of the ankle showed a fracture of the medial and lateral malleoli and a fracture of the lower end of the tibia. The tibia was displaced forward on the talus. The diagnosis of severe fracture dislocation of the ankle joint was made. (a) Is the ankle joint normally a stable joint? If so, what does the stability depend upon? (b) Describe in anatomical terms how this patient's fracture dislocation took place. (c) What type of fracture occurs involving the lateral and medial malleoli—transverse, oblique, or spiral?

17. An 18-year-old girl decided to work as a waitress during the summer vacation to earn part of her college fees. She had never done this type of work before. After working for 8 hours a day, for a week, she found that her feet were swollen and painful, and her legs ached. Using your anatomical knowledge, explain the underlying cause for her discomfort.

18. A 54-year-old man was told by his physician to reduce weight. He was prescribed a diet and was strongly advised to take more exercise. One morning while jogging, he heard a sharp snap and felt a sudden pain in his right lower calf. On examination in the emergency room, the physician noted the upper part of the right calf was swollen and a gap was apparent between the swelling and the right heel. With the patient lying on his back the physician gently squeezed the upper part of the right calf and noted that there was no plantar flexion of the right ankle joint. A diagnosis of rupture of the right Achilles tendon was made. (a) Describe the location of the Achilles tendon. (b) Demonstrate to yourself the movement of plantar flexion of the ankle joint. (c) Explain why the upper part of the right calf was swollen in this patient and why there was a gap below the swelling.

19. A 66-year-old man with a history of diabetes mellitus was admitted to the hospital because of poor circulation of blood in his left leg. As a physician it is important with a patient with circulatory problems to periodically check and document the color and temperature and the quality of the peripheral pulses of both legs. Where would you locate the palpable pulses of the lower limbs? Name the artery at each site.

20. A 58-year-old woman with extensive osteoarthritis of the right hip joint is about to undergo surgery for total hip joint replacement with a prosthesis. As the attending physician you have the responsibility of fully explaining the function of the prosthesis. It should be pointed out that the artificial joint is of the same type as that of the hip. (a) What type of joint is found at the hip? Immediately following the operation the adduction of the patient's affected leg will be maintained with a device to prevent dislocation of the prosthesis. (b) Demonstrate on yourself the movement of adduction of the hip joint. Describe it.

Following the operation the physical therapist will ask the patient to exercise the gluteal and quadriceps muscles so that they will not atrophy from disuse. (c) Name the three gluteal muscles and state their actions. (d) Which member of the quadriceps femoris muscle acts on the hip joint as well as the knee joint?

## NATIONAL BOARD TYPE QUESTIONS

*Answers on page 984*

**In each of the following questions, answer:**
- (a) IF (1) IS CORRECT ONLY
- (b) IF (2) IS CORRECT ONLY
- (c) IF BOTH (1) AND (2) ARE CORRECT, AND
- (d) IF NEITHER (1) NOR (2) IS CORRECT

1. Concerning the gluteal region:
   (1) The gluteus maximus muscle covers the ischial tuberosity when one assumes the sitting position.
   (2) The upper border of the greater trochanter of the femur lies on a line connecting the anterior superior iliac spine to the ischial tuberosity.

2. Concerning the sciatic nerve:
   (1) This large nerve originates from the fourth and fifth lumbar segments of the spinal cord and from the first, second, and third sacral segments of the spinal cord.
   (2) In the buttock, the sciatic nerve lies midway between the greater trochanter of the femur and the ischial tuberosity.

3. Concerning the femoral artery:
   (1) In the thigh, the femoral artery lies midway between the anterior superior iliac spine and the pubic tubercle.
   (2) The femoral artery occupies the medial compartment of the femoral sheath.

4. Concerning the superficial veins of the lower limb:
   (1) The great saphenous vein empties into the popliteal vein through an opening in the deep fascia.
   (2) The dorsal venous arch of the foot is drained on the lateral side by the small saphenous vein, which ascends behind the lateral malleolus.

5. Concerning the dermatomes of the lower limb:
   (1) The dermatome for the lateral surface of the little toe is the first sacral segment of the spinal cord.
   (2) The dermatome for most of the medial surface of the thigh is the third lumbar segment of the cord.

6. Concerning the posterior compartment of the thigh:
   (1) The major blood supply is from the perforating branches of the profunda femoris artery.

   (2) The nerve supply to the muscles is the tibial component of the sciatic nerve, except for the short head of the biceps femoris, which is supplied by the posterior division of the obturator nerve.

7. Concerning the iliotibial tract:
   (1) It receives the insertion of the greater part of the gluteus maximus and the entire insertion of the tensor fasciae latae muscles.
   (2) When these muscles contract, the knee joint is extended.

**In the following questions, answer:**
- (a) IF ONLY (1), (2), AND (3) ARE CORRECT
- (b) IF ONLY (1) AND (3) ARE CORRECT
- (c) IF ONLY (2) AND (4) ARE CORRECT
- (d) IF ONLY (4) IS CORRECT, AND
- (e) IF ALL ARE CORRECT

8. The following nerve(s) innervate(s) at least one muscle, which acts on both the hip and the knee joints.
   (1) Obturator nerve
   (2) Saphenous nerve
   (3) Common peroneal nerve
   (4) Femoral nerve

9. In walking, the hip bone of the suspended leg is raised by which of the following muscles acting on the supported side of the body?
   (1) Obturator externus
   (2) Gluteus medius
   (3) Gluteus maximus
   (4) Gluteus minimus

10. Which of the following muscles are flexors of the thigh?
    (1) Psoas
    (2) Pectineus
    (3) Rectus femoris
    (4) Vastus lateralis

11. The following muscles dorsiflex the foot at the ankle joint.
    (1) Extensor digitorum brevis
    (2) Tibialis anterior
    (3) Peroneus longus
    (4) Extensor hallucis longus

12. The muscles that invert the foot include the:
    (1) Tibialis anterior
    (2) Extensor hallucis longus

(3) Tibialis posterior

(4) Flexor digitorum longus

13. The movement of eversion of the foot takes place mainly in the following joints:
    (1) Ankle joint
    (2) Subtalar joints
    (3) Metatarsophalangeal joint of the little toe
    (4) Calcaneonavicular and calcaneocuboid joints

14. The dorsalis pedis artery enters the sole of the foot:
    (1) By passing around the medial margin of the first metatarsal bone.
    (2) By passing between the third and fourth metatarsal bones.
    (3) By passing between the two heads of the flexor hallucis brevis muscle.
    (4) By passing between the two heads of the first dorsal interosseous muscle.

15. The following structures contribute to the boundaries of the popliteal fossa:
    (1) Semimembranosus muscle
    (2) Plantaris muscle
    (3) Biceps femoris muscle
    (4) Medial head of gastrocnemius muscle

**Multiple choice:**

16. The following structures pass through the greater sciatic foramen, **except** the:
    (a) Superior gluteal artery
    (b) Sciatic nerve
    (c) Obturator internus tendon
    (d) Pudendal nerve
    (e) Inferior gluteal vein

17. The femoral ring is bounded by the following structures, **except** the:
    (a) Femoral vein
    (b) Lacunar ligament
    (c) Superior ramus of pubis
    (d) Femoral artery
    (e) Inguinal ligament

18. A femoral hernia descends through the femoral canal and the **neck** of the hernial sac lies:
    (a) At the saphenous opening.
    (b) Above and medial to the pubic tubercle.
    (c) Below and lateral to the pubic tubercle.
    (d) In the obturator canal.
    (e) Lateral to the iliacus muscle.

19. The following structures pass through the sub-sartorial canal, **except** the:

    (a) Posterior division of the obturator nerve
    (b) Saphenous nerve
    (c) Femoral artery
    (d) Nerve to vastus intermedius
    (e) Femoral vein

20. The floor of the femoral triangle is formed by the following muscles, **except** the:
    (a) Pectineus
    (b) Adductor longus
    (c) Iliacus
    (d) Psoas
    (e) Adductor brevis

21. The peroneal artery is a branch of which artery?
    (a) Anterior tibial artery
    (b) Popliteal artery
    (c) Posterior tibial artery
    (d) Arcuate artery
    (e) Lateral plantar artery

22. Which of the following muscles does **not** insert into the plantar aspect of the foot?
    (a) Flexor digitorum longus
    (b) Peroneus tertius
    (c) Peroneus longus
    (d) Tibialis posterior
    (e) Flexor hallucis longus

23. Which of the following muscles does **not** arise from the calcaneum?
    (a) Flexor digitorum brevis
    (b) Extensor digitorum brevis
    (c) Quadratus plantae
    (d) Flexor hallucis brevis
    (e) Abductor hallucis

24. Which statement is **not** true of the ankle joint?
    (a) It is strengthened by the deltoid (medial collateral ligament).
    (b) It is a hinge joint.
    (c) It is formed by the articulation of the talus and the distal ends of the tibia and the fibula.
    (d) It is most stable in the fully plantar flexed position.
    (e) It is a synovial joint.

25. "Unlocking" of the knee joint to permit flexion is caused by the action of which muscle?
    (a) Vastus medialis
    (b) Articularis genu
    (c) Gastrocnemius
    (d) Biceps femoris
    (e) Popliteus

26. In the adult, the chief arterial supply to the head of the femur is from the:
    (a) Superficial circumflex iliac artery
    (b) Obturator artery
    (c) Branches from the medial and lateral circumflex femoral arteries
    (d) Deep external pudendal artery
    (e) Inferior gluteal artery

**Match the skin area on the left with the appropriate lymphatic drainage listed on the right:**

27. Ball of the big toe

28. Medial side of the knee joint

29. Buttock

30. Calf

    (a) Vertical group of superficial inguinal nodes
    (b) Popliteal nodes
    (c) Horizontal group of superficial inguinal nodes
    (d) Axillary nodes
    (e) None of the above

**Match the statement on the left with the appropriate ligament on the right:**

31. Hyperextension of the hip joint is prevented by _____ .
32. The _____ prevents dislocation of the femur backward at the knee joint.
33. The _____ limits abduction of the tibia at the knee joint.
34. The _____ is attached to the head of the fibula.

    (a) Posterior cruciate ligament
    (b) Ischiofemoral ligament
    (c) Anterior cruciate ligament
    (d) Lateral collateral ligament of the knee joint
    (e) None of the above

**Match the bones on the left with the appropriate arch of the foot listed on the right:**

35. Calcaneum
36. Talus
37. Lateral cuneiform
38. Base of second metatarsal
39. Cuboid
40. Sesamoid bones under the head of the first metatarsal bone

    (a) Medial longitudinal arch only
    (b) Medial and lateral longitudinal arches only
    (c) Transverse arch only
    (d) Medial longitudinal and transverse arches
    (e) Lateral longitudinal and transverse arches

# 11. The Head and Neck

The head and neck region of the body is one in which a large number of important structures are compressed into a relatively small area. It is a very interesting region, since it contains the brain, the special sense organs, the cranial nerves, and branches of the cervical plexus.

## SURFACE ANATOMY
### Surface Landmarks of the Head

#### Nasion

This is the depression in the midline at the root of the nose (Fig. 11-1).

#### Inion

The inion, or *external occipital protuberance*, is a bony prominence in the middle of the squamous part of the occipital bone (Fig. 11-1). It lies in the midline at the junction of the head and neck and gives attachment to the ligamentum nuchae, which is a large ligament that runs down the back of the neck, connecting the skull to the spinous processes of the cervical vertebrae. A line joining the nasion to the inion over the superior aspect of the head would indicate the position of the underlying *falx cerebri*, the *superior sagittal sinus*, and the *longitudinal cerebral fissure*, which separates the right and left cerebral hemispheres.

#### Vertex

This is the highest point on the skull in the sagittal plane (Fig. 11-1).

#### Anterior Fontanelle

In the baby, the anterior fontanelle lies between the two halves of the frontal bone in front and the two parietal bones behind (Fig. 11-1). It is usually not palpable after 18 months.

#### Posterior Fontanelle

In the baby, the posterior fontanelle lies between the squamous part of the occipital bone and the

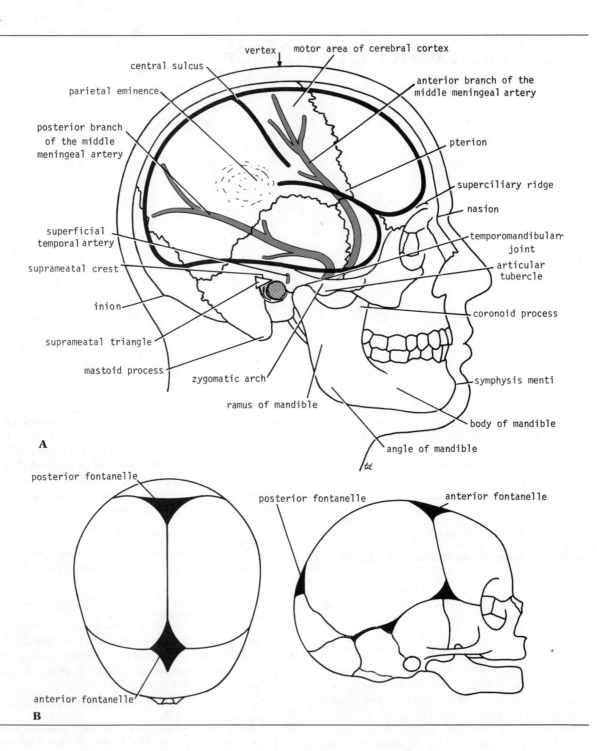

vertex

motor area of cerebral cortex

central sulcus

parietal eminence

anterior branch of the middle meningeal artery

posterior branch of the middle meningeal artery

pterion

superciliary ridge

nasion

superficial temporal artery

temporomandibular joint

suprameatal crest

articular tubercle

inion

coronoid process

suprameatal triangle

mastoid process

zygomatic arch

symphysis menti

ramus of mandible

body of mandible

angle of mandible

A

posterior fontanelle

posterior fontanelle

anterior fontanelle

anterior fontanelle

B

posterior borders of the two parietal bones (Fig. 11-1). It is usually closed by the end of the first year.

## Superciliary Ridges

These are two prominent ridges on the frontal bones above the upper margin of the orbit (Fig. 11-1). Deep to these ridges on either side of the midline lie the *frontal air sinuses.*

## Superior Nuchal Line

This is a slightly curved ridge that runs laterally from the external occipital protuberance to the mastoid process of the temporal bone. It gives attachment to the trapezius and sternocleidomastoid muscles.

## Mastoid Process of the Temporal Bone

The mastoid process projects downward and forward from behind the ear (Figs. 11-1 and 11-4). It is undeveloped in the newborn child and grows only as the result of the pull of the sternocleidomastoid, as the child moves his head. It may be recognized as a bony projection at the end of the second year.

## Auricle and External Auditory Meatus

These structures lie in front of the mastoid process (Fig. 11-53). The external auditory meatus is about 1 inch (2.5 cm) long and forms an S-shaped curve. In order to examine the outer surface of the tympanic membrane in the adult with an otoscope, the tube may be straightened by pulling the auricle upward and backward. In small children, the auricle is pulled straight back or downward and backward.

**Fig. 11-1. (A) Right side of head, showing relations of middle meningeal artery and brain to surface of skull. (B) Superior aspect and right side of neonatal skull. Note positions of anterior and posterior fontanelles.**

## Tympanic Membrane

The tympanic membrane is normally pearly gray in color and is concave toward the meatus (Fig. 11-53). The most depressed part of the concavity is called the *umbo* and is caused by the attachment of the handle of the malleus on its medial surface.

## Parietal Eminence

This structure on the lateral surface of the skull can be felt about 2 inches (5 cm) above the auricle. It lies close to the lower end of the central cerebral sulcus of the brain (Fig. 11-1).

## Zygomatic Arch

The zygomatic arch extends forward in front of the ear and ends in front in the zygomatic bone (Fig. 11-1). Above the zygomatic arch is the *temporal fossa,* which is filled with the *temporalis muscle.* Attached to the lower margin of the zygomatic arch is the masseter muscle. Contraction of both the temporalis and masseter muscles (Fig. 11-26) may be felt by clenching the teeth.

## Superficial Temporal Artery

The pulsations of the superficial temporal artery may be felt as it crosses the zygomatic arch, immediately in front of the auricle (Fig. 11-1).

## Pterion

This is the point where the greater wing of the sphenoid meets the antero-inferior angle of the parietal bone. Lying 1½ inches (4 cm) above the midpoint of the zygomatic arch (Fig. 11-1) it is not marked by an eminence or a depression, but it is important since beneath it lies the *anterior branch of the middle meningeal artery.*

Above and behind the external auditory meatus, deep to the auricle, may be felt a small depression, the *suprameatal triangle* (Fig. 11-1). This is bounded behind by a line drawn vertically upward from the posterior margin of the external auditory meatus, above by the *suprameatal crest* of the temporal bone, and below by the external auditory meatus. The bony floor of the triangle forms the lateral wall of the *mastoid antrum.*

## Temporomandibular Joint

This joint can be easily palpated in front of the auricle (Fig. 11-1). Note that as the mouth is opened, the head of the mandible rotates and moves forward below the tubercle of the zygomatic arch.

## Anterior Border of the Ramus of the Mandible

This may be felt deep to the masseter muscle. The *coronoid process* of the mandible may be felt with the finger inside the mouth, and the *pterygomandibular ligament* may be palpated as a tense band on its medial side (Fig. 11-29).

## Posterior Border of the Ramus of the Mandible

This is overlapped above by the parotid gland (Fig. 11-26), but below, it is easily felt through the skin. The *outer surface of the ramus of the mandible* is covered by the masseter muscle and can be felt on deep palpation when this muscle is relaxed.

## Body of the Mandible

This is best examined by having one finger inside the mouth and another on the outside. Thus it is possible to examine the mandible from the *symphysis menti* as far backward as the angle of the mandible (Fig. 11-1).

## Facial Artery

The pulsations of the facial artery may be felt as it crosses the lower margin of the body of the mandible, at the anterior border of the masseter (Fig. 11-5).

## Anterior Border of the Masseter

This may be easily felt by clenching the teeth.

## Parotid Duct

The parotid duct runs forward from the parotid gland 1 fingerbreadth below the zygomatic arch (Fig. 11-5). It can be rolled beneath the examining finger at the anterior border of the masseter as it turns medially and opens into the mouth opposite the upper second molar tooth (Fig. 11-57).

## Orbital Margin

The orbital margin is formed by the frontal, zygomatic, and maxillary bones (Fig. 11-48).

## Supraorbital Notch

If present, this can be felt at the junction of the medial and intermediate thirds of the upper margin of the orbit. It transmits the *supraorbital nerve*, which can be rolled against the bone (Fig. 11-48).

## Infraorbital Foramen

This lies 5 mm below the lower margin of the orbit (Fig. 11-34), on a line drawn downward from the supraorbital notch to the interval between the two lower premolar teeth.

## Infraorbital Nerve

The infraorbital nerve emerges from the foramen and supplies the skin of the face.

## Maxillary Air Sinus

This is situated within the maxillary bone and lies below the infraorbital foramen on each side (Fig. 11-63).

## Frontal Air Sinus

This is situated within the frontal bone and lies deep to the superciliary ridge on each side (Fig. 11-63).

## **Surface Landmarks of the Neck**

*In the midline anteriorly,* the following structures may be palpated from above downward.

## Symphysis Menti

The lower margin of the symphysis menti may be felt where the two halves of the body of the mandible unite in the midline (Figs. 11-2 and 11-3).

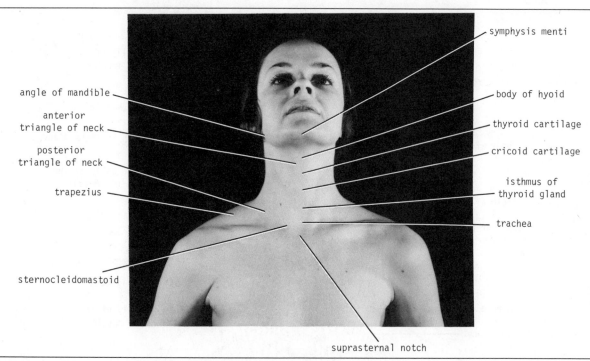

symphysis menti

angle of mandible

body of hyoid

anterior
triangle of neck

thyroid cartilage

posterior
triangle of neck

cricoid cartilage

isthmus of
thyroid gland

trapezius

trachea

sternocleidomastoid

suprasternal notch

**Fig. 11-2. Anterior view of head and neck of a 29-year-old female. Note that the atlanto-occipital joints and the cervical part of the vertebral column are partially extended for full exposure of the front of the neck.**

## Submental Triangle

The submental triangle lies between the symphysis menti and the body of the hyoid bone (Fig. 11-10). It is bounded anteriorly by the midline of the neck, laterally by the *anterior belly of the digastric muscle,* and inferiorly by the body of the hyoid bone. The floor is formed by the *mylohyoid muscle.* The *submental lymph nodes* are located in this triangle.

## Body of the Hyoid Bone

This lies opposite the third cervical vertebra (Fig. 11-44).

## Thyrohyoid Membrane

The thyrohyoid membrane fills in the interval between the hyoid bone and the thyroid cartilage (Fig. 11-3).

## Upper Border of the Thyroid Cartilage

This notched structure lies opposite the fourth cervical vertebra (Figs. 11-2 and 11-44).

## Cricothyroid Ligament

This structure fills in the interval between the cricoid cartilage and the thyroid cartilage (Fig. 11-3).

## Cricoid Cartilage

An important landmark in the neck (Fig. 11-2), this lies at the level of the sixth cervical vertebra; at the junction of the larynx with the trachea; at the level of the junction of the pharynx with the esophagus; at the level of the middle cervical sympathetic ganglion; and at the level where the inferior thyroid artery enters the thyroid gland (Fig. 11-44).

## Cricotracheal Ligament

This structure fills in the interval between the cricoid cartilage and the first ring of the trachea (Fig. 11-64).

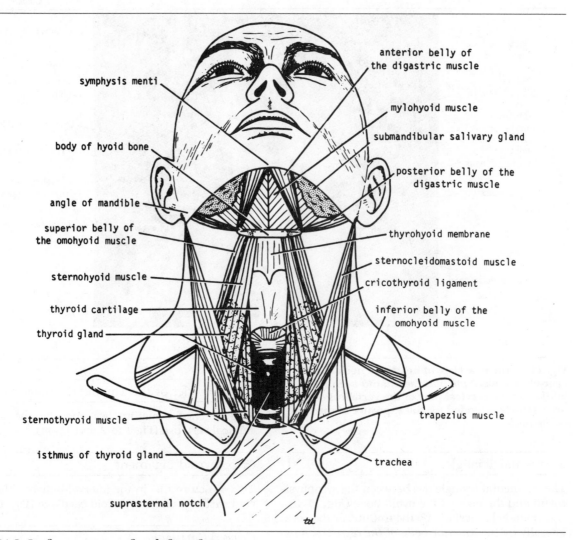

symphysis menti

anterior belly of
the digastric muscle

mylohyoid muscle

submandibular salivary gland

body of hyoid bone

posterior belly of the
digastric muscle

angle of mandible

superior belly of
the omohyoid muscle

thyrohyoid membrane

sternocleidomastoid muscle

sternohyoid muscle

cricothyroid ligament

thyroid cartilage

inferior belly of the
omohyoid muscle

thyroid gland

sternothyroid muscle

trapezius muscle

isthmus of thyroid gland

trachea

suprasternal notch

**Fig. 11-3. Surface anatomy of neck from front.**

## First Ring of the Trachea

This can be felt by gentle palpation just above the isthmus of the thyroid gland.

## Isthmus of the Thyroid Gland

This lies in front of the second, third, and fourth rings of the trachea (Figs. 11-2 and 11-3).

## Inferior Thyroid Veins

The inferior thyroid veins lie in front of the fifth, sixth, and seventh rings of the trachea (Fig. 11-17).

## Thyroidea Ima Artery

When present, this artery ascends in front of the trachea to the isthmus of the thyroid gland, from the brachiocephalic artery (Fig. 11-17).

## Jugular Arch

This vein connects the two anterior jugular veins just above the suprasternal notch (Fig. 11-44).

## Suprasternal Notch

This can be felt between the anterior ends of the clavicles (Fig. 11-2). It is the superior border of the manubrium sterni and lies opposite the lower border of the body of the second thoracic vertebra.

It should be remembered that in the adult the trachea may measure as much as 1 inch (2.5 cm) in diameter, whereas in a baby it may be narrower than a pencil. It should also be remembered that in young children the thymus gland may extend above the suprasternal notch as far as the isthmus of the thyroid gland, and the brachiocephalic artery and the left brachiocephalic vein may protrude above the suprasternal notch.

*In the midline posteriorly,* the following structures can be palpated from above downward.

## External Occipital Protuberance

The external occipital protuberance lies in the midline at the junction of the head and neck (Fig. 11-5). If the index finger is placed on the skin in the midline, it can be drawn downward in the *nuchal*

*groove.* The first spinous process to be felt is that of the *seventh cervical vertebra* (*vertebra prominens*). Cervical spines 1–6 are covered by the *ligamentum nuchae.*

## Sternocleidomastoid Muscle

On the side of the neck, the sternocleidomastoid can be palpated throughout its length as it passes upward from the sternum and clavicle to the mastoid process (Figs. 11-4 and 11-5). The muscle can be made to stand out by asking the patient to approximate his ear to the shoulder of the same side and at the same time rotate his head so that his face looks upward toward the opposite side. If the movement is carried out against resistance, the muscle will be felt to contract, and its anterior and posterior borders will be defined.

The sternocleidomastoid serves to divide the neck into anterior and posterior triangles. The *anterior* triangle of the neck is bounded by the body of the mandible, the sternocleidomastoid, and the midline (Figs. 11-9 and 11-10). The *posterior triangle* is bounded by the anterior border of the trapezius, the sternocleidomastoid, and the clavicle (Figs. 11-9 and 11-10).

## Trapezius Muscle

The anterior border of the trapezius muscle (Fig. 11-2) may be felt by asking the patient to shrug his shoulders. It will be seen to extend from the superior nuchal line of the occipital bone, downward and forward to the posterior border of the lateral third of the clavicle.

## Platysma Muscle

The platysma can be seen as a sheet of muscle by asking the patient to clench the jaws firmly. The muscle extends from the body of the mandible downward over the clavicle onto the anterior thoracic wall (Fig. 11-25).

## Root of the Neck

Here are the *suprasternal notch* in the midline anteriorly (see above) and the clavicles. Each *clavicle* is subcutaneous throughout its entire length and can be easily palpated (Figs. 11-5 and 11-9). It ar-

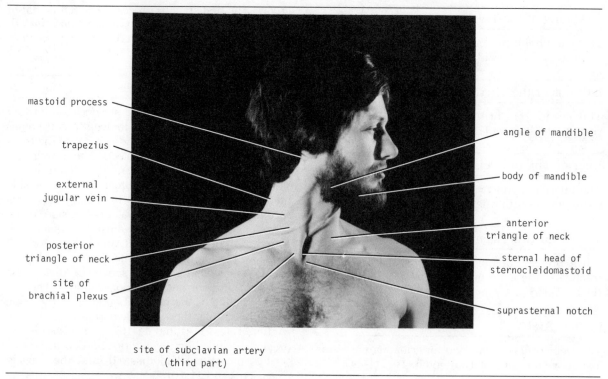

mastoid process

trapezius

external
jugular vein

posterior
triangle of neck

site of
brachial plexus

angle of mandible

body of mandible

anterior
triangle of neck

sternal head of
sternocleidomastoid

suprasternal notch

site of subclavian artery
(third part)

**Fig. 11-4. Anterior view of neck of a 27-year-old male. Note that the head has been laterally rotated to the left at the atlanto-axial joints and at the joints of the cervical part of the vertebral column.**

ticulates at its lateral extremity with the acromion process of the scapula. At the medial end of the clavicle, the *sternoclavicular joint* can be identified.

## Anterior Triangle of the Neck

The *isthmus of the thyroid gland* lies in front of the second, third, and fourth rings of the trachea (Figs. 11-2 and 11-3). The *lateral lobes of the thyroid gland* may be palpated deep to the sternocleidomastoid muscles. This is most easily carried out by standing behind the seated patient and asking the patient to flex the neck forward and so relax the overlying muscles. The observer can then examine both lobes simultaneously with the tips of the fingers of both hands.

## Carotid Sheath

The carotid sheath, which contains the *carotid arteries*, the *internal jugular vein*, the *vagus nerve*, and the *deep cervical lymph nodes*, can be marked out by a line joining the sternoclavicular joint to a point midway between the tip of the mastoid process and the angle of the mandible. At the level of the upper border of the thyroid cartilage, the *common carotid artery* bifurcates into the *internal* and *external carotid arteries* (Fig. 11-5). The pulsations of these arteries can be felt at this level.

## Posterior Triangle of the Neck

Here the spinal part of the *accessory nerve* is relatively superficial as it emerges from the posterior border of the sternocleidomastoid and runs downward and backward to pass beneath the anterior border of the trapezius (Fig. 11-5). The course of this nerve may be indicated as follows: Draw a line from the angle of the mandible to the tip of the

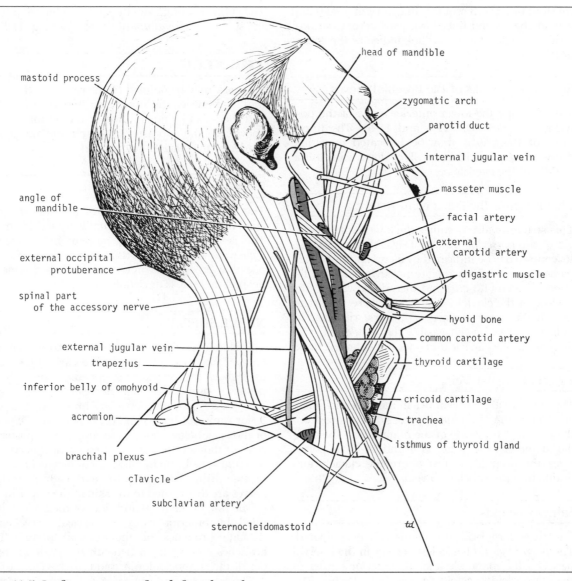

mastoid process

head of mandible

zygomatic arch

parotid duct

internal jugular vein

masseter muscle

angle of
mandible

facial artery

external
carotid artery

external occipital
protuberance

digastric muscle

spinal part
of the accessory nerve

hyoid bone

common carotid artery

external jugular vein

thyroid cartilage

trapezius

inferior belly of omohyoid

cricoid cartilage

acromion

trachea

isthmus of thyroid gland

brachial plexus

clavicle

subclavian artery

sternocleidomastoid

td

**Fig. 11-5. Surface anatomy of neck from lateral aspect.**

mastoid process. Bisect this line at right angles and extend the second line downward across the posterior triangle; the second line indicates the course of the nerve.

### Roots and Trunks of the Brachial Plexus

These occupy the lower anterior angle of the posterior triangle (Figs. 11-4 and 11-5). The upper limit of the plexus may be indicated by a line drawn from the cricoid cartilage downward to the middle of the clavicle.

### Third Part of the Subclavian Artery

This structure also occupies the lower anterior angle of the posterior triangle (Figs. 11-4 and 11-5). Its course may be indicated by a curved line, which passes upward from the sternoclavicular joint for about ½ inch (1.3 cm) and then downward to the middle of the clavicle. It is here, where the artery lies on the upper surface of the first rib, that its pulsations can be felt easily. The *subclavian vein* lies behind the clavicle and does not enter the neck.

### External Jugular Vein

The external jugular vein lies in the superficial fascia deep to the platysma. It passes downward from the region of the angle of the mandible to the middle of the clavicle (Figs. 11-4 and 11-5). It perforates the deep fascia just above the clavicle and drains into the subclavian vein.

### Salivary Glands

The three large salivary glands can be palpated. The *parotid gland* lies below the ear in the interval between the mandible and the anterior border of the sternocleidomastoid muscle (Fig. 11-26). The surface marking of the parotid duct is given on page 702.

The *submandibular gland* may be divided into superficial and deep parts. The superficial part lies beneath the lower margin of the body of the mandible (Fig. 11-33). The deep part of the submandibular gland, the *submandibular duct*, and the *sublingual gland* may be palpated through the mucous membrane covering the floor of the mouth in the interval between the tongue and the lower jaw.

The submandibular duct opens into the mouth on the side of the *frenulum of the tongue* (Fig. 11-57).

## THE NECK

The neck may be defined as the region of the body that lies between the lower margin of the mandible and the superior nuchal line of the occipital bone above and the suprasternal notch and the upper border of the clavicle below.

### SKIN

The natural lines of cleavage of the skin are constant and run almost horizontally around the neck. This is important clinically, since an incision along a cleavage line will heal as a narrow scar, whereas one that crosses the lines will heal as a wide or heaped-up scar. (For details, see p. 5.)

### NERVE SUPPLY

The skin overlying the trapezius muscle on the back of the neck, and that of the back of the scalp as high as the vertex, is supplied segmentally by posterior rami of cervical nerves 2–5 (Fig. 11-6). The *greater occipital nerve* is a branch of the posterior ramus of the second cervical nerve. The skin of the back below the neck is supplied by posterior rami of thoracic nerves. The first cervical nerve has no cutaneous branch, and the cutaneous branches of the fifth, sixth, seventh, and eighth cervical nerves are distributed to the skin of the upper limb.

The skin of the front and sides of the neck is supplied by anterior rami of cervical nerves 2–4 through branches of the cervical plexus. The branches emerge from beneath the posterior border of the sternocleidomastoid muscle (Fig. 11-6).

The *lesser occipital nerve* (C2) hooks around the accessory nerve and ascends along the posterior border of the sternocleidomastoid muscle, to supply the skin over the lateral part of the occipital region and the medial surface of the auricle (Fig. 11-6).

The *great auricular nerve* (C2 and 3) ascends across the sternocleidomastoid muscle in company with the external jugular vein. On reaching the parotid gland it divides into branches that supply the skin over the angle of the mandible, the parotid

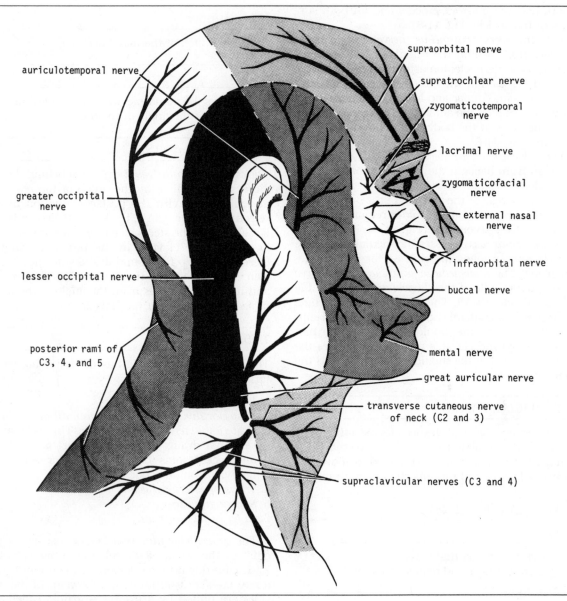

**Fig. 11-6. Sensory nerve supply to skin of head and neck. Note that the skin over the angle of the jaw is supplied by the great auricular nerve (C2 and 3) and not by branches of the trigeminal nerve.**

gland, and the mastoid process and on both surfaces of the auricle (Fig. 11-6).

The *transverse cutaneous nerve* (C2 and 3) emerges from behind the middle of the posterior border of the sternocleidomastoid muscle. It passes forward across that muscle deep to the platysma and divides into upper and lower branches, which supply the skin on the anterior and lateral surfaces of the neck, from the body of the mandible to the sternum (Fig. 11-6).

The *supraclavicular nerves* (C3 and 4), having emerged from beneath the posterior border of the sternocleidomastoid muscle, descend across the posterior triangle of the neck and pass onto the anterior thoracic wall and shoulder region, down to the level of the second rib (Fig. 11-6). The *medial supraclavicular nerve* crosses the medial end of the clavicle and supplies the skin as far as the meian plane. The *intermediate supraclavicular nerve* crosses the middle of the clavicle and supplies the skin of the chest wall. The *lateral supraclavicular nerve* crosses the lateral end of the clavicle and supplies the skin over the shoulder and the upper half of the deltoid muscle; this nerve also supplies the posterior aspect of the shoulder as far down as the spine of the scapula.

## SUPERFICIAL FASCIA

The superficial fascia of the neck forms a thin layer that encloses the platysma muscle. Also embedded in it are the cutaneous nerves referred to above, the superficial veins, and the superficial lymph nodes.

## PLATYSMA (FIG. 11-25)

### Origin

From the deep fascia that covers the upper part of the pectoralis major and deltoid muscles.

### Insertion

It passes upward into the neck as a thin muscular sheet embedded in the superficial fascia. It is inserted into the lower margin of the body of the mandible; some of the posterior fibers enter the face and blend with the muscle at the angle of the mouth. Below the chin, some of the anterior fibers interlace with the muscle fibers of the opposite side.

### Nerve Supply

Cervical branch of the facial nerve.

### Action

It depresses the mandible and also draws down the lower lip and the angle of the mouth.

## SUPERFICIAL VEINS

The *external jugular vein* begins just behind the angle of the mandible by the union of the posterior auricular vein with the posterior division of the retromandibular vein (Fig. 11-7). It descends obliquely across the sternocleidomastoid muscle and, just above the clavicle in the posterior triangle, pierces the deep fascia and drains into the subclavian vein (Fig. 11-16). It varies considerably in size, and its course extends from the angle of the mandible to the middle of the clavicle.

### Tributaries

The external jugular vein has the following tributaries:

1. *Posterior auricular vein.*
2. *Posterior division of the retromandibular vein.*
3. *Posterior external jugular vein.* This is a small vein that drains the posterior part of the scalp and neck. It joins the external jugular vein about halfway along its course.
4. *Transverse cervical vein.*
5. *Suprascapular vein.*
6. *Anterior jugular vein.*

The *anterior jugular vein* begins just below the chin, by the union of several small veins (Fig. 11-7). It runs down the neck close to the midline. Just above the suprasternal notch, the veins of the two sides are united by a transverse trunk, called the *jugular arch.* The vein then turns sharply laterally and passes deep to the sternocleidomastoid muscle to drain into the external jugular vein.

## SUPERFICIAL LYMPH NODES

The *superficial cervical lymph nodes* lie along the external jugular vein superficial to the sternocleidomastoid muscle (Fig. 11-19). They receive lym-

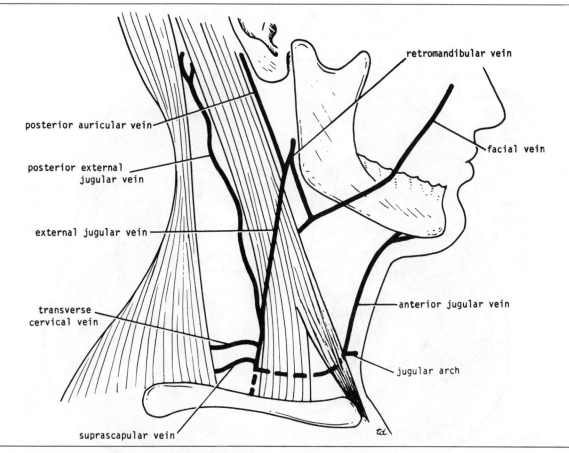

**Fig. 11-7. Major superficial veins of face and neck.**

phatic vessels from the occipital and mastoid lymph nodes (see p. 742) and drain into the deep cervical lymph nodes.

## DEEP CERVICAL FASCIA

The deep cervical fascia consists of areolar tissue that supports the muscles, vessels, and viscera of the neck (Fig. 11-8). In certain areas it is condensed to form well-defined fibrous sheets called the *investing layer*, the *pretracheal layer*, and the *prevertebral* layer. It is also condensed around the carotid vessels to form the *carotid sheath*.

The *investing layer of deep cervical fascia* completely encircles the neck, splitting to enclose the sternocleidomastoid and trapezius muscles; it is at-

tached posteriorly to the ligamentum nuchae (Fig. 11-8). It roofs over the anterior and posterior triangles of the neck.

Superiorly, the fascia is attached in front to the hyoid bone, and above this, it splits to enclose the submandibular salivary gland and is attached to the lower border of the mandible. There it splits to enclose the parotid gland, which it provides with a strong sheath (Fig. 11-26). Above the parotid it is attached to the zygomatic arch and the base of the skull. Behind the ear the fascia is attached to the mastoid process, the superior nuchal line of the occipital bone, and the external occipital protuberance. Between the angle of the mandible and the styloid process of the temporal bone, the fascial layer is thickened to form the *stylomandibular ligament* (Fig. 11-30).

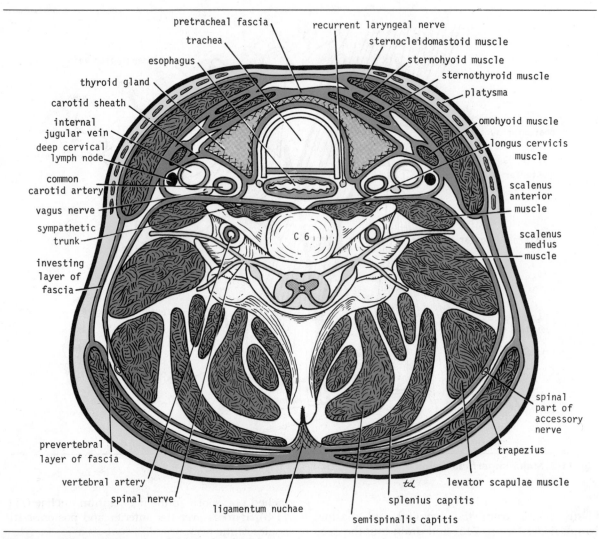

**Fig. 11-8. Cross section of neck at the level of the sixth cervical vertebra.**

Inferiorly, the fascial layer is attached to the acromion, the clavicle, and the manubrium sterni. Over the lower part of the anterior triangle, the investing fascia splits into two layers, which are attached to the anterior and posterior margins of the upper border of the manubrium. Between these two layers is a slitlike space, called the *suprasternal space* (Fig. 11-44). It contains the jugular arch, connective tissue, and sometimes a lymph node.

Over the lower part of the posterior triangle, the investing layer also splits into two layers, which are attached to the anterior and posterior borders of the clavicle. The deeper layer encloses the inferior belly of the omohyoid muscle (Fig. 11-10) and binds it down to the clavicle.

The thin *pretracheal layer of deep cervical fascia* is attached above to the thyroid and cricoid cartilages. Below, it extends into the thorax and blends with the fibrous pericardium. Laterally, it blends with the carotid sheath (see below) and with the investing layer of deep cervical fascia beneath the sternocleidomastoid muscle (Fig. 11-8). It completely surrounds the thyroid gland, forming a sheath for it, and binds the gland to the larynx (Fig. 11-17). It encloses the parathyroid glands and invests the infrahyoid muscles.

The *prevertebral layer of deep cervical fascia* covers the prevertebral muscles, namely, the longus capitis and longus cervicis (Fig. 11-8). It passes around the neck to be attached to the ligamentum nuchae. In the posterior triangle, it forms the fascial floor and here covers the scalenus anterior, the scalenus medius, the levator scapulae, the splenius capitis, and the semispinalis capitis. Superiorly, it is attached to the base of the skull; inferiorly, it enters the thorax and blends with the anterior longitudinal ligament of the vertebral column. The interval between the pharynx and the prevertebral fascia is called the *retropharyngeal space.*

As the subclavian artery and the brachial plexus emerge in the interval between the scalenus anterior and the scalenus medius muscles, they carry with them a sheath of the fascia, which extends into the axilla and is called the *axillary sheath.*

All the anterior rami of the cervical nerves that emerge in the interval between the scalenus anterior and scalenus medius lie at first deep to the prevertebral fascia. The *phrenic nerve*, the *dorsal scapular nerve*, and the *nerve to the serratus anterior* are bound down to the underlying muscles by this fascia. The spinal part of the *accessory nerve* lies superficial to the fascia.

The *carotid sheath* is a condensation of deep fascia, in which are embedded the common and internal carotid arteries, the internal jugular vein, and the vagus nerve (Fig. 11-8). The deep cervical group of lymph nodes form a chain along the internal jugular vein and are also embedded in the carotid sheath. The carotid sheath blends in front with the pretracheal and investing layers of deep fascia, and behind, with the prevertebral layer of deep fascia.

## The Triangles of the Neck

For purposes of description, the neck is divided into anterior and posterior triangles by the sternocleidomastoid muscle; the anterior triangle lies in front of the muscle and the posterior triangle lies behind it (Figs. 11-9 and 11-10).

## STERNOCLEIDOMASTOID (Fig. 11-11)

### Origin

By a rounded tendon from the front of the upper part of the manubrium sterni and by a muscular head from the medial third of the upper surface of the clavicle.

### Insertion

The two heads join one another, and the muscle is inserted into the mastoid process of the temporal bone and the lateral part of the superior nuchal line of the occipital bone.

### Nerve Supply

Spinal part of accessory nerve and the anterior rami of the second and third cervical nerves. The spinal part of the accessory nerve pierces the deep surface of the muscle and emerges from its posterior border. The cervical nerves are believed to be sensory (proprioceptive).

### Action

Both muscles acting together extend the head at the atlanto-occipital joint and flex the cervical part of the vertebral column. The contraction of one muscle pulls the ear down to the tip of the shoulder on the same side and rotates the head so that the face looks upward to the opposite side (i.e., it pulls the mastoid process of the same side down toward the sternum).

If the head is fixed by contracting the pre- and postvertebral muscles, the two sternocleidomastoid muscles can act as accessory muscles of inspiration.

### *Posterior Triangle of the Neck*

The posterior triangle of the neck is bounded anteriorly by the posterior border of the sternocleidomastoid, posteriorly by the anterior border of the trapezius, and inferiorly by the middle third of the clavicle (Figs. 11-9, 11-10, and 11-11). The triangle is covered by skin, superficial fascia, platysma, and the investing layer of deep fascia. Running across the triangle in this covering are the supraclavicular nerves.

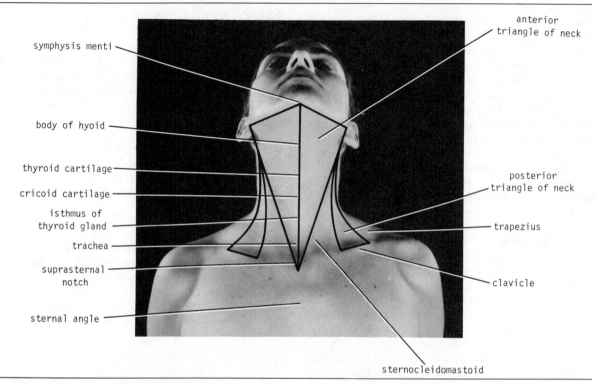

symphysis menti

body of hyoid

thyroid cartilage

cricoid cartilage

isthmus of
thyroid gland

trachea

suprasternal
notch

sternal angle

anterior
triangle of neck

posterior
triangle of neck

trapezius

clavicle

sternocleidomastoid

**Fig. 11-9. Anterior view of neck of a 29-year-old female. Important surface landmarks are shown, and the boundaries of the anterior and posterior triangles are outlined.**

The muscular floor of the triangle is covered by the prevertebral layer of deep fascia. It is formed from above downward by the semispinalis capitis, splenius capitis, levator scapulae, and scalenus medius. A small part of the scalenus anterior may be present, but it is usually overlapped and hidden by the sternocleidomastoid muscle.

The inferior belly of the omohyoid subdivides the posterior triangle into a large occipital triangle above and a small supraclavicular triangle below.

## OMOHYOID (FIG. 11-11)

### Origin and Insertion

This muscle has an inferior belly, an intermediate tendon, and a superior belly. The *inferior belly* arises from the upper margin of the scapula and the suprascapular ligament. The inferior belly is a narrow, flat muscle, which passes upward and forward across the lower part of the posterior triangle of the neck. It passes deep to the sternocleidomastoid muscle and ends in the intermediate tendon. The *intermediate tendon* is held in position by a loop of deep fascia, which slings the tendon to the clavicle and the first rib. The *superior belly* ascends almost vertically in the anterior triangle of the neck and is inserted into the lower border of the body of the hyoid bone.

### Nerve Supply

Ansa cervicalis (C1, 2, and 3).

### Action

Depresses the hyoid bone.

### OCCIPITAL TRIANGLE

The occipital triangle lies below the occipital bone of the skull (Fig. 11-10). It is bounded anteriorly by

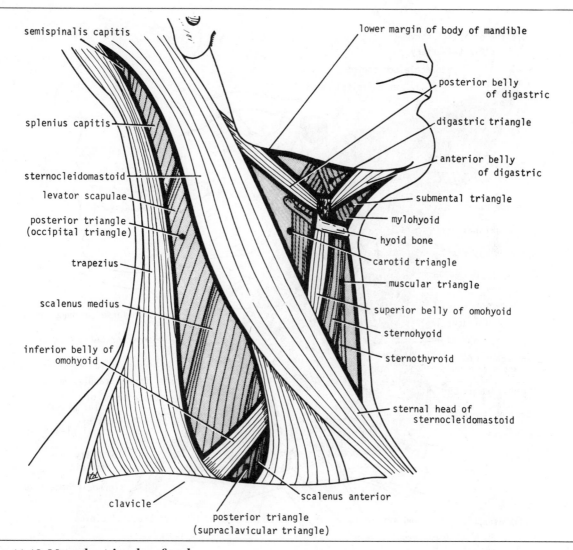

**Fig. 11-10. Muscular triangles of neck.**

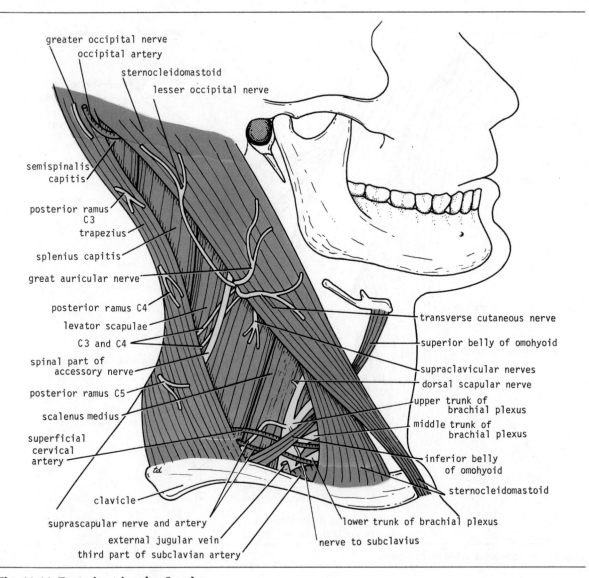

greater occipital nerve
occipital artery
sternocleidomastoid
lesser occipital nerve

semispinalis
capitis

posterior ramus
C3
trapezius

splenius capitis

great auricular nerve

posterior ramus C4

levator scapulae

C3 and C4

spinal part of
accessory nerve

posterior ramus C5

scalenus medius

superficial
cervical
artery

clavicle

suprascapular nerve and artery

external jugular vein

third part of subclavian artery

transverse cutaneous nerve

superior belly of omohyoid

supraclavicular nerves

dorsal scapular nerve

upper trunk of
brachial plexus

middle trunk of
brachial plexus

inferior belly
of omohyoid

sternocleidomastoid

lower trunk of brachial plexus

nerve to subclavius

**Fig. 11-11. Posterior triangle of neck.**

the posterior border of the sternocleidomastoid, posteriorly by the anterior border of the trapezius, and inferiorly by the inferior belly of the omohyoid.

## SUPRACLAVICULAR TRIANGLE

The supraclavicular triangle lies above the middle third of the clavicle (Fig. 11-10). It is bounded anteriorly by the posterior border of the sternocleidomastoid, superiorly by the inferior belly of the omohyoid, and inferiorly by the clavicle.

## CONTENTS OF THE POSTERIOR TRIANGLE

### Arteries

Subclavian artery (third part), superficial cervical artery, suprascapular artery, occipital artery.

### Veins

External jugular vein and its tributaries; subclavian vein (occasionally).

### Nerves

Brachial plexus, spinal part of accessory nerve, branches of cervical plexus.

## THIRD PART OF THE SUBCLAVIAN ARTERY

The subclavian artery, for purposes of description, is divided into three parts by the scalenus anterior muscle, which crosses the artery anteriorly (Fig. 11-18). The first part of the artery extends from its origin (see p. 740) to the medial border of the scalenus anterior. The second part lies posterior to this muscle, and the third part extends from the lateral border of the scalenus anterior to the outer border of the first rib; here, the subclavian artery becomes the axillary artery.

The third part of the subclavian artery enters the antero-inferior angle of the posterior triangle (Fig. 11-11) and disappears behind the middle of the clavicle. Together with the brachial plexus of nerves, it carries with it a sheath of fascia, the *axillary sheath*, derived from the prevertebral layer of deep cervical fascia.

### Branches

The third part of the subclavian artery usually has no branches.*

### Relations

#### Anteriorly

The skin, superficial fascia, platysma, supraclavicular nerves, investing layer of deep cervical fascia, external jugular vein and its tributaries (the transverse cervical, suprascapular, and anterior jugular veins), suprascapular artery, and clavicle. The nerve to the subclavius crosses the artery.

At first, the artery is covered by the sternocleidomastoid muscle. In the intermediate part of its course, it is comparatively superficial, and its pulsations can easily be felt. The terminal part of the artery descends behind the clavicle and the subclavius muscle (Fig. 11-11). The subclavian vein lies below and in front of the artery.

#### Posteriorly

The lower trunk of the brachial plexus and the scalenus medius (Fig. 11-11).

#### Superiorly

The upper and middle trunks of the brachial plexus.

#### Inferiorly

The upper surface of the first rib (Fig. 11-18).

## SUPERFICIAL CERVICAL ARTERY

The superficial cervical artery is a branch of the thyrocervical trunk, which is a branch of the first part of the subclavian artery (Figs. 11-11 and 11-18). It runs across the lower part of the posterior triangle in front of the trunks of the brachial plexus and disappears deep to the trapezius muscle.

---

*Occasionally the superficial cervical or the suprascapular artery, or both, arise from the third part of the subclavian artery.

## SUPRASCAPULAR ARTERY

The suprascapular artery is also a branch of the thyrocervical trunk (Fig. 11-18). It runs across the lower part of the posterior triangle, inferior to the superficial cervical artery. It passes in front of the third part of the subclavian artery and the trunks of the brachial plexus (Fig. 11-11). It then follows the suprascapular nerve into the supraspinous fossa and takes part in the arterial anastomosis around the scapula.

## OCCIPITAL ARTERY

The occipital artery is a branch of the external carotid artery (Fig. 11-15). It enters the posterior triangle at its superior angle, appearing between the sternocleidomastoid and the trapezius muscles (Fig. 11-11). The artery then ascends in a tortuous course over the back of the scalp, accompanied by the greater occipital nerve.

## EXTERNAL JUGULAR VEIN

The external jugular vein is an important superficial vein in the neck and is described fully on page 710.

## SUBCLAVIAN VEIN

The subclavian vein lies below and in front of the third part of the subclavian artery (Fig. 11-18), behind the clavicle; it does not usually enter the posterior triangle of the neck. Occasionally, however, it bulges upward above the clavicle. It lies on the upper surface of the first rib in front of the subclavian artery and is a continuation of the axillary vein.

## BRACHIAL PLEXUS

The brachial plexus is formed from the anterior rami of the fifth, sixth, seventh, and eighth cervical nerves and from the first thoracic nerve (Fig. 11-12). It lies in the anteroinferior angle of the posterior triangle (Fig. 11-11). For purposes of description, the plexus may be divided into the roots, the trunks, the divisions, and the cords. (See Fig. 9-22).

The *roots of the brachial plexus* enter the posterior triangle of the neck by emerging through the interval between the scalenus anterior and scalenus medius muscles (Fig. 11-18). Together with the subclavian artery, the plexus acquires a sheath, the *axillary sheath*, which is derived from the prevertebral layer of deep cervical fascia.

The *trunks of the brachial plexus* are formed as follows: The fifth and sixth cervical roots quickly unite to form the *upper trunk* of the plexus. The seventh cervical root continues as the *middle trunk* of the plexus. The eighth cervical and first thoracic roots unite to form the *lower trunk* of the plexus, which comes to lie behind the third part of the subclavian artery (Fig. 11-11).

The *divisions of the brachial plexus* are formed by each trunk's dividing into anterior and posterior branches (Fig. 11-12). This takes place in the supraclavicular triangle.

The *cords of the brachial plexus* are formed as follows: The *lateral cord* is formed by the union of the anterior divisions of the upper and middle trunks (Fig. 11-12). The *posterior cord* is formed by the union of the posterior divisions of the upper, middle, and lower trunks. The *medial cord* is formed from the anterior division of the lower trunk.

The cords of the plexus leave the posterior triangle by descending behind the clavicle and entering the axilla. The arrangement and branches of the brachial plexus in the axilla are described on page 428.

### *Branches of the Brachial Plexus in the Posterior Triangle of the Neck*

#### Branches from the Roots

The *dorsal scapular nerve* arises from the fifth cervical root (Fig. 11-11). It pierces the scalenus medius and supplies the levator scapulae and the rhomboid muscles.

The *long thoracic nerve* arises from the fifth, sixth, and seventh cervical roots (Fig. 11-12). After its fibers of origin join, the nerve descends behind the brachial plexus and the subclavian vessels and crosses the outer border of the first rib, to enter the axilla. It supplies the serratus anterior muscle.

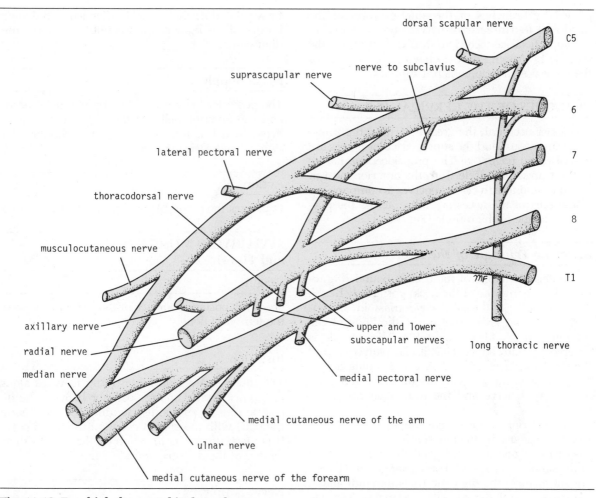

**Fig. 11-12. Brachial plexus and its branches.**

## Branches from the Trunks

The *suprascapular nerve* arises from the upper trunk of the brachial plexus (Fig. 11-11). It passes laterally and downward, accompanied by the suprascapular vessels. It enters the supraspinous fossa of the scapula through the suprascapular notch. It supplies the supraspinatus and infraspinatus muscles.

The *nerve to the subclavius* arises from the upper trunk of the brachial plexus (Fig. 11-11). It passes downward in front of the brachial plexus and the third part of the subclavian artery. It leaves the posterior triangle by descending behind the clavicle and in front of the subclavian vein, to supply the subclavius muscle. The importance of this nerve clinically is that it may contain *accessory phrenic fibers*, which join the phrenic nerve in the superior mediastinum. (See p. 120.)

## ACCESSORY NERVE (SPINAL PART)

The spinal part of the accessory nerve enters the posterior triangle by emerging from beneath the posterior border of the sternocleidomastoid muscle (Fig. 11-11). It runs downward and laterally across the posterior triangle on the levator scapulae muscle, but is separated from the muscle by the prevertebral layer of deep cervical fascia. It is accompanied by branches of the anterior rami of the third and fourth cervical nerves and is related to the superficial cervical lymph nodes. It leaves the triangle by passing deep to the anterior border of the trapezius muscle, which it supplies.

## BRANCHES OF THE CERVICAL PLEXUS

The lesser occipital, the great auricular, the transverse cutaneous, and the supraclavicular nerves are described on page 708. The proprioceptive nerves to the trapezius muscle from the anterior rami of the third and fourth cervical nerves accompany the spinal part of the accessory nerve across the posterior triangle to the muscle (Fig. 11-11).

## *Anterior Triangle of the Neck*

The anterior triangle of the neck is bounded anteriorly by the midline of the neck, posteriorly by the anterior border of the sternocleidomastoid, and superiorly by the lower margin of the body of the mandible (Figs. 11-9 and 11-13). The triangle is covered by skin, superficial fascia, platysma, and the investing layer of deep fascia. Running across the triangle in this covering are the cervical branch of the facial nerve and the transverse cutaneous nerve.

The anterior triangle may be subdivided into smaller triangles by the anterior and posterior bellies of the digastric muscle and the superior belly of the omohyoid muscle. These triangles are called the submental, the digastric (or submandibular), the carotid, and the muscular triangles (Fig. 11-10).

## DIGASTRIC MUSCLE (FIG. 11-13)

### Origin and Insertion

This muscle has a posterior belly, an intermediate tendon, and an anterior belly. *The posterior belly* arises from the medial surface of the mastoid process of the temporal bone, passes downward and forward across the carotid sheath, and ends in the intermediate tendon. The *intermediate tendon* pierces the stylohyoid insertion and is held in position by a loop of deep fascia, which binds the tendon down to the junction of the body and greater cornu of the hyoid bone. The *anterior belly* runs

forward and medially and is attached to the lower border of the body of the mandible, near the median plane.

### Nerve Supply

The posterior belly is supplied by the facial nerve, and the anterior belly by the nerve to the mylohyoid, which is a branch of the mandibular division of the trigeminal nerve.

### Action

Depresses the mandible or elevates the hyoid bone.

## STYLOHYOID MUSCLE (FIGS. 11-13 and 11-14)

### Origin

From the posterior surface of the styloid process of the temporal bone.

### Insertion

The muscle passes downward and forward along the upper border of the posterior belly of the digastric muscle and is inserted into the junction of the body with the greater cornu of the hyoid bone. It is pierced near its insertion by the intermediate tendon of the digastric muscle.

### Nerve Supply

Facial nerve.

### Action

Elevates the hyoid bone.

## SUBMENTAL TRIANGLE

The submental triangle lies below the chin and is bounded anteriorly by the midline of the neck, laterally by the anterior belly of the digastric, and inferiorly by the body of the hyoid bone. The floor of the triangle is formed by the mylohyoid muscle. It contains the submental lymph nodes and the beginning of the anterior jugular vein.

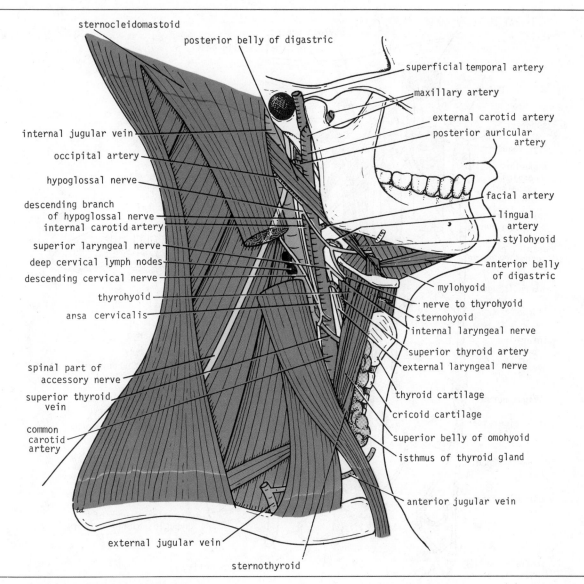

**Fig. 11-13. Anterior triangle of neck.**

sternocleidomastoid

posterior belly of digastric

superficial temporal artery

maxillary artery

external carotid artery

posterior auricular artery

internal jugular vein

occipital artery

hypoglossal nerve

descending branch of hypoglossal nerve

internal carotid artery

superior laryngeal nerve

deep cervical lymph nodes

descending cervical nerve

thyrohyoid

ansa cervicalis

spinal part of accessory nerve

superior thyroid vein

common carotid artery

facial artery

lingual artery

stylohyoid

anterior belly of digastric

mylohyoid

nerve to thyrohyoid

sternohyoid

internal laryngeal nerve

superior thyroid artery

external laryngeal nerve

thyroid cartilage

cricoid cartilage

superior belly of omohyoid

isthmus of thyroid gland

anterior jugular vein

external jugular vein

sternothyroid

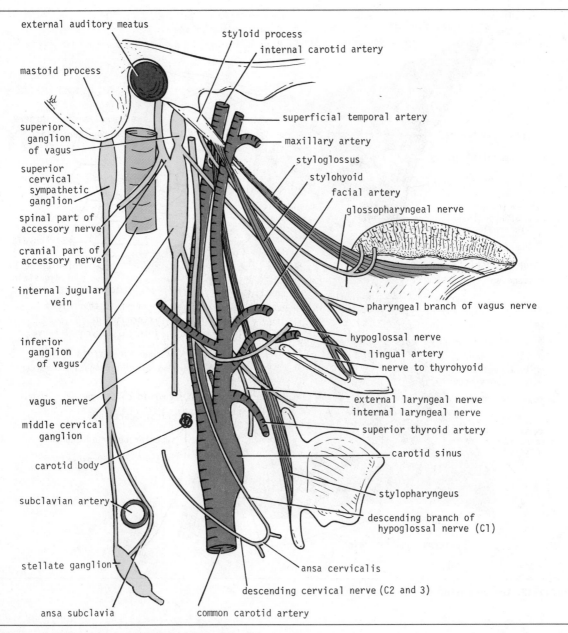

external auditory meatus
styloid process
internal carotid artery
mastoid process
superior ganglion of vagus
superior cervical sympathetic ganglion
spinal part of accessory nerve
cranial part of accessory nerve
internal jugular vein
inferior ganglion of vagus
vagus nerve
middle cervical ganglion
carotid body
subclavian artery
stellate ganglion
ansa subclavia
common carotid artery
descending cervical nerve (C2 and 3)
ansa cervicalis
descending branch of hypoglossal nerve (C1)
stylopharyngeus
carotid sinus
superior thyroid artery
internal laryngeal nerve
external laryngeal nerve
nerve to thyrohyoid
lingual artery
hypoglossal nerve
pharyngeal branch of vagus nerve
glossopharyngeal nerve
facial artery
stylohyoid
styloglossus
maxillary artery
superficial temporal artery

**Fig. 11-14. Styloid muscles, vessels, and nerves of neck.**

## DIGASTRIC TRIANGLE

The digastric triangle lies below the body of the mandible (Fig. 11-10). It is bounded anteriorly by the anterior belly of the digastric and posteriorly by the posterior belly of the digastric and the stylohyoid muscles. It is bounded above by the lower border of the body of the mandible. The floor of the triangle is formed by the mylohyoid and hyoglossus muscles and the superior constrictor muscle of the pharynx.

The anterior part of the triangle contains the submandibular salivary gland (Fig. 11-3), with the facial artery deep to it, and the facial vein and submandibular lymph nodes superficial to it. The hypoglossal nerve runs on the hyoglossus muscle deep to the gland (Fig. 11-13); the nerve and vessels to the mylohyoid muscle run on the inferior surface of this muscle (Fig. 11-27).

In the posterior part of the triangle lies the carotid sheath, with the carotid arteries, internal jugular vein, and vagus nerve (Figs. 11-8 and 11-13). The stylopharyngeus muscle and the glossopharyngeal nerve are deeply placed here (Fig. 11-28). The lower part of the parotid gland projects into the triangle.

## CAROTID TRIANGLE

The carotid triangle lies behind the hyoid bone. It is bounded superiorly by the posterior belly of the digastric, inferiorly by the superior belly of the omohyoid, and posteriorly by the anterior border of the sternocleidomastoid muscle (Fig. 11-10). Its floor is formed by portions of the thyrohyoid, hyoglossus, and middle and inferior constrictor muscles of the pharynx.

The triangle contains the carotid sheath, with the common carotid artery dividing within the triangle into internal and external carotid arteries; numerous branches of the external carotid artery; the internal jugular vein and its tributaries; the hypoglossal nerve with its descending branch; the internal and external laryngeal nerves; the accessory and vagus nerves; and part of the chain of deep cervical lymph nodes (Fig. 11-13).

The internal jugular vein tends to be overlapped by the anterior border of the sternocleidomastoid muscle and is therefore hidden from view.

## MUSCULAR TRIANGLE

The muscular triangle lies below the hyoid bone. It is bounded anteriorly by the midline of the neck, superiorly by the superior belly of the omohyoid, and inferiorly by the anterior border of the sternocleidomastoid muscle (Fig. 11-10). Its floor is formed by the sternohyoid and sternothyroid muscles. Beneath the floor lie the thyroid gland, the larynx, the trachea, and the esophagus (Fig. 11-13).

## STERNOHYOID (FIGS. 11-3 and 11-13)

### Origin

From the posterior surface of the manubrium sterni and the adjoining portion of the clavicle.

### Insertion

The muscle runs upward and medially and is inserted into the medial part of the lower border of the body of the hyoid bone.

### Nerve Supply

Ansa cervicalis (C1, 2, and 3).

### Action

Depresses the hyoid bone.

## OMOHYOID

The omohyoid muscle is described on page 714.

## STERNOTHYROID (FIG. 11-13)

### Origin

From the posterior surface of the manubrium sterni.

### Insertion

The muscle runs upward deep to the sternohyoid, covering the lateral lobe of the thyroid gland. It is inserted into the oblique line on the lamina of the thyroid cartilage.

## Nerve Supply

Ansa cervicalis (C1, 2, and 3).

## Action

Depresses the larynx.

## THYROHYOID (FIG. 11-13)

### Origin

From the oblique line on the lamina of the thyroid cartilage.

### Insertion

The muscle runs upward over the thyrohyoid membrane and is inserted into the lower border of the body of the hyoid bone.

### Nerve Supply

The first cervical nerve via a branch of the hypoglossal nerve.

### Action

Depresses the hyoid bone or elevates the larynx.

## Main Arteries of the Neck (Fig. 11-15)

### COMMON CAROTID ARTERY

The right common carotid artery arises from the brachiocephalic artery behind the right sternoclavicular joint (Fig. 11-18). The left artery arises from the arch of the aorta in the superior mediastinum. (See p. 117.) The common carotid artery runs upward and backward through the neck, from the sternoclavicular joint to the upper border of the thyroid cartilage, where it divides into the external and internal carotid arteries (Fig. 11-13). At its point of division, the terminal part of the common carotid artery or the beginning of the internal carotid artery shows a localized dilatation, called the *carotid sinus* (Fig. 11-14). The tunica media of the sinus is thinner than elsewhere, but the adventitia is relatively thick and contains numerous nerve endings derived from the glossopharyngeal nerve. The carotid sinus serves as a reflex pressoreceptor mechanism, which assists in the regulation of the blood pressure in the cerebral arteries.

The *carotid body* is a small, reddish-brown structure, which lies posterior to the point of bifurcation of the common carotid artery (Fig. 11-14). It is innervated by the glossopharyngeal nerve and is a chemoreceptor, being sensitive to excess carbon dioxide and reduced oxygen tension in the blood. Such a stimulus reflexly produces a rise in blood pressure and heart rate and an increase in respiratory movements.

The common carotid artery is embedded in the carotid sheath throughout its course and is closely related to the internal jugular vein and vagus nerve (Fig. 11-8). Apart from the two terminal branches, the common carotid artery gives off no branches.

### Relations

#### Anterolaterally

The skin, superficial fascia, platysma, investing layer of deep cervical fascia, sternocleidomastoid, sternohyoid, sternothyroid, and superior belly of the omohyoid. The descendens hypoglossi and the ansa cervicalis are embedded in the anterior wall of the carotid sheath (Fig. 11-13). The superior and middle thyroid veins and the anterior jugular vein cross the artery.

#### Posteriorly

The transverse processes of the lower four cervical vertebrae, the longus capitis and longus colli muscles, and the origin of the scalenus anterior muscle; the sympathetic trunk (Fig. 11-18). In the lower part of the neck are the vertebral vessels and the inferior thyroid artery, as it bends medially to supply the thyroid gland. On the left side, the thoracic duct arches laterally and crosses behind the artery (Fig. 11-18). Behind its termination is the carotid body.

#### Medially

The larynx and pharynx, and below these, the trachea and esophagus (Fig. 11-8). The lobe of the thyroid gland, the inferior thyroid artery, and the recurrent laryngeal nerve also lie medially.

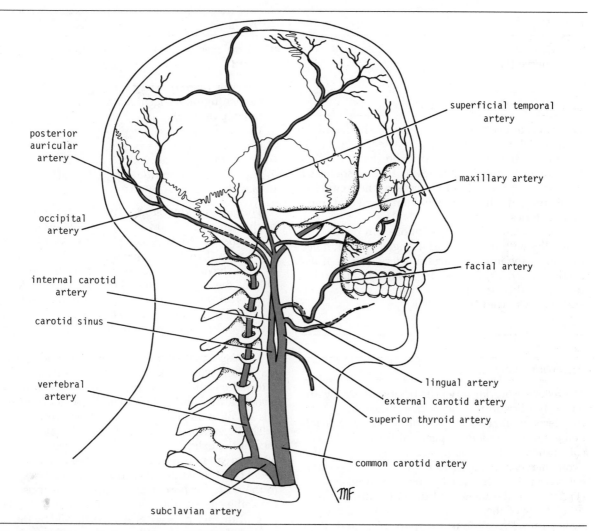

posterior
auricular
artery

occipital
artery

internal carotid
artery

carotid sinus

vertebral
artery

subclavian artery

superficial temporal
artery

maxillary artery

facial artery

lingual artery

external carotid artery

superior thyroid artery

common carotid artery

**Fig. 11-15. Main arteries of the head and neck. Note that for clarity, the thyrocervical trunk, the costocervical trunk, and the internal thoracic artery, branches of the subclavian artery, have not been shown.**

## Laterally

The internal jugular vein, and posterolaterally, the vagus nerve (Fig. 11-8).

## EXTERNAL CAROTID ARTERY

The external carotid artery is one of the terminal branches of the common carotid artery (Fig. 11-8). It begins at the level of the upper border of the thyroid cartilage and terminates in the substance of the parotid gland behind the neck of the mandible by dividing into the superficial temporal and maxillary arteries.

At its origin, where its pulsations can be felt, the artery lies within the carotid triangle. At first, it lies medial to the internal carotid artery, but as it ascends in the neck, it passes backward and laterally. It is crossed by the posterior belly of the digastric and the stylohyoid (Fig. 11-13).

## *Relations*

### Anterolaterally

The artery is overlapped at its beginning by the anterior border of the sternocleidomastoid. Above this level the artery is comparatively superficial, being covered by skin, superficial fascia, the cervical branch of the facial nerve, the transverse cutaneous nerve, and the investing layer of deep cervical fascia. It is crossed by the hypoglossal nerve (Fig. 11-13), the facial and lingual veins, the posterior belly of the digastric muscle, and the stylohyoid muscles. Within the parotid gland it is crossed by the facial nerve (Fig. 11-26). The internal jugular vein first lies lateral to the artery and then posterior to it.

### Medially

The wall of the pharynx, the styloid process, and the internal carotid artery. The stylopharyngeus muscle, the glossopharyngeal nerve, and the pharyngeal branch of the vagus pass between the external and internal carotid arteries (Fig. 11-14).

For the relations of the external carotid artery in the parotid gland, see page 758.

## *Branches*

The branches of the external carotid artery are as follows:

1. Superior thyroid artery.
2. Ascending pharyngeal artery.
3. Lingual artery.
4. Facial artery.
5. Occipital artery.
6. Posterior auricular artery.
7. Superficial temporal artery.
8. Maxillary artery.

The *superior thyroid artery* arises from the front of the external carotid artery near its origin (Figs. 11-13 and 11-15). It passes almost vertically downward with the vein. It runs superficial and parallel to the external laryngeal nerve to reach the upper pole of the thyroid gland (Fig. 11-17). It gives off: (1) a *branch to the sternocleidomastoid muscle* and (2) the *superior laryngeal artery*, which pierces the thyrohyoid membrane with the internal laryngeal nerve.

The *ascending pharyngeal artery* is a long, slender vessel, which arises on the deep surface of the external carotid artery. It ascends between the internal carotid and the pharynx. It gives off numerous small branches to adjacent structures.

The *lingual artery* arises from the anterior surface of the external carotid artery, opposite the tip of the greater cornu of the hyoid bone (Figs. 11-13 and 11-15). It loops upward and then passes deep to the posterior border of the hyoglossus muscle to enter the submandibular region. (See p. 782.) The loop of the artery is crossed superficially by the hypoglossal nerve.

The *facial artery* arises from the anterior surface of the external carotid artery, just above the level of the tip of the greater cornu of the hyoid bone (Figs. 11-13 and 11-15). It arches upward deep to the posterior belly of the digastric muscle and reaches the posterior part of the submandibular salivary gland. (See p. 781.)

The *occipital artery* arises from the posterior surface of the external carotid artery, opposite the facial artery (Figs. 11-13 and 11-15). Close to its origin it is crossed by the hypoglossal nerve. It passes upward under cover of the lower border of the posterior belly of the digastric muscle and finally reaches the back of the scalp by entering the apical region of the posterior triangle, between the sternocleidomastoid and the trapezius (Fig. 11-11). Its terminal part accompanies branches of the greater occipital nerve, to supply the back of the scalp.

The *posterior auricular artery* arises from the posterior surface of the external carotid artery, at the level of the upper border of the posterior belly of the digastric (Figs. 11-13 and 11-15). It passes upward and backward along the upper border of the posterior belly of the digastric, deep to the parotid gland, and reaches the groove between the auricle and the back of the scalp.

The *superficial temporal artery* (Figs. 11-13 and 11-15) is described on page 746.

The *maxillary artery* (Figs. 11-13 and 11-15) is described on page 768.

## INTERNAL CAROTID ARTERY

The internal carotid artery is one of the terminal branches of the common carotid artery (Figs. 11-13 and 11-15). It begins at the level of the upper border of the thyroid cartilage and ascends in the neck to the base of the skull. It enters the cranial cavity through the carotid canal in the petrous part of the temporal bone. (See p. 787.) It lies embedded in the carotid sheath with the internal jugular vein and vagus nerve. At its beginning it lies superficially in the carotid triangle. After ascending deep to the posterior belly of the digastric, it lies deep to the parotid gland, the styloid process, and the muscles attached to it (Figs 11-14 and 11-26). *The internal carotid artery gives off no branches in the neck.*

### Relations

#### Anterolaterally

*Below the digastric* lie the skin, superifical fascia, platysma, transverse cutaneous nerve, investing layer of deep cervical fascia, anterior border of the sternocleidomastoid, lingual and facial veins, hypoglossal nerve, descending branch of hypoglossal nerve, and occipital artery (Fig. 11-13).

*Above the digastric* lie the posterior auricular artery, stylohyoid muscle, styloid process, stylopharyngeus muscle, glossopharyngeal nerve, pharyn-

geal branch of the vagus, and parotid gland and its contents, namely, the facial nerve, retromandibular vein, and external carotid artery (Figs. 11-14 and 11-26).

#### Posteriorly

The superior laryngeal nerve, cervical part of the sympathetic trunk (Fig. 11-14), longus capitis muscle, and transverse processes of the upper three cervical vertebrae.

#### Medially

The pharyngeal wall and superior laryngeal nerve.

#### Laterally

The internal jugular vein and the vagus nerve.

## Main Veins of the Neck (Fig. 11-16)

The main veins of the neck that lie superficial to the deep fascia of the neck, namely, the external and anterior jugular veins, have been described previously.

### INTERNAL JUGULAR VEIN

The internal jugular vein receives blood from the brain, from the face, and from the neck. It begins at the jugular foramen in the skull as a continuation of the sigmoid sinus. (See p. 795.) It descends through the neck in the carotid sheath and unites with the subclavian vein behind the medial end of the clavicle, to form the brachiocephalic vein (Figs. 11-16 and 11-18).

The vein has a dilatation at its upper end, called the *superior bulb*, and another near its termination, called the *inferior bulb*. Directly above the inferior bulb is a bicuspid valve.

### Relations

#### Anterolaterally

The skin, superficial fascia, platysma, transverse cutaneous nerve, investing layer of deep cervical fascia, sternocleidomastoid, and parotid salivary

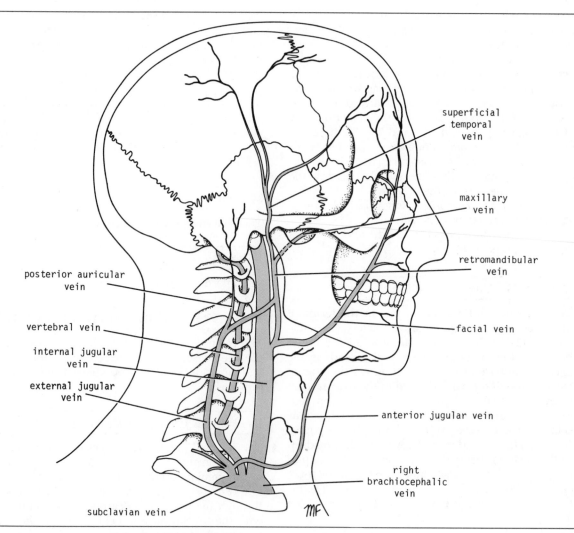

**Fig. 11-16. Main veins of the head and neck.**

gland. Its lower part is covered by the sternothyroid, sternohyoid, and omohyoid muscles, which intervene between the vein and the sternocleidomastoid (Fig. 11-13). The ansa cervicalis crosses the vein. Higher up, it is crossed by the stylohyoid, the posterior belly of the digastric, the posterior auricular and occipital arteries, and the spinal part of the accessory nerve. The styloid process and the stylopharyngeus muscles separate the vein from the parotid gland. The chain of deep cervical lymph nodes runs alongside the vein.

## Posteriorly

The transverse processes of the cervical vertebrae, levator scapulae, scalenus medius, scalenus anterior, cervical plexus, phrenic nerve, thyrocervical trunk, vertebral vein, and the first part of the subclavian artery (Fig. 11-18). On the left side it passes in front of the thoracic duct.

### Medially

Above lie the internal carotid artery and the ninth, tenth, eleventh, and twelfth cranial nerves. Below lie the common carotid artery and the vagus nerve.

### Tributaries

The *inferior petrosal sinus*, which assists in draining the cavernous sinus, leaves the skull through the anterior part of the jugular foramen and joins the internal jugular vein at or below the superior bulb (Figs. 11-41 and 11-56).

The *facial vein*, having left the face and crossed superficially over the submandibular salivary gland, is joined by the anterior division of the retromandibular vein (Fig. 11-7). The vein then crosses the hypoglossal nerve, the loop of the lingual artery, and the external and internal carotid arteries to join the internal jugular vein.

The *pharyngeal veins* drain the pharyngeal venous plexus and join the facial, lingual, or internal jugular vein.

The *lingual vein* joins the facial vein or drains into the internal jugular vein.

The *superior thyroid vein* leaves the superior pole of the thyroid gland and drains into the facial or internal jugular vein (Fig. 11-17).

The *middle thyroid vein* leaves the lobe of the thyroid gland and drains into the internal jugular vein at the level of the cricoid cartilage (Fig. 11-17).

Occasionally, the *occipital vein* accompanies the occipital artery and drains into the internal jugular vein. More often, it joins the vertebral or posterior auricular veins.

### DEEP CERVICAL LYMPH NODES

The deep cervical lymph nodes form a chain along the course of the internal jugular vein (Fig. 11-13). They are embedded in the fascia of the carotid sheath, and the majority lie on the anterolateral aspect of the internal jugular vein. They receive afferent lymph vessels from neighboring anatomical structures and from all the other groups of lymph nodes in the head and neck (Fig. 11-19). The efferent lymph vessels from the nodes join to form the *jugular lymph trunk*. This vessel drains into either the thoracic duct, the right lymphatic duct, or the subclavian lymph trunk, or it may drain independently into the brachiocephalic vein.

## Main Nerves of the Neck

### VAGUS NERVE (TENTH CRANIAL NERVE)

The vagus nerve is composed of both motor and sensory fibers. It originates in the medulla oblongata and leaves the skull through the middle of the jugular foramen in company with the ninth and eleventh cranial nerves. (See p. 795.) The vagus nerve possesses two sensory ganglia, a rounded *superior ganglion*, which is situated on the nerve within the jugular foramen, and a cylindrical *inferior ganglion*, which lies on the nerve just below the foramen (Fig. 11-14). Below the inferior ganglion the cranial part of the accessory nerve joins the vagus nerve and is distributed mainly in its pharyngeal and recurrent laryngeal branches.

The vagus nerve passes vertically down the neck within the carotid sheath, lying at first between the internal jugular vein and the internal carotid artery and then between the vein and the common carotid artery (Fig. 11-8). At the root of the neck the nerve accompanies the common carotid artery and lies anterior to the first part of the subclavian artery (Fig. 11-18). The further course of the vagus nerve in the thorax is described on page 118. (See also Fig. 11-87.)

### Branches of the Vagus Nerve in the Neck

The *meningeal branch* of the vagus nerve in the neck arises from the superior ganglion and supplies the dura mater in the posterior fossa of the skull.

The *auricular branch* arises from the superior ganglion and passes through a bony canal of the skull, to emerge behind the external auditory meatus. It supplies the medial surface of the auricle, the floor of the external auditory meatus, and the adjacent part of the tympanic membrane.

The *pharyngeal branch* arises from the inferior ganglion and contains motor fibers from the cranial part of the accessory nerve (Fig. 11-14). It passes forward between the internal and external carotid arteries, to reach the pharyngeal wall. It joins branches from the glossopharyngeal nerve and the sympathetic trunk, to form the *pharyngeal*

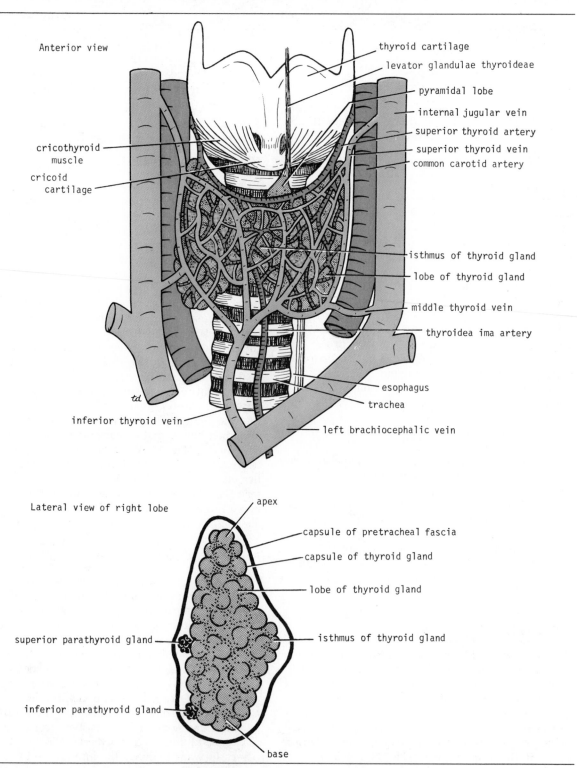

Anterior view

thyroid cartilage
levator glandulae thyroideae
pyramidal lobe
internal jugular vein
superior thyroid artery
superior thyroid vein
common carotid artery

cricothyroid muscle
cricoid cartilage

isthmus of thyroid gland
lobe of thyroid gland
middle thyroid vein
thyroidea ima artery

esophagus
trachea

inferior thyroid vein
left brachiocephalic vein

td

Lateral view of right lobe

apex
capsule of pretracheal fascia
capsule of thyroid gland
lobe of thyroid gland

isthmus of thyroid gland

superior parathyroid gland

inferior parathyroid gland

base

**Fig. 11-17. Thyroid gland; its blood supply and venous drainage.**

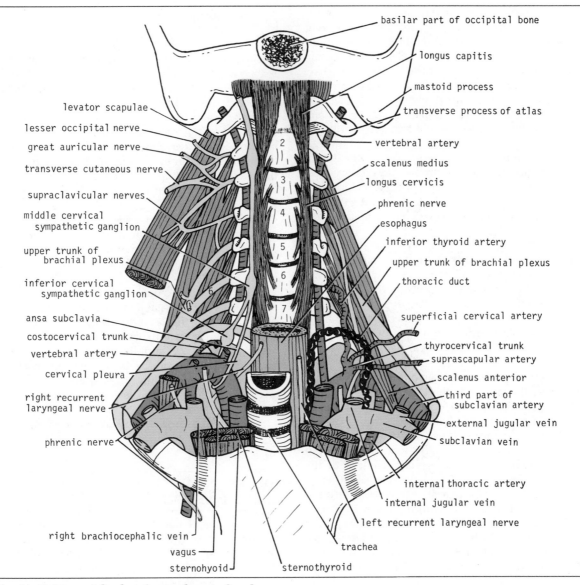

**Fig. 11-18. Prevertebral region and root of neck.**

*plexus*. The pharyngeal nerve supplies all the muscles of the pharynx except the stylopharyngeus (glossopharyngeal nerve) and all the muscles of the soft palate except the tensor veli palatini (mandibular division of the trigeminal nerve).

The *superior laryngeal nerve* arises from the inferior ganglion and runs downward and medially behind the internal carotid artery (Figs. 11-13 and 11-14). It divides into internal and external laryngeal nerves.

The *internal laryngeal nerve* pierces the thyrohyoid membrane, along with the internal laryngeal artery (Fig. 11-13). It is a sensory nerve, which supplies the floor of the piriform fossa and the mucous membrane of the larynx down as far as the vocal folds.

The *external laryngeal nerve* is a fine nerve, which descends in company with the superior thyroid artery (Fig. 11-13). It passes deep to the thyroid gland and supplies the cricothyroid muscle.

*Two* or *three cardiac branches* arise from the vagus as it descends through the neck. They join or accompany the cardiac branches of the sympathetic trunk and end in the cardiac plexus in the thorax.

The *recurrent laryngeal nerve* arises on the right side from the vagus, as the latter crosses the first part of the subclavian artery (Fig. 11-18). It hooks backward and upward behind the artery and ascends in the groove between the trachea and the esophagus. It passes deep to the lobe of the thyroid gland and comes into close relationship with the inferior thyroid artery. The nerve crosses either in front of or behind the artery or may pass between its branches. It passes beneath the lower border of the inferior constrictor muscle and supplies all the muscles of the larynx except the cricothyroid, which is supplied by the external laryngeal branch of the superior laryngeal nerve. The nerve also supplies the mucous membrane of the larynx below the vocal folds, and the mucous membrane of the upper part of the trachea.

On the left side, the recurrent laryngeal nerve arises from the vagus as the latter crosses the arch of the aorta in the thorax. It hooks around beneath the arch behind the ligamentum arteriosum (Fig. 3-16) and ascends into the neck in the groove between the trachea and the esophagus (See p. 737.)

See also the summary of cranial nerves, Table 11-3, page 812.

## ACCESSORY NERVE (ELEVENTH CRANIAL NERVE)

The accessory nerve is composed of motor fibers. It is formed by the union of cranial and spinal roots. (See p. 811.) The cranial root is smaller and arises in the medulla oblongata. The spinal roots arise from the upper five cervical segments of the spinal cord. The spinal roots unite to form a trunk that ascends in the vertebral canal to enter the skull through the foramen magnum (Fig. 11-88). Both the cranial and spinal roots come together and pass through the middle of the jugular foramen. (See p. 795.)

The *cranial root* now separates from the spinal root and joins the vagus at is inferior ganglion (Fig. 11-14). It is distributed mainly in the pharyngeal and recurrent laryngeal branches of the vagus nerve.

The *spinal root* runs downward and laterally and crosses the internal jugular vein (Fig. 11-14). It passes beneath the posterior belly of the digastric and reaches the deep surface of the sternocleidomastoid, which it supplies. The nerve emerges above the middle of the posterior border of the sternocleidomastoid and crosses the posterior triangle of the neck on the levator scapulae. (See p. 719.)

See also the summary of cranial nerves, Table 11-3, page 812.

## HYPOGLOSSAL NERVE (TWELFTH CRANIAL NERVE)

The hypoglossal nerve is the motor nerve to the tongue muscles. It arises in the medulla oblongata and leaves the skull through the hypoglossal canal in the occipital bone. (See p. 794.) It now comes into close relationship with the ninth, tenth, and eleventh cranial nerves, the internal carotid artery, and the internal jugular vein. It descends between the internal carotid artery and the internal jugular vein until it reaches the lower border of the posterior belly of the digastric muscle, where it turns forward and medially (Fig. 11-13). The nerve loops around the occipital artery and crosses the internal and external carotid arteries and the loop of the lingual artery. Here, it is crossed by the facial vein. It passes forward and upward, deep to the intermediate tendon of the digastric muscle, the stylohyoid, and the posterior margin of the mylohyoid muscle. Its further course in the submandibular region is described on page 781. See also Figure 11-88B.

In the upper part of its course, the hypoglossal nerve is joined by a small branch from the cervical plexus (C1 and sometimes C2). This branch later leaves the hypoglossal nerve as its *descending branch*, the *nerve to the thyrohyoid* and the *nerve to the geniohyoid*.

## Branches

The *meningeal* branch of the hypoglossal nerve arises from the nerve as it traverses the hypoglossal canal. It supplies the meninges in the posterior cranial fossa.

The *descending branch*, which is composed of C1 fibers, arises from the hypoglossal nerve as it curves forward below the posterior belly of the digastric (Fig. 11-13). It descends in front of the internal and common carotid arteries, embedded in the carotid sheath. It is joined by the descending cervical nerve (C2 and 3) from the cervical plexus, to form a loop, called the *ansa cervicalis* (Fig. 11-14). Branches from the loop supply the omohyoid, the sternohyoid, and the sternothyroid muscles.

The *nerve to the thyrohyoid*, which is composed of C1 fibers, arises from the hypoglossal nerve as it passes deep to the mylohyoid muscle (Fig. 11-13). It descends across the greater cornu of the hyoid bone and supplies the thyrohyoid muscle.

The *muscular branches* to the tongue are described on page 781. The nerve supply to the geniohyoid muscle, which is composed of C1 fibers, is given off on the side of the tongue.

See the summary of cranial nerves, Table 11-3, page 812.

# Cervical Part of the Sympathetic Trunk

The cervical part of the sympathetic trunk extends upward to the base of the skull and below to the neck of the first rib, where it becomes continuous with the thoracic part of the sympathetic trunk. It lies directly behind the internal and common carotid arteries (i.e., medial to the vagus) and is embedded in deep fascia between the carotid sheath and the prevertebral layer of deep fascia (Fig. 11-8).

The sympathetic trunk possesses three ganglia: the superior, middle, and inferior cervical ganglia.

## SUPERIOR CERVICAL GANGLION

The *superior cervical ganglion* is large and lies immediately below the skull (Fig. 11-14).

### Branches

1. The *internal carotid nerve*, consisting of postganglionic fibers, ascends from the upper pole of the ganglion and accompanies the internal carotid artery into the carotid canal in the temporal bone. It divides into branches around the artery to form the *internal carotid plexus.*
2. *Gray rami communicantes* to the upper four anterior rami of the cervical nerves.
3. *Arterial branches* to the common and external carotid arteries. These branches form a plexus around the arteries and are distributed along the branches of the external carotid artery.
4. *Cranial nerve branches*, which join the ninth, tenth, and twelfth cranial nerves.
5. *Pharyngeal branches*, which unite with the pharyngeal branches of the glossopharyngeal and vagus nerves to form the pharyngeal plexus.
6. The *superior cardiac branch*, which descends in the neck behind the common carotid artery. It ends in the cardiac plexus in the thorax. (See p. 111.)

## MIDDLE CERVICAL GANGLION

The middle cervical ganglion is small and lies at the level of the cricoid cartilage (Fig. 11-18). It is related to the loop of the inferior thyroid artery.

### Branches

1. *Gray rami communicantes* to the anterior rami of the fifth and sixth cervical nerves.
2. *Thyroid branches*, which pass along the inferior thyroid artery to the thyroid gland.
3. The *middle cardiac branch*, which descends in the neck behind the common carotid artery. It ends in the cardiac plexus in the thorax. (See p. 111.)

## INFERIOR CERVICAL GANGLION

The inferior cervical ganglion in the majority of subjects is fused with the first thoracic ganglion to form the *stellate ganglion*. It lies in the interval between the transverse process of the seventh cervical vertebra and the neck of the first rib, behind the vertebral artery (Fig. 11-18).

### Branches

1. *Gray rami communicantes* to the anterior rami of the seventh and eighth cervical nerves.
2. *Arterial branches* to the subclavian and vertebral arteries.
3. The *inferior cardiac branch*, which descends behind the subclavian artery to join the cardiac plexus in the thorax. (See p. 111.)

The part of the sympathetic trunk connecting the middle cervical ganglion to the inferior or stellate ganglion is represented by two or more nerve bundles. The most anterior bundle crosses in front of the first part of the subclavian artery and then turns upward behind it. This anterior bundle is referred to as the *ansa subclavia* (Figs. 11-14 and 11-18).

## Cervical Plexus

The cervical plexus is formed by the anterior rami of the first four cervical nerves. The rami are joined by connecting branches, which form loops that lie in front of the origins of the levator scapulae and the scalenus medius muscles (Fig. 11-18). The plexus is covered in front by the prevertebral layer of deep cervical fascia and is related to the internal jugular vein within the carotid sheath.

### Branches

1. *Cutaneous.* Lesser occipital, greater auricular, transverse cutaneous, and supraclavicular. These have been fully described on page 708.
2. *Muscular branches to the neck muscles.* Prevertebral muscles, sternocleidomastoid (proprioceptive, C2 and 3), levator scapulae (C3 and 4), and trapezius (proprioceptive, C3 and 4).
   A branch from C1 joins the hypoglossal nerve.

Some of these C1 fibers later leave the hypoglossal as the descending branch, which unites with the *descending cervical nerve* (C2 and 3), to form the *ansa cervicalis* (Fig. 11-14). The first, second, and third cervical nerve fibers within the *ansa cervicalis* supply the omohyoid, sternohyoid, and sternothyroid muscles. Other C1 fibers within the hypoglossal nerve leave it as the nerve to the thyrohyoid and geniohyoid.
3. *Nerve supply to the diaphragm.* Phrenic nerve.

## PHRENIC NERVE

The phrenic nerve is the *only motor nerve supply to the diaphragm*. It also contains sensory fibers and sympathetic fibers. Although some of the sensory fibers are proprioceptive fibers for the muscle of the diaphragm, the majority supply the pleura and peritoneum covering the upper and lower surfaces of the central part of the diaphragm. Other sensory fibers supply the mediastinal pleura and the pericardium (See p. 120.)

The phrenic nerve arises from the third, fourth, and fifth cervical nerves of the cervical plexus. The roots of the phrenic nerve unite at the lateral border of the scalenus anterior muscle at the level of the cricoid cartilage (Fig. 11-18). The nerve then runs vertically downward across the front of the scalenus anterior, behind the prevertebral layer of deep fascia. Because of the obliquity of the scalenus anterior muscle, the nerve crosses the muscle from its lateral to its medial border. The phrenic nerve enters the thorax by passing in front of the subclavian artery and behind the beginning of the brachiocephalic vein. As it descends, it crosses the internal thoracic artery from the lateral to the medial side.

The further course of the phrenic nerve in the thorax is described on page 120.

### Relations in the Neck

#### Anteriorly

The prevertebral layer of deep fascia, the internal jugular vein, the superficial cervical and suprascapular arteries, and, on the left side, the thoracic duct; the beginning of the brachiocephalic vein (Fig. 11-18).

## Posteriorly

The scalenus anterior, the subclavian artery, and the cervical dome of pleura.

The *accessory phrenic nerve* is described on page 900.

# Viscera of the Neck

## THYROID GLAND

The thyroid gland consists of right and left lobes connected by a narrow isthmus (Fig. 11-17). It is a very vascular organ, surrounded by a sheath derived from the pretracheal layer of deep fascia. The sheath attaches the gland to the larynx and the trachea.

Each lobe is pear-shaped, with it apex being directed upward as far as the oblique line on the lamina of the thyroid cartilage; its base lies below at the level of the fourth or fifth tracheal ring.

The *isthmus* extends across the midline in front of the second, third, and fourth tracheal rings (Fig. 11-17). A *pyramidal lobe* is often present, and it projects upward from the isthmus, usually to the left of the midline. A fibrous or muscular band frequently connects the pyramidal lobe to the hyoid bone; if it is muscular, it is referred to as the *levator glandulae thyroideae* (Fig. 11-17).

### *Relations of the Lobes*

#### Anterolaterally

The sternothyroid, superior belly of the omohyoid, sternohyoid, and anterior border of the sternocleidomastoid (Fig. 11-8).

#### Posterolaterally

The carotid sheath with the common carotid artery, internal jugular vein, and vagus nerve (Fig. 11-8).

#### Medially

The larynx, trachea, inferior constrictor of the pharynx, and esophagus. Associated with these structures are the cricothyroid muscle and its nerve supply, the external laryngeal nerve. In the groove between the esophagus and the trachea is the recurrent laryngeal nerve (Fig. 11-8).

The rounded posterior border of each lobe is related posteriorly to the superior and inferior parathyroid glands (Fig. 11-17) and the anastomosis between the superior and inferior thyroid arteries.

### *Relations of the Isthmus*

#### Anteriorly

The sternothyroids, sternohyoids, anterior jugular veins, fascia, and skin.

#### Posteriorly

The second, third, and fourth rings of the trachea.

The terminal branches of the superior thyroid arteries anastomose along its upper border.

### *Blood Supply*

The *arteries* to the thyroid gland are (1) the superior thyroid artery, (2) the inferior thyroid artery, and sometimes (3) the thyroidea ima. The arteries anastomose profusely with one another over the surface of the gland.

The *superior thyroid artery*, a branch of the external carotid artery, descends to the upper pole of each lobe, accompanied by the external laryngeal nerve (Fig. 11-13).

The *inferior thyroid artery*, a branch of the thyrocervical trunk, ascends behind the gland to the level of the cricoid cartilage (Fig. 11-18). It then turns medially and downward, to reach the posterior border of the gland. The recurrent laryngeal nerve crosses either in front of or behind the artery or may pass between its branches.

The *thyroidea ima*, if present, may arise from the brachiocephalic artery or the arch of the aorta. It ascends in front of the trachea to the isthmus (Fig. 11-17).

The *veins* from the thyroid gland are (1) the superior thyroid, which drains into the internal jugular vein; (2) the middle thyroid, which drains into the internal jugular vein; and (3) the inferior thyroid (Fig. 11-17). The latter vein receives its tributaries from the isthmus and the lower poles of the gland. The inferior thyroid veins of the two sides anastomose with one another as they descend in front of the trachea. They drain into the left brachiocephalic vein in the thorax.

## Lymphatic Drainage

The lymph from the thyroid gland drains mainly laterally into the deep cervical lymph nodes. A few lymph vessels descend to the paratracheal nodes.

## Development of the Thyroid Gland

The thyroid gland develops as an endodermal outgrowth from the midline of the floor of the pharynx, between the *tuberculum impar* and the *copula*. Later, this thickening becomes a diverticulum called the thyroglossal duct. As development continues, the duct elongates and its distal end becomes bilobed. The duct becomes a solid cord and migrates down the neck, passing either anterior, through, or posterior to the developing hyoid bone. By the seventh week it reaches its final position in relation to the larynx and the trachea. Meanwhile, the solid cord connecting the thyroid gland to the tongue breaks up and disappears. The site of origin of the thyroglossal duct on the tongue remains as a pit called the *foramen cecum*. As a result of epithelial proliferation, the bilobed terminal swellings of the thyroglossal duct expand to form the thyroid gland.

## Congenital Anomalies

*Agenesis of the thyroid gland* may occur and is the commonest cause of cretinism.

*Incomplete descent* of the thyroid gland may occur, and the thyroid may be found at any point between the base of the tongue and the trachea. *Lingual thyroid* is the commonest form of incomplete descent.

A *thyroglossal cyst* may appear in childhood or adolescence or in young adults. It is due to a persistence of a segment of the thyroglossal duct. Such a cyst occurs in the midline of the neck at any point along the thyroglossal tract.

## PARATHYROID GLANDS

The parathyroid glands are yellowish-brown, ovoid bodies, measuring about 6 mm long in their greatest diameter. They are usually four in number and are intimately related to the posterior border of the thyroid gland, lying within its fascial capsule (Fig. 11-17).

The two *superior parathyroid glands* are the more constant in position and lie at the level of the middle of the posterior border of the thyroid gland.

The two *inferior parathyroid glands* usually lie close to the inferior poles of the thyroid gland. They may lie within the fascial sheath, embedded in the thyroid substance, or outside the fascial sheath. Sometimes they are found some distance caudal to the thyroid gland, in association with the inferior thyroid veins; or they may even reside in the superior mediastinum.

## Blood Supply

The arterial supply to the parathyroid glands is from the superior and inferior thyroid arteries.

## TRACHEA

The trachea is a mobile cartilaginous and membranous tube (Fig. 11-44). It commences at the lower border of the cricoid cartilage of the larynx and extends downward in the midline of the neck (Fig. 11-44 and 11-64). In the thorax it ends by dividing into two main bronchi at the level of the disc between the fourth and fifth thoracic vertebrae. (See p. 89.)

## Relations in the Neck

### Anteriorly

The skin, fascia, isthmus of the thyroid gland (in front of the second, third, and fourth rings), inferior thyroid veins, jugular arch, thyroidea ima artery (if present), and left brachiocephalic vein in the child. It is overlapped by the sternothyroids and sternohyoids (Fig. 11-8).

### Posteriorly

The right and left recurrent laryngeal nerves, the esophagus, and the vertebral column (Fig. 11-8).

### Laterally

The lobes of the thyroid gland (down as far as the fifth or sixth ring) and the carotid sheath.

### *Blood Supply in the Neck*

The blood supply of the trachea in the neck is derived mainly from the inferior thyroid arteries.

### *Lymphatic Drainage in the Neck*

The lymph vessels drain into the pretracheal and paratracheal lymph nodes.

### *Nerve Supply in the Neck*

The nerve supply is from the vagi, the recurrent laryngeal nerves, and the sympathetic trunks.

### ESOPHAGUS

The esophagus is a muscular tube about 10 inches (25 cm) long, extending from the pharynx to the stomach (Figs. 11-44 and 11-59). It begins at the level of the cricoid cartilage, opposite the body of the sixth cervical vertebra. It commences in the midline, but as it descends through the neck, it inclines to the left side (Fig. 11-8). Its further course in the thorax is described on page 121.

### *Relations in the Neck*

#### Anteriorly

The trachea; the recurrent laryngeal nerves ascend, one on each side, in the groove between the trachea and the esophagus (Fig. 11-8).

#### Posteriorly

The prevertebral layer of deep cervical fascia, the longus colli, and the vertebral column (Fig. 11-8).

### Laterally

On each side lie the lobe of the thyroid gland and the carotid sheath (Fig. 11-8). On the left side the thoracic duct ascends along the left margin for a short distance (Fig. 11-18).

### *Blood Supply in the Neck*

The *arteries* of the esophagus in the neck are derived from the inferior thyroid arteries. The *veins* drain into the inferior thyroid veins.

### *Lymphatic Drainage in the Neck*

The lymph vessels drain into the deep cervical lymph nodes.

### *Nerve Supply in the Neck*

The nerves are derived from the recurrent laryngeal nerves and from the sympathetic trunks.

## The Root of the Neck

The root of the neck may be defined as the area of the neck immediately above the inlet into the thorax (Fig. 11-18).

### SCALENUS ANTERIOR (FIG. 11-18)

#### Origin

From the anterior tubercles of the transverse processes of the third, fourth, fifth, and sixth cervical vertebrae.

#### Insertion

The fibers pass downward and laterally, to be inserted into the scalene tubercle on the inner border of the first rib and into a ridge on the upper surface of the first rib.

#### Nerve Supply

From the anterior rami of the fourth, fifth, and sixth cervical nerves.

## Action

It assists in elevating the first rib. When acting from below, it laterally flexes and rotates the cervical part of the vertebral column.

The muscle is an important landmark in the neck, and for this reason its relations should be understood.

### Relations

#### Anteriorly

The prevertebral layer of deep cervical fascia, which binds the phrenic nerve down to the anterior surface of the muscle; the superficial cervical and suprascapular arteries, which cross the phrenic nerve; and the internal jugular and subclavian veins (Fig. 11-18).

#### Posteriorly

The subclavian artery, brachial plexus, and cervical dome of the pleura.

#### Medially

The vertebral artery and vein, inferior thyroid artery, thyrocervical trunk, sympathetic trunk, and, on the left side, thoracic duct.

#### Laterally

The roots of the phrenic nerve unite at the lateral border of the muscle at the level of the cricoid cartilage, before the nerve starts to descend on its anterior surface. The roots of the brachial plexus and the subclavian artery emerge from behind the lateral border of the muscle, to enter the posterior triangle of the neck.

## SCALENUS MEDIUS (FIG. 11-18)

### Origin

From the transverse process of the atlas and the posterior tubercles of the transverse processes of the next five cervical vertebrae.

### Insertion

The muscle passes downward and laterally and is inserted into the upper surface of the first rib behind the groove for the subclavian artery. The muscle lies behind the roots of the brachial plexus and behind the subclavian artery.

### Nerve Supply

Branches from the anterior rami of the cervical nerves.

### Action

It assists in elevating the first rib. When acting from below, it laterally flexes and rotates the cervical part of the vertebral column.

## SCALENUS POSTERIOR

The scalenus posterior muscle may be absent or blended with the scalenus medius.

### Origin

From the posterior tubercles of the transverse processes of the lower cervical vertebrae.

### Insertion

It is inserted into the outer surface of the second rib.

### Nerve Supply

Branches from the anterior rami of the lower cervical nerves.

### Action

It elevates the second rib. When active from below, it laterally flexes the cervical part of the vertebral column.

For a summary of muscles of the neck, their nerve supply, and their action, see Table 11-1.

Table 11-1. Muscles of the Neck

| Name of muscle | Origin | Insertion | Nerve supply | Action |
|---|---|---|---|---|
| Platysma | Deep fascia over pectoralis major and deltoid | Body of mandible and angle of mouth | Facial nerve cervical branch | Depresses mandible and angle of mouth |
| Sternocleidomastoid | Manubrium sterni and medial third of clavicle | Mastoid process of temporal bone and occipital bone | Spinal part of accessory nerve and C2 and 3 | Two muscles acting together extend head and flex neck; one muscle rotates head to opposite side |
| Digastric | | | | |
|   Posterior belly | Mastoid process of temporal bone | Intermediate tendon is held to hyoid by fascial sling | Facial nerve | Depresses mandible or elevates hyoid bone |
|   Anterior belly | Body of mandible | | Nerve to mylohyoid | |
| Stylohyoid | Styloid process | Body of hyoid bone | Facial nerve | Elevates hyoid bone |
| Mylohyoid | Mylohyoid line of body of mandible | Body of hyoid bone and fibrous raphe | Inferior alveolar nerve | Elevates floor of mouth and hyoid bone or depresses mandible |
| Geniohyoid | Inferior mental spine of mandible | Body of hyoid bone | First cervical nerve | Elevates hyoid bone or depresses mandible; depresses hyoid bone |
| Sternohyoid | Manubrium sterni and clavicle | Body of hyoid bone | Ansa cervicalis; C1, 2, and 3 | |
| Sternothyroid | Manubrium sterni | Oblique line lamina of thyroid cartilage | Ansa cervicalis; C1, 2, and 3 | Depresses larynx |
| Thyrohyoid | Oblique line lamina of thyroid cartilage | Lower border of body of hyoid bone | First cervical nerve | Depresses hyoid bone or elevates larynx |
| Omohyoid | | | | |
|   Inferior belly | Upper margin of scapula and suprascapular ligament | Intermediate tendon is held to clavicle and first rib by fascial sling | Ansa cervicalis; C1, 2, and 3 | Depresses hyoid bone |
|   Superior belly | Lower border body of hyoid bone | | | |
| Scalenus anterior | Transverse processes of third, fourth, fifth, and sixth cervical vertebrae | First rib | C4, 5, and 6 | Elevates first rib; laterally flexes and rotates cervical part of vertebral column |
| Scalenus medius | Transverse processes of upper six cervical vertebrae | First rib | Anterior rami of cervical nerves | Elevates first rib; laterally flexes and rotates cervical part of vertebral column |
| Scalenus posterior | Transverse processes of lower cervical vertebrae | Second rib | Anterior rami of cervical nerves | Elevates second rib; laterally flexes and rotates cervical part of vertebral column |

## SUBCLAVIAN ARTERY

The *right subclavian artery* arises from the brachiocephalic artery, behind the right sternoclavicular joint (Figs. 11-15 and 11-18). It passes upward and laterally as a gentle curve behind the scalenus anterior muscle, and at the outer border of the first rib, it becomes continuous with the axillary artery.

The *left subclavian artery* arises from the arch of the aorta, behind the left common carotid artery. (See p. 117.) It ascends to the root of the neck and then arches laterally in a manner similar to that of the right subclavian artery.

For purposes of description, it is usual to divide the subclavian artery into three parts by the presence of the scalenus anterior muscle.

### First Part of the Subclavian Artery

The first part of the subclavian artery extends from its origin to the medial border of the scalenus anterior (Fig. 11-18).

### Relations

#### Anteriorly

From medial to lateral are the common carotid artery, the ansa subclavia, the vagus nerve, the internal jugular and vertebral veins, and on the left side, the phrenic nerve. In addition, there are the cardiac branches of the vagus and sympathetic nerves.

#### Posteriorly

The dome of the cervical pleura, the apex of the lung, the ansa subclavia, and, on the right side, the right recurrent laryngeal nerve.

The branches of the first part of the subclavian artery are as follows:

### Vertebral Artery

The vertebral artery arises from the upper margin of the subclavian artery and ascends in the neck between the longus colli and the scalenus anterior muscles (Fig. 11-18). It passes in front of the transverse process of the seventh cervical vertebra and enters the foramen in the transverse process of the *sixth* cervical vertebra (Fig. 11-15). It then ascends

through the foramina in the transverse processes of the upper six cervical vertebrae. Having emerged from the transverse process of the atlas, it curves backward behind the lateral mass of the atlas. It then passes medially, pierces the dura mater, and enters the vertebral canal. The vertebral artery then ascends into the skull through the foramen magnum. The further course of the artery is described on page 807.

### Relations of the Vertebral Artery

*Anteriorly.* The common carotid artery, vertebral vein, inferior thyroid artery; on the left side, it is crossed by the thoracic duct (Figs. 11-8 and 11-18).

*Posteriorly.* The transverse process of the seventh cervical vertebra, the cervicothoracic sympathetic ganglion (stellate ganglion), and the anterior rami of the seventh and eighth cervical nerves.

As the artery ascends through the foramina in the transverse processes, it lies in front of the anterior rami of the cervical nerves (Fig. 11-8).

### Branches of the Vertebral Artery in the Neck

Spinal and muscular branches arise from the vertebral artery. The spinal branches enter the vertebral canal through the intervertebral foramina.

### Thyrocervical Trunk

The thyrocervical trunk is a wide, short trunk that arises from the front of the first part of the subclavian artery, at the medial border of the scalenus anterior muscle (Fig. 11-18). It gives off three branches: (1) the inferior thyroid, (2) the superficial cervical, and (3) the suprascapular arteries.

The *inferior thyroid artery* ascends along the medial border of the scalenus anterior to the level of the cricoid cartilage (Fig. 11-18). It then turns medially and downward, passing behind the vagus and the common carotid artery and in front of the vertebral artery and vein and the sympathetic trunk. It then reaches the posterior border of the thyroid gland and is closely related to the recurrent laryngeal nerve.

The *superficial cervical* and *suprascapular arteries* pass laterally across the scalenus anterior and the phrenic nerve (Fig. 11-18); their further course is described on pages 717 and 718.

## Internal Thoracic Artery

The internal thoracic artery arises from the lower border of the first part of the subclavian artery (Fig. 11-18). It enters the thorax by descending behind the first costal cartilage in front of the pleura. It is crossed obliquely by the phrenic nerve, from the lateral to the medial side. The further course of the artery in the thorax is described on page 74.

## Second Part of the Subclavian Artery

The second part of the subclavian artery lies behind the scalenus anterior muscle.

## Relations

### Anteriorly

The scalenus anterior muscle.

### Posteriorly

The dome of the cervical pleura and the apex of the lung (Fig. 11-18).

The branch of the second part of the subclavian artery is as follows:

### Costocervical Trunk

The costocervical trunk arises from the back of the second part of the subclavian artery and runs backward above the pleura to the neck of the first rib (Fig. 11-18). Here, it divides into the superior intercostal and deep cervical arteries.

The *superior intercostal artery* gives rise to the posterior intercostal arteries of the first and second intercostal spaces. (See p. 69.)

The *deep cervical artery* passes backward between the transverse process of the seventh cervical vertebra and the neck of the first rib. It supplies the muscles of the back of the neck.

## Third Part of the Subclavian Artery

The third part of the subclavian artery extends from the lateral border of the scalenus anterior to the outer border of the first rib (Fig. 11-18). It is described on page 717.

## SUBCLAVIAN VEIN

The subclavian vein begins at the outer border of the first rib as a continuation of the axillary vein (Fig. 11-18). At the medial border of the scalenus anterior it joins the internal jugular vein, to form the brachiocephalic vein.

## Relations

### Anteriorly

The clavicle and the subclavius muscle.

### Posteriorly

The scalenus anterior muscle and the phrenic nerve.

### Inferiorly

The upper surface of the first rib.

## Thoracic Duct

The thoracic duct begins in the abdomen at the upper end of the cisterna chyli. (See p. 261.) It enters the thorax through the aortic opening in the diaphragm and ascends through the posterior mediastinum, inclining gradually to the left. (See p. 118.) On reaching the superior mediastinum, it is found passing upward along the left margin of the esophagus. At the root of the neck, it continues to ascend along the left margin of the esophagus until it reaches the level of the transverse process of the seventh cervical vertebra. Here, it bends laterally behind the carotid sheath and in front of the vertebral vessels (Fig. 11-18). On reaching the medial border of the scalenus anterior, it turns downward in front of the left phrenic nerve and the first part of the subclavian artery and drains into the beginning of the left brachiocephalic vein. It may, however, end in the terminal part of the subclavian or internal jugular veins.

## Lymphatic Drainage of the Head and Neck

The lymph nodes in the head and neck are made up of a number of regional groups and a terminal group (Fig. 11-19). The regional groups comprise (1) the occipital, (2) the retroauricular (mastoid), (3) the parotid, (4) the buccal (facial), (5) the submandibular, (6) the submental, (7) the anterior cervical, (8) the superficial cervical, (9) the retropharyngeal, (10) the laryngeal, and (11) the tracheal.

The terminal group of nodes receives all the lymph vessels of the head and neck, either directly or indirectly, via one of the regional groups. The terminal group is closely related to the carotid sheath, and, in particular, to the internal jugular vein, and is referred to as the deep cervical group.

### REGIONAL GROUPS OF LYMPH NODES

The *occipital lymph nodes* are situated over the occipital bone at the apex of the posterior triangle of the neck (Fig. 11-19). They receive lymph from the back of the scalp. The efferent lymph vessels drain into the deep cervical lymph nodes.

The *retroauricular (mastoid) lymph nodes* are situated over the lateral surface of the mastoid process of the temporal bone (Fig. 11-19). They receive lymph from a strip of scalp above the auricle and from the posterior wall of the external auditory meatus. The efferent lymph vessels drain into the deep cervical lymph nodes.

The *parotid lymph nodes* are situated on or within the parotid salivary gland (Figs. 11-19 and 11-26). They receive lymph from a strip of scalp above the parotid salivary gland, from the lateral surface of the auricle and the anterior wall of the external auditory meatus, and from the lateral parts of the eyelids. The nodes that are deeply placed in the parotid salivary gland also receive lymph from the middle ear. The efferent lymph vessels drain into the deep cervical lymph nodes.

The *buccal (facial) lymph nodes* are situated over the buccinator muscle, close to the facial vein (Fig. 11-19). They lie along the course of lymph vessels that ultimately drain into the submandibular nodes.

The *submandibular lymph nodes* are situated on the superficial surface of the submandibular salivary gland, beneath the investing layer of deep cervical fascia (Fig. 11-19). They may be palpated just below the lower border of the body of the mandible. They receive lymph from a wide area, including the front of the scalp; the nose and adjacent cheek; the upper lip and the lower lip (except the center part); the frontal, maxillary, and ethmoid air sinuses; the upper and lower teeth (except the lower incisors); the anterior two-thirds of the tongue (except the tip); the floor of the mouth and vestibule; and the gums. The efferent lymph vessels drain into the deep cervical lymph nodes.

The *submental lymph nodes* are situated in the submental triangles between the anterior bellies of the digastric muscles (Fig. 11-19). They receive lymph from the tip of the tongue, the floor of the mouth beneath the tip of the tongue, the incisor teeth and the associated gums, the center part of the lower lip, and the skin over the chin. The efferent lymph vessels drain into the submandibular and deep cervical lymph nodes.

The *anterior cervical lymph nodes* are situated along the course of the anterior jugular veins (Fig. 11-19). They receive lymph from the skin and superficial tissues of the front of the neck. The efferent lymph vessels drain into the deep cervical lymph nodes.

The *superficial cervical lymph nodes* are situated along the course of the external jugular vein (Fig. 11-19). They receive lymph from the skin over the angle of the jaw, the skin over the apex of the parotid salivary gland, and the lobe of the ear. The efferent lymph vessels drain into the deep cervical lymph nodes.

The *retropharyngeal lymph nodes* are situated in the retropharyngeal space, in the interval between the pharyngeal wall and the prevertebral fascia. They receive lymph from the nasal part of the pharynx, the auditory tube, and the upper part of the cervical vertebral column. The efferent lymph vessels drain into the deep cervical lymph nodes.

The *laryngeal lymph nodes* are situated in front of the larynx on the cricothyroid ligament (Fig. 11-19). One or two small nodes may be found in front of the thyrohyoid membrane. They receive lymph from adjacent structures, and their efferent vessels drain into the deep cervical lymph nodes.

The *tracheal lymph nodes* (Fig. 11-19) are situated lateral to the trachea (paratracheal nodes)

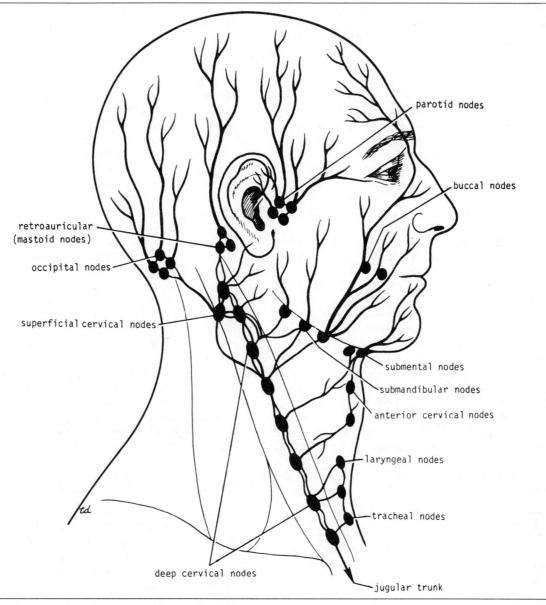

**Fig. 11-19. Lymphatic drainage of head and neck.**

and in front of the trachea (pretracheal nodes). They receive lymph from neighboring structures, including the thyroid gland. The efferent lymph vessels drain into the deep cervical lymph nodes.

## DEEP CERVICAL LYMPH NODES

The deep cervical lymph nodes form a chain along the course of the internal jugular vein, from the skull to the root of the neck (Fig. 11-19). They are embedded in the fascia of the carotid sheath and the tunica adventitia of the internal jugular vein; the majority lie on the anterolateral aspect of the internal jugular vein. Two of the nodes are often referred to clinically and are called the *jugulodigastric node* and the *jugulo-omohyoid node*. The *jugulodigastric node* lies just below the posterior belly of the digastric muscle and is placed just below and behind the angle of the mandible. It is chiefly concerned with the lymph drainage of the tonsil and the tongue.

The *jugulo-omohyoid node* is related to the intermediate tendon of the omohyoid muscle and is associated mainly with the lymph drainage of the tongue.

The deep cervical lymph nodes receive lymph from neighboring anatomical structures and from all the other regional lymph nodes in the head and neck. The efferent lymph vessels join to form the *jugular lymph trunk*. This vessel drains into the thoracic duct or the right lymphatic duct. Alternatively, it may drain into the subclavian lymph trunk or independently into the brachiocephalic vein.

## THE SCALP

The scalp consists of five layers, the first three of which are intimately bound together and move as a unit (Fig. 11-20). In order to assist one in memorizing the names of the five layers of the scalp, use each letter of the word SCALP to denote the layer of the scalp.

1. *Skin*, which is thick and hair-bearing and contains numerous sebaceous glands.
2. *Connective tissue beneath the skin*, which is fibro-fatty, the fibrous septa uniting the skin to the underlying aponeurosis of the occipitofrontalis muscle (Fig. 11-20). It is in this layer that

numerous arteries and veins ramify. The arteries are branches of the external and internal carotid arteries, and a free anastomosis takes place between them.

3. *Aponeurosis (epicranial)*, which is a thin, tendinous sheet that unites the occipital and frontal bellies of the occipitofrontalis muscle (Figs. 11-20 and 11-25). The lateral margins of the aponeurosis are attached to the temporal fascia.

The *subaponeurotic space* is the potential space beneath the epicranial aponeurosis. It is limited in front and behind by the origins of the occipitofrontalis muscle, and it extends laterally as far as the attachment of the aponeurosis to the temporal fascia.

4. *Loose areolar tissue*, which occupies the subaponeurotic space (Fig. 11-20) and loosely connects the epicranial aponeurosis to the periosteum of the skull (the pericranium). The areolar tissue contains a few small arteries, but it also contains some important *emissary veins*. The emissary veins are valveless and connect the superficial veins of the scalp with the *diploic veins* of the skull bones and with the intracranial venous sinuses (Fig. 11-20).

5. *Pericranium*, which is the periosteum covering the outer surface of the skull bones. It is important to remember that at the sutures between individual skull bones, the pericranium becomes continuous with the periosteum (endosteum) on the inner surface of the skull bones (Fig. 11-20).

## OCCIPITOFRONTALIS (FIG. 11-25)

### Origin

It consists of four bellies, two occipital and two frontal, connected by an aponeurosis. Each *occipital belly* arises from the highest nuchal line on the occipital bone and passes forward, to be attached to the aponeurosis.

**Fig. 11-20. (A) Coronal section of upper part of head showing layers of scalp, sagittal suture of skull, falx cerebri, superior and inferior sagittal venous sinuses, arachnoid granulations; emissary veins, and relation of cerebral blood vessels to subarachnoid space. (B) Sensory nerve supply and arterial supply to the scalp.**

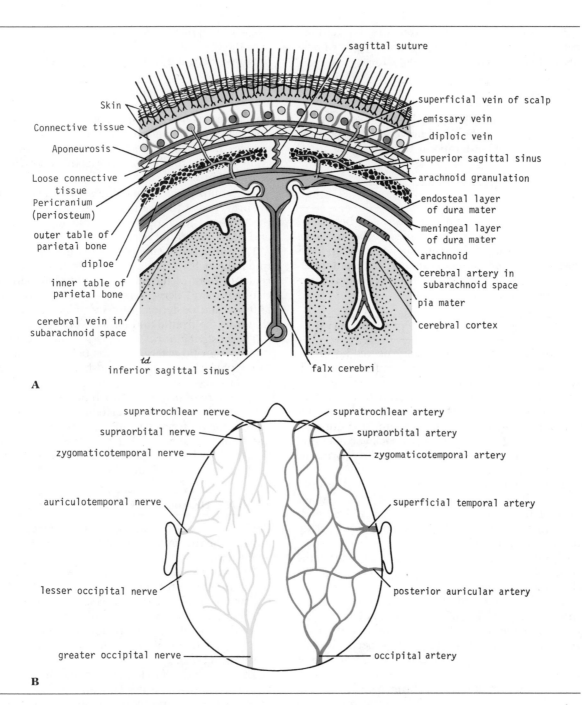

Skin

Connective tissue

Aponeurosis

Loose connective tissue

Pericranium (periosteum)

outer table of parietal bone

diploe

inner table of parietal bone

cerebral vein in subarachnoid space

sagittal suture

superficial vein of scalp

emissary vein

diploic vein

superior sagittal sinus

arachnoid granulation

endosteal layer of dura mater

meningeal layer of dura mater

arachnoid

cerebral artery in subarachnoid space

pia mater

cerebral cortex

*td*

inferior sagittal sinus

falx cerebri

**A**

supratrochlear nerve

supraorbital nerve

zygomaticotemporal nerve

auriculotemporal nerve

lesser occipital nerve

greater occipital nerve

supratrochlear artery

supraorbital artery

zygomaticotemporal artery

superficial temporal artery

posterior auricular artery

occipital artery

**B**

Each *frontal belly* arises from the skin and superficial fascia of the eyebrow and passes backward, to be attached to the aponeurosis.

## Nerve Supply

The occipital belly is supplied by the posterior auricular branch of the facial nerve; the frontal belly is supplied by the temporal branch of the facial nerve.

## Action

The first three layers of the scalp can be moved forward or backward, the loose areolar tissue of the fourth layer of the scalp allowing the aponeurosis to move on the pericranium. The frontal bellies can raise the eyebrows in expressions of surprise or horror.

## SENSORY NERVE SUPPLY OF THE SCALP

The main trunks of the sensory nerves lie in the superficial fascia. Moving laterally from the midline anteriorly, the following nerves are present.

The *supratrochlear nerve*, a branch of the ophthalmic division of the trigeminal nerve, winds around the superior orbital margin and supplies the scalp (Fig. 11-20). It passes backward close to the median plane and reaches nearly as far as the vertex of the skull.

The *supraorbital nerve*, a branch of the ophthalmic division of the trigeminal nerve, winds around the superior orbital margin and ascends over the forehead (Fig. 11-20). It supplies the scalp as far backward as the vertex.

The *zygomaticotemporal nerve*, a branch of the maxillary division of the trigeminal nerve, supplies the scalp over the temple (Fig. 11-20).

The *auriculotemporal nerve*, a branch of the mandibular division of the trigeminal nerve, ascends over the side of the head from in front of the auricle (Fig. 11-20). Its terminal superficial temporal branches supply the skin over the temporal region.

The *lesser occipital nerve*, a branch of the cervical plexus (C2), supplies the scalp over the lateral part of the occipital region (Fig. 11-20). It also supplies the skin over the medial surface of the auricle.

The *greater occipital nerve*, a branch of the posterior ramus of the second cervical nerve, ascends over the back of the scalp and supplies the skin as far forward as the vertex of the skull (Fig. 11-20).

## ARTERIAL SUPPLY OF THE SCALP

The scalp has a very rich supply of blood and, for this reason, the smallest cut bleeds profusely. The arteries lie in the superficial fascia. Moving laterally from the midline anteriorly, the following arteries are present:

The *supratrochlear* and the *supraorbital arteries*, branches of the ophthalmic artery, ascend over the forehead in company with the supratrochlear and the supraorbital nerves (Fig. 11-20).

The *superficial temporal artery*, the smaller terminal branch of the external carotid artery, ascends in front of the auricle in company with auriculotemporal nerve (Fig. 11-20). It divides into anterior and posterior branches, which supply the skin over the frontal and temporal regions.

The *posterior auricular artery*, a branch of the external carotid artery, ascends behind the auricle to supply the scalp above and behind the auricle (Fig. 11-20).

The *occipital artery*, a branch of the external carotid artery, ascends in a tortuous course from the apex of the posterior triangle, in company with the greater occipital nerve (Fig. 11-20). It supplies the skin over the back of the scalp and reaches as high as the vertex of the skull.

## VENOUS DRAINAGE OF THE SCALP

The *supratrochlear* and the *supraorbital veins* unite at the medial margin of the orbit to form the facial vein.

The *superficial temporal vein* unites with the maxillary vein in the substance of the parotid gland to form the retromandibular vein (Fig. 11-26).

The *posterior auricular vein* unites with the posterior division of the retromandibular vein, just below the parotid gland, to form the external jugular vein (Fig. 11-16).

The *occipital vein* drains into the suboccipital venous plexus, which lies beneath the floor of the upper part of the posterior triangle; the plexus in turn drains into the vertebral veins. Occasionally, the occipital vein drains forward into the internal jugular vein.

The veins of the scalp freely anastomose with one another and are connected to the diploic veins and the intracranial venous sinuses by the valveless *emissary veins* (Fig. 11-20).

# THE FACE

## Skin of the Face

The skin of the face possesses numerous sweat and sebaceous glands. It is connected to the underlying bones by loose connective tissue, in which are embedded the muscles of facial expression. *There is no deep fascia in the face.*

Wrinkle lines of the face result from the repeated folding of the skin perpendicular to the long axis of the underlying contracting muscles, coupled with the loss of youthful skin elasticity. Surgical scars of the face are less conspicuous if they follow the wrinkle lines.

## Development of the Face

Early in development, the face of the embryo is represented by an area bounded cranially by the neural plate, caudally by the pericardium, and laterally by the mandibular process of the first pharyngeal arch on each side (Fig. 11-21). In the center of this area is a depression in the ectoderm known as the *stomodeum.* In the floor of the depression is the *buccopharyngeal membrane.* By the fourth week, the buccopharyngeal membrane breaks down so that the stomodeum communicates with the foregut.

The further development of the face is dependent on the coming together and fusion of a number of important processes, namely, the *frontonasal process*, the *maxillary processes*, and the *mandibular processes* (Fig. 11-21). The frontonasal process begins as proliferation of mesenchyme on the ventral surface of the developing brain, and this grows toward the stomodeum. Meanwhile, the maxillary process grows out from the upper end of each first arch and passes medially, forming the lower border of the developing orbit. The mandibular processes of the first arches now approach one another in the midline below the stomodeum and fuse to form the lower jaw and lower lip (Fig. 11-21).

The *olfactory pits* appear as depressions in the lower edge of the advancing frontonasal process,

dividing it into a *medial nasal process* and two *lateral nasal processes.* With further development, the maxillary processes grow medially and fuse with the lateral nasal processes and with the medial nasal process (Fig. 11-21). The medial nasal process forms the *philtrum* of the upper lip and the *premaxilla.* The maxillary processes extend medially, forming the upper jaw and the cheek, and finally bury the premaxilla and fuse in the midline. The various processes that ultimately form the face unite during the second month.

The *upper lip* is formed by the growth medially of the maxillary processes of the first pharyngeal arch on each side. Ultimately, the maxillary processes meet in the midline and fuse with each other and with the medial nasal process (Fig. 11-21). Thus the lateral parts of the upper lip are formed from the maxillary processes, and the medial part, or philtrum from the medial nasal process, with contributions from the maxillary processes.

The *lower lip* is formed from the mandibular process of the first pharyngeal arch on each side (Fig. 11-21). These processes grow medially below the stomodeum, and fuse in the midline to form the entire lower lip.

The area of skin overlying the frontonasal process and its derivatives receives its sensory nerve supply from the ophthalmic division of the trigeminal nerve, while the maxillary division of the trigeminal nerve supplies the area of skin overlying the maxillary process. The area of skin overlying the mandibular process is supplied by the mandibular division of the trigeminal nerve.

The muscles of facial expression are derived from mesenchyme of the second pharyngeal arch. The nerve supply of these muscles is the nerve of the second pharyngeal arch, namely, the seventh cranial nerve.

## CONGENITAL ANOMALIES

*Cleft upper lip* may be confined to the lip or may be associated with a cleft palate. The anomaly is usually *unilateral cleft lip* and is caused by failure of the maxillary process to fuse with the medial nasal process (Fig. 11-22). *Bilateral cleft lip* is caused by the failure of both maxillary processes to fuse with the medial nasal process, which then remains as a central flap of tissue.

*Oblique facial cleft* is a rare condition in which

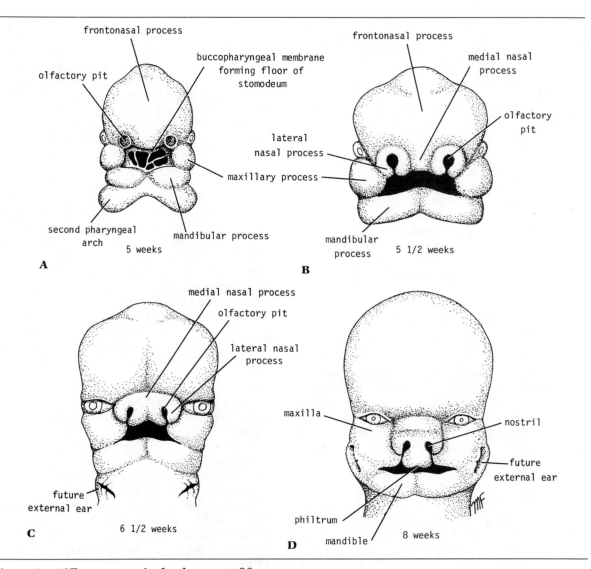

**Fig. 11-21. Different stages in development of face.**

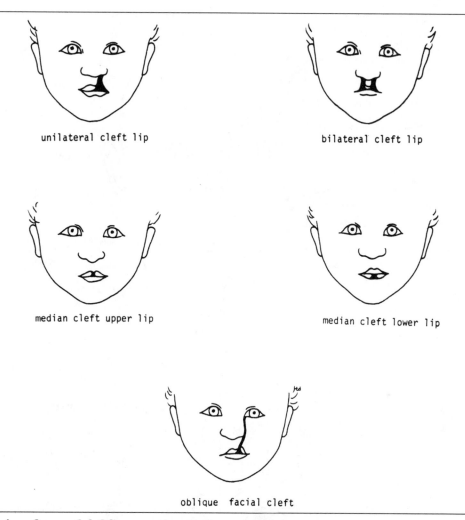

unilateral cleft lip

bilateral cleft lip

median cleft upper lip

median cleft lower lip

oblique facial cleft

**Fig. 11-22. Various forms of cleft lip.**

the cleft lip on one side extends to the medial margin of the orbit (Fig. 11-22). This is caused by the failure of the maxillary process to fuse with the lateral and medial nasal processes.

*Cleft lower lip* is extremely rare. It is exactly central and is caused by incomplete fusion of the mandibular processes (Fig. 11-22).

## Sensory Nerves of the Face

The skin of the face is supplied by branches of the three divisions of the trigeminal nerve, except for the small area over the angle of the mandible and the parotid gland (Fig. 11-23), which is supplied by the great auricular nerve (C2 and 3). The overlap of the three divisions of the trigeminal nerve is slight in comparison to the considerable overlap of adjacent dermatomes of the trunk and limbs. The ophthalmic nerve supplies the region developed from the frontonasal process; the maxillary nerve serves the region developed from the maxillary process of the first pharyngeal arch; and the mandibular nerve serves the region developed from the mandibular process of the first pharyngeal arch.

It should be pointed out that these nerves not only supply the skin of the face, but also supply

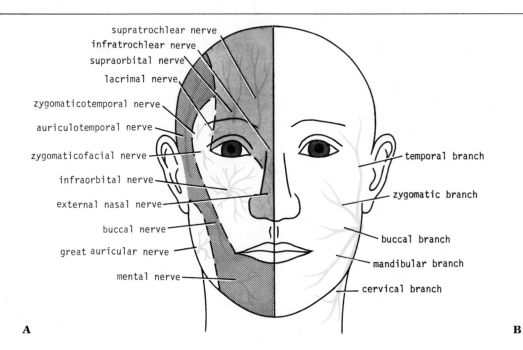

supratrochlear nerve
infratrochlear nerve
supraorbital nerve
lacrimal nerve
zygomaticotemporal nerve
auriculotemporal nerve
zygomaticofacial nerve
infraorbital nerve
external nasal nerve
buccal nerve
great auricular nerve
mental nerve

temporal branch
zygomatic branch
buccal branch
mandibular branch
cervical branch

A

B

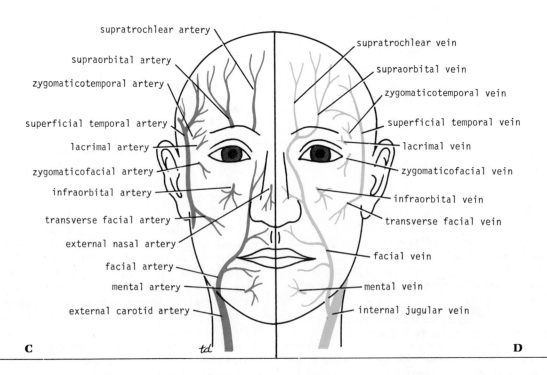

supratrochlear artery
supraorbital artery
zygomaticotemporal artery
superficial temporal artery
lacrimal artery
zygomaticofacial artery
infraorbital artery
transverse facial artery
external nasal artery
facial artery
mental artery
external carotid artery

supratrochlear vein
supraorbital vein
zygomaticotemporal vein
superficial temporal vein
lacrimal vein
zygomaticofacial vein
infraorbital vein
transverse facial vein
facial vein
mental vein
internal jugular vein

C

D

proprioceptive fibers to the underlying muscles of facial expression. They are, in addition, the sensory nerve supply to the mouth, teeth, nasal cavities, and paranasal air sinuses.

## OPHTHALMIC NERVE

The ophthalmic nerve supplies the skin of the forehead, the upper eyelid, the conjunctiva, and the side of the nose down to and including the tip. Five branches of the nerve pass to the skin.

1. The *lacrimal nerve* supplies the skin and conjunctiva of the lateral part of the upper eyelid (Fig. 11-23).
2. The *supraorbital nerve* winds around the upper margin of the orbit at the supraorbital notch (Fig. 11-23). It divides into lateral and medial branches and supplies the skin and conjunctiva on the central part of the upper eyelid; it also supplies the skin of the forehead.
3. The *supratrochlear nerve* winds around the upper margin of the orbit medial to the supraorbital nerve (Fig. 11-23). It divides into branches that supply the skin and conjunctiva on the medial part of the upper eyelid and the skin over the lower part of the forehead, close to the median plane.
4. The *infratrochlear nerve* leaves the orbit below the pulley of the superior oblique muscle. It supplies the skin and conjunctiva on the medial part of the upper eyelid and the adjoining part of the side of the nose (Fig. 11-23).
5. The *external nasal nerve* leaves the nose by emerging between the nasal bone and the upper nasal cartilage. It supplies the skin on the side of the nose down as far as the tip (Fig. 11-23).

## MAXILLARY NERVE

The maxillary nerve supplies the skin on the posterior part of the side of the nose, the lower eyelid, the cheek, the upper lip, and the lateral side of the orbital opening. Three branches of the nerve pass to the skin.

1. The *infraorbital nerve* is a direct continuation of the maxillary nerve. It enters the orbit and appears on the face through the infraorbital foramen. It immediately divides into numerous small branches, which radiate out from the foramen and supply the skin of the lower eyelid and cheek, the side of the nose, and the upper lip (Fig. 11-23).
2. The *zygomaticofacial nerve* passes onto the face through a small foramen on the lateral side of the zygomatic bone. It supplies the skin over the prominence of the cheek (Fig. 11-23).
3. The *zygomaticotemporal nerve* emerges in the temporal fossa through a small foramen on the posterior surface of the zygomatic bone. It supplies the skin over the temple (Fig. 11-23).

## MANDIBULAR NERVE

The mandibular nerve supplies the skin of the lower lip, the lower part of the face, the temporal region, and part of the auricle. It then passes upward to the side of the scalp. Three branches of the nerve pass to the skin.

1. The *mental nerve* emerges from the mental foramen of the mandible and supplies the skin of the lower lip and chin (Fig. 11-23).
2. The *buccal nerve* emerges from beneath the anterior border of the masseter muscle and supplies the skin over a small area of the cheek (Fig. 11-23).
3. The *auriculotemporal nerve* ascends from the upper border of the parotid gland between the superficial temporal vessels and the auricle. It supplies the skin of the auricle, the external auditory meatus, the outer surface of the tympanic membrane, and the skin of the scalp above the auricle (Fig. 11-23).

## Arterial Supply of the Face

The face receives a rich blood supply from two main vessels, the facial and the superficial temporal arteries. These are supplemented by a number of small arteries that accompany the sensory nerves of the face.

The *facial artery* is the highest of the three branches that arise from the anterior aspect of the external carotid artery (Figs. 11-13 and 11-15).

**Fig. 11-23. (A) Sensory nerve supply to skin of face. (B) Branches of seventh cranial nerve to muscles of facial expression. (C) Arterial supply of face. (D) Venous drainage of face.**

Having arched upward and over the submandibular salivary gland, it curves around the inferior margin of the body of the mandible at the anterior border of the masseter muscle. It is here that the pulse can be easily felt (Fig. 11-5). It runs upward in a tortuous course toward the angle of the mouth and is covered by the platysma and the risorius muscles. It then ascends deep to the zygomaticus muscles and the levator labii superioris muscle and runs along the side of the nose to the medial angle of the eye, where it anastomoses with the terminal branches of the ophthalmic artery (Fig. 11-23).

### Branches

1. The *submental artery* arises from the facial artery as the latter is curving around the lower border of the body of the mandible. It runs forward along the lower border of the mandible and supplies the skin of the chin and lower lip.
2. The *inferior labial artery* arises near the angle of the mouth. It runs medially in the lower lip and anastomoses with its fellow of the opposite side.
3. The *superior labial artery* arises near the angle of the mouth. It runs medially in the upper lip and gives branches to the septum and ala of the nose.
4. The *lateral nasal artery* arises from the facial artery as the latter ascends alongside the nose. It supplies the skin on the side and dorsum of the nose.

The *superficial temporal artery* (Fig. 11-23), the smaller terminal branch of the external carotid artery, commences in the parotid gland. It ascends in front of the auricle to supply the scalp. (See p. 746.)

The *transverse facial artery*, a branch of the superficial temporal artery, arises within the parotid gland. It runs forward across the cheek just above the parotid duct (Fig. 11-23).

The *supraorbital* and *supratrochlear arteries*, branches of the ophthalmic artery, supply the skin of the forehead (Fig. 11-23).

## Venous Drainage of the Face

The *facial vein* is formed at the medial angle of the eye by the union of the supraorbital and supratrochlear veins (Fig. 11-23). It is connected to the superior ophthalmic vein directly through the supraorbital vein. By means of the superior ophthalmic vein, the facial vein is connected to the cavernous sinus (Fig. 11-41); this connection is of great clinical importance, since it provides a pathway for the spread of infection from the face to the cavernous sinus. The facial vein descends behind the facial artery to the lower margin of the body of the mandible. It crosses superficial to the submandibular gland and is joined by the anterior division of the retromandibular vein. The facial vein ends by draining into the internal jugular vein.

### Tributaries

The facial vein receives tributaries that correspond to the branches of the facial artery. It is joined to the pterygoid venous plexus by the *deep facial vein*, and to the cavernous sinus by the superior ophthalmic vein.

The *transverse facial vein* joins the superficial temporal vein within the parotid gland.

## Lymphatic Drainage of the Face

The forehead and the anterior part of the face drain into the submandibular lymph nodes (Fig. 11-24). A few buccal lymph nodes may be present along the course of these lymph vessels. The lateral part of the face, including the lateral parts of the eyelids, is drained by lymph vessels that end in the parotid lymph nodes. The central part of the lower lip and the skin of the chin is drained into the submental lymph nodes.

## Bones of the Face

The bones that form the front of the skull are shown diagrammatically in Figure 11-24. The superior orbital margins and the area above them are formed by the *frontal bone*, which contains the *frontal air sinuses*. The lateral orbital margin is formed by the *zygomatic bone*, and the inferior orbital margin by the *zygomatic bone* and the *maxilla*. The medial orbital margin is formed above the maxillary process of the *frontal bone* and below by the frontal process of the *maxilla*.

The root of the nose is formed by the *nasal bones*, which articulate below with the maxilla and above with the frontal bones. Anteriorly, the nose

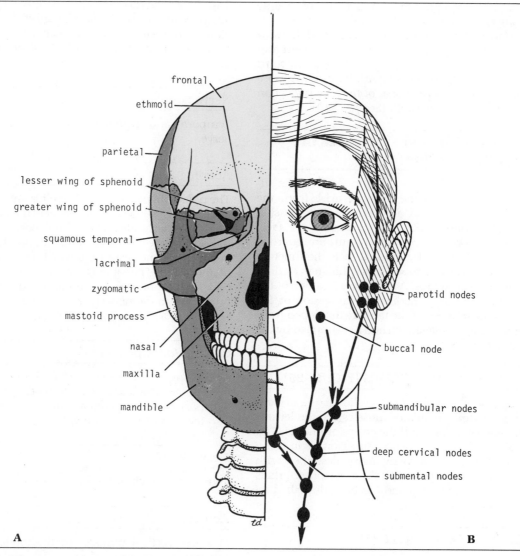

**Fig. 11-24. (A) Bones of front of skull and (B) lymphatic drainage of face.**

is completed by upper and lower plates of hyaline cartilage and small cartilages of the ala nasi.

The important central bone of the middle third of the face is the maxilla, containing its teeth and the maxillary air sinus. The bone of the lower third of the face is the mandible, with its teeth. A more detailed account of the bones of the face is given in the discussion on the skull. (See p. 782.)

## Muscles of the Face (Muscles of Facial Expression)

The muscles of the face are embedded in the superifical fascia, and the majority arise from the bones of the skull and are inserted into the skin (Figs. 11-34 and 11-35). The orifices of the face, namely, the orbit, nose, and mouth, are guarded by the eyelids, nostrils, and lips. It is the function of the facial muscles to serve as sphincters or dilators of these structures. A secondary function of the facial muscles is to modify the expression of the face. All the muscles of the face are developed from the second pharyngeal arch and are supplied by the facial nerve.

### MUSCLES OF THE EYELIDS (FIG. 11-25)

The sphincter muscle of the eyelids is the orbicularis oculi, and the dilator muscles are the levator palpebrae superioris and the occipitofrontalis (Fig. 11-25). The levator palpebrae superioris is in fact a muscle of the orbital cavity and is described with these muscles on page 816. The occipitofrontalis forms part of the scalp and has been described on page 744.

#### *Orbicularis Oculi (Fig. 11-25)*

##### Origin and Insertion

The *palpebral part*, which is confined to the eyelids, arises from the medial palpebral ligament, arches across both lids in front of the tarsal plates, and is inserted into the lateral palpebral raphe. Some of the fibers arise from the lacrimal bone and from the fascia covering the lacrimal sac; they pass laterally behind the lacrimal sac and enter both eyelids.

The *orbital part*, which lies flat on the surface of the orbital margin on the forehead and cheek,

arises from the medial end of the medial palpebral ligament and from the adjoining bone. The fibers pass laterally as a series of concentric loops, there being no interruption on the lateral side of the orbit.

##### Nerve Supply

Temporal and zygomatic branches of the facial nerve.

##### Action

The palpebral part closes the eyelids and dilates the lacrimal sac; it also directs the lacrimal puncta into the lacus lacrimalis. The orbital portion pulls on the skin of the forehead, temple, and cheek like a purse string and draws it toward the medial angle of the orbit. The skin is thrown into prominent folds, which overlap the eyelids and add further protection to the underlying eye. The movement is referred to as "screwing up the eye."

#### *Corrugator Supercilii*

The corrugator supercilii is a small muscle that lies deep to the orbicularis oculi. It arises from the medial part of the superciliary arch and is inserted into the skin of the eyebrow.

##### Nerve Supply

Temporal branch of the facial nerve.

##### Action

Produces vertical wrinkles of the forehead, as in frowning.

### MUSCLES OF THE NOSTRILS

The sphincter muscle is the compressor naris, and the dilatator is the dilatator naris.

#### *Compressor Naris (Fig. 11-25)*

##### Origin

From the frontal process of the maxilla.

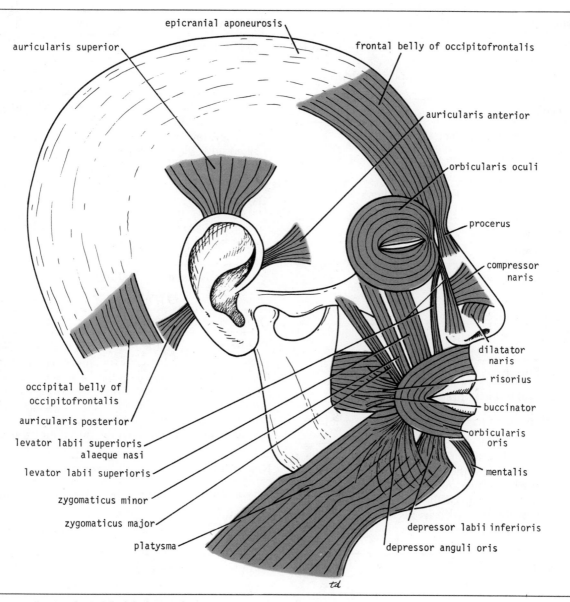

**Fig. 11-25. Muscles of facial expression.**

### Insertion

The fibers pass medially into a thin aponeurosis, which crosses the bridge of the nose and is continuous with the muscle of the opposite side.

### Nerve Supply

Buccal branch of the facial nerve.

### Action

Compresses the mobile nasal cartilages.

## *Dilatator Naris (Fig. 11-25)*

### Origin

From the maxilla.

### Insertion

Into the ala of the nose.

### Nerve Supply

The buccal branch of the facial nerve.

### Action

Pulls the ala laterally, thus widening the nasal aperture. In addition, the mobile alar cartilages can be elevated slightly, and this is associated with wrinkling of the skin of the nose. This movement is produced by the procerus and the levator labii superioris alaeque nasi.

## *Procerus (Fig. 11-25)*

### Origin

This is continuous with the medial margin of the frontal belly of the occipitofrontalis. It arises from the lower part of the nasal bone and the lateral nasal cartilage.

### Insertion

The skin between the eyebrows.

### Nerve Supply

Buccal branch of the facial nerve.

### Action

Pulls down the medial end of the eyebrow and wrinkles the skin of the nose.

   The levator labii superioris alaeque nasi is described with the muscles of the lips.

## MUSCLES OF THE LIPS AND CHEEKS

The sphincter muscle is the orbicularis oris. The dilator muscles consist of a series of small muscles that radiate out from the lips.

### *Sphincter Muscle of the Lips*

#### *Orbicularis Oris (Fig. 11-25)*

##### Origin and Insertion

The fibers encircle the oral orifice within the substance of the lips. Some of the fibers arise near the midline from the maxilla above and the mandible below. Other fibers arise from the deep surface of the skin and pass obliquely to the mucous membrane lining the inner surface of the lips. Many of the fibers are derived from the buccinator muscle.

##### Nerve Supply

Buccal and mandibular branches of the facial nerve.

##### Action

Compresses the lips together.

### *Dilator Muscles of the Lips (Fig. 11-25)*

The dilator muscles radiate out from the lips, and their action is to separate the lips; this movement is usually accompanied by separation of the jaws.

   The muscles arise from the bones and fascia around the oral aperture and converge, to be inserted into the substance of the lips. Traced from the side of the nose to the angle of the mouth and then below the oral aperture, the muscles are named as follows:

1. Levator labii superioris alaeque nasi.
2. Levator labii superioris.
3. Zygomaticus minor.
4. Zygomaticus major.
5. Levator anguli oris (deep to the zygomatic muscles).
6. Risorius.
7. Depressor anguli oris.
8. Depressor labii inferioris.
9. Mentalis.

## Nerve Supply

Buccal and mandibular branches of the facial nerve.

## MUSCLE OF THE CHEEK

### Buccinator (Figs. 11-25 and 11-29)

### Origin

From the outer surface of the alveolar margins of the maxilla and mandible opposite the molar teeth, and from the pterygomandibular ligament.

### Insertion

The muscle fibers pass forward, forming the muscle layer of the cheek. The muscle is pierced by the parotid duct. At the angle of the mouth the central fibers decussate, those from below entering the upper lip and those from above entering the lower lip; the highest and lowest fibers continue into the upper and lower lips, respectively, without intersecting. The buccinator muscle thus blends and forms part of the orbicularis oris muscle.

### Nerve Supply

Buccal branch of the facial nerve.

### Action

Compresses the cheeks and lips against the teeth.

## Facial Nerve

As the facial nerve runs forward within the substance of the parotid salivary gland (see p. 760), it divides into its five terminal branches (Fig. 11-23).

1. The *temporal branch* emerges from the upper border of the gland and supplies the anterior and superior auricular muscles, the frontal belly of the occipitofrontalis, the orbicularis oculi, and the corrugator supercilii.
2. The *zygomatic branch* emerges from the anterior border of the gland and supplies the orbicularis oculi.
3. The *buccal branch* emerges from the anterior border of the gland below the parotid duct and supplies the buccinator muscle and the muscles of the upper lip and nostril.
4. The *mandibular branch* emerges from the anterior border of the gland and supplies the muscles of the lower lip.
5. The *cervical branch* emerges from the lower border of the gland and passes forward in the neck below the mandible to supply the platysma muscle; it may cross the lower margin of the body of the mandible to supply the depressor anguli oris muscle.

The facial nerve is the nerve of the second pharyngeal arch and supplies all the muscles of facial expression. It *does not supply the skin*, but its branches communicate with branches of the trigeminal nerve. It is believed that the proprioceptive nerve fibers of the facial muscles leave the facial nerve in these communicating branches and pass to the central nervous system via the trigeminal nerve. A summary of the origin and distribution of the facial nerve is given on page 891. See also Figure 11-85.

## PAROTID REGION

The parotid region comprises the parotid salivary gland and the structures immediately related to it.

## Parotid Gland

The paired parotid glands, together with the paired submandibular and sublingual glands and the numerous small glands scattered throughout the mouth cavity, constitute the salivary glands.

## TYPE AND POSITION OF GLAND

The parotid gland is the largest of the salivary glands and is composed almost entirely of serous acini. It is situated below the external auditory meatus and lies in a deep hollow behind the ramus of the mandible and in front of the sternocleidomastoid (Fig. 11-26).

## SHAPE AND LOBES OF THE GLAND; PROCESSES OF THE GLAND

As seen from the superficial surface, the parotid gland is roughly wedge-shaped, with its base above and its apex behind the angle of the mandible (Fig. 11-26). If cut across in a horizontal plane, it would also be found to be wedge-shaped, with its base in the lateral position and its apex against the pharyngeal wall.

During development, the parotid gland covers the lateral surface of the facial nerve. As development proceeds, the deep part of the gland extends medially between the branches of the facial nerve. In the fully formed gland, the facial nerve may be said to divide the parotid gland into *superficial* and *deep parts*, or *lobes* (Fig. 11-26).

The superior margin of the gland extends upward behind the temporomandibular joint into the posterior part of the mandibular fossa. This part of the gland is called the *glenoid process.*

The anterior margin of the gland extends forward superficial to the masseter muscle to form the *facial process.* A small part of the facial process may be separate from the main gland and is called the *accessory part of the gland* (Fig. 11-26).

The deep part of the gland may extend forward between the medial pterygoid muscle and the ramus of the mandible to form the *pterygoid process.*

## CAPSULES OF THE GLAND

The parotid gland is a lobulated, yellowish mass surrounded by a connective-tissue capsule. In addition, the gland is enclosed in a dense fibrous capsule derived from the investing layer of deep cervical fascia (Fig. 11-26). At the lower extremity of the gland the fascia splits into two layers; above,

the superficial layer is attached to the zygomatic arch, and the deep layer is attached to the tympanic part of the temporal bone.

A portion of the fascia, which extends between the styloid process and the angle of the mandible, is called the *stylomandibular ligament,* and this intervenes between the parotid and submandibular salivary glands (Figs. 11-26 and 11-30).

## PAROTID DUCT

The parotid duct emerges from the facial process of the gland and passes forward over the lateral surface of the masseter muscle. At the anterior border of the muscle it turns sharply medially and pierces the buccal pad of fat and the buccinator muscle (Fig. 11-26). It then passes forward for a short distance between the muscle and the mucous membrane and finally opens into the vestibule of the mouth upon a small papilla, opposite the upper second molar tooth (Fig. 11-57). The oblique passage of the duct forward between the mucous membrane and the buccinator serves as a valvelike mechanism and prevents the inflation of the duct system during violent blowing (e.g., as in glass blowing or trumpet playing). The accessory part of the gland is drained by a small duct that opens into the upper border of the parotid duct.

The parotid duct runs forward from the parotid gland 1 fingerbreadth below the zygomatic arch (Fig. 11-26). If the masseter is tightly contracted, the parotid duct can be palpated and rolled on the firm anterior edge of this muscle. The accessory part of the parotid gland and the transverse facial artery lie above the duct.

## STRUCTURES WITHIN THE PAROTID GLAND

The structures within the parotid gland, from lateral to medial, are (1) the facial nerve, (2) the retromandibular vein, and (3) the external carotid artery. Some members of the parotid group of lymph nodes are also located within the gland (Fig. 11-26).

**Fig. 11-26. Parotid gland and its relations. (A) Lateral surface of gland and course of parotid duct. (B) Horizontal section of parotid gland.**

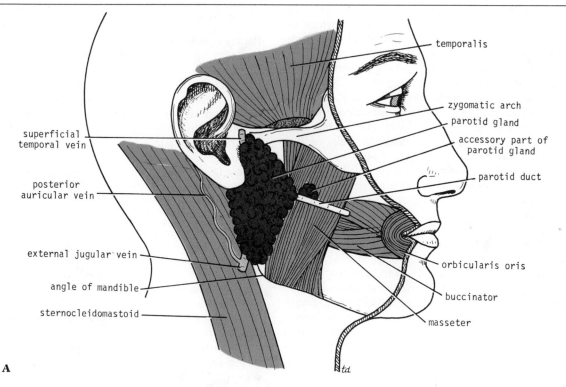

**A**

temporalis

zygomatic arch

parotid gland

accessory part of
parotid gland

parotid duct

orbicularis oris

buccinator

masseter

superficial
temporal vein

posterior
auricular vein

external jugular vein

angle of mandible

sternocleidomastoid

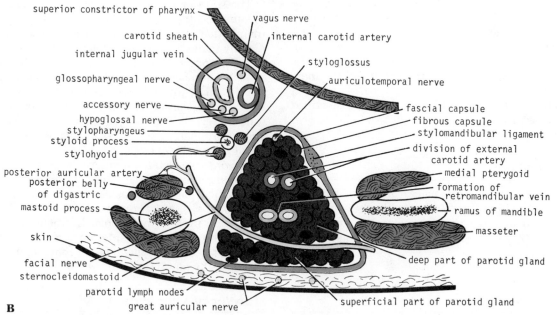

**B**

superior constrictor of pharynx

carotid sheath

internal jugular vein

glossopharyngeal nerve

accessory nerve

hypoglossal nerve

stylopharyngeus

styloid process

stylohyoid

posterior auricular artery

posterior belly
of digastric

mastoid process

skin

facial nerve

sternocleidomastoid

parotid lymph nodes

great auricular nerve

vagus nerve

internal carotid artery

styloglossus

auriculotemporal nerve

fascial capsule

fibrous capsule

stylomandibular ligament

division of external
carotid artery

medial pterygoid

formation of
retromandibular vein

ramus of mandible

masseter

deep part of parotid gland

superficial part of parotid gland

The *facial nerve* emerges from the stylomastoid foramen and enters the posteromedial surface of the gland (Fig. 11-26). It passes forward within the gland, superficial to the retromandibular vein and the external carotid artery, and divides into its five terminal branches. The branches of nerve leave the gland on its anteromedial surface (Fig. 11-23).

Branches of the facial nerve immediately before it enters the parotid gland are (1) a *muscular branch*, which supplies the posterior belly of the digastric and the stylohyoid (Fig. 11-26) and (2) the *posterior auricular nerve*, which ascends behind the ear and supplies the posterior and superior auricular muscles and the occipital belly of the occipitofrontalis.

Branches of the facial nerve within the parotid gland are the five terminal branches described on page 757.

The *retromandibular vein* is formed within the parotid gland by the union of the superficial temporal and the maxillary veins. It divides into anterior and posterior divisions, which leave the lower border of the gland. The anterior division joins the facial vein, and the posterior division unites with the posterior auricular vein to form the external jugular vein (Fig. 11-16).

The *external carotid artery*, having left the carotid triangle by passing deep to the posterior belly of the digastric and the stylohyoid, ascends and enters the substance of the parotid gland (Fig. 11-13). At the level of the neck of the mandible, it divides into the superficial temporal artery and the maxillary artery.

The *parotid group of lymph nodes* are described on page 742.

## Blood Supply

The external carotid artery and its terminal branches within the gland, namely, the superficial temporal and the maxillary arteries, supply the gland. The veins drain into the retromandibular vein.

## Lymph Drainage

The lymph vessels drain into the parotid lymph nodes and the deep cervical lymph nodes.

## Nerve Supply

Parasympathetic secretomotor fibers from the inferior salivary nucleus of the ninth cranial nerve supply the parotid gland. The nerve fibers pass to the otic ganglion via the tympanic branch of the ninth cranial nerve and the lesser petrosal nerve. Postganglionic parasympathetic fibers reach the parotid gland via the auriculotemporal nerve, which lies in contact with the deep surface of the gland.

Postganglionic sympathetic fibers reach the gland as a plexus of nerves around the external carotid artery.

## RELATIONS OF THE PAROTID GLAND

The structures that are intimately related to the deep surface of the gland are sometimes collectively referred to as the *parotid bed*.

For purposes of description, the relations may be divided into superficial, superior, posteromedial, and anteromedial.

The *superficial relations* are the parotid lymph nodes, superficial fascia, great auricular nerve, and skin (Fig. 11-26).

The *superior relations* are the external auditory meatus and the posterior surface of the temporomandibular joint (Fig. 11-26). The glenoid lobe is directly related to the auriculotemporal nerve.

The *posteromedial relations* are the mastoid process, the sternocleidomastoid, the posterior belly of the digastric, the styloid process and its attached muscles, the carotid sheath with the internal carotid artery, the internal jugular vein, and the vagus, glossopharyngeal, accessory, hypoglossal, and facial nerves (Fig. 11-26).

The *anteromedial relations* are the posterior border of the ramus of the mandible, the temporomandibular joint, the masseter, the medial pterygoid muscle, the terminal branches of the facial nerve, and the stylomandibular ligament (Fig. 11-26). At the union of the anteromedial and posteromedial surfaces, the gland lies in contact with the pharyngeal wall (Fig. 11-26).

## *Muscles of Mastication*

These consist of the masseter, temporalis, lateral pterygoid, and medial pterygoid muscles.

## MASSETER (FIG. 11-26)

### Origin

From the lower border and medial surface of the zygomatic arch.

### Insertion

Its fibers run downward and backward to be attached to the lateral aspect of the ramus of the mandible.

### Nerve Supply

Mandibular division of the trigeminal nerve.

### Action

Raises the mandible to occlude the teeth in mastication.

The temporalis, lateral pterygoid, and medial pterygoid muscles are described in the next section.

## THE TEMPORAL AND INFRATEMPORAL FOSSAE

The temporal region is situated on the side of the head. The temporal fossa is bounded above by the superior temporal line on the side of the skull, in front by the frontal process of the zygomatic bone, and below by the zygomatic arch (Fig. 11-27).

The infratemporal region is situated beneath the base of the skull, between the pharynx and the ramus of the mandible. The infratemporal fossa is bounded in front by the posterior surface of the maxilla, behind by the styloid process, above by the infratemporal surface of the greater wing of the sphenoid, medially by the lateral pterygoid plate, and laterally by the ramus of the mandible (Fig. 11-28).

The temporal and infratemporal fossae communicate with each other deep to the zygomatic arch.

## Contents of the Temporal Fossa

The temporal fossa is occupied by the temporalis muscle and its covering fascia, the deep temporal nerves and vessels, the auriculotemporal nerve, and the superficial temporal artery.

## TEMPORALIS (FIG. 11-27)

### Origin

The muscle is fan-shaped and arises from the bony floor of the temporal fossa and from the deep surface of the temporal fascia.

### Insertion

The muscle fibers converge to a tendon, which passes deep to the zygomatic arch and is inserted on the coronoid process of the mandible and the anterior border of the ramus of the mandible.

### Nerve Supply

Deep temporal nerves, which are branches of the anterior division of the mandibular division of the trigeminal nerve.

### Action

The anterior and superior fibers elevate the mandible; the posterior fibers retract the mandible.

The *temporal fascia* covers the temporal muscle above the zygomatic arch. It is attached above to the superior temporal line and below to the upper border of the zygomatic arch. Its deep surface gives origin to the temporalis muscle.

The *deep temporal nerves* are two in number and arise from the anterior division of the mandibular nerve. (See p. 767.) They emerge from the upper border of the lateral pterygoid muscle and enter the deep surface of the temporalis muscle (Fig. 11-27). The deep temporal arteries, also two in number, are branches of the maxillary artery. They accompany the nerves and supply the temporalis muscle.

The *auriculotemporal nerve*, a branch of the posterior division of the mandibular nerve (see p. 767), emerges from the upper border of the parotid gland behind the temporomandibular joint (Fig. 11-27). It crosses the root of the zygomatic arch behind the superficial temporal artery and in front of the auricle. It is distributed to the skin of the

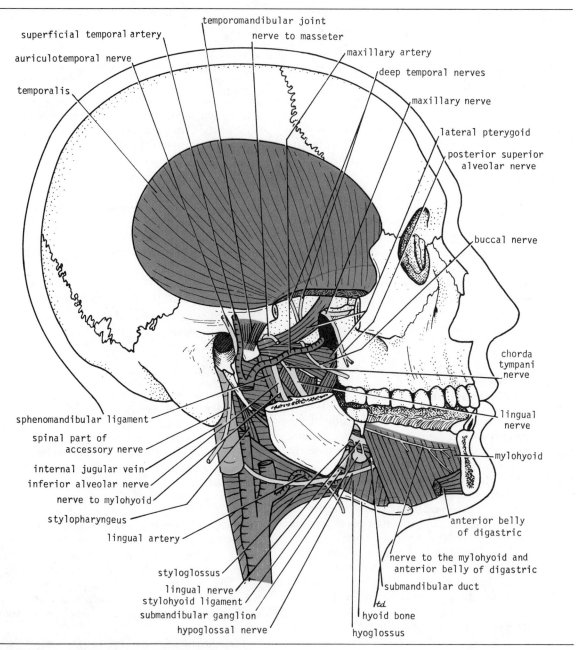

**Fig. 11-27. Infratemporal and submandibular regions. Parts of zygomatic arch; ramus and body of mandible have been removed to display deeper structures.**

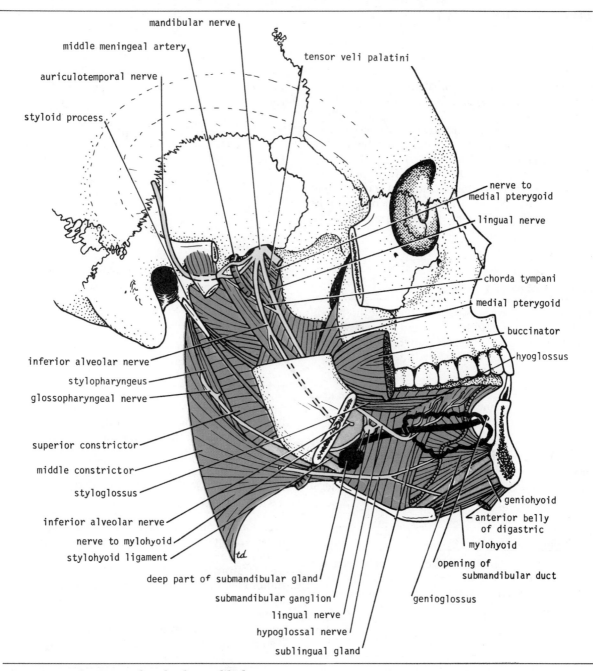

**Fig. 11-28. Infratemporal and submandibular regions. Parts of zygomatic arch; ramus and body of mandible have been removed. Mylohyoid and lateral pterygoid muscles have also been removed to display deeper structures. Outline of sublingual gland is shown as solid black wavy line.**

auricle, the external auditory meatus, and the scalp over the temporal region. (See p. 746.)

The *superficial temporal artery*, a terminal branch of the external carotid artery, emerges from the upper border of the parotid gland behind the temporomandibular joint (Fig. 11-27). It crosses the root of the zygomatic arch in front of the auriculotemporal nerve and the auricle. It is here that its pulsations may be easily felt. It ascends onto the scalp and divides into anterior and posterior divisions. (See p. 746.)

# Contents of the Infratemporal Fossa

The infratemporal fossa is occupied by the lateral and medial pterygoid muscles, the branches of the mandibular nerve, the otic ganglion, the chorda tympani, the maxillary artery, and the pterygoid venous plexus.

## LATERAL PTERYGOID (FIG. 11-27)

### Origin

The *upper head* arises from the infratemporal surface of the greater wing of the sphenoid. The *lower head* arises from the lateral surface of the lateral pterygoid plate.

### Insertion

The two heads converge as they pass backward and are inserted into the front of the neck of the mandible and the articular disc of the temporomandibular joint.

### Nerve Supply

From the anterior division of the mandibular division of the trigeminal nerve.

### Action

Pulls the neck of the mandible forward with the articular disc during the process of opening the mouth. Acting with the medial pterygoid of the same side, it pulls the neck of the mandible forward with the articular disc, causing the jaw to ro-

tate around the opposite condyle, as in the movement of chewing.

## MEDIAL PTERYGOID (FIG. 11-27)

### Origin

The *superficial head* arises from the tuberosity of the maxilla. The *deep head* arises from the medial surface of the lateral pterygoid plate.

### Insertion

The fibers run downward, backward, and laterally and are inserted into the medial surface of the angle of the mandible.

### Nerve Supply

Mandibular division of the trigeminal nerve.

### Action

Assists in elevating the mandible.

The muscles of the head, their nerve supply, and their actions are summarized in Table 11-2.

## MANDIBULAR DIVISION OF THE TRIGEMINAL NERVE

The sensory and motor roots of the mandibular nerve emerge from the skull through the foramen ovale in the greater wing of the sphenoid bone (Fig. 11-28). Immediately below the foramen, the small motor root of the mandibular nerve unites with the large sensory root. The mandibular nerve now descends between the tensor veli palatini medially and the lateral pterygoid laterally and divides into a small anterior and a large posterior division.

### Branches from the Main Trunk

1. A *meningeal branch*, which enters the skull through the foramen ovale.
2. The *nerve to the medial pterygoid.* This is a small branch that supplies the medial pterygoid muscle (Fig. 11-28). It gives off two branches, which pass without interruption through the otic ganglion (see below), to supply the tensor tympani and the tensor veli palatini.

Table 11-2. Muscles of the Head

| Name of muscle | Origin | Insertion | Nerve supply | Action |
|---|---|---|---|---|
| MUSCLE OF SCALP | | | | |
| Occipitofrontalis Occipital belly | Highest nuchal line of occipital bone | Epicranial aponeurosis | Facial nerve | Moves scalp on skull and raises eyebrows |
| Frontal belly | Skin and superficial fascia of eyebrows | | | |
| MUSCLES OF FACIAL EXPRESSION | | | | |
| Orbicularis oculi Palpebral part | Medial palpebral ligament | Lateral palpebral raphe | Facial nerve | Closes eyelids and dilates lacrimal sac |
| Orbital part | Medial palpebral ligament and adjoining bone | Loops return to origin | Facial nerve | Throws skin around orbit into folds to protect eyeball |
| Corrugator supercilii | Superciliary arch | Skin of eyebrow | Facial nerve | Vertical wrinkles of forehead, as in frowning |
| Compressor nasi | Frontal process of maxilla | Aponeurosis of bridge of nose | Facial nerve | Compresses mobile nasal cartilages |
| Dilator naris | Maxilla | Ala of nose | Facial nerve | Widens nasal aperture |
| Procerus | Nasal bone | Skin between eyebrows | Facial nerve | Wrinkles skin of nose |
| Orbicularis oris | Maxilla, mandible and skin | Encircles oral orifice | Facial nerve | Compresses lips together |
| DILATOR MUSCLES OF LIPS | | | | |
| Levator labii superioris alaeque nasi | | | | |
| Levator labii superioris | | | | |
| Zygomaticus minor | Arise from bones and fascia around oral aperture and insert into substance of lips | | Facial nerve | Separate lips |
| Zygomaticus major | | | | |
| Levator anguli oris | | | | |
| Risorius | | | | |
| Depressor anguli oris | | | | |
| Depressor labii inferioris | | | | |
| Mentalis | | | | |
| Buccinator | Outer surface of alveolar margins of maxilla and mandible and pterygomandibular ligament | | Facial nerve | Compresses cheeks and lips against teeth |
| Platysma | See Table 11-1 | | | |

Table 11-2 (continued)

| Name of muscle | Origin | Insertion | Nerve supply | Action |
|---|---|---|---|---|
| Platysma | See Table 11-1 | | | |
| MUSCLES OF MASTICATION | | | | |
| Masseter | Zygomatic arch | Lateral surface ramus of mandible | Mandibular division of trigeminal nerve | Elevates mandible to occlude teeth |
| Temporalis | Floor of temporal fossa | Coronoid process of mandible | Mandibular division of trigeminal nerve | Anterior and superior fibers elevate mandible; posterior fibers retract mandible |
| Lateral pterygoid (two heads) | Greater wing of sphenoid and lateral pterygoid plate | Neck of mandible and articular disc | Mandibular division of trigeminal nerve | Pulls neck of mandible forward |
| Medial pterygoid | Tuberosity of maxilla and lateral pterygoid plate | Medial surface of angle of mandible | Mandibular division of trigeminal nerve | Elevates mandible |

## Branches from the Anterior Division

The anterior division of the trigeminal nerve gives off three motor branches and *one sensory branch, the buccal nerve.*

1. The *masseteric nerve* runs laterally, above the lateral pterygoid and in front of the temporomandibular joint, to supply the deep surface of the masseter muscle (Fig. 11-27).
2. The *two deep temporal nerves* run upward above the upper border of the lateral pterygoid and enter the deep surface of the temporalis muscle (Fig. 11-27).
3. The *nerve to the lateral pterygoid muscle* enters the deep surface of the muscle.
4. The *buccal nerve* is a sensory nerve that runs forward between the two heads of the lateral pterygoid (Fig. 11-27). It emerges on the cheek from beneath the anterior border of the masseter and supplies the skin over the cheek and the mucous membrane lining the cheek. (*It does not supply the buccinator muscle*, which is supplied by the buccal branch of the facial nerve.)

## Branches from the Posterior Division

The posterior division of the trigeminal nerve gives off two sensory branches and *one branch containing motor fibers, the inferior alveolar nerve.*

1. The *auriculotemporal nerve* arises by two roots, which embrace the middle meningeal artery (Fig. 11-28). The nerve runs backward between the neck of the mandible and the sphenomandibular ligament. It then ascends behind the temporomandibular joint, close to the parotid gland and in company with the superficial temporal vessels. Its further course in the scalp is described on page 746. It receives postganglionic parasympathetic secretomotor fibers from the otic ganglion (see below), which it conveys to the parotid gland.

### Branches

It gives sensory branches to the skin of the auricle, the external auditory meatus, the tympanic membrane, the parotid gland, the temporomandibular joint, and temporal branches to the skin of the scalp.

2. The *lingual nerve* runs downward on the medial surface of the lateral pterygoid and then on the lateral surface of the medial pterygoid (Figs. 11-27 and 11-28). It emerges at the lower border of the lateral pterygoid, anterior to the inferior alveolar nerve. It continues its course downward and forward between the ramus of the mandible and the medial pterygoid. The nerve now passes forward and medially (Fig. 11-28) beneath the lower border of the superior constrictor muscle of the pharynx, related medially to the styloglossus muscle and laterally to the lower third molar tooth. It then runs forward on the lateral surface of the hyoglossus in the submandibular region. (See p. 780.) At the lower border of the lateral pterygoid muscle, it is joined by the *chorda tympani nerve* (Fig. 11-27); in addition, it frequently receives a branch from the inferior alveolar nerve.

### Branches

The lingual nerve gives off no branches in the infratemporal fossa.

3. The *inferior alveolar nerve* which is made up of motor and sensory nerve fibers, descends deep to the lateral pterygoid muscle and then onto the lateral surface of the sphenomandibular ligament (Fig. 11-27). It then enters the mandibular canal through the mandibular foramen and runs forward below the teeth. Having supplied the teeth of the lower jaw, it emerges through the mental foramen to supply the skin of the face. (See p. 751.)

### Branches

The *mylohyoid nerve* arises from the inferior alveolar nerve just above the mandibular foramen (Fig. 11-27). It pierces the sphenomandibular ligament and runs forward on the medial surface of the body of the mandible below the mylohyoid line. It lies superficial to the mylohyoid muscle (Fig. 11-27), which it supplies; it also supplies the anterior belly of the digastric muscle.

A *communicating branch*, which joins the lingual nerve, is frequently given off by the inferior alveolar nerve.

## OTIC GANGLION

The otic ganglion is a small parasympathetic ganglion that is functionally associated with the glossopharyngeal nerve. It is situated just below the foramen ovale, between the mandibular nerve and the tensor veli palatini. The ganglion adheres to the nerve to the medial pterygoid, but functionally it is completely separate from it.

The preganglionic parasympathetic fibers originate in the inferior salivary nucleus of the glossopharyngeal nerve. They leave the glossopharyngeal nerve by its tympanic branch and then pass via the tympanic plexus and the lesser petrosal nerve to the otic ganglion. Here, the fibers synapse, and the postganglionic fibers leave the ganglion and join the auriculotemporal nerve. They are conveyed by this nerve to the parotid gland and serve as secretomotor fibers.

Postganglionic sympathetic fibers from the plexus around the middle meningeal artery pass through the ganglion without interruption. They join the auriculotemporal nerve and are conveyed to the parotid gland, where they supply the blood vessels of the gland.

## CHORDA TYMPANI

The chorda tympani is a branch of the facial nerve in the temporal bone. (See p. 836 and also Fig. 11-85.) It enters the infratemporal fossa through the petrotympanic fissure and runs downward and forward to join the lingual nerve (Fig. 11-28). The nerve carries secretomotor preganglionic parasympathetic fibers to the submandibular and sublingual salivary glands. (See p. 780.) It also carries taste fibers from the anterior two-thirds of the tongue and the floor of the mouth. The cell bodies of the taste fibers are situated in the sensory geniculate ganglion of the facial nerve (see p. 836), and they end by synapsing with cells of the nucleus solitarius in the pons of the brain.

## MAXILLARY ARTERY

The maxillary artery is the larger terminal branch of the external carotid artery (see p. 760) in the parotid gland. It arises behind the neck of the mandible and runs forward medial to it, to reach the lower border of the lateral pterygoid muscle (Fig. 11-27). The artery then passes upward and forward, either superficial or deep to the lower head of the lateral pterygoid. It then leaves the infratemporal fossa by entering the pterygopalatine fossa. Its further course is described on page 839.

### Branches in the Infratemporal Fossa

1. The *inferior alveolar artery* follows the inferior alveolar nerve.
2. The *middle meningeal artery*. This is an important artery, which passes upward deep to the lateral pterygoid muscle behind the mandibular nerve (Fig. 11-28). It is embraced by the two roots of origin of the auriculotemporal nerve. The artery enters the skull through the foramen spinosum (Figs. 11-28 and 11-29).
3. Small branches, which supply the lining of the external auditory meatus and the tympanic membrane.
4. Numerous small muscular branches, which supply the muscles of mastication.

## PTERYGOID VENOUS PLEXUS

The pterygoid venous plexus is a venous network that surrounds the lateral pterygoid muscle. It receives veins that correspond to branches of the maxillary artery, and its posterior end is drained by the maxillary vein. It communicates with the facial vein through the *deep facial vein*.

## MAXILLARY VEIN

The maxillary vein is a short vessel that drains the posterior end of the pterygoid venous plexus. It runs backward with the maxillary artery on the medial side of the neck of the mandible and joins the superficial temporal vein within the parotid gland, to form the *retromandibular vein* (Fig. 11-26).

# Temporomandibular Joint

### Articulation

This occurs between the articular tubercle and the anterior portion of the mandibular fossa of the temporal bone above and the head (condyloid pro-

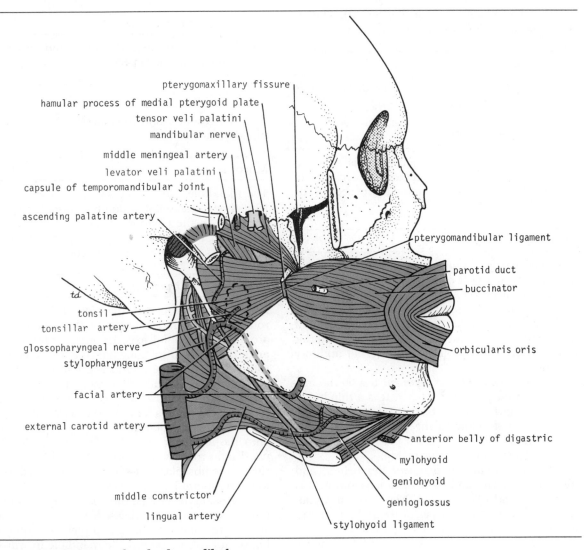

pterygomaxillary fissure

hamular process of medial pterygoid plate

tensor veli palatini

mandibular nerve

middle meningeal artery

levator veli palatini

capsule of temporomandibular joint

ascending palatine artery

td

tonsil

tonsillar artery

glossopharyngeal nerve

stylopharyngeus

facial artery

external carotid artery

middle constrictor

lingual artery

pterygomandibular ligament

parotid duct

buccinator

orbicularis oris

anterior belly of digastric

mylohyoid

geniohyoid

genioglossus

stylohyoid ligament

**Fig. 11-29. Infratemporal and submandibular regions, showing courses of facial and lingual arteries.**

cess) of the mandible below (Fig. 11-30). The articular surfaces are covered with fibrocartilage.

## Type

Synovial. The articular disc divides the joint into upper and lower cavities (Fig. 11-31).

## Capsule

The capsule surrounds the joint and is attached above to the articular tubercle and the margins of the mandibular fossa, and below to the neck of the mandible.

## Ligaments

The *lateral temporomandibular ligament* strengthens the lateral aspect of the capsule, and its fibers run downward and backward from the tubercle on the root of the zygoma to the lateral surface of the neck of the mandible (Fig. 11-30). This ligament limits the movement of the mandible in a posterior direction and thus protects the external auditory meatus.

The *sphenomandibular ligament* lies on the medial side of the joint (Fig. 11-30). It is a thin band, which is attached above to the spine of the sphenoid bone and below to the lingula of the mandibular foramen. It represents the remains of the first pharyngeal arch in this region.

The *stylomandibular ligament* lies behind and medial to the joint and some distance from it. It is merely a band of thickened deep cervical fascia, which extends from the apex of the styloid process to the angle of the mandible (Fig. 11-30).

The *articular disc* divides the joint into upper and lower cavities (Fig. 11-31). It is an oval plate of fibrocartilage, which is attached circumferentially to the capsule. It is also attached in front to the tendon of the lateral pterygoid muscle, and by fibrous bands to the head of the mandible. These bands ensure that the disc moves forward and backward with the head of the mandible during protraction and retraction of the mandible. The upper surface of the disc is concavoconvex from before backward to fit the shape of the articular tubercle and the mandibular fossa; the lower surface is concave to fit the head of the mandible.

## Synovial Membrane

This lines the capsule in the upper and lower cavities of the joint (Fig. 11-31).

## Nerve Supply

Auriculotemporal and masseteric branches of the mandibular nerve.

## Movements

The mandible can be depressed or elevated, protruded or retracted. Rotation may also occur, as in chewing. In the position of rest, the teeth of the upper and lower jaws are slightly apart. On closure of the jaws, the teeth come into contact.

### Depression of the Mandible

As the mouth is opened, the head of the mandible rotates on the undersurface of the articular disc around a horizontal axis. In order to prevent the angle of the jaw impinging unnecessarily on the parotid gland and the sternocleidomastoid muscle, the mandible is pulled forward. This is accomplished by the contraction of the lateral pterygoid muscle, which pulls forward the neck of the mandible and the articular disc, so that the latter moves onto the articular tubercle (Fig. 11-31). The forward movement of the disc is limited by the tension of the fibroelastic tissue, which tethers the disc to the temporal bone posteriorly.

Depression of the mandible is brought about by the contraction of the digastrics, the geniohyoids, and the mylohyoids; the lateral pterygoids play an important role by pulling the mandible forward.

### Elevation of the Mandible

The movements in depression of the mandible are reversed. First, the head of the mandible and the disc move backward, and then the head rotates on the lower surface of the disc.

Elevation of the mandible is brought about by the contraction of the temporalis, the masseter, and the medial pterygoids. The head of the mandible is pulled backward by the posterior fibers of the temporalis. The articular disc is pulled back-

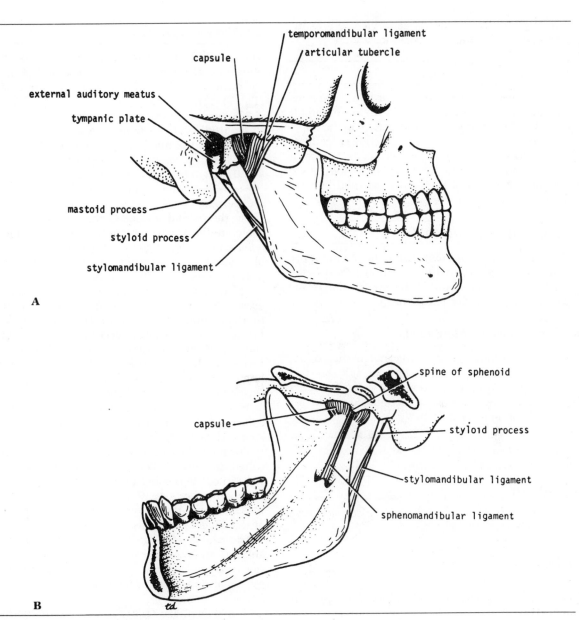

**Fig. 11-30. Temporomandibular joint as seen from (A) lateral aspect and (B) medial aspect.**

ward by the fibroelastic tissue, which tethers the disc to the temporal bone posteriorly.

### Protrusion of the Mandible

The articular disc is pulled forward onto the anterior tubercle, carrying the head of the mandible with it. All movement thus takes place in the upper cavity of the joint. In protrusion, the lower teeth are drawn forward over the upper teeth, which is brought about by the contraction of the lateral pterygoid muscles of both sides, assisted by both medial pterygoids.

### Retraction of the Mandible

The articular disc and the head of the mandible are pulled backward into the mandibular fossa. Retraction is brought about by the contraction of the posterior fibers of the temporalis.

### Lateral Chewing Movements

These are accomplished by alternately protruding and retracting the mandible on each side. For this to take place a certain amount of rotation occurs and the muscles responsible on both sides work alternately and not in unison.

### Important Relations of the Temporomandibular Joint

#### Anteriorly

The mandibular notch and the masseteric nerve and artery (Fig. 11-27).

#### Posteriorly

The tympanic plate of the external auditory meatus (Fig. 11-30) and the glenoid process of the parotid gland.

#### Laterally

The parotid gland, fascia, and skin (Fig. 11-26).

#### Medially

The maxillary artery and vein and the auriculotemporal nerve.

## THE MANDIBLE

The mandible consists of a horseshoe-shaped *body* and a pair of *rami*. The body of the mandible meets the ramus on each side at the *angle of the mandible* (Fig. 11-32).

The *body of the mandible*, on its external surface in the midline, has a faint ridge indicating the line of fusion of the two halves during development at the *symphysis menti*. The *mental foramen* can be seen below the second premolar tooth; it transmits the terminal branches of the inferior alveolar nerve and vessels.

On the medial surface of the body of the mandible in the median plane are seen the *mental spines*; these give origin to the genioglossus muscles above and the geniohyoid muscles below (Fig. 11-32). The *mylohyoid line* can be seen as an oblique ridge, which runs backward and laterally from the area of the mental spines to an area below and behind the third molar tooth. The *submandibular fossa*, for the superficial part of the submandibular salivary gland, lies below the posterior part of the mylohyoid line. The *sublingual fossa*, for the sublingual gland, lies above the anterior part of the mylohyoid line (Fig. 11-32).

The upper border of the body of the mandible is called the *alveolar* part; in the adult it contains sixteen sockets for the roots of the teeth.

The lower border of the body of the mandible is called the *base*. The *digastric fossa* is a small, roughened depression on the base, or behind the base, on either side of the symphysis menti (Fig. 11-32). It is in these fossae that the anterior bellies of the digastric muscles are attached.

The *ramus of the mandible* is vertically placed and has an anterior *coronoid process* and a posterior *condyloid process*, or *head*; the two processes are separated by the *mandibular notch* (Fig. 11-32).

On the lateral surface of the ramus are markings for the attachment of the masseter muscle. On the medial surface is the *mandibular foramen*, for the

**Fig. 11-31. Temporomandibular joint (A) with mouth closed and (B) with mouth open. Note position of head of mandible and articular disc in relation to articular tubercle in each case.**

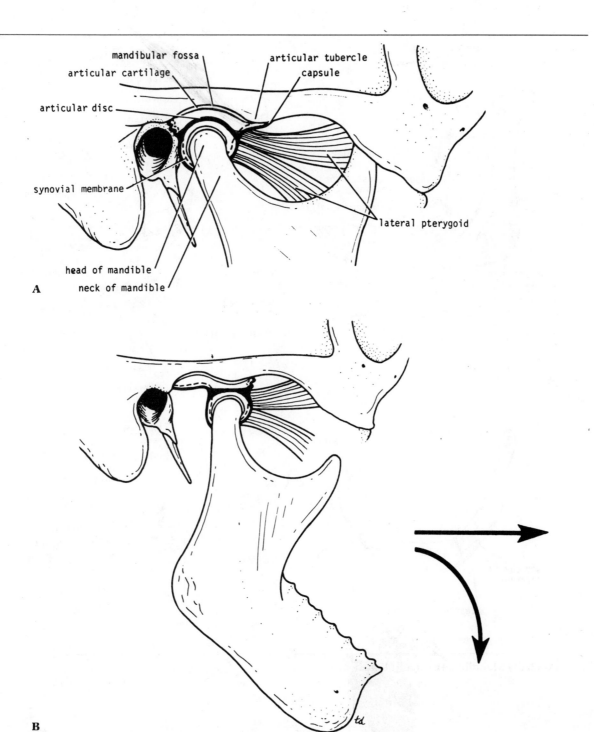

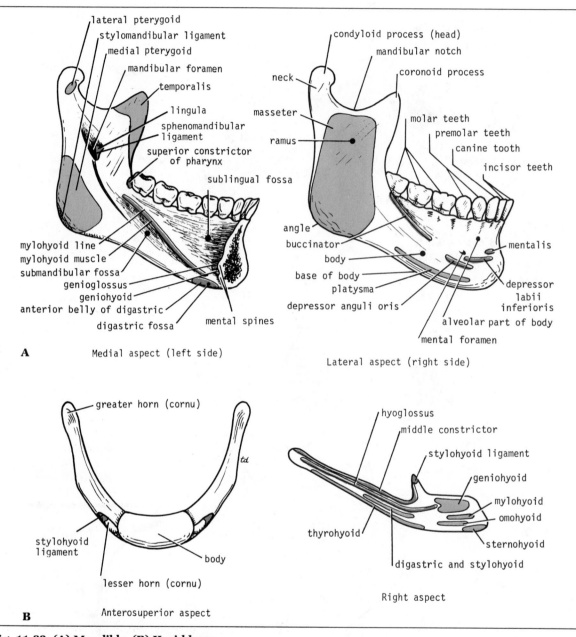

lateral pterygoid
stylomandibular ligament
medial pterygoid
mandibular foramen
temporalis
lingula
sphenomandibular ligament
superior constrictor of pharynx
sublingual fossa
mylohyoid line
mylohyoid muscle
submandibular fossa
genioglossus
geniohyoid
anterior belly of digastric
digastric fossa
mental spines

**A**    Medial aspect (left side)

condyloid process (head)
mandibular notch
coronoid process
neck
masseter
ramus
molar teeth
premolar teeth
canine tooth
incisor teeth
angle
buccinator
body
base of body
platysma
depressor anguli oris
mentalis
depressor labii inferioris
alveolar part of body
mental foramen

Lateral aspect (right side)

greater horn (cornu)
td
stylohyoid ligament
body
lesser horn (cornu)

**B**    Anterosuperior aspect

hyoglossus
middle constrictor
stylohyoid ligament
geniohyoid
mylohyoid
omohyoid
thyrohyoid
sternohyoid
digastric and stylohyoid

Right aspect

**Fig. 11-32. (A) Mandible. (B) Hyoid bone.**

inferior alveolar nerve and vessels. In front of the foramen is a projection of bone, called the *lingula*, for the attachment of the *sphenomandibular ligament* (Fig. 11-32). The foramen leads into the *mandibular canal*, which opens on the lateral surface of the body of the mandible at the *mental foramen*. (See above.) The *incisive canal* is a continuation forward of the mandibular canal beyond the mental foramen and below the incisor teeth.

The *coronoid process* receives on its medial surface the attachment of the temporalis muscle. Below the *condyloid process*, or *head*, is a short *neck* (Fig. 11-32).

The important muscles and ligaments that are attached to the mandible are shown in Figure 11-32.

## The Hyoid Bone

The hyoid bone is U-shaped and consists of a body and two greater and two lesser cornua (Fig. 11-32). It is attached to the skull by the stylohyoid ligament and to the thyroid cartilage by the thyrohyoid membrane. It is very mobile and lies in the neck just above the larynx and below the mandible. The hyoid bone forms a base for the tongue and is suspended in position by muscles that connect it to the mandible, to the styloid process of the temporal bone, to the thyroid cartilage, to the sternum, and to the scapula.

The important muscles that are attached to the hyoid bone are shown in Figure 11-32.

## SUBMANDIBULAR REGION

The submandibular region lies under cover of the body of the mandible, between the mandible and the hyoid bone. It contains the following structures:

### Muscles

Digastric, mylohyoid, hyoglossus, geniohyoid, genioglossus, and styloglossus.

### Salivary Glands

Submandibular and sublingual glands.

### Nerves

Lingual, glossopharyngeal, hypoglossal.

### Parasympathetic Ganglion

Submandibular ganglion.

### Blood Vessels

Facial artery and vein and lingual artery and vein.

### Lymph Nodes

Submandibular group of lymph nodes.

## DIGASTRIC MUSCLE (FIG. 11-33)

The digastric muscle is described on page 720.

## MYLOHYOID (FIGS. 11-27 and 11-33)

### Origin

This is a flat, triangular sheet of muscle, which arises from the whole length of the mylohyoid line of the mandible.

### Insertion

The fibers run downward and forward. The posterior fibers are inserted into the body of the hyoid bone; the anterior fibers are inserted into a fibrous raphe, which extends from the symphysis menti to the body of the hyoid bone.

### Nerve Supply

Mylohyoid branch of the inferior alveolar nerve.

### Action

The two mylohyoid muscles form a muscular sheet, which supports the tongue and the floor of the mouth. When the mandible is fixed, they elevate the floor of the mouth and the hyoid bone during the first stage of swallowing. When the hyoid bone is fixed, it assists in the depression of the mandible and the opening of the mouth.

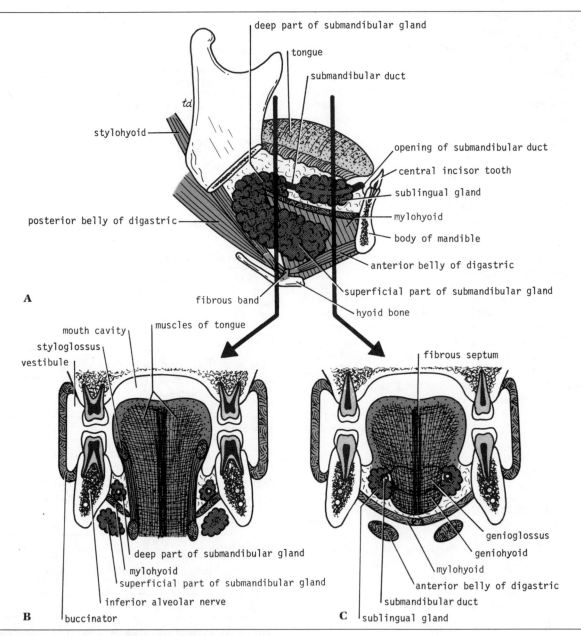

Fig. 11-33. (A) Submandibular and sublingual salivary gland (lateral view). (B) Coronal section through superficial and deep parts of submandibular salivary glands. (C) Coronal section (anterior to B) through sublingual salivary glands and ducts of submandibular salivary glands.

## HYOGLOSSUS (FIGS. 11-27 and 11-28)

### Origin

From the upper border of the body and greater cornu of the hyoid bone.

### Insertion

The muscle is quadrilateral in shape and runs upward deep to the mylohyoid muscle to enter the side of the tongue. It ends when its fibers mix with those of other muscles of the tongue.

### Nerve Supply

Hypoglossal nerve.

### Action

Depresses the tongue.

## GENIOHYOID (FIGS. 11-28 and 11-29)

### Origin

From the inferior mental spine, behind the symphysis menti of the mandible.

### Insertion

It is a narrow muscle, which lies above the mylohyoid; it is inserted onto the anterior surface of the body of the hyoid bone. Its medial surface lies in contact with the corresponding muscle of the opposite side.

### Nerve Supply

First cervical nerve through the hypoglossal nerve. (See p. 733.)

### Action

Elevates the hyoid bone and draws it forward; or depresses the mandible.

## GENIOGLOSSUS (FIG. 11-28)

### Origin

From the superior mental spine, behind the symphysis menti of the mandible.

### Insertion

This is a fan-shaped muscle, which widens out as it extends backward into the tongue. The superior fibers pass to the tip of the tongue, and middle fibers pass to the dorsum of the tongue, and a few of the inferior fibers are attached to the body of the hyoid bone.

### Nerve Supply

Hypoglossal nerve.

### Action

Draws the tongue forward and protrudes the tip so that it points to the opposite side. The two muscles acting in unison protrude the tongue in the midline. The muscles also depress the tongue.

## STYLOGLOSSUS (FIGS. 11-14 and 11-28)

### Origin

From the tip of the styloid process.

### Insertion

The fibers run downward and forward on the lateral surface of the superior constrictor muscle. On reaching the interval between the superior and middle constrictor muscles, the styloglossus passes forward to enter the side of the tongue.

### Nerve Supply

Hypoglossal nerve.

### Action

Draws the tongue upward and backward.

## SUBMANDIBULAR GLAND

The paired submandibular glands, together with the paired sublingual and parotid glands and the numerous small glands scattered throughout the mouth cavity, constitute the salivary glands.

### Type and Parts of the Gland

The submandibular gland is a large salivary gland and is composed of a mixture of serous and mucous acini, the former predominating. It lies partly under cover of the body of the mandible and is made up of a large superficial part and a small deep part, which are continuous with each other around the posterior border of the mylohyoid muscle (Fig. 11-33).

The *superficial part of the gland* lies in the digastric triangle, reaching upward under cover of the body of the mandible. Posteriorly, it is separated from the parotid gland by the stylomandibular ligament.

### Relations
#### Anteriorly

The anterior belly of the digastric (Fig. 11-33).

#### Posteriorly

The stylohyoid, the posterior belly of the digastric, the stylomandibular ligament, and the parotid gland.

#### Medially

The mylohyoid, and the mylohyoid nerve and vessels, the hyoglossus, and the lingual and hypoglossal nerves (Fig. 11-27).

#### Laterally

The gland lies in contact with the submandibular fossa on the medial surface of the mandible. Inferolaterally, it is covered by the investing layer of deep cervical fascia, the platysma muscle, and skin. It is crossed by the cervical branch of the facial nerve and facial vein. The submandibular lymph nodes also lie lateral to it.

The facial artery is related to the posterior and superior aspects of the superficial part of the gland (Fig. 11-29).

The *deep part of the gland* extends forward in the interval between the mylohyoid below and laterally and the hyoglossus and styloglossus medially (Fig. 11-33). Its posterior end is continuous with the superficial part of the gland around the posterior border of the mylohyoid muscle; its anterior end reaches as far as the sublingual gland.

### Relations
#### Anteriorly

The sublingual gland (Fig. 11-33).

#### Posteriorly

The stylohyoid, the posterior belly of the digastric, the stylomandibular ligament, and the parotid gland.

#### Medially

The hyoglossus and styloglossus (Fig. 11-33).

#### Laterally

The mylohyoid muscle and the superficial part of the gland (Fig. 11-33).

#### Superiorly

It is related superiorly to the lingual nerve and the submandibular ganglion; it is covered by the mucous membrane of the floor of the mouth.

#### Inferiorly

The hypoglossal nerve.

### Capsules of the Gland

The submandibular gland is a lobulated mass, surrounded by a connective-tissue capsule. In addition, the gland is partly enclosed in a dense fibrous capsule derived from the investing layer of deep cervical fascia; one layer of this fascia passes up from the hyoid bone, covering the inferolateral surface of the gland, and is attached to the lower border of the body of the mandible. Another, deeper layer passes up from the hyoid bone, covering the medial surface of the gland, and is attached to the mylohyoid line. A portion of the fascia that extends between the styloid process and the angle of the mandible is called the *stylomandibular ligament* and intervenes between the parotid and submandibular glands.

## Submandibular Duct

The submandibular duct emerges from the anterior end of the deep part of the gland (Fig. 11-33). It passes forward along the side of the tongue, beneath the mucous membrane of the floor of the mouth. It is crossed laterally by the lingual nerve and then lies between the sublingual gland and the genioglossus muscle (Fig. 11-28). It opens into the mouth on the summit of a small papilla, which is situated at the side of the frenulum of the tongue (Fig. 11-57).

Clinically, it is important to remember that the submandibular duct and the deep part of the gland can be readily palpated through the mucous membrane of the floor of the mouth alongside the tongue. Saliva can usually be seen emerging from the orifice of the duct.

## Blood Supply

The arteries are branches of the facial and lingual arteries. The veins drain into the facial and lingual veins.

## Lymph Drainage

The lymph vessels drain into the submandibular and deep cervical lymph nodes.

## Nerve Supply

Parasympathetic secretomotor supply from the superior salivary nucleus of the seventh cranial nerve. The nerve fibers pass to the submandibular ganglion and other small ganglia close to the duct via the chorda tympani nerve and the lingual nerve. Postganglionic parasympathetic fibers reach the gland either directly or along the duct.

Postganglionic sympathetic fibers reach the gland as a plexus of nerves around the facial and lingual arteries.

## SUBLINGUAL GLAND

The sublingual gland is the smallest of the three main salivary glands (Fig. 11-33). It lies beneath the mucous membrane of the floor of the mouth, close to the midline. It contains both serous and mucous acini, the latter predominating.

## Relations

### Anteriorly

The gland of the opposite side.

### Posteriorly

The deep part of the submandibular gland (Fig. 11-33).

### Medially

The genioglossus muscle, the lingual nerve, and the submandibular duct (Fig. 11-28).

### Laterally

The gland is related laterally to the sublingual fossa of the medial surface of the mandible.

### Superiorly

The mucous membrane of the floor of the mouth, which is elevated by the gland to form the *sublingual fold* (Fig. 11-57).

### Inferiorly

The gland is supported by the mylohyoid muscle (Fig. 11-33).

## Sublingual Ducts

The sublingual ducts are eight to twenty in number. The majority open into the mouth on the summit of the sublingual fold (Fig. 11-57), but a few may open into the submandibular duct.

## Blood Supply

The gland is supplied by branches of the facial and lingual arteries. The veins drain into the facial and lingual veins.

## Lymph Drainage

The lymph vessels drain into the submandibular and the deep cervical lymph nodes.

## Nerve Supply

Parasympathetic secretomotor supply from the superior salivary nucleus of the seventh cranial nerve. The nerve fibers pass to the submandibular ganglion via the chorda tympani nerve and the lingual nerve. Postganglionic parasympathetic fibers pass to the gland via the lingual nerve.

Postganglionic sympathetic fibers reach the gland as a plexus of nerves around the facial and lingual arteries.

## LINGUAL NERVE

The lingual nerve is a branch of the posterior division of the mandibular nerve in the infratemporal fossa (See p. 767.) It enters the submandibular region by passing forward and medially beneath the lower border of the superior constrictor muscle of the pharynx, related medially to the styloglossus muscle and laterally to the *lower third molar tooth* (Fig. 11-27). The nerve then passes forward on the lateral surface of the hyoglossus (Fig. 11-28), above the deep part of the submandibular gland; at the anterior margin of the hyoglossus, it passes downward, crossing the lateral surface of the submandibular duct, and then, winding below it, passes upward and forward on its medial side (Fig. 11-28). The nerve now lies under cover of the sublingual gland on the lateral surface of the genioglossus muscle. It ends by dividing into terminal branches, which supply the mucous membrane of the anterior two-thirds of the tongue and the floor of the mouth.

The lingual nerve is joined by the chorda tympani nerve in the infratemporal fossa. (See p. 768.)

## Branches

1. *Ganglionic branches.* Parasympathetic secretomotor fibers from the superior salivary nucleus of the seventh cranial nerve join the lingual nerve from the chorda tympani (Fig. 11-28). The preganglionic fibers leave the lingual nerve, and the majority pass to the submandibular ganglion; others pass to other smaller ganglia in the region. Postganglionic fibers pass to the submandibular salivary gland. Those fibers passing to the sublingual gland do so directly, or reach the gland via branches of the lingual nerve.
2. *Sensory branches.* The terminal branches of the lingual nerve are distributed to the mucous membrane of the anterior two-thirds of the tongue and the floor of the mouth and the lingual surface of the gums (the circumvallate papillae are supplied by the glossopharyngeal nerve). Common sensation passes to the central nervous system via the mandibular and trigeminal nerves. The special taste fibers leave the lingual nerve in the chorda tympani and reach the central nervous system in the facial nerve.
3. *Communicating branches.* These connect the nerve to the hypoglossal nerve on the side of the tongue.

## SUBMANDIBULAR GANGLION

The submandibular ganglion is a small parasympathetic ganglion, situated on the lateral surface of the hyoglossus muscle (Fig. 11-28). It lies below the lingual nerve and above the submandibular duct. The ganglion is connected to the lingual nerve by several branches; other branches pass to the submandibular salivary gland.

Preganglionic parasympathetic fibers reach the ganglion from the superior salivary nucleus of the seventh cranial nerve via the chorda tympani and lingual nerves. The preganglionic fibers synapse within the ganglion, and the postganglionic fibers pass to the submandibular salivary gland, to which they are secretomotor. Other postganglionic secretomotor fibers pass to the sublingual salivary gland and small salivary glands in the mouth. They do so by returning to the lingual nerve and traveling with the branches of the nerve to reach the glands.

A few postganglionic sympathetic fibers pass without interruption through the ganglion.

Smaller parasympathetic ganglia are also present in this region. They have similar connections and are found along the submandibular duct or in the substance of the submandibular gland.

## GLOSSOPHARYNGEAL NERVE

The glossopharyngeal nerve descends in the neck within the carotid sheath. (See p. 722.) It then winds forward around the stylopharyngeus muscle and passes between the superior and middle constrictor muscles (Fig. 11-28). The lingual branch of the nerve (see below) enters the submandibular region.

### Branches

1. The *tympanic branch* is described on page 836.
2. The *carotid branch* to the carotid sinus and carotid body is described on page 724.
3. The *muscular branch* supplies the stylopharyngeus muscle (Fig. 11-28).
4. The *pharyngeal branches* unite on the outer surface of the middle constrictor muscle with the pharyngeal branch of the vagus and the pharyngeal branch of the sympathetic trunk, to form the *pharyngeal plexus.* By means of these branches, the glossopharyngeal nerve gives sensory fibers to the mucous membrane of the pharynx, tonsil, and soft palate.
5. The *lingual branch* supplies general sensory and special taste fibers to the mucous membrane of the posterior third of the tongue and the circumvallate papillae region of the anterior part of the tongue. (See p. 842.) The lingual branch enters the tongue below the styloglossus muscle (Fig. 11-28).

## HYPOGLOSSAL NERVE

The hypoglossal nerve descends in the carotid sheath in front of the vagus nerve. On reaching the lower border of the posterior belly of the digastric (Fig. 11-13), it loops forward around the occipital artery. (See p. 732.) It now crosses the loop of the lingual artery, just above the tip of the greater cornu of the hyoid bone, and runs forward on the surface of the hyoglossus (Fig. 11-14). Having passed deep to the posterior belly of the digastric and the stylohyoid, it runs superior to the posterior margin of the mylohyoid between the mylohyoid and the hyoglossus (Fig. 11-27). It lies below the deep part of the submandibular gland, the submandibular duct, and the lingual nerve (Fig. 11-28). At the anterior margin of the hyoglossus, it curves upward toward the tip of the tongue, supplying branches to the muscles.

### Branches in the Submandibular Region

1. The *nerve to the thyrohyoid* (first cervical nerve fibers; see p. 733).
2. The *nerve to the geniohyoid* (first cervical nerve fibers; see p. 733).
3. *Muscular branches* to all the muscles of the tongue except the palatoglossus, which is supplied by the pharyngeal plexus. It thus supplies the styloglossus, the hyoglossus, the genioglossus, and the intrinsic muscles of the tongue.
4. *Communicating branch.* The hypoglossal nerve communicates with the lingual nerve on the side of the tongue.

## FACIAL ARTERY

The facial artery arises from the external carotid artery just above the tip of the greater cornu of the hyoid bone. (See p. 726.) Having passed deep to the posterior belly of the digastric and the stylohyoid, it arches upward (Fig. 11-29) and grooves the posterior border of the submandibular salivary gland. It then turns downward between the gland and the medial surface of the mandible. On reaching the lower border of the body of the mandible, it hooks around it, to enter the face at the anterior margin of the masseter muscle. Its further course in the face has been described on page 751.

### Branches in the Submandibular Region

1. *Ascending palatine artery.* This ascends alongside the pharynx, to reach the base of the skull (Fig. 11-29).
2. *Tonsillar artery.* This artery perforates the superior constrictor muscle and provides the tonsil

with its main blood supply (Figs. 11-29 and 11-60).

3. *Glandular arteries* supply the submandibular salivary gland.
4. *Submental artery.* This artery runs forward along the lower border of the body of the mandible to supply the region of the chin and lower lip.

## FACIAL VEIN

The facial vein leaves the face by crossing the lower margin of the body of the mandible behind the facial artery. It is joined by the anterior division of the retromandibular vein (Fig. 11-16), crosses the loop of the lingual artery, and drains into the internal jugular vein.

## LINGUAL ARTERY

The lingual artery arises from the external carotid artery just below the tip of the greater cornu of the hyoid bone. (See p. 726.) It runs forward, forming an upward loop, which is crossed by the hypoglossal nerve (Fig. 11-27). It then passes deep to the posterior margin of the hyoglossus, proceeds forward deep to that muscle, and finally turns upward to supply the tip of the tongue (Fig. 11-28).

### Branches

1. The *dorsal lingual branches* are two or three in number and ascend to the dorsum of the tongue.
2. The *sublingual artery* supplies the sublingual salivary gland and neighboring structures.

## LINGUAL VEINS

The dorsal lingual veins join to form the lingual vein. A deep lingual vein joins the sublingual vein to form a vein that follows the hypoglossal nerve. Ultimately, all the veins drain into the internal jugular vein.

## LYMPH NODES

The submandibular lymph nodes are situated on the superficial surface of the submandibular salivary gland (Fig. 11-19). They are fully described on page 742.

## THE SKULL

The skull is composed of a number of separate bones united at immobile joints called *sutures*. The connective tissue between the bones is called a *sutural ligament*. The mandible is an exception to this rule, for it is united to the skull by the mobile temporomandibular joint. (See p. 768.)

The bones of the skull may be divided into those of the *cranium* and those of the face. The *vault* is the upper part of the cranium, and the *base of the skull* is the lowest part of the cranium (Figs. 11-34 and 11-35).

The skull bones are made up of *external* and *internal* tables of compact bone, separated by a layer of spongy bone, called the *diploë* (Fig. 11-20). The internal table is thinner and more brittle than the external table. The bones are covered on the outer and inner surfaces with periosteum; the outer layer is referred to as the *pericranium* and the inner covering as the *endocranium*.

The *cranium* consists of the following bones, two of which are paired (Figs. 11-34 through 11-36):

| | |
|---|---|
| Frontal bone | 1 |
| Parietal bones | 2 |
| Occipital bone | 1 |
| Temporal bones | 2 |
| Sphenoid bone | 1 |
| Ethmoid bone | 1 |

The *facial bones* consist of the following, two of which are single:

| | |
|---|---|
| Zygomatic bones | 2 |
| Maxillae | 2 |
| Nasal bones | 2 |
| Lacrimal bones | 2 |
| Vomer | 1 |
| Palatine bones | 2 |
| Inferior conchae | 2 |
| Mandible | 1 |

It is unnecessary for a student of medicine to know the detailed structure of each individual skull bone. However, he should be familiar with the skull as a whole and should have a dried skull available for reference as he is reading the following description.

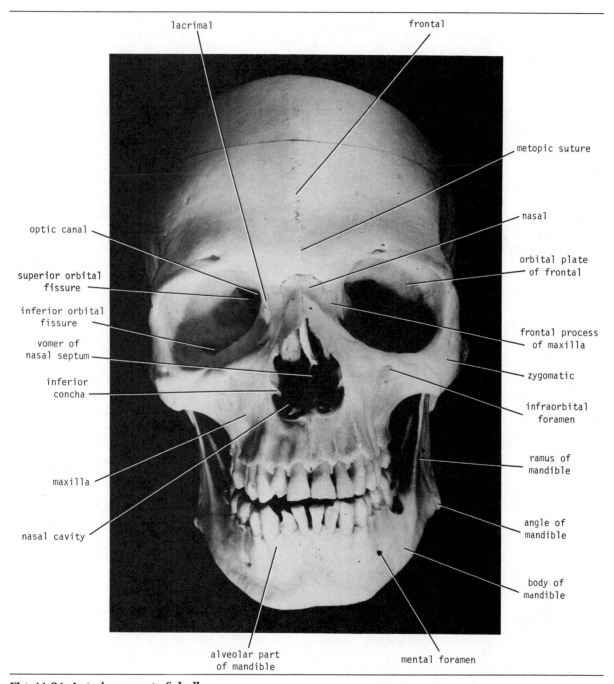

lacrimal

frontal

metopic suture

optic canal

nasal

superior orbital
fissure

orbital plate
of frontal

inferior orbital
fissure

frontal process
of maxilla

vomer of
nasal septum

zygomatic

inferior
concha

infraorbital
foramen

maxilla

ramus of
mandible

nasal cavity

angle of
mandible

body of
mandible

alveolar part
of mandible

mental foramen

**Fig. 11-34. Anterior aspect of skull.**

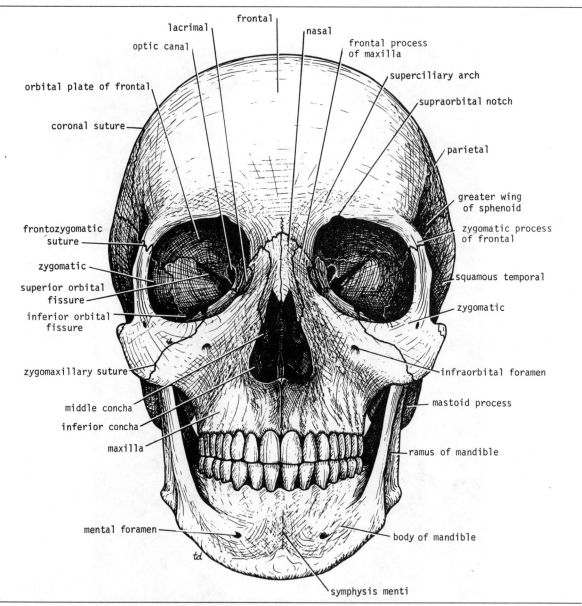

**Fig. 11-35. Bones of the anterior aspect of skull.**

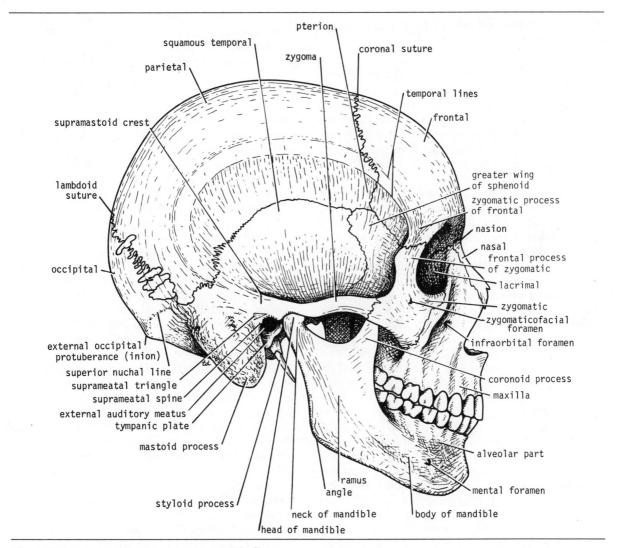

**Fig. 11-36. Bones of the lateral aspect of skull.**

## Anterior View of Skull (Figs. 11-34 and 11-35)

The *frontal bone* curves downward to make the upper margins of the orbits. The *superciliary arches* can be seen on either side, and the *supraorbital notch*, or *foramen*, can be recognized. Medially, the frontal bone articulates with the frontal processes of the maxillae and with the nasal bones. Laterally, the frontal bone articulates with the zygomatic bone.

The *orbital margins* are bounded by the frontal bone superiorly, the zygomatic bone laterally, the maxilla inferiorly, and the processes of the maxilla and frontal bone medially.

Within the *frontal bone*, just above the orbital margins, are two hollow spaces lined with mucous membrane, called the *frontal air sinuses*.

The two nasal bones form the bridge of the nose. Their lower borders, with the maxillae, make the *anterior nasal aperture*. The nasal cavity is divided into two by the bony nasal septum, which is largely formed by the vomer. The *superior* and *middle conchae* jut into the nasal cavity from the *ethmoid* on each side; the inferior conchae are separate bones.

The two *maxillae* form the upper jaw, the anterior part of the hard palate, part of the lateral walls of the nasal cavities, and part of the floors of the orbital cavities. The two bones meet in the midline at the *intermaxillary suture* and form the lower margin of the nasal aperture. Below the orbit the maxilla is perforated by the *infraorbital foramen*. The *alveolar process* projects downward and, together with the fellow of the opposite side forms the *alveolar arch*, which carries the upper teeth. Within each maxilla is a large pyramid-shaped cavity lined with mucous membrane, called the *maxillary sinus*.

The *zygomatic bone* forms the prominence of the cheek and part of the lateral wall and floor of the orbital cavity. Medially, it articulates with the maxilla, and laterally, it articulates with the zygomatic process of the temporal bone to form the *zygomatic arch*. The zygomatic bone is perforated by two foramina for the zygomaticofacial and zygomaticotemporal nerves.

The *mandible*, or lower jaw, consists of a horizontal *body* and two vertical *rami* (for details, see p. 772); the body joins the ramus at the *angle of the mandible*. The *mental foramen* opens onto the anterior surface of the body of the mandible, below the second premolar tooth. The upper border of the mandible, the alveolar part, carries the lower teeth.

## Lateral View of Skull (Fig. 11-36)

The *frontal bone* forms the anterior part of the side of the skull and articulates with the parietal bone at the *coronal suture*.

The *parietal bones* form the sides and roof of the cranium and articulate with each other in the midline at the *sagittal suture*. They articulate with the occipital bone behind, at the *lambdoid suture*.

The skull is completed from the side by the squamous part of the *occipital bone*; parts of the *temporal bone*, namely, the *squamous, tympanic, mastoid process, styloid process,* and *zygomatic process*; and the *greater wing of the sphenoid*. The ramus and body of the mandible lie inferiorly.

Note that the thinnest part of the lateral wall of the skull is where the antero-inferior corner of the parietal bone articulates with the greater wing of the sphenoid; this point is referred to as the *pterion*.

Clinically, the pterion is a very important area, since it overlies the anterior division of the *middle meningeal artery* and *vein*. On the outer surface of the skull, the pterion lies about 1 inch (2.5 cm) behind the frontal process of the zygomatic bone and about 1½ inches (4 cm) above the zygomatic arch.

Identify the *superior* and *inferior temporal lines*, which begin as a single line from the posterior margin of the zygomatic process of the frontal bone and diverge as they arch backward. The *temporal fossa* lies below the inferior temporal line. The lower limit of the temporal fossa is the *infratemporal crest* of the greater wing of the sphenoid, which is level with the upper border of the zygomatic arch.

The *infratemporal fossa* lies below the infratemporal crest on the greater wing of the sphenoid. The *pterygomaxillary fissure* is a vertical fissure that lies within the fossa between the pterygoid process of the sphenoid bone and back of the maxilla. It leads medially into the *pterygopalatine fossa*.

The *inferior orbital fissure* is a horizontal fissure between the greater wing of the sphenoid bone and the maxilla. It leads forward into the orbit.

The *pterygopalatine fossa* is a small space behind and below the orbital cavity. It communicates laterally with the infratemporal fossa through the pterygomaxillary fissure, medially with the nasal cavity through the *sphenopalatine foramen*, superiorly with the skull through the *foramen rotundum*, and anteriorly with the orbit through the *inferior orbital fissure*.

## Posterior View of Skull (Fig. 11-37)

The posterior parts of the two parietal bones with the intervening *sagittal suture* are seen above. Below, the parietal bones articulate with the squamous part of the occipital bone at the *lambdoid suture*. On each side the occipital bone articulates with the temporal bone. In the midline of the occipital bone is a roughened elevation called the *external occipital protuberance*, which gives attachment to muscles and the ligamentum nuchae. On either side of the protuberance the *superior nuchal lines* extend laterally toward the temporal bone.

## Superior View of Skull (Fig. 11-37)

Anteriorly, the frontal bone articulates with the two parietal bones at the *coronal suture*. Occasionally, the two halves of the frontal bone fail to fuse, leaving a midline *metopic suture*. Behind, the two parietal bones articulate in the midline at the *sagittal suture*. The center of the parietal bone forms a small eminence, the *parietal eminence*.

## Inferior View of Skull (Fig. 11-38)

If the mandible is discarded, the anterior part of this aspect of the skull is seen to be formed by the *hard palate*.

The *palatal processes of the maxillae* and the *horizontal plates of the palatine bones* can be identified. In the midline anteriorly is the *incisive fossa* and *foramen*. Posterolaterally are the *greater* and *lesser palatine foramina*.

Above the posterior edge of the hard palate are the *choanae* (posterior nasal apertures). These are separated from each other by the posterior margin of the vomer and are bounded laterally by the medial pterygoid plates of the sphenoid bone. The inferior end of the *medial pterygoid plate* is prolonged as a curved spike of bone, the *pterygoid hamulus*. The superior end widens to form the *scaphoid fossa*.

Posterolateral to the *lateral pterygoid plate*, the greater wing of the sphenoid is pierced by the large *foramen ovale* and the small *foramen spinosum*. Posterolateral to the foramen spinosum is the *spine of the sphenoid*. Above the medial border of the scaphoid fossa, the sphenoid bone is pierced by the *pterygoid canal*.

Behind the spine of the sphenoid, in the interval between the greater wing of the sphenoid and the petrous part of the temporal bone, there is a groove for the cartilaginous part of the *auditory tube*. The opening of the bony part of the tube can be identified.

The *mandibular fossa* of the temporal bone and the *articular tubercle* form the upper articular surfaces for the temporomandibular joint. Separating the mandibular fossa from the tympanic plate posteriorly is the *squamotympanic fissure*, through the medial end of which (petrotympanic fissure) the chorda tympani exits from the tympanic cavity.

The *styloid process* of the temporal bone projects downward and forward from its inferior aspect. The opening of the *carotid canal* can be seen on the inferior surface of the petrous part of the temporal bone.

The medial end of the petrous part of the temporal bone is irregular and, together with the basilar part of the occipital bone and the greater wing of the sphenoid, forms the *foramen lacerum*. During life, the foramen lacerum is closed with fibrous tissue, and only a few very small vessels pass through this foramen from the cavity of the skull to the exterior.

The *tympanic plate*, which forms part of the temporal bone, is C-shaped on section and forms the bony part of the *external auditory meatus*. While examining this region, identify the *suprameatal crest* on the lateral surface of the squamous part of the temporal bone, the *suprameatal triangle*, and the *suprameatal spine*.

In the interval between the styloid and mastoid processes, the *stylomastoid foramen* can be seen. Medial to the styloid process, the petrous part of the temporal bone has a deep notch, which, together with a shallower notch on the occipital bone, forms the *jugular foramen*.

Behind the posterior apertures of the nose and in front of the foramen magnum are the sphenoid bone and the basilar part of the occipital bone. The *pharyngeal tubercle* is a small prominence on the

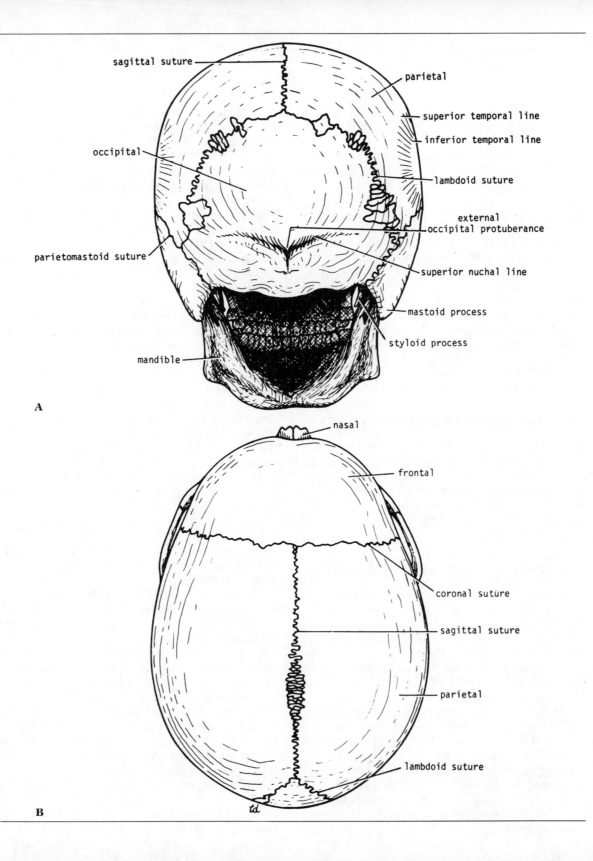

**A**

sagittal suture

parietal

superior temporal line

inferior temporal line

occipital

lambdoid suture

external
occipital protuberance

superior nuchal line

parietomastoid suture

mastoid process

styloid process

mandible

**B**

nasal

frontal

coronal suture

sagittal suture

parietal

lambdoid suture

*td*

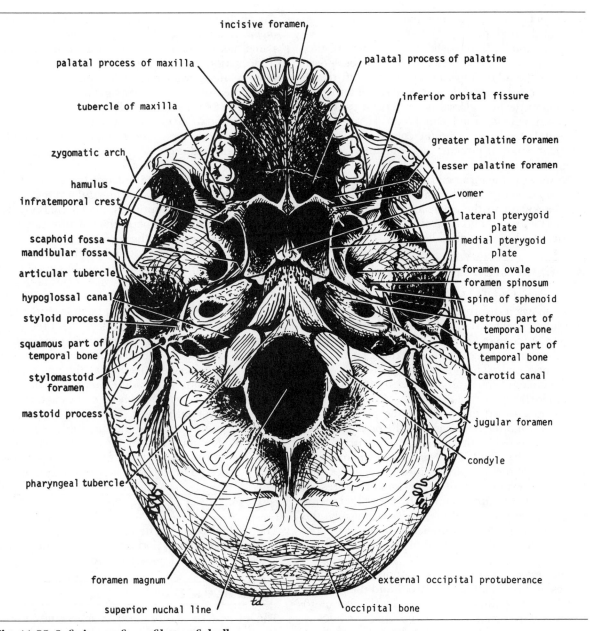

incisive foramen

palatal process of maxilla

palatal process of palatine

tubercle of maxilla

inferior orbital fissure

greater palatine foramen

zygomatic arch

lesser palatine foramen

hamulus

vomer

infratemporal crest

lateral pterygoid plate

scaphoid fossa

medial pterygoid plate

mandibular fossa

foramen ovale

articular tubercle

foramen spinosum

hypoglossal canal

spine of sphenoid

styloid process

petrous part of temporal bone

squamous part of temporal bone

tympanic part of temporal bone

stylomastoid foramen

carotid canal

mastoid process

jugular foramen

condyle

pharyngeal tubercle

external occipital protuberance

foramen magnum

occipital bone

superior nuchal line

**Fig. 11-38. Inferior surface of base of skull.**

**Fig. 11-37. Bones of the skull viewed from (A) posterior aspect and (B) superior aspect.**

undersurface of the basilar part of the occipital bone in the midline.

The *occipital condyles* should be identified; they articulate with the superior aspect of the lateral mass of the atlas. Superior to the summit of the occipital condyle is the *hypoglossal canal* for transmission of the hypoglossal nerve. This must not be confused with a small canal that is sometimes present behind the occipital condyle, called the *condylar canal*. If present, this transmits an emissary vein.

Posterior to the foramen magnum in the midline is the *external occipital crest*, which runs upward and backward to the external occipital protuberance. The *inferior* and *superior nuchal lines* should be identified as they curve laterally on each side.

## Neonatal Skull (Fig. 11-40)

The neonatal skull, when compared with the adult skull, shows a disproportionately large size of the cranium relative to the face. In childhood, the growth of the mandible, the maxillary sinuses, and the alveolar processes of the maxillae results in a great increase in length of the face.

The bones of the skull are smooth and unilaminar, there being no diploë present. Most of the skull bones are ossified at birth, but the process is incomplete, and the bones are mobile on each other, being connected by fibrous tissue or cartilage. The bones of the vault are ossified in membrane; the bones of the base are ossified in cartilage. The bones of the vault are not closely knit at sutures, as in the adult, but are separated by unossified membranous intervals called *fontanelles*. Clinically, the anterior and posterior fontanelles are most important and are easily examined in the midline of the vault.

The *anterior fontanelle* is bounded by the two halves of the frontal bone in front and the two parietal bones behind (Fig. 11-40). By the age of 18 months it is very small and no longer clinically palpable. The *posterior fontanelle* is bounded by the two parietal bones in front and the occipital bone behind. By the end of the first year it can no longer be palpated.

The *tympanic part of the temporal bone* is merely a C-shaped ring at birth, compared with a C-shaped curved plate in the adult. This means that the external auditory meatus is almost entirely cartilaginous in the newborn, and the *tympanic membrane* is nearer the surface. Although the tympanic membrane is nearly as large as in the adult, it faces more inferiorly. During childhood the tympanic plate grows laterally, forming the bony part of the meatus, and the tympanic membrane comes to face more directly laterally.

The *mastoid process* is not present at birth (Fig. 11-40) and develops later in response to the pull of the sternocleidomastoid muscle when the child moves his head.

At birth, the mastoid antrum lies about 3 mm deep to the floor of the *suprameatal triangle*. As the growth of the skull continues, the lateral bony wall thickens, so that at puberty the antrum may lie as much as 15 mm from the surface.

The mandible has right and left halves at birth, united in the midline with fibrous tissue. The two halves fuse at the *symphysis menti* by the end of the first year.

The *angle of the mandible* at birth is obtuse (Fig. 11-40), the head being placed level with the upper margin of the body, and the coronoid process lying at a superior level to the head. It is only after eruption of the permanent teeth that the angle of the mandible assumes the adult shape and the head and neck grow, so that the head comes to lie higher than the coronoid process.

In old age, the size of the mandible is reduced when the teeth are lost. As the alveolar part of the bone becomes smaller, the ramus becomes oblique in position so that the head is bent posteriorly.

## The Cranial Cavity

The cranial cavity contains the brain and its surrounding meninges, portions of the cranial nerves, arteries, veins, and venous sinuses.

## Vault of the Skull

The internal surface of the vault shows the coronal, sagittal, and lambdoid sutures. It is interesting to note that they are less tortuous on the inside than on the outside of the skull. In the midline a shallow sagittal groove is present that lodges the *superior sagittal sinus*. On each side of the groove are a number of small pits, called *granular pits*, which lodge the *lateral lacunae* and *arachnoid granulations*. (See p. 800.) A small foramen is commonly

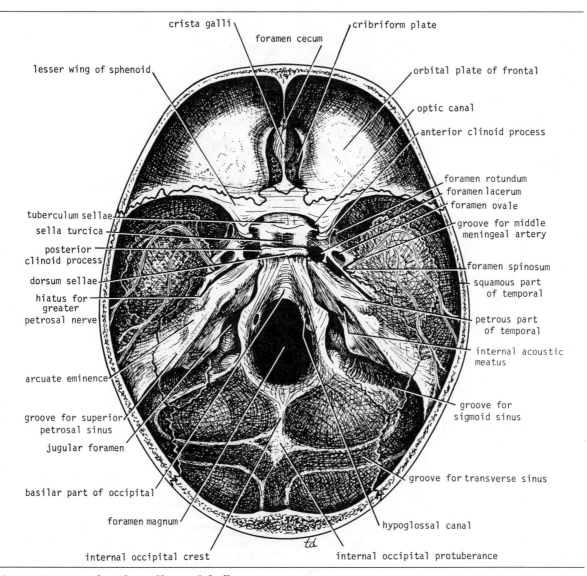

crista galli

cribriform plate

foramen cecum

lesser wing of sphenoid

orbital plate of frontal

optic canal

anterior clinoid process

foramen rotundum

foramen lacerum

foramen ovale

tuberculum sellae

groove for middle
meningeal artery

sella turcica

posterior
clinoid process

foramen spinosum

squamous part
of temporal

dorsum sellae

hiatus for
greater
petrosal nerve

petrous part
of temporal

internal acoustic
meatus

arcuate eminence

groove for superior
petrosal sinus

groove for
sigmoid sinus

jugular foramen

basilar part of occipital

groove for transverse sinus

foramen magnum

hypoglossal canal

internal occipital crest

internal occipital protuberance

**Fig. 11-39. Internal surface of base of skull.**

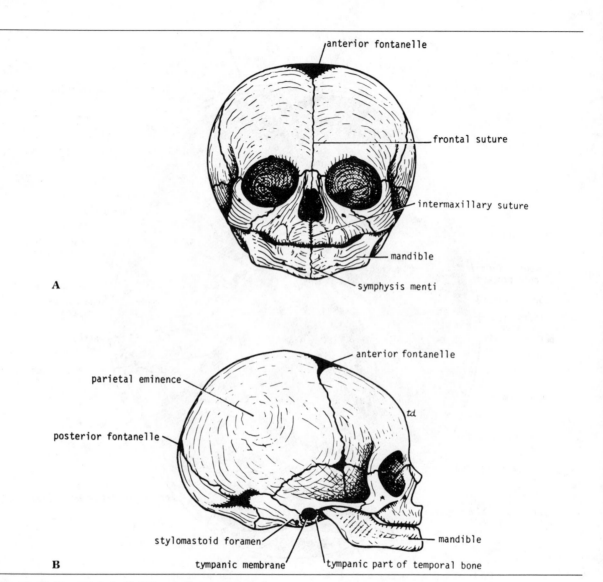

**Fig. 11-40. Neonatal skull: (A) anterior aspect and (B) lateral aspect.**

present, which perforates each parietal bone at the side of the sagittal groove; it transmits an emissary vein from the superior sagittal sinus. A number of narrow grooves are present for the anterior and posterior divisions of the *middle meningeal vessels* as they pass up the side of the skull to the vault.

## Base of the Skull (Fig. 11-39)

The interior of the base of the skull, for ease of description, is conveniently divided into three cranial fossae: anterior, middle, and posterior.

### ANTERIOR CRANIAL FOSSA

The anterior cranial fossa lodges the frontal lobes of the cerebral hemispheres. It is bounded anteriorly by the inner surface of the frontal bone and in the midline there is a crest for the attachment of the *falx cerebri.* Its posterior boundary is the sharp lesser wing of the sphenoid, which articulates laterally with the frontal bone and meets the anterior inferior angle of the parietal bone, or pterion. The medial end of the lesser wing of the sphenoid forms the *anterior clinoid process* on each side, which gives attachment to the *tentorium cerebelli.* The median part of the anterior cranial fossa is limited posteriorly by the groove for the optic chiasma.

The floor of the fossa is formed by the ridged orbital plates of the frontal bone laterally and by the *cribriform plate* of the ethmoid medially (Fig. 11-39). The *crista galli* is a sharp upward projection of the ethmoid bone in the midline, for the attachment of the falx cerebri. Between the crista galli and the crest of the frontal bone is a small aperture, the *foramen cecum,* for the transmission of a small vein from the nasal mucosa to the superior sagittal sinus. Alongside the crista galli is a narrow slit in the cribriform plate for the passage of the *anterior ethmoidal nerve* into the nasal cavity. The upper surface of the cribriform plate supports the *olfactory bulbs,* and the small perforations in the cribriform plate are for the *olfactory nerves.*

### MIDDLE CRANIAL FOSSA

The middle cranial fossa consists of a small median part and expanded lateral parts (Fig. 11-39).

The median raised part is formed by the body of the sphenoid, and the expanded lateral parts form concavities on either side, which lodge the *temporal lobes* of the *cerebral hemispheres.*

It is bounded anteriorly by the lesser wings of the sphenoid and posteriorly by the superior borders of the petrous parts of the temporal bones. Laterally lie the squamous parts of the temporal bones, the greater wings of the sphenoid, and the parietal bones.

The floor of each lateral part of the middle cranial fossa is formed by the greater wing of the sphenoid and the squamous and petrous parts of the temporal bone.

Anteriorly, the *optic canal* transmits the optic nerve and the ophthalmic artery, a branch of the internal carotid artery, to the orbit. The *superior orbital fissure,* which is a slitlike opening between the lesser and greater wings of the sphenoid, transmits the lacrimal, the frontal, the trochlear, the oculomotor, the nasociliary, and the abducent nerves, together with the superior ophthalmic vein. The sphenoparietal venous sinus runs medially along the posterior border of the lesser wing of the sphenoid and drains into the cavernous sinus.

The *foramen rotundum,* which is situated behind the medial end of the superior orbital fissure, perforates the greater wing of the sphenoid and transmits the maxillary nerve from the trigeminal ganglion to the pterygopalatine fossa.

The *foramen ovale* lies posterolateral to the foramen rotundum (Fig. 11-39). It perforates the greater wing of the sphenoid and transmits the large sensory root and small motor root of the trigeminal nerve to the infratemporal fossa; the lesser petrosal nerve also passes through it.

The small *foramen spinosum* lies posterolateral to the foramen ovale and also perforates the greater wing of the sphenoid. The foramen transmits the middle meningeal artery from the infratemporal fossa (see p. 768) into the cranial cavity. The artery then runs forward and laterally in a groove on the upper surface of the squamous part of the temporal bone and the greater wing of the sphenoid (Fig. 11-49). After a short distance the artery divides into anterior and posterior branches. The anterior branch passes forward and upward to the anterior inferior angle of the parietal bone (Fig. 11-1). Here, the bone is deeply grooved or tunneled by the artery for a short distance before it runs backward and upward on the parietal bone. It is at this site that the artery may be damaged following a blow to the side of the head. The posterior branch passes backward and upward across the squamous part of the temporal bone to reach the parietal bone.

The large and irregularly shaped *foramen lacerum* lies between the apex of the petrous part of the temporal bone and the sphenoid bone (Fig. 11-39). The inferior opening of the foramen lacerum in life is filled by cartilage and fibrous tissue, and only very small blood vessels pass through this tissue from the cranial cavity to the neck.

The *carotid canal* opens into the side of the foramen lacerum above the closed inferior opening. The internal carotid artery enters the foramen through the carotid canal and immediately turns upward to reach the side of the body of the sphenoid bone. Here, the artery turns forward in the cavernous sinus to reach the region of the anterior clinoid process. At this point the internal carotid artery turns vertically upward, medial (Fig. 11-49) to the anterior clinoid process, and emerges from the cavernous sinus. (See p. 807.)

Lateral to the foramen lacerum is an impression on the apex of the petrous part of the temporal bone for the *trigeminal ganglion*. On the anterior surface of the petrous bone are two grooves for nerves; the largest medial groove is for the *greater petrosal nerve*, a branch of the facial nerve; the smaller lateral groove is for the *lesser petrosal nerve*, a branch of the tympanic plexus. The greater petrosal nerve enters the foramen lacerum deep to the trigeminal ganglion and joins the *deep petrosal nerve* (sympathetic fibers from around the internal carotid artery), to form the *nerve of the pterygoid canal*. The lesser petrosal nerve passes forward to the foramen ovale.

The abducent nerve bends sharply forward across the apex of the petrous bone, medial to the trigeminal ganglion. It is here that it leaves the posterior cranial fossa and enters the cavernous sinus.

The *arcuate eminence* is a rounded eminence found on the anterior surface of the petrous bone and is caused by the underlying *superior semicircular canal*.

The *tegmen tympani* is a thin plate of bone, which is a forward extension of the petrous part of the temporal bone and adjoins the squamous part of the bone (Fig. 11-39). From behind forward, it forms the roof of the mastoid antrum, the tympanic cavity, and the auditory tube. It is important to realize that this thin plate of bone is the only major barrier that separates infection in the tympanic cavity from the temporal lobe of the cerebral hemisphere (Fig. 11-56).

The median part of the middle cranial fossa is formed by the body of the sphenoid bone (Fig. 11-39). In front is the *sulcus chiasmatis*, which is related to the optic chiasma and leads laterally to the *optic canal* on each side. Posterior to the sulcus is an elevation, the *tuberculum sellae*. Behind the elevation is a deep depression, the *sella turcica*, which lodges the *hypophysis cerebri*. The sella turcica is bounded posteriorly by a square plate of bone called the *dorsum sellae*. The superior angles of the dorsum sellae have two tubercles, called the *posterior clinoid processes*, which give attachment to the fixed margin of the tentorium cerebelli.

The cavernous sinus is directly related to the side of the body of the sphenoid (Figs. 11-41 and 11-42). It carries in its lateral wall the third and fourth cranial nerves and the ophthalmic and maxillary divisions of the fifth cranial nerve (Fig. 11-43). The internal carotid artery and the sixth cranial nerve pass forward through the sinus.

## POSTERIOR CRANIAL FOSSA

The posterior cranial fossa is very deep and lodges the parts of the hindbrain, namely, the *cerebellum*, *pons*, and *medulla oblongata*. Anteriorly, the fossa is bounded by the superior border of the petrous part of the temporal bone, and posteriorly it is bounded by the internal surface of the squamous part of the occipital bone (Fig. 11-39). The floor of the posterior fossa is formed by the basilar, condylar, and squamous parts of the occipital bone and the mastoid part of the temporal bone.

The roof of the fossa is formed by a fold of dura, the *tentorium cerebelli*, which intervenes between the cerebellum below and the occipital lobes of the cerebral hemispheres above (Fig. 11-42).

The *foramen magnum* occupies the central area of the floor and transmits the medulla oblongata and its surrounding meninges, the ascending spinal parts of the accessory nerves, and the two vertebral arteries.

The *hypoglossal canal* is situated above the an-

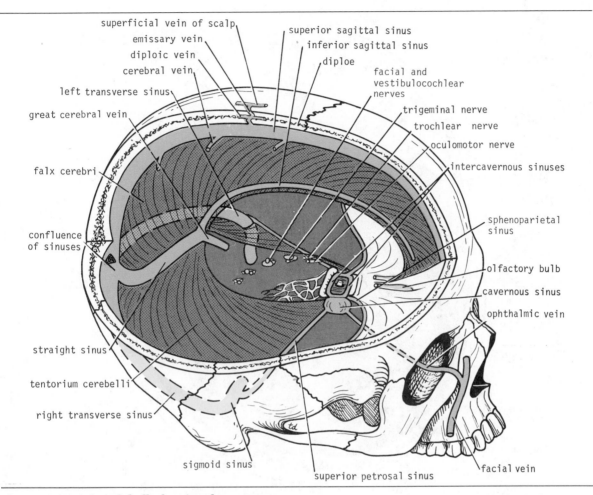

superficial vein of scalp
emissary vein
diploic vein
cerebral vein
left transverse sinus
great cerebral vein
falx cerebri
confluence of sinuses
straight sinus
tentorium cerebelli
right transverse sinus
sigmoid sinus
superior sagittal sinus
inferior sagittal sinus
diploe
facial and vestibulocochlear nerves
trigeminal nerve
trochlear nerve
oculomotor nerve
intercavernous sinuses
sphenoparietal sinus
olfactory bulb
cavernous sinus
ophthalmic vein
facial vein
superior petrosal sinus

**Fig. 11-41. Interior of skull, showing dura mater and its contained venous sinuses. Note connections of veins of scalp with veins of face and venous sinuses.**

terolateral boundary of the foramen magnum (Fig. 11-39) and transmits the *hypoglossal nerve.*

The *jugular foramen* lies between the lower border of the petrous part of the temporal bone and the condylar part of the occipital bone. It transmits the following structures from before backward: the *inferior petrosal sinus,* the *ninth, tenth,* and *eleventh cranial nerves,* and the large *sigmoid sinus.* The inferior petrosal sinus descends in the groove on the lower border of the petrous part of the temporal bone, to reach the foramen. The sigmoid sinus turns down through the foramen to become the *internal jugular vein.*

The *internal acoustic meatus* pierces the posterior surface of the petrous part of the temporal bone. It transmits the vestibulocochlear nerve and the motor and sensory roots of the facial nerve.

The *condylar canal* is sometimes present just lateral to the foramen magnum, and an emissary vein passes through it.

The *internal occipital crest* runs upward in the midline posteriorly from the foramen magnum to the *internal occipital protuberance;* to it is attached the small *falx cerebelli* over the *occipital sinus.*

On each side of the internal occipital protuberance there is a wide groove for the *transverse sinus*

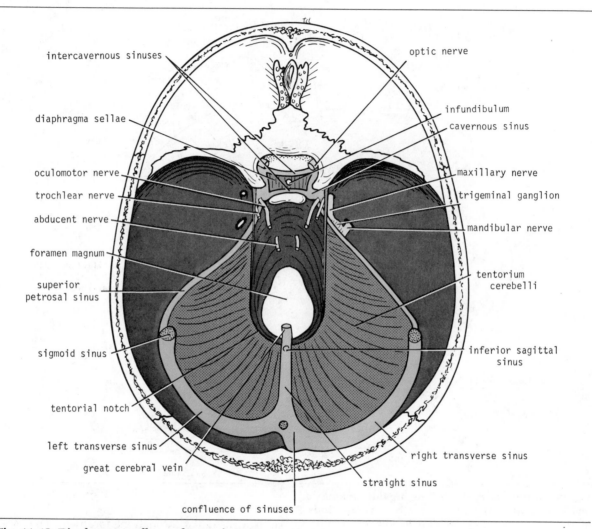

intercavernous sinuses

optic nerve

diaphragma sellae

infundibulum

cavernous sinus

oculomotor nerve

maxillary nerve

trochlear nerve

trigeminal ganglion

abducent nerve

mandibular nerve

foramen magnum

superior petrosal sinus

tentorium cerebelli

sigmoid sinus

inferior sagittal sinus

tentorial notch

left transverse sinus

right transverse sinus

great cerebral vein

straight sinus

confluence of sinuses

**Fig. 11-42. Diaphragma sellae and tentorium cerebelli. Note position of venous sinuses.**

**Fig. 11-43. (A) Forebrain has been removed, leaving midbrain, hypophysis cerebri, and internal carotid and basilar arteries in position. (B) Sagittal section through sella turcica, showing hypophysis cerebri. (C) Coronal section through body of sphenoid, showing hypophysis cerebri and cavernous sinuses. Note position of cranial nerves.**

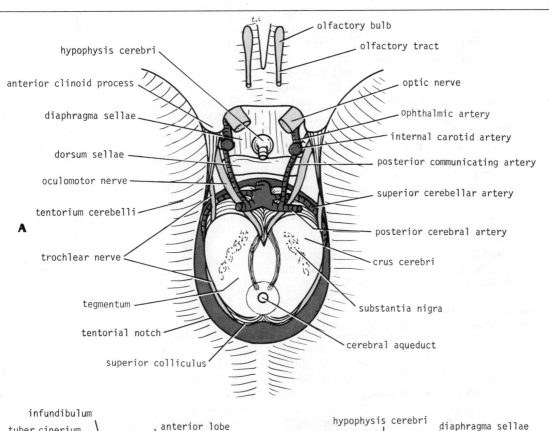

**A**

olfactory bulb

olfactory tract

hypophysis cerebri

anterior clinoid process

diaphragma sellae

dorsum sellae

oculomotor nerve

tentorium cerebelli

trochlear nerve

tegmentum

tentorial notch

superior colliculus

optic nerve

ophthalmic artery

internal carotid artery

posterior communicating artery

superior cerebellar artery

posterior cerebral artery

crus cerebri

substantia nigra

cerebral aqueduct

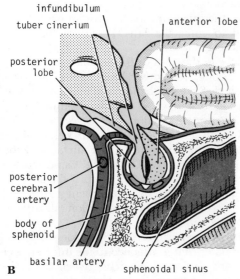

**B**

infundibulum

tuber cinerium

anterior lobe

posterior lobe

posterior cerebral artery

body of sphenoid

basilar artery

sphenoidal sinus

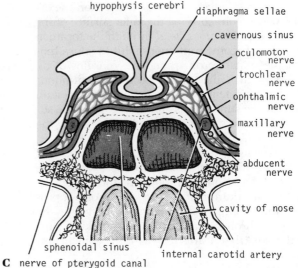

**C**

hypophysis cerebri

diaphragma sellae

cavernous sinus

oculomotor nerve

trochlear nerve

ophthalmic nerve

maxillary nerve

abducent nerve

cavity of nose

internal carotid artery

sphenoidal sinus

nerve of pterygoid canal

(Fig. 11-39). This groove sweeps around on either side, on the internal surface of the occipital bone, to reach the posterior inferior angle or corner of the parietal bone. The groove now passes onto the mastoid part of the temporal bone, and it is here that the transverse sinus becomes the *sigmoid sinus.* The *superior petrosal sinus* runs backward along the upper border of the petrous bone in a narrow groove and drains into the sigmoid sinus. As the sigmoid sinus descends to the jugular foramen, it deeply grooves the back of the petrous bone and the mastoid part of the temporal bone. Here, it lies directly posterior to the mastoid antrum.

## The Meninges

The brain and spinal cord are surrounded by three membranes, or meninges: the dura mater, the arachnoid mater, and the pia mater.

### DURA MATER OF THE BRAIN

The dura mater is conventionally described as two layers, the endosteal layer and the meningeal layer (Fig. 11-20). These are closely united except along certain lines, where they separate to form venous sinuses.

The *endosteal layer* is nothing more than the ordinary periosteum covering the inner surface of the skullbones. *It does not extend* through the foramen magnum to become continuous with the dura mater of the spinal cord. Around the margins of all the foramina in the skull it becomes continuous with the periosteum on the outside of the skull bones. At the sutures it is continuous with the sutural ligaments. It is most strongly adherent to the bones over the base of the skull.

The *meningeal layer* is the dura mater proper. It is a dense, strong fibrous membrane covering the brain and is continuous through the foramen magnum with the dura mater of the spinal cord. It provides tubular sheaths for the cranial nerves as the latter pass through the foramina in the skull. Outside the skull the sheaths fuse with the epineurium of the nerves.

The meningeal layer sends inward four septa, which divide the cranial cavity into freely communicating spaces lodging the subdivisions of the brain. The function of these septa is to restrict the rotatory displacement of the brain.

The *falx cerebri* is a sickle-shaped fold of dura mater that lies in the midline between the two cerebral hemispheres (Figs. 11-41 and 11-44). Its narrow end in front is attached to the internal frontal crest and the crista galli. Its broad posterior part blends in the midline with the upper surface of the tentorium cerebelli. The superior sagittal sinus runs in its upper fixed margin; the inferior sagittal sinus runs in its lower concave free margin; and the straight sinus runs along its attachment to the tentorium cerebelli.

The *tentorium cerebelli* is a crescent-shaped fold of dura mater that roofs over the posterior cranial fossa (Figs. 11-41 and 11-42). It covers the upper surface of the cerebellum and supports the occipital lobes of the cerebral hemispheres. In front there is a gap, the *tentorial notch,* for the passage of the midbrain (Fig. 11-43), thus producing an inner free border and an outer attached or fixed border. The fixed border is attached to the posterior clinoid processes, the superior borders of the petrous bones, and the margins of the grooves for the transverse sinuses on the occipital bone. The free border runs forward at its two ends, crosses the attached border, and is affixed to the anterior clinoid process on each side. At the point where the two borders cross, the third and fourth cranial nerves pass forward to enter the lateral wall of the cavernous sinus (Fig. 11-43).

Close to the apex of the petrous part of the temporal bone, the lower layer of the tentorium is pouched forward beneath the superior petrosal sinus, to form a recess for the trigeminal nerve and the trigeminal ganglion.

The falx cerebri and the falx cerebelli are attached to the upper and lower surfaces of the tentorium, respectively. The straight sinus runs along its attachment to the falx cerebri, the superior petrosal sinus along its attachment to the petrous bone, and the transverse sinus along its attachment to the occipital bone (Fig. 11-42).

The *falx cerebelli* is a small, sickle-shaped fold of dura mater that is attached to the internal occipital crest and projects forward between the two cerebellar hemispheres. Its posterior fixed margin contains the occipital sinus.

The *diaphragma sellae* is a small circular fold of

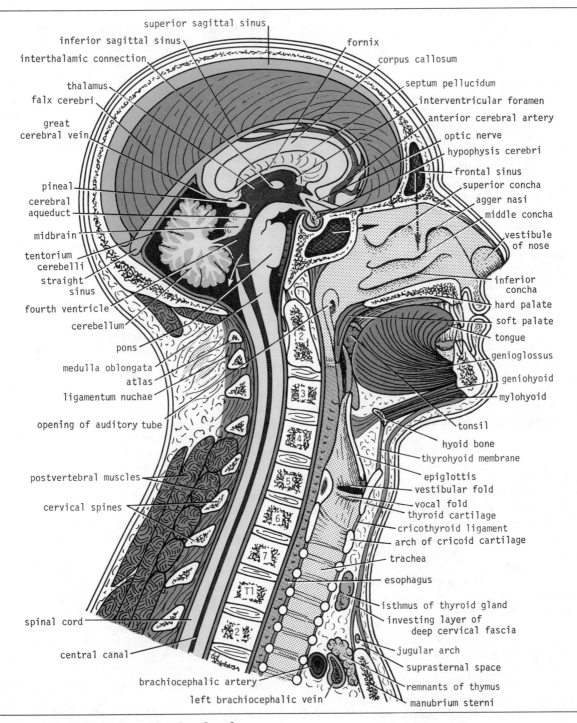

**Fig. 11-44. Sagittal section of head and neck.**

dura mater that forms the roof for the sella turcica (Fig. 11-42). A small opening in its center allows passage of the stalk of the hypophysis cerebri (Fig. 11-43).

## Dural Nerve Supply

Branches of the trigeminal, vagus, and the first three cervical nerves and branches from the sympathetic system pass to the dura.

There are numerous sensory endings in the dura. The dura is sensitive to stretching, which produces the sensation of headache. Stimulation of the sensory endings of the trigeminal nerve above the level of the tentorium cerebelli produces referred pain to an area of skin on the same side of the head. Stimulation of the dural endings below the level of the tentorium produces referred pain to the back of the neck and back of the scalp along the distribution of the greater occipital nerve.

## Dural Arterial Supply

Numerous arteries supply the dura mater from the internal carotid, maxillary, ascending pharyngeal, occipital, and vertebral arteries. From the clinical standpoint, the most important is the middle meningeal artery, which is commonly damaged in head injuries.

The *middle meningeal artery* arises from the maxillary artery in the infratemporal fossa. (See p. 768.) It enters the cranial cavity and runs forward and laterally in a groove on the upper surface of the squamous part of the temporal bone (Fig. 11-49). In order to enter the cranial cavity, it passes through the foramen spinosum to *lie between the meningeal and endosteal* layers of dura. Its further course in the middle cranial fossa is described on page 793. The anterior (frontal) branch deeply grooves or tunnels the antero-inferior angle of the parietal bone, and its course corresponds roughly to the line of the underlying precentral gyrus of the brain. The posterior (parietal) branch curves backward and supplies the posterior part of the dura mater.

The *meningeal veins* lie in the endosteal layer of dura. The middle meningeal vein follows the branches of the middle meningeal artery and drains into the pterygoid venous plexus or the sphenoparietal sinus. The veins lie lateral to the arteries.

## ARACHNOID MATER OF THE BRAIN

The arachnoid mater is a delicate impermeable membrane covering the brain and lying between the pia mater internally and the dura mater externally (Fig. 11-20). It is separated from the dura by a potential space, the *subdural space*, and from the pia by the *subarachnoid space*, which is filled with *cerebrospinal fluid*.

The arachnoid bridges over the sulci on the surface of the brain, and in certain situations the arachnoid and pia are widely separated to form the subarachnoid cisternae. The *cisterna cerebellomedullaris* lies between the inferior surface of the cerebellum and the roof of the fourth ventricle. The *cisterna pontis* lies on the anterior surface of the pons and the medulla oblongata. The *cisterna interpeduncularis* lies between the two cerebral peduncles. All the cisternae are in free communication with each other and with the remainder of the subarachnoid space.

In certain areas the arachnoid projects into the venous sinuses to form *arachnoid villi*. The arachnoid villi are most numerous along the superior sagittal sinus. Aggregations of arachnoid villi are referred to as *arachnoid granulations* (Fig. 11-20). Arachnoid villi serve as sites where the cerebrospinal fluid diffuses into the bloodstream.

The arachnoid is connected to the pia mater across the fluid-filled subarachnoid space by delicate strands of fibrous tissue.

It is important to remember that structures passing to and from the brain to the skull or its foramina must pass through the subarachnoid space. All the cerebral arteries and veins lie in the space, as do the cranial nerves (Fig. 11-20). The arachnoid fuses with the epineurium of the nerves at their point of exit from the skull. In the case of the optic nerve, the arachnoid forms a sheath for the nerve, which extends into the orbital cavity through the optic canal and fuses with the sclera of the eyeball (Fig. 11-52). Thus, the subarachnoid space extends around the optic nerve as far as the eyeball. (See p. 823.)

The *cerebrospinal fluid* is produced by the *cho-*

*roid plexuses* within the lateral, third, and fourth ventricles of the brain. It escapes from the ventricular system of the brain through the three foramina in the roof of the fourth ventricle and so enters the subarachnoid space. It now circulates both upward over the surfaces of the cerebral hemispheres and downward around the spinal cord. The spinal subarachnoid space extends down as far as the *second sacral vertebra* (see Fig. 12-7). Eventually, the fluid enters the bloodstream by passing into the arachnoid villi and diffusing through their walls.

In addition to removing waste products associated with neuronal activity, the cerebrospinal fluid provides a fluid medium in which the brain floats. This mechanism effectively protects the brain from trauma.

## PIA MATER OF THE BRAIN

The pia mater is a vascular membrane which closely invests the brain, covering the gyri and descending into the deepest sulci (Fig. 11-20). It extends over the cranial nerves and fuses with their epineurium. The cerebral arteries entering the substance of the brain carry a sheath of pia with them.

The pia mater forms the *tela choroidea* of the roof of the third and fourth ventricles of the brain, and it fuses with the ependyma to form the choroid plexuses in the lateral, third, and fourth ventricles of the brain.

## THE VENOUS BLOOD SINUSES

The venous sinuses of the cranial cavity are situated between the layers of the dura mater (Fig. 11-20). They are lined by endothelium, and their walls are devoid of muscular tissue. They contain no valves. They receive tributaries from the various parts of the brain, from the diploë, from the orbit, and from the internal ear.

The *superior sagittal sinus* occupies the upper fixed border of the falx cerebri (Fig. 11-41). It begins in front at the foramen cecum, where it occasionally receives a vein from the nasal cavity. It runs backward, grooving the vault of the skull, and at the internal occipital protuberance it deviates to one or the other side (usually the right) and be-

comes continuous with the corresponding transverse sinus. The sinus communicates through small openings with two or three irregularly shaped *venous lacunae* on each side. Numerous arachnoid villi and granulations project into the lacunae, which also receive the diploic and meningeal veins (Fig. 11-20).

The superior sagittal sinus receives in its course the *superior cerebral veins*. At the internal occipital protuberance it is dilated to form the *confluence of the sinuses* (Fig. 11-41). Here, the superior sagittal sinus usually becomes continuous with the right transverse sinus; it is connected to the opposite transverse sinus and receives the *occipital sinus*.

The *inferior sagittal sinus* occupies the free lower margin of the falx cerebri. It runs backward and joins the *great cerebral vein* at the free margin of the tentorium cerebelli, to form the *straight sinus* (Fig. 11-41). It receives a few cerebral veins from the medial surface of the cerebral hemisphere.

The *straight sinus* occupies the line of junction of the falx cerebri with the tentorium cerebelli (Fig. 11-41). It is formed by the union of the inferior sagittal sinus with the great cerebral vein. It ends by turning to the left (sometimes to the right), to form the transverse sinus.

The *transverse sinuses* are paired structures and begin at the internal occipital protuberance (Figs. 11-41 and 11-42). The right sinus is usually continuous with the superior sagittal sinus, and the left is continuous with the straight sinus. Each sinus occupies the attached margin of the tentorium cerebelli, grooving the occipital bone and the posteroinferior angle of the parietal bone. They receive the superior petrosal sinuses, inferior cerebral and cerebellar veins, and diploic veins. They end by turning downward as the sigmoid sinuses (Fig. 11-42).

The *sigmoid sinuses* are a direct continuation of the transverse sinuses. Each sinus turns downward and medially and grooves the mastoid part of the temporal bone (Fig. 11-42). Here it lies behind the mastoid antrum. The sinus then turns forward and then downward through the posterior part of the jugular foramen, to become continuous with the superior bulb of the internal jugular vein (Fig. 11-56).

The *occipital sinus* is a small sinus occupying the attached margin of the falx cerebelli. It commences near the foramen magnum, where it communicates with the vertebral veins and drains into the confluence of sinuses.

The *cavernous sinuses* are situated in the middle cranial fossa on each side of the body of the sphenoid bone (Fig. 11-41). Numerous trabeculae cross their interior, giving them a spongy appearance, hence the name. Each sinus extends from the superior orbital fissure in front to the apex of the petrous part of the temporal bone behind.

The internal carotid artery, surrounded by its sympathetic nerve plexus, runs forward through the sinus (Fig. 11-43). The abducent nerve also passes through the sinus. The internal carotid artery and the nerves are separated from the blood by an endothelial covering.

The third and fourth cranial nerves, and the ophthalmic and maxillary divisions of the trigeminal nerve run forward in the lateral wall of the sinus (Fig. 11-43). They lie between the endothelial lining and the dura mater. The *tributaries* are the superior and inferior ophthalmic veins, the cerebral veins, the sphenoparietal sinus, and the central vein of the retina.

The sinus drains posteriorly into the superior and inferior petrosal sinuses, and inferiorly into the pterygoid venous plexus.

The two sinuses communicate with one another by means of the *anterior* and *posterior intercavernous sinuses*, which run in the diaphragma sellae in front and behind the stalk of the hypophysis cerebri (Fig. 11-42). Each sinus has an important communication with the facial vein through the superior ophthalmic vein.

The *superior* and *inferior petrosal sinuses* are small sinuses situated on the superior and inferior borders of the petrous part of the temporal bone on each side (Fig. 11-41). Each superior sinus drains the cavernous sinus into the transverse sinus, and each inferior sinus drains the cavernous sinus into the internal jugular vein.

## HYPOPHYSIS CEREBRI

The hypophysis cerebri, or pituitary gland, is a small oval structure attached to the undersurface of the brain by the *infundibulum* (Fig. 11-43). It is well protected by virtue of its location in the sella turcica of the sphenoid bone. Because the hormones produced by the gland influence the activities of many other endocrine glands, the hypophysis cerebri is often referred to as the master endocrine gland. For this reason, it is of very great importance and vital to life.

The pituitary gland is divided into an *anterior lobe*, or *adenohypophysis*, and a *posterior lobe*, or *neurohypophysis*. The anterior lobe is subdivided into the *pars anterior* (sometimes called the pars distalis) and the *pars intermedia*, which may be separated by a cleft that is a remnant of an embryonic pouch. A projection from the pars anterior, the *pars tuberalis*, extends up along the anterior and lateral surfaces of the pituitary stalk.

### Relations

#### Superiorly

The diaphragma sellae, which has a central aperture that allows the passage of the infundibulum. The diaphragma sellae separates the anterior lobe from the optic chiasma.

#### Inferiorly

The body of the sphenoid, with its sphenoid air sinuses.

#### Laterally

The cavernous sinus and its contents.

#### Posteriorly

The dorsum sellae, the basilar artery, and the pons.

### Blood Supply

The arteries are derived from the *superior* and *inferior hypophyseal arteries*, branches of the internal carotid artery. The veins drain into the intercavernous sinuses.

## PARTS OF THE BRAIN

For a detailed description of the gross structure of the brain, a textbook of neuroanatomy should be consulted. In the following account, only the main parts of the brain are described.

| *Major Parts of the Brain* | *Cavities of the Brain* |
|---|---|
| Forebrain— Cerebrum | Right and left lateral ventricles |
| Diencephalon | Third ventricle |
| Midbrain | Cerebral aqueduct |
| Hindbrain— Pons, Medulla oblongata, Cerebellum | Fourth ventricle and central canal |

The brain is that part of the central nervous system that lies inside the cranial cavity. It is continuous with the spinal cord through the foramen magnum.

## Cerebrum

The *cerebrum* is the largest part of the brain and consists of two *cerebral hemispheres*, connected by a mass of white matter called the *corpus callosum* (Fig. 11-44). Each hemisphere extends from the frontal to the occipital bones, above the anterior and middle cranial fossae, and posteriorly, above the tentorium cerebelli. The hemispheres are separated by a deep cleft, the *longitudinal fissure*, into which projects the *falx cerebri* (Fig. 11-44).

The surface layer of each hemisphere is called the *cortex* and is composed of *gray matter* (Fig. 11-20). The cerebral cortex is thrown into folds, or *gyri*, separated by fissures, or *sulci*. By this means the surface area of the cortex is greatly increased. A number of the large sulci conveniently subdivide the surface of each hemisphere into *lobes*. The lobes are named for the bones of the cranium under which they lie (Fig. 11-45).

The *frontal lobe* is situated in front of the *central sulcus* (Fig. 11-45) and above the *lateral sulcus*. The *parietal lobe* is situated behind the central sulcus and above the lateral sulcus. The *occipital lobe* lies below the *parieto-occipital sulcus*. Below the lateral sulcus is situated the *temporal lobe*.

The *precentral gyrus* lies immediately anterior to the central sulcus and is known as the *motor area* (Fig. 11-45). The large motor nerve cells in this area control voluntary movements on the opposite side of the body. The majority of the nerve fibers cross over to the opposite side in the medulla oblongata as they descend to the spinal cord.

In the motor area the body is represented in an inverted position, with the nerve cells controlling the movements of the feet located in the upper part, and those controlling movements of the face and hands in the lower part (Fig. 11-45).

The *postcentral gyrus* lies immediately posterior to the central sulcus and is known as the *sensory area* (Fig. 11-45). The small nerve cells in this area receive and interpret sensations of pain, temperature, touch, and pressure from the opposite side of the body.

The *superior temporal gyrus* lies immediately below the lateral sulcus (Fig. 11-45). The middle of this gyrus is concerned with the reception and interpretation of sound and is known as the *auditory area*.

*Broca's area*, or the *motor speech area*, lies just above the lateral sulcus (Fig. 11-45). It controls the movements employed in speech. It is dominant in the left hemisphere in right-handed persons and dominant in the right hemisphere in left-handed persons.

The *visual area* is situated on the posterior pole and medial aspect of the cerebral hemisphere in the region of the *calcarine sulcus* (Fig. 11-45). It is the receiving area for visual impressions.

The cavity present within each cerebral hemisphere is called the *lateral ventricle*. The lateral ventricles communicate with the third ventricle through the *interventricular foramina* (Fig. 11-44).

## Diencephalon

The diencephalon is almost completely hidden from the surface of the brain. It consists of a dorsal *thalamus* (Fig. 11-44) and a ventral *hypothalamus*. The thalamus is a large mass of gray matter that lies on either side of the third ventricle. It is the

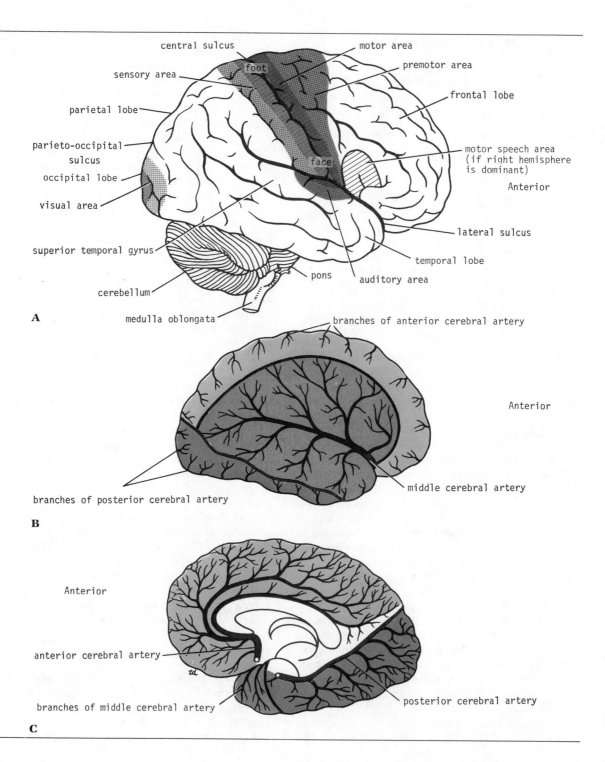

central sulcus

motor area

sensory area

premotor area

foot

parietal lobe

frontal lobe

parieto-occipital
sulcus

motor speech area
(if right hemisphere
is dominant)

occipital lobe

face

Anterior

visual area

superior temporal gyrus

lateral sulcus

cerebellum

temporal lobe

pons

auditory area

medulla oblongata

**A**

branches of anterior cerebral artery

Anterior

branches of posterior cerebral artery

middle cerebral artery

**B**

Anterior

anterior cerebral artery

branches of middle cerebral artery

posterior cerebral artery

**C**

great relay station on the afferent sensory pathway to the cerebral cortex.

The hypothalamus forms the lower part of the lateral wall and floor of the third ventricle. The following structures are found in the floor of the third ventricle from before backward: the *optic chiasma* (Fig. 11-46), the *tuber cinereum* and the *infundibulum*, the *mammillary bodies*, and the *posterior perforated substance*.

## Midbrain

The midbrain is the narrow part of the brain that passes through the tentorial notch and connects the forebrain to the hindbrain (Fig. 11-44).

The midbrain comprises two lateral halves, called the *cerebral peduncles*; each of these is divided into an anterior part, the *crus cerebri*, and a posterior part, the *tegmentum*, by a pigmented band of gray matter, the *substantia nigra* (Fig. 11-43). The narrow cavity of the midbrain is the *cerebral aqueduct*, which connects the third and fourth ventricles. The *tectum* is the part of the midbrain posterior to the cerebral aqueduct; it has four small surface swellings, namely, the *two superior* (Fig. 11-43) and *two inferior colliculi*. The colliculi are deeply placed between the cerebellum and the cerebral hemispheres.

The *pineal body* is a small glandular structure that lies between the superior colliculi (Fig. 11-44). It is attached by a stalk to the region of the posterior wall of the third ventricle. A small recess of the ventricles, called the *pineal recess*, extends into the base of the stalk. The pineal commonly calcifies in middle age, and thus it may be visualized in radiographs.

Fig. 11-45. (A) Right side of brain, showing some important localized areas of cerebral function. Note that the motor speech area is most commonly located in the left rather than the right cerebral hemisphere. (B) Lateral surface of cerebral hemisphere, showing areas supplied by cerebral arteries. In this and next diagram, areas colored blue are supplied by anterior cerebral artery; those colored red, by middle cerebral artery; and those colored green, by posterior cerebral artery. (C) Medial surface of cerebral hemisphere, showing areas supplied by cerebral arteries.

## Hindbrain

The *pons* is situated on the anterior surface of the cerebellum below the midbrain and above the medulla oblongata (Fig. 11-44). It is composed mainly of nerve fibers, which connect the two halves of the cerebellum. It also contains ascending and descending fibers connecting the forebrain, the midbrain, and the spinal cord. Some of the nerve cells within the pons serve as relay stations, while others form cranial nerve nuclei.

The *medulla oblongata* is conical in shape and connects the pons above to the spinal cord below (Fig. 11-44). A *median fissure* is present on the anterior surface of the medulla, and on each side of this is a swelling, called the *pyramid* (Fig. 11-46). The pyramids are composed of bundles of nerve fibers that originate in large nerve cells in the precentral gyrus of the cerebral cortex. The pyramids taper below, and here the majority of the descending fibers cross over to the opposite side, forming the *decussation of the pyramids*.

Posterior to the pyramids are the *olives*, which are oval elevations produced by the underlying *olivary nuclei* (Fig. 11-46). Behind the olives are the *inferior cerebellar penduncles*, which connect the medulla to the cerebellum.

On the posterior surface of the inferior part of the medulla oblongata are the *gracile* and *cuneate tubercles*, produced by the medially placed underlying *nucleus gracilis* and the laterally placed underlying *nucleus cuneatus*.

The *cerebellum* lies within the posterior cranial fossa beneath the tentorium cerebelli (Fig. 11-44). It is situated posterior to the pons and the medulla oblongata. It consists of two hemispheres connected by a median portion, the *vermis*. The cerebellum is connected to the midbrain by the *superior cerebellar peduncles*, to the pons by the *middle cerebellar peduncles*, and to the medulla by the *inferior cerebellar peduncles*.

The surface layer of each cerebellar hemisphere, called the *cortex*, is composed of gray matter. The cerebellar cortex is thrown into folds, or *folia*, separated by closely set transverse fissures. Certain masses of gray matter are found in the interior of the cerebellum, embedded in the white matter; the largest of these is known as the *dentate nucleus*.

The cerebellum plays an important role in the

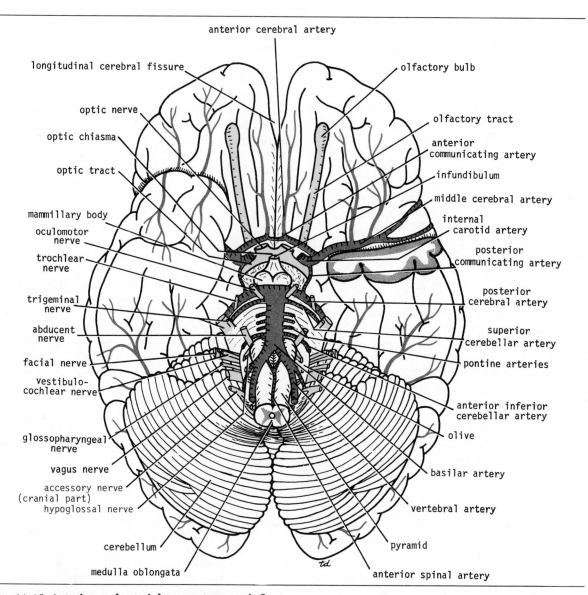

**Fig. 11-46. Arteries and cranial nerves seen on inferior surface of brain. In order to show the course of the middle cerebral artery, the anterior pole of the left temporal lobe has been removed.**

control of muscle tone and the coordination of muscle movement on the same side of the body.

The cavity of the hind brain is the fourth ventricle (Fig. 11-44). This is bounded in front by the pons and the medulla oblongata, and behind by the *superior* and *inferior medullary vela* and the cerebellum. The fourth ventricle is connected above to the third ventricle by the cerebral aqueduct, and below it is continuous with the central canal of the spinal cord. It communicates with the subarachnoid space through three openings in the lower part of the roof, a median and two lateral openings.

## Ventricles of the Brain

The ventricles of the brain consist of the two lateral ventricles, the third ventricle, and the fourth ventricle. The lateral ventricles are in communication with the third ventricle through the *interventricular foramina* (Fig. 11-44); the third ventricle communicates with the fourth ventricle by the cerebral aqueduct. The ventricles are filled with cerebrospinal fluid, which is produced by the *choroid plexuses* of the two lateral ventricles, the third ventricle, and the fourth ventricle. The cerebrospinal fluid escapes from the ventricular system through the three foramina in the roof of the fourth ventricle and enters the subarachnoid space. The circulation of the cerebrospinal fluid in the subarachnoid space and the fluid's ultimate absorption into the bloodstream is described on page 800.

## Blood Supply of the Brain

### ARTERIES OF THE BRAIN

The brain is supplied by the two internal carotid and the two vertebral arteries. The four arteries anastomose on the inferior surface of the brain and form the *circulus arteriosus*.

### INTERNAL CAROTID ARTERY

The internal carotid artery emerges from the cavernous sinus on the medial side of the anterior clinoid process by perforating the dura mater. (See p. 794.) It then enters the subarachnoid space by piercing the arachnoid mater and turns backward to the region of the anterior perforated substance of the brain, at the medial end of the lateral cerebral sulcus. Here, it divides into the anterior and middle cerebral arteries (Fig. 11-46).

### *Branches of the Cerebral Portion of the Internal Carotid Artery*

1. The *ophthalmic artery* arises as the internal carotid artery emerges from the cavernous sinus (Fig. 11-49). It enters the orbit through the optic canal, below and lateral to the optic nerve.
2. The *posterior communicating artery* runs backward to join the posterior cerebral artery (Fig. 11-46).
3. The *choroidal artery*, a small branch, passes backward, enters the inferior horn of the lateral ventricle, and ends in the choroid plexus.
4. The *anterior cerebral artery* runs forward and medially and enters the longitudinal fissure of the cerebrum (Fig. 11-46). It is joined to the fellow of the opposite side by the *anterior communicating artery*. It curves backward over the corpus callosum, and its *cortical branches* supply all the medial surface of the cerebral cortex as far back as the parieto-occipital sulcus (Fig. 11-45). They also supply a strip of cortex about 1 inch (2.5 cm) wide on the adjoining lateral surface. The anterior cerebral artery thus supplies the "leg area" of the precentral gyrus. A number of *central branches* pierce the brain substance and supply the deep masses of gray matter within the cerebral hemisphere.
5. The *middle cerebral artery*, the largest branch of the internal carotid, runs laterally in the lateral cerebral sulcus (Fig. 11-46). *Cortical branches* supply the entire lateral surface of the hemisphere, except for the narrow strip supplied by the anterior cerebral artery, the occipital pole, and the inferolateral surface of the hemisphere, which are supplied by the posterior cerebral artery. This artery thus supplies all the motor area except the "leg area." *Central branches* enter the anterior perforated substance and supply the deep masses of gray matter within the cerebral hemisphere.

### Vertebral Artery

The vertebral artery, a branch of the first part of the subclavian artery (Fig. 11-15), ascends through the foramina in the transverse processes of the upper six cervical vertebrae. (See p. 740.) It enters the skull through the foramen magnum and passes upward, forward, and medially on the medulla oblongata (Fig. 11-46). At the lower border of the pons it joins the vessel of the opposite side to form the *basilar artery*.

## Cranial Branches of the Vertebral Artery

1. *Meningeal arteries.*
2. *Anterior* and *posterior spinal arteries.*
3. *Posterior inferior cerebellar artery.*
4. *Medullary arteries.*

### Basilar Artery

The basilar artery, formed by the union of the two vertebral arteries, ascends in a groove on the anterior surface of the pons (Fig. 11-46). At the upper border of the pons it divides into the two posterior cerebral arteries.

## Branches

1. It gives off branches to the pons, cerebellum, and internal ear.
2. The posterior cerebral arteries.

The *posterior cerebral artery* on each side curves laterally and backward around the midbrain (Fig. 11-46). *Cortical branches* supply the inferolateral surface of the temporal lobe and the lateral and medial surfaces of the occipital lobe (Fig. 11-45). It thus supplies the visual cortex. *Central branches* pierce the brain substance and supply (1) the deep masses of gray matter within the cerebral hemisphere and (2) the midbrain.

The *circulus arteriosus* lies in the interpeduncular fossa at the base of the brain. It is formed by the anastomosis between the two internal carotid arteries and the two vertebral arteries (Fig. 11-46). The anterior communicating, the anterior cerebral, the internal carotid, the posterior communicating, the posterior cerebral, and the basilar arteries all contribute to the circle. The circulus arteriosus allows blood that enters by either internal carotid or vertebral arteries to be distributed to any part of both cerebral hemispheres. Cortical and central branches arise from the circle and supply the brain substance.

## VEINS OF THE BRAIN

The veins of the brain have no muscular tissue in their very thin walls and they possess no valves. They emerge from the brain and lie in the subarachnoid space. The veins pierce the arachnoid mater and the meningeal layer of the dura and drain into the cranial venous sinuses (Fig. 11-20). There are cerebral and cerebellar veins and veins of the brain stem. The *great cerebral vein* is formed by the union of the two *internal cerebral veins* and drains into the straight sinus (Fig. 11-41).

# Cranial Nerves

There are twelve pairs of cranial nerves, which are numbered and named from before backward, as follows:

1. Olfactory (sensory).
2. Optic (sensory).
3. Oculomotor (motor).
4. Trochlear (motor).
5. Trigeminal (mixed).
6. Abducent (motor).
7. Facial (mixed).
8. Vestibulocochlear (sensory).
9. Glossopharyngeal (mixed).
10. Vagus (mixed).
11. Accessory (motor).
12. Hypoglossal (motor).

The nerves emerge from the brain and are transmitted through foramina in the base of the skull.

## OLFACTORY NERVE

The nerve fibers of the olfactory nerve originate as the central processes of the *olfactory receptor nerve cells* in the mucous membrane of the upper part of the nose (above the superior concha). (See p. 855.) Bundles of the nerve fibers pierce the cribriform plate of the ethmoid bone and end in the *olfactory bulb* in the anterior cranial fossa (Fig. 11-41). Emerging from the posterior end of the olfactory bulb is a white band, the *olfactory tract* (Fig. 11-46), which passes backward and is attached to the cerebrum in the region of the anterior perforated substance. It ends by dividing into *medial* and *lateral olfactory striae.*

## OPTIC NERVE

The optic nerve is about 1.6 inches (4 cm) long. It leaves the orbital cavity by passing through the optic canal in company with the ophthalmic artery and then enters the cranial cavity (Fig. 11-49). Within the orbit the nerve is surrounded by the three meninges, the dura, arachnoid, and pia mater, which carry with them an extension of the subarachnoid space (Fig. 11-52). The nerves on both sides now join one another to form the *optic chiasma* (Fig. 11-49). Here, the nerve fibers that arise from the nasal half of the retina cross over to the opposite side; the fibers from the temporal half of the retina remain on the same side.

The upper surface of the optic chiasma is attached to the *lamina terminalis* of the brain, and its lower surface is separated from the hypophysis cerebri by the diaphragma sellae.

The *optic tract* emerges from the posterolateral angle of the optic chiasma and passes backward around the lateral side of the midbrain to reach the *lateral geniculate body.* A small number of fibers, subserving pupillary and ocular reflexes, bypass the lateral geniculate body and go directly to the *pretectal nucleus* and the *superior colliculus.* From the lateral geniculate body the *optic radiation* curves backward to the occipital visual cortex.

## OCULOMOTOR NERVE

The oculomotor nerve supplies all the muscles of the orbit except the superior oblique and the lateral rectus; it also supplies the sphincter pupillae and the ciliary muscle with parasympathetic fibers.

The oculomotor nerve emerges from the anterior aspect of the midbrain medial to the cerebral peduncle (Fig. 11-43). It pierces the arachnoid and dura and runs forward in the lateral wall of the cavernous sinus. Here, it divides into a superior and an inferior ramus, which enter the orbit through the superior orbital fissure. (See p. 823.)

## TROCHLEAR NERVE

The trochlear nerve, the most slender of the cranial nerves, supplies the superior oblique muscle in the orbit. The nerve emerges from the posterior surface of the midbrain (Fig. 11-43), just below the inferior colliculi. It then curves forward around the lateral side of the cerebral peduncle (Fig. 11-46). After piercing the arachnoid and dura mater, it runs forward in the lateral wall of the cavernous sinus, lying slightly below the oculomotor nerve.

The trochlear nerve enters the orbit through the superior orbital fissure. (See p. 823.)

## TRIGEMINAL NERVE

The trigeminal nerve is the largest cranial nerve. It supplies the sensory fibers to the skin of the scalp, the face, the mouth, the teeth, the nasal cavity, and the paranasal air sinuses and supplies motor fibers to the muscles of mastication (and the tensor veli palatini and tensor tympani muscles).

The trigeminal nerve emerges from the anterior surface of the pons by a large sensory and a small motor root, the motor root lying medial to the sensory root (Fig. 11-46). The nerve passes forward out of the posterior cranial fossa, below the superior petrosal sinus, and carries with it a pouch derived from the meningeal layer of dura mater. On reaching the depression on the apex of the petrous part of the temporal bone in the middle cranial fossa, the large sensory root expands to form the trigeminal ganglion (Fig. 11-49). The trigeminal ganglion is crescentic in shape and lies within the pouch of dura mater, called the *trigeminal cave.* The motor root of the trigeminal nerve is situated below the sensory ganglion and is completely separate from it. The ophthalmic, maxillary, and mandibular nerves arise from the anterior border of the ganglion (Fig. 11-49).

The *ophthalmic nerve* is purely sensory and is the smallest division of the trigeminal nerve. It pierces the dura mater and runs foward in the lateral wall of the cavernous sinus below the oculomotor and trochlear nerves (Figs. 11-43 and 11-49). It divides into three branches, the *lacrimal, frontal,* and *nasociliary* nerves, which enter the orbital cavity through the superior orbital fissure. (See pp. 823 and 825.)

The *maxillary nerve* is purely sensory. It pierces the dura mater and runs forward along the lower part of the lateral wall of the cavernous sinus (Figs. 11-43 and 11-49). It leaves the skull through the foramen rotundum, to enter the pterygopalatine fossa. (See p. 838.)

The *mandibular nerve* is motor and sensory and is the largest division of the trigeminal nerve. The large sensory root leaves the lateral part of the trigeminal ganglion, pierces the dura mater, and passes almost at once through the foramen ovale (Fig. 11-49). The small motor root passes beneath the ganglion, then through the foramen ovale. Immediately after emerging from the foramen, the motor root joins the sensory root. The course and branches of the mandibular nerve in the infratemporal fossa are described on page 764.

## ABDUCENT NERVE

The abducent nerve is a motor nerve and supplies the lateral rectus muscle of the eyeball. It emerges from the anterior surface of the brain, in the groove between the lower border of the pons and the medulla oblongata (Fig. 11-46). It lies at first in the posterior cranial fossa (Fig. 11-42). It then pierces the dura mater lateral to the dorsum sellae and turns sharply forward, crossing the superior border of the petrous part of the temporal bone. Having entered the cavernous sinus, it runs forward below and lateral to the internal carotid artery (Fig. 11-43). It enters the orbital cavity through the superior orbital fissure. (See p. 825.)

## FACIAL NERVE

The facial nerve has a medial motor root and a lateral sensory root, the *nervus intermedius*. The motor root supplies the muscles of the face, the scalp and auricle, the buccinator, the platysma, the stapedius, the stylohyoid, and the posterior belly of the digastric. The sensory root carries taste fibers from the anterior two-thirds of the tongue, the floor of the mouth, and the soft palate. It also conveys parasympathetic secretomotor fibers to the submandibular and sublingual salivary glands, the lacrimal gland, and the glands of the nose and palate.

The two roots of the facial nerve emerge from the anterior surface of the brain in the groove between the lower border of the pons and the medulla oblongata (Fig. 11-46). They pass laterally and forward in the posterior cranial fossa with the vestibulocochlear nerve to the opening of the internal acoustic meatus (Fig. 11-54). At the bottom of

the meatus, the nerve enters the facial canal and runs laterally above the vestibule of the labyrinth (Fig. 11-56) until it reaches the medial wall of the tympanic cavity. (See p. 836.)

## VESTIBULOCOCHLEAR NERVE

The vestibulocochlear nerve consists of two sets of sensory fibers: *vestibular* and *cochlear*. The vestibular fibers, which are concerned with equilibrium, represent the central processes of nerve cells of the *vestibular ganglion*. This is located in the outer part of the internal acoustic meatus.

The cochlear fibers, which are concerned with hearing, represent the central processes of nerve cells of the *spiral ganglion of the cochlea.*

The two parts of the nerve leave the anterior surface of the brain in the groove between the lower border of the pons and the medulla oblongata (Fig. 11-46). They cross the posterior cranial fossa and enter the internal acoustic meatus lying inferior to the facial nerve (Fig. 11-54).

## GLOSSOPHARYNGEAL NERVE

The glossopharyngeal nerve is a motor and sensory nerve. The motor fibers supply the stylopharyngeus muscle; parasympathetic secretomotor fibers supply the parotid salivary gland. The sensory fibers (including the taste fibers) pass to the posterior third of the tongue and the pharynx.

The glossopharyngeal nerve emerges from the anterior surface of the upper part of the medulla oblongata by three or four rootlets along the groove between the olive and the inferior cerebellar peduncle (Fig. 11-46). It passes forward and laterally beneath the cerebellum in the posterior cranial fossa and leaves the skull by passing downward through the central part of the jugular foramen. Its further course in the neck is described on page 781.

The *superior* and *inferior sensory glossopharyngeal ganglia* are situated on the nerve as it passes through the jugular foramen.

## VAGUS NERVE

The vagus nerve is composed of motor and sensory fibers. It supplies the heart and the major part of the respiratory and intestinal tracts.

The vagus nerve emerges from the anterior surface of the upper part of the medulla oblongata by eight or ten rootlets along the groove between the olive and the inferior cerebellar peduncle (Fig. 11-46). It lies below the glossopharyngeal nerve. The nerve passes laterally beneath the cerebellum in the posterior cranial fossa and leaves the skull through the central part of the jugular foramen. Its further course in the neck is described on page 729.

The *superior vagal sensory ganglion* is situated on the nerve as it passes through the jugular foramen. The *inferior vagal sensory ganglion* lies on the nerve a short distance below the foramen.

## ACCESSORY NERVE

The accessory nerve is a motor nerve. It consists of a small cranial root, which is distributed through the branches of the vagus nerve to the muscles of the soft palate, pharynx, and larynx, and a large spinal root, which innervates the sternocleidomastoid and trapezius muscles.

The *cranial root* emerges from the anterior surface of the upper part of the medulla oblongata by four or five rootlets along the groove between the olive and the inferior cerebellar peduncle (Fig. 11-46). It lies below the vagus nerve. The nerve runs laterally beneath the cerebellum in the posterior cranial fossa and joins the spinal root.

The *spinal root* arises from nerve cells in the anterior gray column of the upper five segments of the cervical part of the spinal cord. The nerve fibers emerge on the lateral surface of the spinal cord and form a nerve trunk. The nerve ascends alongside the spinal cord and enters the skull through the foramen magnum; it then turns laterally to join the cranial root.

The two roots are united for a short distance. They pass through the jugular foramen, and then the cranial portion separates from the spinal root and becomes adherent to the inferior ganglion of the vagus (Fig. 11-14). The fibers of the cranial root are distributed chiefly in the pharyngeal and recurrent laryngeal branches of the vagus.

The spinal root runs backward and laterally, crossing the internal jugular vein to reach the upper part of the sternocleidomastoid muscle. Its further course in the neck is described on page 781.

## HYPOGLOSSAL NERVE

The hypoglossal nerve is the motor nerve supply to the muscles of the tongue. The nerve emerges as a number of rootlets on the anterior surface of the medulla oblongata, in the groove between the pyramid and the olive (Fig. 11-46). The rootlets run laterally in the posterior cranial fossa and leave the skull through the hypoglossal canal. On emerging from the canal, the rootlets unite to form the nerve trunk. Its further course in the neck is described on page 719.

The cranial nerves, their component parts, their function, and the openings through which they exit from the skull are summarized in Table 11-3.

# THE ORBITAL REGION

The orbits are a pair of bony cavities that contain the eyeballs, their associated muscles, nerves, vessels, and fat, and most of the lacrimal apparatus. The orbital opening is guarded by two thin, movable folds, the eyelids, which are situated in front of the eye.

## EYELIDS

The eyelids are placed in front of the eye, which is protected from injury and excessive light by their closure (Fig. 11-47). The upper eyelid is larger and more mobile than the lower, and they meet each other at the *medial* and *lateral angles*. The *palpebral fissure* is the elliptical opening between the eyelids and is the entrance into the conjunctival sac. When the eye is closed, the upper eyelid completely covers the cornea of the eye. When the eye is open and looking straight ahead, the upper lid just covers the upper margin of the cornea. The lower lid lies just below the cornea when the eye is open and rises only slightly when the eye is closed.

The superficial surface of the eyelids is covered by skin, and the deep surface is covered by a mucous membrane, called the *conjunctiva*. The *eyelashes*, which are short, curved hairs, are present on the free edges of the eyelids (Fig. 11-47). They are arranged in double or triple rows at the mucocutaneous junction. The sebaceous glands (glands of Zeis) open directly into the eyelash follicles. The *ciliary glands* (glands of Moll) are mod-

Table 11-3. Cranial Nerves

| Name | Components | Function | Opening in skull |
|---|---|---|---|
| I. Olfactory | Sensory | Smell | Openings in cribriform plate of ethmoid |
| II. Optic | Sensory | Vision | Optic canal |
| III. Oculomotor | Motor | Lifts upper eyelid, turns eyeball upward, downward, and medially; constricts pupil; accommodates eyes | Superior orbital fissure |
| IV. Trochlear | Motor | Assists in turning eyeball downward and laterally | Superior orbital fissure |
| V. Trigeminal<br>Ophthalmic division | Sensory | Cornea, skin of forehead, scalp, eyelids; also mucous membrane of paranasal sinuses and nasal cavity | Superior orbital fissure |
| Maxillary division | Sensory | Skin of face over maxilla; teeth of upper jaw; mucous membrane of nose, the maxillary air sinus, and palate | Foramen rotundum |
| Mandibular division | Motor | Muscles of mastication, mylohyoid, anterior belly of digastric, tensor veli palatini, and tensor tympani | Foramen ovale |
| | Sensory | Skin of cheek, skin over mandible and side of head, teeth of lower jaw and temporomandibular joint; mucous membrane of mouth and anterior two-thirds of tongue | |
| VI. Abducent | Motor | Lateral rectus muscle: turns eyeball laterally | Superior orbital fissure |
| VII. Facial | Motor | Muscles of face, the cheek and scalp, stapedius muscle of middle ear, stylohyoid, and posterior belly of digastric | Internal acoustic meatus, facial canal, stylomastoid foramen |
| | Sensory | Taste from anterior two-thirds of tongue, floor of mouth and soft palate | |
| | Secretomotor parasympathetic | Submandibular and sublingual salivary glands, the lacrimal gland, and glands of nose and palate | |
| VIII. Vestibulocochlear<br>Vestibular | Sensory | Position and movement of head | Internal acoustic meatus |
| Cochlear | Sensory | Hearing | |

Table 11-3 (continued)

| Name | Components | Function | Opening in skull |
|------|-----------|----------|------------------|
| IX. Glossopharyngeal | Motor | Stylopharyngeus muscle: assists swallowing | |
| | Secretomotor parasympathetic | Parotid salivary gland | Jugular foramen |
| | Sensory | General sensation and taste from posterior third of tongue and pharynx; carotid sinus and carotid body | |
| X. Vagus | Motor | Constrictor muscles of pharynx and intrinsic muscles of larynx; involuntary muscle of bronchi, heart, alimentary tract from pharynx to distal one-third of transverse colon; liver and pancreas | Jugular foramen |
| | Sensory | Taste from epiglottis and vallecula and afferent fibers from structures named above | |
| XI. Accessory | | | |
| Cranial root | Motor | Muscles of soft palate, pharynx, and larynx | Jugular foramen |
| Spinal root | Motor | Sternocleidomastoid and trapezius muscles | |
| XII. Hypoglossal | Motor | Muscles of tongue controlling its shape and movement (except palatoglossus) | Hypoglossal canal |

ified sweat glands that open separately between adjacent lashes. The *tarsal glands* are long modified sebaceous glands, which pour their oily secretion onto the margin of the lid; their openings lie behind the eyelashes (Fig. 11-47). This oily material prevents the overflow of tears and helps to make the closed eyelids airtight.

The lateral angle of the palpebral fissure is more acute than the medial and lies directly in contact with the eyeball. The more rounded medial angle is separated from the eyeball by a small space, the *lacus lacrimalis*, in the center of which is a small, reddish-yellow elevation, the *caruncula lacrimalis* (Fig. 11-47). A reddish semilunar fold, called the *plica semilunaris*, lies on the lateral side of the caruncle.

Near the medial angle of the eye, the eyelashes and the tarsal glands stop abruptly, and there is a small elevation, the *papilla lacrimalis*. On the summit of the papilla is a small hole, the *punctum lacrimale* which leads into the *canaliculus lacrimalis* (Fig. 11-47). The papilla lacrimalis projects into the lacus, and the punctum and canaliculus serve to carry tears down into the nose. (See p. 815.)

The *conjunctiva* is a thin mucous membrane that lines the eyelids and is reflected at the *superior* and *inferior fornices* onto the anterior surface of the eyeball (Fig. 11-47). Its epithelium is continuous with that of the cornea. The upper lateral part of the superior fornix is pierced by the ducts of the lacrimal gland. (See below.) The conjunctiva thus forms a potential space, the *conjunctival sac*, which is

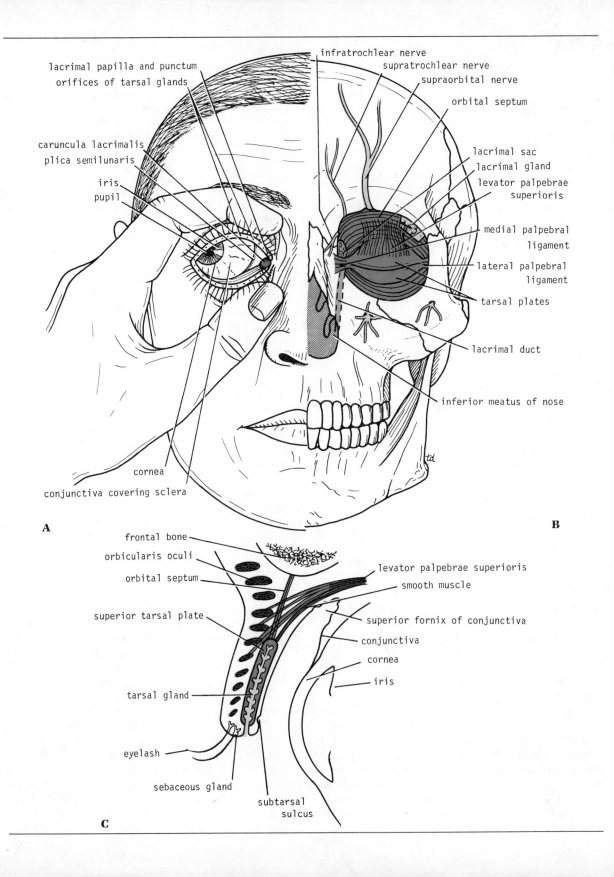

lacrimal papilla and punctum
orifices of tarsal glands

infratrochlear nerve
supratrochlear nerve
supraorbital nerve
orbital septum

caruncula lacrimalis
plica semilunaris

lacrimal sac
lacrimal gland
levator palpebrae
superioris

iris
pupil

medial palpebral
ligament

lateral palpebral
ligament

tarsal plates

lacrimal duct

inferior meatus of nose

cornea

conjunctiva covering sclera

A

B

frontal bone
orbicularis oculi
orbital septum

levator palpebrae superioris
smooth muscle

superior tarsal plate

superior fornix of conjunctiva
conjunctiva
cornea
iris

tarsal gland

eyelash

sebaceous gland

subtarsal
sulcus

C

open at the palpebral fissure.

Beneath the eyelid is a groove, the *subtarsal sulcus*, which runs close to and parallel with the margin of the lid (Fig. 11-47). The sulcus tends to trap small foreign particles introduced into the conjunctival sac and is thus clinically important.

The fibrous framework of the eyelids is formed by a membranous sheet, the *orbital septum* (Fig. 11-47). This is attached to the orbital margin, where it is continuous with the periosteum. The orbital septum is thickened at the margins of the lids to form the *tarsal plates*. These are crescent-shaped laminae of dense fibrous tissue, and the superior tarsal plate is the larger. The lateral ends of the plates are attached by a band, the *lateral palpebral ligament*, to a bony tubercle just within the orbital margin. The medial ends of the plates are attached by a band, the *medial palpebral ligament*, to the crest of the lacrimal bone (Fig. 11-47). The tarsal glands are embedded in the posterior surface of the tarsal plates.

The superficial surface of the tarsal plates and the orbital septum are covered by the palpebral fibers of the *orbicularis oculi muscle*. (See p. 754.) The aponeurosis of insertion of the *levator palpebrae superioris muscle* pierces the orbital septum, to reach the anterior surface of the superior tarsal plate and the skin (Fig. 11-47).

## Movements of the Eyelids

The position of the eyelids at rest depends on the tone of the orbicularis oculi and the levator palpebrae superioris muscles and the position of the eyeball. The eyelids are closed by the contraction of the orbicularis oculi and the relaxation of the levator palpebrae superioris muscles. The eye is opened by the levator palpebrae superioris raising the upper lid. On looking upward, the levator palpebrae superioris contracts, and the upper lid moves with the eyeball. On looking downward, both lids move, the upper lid continues to cover the upper part of the cornea, and the lower lid is pulled downward slightly by the conjunctiva, which is attached to the sclera and the lower lid.

## LACRIMAL APPARATUS

The lacrimal gland consists of a large *orbital part* and a small *palpebral part*, which are continuous with each other around the lateral edge of the aponeurosis of the levator palpebrae superioris. It is situated above the eyeball in the anterior and upper part of the orbit posterior to the orbital septum (Fig. 11-47). About twelve ducts open from the lower surface of the gland into the lateral part of the superior fornix of the conjunctiva.

## Nerve Supply of Lacrimal Gland

The parasympathetic secretomotor nerve supply is derived from the *lacrimal nucleus* of the facial nerve. The preganglionic fibers reach the pterygopalatine ganglion (sphenopalatine ganglion) via the nervus intermedius and its great petrosal branch and via the nerve of the pterygoid canal. The postganglionic fibers leave the ganglion and join the maxillary nerve. They then pass into its zygomatic branch and the zygomaticotemporal nerve. They reach the lacrimal gland within the lacrimal nerve.

The sympathetic postganglionic fibers travel in the internal carotid plexus, the deep petrosal nerve, the nerve of the pterygoid canal, the maxillary nerve, the zygomatic nerve, the zygomaticotemporal nerve, and finally the lacrimal nerve.

The *tears* circulate across the cornea and accumulate in the *lacus lacrimalis*. From here, the tears enter the *canaliculi lacrimales* through the *puncta lacrimalia*. The canaliculi lacrimales pass medially and open into the *lacrimal sac* (Fig. 11-47). This lies in the lacrimal groove behind the medial palpebral ligament and is the upper blind end of the nasolacrimal duct.

The *nasolacrimal duct* is about ½ inch (1.3 cm) long and emerges from the lower end of the lacrimal sac (Fig. 11-47). The duct descends downward, backward, and laterally in an osseous canal and opens into the inferior meatus of the nose. The

**Fig. 11-47. (A) Right eye, with eyelids separated to show openings of tarsal glands, plica semilunaris, caruncula lacrimalis, and puncta lacrimalia. (B) Left eye, showing superior and inferior tarsal plates and lacrimal gland, sac, and duct. Note that a small window has been cut in the orbital septum to show underlying lacrimal gland and fat (yellow). (C) Sagittal section through upper eyelid, and superior fornix of conjunctiva. Note presence of smooth muscle in levator palpebrae superioris.**

opening is guarded by a fold of mucous membrane known as the *lacrimal fold*. This prevents air from being forced up the duct into the lacrimal sac on blowing the nose.

# THE ORBIT

The orbit is a pyramid-shaped cavity with its base in front and its apex behind (Fig. 11-48). The *orbital margin* is formed above by the frontal bone, which is notched or canalized for the passage of the supraorbital nerve and vessels. The lateral margin is formed by the processes of the frontal and zygomatic bones. The inferior margin is formed by the zygomatic bone and the maxilla. The medial margin is formed by the processes of the maxilla and the frontal bone.

The *roof* of the orbit is formed by the orbital plate of the frontal bone, which separates the orbital cavity from the anterior cranial fossa. The *lateral wall* is composed of the zygomatic bone and the greater wing of the sphenoid (Fig. 11-48). The *floor* is formed by the orbital plate of the maxilla, which separates the orbital cavity from the maxillary sinus. The *medial wall* consists, from before backward, of the frontal process of the maxilla, the lacrimal bone, the orbital plate of the ethmoid (which separates the orbital cavity from the ethmoid sinuses), and the body of the sphenoid.

## OPENINGS INTO THE ORBITAL CAVITY

The main *orbital opening* lies anteriorly and is bounded by the orbital margin (Fig. 11-48). The *supraorbital notch*, or *canal*, is situated on the superior orbital margin. The *infraorbital groove* and *canal* lie on the floor of the orbit. Anteriorly on the medial part of the floor lies the *nasolacrimal canal*.

Posteriorly is the *inferior orbital fissure*, which leads from the pterygopalatine fossa and the infratemporal fossa into the orbital cavity. The *superior orbital fissure* leads from the middle cranial fossa into the orbit (Fig. 11-48). The *optic canal* also leads forward from the middle fossa into the orbit.

On the lateral wall are two small openings for the *zygomaticotemporal* and *zygomaticofacial nerves*.

On the medial wall along the upper margin of the ethmoid bone are the *anterior* and *posterior ethmoidal foramina*.

## ORBITAL FASCIA

The orbital fascia is the periosteum of the bones that form the walls of the orbit. It is loosely attached to the bones and is continuous through the foramina and fissures with the periosteum covering the outer surfaces of the bones. In the case of the superior orbital fissure, the optic canal, and the anterior ethmoidal canal, it becomes continuous with the endosteal layer of the dura mater. The *muscle of Müller*, or *orbitalis muscle*, is a thin layer of smooth muscle that bridges the inferior orbital fissure. It is supplied by sympathetic nerves, and its function is unknown.

# Muscles of the Orbit

The muscles of the orbit are the levator palpebrae superioris, the four recti, and the two oblique muscles.

## LEVATOR PALPEBRAE SUPERIORIS (FIGS. 11-47 to 11-49)

### Origin

From the undersurface of the lesser wing of the sphenoid bone, above and in front of the optic canal.

### Insertion

It is a flat muscle, which widens as it passes forward. It ends anteriorly in a wide aponeurosis, which splits into two lamellae. The superior lamella is inserted into the anterior surface of the superior tarsal plate and into the skin of the upper lid. The *inferior lamella contains smooth muscle fibers*, which are attached to the upper margin of the superior tarsal plate.

**Fig. 11-48. (A) Right eyeball exposed from front. (B) Muscles and nerves of left orbit seen from in front. (C) Bones forming walls of right orbit. (D) Optic canal and superior and inferior orbital fissures on left side.**

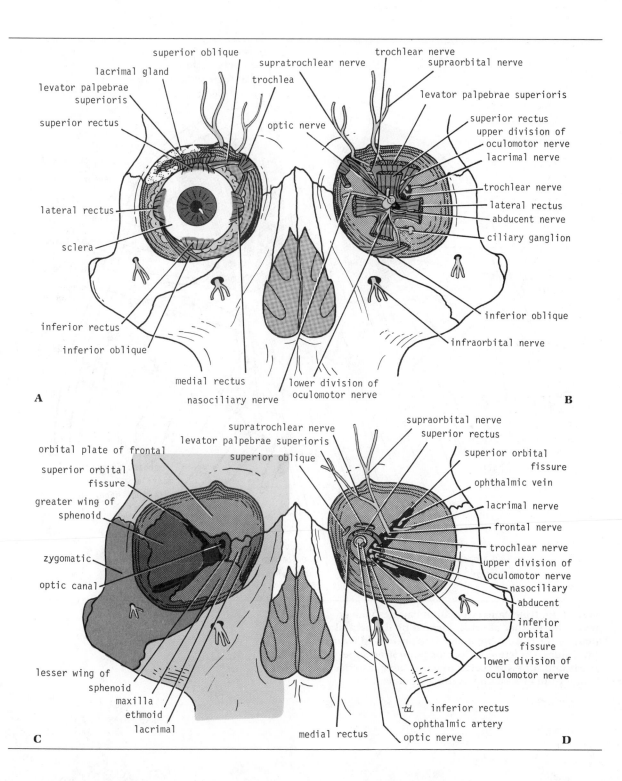

superior oblique
lacrimal gland
levator palpebrae
superioris
superior rectus
trochlea
supratrochlear nerve
trochlear nerve
supraorbital nerve
optic nerve
levator palpebrae superioris
superior rectus
upper division of
oculomotor nerve
lacrimal nerve
trochlear nerve
lateral rectus
lateral rectus
abducent nerve
sclera
ciliary ganglion
inferior rectus
inferior oblique
inferior oblique
infraorbital nerve
medial rectus
lower division of
oculomotor nerve
nasociliary nerve

**A**

**B**

supratrochlear nerve
levator palpebrae superioris
superior oblique
supraorbital nerve
superior rectus
orbital plate of frontal
superior orbital
fissure
greater wing of
sphenoid
superior orbital
fissure
ophthalmic vein
lacrimal nerve
frontal nerve
trochlear nerve
upper division of
oculomotor nerve
nasociliary
abducent
zygomatic
optic canal
inferior
orbital
fissure
lesser wing of
sphenoid
maxilla
ethmoid
lacrimal
lower division of
oculomotor nerve
medial rectus
inferior rectus
ophthalmic artery
optic nerve

**C**

**D**

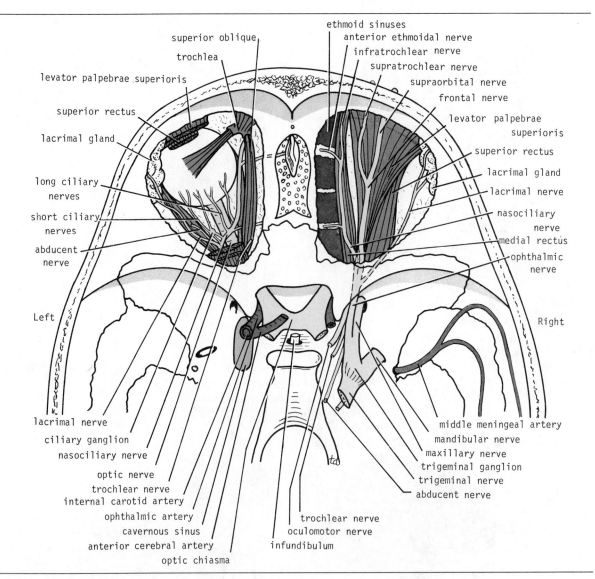

**Fig. 11-49. Right and left orbital cavities viewed from above. Roof of orbit, formed by the orbital plate of the frontal bone, has been removed from both sides. On left side, levator palpebrae superioris and superior rectus muscles have also been removed to expose underlying structures.**

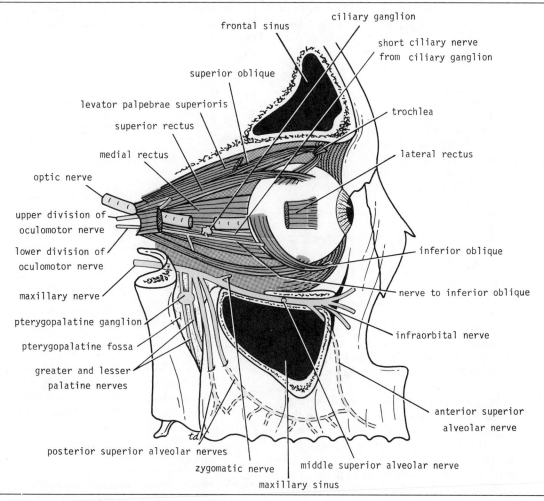

**Fig. 11-50. Muscles and nerves of right orbit viewed from lateral side. Maxillary nerve and pterygopalatine ganglion are also shown.**

## Nerve Supply

The superior ramus of the oculomotor nerve. The smooth muscle fibers are innervated by sympathetic nerves from the superior cervical sympathetic ganglion.

## Action

The levator palpebrae superioris raises the upper lid. Stimulation of the sympathetic innervation re-

sults in further elevation of the lid. Division of the cervical sympathetic paralyzes the smooth muscle and causes drooping of the upper lid *(ptosis)*.

## THE RECTI (FIGS. 11-48 to 11-50)

### Origin

The four recti arise from a fibrous ring, called the *common tendinous ring* (Fig. 11-48). It is a thickening of the periosteum. The ring surrounds the optic canal and bridges the superior orbital fissure. The superior rectus arises from the upper part of the ring, the inferior rectus from the lower part of the ring, and the medial rectus from the medial part. The lateral rectus arises by two heads from the lateral part of the ring.

## Insertion

As each rectus muscle passes forward, it becomes wider and diverges from its neighbor. Together, they form a muscular cone that encloses the optic nerve and the posterior part of the eyeball. The tendon of each muscle pierces the fascial sheath of the eyeball (see below) and is inserted into the sclera about 6 mm behind the margin of the cornea.

## Nerve Supply

The superior, inferior, and medial recti are supplied by the oculomotor nerve; the lateral rectus is supplied by the abducent nerve.

## Actions

The lateral rectus rotates the eyeball so that the cornea looks laterally (Fig. 11-51). The medial rectus rotates the eyeball so that the cornea looks medially.

Because the superior and inferior recti are inserted on the medial side of the vertical axis of the eyeball, they not only raise and depress the cornea, respectively, but *rotate it medially* (Fig. 11-51). For the superior rectus muscle to raise the cornea directly upward, the inferior oblique must assist; and for the inferior rectus to depress the cornea directly downward, the superior oblique must assist. (See action of oblique muscles.)

## SUPERIOR OBLIQUE (FIGS. 11-48 to 11-50)

### Origin

From the body of the sphenoid bone, above and medial to the optic canal.

### Insertion

Its rounded belly passes forward and gives way to a slender tendon, which passes through a fibrocartilaginous pulley attached to the frontal bone. The tendon now turns backward and laterally, pierces the fascial sheath of the eyeball, and is inserted into the sclera beneath the superior rectus. It is attached to the sclera behind the coronal equator of the eyeball, and the line of pull of the tendon passes medial to the vertical axis.

## Nerve Supply

Trochlear nerve.

## Action

The superior oblique rotates the eyeball (Fig. 11-51) so that the cornea looks (1) downward and (2) laterally. (See action with inferior rectus muscle.)

## INFERIOR OBLIQUE (FIGS. 11-48 and 11-50)

### Origin

From the anterior part of the floor of the orbit.

### Insertion

The narrow muscle passes backward and laterally below the inferior rectus. It is inserted into the sclera behind the coronal equator, and the line of pull of the tendon passes medial to the vertical axis.

### Nerve Supply

Inferior ramus of the oculomotor nerve.

### Action

The inferior oblique (Fig. 11-51) rotates the eyeball so that the cornea looks (1) upward and (2) laterally. (See action with superior rectus muscle.)

The extraocular muscles, their nerve supply, and their actions are summarized in Table 11-4.

**Fig. 11-51. Actions of four recti and two oblique muscles of right orbit,** *assuming that each muscle is acting alone.* **Position of pupil in relation to vertical and horizontal planes should be noted in each case. The actions of the superior and inferior recti and the oblique muscles in the living intact eye are tested clinically as described on page 822.**

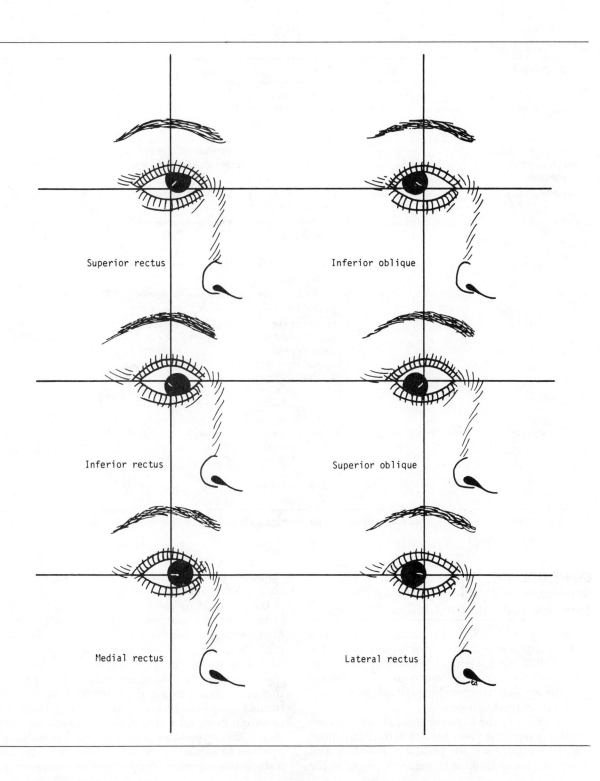

Table 11-4. Extraocular Muscles

| Name of muscle | Origin | Insertion | Nerve supply | Action |
|---|---|---|---|---|
| Levator palpebrae superioris | Lesser wing of sphenoid | Anterior surface and upper border of superior tarsal plate | | |
| Voluntary portion Involuntary portion | | | Oculomotor nerve Sympathetic nerves | Raises upper eyelid |
| Superior rectus | Common tendinous ring | Sclera 6 mm behind corneal margin | Oculomotor nerve | Raises and medially rotates cornea |
| Inferior rectus | Common tendinous ring | Sclera 6 mm behind corneal margin | Oculomotor nerve | Depresses cornea and medially rotates cornea |
| Lateral rectus | Common tendinous ring | Sclera 6 mm behind corneal margin | Abducent nerve | Moves cornea laterally |
| Medial rectus | Common tendinous ring | Sclera 6 mm behind corneal margin | Oculomotor nerve | Moves cornea medially |
| Superior oblique | Body of sphenoid | By way of pulley and attached to sclera behind coronal equator of eyeball; line of pull of tendon passes medial to vertical axis | Trochlear nerve | Moves cornea downward and laterally |
| Inferior oblique | Anterior part of floor of orbit | Attached to sclera behind coronal equator; line of pull of tendon passes medial to vertical axis | Oculomotor nerve | Moves cornea upward and laterally |

The actions of the superior and inferior recti and the oblique muscles in the living intact eye are tested clinically as described in text below.

## Clinical Testing for the Actions of the Superior and Inferior Recti and the Superior and Inferior Oblique Muscles

Since the actions of the superior and inferior recti and the superior and inferior oblique muscles are complicated when a patient is asked to look vertically upward or vertically downward, the physician tests the eye movements where the single action of each muscle predominates.

The origins of the superior and inferior recti are situated about 25 degrees medial to their insertions and, therefore, when the patient is asked to turn the cornea laterally, these muscles are placed in the optimum position to raise the cornea (superior rectus) or lower it (inferior rectus).

Using the same rationale, the superior and the inferior oblique muscles may be tested. The pulley of the superior oblique and the origin of the inferior oblique muscles lie medial and anterior to their insertions. The physician tests the action of these muscles by asking the patient first to look medially, thus placing these muscles in the optimum position to lower the cornea (superior oblique), or to raise it (inferior oblique). In other words, when you ask a patient to look medially and downward at the tip of his or her nose, you are testing the superior oblique at its best position.

Conversely, by asking the patient to look medially and upward, you are testing the inferior oblique at its best position.

Since the lateral and medial recti are simply placed relative to the eyeball, by asking the patient to turn his or her cornea directly laterally tests the lateral rectus and by turning the cornea directly medially tests the medial rectus.

## FASCIAL SHEATH OF THE EYEBALL

The fascial sheath is a thin membrane, which surrounds the eyeball from the optic nerve to the corneoscleral junction (Fig. 11-52). It separates the eyeball from the orbital fat and provides it with a socket for free movement. It is perforated by the tendons of the orbital muscles and is reflected onto each of them as a tubular sheath. The sheaths for the tendons of the medial and lateral recti are attached to the medial and lateral walls of the orbit by triangular ligaments called the *medial* and *lateral check ligaments.* The lower part of the fascial sheath, which passes beneath the eyeball and connects the check ligaments, is thickened and serves to suspend the eyeball; it is called the *suspensory ligament of the eye* (Fig. 11-52). By this means the eye is suspended from the medial and lateral walls of the orbit, as if in a hammock.

## Nerves of the Orbit

### OPTIC NERVE

The optic nerve enters the orbit from the middle cranial fossa by passing through the optic canal (Fig. 11-49). It is accompanied by the ophthalmic artery, which lies on its lower lateral side. The nerve is surrounded by sheaths of pia mater, arachnoid mater, and dura mater (Fig. 11-52). It runs forward and laterally within the cone of the recti muscles and pierces the sclera at a point medial to the posterior pole of the eyeball. Here, the meninges fuse with the sclera, so that the subarachnoid space with its contained cerebrospinal fluid extends forward from the middle cranial fossa, around the optic nerve, and through the optic canal, as far as the eyeball. A rise in pressure of the cerebrospinal fluid within the cranial cavity will therefore be transmitted to the back of the eyeball.

## LACRIMAL NERVE

The lacrimal nerve arises from the ophthalmic division of the trigeminal nerve in the lateral wall of the cavernous sinus. (See p. 809.) It is a slender nerve and enters the orbit through the upper part of the superior orbital fissure (Fig. 11-48). It passes forward along the upper border of the lateral rectus muscle (Fig. 11-49). It is joined by a branch of the zygomaticotemporal nerve, which later leaves it to enter the lacrimal gland (parasympathetic secretomotor fibers). The lacrimal nerve ends by supplying the skin of the lateral part of the upper lid.

## FRONTAL NERVE

The frontal nerve arises from the ophthalmic division of the trigeminal nerve in the lateral wall of the cavernous sinus. (See p. 809.) It enters the orbit through the upper part of the superior orbital fissure (Fig. 11-48) and passes forward on the superior surface of the levator palpebrae superioris, between that muscle and the roof of the orbit (Fig. 11-49). Just before it reaches the orbital margin, it divides into the *supratrochlear* and *supraorbital nerves.* The small supratrochlear nerve passes above the pulley for the superior oblique muscle and winds around the upper margin of the orbital cavity, to supply the skin of the forehead. The larger supraorbital nerve passes through the supraorbital notch, or foramen, and supplies the skin of the forehead lateral to the area supplied by the supratrochlear nerve. (See p. 751.) The frontal nerve also supplies the mucous membrane of the frontal air sinus.

## TROCHLEAR NERVE

The trochlear nerve leaves the lateral wall of the cavernous sinus (see p. 809) and enters the orbit through the upper part of the superior orbital fissure (Fig. 11-48). It runs forward and medially across the origin of the levator palpebrae superioris and enters the superior oblique muscle (Fig. 11-49).

## OCULOMOTOR NERVE

The *superior ramus* of the oculomotor nerve leaves the lateral wall of the cavernous sinus (see p. 809)

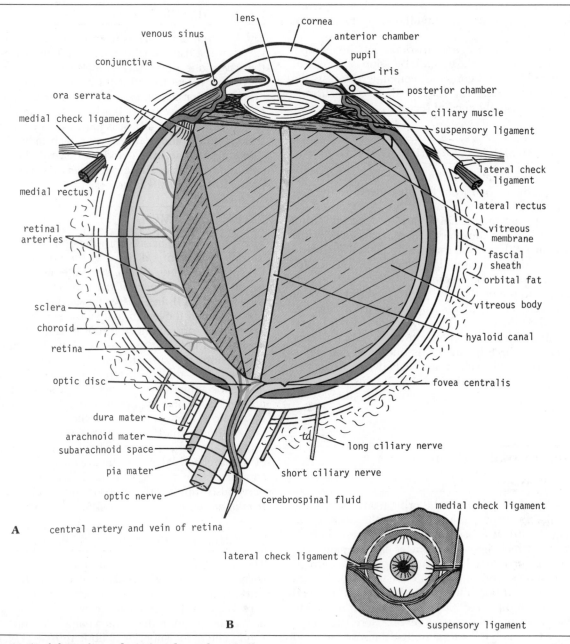

**Fig. 11-52. (A) Horizontal section through eyeball and optic nerve. Note that central artery and vein of retina cross subarachnoid space to reach optic nerve. (B) Check ligaments and suspensory ligament of the eyeball.**

and enters the orbit through the lower part of the superior orbital fissure, within the tendinous ring (Fig. 11-48). It supplies the superior rectus muscle, then pierces it, and supplies the overlying levator palpebrae superioris muscle (Fig. 11-50).

The *inferior ramus* of the oculomotor nerve enters the orbit in a similar manner and gives off branches to the inferior rectus, the medial rectus, and the inferior oblique muscles. The nerve to the inferior oblique gives off a branch (Fig. 11-50) that passes to the ciliary ganglion and carries parasympathetic fibers. (See below.)

## NASOCILIARY NERVE

The nasociliary nerve arises from the ophthalmic division of the trigeminal nerve in the lateral wall of the cavernous sinus. (See p. 809.) It enters the orbit through the lower part of the superior orbital fissure, within the tendinous ring (Fig. 11-48). It crosses above the optic nerve with the ophthalmic artery, to reach the medial wall of the orbital cavity. It then runs forward along the upper margin of the medial rectus muscle and ends by dividing into the *anterior ethmoidal* and *infratrochlear nerves* (Fig. 11-49).

### Branches

1. The *communicating branch to the ciliary ganglion.* This is a sensory nerve. The sensory fibers from the eyeball pass to the ciliary ganglion via the short ciliary nerves, pass through the ganglion without interruption, and then join the nasociliary nerve by means of the communicating branch.
2. The long *ciliary nerves,* two or three in number, arise from the nasociliary nerve as it crosses the optic nerve (Fig. 11-49). They contain sympathetic fibers for the dilator pupillae muscle. The nerves pass forward with the short ciliary nerves and pierce the sclera of the eyeball close to the optic nerve. They continue forward between the sclera and the choroid, to reach the iris.
3. The *posterior ethmoidal nerve* passes through the posterior ethmoidal foramen, to supply the ethmoidal and sphenoidal air sinuses (Fig. 11-49).
4. The *infratrochlear nerve* passes forward below the pulley of the superior oblique muscle and supplies the skin of the medial part of the upper eyelid and the adjacent part of the nose (Fig. 11-47).
5. The *anterior ethmoidal nerve* passes through the anterior ethmoidal foramen and enters the anterior cranial fossa on the upper surface of the cribriform plate of the ethmoid (Fig. 11-49). It crosses the cribriform plate and enters the nasal cavity through a slitlike opening alongside the crista galli. After supplying an area of mucous membrane, it appears on the face as the external nasal branch at the lower border of the nasal bone (Fig. 11-23). It supplies the skin of the nose down as far as the tip. (See p. 751.)

## ABDUCENT NERVE

The abducent nerve leaves the cavernous sinus (see p. 810) and enters the orbit through the lower part of the superior orbital fissure, within the tendinous ring (Fig. 11-48). It runs forward and supplies the lateral rectus muscle.

## CILIARY GANGLION

The ciliary ganglion is about the size of a pinhead (Fig. 11-50). It is a parasympathetic ganglion and is situated in the posterior part of the orbit on the lateral side of the optic nerve. It receives its preganglionic parasympathetic fibers from the oculomotor nerve via the nerve to the inferior oblique. The postganglionic fibers leave the ganglion in the *short ciliary nerves,* which pass forward to the back of the eyeball.

A number of sympathetic fibers pass from the internal carotid plexus into the orbit and run through the ganglion without interruption.

## Blood Vessels of the Orbit

### OPHTHALMIC ARTERY

The ophthalmic artery is a branch of the internal carotid artery after that vessel emerges from the cavernous sinus. (See p. 807.) It passes forward through the optic canal below and lateral to the optic nerve (Fig. 11-49). It runs forward, at first lateral to the optic nerve, then crosses above it obliquely to reach the medial wall of the orbit. It crosses the optic nerve with the nasociliary nerve. It now gives off numerous branches, some of which accompany the nerves in the orbital cavity.

## Branches

1. The *central artery of the retina* is a small branch that pierces the meningeal sheaths of the optic nerve to gain entrance to the nerve (Fig. 11-52). It runs in the substance of the optic nerve and enters the eyeball at the center of the *optic disc*. Here, it divides into branches, which may be studied in a patient through an ophthalmoscope. The branches are end arteries.
2. *The muscular branches.*
3. The *ciliary arteries.* These may be divided into anterior and posterior groups. The former group enters the eyeball near the corneoscleral junction; the latter group enters near the optic nerve.
4. The *lacrimal artery* to the lacrimal gland.
5. The *supratrochlear* and *supraorbital arteries* are distributed to the skin of the forehead. (See p. 752.)

## OPHTHALMIC VEINS

The *superior ophthalmic vein* communicates in front with the facial vein (Fig. 11-41). The *inferior ophthalmic vein* communicates through the inferior orbital fissure with the pterygoid venous plexus. Both veins pass backward through the superior orbital fissure and drain into the cavernous sinus.

## LYMPHATIC VESSELS

There are no lymphatic vessels or nodes in the orbital cavity.

## THE EYE

The eyeball (Fig. 11-52) is embedded in orbital fat, but is separated from it by the fascial sheath of the eyeball. (See p. 823.) The eyeball consists of three coats, which, from without inward, are: (1) the fibrous coat, (2) the vascular pigmented coat, and (3) the nervous coat.

## COATS OF THE EYEBALL

### Fibrous Coat

The fibrous coat is made up of a posterior opaque part, the sclera, and an anterior transparent part, the cornea (Fig. 11-52). The *sclera* is composed of dense fibrous tissue and is white in color. Posteriorly, it is pierced by the optic nerve and is fused with the dural sheath of that nerve. The *lamina cribrosa* is the area of the sclera that is pierced by the nerve fibers of the optic nerve. It is a relatively weak area and can be made to bulge into the eyeball by a rise of cerebrospinal fluid pressure in the tubular extension of the subarachnoid space, which surrounds the optic nerve. If the intraocular pressure should rise, the lamina cribrosa will bulge outward, producing a cupped disc, as seen through the ophthalmoscope.

The sclera is also pierced by the ciliary arteries and nerves and their associated veins, the venae *vorticosae.* The sclera is directly continuous in front with the cornea at the *corneoscleral junction*, or *limbus.*

The transparent *cornea* is largely responsible for the refraction of the light entering the eye (Fig. 11-52). It consists of the following layers, from before backward: (1) the corneal epithelium, which is continuous with that of the conjunctiva; (2) the substantia propria, composed of transparent connective tissue; (3) the posterior elastic lamina; and (4) the endothelium, which is in contact with the aqueous humor.

### Vascular Pigmented Coat

The vascular pigmented coat consists, from behind forward, of the choroid, the ciliary body, and the iris.

The *choroid* is composed of an outer pigmented layer and an inner, highly vascular layer.

The *ciliary body* is continuous posteriorly with the choroid, and anteriorly it lies behind the peripheral margin of the iris (Fig. 11-52). It is composed of (1) the ciliary ring, (2) the ciliary processes, and (3) the ciliary muscle.

The *ciliary ring* is the posterior part of the body, and its surface has shallow grooves, the *ciliary striae.*

The *ciliary processes* are radially arranged folds,

or ridges, to the posterior surfaces of which are connected the suspensory ligaments of the lens.

The *ciliary muscle* (Fig. 11-52) is composed of meridianal and circular fibers of smooth muscle. The meridianal fibers run backward from the region of the corneoscleral junction to the ciliary processes. The circular fibers are fewer in number and lie internal to the meridianal fibers.

## Nerve Supply of Ciliary Muscle

The parasympathetic fibers from the oculomotor nerve. After synapsing in the ciliary ganglion, the postganglionic fibers pass forward to the eyeball in the short ciliary nerves.

## Action

Contraction of the ciliary muscle, especially the meridianal fibers, pulls the ciliary body forward. This relieves the tension in the suspensory ligament, and the elastic lens becomes more convex. This increases the refractive power of the lens.

The *iris* is a thin, contractile, pigmented diaphragm with a central aperture, the *pupil* (Fig. 11-52). It is suspended in the aqueous humor between the cornea and the lens. The periphery of the iris is attached to the anterior surface of the ciliary body. It divides the space between the lens and the cornea into an *anterior* and a *posterior chamber*.

The muscle fibers of the iris are involuntary and consist of circular and radiating fibers. The circular fibers form the *sphincter pupillae* and are arranged around the margin of the pupil. The radial fibers form the *dilator pupillae* and consist of a thin sheet of radial fibers that lie close to the posterior surface.

## Nerve Supply

*The sphincter pupillae* is supplied by parasympathetic fibers from the oculomotor nerve. After synapsing in the ciliary ganglion, the postganglionic fibers pass forward to the eyeball in the short ciliary nerves.

The *dilator pupillae* is supplied by sympathetic fibers, which pass forward to the eyeball in the long ciliary nerves.

## Action

The sphincter pupillae constricts the pupil in the presence of bright light and during accommodation. The dilator pupillae dilates the pupil in the presence of light of low intensity or in the presence of excessive sympathetic activity such as occurs in fright.

## *Nervous Coat: The Retina*

The retina consists of an *outer pigmented layer* and an *inner nervous layer*. Its outer surface is in contact with the choroid, and its inner surface is in contact with the vitreous body (Fig. 11-52). The posterior three-quarters of the retina is the receptor organ. Its anterior edge forms a wavy ring, the *ora serrata*, and it is here that the nervous tissues end. The anterior part of the retina is nonreceptive and consists merely of pigment cells, with a deeper layer of columnar epithelium. This anterior part of the retina covers the ciliary processes and the back of the iris.

At the center of the posterior part of the retina is an oval yellowish area, the *macula lutea*, which is the area of the retina for the most distinct vision. It has a central depression, the *fovea centralis* (Fig. 11-52).

The optic nerve leaves the retina about 3 mm to the medial side of the macula lutea by the optic disc. The *optic disc* is slightly depressed at its center, where it is pierced by the *central artery of the retina*. At the optic disc there is a complete absence of *rods* and *cones*, so that it is insensitive to light and is referred to as the *"blind spot."* On ophthalmoscopic examination, the optic disc is seen to be pale pink in color, much paler than the surrounding retina.

## CONTENTS OF THE EYEBALL

The contents of the eyeball consist of the refractive media, the aqueous humor, the vitreous body, and the lens.

## *Aqueous Humor*

The aqueous humor is a clear fluid that fills the anterior and posterior chambers of the eyeball (Fig. 11-52). It is believed to be a secretion or transudate from the ciliary processes, from which it enters the posterior chamber. It then flows into the anterior chamber through the pupil and is drained away through the spaces at the iridocorneal angle into the *canal of Schlemm.* Obstruction to the draining of the aqueous humor results in a rise in intraocular pressure, called *glaucoma.* This may produce degenerative changes in the retina, with consequent blindness.

The function of the aqueous humor is to support the wall of the eyeball by exerting internal pressure. It also nourishes the lens and removes the products of metabolism; these functions are important because the lens does not possess a blood supply.

### Vitreous Body

The vitreous body fills the eyeball behind the lens (Fig. 11-52). It is a transparent gel enclosed by the *vitreous membrane.* The *hyaloid canal* is a narrow channel that runs through the vitreous body from the optic disc to the posterior surface of the lens; in the fetus, it is filled by the hyaloid artery, which disappears before birth.

In front, in the region of the margin of the lens, the vitreous membrane is thickened and consists of two layers. The posterior layer covers the vitreous body; the anterior layer consists of a series of delicate, radially arranged fibers. Collectively, the fibers form the *suspensory ligament of the lens;* they are attached laterally to the ciliary processes and centrally to the capsule of the lens in the region of the equator (Fig. 11-52).

No blood vessels are found in the vitreous body. The function of the vitreous body is to contribute slightly to the magnifying power of the eye. It supports the posterior surface of the lens and assists in holding the neural part of the retina against the pigmented part of the retina.

### Lens

The lens (Fig. 11-52) is a transparent, biconvex body enclosed in a transparent capsule. It is situated behind the iris and in front of the vitreous body and is encircled by the ciliary processes.

The lens consists of (1) an elastic *capsule,* which envelops the structure; (2) a *cuboidal epithelium,* which is confined to the anterior surface of the lens; and (3) *lens fibers,* which are formed from the cuboidal epithelium at the equator of the lens. The lens fibers make up the bulk of the lens.

The elastic lens capsule is under tension, causing the lens constantly to endeavor to assume a globular rather than a disc shape. The equatorial region, or circumference, of the lens is attached to the ciliary processes of the ciliary body by the suspensory ligament. The pull of the radiating fibers of the suspensory ligament tends to keep the elastic lens flattened, so that the eye may be focused on distant objects.

To accommodate the eye for close objects, the ciliary muscle contracts and pulls the ciliary body forward and inward, so that the radiating fibers of the suspensory ligament are relaxed. This allows the elastic lens to assume a more globular shape.

With advancing age the lens becomes denser and less elastic, and, as a result, the ability to accommodate is lessened (presbyopia). This disability may be overcome by the use of an additional lens in the form of glasses to assist the eye in focusing on nearby objects.

## THE EAR

The ear may be divided into the external ear, the middle ear, or tympanic cavity, and the internal ear, or labyrinth, the last containing the organs of hearing and of balance.

## External Ear

The external ear consists of the auricle and the external auditory meatus.

The *auricle* has a characteristic shape (Fig. 11-53A) and serves to collect air vibrations. It consists of a thin plate of elastic cartilage covered by skin. It possesses both extrinsic and intrinsic muscles, which are supplied by the facial nerve.

The *external auditory meatus* is a sinuous tube

**Fig. 11-53. (A) Different parts of auricle of external ear. Arrow indicates direction that auricle should be pulled to straighten external auditory meatus prior to insertion of otoscope in the adult. (B) External and middle portions of right ear, viewed from in front. (C) Right tympanic membrane as seen through otoscope.**

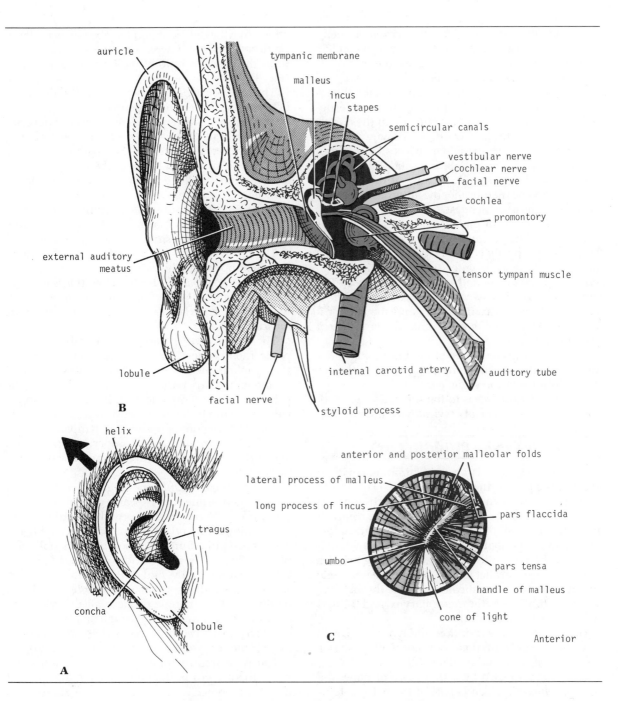

A
helix
tragus
concha
lobule

B
auricle
tympanic membrane
malleus
incus
stapes
semicircular canals
vestibular nerve
cochlear nerve
facial nerve
cochlea
promontory
external auditory meatus
tensor tympani muscle
lobule
facial nerve
styloid process
internal carotid artery
auditory tube

C
anterior and posterior malleolar folds
lateral process of malleus
long process of incus
pars flaccida
umbo
pars tensa
handle of malleus
cone of light
Anterior

that leads from the auricle to the tympanic membrane (Figs. 11-53 and 11-54). It serves to conduct sound waves from the auricle to the tympanic membrane. In the adult it measures about 1 inch (2.5 cm) long and may be straightened for the insertion of an otoscope by pulling the auricle upward and backward. In the young child the auricle is pulled straight backward, or downward and backward. The meatus is narrowest about 5 mm from the tympanic membrane, and this narrow area is called the *isthmus*. Because of the obliquity of the tympanic membrane, the anteroinferior wall of the meatus is the longest.

The framework of the outer third of the meatus is elastic cartilage, and the inner two-thirds is bone, formed by the tympanic plate. The meatus is lined by skin, and its outer third is provided with *hairs* and *sebaceous* and *ceruminous glands*. The latter are modified sweat glands, which secrete a yellowish-brown wax. The hairs and the wax provide a sticky barrier that prevents the entrance of foreign bodies.

The *sensory nerve* supply of the lining skin is derived from the auriculotemporal nerve and the auricular branch of the vagus nerve.

The *lymph drainage* is to the superficial parotid, mastoid, and superficial cervical lymph nodes.

## Tympanic Cavity (Middle Ear)

The tympanic cavity is an air-containing cavity in the petrous part of the temporal bone (Fig. 11-54) and is lined with mucous membrane. It contains the auditory ossicles, whose function is to transmit the vibrations of the tympanic membrane (eardrum) to the perilymph of the internal ear. It is a narrow, oblique, slitlike cavity, whose long axis lies approximately parallel to the plane of the tympanic membrane. It communicates in front through the auditory tube with the nasopharynx and behind with the mastoid antrum.

For purposes of description, the tympanic cavity is said to have a roof, floor, anterior wall, posterior wall, lateral wall, and medial wall.

The *roof* is formed by a thin plate of bone, the *tegmen tympani*, which is part of the petrous temporal bone (Figs. 11-55 and 11-56). It separates the tympanic cavity from the meninges and the temporal lobe of the brain in the middle cranial fossa.

The *floor* is formed by a thin plate of bone,

which may be deficient and may be partly replaced by fibrous tissue. It separates the tympanic cavity from the superior bulb of the internal jugular vein (Fig. 11-56).

The *anterior wall* is formed below by a thin plate of bone that separates the tympanic cavity from the internal carotid artery and its surrounding sympathetic plexus (Fig. 11-56). At the upper part of the anterior wall there are the openings into two canals. The lower and larger of these leads into the auditory tube, and the upper and smaller is the entrance into the canal for the tensor tympani muscle (Fig. 11-55). The thin, bony septum, which separates the canals, is prolonged backward on the medial wall, where it forms a shelflike projection.

The *posterior wall* has in its upper part a large, irregular opening, the *aditus to the mastoid antrum* (Figs. 11-55 and 11-56). Below this is a small, hollow, conical projection, the *pyramid*, from whose apex emerges the tendon of the *stapedius muscle*.

The *lateral wall* is largely formed by the tympanic membrane (Figs. 11-53 and 11-55). Above the tympanic membrane the lateral wall is formed by the squamous part of the temporal bone. The presence of the tympanic membrane serves to divide the tympanic cavity artificially into two regions. The region opposite the tympanic membrane is the *tympanic cavity* proper, and the region above the level of the tympanic membrane is the *epitympanic recess*.

The *tympanic membrane* (Fig. 11-53) is a thin, fibrous membrane that is pearly gray in color. It is covered on the outer surface with stratified squamous epithelium and on the inner surface with low columnar epithelium. The membrane is obliquely placed, facing downward, forward, and laterally. It is concave laterally, and at the depth of the concavity is a small depression, the *umbo*, produced by the tip of the handle of the malleus. When the membrane is illuminated through an otoscope, the concavity produces a "cone of light," which radiates anteriorly and inferiorly from the umbo.

The tympanic membrane is circular and measures about 1 cm in diameter. The greater part of its circumference is thickened, and it is slotted into a groove in the bone. The groove, or *tympanic sulcus*, is deficient superiorly, which forms a notch. From the sides of the notch, two bands, termed the *anterior* and *posterior malleolar folds*, pass to the

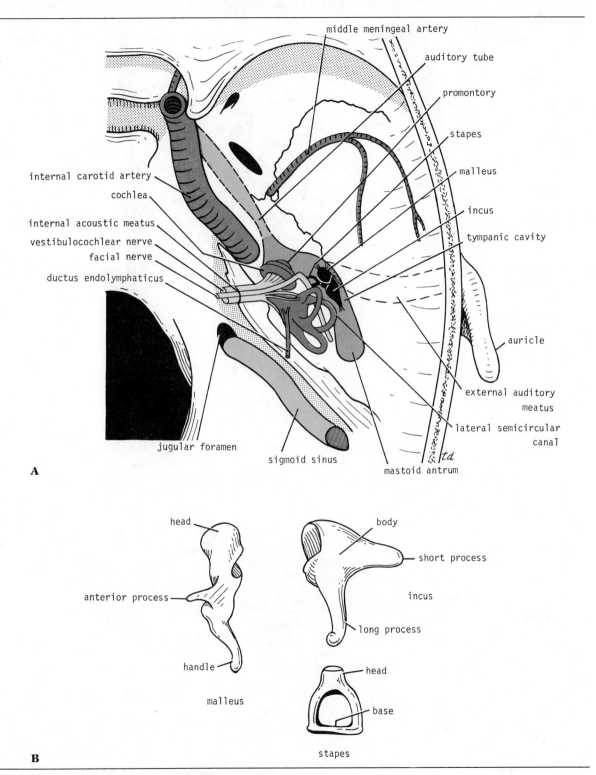

middle meningeal artery

auditory tube

promontory

stapes

malleus

incus

tympanic cavity

internal carotid artery

cochlea

internal acoustic meatus

vestibulocochlear nerve

facial nerve

ductus endolymphaticus

auricle

external auditory meatus

lateral semicircular canal

jugular foramen

sigmoid sinus

mastoid antrum

*td*

**A**

head

body

short process

anterior process

incus

long process

handle

malleus

head

base

stapes

**B**

**Fig. 11-54. (A) Parts of right ear in relation to temporal bone, as viewed from above. (B) The auditory ossicles.**

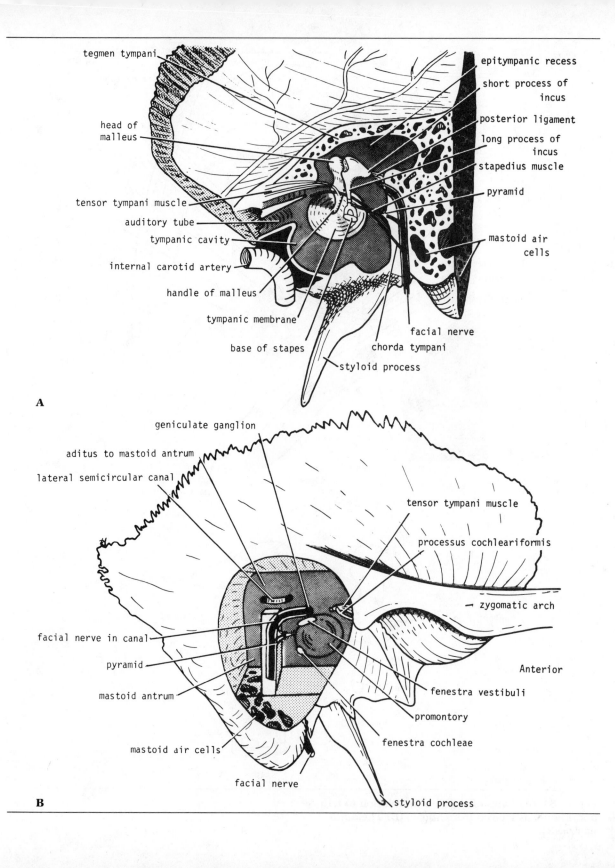

tegmen tympani

epitympanic recess

short process of
incus

head of
malleus

posterior ligament

long process of
incus

stapedius muscle

pyramid

tensor tympani muscle

auditory tube

tympanic cavity

internal carotid artery

mastoid air
cells

handle of malleus

tympanic membrane

base of stapes

facial nerve

chorda tympani

styloid process

**A**

geniculate ganglion

aditus to mastoid antrum

lateral semicircular canal

tensor tympani muscle

processus cochleariformis

zygomatic arch

facial nerve in canal

pyramid

Anterior

mastoid antrum

fenestra vestibuli

promontory

mastoid air cells

fenestra cochleae

facial nerve

styloid process

**B**

lateral process of the malleus. The small triangular area on the tympanic membrane that is bounded by the folds is slack and is called the *pars flaccida* (Fig. 11-53). The remainder of the membrane is tense and is called the *pars tensa*. The handle of the malleus is bound down to the inner surface of the tympanic membrane by the mucous membrane.

The tympanic membrane is extremely sensitive to pain and is innervated on its outer surface by the auriculotemporal nerve and the auricular branch of the vagus.

The *medial wall* is formed by the lateral wall of the inner ear. The greater part of the wall shows a rounded projection, called the *promontory*, which results from the underlying first turn of the cochlea (Figs. 11-53 and 11-55). Above and behind the promontory lies the *fenestra vestibuli*, which is oval in shape and closed by the footpiece, or base, of the stapes. On the medial side of the window is the perilymph of the scala vestibuli of the internal ear. Below the posterior end of the promontory lies the *fenestra cochleae*, which is round in shape and closed by the *secondary tympanic membrane*. On the medial side of this window is the perilymph of the blind end of the scala tympani. (See p. 837.)

The bony shelf derived from the anterior wall first separates the canal for the tensor tympani from the auditory tube, and then extends backward on the medial wall above the promontory and above the fenestra vestibuli. It supports the tensor tympani muscle. Its posterior end is curved upward and forms a pulley, the *processus cochleariformis*, around which the tendon of the tensor tympani bends laterally to reach its insertion on the handle of the malleus (Fig. 11-56).

A rounded ridge runs horizontally backward above the promontory and the fenestra vestibuli and is known as the *prominence of the facial nerve canal*. On reaching the posterior wall, it curves downward behind the pyramid; it forms a ridge on the medial wall of the aditus of the mastoid antrum.

**Fig. 11-55. (A) Lateral wall of right middle ear viewed from medial side. Note position of ossicles and mastoid antrum. (B) Medial wall of right middle ear viewed from lateral side. Note position of facial nerve in its bony canal.**

## Auditory Ossicles

The auditory ossicles are the malleus, incus, and stapes (Figs. 11-54 and 11-55).

The *malleus* is the largest ossicle and possesses a head, a neck, a long process or handle, an anterior process, and a lateral process.

The *head* is rounded in shape and lies within the epitympanic recess. It articulates posteriorly with the incus. The *neck* is the constricted part below the head. The *handle* passes downward and backward and is firmly attached to the medial surface of the tympanic membrane. It can be seen through the tympanic membrane on otoscopic examination. The *anterior process* is a spicule of bone that is connected to the anterior wall of the tympanic cavity by a ligament. The lateral process projects laterally and is attached to the anterior and posterior malleolar folds of the tympanic membrane.

The *incus* possesses a large body and two processes (Fig. 11-54).

The *body* is rounded and compressed laterally. It lies within the epitympanic recess and articulates anteriorly with the head of the malleus.

The *long process* descends behind and parallel to the handle of the malleus. Its lower end bends medially and articulates with the head of the stapes. Its shadow on the tympanic membrane can sometimes be recognized on otoscopic examination.

The *short process* projects backward and is attached to the posterior wall of the tympanic cavity by a ligament.

The *stapes* has a head, a neck, two limbs, and a base (Fig. 11-54).

The *head* is small and articulates with the long process of the incus. The *neck* is narrow and receives the insertion of the stapedius muscle. The *two limbs* diverge from the neck and are attached to the oval *base*. The edge of the base is attached to the margin of the fenestra vestibuli by a ring of fibrous tissue, the *anular ligament*.

## MUSCLES OF THE OSSICLES

### Tensor Tympani (Figs. 11-53 and 11-56)

#### Origin

From the cartilage of the auditory tube and the bony walls of its own canal.

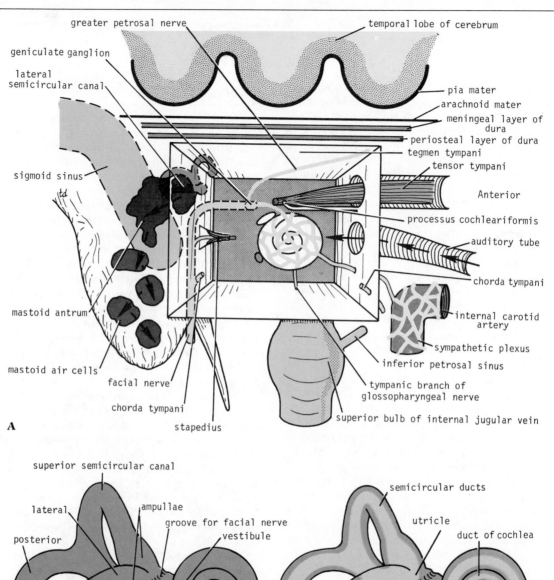

**A**

greater petrosal nerve

temporal lobe of cerebrum

geniculate ganglion

lateral
semicircular canal

pia mater

arachnoid mater

meningeal layer of
dura

periosteal layer of dura

tegmen tympani

tensor tympani

sigmoid sinus

Anterior

td

processus cochleariformis

auditory tube

chorda tympani

mastoid antrum

internal carotid
artery

sympathetic plexus

inferior petrosal sinus

mastoid air cells

facial nerve

chorda tympani

tympanic branch of
glossopharyngeal nerve

superior bulb of internal jugular vein

stapedius

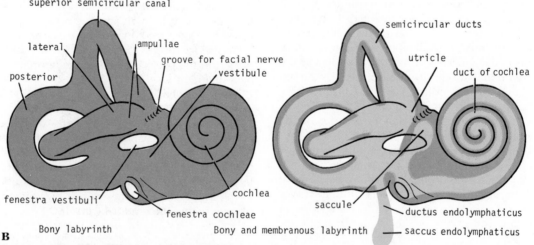

superior semicircular canal

semicircular ducts

lateral

ampullae

utricle

duct of cochlea

posterior

groove for facial nerve

vestibule

fenestra vestibuli

cochlea

saccule

ductus endolymphaticus

fenestra cochleae

saccus endolymphaticus

**B**  Bony labyrinth

Bony and membranous labyrinth

Fig. 11-56. (A) Diagrammatic representation of
middle ear and its relations. (B) Bony and membranous labyrinths.

## Insertion

The slender muscle passes backward and ends in a rounded tendon, which turns laterally around the processus cochleariformis and is inserted into the root of the handle of the malleus.

## Nerve Supply

A branch from the nerve to the medial pterygoid muscle, which is a branch of the mandibular division of the trigeminal nerve.

## Action

Reflexly damps down the vibrations of the malleus by making the tympanic membrane more tense.

## *Stapedius (Fig. 11-56)*

### Origin

From the internal walls of the hollow pyramid.

### Insertion

The tendon emerges from the apex of the pyramid and is inserted into the back of the neck of the stapes.

### Nerve Supply

From the facial nerve, which lies behind the pyramid.

### Action

Reflexly damps down the vibrations of the stapes by pulling on the neck of that bone.

## MOVEMENTS OF THE AUDITORY OSSICLES

The malleus and incus rotate on an anteroposterior axis that runs through (1) the ligament connecting the anterior process of the malleus to the anterior wall of the tympanic cavity; (2) the anterior process of the malleus and the short process of the incus; and (3) the ligament connecting the short process of the incus to the posterior wall of the tympanic cavity.

When the tympanic membrane moves medially, the handle of the malleus also moves medially. The head of the malleus and the head of the incus move laterally. The long process of the incus moves medially with the stapes. The base of the stapes is pushed medially in the fenestra vestibuli, and the motion is communicated to the perilymph in the scala vestibuli. Liquid being incompressible, the perilymph causes an outward bulging of the secondary tympanic membrane in the fenestra cochleae at the lower end of the scala tympani. The above movements are reversed if the tympanic membrane moves laterally. Excessive lateral movements of the head of the malleus cause a temporary separation of the articular surfaces between the malleus and incus, so that the base of the stapes is not pulled laterally out of the fenestra vestibuli.

It is interesting to note that during the passage of the vibrations from the tympanic membrane to the perilymph and via the small ossicles, the leverage increases at a rate of 1.3 to 1. Moreover, the area of the tympanic membrane is about 17 times greater than that of the base of the stapes, causing the effective pressure on the perilymph to increase by a total of 22 to 1.

## AUDITORY TUBE

The auditory tube extends from the anterior wall of the tympanic cavity downward, forward, and medially to the nasal pharynx (Fig. 11-53). Its posterior third is bony, and its anterior two-thirds is cartilaginous. It joins the nasal pharynx by passing over the upper border of the superior constrictor muscle (Fig. 11-58). It serves to equalize air pressures in the tympanic cavity and the nasal pharynx.

## MASTOID ANTRUM

The mastoid antrum lies behind the tympanic cavity in the petrous part of the temporal bone (Fig. 11-54). It communicates with the epitympanic recess by the aditus (Fig. 11-55). It is rounded in shape and may be as large as 1 cm in diameter.

The *anterior wall* is related to the tympanic cavity and contains the aditus to the mastoid antrum (Fig. 11-56).

The *posterior wall* separates the antrum from the sigmoid venous sinus and the cerebellum (Fig. 11-56).

The *lateral wall* is 1.5 cm thick and forms the floor of the suprameatal triangle. (See p. 701.)

The *medial wall* is related to the posterior semicircular canal (Fig. 11-56).

The *superior wall* is the thin plate of bone, the tegmen tympani, which is related to the meninges of the middle cranial fossa and the temporal lobe of the brain (Fig. 11-56).

The *inferior wall* is perforated with holes, through which the antrum communicates with the mastoid air cells (Fig. 11-56).

## MASTOID AIR CELLS

The mastoid process begins to develop during the second year of life. The mastoid air cells are a series of communicating cavities within the process, which are continuous above with the antrum and the tympanic cavity (Fig. 11-56). They are lined with mucous membrane. Their degree of development shows considerable individual variation, and the greatest growth occurs at puberty.

## FACIAL NERVE

On reaching the bottom of the internal acoustic meatus (see p. 810), the facial nerve enters the facial canal (Fig. 11-54). The nerve runs laterally above the vestibule of the internal ear, until it reaches the medial wall of the tympanic cavity. Here, the nerve expands to form the sensory *geniculate ganglion* (Figs. 11-55 and 11-56). The nerve then bends sharply backward above the promontory.

On arriving at the posterior wall of the tympanic cavity, it curves downward on the medial side of the aditus of the mastoid antrum (Fig. 11-56). It descends in the posterior wall of the tympanic cavity, behind the pyramid, and finally emerges through the stylomastoid foramen. Its further course in the neck is described on page 757.

### Important Branches of the Intrapetrous Part of the Facial Nerve

1. The *greater petrosal nerve* arises from the facial nerve at the geniculate ganglion (Fig. 11-56). It contains preganglionic parasympathetic fibers, which pass to the pterygopalatine ganglion and are there relayed through the zygomatic and lacrimal nerves to the lacrimal gland; other postganglionic fibers pass through the nasal and palatine nerves to the glands of the mucous membrane of the nose and palate. It also contains many taste fibers from the mucous membrane of the palate.

   The nerve emerges on the superior surface of the petrous part of the temporal bone and runs forward in a groove. It runs below the trigeminal ganglion and enters the foramen lacerum. (See p. 794.) It is here joined by the deep petrosal nerve from the sympathetic plexus on the internal carotid artery and forms the *nerve of the pterygoid canal*. This passes forward and enters the pterygopalatine fossa, where it ends in the pterygopalatine ganglion.

2. The *nerve to the stapedius* arises from the facial nerve as it descends in the facial canal behind the pyramid (Fig. 11-56). It supplies the muscle within the pyramid.

3. The *chorda tympani* arises from the facial nerve just above the stylomastoid foramen (Fig. 11-55). It enters the tympanic cavity close to the posterior border of the tympanic membrane. It then runs forward over the tympanic membrane and crosses the root of the handle of the malleus (Fig. 11-55). It lies in the interval between the mucous membrane and the fibrous layers of the tympanic membrane. The nerve leaves the tympanic cavity through the petrotympanic fissure and enters the infratemporal fossa, where it joins the lingual nerve. (See p. 768.)

The chorda tympani contains many taste fibers from the mucous membrane covering the anterior two-thirds of the tongue (not the vallate papillae) and the floor of the mouth. The taste fibers are the peripheral processes of the cells in the geniculate ganglion.

The nerve also contains preganglionic parasympathetic secretomotor fibers, which reach the submandibular ganglion and are there relayed to the submandibular and sublingual salivary glands.

## TYMPANIC NERVE

The tympanic nerve arises from the glossopharyngeal nerve, just below the jugular foramen. (See p.

781.) It passes through the floor of the tympanic cavity and grooves the surface of the promontory (Fig. 11-56). Here it splits up into branches, which form the *tympanic plexus.* The tympanic plexus supplies the lining of the tympanic cavity and gives off the lesser petrosal nerve.

The *lesser petrosal nerve* contains secretomotor fibers for the parotid gland. (See p. 760.) It passes through a small opening on the anterior surface of the petrous part of the temporal bone and leaves the skull through the foramen ovale. The nerve then joins the otic ganglion.

## The Internal Ear, or Labyrinth

The labyrinth is situated in the petrous part of the temporal bone, medial to the middle ear (Fig. 11-54). It consists of (1) the bony labyrinth, comprising a series of cavities within the bone, and (2) the membranous labyrinth, comprising a series of membranous sacs and ducts contained within the bony labyrinth. For a detailed description of the microscopic structure of the labyrinth, a textbook of histology should be consulted.

The *bony labyrinth* consists of three parts: the vestibule, the semicircular canals, and the cochlea (Fig. 11-56). These are cavities situated in the substance of dense bone. They are lined by endosteum and contain a clear fluid, the *perilymph,* in which is suspended the membranous labyrinth.

The *vestibule,* the central part of the bony labyrinth, lies posterior to the cochlea and anterior to the semicircular canals. In its lateral wall are the *fenestra vestibuli,* which is closed by the base of the stapes and its anular ligament, and the *fenestra cochleae,* which is closed by the *secondary tympanic membrane.* Lodged within the vestibule are the *saccule* and *utricle* of the membranous labyrinth (Fig. 11-56).

The *semicircular canals* open into the posterior part of the vestibule; there are three: *superior, posterior,* and *lateral.* Each canal has a swelling at one end called the *ampulla.* The canals open into the vestibule by five orifices, one of which is common to two of the canals. Lodged within the canals are the *semicircular ducts* (Fig. 11-56).

The superior semicircular canal is vertical and placed at right angles to the long axis of the petrous bone. The posterior canal is also vertical, but it is placed parallel with the long axis of the petrous

bone. The lateral canal is set in a horizontal position, and it lies in the medial wall of the aditus to the mastoid antrum, above the facial nerve canal.

The *cochlea* opens into the anterior part of the vestibule (Fig. 11-56). Basically, it consists of a central pillar, the *modiolus,* around which a hollow bony tube makes two and one-half spiral turns. Each successive turn is of decreasing radius, so that the whole structure is cone-shaped. The apex faces anterolaterally and the base posteromedially. The first basal turn of the cochlea is responsible for the promontory seen on the medial wall of the tympanic cavity.

The modiolus has a broad base, which is situated at the bottom of the internal acoustic meatus. It is perforated by branches of the cochlear nerve. A spiral ledge, the *spiral lamina,* winds around the modiolus and projects into the interior of the canal and partially divides it. The *basilar membrane* stretches from the free edge of the spiral lamina to the outer bony wall, thus dividing the cochlear canal into the *scala vestibuli* above and the *scala tympani* below. The perilymph within the scala vestibuli is separated from the tympanic cavity by the base of the stapes and the anular ligament, at the fenestra vestibuli. The perilymph in the scala tympani is separated from the tympanic cavity by the secondary tympanic membrane at the fenestra cochleae.

The *membranous labyrinth* is lodged within the bony labyrinth (Fig. 11-56). It is filled with endolymph and surrounded by perilymph. It consists of the utricle and saccule, which are lodged in the bony vestibule; the three semicircular ducts, which lie within the bony semicircular canals; and the duct of the cochlea, which lies within the bony cochlea. All these structures freely communicate with one another.

The *utricle* is the larger of the two vestibular sacs. It is indirectly connected to the saccule and the *ductus endolymphaticus* by the *ductus utriculosaccularis.*

The *saccule* is globular in shape and connected to the utricle, as described previously. The ductus endolymphaticus, after being joined by the ductus utriculosaccularis, passes on to end in a small blind pouch, the *saccus endolymphaticus* (Fig. 11-56). This lies beneath the dura on the posterior surface of the petrous part of the temporal bone.

Located on the walls of the utricle and saccule

are specialized sensory receptors, which are sensitive to the orientation of the head to gravity or other acceleration forces.

The *semicircular ducts*, although much smaller in diameter than the semicircular canals, have the same configuration. They are arranged at right angles to each other, so that all three planes are represented. Whenever the head begins or ceases to move, or whenever a movement of the head accelerates or decelerates, the endolymph in the semicircular ducts will change its speed of movement relative to that of the walls of the semicircular ducts. This change is detected in the sensory receptors in the ampullae of the semicircular ducts.

The *duct of the cochlea* is triangular in cross section and is connected to the saccule by the *ductus reuniens.* The highly specialized epithelium that lies on the *basilar membrane* forms the spiral organ of Corti, and contains the sensory receptors for hearing. For a detailed description of the spiral organ, a textbook of histology should be consulted.

## VESTIBULOCOCHLEAR NERVE

On reaching the bottom of the internal acoustic meatus (see p. 810), the nerve divides into vestibular and cochlear portions (Fig. 11-53).

The *vestibular nerve* is expanded to form the *vestibular ganglion.* The branches of the nerve then pierce the lateral end of the internal acoustic meatus and gain entrance to the membranous labyrinth, where they supply the utricle, and saccule, and the ampullae of the semicircular ducts.

The *cochlear nerve* divides into branches, which enter foramina at the base of the modiolus. The sensory ganglion of this nerve takes the form of an elongated *spiral ganglion* that is lodged in a canal winding around the modiolus, in the base of the spiral lamina. The peripheral branches of this nerve pass from the ganglion to the spiral organ of Corti.

## MAXILLARY NERVE

The maxillary nerve arises from the trigeminal ganglion in the middle cranial fossa. (See p. 809.) It passes forward along the lower part of the lateral wall of the cavernous sinus and leaves the skull through the foramen rotundum, to enter the pterygopalatine fossa (Fig. 11-49). The nerve crosses the upper part of the fossa and enters the orbit by passing through the inferior orbital fissure (Fig. 11-50).

The nerve is now called the *infraorbital nerve,* and it runs forward on the floor of the orbit, first in the infraorbital groove and then in the infraorbital canal. It appears on the face by emerging through the infraorbital foramen. (See p. 751.)

### Branches

1. A *meningeal branch* in the middle cranial fossa.
2. The *ganglionic branches* are two short nerves that hold up the pterygopalatine ganglion in the pterygopalatine fossa (Fig. 11-50). They contain sensory fibers that without interruption have passed through the ganglion from the nose, palate, and pharynx. They also contain postganglionic parasympathetic fibers that are going to the lacrimal gland.
3. The *posterior superior alveolar nerve* arises in the pterygopalatine fossa. It passes downward on the back of the maxilla and pierces its posterior surface (Fig. 11-50). It supplies the maxillary sinus, the upper molar teeth, and the adjoining parts of the gum and cheek.
4. The *zygomatic nerve* arises in the pterygopalatine fossa and enters the orbit through the inferior orbital fissure (Fig. 11-50). It ascends on the lateral wall of the orbit and divides into the zygomaticotemporal and zygomaticofacial nerves (see p. 751), which are distributed to the skin of the face.
5. The *middle superior alveolar nerve* arises from the infraorbital nerve as it lies in the infraorbital groove (Fig. 11-50). It descends in the lateral wall of the maxillary sinus and supplies the upper premolar teeth and the adjoining parts of the gum and cheek.
6. The *anterior superior alveolar nerve* arises from the infraorbital nerve as it lies in the infraorbital canal (Fig. 11-50). It descends in the anterior wall of the maxillary sinus to supply the upper canine and incisor teeth. A small terminal branch supplies part of the lateral wall and floor of the nose.

## PTERYGOPALATINE GANGLION

The pterygopalatine ganglion is a parasympathetic ganglion that is deeply placed in the pterygopalatine fossa (Fig. 11-50).

The preganglionic secretomotor fibers arise in the *lacrimal nucleus* of the facial nerve. They run in the sensory root of the facial nerve, then in its greater petrosal branch, and then in the nerve of the pterygoid canal, which enters the posterior surface of the ganglion.

The postganglionic fibers reach the maxillary nerve by one of its ganglionic branches. They then run in the zygomatic nerve, the zygomaticotemporal nerve, and the lacrimal nerve, to reach the lacrimal gland. Other postganglionic fibers run in the palatine nerves and nasal nerves to the palatine and nasal glands.

Sympathetic postganglionic fibers reach the ganglion via the internal carotid plexus, the deep petrosal nerve, and the nerve of the pterygoid canal. They pass without interruption through the ganglion and emerge in the orbital branches of the ganglion. They supply the orbitalis muscle.

### Branches

These are composed mainly of sensory fibers derived from the maxillary nerve. They reach the ganglion by way of the ganglionic branches of the nerve.

1. *Orbital branches* enter the orbit through the inferior orbital fissure.
2. The *greater* and *lesser palatine nerves* supply the mucous membrane of the palate, tonsil, and nasal cavity (Fig. 11-50).
3. The *nasal branches* enter the nose through the sphenopalatine foramen and supply the mucous membrane of the nasal cavity.
4. The *pharyngeal branch* supplies the mucous membrane of the roof of the nasal part of the pharynx.

### MAXILLARY ARTERY

The maxillary artery leaves the infratemporal fossa (Fig. 11-27) by passing through the pterygomaxillary fissure into the pterygopalatine fossa. (See p.

768.) Here, it splits up into branches, which accompany the branches of the maxillary nerve.

### Branches

1. The *posterior superior alveolar artery* descends on the posterior surface of the maxilla and supplies the maxillary sinus and the molar and premolar teeth.
2. The *infraorbital artery* enters the orbital cavity through the inferior orbital fissure. It ends by emerging on the face with the infraorbital nerve.
3. The *greater palatine artery* descends through the greater palatine canal with the greater palatine nerve. It is distributed to the mucous membrane covering the oral surface of the hard palate.
4. The *pharyngeal branch* passes backward to supply the mucous membrane of the root of the nasopharynx.
5. The *sphenopalatine* artery passes through the sphenopalatine foramen into the nasal cavity. It supplies the mucous membrane of the nasal cavity. One long septal branch descends to the incisive canal and supplies the mucous membrane on the undersurface of the hard palate, behind the incisor teeth.

## THE MOUTH

The mouth extends from the lips to the oropharyngeal isthmus, i.e., the junction of the mouth with the pharynx. It is subdivided into the *vestibule*, which lies between the lips and cheeks externally and the gums and teeth internally, and the *mouth cavity proper*, which lies within the alveolar arches, gums, and teeth (Fig. 11-33).

The *vestibule* is a slitlike space that communicates with the exterior through the *oral fissure*. When the jaws are closed, it communicates with the mouth proper behind the third molar tooth on each side. Superiorly and inferiorly, the vestibule is limited by the reflection of the mucous membrane from the lips and cheeks onto the gums. The *cheek* forms the lateral wall of the vestibule. It is made up of the buccinator muscle (see p. 757) and is lined by mucous membrane. The buccinator is covered on the outside by the buccopharyngeal fascia. Opposite the upper second molar tooth a small papilla is present on the mucous membrane, marking the opening of the duct of the parotid salivary gland (Fig. 11-57).

The *mouth proper* has a roof, which is formed by the hard palate in front and the soft palate behind. The floor is formed largely by the anterior two-thirds of the tongue and by the reflection of the mucous membrane from the sides of the tongue to the gum on the mandible. In the midline, a fold of mucous membrane called the *frenulum of the tongue* connects the undersurface of the tongue to the floor of the mouth (Fig. 11-57). On each side of the frenulum there is a small papilla, on the summit of which is the *orifice of the duct of the submandibular gland*. From the papilla, a rounded ridge of mucous membrane extends backward and laterally. It is produced by the underlying *sublingual gland* and is called the *sublingual fold* (Fig. 11-57).

## SENSORY NERVE SUPPLY OF THE MUCOUS MEMBRANE OF THE MOUTH

The *roof* is supplied by the greater palatine and nasopalatine nerves. The nerve fibers travel in the maxillary nerve.

The *floor* is supplied by the lingual nerve, a branch of the mandibular nerve. The taste fibers travel in the chorda tympani nerve, a branch of the facial nerve.

The *cheek* is supplied by the buccal nerve, a branch of the mandibular nerve.

## Teeth

There are two sets of teeth, which make their appearance at different times of life. The first set, called the *deciduous teeth*, is temporary. The second set is called the *permanent teeth*.

The *deciduous teeth* are twenty in number; four incisors, two canines, and four molars in each jaw. They begin to erupt at about the sixth month after birth and have all erupted by the end of the second year. The approximate times of eruption are as follows:

| | |
|---|---|
| Central incisors | 6–8 months |
| Lateral incisors | 8–10 months |
| First molars | 1 year |
| Canines | 18 months |
| Second molars | 2 years |

The teeth of the lower jaw usually appear before those of the upper jaw.

The *permanent teeth* are thirty-two in number, including four incisors, two canines, four premolars, and six molars in each jaw (Fig. 11-32). They begin to erupt at the sixth year. However, the last tooth to erupt is the third molar, and this may take place between the seventeenth and thirtieth years. The approximate times of eruption are as follows:

| | |
|---|---|
| First molars | 6 years |
| Central incisors | 7 years |
| Lateral incisors | 8 years |
| First premolars | 9 years |
| Second premolars | 10 years |
| Canines | 11 years |
| Second molars | 12 years |
| Third molars (wisdom teeth) | 17–30 years |

The teeth of the lower jaw usually appear before those of the upper jaw.

## Tongue

The tongue is a mass of striated muscle covered with mucous membrane (Figs. 11-28 and 11-33). Its anterior two-thirds lies in the mouth, and its posterior third lies in the pharynx (Fig. 11-44). The muscles attach the tongue to the styloid process and the soft palate above, and to the mandible and the hyoid bone below. The tongue is divided into right and left halves by a median *fibrous septum*.

## MUSCLES OF THE TONGUE

The muscles of the tongue are divided into two types: (1) intrinsic and (2) extrinsic.

The *intrinsic muscles* are confined to the tongue and are not attached to bone. They consist of longitudinal, transverse, and vertical fibers.

### Nerve Supply

Hypoglossal nerve.

**Fig. 11-57. (A) Cavity of mouth. Cheek on left side of face has been cut away to show buccinator muscle and parotid duct. (B) Undersurface of tongue.**

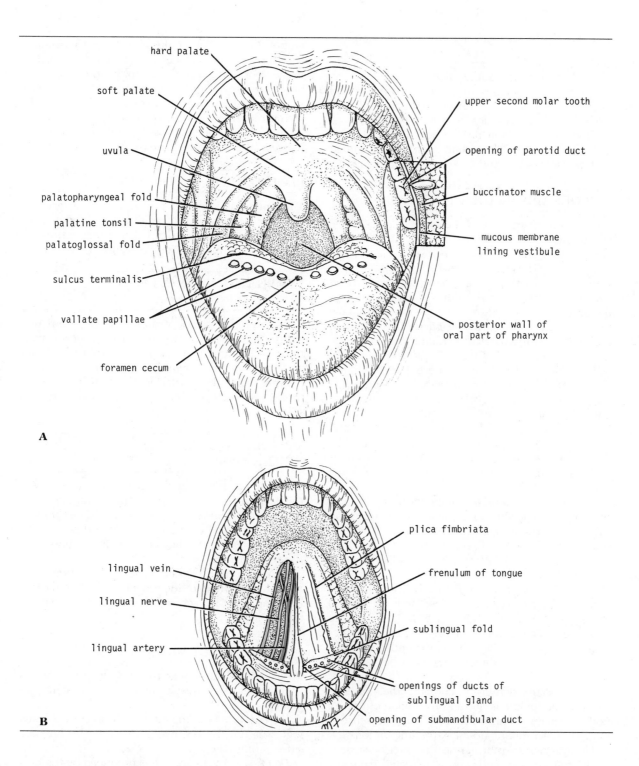

hard palate

soft palate

uvula

palatopharyngeal fold

palatine tonsil

palatoglossal fold

sulcus terminalis

vallate papillae

foramen cecum

upper second molar tooth

opening of parotid duct

buccinator muscle

mucous membrane lining vestibule

posterior wall of oral part of pharynx

**A**

plica fimbriata

lingual vein

frenulum of tongue

lingual nerve

sublingual fold

lingual artery

openings of ducts of sublingual gland

opening of submandibular duct

**B**

## Action

They alter the shape of the tongue.

The *extrinsic muscles* are attached to bones and the soft palate. They are the genioglossus, the hyoglossus, and the styloglossus, which have been described in the section on the submandibular region. (See p. 777.) The palatoglossus is associated with the soft palate and is described on page 851.

## MUCOUS MEMBRANE OF THE TONGUE

The mucous membrane of the upper surface of the tongue may be divided into anterior and posterior parts by a V-shaped sulcus, the *sulcus terminalis* (Fig. 11-57). The apex of the sulcus projects backward and is marked by a small pit, the *foramen cecum*. The sulcus serves to divide the tongue into the anterior two-thirds, or oral part, and the posterior third, or pharyngeal part. The foramen cecum is an embryological remnant and marks the site of the upper end of the thyroglossal duct.

Three types of papillae are present on the upper surface of the anterior two-thirds of the tongue: (1) the filiform papillae, (2) the fungiform papillae, and (3) the vallate papillae.

The *filiform papillae* are very numerous and cover the anterior two-thirds of the tongue on its upper surface. They form small conical projections and are whitish in color due to the thickness of the cornified epithelium.

The *fungiform papillae*, which are much less numerous than the filiform papillae, are scattered on the sides and the apex of the tongue. They are mushroom-shaped and possess a vascular connective tissue core, which imparts a reddish tinge to the papillae.

The *vallate papillae* are ten to twelve in number and are situated in a row immediately in front of the sulcus terminalis (Fig. 11-57). Each papilla measures about 2 mm in diameter and protrudes slightly from the surface. It is surrounded by a circular furrow, in the walls of which lie the taste buds.

The mucous membrane covering the posterior third of the tongue is devoid of papillae but has a nodular irregular surface caused by the presence of underlying lymphatic nodules, the *lingual tonsil.*

The mucous membrane on the inferior surface of the tongue is smooth and is reflected from the tongue to the floor of the mouth. In the midline anteriorly, the underface of the tongue is connected to the floor of the mouth by a fold of mucous membrane, the *frenulum of the tongue.* On the lateral side of the frenulum, the deep lingual vein can be seen through the mucous membrane. Lateral to the lingual vein, the mucous membrane forms a fringed fold called the *plica fimbriata* (Fig. 11-57).

## BLOOD SUPPLY OF THE TONGUE

The tongue is supplied by the lingual artery, the tonsillar branch of the facial artery, and the ascending pharyngeal artery. The veins ultimately drain into the internal jugular vein.

## LYMPHATIC DRAINAGE OF THE TONGUE

The tip of the tongue drains into the submental lymph nodes. The remainder of the anterior two-thirds of the tongue drains into the submandibular and deep cervical lymph nodes on both sides. Lymph from the posterior third of the tongue drains into the deep cervical lymph nodes on both sides.

## SENSORY INNERVATION OF THE TONGUE

The mucous membrane covering the anterior two-thirds of the tongue is supplied by the lingual nerve for ordinary sensations. Taste fibers from the anterior two-thirds of the tongue, excluding the vallate papillae, run in the chorda tympani branch of the facial nerve.

General sensation and taste appreciation from the posterior third of the tongue, including the vallate papillae, are served by the glossopharyngeal nerve.

## MOVEMENTS OF THE TONGUE

*Protrusion* of the tongue may be brought about by the genioglossus muscles on both sides acting together.

*Retraction* of the tongue is produced by the styloglossus and by the hyoglossus muscles on both sides acting together.

*Depression* of the tongue is produced by the hyoglossus and the genioglossus muscles on both sides acting together.

*Retraction* and *elevation* of the posterior third of the tongue is produced by the styloglossus and palatoglossus muscles on both sides acting together. The shape of the tongue is modified by the action of its intrinsic muscles.

## THE PHARYNX

The pharynx is situated behind the nasal cavities, the mouth, and the larynx (Fig. 11-44). It is somewhat funnel-shaped, its upper, wide end lying under the skull, and its lower, narrow end becoming continuous with the esophagus opposite the sixth cervical vertebra. The pharynx has a musculomembranous wall, which is deficient anteriorly. Here, it is replaced by the posterior nasal apertures, the oropharyngeal isthmus (opening into the mouth), and the inlet of the larynx.

The wall of the pharynx has three layers: (1) mucous, (2) fibrous, and (3) muscular.

The *mucous membrane* is continuous with that of the nasal cavities, the mouth, and the larynx (Fig. 11-44). By means of the auditory tubes, it is also continuous with that of the tympanic cavity. The upper part is lined by ciliated columnar epithelium, the lower part by stratified squamous epithelium; where the two areas come together, there is a transitional zone.

The *fibrous layer* lies between the mucous membrane and the muscle layer. It is thicker above, where it is strongly connected to the base of the skull (Fig. 11-59). Below, it becomes continuous with the submucous coat of the esophagus.

*The muscular layer* consists of the superior, middle, and inferior constrictor muscles, whose fibers run in a more or less circular direction (Fig. 11-58), and the stylopharyngeus and salpingopharyngeus muscles, whose fibers run in a more or less longitudinal direction.

### SUPERIOR CONSTRICTOR MUSCLE (FIGS. 11-58 and 11-59)

#### Origin

From the lower part of the posterior border of the medial pterygoid plate, the pterygoid hamulus, the pterygomandibular ligament, the posterior end of the mylohyoid line on the mandible, and the side of the tongue.

#### Insertion

The upper fibers curve medially and upward and are attached to the pharyngeal tubercle of the occipital bone. The middle fibers are inserted into a median fibrous raphe on the posterior wall. The lower fibers curve medially and downward and join the fibrous raphe; they are overlapped by the middle constrictor muscle.

#### Nerve Supply

Pharyngeal plexus.

#### Action

See under inferior constrictor muscle.

### MIDDLE CONSTRICTOR MUSCLE (FIGS. 11-58 and 11-59)

#### Origin

From the lower part of the stylohyoid ligament and from the lesser and greater cornua of the hyoid bone.

#### Insertion

The fibers radiate medially and are inserted into a median fibrous raphe on the posterior wall of the pharynx. The superior fibers overlap the lateral surface of the superior constrictor muscle, and the inferior fibers are overlapped laterally by the inferior constrictor muscle.

#### Nerve Supply

Pharyngeal plexus.

#### Action

See under inferior constrictor muscle.

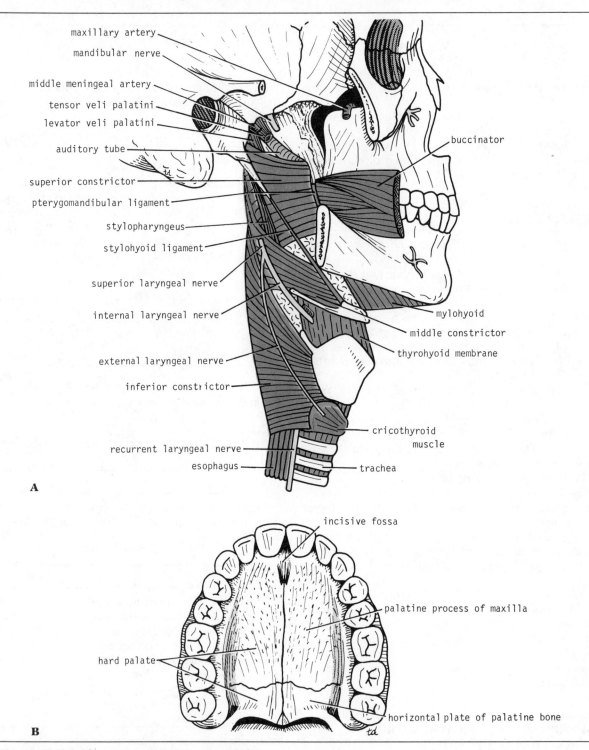

maxillary artery
mandibular nerve
middle meningeal artery
tensor veli palatini
levator veli palatini
auditory tube
superior constrictor
pterygomandibular ligament
stylopharyngeus
stylohyoid ligament
superior laryngeal nerve
internal laryngeal nerve
external laryngeal nerve
inferior constrictor
recurrent laryngeal nerve
esophagus

buccinator
mylohyoid
middle constrictor
thyrohyoid membrane
cricothyroid muscle
trachea

A

incisive fossa
palatine process of maxilla
hard palate
horizontal plate of palatine bone

B

**Fig. 11-58. (A) Three constrictor muscles of pharynx. Superior and recurrent laryngeal nerves are also shown. (B) Hard palate.**

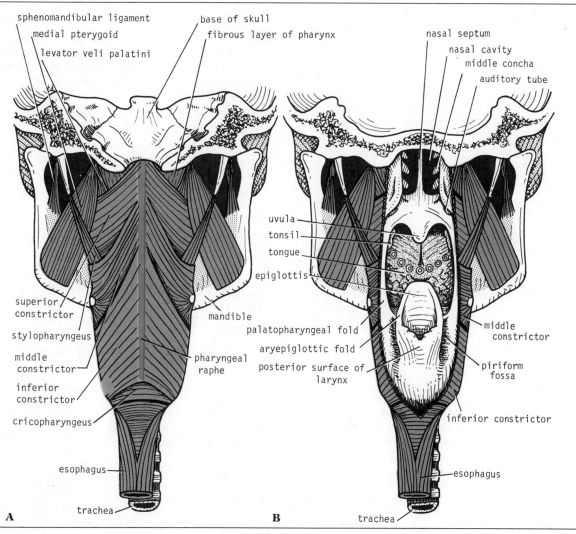

**Fig. 11-59. Pharynx seen from behind. (A) Note three constrictor muscles and position of stylopharyngeus muscles. (B) Greater part of posterior wall of pharynx has been removed to display nasal, oral, and laryngeal parts of pharynx.**

## INFERIOR CONSTRICTOR MUSCLE (FIGS. 11-58 and 11-59)

### Origin

From the oblique line on the lamina of the thyroid cartilage, from a tendinous band connecting the thyroid cartilage to the cricoid cartilage, and from the side of the cricoid cartilage.

### Insertion

The superior fibers curve upward and medially and overlap the lateral surface of the middle constrictor muscle. The lower fibers run horizontally medially and are continuous with the circular fibers of the esophagus below. All the fibers are inserted into a median fibrous raphe on the posterior wall of the pharynx.

### Nerve Supply

Pharyngeal plexus.

### Action

During the process of swallowing, the upper fibers of the superior constrictor muscle contract and pull the posterior pharyngeal wall forward, thus aiding the soft palate in closing off the upper part of the pharynx (nasal part of pharynx).

The successive contraction of the superior, middle, and inferior constrictor muscles propels the bolus of food down into the esophagus.

The lowest fibers of the inferior constrictor muscle (Fig. 11-59), sometimes referred to as the *cricopharyngeus muscle*, are believed to exert a sphincteric effect on the lower end of the pharynx, preventing the entry of air into the esophagus between the acts of swallowing.

## STYLOPHARYNGEUS MUSCLE (FIGS. 11-58 and 11-59)

### Origin

From the medial side of the base of the styloid process of the temporal bone.

### Insertion

The cylindrical muscle descends across the internal carotid artery on the lateral surface of the superior constrictor muscle. It enters the pharyngeal wall by passing between the superior and middle constrictor muscles and is inserted with the palatopharyngeus into the posterior border of the thyroid cartilage.

### Nerve Supply

Glossopharyngeal nerve.

### Action

Elevates the larynx and pharynx during the act of swallowing.

## SALPINGOPHARYNGEUS MUSCLE (FIG. 11-60)

### Origin

From the lower part of the cartilage of the auditory tube.

### Insertion

The fibers pass downward and blend with the palatopharyngeus muscle.

### Nerve Supply

Pharyngeal plexus.

### Action

Assists in elevating the pharynx.

## PALATOPHARYNGEUS MUSCLE (FIG. 11-60)

The palatopharyngeus muscle is described with the soft palate. (See p. 851.)

## Interior of the Pharynx

For purposes of description the pharynx is divided into three parts: nasal, oral, and laryngeal.

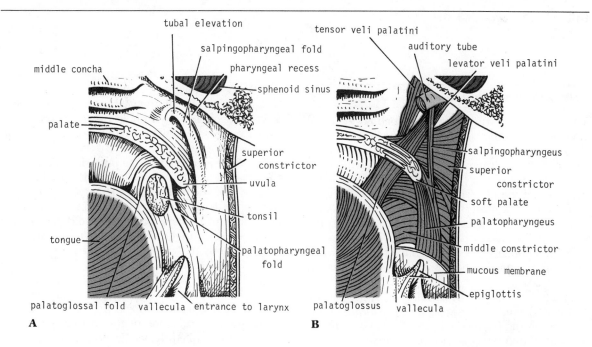

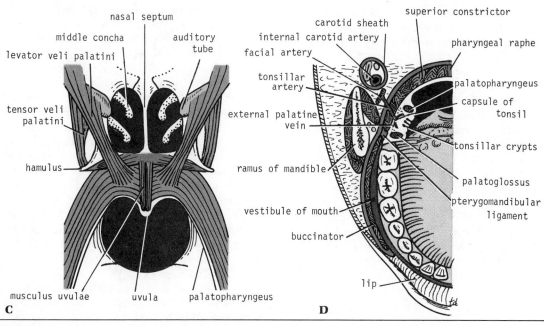

Fig. 11-60. (A) Junction of nose with nasal part of
pharynx and mouth with oral part of pharynx.
Note position of tonsil and opening of auditory
tube. (B) Muscles of soft palate and upper part of
pharynx. (C) Muscles of soft palate as seen from behind. (D) Horizontal section through mouth and
oral part of pharynx, showing relations of tonsil.

## *Nasal Part of the Pharynx*

The nasal part of the pharynx lies behind the nasal cavities, above the soft palate (Fig. 11-44). When the soft palate is raised and the posterior wall of the pharynx is drawn forward, as in swallowing, the nasal part of the pharynx is shut off from the oral part of the pharynx. It has a roof, a floor, an anterior wall, a posterior wall, and lateral walls.

The *roof* is supported by the body of the sphenoid and the basilar part of the occipital bone. A collection of lymphoid tissue, called the *pharyngeal tonsil*, is present in the submucosa of this region (Fig. 11-62).

The *floor* is formed by the sloping upper surface of the soft palate. The *pharyngeal isthmus* is the opening in the floor between the free edges of the soft palate and the posterior pharyngeal wall. During the act of swallowing, this communication between the nasal and oral parts of the pharynx is closed by the elevation of the soft palate and the pulling forward of the posterior wall of the pharynx.

The *anterior wall* is formed by the posterior nasal apertures, separated by the posterior edge of the nasal septum (Fig. 11-59).

The *posterior wall* forms a continuous sloping surface with the roof. It is supported by the anterior arch of the atlas (Fig. 11-62).

The *lateral wall*, on each side, has the pharyngeal opening of the auditory tube. Since the tube is directed forward, medially, and downward, the posterior margin of the tube forms an elevation, called the *tubal elevation* (Fig. 11-60). The salpingopharyngeus muscle, which is attached to the lower margin of the tube, produces a vertical fold of mucous membrane called the *salpingopharyngeal fold*. The *pharyngeal recess* is a small depression in the lateral wall behind the tubal elevation (Fig. 11-60). A collection of lymphoid tissue in the submucosa behind the opening of the auditory tube is called the *tubal tonsil*.

## *Oral Part of the Pharynx*

The oral part of the pharynx lies behind the mouth cavity and extends from the soft palate to the upper border of the epiglottis. It has a roof, a floor, an anterior wall, a posterior wall, and lateral walls (Fig. 11-44).

The *roof* is formed by the undersurface of the soft palate and the pharyngeal isthmus (see above). Small collections of lymphoid tissue are present in the submucosa on the undersurface of the soft palate.

The *floor* is formed by the posterior one-third of the tongue (which is almost vertical) and the interval between the tongue and the anterior surface of the epiglottis. The mucous membrane covering the posterior third of the tongue is devoid of papillae, but is irregular in appearance, due to the presence of the underlying lymphoid tissue, the *lingual tonsil* (Fig. 11-62). The mucous membrane is reflected from the tongue onto the epiglottis. In the midline is an elevation, called the *median glossoepiglottic fold*, and two *lateral glossoepiglottic folds*. The depression on each side of the median glossoepiglottic fold is called the *vallecula* (Fig. 11-60).

The *anterior wall* opens into the mouth through the oropharyngeal isthmus. Below this opening is the pharyngeal part of the tongue (Fig. 11-59).

The *posterior wall* is supported by the body of the second cervical vertebra and the upper part of the body of the third cervical vertebra (Fig. 11-44).

The *lateral walls* on each side have the palatoglossal and the palatopharyngeal arches or folds and the palatine tonsils between them (Fig. 11-60).

The *palatoglossal arch* is a fold of mucous membrane covering the underlying palatoglossus muscle. (See p. 851.) The interval between the two palatoglossal arches marks the boundary between the mouth and the oral pharynx and is called the *oropharyngeal isthmus* (Fig. 11-60).

The *palatopharyngeal arch* is a fold of mucous membrane on the lateral wall of the oral part of the pharynx behind the palatoglossal arch (Fig. 11-60). It covers the underlying palatopharyngeus muscle. (See p. 851.)

The *tonsillar sinus* is a triangular recess on the lateral wall of the oral pharynx between the palatoglossal arch in front and the palatopharyngeal arch behind. It is occupied by the palatine tonsil.

## PALATINE TONSILS

The palatine tonsils are two masses of lymphoid tissue (Fig. 11-60) located in the lateral walls of the oral part of the pharynx in the tonsillar sinuses. Each tonsil is covered by mucous membrane, and its free medial surface projects into the cavity of the pharynx. The surface is pitted by numerous small openings, which lead into the *tonsillar crypts.* The upper part of the medial surface of the tonsil has a deep *intratonsillar cleft.* The tonsil is covered on its lateral surface by a layer of fibrous tissue, called the *capsule* (Fig. 11-60).

The tonsil reaches its maximum size during early childhood, but after puberty it diminishes considerably in size.

## Relations of the Palatine Tonsil

### Anteriorly

The palatoglossal arch, beneath which it may extend for a short distance.

### Posteriorly

The palatopharyngeal arch.

### Superiorly

The soft palate. Here, the tonsil becomes continuous with the lymphoid tissue on the undersurface of the soft palate.

### Inferiorly

The posterior third of the tongue. Here, the palatine tonsil becomes continuous with the lingual tonsil.

### Medially

The cavity of the oral part of the pharynx.

### Laterally

The capsule is separated from the superior constrictor muscle by loose areolar tissue (Fig. 11-60). The external palatine vein descends from the soft palate in this loose connective tissue to join the pharyngeal venous plexus. Lateral to the superior constrictor muscle lie the styloglossus muscle and the loop of the facial artery. The internal carotid artery lies 1 inch (2.5 cm) behind and lateral to the tonsil.

The *arterial supply* to the tonsil is the tonsillar artery, a branch of the facial artery.

The *veins* pierce the superior constrictor muscle and join the external palatine, the pharyngeal, or the facial veins.

The *lymphatic vessels* join the upper deep cervical lymph nodes. The most important node of this group is the jugulodigastric node, which lies below and behind the angle of the mandible.

## Laryngeal Part of the Pharynx

The laryngeal part of the pharynx lies behind the opening into the larynx and the posterior surface of the larynx. It extends between the upper border of the epiglottis and the lower border of the cricoid cartilage. It has an anterior wall, a posterior wall, and lateral walls.

The *anterior wall* is formed by the inlet of the larynx and by the mucous membrane covering the posterior surface of the larynx (Fig. 11-59).

The *posterior wall* is supported by the bodies of the third, fourth, fifth, and sixth cervical vertebrae.

The *lateral wall* is supported by the thyroid cartilage and the thyrohyoid membrane. A small groove in the mucous membrane, called the *piriform fossa,* is situated on each side of the laryngeal inlet (Fig. 11-59). It leads obliquely downward and backward from the region of the back of the tongue to the esophagus. The piriform fossa is bounded medially by the aryepiglottic fold and laterally by the lamina of the thyroid cartilage and the thyrohyoid membrane.

The *nerve supply* of the pharynx is mainly from the pharyngeal plexus; the latter is formed from branches of the glossopharyngeal, vagus, and sympathetic nerves.

The *motor nerve supply* is derived from the cranial part of the accessory nerve, which, via the branch of the vagus to the pharyngeal plexus, supplies all the muscles of the pharynx except the stylopharyngeus, which is supplied by the glossopharyngeal nerve.

The *sensory nerve supply* of the mucous membrane of the nasal part of the pharynx is mainly from the maxillary nerve. The mucous membrane of the oral pharynx is mainly supplied by the glossopharyngeal nerve. The mucous membrane around the entrance into the larynx is supplied by the internal laryngeal branch of the vagus nerve.

The *arterial supply* of the pharynx is derived from branches of the ascending pharyngeal, the ascending palatine, the facial, the maxillary, and the lingual arteries.

*Veins* drain into the pharyngeal venous plexus, which in turn drains into the internal jugular vein.

The *lymphatic vessels* from the pharynx drain either directly into the deep cervical lymph nodes or indirectly via the retropharyngeal or paratracheal nodes.

## THE PALATE

The palate forms the roof of the mouth. It may be divided into two parts: (1) the hard palate in front and (2) the soft palate behind.

The *hard palate* is formed by the palatine processes of the maxillae and the horizontal plates of the palatine bones (Fig. 11-58). It is bounded by the alveolar arches, and behind it is continuous with the soft palate. It forms the floor of the nasal cavities.

The undersurface of the hard palate is covered with mucoperiosteum and possesses a median ridge, on either side of which the mucous membrane shows corrugations. The mucous membrane is covered by stratified squamous epithelium and possesses numerous mucous glands in its posterior part.

The *soft palate* is a mobile fold attached to the posterior border of the hard palate (Fig. 11-60). It is covered on its upper and lower surfaces by mucous membrane and contains an aponeurosis, muscle fibers, lymphoid tissue, glands, vessels, and nerves. Its free posterior border presents in the midline a conical projection called the *uvula*. The soft palate is continuous at the sides with the lateral wall of the pharynx.

The soft palate is composed of (1) mucous membrane, (2) palatine aponeurosis, and (3) muscles.

The *mucous membrane* covers the upper and lower surfaces of the soft palate. It is covered mainly with stratified squamous epithelium. Numerous mucous glands are present on both surfaces, and collections of lymphoid tissue are found in the submucosa.

The *palatine aponeurosis* is a fibrous sheet attached to the posterior border of the hard palate. It is the expanded tendon of the tensor veli palatini, and it splits to enclose the musculus uvulae.

The *muscles* of the soft palate are the tensor veli palatini, the levator veli palatini, the palatoglossus, the palatopharyngeus, and the musculus uvulae.

### Tensor Veli Palatini (Fig. 11-60)

#### Origin

From the spine of the sphenoid, the lateral side of the auditory tube, and the scaphoid fossa.

#### Insertion

The fibers converge as they descend and form a narrow tendon, which turns medially around the pterygoid hamulus. The tendon pierces the origin of the buccinator muscle and, together with the tendon of the muscle of the opposite side, expands to form the palatine aponeurosis.

#### Nerve Supply

From the nerve to the medial pterygoid muscle from the mandibular division of the trigeminal nerve.

#### Action

The two muscles tighten the soft palate so that it may be moved upward or downward as a tense sheet.

### Levator Veli Palatini (Fig. 11-60)

#### Origin

From the undersurface of the petrous part of the temporal bone and from the medial surface of the cartilage of the auditory tube.

#### Insertion

The muscle descends medial to the upper border of the superior constrictor muscle and is inserted into the upper surface of the palatine aponeurosis.

#### Nerve Supply

Pharyngeal plexus.

## Action

Raises the soft palate.

## *Palatoglossus (Fig. 11-60)*

### Origin

From the undersurface of the palatine aponeurosis, where it is continuous with the muscle of the opposite side.

### Insertion

It passes downward and forward beneath the mucous membrane of the lateral wall of the pharynx, where it forms the palatoglossal arch. Having passed in front of the tonsil, it is inserted into the side of the tongue.

### Nerve Supply

Pharyngeal plexus.

### Action

Pulls the root of the tongue upward and backward. Both muscles contracting together cause the palatoglossal arches to approach the midline, and thus the opening (oropharyngeal isthmus) between the oral pharynx and the mouth is narrowed.

## *Palatopharyngeus (Fig. 11-60)*

### Origin

From the posterior border of the hard palate and from the palatine aponeurosis.

### Insertion

It passes downward and backward beneath the mucous membrane of the lateral wall of the pharynx, where it forms the palatopharyngeal arch. It passes behind the tonsil and is inserted into the posterior border of the lamina of the thyroid cartilage. It forms part of the longitudinal muscle of the pharyngeal wall.

### Nerve Supply

Pharyngeal plexus.

### Action

Pulls the wall of the pharynx upward. Acting together, they pull the palatopharyngeal arches toward the midline.

## *Musculus Uvulae (Fig. 11-60)*

### Origin

From the posterior border of the hard palate and the palatine aponeurosis.

### Insertion

The muscle is inserted into the mucous membrane of the uvula.

### Nerve Supply

Pharyngeal plexus.

### Action

Pulls up the uvula.

## MOVEMENTS OF THE SOFT PALATE

The pharyngeal isthmus (communicating channel between the nasal and oral parts of the pharynx) is closed by raising the soft palate. Closure occurs during the production of explosive consonants in speech.

The soft palate is raised by the contraction of the levator veli palatini on each side. At the same time, the upper fibers of the superior constrictor muscle contract and pull the posterior pharyngeal wall forward. The palatopharyngeus muscles on both sides also contract, so that the palatopharyngeal arches are pulled medially, like side curtains. By this means the nasal part of the pharynx is closed off from its oral part.

## CLEFT PALATE

Cleft palate is commonly associated with cleft upper lip. All degrees of cleft palate occur and are caused by failure of the palatal processes of the maxilla to fuse with each other in the midline; in severe cases these processes also fail to fuse with the primary palate (premaxilla) (Fig. 11-61). The first degree of severity is cleft uvula, and the second degree is ununited palatal processes. The third degree is ununited palatal processes and a cleft on one side of the primary palate. This type is usually associated with unilateral cleft lip. The fourth degree of severity, which is rare, consists of ununited palatal processes and a cleft on both sides of the primary palate. This type is usually associated with bilateral cleft lip. A very rare form may occur in which there is a bilateral cleft lip and failure of the primary palate to fuse with the palatal processes of the maxilla on each side.

## THE MECHANISM OF SWALLOWING

After food enters the mouth, it is usually broken down by the grinding action of the teeth and is mixed with saliva. The food is repeatedly passed between the opposing teeth as the result of the movements of the tongue and the "trampoline-like" action of the buccinator muscles of the cheeks. The thoroughly mixed food is now formed into a bolus on the dorsum of the tongue and pushed upward and backward against the undersurface of the hard palate. This is brought about by the contraction of the styloglossus muscles on both sides, which pull the root of the tongue upward and backward. The contraction of the palatoglossus muscles now squeezes the bolus backward into the oral part of the pharynx. The process of swallowing is an involuntary act from this point onward.

The nasal part of the pharynx is now shut off from the oral part of the pharynx by the elevation of the soft palate (see above), the pulling forward of the posterior pharyngeal wall by the upper fibers of the superior constrictor muscle, and the contraction of the palatopharyngeus muscles.

The larynx and laryngeal part of the pharynx are now pulled upward by the contraction of the stylopharyngeus, salpingopharyngeus, thyrohyoid, and palatopharyngeus muscles. The main part of

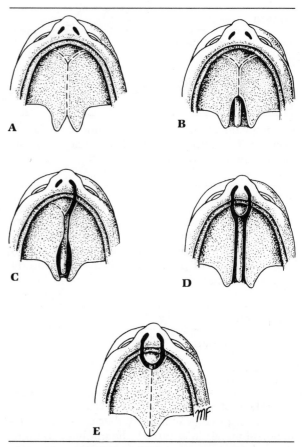

**Fig. 11-61. Different forms of cleft palate: (A) cleft uvula, (B) cleft soft and hard palate, (C) total unilateral cleft palate and cleft lip, (D) total bilateral cleft palate and cleft lip, and (E) bilateral cleft lip and jaw.**

the larynx is thus elevated to the posterior surface of the epiglottis, and the entrance into the larynx is closed. (See p. 863.)

The bolus moves downward over the epiglottis, the closed entrance into the larynx, and reaches the lower part of the pharynx as the result of the successive contraction of the superior, middle, and inferior constrictor muscles. Some of the food slides down the grooves on either side of the entrance into the larynx, i.e., down through the piriform fossae.

Finally, the lower fibers of the inferior constrictor muscle (cricopharyngeus muscle) relax, and the bolus enters the esophagus.

# THE NOSE

The nose consists of the external nose and the nasal cavity.

The *external nose* has a free tip and is attached to the forehead by the *root*, or *bridge*, of the nose. The external orifices of the nose are the two *nostrils*, or *nares* (Fig. 11-62). Each nostril is bounded laterally by the *ala* and medially by the *nasal septum*.

The framework of the external nose is made up above by the nasal bones, the frontal processes of the maxillae, and the nasal part of the frontal bone. Below, the framework is formed of plates of hyaline cartilage, which include the *upper* and *lower nasal cartilages* and the *septal cartilage*. The muscles acting on the external nose are described on page 754.

The skin over the dorsum and sides of the nose is thin and contains a large number of sebaceous glands.

The *nasal cavity* extends from the nostrils, or nares, in front to the *choanae* behind (Fig. 11-62). It is divided into right and left halves by the nasal *septum*. Each half has a floor, a roof, and a lateral wall and medial wall.

The *floor* is formed by the palatine process of the maxilla and the horizontal plate of the palatine bone, i.e., the upper surface of the hard palate.

The *roof* is narrow and is formed from behind forward by the body of the sphenoid, the cribriform plate of the ethmoid, the frontal bone, the nasal bone, and the nasal cartilages.

The *lateral wall* is marked by three projections called the *superior, middle,* and *inferior nasal conchae.* The area below each concha is referred to as a *meatus* (Fig. 11-62).

The *spheno-ethmoidal recess* is a small area of the nose that lies above the superior concha and in front of the body of the sphenoid bone. It receives the opening of the *sphenoidal air sinus* (Fig. 11-62).

The *superior meatus* lies below and lateral to the superior concha. It receives the openings of the *posterior ethmoidal sinuses* (Fig. 11-62).

The *middle meatus* lies below and lateral to the middle concha. It has on its lateral wall a rounded prominence, the *bulla ethmoidalis,* caused by the bulging of the underlying middle ethmoidal air sinuses, which open on its upper border (Fig. 11-62).

A curved cleft, the *hiatus semilunaris,* lies immediately below the bulla. The anterior end of the hiatus leads into a funnel-shaped channel called the *infundibulum.* The *maxillary sinus* opens into the middle meatus via the hiatus semilunaris. The *frontal sinus* opens into and is continuous with the infundibulum. The *anterior ethmoidal sinuses* also open into the infundibulum.

The middle meatus is continuous in front with a depression called the *atrium.* The atrium is limited above by a ridge, the *agger nasi* (Fig. 11-62). Below and in front of the atrium, and just within the nostril, is the *vestibule.* This is lined by modified skin and possesses short, curved hairs, or *vibrissae.*

The *inferior meatus* lies below and lateral to the inferior concha and receives the opening of the *nasolacrimal duct* (Fig. 11-62). A fold of mucous membrane forms an imperfect valve, which guards the opening of the duct.

The *medial wall,* or *nasal septum,* is an osteocartilaginous partition covered by adherent mucous membrane. The upper part is formed by the vertical plate of the ethmoid, and its posterior part is formed by the vomer. The anterior portion is formed by the septal cartilage. Only rarely does it lie in the median plane.

The *mucous membrane* lines the nasal cavities with the exception of the vestibules, which are lined by modified skin. There are two types of mucous membrane: (1) olfactory and (2) respiratory.

The *olfactory mucous membrane* lines the upper surface of the superior concha and the sphenoethmoidal recess. It also lines a corresponding area on the nasal septum and lines the roof. Its function is the reception of olfactory stimuli, and for this purpose it possesses specialized olfactory nerve cells. The central axons of these cells (the *olfactory nerve fibers*) pass through the openings in the cribriform plate of the ethmoid and end in the *olfactory bulbs.* (See p. 808.) There are also supporting and basal cells in the mucous membrane. The surface of the mucous membrane is kept moist by the secretions of numerous serous glands.

The *respiratory mucous membrane* lines the lower part of the nasal cavities. Its function is to warm, moisten, and clean the inspired air. The warming process is accomplished by the presence of a plexus of veins in the submucous connective tissue. The moisture is derived from the abundant production of mucus secreted by glands and goblet

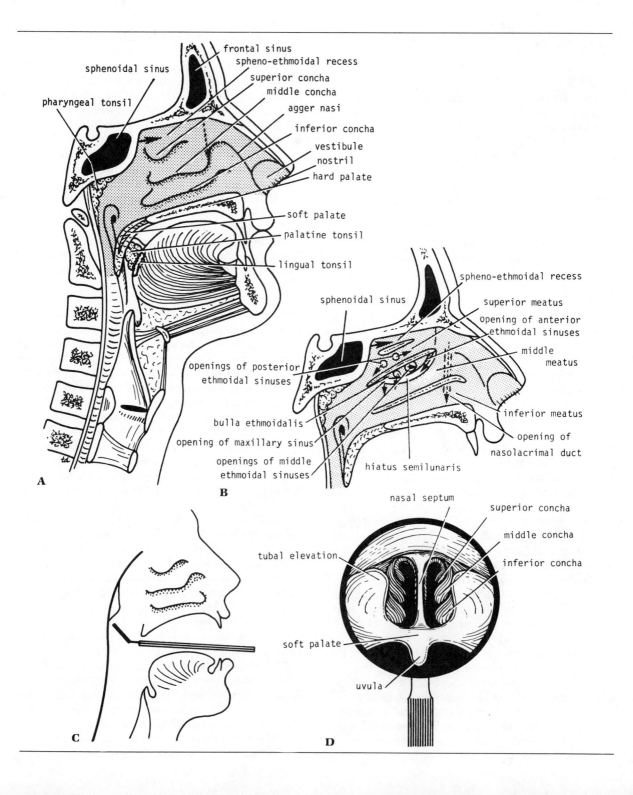

frontal sinus
sphenoidal sinus
spheno-ethmoidal recess
superior concha
middle concha
pharyngeal tonsil
agger nasi
inferior concha
vestibule
nostril
hard palate
soft palate
palatine tonsil
lingual tonsil

spheno-ethmoidal recess
superior meatus
sphenoidal sinus
opening of anterior ethmoidal sinuses
middle meatus
openings of posterior ethmoidal sinuses
inferior meatus
opening of nasolacrimal duct
bulla ethmoidalis
opening of maxillary sinus
openings of middle ethmoidal sinuses
hiatus semilunaris

A

B

nasal septum
superior concha
middle concha
inferior concha
tubal elevation
soft palate
uvula

C

D

cells. Inspired dust particles are removed from the air by the moist, sticky surface of the mucous membrane. The contaminated mucus is continually being moved backward by the ciliary action of the columnar ciliated epithelium that covers the surface. On reaching the pharynx, the mucus is swallowed.

The *nerve supply* to the nasal cavity is as follows:

The *olfactory nerves* arise from the special olfactory cells in the olfactory mucous membrane described above. They ascend through the cribriform plate, to reach the olfactory bulbs. (See p. 808.)

The *nerves of ordinary sensation* are derived from the ophthalmic and maxillary divisions of the trigeminal nerve. The nerve supply to the anterior part of the nasal cavity comes from the anterior ethmoidal nerve. The nerve supply to the posterior part of the nasal cavity comes from the nasal, nasopalatine, and palatine branches of the pterygopalatine ganglion.

The *arterial supply* to the nasal cavity is derived mainly from branches of the maxillary artery. The most important branch is the sphenopalatine artery, which enters the nasal cavity through the pterygopalatine foramen. The sphenopalatine artery anastomoses with the septal branch of the superior labial branch of the facial artery in the region of the vestibule. This is a very common site of bleeding from the nose *(epistaxis)*.

The *veins* form a rich plexus in the submucosa. The plexus is drained by veins that accompany the arteries.

The *lymphatic vessels* draining the vestibule end in the submandibular nodes. The remainder of the nasal cavity is drained by vessels that pass to the upper deep cervical nodes.

## THE PARANASAL SINUSES

The paranasal sinuses are cavities found in the interior of the maxilla, frontal, sphenoid, and ethmoid bones (Figs. 11-62 and 11-63). They are lined with mucoperiosteum and filled with air; they communicate with the nasal cavity through relatively small apertures. The maxillary and sphenoidal sinuses are present in a rudimentary form at birth; they enlarge appreciably after the eighth year and become fully formed in adolescence.

The mucus produced by the glands in the mucous membrane is moved into the nose by ciliary action of the columnar cells. Drainage of the mucus is also achieved by the syphon action created during the blowing of the nose. The function of the sinuses is to act as resonators to the voice; they also reduce the weight of the skull. When the apertures of the sinuses are blocked, or they become filled with fluid, the quality of the voice is markedly changed.

The *maxillary sinus* is located within the body of the maxilla (Fig. 11-63). It is pyramidal in shape, with the base forming the lateral wall of the nose, and the apex in the zygomatic process of the maxilla. The roof is formed by the floor of the orbit, while the floor is formed by the alveolar process. The roots of the first and second premolars and of the third molar, and sometimes the root of the canine, project up into the sinus. The roots of the teeth are enclosed in a thin layer of compact bone; occasionally, this is absent, and the root is in contact with the mucous membrane of the sinus. It is easy to understand how extraction of a tooth may result in a fistula, or how an infected tooth may produce sinusitis.

The maxillary sinus opens into the middle meatus of the nose through the hiatus semilunaris (Fig. 11-63). Unfortunately, this opening lies high up on the medial wall of the sinus, so that the sinus readily accumulates fluid. Since the frontal and anterior ethmoidal sinuses drain into the infundibulum, which in turn drains into the hiatus semilunaris, the chance that infection may spread from these sinuses into the maxillary sinus is great.

The mucous membrane of the maxillary sinus is supplied by the superior alveolar and infraorbital nerves.

The *frontal sinuses*, two in number, are contained within the frontal bone (Fig. 11-62). They are separated from each other by a bony septum, which frequently deviates from the medial plane. Each sinus is roughly triangular in shape, extending upward above the medial end of the eyebrow and backward into the medial part of the roof of the orbit.

**Fig. 11-62. (A) Sagittal section through nose, mouth, larynx, and pharynx. (B) Lateral wall of nose and nasal part of pharynx. (C) Position of mirror in posterior rhinoscopy. (D) Structures seen in posterior rhinoscopy.**

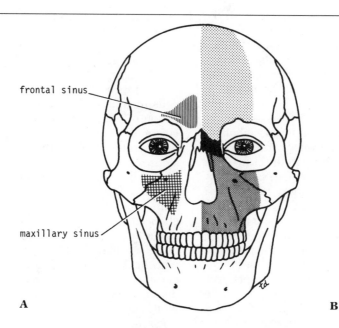

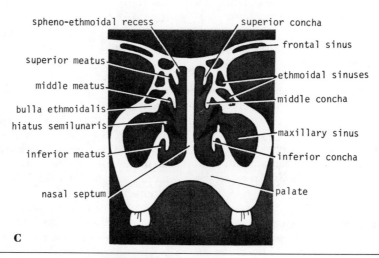

**Fig. 11-63. (A)** Bones of face, showing positions of
frontal and maxillary sinuses. **(B)** Regions where
pain is experienced in sinusitis. (Lightly dotted
area in frontal sinusitis; solid area in spheno-ethmoidal sinusitis; and heavily dotted area in maxillary sinusitis.) **(C)** Coronal section through nasal
cavity, showing the frontal, ethmoidal, and maxillary sinuses.

Table 11-5. The Paranasal Sinuses and Their Site of Drainage into the Nose

| Name of sinus | Site of drainage |
| --- | --- |
| Maxillary sinus | Middle meatus through hiatus semilinaris |
| Frontal sinus | Middle meatus via infundibulum |
| Sphenoid sinuses | Sphenoethmoidal recess |
| Ethmoidal sinuses | |
|   Anterior group | Infundibulum and into middle meatus |
|   Middle group | Middle meatus on or above bulla ethmoidalis |
|   Posterior group | Superior meatus |

Note that maxillary and sphenoidal sinuses are present in rudimentary form at birth and enlarge appreciably after the eighth year and are fully formed in adolescence.

Each frontal sinus opens into the middle meatus of the nose through the infundibulum (Fig. 11-62). The mucous membrane is supplied by the supraorbital nerve.

The *sphenoidal sinuses*, two in number, lie within the body of the sphenoid bone (Fig. 11-62). Each sinus opens into the sphenoethmoidal recess above the superior concha. The mucous membrane is supplied by the posterior ethmoidal nerves.

The *ethmoidal sinuses* are contained within the ethmoid bone, between the nose and the orbit (Fig. 11-63). They are separated from the latter by a thin plate of bone, so that infection may readily spread from the sinuses into the orbit. They can be divided into three groups: anterior, middle, and posterior. The anterior group opens into the infundibulum, the middle group opens into the middle meatus, on or above the bulla ethmoidalis, and the posterior group opens into the superior meatus. The mucous membrane is supplied by the anterior and posterior ethmoidal nerves. See Table 11-5.

## THE LARYNX

The larynx is a specialized organ that provides a protective sphincter at the inlet of the air passages and is responsible for voice production. Above, it opens into the laryngeal part of the pharynx, and below, it is continuous with the trachea.

The framework of the larynx is made up of car-

tilages, which are connected by membranes and ligaments and moved by muscles. It is lined by mucous membrane.

The *thyroid cartilage* (Fig. 11-64) consists of two laminae of hyaline cartilage meeting in the midline in the prominent V angle of the Adam's apple. The posterior border of each lamina is drawn upward into a *superior cornu* and downward into an *inferior cornu*. On the outer surface of each lamina is an *oblique line* for the attachment of the sternothyroid, the thyrohyoid, and the inferior constrictor muscles.

The *cricoid cartilage* is formed from a complete ring of hyaline cartilage (Fig. 11-64). It is shaped like a signet ring and lies below the thyroid cartilage. It has a narrow anterior *arch* and a broad posterior *lamina*. On each side of the lateral surface there is a circular facet for articulation with the inferior cornu of the thyroid cartilage. On each side of the upper border there is an articular facet for articulation with the base of the arytenoid cartilage. All these joints are synovial joints.

The *arytenoid cartilages* are small, two in number, and pyramidal in shape (Fig. 11-64). They are situated at the back of the larynx, on the lateral part of the upper border of the lamina of the cricoid cartilage. Each cartilage has an *apex* above and a *base* below. The apex supports the corniculate cartilage. The base articulates with the cricoid cartilage. Two processes project from the base. The *vocal process* projects horizontally forward and gives attachment to the vocal ligament. The *muscular process* projects laterally and gives attachment to the posterior and lateral cricoarytenoid muscles.

The *corniculate cartilages* (Fig. 11-64) are two small nodules that articulate with the apices of the arytenoid cartilages and give attachment to the aryepiglottic folds. (See below.)

The *cuneiform cartilages* are two small, rod-shaped pieces of cartilage placed so that one is in each aryepiglottic fold. They serve as supports for the folds (Fig. 11-64).

The *epiglottis* is a leaf-shaped elastic cartilage situated behind the root of the tongue (Fig. 11-64). It is connected in front to the body of the hyoid bone, and by its stalk to the back of the thyroid cartilage. The sides of the epiglottis are connected to the arytenoid cartilages by the aryepiglottic folds. The upper edge of the epiglottis is free, and

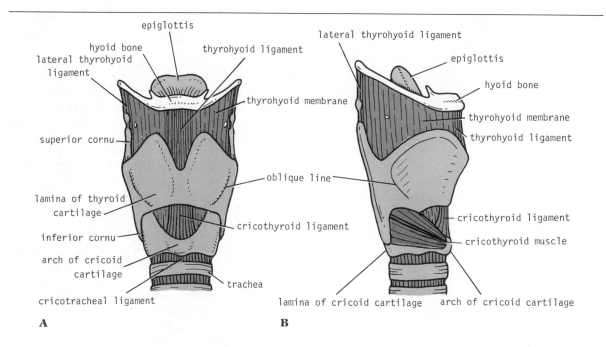

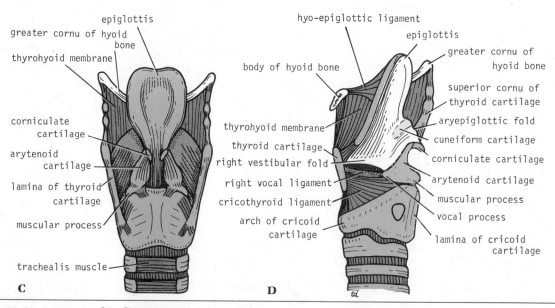

**Fig. 11-64. Larynx and its ligaments: (A) from the front, (B) from the lateral aspect, (C) from behind. (D) Left lamina of thyroid cartilage has been removed to display interior of larynx.**

the covering of mucous membrane is reflected onto the posterior surface of the tongue. Here, there are a *median glossoepiglottic fold* and *lateral pharyngoepiglottic folds*. The *valleculae* are depressions of mucous membrane present on either side of the glossoepiglottic fold.

## Membranes and Ligaments of the Larynx

The *thyrohyoid membrane* connects the upper margin of the thyroid cartilage below to the upper margin of the posterior surface of the body and greater cornu of the hyoid bone above (Fig. 11-64). In the midline the membrane is thickened to form the *median thyrohyoid ligament*; the posterior borders are thickened to form the *lateral thyrohyoid ligaments*. On each side the membrane is pierced by the superior laryngeal vessels and the internal laryngeal nerve.

The *cricotracheal ligament* connects the lower margin of the cricoid cartilage to the first ring of the trachea (Fig. 11-64).

The *fibroelastic membrane* of the larynx lies beneath the mucous membrane lining the larynx. The upper portion of the membrane is called the *quadrangular membrane*, and it extends between the epiglottis and the arytenoid cartilages (Fig. 11-65). Its lower margin forms the *vestibular ligaments*. The lower part of the fibroelastic membrane is called the *cricothyroid ligament*. The anterior part of the cricothyroid ligament is thick and connects the cricoid cartilage to the lower margin of the thyroid cartilage (Fig. 11-64). The lateral part of the ligament is thin and is attached below to the upper margin of the cricoid cartilage. The superior margin of the ligament, instead of being attached to the lower margin of the thyroid cartilage, ascends within the thyroid cartilage on its medial surface. Its upper margin is thickened and forms the important *vocal ligament* on each side (Fig. 11-64). The anterior end of each vocal ligament is attached to the deep surface of the thyroid cartilage. The posterior end is attached to the vocal process of the arytenoid cartilage.

The *hyoepiglottic ligament* attaches the epiglottis to the hyoid bone. The *thyroepiglottic ligament* attaches the epiglottis to the thyroid cartilage.

## Inlet of the Larynx

The inlet of the larynx looks backward and upward into the laryngeal part of the pharynx (Fig. 11-62). The opening is bounded in front by the upper margin of the epiglottis; laterally, by the aryepiglottic fold of mucous membrane, which connects the epiglottis to the arytenoid cartilage; and posteriorly and below, by the mucous membrane stretching between the arytenoid cartilages. The corniculate cartilage on the apex of the arytenoid cartilage and the small bar, the cuneiform cartilage, produce a small elevation on the upper border of each aryepiglottic fold.

## Cavity of the Larynx

The cavity of the larynx extends from the inlet to the lower border of the cricoid cartilage. It may be divided into three parts: (1) the upper part, or vestibule, (2) the middle part, and (3) the lower part.

The *vestibule of the larynx* extends from the inlet to the vestibular folds (Fig. 11-65). The latter are two thick folds of mucous membrane that cover the vestibular ligaments. The vestibule has an anterior, posterior, and lateral wall.

The anterior wall is formed by the posterior surface of the epiglottis, which is covered by mucous membrane. The posterior wall is formed by the arytenoid cartilages and the interarytenoid fold of mucous membrane, containing the transverse arytenoid muscle. The lateral walls are formed by the aryepiglottic folds, which contain the aryepiglottic muscle.

Below, the vestibule is narrowed by the pink vestibular folds, which project medially. The *rima vestibuli* is the gap between the vestibular folds. The *vestibular ligament*, which lies within each vestibular fold, is the thickened lower edge of the quadrangular membrane. (See above.) The ligament stretches from the thyroid cartilage to the side of the arytenoid cartilage.

The *middle part of the larynx* extends from the level of the vestibular folds to the level of the vocal folds. The *vocal folds* are white in color and contain the *vocal ligaments* (Fig. 11-65). Each vocal ligament is the thickened upper edge of the cricothyroid ligament. (See above.) It stretches from the thyroid cartilage in front to the vocal process of the

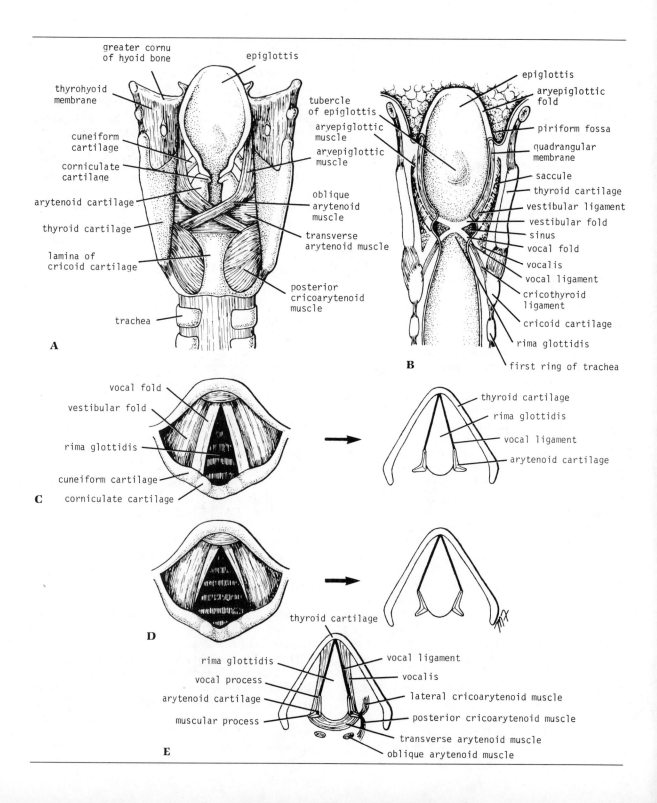

**A**

greater cornu of hyoid bone
epiglottis
thyrohyoid membrane
cuneiform cartilage
corniculate cartilage
arytenoid cartilage
thyroid cartilage
lamina of cricoid cartilage
trachea
tubercle of epiglottis
aryepiglottic muscle
arvepiglottic muscle
oblique arytenoid muscle
transverse arytenoid muscle
posterior cricoarytenoid muscle

**B**

epiglottis
aryepiglottic fold
piriform fossa
quadrangular membrane
saccule
thyroid cartilage
vestibular ligament
vestibular fold
sinus
vocal fold
vocalis
vocal ligament
cricothyroid ligament
cricoid cartilage
rima glottidis
first ring of trachea

**C**

vocal fold
vestibular fold
rima glottidis
cuneiform cartilage
corniculate cartilage

thyroid cartilage
rima glottidis
vocal ligament
arytenoid cartilage

**D**

**E**

thyroid cartilage
rima glottidis
vocal process
arytenoid cartilage
muscular process
vocal ligament
vocalis
lateral cricoarytenoid muscle
posterior cricoarytenoid muscle
transverse arytenoid muscle
oblique arytenoid muscle

arytenoid cartilage behind. The *rima glottidis* is the gap between the vocal folds in front and the vocal processes of the arytenoid cartilages behind.

Between the vestibular and vocal folds on each side is a small recess, called the *sinus of the larynx*. It is lined with mucous membrane, and from it, a small diverticulum, called the *saccule of the larynx*, passes upward between the vestibular fold and the thyroid cartilage (Fig. 11-65).

The *lower part of the larynx* extends from the level of the vocal folds to the lower border of the cricoid cartilage. Its walls are formed by the inner surface of the cricothyroid ligament and the cricoid cartilage.

The *mucous membrane of the larynx* lines the cavity and is covered with ciliated columnar epithelium. On the vocal folds, however, where the mucous membrane is subject to repeated trauma during phonation, the mucous membrane is covered with stratified squamous epithelium. There are many mucous glands contained within the mucous membrane, and they are especially numerous in the saccules. Here, the secretion pours down onto the upper surface of the vocal folds and lubricates them during phonation.

## Muscles of the Larynx

The muscles may be divided into two groups: (1) extrinsic and (2) intrinsic.

The *extrinsic muscles* may be divided into two opposing groups, the elevators of the larynx and the depressors of the larynx. The larynx moves up during the act of swallowing and down following the act. Since the hyoid bone is attached to the thyroid cartilage by the thyrohyoid membrane, it follows that movements of the hyoid bone are accompanied by movements of the larynx.

### ELEVATORS OF THE LARYNX

The elevators of the larynx include the digastric, the stylohyoid, the mylohyoid, and the geniohyoid

Fig. 11-65. (A) Muscles of larynx as seen from behind. (B) Coronal section through larynx. (C) Rima glottidis partially open as in quiet breathing. (D) Rima glottidis wide open as in deep breathing. (E) Muscles that move vocal ligaments.

muscles. The stylopharyngeus, the salpingopharyngeus, and the palatopharyngeus, which are inserted into the posterior border of the lamina of the thyroid cartilage, also elevate the larynx.

### DEPRESSORS OF THE LARYNX

The depressors of the larynx include the sternothyroid, the sternohyoid, and the omohyoid muscles. The action of these muscles is assisted by the elastic recoil of the trachea.

The *intrinsic muscles* may be divided into two groups: those that control the inlet into the larynx and those that move the vocal folds.

### MUSCLE CONTROLLING THE LARYNGEAL INLET

*Oblique Arytenoid (Fig. 11-65)*

Origin

From the muscular process of the arytenoid cartilage.

Insertion

Into the apex of the *opposite* arytenoid cartilage. Some of the fibers continue beyond the apex of the arytenoid cartilage and reach the epiglottis via the aryepiglottic fold. The latter fibers form the *aryepiglottic muscles*.

Nerve Supply

Recurrent laryngeal nerve.

Action

The two muscles contracting together serve as a sphincter to the laryngeal inlet. They approximate the arytenoid cartilages to one another and draw them forward to the epiglottis.

The laryngeal inlet opens as the result of a relaxation of the oblique arytenoid muscle and the elastic recoil of the ligaments of the joints of the arytenoid cartilages and the cricoid cartilage.

## MUSCLES CONTROLLING THE MOVEMENTS OF THE VOCAL FOLDS

The vocal folds may be tightened or they may be relaxed. They may be adducted or they may be abducted. The following muscles perform these actions.

### Cricothyroid (Tensor) (Fig. 11-64)

#### Origin

From the side of the cricoid cartilage.

#### Insertion

The muscle is triangular in shape. The upper fibers pass upward and backward and are inserted onto the lower border of the lamina of the thyroid cartilage. The lower fibers run backward and are inserted onto the anterior border of the inferior cornu of the thyroid cartilage.

#### Nerve Supply

External laryngeal nerve.

#### Action

The vocal ligaments are tensed and elongated by increasing the distance between the angle of the thyroid cartilage and the vocal processes of the arytenoid cartilages. This is brought about by the muscle (1) pulling the thyroid cartilage forward and (2) tilting the lamina of the cricoid cartilage backward with the attached arytenoid cartilages.

### Thyroarytenoid (Relaxor) (Fig. 11-65)

#### Origin

From the inner surface of the angle of the thyroid cartilage.

#### Insertion

The fibers lie lateral to the vocal ligament and are inserted onto the anterolateral surface of the arytenoid cartilage. Some of the fibers run alongside the vocal ligament and are attached to the vocal

process of the arytenoid cartilage. The latter fibers are called the *vocalis* muscle.

#### Nerve Supply

Recurrent laryngeal nerve.

#### Action

Pulls the arytenoid cartilage forward toward the thyroid cartilage and thus shortens and relaxes the vocal ligament.

### Lateral Cricoarytenoid (Adductor) (Fig. 11-65)

#### Origin

From the upper border of the arch of the cricoid cartilage.

#### Insertion

Into the muscular process of the arytenoid cartilage.

#### Nerve Supply

Recurrent laryngeal nerve.

#### Action

Pulls the muscular process of the arytenoid cartilage forward, causing rotation of the arytenoid, so that the vocal process moves medially, and the vocal folds is adducted.

### Transverse Arytenoid (Fig. 11-65)

#### Origin

From the back and medial surface of the arytenoid cartilage.

#### Insertion

The muscle fibers bridge the interval between the arytenoid cartilages. The fibers are attached to the back and medial surface of the opposite arytenoid cartilage.

## Nerve Supply

Recurrent laryngeal nerve.

## Action

Approximates the arytenoid cartilages and closes the posterior part of the rima glottidis.

### *Posterior Cricoarytenoid (Abductor) (Fig. 11-65)*

### Origin

From the back of the lamina of the cricoid cartilage.

### Insertion

The fibers pass upward and laterally, to be inserted into the muscular process of the arytenoid cartilage.

### Nerve Supply

Recurrent laryngeal nerve.

### Action

Pulls the muscular process of the arytenoid cartilage backward, causing rotation of the arytenoid, so that the vocal process moves laterally, and the vocal fold is abducted.

## SPHINCTERIC FUNCTION OF THE LARYNX

There are two sphincters in the larynx: (1) at the inlet and (2) at the rima glottidis.

The sphincter at the inlet is used only during swallowing. As the bolus of food is passed backward between the tongue and the hard palate, the larynx is pulled up beneath the back of the tongue. The inlet of the larynx is narrowed by the action of the oblique arytenoid and aryepiglottic muscles. The epiglottis is pushed backward by the tongue and serves as a cap over the laryngeal inlet. The bolus of food, or fluids, now enters the esophagus by passing over the epiglottis or moving down the grooves on either side of the laryngeal inlet, the *piriform fossae.*

In coughing or sneezing, the rima glottidis serves as a sphincter. After inspiring, the vocal folds are adducted, and the muscles of expiration are made to contract strongly. As a result, the intrathoracic pressure rises, whereupon the vocal folds are suddenly abducted. The sudden release of the compressed air will often dislodge foreign particles or mucus from the respiratory tract and carry the material up into the pharynx. Here, they are either swallowed or expectorated.

In abdominal straining associated with micturition, defecation, and parturition, the air is often held temporarily in the respiratory tract by closing the rima glottidis. After deep inspiration the rima glottidis is closed. The muscles of the anterior abdominal wall now contract, and the upward movement of the diaphragm is prevented by the presence of compressed air within the respiratory tract. After a prolonged effort the person often releases some of the air by momentarily opening the rima glottidis, producing a grunting sound.

## VOICE PRODUCTION IN THE LARYNX

The intermittent release of expired air between the adducted vocal folds results in their vibration and in the production of sound. The *frequency,* or *pitch,* of the voice is determined by changes in the length and tension of the vocal ligaments. The *quality* of the voice depends on the resonators above the larynx, namely, the pharynx, mouth, and paranasal sinuses. The quality is controlled by the muscles of the soft palate, tongue, floor of the mouth, cheeks, lips, and jaws. *Normal speech* depends on the modification of the sound into recognizable consonants and vowels by the use of the tongue, teeth, and lips. Vowel sounds are usually purely oral with the soft palate raised; that is to say, the air is channeled through the mouth rather than the nose. The physician tests the mobility of the soft palate by asking the patient to say "ah" with the mouth open.

*Speech* involves the intermittent release of expired air between the adducted vocal folds. *Singing* a note requires a more prolonged release of the expired air between the adducted vocal folds. In *whispering,* the vocal folds are adducted, but the arytenoid cartilages are separated; the vibrations are given to a constant stream of expired air that passes through the posterior part of the rima glottidis.

## Movements of the Vocal Folds with Respiration

In quiet respiration, the rima glottidis is triangular in shape, with the apex in front (Fig. 11-65C). With forced inspiration, the rima glottidis assumes a diamond shape due to the lateral rotation of the arytenoid cartilages (Fig. 11-65D).

## NERVE SUPPLY OF THE LARYNX

The *sensory nerve supply* to the mucous membrane of the larynx above the vocal folds is from the internal laryngeal branch of the superior laryngeal branch of the vagus. Below the level of the vocal folds the mucous membrane is supplied by the recurrent laryngeal nerve.

The *motor nerve supply* to the intrinsic muscles of the larynx is the recurrent laryngeal nerve, except for the cricothyroid muscle, which is supplied by the external laryngeal branch of the superior laryngeal branch of the vagus.

## BLOOD SUPPLY AND LYMPHATIC DRAINAGE OF THE LARYNX

The arterial supply to the upper half of the larynx is from the superior laryngeal branch of the superior thyroid artery. The lower half of the larynx is supplied by the inferior laryngeal branch of the inferior thyroid artery.

The lymphatic vessels drain into the deep cervical group of nodes.

## CERVICAL PART OF THE VERTEBRAL COLUMN

The cervical part of the vertebral column shows a forward convexity and is made up of seven vertebrae.

A *typical cervical vertebra* has the following characteristics (Fig. 11-66): In each transverse process there is a foramen transversarium for the vertebral vessels. (Note the vertebral artery only passes through transverse processes C 1–6.) The

spines are small and bifid. The body is small and broader from side to side than from before backward; there are small synovial joints on each side. The vertebral foramen is large and triangular in shape. The superior articular processes have small, flat articular facets that face backward and upward; the inferior articular processes have facets that face downward and forward.

The first, second, and seventh cervical vertebrae are atypical.

The *first cervical vertebra*, or *atlas*, has no body and no spinous process (Fig. 11-66). It is merely a ring of bone consisting of anterior and posterior arches and a lateral mass on each side. Each lateral mass has articular surfaces on its upper and lower aspects. The bone articulates above with the occipital condyles, forming the *atlanto-occipital joints*, where nodding movements of the head take place. Below, the bone articulates with the axis, forming the *atlanto-axial joints*, where rotation movements of the head take place.

The *second cervical vertebra*, or *axis*, has a peglike *odontoid process*, which surmounts the body and represents the body of the atlas, which has fused with the axis (Fig. 11-66).

The *seventh cervical vertebra*, or *vertebra prominens*, is so named because it has the longest spinous process. The spine is not bifid. The transverse process is large, but the foramen transversarium is small and does not transmit the vertebral artery (Fig. 11-66).

## VERTEBRAL ARTERY

The vertebral artery arises from the upper border of the first part of the subclavian artery. (See p. 740.) It ascends in front of the anterior ramus of the eighth cervical nerve, the transverse process of the seventh cervical vertebra, and the anterior ramus of the seventh cervical nerve. It enters the foramen transversarium of the sixth cervical vertebra (Fig. 11-15). The artery now ascends through the transverse processes of the upper cervical vertebrae in front of the anterior rami of the upper cervical nerves. Having emerged from the atlas, it winds backward behind the lateral mass of that bone, pierces the dura mater, and ascends vertically into the skull through the foramen magnum. The further course of this artery is described on page 807.

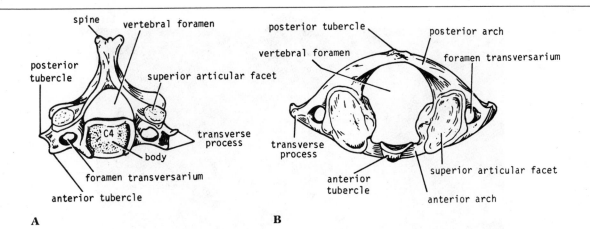

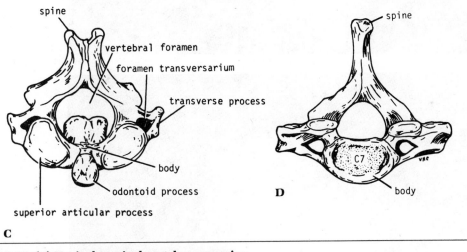

**Fig. 11-66. (A)** Typical cervical vertebra, superior aspect. **(B)** Atlas, or first cervical vertebra, superior aspect. **(C)** Axis, or second cervical vertebra, from above and behind. **(D)** Seventh cervical vertebra, superior aspect. The foramen transversarium forms a passage for the vertebral vein but *not* for the vertebral artery.

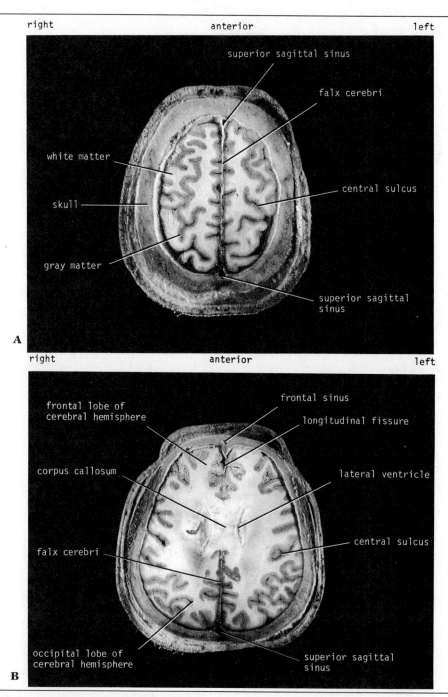

**Fig. 11-67.** (A) Cross section of head a short distance beneath the vault of the skull, viewed from below; (B) cross section of head at level of corpus callosum, viewed from below.

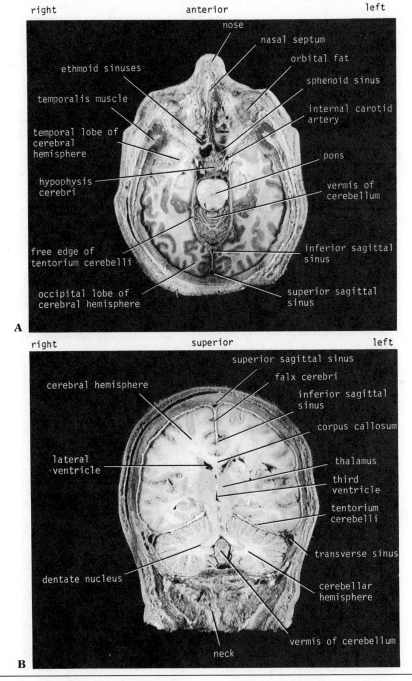

**Fig. 11-68. (A) Cross section of head viewed from below; (B) coronal section of head and upper part of neck.**

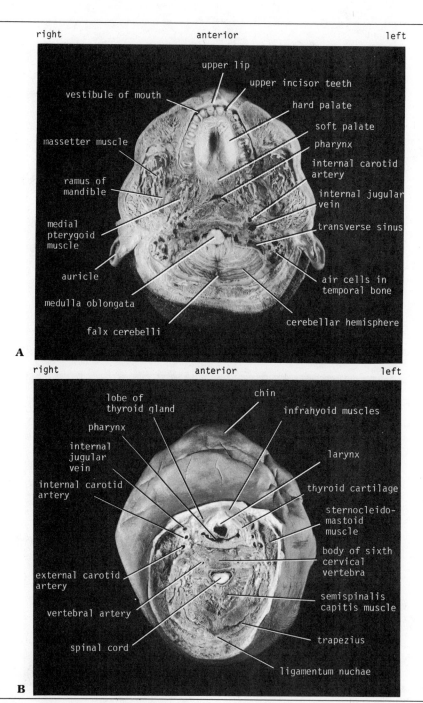

**Fig. 11-69. (A) Cross section of head just below the level of the hard palate, viewed from below; (B) cross section of neck at the level of the sixth cervical vertebra, viewed from below.**

## CERVICAL SPINAL NERVES

The cervical spinal nerves, having emerged from the intervertebral foramina, divide into anterior and posterior rami (Fig. 11-8). The anterior rami pass laterally behind the vertebral artery and run behind the longus capitis above and the scalenus anterior below. The formation of the cervical plexus is described on page 734. The posterior rami run backward around the articular processes and are distributed to the muscles and skin of the back. The posterior ramus of the first cervical nerve does not reach the skin.

Before studying the radiographic appearance of the head and neck, the student is encouraged to examine photographs of sections of the head and neck (see Figs. 11-67 to 11-69).

# RADIOGRAPHIC APPEARANCE OF THE HEAD AND NECK

Radiological examination of the head and neck concentrates mainly on the bony structures, since the brain, muscles, tendons, and nerves blend into a homogeneous mass. However, a few normal structures within the skull become calcified in the adult, and the displacement of such structures may indirectly give evidence of a pathological condition. The pineal gland, for example, is calcified in 50 percent of normal adults. It lies in the midline. The falx cerebri and the choroid plexuses also become calcified frequently. The brain may be studied by injecting contrast media into the subarachnoid space (e.g., *pneumoencephalography*) or into the ventricular system *(ventriculography)*. An indirect method, which is commonly used, is the injection of contrast media into the arterial system leading to the brain *(cerebral arteriogram)*. However, these studies are specialized and considerable experience is required both to perform them and to interpret the results. Fortunately, the introduction of the technique of computerized axial tomography has provided physicians with a safe and accurate method of studying the intracranial contents.

## Radiographic Appearance of the Skull

The selected position of the skull relative to the film cassette will depend on the anatomical area that one wishes to demonstrate. In this text the appearance seen on a straight postero-anterior view and on a lateral view will be described. Routine postero-anterior and lateral views of the skull for the study of the paranasal sinuses will also be described.

The *straight postero-anterior view* of the skull (Fig. 11-70) is taken with the forehead and nose against the film cassette and the X-ray tube positioned behind the head, perpendicular to the film and in line with the external auditory meatus and the palpebral fissure. Unfortunately, in this position the petrous parts of the temporal bones are superimposed on the lower halves of the orbits.

The different parts of the vault of the skull are visible and the sagittal, coronal, and lambdoid sutures may be seen (Fig. 11-71). The frontal sinuses, the upper and lower margins of the orbit, the nasal septum and the conchae, the maxillary sinuses, and the maxillary teeth can be identified.

The rami and body of the mandible are easily recognized. The sphenoidal and ethmoidal air sinuses produce a composite shadow.

The *lateral view of the skull* (Fig. 11-72) is taken with the sagittal plane of the skull parallel with the film cassette. The X-ray tube is centered over the region of the sella turcica.

The different parts of the bones of the vault and base of the skull are well shown (Fig. 11-73). The zygomatic and maxillary bones are superimposed on each other and are not clear. The coronal, squamosal (between the squamous part of the temporal bone and the parietal bone), and lambdoid sutures can be recognized. The inner and outer tables of the skull bones and the intervening diploë can be seen. Depressions on the inner table are commonly seen in children and are produced by the underlying cerebral convolutions.

The grooves produced by the anterior and posterior branches of the middle meningeal vessels may be seen running posteriorly across the parietal bones. A wide groove for the transverse sinus may also be identified as it crosses the occipital bone. Diploic vessels may be recognized as branching dark lines.

The pineal body, if calcified, can be seen as a small shadow above and behind the external auditory meatus.

Anteriorly, the frontal air sinuses are clearly shown superimposed on one another. Behind them

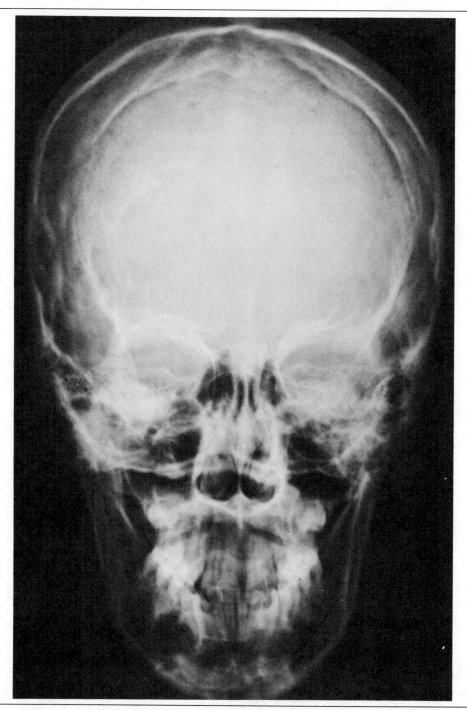

**Fig. 11-70. Postero-anterior radiograph of the skull.**

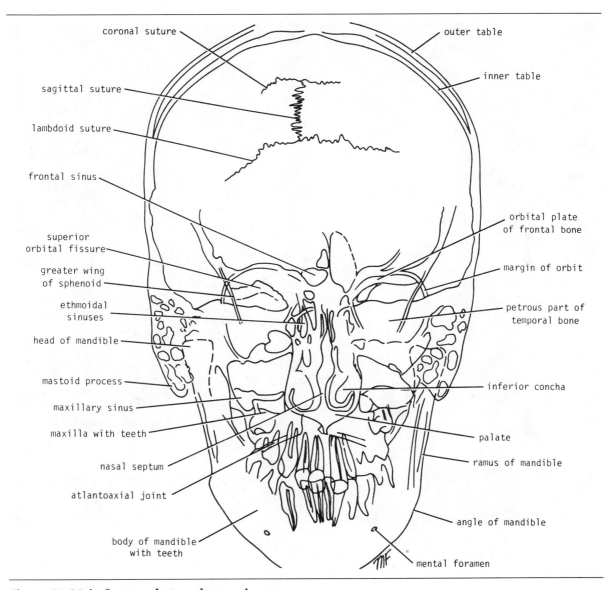

coronal suture

sagittal suture

lambdoid suture

frontal sinus

superior orbital fissure

greater wing of sphenoid

ethmoidal sinuses

head of mandible

mastoid process

maxillary sinus

maxilla with teeth

nasal septum

atlantoaxial joint

body of mandible with teeth

outer table

inner table

orbital plate of frontal bone

margin of orbit

petrous part of temporal bone

inferior concha

palate

ramus of mandible

angle of mandible

mental foramen

**Fig. 11-71. Main features that can be seen in postero-anterior radiograph of skull in Figure 11-70.**

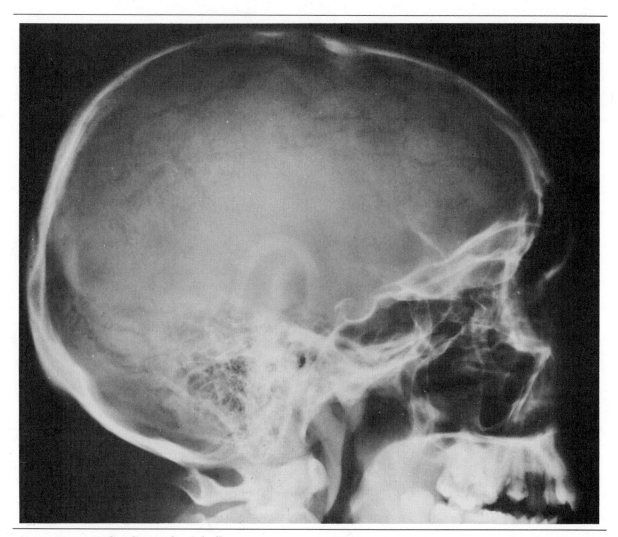

**Fig. 11-72. Lateral radiograph of skull.**

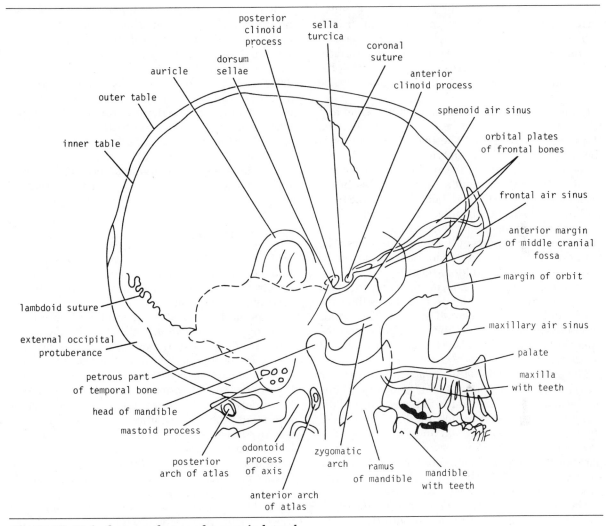

**Fig. 11-73. Main features that can be seen in lateral radiograph of skull in Figure 11-72.**

the two orbital plates of the frontal bones, which form the roofs of the orbits, can be demonstrated. Behind these are the lesser wings of the sphenoid, the anterior clinoid processes, and the sella turcica. The curved lines of the greater wings of the sphenoid and the sphenoidal air sinuses should also be recognized.

Behind the sella turcica, the dorsum sellae and the posterior clinoid processes are clearly seen (Figs. 11-72 and 11-73). The two petrous parts of the temporal bones are superimposed and form a dense shadow between the middle and posterior cranial fossae. Translucent areas formed by the external auditory meatus and, behind them, the mastoid air cells can be identified. The auricle of the external ear frequently produces a curved shadow above the petrous parts of the temporal bones. The temporomandibular joint can be recognized in front of the external auditory meatus.

The nasal bones, the cribriform plate, the hard palate, the maxillary air sinus, and the teeth of the upper and lower jaws can all be seen. The ramus and body of the mandible, the hyoid bone, and the upper part of the cervical vertebral column should be identified.

The *postero-anterior view of the skull to visualize the paranasal sinuses* (Fig. 11-74) is taken with the forehead and nose against the film cassette and the X-ray tube positioned behind the head, but tilted slightly caudally. The frontal and ethmoid sinuses are well shown, but unfortunately the petrous parts of the temporal bones obscure the maxillary sinuses (Fig. 11-75). The ethmoid bones are also superimposed on the sphenoidal sinuses.

The *lateral view of the skull to visualize the paranasal sinuses* (Fig. 11-76) is taken with the patient positioned in exactly the same manner as for a routine lateral radiograph. The sphenoidal and frontal air sinuses are well shown (Fig. 11-77). The ethmoidal and maxillary sinuses are also seen, but the bony trabeculae somewhat obscure the view.

## Cerebral Arteriography

The technique of cerebral arteriography is used in the detection of abnormalities of the cerebral arteries and in the detection and localization of space-occupying lesions such as tumors, blood clots, or abscesses. With the patient under general anesthesia and in the supine position, the head is

centered on a radiographic apparatus that will take repeated radiographs at 2-second intervals. Both anteroposterior and lateral projections are obtained. A radiopaque medium is rapidly injected into the lumen of the common carotid or vertebral arteries. As the radiopaque material is introduced, a series of films are exposed. By this means the cerebral arteries can be demonstrated and their position and patency determined (Figs. 11-78 to 11-81). Unfortunately this technique is not without risk, since the insertion of a needle through the wall of an artery or the manipulation of a catheter within its lumen may dislodge an atheromatous plaque, leading to cerebral embolism.

## Computerized Axial Tomography

A new technique, known as computerized axial tomography, has been introduced for the detection of intracranial lesions. It is safe and provides very accurate information.

Computerized axial tomography relies on the same physics as conventional X-rays, in that structures are distinguished from one another by their ability to absorb energy from X-rays. The beams of X-rays, having passed through the region of the body under consideration, are collected by a special X-ray detector. The information is fed to a computer that processes the information, which is then displayed as a reconstructed picture on a television-like screen. Essentially, the observer sees an image of a thin slice through, for example, the head, which may then be photographed for later examination (Figs. 11-82 and 11-83). The procedure is quick, lasting only a few seconds for each slice, and most patients require no sedation.

## Radiographic Appearance of the Cervical Vertebral Column

The views commonly used to visualize the cervical vertebral column are (1) the anteroposterior and (2) the lateral view.

The *anteroposterior view* is taken with the patient in the supine position, with the film cassette behind the head and neck. The X-ray tube is centered on the prominent anterior border of the thyroid cartilage. In this view, the body of the mandible obscures the upper three cervical vertebrae, unless the mandibular shadow is blurred by the patient's opening and closing the mouth on request during exposure. Alternatively, the X-ray tube can be centered over the open mouth to obtain a good view of the first three cervical vertebrae.

The bodies and spines of the vertebrae may be visualized. The different parts of the transverse processes, the foramen transversarium, and the articular processes are superimposed. It is usually possible to identify the pedicles.

With the X-ray tube centered over the open mouth, the atlanto-axial articulation can be visualized, with the odontoid process lying between the lateral masses of the atlas.

The *lateral view* (Fig. 11-84) is taken with the film cassette situated parallel to the sagittal plane of the vertebral column. The shoulders are lowered as much as possible, so that the seventh cervical vertebra is not obscured. The X-ray tube is directed toward the side of the neck, at a right angle to the film.

The anterior and posterior arches of the atlas are well shown, and the odontoid process and body of the axis can be seen. The spinous processes and the joint surfaces of the articular processes are clearly visualized. The anterior and posterior surfaces of the vertebral bodies and the posterior surface of the vertebral canal are well shown. The transverse processes can be identified, but they are superimposed on the bodies of the vertebrae.

## Summary of the Courses and Distribution of the Cranial Nerves

There are twelve pairs of cranial nerves, which leave the brain and pass through foramina in the skull. All the nerves are distributed in the head and neck except the tenth, which also supplies structures in the thorax and abdomen. (See Table 11-3.)

The cranial nerves are named as follows:

1. Olfactory.
2. Optic.
3. Oculomotor.
4. Trochlear.
5. Trigeminal.
6. Abducent.
7. Facial.
8. Vestibulocochlear.
9. Glossopharyngeal.
10. Vagus.
11. Accessory.
12. Hypoglossal.

The olfactory, optic, and vestibulocochlear nerves are entirely sensory; the oculomotor, trochlear, abducent, accessory, and hypoglossal nerves are entirely motor; and the remaining nerves are mixed.

## OLFACTORY NERVES

The olfactory nerves, or nerves of smell, arise from *olfactory receptor nerve cells* in the olfactory mucous membrane. The olfactory mucous membrane is situated in the upper part of the nasal cavity above the level of the superior concha (Fig. 11-85). Bundles of these olfactory nerve fibers pass through the openings of the cribriform plate of the ethmoid bone to enter the *olfactory bulb* in the cranial cavity. The olfactory bulb is connected to the olfactory area of the cerebral cortex by the *olfactory tract.*

## OPTIC NERVE

The optic nerve, or nerve of sight, is composed of the axons of the cells of the *ganglionic layer* of the retina. The optic nerve emerges from the back of the eyeball and leaves the orbital cavity through the optic canal to enter the cranial cavity. The optic nerve then unites with the optic nerve of the opposite side to form the optic chiasma (Fig. 11-85).

In the chiasma, the fibers from the medial half of each retina cross the midline and enter the *optic tract* of the opposite side, while the fibers from the lateral half of each retina pass posteriorly in the optic tract of the same side. Most of the fibers of the optic tract terminate by synapsing with nerve cells in the *lateral geniculate body* (Fig. 11-85). A few fibers pass to the pretectal nucleus and the superior colliculus and are concerned with light reflexes.

The axons of the nerve cells of the lateral geniculate body pass posteriorly as the *optic radiation* and terminate in the *visual cortex* of the cerebral hemisphere (Fig. 11-85).

## OCULOMOTOR NERVE

The oculomotor nerve emerges on the anterior surface of the midbrain (Fig. 11-86). It passes forward

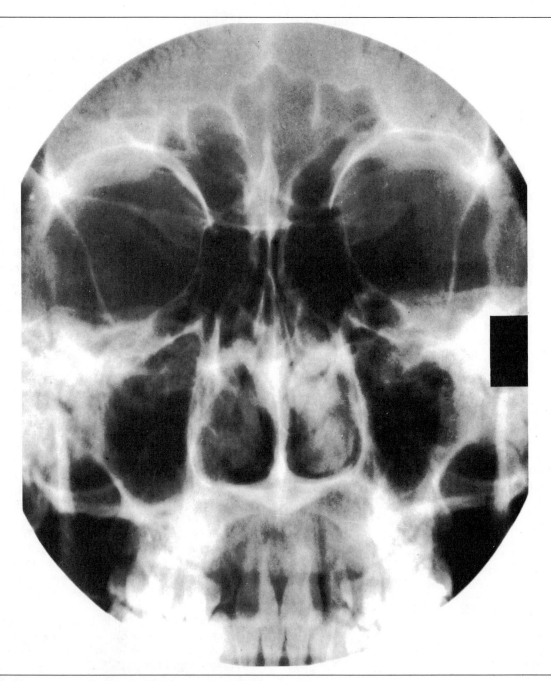

**Fig. 11-74. Postero-anterior radiograph of skull for paranasal sinuses.**

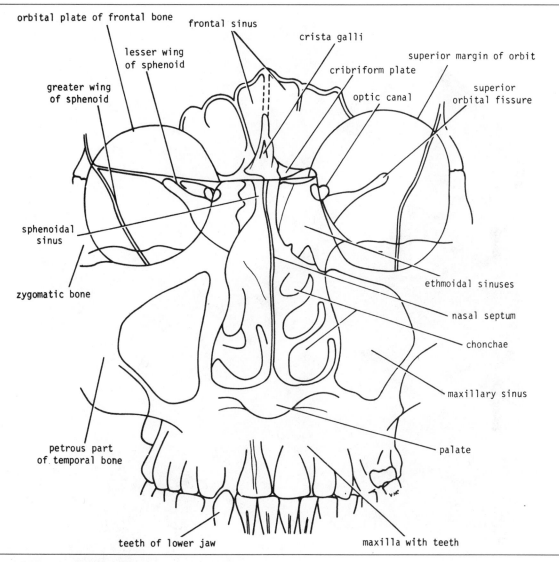

**Fig. 11-75. Main features that can be seen in pos-
tero-anterior radiograph of skull in Figure 11-74.**

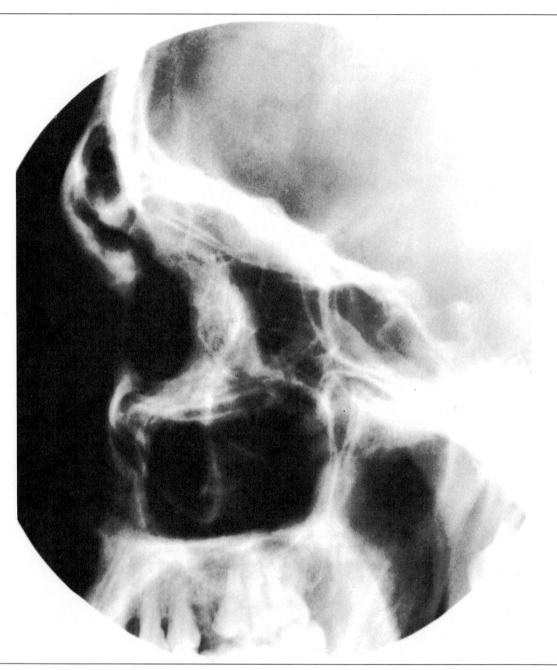

**Fig. 11-76. Lateral radiograph of skull for para-
nasal sinuses.**

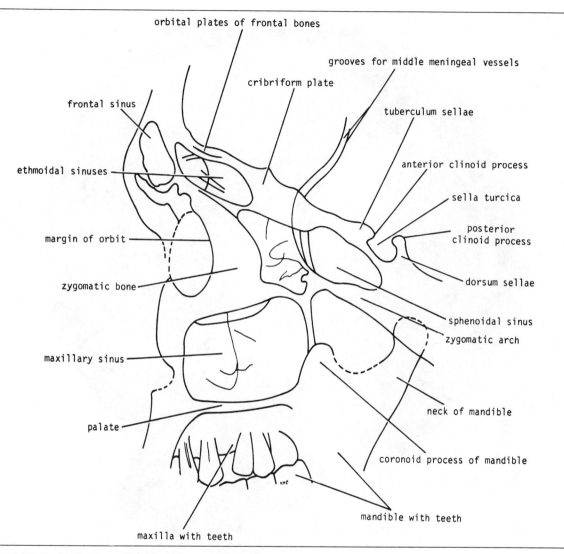

**Fig. 11-77. Main features that can be seen in lateral radiograph of skull in Figure 11-76.**

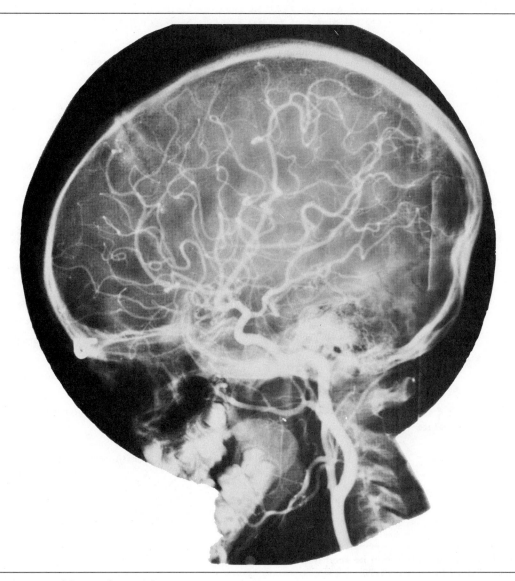

Fig. 11-78. Lateral internal carotid arteriogram. (From R. S. Snell and A. C. Wyman, *An Atlas of Normal Radiographic Anatomy.* Boston: Little, Brown, 1976.)

Fig. 11-79. Main features that can be seen in arteriogram in Figure 11-78.

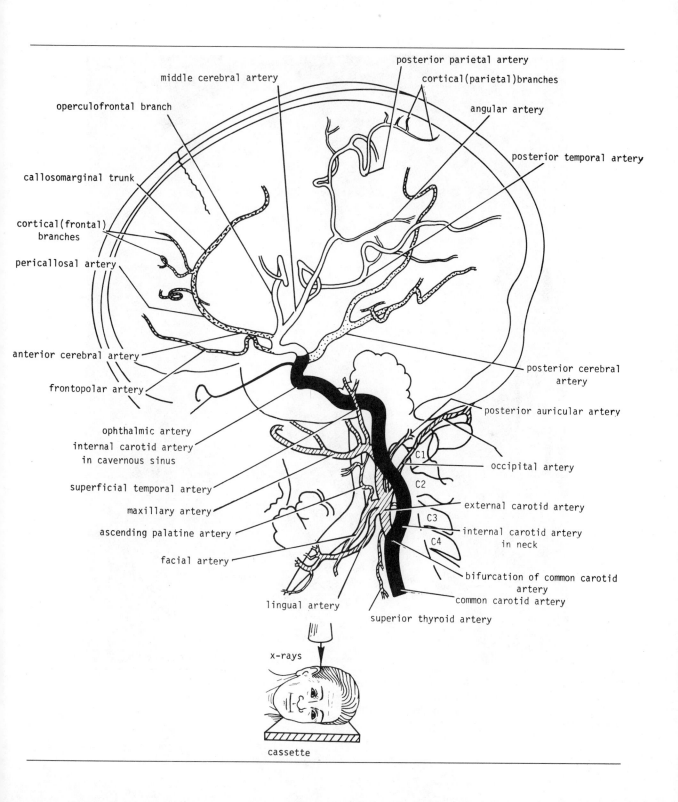

middle cerebral artery

posterior parietal artery

cortical(parietal)branches

angular artery

operculofrontal branch

posterior temporal artery

callosomarginal trunk

cortical(frontal) branches

pericallosal artery

posterior cerebral artery

anterior cerebral artery

posterior auricular artery

frontopolar artery

ophthalmic artery

internal carotid artery in cavernous sinus

occipital artery

superficial temporal artery

external carotid artery

maxillary artery

internal carotid artery in neck

ascending palatine artery

facial artery

bifurcation of common carotid artery

common carotid artery

lingual artery

superior thyroid artery

C1

C2

C3

C4

x-rays

cassette

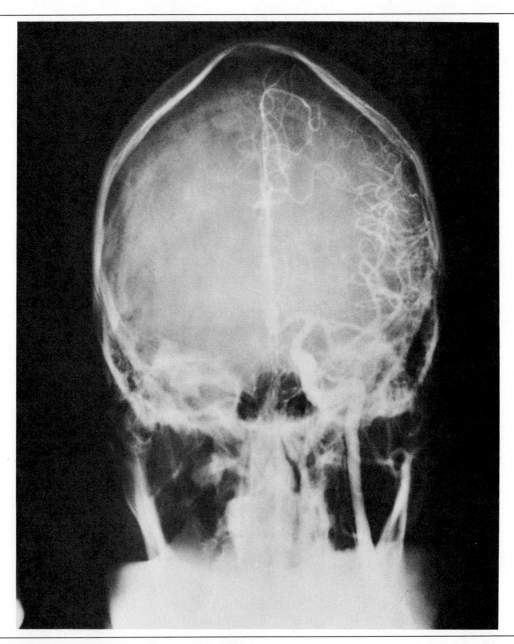

Fig. 11-80. Anteroposterior internal carotid arteriogram. (From R. S. Snell and A. C. Wyman, *An Atlas of Normal Radiographic Anatomy.* Boston: Little, Brown, 1976.)

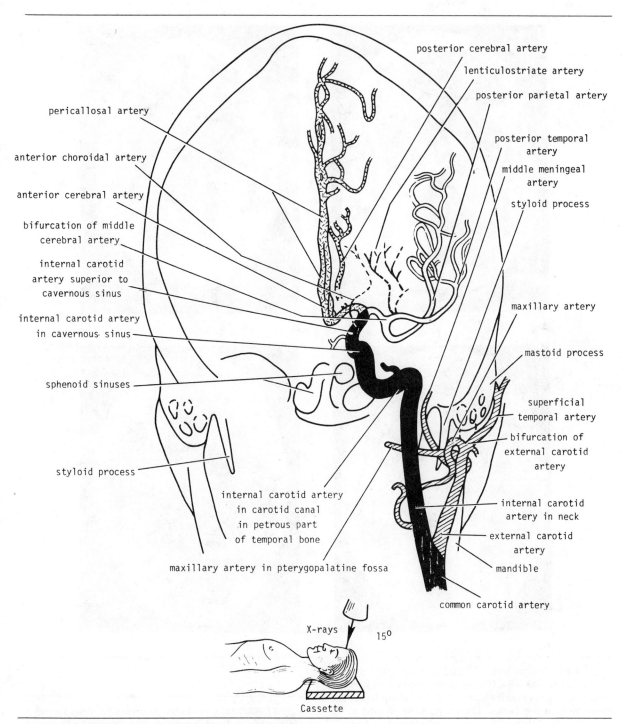

posterior cerebral artery

lenticulostriate artery

posterior parietal artery

posterior temporal artery

middle meningeal artery

styloid process

maxillary artery

mastoid process

superficial temporal artery

bifurcation of external carotid artery

internal carotid artery in neck

external carotid artery

mandible

common carotid artery

pericallosal artery

anterior choroidal artery

anterior cerebral artery

bifurcation of middle cerebral artery

internal carotid artery superior to cavernous sinus

internal carotid artery in cavernous sinus

sphenoid sinuses

styloid process

internal carotid artery in carotid canal in petrous part of temporal bone

maxillary artery in pterygopalatine fossa

X-rays     15°

Cassette

**Fig. 11-81. Main features that can be seen in arteriogram in Figure 11-80.**

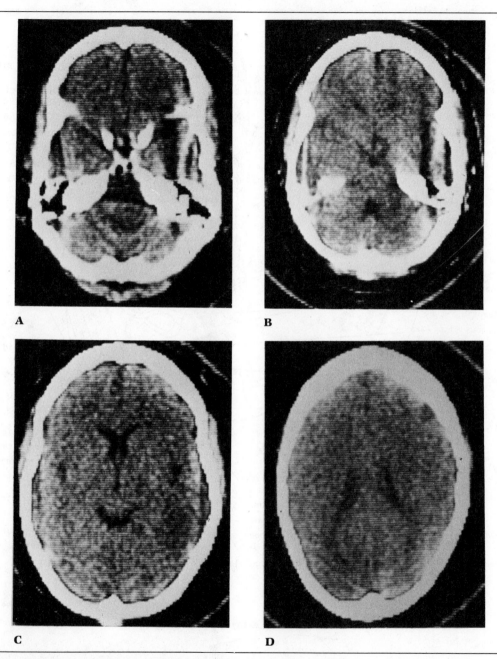

A

B

C

D

Fig. 11-82. Computerized axial tomography of the adult brain. (A), (B), (C), and (D) represent serial cuts taken progressively through the skull from the base toward the vertex. (From R. S. Snell and A. C. Wyman, *An Atlas of Normal Radiographic Anatomy.* Boston: Little, Brown, 1976.)

Fig. 11-83. Main features that can be seen in the computerized axial tomograms shown in Figure 11-82.

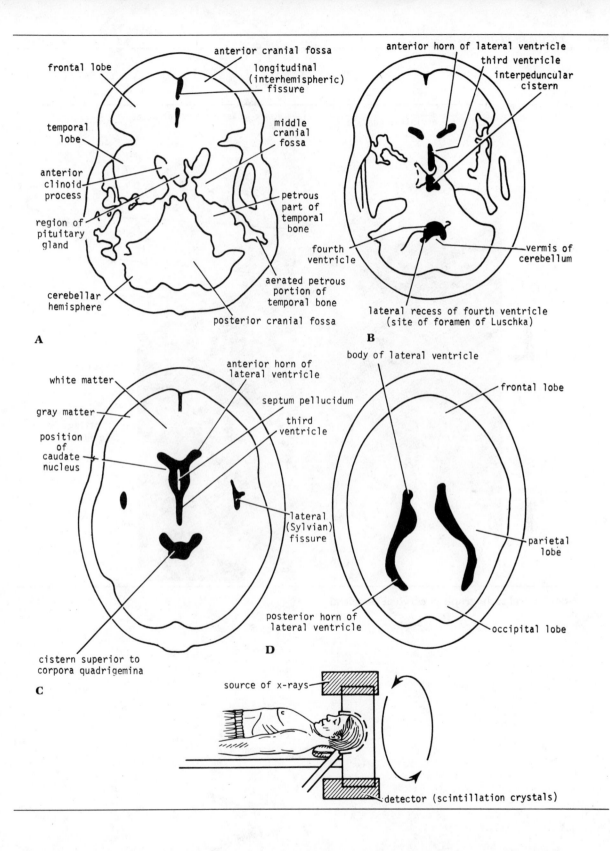

**A**
- frontal lobe
- anterior cranial fossa
- longitudinal (interhemispheric) fissure
- temporal lobe
- middle cranial fossa
- anterior clinoid process
- region of pituitary gland
- petrous part of temporal bone
- cerebellar hemisphere
- aerated petrous portion of temporal bone
- posterior cranial fossa

**B**
- anterior horn of lateral ventricle
- third ventricle
- interpeduncular cistern
- fourth ventricle
- vermis of cerebellum
- lateral recess of fourth ventricle (site of foramen of Luschka)

**C**
- white matter
- anterior horn of lateral ventricle
- gray matter
- septum pellucidum
- third ventricle
- position of caudate nucleus
- lateral (Sylvian) fissure
- cistern superior to corpora quadrigemina

**D**
- body of lateral ventricle
- frontal lobe
- parietal lobe
- posterior horn of lateral ventricle
- occipital lobe

- source of x-rays
- detector (scintillation crystals)

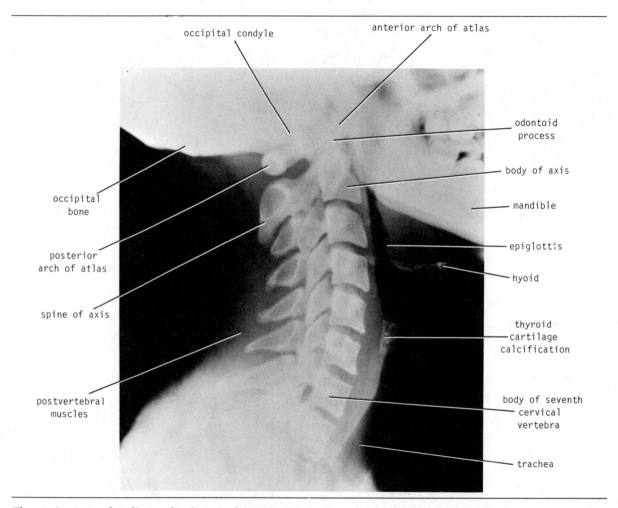

occipital condyle

anterior arch of atlas

odontoid process

body of axis

mandible

epiglottis

hyoid

thyroid cartilage calcification

body of seventh cervical vertebra

trachea

occipital bone

posterior arch of atlas

spine of axis

postvertebral muscles

**Fig. 11-84. Lateral radiograph of cervical vertebral column.**

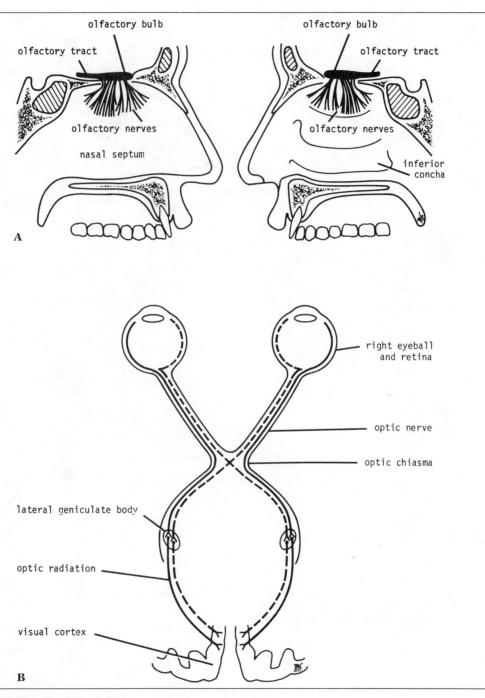

**Fig. 11-85. (A) Distribution of olfactory nerves on nasal septum and lateral wall of nose. (B) The optic nerve and its connections.**

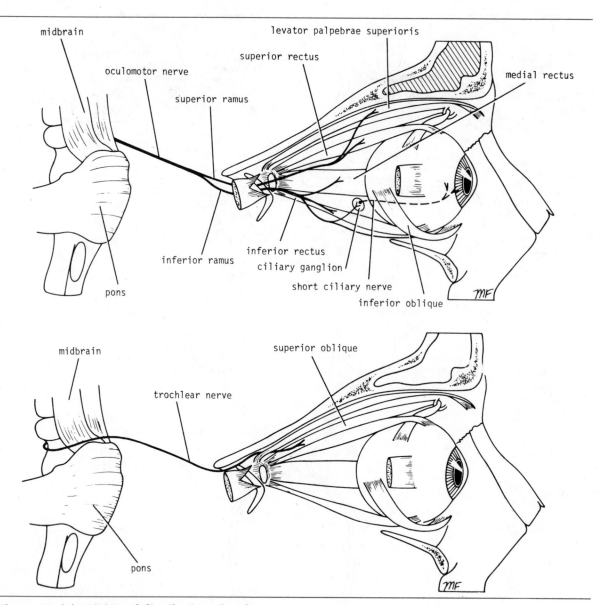

**Fig. 11-86. (A) Origin and distribution of oculomotor nerve. (B) Origin and distribution of trochlear nerve.**

in the middle cranial fossa in the lateral wall of the cavernous sinus. Here, it divides into a *superior* and an *inferior ramus*, which enter the orbital cavity through the superior orbital fissure. The superior and inferior rami of the oculomotor nerve supply the following extrinsic muscles of the eye: the levator palpebrae superioris, superior rectus, medial rectus, inferior rectus, and inferior oblique (Fig. 11-86). The oculomotor nerve also supplies two groups of intrinsic muscles, namely, the constrictor pupillae of the iris and the ciliary muscles. This nerve is therefore responsible for lifting the upper eyelid; turning the eye upward, downward, and medially; constricting the pupil; and allowing accommodation of the eye.

## TROCHLEAR NERVE

The trochlear nerve, the most slender of the cranial nerves, leaves the posterior surface of the midbrain and immediately decussates with the nerve of the opposite side (Fig. 11-86). The trochlear nerve passes forward through the middle cranial fossa in the lateral wall of the cavernous sinus. Having entered the orbital cavity through the superior orbital fissure, it supplies the superior oblique muscle of the eyeball. This nerve therefore assists in turning the eye downward and laterally.

## TRIGEMINAL NERVE

The trigeminal nerve, the largest of the cranial nerves, leaves the anterior aspect of the pons as a small *motor root* and a large *sensory root*. The nerve passes forward from the posterior cranial fossa to reach the apex of the petrous part of the temporal bone in the middle cranial fossa. Here, the large sensory root expands to form the *trigeminal ganglion* (Fig. 11-87). The motor root of the trigeminal nerve is situated below the sensory ganglion and is completely separate from it. The ophthalmic, maxillary, and mandibular nerves arise from the anterior border of the ganglion (Fig. 11-87).

The *ophthalmic nerve* is purely sensory (Fig. 11-87). It runs forward in the lateral wall of the cavernous sinus in the middle cranial fossa and divides into three branches, the *lacrimal, frontal,* and *nasociliary nerves,* which enter the orbital cavity through the superior orbital fissure. The nerves are distributed to the cornea of the eye, the skin of the forehead and scalp, the eyelids, the mucous membrane of the paranasal sinuses, and the nasal cavity.

The *maxillary nerve* is purely sensory (Fig. 11-87). It leaves the skull through the foramen rotundum and is eventually distributed to the skin of the face overlying the maxilla, the teeth of the upper jaw, the mucous membrane of the nose, the maxillary air sinus, and the palate.

The *mandibular nerve* is motor and sensory (Fig. 11-87). The sensory root leaves the trigeminal ganglion and passes out of the skull through the foramen ovale. The motor root of the trigeminal nerve also leaves the skull through the same foramen and joins the sensory root to form the trunk of the mandibular nerve. The sensory fibers of the mandibular nerve supply the skin of the cheek and the skin over the mandible and on the side of the head. They also supply the temporomandibular joint and the teeth of the lower jaw, the mucous membrane of the cheek, the floor of the mouth, and the anterior part of the tongue.

The motor fibers of the mandibular nerve supply the muscles of mastication; the mylohyoid muscle, which forms the floor of the mouth; the anterior belly of the digastric muscle; the tensor veli palatini of the soft palate; and the tensor tympani of the middle ear.

The trigeminal nerve is thus the main sensory nerve of the head and innervates the muscles of mastication. It also tenses the soft palate and the tympanic membrane.

## ABDUCENT NERVE

This small nerve emerges from the anterior surface of the hindbrain between the pons and the medulla oblongata (Fig. 11-87). It passes forward through the cavernous sinus in the middle cranial fossa and enters the orbit through the superior orbital fissure (Fig. 11-87). The abducent nerve supplies the lateral rectus muscle and is, therefore, responsible for turning the eye laterally.

## FACIAL NERVE

The facial nerve emerges as two roots from the anterior surface of the hindbrain between the pons and the medulla oblongata. The roots pass laterally in the posterior cranial fossa with the vestibulo-

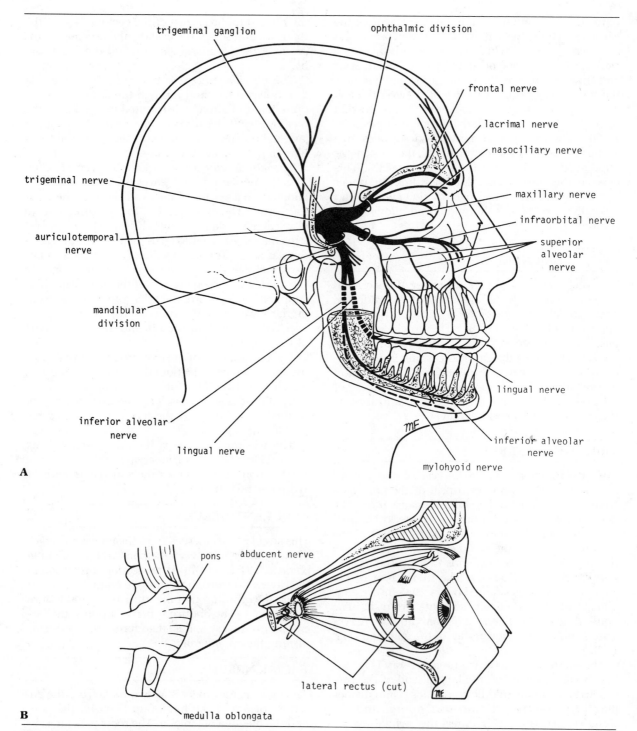

**Fig. 11-87. (A) Distribution of trigeminal nerve. (B) Origin and distribution of abducent nerve.**

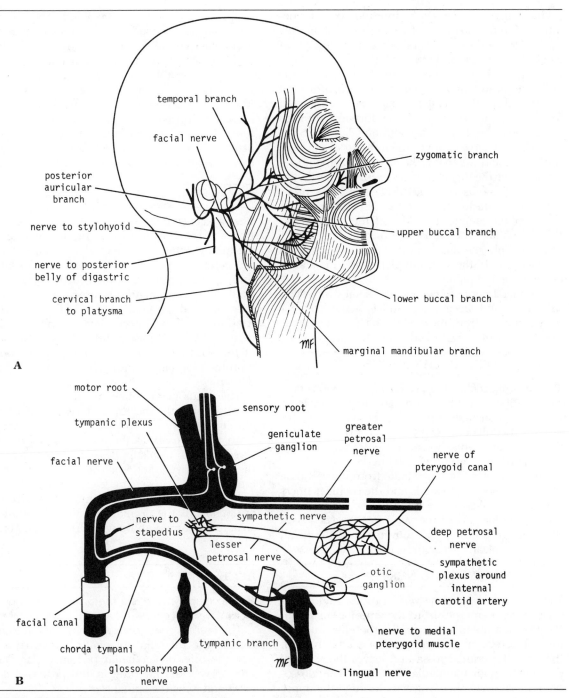

**A**

**B**

**Fig. 11-88. (A) Distribution of facial nerve. (B) Branches of facial nerve within petrous part of temporal bone; the taste fibers are shown in white. The glossopharyngeal nerve is also shown.**

cochlear nerve and enter the internal acoustic meatus in the petrous part of the temporal bone. At the bottom of the meatus the nerve enters the facial canal that runs laterally through the inner ear. The facial nerve then becomes related to the middle ear and the aditus to the tympanic antrum, and emerges from the canal through the stylomastoid foramen. The nerve now passes forward through the parotid gland to its distribution (Fig. 11-88).

The facial nerve supplies the muscles of the face, the cheek, and the scalp, the stylohyoid, the posterior belly of the digastric muscles of the neck, and the stapedius muscle of the middle ear. The sensory root carries taste fibers from the anterior two-thirds of the tongue, the floor of the mouth, and the soft palate. The parasympathetic secretomotor fibers supply the submandibular and sublingual salivary glands, the lacrimal gland, and the glands of the nose and palate.

The facial nerve thus controls facial expression, salivation, and lacrimation and is a pathway for taste sensation from the anterior part of the tongue and floor of the mouth and from the soft palate.

## VESTIBULOCOCHLEAR NERVE

The vestibulocochlear nerve consists of two sets of sensory fibers, vestibular and cochlear. They leave the anterior surface of the brain between the pons and the medulla oblongata (Fig. 11-89). They cross the posterior cranial fossa and enter the internal acoustic meatus with the facial nerve.

The vestibular fibers originate from the vestibule and the semicircular canals and the cochlear fibers from the cochlea of the internal ear (Fig. 11-89). The vestibular part of the nerve is concerned with the sense of position and movement of the head, and the cochlear part with hearing.

## GLOSSOPHARYNGEAL NERVE

The glossopharyngeal nerve is a motor and a sensory nerve. It emerges from the anterior surface of the medulla oblongata between the olive and the inferior cerebellar peduncle. It passes laterally in the posterior cranial fossa and leaves the skull by passing through the jugular foramen. The glossopharyngeal nerve then descends through the upper part of the neck to the back of the tongue (Fig. 11-89). The motor fibers supply the stylopharyngeus muscle; the parasympathetic secretomotor fibers

supply the parotid salivary gland. The sensory fibers, which are concerned with general sensation and taste, pass to the posterior third of the tongue and the pharynx; they also innervate the carotid sinus and carotid body.

The glossopharyngeal nerve thus assists swallowing and promotes salivation. It also conducts sensation from the pharynx and the back of the tongue and carries impulses, which influence the arterial blood pressure and respiration, from the carotid sinus and carotid body.

## VAGUS NERVE

The vagus nerve is composed of motor and sensory fibers. It emerges from the anterior surface of the medulla oblongata between the olive and the inferior cerebellar peduncle. The nerve passes laterally through the posterior cranial fossa and leaves the skull through the jugular foramen. The vagus nerve then descends through the neck alongside the carotid arteries and internal jugular vein within the carotid sheath. It passes through the mediastinum of the thorax (Fig. 11-90), pierces the diaphragm with the esophagus, and terminates within the abdomen.

The vagus nerve innervates the heart and great vessels within the thorax, the larynx, trachea, bronchi, and lungs, and much of the alimentary tract from the pharynx to the distal part of the transverse colon. It also supplies glands associated with the alimentary tract, such as the liver and pancreas.

The vagus nerve has the most extensive distribution of all the cranial nerves and supplies the structures named above with afferent and efferent fibers.

## ACCESSORY NERVE

The accessory nerve is a motor nerve. It consists of a cranial root (part) and a spinal root (part).

The cranial root emerges from the anterior surface of the medulla oblongata between the olive and the inferior cerebellar peduncle (Fig. 11-91). The nerve runs laterally in the posterior cranial fossa and joins the spinal root.

The spinal root arises from nerve cells in the anterior gray column (horn) of the upper five segments of the cervical part of the spinal cord. The

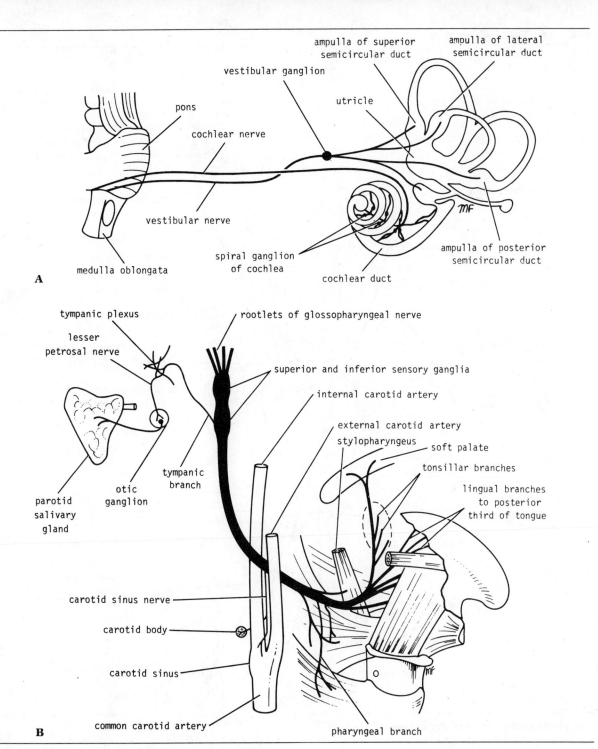

**Fig. 11-89. (A) Origin and distribution of vestibulo-cochlear nerve. (B) Distribution of glossopharyn-geal nerve.**

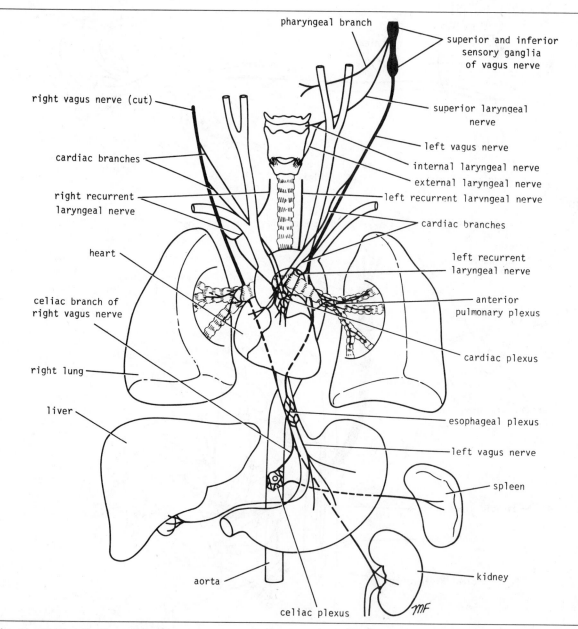

**Fig. 11-90. Distribution of vagus nerve.**

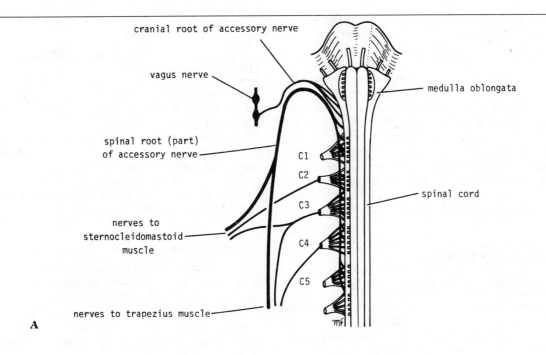

A

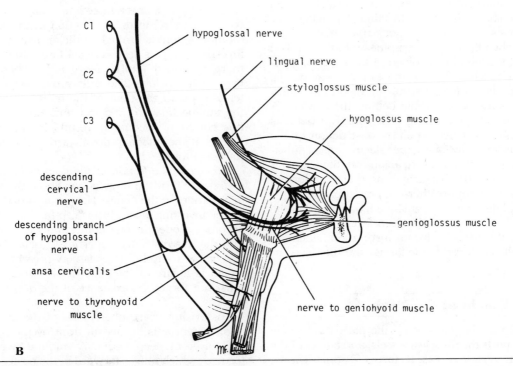

B

**Fig. 11-91. (A) Origin and distribution of accessory nerve. (B) Distribution of hypoglossal nerve.**

nerve ascends alongside the spinal cord (Fig. 11-91) and enters the skull through the foramen magnum. It then turns laterally to join the cranial root.

The two roots unite and leave the skull through the jugular foramen. The roots then separate and the cranial root joins the vagus nerves and is distributed in its branches to the muscles of the soft palate and pharynx (via the pharyngeal plexus), and to the muscles of the larynx (except the cricothyroid muscle). The spinal root supplies the sternocleidomastoid and trapezius muscles.

The accessory nerve thus brings about movements of the soft palate, pharynx, and larynx, and controls the movements of the sternocleidomastoid and trapezius muscles, two large muscles in the neck.

## HYPOGLOSSAL NERVE

The hypoglossal nerve is a motor nerve. It emerges on the anterior surface of the medulla oblongata between the pyramid and the olive, crosses the posterior cranial fossa, and leaves the skull through the hypoglossal canal. The nerve then passes downward and forward in the neck to reach the tongue (Fig. 11-91). The hypoglossal nerve innervates the muscles of the tongue and thus controls the shape and movements of the tongue.

## CLINICAL NOTES

### Superficial Veins of the Neck

The *external jugular vein* descends in the neck in the superficial fascia. It runs from the angle of the mandible over the sternocleidomastoid and pierces the investing layer of deep cervical fascia a fingerbreadth above the midpoint of the clavicle, to drain into the subclavian vein. In clinical practice this vein serves as a useful venous manometer. Normally, when the patient is supine, with the head on pillows, the level of the blood in the external jugular veins reaches about a third of the way up the neck. As the patient sits up, the blood level falls, until it is no longer visible behind the clavicle. A rise in venous pressure may be found in patients with right-sided heart failure, obstruction of the superior vena cava by a neoplasm, raised intrathoracic pressure, or overtransfusion. Singers tend to have enlarged superficial veins, due to prolonged periods of raised intrathoracic pressure.

Should the external jugular vein be severed at the point where it pierces the deep fascia, air may be drawn into the vein during inspiration, since the vein wall is tethered to the fascia, and the vein cannot collapse.

### Deep Fascia of the Neck

The deep fascia lies beneath the platysma muscle and supports the muscles, vessels, and nerves of the neck. In certain areas it forms distinct sheets, which are called (1) the investing, (2) the pretracheal, and (3) the prevertebral layers. These have been described in detail on page 711.

These fascial layers are easily recognizable to the surgeon at operation. Their clinical importance stems from the fact that they may determine the direction of the spread of pus, or may limit the spread of pus, in neck infections. *Ludwig's angina*, which is a streptococcal infection in the submandibular region, is limited by the attachment of the investing layer of deep cervical fascia to the lower margin of the mandible above and the body of the hyoid bone below. The inflammatory edema extends beneath the floor of the mouth and pushes the tongue forward and upward. Failure to treat this conditon adequately results in the eventual spread of the infection downward, with edema of the vocal folds.

*Tuberculous infection of the deep cervical lymph nodes* may result in liquefaction and destruction of one or more of the nodes. The pus is at first limited by the investing layer of the deep fascia. Later, this becomes eroded at one point, and the pus passes into the less restricted superficial fascia. A dumbbell or collar-stud abscess is now present. The clinician is aware of the superficial abscess, but he must not forget the existence of the deeply placed abscess.

Pus arising from tuberculosis of the upper cervical vertebrae is limited in front by the prevertebral layer of deep fascia. A midline swelling is formed, which bulges forward in the posterior wall of the pharynx. The pus then tracks laterally and downward behind the carotid sheath, to reach the posterior triangle. Here, the fascia, which forms a covering to the muscular floor of the triangle, is weaker, and the abscess points behind the sterno-cleidomastoid. Rarely, the abscess may track downward behind the prevertebral fascia, to reach the superior and posterior mediastina in the thorax.

It is important to distinguish this condition from an abscess involving the *retropharyngeal lymph nodes.* These nodes lie in front of the prevertebral layer of fascia, but behind the buccopharyngeal fascia, which covers the outer surface of the constrictor muscles. Such an abscess usually points on the posterior pharyngeal wall and, if untreated, ruptures into the pharyngeal cavity.

## Sternocleidomastoid Muscle

### Congenital Torticollis

It is now generally accepted that most cases of congenital torticollis are a result of excessive stretching of the sternocleidomastoid muscle during a difficult labor. Hemorrhage occurs into the muscle and may be detected as a small rounded "tumor" during the early weeks after birth. Later, this becomes invaded by fibrous tissue, which contracts and shortens the muscle. The mastoid process is thus pulled down toward the sternoclavicular joint of the same side; the cervical spine is flexed; and the face looks upward to the opposite side. If left untreated, asymmetrical growth changes will occur in the face, and the cervical vertebrae may become wedge-shaped.

### Spasmodic Torticollis

This condition, which results from repeated chronic contractions of the sternocleidomastoid and trapezius muscles, is usually psychogenic in origin. Section of the spinal part of the accessory nerve may be necessary in severe cases.

## Posterior Triangle of the Neck

The *spinal part of the accessory nerve* emerges from behind the middle of the posterior border of the sternocleidomastoid muscle. It crosses the posterior triangle on the levator scapulae muscle in a relatively superficial position. It leaves the triangle by passing deep to the anterior border of the trapezius muscle, which it supplies.

This nerve may be injured at operation or from penetrating wounds. The trapezius muscle is paralyzed; the muscle will show wasting; and the shoulder will drop. The patient will experience difficulty in elevating the arm above the head, having abducted it to a right angle by using the deltoid muscle.

The *roots* and *trunks of the brachial plexus* occupy the anterior-inferior angle of the posterior triangle of the neck. Incomplete lesions may result from stab or bullet wounds, traction, or pressure injuries. The clinical findings in the Erb-Duchenne and the Klumpke's lesions are fully described on page 527.

It will be remembered that the axillary sheath, formed from the prevertebral layer of deep cervical fascia, encloses the brachial plexus and the axillary artery. A *brachial plexus nerve block* can easily be obtained by closing the distal part of the sheath in the axilla with finger pressure, inserting a syringe needle into the proximal part of the sheath, and then injecting a local anesthetic. The anesthetic solution is massaged along the sheath, producing a nerve block. The syringe needle may be inserted into the axillary sheath in the lower part of the posterior triangle of the neck or in the axilla.

At the root of the neck, the brachial plexus and the *subclavian artery* enter the posterior triangle through a narrow muscular-bony triangle. The boundaries of the narrow triangle are formed in front by the scalenus anterior, behind by the scalenus medius, and below by the first rib. In the presence of a *cervical rib* (see p. 76), the first thoracic nerve and the subclavian artery will be raised and angulated as they pass over the rib. Partial or complete occlusion of the artery causes ischemic muscle pain in the arm, which is worsened by exercise, Rarely, pressure on the first thoracic nerve causes symptoms of pain in the forearm and hand and wasting of the small muscles of the hand.

In severe traumatic accidents to the upper limb involving laceration of the brachial or axillary arteries, it is important to remember that the hemorrhage may be stopped by exerting strong pressure downward and backward on the third part of the subclavian artery. The use of a blunt object to exert the pressure is of great help, and the artery is compressed against the upper surface of the first rib.

The *cervical dome of the pleura* and the *apex of the lung* extend up into the root of the neck on each side. Covered by the suprapleural membrane, they lie behind the subclavian artery. A penetrating wound above the medial end of the clavicle may involve the apex of the lung.

## Anterior Triangle of the Neck

The *platysma* lies in the superficial fascia that roofs over the anterior triangle. Functionally, it is an unimportant muscle. However, the surgeon must take great care to suture the cut edges of this muscle when sewing up incisions, since failure to do so will result in stretching of the skin scar.

The *nerve supply to the platysma*, the cervical branch of the facial nerve, emerges from the lower end of the parotid gland and travels to the platysma muscle; it then sometimes crosses the lower border of the mandible, to supply the depressor anguli oris. In operations on the face or upper part of the neck, accidental section of this nerve may result in distortion of the shape of the mouth.

The bifurcation of the *common carotid artery* into the internal and external carotid arteries can be easily palpated just beneath the anterior border of the sternocleidomastoid muscle at the level of the superior border of the thyroid cartilage. This is a convenient site to take the *carotid pulse.*

In cases of *carotid sinus hypersensitivity,* pressure on one or both carotid sinuses may cause excessive slowing of the heart rate, a fall in blood pressure, and cerebral ischemia with fainting.

Extensive *arteriosclerosis of the internal carotid artery* in the neck may cause visual impairment or blindness in the eye on the side of the lesion, due to insufficient blood flow through the retinal artery. There may also be motor paralysis and sensory loss on the opposite side of the body caused by insufficient blood flow through the middle cerebral artery.

The *midline structures in the neck* should be readily recognized as one passes an examining finger down the neck from the chin to the suprasternal notch. (For details, see p. 702.) The physician commonly forgets that an enlarged submental lymph node may be due to a pathological condition

anywhere between the tip of the tongue and the point of the chin.

The *trachea* can be readily felt below the larynx. As it descends, it becomes deeply placed and may lie as much as 1½ inches (4 cm) from the surface at the suprasternal notch. Remember that in the adult it may measure as much as 1 inch (2.5 cm) in diameter, but in a 3-year-old child it may measure only ⅕ inch (0.5 cm) in diameter. The trachea is a mobile elastic tube and is easily displaced by the enlargement of adjacent organs or the presence of tumors. Remember also that lateral displacement of the cervical part of the trachea may be due to a pathological lesion in the thorax.

*Tracheotomy* is a commonly performed procedure in which a tube is inserted into the lumen of the trachea through the front of the neck to provide an airway. It may be carried out at two sites in patients with laryngeal obstruction. *High tracheotomy* is performed in the interval between the cricoid cartilage and the isthmus of the thyroid gland.[†] The trachea is steadied by extending the neck over a sandbag. A vertical incision is made through the following tissues: (1) skin; (2) superficial fascia (beware of the anterior jugular veins, which lie on either side of the midline close together in the plane); (3) investing layer of deep cervical fascia; (4) pretracheal fascia (separate the sternohyoid muscles and incise the fascia); and (5) the trachea, which is incised through a vertical incision after retracting the isthmus of the thyroid gland downward.

Avoid the following anatomical pitfalls: (1) In patients with chronic respiratory obstruction, the anterior jugular veins are distended with blood and are often touching in the midline. (2) Stretch and steady the trachea by extending the neck. The trachea is very mobile in the living and may slip to one side, causing you to plunge your knife into the esophagus. (3) If you have to cut the isthmus of the thyroid gland, be sure to ligate the terminal branches of the superior thyroid artery, which anastomose with one another along the upper border of the isthmus. (4) Remember that in small children the trachea is the size of a pencil. An over-

---

[†]In an emergency it is sometimes safer to incise the cricothyroid ligament or membrane. By so doing you avoid the possibility of damaging the isthmus of the thyroid gland.

enthusiastic surgeon may easily plunge his scalpel through the trachea into the esophagus. (5) Make sure the tracheotomy tube is inserted into the lumen of the trachea and does not slide down among the infrahyoid muscles.

*Low tracheotomy* is a more difficult operation. The trachea is opened below the isthmus of the thyroid gland and above the suprasternal notch. A vertical incision is made through the following tissues: (1) skin; (2) superficial fascia; (3) investing layer of deep fascia (the suprasternal space is opened and the jugular arch is found and divided between ligatures); (4) connective tissue containing the inferior thyroid veins and possibly the thyroidea ima artery; (5) pretracheal fascia; and (6) trachea. In incising the trachea, the brachiocephalic vessels must be avoided. In children the thymus may obscure the trachea.

*Self-inflicted cut throat wounds* often fail, since the suicidal patient extends his neck when he makes the wound, and the carotid sheath, with its large vessels, is retracted deeply under cover of the sternocleidomastoid. Unaware of this fact, a suicidal patient often has to make several incisions before the great vessels of the neck are sectioned. The common sites for the wound are immediately above and below the hyoid bone.

The *thyroid gland* is invested in a sheath derived from the pretracheal fascia. This tethers the gland to the larynx and the trachea and explains why the thyroid gland follows the movements of the larynx in swallowing. The close relationship between the trachea and the lobes of the thyroid gland commonly results in pressure on the trachea in patients with pathological enlargement of the thyroid.

The attachment of the sternothyroid muscles to the thryoid cartilage effectively binds down the thyroid gland to the larynx and limits upward expansion of the gland. There being no limitation to downward expansion, it is not uncommon for a pathologically enlarged thyroid gland to extend downward behind the sternum. A retrosternal *goiter* (any abnormal enlargement of the thyroid gland) may compress the trachea and cause dangerous dyspnea; it may also cause severe venous compression.

It should be remembered that the two main arteries supplying the thyroid gland are closely related to important nerves that may be damaged during thyroidectomy operations. The *superior thyroid artery* on each side is related to the external laryngeal nerve, which supplies the cricothyroid muscle. The terminal branches of the *inferior thyroid artery* on each side are related to the recurrent laryngeal nerve. Damage to the external laryngeal nerve results in an inability to tense the vocal folds and in hoarseness. For the results of damage to the recurrent laryngeal nerve, see page 910.

The *parathyroid glands* are usually four in number and are closely related to the posterior surface of the thyroid gland. In partial thyroidectomy operations the posterior part of the thyroid gland is left undisturbed, so that the parathyroid glands are not damaged. The development of the inferior parathyroid glands is closely associated with the thymus. For this reason it is not uncommon for the surgeon to find the inferior parathyroid glands in the superior mediastinum because they have been pulled down into the thorax by the thymus.

As elsewhere in the body, a knowledge of the *lymphatic drainage* of an organ or region is of great clinical importance. Examination of a patient may reveal an enlarged lymph node. It is the physician's responsibility to determine the cause and be knowledgeable about the area of the body that drains its lymph into a particular node. For example, an enlarged submandibular node may be caused by a pathological condition in the scalp, the face, the maxillary sinus, or the tongue. An infected tooth of the upper or lower jaw may be responsible. Often a physician has to search systematically the various areas known to drain into a node to discover the cause.

*Lymph nodes* in the neck should be examined from behind the patient. The examination is made easier by asking the patient to flex the neck slightly to reduce the tension of the muscles. The groups of nodes should be examined in a definite order to avoid omitting any.

Following the identification of enlarged lymph nodes, possible sites of infection or neoplastic growth should be examined, including the face, scalp, tongue, mouth, tonsil, and pharynx.

In the head and neck, all the lymph ultimately drains into the *deep cervical group of nodes.* Secondary carcinomatous deposits in these nodes are very common. The primary growth may be easy to find. On the other hand, there are certain anatomical sites where the primary growth may be small and overlooked; for example, in the larynx, the pharynx, the cervical part of the esophagus, and the external auditory meatus. The bronchi, breast, and stomach are sometimes the site of the primary tumor. In these cases the secondary growth has spread far beyond the local lymph nodes.

When there are cervical metastases, the surgeon usually decides to perform a *block dissection of the cervical nodes.* This procedure involves the removal en bloc of the internal jugular vein, the fascia, the lymph nodes, and the submandibular salivary gland. The aim of the operation is removal of all the lymphatic tissues on the affected side of the neck. The carotid arteries and the vagus nerve are carefully preserved. It is often necessary to sacrifice the hypoglossal and vagus nerves, which may be involved in the cancerous deposits. In patients with bilateral spread, a bilateral block dissection may be necessary. An interval of 3 to 4 weeks is necessary before removing the second internal jugular vein.

The *phrenic nerve*, which arises from the anterior rami of the third, fourth, and fifth cervical nerves, is of considerable clinical importance, for it is the sole nerve supply to the muscle of the diaphragm. (See p. 734.) Each phrenic nerve supplies the corresponding half of the diaphragm, and the nerve is often cut or crushed in the neck to paralyze the diaphragm and immobilize the lung in patients with lung tuberculosis. The paralyzed half of the diaphragm relaxes and is pushed up into the thorax by the positive abdominal pressure. It is mainly the lower lobe of the lung that is collapsed and rested by this procedure.

In about one-third of persons there is an *accessory phrenic nerve*. The root from the fifth cervical nerve may be incorporated in the nerve to the subclavius and may join the main phrenic nerve trunk in the thorax. Clearly, in these patients the accessory phrenic nerve must also be severed if complete paralysis of the diaphragm on one side is to be achieved.

The *cervical part of the sympathetic trunk* has been described on page 733. There are three cervical ganglia; superior, middle, and inferior. The inferior ganglion is most commonly fused with the first thoracic sympathetic ganglion to form the stellate ganglion.

The sympathetic innervation of the upper limb is as follows: The preganglionic fibers leave the spinal cord in the second to the eighth thoracic nerves. On reaching the sympathetic trunk via the white rami, they ascend within the trunk and are relayed in the second thoracic, stellate, and middle cervical ganglia. Postganglionic fibers then join the roots of the brachial plexus as gray rami. Sympathectomy of the upper limb is a relatively common procedure for the treatment of arterial insufficiency. From this information it is clear that the stellate and the second thoracic ganglia should be removed to block the sympathetic pathway to the arm completely.

Unfortunately, the removal of the stellate ganglion also removes the sympathetic nerve supply to the head and neck on that side. This not only produces vasodilatation of the skin vessels but also anhidrosis, nasal congestion, and *Horner's syndrome*. Horner's syndrome comprises (1) constriction of the pupil, (2) drooping of the upper lid, and (3) enophthalmos. For this reason the stellate ganglion is usually left intact in sympathectomies of the upper limb.

*Stellate ganglion block* is performed by first palpating the large anterior tubercle (carotid tubercle) of the transverse process of the sixth cervical vertebra, which lies about a fingerbreadth lateral to the cricoid cartilage. The carotid sheath and the sternocleidomastoid muscle are pushed laterally and the needle of the anesthetic syringe is inserted through the skin over the tubercle. The local anesthetic is then injected beneath the prevertebral layer of deep cervical fascia. This procedure will effectively block the ganglion and its rami communicantes.

## Scalp

The structure of the scalp is described on page 744. It is important to realize that the skin, the subcutaneous tissue, and the epicranial aponeurosis are closely united to one another and are separated from the periosteum by loose areolar tissue.

The skin of the scalp possesses numerous sebaceous glands, the ducts of which are prone to infection and damage by combs. For this reason, *sebaceous cysts* of the scalp are common.

The *scalp has a profuse blood supply* to nourish the hair follicles. Even a small laceration of the scalp may cause a severe loss of blood. It is often difficult to stop the bleeding of a scalp wound because the arterial walls are attached to fibrous septa in the subcutaneous tissue and are unable to contract or retract to allow blood clotting to take place. Local pressure applied to the scalp is the only satisfactory method of stopping the bleeding.

In automobile accidents it is quite common for large areas of the scalp to be cut off the head as a person is projected forward through the windshield. Because of the profuse blood supply, it is often possible to replace large areas of scalp that are only hanging to the skull by a narrow pedicle. Suture them in place, and necrosis will not occur.

The tension of the *epicranial aponeurosis*, produced by the tone of the occipitofrontalis muscles, is important in all deep wounds of the scalp. If the aponeurosis has been divided, the wound will gape open. For satisfactory healing to take place, the opening in the aponeurosis must be closed with sutures.

Often a wound caused by a blunt objects such as a baseball bat closely resembles an incised wound. This is because the scalp is split against the unyielding skull, and the pull of the occipitofrontalis muscles causes a gaping wound. This anatomical fact may be of considerable forensic importance.

*Infections of the scalp* tend to remain localized and are usually painful. This is due to the abundant fibrous tissue in the subcutaneous layer.

Occasionally an infection of the scalp spreads by the emissary veins, which are valveless, to the skull bones, causing *osteomyelitis*. Infected blood in the diploic veins may travel by the emissary veins farther into the venous sinuses and produce *venous sinus thrombosis*.

Blood or puss may collect in the potential space beneath the epicranial aponeurosis. It tends to spread over the calvaria, being limited in front by the orbital margin, behind by the nuchal lines, and laterally by the temporal lines. On the other hand, subperiosteal blood or pus is limited to one bone due to the attachment of the periosteum to the sutural ligaments.

## Face

The *facial skin* receives its sensory nerve supply from the three divisions of the trigeminal nerve. Remember that a small area of skin over the angle of the jaw is supplied by the great auricular nerve (C2 and 3). *Trigeminal neuralgia* is a relatively common condition, in which the patient experiences excruciating pain in the distribution of the mandibular or maxillary division, with the ophthalmic division usually escaping. A physician should be able to map out accurately on a patient's face the distribution of each of the divisions of the trigeminal nerve.

The *facial muscles* are innervated by the facial nerve. Damage to the facial nerve in the internal acoustic meatus (by a tumor), in the middle ear (by infection or operation), in the facial nerve canal (perineuritis, Bell's palsy), or in the parotid gland (by a tumor) or due to lacerations of the face will cause distortion of the face, with drooping of the lower eyelid, and the angle of the mouth will sag on the affected side. This is essentially a lower motor neuron lesion. An upper motor neuron lesion (involvement of the pyramidal tracts) will leave the upper part of the face normal, since the neurons supplying this part of the face receive corticobulbar fibers from both cerebral cortices.

The *blood supply to the skin of the face* is profuse, so that it is rare in plastic surgery for skin flaps to necrose in this region. The superficial temporal artery, as it crosses the zygomatic arch in front of the ear, and the facial artery, as it winds around the lower margin of the mandible level with the anterior border of the masseter, are commonly used by the anesthetist to take the patient's pulse.

The area of facial skin bounded by the nose, the eye, and the upper lid is a potentially dangerous zone to have an infection. For example, a boil in this region may cause thrombosis of the facial vein, with spread of organisms through the inferior ophthalmic veins to the cavernous sinus. The resulting *cavernous sinus thrombosis* may be fatal unless adequately treated with antibiotics.

The *congenital anomalies of cleft lip* and cleft palate have been described in detail on page 747.

The *parotid duct*, which is a comparatively superficial structure on the face, may be damaged in injuries to the face, or may be inadvertently cut during surgical operations on the face. The duct is about 2 inches (5 cm) long and passes forward across the masseter about a fingerbreadth below the zygomatic arch. It then pierces the buccinator muscle, to enter the mouth opposite the upper second molar tooth.

The *parotid salivary gland* consists essentially of

superfical and deep parts, and the important facial nerve lies in the interval between these parts. A benign parotid neoplasm rarely, if ever, causes facial palsy. A malignant tumor of the parotid is usually highly invasive and quickly involves the facial nerve, causing unilateral facial paralysis.

The parotid gland may become acutely inflamed as the result of retrograde bacterial infection from the mouth via the parotid duct. The gland may also become infected via the bloodstream, as in *mumps*. In both cases the gland is swollen; it is painful because the fascial capsule derived from the investing layer of deep cervial fascia is strong and limits the swelling of the gland. The swollen glenoid process, which extends medially behind the temporomandibular joint, is responsible for the pain experienced in *acute parotitis* when eating.

*Frey's syndrome* is an interesting complication that sometimes develops following penetrating wounds of the parotid gland. When the patient eats, beads of perspiration appear on the skin covering the parotid. This condition is due to damage to the auriculotemporal and great auricular nerves. During the process of healing, the parasympathetic secretomotor fibers in the auriculotemporal nerve grow out and join the distal end of the great auricular nerve. Eventually, these fibers reach the sweat glands in the facial skin. By this means, a stimulus intended for saliva production produces sweat secretion instead.

The *temporomandibular joint* lies immediately in front of the external auditory meatus. Fortunately, the great strength of the lateral temporomandibular ligament prevents the head of the mandible from passing backward and fracturing the tympanic plate when a severe blow falls on the chin.

The *articular disc* may become partially detached from the capsule, and this results in its movement becoming noisy and producing an audible click during movements at the joint.

*Dislocation of the temporomandibular joint* sometimes occurs when the mandible is depressed. In this movement the head of the mandible and the articular disc both move forward until they reach the summit of the articular tubercle. In this position the joint is unstable, and a minor blow on the chin, or a sudden contraction of the lateral pterygoid muscles, as in yawning, may be sufficient to pull the disc forward beyond the summit. In bilateral cases the mouth is fixed in an open position, and both heads of the mandible lie in front of the articular tubercles. Reduction of the dislocation is easily achieved by pressing the gloved thumbs downward on the lower molar teeth and pushing the jaw backward. The downward pressure overcomes the tension of the temporalis and masseter muscles, and the backward pressure overcomes the spasm of the lateral pterygoid muscles.

A *"mandibular nerve block"* is used to extract teeth from the lower jaw. With the patient's mouth open, the anterior margin of the ramus of the mandible is palpated, and the pterygomandibular ligament is felt. The syringe needle is inserted through the mucous membrane just lateral to the ligament, and the inferior alveolar nerve is infiltrated with a local anesthetic solution.

The *lingual nerve* passes forward into the submandibular region from the infratemporal fossa by running beneath the origin of the superior constrictor muscle, which is attached to the posterior border of the mylohyoid line on the mandible. Here, it is closely related to the last molar tooth and is liable to be damaged in cases of clumsy extraction of an impacted third molar. It should be remembered that the lingual nerve supplies the mucous membrane covering the anterior two-thirds of the tongue and the floor of the mouth with ordinary sensation. It also supplies taste through the fibers derived from the chorda tympani.

The *submandibular salivary gland* is a common site of calculus formation. This condition is rare in the other salivary glands. The presence of a tense swelling below the body of the mandible, which is greatest before or during a meal and is reduced in size or absent between meals, is diagnostic of the condition. Examination of the floor of the mouth will reveal absence of ejection of saliva from the orifice of the duct of the affected gland. Frequently, the stone can be palpated in the duct, which lies below the mucous membrane of the floor of the mouth.

The *submandibular lymph nodes* are commonly enlarged as the result of a pathological condition of the scalp, face, maxillary sinus, or mouth cavity. One of the commonest causes of painful enlargement of these nodes is acute infections of the teeth.

The *sublingual salivary gland*, which lies beneath the sublingual fold of the floor of the mouth, opens into the mouth by numerous small ducts.

Blockage of one of these ducts is believed to be the cause of *cysts* under the tongue.

# Skull

The *neonatal skull* presents a number of anatomical features that are clinically important. Palpation of the *fontanelles* enables the physician to determine (1) the progress of growth in the surrounding bones; (2) the degree of hydration of the baby (e.g., if the fontanelles are depressed below the surface, the baby is dehydrated); and (3) the state of the intracranial pressure (a bulging fontanelle would indicate a raised intracranial pressure).

Blood may be obtained from a baby or given to a baby by inserting a needle through the anterior fontanelle in the midline into the superior sagittal sinus. Samples of *cerebrospinal fluid* can be obtained by passing a long needle obliquely through the anterior fontanelle into the subarachnoid space or even into the lateral ventricle.

Clinically, it is usually not possible to palpate the anterior fontanelle after 18 months, since the frontal and parietal bones have enlarged to close the gap.

At birth, the *tympanic membrane* faces more downward and less laterally than in maturity; when examined with the otoscope it therefore lies more obliquely in the infant than in the adult.

In the newborn infant, the *mastoid process* is not developed, and the *facial nerve*, as it emerges from the stylomastoid foramen, is very close to the surface. Thus it may be damaged by forceps in a difficult delivery.

## *Fractures of the Skull*

Fractures of the skull are very common in the adult, but much less so in the young child. In the infant skull, the bones are more resilient than in the adult skull, and they are separated by fibrous sutural ligaments. In the adult, the inner table of the skull is particularly brittle. Moreover, the sutural ligaments begin to ossify during middle age.

The type of fracture that occurs in the skull will depend on the age of the patient, the severity of the blow, and the area of skull receiving the trauma. The *adult skull* may be likened to an eggshell in that it possesses a certain limited resilience beyond which it splinters. A severe, localized blow will produce a local indentation, often accompanied by splintering of the bone. Blows to the vault often result in a series of linear fractures, which radiate out through the thin areas of bone. The petrous parts of the temporal bones and the occipital crests strongly reinforce the base of the skull and tend to deflect linear fractures.

In the *young child,* the skull may be likened to a table-tennis ball in that a localized blow produces a depression without splintering. This common type of circumscribed lesion is referred to as a *"pond" fracture.*

In *fractures of the anterior cranial fossa,* the cribriform plate of the ethmoid bone may be damaged. This usually results in tearing of the overlying meninges and underlying mucoperiosteum. The patient will have bleeding from the nose *(epistaxis)* and leakage of cerebrospinal fluid into the nose *(cerebrospinal rhinorrhea).* Fractures involving the orbital plate of the frontal bone will result in hemorrhage beneath the conjunctiva and into the orbital cavity, causing *exophthalmos.* The frontal air sinus may be involved with hemorrhage into the nose.

*Fractures of the middle cranial fossa* are common, since this is the weakest part of the base of the skull. Anatomically, this weakness is due to the presence of numerous foramina and canals in this region; the cavities of the middle ear and the sphenoidal air sinuses are particularly vulnerable. The leakage of cerebrospinal fluid and blood from the external auditory meatus is common. The seventh and eighth cranial nerves may be involved as they pass through the petrous part of the temporal bone. The third, fourth, and sixth cranial nerves may be damaged if the lateral wall of the cavernous sinus is torn. Blood and cerebrospinal fluid may leak into the sphenoidal air sinuses and then into the nose.

In *fractures of the posterior cranial fossa,* blood may escape into the nape of the neck deep to the postvertebral muscles. Some days later, it tracks between the muscles and appears in the posterior triangle, close to the mastoid process. The mucous membrane of the roof of the nasopharynx may be torn, and blood may escape there. In fractures involving the jugular foramen, the ninth, tenth, and eleventh cranial nerves may be damaged. The strong bony walls of the hypoglossal canal usually protect the hypoglossal nerve from injury.

## Fractures of Facial Bones

Signs of fractures of the facial bones include deformity, ocular displacement, or abnormal movement accompanied by crepitation and malocclusion of the teeth. Anesthesia or paresthesia of the facial skin will follow fracture of bones through which branches of the trigeminal nerve pass to the skin.

*Fractures of the nasal bones* are very common. Although the majority are simple fractures and are reduced under local anesthesia, some are associated with severe injuries to the nasal septum, which require careful treatment under general anesthesia.

*Fractures of the maxilla* commonly result from a direct anteroposterior blow to the face. Malocclusion of the teeth, enophthalmos, and anesthesia of the cheek and upper lip are frequent physical findings.

The *zygoma* or *zygomatic arch* may be fractured by a blow to the side of the face. Although it may occur as an isolated fracture, as from a blow from a clenched fist, it may be associated with multiple other fractures of the face, as often seen in automobile accidents.

*Fractures of the mandible* are the most common fractures of the face and they are usually bilateral. The neck, body, angle, symphysis, and ramus are sites of fracture, in decreasing order of frequency.

## Injuries of the Brain

Injuries of the brain are produced by displacement and distortion of the neuronal tissues at the moment of impact. The brain may be likened to a log soaked with water floating submerged in water. The brain is floating in the cerebrospinal fluid in the subarachnoid space and is capable of a certain amount of anteroposterior movement, which is limited by the attachment of the superior cerebral veins to the superior sagittal sinus. Lateral displacement of the brain is limited by the falx cerebri. The tentorium cerebelli and the falx cerebelli also restrict displacement of the brain.

It follows from these anatomical facts that blows on the front or back of the head lead to displacement of the brain, which may produce severe cerebral damage, stretching and distortion of the brain stem, and stretching and even tearing of the

commissures of the brain. The terms *concussion*, *contusion*, and *laceration* are used clinically to describe the degrees of brain injury.

Blows on the side of the head produce less cerebral displacement, and the injuries to the brain consequently tend to be less severe.

*Intracranial hemorrhage* may result from trauma or cerebral vascular lesions. Four varieties will be considered here: (1) extradural, (2) subdural, (3) subarachnoid, and (4) cerebral.

*Extradural hemorrhage* results from the injuries to the meningeal arteries or veins. The commonest artery to be damaged is the anterior division of the middle meningeal artery. A comparatively minor blow to the side of the head, resulting in fracture of the skull in the region of the anterior inferior portion of the parietal bone, may sever the artery. The arterial or venous injury is especially liable to occur if the artery and vein enter a bony canal in this region. Bleeding occurs and strips up the meningeal layer of dura from the internal surface of the skull. The intracranial pressure rises, and the enlarging blood clot exerts local pressure on the underlying motor area in the precentral gyrus. Blood also passes outward through the fracture line, to form a soft swelling under the temporalis muscle.

In order to stop the hemorrhage, the torn artery or vein must be ligated or plugged. The burr hole through the skull wall should be placed about 1⅕ to 1½ inches (3–4 cm) above the midpoint of the zygomatic arch.

*Subdural hemorrhage* results from tearing of the superior cerebral veins at their point of entrance into the superior sagittal sinus. The cause is usually a blow on the front or the back of the head, causing excessive anteroposterior displacement of the brain within the skull.

This condition, which is much more common than middle meningeal hemorrhage, can be produced by a sudden minor blow. Once the vein is torn, blood under low pressure begins to accumulate in the potential space between the dura and the arachnoid. In about half the cases the condition is bilateral.

Acute and chronic forms of the clinical condition occur, depending on the speed of accumulation of fluid in the subdural space. For example, if the patient starts to vomit, the venous pressure will rise as the result of a rise in the intrathoracic pressure. Under these circumstances the extradural blood clot will increase rapidly in size and produce acute symptoms. In the chronic form, over a course of several months, the small blood clot will attract fluid by osmosis, so that a hemorrhagic cyst is formed, which gradually expands and produces pressure symptoms. In both forms the blood clot must be removed through burr holes in the skull.

*Subarachnoid hemorrhage* results from leakage or rupture of a congenital aneurysm on the cerebral arterial circle or, less commonly, from an angioma. The symptoms, which are sudden in onset, will include severe headache, stiffness of the neck, and loss of consciousness. The diagnosis is established by withdrawing heavily blood-stained cerebrospinal fluid through a lumbar puncture.

*Cerebral hemorrhage* is generally due to rupture of the thin-walled *lenticulostriate artery*, a branch of the middle cerebral artery. The hemorrhage involves the vital corticobulbar and corticospinal fibers in the internal capsule and produces hemiplegia on the opposite side of the body. The patient immediately loses consciousness, and the paralysis is evident when consciousness is regained.

*Intracranial hemorrhage in the infant* may occur during birth and may result from excessive molding of the head. Bleeding may occur from the cerebral veins or the venous sinuses. Excessive anteroposterior compression of the head often tears the anterior attachment of the falx cerebri from the tentorium cerebelli. Bleeding then takes place from the *great cerebral veins*, the *straight sinus*, or the *inferior sagittal sinus*.

## Cranial Nerves

The systematic examination of the twelve cranial nerves is an important part of the examination of every neurological patient. It may reveal a lesion of a cranial nerve nucleus or its central connections, or it may show an interruption of the lower motor neurons.

The *olfactory nerve* can be tested by applying substances with different odors to each nostril in turn. It should be remembered that food flavors depend on the sense of smell and not on the sense of taste. Fractures of the anterior cranial fossa or cerebral tumors of the frontal lobes may produce lesions of the olfactory nerves, with consequent loss of the sense of smell *(anosmia)*.

The *optic nerve* is evaluated by first asking the patient whether or not any changes in eyesight have been noted. The acuity of vision is then tested by using charts with lines of print of varying size. The retinas and optic discs should then be examined with an ophthalmoscope. When examining the optic disc, it should be remembered that the intracranial subarachnoid space extends forward around the optic nerve to the back of the eyeball. The retinal artery and vein run in the optic nerve and cross the subarachnoid space of the nerve sheath a short distance behind the eyeball. A rise in cerebrospinal fluid pressure in the subarachnoid space will compress the thin walls of the retinal vein as it crosses the space. This will result in congestion of the retinal veins, edema of the retina, and bulging of the optic disc *(papilledema)*.

The visual fields should then be tested. The patient is asked to gaze straight ahead at a fixed object with the eye under test, the opposite eye being covered. A small object is then moved in an arc around the periphery of the field of vision, and the patient is asked whether or not he can see the object. It is important not to miss loss or impairment of vision in the central area of the field *(central scotoma)*.

Blindness in one-half of each visual field is called *hemianopia*. Lesions of the optic tract and optic radiation produce the same hemianopia for both eyes, i.e., *homonymous hemianopia*. *Bitemporal hemianopia* is a loss of the lateral halves of the fields of vision of both eyes (i.e., loss of function of medial half of both retinas.) This condition is most commonly produced by a tumor of the pituitary gland exerting pressure on the optic chiasma.

The *oculomotor, trochlear,* and *abducent nerves* innervate the muscles that move the eyeball. The oculomotor nerve supplies all the orbital muscles except the superior oblique and the lateral rectus. It also supplies the levator palpebrae superioris and the smooth muscles concerned with accommodation, namely, the sphincter pupillae and the ciliary muscle. The trochlear nerve supplies the superior oblique muscle, and the abducent nerve supplies the lateral rectus.

To examine the ocular muscles, the patient's head is fixed, and he is asked to move the eyes in turn to the left, to the right, upward, and downward, as far as possible in each direction.

In complete *third nerve paralysis* the eye cannot be moved upward, downward, or inward. The eye looks laterally (*external strabismus*) due to the activity of the lateral rectus and downward due to the activity of the superior oblique. The patient sees double (*diplopia*). There is drooping of the upper eyelid (ptosis) due to paralysis of the levator palpebrae superioris. The pupil is fixed and dilated due to the paralysis of the sphincter pupillae and the unopposed action of the dilator pupillae (supplied by the sympathetic).

In *fourth nerve paralysis* the patient complains of double vision on looking straight downward. This is because the superior oblique is paralyzed, and the eye turns medially as well as downward.

In *sixth nerve paralysis* the patient cannot turn the eyeball laterally. When looking straight ahead, the lateral rectus is paralyzed, and the unopposed medial rectus pulls the eyeball medially, causing *internal strabismus*.

The *trigeminal nerve* has sensory and motor roots. The sensory root passes to the trigeminal ganglion, from which emerge the ophthalmic, maxillary, and mandibular divisions. The motor root joins the mandibular division.

The sensory function may be tested by using cotton and a pin over each area of the face supplied by the divisions of the trigeminal nerve. In lesions of the ophthalmic division, the cornea will be insensitive to touch.

The motor function may be tested by asking the patient to clench his teeth. The masseter and the temporalis muscles can be palpated and felt to contract.

The *facial nerve* supplies the muscles of facial expression, supplies the anterior two-thirds of the tongue with taste fibers, and is secretomotor to the lacrimal, submandibular, and sublingual glands.

The anatomical relationship of this nerve to other structures enables a physician to localize lesions of the nerve accurately. If the sixth and seventh nerves are not functioning, this would suggest a lesion within the pons of the brain. If the eighth and seventh nerves are not functioning, this would suggest a lesion in the internal acoustic meatus. If the patient is excessively sensitive to sound in one ear, the lesion probably involves the nerve to the stapedius. Loss of taste over the anterior two-thirds of the tongue implies that the seventh nerve is damaged proximal to the point where it gives off the chorda tympani.

To test the facial nerve, the patient is asked to show the teeth by separating the lips with the teeth clenched, and then to close the eyes. Taste on each half of the anterior two-thirds of the tongue can be tested with sugar, salt, vinegar, and quinine for the sweet, salt, sour, and bitter sensations.

It should be remembered that the part of the facial nerve nucleus that controls the muscles of the upper part of the face receives corticobulbar fibers from both cerebral cortices. Therefore, in patients with an upper motor neuron lesion, only the muscles of the lower part of the face will be paralyzed. However, in patients with a lower motor neuron lesion, all the muscles on the affected side of the face will be paralyzed. The lower eyelid will droop, and the angle of the mouth will sag. Tears will flow over the lower eyelid, and saliva will dribble from the corner of the mouth. The patient will be unable to close the eye and cannot expose the teeth fully on the affected side.

The *vestibulocochlear nerve* innervates the utricle and saccule, which are sensitive to static changes in equilibrium; the semicircular canals, which are sensitive to changes in dynamic equilibrium; and the cochlea, which is sensitive to sound.

Disturbances of vestibuar function include dizziness (*vertigo*) and *nystagmus*. The latter is an uncontrollable pendular movement of the eyes. Disturbances of cochlear function reveal themselves as deafness and ringing in the ears (*tinnitus*). The patient's ability to hear a voice or a tuning fork should be tested, with each ear tested separately.

The *glossopharyngeal nerve* supplies the stylopharyngeus muscle and sends secretomotor fibers to the parotid gland. Sensory fibers innervate the posterior one-third of the tongue.

The integrity of this nerve may be evaluated by testing the patient's sensation of taste on the posterior third of the tongue.

The *vagus nerve* innervates many important organs, but the examination of this nerve depends upon testing the function of the branches to the pharynx, soft palate, and larynx. The pharyngeal reflex may be tested by touching the lateral wall of the pharynx with a spatula. This should immediately cause the patient to gag, i.e., the pharyngeal muscles will contract.

The innervation of the soft palate may be tested by asking the patient to say "ah." Normally, the soft palate rises and the uvula moves backward in the midline.

All the muscles of the larynx are supplied by the recurrent laryngeal branch of the vagus, except the cricothyroid muscle, which is supplied by the external laryngeal branch of the superior laryngeal branch of the vagus. Hoarseness or absence of the voice may occur. Laryngoscopic examination may reveal abductor paralysis. (See p. 910.)

The *accessory nerve* supplies the sternocleidomastoid and the trapezius muscles by means of its spinal part. The patient should be asked to rotate the head to one side against resistance, causing the sternocleidomastoid of the opposite side to come into action. Then the patient should be asked to shrug the shoulders, causing the trapezius muscles to come into action.

The *hypoglossal nerve* supplies the muscles of the tongue. The patient is asked to put out the tongue, and if there is a lesion of the nerve, it will be noted that the tongue deviates toward the paralyzed side. This can be explained as follows. One of the genioglossus muscles, which pull the tongue forward, is paralyzed on the affected side. The other, normal genioglossus muscle pulls the unaffected side of the tongue forward, leaving the paralyzed side of the tongue stationary. The result is the tip of the tongue's deviation toward the paralyzed side. In patients with long-standing paralysis, the muscles on the affected side are wasted, and the tongue is wrinkled on that side.

# Eye

A nonpenetrating blow to the eye may cause herniation of the orbital contents downward through a fracture in the bony orbital floor into the maxillary sinus. The infraorbital nerve may be damaged as it passes through the infraorbital canal. The effect of lesions of the third, fourth, and sixth cranial nerves on movements of the eyeball are dealt with on page 905. Many cases of strabismus are nonparalytic and are due to an imbalance in the action of opposing muscles. This type of strabismus is known as *concomitant strabismus* and is common in infancy.

The *pupillary reflexes*, i.e., the reaction of the pupils to light and accommodation, are dependent on the integrity of nervous pathways. In *the direct light reflex*, the normal pupil reflexly contracts when a light is shone into the patient's eye. The nervous impulses pass from the retina along the optic nerve to the optic chiasma and then along the optic tract. Before reaching the lateral geniculate body, the fibers concerned with this reflex leave the tract and pass to the oculomotor nuclei on both sides via the pretectal nuclei. From the parasympathetic part of the nucleus, efferent fibers leave the midbrain in the oculomotor nerve and reach the ciliary ganglion via the nerve to the inferior oblique. Postganglionic fibers pass to the constrictor pupillae muscles via the short ciliary nerves.

The *consensual light reflex* is tested by shining the light in one eye and noting the contraction of the pupil in the opposite eye. This reflex is possible because the afferent pathway just described travels to the parasympathetic nuclei of both oculomotor nerves.

The *accommodation reflex* is the contraction of the pupil that occurs when a person suddenly focuses on a near object after having focused on a distant object. The nervous impulses pass from the retina via the optic nerve, the optic chiasma, the optic tract, the lateral geniculate body, the optic radiation, and the cerebral cortex of the occipital lobe of the brain. The efferent pathway passes to the parasympathetic nucleus of the oculomotor nerve. From there, the efferent impulses reach the constrictor pupillae via the oculomotor nerve, the ciliary ganglion, and the short ciliary nerves.

# Ear

Otoscopic examination of the tympanic membrane is facilitated by first straightening the external auditory meatus by gently pulling the auricle upward and backward in the adult, and straight backward or backward and downward in the infant. Normally, the tympanic membrane is pearly gray in color and concave.

Pathogenic organisms may gain entrance to the tympanic cavity by ascending through the auditory tube from the nasal part of the pharynx. Acute infection of the tympanic cavity (*otitis media*) produces bulging and redness of the tympanic membrane.

Inadequate treatment of otitis media may result in the spread of the infection into the mastoid antrum and the mastoid air cells (acute mastoiditis). Acute mastoiditis may be followed by the further spread of the organisms beyond the confines of the middle ear. The meninges and the temporal lobe of the brain lie superiorly. A spread of the infection in this direction could produce a meningitis and a cerebral abscess in the temporal lobe. Beyond the medial wall of the tympanic cavity lie the facial nerve and the internal ear. A spread of the infection in this direction may cause a facial nerve palsy and *labyrinthitis* with *vertigo.* The posterior wall of the mastoid antrum is related to the sigmoid venous sinus. If the infection spreads in this direction, a thrombosis in the sigmoid sinus may well take place. These various complications emphasize the importance of knowing the anatomy of this region.

## Mouth

The mouth is one of the important areas of the body that the physician is called upon to examine. Needless to say, he must be able to recognize all the structures visible in the mouth and be familiar with the normal variations in the color of the mucous membrane covering underlying structures. The sensory nerve supply and lymphatic drainage of the mouth cavity should be known. The close relation of the lingual nerve to the lower third molar tooth should be remembered. The close relation of the submandibular duct to the floor of the mouth may enable one to palpate a calculus in cases of periodic swelling of the submandibular salivary gland.

### Laceration of the Tongue

A wound of the tongue is often caused by a blow on the chin when the tongue is partly protruded from the mouth. It may also occur when a patient accidentally bites the tongue while eating, during recovery from an anesthetic, or during an epileptic attack. Bleeding is halted by grasping the tongue between the finger and thumb posterior to the laceration, thus occluding the branches of the lingual artery.

## Pharynx

At the junction of the mouth with the oral part of the pharynx, and the nose with the nasal part of the pharynx, there are collections of lymphoid tissue of considerable clinical importance. The pala-

tine tonsils and the nasopharyngeal tonsils are the most important.

The *palatine tonsils* reach their maximum normal size in early childhood. After puberty, together with other lymphoid tissues in the body, they gradually atrophy. The palatine tonsils are a common site of infection, producing the characteristic sore throat and pyrexia. The deep cervical lymph node situated below and behind the angle of the mandible, which drains lymph from this organ, is usually enlarged and tender. Recurrent attacks of tonsillitis are best treated by tonsillectomy. Following tonsillectomy, the external palatine vein, which lies lateral to the tonsil, may be the source of troublesome postoperative bleeding.

A *peritonsillar abscess (quinsy)* is caused by spread of infection from the palatine tonsil to the loose connective tissue outside the capsule.

The *nasopharyngeal tonsil* or *pharyngeal tonsil,* consists of a collection of lymphoid tissue beneath the epithelium of the roof of the nasal part of the pharynx. Like the palatine tonsil, it is largest in early childhood and starts to atrophy after puberty.

Excessive hypertrophy of the lymphoid tissue, usually associated with infection, causes the pharyngeal tonsils to become enlarged; they are then commonly referred to as *adenoids.* Marked hypertrophy blocks the posterior nasal openings and causes the patient to snore loudly at night and to breathe through the open mouth. The close relationship of the infected lymphoid tissue to the auditory tube may be the cause of deafness and recurrent otitis media. Adenoidectomy is the treatment of choice in cases of hypertrophied adenoids with infection.

The nasal part of the pharynx may be viewed clinically by a mirror passed through the mouth (Fig. 11-62).

The *piriform fossa* is a recess of mucous membrane situated on either side of the entrance of the larynx. It is bounded medially by the aryepiglottic folds and laterally by the thyroid cartilage. Clinically it is important, since it is a common site for the lodging of sharp ingested bodies such as fish bones. The presence of such a foreign body immediately causes the patient to gag violently. Once the object has become jammed, it is difficult for the patient to remove it without a physician's assistance.

## Pharyngeal Pouch

An examination of the lower part of the posterior surface of the inferior constrictor muscle reveals a potential gap between the upper oblique and the lower horizontal fibers *(cricopharyngeus)*. This area is marked by a dimple in the lining mucous membrane. It is believed that the function of the cricopharyngeus is to prevent the entry of air into the esophagus. Should the cricopharyngeus fail to relax during swallowing, the internal pharyngeal pressure may rise and force the mucosa and submucosa of the dimple posteriorly, to produce a diverticulum. Once the diverticulum has been formed, it may gradually enlarge and fill with food with each meal. Unable to expand posteriorly because of the vertebral column, it turns downward, usually on the left side. The presence of the pouch filled with food causes difficulty in swallowing *(dysphagia)*.

## Nose

The walls of the nasal cavity have been fully described on page 853. Fractures involving the nasal bones are very common. Blows directed from the front may cause one or both nasal bones to be displaced downward and inward. Lateral fractures also occur in which one nasal bone is driven inward and the other outward; the nasal septum is usually involved.

*Infection of the nasal cavity* can spread in a number of directions. The paranasal sinuses are especially prone to infection. Organisms may spread via the nasal part of the pharynx and the auditory tube to the middle ear. It is possible for organisms to ascend to the meninges of the anterior cranial fossa, along the sheaths of the olfactory nerves through the cribriform plate, and produce meningitis.

*Foreign bodies* in the nose are common in children. The presence of the nasal septum and the existence of the folded, shelflike conchae make impaction and retention of balloons, peas, and small toys relatively easy.

*Epistaxis*, or bleeding from the nose, is a frequent condition. The most common cause is nosepicking. The bleeding may be arterial or venous, and the majority of episodes occur on the antero-inferior portion of the septum and involve the septal branches of the sphenopalatine and facial vessels.

*Examination of the nasal cavity* may be carried out by inserting a speculum through the external nares or by means of a mirror in the pharynx. In the latter case the choanae and the posterior border of the septum can be visualized (Fig. 11-62).

It should be remembered that the nasal septum is rarely situated in the midline. A severely deviated septum may interfere with drainage of the nose and the paranasal sinuses.

## Paranasal Sinuses

Infection of the paranasal sinuses is a common complication of nasal infections. Rarely, the cause of maxillary sinusitis is extension from an apical dental abscess. The frontal, ethmoidal, and maxillary sinuses may be palpated clinically for areas of tenderness. The frontal sinus may be examined by pressing the finger upward beneath the medial end of the superior orbital margin. It is here that the floor of the frontal sinus is closest to the surface.

The ethmoidal sinuses can be palpated by pressing the finger medially against the medial wall of the orbit. The maxillary sinus can be examined for tenderness by pressing the finger against the anterior wall of the maxilla below the inferior orbital margin; pressure over the infraorbital nerve may reveal increased sensitivity.

Directing the beam of a flashlight either through the roof of the mouth or through the cheek in a darkened room will often enable a physician to determine whether or not the maxillary sinus is full of inflammatory fluid rather than air. This method of transillumination is simple and effective. Radiological examination of the sinuses is also most helpful in making a diagnosis. One should always compare the clinical findings of each sinus on the two sides of the body.

The frontal sinus is innervated by the supraorbital nerve, which also supplies the skin of the forehead and scalp as far back as the vertex. It is therefore not surprising that patients with frontal sinusitis have pain referred over this area. The maxillary sinus is innervated by the infraorbital nerve and, in this case, pain is referred to the upper jaw, including the teeth.

The frontal sinus drains into the hiatus semilunaris, via the infundibulum, close to the orifice of the maxillary sinus on the lateral wall of the nose. It is thus not unexpected to find that a patient with frontal sinusitis nearly always has a maxillary sinusitis. The maxillary sinus is particularly prone to infection, since its drainage orifice through the hiatus semilunaris is badly placed near the roof of the sinus. In other words, the sinus has to fill up with fluid before it can effectively drain with the person in the upright position. The relation of the apices of the roots of the teeth in the maxilla to the floor of the maxillary sinus has already been emphasized.

## Larynx

The muscles of the larynx are innervated by the recurrent laryngeal nerves, with the exception of the cricothryoid muscle, which is supplied by the external laryngeal nerve. Both these nerves are vulnerable during operations on the thyroid gland because of the close relationship between them and the arteries of the gland. The left recurrent laryngeal nerve may be involved in a bronchial or esophageal carcinoma or in secondary metastatic deposits in the mediastinal lymph nodes. The right and left recurrent laryngeal nerves may be damaged by malignant involvement of the deep cervical lymph nodes.

*Section of the external laryngeal nerve* produces weakness of the voice, since the vocal fold cannot be tensed. The cricothyroid muscle is paralyzed.

*Unilateral complete section of the recurrent laryngeal nerve* results in the vocal fold on the affected side assuming the position midway between abduction and adduction. It lies just lateral to the midline. Speech is not greatly affected, since the other vocal fold compensates to some extent and moves toward the affected vocal fold.

*Bilateral complete section of the recurrent laryngeal nerve* results in both vocal folds assuming the position midway between abduction and adduction. Breathing is impaired, since the rima glottidis is partially closed and speech is lost.

*Unilateral partial section of the recurrent laryngeal nerve* results in a greater degree of paralysis of the abductor muscles than of the adductor muscles. The affected vocal fold assumes the adducted

midline position. This phenomenon has not been explained satisfactorily. It must be assumed that the abductor muscles receive a greater number of nerves than the adductor muscles, and thus partial damage of the recurrent laryngeal nerve results in damage to relatively more nerve fibers to the abductor muscles. Another possibility is that the nerve fibers to the abductor muscles are traveling in a more exposed position in the recurrent laryngeal nerve and are therefore more prone to be damaged.

*Bilateral partial section of the recurrent laryngeal nerve* results in bilateral paralysis of the abductor muscles and the drawing together of the vocal folds. Acute breathlessness (dyspnea) and stridor follow, and tracheotomy is necessary.

The *mucous membrane of the larynx* is loosely attached to the underlying structures by submucous connective tissue. In the region of the vocal folds, however, the mucous membrane is firmly attached to the vocal ligaments. This fact is of clinical importance in cases of edema of the larynx. The accumulation of tissue fluid causes the mucous membrane above the rima glottidis to swell and encroach on the airway. In very severe cases a tracheotomy may be necessary.

The interior of the larynx may be inspected indirectly through a laryngeal mirror passed through the open mouth into the oral pharynx. A more satisfactory method is the direct method using the *laryngoscope*. The neck is brought forward on a pillow and the head is fully extended at the atlanto-occipital joint. The illuminated instrument can then be introduced into the larynx over the back of the tongue. The valleculae, the piriform fossae, the epiglottis, and the aryepiglottic folds are clearly seen. The two elevations produced by the corniculate and cuneiform cartilages can be recognized. Within the larynx, the vestibular folds and the vocal folds can be seen. The former are fixed, widely separated, and reddish in color; the latter move with respiration and are white in color. With quiet breathing, the rima glottidis is triangular in shape, with the apex in front. With deep inspiration, the rima glottidis assumes a diamond shape due to the lateral rotation of the arytenoid cartilages.

If the patient is asked to breathe deeply, the vocal folds become widely abducted, and the inside of the trachea can be seen.

# CLINICAL PROBLEMS

*Answers on page 975*

1. An 8-year-old girl was examined by a pediatrician and found to have a painless, superficial, fluctuant swelling below and behind the angle of the jaw on the right side. The skin over the swelling was cool to touch and showed no redness. Careful palpation of the neck revealed two tender, firm lumps matted together beneath the anterior border of the sternocleidomastoid on the right side. The left side of the neck was normal. Examination of the palatine tonsils showed moderate hypertrophy on both sides, and a few pustules caused by pus exuding from the tonsillar crypts on the right side. In the absence of signs of acute infection, a diagnosis of tuberculous cervical lymphadenitis was made. Using your knowledge of anatomy, name the group of lymph nodes involved in the disease. What anatomical structures would tend to limit the spread of this disease in the neck? Since the lymph node showing the most advanced stage of the disease was situated below and behind the mandible, which organ in the oral part of the pharynx was most likely to have served as the portal of entry to the tubercle bacilli?

2. A 25-year-old woman visited her physician complaining of a swelling on the front of her neck and breathlessness. On examination, a small, solitary swelling of firm consistency was found to the left of the midline of the neck. The swelling was not attached to the skin, but moved upward on swallowing. On careful palpation, it was found to be continuous with the lower pole of the left lobe of the thyroid gland. A diagnosis of adenoma of the thyroid gland was made. From your knowledge of anatomy, explain why the tumor moved upward when the patient swallowed. What structure in the neck was being pressed upon by the adenoma to cause breathlessness? Which lymph nodes would you examine for metastases if you suspected a malignant tumor?

3. A 10-year-old boy was taken to a pediatrician because his parents had noticed that he held his head inclined to one side. On examination, the cervical part of the vertebral column was found to be held in a position of slight flexion.

The anterior border of the sternocleidomastoid muscle on the right side was more prominent than on the left side; the right ear was held nearer the shoulder on the right side than normal; and the boy tended to hold his head inclined so that he looked upward to the left. A diagnosis of congenital torticollis was made. Which muscle is responsible for this condition?

4. A young, enthusiastic surgeon decided to incise an abscess situated in the middle of the posterior triangle of the neck on the right side. On making the incision, he found that the interior of the abscess was extensive, and he felt that he should explore more deeply. On recovering from the operation, the patient thanked the surgeon profusely for ridding her of the abscess, which had healed up nicely. There was one thing, however, that she could not understand. She could no longer raise her right hand above her head to brush her hair. Using your knowledge of anatomy, explain this patient's disability.

5. A 65-year-old man was examined by his physician and found to have right-sided heart failure. As the patient lay propped up on pillows in bed, his physician noticed that the blood in the external jugular veins could be easily seen, rising nearly as high as the angle of the mandible. What is the significance of the blood level in the external jugular vein? What is the surface marking of this vein? Into what does the external jugular vein drain?

6. A 70-year-old man visited his physician complaining of a small swelling below his chin. On examination, a single small, hard swelling could be palpated in the submental triangle. It was mobile on the deep tissues and not tethered to the skin. A diagnosis of a malignant secondary deposit in a submental lymph node was made. Where would you look for the primary carcinoma? Where do the submental lymph nodes drain?

7. A physician commonly uses an electrocardiogram to evaluate patients with cardiac disease. In patients suspected of arrhythmias, the carotid sinus is often pressed upon while the electrocardiogram is being taken. Unilateral pres-

sure on the carotid sinus is sometimes used to stop ventricular tachycardia of atrial origin. Using your knowledge of anatomy, explain how pressure on the carotid sinus can influence the heart.

8. A 45-year-old man was referred to a surgeon because of a pigmented skin lesion in the right temporal region. A diagnosis of malignant melanoma had been made. The surgeon explained to the patient the seriousness of the condition. Although there were no secondary metastases in the regional lymph nodes, it was decided to perform a block dissection of the lymph nodes on the right side of the neck and to remove the internal jugular vein on that side as well. Which group of lymph nodes receives the lymph from the skin of the right temporal region? Why is it necessary to remove the internal jugular vein when performing a block dissection of the deep cervical lymph nodes? Which important nerves are related to the deep cervical lymph nodes and may have to be sacrificed in a block dissection of the neck?

9. A 35-year-old woman has a partial thyroidectomy for the treatment of thyrotoxicosis. During the operation a ligature slipped off the right superior thyroid artery. In order to stop the hemorrhage, the surgeon blindly grabbed for the artery with artery forceps. The operation was completed without further incident. The next morning the surgeon noticed that the patient spoke with a husky voice. Using your knowledge of anatomy, state what is likely to have happened to the patient. What is the condition of the right vocal fold? Where does the superior thryoid artery originate? Into what vessel does the superior thyroid vein drain?

10. Following a total thyroidectomy for carcinoma of the thyroid gland, a 55-year-old male patient noticed tingling and numbness of the fingers, toes, and lips. Painful cramp of the hands and feet was also experienced. Strong muscle spasms producing adduction of the thumb, flexion of the wrist and metacarpophalangeal joints, extension of the phalangeal joints, and plantar flexion of the feet also occurred. Laboratory examination of the blood revealed a blood calcium level of 4 mg per 100 ml. Which organ was damaged during the total thyroidectomy?

11. A 55-year-old woman visited her physician complaining of difficulty in swallowing. She stated that she had first noticed the condition 3 months previously and that it had become progressively worse. She now found it difficult to swallow milk puddings. During the last month she had lost 28 pounds. On questioning, the patient said she felt the obstruction was at the root of the neck (i.e., just above the upper border of the manubrium sterni). On examination of the neck, a hard, fixed lump was felt deep to the anterior border of the right sternocleidomastoid muscle. The lump was considered to be a deep cervical lymph node, which was enlarged due to a secondary carcinomatous deposit. From the history, and using your anatomical knowledge, make your diagnosis.

12. A 53-year-old woman visited her physician because of a dull, aching pain in the forearm on the left side. The discomfort was made worse by exercising the arm, especially in the elevated position. The pain was relieved by rest. She noticed that her left hand was sometimes colder than the one on the right, and when held above the head, became white. When the left arm was held by the side for any length of time, the hand became blue, especially in cold weather. On examination, the radial pulse was found to be absent on the left and normal on the right. The brachial arterial pulse was weak on the left, but normal on the right. The pulsations of the subclavian arteries were normal on both sides of the neck. It was possible to produce the color changes described by the patient by suitable positioning of the left arm. Using your knowledge of anatomy, state what possible structure or structures in the neck could produce these signs and symptoms.

13. A 65-year-old woman visited her physician because she had noticed a painful swelling behind her right ear. She had first noticed the swelling 4 days previously, and since that time it had progressively enlarged. On examination, three discrete, tender swellings were found over the right mastoid process. They were not attached to the bone and were not tethered to the skin. The swellings did not fluctuate, but the skin over them was redder and warmer than normal. The patient was wearing a new Easter hat and was disinclined to remove it. The over-

worked physician assumed that the lymph nodes were enlarged because of infection, but he was unable to determine the site of entry of the organisms from his cursory examination. He prescribed a wide-spectrum antibiotic and told the patient to return in a week. Comment on the treatment of this patient.

14. A 46-year-old man was seen in the emergency room after being knocked down in a street brawl. He had received a blow on the head with an empty bottle. On examination, the patient was found to be conscious and had a large doughlike swelling over the back of the head. The skin was intact, and the swelling fluctuated on palpation. No other abnormal signs were present. A diagnosis of hematoma of the scalp was made. The patient was sent to the radiology department for routine antero-posterior and lateral radiographs of the skull. Name the layer of the scalp in which the blood clot was situated (a) if the swelling was extensive, but did not pass inferiorly beyond the temporal lines laterally, the orbital margin in front, and the nuchal lines behind; (b) if the swelling was limited to the area occupied by an underlying skull bone; and (c) if the swelling was small, superficial, and with no fluctuation. Why was it necessary to X-ray the skull?

15. A 26-year-old man was admitted to the hospital as an emergency case. He had a severe infection of the scalp. Why are such infections potentially dangerous?

16. A 16-year-old boy fell off his bicycle and cut his scalp. The wound was less than ½ inch (1.3 cm) long, and within a few moments he was sitting in a pool of blood. Why does the scalp bleed so profusely from a small wound? Why does a deep wound of the scalp always require sutures?

17. A 17-year-old girl visited her dermatologist because of severe acne of the face. On examination, it was found that a small abscess was present on the side of the nose. The patient was given antibiotics and warned not to press the abscess. Why is it dangerous to squeeze, prick, or incise a boil in the area between the eye and the upper lip or between the eye and the side of the nose?

18. A 7-year-old boy was suffering from a severe infection of the right middle ear (otitis media).

Within the course of a week, the infection had spread to the mastoid antrum and the mastoid air cells (mastoiditis). The organisms did not respond to antibiotics, so the surgeon decided to perform a radical mastoid operation. Following the operation it was noticed that the boy's face was distorted. The mouth was drawn upward to the left, and he was unable to close his right eye. Saliva tended to accumulate in his right cheek and dribble from the corner of his mouth. What structure had been damaged during the operation?

19. A 70-year-old man who was suffering from hypertension suddenly collapsed, with hemorrhage into the corticospinal and corticobulbar fibers (upper motor neurons) on the right side. On recovering consciousness, he was found to be paralyzed on the left side of his body. What facial changes would you expect to find in this patient?

20. A 43-year-old woman visited her physician complaining of severe intermittent pain of short duration on the right side of her face. The pain was precipitated by brushing the teeth or drinking cold fluids. On examination, no abnormal physical signs were found. When asked to point out on her face the area where the pain was experienced, the patient mapped out the skin area innervated by the mandibular division of the trigeminal nerve. A diagnosis of trigeminal neuralgia was made. Which area of facial skin is innervated by the mandibular division of the trigeminal nerve? Which area of facial skin is not innervated by the trigeminal nerve?

21. A 60-year-old woman visited her physician with a swelling over the parotid gland on the right side. She stated that she had first noticed the swelling 3 months previously, and since that time it had rapidly increased in size. Recently she had noticed that the right side of her face "felt weak," and she could no longer whistle for her dog. On examination, a hard swelling deeply attached to the parotid gland was found. On testing the facial muscles, it was found that the muscles on the right side were weaker than those on the left side. What is the connection between the parotid swelling and the right-sided facial weakness? Where does the parotid duct open into the mouth?

22. A 10-year-old boy was playing darts with his brother. He bent down to pick up a fallen dart when another dart fell from the dart board and hit him on the side of the face. The dart penetrated the side of the face and entered the parotid gland. The small wound healed well, without infection. Six months later the boy's mother noticed that just before mealtimes the boy began to sweat profusely on the facial skin close to the old dart wound. What is wrong with this patient?

23. An exhausted medical student decided to brush up his gross anatomy by attending a lecture given by an old and revered visiting professor. After 45 minutes the lecture began to bore him, and his mind began to wander. He could not forget the attractive brunette nurse in the surgical clinic whom he had dated the previous evening. After 5 more minutes he found he just could not keep his eyes open. When would this lecture end? Just then, he involuntarily opened his mouth wide and yawned. To his great consternation he could not close his mouth. His jaw was stuck. What is your diagnosis?

24. A 35-year-old man was struck a blow on the lower jaw with a fist. He was seen in the emergency room with blood-stained saliva trickling from the mouth and obvious difficulty in talking and swallowing. On examination, the teeth of the lower jaw were found to be no longer in correct alignment. What is your diagnosis?

25. A radiograph of a patient's mouth revealed a stone in the right submandibular duct. Where is it possible to palpate the submandibular duct? Where does the duct open into the mouth? Where is the lumen of the submandibular duct narrowest?

26. A 6-month-old infant is suspected of having a raised intracranial pressure. What simple method of physical examination would verify this?

27. While playing baseball, a player was struck on the right side of the head with the ball. The player fell to the ground, but did not lose consciousness. After resting for an hour and then getting up, he was seen to be confused and irritable. Later, he staggered and fell to the floor. On questioning, he was seen to be drowsy, and twitching of the lower left half of his face and left arm was noted. A diagnosis of extradural hemorrhage was made. Which artery is likely to have been damaged? What is responsible for the drowsiness and muscle twitching?

28. Following an automobile accident, a patient was found to have an internal (medial) strabismus of the right eye. Which cranial nerve was damaged? Which muscle was paralyzed?

29. On examination, a patient is found to have a bitemporal hemianopia. An enlargement of which anatomical structure is likely to cause this condition?

30. A patient is suspected of having a lesion of the glossopharyngeal nerve. How would you test the integrity of this cranial nerve?

31. A 25-year-old woman was found to be suffering from thyrotoxicosis. Not only was she complaining of nervousness, excessive sweating, and loss of weight, but she also had bilateral exophthalmos. What anatomical structures are believed to be responsible for the exophthalmos?

32. A patient comes to you complaining of a foreign body in the eye. The foreign body has been in the conjunctival sac for several hours, and the patient has failed to remove it. Where would you look for a foreign body that has been floating around the conjunctival sac?

33. An 11-year-old girl is suffering from an acute infection of the right middle ear (acute otitis media). Trace the path taken by the pathogenic organisms from the nasal part of the pharynx to the middle ear. What anatomical structures are likely to be involved if the infection spreads beyond the confines of the middle ear?

34. Following a severe cold, a patient complained of a frontal headache and a dull, aching pain of the right side of the face. What anatomical structures are likely to become secondarily infected from the nose? Explain the distribution of the pain.

35. While sitting in a restaurant which specializes in seafood, a physician suddenly became aware of a man standing up at a nearby table who was choking violently and trying desperately to get his finger into the back of his mouth. He was gasping that he had a fish bone stuck in his throat. Name the anatomical site in the throat in which fish bones are very liable to lodge. What is the sensory nerve supply to the mucous membrane lining this area?

36. Following a partial thyroidectomy in which difficulty was experienced in tying the left inferior thyroid artery, the patient was found to have unilateral partial section of the left recurrent laryngeal nerve. Describe the position assumed by the left vocal fold.

37. A 49-year-old woman was found on ophthalmoscopic examination to have edema of both optic discs (bilateral papilledema) and congestion of the retinal veins. The cause of the condition was found to be a rapidly growing intracranial tumor. Using your knowledge of anatomy, explain the papilledema. Why does the patient exhibit bilateral papilledema?

38. A 29-year-old woman was struck in the face with a baseball while playing with her son. X-ray examination revealed multiple fractures of the bones around the orbit. Name the bones that form the orbital margin.

39. A 23-year-old woman complained that recently, when she had a cold, she took an airplane flight, during which she became deaf and both ears started to ache. She told the flight attendant of her discomfort and was advised to suck on a candy. After a few moments the woman experienced a popping sensation in both ears and her hearing returned. Her earache immediately disappeared. Using your knowledge of anatomy, can you explain this common complaint of air travelers?

40. A 70-year-old man, who suffered from severe bronchitis and emphysema, slipped on an icy sidewalk and sustained a fracture of the distal third of the right radius. On examination in the emergency room, the physician decided, in view of the patient's chest condition, to reduce the fracture under a local anesthetic. Where can one perform a brachial plexus nerve block? Name the important layer of fascia that allows one to carry out this nerve block procedure.

## NATIONAL BOARD TYPE QUESTIONS

*Answers on page 984*

**In each of the following questions, answer:**
  (a) if (1) is correct only
  (b) if (2) is correct only
  (c) if both (1) and (2) are correct, and
  (d) if neither (1) nor (2) is correct

1. (1) The temporomandibular joint cannot be felt from the surface, because it is covered by the masseter muscle.
   (2) The mastoid process of the temporal bone cannot be palpated in the newborn.

2. (1) The anterior division of the middle meningeal artery lies beneath the anterior inferior angle of the parietal bone, i.e., 1½ inches above the midpoint of the zygomatic arch.
   (2) The anterior fontanelle can easily be palpated in a baby and lies between the squamous part of the temporal bone, the parietal bone, and the greater wing of the sphenoid.

3. (1) The deep cervical lymph nodes are situated in the neck along a line that extends from the midpoint between the tip of the mastoid process and the angle of the mandible down to the sternoclavicular joint.

   (2) The external jugular vein runs down the neck from the angle of the jaw to the middle of the clavicle.

4. (1) The parotid duct opens into the vestibule of the mouth opposite the third lower molar tooth.
   (2) The isthmus of the thyroid gland lies anterior to the second, third, and fourth rings of the trachea.

5. Concerning the face:
   (1) A boil on the side of the nose is potentially dangerous because of the possibility of spread of infection from the facial vein to the cavernous venous sinus via the superior ophthalmic vein.
   (2) The entire skin of the face is innervated by the three divisions of the trigeminal nerve.

6. Concerning the posterior triangle of the neck:
   (1) The spinal part of the accessory nerve crosses the triangle on the superficial surface of the splenius capitis muscle.
   (2) The roots of the brachial plexus emerge into the triangle between the scalenus anterior and scalenus medius muscles.

7. Concerning the carotid sheath:
   (1) The sheath is formed from the investing, pretracheal, and prevertebral layers of the deep cervical fascia.
   (2) The sheath contains the three carotid arteries, the internal jugular vein, the vagus nerve, and the deep cervical lymph nodes.
8. Concerning the root of the neck:
   (1) The cervical plexus is formed from the posterior rami of the cervical spinal nerves.
   (2) The vertebral artery ascends the neck through the foramen transversaria of the upper seven cervical vertebrae.
9. Concerning the thyroid gland:
   (1) The superior thyroid artery is closely associated with the external laryngeal nerve.
   (2) The inferior thyroid veins drain into the internal jugular veins.
10. Concerning the salivary glands:
   (1) The parotid salivary gland is innervated by secretomotor fibers that travel with the auriculotemporal nerve.
   (2) The deep part of the submandibular salivary gland lies between the mylohyoid muscle and the lower border of the body of the mandible.

**Multiple Choice:**
11. Which of the following muscles is primarily responsible for protrusion of the tongue?
   (a) Styloglossus
   (b) Palatoglossus
   (c) Hypoglossus
   (d) Genioglossus
   (e) None of the above
12. The following muscles of the pharynx receive their motor innervation from the pharyngeal plexus via the cranial part of the accessory nerve, **except** the:
   (a) Superior constrictor
   (b) Palatopharyngeus
   (c) Stylopharyngeus
   (d) Middle constrictor
   (e) Salpingopharyngeus
13. Which of the following muscles serve to elevate the soft palate during swallowing?
   (a) Tensor veli palatini
   (b) Palatoglossus
   (c) Palatopharyngeus

   (d) Levator veli palatini
   (e) Salpingopharyngeus
14. Which of the following muscles partially inserts on the articular disc of the temporomandibular joint?
   (a) Medial pterygoid
   (b) Anterior fibers of temporalis
   (c) Masseter
   (d) Posterior fibers of temporalis
   (e) Lateral pterygoid
15. All the following statements concerning cervical vertebrae are correct, **except:**
   (a) Each transverse process has a foramen
   (b) The atlas has no spinous process and no body
   (c) None of the spinous processes can be palpated
   (d) Most of the cervical vertebrae have bifid spines
   (e) There is a synovial joint between the odontoid process of the axis and the anterior arch of the atlas
16. Assuming the patient's eyesight is normal, in which cranial nerve is there likely to be a lesion when the direct and consensual light reflexes are absent?
   (a) Trochlear nerve
   (b) Optic nerve
   (c) Abducent nerve
   (d) Oculomotor nerve
   (e) Trigeminal nerve
17. A patient is unable to taste a piece of sugar placed on the anterior part of the tongue. Which cranial nerve is likely to have a lesion?
   (a) Hypoglossal
   (b) Vagus
   (c) Glossopharyngeal
   (d) Facial
   (e) Maxillary division of trigeminal
18. On asking a patient to say "ah," the uvula is seen to be drawn upward to the right. Which cranial nerve is likely to be damaged?
   (a) Left glossopharyngeal nerve
   (b) Right hypoglossal
   (c) Left accessory nerve
   (d) Right vagus
   (e) Right trigeminal nerve
19. When testing the sensory innervation of the face, it is important to remember that the skin of the tip of the nose is supplied by the:

(a) Zygomatic branch of the facial nerve
(b) Maxillary division of the trigeminal nerve
(c) Ophthalmic division of the trigeminal nerve
(d) External nasal branch of the facial nerve
(e) Buccal branch of the mandibular division of the trigeminal nerve

20. All of the following spatial relations to the scalenus anterior muscle are correct, **except:**
    (a) The origin of the vertebral artery is medial to it
    (b) The roots of the brachial plexus are posterior to it
    (c) The third part of the subclavian artery is lateral to it
    (d) The subclavian vein is posterior to it
    (e) The phrenic nerve is anterior to it

**In the following questions, answer:**
    (a) IF ONLY (1), (2), AND (3) ARE CORRECT
    (b) IF (1) AND (3) ARE CORRECT
    (c) IF ONLY (2) AND (4) ARE CORRECT
    (d) IF ONLY (4) IS CORRECT, AND
    (e) IF ALL ARE CORRECT

21. Concerning the stellate ganglion:
    (1) It is formed from a fusion of the inferior cervical ganglion with the first thoracic ganglion.
    (2) It has white and gray rami communicans, which pass to spinal nerves.
    (3) It lies in the interval between the transverse process of the seventh cervical vertebra and the neck of the first rib.
    (4) It is situated in front of the vertebral artery.

22. The chorda tympani contains which of the following types of nerve fibers?
    (1) Sympathetic preganglionic
    (2) Special sensory (taste)
    (3) General sensation (touch and pressure)
    (4) Parasympathetic preganglionic

23. Concerning the pituitary gland (hypophysis cerebri):
    (1) It is deeply placed within the sella turcica of the skull.
    (2) It is separated from the optic chiasma by the diaphragma sellae.
    (3) The sphenoid sinus lies inferior to it.
    (4) It receives its arterial supply from the internal carotid artery.

24. Concerning the cervical part of the esophagus:
    (1) The lumen is narrowed at the junction with the pharynx.
    (2) The lymph drains into the deep cervical lymph nodes.
    (3) The sensory nerve supply is the recurrent laryngeal nerve.
    (4) It is the site of an important portal-systemic anastomosis.

25. Concerning the submandibular lymph nodes:
    (1) They are situated on the superficial surface of the submandibular salivary gland.
    (2) They drain the tip of the tongue.
    (3) They drain the skin of the scalp covering the forehead.
    (4) They drain the tonsil.

26. Which of the following statements are true regarding the parotid salivary gland?
    (1) The facial nerve passes through it, dividing the gland into superficial and deep parts.
    (2) The retromandibular vein is formed within it by the union of the superficial temporal vein and the maxillary vein.
    (3) The secretomotor nerve supply is derived from the glossopharyngeal nerve.
    (4) The parotid duct pierces the buccinator muscle to enter the mouth.

27. Which of the following statements concerning the tongue is/are correct?
    (1) The intrinsic muscles of the tongue are innervated by the hypoglossal nerve.
    (2) The taste buds on the vallate papillae are innervated by the glossopharyngeal nerve.
    (3) The posterior third of the tongue forms part of the anterior wall of the oral pharynx.
    (4) The foliate papillae are confined to the posterior one-third of the dorsum of the tongue.

**Match the muscles of the orbital cavity listed on the left with the appropriate nerve supply listed on the right:**

28. Levator palpebrae superioris
29. Inferior oblique
30. Lateral rectus
31. Superior oblique
32. Orbicularis oculi

(a) Facial nerve
(b) Trochlear nerve
(c) Trigeminal nerve
(d) Oculomotor nerve
(e) Abducent nerve

**Match the cranial nerves listed on the left with the appropriate openings in the skull listed on the right, through which each nerve exits from the cranial cavity:**

33. Mandibular division of the trigeminal

34. Vagus

35. Abducent

36. Ophthalmic division of the trigeminal

37. Maxillary division of the trigeminal

38. Oculomotor

(a) Superior orbital fissure
(b) Foramen rotundum
(c) Foramen ovale
(d) Jugular foramen
(e) None of the above

**Match the foramen in the skull on the left with the bone in which it is located on the right:**

39. Optic canal
40. Carotid canal
41. Foramen spinosum
42. Hypoglossal canal
43. Foramen rotundum
44. Facial nerve canal
45. Foramen magnum

(a) Sphenoid
(b) Occipital
(c) Temporal
(d) Frontal
(e) None of the above

# 12. The Back

The back, which extends from the skull to the tip of the coccyx, may be defined as the posterior surface of the trunk. Superimposed on the upper part of the posterior surface of the thorax are the scapulae and the muscles that connect the scapulae to the trunk.

## SURFACE ANATOMY

The entire posterior aspect of the patient should be examined from head to foot, and the arms should hang loosely at the side. All clothes and the shoes should be removed.

*In the midline* the following structures can be palpated from above downward.

### External Occipital Protuberance

This lies at the junction of the head and neck (Fig. 12-2). If the index finger is placed on the skin in the midline, it can be drawn downward from the protuberance in the *nuchal groove.*

### Cervical Vertebrae

The first spinous process to be felt (Fig. 12-1). is that of the *seventh cervical vertebra (vertebra prominens).* Cervical spines 1–6 are covered by the *ligamentum nuchae,* a large ligament that runs down the back of the neck connecting the skull to the spinous processes of the cervical vertebrae.

### Thoracic and Lumbar Vertebrae

The nuchal groove is continuous below with a furrow that runs down the middle of the back over the tips of the *spines of the thoracic* and *lumbar vertebrae.* The spines of all the thoracic and the upper four lumbar vertebrae can be individually palpated. The most prominent spine is that of the first thoracic vertebra; the others may be easily recognized when the trunk is bent forward.

### Sacrum

The *spines of the sacrum* are fused with each other in the midline to form the *median sacral crest.* The

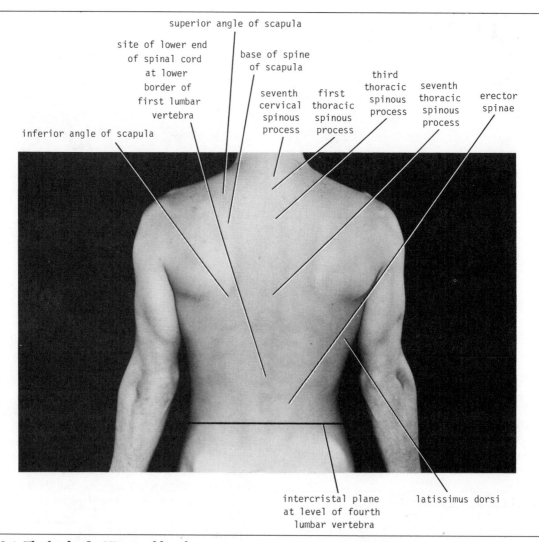

superior angle of scapula

site of lower end
of spinal cord
at lower
border of
first lumbar
vertebra

base of spine
of scapula

seventh
cervical
spinous
process

first
thoracic
spinous
process

third
thoracic
spinous
process

seventh
thoracic
spinous
process

erector
spinae

inferior angle of scapula

intercristal plane
at level of fourth
lumbar vertebra

latissimus dorsi

**Fig. 12-1. The back of a 27-year-old male.**

crest can be felt beneath the skin in the uppermost part of the natal cleft between the buttocks.

The *sacral hiatus* is situated on the posterior aspect of the lower end of the sacrum, and it is here that the *extradural space* terminates. The hiatus lies about 2 inches (5 cm) above the tip of the coccyx and beneath the skin of the natal cleft.

## Coccyx

The inferior surface and tip of the coccyx can be palpated in the natal cleft about 1 inch (2.5 cm) behind the anus (Fig. 12-2). The anterior surface of the coccyx may be palpated with a gloved finger in the anal canal.

## Upper Lateral Part of Thorax

This is covered by the scapula and its associated muscles. The scapula lies posterior to the first to the seventh ribs (Figs. 12-1 and 12-2).

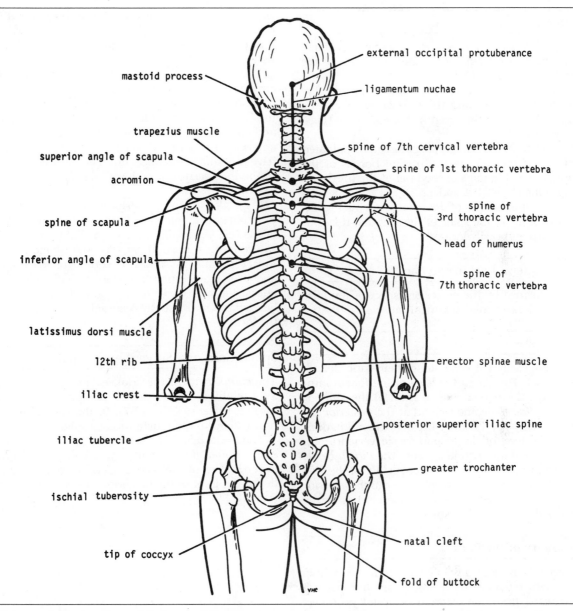

mastoid process

external occipital protuberance

ligamentum nuchae

trapezius muscle

spine of 7th cervical vertebra

superior angle of scapula

spine of 1st thoracic vertebra

acromion

spine of
3rd thoracic vertebra

spine of scapula

head of humerus

inferior angle of scapula

spine of
7th thoracic vertebra

latissimus dorsi muscle

erector spinae muscle

12th rib

iliac crest

posterior superior iliac spine

iliac tubercle

greater trochanter

ischial tuberosity

natal cleft

tip of coccyx

fold of buttock

**Fig. 12-2. Posterior view of skeleton, showing sur-
face markings on the back.**

## Scapula

The *medial border* forms a prominent ridge, which ends above at the superior angle and below at the inferior angle (Fig. 12-1).

The *superior angle* can be palpated opposite the first thoracic spine and the *inferior angle* can be palpated opposite the seventh thoracic spine (Figs. 12-1 and 12-2).

The *crest of the spine of the scapula* can be palpated and traced medially to the medial border of the scapula, which it joins at the level of the third thoracic spine (Figs. 12-1 and 12-2).

The *acromion process of the scapula* forms the lateral extremity of the spine of the scapula. It is subcutaneous and easily located.

## Lower Lateral Part of the Back

This is formed by the posterior aspect of the false pelvis and its associated gluteal muscles.

## Iliac Crests

These are easily palpable along their entire length (Fig. 12-2). They lie at the level of the fourth lumbar spine. Each crest ends in front at the *anterior superior iliac spine* and behind at the *posterior superior iliac spine*; the latter lies beneath a skin dimple at the level of the second sacral vertebra and the middle of the sacroiliac joint. The *iliac tubercle* is a prominence felt on the outer surface of the iliac crest about 2 inches (5 cm) posterior to the anterior superior iliac spine. The iliac tubercle lies at the level of the fifth lumbar spine.

## Symmetry of the Back

Observe the back as a whole and compare the two sides with reference to an imaginary line passing downward from the external occipital protuberance to the natal cleft.

The *posterior vertebral musculature*, which mainly controls the movements of the vertebral column and maintains the postural curves of the column, can be palpated. The muscles are large and lie on either side of the spines of the vertebrae (Figs. 12-1, 12-2, and 12-9). They should be examined with the flat of the hand. If they exhibit normal tone, they are firm to the touch. A spastic mus-

cle feels harder than normal; it is also shorter than normal, which produces a concavity of the vertebral column on the side of the muscular contraction.

The *curves of the vertebral column* may be examined by inspecting the lateral contour of the back. Normally, the posterior surface is concave in the cervical region, convex in the thoracic region, and concave in the lumbar region (Fig. 12-3). The anterior surface of the sacrum and coccyx together have an anterior concavity. The lumbar region meets the sacrum at a sharp angle, the *lumbosacral angle.*

Inspection of the posterior surface of the back, with particular reference to the vertical alignment of the vertebral spines, reveals a slight lateral curvature in most normal persons. Right-handed persons, especially those whose work involves extreme and prolonged muscular effort, usually exhibit a lateral thoracic curve to the right; if they are left-handed, there is usually a lateral thoracic curve to the left.

## Spinal Cord and Subarachnoid Space

The *spinal cord* in the adult extends down to the level of the lower border of the spine of the first lumbar vertebra (Fig. 12-7). In the young child it may extend to the fourth lumbar spine.

The *subarachnoid space* (Fig. 12-4), with the *cerebrospinal fluid* it contains, extends down to the lower border of the second sacral vertebra, which lies at the level of the posterior superior iliac spine (Fig. 12-7).

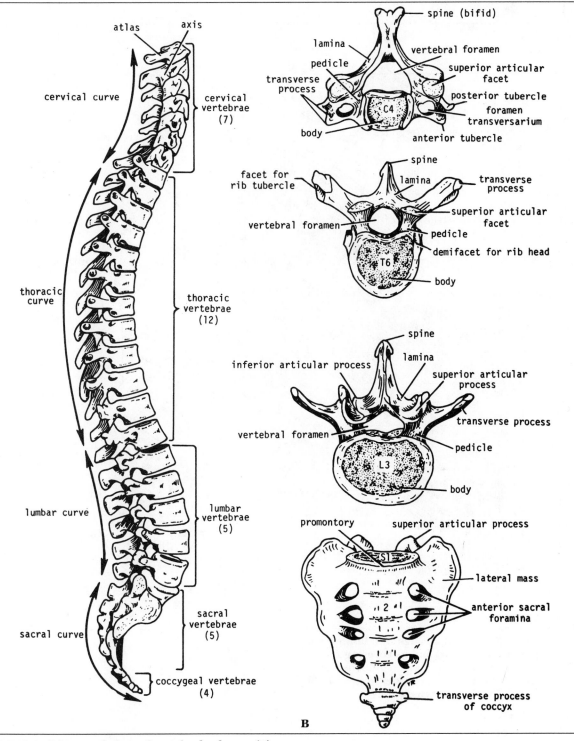

**Fig. 12-3. (A) Lateral view of vertebral column. (B) General features of different kinds of vertebrae.**

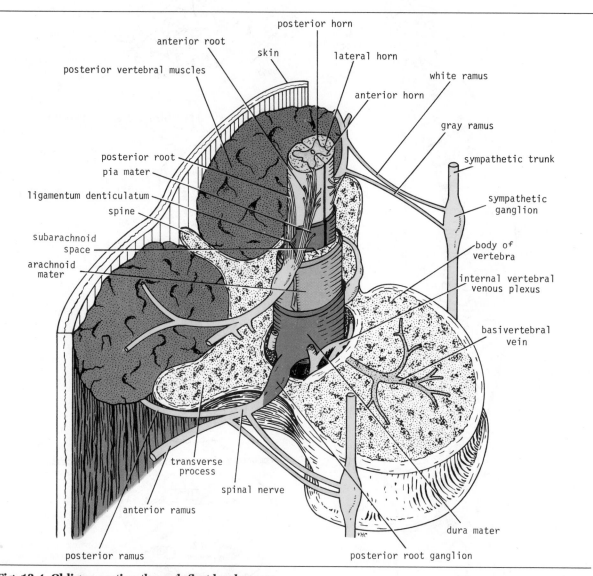

**Fig. 12-4. Oblique section through first lumbar vertebra, showing spinal cord and its covering membranes. Note relationship between spinal nerve and sympathetic trunk on each side. Note, also, important internal vertebral venous plexus.**

# THE VERTEBRAL COLUMN

The vertebral column is the central pillar of the body. It serves to protect the spinal cord and supports the weight of the head and trunk, which it transmits to the hip bones and the lower limbs. It is a flexible structure made up of irregular bones called *vertebrae*, separated by fibrocartilaginous discs called *intervertebral discs*. The intervertebral discs form one quarter of the length of the column.

The vertebrae are arranged in the following groups:

Cervical    (7)
Thoracic    (12)
Lumbar    (5)
Sacral    (5, fused to form the sacrum)
Coccygeal    (4; the lower 3 are commonly fused)

## General Characteristics of a Vertebra

Although vertebrae show regional differences, they all possess a common pattern (Fig. 12-3).

A *typical vertebra* consists of a rounded *body* anteriorly and a *vertebral arch* posteriorly. These enclose a space called the *vertebral foramen*, through which run the the spinal cord and its coverings. The vertebral arch consists of a pair of cylindrical *pedicles*, which form the sides of the arch, and a pair of flattened *laminae*, which complete the arch posteriorly.

The vertebral arch gives rise to seven processes, one spinous, two transverse, and four articular (Fig. 12-3).

The *spinous process*, or *spine*, is directed posteriorly from the junction of the two laminae. The *transverse processes* are directed laterally from the junction of the laminae and the pedicles. Both the spinous and transverse processes serve as levers and receive attachments of muscles and ligaments.

The *articular processes* are vertically arranged and consist of two superior and two inferior processes. They arise from the junction of the laminae and the pedicles, and their articular surfaces are covered with hyaline cartilage. The superior articular processes of one vertebral arch articulate with the inferior articular processes of the arch above, forming a synovial joint.

The pedicles are notched on their upper and lower borders, forming the *superior* and *inferior vertebral notches*. The superior notch of one vertebra and the inferior notch of an adjacent vertebra together form an *intervertebral foramen*. These foramina in an articulated skeleton serve to transmit the spinal nerves and blood vessels.

## CERVICAL VERTEBRAE

A *typical cervical vertebra* has the following characteristics (Fig. 12-3):

Each transverse process possesses a *foramen transversarium* for the vertebral vessels (note that the vertebral artery passes only through transverse processes C1–6). The spines are small and bifid. The body is small and broader from side to side than from front to back; there are small synovial joints on each side. The vertebral foramen is large and triangular in shape. The superior articular processes have small, flat articular facets, which face backward and upward; the inferior articular processes have facets that face downward and forward.

The first, second, and seventh cervical vertebrae are atypical.

The *first cervical vertebra*, or *atlas*, has no body and no spinous process (Fig. 11-66). It is merely a ring of bone consisting of anterior and posterior arches and a lateral mass on each side. Each lateral mass has articular surfaces on its upper and lower aspects. The bone articulates above with the occipital condyles, forming the *atlanto-occipital joints*. Below, the bone articulates with the axis, forming the *atlanto-axial joints*.

The *second cervical vertebra*, or *axis*, has a peglike odontoid process, which surmounts the body and represents the body of the atlas, which has fused with the axis (Fig. 11-66).

The *seventh cervical vertebra*, or *vertebra prominens*, is so named because it has the longest spinous process (Fig. 11-66). The spine is not bifid. The transverse process is large, but the foramen transversarium is small and transmits the vertebral vein or veins.

## THORACIC VERTEBRAE

The thoracic vertebrae increase in size from above downward. The body is heart-shaped (Fig. 12-3). The vertebral foramen is relatively small and circular. The spines are long and inclined downward.

Costal facets are present on the sides of the bodies where the heads of the ribs articulate, and on the transverse processes for articulation with the tubercles of the ribs (T11 and 12 have no facets on the transverse processes.) The superior articular processes bear facets that face backward and laterally, while the facets on the inferior articular processes face forward and medially. The inferior articular processes of the twelfth vertebra face laterally, as do those of the lumbar vertebrae.

## LUMBAR VERTEBRAE

The body of each lumbar vertebra is massive and kidney-shaped (Figs. 12-3 and 12-4). The pedicles are strong and directed backward. The laminae are thick, and the vertebral foramina are triangular in shape. The transverse processes are long and slender. The spinous process is short, flat, and quadrangular in shape (Fig. 12-5) and projects directly backward. The articular surfaces of the superior articular processes face medially, and those of the inferior articular processes face laterally.

The lumbar vertebrae have no facets for articulation with ribs and no foramina in the transverse processes.

## SACRUM

The sacrum (Fig. 12-3) consists of five rudimentary vertebrae fused together to form a wedge-shaped bone, which is concave anteriorly. The upper border, or base, of the bone articulates with the fifth lumbar vertebra. The narrow inferior border articulates with the coccyx. Laterally, the sacrum articulates with the two innominate, or hip, bones to form the sacroiliac joints (Fig. 6-5). The anterior and upper margin of the first sacral vertebra bulges forward as the posterior margin of the pelvic inlet and is known as the *sacral promontory.*

The vertebral foramina are present and form the *sacral canal.* The laminae of the fifth sacral vertebra, and sometimes those of the fourth also, fail to meet in the midline, forming the *sacral hiatus* (Fig. 6-9).

The anterior and posterior surfaces of the sacrum have four foramina on each side for the passage of the anterior and posterior rami of the upper four sacral nerves.

## COCCYX

The coccyx consists of four vertebrae fused together to form a small triangular bone, which articulates at its base with the lower end of the sacrum (Fig. 12-3). The first coccygeal vertebra commonly is not fused, or is incompletely fused, with the second vertebra.

## INTERVERTEBRAL DISCS

The intervertebral discs are responsible for one-quarter of the length of the vertebral column (Fig. 12-5). They are thickest in the cervical and lumbar regions, where the movements of the vertebral column are greatest. They may be regarded as semi-elastic discs, which lie between the rigid bodies of adjacent vertebrae (Fig. 12-6). Their physical characteristics permit them to serve as shock absorbers when the load on the vertebral column is suddenly increased, as when one is jumping from a height. Their elasticity allows the rigid vertebrae to move one upon the other. Unfortunately, their resilience is gradually lost with advancing age.

Each disc consists of a peripheral part, the anulus fibrosus, and a central part, the nucleus pulposus (Fig. 12-6). The *anulus fibrosus* is composed of fibrocartilage, in which the collagen fibers are arranged in concentric lamellae. The collagen bundles pass obliquely between adjacent vertebral bodies, and their inclination is reversed in alternate lamellae (Fig. 12-13C). The more peripheral fibers are strongly attached to the anterior and posterior longitudinal ligaments of the vertebral column.

The *nucleus pulposus* in the child is an ovoid mass of gelatinous material containing a large amount of water, a small number of collagen fibers, and a few cartilage cells. It is normally under pressure and situated slightly nearer to the posterior than to the anterior margin of the disc.

The upper and lower surfaces of the bodies of adjacent vertebrae that abut onto the disc are covered with thin plates of hyaline cartilage.

**Fig. 12-5. Sagittal section through lumbar part of vertebral column in position of flexion. Note that spines and laminae are well separated in this position, enabling one to introduce lumbar puncture needle into subarachnoid space.**

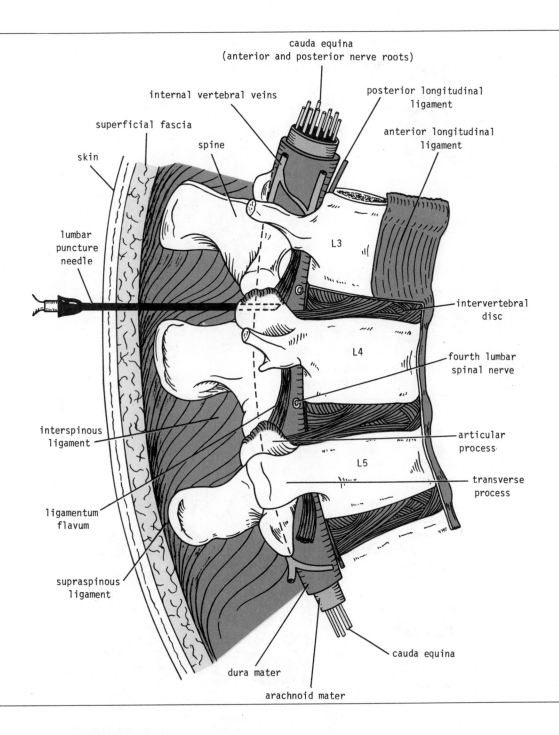

cauda equina
(anterior and posterior nerve roots)

internal vertebral veins

posterior longitudinal
ligament

anterior longitudinal
ligament

superficial fascia

spine

skin

L3

lumbar
puncture
needle

intervertebral
disc

L4

fourth lumbar
spinal nerve

interspinous
ligament

articular
process

L5

transverse
process

ligamentum
flavum

supraspinous
ligament

dura mater

cauda equina

arachnoid mater

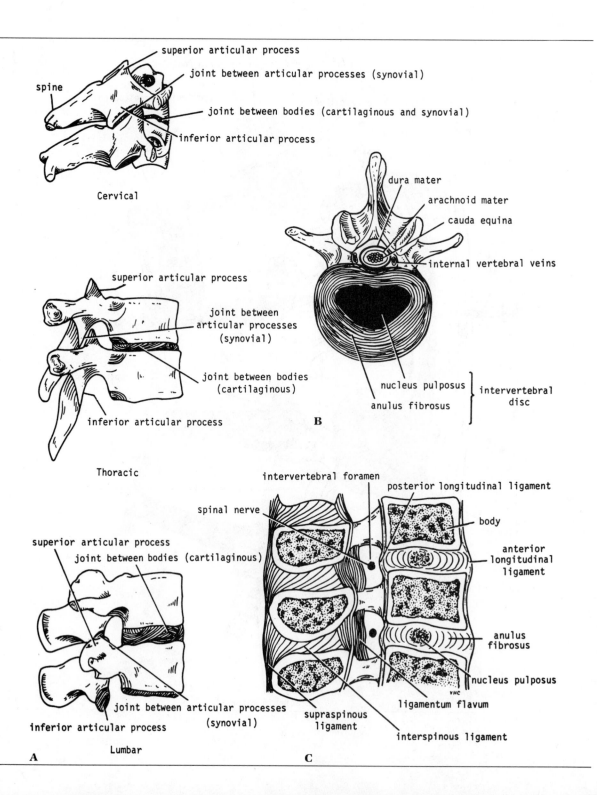

superior articular process

joint between articular processes (synovial)

joint between bodies (cartilaginous and synovial)

inferior articular process

spine

Cervical

superior articular process

joint between
articular processes
(synovial)

joint between bodies
(cartilaginous)

inferior articular process

Thoracic

dura mater

arachnoid mater

cauda equina

internal vertebral veins

nucleus pulposus

anulus fibrosus

intervertebral
disc

B

superior articular process

joint between bodies (cartilaginous)

inferior articular process

joint between articular processes
(synovial)

A          Lumbar

intervertebral foramen

spinal nerve

posterior longitudinal ligament

body

anterior
longitudinal
ligament

anulus
fibrosus

nucleus pulposus

ligamentum flavum

supraspinous
ligament

interspinous ligament

C

The semifluid nature of the nucleus pulposus allows it to change shape and permits one vertebra to rock forward or backward on another, as in flexion and extension of the vertebral column.

A sudden increase in the compression load on the vertebral column causes the semifluid nucleus pulposus to become flattened. The outward thrust of the nucleus is accommodated by the resilience of the surrounding anulus fibrosus. Sometimes, the outward thrust is too great for the anulus fibrosus and it ruptures, allowing the nucleus pulposus to herniate (Fig. 12-15).

With advancing age the water content of the nucleus pulposus diminishes and is replaced by fibrocartilage. In old age the discs are thin and less elastic, and it is no longer possible to distinguish the nucleus from the anulus.

No discs are found between the first two cervical vertebrae or in the sacrum or coccyx.

## Joints of the Vertebral Column

With the exception of the first two cervical vertebrae, the remainder of the mobile vertebrae articulate with each other by means of cartilaginous joints between their bodies and by synovial joints between their articular processes (Fig. 12-6).

### JOINTS BETWEEN TWO VERTEBRAL BODIES

The upper and lower surfaces of the bodies of adjacent vertebrae are covered by thin plates of hyaline cartilage. Sandwiched between the plates of hyaline cartilage is an intervertebral disc of fibrocartilage (Figs. 12-5 and 12-6). The collagen fibers of the disc strongly unite the bodies of the two vertebrae. (The structure of an intervertebral disc is fully described on p. 926.)

**Fig. 12-6. (A) Joints in cervical, thoracic, and lumbar regions of vertebral column. (B) Third lumbar vertebra seen from above, showing relationship between intervertebral disc and cauda equina. (C) Sagittal section through three lumbar vertebrae, showing ligaments and intervertebral discs. Note relationship between emerging spinal nerve in an intervertebral foramen and the intervertebral disc.**

In the lower cervical region, small synovial joints are present at the sides of the intervertebral disc between the upper and lower surfaces of the bodies of the vertebrae.

### Ligaments

The *anterior* and *posterior longitudinal ligaments* run as continuous bands down the anterior and posterior surfaces of the vertebral column from the skull to the sacrum (Figs. 12-5 and 12-6). The anterior ligament is wide and is strongly attached to the front and sides of the margins of the vertebral bodies and to the intervertebral discs. The posterior ligament is weak and narrow and is attached to the posterior borders of the discs.

### JOINTS BETWEEN TWO VERTEBRAL ARCHES

The joints between two vertebral arches consist of synovial joints between the superior and inferior articular processes of adjacent vertebrae (Fig. 12-6). The articular facets are covered with hyaline cartilage, and the joints are surrounded by a capsular ligament.

### Ligaments

The *supraspinous ligament* connects the tips of the vertebral spines (Fig. 12-6). The *interspinous ligament* runs between adjacent spines. The *ligamentum flavum* connects adjacent laminae. In the cervical region the supraspinous and interspinous ligaments are greatly thickened to form the very thick *ligamentum nuchae.* The latter extends from the spine of the seventh cervical vertebra to the external occipital protuberance, its anterior border being strongly attached to the cervical spines in between.

### ATLANTO-OCCIPITAL JOINTS

The atlanto-occipital joints are synovial joints that are formed between the occipital condyles, which are found on either side of the foramen magnum above and the facets on the superior surfaces of the lateral masses of the atlas below (Fig. 12-7).

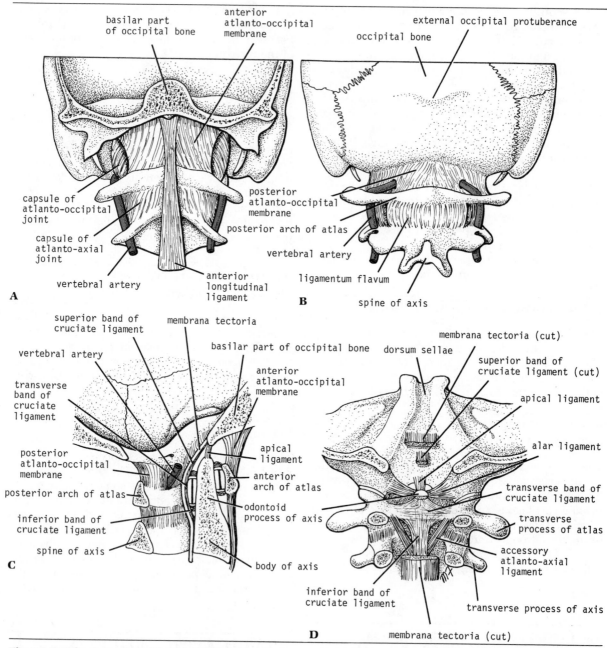

**Fig. 12-7. Atlanto-occipital joints: (A) anterior view; (B) posterior view. Atlanto-axial joints: (C) sagittal section; (D) posterior view; note that the posterior arch of the atlas and the laminae and spine of the axis have been removed.**

## Ligaments

The *anterior atlanto-occipital membrane*, a continuation of the anterior longitudinal ligament, connects the anterior arch of the atlas to the anterior margin of the foramen magnum.

The *posterior atlanto-occipital membrane*, which is similar to the ligamentum flavum, connects the posterior arch of the atlas to the posterior margin of the foramen magnum.

## ATLANTO-AXIAL JOINTS

The atlanto-axial joints are three synovial joints, one of which is between the odontoid process and the anterior arch of the atlas, while the other two are between the lateral masses of the bones (Fig. 12-7).

## Ligaments

The *apical ligament* is a median-placed structure connecting the apex of the odontoid process to the anterior margin of the foramen magnum.

The *alar ligaments* lie one on each side of the apical ligament and connect the odontoid process to the medial sides of the occipital condyles.

The *cruciate ligament* consists of a strong transverse part and a weak vertical part. The transverse part is attached on each side to the inner aspect of the lateral mass of the atlas and binds down the odontoid process to the anterior arch of the atlas. The vertical part runs from the posterior surface of the body of the axis to the anterior margin of the foramen magnum.

The *membrana tectoria* is an upward continuation of the posterior longitudinal ligament. It is attached above to the occipital bone just within the foramen magnum. It covers the posterior surface of the odontoid process and the apical, alar, and cruciate ligaments.

## Curves of the Vertebral Column

### CURVES IN THE SAGITTAL PLANE

In the fetus the vertebral column has one continuous anterior concavity. As development proceeds, the lumbosacral angle appears. After birth, when the child becomes able to raise his head and keep it poised on the vertebral column, the cervical part of the vertebral column becomes concave posteriorly (Fig. 12-8). Toward the end of the first year, when the child begins to stand upright, the lumbar part of the vertebral column becomes concave posteriorly. The development of these secondary curves is largely due to modification in the shape of the intervertebral discs.

In the adult in the standing position (Fig. 12-8), the vertebral column therefore exhibits in the sagittal plane the following regional curves: cervical—posterior concavity, thoracic—posterior convexity, lumbar—posterior concavity, and sacral—posterior convexity. During the later months of pregnancy, with the increase in size and weight of the fetus, women tend to increase the posterior lumbar concavity in an attempt to preserve their center of gravity. In old age the intervertebral discs atrophy, resulting in a loss of height and a gradual return of the vertebral column to a continuous anterior concavity.

### CURVES IN THE CORONAL PLANE

In late childhood it is quite common to find the development of minor lateral curves in the thoracic region of the vertebral column. This is normal and is usually due to the predominant use of one of the upper limbs. For example, right-handed persons will often have a slight right-sided thoracic convexity. Slight compensatory curves are always present above and below such a curvature.

## Movements of the Vertebral Column

As has been seen in the previous sections, the vertebral column consists of a number of separate vertebrae accurately positioned one upon the other and separated by intervertebral discs. The vertebrae are held in position relative to one another by strong ligaments that severely limit the degree of movement possible between adjacent vertebrae. Nevertheless, the summation of all these movements gives the vertebral column as a whole a remarkable degree of mobility.

The following movements are possible: flexion, extension, lateral flexion, rotation, and circumduction:

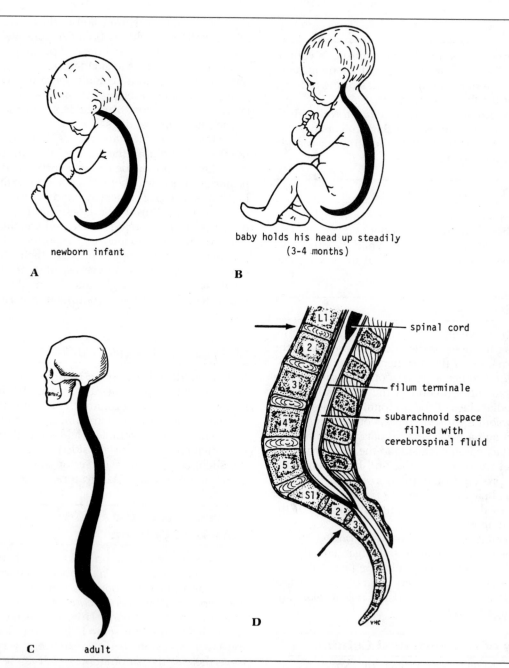

**Fig. 12-8. (A, B, C) Curves of vertebral column at different ages. Note that in adult (D), lower end of spinal cord lies at level of lower border of body of first lumbar vertebra (arrow), and subarachnoid space ends at lower border of body of second sacral vertebra (arrow).**

*Flexion* is a forward movement, and *extension* is a backward movement. They are both extensive in the cervical and lumbar regions, but restricted in the thoracic region.

*Lateral flexion* is the bending of the body to one or the other side. It is extensive in the cervical and lumbar regions, but restricted in the thoracic region.

*Rotation* is a twisting of the vertebral column. This is most extensive in the lumbar region.

*Circumduction* is a combination of all these movements.

The type and range of movements possible in each region of the column is largely dependent on the thickness of the intervertebral discs and the shape and direction of the articular processes. In the thoracic region, the ribs, the costal cartilages, and the sternum severely restrict the range of movement.

The *atlanto-occipital joints* permit extensive flexion and extension of the head. The *atlanto-axial joints* allow a wide range of rotation of the atlas and thus of the head on the axis.

The vertebral column is moved by numerous muscles, many of which are attached directly to the vertebrae, while others, such as the sternocleidomastoid and the abdominal wall muscles, are attached to the skull or to the ribs or fasciae.

*In the cervical region*, flexion is produced by the longus cervicis, the scalenus anterior, and the sternocleidomastoid muscles. Extension is produced by the postvertebral muscles (see below). Lateral flexion is produced by the scalenus anterior and medius and the trapezius and sternocleidomastoid muscles. Rotation is produced by the sternocleidomastoid on one side and the splenius on the other side.

*In the thoracic region*, rotation is produced by the semispinalis and rotatores muscles, assisted by the oblique muscles of the anterolateral abdominal wall.

*In the lumbar region*, flexion is produced by the rectus abdominis and the psoas muscles. Extension is produced by the postvertebral muscles. Lateral flexion is produced by the postvertebral muscles, the quadratus lumborum, and the oblique muscles of the anterolateral abdominal wall. The psoas may also play a part in this movement. Rotation is produced by the rotatores muscles and the oblique muscles of the anterolateral abdominal wall.

# MUSCLES OF THE BACK

The muscles of the back may be divided into three main groups: (1) the superficial muscles associated with the shoulder girdle, (2) the intermediate muscles involved with respiration, and (3) the deep muscles belonging to the vertebral column.

## Superficial Muscles

These muscles belong to the upper limb and are the trapezius, latissimus dorsi, levator scapulae, and the rhomboid minor and major. They are described in Chapter 9.

## Intermediate Muscles

These muscles are associated with respiration and are the serratus posterior superior, serratus posterior inferior, and the levatores costarum. They are described with the thorax in Chapter 2.

## Deep Muscles of the Back (Postvertebral Muscles)

In the standing position the line of gravity (Fig. 12-9) passes through the odontoid process of the axis, behind the centers of the hip joints, and in front of the knee and ankle joints. It follows that when the body is in this position, the greater part of its weight falls in front of the vertebral column. It is therefore not surprising to find that the postvertebral muscles of the back are well developed in man. The postural tone of these muscles is the major factor responsible for the maintenance of the normal curves of the vertebral column.

The deep muscles of the back form a broad, thick column of muscle tissue, which occupies the hollow on each side of the spinous processes (Fig. 12-9). They extend from the sacrum to the skull and lie beneath the thoracolumbar fascia. It must be realized that this complicated muscle mass is composed of many separate muscles of varying length. Each individual muscle may be regarded as a string, which, when pulled on, causes one or several vertebrae to be extended or rotated on the vertebra below. Since the origins and insertions of the different groups of muscles overlap, entire regions of the vertebral column can be made to move smoothly.

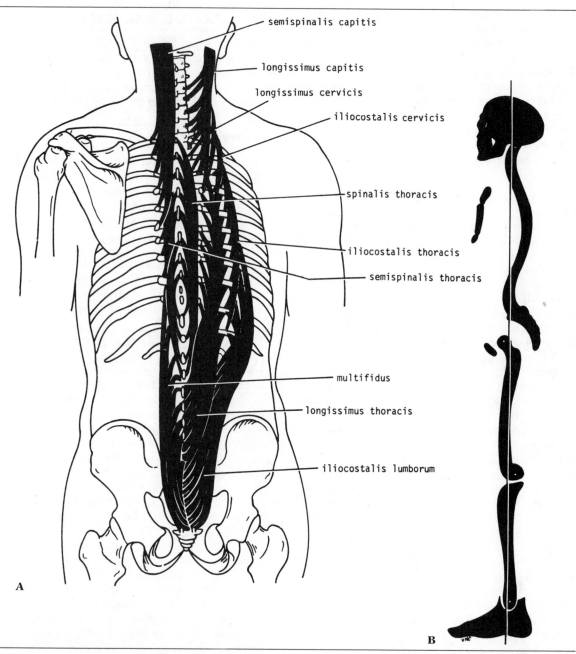

semispinalis capitis

longissimus capitis

longissimus cervicis

iliocostalis cervicis

spinalis thoracis

iliocostalis thoracis

semispinalis thoracis

multifidus

longissimus thoracis

iliocostalis lumborum

A

B

**Fig. 12-9. (A) Arrangement of deep muscles of back. (B) Lateral view of skeleton, showing line of gravity. Since greater part of body weight lies anterior to vertebral column, deep muscles of back are important in maintaining normal postural curves of vertebral column in standing position.**

The spines and transverse processes of the vertebrae serve as levers that facilitate the muscle actions. The muscles of longest length lie superficially and run vertically from the sacrum to the rib angles, the transverse processes, and the upper vertebral spines (Fig. 12-9). The muscles of intermediate length run obliquely from transverse processes to spines. The shortest and deepest muscle fibers run between spines and between transverse processes of adjacent vertebrae.

The deep muscles of the back may be classified as follows:

*Superficial Vertically Running Muscles*

|                |                      |
|----------------|----------------------|
| Erector spinae | ........... iliocostalis |
|                | ........... longissimus |
|                | ........... spinalis |

*Intermediate Oblique Running Muscles*

|                    |                      |
|--------------------|----------------------|
| Transversospinalis | ........... semispinalis |
|                    | ........... multifidus |
|                    | ........... rotatores |

*Deepest Muscles*
Interspinales
Intertransversarii

A knowledge of the detailed attachments of the various muscles of the back has no practical value to a medical student, and the attachments are therefore omitted in this text.

## SPLENIUS

The splenius is a detached part of the deep muscles of the back. It consists of two parts. The *splenius capitis* arises from the lower part of the ligamentum nuchae and the upper four thoracic spines and is inserted into the superior nuchal line of the occipital bone and the mastoid process of the temporal bone.

The *splenius cervicis* has a similar origin, but is inserted into the transverse processes of the upper cervical vertebrae.

## Nerve Supply

All the deep muscles of the back are innervated by the posterior rami of the spinal nerves.

## Deep Fascia of the Back (Thoracolumbar Fascia)

The lumbar part of the deep fascia is situated in the interval between the iliac crest and the twelfth rib. It forms a strong aponeurosis and laterally gives origin to the middle fibers of the transversus and the upper fibers of the internal oblique muscle. (See p. 154.)

Medially, the lumbar part of the deep fascia splits into three lamellae. The posterior lamella covers the deep muscles of the back and is attached to the lumbar spines. The middle lamella passes medially, to be attached to the tips of the transverse processes of the lumbar vertebrae; it lies in front of the deep muscles of the back and behind the quadratus lumborum. The anterior lamella passes medially and is attached to the anterior surface of the transverse processes of the lumbar vertebrae; it lies in front of the quadratus lumborum muscle.

In the thoracic region, the deep fascia is attached medially to the vertebral spines and laterally to the angles of the ribs. It covers the posterior surface of the deep muscles of the back.

In the cervical region, the deep fascia is very much thinner and of no special importance.

## Blood Vessels of the Back

### ARTERIES

The following arteries supply the structures of the back.

*In the cervical region,* branches arise from the occipital artery, a branch of the external carotid; from the vertebral artery, a branch of the subclavian; from the deep cervical artery, a branch of the costocervical trunk; from the ascending cervical artery, a branch of the inferior thyroid artery.

*In the thoracic region,* branches arise from the posterior intercostal arteries and *in the lumbar region,* from the subcostal and lumbar arteries. *In the sacral region,* branches arise from the iliolumbar and lateral sacral arteries, branches of the internal iliac artery.

### VEINS

The veins draining the structures of the back form complicated plexuses extending along the vertebral column from the skull to the coccyx. The veins may be divided into (1) those that lie external to the vertebral column and surround it and form the *external vertebral venous plexus* and (2) those that lie within the vertebral canal and form the *internal vertebral venous plexus* (Fig. 12-4). There is free communication between these plexuses and the veins in the neck, thorax, abdomen, and pelvis. Above, they communicate through the foramen magnum with the occipital and basilar venous sinuses within the cranial cavity. The internal vertebral plexus lies within the vertebral canal, but outside the dura mater of the spinal cord. It is embedded in areolar tissue and receives tributaries from the vertebrae by way of the *basivertebral veins* (Fig. 12-4) and from the meninges and spinal cord. The internal plexus is drained by the *intervertebral veins*, which pass outward with the spinal nerves through the intervertebral foramina. Here, they are joined by tributaries from the external vertebral plexus and in turn drain into the vertebral, intercostal, lumbar, and lateral sacral veins.

The external and internal vertebral plexuses form a capacious venous network whose walls are thin and whose channels have incompetent valves or are valveless. Free venous blood flow may therefore take place between the skull, the neck, the thorax, the abdomen, the pelvis, and the vertebral plexuses, with the direction of flow depending on the pressure differences that exist at any given time between the regions. This fact is of considerable clinical significance.

## Lymphatic Drainage of the Back

The deep lymphatic vessels follow the veins and drain into the deep cervical, posterior mediastinal, lateral aortic, and sacral nodes. The lymphatic vessels from the skin of the neck drain into the cervical nodes; those from the trunk above the iliac crests drain into the axillary nodes; and those from below the level of the iliac crests drain into the superficial inguinal nodes. (See p. 182.)

## Nerves of the Back

The skin and muscles of the back are supplied in a segmental manner by the posterior rami of the thirty-one pairs of spinal nerves. The posterior rami of the first, sixth, seventh, and eighth cervical nerves and the fourth and fifth lumbar nerves supply the deep muscles of the back and do not supply the skin. The posterior ramus of the second cervical nerve (the *greater occipital nerve*) ascends over the back of the head and supplies the skin of the scalp.

The posterior rami run downward and laterally and supply a band of skin at a lower level than the intervertebral foramen from which they emerge. Considerable overlap of skin areas supplied occurs, so that section of a single nerve causes diminished, but not total, loss of sensation. Each posterior ramus divides into a medial and a lateral branch, the first cervical and the fourth and fifth sacral and the first coccygeal nerves excepted.

# SPINAL CORD AND MENINGES
## Spinal Cord

The spinal cord is a white structure that begins above at the foramen magnum, where it is continuous with the medulla oblongata of the brain. It terminates below in the adult at the level of the lower border of the first lumbar vertebra (Fig. 12-8). In the young child it is relatively longer and ends at the upper border of the third lumbar vertebra. The spinal cord is roughly cylindrical in shape. However, in the cervical region, where it gives origin to the brachial plexus, and in the lower thoracic and lumbar regions, where it gives origin to the lumbosacral plexus, there are fusiform enlargements, called the *cervical* and *lumbar enlargements*.

Inferiorly, the spinal cord tapers off into the *conus medullaris*, from the apex of which a prolongation of the pia mater, the *filum terminale*, descends to be attached to the back of the coccyx (Figs. 12-8 and 12-10). The cord possesses in the midline anteriorly a deep longitudinal fissure, the *anterior median fissure*, and on the posterior surface a shallow furrow, the *posterior median sulcus*.

Along the whole length of the spinal cord are attached thirty-one pairs of spinal nerves by the *an-*

Fig. 12-10. **(A) Lower end of spinal cord and cauda equina. (B) Section through thoracic part of spinal cord, showing anterior and posterior roots of spinal nerves and meninges. (C) Transverse section through spinal cord, showing meninges and position of cerebrospinal fluid.**

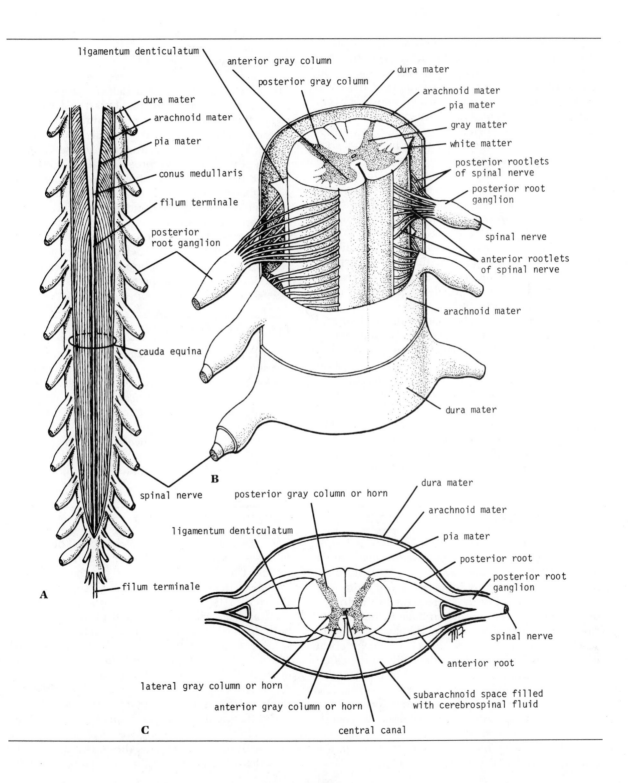

*terior*, or *motor*, *roots* and the *posterior*, or *sensory*, *roots* (Fig. 12-10). Each root is attached to the cord by a series of rootlets, which extend the whole length of the corresponding segment of the cord. Each posterior nerve root possesses a posterior root ganglion, the cells of which give rise to peripheral and central nerve fibers.

The spinal nerve roots pass from the spinal cord to the level of their respective intervertebral foramina, where they unite to form a *spinal nerve*. Having emerged from the intervertebral foramen, each spinal nerve immediately divides into its *anterior* and *posterior rami*, each containing both motor and sensory fibers. Because of the disproportionate growth in length of the vertebral column during development, as compared with that of the spinal cord, the length of the roots increases progressively from above downward. In the upper cervical region the spinal nerve roots are short and run almost horizontally, but the roots of the lumbar and sacral nerves below the level of the termination of the cord (lower border of the first lumbar vertebra in the adult) form a vertical leash of nerves around the filum terminale. These lower nerve roots together are called the *cauda equina* (Fig. 12-10).

## Meninges

The spinal cord, like the brain, is surrounded by three meninges: the dura mater, the arachnoid mater, and the pia mater.

### DURA MATER OF THE SPINAL CORD

The dura mater is a dense, strong, fibrous membrane that encloses the spinal cord and cauda equina (Figs. 12-4 and 12-10). It is continuous above through the foramen magnum with the meningeal layer of dura covering the brain. Inferiorly, it ends on the filum terminale at the level of the lower border of the second sacral vertebra (Fig. 12-8). The dural sheath lies loosely in the vertebral canal and is separated from the walls of the canal by the *extradural space*. This contains loose areolar tissue and the internal vertebral venous plexus. The dura mater extends along each nerve root and becomes continuous with connective tissue surrounding each spinal nerve (*epineurium*). The inner surface of the dura mater is in contact with the arachnoid mater (Fig. 12-10).

### ARACHNOID MATER OF THE SPINAL CORD

The arachnoid mater is a delicate impermeable membrane covering the spinal cord and lying between the pia mater internally and the dura mater externally (Fig. 12-4). It is separated from the pia mater by a wide space, the *subarachnoid space*, which is filled with *cerebrospinal fluid* (Figs. 12-8 and 12-10). The subarachnoid space is crossed by a number of fine strands of connective tissue. The arachnoid mater is continuous above through the foramen magnum with the arachnoid covering the brain. Inferiorly, it ends on the filum terminale at the level of the lower border of the second sacral vertebra (Fig. 12-8). The arachnoid mater is continued along the spinal nerve roots, forming small lateral extensions of the subarachnoid space.

### PIA MATER OF THE SPINAL CORD

The pia mater is a vascular membrane that closely invests the spinal cord (Fig. 12-4). The pia mater is thickened on either side between the nerve roots to form the *ligamentum denticulatum*, which passes laterally to adhere to the arachnoid and dura. It is by this means that the spinal cord is suspended in the middle of the dural sheath. The pia mater extends along each nerve root and becomes continuous with the connective tissue surrounding each spinal nerve (Fig. 12-10).

### CEREBROSPINAL FLUID

The cerebrospinal fluid is produced by the *choroid plexuses*, within the lateral, third, and fourth ventricles of the brain. It escapes from the ventricular system of the brain through the three foramina in the roof of the fourth ventricle and so enters the subarachnoid space. (See p. 800.) It now circulates both upward over the surface of the cerebral hemispheres and downward around the spinal cord. The spinal part of the subarachnoid space extends down as far as the lower border of the second sacral vertebra (Fig. 12-8). Eventually, the fluid enters the bloodstream by passing into the *arachnoid villi* and diffusing through their walls.

In addition to removing waste products associated with neuronal activity, the cerebrospinal fluid provides a fluid medium that surrounds the spinal cord. This fluid, together with the bony and ligamentous walls of the vertebral canal, effectively protects the spinal cord from trauma.

## Blood Supply of the Spinal Cord

The *posterior spinal arteries*, which arise either directly or indirectly from the vertebral arteries, divide to form two descending branches, which run down the side of the spinal cord, one behind and one in front of the attachments of the posterior spinal nerve roots. The *anterior spinal arteries*, which arise from the vertebral arteries, unite to form a single artery, which runs down within the anterior median fissure.

The posterior and anterior spinal arteries are reinforced by *radicular arteries*, which enter the vertebral canal through the intervertebral foramina.

The *veins* of the spinal cord drain into the internal vertebral venous plexus.

## RADIOGRAPHIC APPEARANCES OF THE VERTEBRAL COLUMN

### Cervical Region

The views commonly used are (1) the anteroposterior and (2) the lateral.

The *anteroposterior* view is taken with the patient in the supine position. The film cassette is placed behind the head and the neck, and the X-ray tube is centered over the front of the thyroid cartilage. The atlanto-axial articulation may be demonstrated by asking the patient to keep the mandible in motion while the film is being exposed, or by directing the X-ray tube through the open mouth. By using the latter method, the entire length of the odontoid process may be visualized lying between the lateral masses of the atlas.

Below the level of the third cervical vertebra, the bodies of the vertebrae are well shown and the spines are clearly seen. The laminae can be identified. The transverse processes, the foramina transversaria, and the articular processes overlap one another and are difficult to distinguish separately. The lumen of the trachea may be seen as a tubular transradiancy that narrows at the upper end, where it becomes continuous with the cavity of the larynx.

The *lateral view* is taken with the patient sitting up and the shoulders dropped, so that the seventh cervical vertebra may be demonstrated. The film cassette is placed at the side of the neck in the parasagittal plane. The X-ray tube is directed at the side of the neck at right angles to the long axis of the vertebral column and the film.

The atlanto-occipital joint is difficult to make out. The anterior and posterior arches of the atlas are well shown (Fig. 12-11), and the body of the axis is easily identified. The odontoid process of the axis extends upward, close to the posterior margin of the anterior arch of the atlas. The articular processes are well shown, and the spinous processes can be clearly seen. The transverse processes are difficult to make out since they are superimposed on the vertebral bodies. The intervertebral disc spaces between the bodies of adjacent vertebrae are easily defined and are of equal height.

The anterior and posterior surfaces of the vertebral bodies and the posterior wall of the vertebral canal form smooth curved lines that are roughly parallel (Fig. 12-11).

### Thoracic Region

The views commonly used are (1) the anteroposterior and (2) the lateral.

The *anteroposterior view* is taken with the patient in the supine position. The film cassette is placed behind the thorax, and the X-ray tube is centered over the front of the sternum.

Because of the curvature of the thoracic part of the vertebral column, the upper and lower margins of the bodies of adjacent vertebrae overlap. The spinous processes and laminae are superimposed on the bodies. The transverse processes can be identified, but they are obscured by the heads and necks of the ribs. Note that the first rib and the tenth, eleventh, and twelfth ribs on each side articulate only with the bodies of the first, tenth, eleventh, and twelfth thoracic vertebrae, respectively; all the other ribs articulate with two vertebrae.

The pedicles are clearly seen as ovoid structures that are superimposed on the lateral parts of the bodies.

The transradiant trachea and the heart shadow are superimposed on the thoracic vertebrae.

The *lateral view* is taken with the patient lying on his side, with the arms stretched above his head.

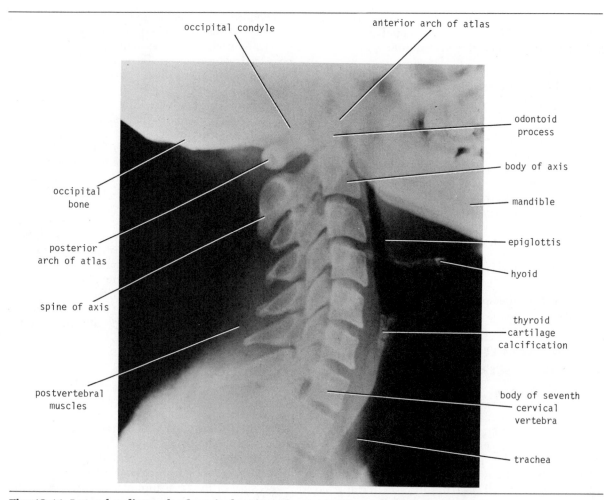

occipital condyle

anterior arch of atlas

occipital bone

posterior arch of atlas

spine of axis

postvertebral muscles

odontoid process

body of axis

mandible

epiglottis

hyoid

thyroid cartilage calcification

body of seventh cervical vertebra

trachea

**Fig. 12-11. Lateral radiograph of cervical region of vertebral column.**

If it is desirable to demonstrate the postural curves, the patient assumes the standing position. The film cassette is placed against the side of the thorax, and the X-ray tube is directed laterally through the vertebral column at right angles to the film.

The rectangular vertebral bodies and the intervertebral disc spaces are clearly seen, even though the ribs and lungs are superimposed on them. The upper four vertebrae are obscured by the shadows of the shoulder girdle.

The pedicles and intervertebral foramina are well demonstrated. However, the spinous processes, the laminae, the transverse processes, and the ribs are superimposed on one another, and their detail is obscured. The vertebral canal is well shown.

### Lumbosacral Region

The views commonly used are (1) the anteroposterior and (2) the lateral.

The *anteroposterior view* is taken with the patient in the supine position. The film cassette is placed behind the lumbar region and buttocks, and the X-ray tube is centered over the umbilicus. To diminish the distortion produced by the lumbar curvature, the patient can be asked to flex his knees and hips, which may straighten the lumbar curvature to some extent.

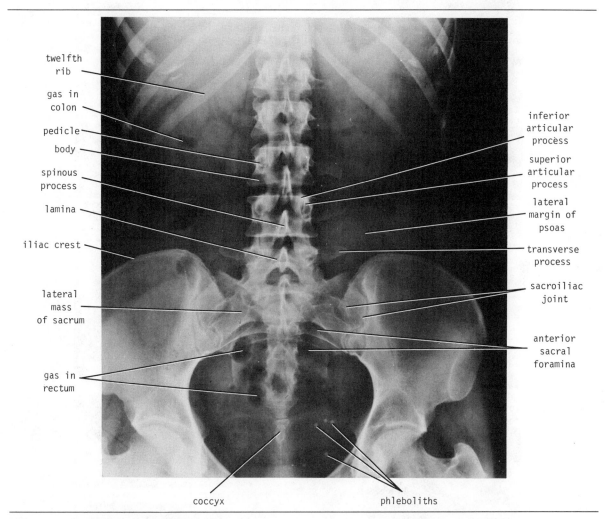

twelfth
rib

gas in
colon

pedicle

body

spinous
process

lamina

iliac crest

lateral
mass
of sacrum

gas in
rectum

coccyx

inferior
articular
procèss

superior
articular
process

lateral
margin of
psoas

transverse
process

sacroiliac
joint

anterior
sacral
foramina

phleboliths

**Fig. 12-12. Anteroposterior radiograph of lower thoracic, lumbar, and sacral regions of vertebral column.**

The bodies, transverse processes, spinous processes, laminae, and intervertebral disc spaces are clearly seen (Fig. 12-12). The pedicles produce ovoid shadows, and the articular processes and posterior intevertebral joints can be delineated.

Because of the obliquity of the sacroiliac joint, it is visualized as two lines, the lateral one corresponding to the anterior margin and the medial one to the posterior margin (Fig. 12-12). The lower segments of the sacrum and the coccyx are tilted posteriorly and are usually overlapped by the sym-

physis pubis. In addition, the presence of gas and fecal material in the rectum and sigmoid colon commonly obscures the sacrum. In order to demonstrate the sacrum in a more direct anteroposterior view, the X-ray tube may be tilted.

The *lateral view* is taken with the patient lying on his side. If it is desirable to demonstrate the postural curves, the patient assumes the standing position. The film cassette is placed against the side of the lumbar region, and the X-ray tube is directed laterally through the lumbar part of the vertebral column at right angles to the film.

The large vertebral bodies, the intervertebral disc spaces, and the intervertebral foramina are

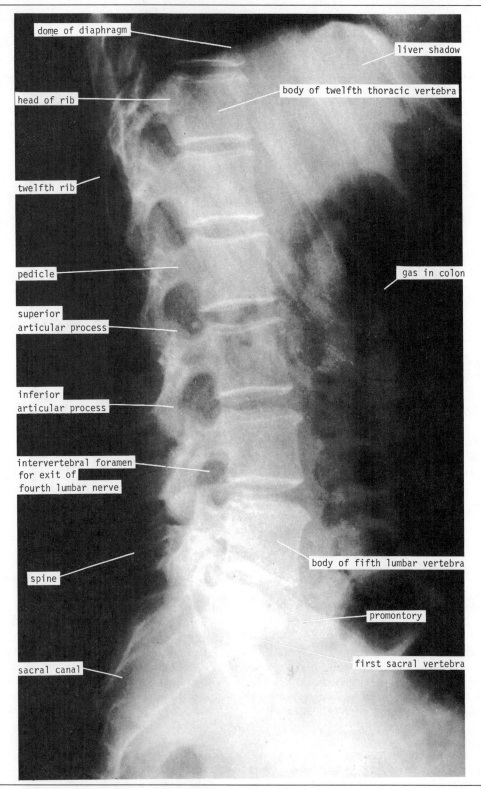

Fig. 12-13. Lateral radiograph of lower thoracic,
lumbar, and sacral regions of vertebral column.

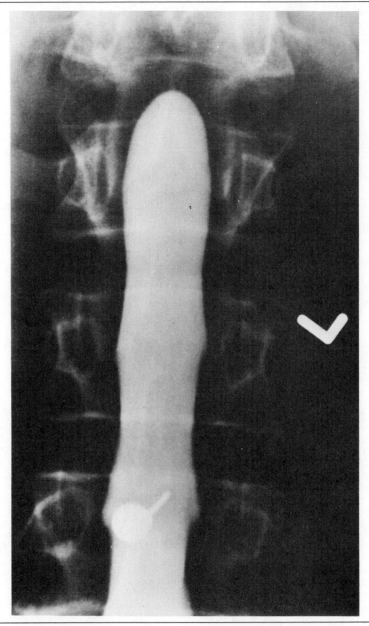

**Fig. 12-14. Posteroanterior myelogram of the lumbar region.**

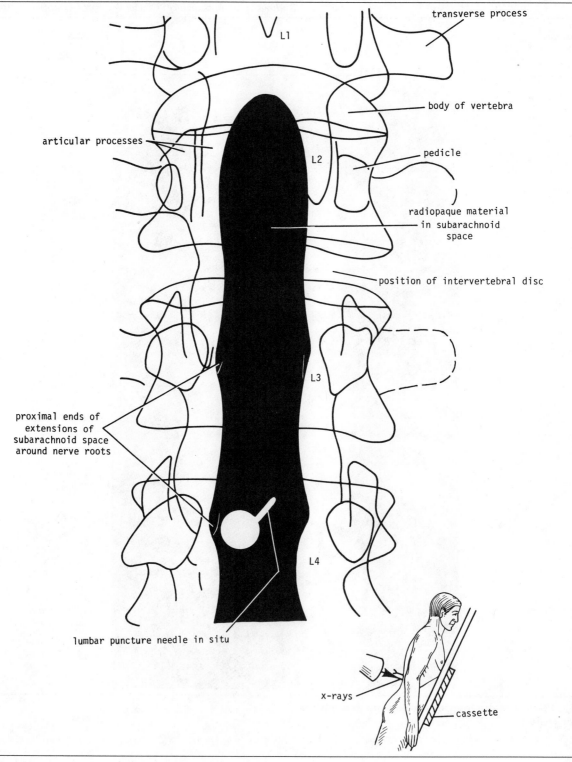

**Fig. 12-15. Main features that can be seen in myelogram in Figure 12-14.**

clearly seen (Fig. 12-13). The pedicles, the articular processes, and the spinous processes are easily visualized. The transverse processes can be identified, but they are superimposed on the sides of the preceding structures. The anterior and posterior surfaces of the vertebral bodies and the posterior wall of the vertebral canal form smooth curved lines that are roughly parallel.

Occasionally, the fifth lumbar vertebra is partly or completely fused with the first sacral vertebra. Not infrequently, the first sacral vertebra is separate from the remainder of the sacrum and has the appearance of a sixth lumbar vertebra.

The *sacrum* on lateral view shows the promontory, the sacral canal, and the fused sacral bodies and spinous processes (Fig. 12-13). Note the localized anterior angulation between the body of the fifth lumbar vertebra and the first sacral vertebra.

## Coccyx

The coccyx is not well shown on routine anteroposterior and lateral radiographs. This is due to its oblique position relative to the film and to the presence of gas and feces in the rectum and sigmoid colon. These difficulties may be partially overcome by tilting the X-ray tube and evacuating the contents of the rectum and sigmoid colon.

## Radiographic Appearances of the Spinal Subarachnoid Space

The subarachnoid space may be studied radiographically by the injection of contrast media into the subarachnoid space by lumbar puncture. Iodized oil has been used with success. This technique is referred to as *myelography* (Figs. 12-14 and 12-15).

If the patient is sitting in the upright position, the oil sinks to the lower limit of the subarachnoid space at the level of the lower border of the second sacral vertebra. By placing the patient on a tilting table, the oil can be made to gravitate gradually to higher levels of the vertebral column.

A normal myelogram will show pointed lateral projections at regular intervals at the intervertebral space levels. This appearance is due to the opaque medium filling the lateral extensions of the subarachnoid space around each spinal nerve. The presence of a tumor or a prolapsed intervertebral disc may obstruct the movement of the oil from one region to another when the patient is tilted.

## CLINICAL NOTES
### Examination of the Back

Aches and pains referred to the back region are very common symptoms encountered in clinical practice. A thorough knowledge of the anatomy of the region, together with a conscientious physical examination of the patient, enables one to make a diagnosis in a substantial number of cases. Unfortunately, it must be admitted that the cause of many cases of low back pain is often difficult to determine.

It is important that the whole area of the back and legs be examined and that the shoes be removed. Unequal length of the legs or disease of the hip joints may lead to abnormal curvatures of the vertebral column. The patient should be asked to walk up and down the examination room so that the normal tilting movement of the pelvis may be observed. As one side of the pelvis is raised, a coronal lumbar convexity develops on the opposite side, with a compensatory thoracic convexity on the same side. When a person assumes the sitting position, it will be noted that the normal lumbar curvature becomes flattened, with an increase in the interval between the lumbar spines.

It is useful to remember that the vertebral column consists of a series of small blocks poised one upon the other and that their position is maintained primarily by the normal tone of the muscles, especially of the deep back muscles, which are acting on the vertebrae. Paralysis of the muscles, as in poliomyelitis, may lead to gross deformity, while irritation of nerves may lead to pain, muscle spasm, and postural changes.

While it is easy to remember that the structures of the back are supplied by segmentally arranged posterior rami of the spinal nerves, the anterior rami and the structures they innervate are often overlooked. Chronic prostatitis in the male or chronic cervicitis of the uterus in the female may lead to referred pain in the back. The importance of a thorough pelvic examination in cases of low back pain cannot be overemphasized.

The vertebral column serves to protect the delicate spinal cord, but should this bony and ligamentous structure be severely damaged, uncautious movement of the vertebrae may result in their being the instrument for severing the spinal cord.

The normal range of movement of the different parts of the vertebral column should be tested. In the cervical region, flexion, extension, lateral rotation, and lateral flexion are possible. Remember that about half of the movement we refer to as flexion is carried out at the atlanto-occipital joints. In flexion the patient should be able to touch his chest with his chin, and in extension he should be able to look directly upward. In lateral rotation the patient should be able to place his chin nearly in line with his shoulder. Half of lateral rotation occurs between the atlas and the axis. In lateral flexion the head can normally be tilted 45 degrees to each shoulder. It is important that the shoulder is not raised when this movement is being tested.

In the thoracic region the movements are limited by the presence of the ribs and sternum. Rotation is possible because the articular processes lie on an arc of a circle whose center is within the vertebral body. When testing for rotation, make sure that the patient does not rotate the pelvis.

In the lumbar region, flexion, extension, lateral rotation, and lateral flexion are possible. Flexion and extension are fairly free. Lateral rotation is, however, limited by the interlocking of the articular processes. Lateral flexion in the thoracic and lumbar regions is tested by asking the patient to slide, in turn, each hand down the lateral side of the thigh.

## Abnormal Curves of the Vertebral Column

*Kyphosis* is the term used to describe an exaggeration in the sagittal curvature present in the thoracic part of the vertebral column. It may be due to muscular weakness or structural changes in the vertebral bodies or intervertebral discs. In sickly adolescents, for example, where the muscle tone is poor, long hours of study or work over a low desk may lead to a gently curved kyphosis of the upper thoracic region. The person is said to be "round-shoul-

dered." Crush fractures or tuberculous destruction of the vertebral bodies leads to acute angular kyphosis of the vertebral column. In the aged, *osteoporosis* (abnormal rarefaction of bone) and/or degeneration of the intervertebral discs leads to *senile kyphosis*, involving the cervical, thoracic, and lumbar regions of the column.

*Lordosis* is the term used to describe an exaggeration in the sagittal curvature present in the lumbar region. Lordosis may be due to an increase in the weight of the abdominal contents, as with the gravid uterus or a large ovarian tumor, or it may be due to disease of the vertebral column such as spondylolisthesis. (See p. 949.) The possibility that it is a postural compensation for a kyphosis in the thoracic region or a disease of the hip joint (congenital dislocation) must not be overlooked.

*Scoliosis* is the term used to describe a lateral deviation of the vertebral column. This is most commonly found in the thoracic region and may be due to muscular or vertebral defects. Paralysis of muscles due to poliomyelitis may cause severe scoliosis. The presence of a congenital hemivertebra may cause scoliosis. Very often, scoliosis is compensatory and may be due to a short leg or hip disease.

## Vertebral Venous Plexus

The longitudinal thin-walled, valveless vertebral venous plexus has been fully described on page 935. Because this plexus communicates above with the intracranial venous sinuses and segmentally with the veins of the thorax, abdomen, and pelvis, it is a clinically important structure. Batson has shown experimentally in cadavers and animals that pelvic venous blood enters not only the inferior vena cava but also the vertebral venous plexus and by this route may also enter the skull. This is especially likely to occur if the intra-abdominal pressure is increased. It is interesting to note that the internal vertebral venous plexus is not subject to external pressures when the intra-abdominal pressure rises. In fact, a rise in pressure on the abdominal veins would tend to force the blood backward out of the abdominal cavity into the veins within the vertebral canal. The existence of this venous plexus may explain how carcinoma of the prostate may metastasize to the vertebral column and the cranial cavity. The spread of the malignant cells intravenously would be aided by any activity that increased the intra-abdominal pressure, such as coughing or sneezing.

## Intervertebral Discs

The structure of the intervertebral disc has been described on page 926. The physical properties of the disc allow one vertebra to rock upon another and at the same time serve as a most efficient shock-absorbing mechanism. The resistance of these discs to compression forces is undeniable, as seen, for example, in circus acrobats who can support four or more of their colleagues on their shoulders. Nevertheless, the discs are vulnerable to sudden shocks, particularly if the vertebral column is flexed and the disc is undergoing degenerative changes.

The discs most commonly affected are those in areas where a mobile part of the column joins a relatively immobile part, i.e., the cervicothoracic junction and the lumbosacral junction. In these areas the posterior part of the anulus fibrosus ruptures, and the nucleus pulposus is forced posteriorly like toothpaste out of a tube. This is referred to as a *herniation of the nucleus pulposus.* The herniation may result either in a central protrusion in the midline under the posterior longitudinal ligament of the vertebrae or in a lateral protrusion at the side of the posterior ligament close to the intervertebral foramen (Fig. 12-16). The escape of the nucleus pulposus will produce narrowing of the space between the vertebral bodies, which may be visible on radiography. Slackening of the anterior and posterior longitudinal ligaments results in abnormal mobility of the vertebral bodies, producing local pain and subsequent development of osteoarthritis.

*Cervical disc herniations* are less common than herniations in the lumbar region. The discs most liable to this condition are those between the fifth and sixth or sixth and seventh vertebrae. Lateral protrusions cause pressure on a spinal root. Each spinal root emerges above the corresponding vertebra; thus, the C5–6 disc protrusion compresses the C6 nerve root. Pain is felt near the lower part of the back of the neck and shoulder and along the arm in the distribution of the nerve root involved. Central protrusions may press on the spinal cord and the anterior spinal artery and involve the pyramidal tracts.

*Lumbar disc herniations* are more common than cervical disc herniations (Fig. 12-16). The discs most usually affected are those between the fourth and fifth lumbar vertebrae and between the fifth lumbar vertebra and the sacrum. In the lumbar region the roots of the cauda equina run posteriorly over a number of intervertebral discs (Fig. 12-16). A lateral herniation may press on one or two roots and often involves the nerve root going to the intervertebral foramen just below. The nucleus pulposus occasionally herniates directly backward, and, if it is a large herniation, the whole cauda equina may be compressed, producing paraplegia.

There is usually an initial period of back pain due to the injury to the disc. The back muscles show spasm, especially on the side of the herniation, due to pressure on the spinal nerve root. As a consequence, the vertebral column shows a scoliosis, with its concavity on the side of the lesion. Pain is referred down the leg and foot in the distribution of the affected nerve. Since the sensory roots most commonly pressed upon are the fifth lumbar and the first sacral, pain is usually felt down the back and lateral side of the leg, radiating to the sole of the foot. This condition is often called *sciatica.* In severe cases there may be paresthesia or actual sensory loss.

Pressure on the motor roots causes muscle weakness. Involvement of the fifth lumbar motor root produces weakness of dorsiflexion of the ankle, while pressure on the first sacral motor root causes weakness of plantar flexion, and the ankle jerk may be diminished or absent (Fig. 12-16).

A large, centrally placed protrusion may give rise to bilateral pain and muscle weakness in both legs. Acute retention of urine may also occur.

## Intervertebral Foramina

The intervertebral foramina transmit the spinal nerves and the small segmental arteries and veins, all of which are embedded in areolar tissue. Each foramen is bounded above and below by the pedicles of adjacent vertebrae, in front by the lower part of the vertebral body and by the intervertebral disc, and behind by the articular processes and the joint between them. In this situation the spinal nerve is very vulnerable and may be pressed upon or irritated by disease of the surrounding structures. Herniation of the intervertebral disc, frac-

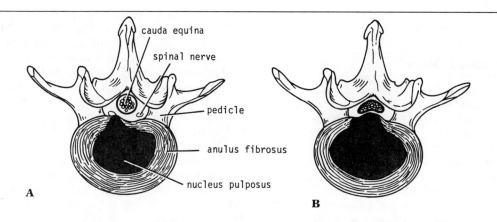

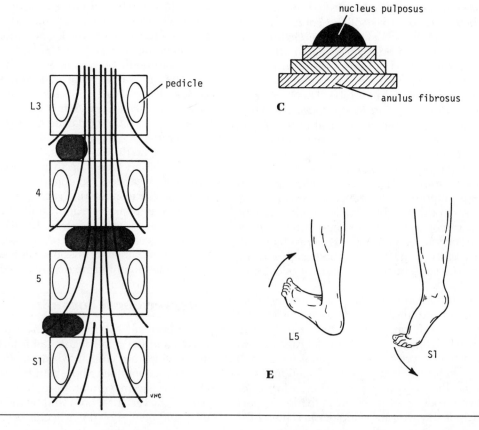

tures of the vertebral bodies, and osteoarthritis involving the joints of the articular processes or the joints between the vertebral bodies may all result in pressure, stretching, or edema of the emerging spinal nerve.

## Dislocations of the Vertebral Column

Dislocations without fracture occur only in the cervical region, since the inclination of the articular processes of the cervical vertebrae permits dislocation to take place without their being fractured. In the thoracic and lumbar regions, dislocations can occur only if the vertically placed articular processes are fractured.

Dislocations commonly occur between the fourth and fifth or fifth and sixth cervical vertebrae, where mobility is greatest. In unilateral dislocations the inferior articular process of one vertebra is forced forward over the anterior margin of the superior articular process of the vertebra below. The spinal nerve on the same side is usually nipped in the intervertebral foramen, producing severe pain. Fortunately, the large size of the vertebral canal allows the spinal cord to escape damage in most cases.

Bilateral cervical dislocations are almost always associated with severe injury to the spinal cord. Death occurs immediately if the upper cervical vertebrae are involved, since the respiratory muscles, including the diaphragm, are paralyzed.

**Fig. 12-16. (A) Posterolateral rupture of anulus fibrosus, permitting nucleus pulposus to herniate backward and exert pressure on a spinal nerve. (B) Posterior rupture of anulus fibrosus, showing nucleus pulposus exerting pressure on spinal nerve roots of cauda equina. (C) Diagrammatic representation of intervertebral disc, showing concentric rings of anulus fibrosus and centrally placed, rounded nucleus pulposus; exploded view. (D) Diagrammatic representation of posterior view of vertebral bodies in lumbar region, showing relationship that might exist between herniated nucleus pulposus and spinal nerve roots. (E) Pressure on L5 motor nerve root produces weakness of dorsiflexion of ankle joint; pressure on S1 motor nerve root produces weakness of plantar flexion of the ankle joint.**

## Fractures of the Vertebral Column

Fractures of the spinous processes, transverse processes, or laminae are caused by direct injury or, in rare cases, by severe muscular activity. Compression fractures of the vertebral bodies are usually caused by an excessive flexion-compression type of injury and take place at the sites of maximum mobility or at the junction of the mobile and fixed regions of the column. It is interesting to note that the body of a vertebra in such a fracture is crushed, while the strong posterior longitudinal ligament remains intact. The vertebral arches remain unbroken and the intervertebral ligaments remain intact, so that vertebral displacement and spinal cord injury do not occur.

*Fracture dislocations* are caused by an excessive flexion-compression type of injury and take place at the sites of maximum mobility or the junction of the mobile and fixed regions of the column. Since the articular processes are fractured and the ligaments are torn, the vertebrae involved are unstable, and the spinal cord is usually severely damaged or severed, with accompanying paraplegia.

## Spondylolisthesis

In spondylolisthesis, the body of a lower lumbar vertebra, usually the fifth, moves forward on the body of the vertebra below and carries with it the whole of the upper portion of the vertebral column. The essential defect is in the pedicles of the migrating vertebra. It is now generally believed that in this condition the pedicles are abnormally formed, and accessory centers of ossification are present that fail to unite. The spine, laminae, and inferior articular processes remain in position, while the remainder of the vertebra, having lost the restraining influence of the inferior articular processes, slips forward. Since the laminae are left behind, the vertebral canal is not narrowed, but the nerve roots may be pressed upon, causing low backache and sciatica. In severe cases the trunk becomes shortened, and the lower ribs contact the iliac crest.

## Sacroiliac Joint

The sacroiliac joint has been described on page 322. The clinical aspects of this joint are referred to again since disease of this joint can cause low back pain and may be confused with disease of the lumbosacral joints. Essentially, the sacroiliac joint is a synovial joint that has irregular elevations on one articular surface that fit into corresponding depressions on the other articular surface. It is a very strong joint and is responsible for the transfer of weight from the vertebral column to the hip bones. The joint is innervated by the lower lumbar and sacral nerves, so that disease in the joint may produce low back pain and sciatica.

The sacroiliac joint is inaccessible to clinical examination. However, a small area located just medial to and below the posterior superior iliac spine is where the joint comes closest to the surface. In disease of the lumbosacral region, movements of the vertebral column in any direction cause pain in the lumbosacral part of the column. In sacroiliac disease, pain is extreme on rotation of the vertebral column and is worst at the end of forward flexion. The latter movement causes pain, since the hamstring muscles hold the innominate bones in position while the sacrum is rotating forward as the vertebral column is flexed.

## Spinal Cord Injuries

The degree of spinal cord injury at different vertebral levels is largely governed by anatomical factors. In the cervical region, dislocation or fracture dislocation is common, but the large size of the vertebral canal often results in the spinal cord escaping severe injury. However, when there is considerable displacement, the cord is sectioned and death occurs immediately. Respiration ceases if the lesion occurs above the segmental origin of the phrenic nerves.

In fracture dislocations of the thoracic region, displacement is often considerable, and the small size of the vertebral canal results in severe injury to the spinal cord.

In fracture dislocations of the lumbar region, two anatomical facts aid the patient. First, the spinal cord in the adult extends only down as far as the level of the lower border of the first lumbar vertebra. Second, the large size of the vertebral foramen in this region gives the roots of the cauda equina ample room. Nerve injury may therefore be minimal in this region.

Injury to the spinal cord may produce partial or complete loss of function at the level of the lesion, and partial or complete loss of function of afferent and efferent nerve tracts below the level of the lesion. The symptoms and signs of spinal shock and paraplegia in flexion and extension are beyond the scope of this book. For further information a textbook of neurology should be consulted.

## Lumbar Puncture

Lumbar puncture may be performed to withdraw a sample of cerebrospinal fluid for examination or to inject drugs to combat infection or induce anesthesia. Fortunately, the spinal cord terminates below at the level of the lower border of the first lumbar vertebra in the adult. (In the infant it may reach as low as the third lumbar vertebra.) The subarachnoid space extends down as far as the lower border of the second sacral vertebra. The lower lumbar part of the vertebral canal is thus occupied by the subarachnoid space, which contains the lumbar and sacral nerve roots and the filum terminale (the cauda equina). A needle introduced into the subarachnoid space in this region usually pushes the nerve roots to one side without causing damage.

With the patient lying on his side with the vertebral column well flexed, the space between adjoining laminae in the lumbar region is opened to a maximum (Fig. 12-5). An imaginary line joining the highest points on the iliac crests passes over the fourth lumbar spine. With a careful aseptic technique and under local anesthesia, the lumbar puncture needle, fitted with a stylet, is passed into the vertebral canal above or below the fourth lumbar spine. The needle will pass through the following anatomical structures before it enters the subarachnoid space: (1) skin, (2) superficial fascia, (3) supraspinous ligament, (4) interspinous ligament, (5) ligamentum flavum, (6) areolar tissue containing the internal vertebral venous plexus, (7) dura mater, and (8) arachnoid mater. The depth to which the needle will have to pass will vary from an inch or less in a child to as much as 4 inches (10 cm) in a fat adult.

As the stylet is withdrawn, a few drops of blood commonly escape. This usually indicates that the point of the needle is situated in one of the veins of the internal vertebral plexus and has not yet reached the subarachnoid space. If the entering needle should stimulate one of the nerve roots of the cauda equina, the patient will experience a fleeting discomfort in one of the dermatomes, or a muscle will twitch, depending on whether a sensory or a motor root was impaled.

The cerebrospinal fluid pressure may be measured by attaching a manometer to the needle. In the recumbent position, the normal pressure is about 120 mm of water.

A block of the subarachnoid space in the vertebral canal, which may be caused by a tumor of the spinal cord or the meninges, may be detected by compressing the internal jugular veins in the neck. This raises the cerebral venous pressure and inhibits the absorption of cerebrospinal fluid in the arachnoid granulations, thus producing a rise in the manometric reading of the cerebrospinal fluid

pressure. If this rise fails to occur, the subarachnoid space is blocked and the patient is said to exhibit a positive *Queckenstedt's sign.*

It is interesting to note that the cerebrospinal fluid pressure normally fluctuates slightly with the heart beat and with each phase of respiration.

## Caudal Anesthesia

Solutions of anesthetics may be injected into the sacral canal through the sacral hiatus. The solutions pass upward in the loose connective tissue and bathe the spinal nerves as they emerge from the dural sheath. Obstetricians use this method of nerve block to relieve the pains of the second stage of labor. Its advantage is that, administered by this method, the anesthetic does not affect the infant.

## CLINICAL PROBLEMS

*Answers on page 979*

1. A 10-year-old boy was examined by a pediatrician because his mother had noticed that his back was abnormally "bent." Questioning revealed that the boy had had a normal childhood, but that at the age of 7 years had had acute osteomyelitis of the lower end of his right femur. Although antibiotics had been given, surgical intervention was necessary. On examination, the child was noted to have a slight lumbar scoliosis with the convexity to the right and a slight thoracic scoliosis with the convexity to the left. With the child standing upright, it was noted that the pelvis was tilted downward on the right side. On examination of the legs, with the child in the supine position, the right leg showed 1 inch (2.5 cm) of shortening that was confined to the thigh. This was revealed by measuring the distance between the anterior superior iliac spine and the adductor tubercle of the femur on each side. An old surgical scar was present on the lower medial side of the right thigh. Using your knowledge of anatomy, explain the possible connection between the scoliosis and the short right leg. Why do you think the right leg was shorter than the left?

2. A 15-year-old girl, measuring 5 feet 11 inches (1.8 m) tall, was examined by a pediatrician because she was "round-shouldered." On ques-

tioning, a history of numerous illnesses, including measles, whooping cough, and pneumonia, was obtained. She said she tired easily and found that her back ached when she worked at her desk at school. Using your knowledge of anatomy, can you explain why this child was round-shouldered? Why did she experience back pain while working at her school desk?

3. In a 17-year-old boy a kyphosis of the thoracic part of the vertebral column suddenly developed. On questioning, he was found to have a history of tuberculosis of the right lung, which had been successfully treated when he was 12 years of age. Knowing that the tubercle bacillus can infect and destroy bone, can you explain the deformity?

4. A 70-year-old man visited his physician complaining of pain in his back and down the outer side of his right leg. On questioning, he gave a history of difficulty in micturition, which had started about 1 month previously. A rectal examination revealed that the prostate was harder and more fixed than normal. A diagnosis of carcinoma of the prostate with metastases in the lower lumbar vertebrae was made. Radiographic examination revealed secondary malignant deposits in the fifth lumbar vertebral body and in both innominate bones. Using your knowledge of anatomy, can you explain

the most likely route taken by the cancer cells as they spread to the innominate bones and lumbar vertebrae? Can you explain the abnormal fixation of the prostate and the right-sided sciatica?

5. A 45-year-old professor of anatomy woke up one morning with a severe pain near the lower part of the back of the neck and right shoulder. The pain was also referred along the outer side of the right upper arm. Movement of the neck caused an increase in the intensity of pain, which was also accentuated by coughing. A lateral radiograph of the neck showed a slight narrowing of the space between the fifth and sixth cervical vertebral bodies. Using your knowledge of anatomy, state which nerve root was involved. Also state the nature of the disease.

6. A medical student offered to move a grand piano for his landlady. He had just finished his final anatomy examinations and was in poor physical shape. Undaunted and being a kindly person, he struggled with the antique monstrosity. Suddenly, he experienced an acute pain in the back, which extended down the back and outer side of his left leg. Immediately realizing what had happened, he withdrew from the operation, leaving the landlady to complete the move. On being examined by an orthopedic surgeon, the student was found to have a slight lumbar scoliosis with the convexity on the right side. The deep muscles of the back in the left lumbar region felt firmer than normal and were tender to touch. When asked to place one finger where the pain hurt most, the student pointed to the left lumbar region opposite the fifth lumbar spine. He also ran his hand down the lateral side of his left leg and ankle and onto the sole of the foot. He said the pain was accentuated by coughing. There was no evidence of muscle weakness, but the left ankle jerk was diminished. With the student supine, flexing the left hip joint with the knee extended caused a marked increase in the pain. A lateral radiograph of the lumbar vertical column revealed nothing abnormal. Using your knowledge of anatomy, make a diagnosis. In anatomical terms, explain the signs and symptoms.

7. An 11-year-old boy was showing off in front of friends and diving into the shallow end of a swimming pool. After one particularly daring dive, he surfaced quickly and climbed out of the pool, holding his head between his hands. He said he had hit the bottom of the pool with his head and now had severe pain in the root of the neck, which was made worse by attempting to move his neck. On examination, the boy held his head rotated to the right and complained of severe pain in the region of the back of the neck and left shoulder. Any attempt to move the neck passively in any direction greatly accentuated the pain, so that all movements were restricted. The deep muscles on the left side of the back of the neck were tender and in spasm. No other neurological signs or symptoms were present. What is your diagnosis? What other tests would you advise to confirm the diagnosis?

8. A 50-year-old coal miner was crouching down at the mine face, servicing a drilling machine. Suddenly, part of the roof of the mine shaft collapsed, and a large rock struck him on the upper part of his back. The miner was dug out alive. Examination by a physician showed an obvious forward displacement of the upper thoracic spines on the sixth thoracic spine. What is your diagnosis? What neurological signs and symptoms would you expect to find? What anatomical factors in the thoracic region determine the degree of injury that may occur in this region?

9. A 75-year-old woman was dusting the top of a high closet while balanced on a chair. She lost her balance and fell to the floor, catching her right lumbar region on the edge of the chair. On examination 3 hours later, a large swollen bruised area was found in the right lumbar region, which was extremely tender to touch. After resting in bed for a week the patient still complained of pain in the right lumbar region on movement. What tests would you, as the examining physician, have performed on this woman at the time of the accident? What is the most likely reason for the persistence of the severe pain?

10. A 25-year-old woman visited her physician complaining of low back pain off and on for the past 5 years, which she said was getting worse. On examination, the patient was found to have a severe lumbar lordosis, with excessive prom-

inence of the first sacral spine. A prominent fold of skin was seen on either side above the iliac crests, and the last ribs appeared to rest on the iliac crests. Using your knowledge of anatomy, make the diagnosis. What are the underlying embryological reasons for the condition?

11. Name, in order, the structures pierced when a lumbar puncture needle is introduced into the subarachnoid space below the fourth lumbar spine.

12. An intelligent woman, aged 45 years, visited her physician complaining of low back pain of 3 months' duration. Prior to this time she had never experienced any aches or pains in the back. In fact, she volunteered that she had always been interested in athletics and was an active member of the local tennis club. A group of medical students was invited to examine and question the patient. Nothing abnormal could be detected.

The physician then asked the patient about her general health and whether or not she had noticed anything else abnormal. She said she was otherwise perfectly fit and had noticed nothing abnormal. The physician then widened the field of his examination. He listened to her chest, examined her thyroid gland, and finally examined both breasts carefully. A large, hard, fixed mass was found in the upper quadrant of the left breast. Several hard enlarged pectoral lymph nodes were also found in the left axilla. A diagnosis of cancer of the left breast was made. What is the connection between cancer of the left breast and low back pain?

# NATIONAL BOARD TYPE QUESTIONS

*Answers on page 984*

**Multiple Choice:**

1. The first cervical vertebra (atlas) has all of the following **except:**
   (a) Lateral masses
   (b) Inferior articular facets
   (c) Anterior arch
   (d) Spinous process
   (e) Superior articular facets

2. The following statements are true of an intervertebral disc **except:**
   (a) During aging, the fluid within the nucleus pulposus is replaced by fibrocartilage.
   (b) The discs are thickest in the lumbar region.
   (c) The atlanto-axial joint possesses no disc.
   (d) The discs play a major role in the development of the curvatures of the vertebral column.
   (e) The nucleus pulposus is most likely to herniate in an anterolateral direction.

3. The cauda equina consists of which of the following components?
   (a) A bundle of posterior roots of lumbar, sacral, and coccygeal spinal nerves.
   (b) The filum terminale.
   (c) A bundle of anterior and posterior roots of lumbar, sacral, and coccygeal spinal nerves.
   (d) A bundle of lumbar, sacral, and coccygeal spinal nerves and the filum terminale.
   (e) A bundle of anterior and posterior roots of lumbar, sacral, and coccygeal spinal nerves and the filum terminale.

4. The spinal cord in the adult ends inferiorly at the level of the:
   (a) L5 vertebra
   (b) L3 vertebra
   (c) S2–3 vertebrae
   (d) T12 vertebra
   (e) L1 vertebra

5. Herniation of the intervertebral disc between the fifth and sixth cervical vertebrae will compress the:
   (a) Fourth cervical nerve root
   (b) Sixth cervical nerve root
   (c) Fifth cervical nerve root
   (d) Seventh and eighth cervical nerve roots
   (e) Seventh cervical nerve root

6. The subarachnoid space ends inferiorly in the adult at the level of:
   (a) The coccyx
   (b) The lower border of L1
   (c) S2–3
   (d) S5
   (e) The promontory of the sacrum

**In each of the following questions, answer:**
  (a) IF (1), (2), AND (3) ONLY ARE CORRECT
  (b) IF (1) AND (3) ONLY ARE CORRECT
  (c) IF (2) AND (4) ONLY ARE CORRECT
  (d) IF (4) ONLY IS CORRECT, AND
  (e) IF ALL ARE CORRECT

7. (1) When one is performing a lumbar puncture, the needle is usually inserted between the laminae of the third and fourth or fourth and fifth lumbar vertebrae.
  (2) Straining during defecation may result in the metastasis of malignant cells from the prostate via the internal vertebral venous plexus to the cranial cavity.
  (3) A dislocation that occurs in the thoracic region of the vertebral column is **more** likely to result in spinal cord damage than one that occurs in the cervical region.
  (4) The sacral hiatus is an aperture situated at the inferior end of the anterior surface of the sacrum.

8. (1) The intervertebral foramina are bounded above and below by the pedicles of adjacent vertebrae, in front by the vertebral bodies and intervertebral discs and behind by the articular processes.
  (2) The atlanto-axial joints permit flexion and extension of the head.
  (3) Injection of an anesthetic into the sacral canal (caudal anesthesia) can be used to block pain and sensation from the cervix, vagina, and perineum during childbirth.
  (4) The vertebral artery ascends the neck through the foramen transversaria of all the cervical vertebrae.

9. The bodies of adjacent vertebrae are joined together by:
  (1) The ligamenta flavum.
  (2) The anterior longitudinal ligament.
  (3) The interspinous ligaments.
  (4) The posterior longitudinal ligament.

**In each of the following questions, answer:**
  (a) IF (1) IS CORRECT ONLY
  (b) IF (2) IS CORRECT ONLY
  (c) IF BOTH (1) AND (2) ARE CORRECT, AND
  (d) IF NEITHER (1) NOR (2) IS CORRECT

10. Concerning the curves of the vertebral column:
  (1) The early fetal spine has only one curve.
  (2) The aged gradually lose their secondary curves.

11. Concerning the functions of the vertebral column:
  (1) It protects the spinal cord.
  (2) Throughout life, the marrow of the vertebral body has a hemopoetic function.

12. Concerning the atlanto-axial joints:
  (1) The apical ligament connects the apex of the odontoid process to the anterior arch of the atlas.
  (2) The transverse part of the cruciate ligament is attached on each side to the inner aspect of the lateral mass of the atlas and binds the odontoid process to the anterior arch of the atlas.

# Answers to Clinical Problems

1. See Figure A.

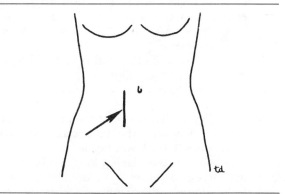

**Fig. A. Position of abdominal incision.**

2. Self-demonstration.
3. Self-demonstration.
4. Self-demonstration.
5. The virus of poliomyelitis attacks and destroys the motor anterior horn cells of the spinal cord.

This resulted in this case in paralysis of the muscles that normally laterally flex the vertebral column on the right side. The muscles on the left side, being now unopposed, slowly caused the left lateral flexion deformity.

6. Self-demonstration.
7. Inversion.
8. Self-demonstration. The patient is unable to separate her fingers.
9. The appendectomy incision was made along one of the lines of cleavage with minimum disruption of the dermal collagen. The gallbladder incision was made at right angles to the lines of cleavage with maximum disruption of the dermal collagen. The surgeon was aware of this effect, but he decided on the vertical incision to permit greater exposure. The safety of the patient must have a higher priority than the cosmetic defects of the future scar.
10. A horizontal incision should be made, if possible, following one of the skin creases in the neck.
11. (a) Regeneration would take place from the epidermal cells of the hair follicles, sebaceous

glands, and sweat glands, and from the epidermal cells at the edges of the burn. (b) Regeneration would take place from the epidermal cells of the ends of the sweat glands that lie in the superficial fascia and from those at the edges of the burn. A split-thickness skin graft is the treatment of choice.

12. Hair follicles and sweat glands.

13. The tumor has destroyed or is interrupting the normal function of the efferent motor fibers and afferent sensory fibers of the lumbar segments of the spinal cord. This would explain the loss of motor function of the muscles on the front of the thigh (the quadriceps femoris), the wasting of these muscles, and the loss of the knee jerk (segments L2–4). The loss of skin sensation is confined to the dermatomes L1, 2, 3, and 4. The absence of muscle tone is due to interruption of the nervous reflex arc upon which muscle tone is based. (See text.)

14. Destruction of the nerve connections between the higher nervous centers and the spinal cord deprives the lower motor neurons of inhibitory impulses. This results in an increased activity of the simple reflex arc upon which skeletal muscle tone is dependent. By this means the tone of the muscle is increased above the normal level.

15. Repeated unaccustomed extension of the wrist has produced traumatic *tenosynovitis* of the synovial sheaths of the long extensor tendons of the fingers.

16. The patient had a traumatic bursitis of the prepatellar bursa as the result of repeated minor trauma applied to the knee while kneeling on a hard surface.

17. A severe tear of a cartilaginous disc never heals. It is usual to operate and remove the affected disc if it is interfering with the normal functioning of the joint.

18. The infection under the nail has spread into the lymphatic vessels and to the draining lymph nodes. The lymphatic vessels were red and inflamed *(lymphangitis)* and could be seen through the skin. The lymph nodes were inflamed *(lymphadenitis)* and were tender and swollen.

19. Coronary arteries are functional end arteries. If a coronary artery should become blocked, the caliber of its anastomosing branches is insufficient to keep the cardiac muscle alive and it undergoes necrosis.

20. A 19-year-old person under normal conditions possesses only yellow nonhematopoietic marrow in the distal bones of the limbs. A satisfactory biopsy specimen of red marrow can be obtained at this age from the sternum or iliac crests.

## Chapter 2

1. Self-examination.

2. Seventh thoracic vertebra.

3. Right atrium.

4. The suprapleural membrane was damaged by the shrapnel and was not repaired at operation. Subsequently, herniation of the cervical pleura and apex of the lung took place, which resulted in the skin above the clavicle bulging upward during forced expiration.

5. The needle was inserted incorrectly and damaged the eighth intercostal nerve. This produced altered sensation (paresthesia) in the eighth thoracic dermatome. Needles should always be inserted close to the upper border of a rib, i.e., as far away from the neurovascular bundle as possible.

6. Palpate the posterior axillary lymph nodes, which lie against the posterior wall of the axilla. The fingers are pressed upward and posteriorly in the armpit. This group of nodes drains the skin of the back above the level of the iliac crests.

7. The thoracic wall of a child is very elastic, and fractures of ribs in children are rare.

8. In old age the costal cartilages may undergo ossification.

9. Normally, the left dome of the diaphragm lies at a lower level than the right dome and usually it lies opposite the lower border of the fifth rib. The possibility that the left lung is collapsed, or that there is a collection of fluid or gas under the left dome of the diaphragm, should be considered in cases like the one presented.

10. The parietal pleura covering the internal surface of the tenth intercostal space is innervated by the tenth intercostal nerve. Under normal conditions a person is not conscious of the parietal pleura. However, should the parietal pleura become involved in a disease process,

the brain assumes incorrectly that the painful stimuli are arising from the dermatome of the tenth thoracic nerve. This phenomenon is known as *referred pain*.

11. The anatomical and physiological changes that take place in the thorax with advancing years are fully described on page 78.

12. Hiccup is the involuntary spasmodic contraction of the diaphragm accompanied by the approximation of the vocal folds and closure of the glottis in the larynx.

---

## Chapter 3

1. This patient, a smoker, has an advanced carcinoma of the bronchus in the upper lobe of the left lung that has metastasized (spread) to the bronchomediastinal lymph nodes. Enlargement of these nodes has resulted in pressure on the left recurrent laryngeal nerve as it passes under the arch of the aorta. Partial injury of the recurrent laryngeal nerve results in paralysis of the abductor muscles of the vocal cords, leaving the adductor muscles unopposed. The left vocal cord was therefore adducted and immobile.

2. This patient has a syphilitic aneurysm (dilatation) of the arch of the aorta. With each cardiac systole the swelling expands and recoils and pushes downward the bifurcation of the trachea and the left main bronchus. Hence the tracheal tug.

3. This patient has a very large goiter (enlargement of the thyroid gland), which extends down the neck into the superior mediastinum. The left lobe of the thyroid is larger than the right. When she falls asleep she has a tendency to flex her neck laterally to the left, which kinks the trachea over the enlarged left lobe of the thyroid in the superior mediastinum. By experience, she has found that she can maintain the patency of the trachea by flexing her neck to the right and supporting her head on pillows. The veins of the neck are congested due to partial obstruction of the brachiocephalic veins by the goiter.

4. Disease of the lung does not cause pain until the parietal pleura is involved. Lung tissue and the visceral pleura are not innervated with pain fibers. The costal parietal pleura is innervated by the intercostal nerves, which have pain endings in the pleura. The boy had pneumonia of the right middle lobe, which later spread to the pleurae, causing pleurisy. Once the parietal pleura was involved he experienced pain he could localize. Movement of the inflamed pleural surfaces against one another, as in deep inspiration or coughing, accentuated the pain.

5. Esophageal stenosis commonly follows damage to the esophageal wall after drinking caustic fluids. The fluid tends to cause the maximum damage where it is briefly held up by normal anatomical constrictions. These are at the beginning where the pharynx joins the esophagus, where the aorta and left bronchus cross the esophagus, and where the esophagus goes through the diaphragm into the stomach. Esophagoscopy revealed that the patient had stenosis at the lower sites, the one at the higher level being the more severe.

6. When the mother's back was turned, the child placed the small pin in his mouth. The child failed to keep still while the mother was looking for the pin. She smacked him, and the shock caused him to take a deep breath and inhale the pin. It rapidly passed down the trachea and into the right main bronchus, where it lodged. The pin was easily recognized on postero-anterior and right-oblique radiographs of the chest and was successfully removed through a bronchoscope. The right principal bronchus is the more vertical and wider of the two principal bronchi. For this reason, foreign bodies tend to be lodged in the right principal bronchus rather than in the left.

7. If, after careful radiographic examination, it is found that a lesion is localized to one or more bronchopulmonary segments, it is certainly possible to resect the area surgically. The surgeon obviously must have a sound knowledge of the bronchopulmonary segments and how they lie in relation to one another.

8. The posterior wall of the left atrium lies in contact with the anterior wall of the esophagus, separated only by the pericardium. Fluoroscopy, or a lateral radiograph of the thorax, following a barium swallow would make possible the determination of the size of the smooth indentation of the anterior wall of the barium-filled esophagus produced by the enlarged atrium.

9. The upper lobe of the left lung is situated above and in front of the oblique fissure. The stethoscope should therefore be placed on the anterior chest wall above the level of the sixth rib. (See p. 57.)

10. (a) No. (b) Contrast media are introduced into the bronchial tree, followed by radiography. This is known as bronchography.

11. The patient has had a left-sided pneumothorax. The air has entered the left pleural cavity as the result of rupture of one of the emphysematous cysts of the left lung. (a) The air in the left pleural cavity displaced the mobile mediastinum over to the right. (b) The left lung collapsed immediately when air entered the left pleural cavity, since the air pressures within the bronchial tree and in the pleural cavity were then equal. The elastic recoil of the lung tissue caused the lung to collapse. (c) Atmospheric pressure.

12. Carcinoma of the esophagus tends to spread via the lymphatic vessels through the esophageal opening in the diaphragm and so enter the celiac lymph nodes. The best chance of removing all the cancerous tissue is therefore to remove (1) the primary tumor and an area of normal adjacent esophagus and (2) the lymphatic vessels and nodes that drain the diseased area, plus the other organs draining into the same nodes.

13. The lower third of the esophagus is the site of a portal-systemic anastomosis between the esophageal veins of the azygos system and the left gastric vein. Portal hypertension due to cirrhosis of the liver means that the portal circulation through the liver is obstructed by fibrous tissue. As the result of this, there is a dilatation of the portal-systemic anastomoses to allow blood to return to the circulation via systemic veins. Many of the dilated veins lie within the mucous membrane and submucosa and become easily damaged by swallowed food. Copious hemorrhage from these veins is difficult to treat and is often terminal.

14. In a newborn baby, the ribs are practically horizontal, so that very little movement of the costal walls takes place on respiration. For the first 2 years of life, a baby relies mainly on the descent of the diaphragm for inspiration. In order to accommodate the abdominal viscera, especially the relatively large liver, there has to be a graduated relaxation of the anterior abdominal wall musculature. The alternate relaxation and contraction of the abdominal wall muscles with the phases of inspiration and expiration were the movements observed by the mother.

15. Asthmatic patients have difficulty in getting air in and out of their lower respiratory tract due to contraction of their bronchial musculature. Inspiration is especially difficult. By grasping the table and chair, the patient was fixing his upper limbs and shoulder girdle so that he could use to better mechanical advantage his accessory muscles of inspiration, namely, the pectoralis major and minor and the serratus anterior muscles.

16. Excessive amounts of pericardial fluid embarrass the action of the heart by preventing adequate filling of the atria. Under local anesthesia, the needle should be inserted to the left of the xiphoid process at an angle of 45 degrees to the skin, so that it passes upward, backward, and to the left. Its passage misses the pleura and lungs because of the cardiac notch.

17. Cardiac pain is sometimes felt over the heart; it is never felt in the heart. It is usually experienced over the middle of the sternum and may radiate down the inside of either arm or up into the neck. The nervous pathways taken by the afferent pain fibers are fully described on page 138.

18. Yes, in selected cases. For details see page 138.

19. The aortic valve is best heard by placing the stethoscope in the second right intercostal space, close to the sternum.

20. Over the apex beat of the heart.

21. Severe obesity restricts the movements of the thoracic cage and abdominal musculature. The presence of large amounts of retroperitoneal and omental fat within the abdomen limits the descent of the diaphragm and decreases the extent of inspiration. As the result of alveolar hypoventilation, moist air stagnates in the small bronchioles and alveoli, and infection is a common complication.

22. In this patient, the spinal cord was severed above the level of the origin of the phrenic nerve (C3, 4, and 5), causing paralysis of the dia-

phragm. All the intercostal muscles, which are supplied by the intercostal nerves (T1–11), were also paralyzed.

## Chapter 4

1. All lower thoracic injuries require a thorough abdominal examination, since the lower thoracic wall forms part of the anterior and lateral walls of the abdomen. The liver, spleen, kidneys, and part of the stomach are protected by the lower part of the thoracic cage. The spleen lies along the inner surface of the left ninth, tenth, and eleventh ribs. Trauma applied in this region may well cause rupture of the spleen, with consequent intra-abdominal hemorrhage and a fall in blood pressure.

2. The fundus of the gallbladder lies against the anterior abdominal wall next to the tip of the right ninth costal cartilage, i.e., where the linea semilunaris crosses the costal margin on the right side. The peritoneum on the undersurface of the diaphragm (central part) is supplied by the phrenic nerve (C3, 4, and 5). In this patient it must be assumed that the inflamed gallbladder was irritating this area of the peritoneum. The pain was referred to the shoulder along the supraclavicular nerves (C3 and 4), which supply the skin of this region.

3. The right hypochondrium is located lateral to the right vertical line (line passing through the midpoint between the anterior superior iliac spine and the symphysis pubis) and above the subcostal plane (line joining the inferior margin of the tenth costal cartilage on each side and lying opposite the third lumbar vertebra). The referred pain is due to the phrenic and supraclavicular nerves' having the same segmental nervous origin in the spinal cord. (See 2.)

4. A ruptured appendix produces a peritonitis, which is localized at first, but may become generalized later. Inflammation of the parietal peritoneum causes pain over the area and reflex spasm of the anterior abdominal muscles. The parietal peritoneum, the abdominal muscles, and the overlying skin are supplied by the same segmental nerves (T11 and 12 and L1). This is essentially a protective mechanism and an attempt by the body to keep that area of the abdomen at rest so that the inflammatory lesion will remain localized.

5. The lymphatic drainage of the skin of the umbilicus is upward to the anterior axillary lymph nodes on both sides and downward to the superficial inguinal nodes on both sides. Today, it would be difficult for a physician to vaccinate an actress in a spot that would not show!

6. The patient's fall ruptured the urethra in the superficial perineal pouch. When he attempted to micturate, the urine extravasated through the superficial perineal pouch beneath the Colles' fascia. The urine then passed over the scrotum and penis under the membranous layer of the superficial fascia and up onto the anterior abdominal wall. It did not pass backward because of the attachment of the fascia to the posterior edge of the perineal membrane, and did not extend into the thighs because of the attachment of the fascia to the fascia lata, just below the inguinal ligament.

7. A sudden unexpected blow on the anterior abdominal wall causes excessive stretching of this structure. In this case the right inferior epigastric artery was ruptured, and bleeding occurred into the rectus sheath. If a person is expecting a blow, he automatically contracts his abdominal muscles and protects the underlying structures.

8. The shrapnel cut the ninth, tenth, and eleventh intercostal nerves just inferior to the costal margin on the right side. The diminished skin sensation was due to the loss of the sensory nerve supply to the ninth, tenth, and eleventh thoracic dermatomes, i.e., a band of skin extending forward to the region of the umbilicus. Extensive portions of the oblique, transversus, and rectus abdominis muscles on the right side were paralyzed. Atrophy of these muscles resulted in loss of support to the abdominal viscera, which consequently sagged forward (visceroptosis).

9. The presence of duodenal contents in the peritoneal cavity causes great irritation of the parietal peritoneum (chemical peritonitis). This produces pain and reflex contraction of the muscles in the upper part of the anterior abdominal wall and interferes with the normal thoracoabdominal rhythm during respiration.

The increased tone of the abdominal muscles is responsible for the boardlike rigidity felt in patients with peritonitis. It is an attempt to keep the area at rest and so diminish the chances of spreading the irritating fluid throughout the peritoneal cavity. (See also 4.)

10. The child had a right indirect inguinal hernia with a preformed sac, due to the failure of the upper part of the processus vaginalis to become obliterated prior to birth. The contents of such a hernia usually consist of omentum and coils of small intestine. With gentle pressure it is possible to push the contents back into the abdominal cavity. The movement of gas in the bowel is responsible for the gurgling sounds. When the child cried, additional abdominal contents were forced down into the hernial sac. An indirect inguinal hernial sac as it emerges from the superficial inguinal ring lies above and medial to the pubic tubercle.

11. Following excessive exertion and an increase in intra-abdominal pressure, a hernial sac was forced down through the left femoral canal. The patient had a left-sided femoral hernia. The neck of a femoral hernial sac is always situated below and lateral to the pubic tubercle.

12. The patient had a right indirect inguinal hernia. The upper part of the processus vaginalis was only partially obliterated prior to birth. The great increase in intra-abdominal pressure that occurred while digging resulted in a coil of small intestine being forced down the processus, opening up the peritoneal sac. By exerting pressure on the body of the hernial sac, the contents were returned to the abdominal cavity. Since the neck of the sac was situated lateral to the inferior epigastric vessels and above and medial to the midpoint of the inguinal ligament, it was possible to press on the deep inguinal ring and prevent the abdominal contents from reentering the sac.

13. The patient had a left-sided direct inguinal hernia. The hernial sac expands on coughing due to the rise in intra-abdominal pressure. Direct inguinal hernias rarely descend into the scrotum. It is due to a weakness of the posterior wall of the inguinal canal medial to the inferior epigastric vessels. There is no preformed sac.

14. The child has an infantile umbilical hernia due to weakness of the scar tissue of the umbilicus.

Treatment is usually not necessary.

15. The patient has a fatty epigastric hernia. This is common in middle-aged manual workers. A small protrusion of extraperitoneal fat is forced forward between the fibers of the linea alba. Later, the fat pulls forward a small peritoneal sac.

16. The commonest sites are (1) in the midline of the abdomen and (2) the flanks. In the midline, the structures traversed by the cannula would be skin, superficial fascia, deep fascia, linea alba, fascia transversalis, extraperitoneal connective tissue, and peritoneum. In the flanks, the following structures would be traversed: skin, superficial fascia, deep fascia, aponeurosis of the external oblique muscle, internal oblique muscle, transversus abdominis muscle, fascia transversalis, extraperitoneal connective tissue, and peritoneum. The linea alba is relatively bloodless and is the preferred site.

17. The boy had an incompletely descended testicle on the right side. The right testicle was situated in the inguinal canal. Surgical treatment was necessary in order to place the testis in the scrotum. The high temperature of the abdominal cavity and inguinal canal inhibits normal spermatogenesis.

18. The boy had a maldescended testis on the left side. Instead of following the gubernaculum down into the scrotum, it passed laterally and came to rest in the superficial fascia in the upper part of the left thigh. A maldescended testis is very prone to injury and should be placed in the scrotum by surgical means.

19. The patient had a right encysted hydrocele of the spermatic cord. This is a cyst in the remnant of the upper part of the processus vaginalis and is connected to the tunica vaginalis by a fibrous strand (a further remnant of the processus). On pulling down the testis and the tunica vaginalis, the cyst was pulled medially by the fibrous strand.

20. The student failed to examine the entire patient and made an erroneous diagnosis. He may have been sidetracked in his thoughts by the patient's stating that he had a poor appetite. One thing is certain—the student had forgotten his anatomy! (a) Malignant disease of the testis metastasizes to the lumbar lymph nodes lying on the transpyloric plane. This is the normal

lymph drainage of the testis. (b) The inguinal lymph nodes are involved only if the tumor spreads locally into the tissues of the scrotum outside the testis.

21. The normal flow of portal venous blood through the liver is impaired in cirrhosis. Portal hypertension develops, and the venous blood returns to the general circulation via the portal-systemic venous anastomoses. The paraumbilical veins link the portal vein to the superficial systemic veins of the anterior abdominal wall; the latter dilate and become varicosed. By applying digital pressure to these veins, it is possible to determine that the direction of blood flow is from the umbilicus outward to the periphery.

22. A paramedian incision can be extended to the full length of the rectus sheath. It does not damage the rectus muscle or its nerve and blood supply. It is closed in layers and this results in a strong repair. The pararectus incision may damage the segmental nerves and can only be extended by cutting the nerves to the rectus muscle. The transrectus incision alway cuts the nerves that supply that part of the rectus muscle medial to the incision. The latter incision should be avoided.

23. The advantages are as follows: (1) The skin wound can be short and made in the lines of cleavage; (2) each muscle layer is split in the direction of its fibers, and practically no muscle damage occurs; (3) the nerve supply to the muscles remains intact. The disadvantage is that the diagnosis must be certain, because it is a difficult incision to enlarge and a wide abdominal exploration through this incision is out of the question.

24. The superficial inguinal lymph nodes.

25. Palpate the upper part of the scrotum between finger and thumb, and you can roll the vas deferens as a cordlike structure. It has a smooth external wall and is firm in consistency. Remember there are two sides and always compare the two.

26. The patient had a tuberculous infection of the lumbar vertebral column with destruction of the bodies of the vertebrae, hence the kyphosis. The tuberculous pus extended laterally and to the right and entered the right psoas fascial sheath. From there, it extended downward into the thigh, producing a swelling above and below the inguinal ligament. Since the pus in each swelling was continuous, pressure could be transmitted from one swelling to the other.

27. The tunica vaginalis covers the front and sides of the testis, and the visceral layer is in direct contact with the testis. Infection causes an excessive accumulation of fluid within the tunica, a condition known as hydrocele.

28. The patient has a cyst of the appendix of the testis, a structure that is derived embryologically from the paramesonephric duct.

29. The patient has a patent urachus. The urachus is the remains of the allantois, which, in the fetus, communicates with the urinary bladder. Normally, the allantois is obliterated prior to birth. Should it remain patent, nothing abnormal occurs unless urinary obstruction develops in the patient in later life (e.g., enlarged prostate). The urine then takes the least line of resistance through the urachus to the umbilicus.

30. The left testicular vein drains into the left renal vein. A malignant tumor of the left kidney could spread along the left renal vein, blocking the exit of the left testicular vein and causing congestion and varicosity of the left pampiniform plexus.

## Chapter 5

1. The sudden increase in pain in the right iliac region would suggest that the appendix suddenly perforated, and the infected intestinal contents gushed out into the peritoneal cavity. The infecting organisms quickly multiplied and spread the inflammation to the parietal peritoneum in the right iliac region, and the pain was intensified. The parietal peritoneum, the muscles of the anterior abdominal wall, and the overlying skin are all supplied by the same segmental nerves (T12 and L1). The irritation of the parietal peritoneum reflexly produced an increase in the tone of the muscles in this region, which, when felt by the examiner's hand, was interpreted as rigidity. This is an attempt by the body to reduce movement in this area and so help localize the infection. If the perforation is a slow process, the greater omentum becomes stuck down to the appendix by inflammatory exudate, and its presence consid-

erably restricts the spread of infection.

2. Some of the gastric contents, on entering the greater sac, passed up under the diaphragm on the upper surface of the liver and occupied the right anterior subphrenic space. The parietal peritoneum on the undersurface of the central part of the diaphragm is supplied by the phrenic nerve (C3, 4, and 5). The supraclavicular nerves, which supply the skin over the shoulder, are derived from the same segments of the spinal cord. This is another example of referred pain.

3. The ulcer was situated on the posterior wall of the first part of the duodenum and had eroded the large gastroduodenal artery, which lies posterior to the duodenum in this part of its course.

4. The following structures lie posterior to the stomach: the lesser sac, the pancreas, the left suprarenal gland, the upper part of the left kidney, the diaphragm, and the spleen. The splenic artery may be eroded by a gastric ulcer.

5. The infected peritoneal fluid tends to drain downward in the paracolic gutters to enter the pelvic cavity if the patient is placed in a half-sitting position. If possible, it is an advantage to drain the fluid away from the subphrenic spaces, where toxins are known to be quickly absorbed into the lymphatics.

6. The infected material lies in the greater peritoneal sac above the transverse colon. (a) If it travels in an upward direction, it may pass into the right posterior subphrenic space between the right kidney and the right lobe of the liver. Alternatively, it may pass in front of the right lobe of the liver to the right of the falciform ligament, to the right anterior subphrenic space. (b) If it travels downward, it will pass around the right margin of the right colic flexure and enter the right lateral paracolic gutter. From here, it will travel vertically downward, cross the pelvic brim, and enter the pelvic cavity.

7. (1) Lesser sac, (2) duodenal fossae, (3) cecal fossae, and (4) intersigmoid fossa.

8. Lymph from the right inferior part of the greater curvature of the stomach drains into nodes lying along the right gastroepiploic artery, the gastroduodenal artery, the root of the hepatic artery, and the nodes around the celiac artery. Lymph from the left superior part of the greater curvature drains into nodes lying along the short gastric and the left gastroepiploic arteries and the splenic artery, and finally drains into nodes around the celiac artery. Cancer of the stomach initially spreads locally by the lymphatic vessels in the mucous membrane, and for this reason the entire stomach must be removed. The best chance one has of removing all the cancer cells is to remove all the lymphatic vessels and nodes that drain the stomach. A total gastrectomy is therefore performed, and this includes removal of the lower end of the esophagus and the first part of the duodenum; the spleen and the gastrosplenic and lienorenal ligaments and their lymph nodes; the splenic vessels; the tail and body of the pancreas, and their associated nodes; and the nodes along the lesser and greater curvatures of the stomach, along with the greater omentum. The continuity of the gut is restored by anastomosing the esophagus to the jejunum.

9. The mucous membrane of the first part of the duodenum is smooth and devoid of plicae circulares. The remainder of the duodenum possesses numerous plicae, which, together with the waves of peristalsis, break up the barium meal, giving it a floccular appearance.

10. The vagus nerves are responsible for the nervous control of gastric secretion, and the antral mucosa is responsible for the production of the hormone gastrin, which exerts an endocrine control on gastric secretion. Section of the vagus nerves (*vagotomy*), combined with resection of the antrum of the stomach, is a common surgical procedure for the treatment of peptic ulcer. The proximal part of the stomach is anastomosed to the jejunum (*gastrojejunostomy*) to restore the continuity of the gut.

11. A small perforation of the duodenum may result in the duodenal contents' running down the right paracolic gutter to the right iliac region. The signs then closely resemble those of a perforated appendix.

12. Gallstones have been known to ulcerate through the posterior wall of the gallbladder into the transverse colon or duodenum. Those that enter the transverse colon are passed through the anal canal without difficulty. A large stone entering the duodenum will become

impacted at the ileocecal junction and produce small-bowel obstruction.

13. A list of the naked-eye differences between the small and large intestine is given on page 238; the differences between the jejunum and ileum are given on page 228.

14. The longitudinal muscle of the colon and cecum is confined to three visible bands, known as the teniae coli. These can be followed to the base of the appendix without difficulty. Here the teniae spread out to enclose the appendix, so that it has a complete coat of longitudinal muscle.

15. The vermiform appendix is supplied by a single artery (often represented as two small arteries), the appendicular artery, which is small and derived from the posterior cecal artery. The gallbladder receives a rich blood supply from the cystic artery and from numerous small vessels from the visceral surface of the liver.

16. The initial pain of appendicitis is a vague discomfort and is referred to the umbilical region. The afferent nerve fibers accompany the sympathetic nerves to the superior mesenteric plexus. They enter the thorax via the splanchnic nerves and enter the spinal cord at the level of the tenth thoracic segment. The tenth thoracic intercostal nerve supplies the skin of the umbilicus, and for this reason pain is referred to the umbilicus. Once the inflammatory process has extended beyond the confines of the appendix and has involved the parietal peritoneum, a severe localized pain is felt in the right iliac region. The parietal peritoneum, the overlying muscles of the anterior abdominal wall, and the covering skin are all supplied by the first lumbar nerve.

17. The terminal arteries to the colon pierce the circular muscle between the teniae coli to reach the mucous membrane. The pathway through the circular muscle is a potentially weak site, and it is here that diverticula occur. The intracolonic pressure (made greater by excessive purgation) forces a small diverticulum of mucous membrane through the weak area *(diverticulosis)*. The absence of muscle in the wall of the diverticulum leads to stasis and infection *(diverticulitis)*.

18. The afferent nerve fibers from the gallbladder and stomach accompany the sympathetic nerves to the celiac plexus. They then run together through the greater splanchnic nerves and enter the spinal cord between the fifth and ninth thoracic segments. The seventh, eighth, and ninth intercostal nerves supply the skin in the epigastric region, so that pain from both organs is referred to this region.

19. Any variation in which the bile duct and the pancreatic duct open by a common orifice into the duodenum is likely to cause this problem. Gallstones are usually associated with infected bile. A stone impacted at the orifice into the duodenum will allow reflux of infected bile along the main pancreatic duct, and pancreatitis will occur.

20. The tail of the pancreas lies within the lienorenal ligament, and its tip is related to the hilus of the spleen. The surgeon has to take extreme care not to damage the tail of the pancreas during a splenectomy.

21. The spleen lies in the left hypochondrium, and its long axis lies along the tenth rib. The spleen has a notched anterior border due to incomplete fusion of its parts during development. The spleen is unable to expand vertically downward because of the presence of the left colic flexure, which is suspended from the diaphragm by the phrenicocolic ligament. Pathological enlargement of the spleen therefore occurs downward and medially, toward the umbilicus.

22. About 10 percent of persons have accessory spleens. These should always be looked for when one is performing splenectomy for such conditions as thrombocytopenic purpura. If an accessory spleen is missed, it will enlarge and take over the functions of the main spleen.

23. The right kidney. It is the only normal kidney that can be palpated. The lower pole may be felt in a thin person at the end of inspiration, when the contracted diaphragm has pushed it down to its lowest level. When the diaphragm relaxes on expiration, the kidney returns to its original position.

24. Both kidneys originate in the pelvis and with development rise up on the posterior abdominal wall until the hili lie opposite the second lumbar vertebra. Occasionally, one of the kidneys fails to reach its normal position.

25. The bridge of renal tissue, which unites the

lower poles of the two kidneys to form the horseshoe, becomes trapped behind the inferior mesenteric artery. The artery arrests the ascent of the kidneys.

26. An aberrant renal artery may cross the pelviureteric junction and obstruct the flow of urine.

27. (a) Spasm of the smooth muscle in the wall of the renal pelvis and ureter as it attempts to move the calculus down the urinary tract. (b) Afferent pain fibers enter the spinal cord in the first and second lumbar segments. The anterior rami of the first lumbar nerves are distributed to the skin in the lumbar region and groin as the iliohypogastric and ilioinguinal nerves. The pain experienced in the front of the thigh was referred along the femoral branch of the genitofemoral nerve (L1 and 2). (c) In front of the tips of the transverse processes of the lumbar vertebrae; in front of the sacroiliac joint; in the region of the spine of the ischium. (d) At the pelviureteric junction, where the ureter crosses the pelvic brim and where it enters the bladder.

28. A case of bifid ureters in which one ureter opens into the urinary tract below the bladder sphincter in the male, or into the vagina in the female.

29. The ileocolic, the right colic, the middle colic, the left colic, and the sigmoid arteries anastomose with one another to form an arterial trunk, which runs along the inner margin of the colon from the ileocolic junction to the junction of the pelvic colon and rectum. This anastomotic vessel is referred to as the marginal artery.

30. Lumbar sympathectomy would remove the influence of the sympathetic vasoconstrictor fibers on the arteries of the lower limb. This would have the effect of opening up the collateral circulation in the region of the arterial obstruction and would improve the blood flow to the distal part of the limb.

31. A small carcinoma of the head of the pancreas was found at operation to be compressing the bile duct. Back pressure along the bile ducts produced dilation of the gallbladder, which could be felt in the region of the tip of the right ninth costal cartilage.

32. (a) The pain is due to the spastic contraction of the muscle of the gallbladder attempting to flush the stone down the bile ducts, and to the distention of the ducts by the stone. (b) The afferent pain fibers from the gallbladder and bile ducts enter the spinal cord between segments T5 and T9. Pain is referred to the epigastrium via the seventh to the ninth intercostal nerves. (c) A variable amount of bile gets past the stone.

33. The patient could be kept alive by dialysis treatment. *Peritoneal dialysis.* In this form of treatment, the dializing fluid flows into the peritoneal cavity and the peritoneum serves as a semipermeable membrane between the fluid and the blood (see p. 281). *Hemodialysis.* Because of the risk of infection, hemodialysis is the more common form of treatment. The patient's arterial blood is pumped through coils of plastic tubing or allowed to flow between sheets of plastic and is then returned to the patient's circulation by way of a vein. In this form of dialysis the thin plastic serves as a semipermeable membrane.

34. After a careful clinical examination and arteriography of the abdominal aorta, a diagnosis of advanced arteriosclerosis in the region of the bifurcation of the abdominal aorta was made. The gradual blockage of the aorta had resulted in insufficient blood reaching both legs, thus causing pain (claudication) on walking. The difficulty with erection of the penis was caused by a lack of blood entering both internal iliac arteries.

35. By the eighth month of pregnancy, the enlarged uterus is an abdominal organ and often compresses the inferior vena cava, interfering with the venous return from the legs. This causes excess tissue fluid to accumulate in the subcutaneous tissues of the feet and ankles (edema) and engorgement of the superficial veins. The high levels of progesterone in the blood that occur in pregnancy also cause the smooth muscle in the wall of the veins to relax, thus permitting the veins to dilate.

## Chapter 6

1. Red bone marrow is readily and safely obtained from the iliac crests at all ages. The iliac crest is subcutaneous along its entire length. The following structures are penetrated by the needle:

(1) skin, (2) superficial fascia, (3) periosteum, and (4) bone.

2. The subarachnoid space ends below at the level of the second sacral vertebra. This lies on the level of the posterior superior iliac spine. The latter is easily found, since it lies beneath a skin dimple immediately above the buttock on each side.

3. Caudal analgesia (anesthesia) is very effective in producing a painless labor provided that it is performed skillfully. The anesthetic solutions are introduced into the sacral canal through the sacral hiatus. Sufficient solution is given so that the nerve roots up as far as T11 and 12 and L1 are blocked. This will make the uterine contractions painless during the first stage of labor. If the nerve fibers of S2, 3, and 4 are also blocked, the perineum will be anesthetized. The distention of the perineum by the fetal head during the second stage of labor will not then be felt. The needle will pierce (1) the skin, (2) the fascia, and (3) the ligaments filling in the sacral hiatus.

4. The swelling was the urinary bladder. In the adult the urinary bladder is a pelvic structure. When it becomes filled, the superior wall rises out of the pelvis and may reach the umbilicus, or in extreme cases, a higher level.

5. The girl was 6 months pregnant. The nonpregnant uterus is a pelvic organ. The fundus of the pregnant uterus rises out of the pelvis, so that it is palpable above the symphysis pubis at 3 months and reaches the umbilicus by about the sixth month.

6. The puborectalis fibers of the levator ani muscle pass around the anorectal junction and sling it up to the back of the body of the pubis. The puborectalis and the internal and external anal sphincters are responsible for anal continence. Division of the puborectalis muscle severely damages the muscular ring at this site and results in incontinence.

7. Normally, it is difficult or impossible to feel the sacral promontory by means of a vaginal examination. The normal diagonal conjugate measures about 5 inches (13 cm). This patient's pelvis was flattened anteroposteriorly, and the sacral promontory projected too far forward. It is very likely that she would have an obstructed labor.

8. This man had a fractured pelvis. An anteroposterior radiograph of the pelvis showed a dislocation of the symphysis pubis and a linear fracture through the lateral part of the sacrum on the right side. The urethra was also damaged. Further examination revealed a tear in the urinary bladder, with leakage of urine into the peritoneal cavity. This man had a full bladder at the time of injury; a full bladder is more prone to injury in these circumstances than is an empty bladder.

9. The expectant mother is told that her pelvic organs are suspended within the pelvis by a muscular sling (the levator ani muscles). If she concentrates, she will find that it is possible to contract these muscles and "pull up her rectum into her body." Once she finds that she can perform this action, she will have no difficulty in repeating the exercise several times a day. It is essential that the tone of the levatores ani muscles be developed during the antenatal period. These muscles become enormously stretched as the child's head passes through the pelvic floor during parturition. If the muscles are healthy and possess good tone, they will spring back in position and once again support the pelvic viscera, the uterus in particular.

10. The coccyx is commonly bruised or fractured in injuries of this type. The bone can be palpated beneath the skin in the natal cleft. A gloved finger in the anal canal can also palpate the anterior surface of this bone.

11. Most fractures of the upper part of the ilium have little displacement of the bone fragments. This is because the iliacus muscle is attached to the inner surface and the gluteal muscles are attached to the outer surface. Splinting the bones is unnecessary because of the attachments of these muscles.

12. Osteomalacia produces softening of the pelvic bones. In many cases the weight of the trunk forces the sacrum forward and reduces the anteroposterior diameter of the pelvis. The pelvis should be carefully examined clinically and by X-ray pelvimetry. If it is found that the pelvis is no longer capacious enough to permit the descent of the baby's head during labor, this fact should be explained to the patient, and the possibility of performing a cesarean section in any future pregnancy should be discussed.

## Chapter 7

1. A loop of pelvic colon and coils of the terminal part of the ileum are normally found in the rectouterine pouch. The nature of the symptoms strongly suggested that the patient had disease of the large bowel. On sigmoidoscopic examination, a papilliferous carcinoma of the lower part of the pelvic colon was seen. The ulcerated surface of the tumor accounted for the blood-stained mucus.

2. This patient had impaled his rectum on the leg of the chair. At operation, a laceration of the anterior wall of the middle of the rectum was found. The pelvic peritoneum was contaminated with rectal contents. The upper one-third of the rectum is covered on the anterior and lateral surfaces by peritoneum; the middle one-third is covered on the anterior surface only by peritoneum; and the lower one-third is devoid of a peritoneal covering.

3. A great variety of foreign bodies have been successfully removed from the rectum. The transverse mucosal folds of the rectal wall and spasm of the anal sphincters often prevent the patient from ridding himself of the object. If it is found impossible to grasp the foreign body from below, the abdomen has to be opened so that the object can be pushed down from above. In this patient, a pretty, conical-shaped vase was delivered per anum, and inscribed on its outside was: "A Gift from Rockport"!

4. As so often occurs, rectal bleeding is assumed by the laymen to be due to hemorrhoids (and often, unfortunately, also by the patient's physician, who fails to examine the rectum). Digital examination of the rectum in this patient revealed a large, hard-based ulcer on the posterior wall of the rectum. It was possible, by careful palpation, to feel extensive induration of the pararectal tissues extending posterolaterally to the sacrum. On examination of the right leg, some weakness of the muscles supplied by the sciatic nerve was found, and the patient indicated that the pain was felt in skin areas supplied by branches of the sciatic nerve. This patient had an advanced carcinoma of the rectum, with involvement of the sacral plexus. At operation, the liver was found to be enlarged, due to the spread of neoplastic cells up the superior rectal, inferior mesenteric, splenic, and portal veins to the liver.

5. The superior wall of the urinary bladder in the adult rises out of the pelvis as the viscus fills with urine. A low, severe blow to the anterior abdominal wall can easily result in rupture of a full bladder. Once the urine escapes into the peritoneal cavity, the bladder can no longer be detected above the pubis, and the desire or the ability to micturate ceases. In this case, the urine accumulated in the rectovesical pouch and could be felt there on rectal examination.

6. Afferent nerves passing from the bladder to the spinal cord travel in company with the parasympathetic and sympathetic fibers that innervate the bladder. They enter the cord at segments S2, 3, and 4 and L1 and 2. From these levels the nerve impulses ascend within the cord to higher centers in the brain, where the stimuli of bladder distention are consciously recognized. (a) A transection of the spinal cord in the midthoracic region would permanently deprive the patient of this information from the bladder. (b) After recovering from spinal shock, the bladder would fill and empty reflexly, and the patient would have an automatic bladder.

7. The patient has a stone in the urinary bladder. As the bladder empties, the bladder wall contracts down on the stone, and this irritates the sensitive mucous membrane and causes pain in the hypogastric region and sometimes a little bleeding. The afferent nerves from the bladder enter the spinal cord at segments S2, 3, and 4 and L1 and 2. The pain is often referred to the penis, which receives its sensory nerve supply from the pudendal nerve (S2, 3, and 4) and the ilioinguinal nerve (L1 and 2). When a patient with a bladder stone is standing or sitting, the stone gravitates to the neck region and irritates the mucous membrane of the trigone, which is especially sensitive. When the patient lies down, the stone rolls backward above the interureteric ridge and leaves the trigone area. For this reason, the pain was relieved by the patient's assuming the supine position.

8. The patient was suffering from benign hypertrophy of the prostate with obstruction of the prostatic urethra. The prostate is an important

anterior relation of the lower part of the rectum and can easily be examined by digital palpation through the anterior rectal wall. As the superior wall of the bladder rises out of the pelvis, the peritoneum is peeled off the posterior surface of the anterior abdominal wall. The surgeon's tube passing through the anterior abdominal wall in the midline just above the pubis would therefore not enter the peritoneal cavity.

9. The prostatic venous plexus is drained into the internal iliac veins. Large valveless veins also connect the plexus to the valveless vertebral veins. On coughing or sneezing, the blood may be forced from the prostatic plexus in the pelvis into the vertebral veins. Dislodged prostatic cancer cells may be carried along this route to the vertebral column. They may also pass up the vertebral plexus to enter the veins of the skull.

10. The median (middle) lobe of the prostate is located between the prostatic urethra and the ejaculatory ducts, just inferior to the sphincter vesicae. Benign hypertrophy of the median lobe results in its upward expansion within the sphincter vesicae. The sphincter can no longer function effectively, and urine continues to dribble into the urethra, giving the patient an intense desire to continue to micturate.

11. The pelvic parts of the ureters in the female can be palpated through the lateral fornices of the vagina as they pass forward close to the cervix to enter the bladder. In this patient, the pelvic part of the right ureter was irregularly thickened, which was suggestive of a tuberculous infection. A pyelogram confirmed that the patient had a tuberculous lesion of the right kidney, and cystoscopic examination showed a tuberculous ulcer of the ureteric orifice in the bladder.

12. The fertilized ovum in this case had become implanted in the right uterine tube. Each uterine tube is situated in the upper free border of the broad ligament of peritoneum. Rupture of the tube due to ectopic gestation nearly always occurs through the peritoneal covering. Hemorrhage takes place into the lower part of the peritoneal cavity. If the hemorrhage occurs suddenly, digital palpation through the posterior fornix reveals only a "doughy fullness" of the rectouterine pouch. If the bleeding is slow, the blood has time to clot, and a mass will be felt in the rectouterine pouch.

13. The long axis of the uterus in the majority of normal women lies at right angles to the long axis of the vagina (anteverted); the body of the uterus is also bent forward on the cervix at the internal os (anteflexion).

14. The lymphatic vessels from the cervix drain mainly into the internal and external iliac nodes.

15. The uterus is mainly supported by the tone of the levatores ani muscles and the ligaments of the visceral pelvic fascia, namely, the transverse cervical, sacrocervical, and pubocervical ligaments.

16. The urethra lies directly in contact with the lower half of the anterior vaginal wall, and the bladder lies in contact with the upper half of the vaginal wall. The bulging downward of the bladder with the anterior vaginal wall is referred to as a *cystocele.*

## Chapter 8

1. This patient had internal hemorrhoids, i.e., varicosities of the tributaries of the superior rectal vein. The varicosities had become pedunculated and remained prolapsed outside the anus after defecation (third degree internal hemorrhoids). The closure of the anal sphincters on the pedicles of the hemorrhoids caused congestion of the mucous membrane and the production of excessive mucus, which was responsible for the perianal moisture and irritation *(pruritus ani).* Abrasion of the hemorrhoids by the fecal masses during defecation was responsible for the bleeding. The patient was treated by hemorrhoidectomy.

2. The forward edge of a hard fecal mass may have caught one of the anal valves and torn it downward as it descended. The mucous membrane lining the anterior and posterior walls of the anal canal at this level is poorly supported by the superficial external sphincter muscle. This may explain why fissures are commonly found on the anterior and posterior anal walls. The mucous membrane of the lower half of the anal canal is innervated by the inferior rectal nerve and is very sensitive to pain. The mucous

membrane of the upper half of the canal is supplied by autonomic afferent fibers and is sensitive only to stretch. The external anal sphincter is supplied by the inferior rectal nerve and is reflexly in a state of spasm, due to the painful afferent impulses arising from the fissure.

3. The anorectal ring of muscle must be kept intact. This structure is composed of the puborectalis part of the levator ani muscle, the internal anal sphincter, and the deep part of the external anal sphincter; the fibers blend together at the anorectal junction.

4. The infection in the base of an anal fissure often tracks laterally through the external anal sphincter to enter the ischiorectal fossa. The fat in the fossa is poorly supplied with blood and very prone to abscess formation. The only structures of importance that cross the ischiorectal fossa are the inferior rectal vessels and nerve.

5. The bulbous part of the urethra, or the membranous part, may be damaged in accidents of this nature. Rupture of the bulbous part of the urethra had occurred in this case, with extensive extravasation of urine into the superficial perineal pouch. The urine had passed forward over the scrotum and penis deep to the membranous layer of the superficial fascia.

6. Just beyond the fossa terminalis of the penile urethra, a fold of mucous membrane projects downward from the roof of the urethra. This fold will sometimes completely obstruct the passage of a catheter if the point is directed toward the roof. With the patient lying in a supine position, hold the penis vertically. Introduce the catheter into the external meatus in such a way that the point is directed first toward the floor of the urethra until this fold is passed.

7. This patient had a cyst of the left greater vestibular gland. Chronic gonococcal infection had resulted in blockage of the duct and retention of the secretion.

8. The vulva is drained into the superficial inguinal lymph nodes.

## Chapter 9

1. The gastroduodenal artery is a relatively large vessel situated behind the first part of the duodenum. It may be eroded by a chronic duodenal ulcer situated on the posterior duodenal wall. The cephalic, basilic, and median cubital veins, and their tributaries, are located in front of the cubital fossa and may be used for transfusion. In the forearm, the cephalic and basilic veins can be seen as they wind around the lateral and medial borders of the forearm, respectively. The cephalic vein lies in a constant position behind the styloid process of the radius, and it is here that it may be exposed through a small skin incision.

2. The patient had an acute bacterial infection under the nail folds *(paronychia)* of the right index finger. The infection had spread into the lymphatic vessels draining the area, and they themselves had become inflamed *(lymphangitis)*. The red streaks were due to localized vasodilatation of blood vessels along the course of the lymphatic vessels in the forearm. The lymphatic vessels from the index finger pass to the dorsum of the hand and then follow the cephalic vein to the infraclavicular group of axillary lymph nodes. These were inflamed, enlarged, and tender *(lymphadenitis)*.

3. The first thoracic dermatome is situated along the medial side of the lower part of the arm and in the elbow region. Although there is considerable overlap of neighboring dermatomes, there would be loss of sensation in the region of the medial epicondyle in this patient.

4. The clavicle is the commonest bone in the body to be fractured. The violent force applied to the clavicle is usually indirect, namely, from the arm through the scapula. (a) Anatomically, the weakest part of the clavicle is the junction of the middle and lateral thirds, and this is where a fracture usually occurs. (b) The lateral bony fragment is displaced downward by the weight of the arm and pulled forward and medially by the pectoral muscles. The medial fragment is elevated by the sternocleidomastoid muscle. (c) Dislocation of the sternoclavicular joint is prevented by the ligaments of the joint, especially the very strong costoclavicular ligament. The very strong coracoclavicular ligament prevents dislocation of the acromioclavicular joint. Needless to say, if the mechanical force was great enough, and the clavicle strong enough, dislocation of one or other of these joints would

occur. (d) The supraclavicular nerves, or a communicating vein between the cephalic and the internal jugular vein, may be damaged with a fractured clavicle.

5. (a) The contracting fibrous tissue of the malignant tumor had pulled the lactiferous ducts and the nipple, elevating the latter above the nipple on the opposite side. (b) The dimpling of the skin was due to the fibrous tissue pulling upon the suspensory ligaments of the breast. (c) The lateral part of the breast is drained into the pectoral or anterior group of axillary lymph nodes, situated just posterior to the lower part of the pectoralis major muscle. The lymph nodes can be most easily felt by asking the patient to put her hand on her hip and press hard; she will then contract her pectoralis major muscle, providing a firm background against which the pectoral group of axillary nodes can be palpated. In this case, the malignant neoplasm had spread to involve these nodes in the left axilla. The medial part of the breast drains into the lymph nodes along the course of the internal thoracic artery.

6. A radial incision should be made to avoid cutting across the ducts of the gland, and also to limit the incision to one lobe of the gland. A transverse incision would cut across the fibrous septa that separate the lobes, and so aid the spread of infection from one lobe to another.

7. Cancer of the breast spreads relatively early via the lymphatic vessels to the regional lymph nodes. Later, it may spread via the bloodstream to distant sites, such as the bones of the vertebral column. The best chance of long survival, provided that the disease is still localized to the regional lymph nodes, is to remove the breast and associated structures, lymph vessels, and lymph nodes en bloc, as described on page 526. The postoperative radiotherapy can be directed to the axilla if it is felt that some malignant cells may have seeded out there; or directed to the internal thoracic nodes, which are difficult to remove surgically. The patient's postoperative anguish at losing a breast can be, in many cases, alleviated by plastic surgery. If this is not possible, the patient can wear a surgical brassiere with a prosthesis.

8. The subclavian artery lies behind the medial part of the clavicle, and at the outer border of the first rib it becomes the axillary artery. The arterial supply to the upper limb can be occluded by applying deep pressure downward and backward, compressing the subclavian artery against the upper surface of the first rib.

9. The radial artery lies in front of the distal third of the shaft of the radius; it is directly in contact with the front of the bone. On its lateral side lies the terminal part of the tendon of the brachioradialis, and on its medial side is the tendon of the flexor carpi radialis muscle. The artery is covered anteriorly by skin and fascia.

10. This patient had an Erb-Duchenne palsy, i.e., a lesion of the fifth and the sixth roots of the brachial plexus. The disability is fully described on page 527.

11. This is a case of Klumpke's palsy, in which the first thoracic nerve has been torn. It should be remembered that this nerve supplies all the small muscles of the hand via the median and ulnar nerves. The condition is fully described on page 527.

12. When removing the fat and lymph nodes from the axilla, the surgeon endeavors to preserve the long thoracic nerve. Sometimes it is cut by accident or has to be sacrificed because of its involvement in cancerous metastatic deposits. In this case, the nerve was purposely removed, and its removal resulted in paralysis of the serratus anterior muscle. Abduction of the shoulder joint to a right angle requires the action of the supraspinatus and deltoid muscles. To raise the arm further above the head requires that the scapula be rotated by the trapezius and the serratus anterior muscles, the latter muscle contributing most to this movement. The paralysis of the serratus anterior explains why the patient experienced difficulty in combing her hair. Another important function of the serratus anterior is to keep the scapula applied to the chest wall. Paralysis of this muscle resulted in "winged" scapula.

13. This patient exhibited a typical radial nerve palsy with wristdrop. The various muscles paralyzed are described on page 529. The radial nerve was presumably damaged in the spiral groove of the humerus when that bone was fractured, or was involved in the callus during the repair process. The surgeon probably waited to see if there was evidence of regener-

ation of the nerve fibers and then, because regeneration was delayed or absent, decided to explore the radial nerve in order to free it from the callus and approximate its proximal and distal ends. In any event, he was obviously unable to improve the situation, possibly because damage to the radial nerve was excessive, and the nerve fibers never regenerated. With the wrist held permanently in the flexed position, the long flexor muscles of the fingers are working at a mechanical disadvantage, and the fingers are unable to grip objects effectively. (Try it on yourself.)

14. The motor and sensory defects and the deformity that follow damage to the radial nerve in the axilla are described on page 531.

15. The axillary nerve gives motor fibers to the deltoid and teres minor muscles. If the shoulder joint is dislocated, it is impossible to test for activity in these muscles. The axillary nerve, however, also supplies the skin covering the lower half of the deltoid muscle, and it is a simple matter to ask the patient if she can feel a pinprick, or light touch with a piece of cotton, over this area. Damage to the axillary nerve produces paresthesia or anesthesia in the skin over the *lower half* of the deltoid muscle.

16. The glass fragment had severed the median nerve as it lay between the tendons of flexor digitorum superficialis and the flexor carpi radialis muscles and under cover of the palmaris longus tendon. The palmar cutaneous branch of the median nerve had also been severed. The effects of median nerve palsy are fully described on page 532.

17. This patient was suffering from the carpal tunnel syndrome, in which there is pressure exerted on the main trunk of the median nerve as it passes beneath the flexor retinaculum. Altered sensation was felt in the skin areas supplied by the digital branches of the median nerve. Although the thenar muscles, which are supplied by the median nerve, did not appear to be weakened, it is clear from her statement about the difficulty she experienced in buttoning up her clothes that the muscles were in fact not acting normally.

18. The patient's old supracondylar fracture of the right humerus had increased the carrying angle

on the right side to such an extent that the ulnar nerve was running around the medial epicondyle like a string around a pulley when the elbow joint was flexed and extended. Repeated friction caused interstitial neuritis of the ulnar nerve and consequent interference with the motor and sensory functions of the nerve. The effects of ulnar nerve palsy are fully described on page 536. The upstroke of writing is produced by flexion of the metacarpophalangeal joint and extension of the interphalangeal joints; both movements are normally carried out by the lumbricals and the interossei, which are supplied by the ulnar nerve.

19. The ulnar artery and nerve of the left hand were transected in front of the flexor retinaculum.

20. The patient had a subcoracoid dislocation of the left shoulder joint. The head of the humerus was dislocated downward through the weakest part of the capsule of the joint. It was then displaced medially in front of the scapula and behind the subscapularis muscle. The greater tuberosity of the humerus no longer displaced the deltoid muscle laterally, and the normal curve of the shoulder was therefore lost. The head of the humerus had come to rest below the coracoid process of the scapula and was responsible for the fullness below the lateral end of the clavicle.

21. This patient had supraspinatus tendinitis. During the middle range of abduction, the tendon of the supraspinatus impinges against the outer border of the acromion. Normally, the large subacromial bursa intervenes and ensures that the movement is relatively free of friction and is painless. In this condition, the bursa has degenerated and the supraspinatus tendon exhibits a localized area of collagen degeneration.

22. The sudden traction on the wrist resulted in the small head of the radius being partially pulled out of the anular ligament. This accident occurs only when the head of the radius is relatively small as compared with the size of the anular ligament and is almost entirely confined to children under the age of 6.

23. (1) The fracture line may deprive the proximal fragment of its arterial supply and result in ischemic necrosis of this fragment. (2) Because

of the articulation of the scaphoid with other bones, the fracture line may enter a joint and be bathed in synovial fluid. The presence of synovial fluid may inhibit union between the bone fragments. (3) The scaphoid is a difficult bone to immobilize because of its position and small size.

24. At the time of the fracture of the humerus, or following the application of the plaster cast, the child had spasm of the brachial artery in the region of the cubital fossa; this was followed by Volkmann's contracture. The anatomical changes in this condition are described on page 542.

25. The patient had Dupuytren's contracture involving the palmar aponeurosis at the base of the ring and little fingers of the left hand. The distal end of the aponeurosis gives four slips to the medial four fingers. Each slip is attached to the base of the proximal phalanx and to the fibrous flexor sheath of each finger. Fibrous contraction of the slip to the ring finger resulted in permanent flexion of the metacarpophalangeal joint.

26. This patient had acute suppurative tenosynovitis of the digital sheath of the index finger of the right hand, following the introduction into the sheath of pathogenic bacteria from the point of the thorn. In this condition, if the tension within the sheath is not relieved, it is likely to rupture at its proximal end, with discharge of pus into the thenar fascial space. If the hand remains untreated, the infection of the thenar space may spread to the midpalmar space and could spread upward into the forearm or downward to the interval between the index and middle fingers. The early administration of antibiotics is the treatment of choice.

27. This patient had an acute infection of the midpalmar fascial space of the right hand. The infected nail penetrated through the skin and the palmar aponeurosis and inoculated the fascial space with pathogenic organisms. The lymphatic drainage of this area is into the network of lymphatic vessels present in the subcutaneous tissue on the dorsum of the hand. For this reason, edema of the loose skin on the back of the hand is common in infections of the palm.

28. This child had a pulp-space infection of the left thumb. The danger here is that the tension within the pulp space will rise and occlude the blood supply to the diaphysis of the terminal phalanx; osteomyelitis of the terminal phalanx may also occur. The presence of a small area of devitalized skin over the center of the pulp would suggest that pus is accumulating within the space and is pointing onto the surface. The lymphatic drainage of the thumb is via vessels that accompany the cephalic vein and drain into the deltopectoral group of axillary nodes.

29. The position of function is described on page 514. Its importance lies in the fact that should some part of the hand become stiff or fixed permanently by adhesions, the patient would have a hand positioned to give the maximum mechanical efficiency.

30. When flexed, all fingers (excluding the thumb) point toward the tubercle of the scaphoid. When a finger is unstable following a fracture, it is tempting to align its long axis parallel to one of the borders of the hand, which is incorrect and will result in malfunction.

31. Absence of the tendon reflexes of the biceps brachii and the triceps muscles of the left arm would indicate the presence of disease in the C5, 6, 7, and 8 segments of the spinal cord or in the motor or sensory nerve fibers passing to or from these muscles.

32. This patient avulsed the insertion of the dorsal extensor expansion into the distal phalanx of the right index finger. The fold of the sheet suddenly caused flexion of the phalanx while the extensor tendons going to the extensor expansion were taut. The last 20 degrees of active extension was lost in the terminal interphalangeal joint, producing a condition known as mallet finger.

33. With fractures of the shaft of the humerus, the displacement of the bone fragments depends on the relation of the site of the fracture to the insertion of the deltoid. In this patient the fracture line was proximal to the insertion. The proximal fragment was adducted by the pectoralis major, latissimus dorsi, and the teres major muscles, and the distal fragment was pulled proximally by the deltoid, biceps, and triceps. The radial nerve, as it lies in the spiral

groove of the humerus, may be damaged in such fractures. It escaped unharmed in this patient.

## Chapter 10

1. This patient sustained damage to the right sciatic nerve as a direct result of the incorrect administration of an intramuscular injection into the right buttock. On questioning, the nurse demonstrated the site where she gave the injections; it lay over the course of the sciatic nerve. The problem was compounded by the fact that the patient received three intramuscular injections into the same site, in the same buttock, each day for 3 weeks. The common peroneal branch of the sciatic nerve was damaged, resulting in the loss of skin sensation in the areas normally supplied by the lateral cutaneous nerve of the calf and the superficial peroneal nerve. The muscles of the anterior and lateral compartments of the leg were partially paralyzed. The unopposed plantar flexors and invertors of the foot caused the patient to hold his right foot in equinovarus.

    All the nurses should be instructed on the extent of the area referred to as the buttock. So many restrict this area to the summit of the buttock, which, in actual fact, overlies the sciatic nerve. Intramuscular injections should be restricted to the upper outer quadrant of the buttock; and alternate buttocks, or other sites for injection, should be used when there are multiple injections extending over many weeks.

2. This little girl had a congenital dislocation of the right hip joint. The left hip joint was normal. When she stood on the right leg and lifted the left leg off the ground, the right hip joint could not act normally as a fulcrum for the pelvis and the contracting gluteus medius and minimus muscles. In fact, the right femoral head was not situated in the acetabulum, but had ridden up onto the gluteal surface of the ilium, due to a failure of the upper border of the acetabulum to develop adequately. As a consequence of this, the gluteus medius and minimus muscles on the right side could not tilt the pelvis, and it sagged downward on the unsupported side. Congenital dislocation of the hip is common in female Italian children. The reason for this is unknown, but a genetic factor is probably responsible. The Trendelenburg's sign is fully explained on page 687.

3. This patient had a malignant melanoma of the right second toe, which had spread by way of the lymphatics to involve the vertical group of superficial inguinal lymph nodes. The fact that the patient ignored the lump for three months worsened the prognosis. The treatment of choice is radical amputation of the toe and complete block dissection and removal of all the inguinal lymph nodes on the right side. The extensive physical examination was necessary because we know that the lymph from a wide area of the body drains into the inguinal lymph nodes. For details, see page 678.

4. This patient had acute intestinal obstruction secondary to a strangulated left femoral hernia. The pain was experienced in the region of the umbilicus, which is the area of the tenth thoracic dermatome. Pain experienced in the small bowel is referred to the umbilical region. (See p. 284.) When, 2 days before, the patient coughed, a loop of ileum was forced down into a preexisting femoral hernial sac. The unyielding nature of the femoral ring resulted in venous congestion of the gut and, later, arterial occlusion, at which point peristalsis ceased (*paralytic ileus*) and intestinal obstruction occurred.

    Such a patient should be operated on immediately. The surgeon must remember the relations of the femoral ring when returning the hernial contents to the abdomen: medially, the sharp edge of the lacunar ligament; laterally, the femoral vein; anteriorly, the inguinal ligament; and posteriorly, the pectineus and superior ramus of the pubis. A femoral hernia is found below and lateral to the pubic tubercle; an inguinal hernia is situated above and medial to the tubercle.

5. This patient had varicose veins of the great saphenous and small saphenous venous systems, which lie in the superficial fascia of the legs. Before operating on these veins, it is imperative to determine whether or not the deep veins of the leg, the venae comitantes, are patent. It is possible that a woman with six children may have experienced thrombosis of her deep veins during or following one of the pregnancies. A

person with deep vein thrombosis depends on the dilated superficial veins to return the blood in the leg to the general circulation, and deep vein thrombosis would be a contraindication to operation. The superficial epigastric, the superficial circumflex iliac, and the superficial external pudendal veins, together with the important perforating veins, must be ligated for a successful result to be obtained. Large varicose veins possess incompetent valves. If the leg is raised above the level of the heart in a supine patient, the varicose veins quickly empty. If the great saphenous vein is now occluded by digital pressure at the saphenous opening and the patient is asked to stand, the veins, when the digital pressure is removed, fill from above and not from below, as they should normally. With the patient standing, a cough will transmit a fluid thrill from the abdomen to the hand examining the veins, because the incompetent valves do not impede the passage of the pressure wave. The venous pump of the leg is described on page 677.

6. This angle is within normal limits in a 5-year-old. It is too great in a 35-year-old man; the condition is called coxa valga, in which adduction of the hip joint is limited. When the angle of the femoral neck is smaller than normal (coxa vara), abduction of the hip joint is limited.

7. The femoral head receives its blood supply from two sources: (1) a small artery that runs with the round ligament and (2) a profuse blood supply from the medial circumflex femoral artery, branches of which ascend the femoral neck beneath the synovial membrane. Fracture of the femoral neck may deprive the femoral head of part or all the blood from source 2, and avascular necrosis will occur. The blood supply to the trochanters is profuse, causing both fragments in a fracture of this region to have an adequate blood supply.

8. The shortening of the leg resulted from the upward pull of the distal fragment by the hamstrings and the quadriceps femoris muscles. The backward rotation of the distal fragment was caused by the pull of the two heads of the gastrocnemius muscles. The muscles responsible for the shortening are very powerful, and prolonged traction to the distal fragment, using

weights connected to a pin driven through the fragment, is required to obtain reduction of such a fracture.

9. (a) Vasodilatation of the collateral arteries can be obtained by preganglionic lumbar sympathectomy (see p. 677), provided that the arteries are free of disease. (b) The muscular and genicular branches of the femoral artery, the perforating branches of the profunda femoris, and the muscular and genicular branches of the popliteal artery anastomose with one another.

10. The sympathetic innervation of the blood vessels of the lower limb is fully described on page 677.

11. This patient was found on examination to have a swelling in front of her patellar tendon on the right side. Repeated unaccustomed trauma to the subcutaneous infrapatellar bursa had produced an inflammatory response in the bursal wall, which resulted in an excessive production of fluid, hence the swelling.

12. As the student fell, the medial semilunar cartilage was drawn laterally within the knee joint. The sudden movement of the knee joint, which occurred on striking the ground, resulted in the grinding of the relatively immobile semilunar cartilage between the medial femoral and tibial condyles. The cartilage split along part of its length, and the detached portion became jammed, like a wedge, between the articular surfaces, limiting further extension, i.e., "locking the joint." The tenderness was experienced over the torn medial semilunar cartilage. The trauma stimulated an excessive production of synovial fluid, which filled the joint cavity and the suprapatellar bursa. The distension of the latter was responsible for the large amount of swelling seen on the front of the joint.

13. The knee jerk depends on the integrity of (1) the afferent neuron, which extends from the stretch receptors in the ligamentum patellae to the spinal cord via the femoral nerve and lumbar plexus; (2) the connection of the afferent neuron to the efferent neuron in the spinal cord at segments L2, 3, and 4; and (3) the efferent neuron, which extends from the spinal cord to the motor end plates of the quadriceps femoris muscle via the lumbar plexus and femoral nerve. Absence of the knee jerk signifies a break

in this simple reflex arc. An exaggerated knee jerk signifies damage to the upper motor neuron, which normally influences the activity of this reflex arc.

The ankle jerk depends on the normal functioning of another simple reflex arc, acting at the level of the first and second sacral segments of the cord.

The Babinski test normally results in plantar flexion of the big toe. In a positive Babinski test, stimulation of the sole is followed by dorsiflexion of the big toe and indicates damage to the upper motor neuron. (Babies normally have a positive Babinski test.)

14. In this patient the anterior cruciate ligament in the right knee had been ruptured. The strength of the knee joint depends primarily on the tone of the quadriceps femoris muscle and secondarily on the ligaments. Operative union of torn cruciate ligaments is considered by many orthopedic surgeons to be unsatisfactory. The loss of this ligament can be compensated for by developing the tone of the quadriceps muscle.

15. In lay language, this girl "sprained" her left ankle. The movement of inversion of the foot normally takes place in the subtalar and transverse tarsal joints. Overinversion places a strain on the lateral ligaments of the ankle joint. In this case, the localized tenderness found below and in front of the lateral malleolus would indicate that some of the fibers of the anterior talofibular ligament had been torn. The resulting hemorrhage was responsible for the swelling in the area. The movements of eversion and inversion of the foot take place in the subtalar and transverse tarsal joints.

16. (a) The ankle is a very stable joint. The body of the talus is held firmly in position in the mortice formed by the lower end of the tibia and the medial and lateral malleoli. This arrangement is further strengthened by a very strong medial or deltoid ligament and a less strong lateral ligament. (b) The mechanics of fracture dislocation of the ankle joint are described on page 690. (c) In a fracture dislocation that involves rotation and overeversion of the foot, there is usually a spiral fracture of the lateral malleolus and a transverse fracture across the medial malleolus. The medial ligament is so strong that it pulls off the medial malleolus.

17. This patient was suffering from acute foot strain. Previously employed in a sedentary occupation, she suddenly exposed the arches of her feet to a tremendous work load. The arches were carrying not only the body weight, but also the weight of loaded trays. Normally, the arches are supported in position by the shape of the bones, by the strong ligaments, and—most important of all in the active foot—by the tone of muscles. Once untrained muscles become fatigued, they stretch. In the case of the medial longitudinal arch, the head of the talus starts to sag down between the calcaneum and the navicular bone. The calcaneonavicular ligament and the other plantar ligaments first become stretched and then may even be torn. At this stage there is pain and swelling of the foot. The tendons of the long muscles also stretch, producing pain in the leg.

The feet should be rested and elevated to eliminate the swelling. Graded physiotherapy to the muscles should then be started. When such a patient returns to work, the hours of duty should be gradually increased so that the muscles are trained to carry the work load. A person who continues to walk on a foot with acute foot strain will end up with permanently flat feet.

18. (a) The Achilles tendon or tendo calcaneus is attached to the posterior surface of the calcaneum. It is the tendon of insertion of the gastrocnemius and soleus muscles. (b) Plantar flexion of the ankle joint is the movement in which the toes point downward. (c) When the tendo calcaneus ruptures, the bellies of the gastrocnemius and soleus retract upward, leaving a gap between the divided ends of the tendo calcaneus.

19. The femoral artery can be palpated at a point midway between the anterior superior iliac spine and the symphysis pubis. The popliteal artery is felt deep in the popliteal fossa, with the deep fascia relaxed and the knee joint flexed. The dorsalis pedis artery is felt on the front of the ankle, between the tendons of the extensor hallucis longus and the extensor digitorum longus. The posterior tibial artery can be felt behind the medial malleolus.

20. (a) The hip joint is a ball-and-socket synovial joint. (b) Adduction of the hip joint is the move-

ment of the lower limb medially; it is limited by contact with the opposite limb. (c) The gluteus maximus extends and laterally rotates the hip joint; through the iliotibial tract it helps to maintain the knee joint in extension. It is also an extensor of the trunk on the thigh. The gluteus medius and gluteus minimus abduct the thigh at the hip joint; the anterior fibers also rotate the thigh medially. (d) The rectus femoris flexes the hip joint and extends the knee joint.

---

## Chapter 11

1. Tuberculous cervical lymphadenitis is much less common now than previously due to adequate pasteurization of cows' milk. The organism commonly gains entrance to the palatine tonsil and spreads to the member of the deep cervical group of lymph nodes that drains the tonsil and is situated below and behind the angle of the jaw. The infection may remain localized to this node for some time or involve other neighboring nodes, which become matted together. Once the disease has resulted in the destruction of the interior of the node, the caseating material liquifies and breaks through the capsule of the node. To begin with, the abscess is localized beneath the investing layer of deep cervical fascia. Later, it erodes through the fascia and produces a large cold abscess beneath the skin. This soon becomes secondarily infected and breaks through the skin, to form a discharging sinus.

2. The thyroid gland is invested by the pretracheal layer of deep cervical fascia, which binds the gland to the larynx. Thus, as the larynx moves upward on swallowing, the thyroid gland and the adenoma move upward also. Each lobe of the thyroid gland is closely related to the trachea. A localized enlargement of the gland, such as an adenoma, often presses on the trachea and partially occludes the lumen, producing dyspnea. The thyroid gland is drained mainly into the deep cervical group of lymph nodes.

3. Congenital torticollis is due to hemorrhage into the sternocleidomastoid muscle during birth. Fibrous infiltration of the blood clot, followed by contracture of the fibrous tissue, results in shortening of the affected muscle. It is only when the neck begins to elongate in childhood that the condition is noticed. In this patient the right sternocleidomastoid was diseased. An understanding of the precise actions of this muscle (see p. 713) will explain the deformity. Failure to correct the deformity leads to asymmetrical changes in the face, which result from the eyes attempting to work on the same horizontal plane. Wedge-shaped deformities of the cervical vertebrae also occur.

4. The spinal part of the accessory nerve crosses the posterior triangle of the neck in a comparatively superficial position. The surface marking of the nerve is as follows: Draw a line from the angle of the mandible to the mastoid process and bisect it at right angles. The latter line, if continued downward and backward, closely follows the course of the nerve. The surgeon had forgotten his anatomy and cut the right spinal part of the accessory nerve, thus paralyzing the right trapezius muscle. In order for a person to raise the hand above the head, the trapezius muscle, assisted by the serratus anterior, must rotate the scapula so that the glenoid cavity faces upward and laterally. (See p. 434.)

5. The external jugular vein can be used clinically as a venous manometer. The zero line is the level of the right atrium in the thorax. Normally, in a person propped up on pillows in bed, the venous pressure is so low that the blood level does not extend above the clavicle. This patient had right-sided heart failure with a backup of venous blood on the right side of the heart. Consequently, the venous pressure was high, and the external jugular vein was engorged with blood throughout its length. The surface marking of this vein is from the angle of the jaw to the midpoint of the clavicle. The external jugular vein drains into the subclavian vein.

6. The submental lymph nodes drain the tip of the tongue, the floor of the mouth in the region of the frenulum of the tongue, the gums and incisor teeth, the middle third of the lower lip, and the skin over the chin. On examination of the inside of the patient's mouth, a small, hard-based carcinomatous ulcer was found on the right side of the tip of the tongue. On question-

ing, the patient said he had noticed the ulcer for about 4 months, but since it did not cause much discomfort, he had ignored it. The submental lymph nodes ultimately drain into the deep cervical group of lymph nodes. In this case they were not yet involved in the spread of the malignant condition.

7. The carotid sinus has a thick tunica adventitia and a thin tunica media. The adventitia contains numerous nerve endings derived from the glossopharyngeal nerve. A rise in pressure within the lumen, due either to a rise in blood pressure or to pressure applied to the neck, will result in an increase in the discharge of afferent impulses from the nerve endings. These will ascend in the glossopharyngeal nerve, reflexly inhibit the cardiac center and the vasomotor center in the medulla, and produce a slowing of the heart and a fall in the arterial blood pressure.

8. The lymph from the temporal region drains into the parotid group of lymph nodes. The efferent lymph from these nodes drains into the deep cervical lymph nodes, which are embedded in the carotid sheath and the tunica adventitia of the internal jugular vein. To ensure that all the deep cervical nodes are removed, it is therefore necessary to remove the internal jugular vein on that side. The spinal part of the accessory nerve and the hypoglossal nerves are carefully preserved in block dissections of the neck. However, if they are involved in secondary cancerous growth, they are sacrificed.

9. A good surgeon, *never, never* blindly grabs at a bleeding artery. Pack the wound and then later, when the bleeding has stopped, remove the pack and clamp the artery under direct vision. In this case the superior laryngeal artery was successfully clamped, but the external laryngeal nerve was also included and severely damaged. As a consequence of this, the cricothyroid muscle on the right side was paralyzed. This muscle normally tenses the vocal fold of the same side. In this patient the right vocal fold was slack, causing hoarseness of the voice. The superior thyroid artery arises from the external carotid artery at a level just below the tip of the greater cornu of the hyoid bone. The superior thyroid vein drains into the internal jugular vein.

10. The patient is suffering from parathyroid tetany due to accidental removal of one or more of the parathyroid glands during the operation of total thyroidectomy. Sometimes the condition occurs following interference with the blood supply to these glands. The function of the parathyroid gland is to secrete a hormone that regulates calcium metabolism and plasma calcium concentration. A low level of plasma calcium results in increased neuromuscular excitability and the clinical syndrome known as *tetany.*

11. The patient had a carcinoma of the cervical part of the esophagus. Radiological examination following the swallowing of a barium emulsion revealed the stenosis produced by the neoplasm. The growth has spread by the lymphatics to the deep cervical lymph nodes.

12. This patient is suffering from vascular insufficiency of the left arm, which is due to partial constriction of the subclavian artery. An anteroposterior radiograph of the neck revealed the presence of a complete cervical rib on the left side. The subclavian artery was found at operation to be angulated as it passed over the rib. A fusiform dilation of the artery distal to the constriction was also noted. Such a dilation is a common finding and may be the site of the formation of blood clots on the intima. Pieces of the thrombus sometimes become detached and form emboli, which may block the brachial artery or one of its branches and so further diminish the vascular supply to the hand.

13. This overworked physician was guilty of negligence and should never have prescribed treatment without thoroughly examining his patient. Of course, the patient did not wish to disturb her Easter bonnet, but the physician should have been firm and had the patient remove it. Further examination would have revealed a large sebaceous cyst on the temporal region of the scalp, which the patient had had for 30 years. She was embarrassed by its presence and always contrived to conceal it beneath her hat. Two weeks previously she had inadvertently stuck her hatpin into the cyst, which had become infected. The organisms had spread by the lymphatics to the regional lymph nodes, hence the lymphadenitis of the mastoid nodes.

14. (a) If the swelling is very extensive, the hematoma is situated beneath the epicranial aponeurosis and is limited only by the attachment of the aponeurosis to the skull. (b) If the swelling is large but restricted to one bone, the hematoma is situated beneath the periosteum of the skull bone and is limited by the attachment of the periosteum to the sutural ligaments. (c) If the swelling is small, superficial, and tense, it probably lies in the subcutaneous tissue and is limited by the fibrous tissue that binds the skin to the epicranial aponeurosis. In all injuries to the head there should be a complete radiological examination; severe fractures of the skull are sometimes missed on clinical examination.

15. The veins of the scalp are connected to the diploic veins of the skull bones, and to the intracranial venous sinuses, by valveless veins, called emissary veins. Thrombosis of the scalp veins secondary to infection could easily result in the spread of pathogenic organisms from the scalp to the skull bones, producing an osteomyelitis, or thrombosis of the venous sinuses, resulting in cerebral edema and possibly death.

16. The tunica adventitia of the numerous arteries supplying the scalp is anchored to the fibrous septa that bind the skin to the epicranial aponeurosis. A sectioned artery is therefore unable to contract or retract to slow the circulation so that clotting may take place. Bleeding may be stopped by applying firm pressure to the wound. A deep wound of the scalp involving the section of the epicranial aponeurosis always gapes open due to the pull of the occipitofrontalis muscles. For this reason the aponeurosis should be sutured.

17. The "danger area" of the face is drained by the facial vein. Interference with a boil may lead to spread of infection and thrombosis of the facial vein. The pathogenic organisms may then spread via the inferior ophthalmic veins to the cavernous sinus and cause thrombosis there. Cavernous sinus thrombosis is a serious condition, resulting in cerebral edema, and in the days before antibiotics it was always fatal.

18. The right facial nerve had been damaged as it lay in the facial nerve canal in the medial wall of the aditus of the tympanic antrum. The unopposed facial muscles on the left side pulled the left corner of the mouth upward and to the left. The right orbicularis oculi was paralyzed; hence the boy could not close his right eye. The right buccinator was paralyzed, resulting in ballooning of the cheek and accumulation of saliva.

19. The muscles of the upper part of the face are controlled by the cerebral cortex on both sides of the brain. The muscles of the lower part of one side of the face are controlled by the cerebral cortex on the opposite side of the brain. This patient had a hemorrhage into the right internal capsule involving the genu and the anterior two-thirds of the posterior limb. The upper part of the face moved normally, but the muscles on the left side of the mouth were paralyzed. (The upper motor neurons coming from the right cerebral cortex cross over at the level of the facial nerve nucleus.) As a result, the unopposed action of the right facial muscles pulled the right corner of the mouth upward.

20. The mandibular division of the trigeminal nerve supplies a strip of skin that extends from the temporal region of the scalp in front of the ear (including the anterior part of the auricle) and it also supplies the skin over the lower jaw, including the lower lip. The skin over the parotid gland and a small area over the angle of the mandible is supplied by the great auricular nerve (C2 and 3).

21. The facial nerve, the retromandibular vein, and the external carotid artery lie within the parotid salivary gland. This patient had a highly invasive carcinoma of the right parotid gland, which quickly involved the right facial nerve, with consequent weakness of the right facial muscles. The method used clinically to test the integrity of the facial nerve is fully described on page 906. A benign tumor of the parotid gland tends not to damage the facial nerve.

22. The patient is suffering from Frey's syndrome. The dart had damaged the parasympathetic secretomotor fibers to the parotid gland and also branches of the great auricular nerve, which supply the overlying skin. On regeneration, some of the parasympathetic fibers had crossed over and joined the distal end of the great auricular nerve. A stimulus normally producing salivation stimulated the sweat glands instead. The best method of treatment is avulsion of the

auriculotemporal nerve, which carries the parasympathetic fibers.

23. The student had dislocated his temporomandibular joints on both sides. When he yawned, his lateral pterygoid muscles reflexly contracted forcibly and pulled the head of the mandible and the articular disc forward over the summit of the articular tubercle in each joint. Reduction is easily performed by pressing the gloved thumbs downward and backward on the last molar teeth.

24. This patient had a bilateral fracture of the mandible. The fracture line went through the socket of the canine tooth on each side, which is the weakest part of the body of the mandible. The fractures were compound, since the mucoperiosteum is invariably torn in fractures of the body of the mandible. The center part of the mandible with the incisor teeth was depressed due to the downward pull of the anterior bellies of the digastric muscles.

25. The submandibular duct may be palpated through the floor of the mouth alongside the tongue. It is often possible to palpate a calculus in the duct in this situation. The duct opens into the mouth at the side of the frenulum of the tongue. The narrowest part of the duct is at the orifice.

26. The anterior and posterior fontanelles of the skull can be easily palpated in the infant. The anterior fontanelle remains open until 18 months, and the posterior fontanelle closes by the end of the first year. In cases of raised intracranial pressure, the fontanelles bulge upward.

27. A minor blow on the side of the head may easily fracture the thin anterior part of the parietal bone or the squamous part of the temporal bone. The anterior branch of the middle meningeal artery commonly enters a bony canal in this region and is sectioned at the time of the fracture. The resulting hemorrhage causes the gradual accumulation of blood under pressure outside the meningeal layer of the dura mater. As the blood clot enlarges, pressure is exerted on the underlying brain, and the symptoms of confusion and irritability become apparent. This is followed later by drowsiness. Pressure on the lower end of the right precentral gyrus

or motor area causes twitching of the facial muscles on the left and later twitching of the left arm muscles. As the blood clot progressively enlarges, the intracranial pressure rises and the patient's condition deteriorates. The accurate placing of a burr hole in the skull (see p. 904) and the tying off of the middle meningeal artery will save the patient's life.

28. Right abducent nerve and right lateral rectus muscle.

29. Bitemporal hemianopia is a loss of both temporal fields of vision and is due to the interruption of the optic nerve fibers derived from the medial halves of both retinae. Pressure on the optic chiasma by a tumor of the pituitary gland is the most common cause of the condition.

30. The glossopharyngeal nerve supplies the mucous membrane of the posterior third of the tongue with taste fibers and those for common sensation. These sensations can easily be tested by using appropriate stimuli.

31. The ocular manifestations of hyperthyroidism are proptosis, retraction of the upper lid, and lid-lag (the upper lid fails to move with the eye as the patient is asked to look downward).

    Exophthalmos was originally believed to be due to the excessive contraction of smooth muscle present in the back of the orbital cavity forcing the orbital contents forward. Now it is believed to be due to a general weakness of the recti muscles resulting from infiltration with fat and lymphoid tissue.

    Retraction of the upper lid and lid-lag may be due to increased tone of the smooth muscle component of the levator palpebrae superioris. It is more likely to be a secondary mechanical defect associated with the exophthalmos.

32. Foreign bodies in the conjunctival sac tend to lodge in the subtarsal sulcus. The eyelid should be everted and the foreign body wiped off with a piece of moist cotton.

33. The pathogenic organisms ascend from the nasal part of the pharynx through the auditory tube into the tympanic cavity. They may then spread posteriorly into the mastoid antrum and the mastoid air cells (acute mastoiditis). They may extend medially from the antrum to the sigmoid sinus, causing venous sinus thrombosis and possibly septicemia. The organisms may

spread superiorly through the tegmen tympani, causing meningitis and possibly later an abscess in the temporal lobe of the brain or in the cerebellum. They may spread medially to involve the facial nerve or the labyrinth of the internal ear (labyrinthitis).

34. The patient was suffering from a frontal and maxillary sinusitis on the right side. The supraorbital nerve supplies not only the frontal sinus but also the skin of the scalp as far back as the vertex. The pain is commonly referred to this area. The maxillary sinus is innervated by the infraorbital nerve, which also supplies the skin of the face. The pain is referred to the skin of the face and also commonly to the upper teeth.

35. The pyriform fossa, which is innervated by the internal laryngeal branch of the superior laryngeal branch of the vagus.

36. The left vocal fold assumes the adducted midline position.

37. The optic nerves are surrounded by sheaths derived from the pia mater, arachnoid mater, and dura mater. There is an extension on the intracranial subarachnoid space forward around the optic nerve to the back of the eyeball. A rise in cerebrospinal fluid pressure caused by an intracranial tumor will compress the thin walls of the retinal vein as it crosses the extension of the subarachnoid space. This will result in congestion of the retinal vein and bulging of the optic disc. Since both subarachnoid extensions are continuous with the intracranial subarachnoid space, both eyes will exhibit papilledema.

38. The bones that form the orbital margin are the frontal, zygomatic, and maxillary bones.

39. A severe cold involving the mucous membrane of the nasopharynx may spread to the auditory tube. Inflammatory edema may cause the mucous membrane of the tube to swell and block the lumen of the tube. As a result, the air in the tympanic cavity is no longer in communication with the nasopharynx. The trapped air is now slowly absorbed into the bloodstream, producing a partial vacuum in the tympanic cavity. The tympanic membrane is sucked medially and its mobility is impaired. The failure of the tympanic membrane to vibrate is responsible for the deafness. The vacuum and the pull on the tympanic membrane causes the ear to ache.

Repeated swallowing while sucking on a candy causes the salpingopharyngeus muscle to pull open the auditory tube, and in many cases allows air to pass the obstruction. The sudden popping sounds in the ear result from air rushing into the tympanic cavity. Once the pressure is equalized on both sides of the tympanic membrane, normal hearing returns and the earache ceases.

40. The procedure for carrying out a brachial plexus nerve block is described on page 897. The important layer of fascia that encloses the axillary vessels and the brachial plexus is the prevertebral layer of deep cervical fascia. This layer of fascia forms the axillary sheath, which confines the anesthetic agent to the area of the brachial plexus.

## Chapter 12

1. Acute staphylococcal infection of the lower end of the right femur at the age of 7 years resulted in inhibition of growth at the lower femoral epiphysis. During the next 3 years this resulted in 1 inch (2.5 cm) of shortening of the right femur. When the boy stood upright, with both feet on the ground, the pelvis tilted downward on the right side. To compensate for the pelvic tilt, a lumbar scoliosis developed, and in order to keep the line of gravity running through the vertebral column, a second compensatory curve developed in the thoracic region.

Osteomyelitis occurring close to an epiphysis may cause diminished growth. At other times, owing to the local hyperemia, it may cause increased growth.

2. This child's father and mother were both over 6 feet (1.8 m) tall. On examination, she was found to be thin, with poor muscular development. On inspection of her back, she was noted to have a gentle kyphosis confined to the upper thoracic region. She was able to correct this partially when asked to straighten her back. On testing her muscular strength, it was found to be generally weak. It was clear that this girl's muscular development had not kept pace with her body growth. This, together with her numerous illnesses, had resulted in the vertebral

column not receiving adequate muscular support. Since she was above average height, her school desk was too low for her, and she was forced to stoop over her books. This further exacerbated the condition. In order to support her back when her muscles were fatigued, she tended to rotate her vertebrae, lock her articular processes, and stretch the intervertebral ligaments. Over a period of time, this caused discomfort and pain. If such a condition is allowed to continue without treatment, structural changes take place in the vertebrae, so that a permanent kyphosis occurs. Graduated muscular exercises, good food, and fresh air, together with a raised desk top, would be the treatment of choice in this patient.

3. This boy had tuberculosis of the right lung, which had become quiescent with treatment. Since he did not follow his physician's advice concerning rest and diet, the disease recently had become active and spread via the bloodstream to the fifth thoracic vertebra. Bone destruction started in the cancellous bone of the vertebral body, close to the intervertebral disc. As the disease progressed, the intervertebral disc was destroyed, and the adjacent vertebra became involved. At this stage, the first vertebral body collapsed, producing an acute angular kyphosis.

4. This patient had advanced carcinoma of the prostate, which had already spread locally beyond the capsules of the gland to involve the sacral plexus on the right side. This explains the fixation of the gland and the right-sided sciatica. In addition, the carcinomatous cells had entered the prostatic venous plexus and had ascended in the internal vertebral venous plexus. Frequent straining on micturition no doubt assisted the process and resulted in the carcinomatous cells reaching the bone marrow of the vertebral body and the innominate bones.

5. This patient had symptoms suggestive of irritation of the right sixth cervical nerve root. The radiograph revealed narrowing of the space between the fifth and sixth cervical vertebral bodies, suggesting a herniation of the nucleus pulposus of the intervertebral disc at this level.

6. Herniation of the nucleus pulposus of the intervertebral disc between the fifth lumbar vertebra and the sacrum was the diagnosis. The herniation occurred on the left side and was relatively small, hence the absence of radiological evidence. The pain occurred in the dermatomes of the fifth lumbar and first sacral segments, and the sensory roots of these segments were presumably involved on the left side. The deep muscles of the back in the left lumbar region were in spasm due to irritation of the motor roots of those segments of the cord. The diminished left ankle jerk could be caused by pressure either on the motor or on the sensory roots of the first sacral nerve. The lumbar scoliosis was produced by the spasm and shortening of the deep muscles of the back in the left lumbar region.

7. The boy's head, on striking the bottom of the pool, forced the neck into excessive flexion. The left inferior articular process of the fifth cervical vertebra was forced over the anterior margin of the left superior articular process of the sixth cervical vertebra, producing a unilateral dislocation, hence the rotation of the head to the right. The tearing of the capsular ligaments and the nipping of the sixth left cervical nerve caused spasm of the neck muscles and extreme pain in the sixth left cervical dermatome. A lateral radiograph of the cervical vertebral column would reveal the dislocation. The large vertebral canal in the cervical region permitted the spinal cord to escape injury.

8. This patient had a severe fracture dislocation between the fifth and sixth thoracic vertebrae. In cases of this nature the spinal cord is invariably severely damaged or even severed. One would expect to find the signs and symptoms of spinal shock and paraplegia. The vertical arrangement of the articular processes and the low mobility of this region because of the thoracic cage mean that a dislocation can occur in this region only if the articular processes are fractured by a great force. The small circular vertebral canal leaves little space around the spinal cord, so that severe cord injuries are certain.

9. Severe direct trauma to the lumbar region should immediately alert the physician to the possibility of damage to the kidney, muscles, and transverse processes of lumbar vertebrae. Anteroposterior and lateral radiographs of the lumbar vertebral column will exclude or con-

firm a fracture. The examination of a 24-hour specimen of urine for blood will exclude or confirm injury to the kidney. The lumbar transverse processes are long and tapered and may be fractured by direct trauma. This may well be the cause of the persistent pain. Damage to muscles such as the erector spinae or quadratus lumborum, followed by fibrous tissue adhesions, is often responsible for protracted pain in the elderly.

10. This patient had spondylolisthesis involving the fifth lumbar vertebra. During the 5 years that she had experienced back pain, the body of the fifth lumbar vertebra had slowly become dislocated in front of the sacrum. The details of this condition, with the embryological explanation, are given on page 949.

11. Skin, superficial fascia, supraspinous ligament, interspinous ligament, ligamentum flavum, areolar tissue containing the internal vertebral venous plexus, dura mater, and arachnoid mater.

12. This patient had advanced carcinoma of the left breast, with lymphatic spread to the pectoral lymph nodes of the left axilla and blood-borne metastases in the bodies of the second and third lumbar vertebrae. On further questioning, the patient admitted knowledge of the lump in the left breast, but said she had been afraid to mention it because she thought that it might indicate cancer. She had first noticed the lump 6 months earlier.

The moral of this case history is that you should always examine your patients thoroughly and completely. You may learn your anatomy by regions, but remember that your patient is a whole person.

# Answers to National Board Type Questions

## Chapter 1

| | | |
|---|---|---|
| 1. (e) | 10. (a) | 18. (d) |
| 2. (c) | 11. (b) | 19. (a) |
| 3. (b) | 12. (c) | 20. (b) |
| 4. (e) | 13. (c) | 21. (d) |
| 5. (b) | 14. (a) | 22. (e) |
| 6. (a) | 15. (e) | 23. (a) |
| 7. (c) | 16. (a) | 24. (b) |
| 8. (b) | 17. (b) | 25. (c) |
| 9. (d) | | |

## Chapter 2

| | | |
|---|---|---|
| 1. (a) | 6. (c) | 11. (d) |
| 2. (c) | 7. (b) | 12. (a) |
| 3. (c) | 8. (c) | 13. (e) |
| 4. (a) | 9. (e) | 14. (b) |
| 5. (d) | 10. (b) | 15. (c) |

## Chapter 3

| | | |
|---|---|---|
| 1. (c) | 8. (d) | 15. (c) |
| 2. (e) | 9. (c) | 16. (d) |
| 3. (b) | 10. (c) | 17. (c) |
| 4. (a) | 11. (e) | 18. (b) |
| 5. (e) | 12. (d) | 19. (c) |
| 6. (c) | 13. (b) | 20. (a) |
| 7. (b) | 14. (c) | |

## Chapter 4

| | | |
|---|---|---|
| 1. (a) | 10. (e) | 19. (c) |
| 2. (d) | 11. (c) | 20. (a) |
| 3. (c) | 12. (d) | 21. (b) |
| 4. (a) | 13. (d) | 22. (b) |
| 5. (d) | 14. (b) | 23. (b) |
| 6. (e) | 15. (c) | 24. (c) |
| 7. (a) | 16. (e) | 25. (b) |
| 8. (b) | 17. (e) | 26. (c) |
| 9. (c) | 18. (d) | |

## Chapter 5

| | | |
|---|---|---|
| 1. (c) | 11. (c) | 21. (c) |
| 2. (b) | 12. (e) | 22. (c) |
| 3. (d) | 13. (a) | 23. (c) |
| 4. (d) | 14. (e) | 24. (e) |
| 5. (c) | 15. (d) | 25. (a) |
| 6. (b) | 16. (b) | 26. (b) |
| 7. (e) | 17. (b) | 27. (d) |
| 8. (d) | 18. (c) | 28. (c) |
| 9. (e) | 19. (d) | 29. (d) |
| 10. (c) | 20. (b) | 30. (a) |

## Chapter 6

| | | |
|---|---|---|
| 1. (d) | 8. (c) | 15. (b) |
| 2. (c) | 9. (a) | 16. (a) |
| 3. (d) | 10. (e) | 17. (d) |
| 4. (c) | 11. (b) | 18. (e) |
| 5. (a) | 12. (d) | 19. (b) |
| 6. (b) | 13. (d) | 20. (e) |
| 7. (c) | 14. (d) | |

## Chapter 7

| | | |
|---|---|---|
| 1. (b) | 8. (b) | 15. (a) |
| 2. (c) | 9. (c) | 16. (b) |
| 3. (d) | 10. (e) | 17. (e) |
| 4. (a) | 11. (e) | 18. (b) |
| 5. (d) | 12. (a) | 19. (b) |
| 6. (c) | 13. (c) | 20. (e) |
| 7. (d) | 14. (b) | |

## Chapter 8

| | | |
|---|---|---|
| 1. (a) | 6. (e) | 11. (b) |
| 2. (a) | 7. (a) | 12. (a) |
| 3. (b) | 8. (b) | 13. (d) |
| 4. (c) | 9. (d) | 14. (e) |
| 5. (e) | 10. (e) | 15. (c) |

## Chapter 9

| | | |
|---|---|---|
| 1. (c) | 7. (c) | 13. (a) |
| 2. (a) | 8. (a) | 14. (a) |
| 3. (d) | 9. (a) | 15. (a) |
| 4. (a) | 10. (c) | 16. (c) |
| 5. (d) | 11. (a) | 17. (d) |
| 6. (c) | 12. (d) | 18. (a) |

| | | |
|---|---|---|
| 19. (e) | 25. (a) | 31. (e) |
| 20. (b) | 26. (a) | 32. (e) |
| 21. (d) | 27. (c) | 33. (d) |
| 22. (b) | 28. (d) | 34. (e) |
| 23. (d) | 29. (d) | 35. (d) |
| 24. (c) | 30. (d) | |

## Chapter 10

| | | |
|---|---|---|
| 1. (b) | 15. (e) | 28. (a) |
| 2. (c) | 16. (c) | 29. (c) |
| 3. (d) | 17. (d) | 30. (b) |
| 4. (b) | 18. (c) | 31. (b) |
| 5. (c) | 19. (d) | 32. (c) |
| 6. (a) | 20. (e) | 33. (e) |
| 7. (c) | 21. (c) | 34. (d) |
| 8. (d) | 22. (b) | 35. (b) |
| 9. (c) | 23. (d) | 36. (a) |
| 10. (a) | 24. (d) | 37. (d) |
| 11. (c) | 25. (e) | 38. (d) |
| 12. (a) | 26. (c) | 39. (e) |
| 13. (c) | 27. (a) | 40. (a) |
| 14. (d) | | |

## Chapter 11

| | | |
|---|---|---|
| 1. (b) | 16. (d) | 31. (b) |
| 2. (a) | 17. (d) | 32. (a) |
| 3. (c) | 18. (c) | 33. (c) |
| 4. (b) | 19. (c) | 34. (d) |
| 5. (a) | 20. (d) | 35. (a) |
| 6. (b) | 21. (a) | 36. (a) |
| 7. (c) | 22. (c) | 37. (b) |
| 8. (d) | 23. (e) | 38. (a) |
| 9. (a) | 24. (a) | 39. (a) |
| 10. (a) | 25. (b) | 40. (c) |
| 11. (d) | 26. (e) | 41. (a) |
| 12. (c) | 27. (a) | 42. (b) |
| 13. (d) | 28. (d) | 43. (a) |
| 14. (e) | 29. (d) | 44. (c) |
| 15. (c) | 30. (e) | 45. (b) |

## Chapter 12

| | | |
|---|---|---|
| 1. (d) | 5. (b) | 9. (c) |
| 2. (e) | 6. (c) | 10. (c) |
| 3. (e) | 7. (a) | 11. (c) |
| 4. (e) | 8. (b) | 12. (b) |

# Index